Handbook of Hydrogen Energy

MECHANICAL and AEROSPACE ENGINEERING

Frank Kreith & Darrell W. Pepper
Series Editors

RECENTLY PUBLISHED TITLES

Air Distribution in Buildings, *Essam E. Khalil*

Alternative Fuels for Transportation, *Edited by Arumugam S. Ramadhas*

Computer Techniques in Vibration, *Edited by Clarence W. de Silva*

Distributed Generation: The Power Paradigm for the New Millennium, *Edited by Anne-Marie Borbely and Jan F. Kreider*

Elastic Waves in Composite Media and Structures: With Applications to Ultrasonic Nondestructive Evaluation, *Subhendu K. Datta and Arvind H. Shah*

Elastoplasticity Theory, *Vlado A. Lubarda*

Energy Audit of Building Systems: An Engineering Approach, *Moncef Krarti*

Energy Conversion, *Edited by D. Yogi Goswami and Frank Kreith*

Energy Management and Conservation Handbook, *Edited by Frank Kreith and D. Yogi Goswami*

The Finite Element Method Using MATLAB®, Second Edition, *Young W. Kwon and Hyochoong Bang*

Fluid Power Circuits and Controls: Fundamentals and Applications, *John S. Cundiff*

Fuel Cells: Principles, Design, and Analysis, *Shripad Revankar and Pradip Majumdar*

Fundamentals of Environmental Discharge Modeling, *Lorin R. Davis*

Handbook of Energy Efficiency and Renewable Energy, *Edited by Frank Kreith and D. Yogi Goswami*

Handbook of Hydrogen Energy, *Edited by S.A. Sherif, D. Yogi Goswami, Elias K. Stefanakos, and Aldo Steinfeld*

Heat Transfer in Single and Multiphase Systems, *Greg F. Naterer*

Heating and Cooling of Buildings: Design for Efficiency, Revised Second Edition, *Jan F. Kreider, Peter S. Curtiss, and Ari Rabl*

Intelligent Transportation Systems: Smart and Green Infrastructure Design, Second Edition, *Sumit Ghosh and Tony S. Lee*

Introduction to Biofuels, *David M. Mousdale*

Introduction to Precision Machine Design and Error Assessment, *Edited by Samir Mekid*

Introductory Finite Element Method, *Chandrakant S. Desai and Tribikram Kundu*

Large Energy Storage Systems Handbook, *Edited by Frank S. Barnes and Jonah G. Levine*

Machine Elements: Life and Design, *Boris M. Klebanov, David M. Barlam, and Frederic E. Nystrom*

Mathematical and Physical Modeling of Materials Processing Operations, *Olusegun Johnson Ilegbusi, Manabu Iguchi, and Walter E. Wahnsiedler*

Mechanics of Composite Materials, *Autar K. Kaw*

Mechanics of Fatigue, *Vladimir V. Bolotin*

Mechanism Design: Enumeration of Kinematic Structures According to Function, *Lung-Wen Tsai*

Mechatronic Systems: Devices, Design, Control, Operation and Monitoring, *Edited by Clarence W. de Silva*

The MEMS Handbook, Second Edition (3 volumes), *Edited by Mohamed Gad-el-Hak*

MEMS: Introduction and Fundamentals

MEMS: Applications

MEMS: Design and Fabrication

Multiphase Flow Handbook, *Edited by Clayton T. Crowe*

Nanotechnology: Understanding Small Systems, Second Edition, *Ben Rogers, Sumita Pennathur, and Jesse Adams*

Nuclear Engineering Handbook, *Edited by Kenneth D. Kok*

Optomechatronics: Fusion of Optical and Mechatronic Engineering, *Hyungsuck Cho*

Practical Inverse Analysis in Engineering, *David M. Trujillo and Henry R. Busby*

Pressure Vessels: Design and Practice, *Somnath Chattopadhyay*

Principles of Solid Mechanics, *Rowland Richards, Jr.*

Principles of Sustainable Energy Systems, Second Edition, *Edited by Frank Kreith with Susan Krumdieck, Co-Editor*

Thermodynamics for Engineers, *Kau-Fui Vincent Wong*

Vibration and Shock Handbook, *Edited by Clarence W. de Silva*

Vibration Damping, Control, and Design, *Edited by Clarence W. de Silva*

Viscoelastic Solids, *Roderic S. Lakes*

Weatherization and Energy Efficiency Improvement for Existing Homes: An Engineering Approach, *Moncef Krarti*

Handbook of Hydrogen Energy

EDITED BY

S.A. Sherif • D. Yogi Goswami
Elias K. Stefanakos • Aldo Steinfeld

CRC Press
Taylor & Francis Group
Boca Raton London New York

CRC Press is an imprint of the
Taylor & Francis Group, an **informa** business

CRC Press
Taylor & Francis Group
6000 Broken Sound Parkway NW, Suite 300
Boca Raton, FL 33487-2742

Printed on acid-free paper
Version Date: 20140530

International Standard Book Number-13: 978-1-4200-5447-7 (Hardback)

Library of Congress Cataloging-in-Publication Data

Handbook of hydrogen energy / edited by S.A. Sherif, D. Yogi Goswami, E.K. (Lee) Stefanakos, Aldo Steinfeld.
pages cm.
Includes bibliographical references and index.
ISBN 978-1-4200-5447-7
1. Hydrogen as fuel. I. Sherif, S. A. II. Goswami, D. Yogi. III. Stefanakos, Elias K. IV. Steinfeld, Aldo.

TP359.H8H37 2014
665.8'1--dc23 2013036876

Visit the Taylor & Francis Web site at
http://www.taylorandfrancis.com

and the CRC Press Web site at
http://www.crcpress.com

Contents

Section V Biological Hydrogen Production

Section VI Solar Hydrogen Production

Section VII Hydrogen Storage, Transportation, Handling, and Distribution

Foreword

Hydrogen energy was proposed nearly four decades ago as a permanent solution to the interrelated global problems of the depletion of fossil fuels and the environmental problems caused by their utilization. It was formally presented at the landmark The Hydrogen Economy Miami Energy (THEME) Conference, March 18–20, 1974, Miami Beach, Florida. It immediately caught the imagination and attention of socially contentious energy and environmental scientists and engineers. Research and development activities ensued around the world in order to develop the technologies needed for the introduction of the hydrogen energy system. It took a quarter of a century to research and develop most of the technologies required.

The hydrogen energy system started making inroads in the energy field early in the twenty-first century. Several types of fuel cells were developed for efficient conversion of hydrogen to electricity, as well as heat. In the United States, Germany, and Japan, solid oxide fuel cells are used to produce electricity and to heat homes and buildings. Hydrogen-fueled forklifts are now replacing battery-powered forklifts in warehouses, since they are much more economical. Several municipalities are experimenting with hydrogen-fueled buses as they are much quieter and cleaner. Major car manufacturers have developed clean and efficient hydrogen cars, which are already being tested in major cities around the world. Construction of hydrogen-fueling stations is also accelerating in several countries and in major cities, especially in Germany, Japan, and California. Car companies have announced that they will offer hydrogen-fueled cars for sale to the public starting 2015. Railway companies are experimenting with hydrogen-fueled locomotives. There are experimental trams running on hydrogen. Many navies are replacing their diesel-fueled submarines with hydrogen-fueled ones. Boeing and Airbus are studying the feasibility of hydrogen-fueled passenger planes. A hydrogen-powered supersonic private plane is also under development.

Clearly, the time has arrived for a handbook of hydrogen energy. I congratulate the editors, S.A. Sherif, D. Yogi Goswami, E. K. (Lee) Stefanakos, and Aldo Steinfield, for seeing the need for such a book and producing it. It covers all aspects of hydrogen energy, including hydrogen production, hydrogen storage and delivery, hydrogen conversion and utilization, and hydrogen safety and sensors. The handbook also provides very useful information, such as inversion curves, physical and thermodynamic tables, properties of storage materials, data on specific heats, and compressibility and temperature–entropy charts.

I strongly recommend this excellent handbook to energy and environmental scientists and engineers, and graduate students, as well as for the pioneers of the hydrogen energy system. It should also be in every library as a reference material.

T. Nejat Veziroğlu
President, International Association for Hydrogen Energy

Editors

Dr. S.A. Sherif is a tenured professor of mechanical and aerospace engineering and is the founding director of the Wayne K. and Lyla L. Masur HVAC Laboratory and the director of the Industrial Assessment Center at the University of Florida. He served as editor in chief, associate editor, subject editor, and member of the editorial board of 15 thermal science journals. He is also an emeritus editor of the *International Journal of Hydrogen Energy*. He is a fellow of the ASME, a fellow of ASHRAE, an associate fellow of AIAA, a member of Commission B-1 on Thermodynamics and Transfer Processes of the International Institute of Refrigeration, and a member of the Advisory Board of Directors of the International Association for Hydrogen Energy. He is a past chair for the ASME Advanced Energy Systems Division, the K-19 Committee on Environmental Heat Transfer of the ASME Heat Transfer Division (2003–2007), the Coordinating Group on Fluid Measurements (1992–1994), and the Fluid Applications and Systems Technical Committee (2008–2010) of the ASME Fluids Engineering Division. He is also a past chair of the Steering Committee of the Intersociety Energy Conversion Engineering Conference (2001–2003), ASHRAE's Standards Project Committee 41.6 on Measurement of Moist Air Properties (1989–1994), and ASHRAE's TC 1.1 on Thermodynamics and Psychrometrics (2012–2013). He also served as a member of the ASME's Energy Resources Board (2001–2003) and was the board's representative to the ASME's International Mechanical Engineering Congress Committee (2003–2006). He served as the head of the Refrigeration Section of ASHRAE (2004–2008), as the technical conference chair of the 2008 ASME Summer Heat Transfer Conference, as the general conference chair of the 2013 ASME Summer Heat Transfer Conference, and as a member of the ASME Frank Kreith Energy Award Selection Committee (2005–2011). He currently serves as the chair of the ASME Heat Transfer Division Executive Committee and as a member of the ASME's Basic Engineering Group Operating Board (2010–2014). He is the recipient of the E.K. Campbell Award of Merit from ASHRAE for "outstanding service and achievement in teaching" in 1997 and a "TIP" teaching award from the University of Florida in 1998. He is the recipient of the Kuwait Prize in Applied Sciences (2001), an ASHRAE Distinguished Service Award (2003), an ASHRAE Exceptional Service Award (2010), an AIAA Best Paper Award (2005), and an ASME Best Paper Award (2005). In 2007, he received a Division III Superior Accomplishment Award from the University of Florida's Office of the Provost, and in 2008 he was elected as an ASHRAE Distinguished Lecturer. He received certificates of appreciation from the ASME Heat Transfer Division, Fluids Engineering Division, Advanced Energy Systems Division, and Solar Energy Division, as well as ASHRAE and AIAA. He served as a NASA Faculty Fellow at the Kennedy Space Center (1993) and the Marshall Space Flight Center (1996 and 1997), as an Air Force Faculty Fellow at the Arnold Engineering Development Center (1994), and as a DOE Faculty Fellow at the Argonne National Laboratory (2002). Dr. Sherif has 110 refereed journal papers, 21 book chapters, 180 conference papers, 100 technical reports, and 2 US patents to his credit.

Dr. D. Yogi Goswami, PhD, PE, is a distinguished university professor and a codirector of the Clean Energy Research Center at the University of South Florida.

Professor Goswami is the editor in chief of the *Solar Energy* journal. Within the field of renewable energy, he has published, as an author or editor, 16 books, 14 book chapters, and more than 300 refereed technical papers. He also holds 18 patents. Some of his inventions have been commercialized and are available worldwide.

Professor Goswami has served as the president of the International Solar Energy Society (ISES), governor of ASME International, senior vice president of ASME, and president of the International Association for Solar Energy Education (IASEE). He has also served as a vice president of ASME, ISES, and IASEE. He currently serves as the chair of the management committee of the ASME-ITI, LLC.

Dr. Goswami is a recipient of the Farrington Daniels Award from ISES (highest award of ISES), Frank Kreith Energy Award and Medal from ASME (highest award of ASME), John Yellott Award for Solar Energy from ASME, the Charles Greely Abbott Award (highest award of the American Solar Energy Society) and Hoyt Clarke Hottel Award of the ASES, and more than 50 awards and certificates from major engineering and scientific societies for his work in the field of renewable energy.

Dr. Goswami is a fellow of the ASME, ASES, ASHRAE, AAAS, and the National Academy of Inventors.

Dr. Elias K. Stefanakos, PhD, PE, currently serves as the professor of electrical engineering and the director of the Clean Energy Research Center (CERC) at the University of South Florida (USF), Tampa, Florida. He served as chairman of the Department of Electrical Engineering at USF for 13 years up to August 2003. He has also served as an associate editor of the *Journal of Solar Energy* (*Photovoltaics*) and coeditor-USA of the *Journal of Asian Electric Vehicles*. He has published over 200 research papers in refereed journals and international conferences in the areas of materials, renewable energy sources and systems, hydrogen and fuel cells, and electric and hybrid vehicles. He has received over $20 million in contracts and grants from agencies such as the National Science Foundation (NSF), US Department of Energy (USDOE), National Aeronautics and Space Administration (NASA), Defense Advanced Research Projects Agency (DARPA), and others. He has also been a consultant to a number of companies and international organizations.

Dr. Aldo Steinfeld is full professor in the Department of Mechanical and Process Engineering at ETH Zurich, where he holds the chair of renewable energy carriers. He also serves as director of the Solar Technology Laboratory at the Paul Scherrer Institute.

Professor Steinfield's research program is aimed at the advancement of the thermal and thermochemical engineering sciences applied to renewable energy technologies. His research interests focus on high-temperature heat/mass transfer phenomena and multiphase reacting flows, with applications in solar power and fuels production, decarbonization and metallurgical processes, CO_2 capture and recycling, and sustainable energy systems. He has pioneered the development of solar reactor technologies for thermochemically producing clean fuels using concentrated solar radiation.

Professor Steinfield served as the editor in chief of the *Journal of Solar Energy Engineering* (2005–2009). He has authored over 240 original articles in refereed journals and filed 24 patents. His contributions to science and education have been recognized with the ASME Calvin W. Rice Award (2006), the UOP/Honeywell Lecturer (2006), the University of Minnesota Founders Lecturer (2007), the John I. Yellott Award (2008), the European Research Council Advanced Grant (2012), and the ASME Heat Transfer Memorial Award (2013). Professor Steinfeld is an ASME fellow and a member of the Swiss Academy of Engineering Sciences.

Contributors

Salvador M. Aceves
Engineering Department
Lawrence Livermore National Laboratory
Livermore, California

Tadeusz Bak
Solar Energy Technologies
University of Western Sydney
Richmond, New South Wales, Australia

Frano Barbir
Department of Thermodynamics, Thermotechnics and Heat Engines
University of Split
Split, Croatia

Halil Berberoğlu
Department of Mechanical Engineering
University of Texas at Austin
Austin, Texas

Chad Blake
National Renewable Energy Laboratory
Golden, Colorado

Robert C. Bowman Jr.
Oak Ridge National Laboratory
Oak Ridge, Tennessee

Geoffrey Bromaghim
Technology Transition Corporation, Ltd.
Newcastle Upon Tyne, United Kingdom

Tom Burdyny
Department of Mechanical Engineering
University of Victoria
Victoria, British Columbia, Canada

Jo-Shu Chang
Department of Chemical Engineering
and
Research Center for Energy Technology and Strategy
National Cheng Kung University
Tainan, Taiwan, People's Republic of China

Paul Charpentier
Department of Chemical and Biochemical Engineering
University of Western Ontario
London, Ontario, Canada

Shing-Der Chen
Department of Chemical Engineering
National Cheng Kung University
Tainan, Taiwan, People's Republic of China

Amgad Elgowainy
Center for Transportation Research
Argonne National Laboratory
Argonne, Illinois

Hesham El Naggar
Department of Civil and Environmental Engineering
University of Western Ontario
London, Ontario, Canada

Said S.E.H. Elnashaie
Department of Chemical and Biological Engineering
University of British Columbia
Vancouver, British Columbia, Canada

and

Department of Chemical Engineering
New Mexico Technical University
Albuquerque, New Mexico

Paul A. Erickson
Department of Mechanical and Aerospace Engineering
University of California, Davis
Davis, California

Francisco Espinosa-Loza
Engineering Department
Lawrence Livermore National Laboratory
Livermore, California

Kamiel S. Gabriel
Faculty of Engineering and Applied Sciences
University of Ontario Institute of Technology
Oshawa, Ontario, Canada

Monterey Gardiner
Office of Fuel Cells
U.S. Department of Energy
Washington, District of Columbia

Nirmal V. Gnanapragasam
Faculty of Engineering and Applied Science
University of Ontario Institute of Technology
Oshawa, Ontario, Canada

D. Yogi Goswami
Clean Energy Research Center
University of South Florida
Tampa, Florida

S. Gursu
OMV Petrol Ofisi AS
Istanbul, Turkey

Hisham Hafez
Department of Civil and Environmental Engineering
University of Western Ontario
London, Ontario, Canada
and
Greenfield Specialty Alcohols
Chatham, Ontario, Canada

Karen Hall
Technology Transition Corporation, Ltd.
Newcastle Upon Tyne, United Kingdom

Gamal Ibrahim
Department of Basic Sciences and Engineering
University of Menoufia
Menoufia, Egypt

Gary G. Ihas
Department of Physics
University of Florida
Gainesville, Florida

Henry V. Kehiaian (Deceased)
ITODYS
University of Paris VII
Paris, France

Elias Ledesma-Orozco
Mechanical Engineering Department
Universidad de Guanajuato
Salamanca, Mexico

Kuo-Shing Lee
Department of Safety Health and Environmental Engineering
Central Taiwan University of Science and Technology
Taichung, Taiwan, People's Republic of China

Dennis Y.C. Leung
Department of Mechanical Engineering
The University of Hong Kong
Pokfulam, Hong Kong, People's Republic of China

Michael K.H. Leung
Department of Mechanical Engineering
The University of Hong Kong
Pokfulam, Hong Kong, People's Republic of China

M. Lewis
Argonne National Laboratory
Argonne, Illinois

Wenxian Li
Solar Energy Technologies
University of Western Sydney
Richmond, New South Wales, Australia

Madhukar Mahishi
Cummins Power Generation
Minneapolis, Minnesota

Anton Meier
Solar Technology Laboratory
Paul Scherrer Institute
Villigen, Switzerland

Marianne Mintz
Center for Transportation Research
Argonne National Laboratory
Argonne, Illinois

George Nakhla
Department of Chemical and Biochemical Engineering
and
Department of Civil and Environmental Engineering
University of Western Ontario
London, Ontario, Canada

Greg F. Naterer
Faculty of Engineering and Applied Science
Memorial University of Newfoundland
St. John's, Newfoundland, Canada

Meng Ni
Department of Mechanical Engineering
The University of Hong Kong
Pokfulam, Hong Kong, People's Republic of China

Janusz Nowotny
Solar Energy Technologies
University of Western Sydney
Penrith, New South Wales, Australia

Gregory L. Olson
HRL Laboratories, LLC
Malibu, California

Guillaume Petitpas
Engineering Department
Lawrence Livermore National Laboratory
Livermore, California

Laurent Pilon
Department of Mechanical and Aerospace Engineering
University of California, Los Angeles
Los Angeles, California

Bale V. Reddy
Faculty of Engineering and Applied Science
University of Ontario Institute of Technology
Oshawa, Ontario, Canada

Carl Rivkin
National Renewable Energy Laboratory
Golden, Colorado

Marc A. Rosen
Faculty of Engineering and Applied Science
University of Ontario Institute of Technology
Oshawa, Ontario, Canada

Andrew Rowe
Department of Mechanical Engineering
University of Victoria
Victoria, British Columbia, Canada

Ganesh D. Saratale
Department of Chemical Engineering
National Cheng Kung University
Tainan, Taiwan, People's Republic of China

Prakash C. Sharma
Department of Physics
Tuskegee University
Tuskegee, Alabama

S.A. Sherif
Department of Mechanical and Aerospace Engineering
University of Florida
Gainesville, Florida

Roger Sierens
Department of Flow, Heat and Combustion Mechanics
Ghent University
Ghent, Belgium

Sesha S. Srinivasan
Department of Physics
Tuskegee University
Tuskegee, Alabama

Elias K. Stefanakos
Clean Energy Research Center
University of South Florida
Tampa, Florida

Aldo Steinfeld
Department of Mechanical and Process Engineering
ETH Zurich
Zurich, Switzerland

and

Solar Technology Laboratory
Paul Scherrer Institute
Villigen, Switzerland

Ned T. Stetson
U.S. Department of Energy
Washington, District of Columbia

Neil S. Sullivan
Department of Physics
University of Florida
Gainesville, Florida

S. Suppiah
Chalk River Laboratories
Atomic Energy of Canada Limited
Chalk River, Ontario, Canada

Hong-Yue (Ray) Tang
Hydrogen Production and Utilization Laboratory
University of California, Davis
Davis, California

K. Mark Thomas
School of Chemical Engineering and Advanced Materials
Newcastle University
Newcastle Upon Tyne, United Kingdom

Sebastian Verhelst
Department of Flow, Heat and Combustian Mechanics
Ghent University
Ghent, Belgium

David R. Vernon
Schatz Energy Research Center
Humboldt State University
Arcata, California

T. Nejat Veziroglu
International Association for Hydrogen Energy
Miami, Florida

Thomas Wallner
Energy Systems Division
Argonne National Laboratory
Argonne, Illinois

Liang-Ming Whang
Department of Environmental Engineering
National Cheng Kung University
Tainan, Taiwan, Republic of China

Emhemmed A.E.A. Youssef
Department of Chemical and Biochemical Engineering
University of Western Ontario
London, Ontario, Canada

1

Hydrogen Economy

S.A. Sherif
University of Florida

Frano Barbir
University of Split

T. Nejat Veziroglu
International Association for Hydrogen Energy

CONTENTS

This chapter presents an overview of how a hydrogen economy can be realized. We will give an overview of the principles of hydrogen energy production, storage, and utilization. We will also discuss different aspects of hydrogen safety. Hydrogen production will cover a whole array of methods including electrolysis, thermolysis, photolysis, thermochemical cycles, and production from biomass. Hydrogen storage will cover all modes of gaseous, liquid, slush, and metal hydride storage. Hydrogen utilization will focus on a large cross section of applications such as fuel cells and catalytic combustion of hydrogen. Details of many of these topics will be presented in the rest of the handbook.

1.1 Introduction

A hydrogen economy is one in which hydrogen is the main energy carrier along with electricity. Hydrogen would be produced from renewable energy sources such as solar or wind energy through water electrolysis. While this is considered an ideal scenario for a hydrogen economy, other possibilities exist. In theory, hydrogen and electricity can satisfy all the energy needs of humankind and form an energy system that would be permanent and independent of energy sources [1–3]. Hydrogen has unique characteristics that make it an ideal energy carrier [4]. These include the fact that (1) it can be produced from and converted into electricity at relatively high efficiencies; (2) its raw material for production is water, which is available in abundance; (3) it is a completely renewable fuel; (4) it can be stored in gaseous form (convenient for large-scale storage), in liquid form (convenient for air and space transportation), or in the form of metal hydrides (convenient for surface vehicles and other relatively small-scale storage requirements); (5) it can be transported over large distances through pipelines or via tankers; (6) it can be converted into other forms of energy in more ways and more efficiently than any other fuel (such as catalytic combustion, electrochemical conversion, and hydriding); and (7) it is environmentally compatible since its production, storage, transportation, and end use do not produce any pollutants (except for small amounts of nitrogen oxides), greenhouse gases, or any other harmful effects on the environment.

In a hydrogen economy, electricity and hydrogen would be produced in large quantities from available energy sources and used in every application where fossil fuels are being used today. This includes large industrial plants as well as small, decentralized units, wherever the primary energy source (solar, nuclear, and even fossil) is available. Electricity would be used directly or transformed into hydrogen. For large-scale storage, hydrogen can be stored underground in ex-mines, caverns, and/or aquifers. Energy transport to the end users, depending on distance and overall economics, would either be in the form of electricity or in the form of hydrogen. Hydrogen would be transported by means of pipelines or supertankers. It would then be used in transportation and industrial, residential, and commercial sectors as a fuel. Some of it would be used to generate electricity (via fuel cells), depending on demand, geographical location, or time of the day.

1.2 Hydrogen Production

Hydrogen is usually found in small amounts mixed with natural gas in crustal reservoirs. However, a few wells have been found to contain large amounts of hydrogen, such as some wells in Kansas that contain 40% hydrogen, 60% nitrogen, and trace amounts of hydrocarbons [5]. Logical sources of hydrogen are hydrocarbon (fossil) fuels (C_XH_Y) and water (H_2O). Presently, hydrogen is mostly being produced from fossil fuels (natural gas, oil, and coal). However, except for the space program, hydrogen is not being used directly as a fuel or an energy carrier. It is being used in refineries to upgrade crude oil (hydrotreating and hydrocracking), in the chemical industry to synthesize various chemical compounds (such as ammonia and methanol), and in metallurgical processes (as a reduction or protection gas). The total annual hydrogen production worldwide in 1996 was about 40 million tons (5.6 EJ) [6]. Less than 10% of that amount was supplied by industrial gas companies; the rest is being produced at consumer-owned and consumer-operated plants (the so-called captive production), such as refineries, and by ammonia and methanol producers. Production of hydrogen as an energy carrier would require an increase in production rates by several orders of magnitude.

To help speed up the introduction of fuel cell–powered vehicles, car manufacturers in collaboration with oil- and fuel-processing companies are intensively working on the development of onboard fuel processors. These devices would enable the use of conventional fuels such as gasoline and diesel, as well as methanol. This would allow the car companies to plan to overcome nonexistence of a hydrogen refueling infrastructure and the difficulties of onboard hydrogen storage. However, development of such compact fuel processors poses several engineering challenges, such as rapid start-up, high efficiency, and very high purity of the produced gas (<100 ppm CO).

The most logical source for large-scale hydrogen production is water. Methods of hydrogen production from water include electrolysis, direct thermal decomposition or thermolysis, thermochemical processes, and photolysis.

1.2.1 Electrolysis

The production of hydrogen by water electrolysis is a mature technology, based on a fundamentally simple process, is very efficient, and does not involve moving parts. It is suitable for large-scale hydrogen production. Typical efficiencies are 72%–82%. Several advanced electrolyzer technologies are being developed such as the advanced alkaline electrolysis (which employs new materials for membranes and electrodes that allow further improvement in efficiency—up to 90% [7,8]), the solid polymer electrolytic (SPE) process (which employs a proton-conducting ion exchange membrane as an electrolyte and as a membrane that separates the electrolysis cell [8,9]), and high-temperature steam electrolysis (which operates between 700°C and 1000°C and which employs oxygen ion–conducting ceramics as electrolyte [10]). An electrolysis plant can operate over a wide range of capacity factors and is convenient for a wide range of operating capacities, which makes this process interesting for coupling with renewable energy sources, particularly with photovoltaics (PVs). The latter generate low-voltage direct current, which is exactly what is required for the electrolysis process. Theoretical and experimental studies on the performance of PV-electrolyzer systems have been performed [11–14]. Examples of experimental PV-electrolysis plants that are currently in operation worldwide include the Solar-Wasserstoff-Bayern pilot plant (Neunburg vorm Wald, Germany) [15], the HYSOLAR project (Saudi Arabia) [16], Schatz Energy Center (Humboldt State University, Arcata, California) [17], Helsinki University of Technology (Helsinki, Finland) [18], and INTA Energy Laboratory (Huelva, Spain) [19].

1.2.2 Direct Thermal Decomposition of Water (Thermolysis)

Water can be split thermally at temperatures above 2000 K [20]. The degree of dissociation is a function of temperature: only 1% at 2000 K, 8.5% at 2500 K, and 34% at 3000 K. The product is a mixture of gases at extremely high temperatures. The main problems in connection with this method are related to materials required for extremely high temperatures, recombination of the reaction products at high temperatures, and separation of hydrogen from the mixture.

1.2.3 Thermochemical Cycles

Production of hydrogen employing thermochemical cycles involves the chemical splitting of water at temperatures lower than those needed for thermolysis, through a series of cyclical chemical reactions that ultimately release hydrogen. Some of the more thoroughly investigated thermochemical process cycles include the following [9,21,22]: sulfuric acid–iodine cycle, hybrid sulfuric acid cycle, hybrid sulfuric acid–hydrogen bromide cycle, calcium bromide–iron oxide cycle (UT-3), and iron–chlorine cycle. Depending on the temperatures at which these processes are occurring, relatively high efficiencies are achievable (40%–50%). However, the problems related to movement of a large mass of materials in chemical reactions, toxicity of some of the chemicals involved, and corrosion at high temperatures remain to be solved in order for these methods to become practical.

1.2.4 Photolysis

Photolysis (or direct extraction of hydrogen from water using only sunlight as an energy source) can be accomplished by employing photobiological systems, photochemical assemblies, or photoelectrochemical cells [23,24]. Intensive research activities are opening new perspectives for photoconversion, where new redox catalysts, colloidal semiconductors, immobilized enzymes, and selected microorganisms could provide means of large-scale solar energy harvesting and conversion into hydrogen.

1.2.5 Hydrogen Production from Biomass

Hydrogen can be obtained from biomass by a pyrolysis/gasification process [25]. The biomass preparation step involves heating of the biomass/water slurry to high temperatures under pressure in a reactor. This process decomposes and partially oxidizes the biomass, producing a gas product consisting of hydrogen, methane, CO_2, CO, and nitrogen. Mineral matter is removed from the bottom of the reactor. The gas stream goes to a high-temperature shift reactor where the hydrogen content is increased. Relatively high-purity hydrogen is produced in the subsequent pressure swing adsorption unit. The whole system is very much similar to a coal gasification plant, with the exception of the unit for pretreatment of the biomass and the design of the reactor. Because of the lower calorific value per unit mass of biomass as compared to coal, the processing facility is larger than that of a comparably sized coal gasification plant.

1.3 Hydrogen Storage

Once hydrogen is produced, it needs to be stored in order to overcome daily and seasonal discrepancies between energy source availability and demand. Several forms of storage are discussed briefly in the following.

1.3.1 Gaseous Hydrogen Storage

Depending on storage size and application, several types of hydrogen storage systems may be available. This includes stationary large storage systems and stationary small storage systems at the distribution, or final user level; mobile storage systems for transport and distribution including both large-capacity devices (such as a liquid hydrogen tanker—bulk carrier) and small systems (such as a gaseous or liquid hydrogen truck trailer); and vehicle tanks to store hydrogen used as fuel for road vehicles. Because of hydrogen's low density, its storage always requires relatively large volumes and is associated with either high pressures (thus requiring heavy vessels) or extremely low temperatures, and/or combination with other materials (much heavier than hydrogen itself). Studies dealing with large underground storage of hydrogen include those reported in Refs. [26–28]. Pressurized gas storage systems are used today in the natural gas business in various sizes and pressure ranges from standard pressure cylinders (50 L, 200 bar) to stationary high-pressure containers (over 200 bar) or low-pressure spherical containers (>30,000 m^3, 12–16 bar). This application range will be similar for hydrogen storage. Storage in vehicular pressurized hydrogen tanks are discussed in Ref. [29].

1.3.2 Metal Hydride Storage

In the presence of some metals and alloys, hydrogen can form metal hydrides. During that process, hydrogen molecules are split and hydrogen atoms are inserted in spaces inside the lattice of the metal or alloy. This creates effective storage comparable to the density of liquid hydrogen. However, when the mass of the metal or alloy is taken into account, then the metal hydride gravimetric storage density becomes comparable to storage of pressurized hydrogen. The best achievable gravimetric storage density is about 0.07 kg of H_2/kg of metal, for a high-temperature hydride such as MgH_2 [30]. During the storage process (charging or absorption), heat is released, which must be removed in order to achieve the continuity of the reaction. During the hydrogen release process (discharging or desorption), heat must be supplied to the storage tank. An advantage of storing hydrogen in hydriding substances is the safety aspect. A serious damage to a hydride tank (such as one that could be caused by a collision) would not pose a fire hazard since hydrogen would remain in the metal structure.

1.3.3 Novel Hydrogen Storage Methods

There are several emerging methods of storing hydrogen that are still under investigation. For example, hydrogen can be physically adsorbed on activated carbon and be *packed* on the surface and inside the carbon structure more densely than if it has been just compressed. Amounts of up to 48 g H_2 per kg of carbon have been reported at 6.0 MPa and 87 K [31]. The adsorption capacity is a function of pressure and temperature; therefore, at higher pressures and/or lower temperatures, even larger amounts of hydrogen can be adsorbed.

For any practical use, relatively low temperatures are needed (less than 100 K). Since the adsorption is a surface process, the adsorption capacity of hydrogen on activated carbon is largely due to the high surface area of the activated carbon, although there are some other carbon properties that affect the ability of activated carbon to adsorb hydrogen. Researchers at the Northeastern University developed a carbon nanotube capable of storing up to 75% of hydrogen by weight [32]. This material is being researched in several laboratories including the National Renewable Energy Laboratory, which is claiming a storage density corresponding to about 10% of the nanotube weight [33]. Hydrogen can also be stored in glass microspheres of approximately 50 μm diameter. The microspheres can be filled with hydrogen by heating them to increase the glass permeability to hydrogen. At room temperature, a pressure of approximately 25 MPa is achieved resulting in storage density of 14% mass fraction and 10 kg H_2/m^3 [34]. At 62 MPa, a bed of glass microspheres can store 20 kg H_2/m^3. The release of hydrogen occurs by reheating the spheres to again increase the permeability. Researchers at the University of Hawaii are investigating hydrogen storage via polyhydride complexes. Complexes have been found that catalyze the reversible hydrogenation of unsaturated hydrocarbons. This catalytic reaction could be the basis for a low-temperature hydrogen storage system with an available hydrogen density greater than 7% [35].

1.3.4 Liquid Hydrogen Storage

Liquid hydrogen's favorable characteristics include its high heating value per unit mass and large cooling capacity due to its high specific heat [36,37]. Liquid hydrogen has some important uses such as in the space program, in high-energy nuclear physics, and in bubble chambers. The transport of hydrogen is vastly more economical when it is in liquid form even though cryogenic refrigeration and special Dewar vessels are required. Although liquid hydrogen can provide a lot of advantages, its uses are restricted in part because liquefying hydrogen by existing conventional methods consumes a large amount of energy (around 30% of its heating value). Liquefying 1 kg of hydrogen in a medium-size plant requires 10–13 kWh of energy (electricity) [37]. In addition, boil-off losses associated with the storage, transportation, and handling of liquid hydrogen can consume up to 40% of its available combustion energy. It is therefore important to search for ways that can improve the efficiency of the liquefiers and diminish the boil-off losses.

1.3.4.1 Hydrogen Liquefaction

The production of liquid hydrogen requires the use of liquefiers that utilize different principles of cooling. In general, hydrogen liquefiers may be classified as conventional, magnetic, or hybrid. Many types of conventional liquefiers exist such as the Linde–Hampson liquefiers, the Linde dual-pressure liquefiers, the Claude liquefiers, the Kapitza liquefiers, the Heylandt liquefiers, and the Collins liquefiers. Conventional liquefiers generally comprise compressors, expanders, heat exchangers, and Joule–Thomson valves. Magnetic liquefiers, on the other hand, utilize the magnetocaloric effect. This effect is based on the principle that some magnetic materials experience a temperature increase upon the application of a magnetic field and a temperature drop upon lifting the magnetic field. The magnetic analog of several conventional liquefiers includes the Brayton liquefiers, the Stirling liquefiers, and the active magnetic regenerative (AMR) liquefier. Additional information on liquid hydrogen production methods can be found in Sherif et al. [39].

1.3.4.2 Liquid Hydrogen Storage

The need for liquid hydrogen is highest in the transportation sector and the space program. Transport of large quantities of hydrogen is usually accomplished by truck tankers of 30–60 m^3 capacity, by rail tank cars of 115 m^3 capacity, and by barge containers of 950 m^3 capacity [39]. Liquid hydrogen storage vessels are usually available in sizes ranging from 1 L Dewar flasks used in laboratory applications to large tanks of 5000 m^3 capacity. The National Aeronautics and Space Administration (NASA) typically uses large tanks of 3800 m^3 capacity (25 m in diameter) [26]. The total boil-off rate from such Dewars is approximately 600,000 LPY (liters per year), which is vented to a burn pond. The contributing mechanisms to boil-off losses in cryogenic hydrogen storage systems are as follows: (1) ortho–para conversion, (2) heat leak (shape and size effect, thermal stratification, thermal overfill, insulation, conduction, radiation, cooldown), (3) sloshing, and (4) flashing. In order to minimize the storage boil-off losses, the conversion rate of ortho-to-para hydrogen should be accelerated with a catalyst that converts the hydrogen during the liquefaction process [40–43]. The use of a catalyst usually results in a larger refrigeration load and consequently in an efficiency penalty primarily because the heat of conversion must be removed. The time for which hydrogen is to be stored usually determines the optimum amount of conversion. For use within a few hours, no conversion is necessary. For example, large-scale use of liquid hydrogen as a fuel for jet aircraft is one of those cases where conversion is not necessary since utilization of the liquid is almost a continuous process and long-term storage is therefore not needed [43]. For some other uses, a partial conversion might be required to create more favorable conditions. It should be noted that for every initial ortho concentration, there exists a unique curve for boil-off of hydrogen with respect to time.

The heat leakage losses are generally proportional to the ratio of surface area to the volume (S/V) of the storage vessel. The most favorable shape is therefore spherical since it has the least S/V ratio. Spherical shape containers have another advantage. They have good mechanical strength since stresses and strains are distributed uniformly. Storage vessels may also be constructed in other shapes such as cylindrical, conical, or any combination of these shapes. Cylindrical vessels are usually required for transportation of liquid hydrogen by trailers or railway cars because of limitations imposed on the maximum allowable diameter of the vessel. For normal highway transportation, the outside diameter of the vessel cannot exceed 2.44 m. From an economics standpoint, cylindrical vessels with either dish, elliptical, or hemispherical heads are very good, and their S/V ratios are only about 10% greater than that of the sphere [44]. Since boil-off losses due to heat leak are proportional to the S/V ratio, the evaporation rate will diminish drastically as the storage tank size is increased. For double-walled, vacuum-insulated, spherical Dewars, boil-off losses are typically 0.3%–0.5% per day for containers having a storage volume of 50 m^3, 0.2% for 103 m^3 tanks, and about 0.06% for 19,000 m^3 tanks [45].

Additional boil-off may be caused by thermal stratification in liquid hydrogen in which case the warmer upper layers evaporate much faster than the bulk liquid [44]. One way of decreasing boil-off losses due to stratification and thermal overfill is by employing high-conductivity plates (conductors) installed vertically in the vessel. The plates produce heat paths of low resistance between the bottom and top of the vessel and can operate most satisfactorily in eliminating temperature gradients and excessive pressures. Another way is to pump the heat out and maintain the liquid at subcooled or saturated conditions. An ideal refrigeration system to perform this task can be an efficient magnetic refrigerator.

The magnetic refrigerator is very suitable for this job because of its relatively higher efficiency, compactness, lower price, and reliability [46–53].

Liquid hydrogen containers are usually of three types: double-jacketed vessels with liquid nitrogen in the outer jacket, superinsulated vessels with either a reflecting powder or multilayer insulation (MLI), and containers with vapor-cooled shields (VCSs) employing superinsulation. Although MLI provides for a low boil-off rate, the addition of a VCS will lower the boil-off losses even further. A VCS is a type of insulation that takes the vapor boil-off and passes it past the tank before being reliquefied or vented. Published data indicate that a reduction of more than 50% in boil-off losses may be achieved for a 100,000 lb liquid hydrogen cryogenic facility with a VCS than without one. Brown [54] showed that locating the VCS at half the distance from the tank to the outer surface of a 4 in. MLI of a 100,000 lb facility would reduce the boil-off by 10%. A dual VCS system on the tank would improve the performance by 40% over a single VCS. Brown [54] also showed that the preferred locations for the inner and outer shields in a dual VCS system are 30% and 66% of the distance from the tank to the outer surface of the MLI.

Another process that leads to boil-off during liquid hydrogen transportation by tankers is sloshing. Sloshing is the motion of liquid in a vessel due to acceleration or deceleration. Due to different types of acceleration and deceleration, there exist different types of sloshing. Acceleration causes the liquid to move to one end and then reflect from that end, thus producing a hydraulic jump. The latter then travels to the other end, thus transforming some of its impact energy to thermal energy. The thermal energy dissipated eventually leads to an increase in the evaporation rate of the liquid [40,44]. The insertion of traverse or antislosh baffles not only restrains the motion of the liquid, thus reducing the impact forces, but also increases the frequency above the natural frequency of the tanker [40,55].

Another source of boil-off is flashing. This problem occurs when liquid hydrogen, at a high pressure (2.4–2.7 atm), is transferred from trucks and railcars to a low-pressure Dewar (1.17 atm). This problem can be reduced if transportation of liquid hydrogen is carried out at atmospheric pressures. Furthermore, some of the low-pressure hydrogen can be captured and reliquefied.

1.3.5 Slush Hydrogen Storage

Slush hydrogen is a mixture of liquid and frozen hydrogen in equilibrium with the gas at the triple point, 13.8 K. The density of the icelike form is about 20% higher than that of the boiling liquid. To obtain the icelike form, one has to remove the heat content of the liquid at 20.3 K until the triple point is reached and then remove the latent heat of fusion. The *cold content* of the icelike form of hydrogen is some 25% higher than that of the saturated vapor at 20.3 K.

Slush hydrogen is of great interest for the space program because of its potential in reducing its physical size and significantly cutting the projected gross liftoff weight [56]. Some of the problems related to liquid hydrogen storage, such as low density, temperature stratification, short holding time due to its low latent heat, hazards associated with high vent rates, and unstable flight conditions caused by sloshing of the liquid in the fuel tank, may be eliminated or reduced if slush hydrogen is used. It is apparent that the use of slush hydrogen should be considered only for cases in which higher density and greater solid content are really needed. This is mainly because the production costs of slush hydrogen now and in the near future are greater than the liquefaction costs of hydrogen due to the larger energy use involved. For additional information on slush hydrogen properties, its production methods, and its applications, the reader is referred to Baker and Matsch [57], Sindt [58], and Voth [59].

1.4 Hydrogen Transport

In a hydrogen economy, it is envisaged that from the production plants and/or storage, hydrogen will be transmitted to consumers by means of underground pipelines (gaseous hydrogen) and/or supertankers (liquid hydrogen). Presently, hydrogen transportation through pipelines is used either in links between nearby production and utilization sites (up to 10 km) or in more extensive networks (roughly 200 km) [60]. Future developments will certainly entail greater flow rates and distances. However, it would be possible to use the existing natural gas pipelines with some modifications. For hydrogen pipelines, it is necessary to use steel less prone to embrittlement by hydrogen under pressure (particularly for very pure hydrogen [>99.5% purity]). Reciprocating compressors used for natural gas can be used for hydrogen without major design modifications. However, special attention must be given to sealing (to avoid hydrogen leaks) and to materials selection for the parts subject to fatigue stress. The use of centrifugal compressors for hydrogen creates more problems due to hydrogen's exceptional lightness.

As a rule, hydrogen transmission through pipelines requires larger-diameter piping and more compression power than natural gas for the same energy throughput. However, due to lower pressure losses in the case of hydrogen, the recompression stations would need to be spaced twice as far apart. In economic terms, most of the studies found that the cost of large-scale transmission of hydrogen is about 1.5–1.8 times that of natural gas transmission. However, transportation of hydrogen over distances greater than 1000 km is more economical than transmission of electricity [61]. Hydrogen in the gas phase is generally transported in pressurized cylindrical vessels (typically at 200 bar) arranged in frames adapted to road transport. The unit capacity of these frames or skids can be as great as 3000 m^3. Hydrogen gas distribution companies also install such frames at the user site to serve as a stationary storage.

1.5 Hydrogen Conversion

Hydrogen as an energy carrier can be converted into useful forms of energy in several ways including combustion in internal combustion engines and jet and rocket engines, combustion with pure oxygen to generate steam, catalytic combustion to generate heat, electrochemical conversion to electricity, and metal hydride conversions.

1.5.1 Combustion in Internal Combustion, Jet, and Rocket Engines

Hydrogen-powered internal combustion engines are on average about 20% more efficient than comparable gasoline engines. The thermal efficiency of an internal combustion engine can be improved by increasing either the compression ratio or the specific heat ratio or both. In hydrogen engines, both ratios are higher than in a comparable gasoline engine due to hydrogen's lower self-ignition temperature and ability to burn in lean mixtures. However, the use of hydrogen in internal combustion engines results in loss of power due to the lower energy content in a stoichiometric mixture in the engine's cylinder. A stoichiometric mixture of gasoline and air, and gaseous hydrogen and air, premixed externally, occupies ~2% and 30% of the cylinder volume, respectively. Under these conditions, the energy of the hydrogen mixture is only 85% of the gasoline mixture, thus resulting in about 15% reduction in power. The power output of a hydrogen engine can be improved by using more advanced fuel injection techniques or liquid hydrogen. For example, if liquid hydrogen is premixed with air, the amount of hydrogen that can be introduced into the combustion chamber can be increased by approximately one-third [62]. One major advantage of hydrogen-powered engines is that they emit far fewer pollutants than comparable gasoline engines. Basically, the only products of hydrogen combustion in air are water vapor and small amounts of nitrogen oxides. Hydrogen has a wide flammability range in air (4%–75% vol.), and therefore high excess air can be utilized more effectively. The formation of nitrogen oxides in hydrogen/air combustion can be minimized with excess air. NO_x emissions can also be lowered by cooling the combustion environment using techniques such as water injection, exhaust gas recirculation, or using liquid hydrogen. The emissions of NO_x in hydrogen engines are typically one order of magnitude smaller than the emissions from comparable gasoline engines primarily because the former run leaner than the latter. Small amounts of unburned hydrocarbons, CO_2, and CO have been detected in hydrogen engines due to lubrication oil [62].

1.5.2 Steam Generation by Hydrogen/Oxygen Combustion

Hydrogen combusted with pure oxygen results in pure steam. This creates temperatures in the flame zone above 3000°C; therefore, additional water has to be injected so that the steam temperature can be regulated at a desired level. Both saturated and superheated vapor can be produced.

A compact hydrogen/oxygen steam generator has been commercially developed by the German Aerospace Research Establishment (DLR) [63]. This generator consists of the ignition, combustion, and evaporation chambers. In the ignition chamber, a combustible mixture of hydrogen and oxygen at a low oxidant/fuel ratio is ignited by means of a spark plug. The rest of the oxygen is added in the combustion chamber to adjust the oxidant/fuel ratio exactly to the stoichiometric one. Water is also injected in the combustion chamber after it has passed through the double walls of the combustion chamber. The evaporation chamber serves to homogenize the steam. The steam's temperature is monitored and controlled. Such a device is close to 100% efficient, since there are no emissions other than steam and little or no thermal losses. Hydrogen steam generators can be used to generate steam for spinning reserve in power plants, for peak-load electricity generation, for industrial steam supply networks, and as a micro steam generator in medical technology and biotechnology [63].

1.5.3 Catalytic Combustion of Hydrogen

Hydrogen and oxygen in the presence of a suitable catalyst may be combined at temperatures significantly lower than flame combustion (from ambient to 500°C). This principle can be used to design catalytic burners and heaters. Catalytic burners require considerably more surface area than conventional flame burners. Therefore, the catalyst is typically dispersed in a porous structure. The reaction rate and resulting temperature are easily controlled by controlling the hydrogen flow rate. The reaction takes place in a reaction zone of the porous catalytic sintered metal cylinders or plates in which hydrogen and oxygen are mixed by diffusion from opposite sides. A combustible mixture is formed only in the reaction zone and assisted with a catalyst (platinum) to burn at low temperatures. The only product of catalytic combustion of hydrogen is water vapor. Due to low temperatures, there are no nitrogen oxides formed. The reaction cannot migrate into the hydrogen supply, since there is no flame and hydrogen concentration is above the higher flammable limit (75%). Possible applications of catalytic burners are in household appliances such as cooking ranges and space heaters. The same principle is also used in hydrogen sensors.

1.5.4 Electrochemical Conversion (Fuel Cells)

Hydrogen can be combined with oxygen without combustion in an electrochemical reaction (reverse electrolysis) and produce direct-current electricity. The device where such a reaction takes place is called a fuel cell. Depending on the type of electrolyte used, there are several types of fuel cells:

- Alkaline fuel cells (AFCs) use concentrated (85 wt.%) KOH as the electrolyte for high-temperature operation (250°C) and less concentrated (35–50 wt.%) for low-temperature operation (<120°C). The electrolyte is retained in a matrix (usually asbestos), and a wide range of electrocatalysts can be used (such as Ni, Ag, metal oxides, and noble metals). This fuel cell is intolerant to CO_2 present in either the fuel or the oxidant [64].
- Polymer electrolyte membrane or proton exchange membrane fuel cells (PEMFCs) use a thin polymer membrane (such as perfluorosulfonated acid polymer) as the electrolyte. The membranes as thin as 12–20 μm have been developed, which are excellent proton conductors. The catalyst is typically platinum with loadings about 0.3 mg/cm^2, or if the hydrogen feed contains minute amounts of CO, Pt–Ru alloys are used. The operating temperature is usually below 100°C, more typically between 60°C and 80°C.
- Phosphoric acid fuel cells (PAFCs) use concentrated phosphoric acid (~100%) as the electrolyte. The matrix used to retain the acid is usually SiC, and the electrocatalyst in both the anode and cathode is platinum black. The operating temperature is typically between 150°C and 220°C [64,65].
- Molten carbonate fuel cells (MCFCs) have the electrolyte composed of a combination of alkali (Li, Na, K) carbonates, which is retained in a ceramic matrix of $LiAlO_2$. Operating temperatures are between 600°C and 700°C where the carbonates form a highly conductive molten salt, with carbonate ions providing ionic conduction. At such high operating temperatures, noble metal catalysts are typically not required [64,65].
- Solid oxide fuel cells (SOFCs) use a solid, nonporous metal oxide, usually Y_2O_3-stabilized ZrO_2, as the electrolyte. The cell operates at 900°C–1000°C where ionic conduction by oxygen ions takes place [64,65].

1.5.5 Energy Conversions Involving Metal Hydrides

Hydrogen's property to form metal hydrides may be used not only for hydrogen storage but also for various energy conversions. When a hydride is formed by the chemical combination of hydrogen with a metal, an element, or an alloy, heat is generated, that is, the process is exothermic. Conversely, in order to release hydrogen from a metal hydride, heat must be supplied. The rate of these reactions increases with an increase in the surface area. Therefore, in general, the hydriding substances are used in powdered form to speed up the reactions. Elements or metals with unfilled shells or subshells are suitable hydriding substances. Metal and hydrogen atoms form chemical compounds by sharing their electrons in the unfilled subshells of the metal atom and the K shells of the hydrogen atoms.

Ideally, for a given temperature, the charging or absorption process and the discharging or desorption process take place at the same constant pressure. However, actually, there is a hysteresis effect, and the pressure is not absolutely constant—for a given temperature, charging pressures are higher than the discharging pressures. The heat generated during the charging process and the heat needed for discharging are functions of the hydriding substance, the hydrogen pressure, and the temperature at which the heat is supplied or extracted. Using different metals and by forming different alloys, different hydriding characteristics can be obtained. In other words, it is possible to make or find hydriding substances that are more suitable for a given application, such as waste heat storage, electricity generation, pumping, hydrogen purification, and isotope separation.

Hydriding substances can be used for electricity storage in two ways. In one of the methods, electricity (direct current) is used to electrolyze the water, and the hydrogen produced is stored in a hydriding substance. When electricity is needed, the hydrogen is released from the hydriding substance by adding heat and used in a fuel cell to produce direct-current electricity. Heat from a fuel cell can be used to release hydrogen from the metal hydride. In the second method, one electrode is covered with a hydriding substance (e.g., titanium–nickel alloy). During the electrolysis of water, the hydriding substance covering the electrode immediately absorbs the hydrogen produced on the surface of the electrode. Then, when electricity is needed, the electrolyzer operates in a reverse mode as a fuel cell producing electricity using the hydrogen released from the metal hydride.

1.6 Hydrogen Safety

Since hydrogen has the smallest molecule, it has a greater tendency to escape through small openings than other liquid or gaseous fuels. Based on properties of hydrogen such as density, viscosity, and diffusion coefficient in air, the propensity of hydrogen to leak through holes or joints of low-pressure fuel lines may be only 1.26–2.8 times faster than a natural gas leak through the same hole (and not 3.8 times faster as frequently assumed based solely on diffusion coefficients). Experiments have indicated that most leaks from residential natural gas lines are laminar [66]. Since natural gas has over three times the energy density per unit volume, the natural gas leak would result in more energy release than a hydrogen leak. For very large leaks from high-pressure storage tanks, the leak rate is limited by the sonic speed. Due to higher sonic velocity (1308 m/s), hydrogen would initially escape much faster than natural gas (sonic velocity of natural gas is 449 m/s).

Some high-strength steels are prone to hydrogen embrittlement. Prolonged exposure to hydrogen, particularly at high temperatures and pressures, can cause the steel to lose strength, eventually leading to failure. However, most other construction, tank, and pipe materials are not prone to hydrogen embrittlement. Therefore, with proper choice of materials, hydrogen embrittlement should not contribute to hydrogen safety risks.

If a leak should occur for whatever reason, hydrogen will disperse much faster than any other fuel, thus reducing the hazard levels. Hydrogen is both more buoyant and more diffusive than gasoline, propane, or natural gas. A hydrogen/air mixture can burn in relatively wide volume ratios, between 4% and 75% of hydrogen in air. The other fuels have much lower flammability ranges, namely, natural gas 5.3%–15%, propane 2.1%–10%, and gasoline 1%–7.8%. However, the range has little practical value. In many actual leak situations, the key parameter that determines if a leak would ignite is the lower flammability limit, and hydrogen's lower flammability limit is four times higher than that of gasoline, 1.9 times higher than that of propane, and slightly lower than that of natural gas. It also has a very low ignition energy (0.02 mJ), about one order of magnitude lower than other fuels. The ignition energy is a function of fuel/air ratio, and for hydrogen, it reaches a minimum at about 25%–30%. At the lower flammability limit, hydrogen ignition energy is comparable with that of natural gas [67]. Its flame velocity is seven times faster than that of natural gas or gasoline. A hydrogen flame would therefore be more likely to progress to a deflagration or even a detonation than other fuels. However, the likelihood of a detonation depends in a complex manner on the exact fuel/air ratio, the temperature, and, particularly, the geometry of the confined space. Hydrogen detonation in the open atmosphere is highly unlikely.

The lower detonability fuel/air ratio for hydrogen is 13%–18%, which is two times higher than that of natural gas and 12 times higher than that of gasoline. Since the lower flammability limit is 4%, an explosion is possible only under the most unusual scenarios; for example, hydrogen would first have to accumulate and reach 13% concentration in a closed space without ignition, and only then an ignition source would have to be triggered. Should an explosion occur, hydrogen has the lowest explosive energy per unit stored energy in the fuel, and a given volume of hydrogen would have 22 times less explosive energy than the same volume filled with gasoline vapor.

A hydrogen flame is nearly invisible, which may be dangerous, because people in the vicinity of a hydrogen flame may not even know there is a fire. This may be remedied by adding some chemicals that will provide the necessary luminosity. The low emissivity of hydrogen flames means that nearby materials and people will be much less likely to ignite and/or be hurt by radiant heat transfer. The fumes and soot from a gasoline fire pose a risk to anyone inhaling the smoke, while hydrogen fires produce only water vapor (unless secondary materials begin to burn).

Liquid hydrogen presents another set of safety issues, such as risk of cold burns and the increased duration of leaked cryogenic fuel. A large spill of liquid hydrogen has some characteristics of a gasoline spill; however, it will dissipate much faster. Another potential danger is a violent explosion of a boiling liquid/expanding vapor in case of a pressure relief valve failure.

In conclusion, hydrogen appears to pose safety risks of the same order of magnitude as gasoline or natural gas. The perception that hydrogen is an unsafe fuel needs to be addressed if a hydrogen economy is to replace the existing fossil fuel–based economy.

1.7 Conclusions

In this chapter, an overview of the hydrogen economy was presented. Many of the aforementioned topics are discussed in much greater detail in the remainder of the handbook.

References

1. Bockris, J.O'M., *Energy: The Solar-Hydrogen Alternative*, Halsted Press, New York, 1975.
2. Bockris, J.O'M. and Veziroglu, T.N., A solar-hydrogen energy system for environmental compatibility, *Environ. Conserv.*, 12(2), 105–118, 1985.
3. Bockris, J.O'M., Veziroglu, T.N., and Smith, D., *Solar Hydrogen Energy: The Power to Save the Earth*, Optima, London, U.K., 1991.
4. Veziroglu, T.N. and Barbir, F., Hydrogen: The wonder fuel, *Int. J. Hydrogen Energy*, 17(6), 391–404, 1992.
5. Goebel, E., Coveney, Jr., R.M., Angino, E.E., Zeller, E.J., and Dreschoff, G.A.M., Geology, composition, isotopes of naturally occurring H_2/N_2 rich gas from wells near Junction City, Kansas, *Oil Gas J.*, 217–222, May 1994.
6. Heydorn, B., *SRI Consulting Chemical Economics Handbook*, SRI, Menlo Park, CA, 1998.
7. Bonner, M., Botts, T., McBreen, J., Mezzina, A., Salzano, F., and Yang, C., Status of advanced electrolytic hydrogen production in the United States and Abroad, *Int. J. Hydrogen Energy*, 9(4), 269–275, 1984.
8. Dutta, S., Technology assessment of advanced electrolytic hydrogen production, *Int. J. Hydrogen Energy*, 15(6), 379–386, 1990.
9. Wendt, H., Water splitting methods, in C.-J. Winter and J. Nitsch (eds.), *Hydrogen as an Energy Carrier*, Springer-Verlag, Berlin/Heidelberg, Germany, pp. 166–238, 1988.
10. Liepa, M.A. and Borhan, A., High-temperature steam electrolysis: Technical and economic evaluation of alternative process designs, *Int. J. Hydrogen Energy*, 11(7), 435–442, 1986.
11. Hancock, Jr., O.G., A photovoltaic-powered water electrolyzer: Its performance and economics, in T.N. Veziroglu and J.B. Taylor (eds.), *Hydrogen Energy Progress V*, Pergamon Press, New York, pp. 335–344, 1984.
12. Carpetis, C., An assessment of electrolytic hydrogen production by means of photovoltaic energy conversion, *Int. J. Hydrogen Energy*, 9(12), 969–992, 1984.
13. Siegel, A. and Schott, T., Optimization of photovoltaic hydrogen production, *Int. J. Hydrogen Energy*, 13(11), 659–678, 1988.
14. Steeb, H., Brinner, A., Bubmann, H., and Seeger, W., Operation experience of a 10 kW PV-electrolysis system in different power matching modes, in T.N. Veziroglu and P.K. Takahashi (eds.), *Hydrogen Energy Progress VIII*, Vol. 2, Pergamon Press, New York, pp. 691–700, 1990.
15. Blank, H. and Szyszka, A., Solar hydrogen demonstration plant in Neunburg vorm Wald, in T.N. Veziroglu, C. Derive, and J. Pottier (eds.), *Hydrogen Energy Progress IX*, Vol. 2, M.C.I., Paris, France, pp. 677–686, 1992.
16. Grasse, W., Oster, F., and Aba-Oud, H., HYSOLAR: The German-Saudi Arabian program on solar hydrogen—5 years of experience, *Int. J. Hydrogen Energy*, 17(1), 1–8, 1992.
17. Lehman, P. and Chamberlain, C.E., Design of a photovoltaic-hydrogen-fuel cell energy system, *Int. J. Hydrogen Energy*, 16(5), 349–352, 1991.
18. Lund, P.D., Optimization of stand-alone photovoltaic system with hydrogen storage for total energy self-sufficiency, *Int. J. Hydrogen Energy*, 16(11), 735–740, 1991.
19. Garcia-Conde, A.G. and Rosa, F., Solar hydrogen production: A Spanish experience, in T.N. Veziroglu, C. Derive, and J. Pottier (eds.), *Hydrogen Energy Progress IX*, Vol. 2, M.C.I., Paris, France, pp. 723–732, 1992.
20. Baykara, S.Z. and Bilgen, E., An overall assessment of hydrogen production by solar water thermolysis, *Int. J. Hydrogen Energy*, 14(12), 881–889, 1989.
21. Engels, H. et al., Thermochemical hydrogen production, *Int. J. Hydrogen Energy*, 12(5), 291–295, 1987.
22. Yalcin, S., A review of nuclear hydrogen production, *Int. J. Hydrogen Energy*, 14(8), 551–561, 1989.
23. Bull, S.R., Hydrogen production by photoprocesses, *Proceedings of International Renewable Energy Conference*, Honolulu, HI, pp. 413–426, 1988.

24. Willner, I. and Steinberger-Willner, B., Solar hydrogen production through photo-biological, photochemical and photoelectrochemical assemblies, *Int. J. Hydrogen Energy*, 13(10), 593–604, 1988.
25. National Hydrogen Association, The hydrogen technology assessment, phase I. A report for NASA, Washington, DC, 1991.
26. Taylor, J.B., Alderson, J.E.A., Kalyanam, K.M., Lyle, A.B., and Phillips, L.A. Technical and economic assessment of methods for the storage of large quantities of hydrogen, *Int. J. Hydrogen Energy*, 11(1), 5–22, 1986.
27. Carpetis, C., Storage, transport and distribution of hydrogen, in C.-J. Winter and J. Nitsch (eds.), *Hydrogen as an Energy Carrier*, Springer-Verlag, Berlin, Germany, pp. 249–289, 1988.
28. Pottier, J.D. and Blondin, E., Mass storage of hydrogen, in Y. Yurum (ed.), *Hydrogen Energy System: Production and Utilization of Hydrogen and Future Aspects*, NATO ASI Series E-295, Kluwer Academic Publishers, Dordrecht, the Netherlands, pp. 167–180, 1995.
29. Mitlitsky, F., Development of an advanced, composite, lightweight, high pressure storage tank for on-board storage of compressed hydrogen, *Proceedings of the Fuel Cells for Transportation TOPTEC: Addressing the Fuel Infrastructure Issue*, Alexandria, VA, SAE, Warrendale, PA, 1996.
30. Veziroglu, T.N., Hydrogen technology for energy needs of human settlements, *Int. J. Hydrogen Energy*, 12(2), 99–129, 1987.
31. Schwartz, J.A. and Amankwah, K.A.G., Hydrogen storage systems, in D.G. Howell (ed.), *The Future of Energy Gases*, U.S. Geological Survey Professional Paper 1570, U.S. Government Printing Office, Washington, DC, pp. 725–736, 1993.
32. Chambers, A., Park, C., Baker, R.T.K., and Rodriguez, N.M., Hydrogen storage in graphite nanofibers, *J. Phys. Chem. B*, 102(22), 4253–4259, 1998.
33. Dillon, A.C., Jones, K.M., and Heben, M.J., Carbon nanotube materials for hydrogen storage, *Proceedings of the 1996 U.S. DOE Hydrogen Program Review*, Vol. II, National Renewable Energy Laboratory, Golden, CO, pp. 747–763, 1996.
34. Rambach, G. and Hendricks, C., Hydrogen transport and storage in engineered glass microspheres, *Proceedings of the 1996 U.S. DOE Hydrogen Program Review*, Vol. II, National Renewable Energy Laboratory, Golden, CO, pp. 765–772, 1996.
35. Jensen, C., Hydrogen storage via polyhydride complexes, *Proceedings of the 1996 U.S. DOE Hydrogen Program Review*, Vol. II, National Renewable Energy Laboratory, Golden, CO, pp. 787–794, 1996.
36. Brewer, G.D., The prospects for liquid hydrogen fueled aircraft, *Int. J. Hydrogen Energy*, 7, 21–41, 1982.
37. Winter, C.J. and Nitsch, J., *Hydrogen as an Energy Carrier*, Springer-Verlag, Berlin, Germany, 1988.
38. Huston, E.L., Liquid and solid storage of hydrogen, *Hydrogen Energy Progress V*, Toronto, Ontario, Canada, 1984.
39. Sherif, S.A., Lordgooei, M., and Syed, M.T., Hydrogen liquefaction, in T.N. Veziroglu (ed.), *Solar Hydrogen Energy System*, Final Technical Report, Clean Energy Research Institute, University of Miami, Coral Gables, FL, pp. C1–C199, August 1989.
40. Hands, B.A., *Cryogenic Engineering*, Academic Press, New York, 1986.
41. Newton, C.L., Hydrogen production, liquefaction and use, *Cryogen. Eng. News*, Part I(8), 50–60, 1967.
42. Newton, C.L., Hydrogen production, liquefaction and use, *Cryogen. Eng. News*, Part II(9), 24–29, 1967.
43. Baker, C.R. and Shaner, R.L., A study of the efficiency of hydrogen liquefaction, *Int. J. Hydrogen Energy*, 3, 321–334, 1978.
44. Scott, R.B., *Cryogenic Engineering*, Van Nostrand Company, Inc., Princeton, NJ, 1962.
45. Ewe, H.H. and Selbach, H.J., The storage of hydrogen, in W.E. Justi (ed.), *A Solar Hydrogen Energy System*, Plenum Press, London, U.K., 1987.

46. Barclay, J.A., Magnetic refrigeration for low-temperature applications, *Proceedings of the Cryocooler Conference*, Boulder, CO, September 1984.
47. Barclay, J.A., Overton, Jr., W.C., and Stewart, W.F., Phase I final report: Magnetic reliquefication of LH_2 storage tank boil-off, Kennedy Space Center, Ref. No. PT-SPD/6037/2511C/000000/04/82, (NASA-Defense Purchase Req. CC-22163B), NASA, Washington, DC, 1983.
48. Barclay, J.A., The theory of an active magnetic regenerative refrigerator, NASA-R-2287, NASA, Washington, DC, 1983.
49. Barclay, J.A. and Steyert, W.A., Materials for magnetic refrigeration between 2 K and 20 K, *Cryogenics*, 22, 73–79, February 1982.
50. Barclay, J.A. and Stewart, W.F., The effect of parasitic refrigeration on the efficiency of magnetic liquefiers, *Proceedings IECEC '82: 17th Intersociety Energy Conversion Engineering Conference*, Los Angeles, CA, p. 1166, August 1982.
51. Barclay, J.A., Use of a ferrofluid as the heat-exchange fluid in a magnetic refrigerator, *J. Appl. Phys.*, 53(4), 2887, 1982.
52. Barclay, J.A. and Sarangi, S., Selection of regenerator geometry for magnetic refrigerator applications, *Proceedings of the ASME Winter Annual Meeting*, New Orleans, LA, December 1984.
53. Barclay, J.A., Stewart, W.F., Overton, W.C., Candler, R.C., and Harkleroad, O.D., Experimental results on a low-temperature magnetic refrigerator, *Cryogenic Engineering Conference*, Boston, MA, August 1985.
54. Brown, N.S., Advanced long term cryogenic storage systems, NASA Marshall Space Flight Center, Huntsville, AL, NASA Conference Publication 2465P, Washington, DC, pp. 7–16, 1986.
55. Barron, R.F., *Cryogenic Systems*, Oxford University Press, Oxford, U.K., 1985.
56. Kandebo, S.W., Researchers explore slush hydrogen as fuel for national aerospace plane, *Aviation Week Space and Technology*, pp. 37–38, June 26, 1989.
57. Baker, C.R. and Matsch, L.C., Production and distribution of liquid hydrogen, *Adv. Pet. Chem. Refin.*, 10, 37–81, 1965.
58. Sindt, C.F., A summary of the characterization study of slush hydrogen, *Cryogenics*, 10, 372–380, October 1970.
59. Voth, R.O., Producing liquid-solid mixtures (slushes) of oxygen or hydrogen using an auger, *Cryogenics*, 25, 511–517, September 1985.
60. Pottier, J.D., Hydrogen transmission for future energy systems, in Y. Yurum (ed.), *Hydrogen Energy System, Utilization of Hydrogen and Future Aspects*, NATO ASI Series E-295, Kluwer Academic Publishers, Dordrecht, the Netherlands, pp. 181–194, 1995.
61. Oney, F., The comparison of pipelines transportation of hydrogen and natural gas, MS thesis, University of Miami, Coral Gables, FL, 1991.
62. Norbeck, J.M., Heffel, J.W., Durbin, T.D., Tabbara, B., Bowden, J.M., and Montano, M.C., *Hydrogen Fuel for Surface Transportation*, SAE, Warrendale, PA, 1996.
63. Sternfeld, H.J. and Heinrich, P., A demonstration plant for the hydrogen/oxygen spinning reserve, *Int. J. Hydrogen Energy*, 14, 703–716, 1989.
64. Kinoshita, K., McLarnon, F.R., and Cairns, E.J., *Fuel Cells: A Handbook*, U.S. Department of Energy, DOE/METC88/6069, Morgantown, WV, 1988.
65. Blumen, L.J.M.J. and Mugerwa, M.N. (eds.), *Fuel Cell Systems*, Plenum Press, New York, 1993.
66. Thomas, C.E., Preliminary hydrogen vehicle safety report, The Ford Motor Company, Contract. No. DE-AC02-94CE50389, U.S. Department of Energy, Washington, DC, 1996.
67. Swain, M.R. and Swain, M.N., A comparison of H_2, CH_4, and C_3H_8 fuel leakage in residential settings, *Int. J. Hydrogen Energy*, 17(10), 807–815, 1992.

Section I

Hydrogen Production Overview

2

Overview of Hydrogen Production

Aldo Steinfeld

ETH Zurich
Paul Scherrer Institute

Pure hydrogen is not found free in nature, but bounded mainly as water and natural gas, and in lower concentration as biomass, fossil fuels, and other hydrogen-containing compounds. Its production from these natural raw materials requires energy-intensive processes. Thus, two crucial aspects are pertinent to the selection of a production methodology, namely, the raw material and the energy source. Both aspects ultimately determine the carbon footprint of hydrogen production.

The production processes can be grouped into electrochemical, thermochemical, photochemical, and their combinations. In electrochemical processes, the energy input is given as electricity, which in turn is generated from various energy sources such as renewables (e.g., wind, solar, geothermal, hydro, biomass) and nonrenewables (e.g., fossil fuels, nuclear). In thermochemical processes, the energy input is given as high-temperature heat, which in turn is generated from various energy sources such as renewables (e.g., concentrated solar thermal, biomass combustion) and nonrenewables (e.g., fossil fuel combustion, nuclear heat). In photochemical processes, the energy input is by the direct absorption of photons of light. Combinations of these energy-intensive processes include photoelectrochemical (e.g., artificial catalytic photosynthesis), photobiological (e.g., photosynthesis by microbial metabolic systems), thermoelectrochemical (e.g., high-temperature electrolysis), and hybrid methods involving multiple steps of the aforementioned processes (e.g., thermochemical and electrochemical splitting cycles).

The carbon footprint of hydrogen production is of utmost importance for its use as a sustainable fuel. The plurality of production processes and associated energy sources is further complicated by the plurality of raw materials. Figure 2.1 lists examples of production processes along with their associated raw materials and energy sources. For the raw materials, the key attribute is the ability to close the material cycle, that is, to regenerate the starting material without emissions discharge to the environment. For the energy sources, the key attribute is renewable or nonrenewable. A most cited example of a production process theoretically approaching zero carbon footprint is the electrolysis of water (Chapter 7) driven by solar, wind, hydropower, or any other renewable electricity source. Competing production processes that also use water as a raw material and a renewable energy resource are solar thermochemical cycles (Chapter 12) and solar photosynthesis (Chapter 13). Thermochemical cycles may also be driven by nuclear process heat (Chapter 8). An example of a production process with a relatively large carbon footprint is the reforming of natural gas driven by heat from fossil fuel combustion (Chapter 3), which presently is the most practiced manufacturing process at an industrial scale because of its economic feasibility. Given the future importance of solid carbonaceous feedstocks such as coal, coke, biomass,

	Processes	Raw Materials	Source of Energy
Electrochemical	Electrolysis	• Water	• Electricity from renewable energy sources (e.g., wind, geothermal, solar, hydro) • Electricity from nonrenewables (e.g., fossil fuels, nuclear)
Thermochemical	Reforming	• Natural gas • Hydrocarbons • + Water	• Combustion of natural gas/syngas • Concentrating solar thermal
	Gasification	• Coal • Carbonaceous materials • Biomass • + Water	• Combustion of coal/biomass/carbonaceous materials/syngas • Concentrating solar thermal
	Decomposition	• Natural gas • Fossil fuel hydrocarbons • Biomethane • Biohydrocarbons	• Natural gas combustion • Concentrating solar thermal
	Thermolysis	• Water	• Concentrating solar thermal
	Thermochemical cycles	• Water	• Concentrating solar thermal • Nuclear heat
Photochemical	Photosynthesis	• Water	• Solar radiation, artificial light
	Photobiological	• Microbial (e.g., algae) • + Water	• Solar radiation

FIGURE 2.1
Production processes, raw materials, and sources of energy.

bitumen, and carbon-containing wastes, gasification technologies are developing rapidly. An example of this category of thermochemical processes with a low carbon footprint is the gasification of biomass driven by heat from biomass combustion (Chapter 5) or from concentrated solar energy (Chapter 12). Alternatively, when coal is the raw material and coal combustion the source of process heat, the large carbon footprint may be mitigated by implementing CO_2 capture technologies (Chapter 4). Ultimately, the carbon footprint of hydrogen production is determined via a life cycle assessment (Chapter 14), which evaluates the environmental burdens associated with the production process by identifying and quantifying the energy and materials used and wastes released to the environment and accounts for the entire life cycle of the process encompassing extraction and processing of raw materials, manufacturing, transportation, distribution, recycling, and final disposal.

Sections II through VI present the fundamental principles of the various hydrogen production processes and their associated technologies and systems. It recognizes that transitioning from a fossil fuel–based economy to a sustainable hydrogen-based economy imposes scientific and technical challenges. Further R&D aimed at near-zero carbon footprint is warranted, while emphasizing the interrelation between the production, storage, and final use of hydrogen as a fuel, along with the social and environmental aspects of energy sustainability.

Section II

Hydrogen Production: Fossil Fuels and Biomass

3

Reformation of Hydrocarbon Fuels

Paul A. Erickson
University of California, Davis

Hong-Yue (Ray) Tang
University of California, Davis

David R. Vernon
Humboldt State University

CONTENTS

3.1 Introduction

Hydrogen is not readily available in nature in the unbound molecular form. Thus, hydrogen is not a primary energy source but, like electricity, is an energy carrier and must be converted from other sources of energy. While hydrogen as an energy carrier has low environmental impact at the point of use, there may be significant impacts from the production and distribution of hydrogen. There are many hydrogen production methods ranging from well-developed industrial processes to emerging pathways in both biological and thermochemical pathways. The most commonly used process to generate hydrogen is reformation of fossil fuels. This chapter will discuss the basic principles and state of the art of steam and autothermal reformation (ATR) processes.

Figure 3.1 shows the energy density of some fuels on both mass and volumetric bases. Note that for Figure 3.1, these values include only the fuel itself and not the tank required to hold such a fuel. The low volumetric energy content of hydrogen and the lack of infrastructure for hydrogen refueling present a significant obstacle for enabling hydrogen-fueled systems for power generation. Producing hydrogen via reformation of a liquid fuel for onboard mobile hydrogen applications shows the potential of using hydrogen technologies while avoiding storage difficulties by allowing the storage and transport of a higher-energy-density liquid. While significant progress has been made recently in hydrogen storage issues, reforming high-energy-density liquid hydrocarbons allows for fast refueling, ambient pressure storage, and potentially higher total energy storage capacity in a given volume. For stationary

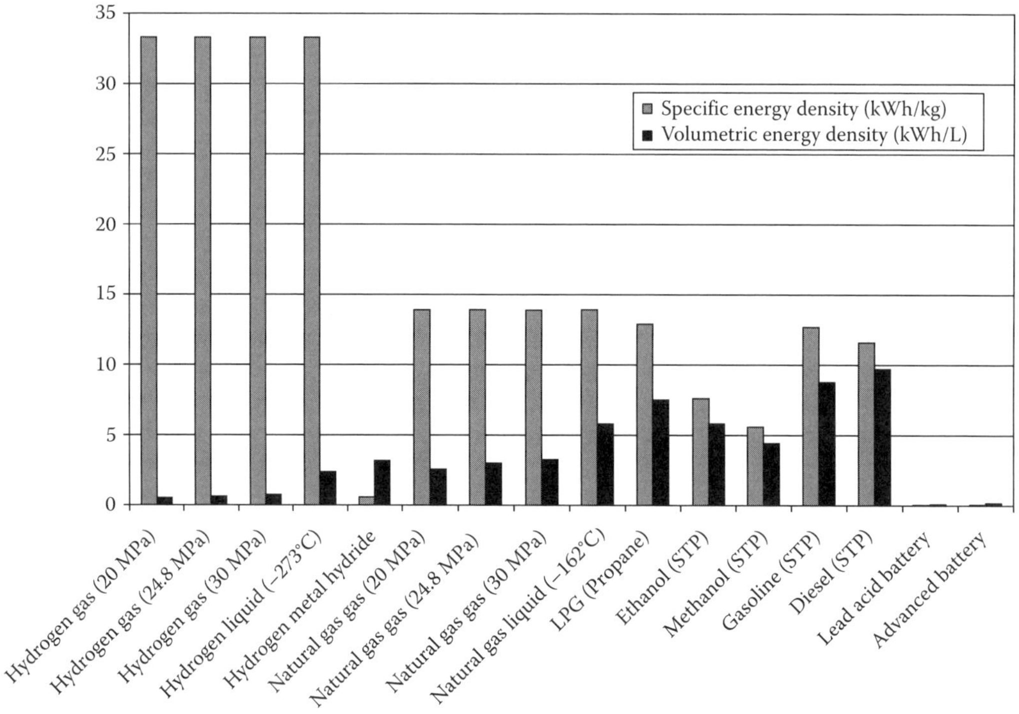

FIGURE 3.1
A comparison of various fuels. Hydrogen has the highest energy per unit weight but the lowest energy per unit volume. Currently, diesel, gasoline, and LPG are common fuel for transportation application.

applications, catalytic reforming of a hydrocarbon feedstock can potentially be combined with carbon dioxide capture and sequestration systems to reduce greenhouse gas emissions.

For about a century, stationary large-scale hydrogen production plants have been providing hydrogen for industrial processes such as for refinery hydrogenation, ammonia production, and gas-to-liquid (GTL) plants, that is, converting methane to methanol.

While large-scale hydrogen production plants are well established, small-scale fuel processors are not as established and are found in the research and development field or in niche markets [1–6]. These potential hydrogen applications include standby and auxiliary power systems, mobile power generation (i.e., auxiliary power units, forklift applications), and stationary distributed power generation.

3.2 What Is Reformation?

Reformation is a chemical process that breaks hydrocarbon molecules into hydrogen from a compounded element to its elemental form. Equation 3.1 represents the typical overall reaction in steam reformation (SR) of a hydrocarbon. In a pure definition, reformation is when the characteristics of a certain fuel

$$C_nH_m + nH_2O \rightarrow nCO + \frac{n+m}{2}H_2 \quad \Delta H\,(298\text{ K}) > 0 \tag{3.1}$$

are reformed into a desirable trait. The SR process shown in Equation 3.1 and other processes like it can be used to change the characteristics of fuels.

Reformation is typically an endothermic reaction due to the hydrogen product having a higher energy state than the hydrocarbon fuel or fuel–water mixture. Thus, energy is needed to volatilize the hydrocarbons and break the C–C and H–C bonds. A heat source, such as an external burner, is used to provide the energy in a typical system. Internal heat sources can also be used by introducing an oxidizer into the fuel–water stream, but this typically will dilute the output stream purity with nitrogen when air is used. A variety of hydrocarbon fuels have been used in reformation studies and have been successfully used in some prototypes and commercial applications: gasoline [7,8], ethanol [4,9], methanol [10–12], dimethyl ether, propane, butane, methane (natural gas), diesel [1,13,14], isooctane, and jet fuel such as JP-8 [15] are some of the examples [5–7,12,16–18]. The reassembly of the elemental hydrocarbon constituents with oxygen, either supplied with the fuel–water mixture or externally added in small amounts, is thermodynamically favored as being exothermic. Thus, with typical reformation of hydrocarbon feedstocks, one has a highly endothermic and rate-limiting fuel steam breakdown followed by a moderately exothermic reassembly through the water–gas shift reaction with the total reaction being endothermic.

Although reformation can proceed unaided by using solely thermal processes at high temperature, the typical reformation system uses a catalyst bed or a series of catalyst beds to shift selectivity toward the desired products (H_2 and CO_2). Inside the fuel processor, the catalyst aids the reformation of the hydrocarbon fuel into a hydrogen-rich gas mixture. The main reformer can be followed by numerous catalytic water–gas shift reactors that bring the H_2 concentrations up and minimize problematic compounds. The conversion, efficiencies, cost, and usable life of the catalysts are strongly dependent on the temperature at which the catalyst is operated; thus, temperature is an important aspect of the control of the fuel processor.

For complex hydrocarbons, dry reforming (or partial oxidation [POX]) is used in prereformers to break apart long-chain hydrocarbons prior to SR. If water or steam is added to the process, then part of the water will be consumed to produce additional hydrogen via the water–gas shift reaction. POX and ATR can use either oxygen or air to produce heat in the overall process. This improves the heat and mass transfer of the catalyst bed by trading off reformate stream purity and results in product dilution when using air as the oxidizer.

3.3 History of Reformation

Historically speaking, reformation is as old as flame itself and is not independent of combustion especially regarding solid or liquid fuel combustion. It is well known that a typical solid or liquid fuel does not burn directly but only in gaseous phase. Thus, the endothermic heating and vaporization as seen with most diffusion flames lead to a local gasification or reformation of the fuel. This gaseous fuel then diffuses until it reaches the oxidizer and then burns to the end products.

Reformation is then an effective short-circuiting of the combustion process and is usually classified in the combustion field as gasification. While it can be argued that reformation has always occurred with combustion and it is not an independent field, reformation as known in literature as a deliberate action had its roots in the late 1700s in solving England's early street lighting problems.

Table 3.1 shows early milestones in the uses of reformation or gasification. In the late 1700s, coal and biomass fuels were not readily applicable to distributed lighting systems because of their traits regarding solid transportability. With the price of whale oil skyrocketing with resource depletion, a crisis in the lighting systems of London was developing. William Murdoch is attributed with devising ways of changing solid fuels such as wood and coal into distributable gaseous fuels by POX or gasification of the solid fuel. One of the early commercial uses of reformation was deriving *town gas* or *syngas* from coal in the early 1800s. This was used for lighting and cooking before natural gas and electricity became more available in the mid-1900s.

TABLE 3.1

Early Milestones of Gasification

Date	Milestone
1792–1794	Scottish scientist William Murdoch produced gas from heating coal for lighting.
1804	Coal gas first patented for lighting.
1813	London and Westminster Gas Light and Coke Company, Great Peter Street, illuminates Westminster Bridge with town gas lights on New Year's Eve using wooden pipes.
1816	Baltimore, Maryland, becomes the first US city to light streets with town gas.
1959	British Gas Council started to replace coal gas with liquid natural gas.

Source: http://www.netl.doe.gov/technologies/coalpower/gasification/basics/3.html.

3.4 Modern Reformation: State of the Art and Technological Barriers

The use of town gas was essentially that of converting a fuel from a solid form to a form that had desirable transport characteristics. Modern reformation also converts liquid, solid, or gaseous fuels into hydrogen that has desirable environmental point-of-use characteristics or allows hydrogen to be used in chemical processing. Reformation can also be used to convert marginally usable low-energy fuels as is found with some biomass sources into a hydrogen gas with benefits in transportability and energy intensity.

The dominant use of hydrogen is in the chemical industry [20]. The primary use of hydrogen is for the production of ammonia and methanol, for hydrotreatment in refineries, and for hydrogenation of unsaturated hydrocarbons. Approximately 90% of the hydrogen used is produced by SR of natural gas and other light refinery hydrocarbons. Other production methods used include gasification and reforming of heavier hydrocarbon fuels, electrolysis, and biological methods.

In a modern reformation facility, natural gas is joined with steam and fed across a catalyst bed. Product gas is collected and fed into various purification steps. The product gas typically carries more chemical energy than the input fuel and steam; thus, the reformation is endothermic, and heat is supplied externally from a combusted fuel.

In a typical methane steam reformer, the reforming process typically requires a temperature from 500°C to 950°C. The reformate, or the product of hydrocarbon reformation, is usually fed into high-temperature and low-temperature water–gas shift reactors to increase the hydrogen concentration and decrease the CO content. The output stream of the water–gas shift reactors is fed into a pressure swing adsorption (PSA) system for purification that gives 99.9% or higher purity. While large-scale hydrogen production has been commercialized for many decades, the current gap to make hydrogen readily available for distributed power generation and transportation applications is dependent on decentralized hydrogen production [21]. Smaller fuel processors for point-of-use application are now actively studied.

In commercial large-scale hydrogen production systems, the steam reformer operates with the catalyst bed temperature typically at 450°C–800°C at 30–45 bar using a Ni-based catalyst held on a ceramic pellet substrate. Cobalt and other noble metals can also be used as catalysts but are generally more expensive. The selection of the catalyst is typically a trade-off in the costs associated with operational life and avoidance of solid carbon (coke) formation, which is also affected by the steam-to-carbon (S/C) ratio. Along with the S/C ratio, other cost factors include flow throughput and pressure drop that reflect operational costs. The catalyst is loaded into high-strength alloy tubes with an outer diameter that ranges from 100 to 150 mm and the length ranges from 10 to 13 m. These tubes must have long creep life and high creep rupture strength due to the high-temperature and high-pressure operation inside the reformer. Central external burners operating at slightly negative gauge pressure provide heat to arrays of steam reformer tubes that can have over 1000°C external wall temperatures. Furnaces holding hundreds of reformer tubes are not unheard of and even small plants might have a 72-tube array in a single furnace. The standard reformer tube is mechanically simple with a header and an orifice at each tube entrance to ensure similar flows in each tube. Temperature for the entire reformate stream is typically monitored by a single temperature sensor at the merged reformate exit. Some operators will monitor tube temperature down the center line of each tube with a multijunction device. Areas of high heat flux can also be noted by observing the dark or cooler regions of the tube from observation points within the furnace. Lifetimes on the

FIGURE 3.2
Georgetown's Test Bed Bus-1 at University of Davis. First demonstrated in April 1994.

order of years with continuous use are expected with the commercial system catalysts and reformer apparatus. Other than methane, various heavy hydrocarbon feedstocks can be used with additional prereforming and suitable catalysts as longer-chain molecules as fuels have the tendency to form carbon at hot spots on the Ni-based catalyst. Ammonia formation and other minor species can also be problematic.

For small-scale hydrogen production, adopting technology from well-understood industrial processes and scale down shows unique challenges [22–24], such as optimizing catalyst operation life and reformer design. In recent years, detailed studies of small-scale reformers for hydrogen production have been published [15,25–27]. Unlike large-scale reformers, small-scale reformers are expected to experience significant transients in normal operation. Frequent start-up and shutdown of reformers can degrade the catalyst and apparatus reducing the life and performance of the system. Reformation systems have been adapted to mobile applications including several iterations of the Georgetown fuel cell buses (1994–2000). The earliest system (1994) is shown in Figure 3.2. Reformation was also used by early fuel cell vehicles including those demonstrated by Daimler (NECAR 3), Toyota (fuel cell electric vehicle [FCEV]), General Motors, and Hyundai motors. An excellent review of small-scale fuel processing for fuel cells and fuel cell vehicles is given in Gunther Kolbs' recent book [28].

3.5 Quantifying Reactor Performance and Parameters

A somewhat standard set of terms and language is used in the reformation industry. Typical parlance includes flow rate in terms of space velocity, stoichiometric terms such as oxygen-to-carbon (O_2/C) ratio and S/C ratio, and output and analysis terms such as conversion, selectivity, and yield.

3.5.1 Flow Rate

Flow rate is typically normalized to an inverse residence time parameter known as space velocity, which can be further distinguished to indicate gaseous species or equivalent liquid species if applicable. Space velocity (sv) is a quasi-nondimensional term that is defined as the reactor volumetric flow rate divided by the reactor volume. This is shown in the following:

$$SV = \frac{\dot{V}}{V} \tag{3.2}$$

where
$\dot{V}$ is the volumetric flow rate
V is the reactor volume

The units are inverse time so as flow rate increases so does space velocity. The units can be done in any time unit but hourly space velocity is commonly used. For gas space velocity, the species temperature and pressure can change the volumetric flow rate at any location in the reactor; thus, nonreacted reactor inlet conditions are often used. Standard temperature and pressure can also be used for normalization. S/C ratio can often change, but as the fuel stream is often the only reactant supplying energy, the space velocity can remove the steam-to-carbon dependence by stating just the volumetric flow rate of the fuel and not the water or steam flow. For fuels that are liquids under standard temperature and pressure, liquid space velocities can be used although the reactants are vaporized and are certainly not reacting in liquid phase. A denotation of LHSV-M would represent a liquid hourly space velocity of methanol, whereas liquid hourly space velocity (LHSV) might be methanol or a premix of methanol and water. GHSV could also be used to represent this flow as gas hourly space velocity at specified inlet conditions and S/C. GHSV-MSTP would mean gas hourly space velocity of methane at standard temperature and pressure, whereas GHSV could mean any or all reactants at standard or another given inlet condition. Another issue with all space velocity terms is the definition of the reactor volume. Those in industry generally prefer to indicate the entire tube including empty locations as it indicates the true length and volume required by the reactor, while those in academic studies typically will only include the volume displaced by the catalyst because the typical research reactor housing is not filled to its full capacity with the catalyst. Those reporting flow values in space velocity should be careful to include a full description of their values to avoid ambiguity or mistranslation of the values.

3.5.2 Stoichiometry

O_2/C ratio and S/C ratio are parameters used to describe the stoichiometry in reformation processes. Those reporting values should clearly define O_2/C ratio nomenclature especially when using air as the oxidizer and/or using fuels with oxygen contained therein (i.e., alcohols). S/C ratio and stoichiometry can also be potentially confused in the O_2/C ratio. Typically, oxygen bound up in compounds will not be counted in the O_2/C ratio, and oxygen is counted in its bound O_2 form and not as an oxygen radical. When using alcohols, it is helpful to note the oxygen–alcohol ratio (i.e., oxygen to methanol [O_2: CH_3OH]) rather than just O_2/C to avoid mistranslation.

3.5.3 Output and Analysis

Outputs of reformation reactors can be equally ambiguous and many output metrics can include all or selected species. Typical output nomenclature includes conversion, selectivity, and yield. These outputs should also be defined carefully for a system as some outputs can be defined by the user and most are specific to a certain reactant or a certain product.

Conversion is defined as the reactant consumed divided by the reactant fed as shown in the following:

$$\text{Conversion} = \frac{\text{Reactant consumed}}{\text{Reactant fed}} \tag{3.3}$$

Conversion is typically given in percent and only indicates how much reactant was consumed. For a single pathway system, this does give insight of the fuel consumption and reactant progression. It does not however give any indication of the product produced. This is an important distinction because in reformation systems, the hydrogen product is typically what is desired. Thus, even though conversion may be high, the hydrogen output could be nonexistent especially with POX reforming or ATR systems that employ an oxidation step. For example, complete combustion implies complete conversion but no hydrogen production.

The selectivity has at least two potential definitions. These are shown in the following:

$$\text{Selectivity} = \frac{\text{Desired product produced}}{\text{Undesired product produced}} \tag{3.4}$$

and

$$\text{Selectivity} = \frac{\text{Desired product produced}}{\text{Reactant consumed}} \times \text{SF} \tag{3.5}$$

where the stoichiometric factor (SF) is used to normalize the selectivity to 100%.

For example, in methanol reformation, the stoichiometric equation is shown in the following:

$$CH_3OH + H_2O \rightarrow CO_2 + 3H_2 \tag{3.6}$$

In this case, if the desired product is hydrogen and the reactant is methanol (CH_3OH), the SF would be 1/3 as 3 mol of hydrogen H_2 is produced per 1 mol CH_3OH reactant. This factor allows normalization to 100%.

The yield shown in the following also uses a similar SF:

$$\text{Yield} = \frac{\text{Desired product produced}}{\text{Reactant fed}} \times \text{SF} \tag{3.7}$$

3.5.4 Reformer Characterization

Space velocity describes the fuel-processing capacity and the volume of the reactor but is insufficient to quantify the properties of the reactor. However inadequate, reactors are often compared using space velocity as a metric. Many researchers speak of a break point in space velocity where the reactor begins to break away from the 100% or previously

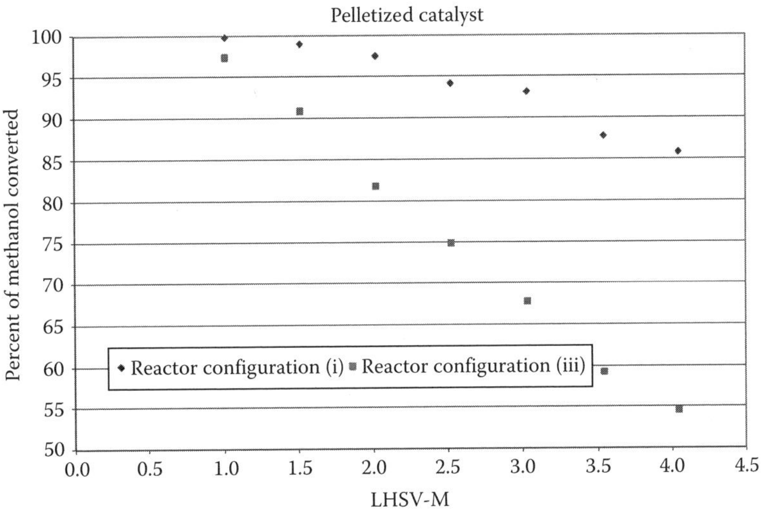

FIGURE 3.3
Two similar reactors with different aspect ratios. (From Davieau, D.D. and Erickson, P.A., *Int. J. Hydrogen Energy*, 32(9), 1192, 2007.)

defined percent conversion or yield. While this is useful to know the limits of performance, the first-order behavior of the reactor is not captured by such characterization. As seen in Figure 3.3 [29], performance can vary greatly for reactors that have the same space velocity. In this figure, the catalyst, flow rate, reactor set point temperature, control scheme, and reactor volumes are identical. The only difference between these data is that the reactor radius of configuration (i) is 0.635 cm (¼ in.) less than that of configuration (iii). This clearly shows that reactor performance cannot be described by space velocity alone.

A better metric is to use the characteristic time constant, which factors additional information such as geometric conditions, heat and mass transfer characteristics, and flow condition and describes the performance directly from the residence time. In addition, the chemical kinetic properties are indirectly captured by the reaction rate coefficient in the residence time distribution (RTD). This is especially important in describing steam reformers as their performances are dependent on the heat transfer characteristic. Since the characteristic time embodies the geometric effect and chemical kinetics, it is more descriptive than using space velocity (SV) [30].

The characteristic time is a modification of the RTD. The overall reformation process based on fuel conversion can be simplified to a first-order reaction. Taking methanol SR as example, using the Arrhenius mechanism, the reaction rate coefficient can be expressed as shown in the following:

$$k(T) = Ae^{E_a/RT} \tag{3.8}$$

Following Fogler's formulation of a segregation model for packed-bed reactors [31], by knowing the reactor's RTD function, $E(t)$, and the reaction rate coefficient from Equation 3.8, the theoretical conversion is shown in the following:

$$\bar{X} = 1 - \int_0^\infty e^{-k(T)t} E(t)\,dt \tag{3.9}$$

By assuming an ideal plug flow reactor (PFR) RTD function as a Dirac function, $E(t) = \partial(t - t_r)$, with an identical residence time t_r for every molecule in the reactor, and substituted into Equation 3.9 yields the following:

$$\bar{X} = 1 - e^{-k(T)t_r} = 1 - e^{\frac{-t_r}{\tau}} \tag{3.10}$$

The first-order characteristic time, τ, represents the time needed to convert 63.2% of the fuel or produce 63.2% of the possible desired product in a particular reactor. Like all first-order systems after five time constants, the reactor should produce near 100% conversion or yield and this characterization captures this behavior. Unlike SV, LHSV-M, or GHSV, characteristic time is descriptive of the actual performance and takes into account catalyst particle size, flow conditions, and other passive enhancements such as baffles and acoustic waves. As characteristic time is based on conversion or yield performance, reactors with the same characteristic time will have the same performance, regardless of temperature, geometry, catalyst, pressure, etc. In other words, given the required characteristic time, there are multiple ways to optimize the reactor design.

3.6 Catalyst Selection

By introducing a catalyst into the reaction, new pathways and acceleration of the reaction occur. An ideal catalyst would not be consumed, but in practice, catalysts do undergo physical and chemical changes. There are three modes of action of catalyst [32]: activity, selectivity, and stability (or degradation behavior).

Activity is a measure of how fast the reaction(s) proceeds in the presence of the catalyst. It is influenced by temperature, concentration of the chemical species, pressure, residence time, and other factors. In reformation, the limiting mechanisms are mass and heat transfer and chemical kinetics and are discussed in a later section. They are fully coupled with one another, and improving the control of the catalyst temperature helps to improve catalyst activity and the overall performance.

The selectivity of the catalyst is the measure of desirable product to the reacted quantity of the feedstock. The realistic selectivity is often less than ideal because of a secondary reaction creating an undesirable by-product that reduces efficiencies. Additional cleanup or removal steps are needed to ensure purity of the output stream. In reformation, CO is undesirable because it can poison the electrode of a proton exchange membrane fuel cell (PEMFC). Carbon formation or coke can also degrade the catalyst, which is the third mode of action of catalysts.

Catalysts help introduce new pathways to reduce the activation energy required to reform a hydrocarbon, provided that the catalyst is active by having sufficient temperature. The reformation is typically an endothermic process; thus, the limiting mechanisms are heat and mass transfer. When a hydrocarbon finds an active site on a catalyst, it breaks down the chemical bonds and reassembles, and it consumes heat and reduces the catalyst and fuel temperature in the process. Heat is added to the system to ensure high catalyst activity to sustain the operation. If the catalyst is highly active, the reformation continues to go forward as long as fuel can find an active site on the catalyst. The porous structure of the catalyst allows fuel to diffuse into the catalyst. After the reactants diffuse into the catalyst and reform, products need to diffuse out from the catalyst to allow the active site to reform the next reactant.

Catalyst degradation is an important issue in catalytic reformation. The chemical, thermal, and mechanical stability of the catalyst determine the operation life of the reactor [32]. Desulfurization is often necessary with most fossil-fuel-based liquids. Care must be taken in controlling the temperature because of the poor heat transfer property of the catalyst or the catalyst bed. A Ni-based catalyst is often used in the high-temperature reformers because it is cheap, active at elevated temperature, and stable. On the other hand, Cu-based catalysts cannot operate at high temperature but have high selectivity toward hydrogen. Care is taken when using a Cu-based catalyst with an exothermic reaction to ensure the integrity of the catalyst. Additionally, other operating conditions such as S:C and O_2:C, where appropriate, should be controlled to minimize carbon formation. The two general types of catalyst substrates are pellets and monolith structures shown in Figure 3.4.

The choice of catalyst depends on the feedstock used. There are wide selections of catalyst formulations for reforming hydrocarbon fuels, and many more are in development. In general, monolith catalysts produce less pressure drop than do pellets. A small amount of catalyst material is washcoated onto a monolith substrate to form the monolithic catalyst. Pellets are usually porous alumina structures with open sites for the catalyst. The advantage of pelletized catalyst over the monolith catalyst is their resistance to poisoning and high amounts of internal surface area.

The reforming catalyst is usually based on nickel/nickel oxide or cobalt composite. They require high temperature to become active and are generally suitable for higher-order hydrocarbon reformation. Lighter hydrocarbons, such as methanol, can be reformed using copper-based medium temperature shift (MTS) [20] catalysts. Many of these formulations used for methanol and ethanol reformation have been used as water–gas shift catalysts for the natural gas reforming industry because of their high selectivity toward hydrogen and CO_2. The catalyst system of copper (Cu) in the presence of zinc oxide (ZnO) supported by alumina (Al_2O_3), derived from industrial catalysts, is the most popular. There have been other catalysts proposed [33–42], but Cu/ZnO/Al_2O_3 remains the primary interest [10,34,43–46]. The low-temperature catalyst allows a less complex heat exchanger and is potentially compatible with phosphoric acid fuel cell (PAFC) or PEMFC applications. Another advantage for reforming at low temperature is the low level of CO formation. The high temperature allows the reverse water–gas shift reaction to consume hydrogen

FIGURE 3.4
Variations of catalysts for steam and ATR. On the left are examples of monolith catalysts. On the right are examples of pelletized catalyst. (Courtesy University of California, Davis, CA.)

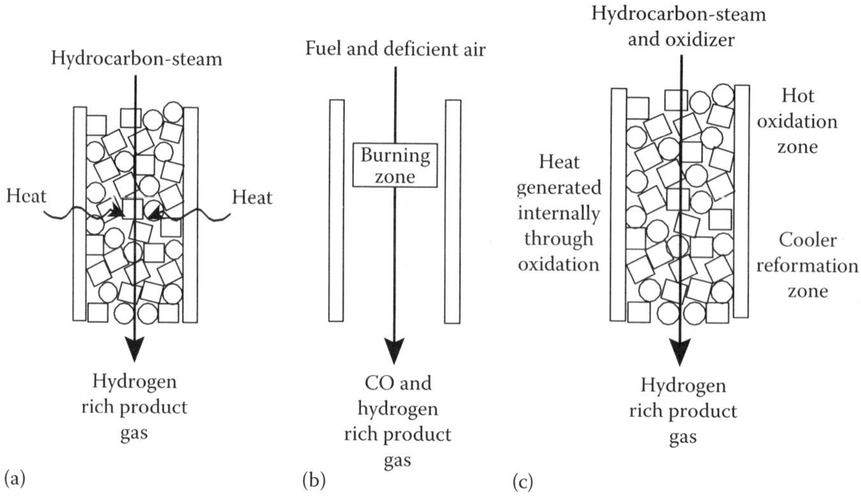

FIGURE 3.5
Graphic representation of (a) SR, (b) POX, and (c) ATR.

even if the catalyst has high selectivity for CO_2 over CO. Other expensive precious metal catalysts are used in high-temperature steam reformers. Rh, Ru, Pt, Pd, and Re supported by porous alumina or magnesium alumina spinel structures are highly active at elevated temperature [47,48]. Current catalyst development is targeting ways to improve selectivity and resistance to degradation by incorporating various oxides and compounds [49], as well as micromachining techniques in fabrication [50–53].

3.7 Types of Reformation

There are currently three major methods for reforming hydrocarbon fuels into hydrogen-rich gas. These are SR, POX reforming, and ATR. Figure 3.5 shows a graphical representation of each process. Today, in large-scale plants, SR is the most common method to obtain hydrogen from a hydrocarbon fuel.

3.8 Steam Reformation

SR is the most widely used method to reform hydrocarbons in large-scale hydrogen production plants. A typical reformer schematic is given in Figure 3.6. This system would be coupled with significantly complex water–gas shift and cleanup devices as well as heat exchangers and preheaters. Specific catalysts are used depending on the hydrocarbon feedstock. The general SR steps are described as fuel breakdown and the water–gas shift reaction and are sometimes followed by methanation as shown in Equations 3.11 through 3.13. All of these reactions can happen at the same physical location, or these may be separated into individual reactors with different catalysts to promote one reaction or another. Higher temperature typically speeds up the rate-limiting steps of fuel breakdown

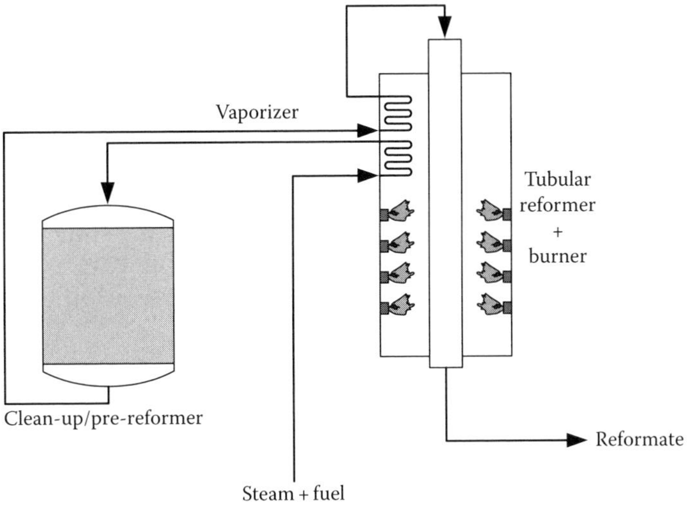

FIGURE 3.6
Schematic of a typical steam reformer.

but higher temperature typically promotes CO and CH_4 production. It is thus typical to have multiple reactors in series each operating at a different temperature with increasing hydrogen purity as the gas progresses through the plant. With heavy hydrocarbons, intermediate species and radicals often are produced and these potentially have carbon or coke formation difficulties:

$$C_nH_m + n\ H_2O \rightarrow n\ CO + \frac{n+m}{2}H_2 \quad \Delta H\ (298\ K) > 0 \tag{3.11}$$

$$CO + H_2O \leftrightarrow CO_2 + H_2 \quad \Delta H\ (298\ K) = -41\ kJ/mol \tag{3.12}$$

$$CO + 3H_2 \rightarrow CH_4 + H_2O \quad \Delta H\ (298\ K) = -206.2\ kJ/mol \tag{3.13}$$

Many hydrocarbons have been used as feedstock for SR. To illustrate some of the key parameters in SR, methanol and ethanol will be described here as examples. The SR kinetics of methanol on a Cu-based catalyst are available in the literature [10,36,54,55]. The exact mechanism of SR is still debated [10,56,57]. As commonly understood, different catalyst formulations promote the reaction in favor of certain pathways that is manifested as selectivity. However complex in application, a simplified reformation process can be expressed for methanol as the overall reaction consisting of methanol decomposition and the water–gas shift reaction. These are shown in Equations 3.4 through 3.6:

$$CH_3OH + H_2O \rightarrow CO_2 + 3H_2 \quad \Delta H\ (298\ K) = +49\ kJ/mol \tag{3.14}$$

$$CH_3OH \rightarrow CO + 2H_2 \quad \Delta H\ (298\ K) = +90.1\ kJ/mol \tag{3.15}$$

$$CO + H_2O \leftrightarrow CO_2 + H_2 \quad \Delta H\ (298\ K) = -41\ kJ/mol \tag{3.16}$$

While the overall reaction is endothermic, the methanol decomposition reaction is endothermic and the faster water–gas shift reaction is exothermic. For methanol SR, 75% dry hydrogen concentration can be produced based on 100% carbon dioxide selectivity as shown in the following:

$$\%H_2 \text{ selectivity} = \frac{\text{mole } H_2}{\text{mole of } H_2 + \text{mole of } CO_2} = \frac{3 \text{ mol}}{3 \text{ mol} + 1 \text{ mol}} = 75\% \quad (3.17)$$

In practice, dry hydrogen concentration is less than 75%. The conversion is strongly dependent on temperature and hence the activity and selectivity of the catalyst used. For a reformer operating at less than ~250 PSI, the pressure effect has been found as less significant [30]. Higher-pressure operation typically allows for higher throughput and better integration with either feedstock or end use such as high-pressure hydrocracking. It should be noted that Equation 3.16 also has a significant backward component that can form CO from the desired products. In the reverse water–gas shift reaction, hydrogen may be consumed to produce CO if the reactor temperature is too high. CO is typically an undesirable by-product when considering operation of a fuel cell system. On the other hand, sufficient heat is needed to obtain high fuel conversion and to assist the forward reaction considering fuel breakdown.

Another key parameter in methanol SR is the S/C ratio (H_2O/CH_3OH, in the case of methanol) [54]. This parameter impacts fuel utilization, efficiencies, and life of the reactor. Sufficient steam is required to achieve full conversion and suppress CO and solid carbon formation. Excessive steam will reduce energy efficiencies because it must be vaporized with the fuel without adding to the hydrogen product. High-temperature steam may also sinter Cu-based catalysts. Insufficient steam results in carbon formation and degraded catalyst performance by means of coking. Experimental work by many researchers has found that an S/C ratio between 1.3 and 1.6 for methanol steam reforming results in higher dry hydrogen yield at the reformer outlet.

Ethanol SR can be carried out with a Ni-based catalyst at about 550°C–650°C. Relatively higher temperature is needed to break the carbon–carbon bond in the fuel. The use of ethanol can favor carbon formation inside the reactor that potentially will degrade catalyst performance. Solid carbon fouls the catalyst by blocking fuel from contacting the active site inside the catalyst structure. The effect of fouling is more pronounced at low temperature or low S/C. It has been reported that an S/C ratio > 1.5 reduces carbon formation but added water has a trade-off in the energy needed for fuel vaporization [58]. In this respect, the ethanol process is quite similar to methanol SR. The following equations show the simplified reactions:

$$CH_3CH_2OH + 3H_2O \rightarrow 2CO_2 + 6H_2 \quad \Delta H\ (298\text{ K}) = 347\text{ kJ/mol} \quad (3.18)$$

$$CH_3CH_2OH + H_2O \rightarrow 2CO + 4H_2 \quad \Delta H\ (298\text{ K}) = 298\text{ kJ/mol} \quad (3.19)$$

$$CO + H_2O \leftrightarrow CO_2 + H_2 \quad \Delta H\ (298\text{ K}) = -41\text{ kJ/mol} \quad (3.20)$$

3.9 Partial Oxidation

POX is an alternative to SR and is generally employed with higher hydrocarbons or if pure oxygen is available [59]. With a lower product concentration of hydrogen, this process sacrifices some efficiency relative to SR but offers rapid dynamic response and compactness. Unfortunately, this process is susceptible to coke formation and must be carried out at high temperatures. POX can be performed with or without a catalyst, but using a catalyst allows for a lower reaction temperature. The POX of methane is described by the chemical reaction shown in the following:

$$CH_4 + \frac{1}{2}O_2 \rightarrow CO + 2H_2 \quad (\Delta H = -36\ \text{kJ/mol}) \tag{3.21}$$

If the oxygen-to-fuel ratio is increased, then the reaction becomes even more exothermic as shown in the following:

$$CH_4 + O_2 \rightarrow CO_2 + 2H_2 \quad (\Delta H = -319\ \text{kJ/mol}) \tag{3.22}$$

These equations illustrate how the amount of heat generated from the POX of methane can be quickly increased from −36 to −319 kJ/mol by simply increasing the amount of oxygen in the reaction, that is, increasing the air flow rate and combusting additional fuel. Therefore, it is possible to reduce reformer start-up times by increasing the temperature rapidly; that is achieved by increasing the air (or oxidant)-to-fuel ratio.

The fast response time of the POX reformer is typically inhibited by the low hydrogen concentration and high CO levels that result. This implies a lagging hydrogen response time because of the potentially necessary water–gas shift and other cleanup systems put in place to clean up the high levels of CO that result from the POX reformer.

POX and catalytic partial oxidation (CPO or CPOX) is similar to a combustion process. With catalyst, CPOX can be carried out at lower temperature. Without catalyst, POX is typically carried out at 1300°C–1500°C to ensure complete conversion. As compared to SR, CPOX has the advantage of short start-up time because of the fast exothermic nature of the reaction. On the other hand, the exothermic nature of the reaction and the heat transfer limitation within the catalyst make it difficult to control the catalyst temperature and reformate composition. Localized hot spots can overheat and sinter the catalyst if present. The product stream is influenced by the process conditions. Temperature, pressure, and O_2/C ratio are some of the process variables. For example, increasing reformer pressure will reduce the hydrogen yield in POX. A POX or CPOX reformer is comparatively compact because it doesn't require external heating to sustain the reaction and lowers the energy required to vaporize the fuel, but it is relatively less efficient than SR and ATR. Experiments on CPOX of various liquid hydrocarbon fuels have been reported by Cheekatamarla [60].

A general equation for POX is shown in the following:

$$C_nH_m + nO_2 \rightarrow nCO + \frac{m}{2}H_2 \quad \Delta H\ (298\ \text{K}) < 0\ \text{kJ/mol} \tag{3.23}$$

POX and CPOX have complex reaction systems. The reaction produces various intermediate species and radicals that decrease the purity of the output stream. They are also potentially coke precursors that cause catalyst deactivation for both precious and nonprecious metal catalysts. Pure oxygen and air have been used as oxidizer in these systems. However, the effect of inert gases such as nitrogen in the POX process can reduce the efficiency of the reformer by lowering the output stream hydrogen concentration. In general, other than O_2/C ratio, there is no control over the chemical species in POX or CPOX. Equations 3.24 and 3.25 show the POX of ethanol, and Equation 3.26 shows the complete oxidation:

$$CH_3CH_2OH + 0.5O_2 \rightarrow 2CO + 3H_2 \quad \Delta H\ (298\ \text{K}) = 57\ \text{kJ/mol} \tag{3.24}$$

$$CH_3CH_2OH + O_2 \rightarrow CO_2 + CO + 3H_2 \quad \Delta H\ (298\ \text{K}) = -226\ \text{kJ/mol} \tag{3.25}$$

$$CH_3CH_2OH + 1.5O_2 \rightarrow 2CO_2 + 3H_2 \quad \Delta H\ (298\ \text{K}) = -509\ \text{kJ/mol} \tag{3.26}$$

In POX or CPOX, CO is one of the products in the reformate stream. This is especially problematic for low-temperature fuel cell applications because the high concentration of CO will poison the electrode. An additional CO oxidation reactor, water–gas shift reactor, or preferential oxidation (PROX) reactor can be used for CO removal. On the other hand, with high-temperature fuel cells, solid oxide fuel cell (SOFC) can utilize CO as a fuel source and it is less sensitive to impurities. Moreover, SOFC operation temperature is near 800°C, which is the typical POX reformate stream temperature. This makes POX and SOFC a good combination for fuel cell power system integration. It should be noted that it is not typically economical to use pure oxygen as oxidizer; thus, air is used.

3.10 Autothermal Reforming

ATR is essentially a combination between SR and CPOX. This is done by bringing the two reforming reactions into close thermal contact or by placing them into a single catalytic reactor. The single catalytic reactor is the most efficient means of heat transfer. ATR has advantages of both SR and POX in that it has potentially high hydrogen product concentration and adequate response to dynamic loads. Ideally, the heat generated from the exothermic POX reaction is used for rapid start-up and supplying the heat needed for the endothermic SR reaction during operation. Once the reactor is at operating temperature, the fuel, steam, and air are all fed into the reactor in the same step. The reactants ignite and form the ideal products of hydrogen and carbon dioxide. It has been found that on noble metal–based catalysts, ATR generally follows equilibrium concentrations in the output gas based on reaction temperature.

With a higher temperature due to the oxidation step, ATR is also capable of reforming multiple fuels, a necessary characteristic if alternative hydrocarbon feedstocks are reformed. Liquid fuels like methanol produced from coal or biomass may contain higher hydrocarbons. For example, at a pulp mill under investigation for hydrogen production from waste, it was found that small amounts of pinenes existed in the methanol fuel used.

In SR systems, the pinenes and other compounds tend to overwhelm any catalyst site until they are reacted. The higher temperatures and oxidation found in ATR may allow faster treatment of the trace levels of higher hydrocarbons. Although the pinenes are relatively small in quantity (about one-tenth of a percent volumetrically for this fuel), their effect on the catalyst can be significant. Similar studies found that trace amounts of heavier oils as found with methanol derived from coal had similar effects in SR [61,62] but could be acceptably dealt with in ATR [63].

ATR is similar to SR with an additional CPOX step. These exothermic steps are fast, and the resulting heat can be used to sustain the SR steps; thus, ATR is termed *autothermal* or thermal neutral. The typical ATR equation of a hydrocarbon is shown in the following:

$$C_nH_m + 0.5mO_2 + 0.5mH_2O \rightarrow mCO + (0.5m + 0.5n)H_2 \quad \Delta H\ (298\ \mathrm{K}) = 0\ \mathrm{kJ/mol} \qquad (3.27)$$

The steps in an ATR process of using methanol as feedstock are shown in Equations 3.28 through 3.33. Equation 3.28 shows the CPOX steps, and combined with the carbon monoxide oxidation step in Equation 3.29, much of the CO will be consumed in these steps:

$$CH_3OH + 0.5O_2 \rightarrow CO + 2H_2 \quad \Delta H\ (298\ \mathrm{K}) = -192\ \mathrm{kJ/mol} \qquad (3.28)$$

$$CO + 0.5O_2 \rightarrow CO_2 \quad \Delta H\ (298\ \mathrm{K}) = -283\ \mathrm{kJ/mol} \qquad (3.29)$$

Another CPOX formulation is by substituting combustion/methanol oxidation steps shown in Equation 3.30. The remaining steps in Equations 3.31 through 3.33 are identical SR steps from the previous section:

$$CH_3OH + 1.5O_2 \rightarrow CO_2 + 2H_2O \quad \Delta H(298\ \mathrm{K}) = -675.4\ \mathrm{kJ/mol} \qquad (3.30)$$

$$CH_3OH + H_2O \rightarrow CO_2 + 3H_2 \quad \Delta H\ (298\ \mathrm{K}) = 50\ \mathrm{kJ/mol} \qquad (3.31)$$

$$CH_3OH \rightarrow CO + 2H_2 \quad \Delta H\ (298\ \mathrm{K}) = 90.1\ \mathrm{kJ/mol} \qquad (3.32)$$

$$CO + H_2O \leftrightarrow CO_2 + H_2 \quad \Delta H\ (298\ \mathrm{K}) = -41\ \mathrm{kJ/mol} \qquad (3.33)$$

The amount of heat generated in the reaction is directly related to the available oxygen in the fuel. If the proper stoichiometry fuel mixture is used, it will result in a thermoneutral condition: a self-sustaining operation as shown in the following:

$$CH_3OH + 0.1O_2 + 0.8H_2O \rightarrow CO_2 + 2.8H_2 \quad \Delta H\ (298\ \mathrm{K}) = 0\ \mathrm{kJ/mol} \qquad (3.34)$$

However, in a typical reactor, heat lost due to conduction is unavoidable. It is necessary to allow additional oxygen to account for this heat lost, as shown in the following:

$$CH_3OH + 0.27O_2 + 1.5H_2O \rightarrow CO_2 + 2.46H_2 + 1.04H_2O \quad \Delta H\ (298\ \mathrm{K}) = -81\ \mathrm{kJ/mol} \qquad (3.35)$$

The O_2/C ratio, or O_2/C, is an important parameter in ATR. Researchers have found O_2/C = 0.2–0.3 to be optimum using a Cu-based catalyst [54,64]. Higher O_2/C ratios will reduce the amount of hydrogen in the output stream and increase the temperature inside the reactor, while lower O_2/C ratios will result in low conversion due to insufficient heat.

The steps in an ATR process of using ethanol as feedstock are described by Equations 3.18 through 3.20, 3.24, and 3.25. The case of complete oxidation of ethanol is described in the following:

$$CH_3CH_2OH + 3O_2 \rightarrow 2CO_2 + 3H_2O \quad \Delta H\ (298\ K) = -1368\ kJ/mol \tag{3.36}$$

In a small fuel processor design, an ATR reactor is compact and has higher heat and power efficiencies when compared to a steam reformer reactor. A typical schematic of the ATR reactor is found in Figure 3.7. The exothermic reactions take place on the surface of the catalyst; thus, they help shorten the warm-up time of the reactor. However, air is typically used to feed into the autothermal reformer; thus, nitrogen dilution will reduce the thermal efficiency and concentration of hydrogen in the reformate output stream. The fast start-up feature has been studied [65,66] for the potential use on fuel cell vehicle [67].

ATR also suffers from mass and heat transfer limitations as in SR but in a different manner. The rate of the exothermic reaction occurring on the catalyst surface is dependent on the rate of reactants mass transfer to and away from active catalyst sites [68,69]. In the case of reforming methanol on Cu-based catalyst, the rate of exothermic reaction is at least two orders of magnitude faster than the endothermic reaction [66,70–72], the heat generated is localized, and a hot spot is created by the heat transfer limitations of the catalyst. This is common in catalytic combustion and often degrades the catalyst by means of sintering. However, a well-designed ATR can take advantage of the exothermic reaction to overcome the heat transfer limitation of the SR step. This reduces the size of the overall reformer while maintaining high H_2 concentration. In addition, the nonlinear relationship between the reaction rate and catalyst temperature has a sharp transition known as the light-off

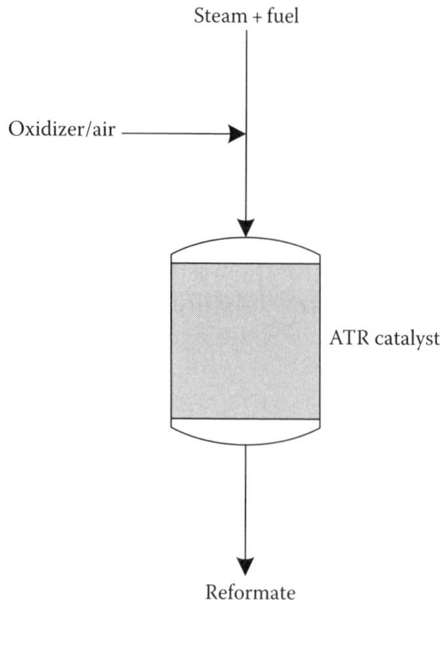

FIGURE 3.7
Schematic of a typical autothermal reformer.

temperature. Operating below the light-off temperature will result in extinction in the combustion process because it is unable to sustain the reaction [31].

The light-off temperature has the following implications. First, selection of a proper O_2/C ratio is based on both stoichiometry and reactor design. Not only sufficient oxygen is needed to sustain the reaction, but it must have excess oxidizer to account for heat lost by the reactor design. Balancing between S/C and O_2/C ratios for a specific catalyst selection can be difficult [65]. Second, if heat lost is significant, excessive oxygen required to sustain the reaction will potentially sinter the catalyst by overheating away from the heat sink. Poor heat transfer of the catalyst will create large temperature gradients, thus degrading (and possibly destroying) the catalyst. Various autothermal methanol reforming reactors have been studied [73,74].

3.11 ATR of Different Fuels

Gasoline and heavy hydrocarbons are typically considered for ATR to produce hydrogen for fuel cells. The advantages of gasoline are its existing fuel infrastructure and high energy density (32.3 MJ/L, based on LHV). However, using gasoline or other fossil petroleum products will not solve the long-term problem of fossil fuel dependence. For theoretical calculations, the optimal temperature to reform gasoline into hydrogen is around 400°C [75]. However, in other studies of gasoline and diesel ATR, temperatures of 600°C–800°C have been used [76,77]. From Sandakane, Saitoh, and Oyama's research [78], a fuel conversion of greater than 97% can be obtained at temperatures in excess of 800°C. High temperatures require expensive materials and longer start-up times for automotive scale reformer systems. The high temperatures required for ATR of gasoline and other higher hydrocarbons make them less than optimal for automotive applications.

Although most work is done on ATR with higher hydrocarbons, focusing on renewably produced alcohols provides a basic understanding of the process. The general form of the ATR reaction of an alcohol fuel using air as the oxidizer, assuming complete conversion of the reactants, is shown in the following:

$$C_nH_mO_p + (2n-2x-p)H_2O + x\left(O_2 + 3.76N_2\right) \rightarrow (2n-2x-p+m/2)H_2 + nCO_2 + 3.76xN_2 \quad (3.37)$$

where $(2n - 2x - p)$ represents the minimum amount of water required in the reaction, and x represents the air-to-fuel ratio. The air-to-fuel ratio and O_2/C are related by a factor of n, the number of carbon atoms in the fuel. The heat of reaction is dependent on the O_2/C and S/C. Note that $(2n - 2x - p)$ is the minimum amount of water required for the reaction and the S/C ratio in practical applications is typically much higher in order to inhibit coke formation.

Methanol is a very attractive fuel for hydrogen production. Like hydrogen, methanol can be produced from multiple feedstocks. Ideally, methanol would be produced renewably, but it may also be produced from abundant coal resources and natural gas. Because it is an oxygenated fuel, it requires lower operating temperatures for effective reformation [75]. Ethanol and higher alcohols have longer carbon chains (two or more carbon atoms) and therefore require higher temperatures and more energy to be reformed. Methanol can be reformed at relatively low temperatures, around 250°C. According to one ATR study, conversions of greater than 90% are possible at temperatures just above 250°C, [79]. Based on thermodynamic equilibrium, the lower the reaction temperature, the lower the CO concentration that can be achieved [80]. This translates into less reformate cleanup and, therefore, a smaller and less complicated fuel-processing system. The lower temperatures also mean that less energy is required to heat the reforming system to operating temperature, so shorter start-up times are possible. It also means that the reformer can be made from a larger variety of materials that potentially reduces manufacturing costs.

In addition to the ease of reformation, methanol can be produced renewably through gasification of biomass or from coal resources. The different feedstocks of methanol result in different methanol purities. For example, coal-derived methanol may include higher hydrocarbons as an impurity [61]. Fortunately, ATR is a proven method of reforming higher hydrocarbons and should be able to cope with varying purities of fuel, depending on the feedstock [62].

The heat of reaction indicates whether a reaction is exothermic (releasing energy) or endothermic (requiring energy) and is defined as the heat of formation of the products minus the heat of formation of the reactants. Since the heat of formation for the oxygen, nitrogen, and hydrogen reactants and products are all zero, the heat of reaction for methanol ATR is simplified as shown in Equation 3.38:

$$\Delta H_r = \Delta H_{f,CO_2} - (1-2x)\Delta H_{f,H_2O(l)} - \Delta H_{f,fuel(l)} \tag{3.38}$$

By evaluating the heat of reaction as a function of O_2/C, x, it is possible to find x_0, the thermoneutral point that produces a net enthalpy change of zero. Plotting the heat of reaction for methanol as a function of x, from $x = 0$ (SR) to $x = 1.5$ (complete combustion), and assuming an ideal reaction, yields Figure 3.8. The stoichiometry of the reaction is simply an $O_2/C = x$ and $S/C = 1–2x$, until $x > 0.5$. At this point, water is no longer consumed in the reaction but rather produced as a product of combustion. It is easy to pinpoint the thermoneutral point that occurs at $x_0 = 0.230$.

The efficiency of a reforming process is defined as the lower heating value of hydrogen produced divided by the lower heating value of fuel consumed as shown in Equation 3.38:

$$\eta = \frac{\text{LHV H}_{2\text{output}}}{\text{LHV fuel}_{\text{input}}} \tag{3.39}$$

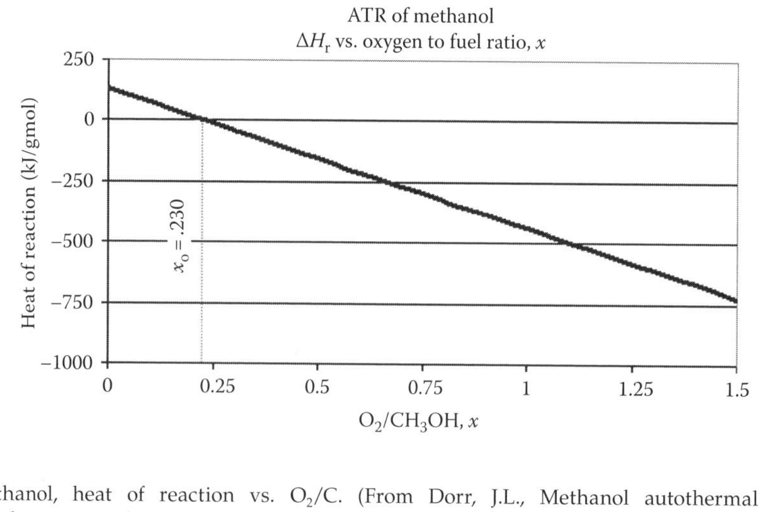

FIGURE 3.8
ATR of methanol, heat of reaction vs. O_2/C. (From Dorr, J.L., Methanol autothermal reformation: Oxygen-to-carbon ratio and reaction progression, Master thesis, University of California, Davis, CA, 2004.)

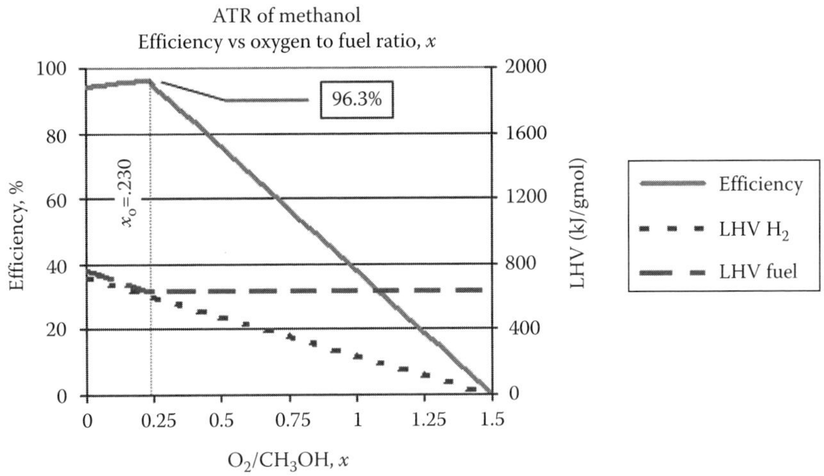

FIGURE 3.9
ATR of methanol, efficiency vs. O_2/C. (From Dorr, J.L., Methanol autothermal reformation: Oxygen-to-carbon ratio and reaction progression, Master thesis, University of California, Davis, CA, 2004.)

Further evaluation of this simple model reveals that the peak efficiency of ATR of methanol occurs at the thermoneutral point, as seen in Figure 3.9. The peak efficiency of ATR of methanol at the thermoneutral point $x_0 = 0.230$ is 96.3%, which is one of the highest theoretical efficiencies of ATR of various hydrocarbon fuels [80]. Therefore, it seems most desirable to operate as close to the thermoneutral point as possible if efficiency is a priority.

It is important to note that below the thermoneutral point ($x < x_0$), the reaction is endothermic and thus requires a heat input. This heat input is assumed to be provided by combusting additional fuel. Also, the amount of energy required to heat the reactants is not included in this efficiency. The amount of excess water in the reaction does play a great role in the amount of heat required to increase the temperature of the reactants. It is also possible to utilize waste heat from the fuel cell, which would effectively increase the efficiency at lower O_2/C ratios, moving the balance toward an SR reaction.

To maximize the overall benefit of hydrogen, renewable energy sources are desired as feedstock. Methanol, although not currently derived from renewable sources but from natural gas, has the potential to be sourced from renewables. Methanol is an attractive choice for many reasons including the basis of carbon-to-hydrogen ratio, fuel processor start-up and operation energies, availability of biorenewable sources, and overall system complexity [62,82–85]. It has been considered as one of the possible feedstocks for hydrogen fuel cells as it can be stored and transported in a liquid form using the existing energy infrastructure and technology with only slight modification [4,9]. Indeed presently, some methanol refueling stations exist for supplying racing fuel across the United States. Methanol can be reformed at relatively low temperatures as compared to other fuels, and the reaction mechanism for carbon formation is less active, which helps to prolong catalyst operation life. An extensive review of methanol SR was done by Palo et al [86]. Ethanol is another promising fuel candidate based on its potential to be carbon neutral [87–90]. Although much debate surrounds the actual carbon neutrality of the fuel, ethanol can be produced through the fermentation of biomass or organic waste materials from agroindustries, forestry residue, and municipal solid waste. Bioethanol is recognized by the automobile industries as an alternative fuel with near-established infrastructure. Hydrogen from ethanol is likewise considered potentially beneficial.

3.12 Limiting Mechanisms in the Reformation Processes

The conversion of fuel and yield of hydrogen are limited by the presence of the physical mechanisms. Heat and mass transfer and chemical kinetics are the major limiting mechanisms in reformation. The effect of these mechanisms is discussed individually in the following sections. A conventional steam reformer has temperature and concentration gradients inside the catalyst bed. Efforts to improve the reactors limiting mechanisms can have significant cost and flow through improvements.

3.12.1 Chemical Kinetics

Catalyst temperature impacts the catalyst activity. Arrhenius behavior as shown in Figure 3.10 for a methanol SR catalyst indicates that activity is exponential with temperature. As typical reformation systems are endothermic, the presence of activity limits is an indication of insufficient heat transfer.

Increasing activity through increased temperature can also damage the catalyst; thus, rugged low activity catalysts may be desired. For example, Ni-based catalysts are typically most active above 800°C. This high-temperature requirement makes a Ni-based catalyst bed difficult to implement for mobile devices because of the long start-up times required. Cu-based catalysts are attractive because they are active at about 260°C. Since the conventional catalyst effectiveness factor is typically less than 5%, being able to control the catalyst bed temperature can potentially improve the overall performance.

3.12.2 Mass Transfer

SR is typically limited by mass and heat transfer [91]; thus, low-cost catalysts are used. Mass transfer includes both external and internal diffusions. Reactant diffusion through the catalyst bed onto a catalyst is known as the external diffusion. Diffusion inside the catalyst pore onto an active site is known as internal diffusion. To improve diffusion, one can increase the catalyst loading and/or use smaller catalyst pellets. Using crushed catalysts is a common practice although this typically increases limitations in heat transfer and

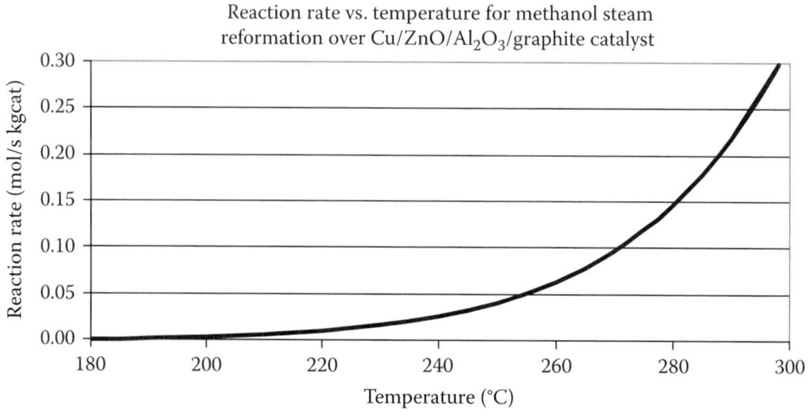

FIGURE 3.10
Reaction rate versus temperature.

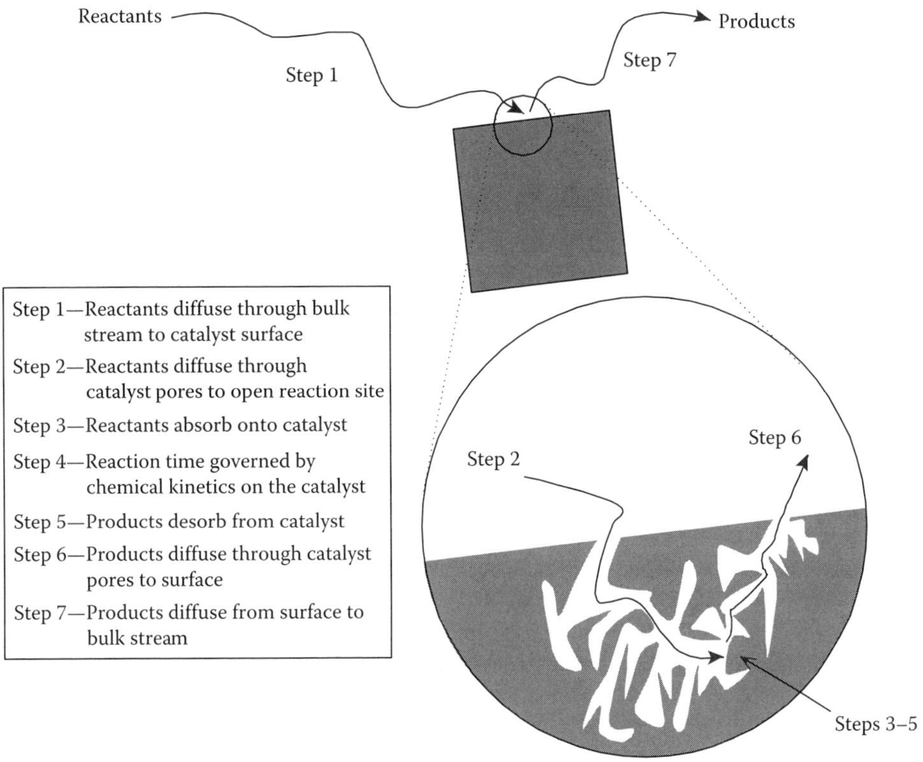
Reactants
Products
Step 1
Step 7
Step 1—Reactants diffuse through bulk stream to catalyst surface
Step 2—Reactants diffuse through catalyst pores to open reaction site
Step 3—Reactants absorb onto catalyst
Step 4—Reaction time governed by chemical kinetics on the catalyst
Step 5—Products desorb from catalyst
Step 6—Products diffuse through catalyst pores to surface
Step 7—Products diffuse from surface to bulk stream
Step 2
Step 6
Steps 3–5

FIGURE 3.11
Mass transfer steps in a typical steam reformer. (From Erickson, P., Enhancing the steam-reformation process with acoustics: An investigation for fuel cell vehicle applications, PhD dissertation, University of Florida, Gainesville, FL, 2002.)

increases pressure drop. Increased catalyst loading will increase the weight of the reactor, while using smaller catalyst pellets will increase flow resistance.

Typical mass transfer steps are given in Figure 3.11.

3.12.3 Heat Transfer

External heat is required by SR as the overall process is endothermic. Inside the reactor, pelletized or crushed catalyst particles are packed randomly with point-to-point contact with each other. Convection is the dominating mode of heat transfer within the reactor. As a consequence, large temperature gradients exist in the region near the reactor wall. Unreacted fuel can potentially flow past in the center region reducing overall fuel conversion. Unreacted fuel may also deactivate the catalyst by accumulating on the catalyst pores blocking active sites causing fouling. Using a small radius reactor can improve heat transfer to the centerline. However, as reactor radius decreases, the reactor-to-catalyst weight ratio increases, and the pressure drop increases rendering inefficient reactor design. Figure 3.12 shows experimental temperature gradients encountered with SR of methanol [93]. Although conduction errors are evident in this figure, experimentation shows that 100°C/cm temperature gradients are frequently encountered near the wall. This result implies that a catalyst might be degraded by sintering at the wall yet simultaneously experience fouling due to relatively cold condensed species forming at the centerline.

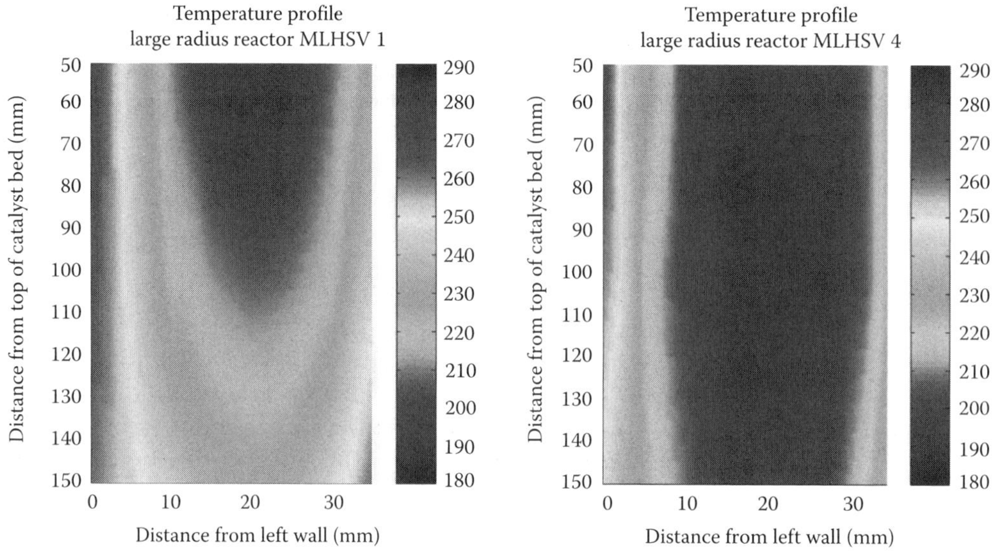

FIGURE 3.12
Example of large temperature gradients in a methanol steam reformer. The temperature differences is 46.4°C from interior wall to centerline, which is a 17.5 mm distance. (From Vernon, D., Understanding the effects of reactor geometry and scaling through temperature profiles in steam-reforming hydrogen production reactors. Master thesis, University of California, Davis, CA, 2006.)

TABLE 3.2
Limiting Mechanisms and Their Effect on Reaction Rate

	Variation of Reaction Rate with:		
Type of Limitation	**Velocity**	**Particle Size**	**Temperature**
External diffusion	$U^{1/2}$	$(dp)^{-3/2}$	~Linear
Internal diffusion	Independent	$(dp)^{-1}$	Exponential
Surface reaction	Independent	Independent	Exponential

Source: Fogler, H.S., *Elements of Chemical Reaction Engineering*, 4th edn., Pearson Education, Inc, 2006.

Table 3.2 shows the overall effects of each limitation and the effect of velocity particle size and temperature on the reaction rate.

3.12.4 Degradation Mechanisms

Reformation can also be limited by catalyst degradation. Degradation is typically classified as attrition, poisoning, fouling, and sintering.

Attrition occurs when the catalyst material is broken off of the substrate and is removed from the reactor. Catalyst and substrate attrition can also lead to physical flow restrictions in the reactor system. This phenomenon is common with frequent temperature cycling in metal housings where the thermal coefficient of expansion and resultant displacement of the housing crushes a pelletized catalyst resulting in a large pressure drop over time.

Poisoning occurs in reformation when a species binds to the reaction site blocking the reaction site from the reactants. Poisoning can be either reversible or nonreversible and may have a temperature effect with higher temperatures inhibiting the poisoning of the

FIGURE 3.13
New monolithic ATR catalyst is shown on the left and a similar cut-away sintered catalyst is shown on the right in this grayscale image.

catalyst. Sulfur and sulfur-containing compounds are especially problematic for typical reformer catalysts and poisoning agents. Deactivation of the catalyst by poisoning can typically be seen as the endothermic active zone of catalyst progresses through the reformer following the flow pattern. Zinc and related species can be used in a sacrificial manner to slow down the poisoning effect in reformers.

Fouling occurs when the active area of a catalyst is physically blocked by species forming on the external or internal surface of the catalyst. When the entrance and exit pores become blocked by condensed species or coke formation, reformer performance can drop significantly due to the induced mass transfer limitation. Because the internal area can be as high as 1600 m^2/g of the catalyst in pelletized catalyst, it is important to keep pore areas clear to avoid mass transfer limitations. Any gas–liquid or gas–solid phase transformations can induce fouling as the liquid or solid blocks the gas reactants from reaching the reaction surface.

Sintering can occur when the catalyst or the substrate changes form. Sintering is typically due to exothermic reactions driving the temperature above the melting point of either the substrate or the catalyst material. In many reformation processes, a ceria substrate is used with a washcoated metal acting as the catalyst material. Typical ceria will melt at 1450°C; thus, high temperatures can lead to pore blocking and absorption of the catalyst through melting of the substrate. It is also important to note that metallic phase transformation and changes can occur far below the melting point of the metal and this phase transformation can change the activity and selectivity of the catalyst. A common rule of thumb is that a certain metal can be used up to 1/3 of its melting point temperature. Structural integrity can also be compromised in extreme sintering cases as shown in Figure 3.13. The need to have fast reactions with heavy complex fuels can be limited by the catalyst's ability to withstand high temperatures and high heat flux.

3.12.5 Controls

Researchers have modeled the SR process [13,83,95,96] and the reformer [82,97–99] in various literature and have also proposed control algorithms [23,100,101] to better control the temperature, but the issue of controlling catalyst temperature has not been fully addressed. A high-level control algorithm sends manipulated inputs to control fuel flow rate and heater power for a given reformer, and the means of getting the heater power to the catalyst is limited by the design of the reformer. It should be noted that these concerns are only dealing with the steady-state control of the reformer temperature and do not address the transients encountered with large flow changes or start-up of the reactor. Slow transient response is a large hurdle for SR. While most current large-scale reformers run continuously for over 12,000 h [20], small-scale reformers will experience frequent fuel rate transients due to changes of demand. Optimizing reactor geometry is one approach to enhance reformer dynamics [102]. However, reformer geometry can impact the characteristic of the heat transfer, thus impacting the control algorithm design. A better approach is to design both the control algorithm and the reformer in parallel.

3.13 Comparison of the Reforming Methods

End use of the hydrogen is perhaps the important aspect when considering types of reformation systems. For example, in combustion applications, high concentrations of CO and unconverted fuel are not typically problematic, yet in fuel cell applications, there are significant requirements for fuel purity. Lower operating temperature fuel cell stacks usually have stricter requirements. Impurity in the hydrogen stream will slowly poison the fuel cell anode over time. CO poisoning is an important issue for PEMFC. High-temperature fuel cell stacks such as molten carbonate fuel cell (MCFC) and SOFC have higher impurity tolerance. They are capable of internal reforming and use CO and some volatile organic compounds (VOCs) as fuel. Selecting a proper fuel cell for a specific application can significantly impact the reformer's technical specifications.

SR has the advantage of a relatively high hydrogen concentration in the product gas, which leads to better utilization by a fuel cell. For example, for SR of methanol, the maximum hydrogen concentration is 75%. The efficiency of the process is also very high, which is desirable for transportation applications. It also has the advantage of a high S/C ratio, which leads to a low instance of coking (solid carbon deposits) without having to raise the reactor temperature. For an onboard steam reformer, the external heat needed for the endothermic reaction is most conveniently provided by simply combusting a portion of the fuel. Therefore, only a fraction of the total fuel used, Y, actually enters the reformer. The remaining fuel, $1 - Y$, is fully combusted to provide the energy required for the desired SR reaction.

In an ideal reactor, there would be 100% heat transfer from the combustion reaction to the SR reaction. However, there are limitations to the efficiency of the heat transfer process and this should be considered when selecting a reforming method. SR is endothermic and therefore an inherently slow process. An SR reactor is most commonly heated externally and has slow dynamic responses that may lead to system degradation. If the load is suddenly decreased and the reactant flow rates are decreased, the reformer heats up and can potentially sinter the catalyst. Alternatively, if the load is suddenly increased, the reformer cannot supply the required hydrogen to the fuel cell stack; consequently, the fuel cell extracts protons from the electrolyte membrane causing irreversible damage. This, coupled with a long start-up time, makes SR a less than optimal choice for onboard reforming or small-scale on-site reforming. It is still a good choice for centralized production of hydrogen on a large scale.

A POX reactor can heat up quickly by simply increasing the flow rate of oxygen to increase the combustion of the incoming fuel; this results in a fast start-up. For the same reason, it is extremely well equipped to handle transient loads. The reactor is much smaller and reformers can be made compact. However, POX produces excess CO and must therefore be accompanied by additional cleanup of the product gas. This increases the size and mass of the fuel-processing system. POX reforming is an exothermic reaction that operates at high temperatures usually above 1000°C. Because there is an absence of water (an S/C ratio of zero), reactors must operate at significantly higher temperatures (1180°C for 2,2,4-trimethylpentane) to avoid coking [103]. The system design must also include a heat exchanger that transfers excess heat to the surroundings. Hot spots may develop as a result of nonuniform mixing and can cause catalyst sintering. Most importantly, the major drawback of POX is the low concentration of hydrogen in the product gas. The addition of air into the reaction dilutes the product gases with nitrogen. For example, when POX is used to reform methanol, the highest theoretical concentration of hydrogen is only 41% (when using air as the source of oxygen), compared to 75% for SR [104]. This directly affects the efficiency of the PEM fuel cell and thus decreases the overall system efficiency, an especially undesirable characteristic for automotive applications.

One potential solution to the drawbacks of SR and POX is to combine the two into ATR. ATR operates ideally at a thermoneutral point, neither consuming nor releasing external energy. This gives ATR a higher efficiency and hydrogen concentration than POX and, at the same time, a better dynamic response than SR and the flexibility to accommodate multiple fuels. Rapid start-up is possible because of the ability to produce heat within the catalyst bed rather than transferring heat from the surroundings. Hot spots are reduced because of the addition of steam in the reforming reaction. This thermal integration lowers the temperature rise potential that is caused by the POX of the fuel and thereby reduces the potential for catalyst sintering. ATR has great potential in applications that require a lightweight, compact reactor capable of reforming multiple fuels [103]. These criteria fit into the needs of the automotive industry, whether for onboard or on-site reforming, and therefore ATR should be considered for transportation applications. ATR can provide a rapid response to hydrogen demand with short start-up times, high efficiencies, and fuel flexibility.

It should be noted that reactant mixing is an important consideration for ATR. Possible mixing schemes include the use of bluff bodies, swirling, and acoustic enhancement. Acoustic enhancement has been proven to improve the capabilities of SR, and there is a reason to believe that it may also be beneficial to ATR. These results have yet to be tested. Other reactant mixing schemes may also enhance ATR reactor performance. Should reactant mixing prove to be beneficial to ATR, it will further reduce the size and weight of the reformer. This in turn would decrease start-up time of the reforming system, one of the most critical aspects of a small-scale reformer.

3.14 Fuel Selection

Cost and environmental reasons are factors in deciding the choice of fuel. Many researchers have built small fuel processors for various hydrocarbon feedstocks [3,4,7,11]. They have shown competitive advantage in implementation, flexibility, and efficiencies [105–108]. Insulation and heat recirculation are important for these reformers to maintain high efficiencies. Coupled with PEMFC, they have demonstrated functional stationary [4,107] and onboard [3,106,109] fuel cell power systems. Although they have presented workable solutions to reform hydrocarbon feedstock, there are a range of issues to be addressed. For example, PEMFCs generally have low CO tolerance at <50 ppm [110–112]; thus, a cleanup system, such as PSA, is necessary to purify the reformed hydrogen stream [3,113–116]. High CO-tolerant electrodes and high-temperature fuel cell operation are also under development to overcome this limitation. Water management [4,105,112] and temperature are also critical to the life and performance of the PEMFC; thus, the temperature and humidity of the hydrogen stream must be regulated [23,24,101,117].

Table 3.3 is an estimate of the hydrogen yield by reforming various hydrocarbons.

Comparing the weight of hydrogen produced per liter of fuel, methanol and ethanol are clearly not competitive with gasoline and diesel. However, both alcohols can be reformed at relatively low temperature and can be made from renewable fuels. The low-temperature reformation process allows less complex reformer design, lower CO selectivity, and good compatibility with low-temperature fuel cell applications. Typical reformer temperatures and catalysts are shown in Table 3.4.

Other fuels such as diesel, gasoline, propane, and logistic fuels such as kerosene and jet fuel have also been used as feedstock. Typically, heavier hydrocarbons will require higher temperature and additional cleanup prior to or after reforming. Desulfurization and pre-reformers are used to ensure high-purity fuel is fed into the fuel processor. Gasoline and

TABLE 3.3

Hydrogen Yield by Reforming Various Hydrocarbons

		SR		POX	
Fuel	**Formula**	**Wt.% H_2**	**g H_2 per L_{fuel}**	**Wt.% H_2**	**g H_2 per L_{fuel}**
Methanol	CH_3OH	19%	150	13%	100
Ethanol	C_2H_5OH	26%	209	22%	168
Methane (LNG)	CH_4	52%	205	38%	151
Gasoline	$C_8H_{15.4}$	43%	301	28%	200
Diesel fuel	$C_{14}H_{25.5}$	42%	357	28%	231

Source: Spiegel, C., *Designing and Building Fuel Cells*, 1st edn., McGraw-Hill, 2007.

TABLE 3.4

Typical Reforming Temperature of Various Hydrocarbons

Fuel	Reformation Temperature (°C)	Catalyst
Glycerol	650–900	Pt
Isooctane/gasoline	650–800	Ru, Ni
Hexadecane/diesel	700–800	Ru, Ni
Natural gas	650–800	Ru, Ni
Methanol	200–300	Cu, Zn, Cr
Ethanol	200–300	Cu, Zn, Cr
Dodecane	450–550	Ni, Rh, Ce
JP-8	>520	Mn, Ni
Propane/*n*-butane	600–800	Pd, Cu, Ni
Methane	>500	Ni
Kerosene/*n*-octane	>500	Ni

diesel have a well-established infrastructure and higher energy storage density for hydrogen. A gasoline reformer can be started and deliver 90% rated hydrogen capacity with less than 50 ppm CO in 60 s [7]. Natural gas or methane is abundant in many parts of the world and is used as feedstock for hydrogen production in industrial settings. It can also be converted to methanol and transported using existing infrastructure. Liquefied petroleum gas (LPG), which has been widely used for cars and buses, is also a potential fuel for mobile reformer systems.

3.15 Internal Reforming in High-Temperature Fuel Cells

Internal reforming refers to reformation taking place at or near the anode of the fuel cell. With SOFC and MCFC, the stack temperature is sufficient to reform low–molecular weight hydrocarbons [119]. Internal reforming has the advantage of reducing cost and complexity since a separated reformer and heat exchangers are not necessary. It is also more energy efficient since less steam is required and heat loss is minimized. The anode catalyst must be able to reform the hydrocarbon fuel and catalyze the oxidation reaction in the high-temperature anode of these fuel cells. The internal reforming systems are favored by Le Chatelier's principle that drives the reaction toward the product side at high temperatures and pressures as the hydrogen produced by reformation is immediately consumed by the fuel cell oxidation reaction.

3.16 Reactor Design

In SR, CPOX, and ATR, maintaining chemical reaction kinetics by activating the catalyst becomes the important aspect in controlling the reformation process. In order to activate the catalyst, sufficient heat must be available for the reaction. Insufficient temperature will deactivate the catalyst, thus reducing efficiencies. It is important to address the need for maintaining proper temperature in reformation processes. For MTS catalysts containing copper, exceeding 300°C generally will make the catalyst unstable and degrade by sintering [35,42,120]. Ni-based catalyst is stable at higher temperatures, but this becomes a tradeoff with the material and operating cost of a high-temperature reformer. Maintaining a uniform temperature profile inside the reactor is ideal but is difficult to achieve. Heat distribution is limited by convection; thus, high-temperature gradients exist inside the catalyst bed. Modifying the reactor design to improve the heat and mass transfer limitation is one possible solution [121–123].

Inside the steam reformer catalyst bed, convection is the primary means to transfer heat from the reactor wall to the catalyst; thus, a small diameter tubular configuration is advantageous. Passive flow baffles can be used inside the reactor to enhance heat transfer [124]. Structured catalysts have also been employed [125]. Figure 3.14 shows how passive flow disturbance can be used to enhance the heat distribution.

Inside the autothermal reformer, an exothermic reaction occurs on the surface of the catalyst in the presence of oxygen; thus, controlling the flow of oxygen can control heat generation. Employing a porous membrane within the catalyst bed to distribute air inside the

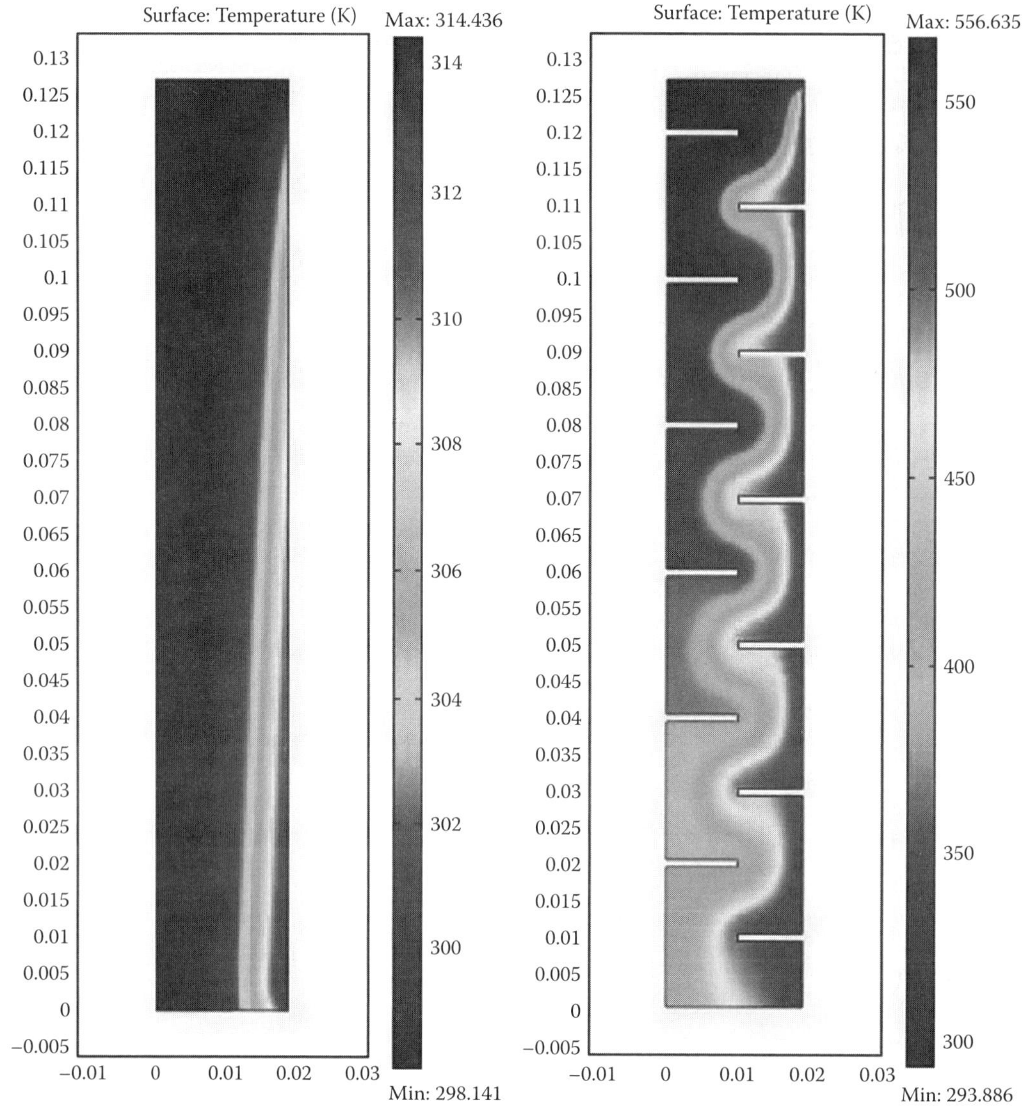

FIGURE 3.14
Simulation of the convective heat transfer inside a large radius cylindrical reactor with and without baffles. The axis of symmetry is on the left side of the domain and heat flux is applied on the right. These two reactors have the same space velocity, but one has dramatically better heat transfer properties. (From Tang, H.-Y., Reactor controller design for steam and autothermal reforming for fuel cell applications, PhD dissertation, University of California, Davis, CA, 2009.)

reactor has been investigated by Lattner [25] and Liu et al. [126]. Other variations of using multiple heating zones [67], dual catalysts [127], and coupling exothermic to endothermic regions of the reactor [128] have also been proposed. These passive methods all intend to optimize energy efficiencies by moving heat from the exothermic reaction region or exhaust to the endothermic reaction region. The various operation conditions can be incorporated into a control algorithm. However, these efforts reduce the flexibility in designing the control algorithm. The best approach is the combination of active control and passive reactor enhancement.

3.17 Reformer Control Issues

Temperature control of the reformer influences fuel conversion and catalyst degradation. The poor heat transfer properties of the catalyst make transferring heat in and out of the catalyst bed difficult as shown in Figure 3.14. In ATR, if excess heat is produced on the catalyst, it will sinter and melt the catalyst. In SR, if insufficient heat is available, unreacted fuel may poison or foul the catalyst. In implementing feedback control, the feedback or the sensor placement location and control variables are not obvious. In ATR, the exothermic reaction typically takes place near the top of the catalyst bed. To control the catalyst temperature, it is necessary to control the available oxidizer in the fuel stream. In SR, the centerline usually is the region with the lowest temperature because convection is the primary mode of heat transfer from the reactor wall to the center. However, using the centerline as the feedback location can produce oscillations in temperature inside the reformer. The high thermal resistance of the catalyst bed potentially creates a long lag in temperature response. Using strictly temperature control can lead to sintering of the catalyst near the reactor wall by excessive heating from the external burner. The situation is more problematic during transients. By the time the centerline temperature has increased, the reactor wall temperature may reach a much higher temperature creating a large temperature gradient. This could reduce the overall efficiencies as heat is not effectively utilized, and the catalyst can be degraded by multiple high-amplitude thermal cycles. A possible solution is to regulate heat flux in the steam reformer to avoid such problems. This is done by combining the ability to control the fuel feed rate and external burner input. The controller will incorporate the heat transfer properties of the reactor and the chemical kinetics of the reformation process to adjust the fuel feed rate and the external heater temperature. This concept allows maintenance of the catalyst in a certain temperature range during transient ramps [129].

3.18 Summary

Direct transport and storage of hydrogen is costly. Hydrogen production at the point of use may be a viable near-term solution. Reforming liquid hydrocarbon fuels for hydrogen production has several advantages for transportation applications: the existing infrastructure for transporting hydrocarbon fuel is well developed; refueling liquid fuel is faster and much more familiar and intuitive than recharging gaseous fuels; and there are wide ranges of possible hydrocarbon feedstocks. SR is the most widely used method of generating hydrogen and produces the highest-quality reformate. POX can respond quickly to transient demand but is associated with low-quality hydrogen. ATR generally combines a higher output gas purity, similar to though not attaining that of SR, with a fast response time, similar to but not attaining that of POX. All of the methods are presently studied in earnest for the small-scale reformers required in mobile and distributed generation applications. In the known and practiced reformation techniques, the resultant reformate is not a pure hydrogen stream; thus, additional cleanup steps are needed for stringent purity applications. If an inert gas, such as nitrogen, is present, it will lower the hydrogen concentration. End use of the hydrogen may change the reformate requirements and hence the fuel processor.

References

1. Amphlett, J.C., Mann, R.F., Peppley, B.A., Roberge, P.R., Rodrigues, A., and Salvador, J.P.: Simulation of a 250 kW diesel fuel processor/PEM fuel cell system, *Journal of Power Sources*, 1998, 71(1–2), 179–184.
2. Han, J., Lee, S.-M., and Chang, H.: Metal membrane-type 25-kW methanol fuel processor for fuel-cell hybrid vehicle, *Journal of Power Sources*, 2002, 112(2), 484–490.
3. Lee, S.H., Han, J., and Lee, K.-Y.: Development of 10-kWe preferential oxidation system for fuel cell vehicles, *Journal of Power Sources*, 2002, 109(2), 394–402.
4. Mathiak, J., Heinzel, A., Roes, J., Kalk, T., Kraus, H., and Brandt, H.: Coupling of a 2.5 kW steam reformer with a 1 kWel PEM fuel cell, *Journal of Power Sources*, 2004, 131(1–2), 112–119.
5. Sopeña, D., Melgar, A., Briceño, Y., Navarro, R.M., Álvarez-Galván, M.C., and Rosa, F.: Diesel fuel processor for hydrogen production for 5 kW fuel cell application, *International Journal of Hydrogen Energy*, 2007, 32(10–11), 1429–1436.
6. Yan, X., Wang, S., Li, X., Hou, M., Yuan, Z., Li, D., Pan, L., Zhang, C., Liu, J., Ming, P., and Yi, B.: A 75-kW methanol reforming fuel cell system, *Journal of Power Sources*, 2006, 162(2), 1265–1269.
7. Ahmed, S., Ahluwalia, R., Lee, S.H.D., and Lottes, S.: A gasoline fuel processor designed to study quick-start performance, *Journal of Power Sources*, 2006, 154(1), 214–222.
8. Docter, A. and Lamm, A.: Gasoline fuel cell systems, *Journal of Power Sources*, 1999, 84(2), 194–200
9. Amphlett, J.C., Creber, K.A.M., Davis, J.M., Mann, R.F., Peppley, B.A., and Stokes, D.M.: Hydrogen production by steam reforming of methanol for polymer electrolyte fuel cells, *International Journal of Hydrogen Energy*, 1994, 19(2), 131–137.
10. Agarwal, V., Patel, S., and Pant, K.K.: H_2 production by steam reforming of methanol over $Cu/ZnO/Al_2O_3$ catalysts: Transient deactivation kinetics modeling, *Applied Catalysis A: General*, 2005, 279(1–2), 155–164.
11. Breen, J.P. and Ross, J.R.H.: Methanol reforming for fuel-cell applications: Development of zirconia-containing Cu-Zn-Al catalysts, *Catalysis Today*, 1999, 51(3–4), 521–533.
12. Palo, D.R., Dagle, R.A., and Holladay, J.D.: Methanol steam reforming for hydrogen production, *Chemical Reviews*, 2007, 107(10), 3992–4021.
13. Ming, Q., Healey, T., Allen, L., and Irving, P.: Steam reforming of hydrocarbon fuels, *Catalysis Today*, 2002, 77(1–2), 51–64.
14. Rosa, F., López, E., Briceño, Y., Sopeña, D., Navarro, R.M., Alvarez-Galván, M.C., Fierro, J.L.G., and Bordons, C.: Design of a diesel reformer coupled to a PEMFC, *Catalysis Today*, 2006, 116(3), 324–333.
15. Roychoudhury, S., Lyubovsky, M., Walsh, D., Chu, D., and Kallio, E.: Design and development of a diesel and JP-8 logistic fuel processor, *Journal of Power Sources*, 2006, 160(1), 510–513.
16. Cutillo, A., Specchia, S., Antonini, M., Saracco, G., and Specchia, V.: Diesel fuel processor for PEM fuel cells: Two possible alternatives (ATR versus SR), *Journal of Power Sources*, 2006, 154(2), 379–385.
17. Hu, J., Wang, Y., VanderWiel, D., Chin, C., Palo, D., Rozmiarek, R., Dagle, R., Cao, J., Holladay, J., and Baker, E.: Fuel processing for portable power applications, *Chemical Engineering Journal*, 2003, 93(1), 55–60.
18. Liu, Z., Mao, Z., Xu, J., Hess-Mohr, N., and Schmidt, V.M.: Modelling of a PEM fuel cell system with propane ATR reforming, *Fuel Cells*, 2006, 6(5), 376–386.
19. National Energy Technology Laboratory, Gasification Systems Technologies; Gasification Basics http://www.netl.doe.gov/technologies/coalpower/gasification/basics/3.html. Accessed on December 7, 2013.
20. Rostrup-Nielsen, J.R. and Rostrup-Nielsen, T.: Large-scale hydrogen production, *CATTECH*, 2002, 6(4), 150–159.
21. *Risø Energy Report 3—Hydrogen and Its Competitors*, Larsen, H., Feidenhans'l and Petersen, L.S. (Eds), Technical University of Denmark, Risø National Laboratory for Sustainable Energy, Holmen Center-Tryk, Holbæk, Denmark, 2004, p. 76.

22. Borup, R.L., Inbody, M.A., Semelsberger, T.A., Tafoya, J.I., and Guidry, D.R.: Fuel composition effects on transportation fuel cell reforming, *Catalysis Today*, 2005, 99(3–4), 263–270.
23. Chen, Y.-H., Yu, C.-C., Liu, Y.-C., and Lee, C.-H.: Start-up strategies of an experimental fuel processor, *Journal of Power Sources*, 2006, 160(2), 1275–1286.
24. Horng, R.-F., Chen, C.-R., Wu, T.-S., and Chan, C.-H.: Cold start response of a small methanol reformer by partial oxidation reforming of hydrogen for fuel cell, *Applied Thermal Engineering*, 2006, 26(11–12), 1115–1124.
25. Lattner, J.R. and Harold, M.P.: Autothermal reforming of methanol: Experiments and modeling, *Catalysis Today*, 2007, 120(1), 78–89.
26. Dams, R.A.J., Hayter, P.R., and Moore, S.C.: *Book Continued Development of Mathematical Model for a Methanol Reformer*, Crown, London, UK, 2001.
27. Radu, R. and Taccani, R.: SIMULINK-FEMLAB integrated dynamic simulation model for a PEM fuel cell system, *Journal of Fuel Cell Science and Technology*, 2006, 3(4), 452–458.
28. Kolb, G.: *Fuel Processing: For Fuel Cells*, Wiley-VCH Verlag GmbH & Co. KGaA, Weinheim, Germany, 2008.
29. Davieau, D.D. and Erickson, P.A.: The effect of geometry on reactor performance in the steam-reformation process, *International Journal of Hydrogen Energy*, 2007, 32(9), 1192–1200.
30. Liao, C.-H. and Erickson, P.A.: Characteristic time as a descriptive parameter in steam reformation hydrogen production processes, *International Journal of Hydrogen Energy*, 2008, 33(6), 1652–1660.
31. Fogler, H.S.: *Elements of Chemical Reaction Engineering*, Pearson Education, Inc, Boston, MA, 4th edn., 2006.
32. Hagen, J.: *Industrial Catalysis—A Practical Approach*, Wiley-VCH Verlag GmbH & Co. KGaA, Weinheim, Germany, 2nd edn., 2006.
33. Catillon, S., Louis, C., and Rouget, R.: Development of new CuO–ZnII/Al_2O_3 catalyst supported on copper metallic foam for the production of hydrogen by methanol steam reforming, *Topics in Catalysis*, 2004, 30–31(1), 463–467.
34. Lindström, B., Pettersson, L.J., and Govind Menon, P.: Activity and characterization of Cu/Zn, Cu/Cr and Cu/Zr on [gamma]-alumina for methanol reforming for fuel cell vehicles, *Applied Catalysis A: General*, 2002, 234(1–2), 111–125.
35. Liu, S., Takahashi, K., Uematsu, K., and Ayabe, M.: Hydrogen production by oxidative methanol reforming on Pd/ZnO, *Applied Catalysis A: General*, 2005, 283(1–2), 125–135.
36. Mastalir, A., Frank, B., Szizybalski, A., Soerijanto, H., Deshpande, A., Niederberger, M., Schomäcker, R., Schlögl, R., and Ressler, T.: Steam reforming of methanol over Cu/ZrO_2/CeO_2 catalysts: A kinetic study, *Journal of Catalysis*, 2005, 230(2), 464–475.
37. Matter, P.H., Braden, D.J., and Ozkan, U.S.: Steam reforming of methanol to H_2 over nonreduced Zr-containing CuO/ZnO catalysts, *Journal of Catalysis*, 2004, 223(2), 340–351.
38. Oguchi, H., Nishiguchi, T., Matsumoto, T., Kanai, H., Utani, K., Matsumura, Y., and Imamura, S.: Steam reforming of methanol over Cu/CeO_2/ZrO_2 catalysts, *Applied Catalysis A: General*, 2005, 281(1–2), 69–73.
39. Papavasiliou, J., Avgouropoulos, G., and Ioannides, T.: Effect of dopants on the performance of CuO-CeO_2 catalysts in methanol steam reforming, *Applied Catalysis B: Environmental*, 2007, 69(3–4), 226–234.
40. Purnama, H., Girgsdies, F., Ressler, T., Schattka, J.H., Caruso, R.A., Schomäcker, R., and Schlögl, R.: Activity and selectivity of a nanostructured CuO/ZrO_2 catalyst in the steam reforming of methanol, *Catalysis Letters*, 2004, 94(1), 61–68.
41. Shan, W., Feng, Z., Li, Z., Zhang, J., Shen, W., and Li, C.: Oxidative steam reforming of methanol on $Ce_{0.9}Cu_{0.1}OY$ catalysts prepared by deposition-precipitation, coprecipitation, and complexation-combustion methods, *Journal of Catalysis*, 2004, 228(1), 206–217.
42. Wang, Z., Wang, W., and Lu, G.: Studies on the active species and on dispersion of Cu in Cu/SiO_2 and Cu/Zn/SiO_2 for hydrogen production via methanol partial oxidation, *International Journal of Hydrogen Energy*, 2003, 28(2), 151–158.

43. Agrell, J., Birgersson, H., Boutonnet, M., Melián-Cabrera, I., Navarro, R.M., and Fierro, J.L.G.: Production of hydrogen from methanol over Cu/ZnO catalysts promoted by ZrO_2 and Al_2O_3, *Journal of Catalysis*, 2003, 219(2), 389–403.
44. Choi, Y. and Stenger, H.G.: Fuel cell grade hydrogen from methanol on a commercial Cu/ZnO/Al_2O_3 catalyst, *Applied Catalysis B: Environmental*, 2002, 38(4), 259–269.
45. Costantino, U., Marmottini, F., Sisani, M., Montanari, T., Ramis, G., Busca, G., Turco, M., and Bagnasco, G.: Cu-Zn-Al hydrotalcites as precursors of catalysts for the production of hydrogen from methanol, *Solid State Ionics*, 2005, 176(39–40), 2917–2922.
46. Agrell, J., Boutonnet, M., Melián-Cabrera, I., and Fierro, J.L.G.: Production of hydrogen from methanol over binary Cu/ZnO catalysts: Part I. Catalyst preparation and characterisation, *Applied Catalysis A: General*, 2003, 253(1), 201–211.
47. Faur Ghenciu, A.: Review of fuel processing catalysts for hydrogen production in PEM fuel cell systems, *Current Opinion in Solid State and Materials Science*, 2002, 6(5), 389–399.
48. Auprêtre, F., Descorme, C., and Duprez, D.: Bio-ethanol catalytic steam reforming over supported metal catalysts, *Catalysis Communications*, 2002, 3(6), 263–267.
49. De Rogatis, L., Montini, T., Casula, M.F., and Fornasiero, P.: Design of Rh@$Ce_{0.2}Zr_{0.8}O_2$-Al_2O_3 nanocomposite for ethanol steam reforming, *Journal of Alloys and Compounds*, 2008, 451(1–2), 516–520.
50. Ha, J.W., Kundu, A., and Jang, J.H.: Poly-dimethylsiloxane (PDMS) based micro-reactors for steam reforming of methanol, *Fuel Processing Technology*, 91(11), 1725–1730.
51. Hwang, S.-M., Kwon, O.J., and Kim, J.J.: Method of catalyst coating in micro-reactors for methanol steam reforming, *Applied Catalysis A: General*, 2007, 316(1), 83–89.
52. Kawamura, Y., Ogura, N., Yamamoto, T., and Igarashi, A.: A miniaturized methanol reformer with Si-based microreactor for a small PEMFC, *Chemical Engineering Science*, 2006, 61(4), 1092–1101.
53. Moreno, A.M. and Wilhite, B.A.: Autothermal hydrogen generation from methanol in a ceramic microchannel network, *Journal of Power Sources*, 195(7), 1964–1970.
54. Agrell, J., Birgersson, H., and Boutonnet, M.: Steam reforming of methanol over a Cu/ZnO/Al_2O_3 catalyst: A kinetic analysis and strategies for suppression of CO formation, *Journal of Power Sources*, 2002, 106(1–2), 249–257.
55. Sahoo, D.R., Vajpai, S., Patel, S., and Pant, K.K.: Kinetic modeling of steam reforming of ethanol for the production of hydrogen over Co/Al_2O_3 catalyst, *Chemical Engineering Journal*, 2007, 125(3), 139–147.
56. Peppley, B.A., Amphlett, J.C., Kearns, L.M., and Mann, R.F.: Methanol-steam reforming on Cu/ZnO/Al_2O_3. Part 1: The reaction network, *Applied Catalysis A: General*, 1999, 179(1–2), 21–29.
57. Choi, Y. and Stenger, H.G.: Kinetics, simulation and optimization of methanol steam reformer for fuel cell applications, *Journal of Power Sources*, 2005, 142(1–2), 81–91.
58. Rabenstein, G. and Hacker, V.: Hydrogen for fuel cells from ethanol by steam-reforming, partial-oxidation and combined auto-thermal reforming: A thermodynamic analysis, *Journal of Power Sources*, 2008, 185(2), 1293–1304.
59. Hoogers, G.: *Fuel Cell Technology Handbook*, CRC Press, Boca Raton, FL, 2003.
60. Cheekatamarla, P.K. and Finnerty, C.M.: Synthesis gas production via catalytic partial oxidation reforming of liquid fuels, *International Journal of Hydrogen Energy*, 2008, 33(19), 5012–5019.
61. Sterchi, P.: The effect of hydrocarbon impurities on the methanol steam-reforming process for fuel cell applications. PhD dissertation, University of Florida, Gainesville, FL, 2001.
62. Yoon, H.C. and Erickson, P.A.: Hydrogen from coal-derived methanol via autothermal reforming processes, *International Journal of Hydrogen Energy*, 2008, 33(1), 57–63.
63. Yoon, H.C.: Comparison of steam and autothermal reforming of methanol for fuel cell applications. PhD dissertation, University of California, Davis, CA, 2008.
64. Velu, S., Suzuki, K., Okazaki, M., Kapoor, M.P., Osaki, T., and Ohashi, F.: Oxidative steam reforming of methanol over CuZnAl(Zr)-oxide catalysts for the selective production of hydrogen for fuel cells: Catalyst characterization and performance evaluation, *Journal of Catalysis*, 2000, 194(2), 373–384.

65. Chan, S.H. and Wang, H.M.: Thermodynamic and kinetic modelling of an autothermal methanol reformer, *Journal of Power Sources*, 2004, 126(1–2), 8–15.
66. Turco, M., Bagnasco, G., Cammarano, C., Senese, P., Costantino, U., and Sisani, M.: Cu/ZnO/Al_2O_3 catalysts for oxidative steam reforming of methanol: The role of Cu and the dispersing oxide matrix, *Applied Catalysis B: Environmental*, 2007, 77(1–2), 46–57.
67. Sundaresan, M., Ramaswamy, S., Moore, R.M., and Hoffman, M.A.: Catalytic burner for an indirect methanol fuel cell vehicle fuel processor, *Journal of Power Sources*, 2003, 113(1), 19–36.
68. Pfefferle, W.C. and Pfefferle, L.D.: Catalytically stabilized combustion, *Progress in Energy and Combustion Science*, 1986, 12(1), 25–41.
69. Pfefferle, L.D. and Pfefferle, W.C.: Catalysis in combustion, *Catalysis Reviews*, 1987, 29(2), 219–267.
70. Otsuka, K., Ina, T., and Yamanaka, I.: The partial oxidation of methanol using a fuel cell reactor, *Applied Catalysis A: General*, 2003, 247(2), 219–229.
71. Sakong, S., Sendner, C., and Groß, A.: Partial oxidation of methanol on Cu(110): Energetics and kinetics, *Journal of Molecular Structure: THEOCHEM*, 2006, 771(1–3), 117–122.
72. Zhou, L., Günther, S., and Imbihl, R.: Low-pressure methanol oxidation over a Cu(110) surface under stationary conditions: (I) reaction kinetics, *Journal of Catalysis*, 2005, 230(1), 166–172.
73. Mu, X., Pan, L., Liu, N., Zhang, C., Li, S., Sun, G., and Wang, S.: Autothermal reforming of methanol in a mini-reactor for a miniature fuel cell, *International Journal of Hydrogen Energy*, 2007, 32(15), 3327–3334.
74. Schildhauer, T.J. and Geissler, K.: Reactor concept for improved heat integration in autothermal methanol reforming, *International Journal of Hydrogen Energy*, 2007, 32(12), 1806–1810.
75. Semelsberger, T.A., Brown, L.F., Borup, R.L., and Inbody, M.A.M.A.: Equilibrium products from autothermal processes for generating hydrogen-rich fuel-cell feeds, *International Journal of Hydrogen Energy*, 2004, 29(10), 1047–1064.
76. Liu, Y., Hayakawa, T., Tsunoda, T., Suzuki, K., Hamakawa, S., Murata, K., Shiozaki, R., Ishii, T., and Kumagai, M.: Steam reforming of methanol over Cu/CeO_2 catalysts studied in comparison with Cu/ZnO and Cu/Zn(Al)O catalysts, *Topics in Catalysis*, 2003, 22(3), 205–213.
77. Simmons, T.: Characteristics of autothermal reforming of a hydrocarbon fuel for hydrogen for fuel cell vehicles. PhD dissertation, University of Florida, Gainesville, FL, 2003.
78. Sadakane, O., Saitoh, K., Oyama, K., Yamauchi, N., and Komatsu, H.: Fuel-cell vehicle fuels: Evaluating the reforming performance of gasoline components, SAE, 2003, 2003-01-0414.
79. Lindström, B., Agrell, J., and Pettersson, L.J.: Combining methanol reforming for hydrogen generation over monolithic catalysts, *Chemical Engineering Journal*, 2003, 93(1), 91–101.
80. Krumpelt, M., Krause, T.R., Carter, J.D., Kopasz, J.P., and Ahmed, S.: Fuel processing for fuel cell systems in transportation and portable power applications, *Catalysis Today*, 2002, 77(1–2), 3–16.
81. Dorr, J.L.: Methanol autothermal reformation: Oxygen-to-carbon ratio and reaction progression. Master thesis, University of California, Davis, CA, 2004.
82. Fukahori, S., Kitaoka, T., Tomoda, A., Suzuki, R., and Wariishi, H.: Methanol steam reforming over paper-like composites of Cu/ZnO catalyst and ceramic fiber, *Applied Catalysis A: General*, 2006, 300(2), 155–161.
83. Patel, S. and Pant, K.K.: Production of hydrogen with low carbon monoxide formation via catalytic steam reforming of methanol, *Journal of Fuel Cell Science and Technology*, 2006, 3(4), 369–374.
84. Wang, J., Li, C.-H., and Huang, T.-J.: Study of partial oxidative steam reforming of methanol over Cu–ZnO/samaria-doped ceria catalyst, *Catalysis Letters*, 2005, 103(3), 239–247.
85. Zhou, S., Yuan, Z., and Wang, S.: Selective CO oxidation with real methanol reformate over monolithic Pt group catalysts: PEMFC applications. *International Journal of Hydrogen Energy*, 2006, 31(7), 924–933.
86. Palo, D.R., Dagle, R.A., and Holladay, J.D.: Methanol steam reforming for hydrogen production, *Chemical Reviews*, 2007, 107(10), 3992–4021.
87. Comas, J., Mariño, F., Laborde, M., and Amadeo, N.: Bio-ethanol steam reforming on Ni/Al_2O_3 catalyst, *Chemical Engineering Journal*, 2004, 98(1–2), 61–68.

88. Giunta, P., Amadeo, N., and Laborde, M.: Simulation of a low temperature water gas shift reactor using the heterogeneous model/application to a PEM fuel cell, *Journal of Power Sources*, 2006, 156(2), 489–496.
89. Mas, V., Kipreos, R., Amadeo, N., and Laborde, M.: Thermodynamic analysis of ethanol/water system with the stoichiometric method, *International Journal of Hydrogen Energy*, 2006, 31(1), 21–28.
90. Wang, W. and Wang, Y.: Thermodynamic analysis of hydrogen production via partial oxidation of ethanol, *International Journal of Hydrogen Energy*, 2008, 33(19), 5035–5044.
91. Yoon, H.C., Otero, J., and Erickson, P.A.: Reactor design limitations for the steam reforming of methanol, *Applied Catalysis B: Environmental*, 2007, 75(3–4), 264–271.
92. Erickson, P.: Enhancing the steam-reformation process with acoustics: An investigation for fuel cell vehicle applications. PhD dissertation, University of Florida, Gainesville, FL, 2002.
93. Vernon, D.: Hydrogen enrichment and thermochemical recuperation in internal combustion engines: An investigation of dilution and inlet temperature effects in the autothermal reformation of ethanol. PhD dissertation, University of California, Davis, CA, 2010.
94. Vernon, D.: Understanding the effects of reactor geometry and scaling through temperature profiles in steam-reforming hydrogen production reactors. Master thesis, University of California, Davis, CA, 2006.
95. Al-Ubaid, A. and Wolf, E.E.: Steam reforming of methane on reduced non-stoichiometric nickel aluminate catalysts, *Applied Catalysis*, 1988, 40, 73–85.
96. Patel, S. and Pant, K.K.: Selective production of hydrogen via oxidative steam reforming of methanol using Cu-Zn-Ce-Al oxide catalysts, *Chemical Engineering Science*, 2007, 62(18–20), 5436–5443.
97. Kundu, A., Shul, Y.G., Kim, D.H., T.S. Zhao, K.D.K., and Trung Van, N.: *Chapter Seven Methanol Reforming Processes: Advances in Fuel Cells*, Elsevier Science, 2007, pp. 419–472.
98. Turco, M., Bagnasco, G., Costantino, U., Marmottini, F., Montanari, T., Ramis, G., and Busca, G.: Production of hydrogen from oxidative steam reforming of methanol: I. Preparation and characterization of $Cu/ZnO/Al_2O_3$ catalysts from a hydrotalcite-like LDH precursor, *Journal of Catalysis*, 2004, 228(1), 43–55.
99. Turco, M., Bagnasco, G., Costantino, U., Marmottini, F., Montanari, T., Ramis, G., and Busca, G.: Production of hydrogen from oxidative steam reforming of methanol: II. Catalytic activity and reaction mechanism on $Cu/ZnO/Al_2O_3$ hydrotalcite-derived catalysts, *Journal of Catalysis*, 2004, 228(1), 56–65.
100. Arcak, M., Gorgun, H., Pedersen, L.M., and Varigonda, S.A.V.S.: A nonlinear observer design for fuel cell hydrogen estimation, *IEEE Transactions on Control Systems Technology*, 2004, 12(1), 101–110.
101. Gorgun, H., Arcak, M., Varigonda, S., and Bortoff, S.A.: Observer designs for fuel processing reactors in fuel cell power systems, *International Journal of Hydrogen Energy*, 2005, 30(4), 447–457.
102. Nummedal, L., Røsjorde, A., Johannessen, E., and Kjelstrup, S.: Second law optimization of a tubular steam reformer, *Chemical Engineering and Processing*, 2005, 44(4), 429–440.
103. Ahmed, S. and Krumpelt, M.: Hydrogen from hydrocarbon fuels for fuel cells, *International Journal of Hydrogen Energy*, 2001, 26(4), 291–301.
104 Lindström, B. and Pettersson, L.J.: Catalytic oxidation of liquid methanol as a heat source for an automotive reformer, *Chemical Engineering and Technology*, 2003, 26(4), 473–478. April 2003. doi:10.1002/ceat.200390071.
105. Lattner, J.R. and Harold, M.P.: Comparison of methanol-based fuel processors for PEM fuel cell systems, *Applied Catalysis B: Environmental*, 2005, 56(1–2), 149–169.
106. Mitchell, W., Bowers, B.J., Garnier, C., and Boudjemaa, F.: Dynamic behavior of gasoline fuel cell electric vehicles, *Journal of Power Sources*, 2006, 154(2), 489–496.
107. Pan, L. and Wang, S.: A compact integrated fuel-processing system for proton exchange membrane fuel cells, *International Journal of Hydrogen Energy*, 2006, 31(4), 447–454.
108. Lattner, J.R. and Harold, M.P.: Comparison of conventional and membrane reactor fuel processors for hydrocarbon-based PEM fuel cell systems, *International Journal of Hydrogen Energy*, 2004, 29(4), 393–417.

109. Bowers, B.J., Zhao, J.L., Ruffo, M., Khan, R., Dattatraya, D., Dushman, N., Beziat, J.-C., and Boudjemaa, F.: Onboard fuel processor for PEM fuel cell vehicles, *International Journal of Hydrogen Energy*, 2007, 32(10–11), 1437–1442.
110. Ahluwalia, R.K., Zhang, Q., Chmielewski, D.J., Lauzze, K.C., and Inbody, M.A.: Performance of CO preferential oxidation reactor with noble-metal catalyst coated on ceramic monolith for on-board fuel processing applications, *Catalysis Today*, 2005, 99(3–4), 271–283.
111. Jiménez, S., Soler, J., Valenzuela, R.X., and Daza, L.: Assessment of the performance of a PEMFC in the presence of CO, *Journal of Power Sources*, 2005, 151, 69–73.
112. Kim, J.-D., Park, Y.-I., Kobayashi, K., Nagai, M., and Kunimatsu, M.: Characterization of CO tolerance of PEMFC by AC impedance spectroscopy, *Solid State Ionics*, 2001, 140(3–4), 313–325.
113. Kim, J.-D., Park, Y.-I., Kobayashi, K., and Nagai, M.: Effect of CO gas and anode-metal loading on H_2 oxidation in proton exchange membrane fuel cell, *Journal of Power Sources*, 2001, 103(1), 127–133.
114. Lee, S.J., Mukerjee, S., Ticianelli, E.A., and McBreen, J.: Electrocatalysis of CO tolerance in hydrogen oxidation reaction in PEM fuel cells, *Electrochimica Acta*, 1999, 44(19), 3283–3293.
115. Pereira, L.G.S., dos Santos, F.R., Pereira, M.E., Paganin, V.A., and Ticianelli, E.A.: CO tolerance effects of tungsten-based PEMFC anodes, *Electrochimica Acta*, 2006, 51(19), 4061–4066.
116. Wagner, N. and Gülzow, E.: Change of electrochemical impedance spectra (EIS) with time during CO-poisoning of the Pt-anode in a membrane fuel cell, *Journal of Power Sources*, 2004, 127(1–2), 341–347.
117. Auckenthaler, T.S.: *Modelling and Control of Three-Way Catalytic Converters*, Swiss Federal Institute of Technology (ETH), Zürich, Switzerland, 2005.
118. Spiegel, C.: *Designing and Building Fuel Cells*, McGraw-Hill Professional Publishing, 1st edn., 2007.
119. Laosiripojana, N. and Assabumrungrat, S.: Catalytic steam reforming of methane, methanol, and ethanol over Ni/YSZ: The possible use of these fuels in internal reforming SOFC, *Journal of Power Sources*, 2007, 163(2), 943–951.
120. Liu, S., Takahashi, K., Uematsu, K., and Ayabe, M.: Hydrogen production by oxidative methanol reforming on Pd/ZnO catalyst: Effects of the addition of a third metal component, *Applied Catalysis A: General*, 2004, 277(1–2), 265–270.
121. Polman, E.A., Der Kinderen, J.M., and Thuis, F.M.A.: Novel compact steam reformer for fuel cells with heat generation by catalytic combustion augmented by induction heating, *Catalysis Today*, 1999, 47(1–4), 347–351.
122. Smith, T.G. and Carberry, J.J.: Design and optimization of a tube-wall reactor, *Chemical Engineering Science*, 1975, 30(2), 221–227.
123. Zanfir, M. and Gavriilidis, A.: Modelling of a catalytic plate reactor for dehydrogenation-combustion coupling, *Chemical Engineering Science*, 2001, 56(8), 2671–2683.
124. Erickson, P.A. and Liao, C.-h.: Statistical validation and an empirical model of hydrogen production enhancement found by utilizing passive flow disturbance in the steam-reformation process, *Experimental Thermal and Fluid Science*, 2007, 32(2), 467–474.
125. Erickson, P.A., Feinstein, J.J., Ralston, M.P., Davieau, D.D., and Sit, I.K.: *Testing of Methane Steam Reforming in a Novel Structured Catalytic Reactor Providing Flow Impingement Heat Transfer*, Proceedings of the AIChe 2010 Annual Meeting, Salt Lake City, Utah, 2010.
126. Liu, S., Li, W., Wang, Y., and Xu, H.: Catalytic partial oxidation of methane to syngas in a fixed-bed reactor with an O_2-distributor: The axial temperature profile and species profile study, *Fuel Processing Technology*, 2008, 89(12), 1345–1350.
127. Zhu, J., Rahuman, M.S.M.M., van Ommen, J.G., and Lefferts, L.: Dual catalyst bed concept for catalytic partial oxidation of methane to synthesis gas, *Applied Catalysis A: General*, 2004, 259(1), 95–100.
128. Patel, K.S. and Sunol, A.K.: Modeling and simulation of methane steam reforming in a thermally coupled membrane reactor, *International Journal of Hydrogen Energy*, 2007, 32(13), 2344–2358.
129. Tang, H.-Y.: Reactor controller design for steam and autothermal reforming for fuel cell applications. PhD dissertation, University of California, Davis, CA, 2009.

4

Hydrogen Production Using Solid Fuels

Nirmal V. Gnanapragasam
University of Ontario Institute of Technology

Bale V. Reddy
University of Ontario Institute of Technology

Marc A. Rosen
University of Ontario Institute of Technology

CONTENTS

Hydrogen is an energy carrier like electricity but with advantageous applications, especially in transportation (through engines and fuel cells). Hydrogen does not occur naturally in large quantities or in high concentrations on Earth. It must be produced from other compounds such as water and hydrocarbons (coal, biomass, solid wastes, natural gas, heavy oils, oil sands, etc).

Solid fuels such as coal, biomass, and solid wastes are increasingly used with water to produce hydrogen, and the energy transfer/conversion processes have been enhanced through recent developments (Gnanapragasam et al., 2010). The search for better uses for solid fuels is gaining prominence rapidly due to environmental imbalances caused by excessive CO_2 emissions from solid fossil fuels in energy conversion processes (IPCC, 2008). The impact of global warming has focused efforts on reducing or capturing emissions of CO_2 and other greenhouse gases from all energy sources.

Producing hydrogen from solid fuels increases their commercial value, especially for coal in the transportation sector. By converting solid fuels into gases and liquids, two major problems associated with solid fuels are addressed: (1) increasing adaptability to gas/liquid intake processes and (2) enabling cost-effective capture of various pollutants during the solid-to-gas/liquid conversion. A third advantage pertaining to certain solid fuels such as solid wastes is their safe and effective disposal (IEA, 2008b). Collective information on these hydrogen production processes is required to assess the expected outcomes of research and development and other improvement efforts now in place and required in the future.

An extended review of selected technological advances in hydrogen production using solid fuels is presented in this chapter, focusing on energy, environment, and economic sustainability. The advantages and problems associated with new developments are assessed based on research and development involving design, integration, and economics.

The process of producing hydrogen from various solid fuels follows several stages of processes, and the sequence is outlined in Figure 4.1. Solid fuels, depending on their natural occurrence and characteristics, are processed (sometimes with water) to use their exergy for deriving hydrogen through various high- and low-temperature processes. Utility processes involve those that pretreat solid fuels in cleaning, drying, and storing before being used in energy conversion processes. Any system that uses solid fuels in producing hydrogen needs oxygen, steam, and electricity for various stages of hydrogen production, and generating them becomes part of the utility processes.

Energy conversion processes primarily involve converting solid fuels into gaseous and liquid forms for further transformation into hydrogen. As listed in section 3 in Figure 4.1,

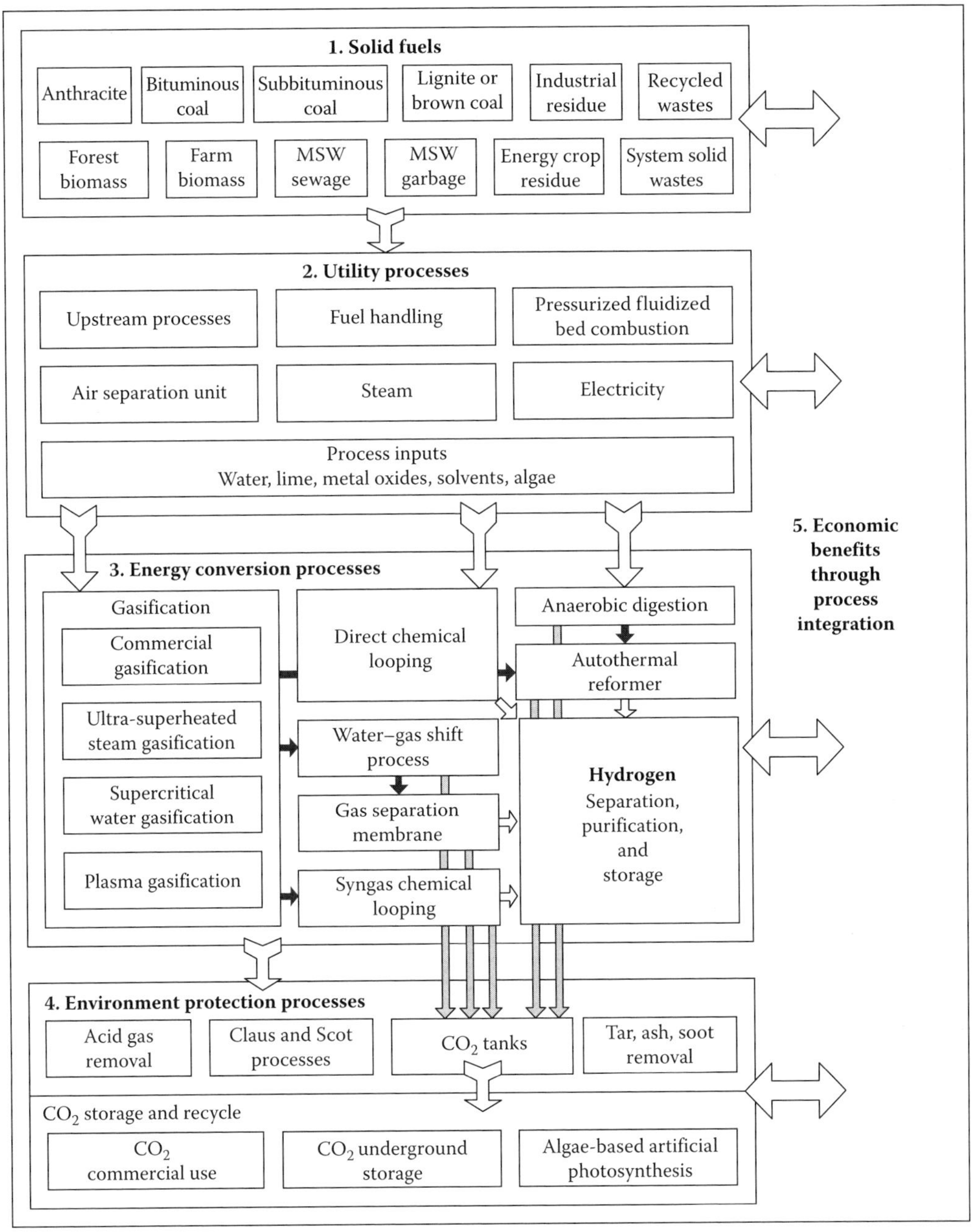

FIGURE 4.1
Organizational flowchart of various processes involved in hydrogen production from solid fuels, representing an outline of subject matter discussed in this chapter.

gasification, chemical looping, and anaerobic digestion are the primary processes that are discussed in this chapter. Although hydrogen storage is listed as part of energy conversion processes, it is of a different nature that involves issues associated with storing hydrogen for longer periods of time. Hydrogen storage is not discussed here but is covered in another chapter of this book.

Processes that enable environmental protection by capturing, controlling, and reducing releases of harmful gaseous and solid elements are discussed to provide an understanding of the energy and costs involved in these processes, which can hinder developments of certain technologies for producing hydrogen using solid fuels. To achieve environmental and economic benefits, energy-coordinated process integration is necessary and is discussed toward the end of this chapter.

4.1 Solid Fuels

Solid fuels supply the energy required to split hydrogen from water through various conversion processes, sometimes yielding additional products. Solid fuels as shown in Figure 4.2 include such hydrocarbons as coal, tar sands, oil shale, and bitumen; renewable organic materials such as biomass/charcoal, agricultural residue, and forest residue; and renewable inorganic materials such as municipal and industrial solid wastes.

4.1.1 Supply and Utilization

Reserves of solid fuels are significantly higher than those for other fossil fuels. Of the current global energy use, 80% is supplied by fossil fuels; 13.5% by renewable sources like solar, wind, and geothermal energy; and 6.5% by nuclear energy (Asif and Muneer, 2007).

Coal is the most significant contributor among fossil fuels to the current global electricity generation, accounting for 40% (IEA, 2008b). The most abundant fossil fuel on the planet, current estimates of global recoverable coal reserves range from 216 years to over 500 years (British Petroleum, 2003) at present usage rates. By the year 2025, it is expected that the United States will require over 250 GW of new electrical generation

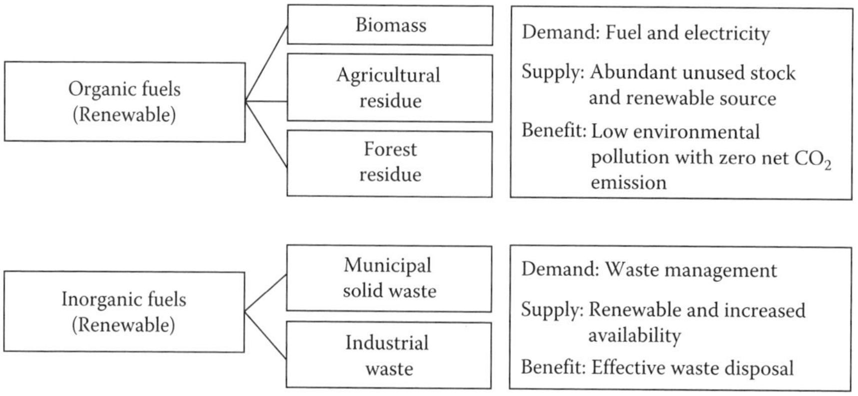

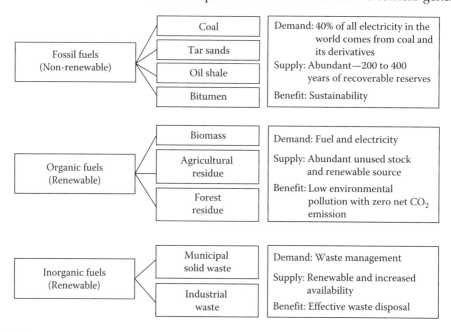

FIGURE 4.2
Types of primary solid fuels with an overview of the demand, supply, and benefit of each.

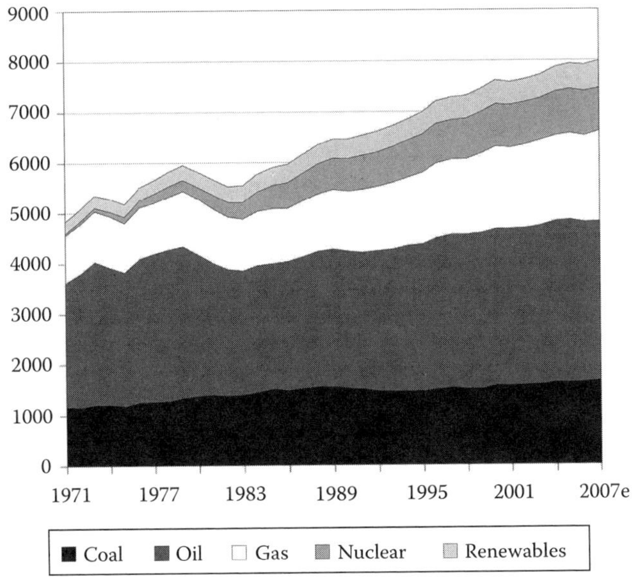

FIGURE 4.3
TPES by fuel in million tons of coal equivalent (Mtce) for member countries of the OECD. (Reprinted from IEA, World energy outlook 2008, International Energy Agency Report, Paris, France, 2008a; IEA, Energy technology perspectives 2008: Scenarios and strategies to 2050, International Energy Agency Report, Paris, France, 2008b.)

capacity even without considering replacing old plants (DOE, 2005). Of this new capacity, the International Energy Agency (IEA) estimates that 80 GW will be met through the construction of coal-based power plants, through advanced cocombustion and cogasification processes. Worldwide installed capacity of coal-based plants is expected to increase by over 40% in the next 20 years, exceeding 1400 GW by 2025 (DOE, 2005). When the hydrogen demand becomes equal to that of oil in the transportation sector or electricity in the power sector, solid fuels especially coal will likely have an important role to play owing to their abundance and low market price (IEA, 2008a). Within the member countries of the Organization for Economic Co-operation and Development (OECD), the nature of coal use relative to other energy sources for the last four decades is shown in Figure 4.3. The rapid increase in nuclear and renewable energy in recent years (after 2001 in Figure 4.3) shows the significant efforts of these countries in curbing greenhouse gas emissions.

Only nonagglomerating coal is considered for the purpose of hydrogen production, especially when using gasification as the primary conversion process. In 2007, hard coal (anthracite and bituminous coal) production increased by 6.5% (or 338.0 Mt) to 5542.9 Mt. Brown coal (subbituminous coal) production increased by 0.9% (or 7.9 Mt) to 945.2 Mt, a little above its 1994 level (IEA, 2008a). Total global coal production increased by 5.6%, well above the 10-year average growth trend of 3.4%. Total global coal consumption increased by 6.2% or 271.7 Mtce in 2007, which follows a 4-year trend of annual increases averaging 6.6%. These numbers suggest an increase in worldwide demand for electricity (which is the largest end use for coal), largely due to the migration of vast numbers of potential consumers to the OECD countries, an increase in living standards across the planet and an increase in global population (IEA, 2008b). The total primary energy supply (TPES) by fuel shares for the year 2007 is shown in Figure 4.4, where it is observed that coal delivers up to 26.4% and still maintains the lead role in electricity generation (IEA, 2009). The calorific values of various coals and biomass based on proximate and ultimate analyses are available elsewhere (Parikh et al., 2007).

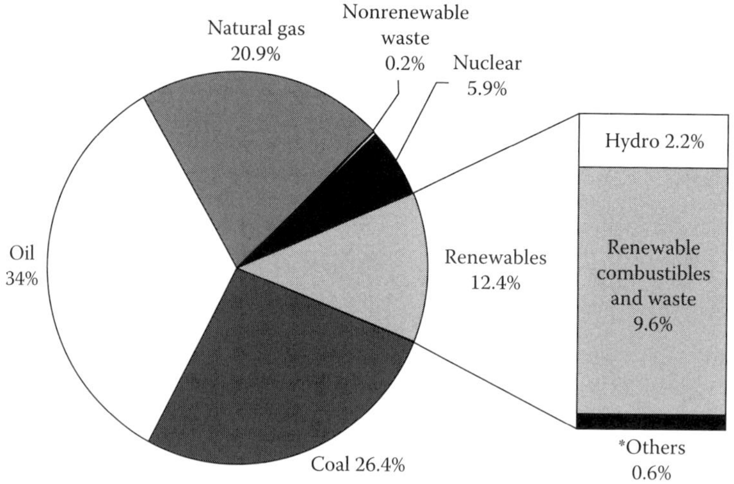

FIGURE 4.4
Breakdown of energy form from the 2007 world TPES. (From International Energy Agency, Renewables information 2009 with 2008 data, 2009. Available at www.iadb.org/intal/intalcdi/PE/2009/03711.pdf, Accessed on December 5, 2013.)

Biomass capable of use in hydrogen production includes agricultural residues, forest residues, energy crop residues, organic solid wastes, and firewood (charcoal). From data gathered by the IEA (2009) shown in Figure 4.4, only 12.4% is contributed by all renewable resources to the TPES in the world. A contribution of 9.6% is directly from renewable solid fuels such as biomass, residues, and municipal solid waste (MSW).

A total of about 36% of energy contributions are from solid fuels in recent years, providing ample opportunity for producing hydrogen using the energy from these solid fuels. Of the 12.4% renewables in Figure 4.4, 77.3% is shown in Figure 4.5 to be from solid biomass, charcoal, and MSW except for 1.1% from landfill gas, far more than any other renewable energy resources. But biomass tends to be bulky, to deteriorate over time, and to be difficult to store and handle. Compared to coal and oil, biomass has a lower energy density (GJ per unit of weight or volume), which makes handling, transport, storage, and combustion more difficult (IEA, 2006). Although the contribution of solid renewables to the TPES has increased by 9% compared to 1990 levels (IEA, 2009), much of it is used for combustion-based energy systems.

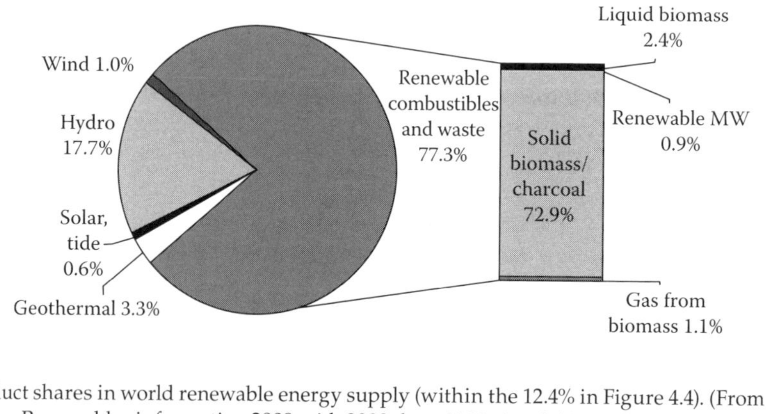

FIGURE 4.5
The 2007 product shares in world renewable energy supply (within the 12.4% in Figure 4.4). (From International Energy Agency, Renewables information 2009 with 2008 data, 2009. Available at www.iadb.org/intal/intalcdi/PE/2009/03711.pdf, Accessed on December 5, 2013.)

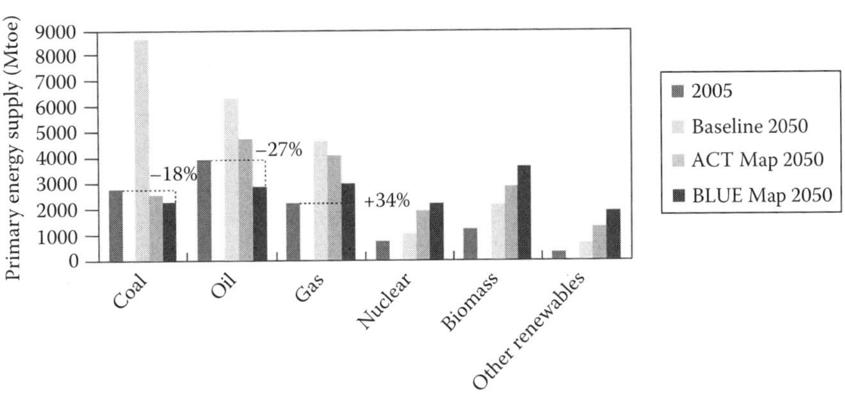

FIGURE 4.6
Comparison of world primary energy supply for year 2005 with values for 2050 using the baseline, ACT map, and BLUE map projections. (Reprinted from IEA, Energy technology perspectives 2008: Scenarios and strategies to 2050, International Energy Agency Report, Paris, France, 2008b.)

The IEA (2008b) recently compared energy resource types, conversion technologies, and associated policies based on three scenarios (used in Figure 4.6) scaled to the year 2050: (1) the ACT map scenario of the IEA, which implies adoption of a *wide range of technologies* with marginal costs of up to USD 50/ton of CO_2 saved when fully commercialized; (2) the BLUE map scenario, which requires deployment of *all technologies* involving costs of up to USD 200/ton of CO_2 saved when fully commercialized; and (3) the baseline scenario, which reflects developments likely to occur with energy and climate policies implemented to date. While the ACT scenarios are demanding, the BLUE scenarios depend on rapid implementation of unprecedented and far-reaching new policies in the energy sector (IEA, 2008b), which will take more time and effort to implement than the ACT scenarios.

From the data in Figure 4.6, coal appears to be the least favorable fuel for the future owing to its CO_2 emissions based on existing commercial-level combustion technologies, which mostly generate electricity. But this trend will change when coal is used for producing hydrogen (Gnanapragasam et al., 2009b), since the downstream processes enable easier capture of CO_2 and, when combined with biomass, can also help in reducing overall CO_2 emissions. Biomass has tremendous potential (as seen from Figure 4.6) for delivering more energy than coal and other resources by 2050 when the BLUE map is implemented effectively. Thus, it appears to be promising to use biomass for hydrogen production along with high-energy-intensity fuels such as coal.

Industrial inorganic solid wastes such as scrap tires (Stringfellow and Beaumant, 2008) and plastics (Aznar et al., 2006) are also being considered as *alternative* fuels through effective gasification processes. The supply and utilization of other solid fuels are discussed in detail elsewhere (IEA, 2006, 2007, 2008c; Asif and Muneer, 2007).

4.1.2 Issues Associated with Solid Fuels

Solid fuels contain carbon and their use in producing hydrogen results in emissions of CO_2, which is currently the most significant greenhouse gas and to which increasing temperatures in the lower atmosphere are attributed (USEPA, 2008). The carbon from solid fossil fuels (shown in Figure 4.2) is of higher concern and requires measures to capture and store underground. Due to the recent necessity to curb CO_2 emissions, combustion processes are becoming of decreasing interest to the energy sector (IPCC, 2008). One option to avoid CO_2

emissions is the use of renewables, but renewable energy sources that can provide base-load electricity—hydro, biomass, and geothermal—together are not anticipated to be able to satisfy even existing electricity demands and are not climate-neutral when operating continuously (Moriarty and Honnery, 2007). Another possible option, coal gasification with carbon capture and sequestration/storage (CCS), entails large energy and cost penalties using commercially proven methods of CO_2 separation (Muradov and Veziroglu, 2008).

In an effort to demonstrate the potential of CO_2-free coal-based power generation, a company in Germany has commissioned and is operating a power plant incorporating a complete CO_2 capture and sequestration facility (Harrabin, 2008). CO_2 capture and separation for gasification (Pennline et al., 2008) involves novel concepts in wet scrubbing with physical sorption, chemical sorption with solid sorbents, and separation by membranes.

Energy crops such as biomass resources are a risk for a sustainable future, since the carbon debt arising from land-use changes can take many years, even centuries, to pay back by using the biomass produced to displace fossil fuels (Fargione et al., 2008). Producing biomass for energy purposes at the expense of either food or fiber supplies, or by increasing deforestation, is of little global benefit (IEA, 2008b).

Other issues regarding solid fuels such as washing, drying, ash removal, and solid pollutant removal are well understood and managed in existing solid fuel–based energy systems and are discussed further elsewhere (de Souza-Santos, 2004; Rezaiyan and Cheremisinoff, 2005; Higman and Burgt, 2008).

4.2 Utility Processes

Utility processes are those that involve preparing solid fuels for actual energy conversion processes where hydrogen production begins in gaseous or liquid forms.

4.2.1 Upstream Processes

Upstream processes (in Figure 4.1) involve cleaning, blending, and upgrading solid fuels to enhance the quality of feedstock, thus improving the efficiency of various downstream conversion processes (CanmetENERGY, 2005) and also simplifying the separation of pollutants associated with solid fuels (NEDO, 2004). Some of the upstream cleaning processes that are in use in Japan and other larger importers of coal include (1) using a cartridge system, where all solid feedstocks are blended to form a uniform mixture containing a standardized composition; (2) treating the feedstock with solvents to clean the fuel of unusable residue; (3) blending of high-sulfur, high-grade coals with low-sulfur, low-grade coals and high-ash biomass (to avoid sintering); and (4) upgrading low-grade solid fuels with pretreatment using heavy oils (NEDO, 2004). These upstream processes likely will eventually be part of every energy conversion facility in the world, due to the long-term environmental and economic benefits and the desire of governments to implement energy efficiency policies (Eldridge et al., 2009) for improving the sustainability of operating existing and future energy systems.

4.2.2 Fuel Handling and Air Separation Unit

Solid fuels arriving at energy transfer plants need temporary storage, drying, crushing/milling, and internal transportation mechanisms. The handling of solid fuels consumes some energy with operation and maintenance costs and is vital to the functioning of systems that produce hydrogen using energy from solid fuels. Due to their availability and widespread use (de Souza-Santos, 2004), solid fuel handling is mature and is not discussed in detail here. Commercially, ProcessBarron (www.processbarron.com) is one of several private companies that specializes in design, manufacture, and commissioning of integrated solid fuel handling equipment for a given energy conversion system.

The nature of integration of such fuel handling equipment within the system should entail limited wastes in all forms (energy, material, and cost). For example, storing the right type of fuel in a feasible environment reduces energy requirements for drying and transportation. Using waste heat from various processes within the system, to dry high-moisture feedstocks before crushing, increases the specific energy output of the fuel and overall system energy efficiency.

Air separation (Kerry, 2006) is crucial to enable both oxyfuel combustion and oxyfuel gasification processes that result in gaseous CO_2 only at the exhaust (after hydrogen separation). This makes the CO_2 underground storage much easier by compressing only CO_2. Removing N_2 from air also increases the residence time of high-temperature synthetic gas (syngas) within the gasifier. When N_2 is not removed from air, the high nitrogen content in the gasifier yields syngas with a low heating value, 4–6 MJ/Nm^3 based on HHV (McKendry, 2002). Oxygen and steam gasification on the other hand may produce a gas with a medium heating value, 10–18 MJ/Nm^3 based on HHV (Hofbauer et al., 2003).

Air separation units (ASUs) consume a considerable amount of electric power, up to 10% of that for the entire system, especially in larger systems using cryogenic separation (Li, 2007), thus requiring a careful assessment of the need for only high-quality processes (Kerry, 2006). Membrane-based gas separation units (ion transport membranes [ITMs]) have been proven to be cheaper to install than cryogenic-based air separation by about 8% (Stiegel, 1999), and they also consume less power and are likely to prevail in most hydrogen-producing systems in the future.

4.2.3 Steam and Electricity: Pressurized Fluidized Bed Combustion

Circulating fluidized bed (CFB) combustors operating at atmospheric pressure are commercially established globally, that is, more than 400 power units are in operation around the world (Kavidass et al., 2000), with approximately 3000 operating CFB boilers in China alone (Yue et al., 2009). Fluidized beds are particularly suited to the combustion of low-quality coals, and most existing circulating fluidized bed combustion/combustor (CFBC) plants burn such materials along with biomass in cofiring mode. Moving to supercritical cycles is a logical step for very large CFB units (IEA, 2008b). A 460 MW supercritical unit is under construction at Lagisza, Poland, with start-up anticipated in 2009. This unit is expected to have a thermal efficiency of 70%. CFB combustor-based power plants using oxygen instead of air (known as oxy fuel combustion) are advantageous for CO_2 capture (Anthony, 2008; Manovic et al., 2008). There are hundreds of atmospheric CFB combustors operating worldwide, including a number of plants as large as 250–300 MW. Designs for larger supercritical CFBC units (600 MW capacity) are being developed (IEA, 2008b) with still higher efficiencies.

Any system that produces hydrogen involves processes that need high-quality steam and electricity, which may be produced on-site with a pressurized fluidized bed combustion unit. This process is of particular interest when considering the energy, environment, and economic benefits through integration and was developed by the US Department of Energy and industry partners (Weinstein and Travers, 2002). The process is still under development but is moving into demonstration and commercial stages owing to its small- and medium-scale capacities and higher operating efficiencies, above 70% (IEA, 2008b).

Pressurized fluidized bed combustion is employed at high pressures in combined cycles, with the boiler exhaust gases routed to generate additional power through heat recovered from the gas turbine exhaust. The combined gas and steam cycle achieves efficiencies up to approximately 44% (DOE, 2007). The first of such units had a capacity of about 80 MW, but two larger units are operating in Karita and Osaki, Japan, the former using supercritical steam (IEA, 2008b).

Incorporating advanced pressurized fluidized bed combustion/combustor (PFBC), with a thermal efficiency exceeding 70%, would enhance the use of nongasifiable solid feedstock, such as certain types of biomass, which then may be cofired with coal in the advanced PFBC to produce process steam and electricity.

4.3 Energy Conversion Processes

Much research on converting solid fuels to hydrogen is in early stages, but developments are expected to accelerate in the near future as the demand for hydrogen increases in various applications. Recent research into processes for converting solid fuels to gaseous forms indicates that such processes have a significant potential for commercialization and should be able to achieve industrial-scale production levels (Gnanapragasam et al., 2010).

Hydrogen derived from coal is slated to be the primary objective in the fuel program of the US Department of Energy (Lior, 2008), and research is ongoing to develop modules for coproducing hydrogen from coal via systems integrated with advanced coal power plants at prices competitive with crude oil (DOE, 2007).

Primary conversion processes in producing hydrogen from solid fuels include direct processes such as gasification, anaerobic digestion, fermentation, and liquefaction (Chmielniak and Sciazko, 2003). There are also direct chemical looping (DCL) processes such as the iron oxide cycle and indirect thermochemical cycles based on copper–chloride and sulfur–iodine. Some of these processes (copper–chloride and sulfur–iodine cycles) are undergoing extensive research for commercial development and implementation (IEA, 2008b) and thus are not discussed here. The process diagrams in this chapter are wherever possible self-explanatory.

Coal, biomass, solid wastes, and oil sand coke (Furimsky, 1998) are potential gasification fuels for syngas production and methanol synthesis (Chmielniak and Sciazko, 2003). Applications of these processes depend on feedstock characteristics, volume of production, and postconversion processes to manage wastes and by-products (Higman and Burgt, 2008). A review of decarbonization processes for fossil fuels ranging from natural gas to coal (Muradov and Veziroglu, 2008) has identified the commercial potential of new technologies, which suggests various uses for solid carbon after CO_2 sequestration.

The average energy and exergy conversion efficiencies of syngas from solid fuel gasification (76% and 75%, respectively) are found to be higher than those for hydrogen (64% and 55%) production from gasification (Bargigli et al., 2004). However, coal-to-syngas conversion generates a significant amount of solid waste, and the material intensity is much higher for syngas than for natural gas and hydrogen (21 and 39 g/g, respectively), indicating a higher load on the environment that should be dealt with carefully.

Technoeconomic comparisons of hydrogen production via steam–CH_4 reforming (SMR), coal and biomass gasification, and water electrolysis have been reported (Langer et al., 2007).

Increases in natural gas prices are observed to make coal gasification as well as biomass gasification competitive, provided its technological barriers are overcome (including feedstock processing and postconversion of products). A comparison of electricity and hydrogen production processes, from coal and natural gas with CO_2 capture, for various technologies and for large-scale and decentralized systems (Damen et al., 2006, 2007), demonstrated that a short-term net power efficiency of 32%–40% is achieved by an integrated gasification combined cycle (IGCC) system with production costs of 4.7–6.3 €ct/kWh.

Coal gasification forms the central element of IGCC systems and has the greatest fuel flexibility of advanced technologies for power generation (Beer, 2007). Current commercial gasification technologies are also well adapted to using biomass and other low-value feedstocks that have high ash residues (Liu and Niksa, 2004). Gasification also permits the control and reduction of gaseous pollutant emissions (Trapp, 2005) and a possible low-cost approach to concentrate CO_2 emissions at high pressure to facilitate underground sequestration.

4.3.1 Gasification

Presently, gasification is the only commercial, large-scale option for converting solids to gases (Rezaiyan and Cheremisinoff, 2005) and one of the cleanest conversion technologies for solid fuels. Academic and industrial research has improved production capabilities and operating efficiencies in recent years. The gasification of carbonaceous, hydrogen-containing fuels is an effective method for thermal hydrogen production (Stiegel and Ramezan, 2006) and is considered a key technology in the transition to a hydrogen economy (Collot, 2006). Gasification converts solid fuels into a syngas comprised mainly of CO, CO_2, H_2, CH_4, H_2O, and other constituents in minor concentrations (Higman and Burgt, 2008). Syngas production offers the possibility of obtaining multiple products that can be used for different applications. Gasification has the highest energy conversion efficiency relative to other solid fuel conversion technologies (Beer, 2000). One particular version of a commercial gasification process is shown in Figure 4.7, as an example of the initial stage of solid fuel conversion.

Producing hydrogen from syngas is a significant step in the clean coal technology roadmap as realized by many countries including Canada (CanmetENERGY, 2005), Japan (NEDO, 2004), and the United States (DEO, 2002). Gasification carries great significance for coal and dry biomass, while for wet biomass and sewage, other conversion processes such as anaerobic digestion and supercritical water gasification (SCWG), respectively, appear advantageous due to the higher moisture content in the feedstock (Mozaffarian et al., 2004).

The British Gas–Lurgi (BGL) gasifier shown in Figure 4.7 is a countercurrent, moving-bed, slagging gasifier operating at pressures of 25 bar or higher (NETL, 2000). The reactor vessel is water cooled and refractory lined. The coal and/or biomass mixture is fed into the top of the gasifier via a lock-hopper system and reacts while moving downward through the gasifier. The coal's ash/mineral matter is removed from the bottom of the gasifier as molten slag through a slag tap, then quenched in water, and removed. Steam and oxygen are injected through tuyere nozzles near the base of the gasifier and react with the coal as the gases move up. This countercurrent action results in a wide temperature difference between the top and the bottom of the gasifier. After gas conversion, the ash and unconverted char from coal/biomass end up as slag and get collected at the bottom of slag lock after cooling. The typical operating temperature for this type of gasifier is from 600°C to 1900°C (Higman and Burgt, 2008). The gasifier can be characterized due to this temperature profile as being divided into drying, devolatilization, gasification, and combustion zones from top to bottom, respectively.

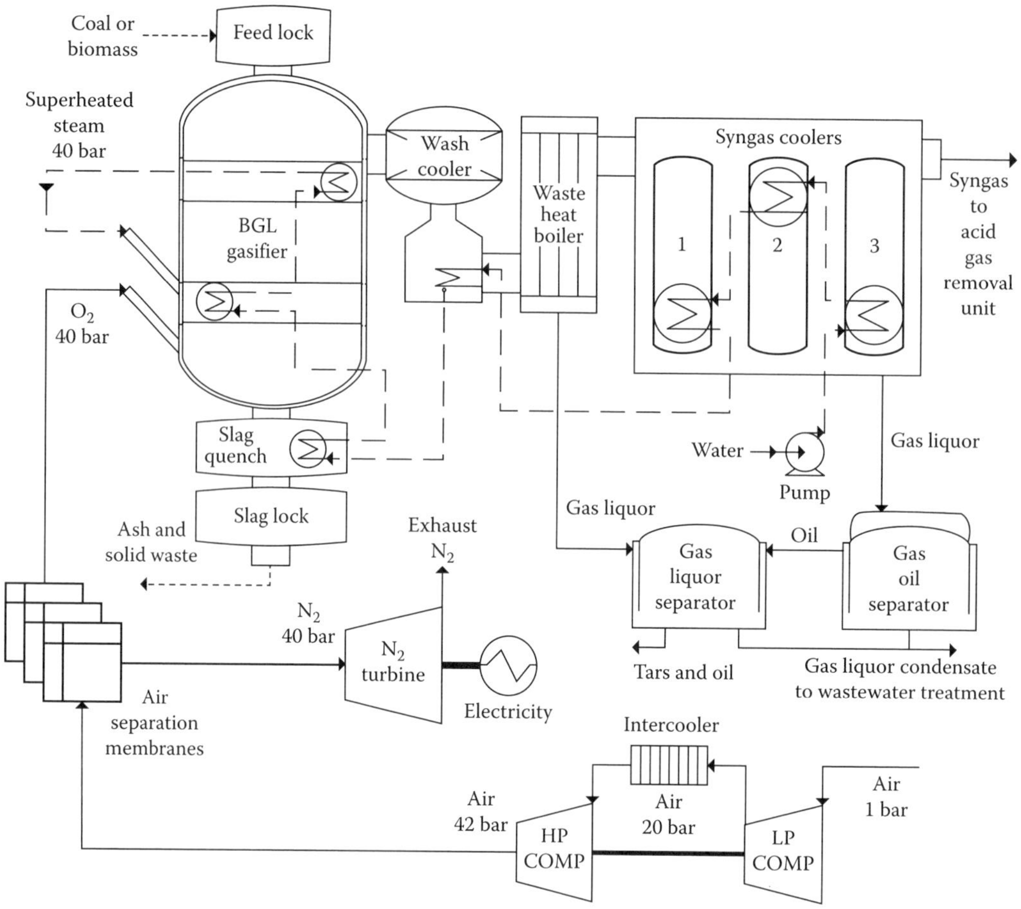

FIGURE 4.7
Industrial-scale gasification process with BGL downdraft gasifier. (Adapted from DOE, Gasification: Reference shelf—System studies, National Energy Technology Laboratory, U.S. Department of Energy, Washington, DC, accessible at: http://www.netl.doe.gov/technologies/coalpower/gasification/pubs/system-studies.html, 2000.)

Experimental and modeling investigations of long-stick wood gasification in a top-lit updraft fixed-bed gasifier describe common problems of many gasifier designs (Saravanakumar et al., 2007). They produce so much tar that the gas cleanup equipment cost is several times the gasifier cost. Fluidized beds typically produce 0.5%–4% tars, while updraft biomass gasifiers produce 10% tars. Both stratified downdraft and top-lit updraft gasifiers produce much lower tar levels, typically 0.1%. For the gasifier, the top-lit updraft mode is found to provide more satisfactory overall performance (NETL, 2000); the BGL gasifier (Figure 4.7) belongs to this category.

4.3.1.1 Gasification Process Design

The two combustion-related factors of greatest impact on gasifier design and operation have been reported to be (1) fuel reactivity and (2) slag flow as it runs down the refractory walls and out of the bottom of the reactor under gravity (Wall, 2007). A gasification product gas with tar content below the targeted limit of 2 g/m^3 can be achieved only under special conditions in terms of gasifier design and operation as well as feedstock characteristics (Corella et al., 2006).

The main chemical reactions that occur inside the gasifier with appropriate energy conditions (Ptasinski et al., 2007; Mountouris et al., 2008) are as follows:

Exothermic combustion reaction:

$$C + O_2 \rightarrow CO_2; \quad \Delta H = -393.5\ \text{kJ/mol} \tag{4.1}$$

Endothermic Boudouard equilibrium process:

$$C + CO_2 \rightarrow 2CO; \quad \Delta H = 172.6\ \text{kJ/mol} \tag{4.2}$$

Endothermic heterogeneous water–gas shift (WGS) reaction:

$$C + H_2O_{(g)} \rightarrow CO + H_2; \quad \Delta H = 131.4\ \text{kJ/mol} \tag{4.3}$$

Exothermic hydrogenation gasification:

$$C + 2H_2 \rightarrow 2CH_4; \quad \Delta H = -74.9\ \text{kJ/mol} \tag{4.4}$$

After formation, these gases tend to react based on local temperature and pressure gradients and nonequilibrium conditions. Depending on the amount of steam available, the following reactions may occur within the gasifier and in the WGS reactors in the H_2 system: Endothermic CH_4 decomposition:

$$CH_4 + H_2O \rightarrow CO + 3H_2; \quad \Delta H = 206.12\ \text{kJ/mol} \tag{4.5}$$

Exothermic WGS reaction:

$$CO + H_2O \rightarrow CO_2 + H_2; \quad \Delta H = -41.2\ \text{kJ/mol} \tag{4.6}$$

At high gasifier temperatures, the sulfur in coal reacts with CH_4 and steam to form H_2S in a two-step process (Patnaik, 2002) as shown in the following discussion; this is removed using stages of separation processes, the descriptions for which can be found in the literature (Garcia et al., 2006):

$$CH_4 + 2S \rightarrow CS_2 + 2H_2 \tag{4.7}$$

$$CS_2 + 2H_2O \rightarrow 2H_2S + CO_2 \tag{4.8}$$

The overall reaction in the gasifier, considering the use of CO_2 as a coal carrier gas and using oxygen instead of air for gasification, may be represented as

$$\begin{aligned} &\dot{n}_{\text{coal}} \overbrace{(y_C C + y_{H_2} H_2 + y_{N_2} N_2 + y_{O_2} O_2 + y_S S + y_{\text{ash}})}^{\text{coal}} + \dot{n}_{O_2} O_2 + \dot{n}_{CO_2} CO_2 + \dot{n}_{\text{st}} \overbrace{(H_2O)}^{\text{steam}} \rightarrow \\ &\dot{n}_{\text{syn}} \underbrace{(y_{CO_2} CO_2 + y_{CO} CO + y_{pH_2} H_2 + y_{CH_4} CH_4 + y_{H_2O} H_2O + y_{pN_2} N_2)}_{\text{Syngas}} + \dot{n}_{H_2S} H_2S + \dot{n}_{\text{ash}} \end{aligned} \tag{4.9}$$

where

y is the molar fraction
n is the molar flow rate
p denotes the product

4.3.1.2 Gasifier Process Optimization

An assessment of the performance of a gasifier operating without air preheating and using a higher-moisture-content fuel (over 10% moisture by weight) demonstrates the benefits of high-temperature air preheat (Young and Pian, 2003). Performance was observed to improve minimally by drying a manure–waste fuel to below 10% moisture. An investigation of the selection of IGCC candidate coals using a pilot-scale gasifier operation showed that high-ash coal (about 17% by weight) yields less than 60% cold-gas efficiency (Yun et al., 2007). To improve the efficiency for such low-reactivity coal, it is suggested that the gasifier design should permit increased reaction time and the option of char recycling and that methods of inducing higher mixing of coal powder with reacting gases be considered. Since all types of coal can be gasified, gasification appears promising for the production of hydrogen from coal. Gasification plants are also advantageous, compared to other coal-based alternatives, for CO_2 capture (Collot, 2006) as mentioned earlier.

An optimization of catalytic steam gasification of biomass at low temperatures for hydrogen production suggests that higher temperatures and steam flow rates increase syngas yield (Moghtaderi, 2007). At 600°C and high steam content (about 90%), hydrogen production can be optimized with the help of a catalyst when using biomass. The status of catalytic gasification of biomass, similarities and differences between dry and wet processes, and challenges for future research and development from both catalysis and process viewpoints have been discussed (van Rossum et al., 2008).

The influences on hydrogen production of gasifier operating temperature, pressure and coal type, and steam-to-carbon ratio that have been examined based on chemical equilibrium (Wang et al., 2006) suggest that the operating pressure in the gasifier be maintained at 20–30 bar for enhanced CO_2 partial pressure and capture efficiency. The appropriate gasifier temperature is between 625°C and 850°C, for which more than 70% hydrogen can be obtained with a production efficiency of 46.45%. The overall system also cogenerates hydrogen and power with near-zero emissions.

An energy analysis of the biomass gasifier showed (Mahishi and Goswami, 2007) that the optimum conditions for hydrogen production occurred at a gasification temperature of 1000 K, a steam–biomass ratio of 3 and an equivalence ratio of 0.1, achieving a 54% energy efficiency for the gasifier. A similar thermodynamic analysis (Gnanapragasam et al., 2009a) of the performance of a gasification process under varying steam-to-coal (S/C) and oxygen-to-coal (O/C) ratios in the gasifier yielded a range of syngas compositions as given in Table 4.1. These values are based on a Gibbs free energy minimization equilibrium model for the gasifier when using anthracite coal with an HHV of 32.85 MJ. From Table 4.1, it is evident that, when 80% S/C and 70% O/C are used, the hydrogen produced is the highest (37 vol.%) and that the remaining CO will be converted to hydrogen and CO_2 during the WGS reaction stage. The amount of CO_2 emission is directly proportional to the carbon in the feedstock.

4.3.1.3 Fluidized Bed Gasifier

A novel fluidized bed gasifier fitted with water-cooled sampling probes for measuring the axial gas concentrations at various gasifier heights has been reported (Ross et al., 2007). Three distinct zones are observed in the gasifier. In the first zone (the bottom third of the distance from the distributor), with a temperature below 950°C, char combustion and gasification reactions dominate. In the second zone, with a temperature between 800°C and 900°C, the devolatilization products from the biomass combine with the char

TABLE 4.1

Simulated Syngas Volumetric Composition (without Nitrogen) for Four S/C Inlet Ratios and Five O/C Inlet Ratios, Considering an Anthracite Feedstock When Gasifier Operates at 20 bar, 1173 K

	Syngas Composition (Volume Fraction)							
S/C (%)	20	40	60	80	40	40	40	40
O/C (%)	70	70	70	70	60	80	90	100
H_2	0.19	0.31	0.35	0.37	0.25	0.31	0.28	0.24
CO	0.74	0.61	0.52	0.46	0.63	0.6	0.59	0.58
CO_2	0	0.05	0.09	0.12	0.04	0.06	0.09	0.12
CH_4	0.07	0.03	0.02	0.01	0.08	0	0	0
H_2O (g)	0	0.01	0.03	0.04	0.01	0.02	0.04	0.05

Source: Adapted from Gnanapragasam, N.V. et al., *Energy Conserv. Manage.*, 50, 1915, 2009a.

gasification reactions. The third zone is the freeboard with a temperature of 800°C and below, where variations in the main gas components occur due to the WGS reaction.

In the case of a fluidized bed gasifier, the fuel is gasified in a bed of small particles fluidized by a suitable gasification medium such as air or steam. The concern for climate change has increased the interest in biomass gasification for which fluidized bed gasifiers are particularly popular, occupying nearly 20% of the market (Basu, 2006). Fluidized bed gasifiers are divided into the following two major types: (1) bubbling and (2) circulating. Depending on the fuel and the application, the gasifier operates at a temperature within the range of 800°C–1000°C and at atmospheric and pressurized conditions. The hot gas from the gasifier passes through a cyclone, which separates most of the solid particles associated with the gas and returns them to the bottom of the gasifier for recirculation. An air preheater located below the cyclone raises the temperature of the gasification air and indirectly controls the temperature inside the gasifier.

4.3.1.4 Ultrasuperheated Steam Gasification

A new method for gasifying carbonaceous materials to syngas comprises the formation of an ultrasuperheated steam (USS) composition containing mainly water vapor, CO_2, and highly reactive free radicals (Lewis, 2007). The USS at temperatures ranging from 1316°C to 2760°C is a clear colorless flame; when it comes into contact with carbonaceous materials (feedstock), rapid gasification occurs to form a syngas. The syngas generated from USS gasification has a higher hydrogen fraction (more than 50%) than other gasification processes (Pei and Kulkarni, 2008). When used within an IGCC system, the overall efficiency is found to be lower, suggesting that USS gasification is more suitable for hydrogen production than power generation.

The USS gasifier is a long cylindrical reactor with a steel casing for the high-pressure (above 30 bar) operation and a ceramic lining (Lewis et al., 2002; Ryu, 2004) to resist the high temperatures of up to 2500°C. The pulverized coal or biomass is fed from the top into a pilot burner as shown in Figure 4.8, which initiates with some propane aided by synthetic air the high-temperature flame, which is a combination of 79% steam and 21% oxygen. The flame disintegrates hydrogen from steam, allowing the syngas to have a higher hydrogen fraction (more than 50%) than other gasification processes (shown in Table 4.1). The reactor

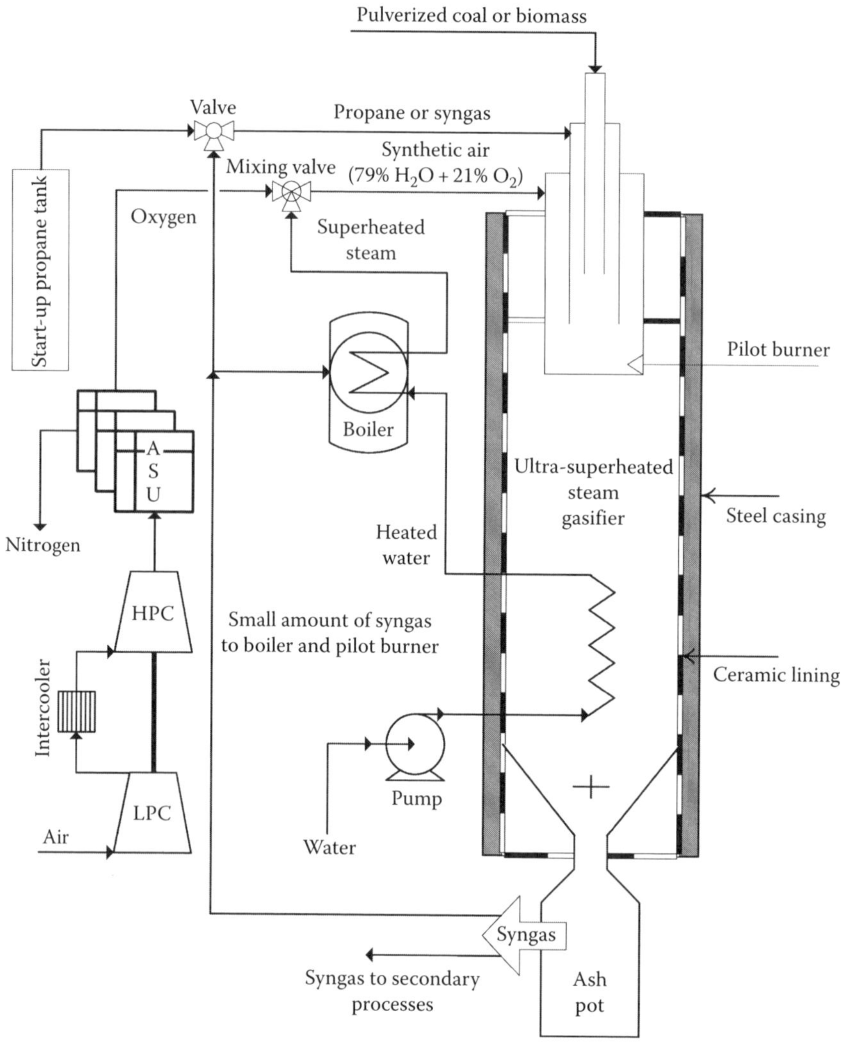

FIGURE 4.8
USS gasification. (Adapted from Lewis, F.M., Generation of an ultra-superheated steam composition and gasification therewith, US Patent No. US 7,229,483 B2, June 12, 2007.)

also generates a tremendous amount of heat, thus enabling self steam generation. The flame is maintained by using some portion of the syngas generated.

The ASU is mandatory for this type of gasifier, since the higher hydrogen percentage depends on the synthetic air composition; thus, oxygen becomes important for high-temperature combustion. This adds to the power requirement for the compressors unless membrane-based separation is used (Stiegel, 1999). The USS process is burner-based, not reaction-based, offering the end user a wide choice of reactor designs. There are no exothermic oxidization reactions within the reactor. Downstream of the USS steam envelope, gasification becomes a steam-only gasification and/or steam reforming process inherently resistant to the production of tars (Sieger and Donovon, 2002). No ash slagging occurs despite very high temperatures (Ryu et al., 2004), which makes the USS process ideal for small-scale hydrogen production.

TABLE 4.2

Experimental Syngas Volumetric Composition for Four Test Cases Involving Two Different USS Compositions with Char and Coal

	Syngas Composition (Volume Fraction of Dry Gas)			
Test Cases	**Wood Char**		**Anthracite Coal**	
Fuel composition (%)	Moisture—1.6 Volatile matter—21.8 Fixed carbon—78.2 Ash—1.4		Moisture—0.9 Volatile matter—11.6 Fixed carbon—81.5 Ash—6.0	
Fuel feed rate (g/s)	2.5		4	
USS composition	**70% H_2O, 30% CO_2**	**80% H_2O, 20% CO_2**	**70% H_2O, 30% CO_2**	**80% H_2O, 20% CO_2**
H_2	0.323	0.37	0.321	0.367
CO	0.254	0.236	0.243	0.203
CO_2	0.4	0.373	0.409	0.402
CH_4	0.02	0.019	0.019	0.014

Source: Adapted from Ryu, C. et al., *J. Energy Inst.*, 77, 46, 2004.

Wood char with a lower higher heating value than that of anthracite (about 2–3 MJ/kg) still is able to produce about 37% hydrogen for about 63% of its weight compared to anthracite, as can be seen from Table 4.2 when USS composition is at 80% steam and 20% CO_2. The data in Table 4.2 are mainly for steam gasification since no oxygen is supplied, thus belonging to a second type of USS process with temperatures ranging from 300°C to 700°C. This particular USS process does not require an ASU, reducing the penalty on the overall energy efficiency.

4.3.1.5 Supercritical Water Gasification

This process is aimed at generating hydrogen from the biogenic feedstock sewage sludge (Gasafi et al., 2007; Li et al., 2009). The process exploits the specific physical and chemical properties of water above its critical point (T = 374°C, P = 221 bar). These properties allow for a nearly complete conversion of the organic substance contained in the feed material to an energy-rich fuel gas containing hydrogen, CO_2, and CH_4. The characteristics of SCWG are examined and compared to other energy conversion technologies by modeling an overall energy system (Yoshida et al., 2003). A SCWG combined cycle is determined to be the most efficient conversion process for biomass with high moisture content. The breakeven point between thermal gasification and SCWG is approximately 40% biomass moisture content.

Gas yields, carbon gasification efficiency, and the total gasification efficiency increase with increasing temperature and reaction time and decreasing feed concentration (Yamaguchi et al., 2009). Hydrogen yield increases from 7% mole fraction to over 30% for an increase of reaction temperature from 600°C to 800°C for the gasifier shown in Figure 4.9. Among the parameters varied, temperature and feed concentration are found to have the most significant effect on the SCWG reaction behaviors. Higher temperatures and lower feed concentrations favor coal utilization for hydrogen production, promoting the reforming reaction of CH_4 and CO (Lu et al., 2008). The SCWG of brown coal requires

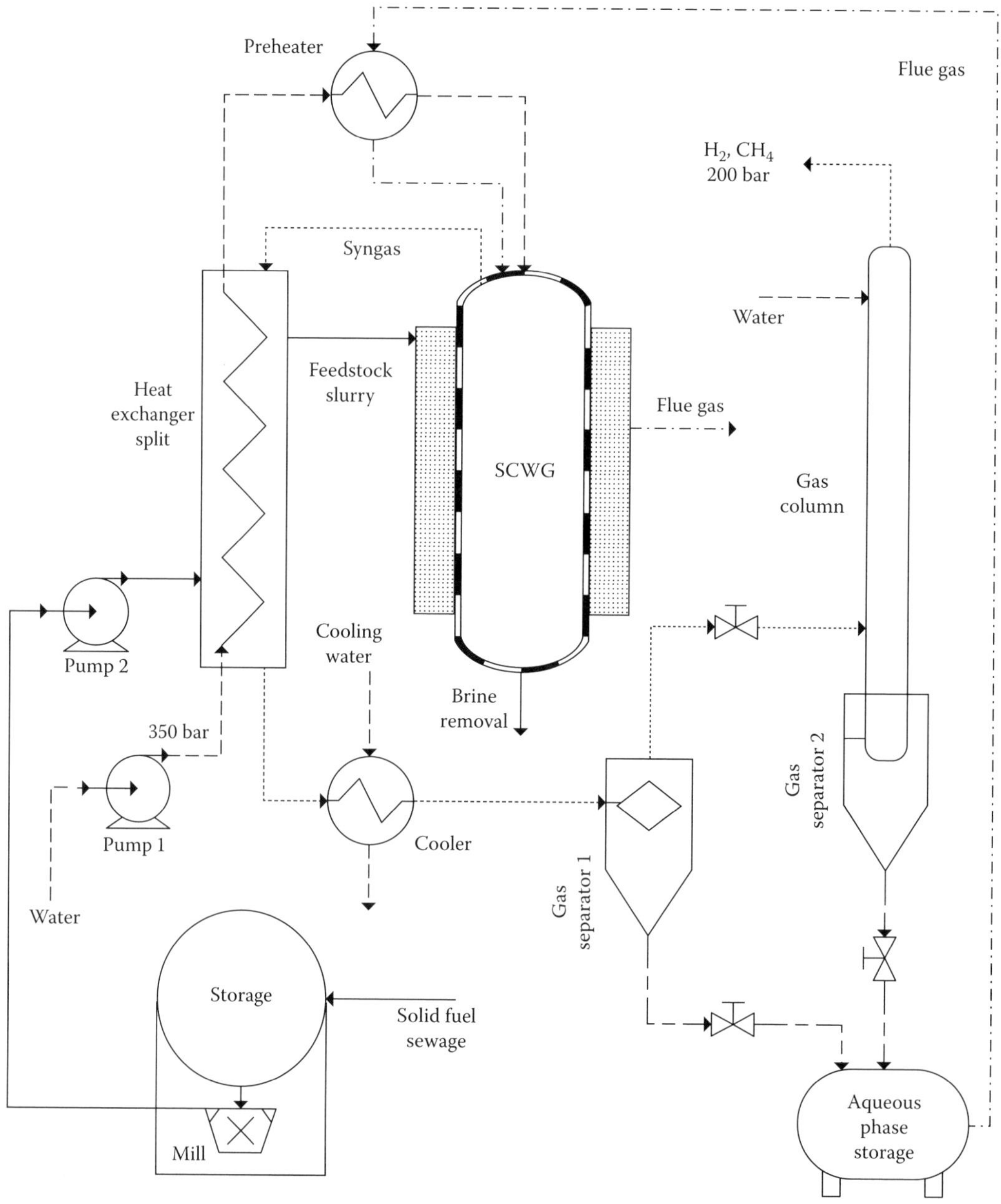

FIGURE 4.9
SCW gasification process. (Adapted from Gasafi, E. et al., *Int. J. Energy Res.*, 31, 346, 2007.)

a catalyst for achieving a high utilization efficiency that leads to a high hydrogen production closer to the theoretical limit.

To improve hydrogen production when using SCWG with biomass, an alkali, such as NaOH, KOH, Na_2CO_3, K_2CO_3, and $Ca(OH)_2$, is used as a catalyst or promoter. Of these, potassium carbonate (K_2CO_3) is found to catalyze gasification of the cellulose (in biomass) with the formation of more hydrogen and less CO (Guo et al., 2007). The activity of K_2CO_3 is higher than that of $Ca(OH)_2$ and K_2CO_3 cannot capture CO_2. A summary of the catalytic mechanism of K_2CO_3 for biomass gasification in supercritical water (SCW) occurs as given in the following (Sinag et al., 2003):

$$K_2CO_3 + H_2O \rightarrow KHCO_3 + KOH \quad (4.10)$$

$$KOH + CO \rightarrow HCOOK \quad (4.11)$$

$$HCOOK + H_2O \rightarrow KHCO_3 + H_2 \quad (4.12)$$

$$2KHCO_3 \rightarrow H_2O + K_2CO_3 + CO_2 \quad (4.13)$$

$$H_2O + CO \leftrightarrow HCOOH \leftrightarrow H_2 + CO_2 \quad (4.14)$$

More hydrogen was also obtained for cellulose gasification with the mixture of K_2CO_3 and $Ca(OH)_2$, and the CO_2 capture is enabled in this particular case. The processes within the SCWG system provide opportunities for waste energy integration, thus enabling a self-sustained process with feedstock delivering the corresponding product of syngas with a higher hydrogen fraction.

4.3.1.6 Plasma Gasification

Plasma gasification is another process for producing a hydrogen-rich syngas. The process has no limitation on feedstock characteristics and smaller requirements of air/oxygen and is a pyrolysis process becoming commercially popular in solid waste management facilities around the world (especially in Canada, Europe, Japan, and the United States). It is mostly used for efficient and clean disposal of municipal solid wastes and garbage (Carabin and Gagnon, 2007). When coal is gasified in a steam and air environment under arc plasma conditions, hydrogen production is enhanced (Galvita et al., 2007). The steam environment is estimated to increase syngas output by 30%–40% for plasma gasification compared to an air environment.

Efforts on process development and energy optimization for plasma gasification of sewage sludge have suggested (Mountouris et al., 2006) that a moisture content (w/w) of about 0.4 produces maximum hydrogen with lower CO, for 0.3 mol/mol of dry waste at a temperature of 1273 K. But at lower moisture content (about 0.1) for almost the same amount of hydrogen produced, an equal amount of CO is also produced, thus enabling higher overall hydrogen production, through the WGS process. Sewage treatment using plasma gasification indicates (Mountouris et al., 2008) that integration of predrying and electrical energy production processes (or hydrogen) not only is self-sufficient from an energy point of view but leads to the availability of additional electrical energy for distribution. The primary roadblock to performance of plasma gasification is the amount of moisture in the feedstock, and the sensitivity is shown in Table 4.3. Correlating with the data from Mountouris et al. (2006), the moisture content of feedstock is not critical for hydrogen production although it is for electricity generation, as seen from the data in Table 4.3. The equilibrium model based on plasma gasification demonstrates that an input of 250 tons/day of sewage sludge with 68% moisture yields a net electric power output of 2.85 MW.

From an energy viewpoint, the ratio of power available for material treatment (after all power losses are subtracted from the arc power) to the total arc power (as shown in

TABLE 4.3

Effect of Feed Moisture Content on Required Oxygen Amount, Energy Values, and Quality of Synthesis Gas

Moisture Content (% w/w)	Oxygen (mol/mol)	Net Thermal (MW)	Net Electrical (MW)	Syngas Heating Value (kWh/Nm³)
Zero gasification energy (when the energy for gasification is self-sustained)				
5.2	0.53	−1.07	4.17	1.12
10.23	0.55	−0.92	4.04	1.03
14.68	0.57	−0.77	3.90	0.96
20.50	0.60	−0.54	3.71	0.86
27.04	0.64	−0.25	3.45	0.74
29.88	0.66	−0.10	3.33	0.69
34.92	0.70	0.19	3.07	0.60
Zero net thermal energy (when net thermal energy is self-sustained)				
Moisture Content (% w/w)	**Oxygen (mol/mol)**	**Gasification Energy (MW)**	**Net Electrical (MW)**	**Syngas Heating Value (kWh/Nm³)**
14.68	0.213	−3.92	2.12	2.14
20.50	0.348	−2.77	2.45	1.52
27.04	0.535	−1.18	2.90	0.95
29.88	0.613	−0.52	3.09	0.78

Source: Adapted from Mountouris, A. et al., *Energy Convers. Manage.*, 49, 2264, 2008.

Figure 4.10) increased with increasing arc power from 0.35–0.41 at arc power 95–100 kW to 0.41–0.46 for arc power higher than 130 kW for wall temperatures of 1100°C–1200°C. The ratio is lower for higher wall temperatures (Van Oost et al., 2009).

Plasma gasification above 1300 K when using biomass appears to produce higher fractions of hydrogen in syngas (Van Oost et al., 2009) than at 800 K, ranging from 28% to 46% by volume. Due to their high costs, plasma systems have been used primarily for the vitrification of high-toxicity waste and mainly at the pilot scale (Moustakas et al., 2008).

As these systems become more accepted and their designs simpler, their use will likely be more extensive. Plasma gas used in gasifiers for vitrification also belongs to this category and can result in a significant waste reduction volume, ranging from about 5:1 for ash input to maximum 50:1 for solid waste (Minutillo et al., 2009). A better solution can be achieved using air as the plasma gas and adding oxygen to sustain the waste gasification. Then, the plasma gasification efficiency is 69%.

4.3.2 Anaerobic Digestion

Anaerobic digestion is similar to SCWG for producing syngas and substitute natural gas (SNG) by converting wet biomass with moisture of about 70–95 wt.% (Mozaffarian et al., 2004). Anaerobic digestion is a biological process that occurs in the absence of air and has been used to convert organic wastes to biogas, that is, a mixture of CH_4 (55–75 vol.%) and CO_2 (25–45 vol.%). During anaerobic digestion, typically 30%–60% of the solid input is converted to biogas. The by-products consist of an undigested residue and various water-soluble substances that may act as nutrients for algae growth and other biodigesters. Depending on the digestion system (wet or dry), the average residence time is between 10 days and 4 weeks. The anaerobic digestion of biomass and organic waste streams has the potential to facilitate energy recovery and sustainability from biodegradable waste (Duerr et al., 2007). With current developments in reformer technologies, hydrogen can be

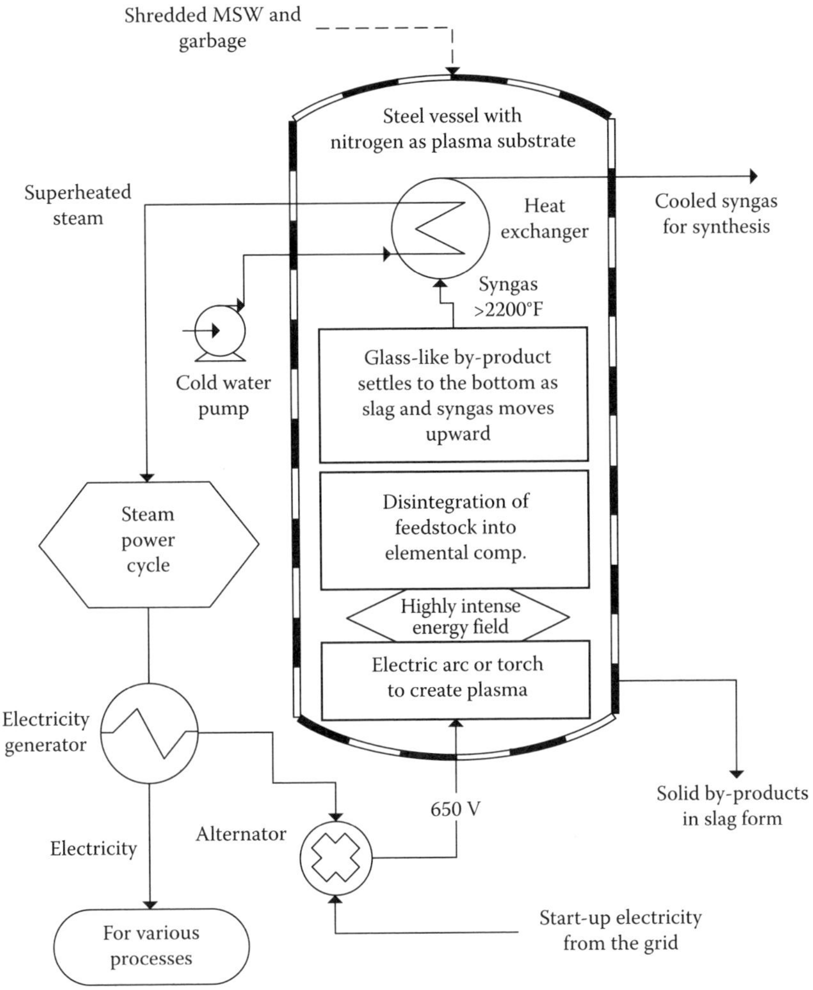

FIGURE 4.10
Plasma gasification process. (Adapted from Mountouris, A. et al., *Energy Convers. Manage.*, 49, 2264, 2008.)

produced from CH_4-derived anaerobic digestion of organic waste material, much of which is currently landfilled through autothermal reforming (ATR).

Factors affecting anaerobic digestion operation and performance of such system as shown in Figure 4.11 include physical composition of the manure, manure type and manure collection method, and frequency of feeding the digester, along with suitable conditions for bacteria growth, for biogas energy production (Brown et al., 2007). Anaerobic bacteria grows and performs best under optimal temperatures in the range of 951°F–1051°F, which has an optimal retention time range of 10–30 days and an optimal pH range of 7–8.5.

The production of biogas through anaerobic digestion offers significant advantages over other waste treatment processes (Ward et al., 2008), including the following:

- Production of less biomass sludge compared to aerobic treatment technologies.
- Success in treating wet wastes of less than 40% dry matter.

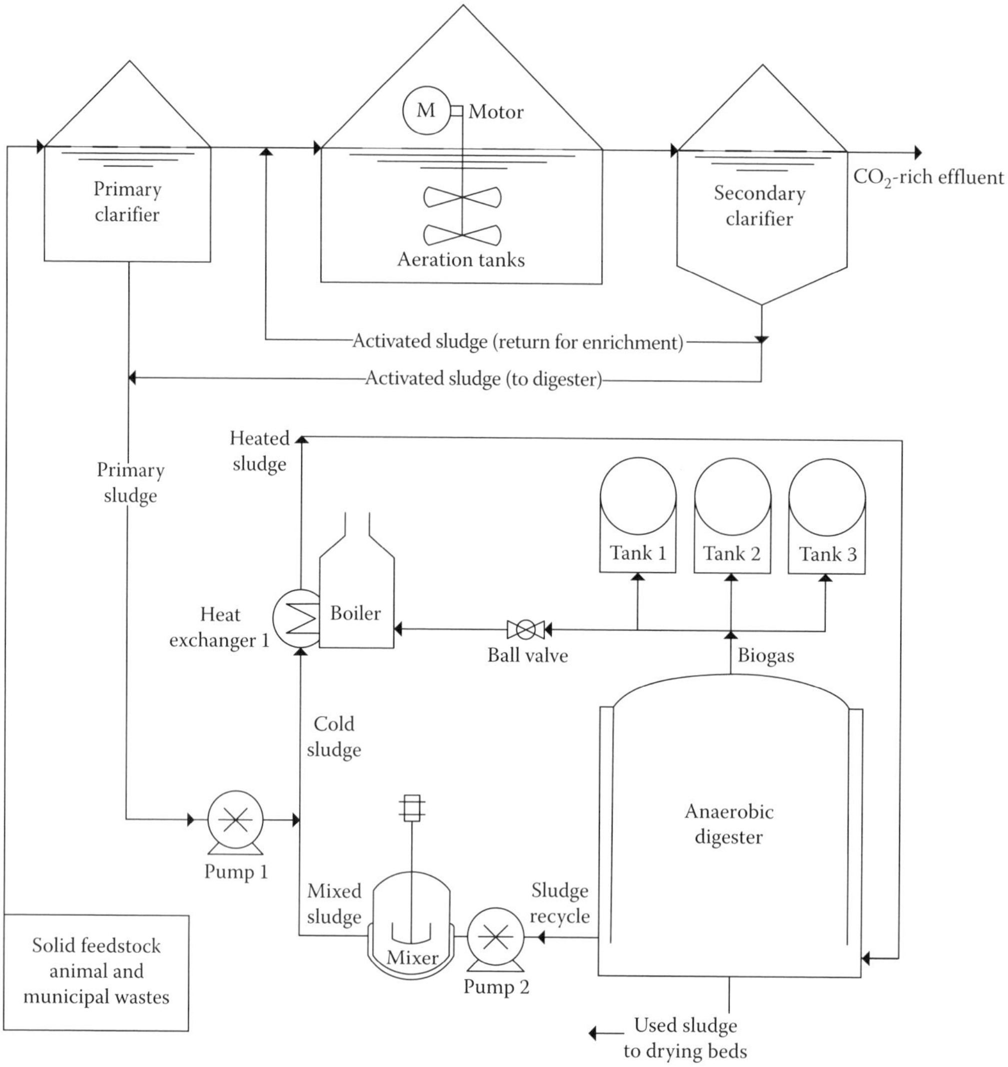

FIGURE 4.11
Anaerobic digestion system and processes. (Adapted from Goodman, B.L., Stabilization: Anaerobic digestion, chapter 7, in *Environmental Engineer's Handbook*, CRC Press/Taylor & Francis Group, LLC, Boca Raton, FL, 1999.)

- More effective pathogen removal. This is especially true for multistage digesters or if a pasteurization step is included in the process.
- Minimal odor emissions, as 99% of volatile compounds are oxidatively decomposed upon combustion, for example, H_2S forms SO_2.
- A high degree of compliance with many national waste strategies, implemented to reduce the amount of biodegradable waste entering landfills.
- A slurry is produced (digestate), which is an improved fertilizer in terms of both its availability to plants and its rheology.
- A source of carbon-neutral energy in the form of biogas.

There are four important biological and chemical stages of anaerobic digestion (Gerardi, 2003):

- Hydrolysis: A chemical reaction where particulates are solubilized and large polymers converted into simpler monomers, when one or more water molecules are split into hydrogen and hydroxide ions.
- Acidogenesis: A biological reaction where simple monomers are converted into volatile fatty acids.
- Acetogenesis: A biological reaction where volatile fatty acids are converted into acetic acid, CO_2, and hydrogen. The acetate is produced by anaerobic bacteria from a variety of energy sources such as hydrogen and carbon.
- Methanogenesis: A biological reaction where acetates are converted into CH_4 and CO_2, while hydrogen is consumed by the formation of CH_4 by microbes known as methanogens.

The material drawn from the anaerobic digester is called sludge or effluent. It is rich in nutrients (ammonia, phosphorus, potassium, and more than a dozen trace elements) and is an excellent soil conditioner and feed for algae growth (Goodman, 1999). Digester sizing and process details are available elsewhere (Goodman, 1999; Gerardi, 2003; Pandey, 2004).

4.3.3 Chemical Looping Processes

Chemical looping combustion (CLC), developed in the mid-1990s, uses metallic oxide as an oxygen carrier for combustion (Ishida and Jin, 1995). The fuel, mostly gases similar in composition to syngas, reduces the metal oxide to the corresponding metal at low temperature in the first reactor. In a second reactor, the metal is oxidized using oxygen in moistened air at high temperature to form the metal oxide, which is recycled to the first reactor. Hydrogen is produced from water in the second reactor. During the reaction in the first reactor, the oxygen in the metal is transferred to the carbon in the fuel forming CO_2 and water (Yu et al., 2003; Jin and Ishida, 2004; Fan et al., 2008). The water is condensed to separate CO_2 and send it to storage. This process exhibits a greater potential for CO_2 separation compared to membrane separation of CO_2.

Experimentation and process modeling and simulation suggest that a maximum coal-to-hydrogen conversion efficiency of 80% can be achieved using coal DCL (Fan et al., 2008; Gnanapragasam et al., 2009b). Some problems with this process include temperature issues relating to metal oxide (very high temperatures destabilize the structure) and sizing the reactor to control reaction rates (Fan et al., 2008).

Exergy analysis allows meaningful efficiencies and losses to be evaluated for energy systems and corresponding design improvements to be identified (Dincer and Rosen, 2007). Such an analysis suggests that the net power efficiency of a power generation system incorporating CLC exceeds that of a conventional power system by reducing combustion irreversibilities (Anheden and Svedberg, 1998). Of the various metal oxides that can be used for the syngas redox process, iron oxide (Fe_2O_3) has been identified as permitting the highest conversion of syngas to combustion products (CO_2 and water) along with a high conversion of steam to hydrogen (Gupta et al., 2007).

4.3.3.1 Syngas Chemical Looping

4.3.3.1.1 Reduction Reactor

The syngas produced from the gasifier contains mostly CO, H_2, CO_2, and CH_4 (Li et al., 2001), and it reduces the metal oxide (Fe_2O_3) to the constituent metal (Fe and FeO). The advantage of using iron oxide (Fe_2O_3) as the oxygen carrier is that it does not involve catalytically dependent reactions (Gupta et al., 2007). The gaseous products are CO_2 and steam. The steam is condensed to obtain sequestration-ready CO_2. The reactions of CO in the syngas with iron oxide and wustite (FeO) are given as follows:

$$Fe_2O_3 + CO \rightarrow 2FeO + CO_2 \tag{4.15}$$

$$FeO + CO \rightarrow Fe + CO_2 \tag{4.16}$$

Similarly, the reactions of hydrogen in the syngas with iron oxide and wustite (FeO) are as follows:

$$Fe_2O_3 + H_2 \rightarrow 2FeO + H_2O \tag{4.17}$$

$$FeO + H_2 \rightarrow Fe + H_2O \tag{4.18}$$

The reactions in Equations 4.15 through 4.18 occur at a pressure of 30 atm and a temperature ranging from 750°C to 900°C within the process shown in Figure 4.12. The particle size for the iron oxide is 2–10 mm. The reduction reactors in the syngas chemical looping (SCL) system are set with inert conditions for CH_4 since the model does not allow CH_4 to react with the iron oxide, thus completing the reactions in Equations 4.15 through 4.18 and enabling the production of hydrogen (Hoffman, 2005). In the SCL system, therefore, the CO_2 stream contains the unreacted CH_4, depending on the inlet conditions. These data are included in the results.

4.3.3.1.2 Oxidation Reactor

This reactor operates at 30 atm and 500°C–700°C to oxidize the metal produced in the reduction reactor using steam. The products are 99% pure hydrogen and magnetite (Fe_3O_4). The reactions are as follows:

$$Fe + H_2O_{(g)} \rightarrow FeO + H_2 \tag{4.19}$$

$$3FeO + H_2O_{(g)} \rightarrow Fe_3O_4 + H_2 \tag{4.20}$$

Both reactions are slightly exothermic and some of the heat may be used for preheating the feedwater to make steam. Hydrogen production using CLC of the syngas is indirect with the use of iron oxide. The actual hydrogen in the syngas is converted to water (Equations 4.17 and 4.18). These oxidation reactions are similar with a principal purpose as that of the SCL system to produce hydrogen as in the case of the DCL system in Figure 4.14.

4.3.3.1.3 Combustion Reactor

The magnetite formed in the oxidation reactor enters the combustion reactor where it reacts with oxygen to form a more stable form of iron oxide III (Fe_2O_3). A significant amount of heat is produced during the oxidation of Fe_3O_4 to Fe_2O_3. The reaction is

$$4Fe_3O_4 + O_2 \rightarrow 6Fe_2O_3 \tag{4.21}$$

The gas composition of exhaust ($N_2 + O_2$ in Figures 4.12 and 4.14) includes the remaining oxygen after the reaction in Equation 4.21 and the corresponding nitrogen.

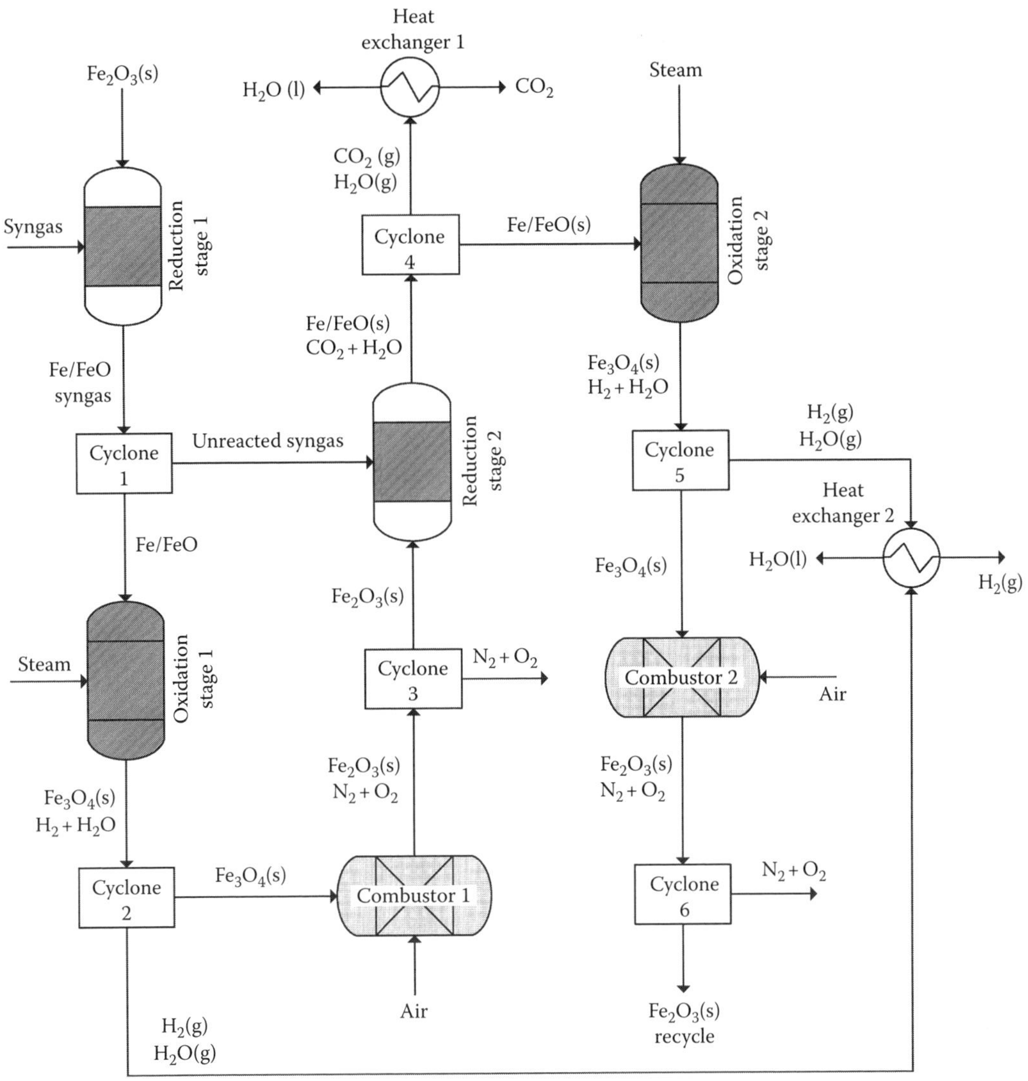

FIGURE 4.12
Syngas-based chemical looping process (SCL). (Adapted from Gnanapragasam, N.V. et al., *Int. J. Hydrogen Energy*, 34, 2606, 2009b.)

Chemical looping with oxygen uncoupling (CLOU) is a novel method to burn solid fuels in gaseous oxygen without the need for an energy-intensive ASU (Mattison et al., 2008). The CO_2 from combustion is inherently separated from the rest of the flue gases. The reaction rate of petroleum coke was found to be approximately 50 times higher when using CLOU compared to the reaction rate of the same fuel with an iron-based oxygen carrier in conventional CLC.

4.3.3.2 Direct Chemical Looping

The DCL process removes the gasification process from the hydrogen production route. Rather, it has a fuel reactor that first combusts some feedstock to generate the temperature required for the endothermic reactions as described in Equations 4.24 through 4.28.

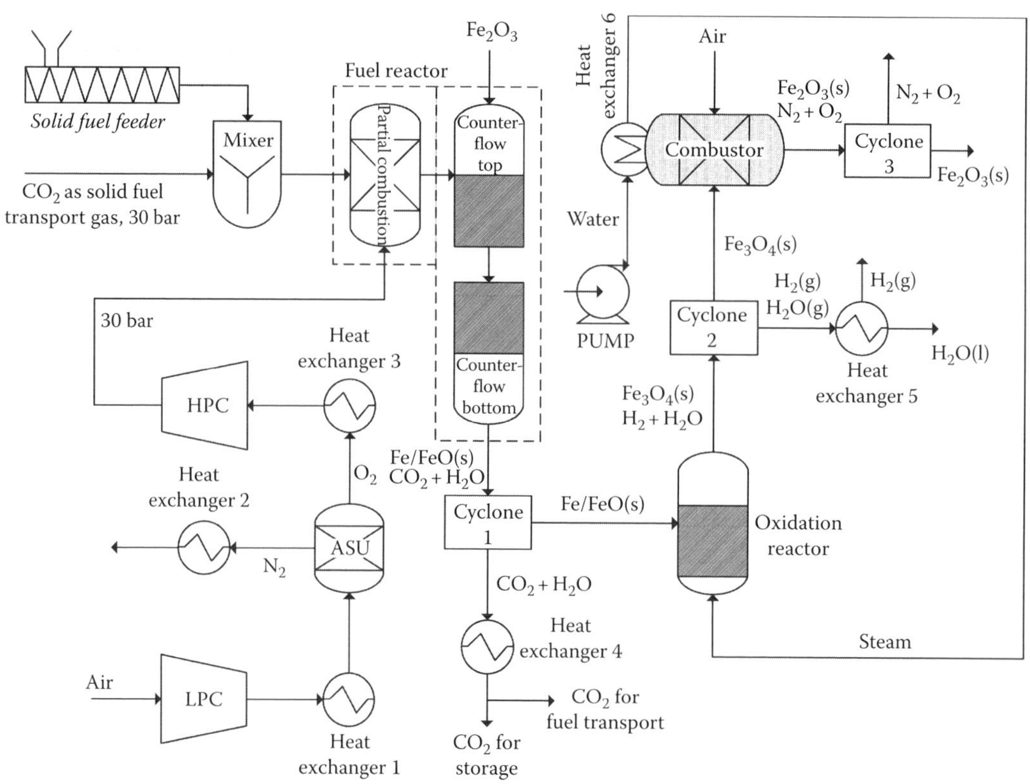

FIGURE 4.13
DCL process without gasification. (Adapted from Gnanapragasam, N.V. et al., *Int. J. Hydrogen Energy*, 34, 2606, 2009b.)

4.3.3.2.1 Fuel Reactor

The principal difference in the DCL system (Figure 4.13) relative to the SCL system (Figure 4.12) is the absence of the gasifier and the addition of a fuel reactor. The latter is an extended form of reduction reactor, and its chemical reactions are discussed as follows. The fuel reactor can be modeled as three separate *RGibbs* reactors (rigorous reaction and multiphase equilibrium based on Gibbs free energy minimization) linked together by restricting products from each of the three reactors: partial combustion, fuel reactor top, and fuel reactor bottom (see Figure 4.13). The reactions and conditions in the fuel reactor, based on Fan et al. (2008) and Mattison et al. (2008), are as follows:

- *Partial combustion*: Coal devolatilization and partial combustion occur (where the formula for coal used here represents Pittsburgh #8 coal, although a different coal similar in composition is used in the calculations) as follows:

$$C_{11}H_{10}O \rightarrow C + CH_4 \quad (4.22)$$

$$C + O_2 \rightarrow CO_2 \quad (4.23)$$

- *Fuel reactor top*: Char gasification and iron oxide reduction occur as follows:

$$2C + O_2 \rightarrow 2CO \quad (4.24)$$

$$C + CO_2 \rightarrow 2CO \quad (4.25)$$

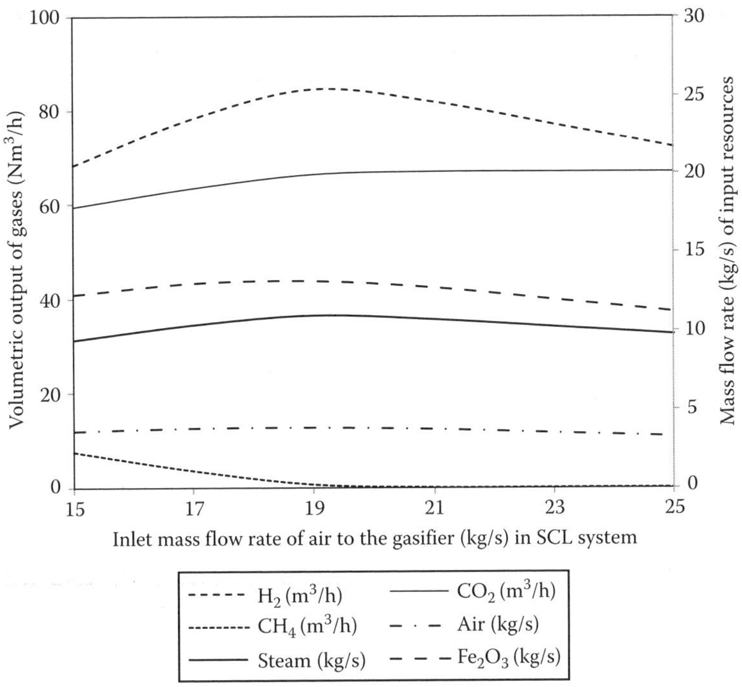

FIGURE 4.14
Variation in gas outputs from the SCL system with inlet mass flow rate of air to the gasifier. The input variations of steam and iron oxide (Fe_2O_3) are also provided. (Adapted from Gnanapragasam, N.V. et al., *Int. J. Hydrogen Energy*, 34, 2606, 2009b.)

$$2C + H_2O \rightarrow CO + H_2 \qquad (4.26)$$

$$CH_4 + 4Fe_2O_3 \rightarrow CO_2 + 2H_2O + 8FeO \qquad (4.27)$$

- *Fuel reactor bottom*: Wustite (FeO) reduction occurs, following the reactions in Equations 4.15 and 4.16.

The overall chemical reaction within the fuel reactor (Fan et al., 2008) is

$$C_{11}H_{10}O + 6.44Fe_2O_3 + 3.34O_2 \rightarrow 11CO_2 + 5H_2O + 12.88Fe \qquad (4.28)$$

The residence time for the coal char in the fuel reactor is between 30 and 90 min (Fan et al., 2008), depending on operating temperature (750°C–900°C) and pressure (1–30 atm). The reaction in Equation 4.27 enables the conversion of CH_4 in the gas stream to CO_2 and H_2 while reducing iron oxide (Fe_2O_3). Thus, in the DCL system, the CO_2 stream may not contain CH_4, unlike the SCL system.

Figures 4.14 and 4.15 show the range (from 15 to 25 kg/s) of air mass flow rate into the gasifier for the SCL system and into the fuel reactor for the DCL system, on the horizontal axis. The distributions of the production of H_2, CO_2, and CH_4 are plotted in Figure 4.14 for the corresponding change in inlet air mass flow rate to the gasifier in the SCL system. In Figure 4.15, a similar distribution is shown for the DCL system, except for CH_4 since

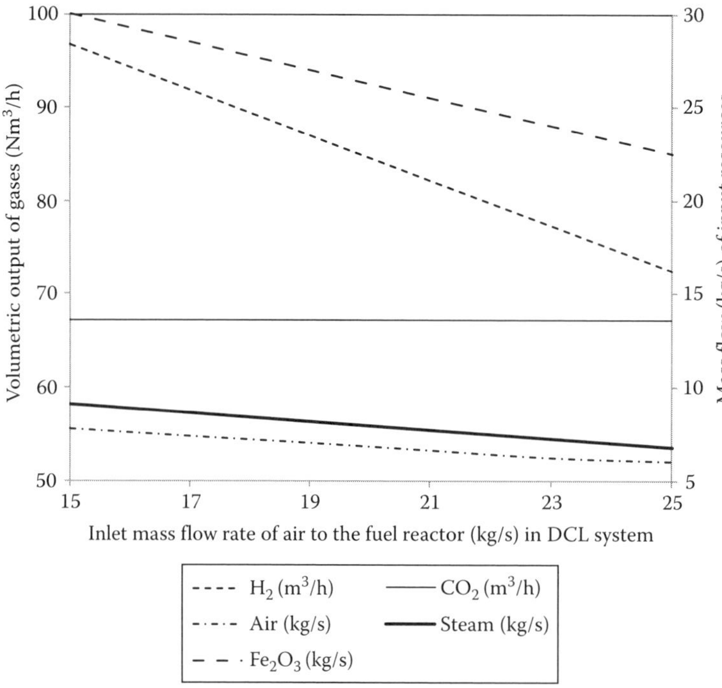

FIGURE 4.15
Variation in gas outputs from the DCL system with inlet mass flow rate of air to the fuel reactor. The input variations of steam and iron oxide (Fe_2O_3) are also provided. (Adapted from Gnanapragasam, N.V. et al., *Int. J. Hydrogen Energy*, 34, 2606, 2009b.)

it is consumed within the process (Equation 4.27) and not produced along with H_2 and CO_2 as in the SCL system. Since CH_4 is not converted to hydrogen or any other form in the SCL system, the H_2 output rate in Figure 4.14 increases with air inlet flow rate until the air inlet flow rate reaches 19 kg/s, when CH_4 is no longer produced in the gasifier of the SCL system. The results in Figures 4.14 and 4.15 are based on simulations of the SCL and DCL systems when anthracite coal is used at a rate of 5 kg/s. Theoretically, the DCL system is better than the SCL system in producing more hydrogen at low oxygen input (Gnanapragasam et al., 2009b), but this observation has not been confirmed at large scales.

4.3.4 Autothermal Reforming

Hydrogen production from coal-derived methanol via ATR has been reported to have fewer trace impurities than other coal-based hydrogen production processes, mainly due to higher operating temperature generated by the partial oxidation step (Yoon and Erickson, 2008) as shown in Figure 4.16. Coal-based methanol has been shown to have higher amounts of trace hydrocarbons than chemical-grade methanol derived from natural gas, so hydrogen production from coal-derived methanol via ATR is feasible considering fuel cell applications.

Natural gas and the thermolysis gases from petroleum and coal can be thermally decomposed to manufacture hydrogen and vapor-deposited carbon materials, in the form of nanoparticles, fibrous materials, or pyrolytic carbon solids (Halloran, 2008). Coal thermolysis can produce hydrogen from coke oven gas and carbon materials from fabricated cokes.

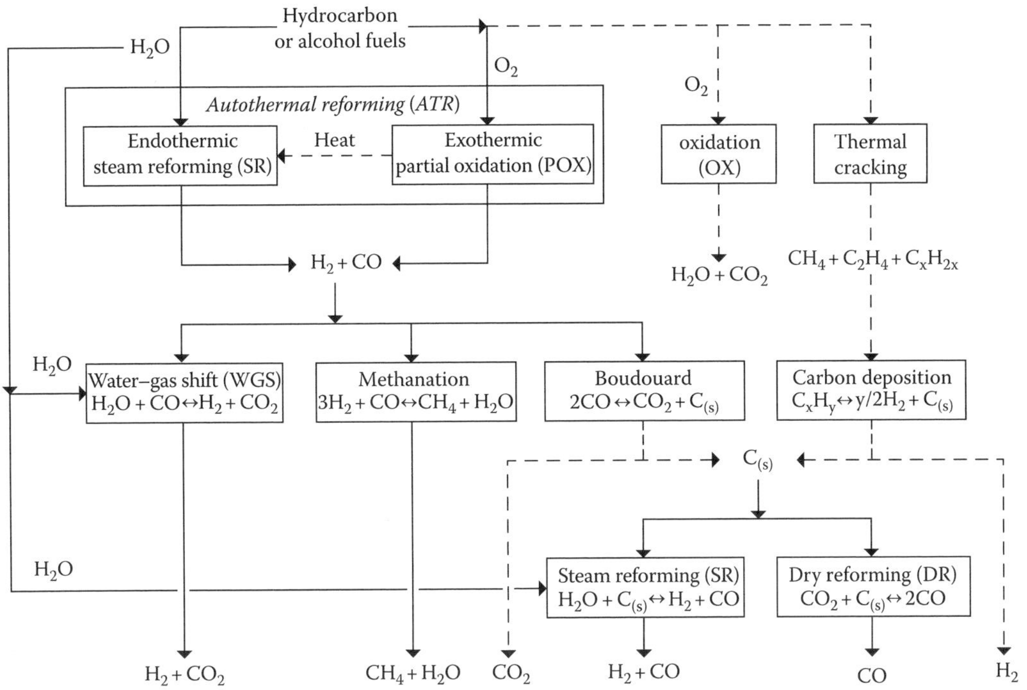

FIGURE 4.16
Flow scheme of possible reactions taking place within ATR process. (From Chen, Y. et al., *Catal. Today*, 116, 334, 2006.)

The combined manufacture of hydrogen and carbon materials would be economical if the market value of the hydrogen and solid carbon materials were greater than the value of coal as a fuel.

ATR uses oxygen and CO_2 or steam in a reaction with CH_4 to form syngas (Xuan et al., 2009). The reaction takes place in a single chamber where the CH_4 is partially oxidized. The reaction is exothermic due to the oxidation. When the ATR uses CO_2, the H_2/CO ratio produced is 1:1; when the ATR uses steam, the H_2/CO ratio produced is 2.5:1.

The exothermic reaction in the ATR process when using CO_2 is

$$2CH_4 + O_2 + CO_2 \rightarrow 3H_2 + 3CO + H_2O \tag{4.29}$$

and when using steam is

$$4CH_4 + O_2 + 2H_2O \rightarrow 10H_2 + 4CO \tag{4.30}$$

The outlet temperature of the syngas is between 950°C and 1100°C, and the outlet pressure can be as high as 100 bar (Mahecha-Botero et al., 2008). The main difference between SMR and ATR is that SMR uses no oxygen. The advantage of ATR is that the H_2/CO can be varied, which is particularly useful for producing some second-generation biofuels, such as dimethyl ether (DME, CH_3OCH_3) that requires a 1:1 H_2/CO ratio. For the generation of hydrogen, the use of steam in the flow of processes in the ATR system is shown in Figure 4.16. The broken lines in Figure 4.16 denote SMR, while the solid lines denote the ATR process.

4.3.5 Hydrogen Separation and Purification Processes

Hydrogen is generated from the gasification process and available in the syngas, but it needs to be separated before and after WGS reaction. Chemical looping does not require physical hydrogen separation since it is a thermochemical process and enables cheaper separation (Andrus, 2009). After hydrogen is separated, it still contains other gaseous elements in smaller amounts that need to be removed for highly purified hydrogen.

4.3.5.1 Water–Gas Shift Process

WGS catalysts have been commercially developed for use by the petrochemical industry (Leach, 1984; Twigg, 1996). Presently, there is renewed interest in the WGS reaction because of its importance in reforming fuels to hydrogen for use in fuel cells. Since biomass gasification yields relatively high CO/H_2 ratios, higher hydrogen contents can be achieved using commercial CO shift catalysts in two fixed-bed reactors operated in series: a high-temperature shift reactor (HTSR) for rapid reaction and a low-temperature shift reactor (LTSR) to shift thermodynamic equilibrium to very low levels of CO (Zhang et al., 2005). The high-temperature shift reaction takes advantage of faster kinetics at elevated temperatures to convert about 75% of the CO into H_2. Conversion is, however, limited by thermodynamic equilibrium, which favors hydrogen formation at low temperatures. Accordingly, the gas is slightly cooled and passed through a second LTSR to convert most of the remaining CO to H_2.

The WGS reactors in Figure 4.17 are arranged to obtain the maximum hydrogen production. The WGS reaction, Equation 4.6, is carried out in two reactors placed in series, including a HTSR and LTSR. These reactors operate at temperatures of about 350°C–400°C

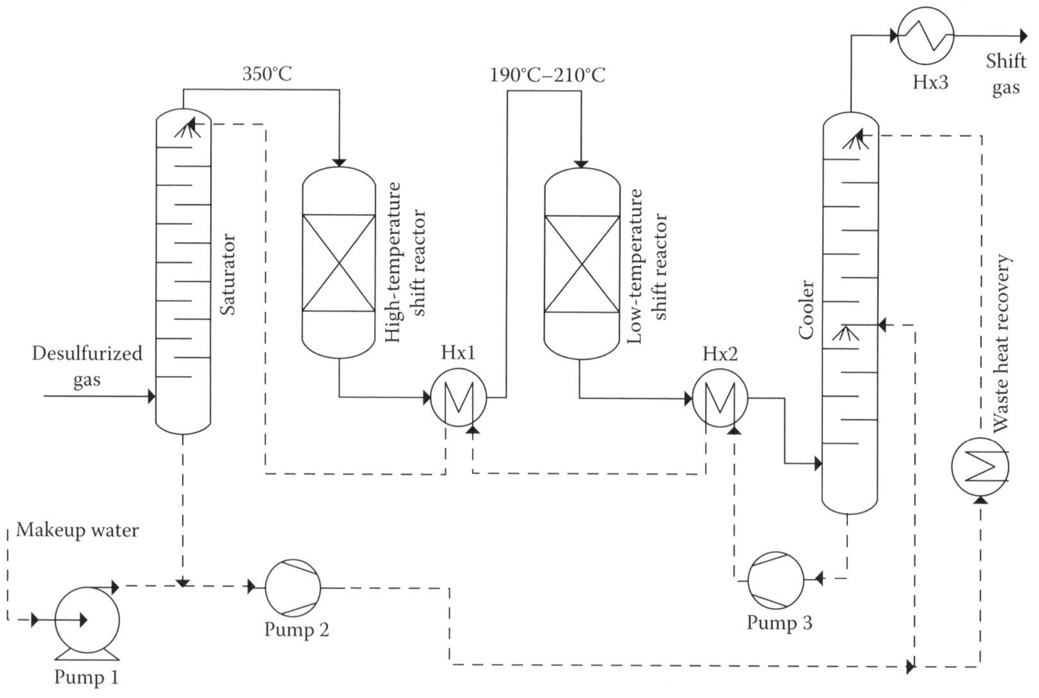

FIGURE 4.17
WGS process. (Adapted from Zhang, R. et al., *Fuel Process. Technol.*, 86, 861, 2005.)

TABLE 4.4

Characteristics of the Syngas from LTSR at Different S_{SHIFT}/C Ratios in the Conditions of Maximum Hydrogen Production

S_{SHIFT}/C	0.6	0.8	1	1.2
S/C	0.7	0.6	0.4	0.2
Molar Composition (%)				
H_2	49.95	48.92	49.58	50.15
CO_2	33.28	32.44	32.39	32.28
CO	0.89	0.55	0.61	0.72
H_2O	14.34	16.61	15.96	15.4
H_2S	0.5	0.49	0.49	0.49
CH_4	0	0	0	0.01
N_2 + Ar	1	1	1	1
Hydrogen Flow Rate (kg/kg coal)				
H_2	0.156	0.159	0.161	0.162

Source: Adapted from Perna, A., *Int. J. Hydrogen Energy*, 33, 2957, 2008.

and 190°C–210°C, respectively (Perna, 2008). The catalysts employed to improve the reaction rate are sulfur tolerant, thus enabling the sulfur removal unit to be downstream of the shift units (Chiesa et al., 2005). The syngas exiting the LTSR is cooled at the operating temperature of the acid gas removal (AGR) unit.

The steam-to-carbon ratio (S_{SHIFT}/C) in the WGS reactors in Table 4.4 is defined as the ratio of the mass flow rate of the feed coal to the mass flow rate of the water input to the shift units, in the range 0.6–1.2. It is assumed that the mass fraction input to the HTSR is 0.67. The syngas converted after the WGS process is similar in composition to Table 4.4.

The catalytic tar process in combination with high- and low-temperature WGS reactors upgrades hydrogen in the raw gas from 5.8–8.8 vol.% to as high as 27–29 vol.% (Zhang et al., 2005). The CO concentration of 13–15 vol.% in the raw gas is reduced to less than 0.5 vol.%. The conversion of CO in the high-temperature WGS reaches 75%–80%, while the CO conversion by the combination of high- and low-temperature WGS reactors exceeds 95% (Mahecha-Boteroa et al., 2009).

4.3.5.2 Gas Separation Membranes

Gas separation using membranes of various kinds, primarily differentiated by the membrane material, is a rapidly evolving field (Li, 2007). The most common commercial materials for membranes include metallic, ceramic, and polymeric substances and, recently, carbon-based nanotubes or pores in compact grid arrangements (Kerry, 2006). It was reported in 2004 that membrane use in power cycles and fuel production using fossil fuels through the integration of membrane technologies has not been fully explored (Bredesen et al., 2004). Since then, several integration works have been reported, including design optimization and feasibility assessments using technoeconomic analyses of the membranes for power and fuel production systems (Amelio et al., 2007; Rezvani et al., 2009). New combustion processes with porous membrane integration have also been recently reviewed (Mujeebu et al., 2009), providing a broader picture of membrane applications in energy conversion processes: mainly air separation and CO_2 and H_2 separation.

Of the current noncryogenic technologies for oxygen separation, membranes such as the ITM provide clean, efficient, and economic means of producing oxygen (Li, 2007). Ionic conducting or ionic and electronic conducting ceramic dense membranes are of particular interest due to their 100% selectivity for oxygen, with three to four orders of magnitude higher flux than the best organic membranes.

Carbon separation membranes (CSMs) are ceramic hollow fiber membranes that provide a good alternative. Ceramic hollow fiber membranes are not commercially available at present, mainly due to the lack of techniques to produce the ceramic membrane with a thin and dense separation layer in a hollow fiber form (Li, 2007). Hollow fiber membranes are frequently produced from polymeric materials and can provide the highest membrane area per unit packing volume, achieving densities as high as 8,000–10,000 m^2/m^3 (Li, 2007). However, this configuration is seldom used in ceramic membranes because of its poor mechanical strength. An early effort at gas separation, via the development of glass hollow fiber membranes (Way and Roberts, 1992), yielded membranes that were too fragile for industrial uses. Many other attempts have been made to develop fabrication techniques for hollow fiber ceramic membranes (Smid et al., 1996; Brinkman et al., 2000; Luyten et al., 2000; Tan et al., 2001; de Jong et al., 2004; Xu et al., 2004; Liu et al., 2006).

A fundamental process for noncryogenic separation of industrial gases is permeation (Kerry, 2006), which involves the diffusion of a substance in solution through a barrier or membrane. Permeability is the capacity of a porous material (membrane) for transmitting a fluid. The membranes used for permeation simulation in this work may be one or a combination of thin, dense, and continuous films formed from cellulose acetate or polymers.

The separation of a component in a gas mixture is carried out in three steps, in which the component dissolves in the membrane wall, diffuses through the membrane material, and is desorbed on the opposite side of the membrane wall. This procedure can be defined by Henry's law of solubility and by Fick's law of diffusivity, while noting that the solubility of a gas in a liquid is proportional to the partial pressure of the gas. These laws can be expressed as follows:

Feed gas solubility in the polymer:

$$z_{iF} = S_i P_F x_i \tag{4.31}$$

Fick's law inside membrane material:

$$\frac{Q_i}{A} = \frac{D_i\left(x_i - y_i\right)}{L_p} \tag{4.32}$$

Permeate solubility in the polymer:

$$z_{iP} = S_i P_P y_i \tag{4.33}$$

The operating performance of any polymeric membrane is characterized by two factors: (1) permeability (which defines productivity, the transport rate for a given species in the feed stream, and, therefore, the cost of the system) and (2) selectivity (which in turn defines recovery or purity of the selected stream). Expressions for permeability and selectivity can be derived from Equations 4.31 through 4.33 (Fleming and Dupuis, 1993). For oxygen separation from air, N_2 is more permeable than O_2 due to the higher molar fraction of N_2 in air, whereas H_2 is more permeable than CO_2 due to the lower molar fraction of CO_2 in the exhaust gas stream after WGS reaction.

4.3.5.3 Pressure Swing Adsorption for Hydrogen Purification

The quality of the hydrogen produced can be a major issue for automotive applications. The pressure swing adsorption (PSA) unit produces hydrogen with 99.999% of purity, operates at above 20 bar, and reaches H_2 separation efficiencies in the range 85%–90%. The PSA purge gas (waste gas) can be used to superheat steam or can be combusted for power generation in a gas turbine combined cycle (Perna, 2008). New hydrogen plants are almost invariably designed using PSA for final hydrogen purification (Hamelinck and Faaij, 2002).

PSA is based on the difference in adsorption behavior between different molecules (Katofsky, 1993). The process separates components of a gas stream by selective adsorption to a solid at high pressure and subsequent desorption at low pressure. This adsorption/desorption is a batch process, but by placing two beds in parallel, it operates nearly continuously. While adsorption takes place in one bed, the other is desorbed (LaCava et al., 1998). First, activated carbon in the set of beds *A* selectively adsorbs nearly all CO_2 and all H_2O. The remaining gas then passes to the second set of beds *B* containing a zeolite molecular sieve, which selectively adsorbs essentially all the remaining compounds and some hydrogen. The overall recovery of hydrogen is increased by recycling some of the desorbed gas from the *B* beds (LaCava et al., 1998). There is a trade-off in that the recycled gas must be recompressed and cooled to near-ambient temperature, adding to capital and operating costs, and a slightly larger PSA unit will also be needed. As with the methanol synthesis loop, some of the recycled gas must be purged to prevent the buildup of CH_4 and other nonhydrogen gases. Recovery rates of over 90% are achievable, and the product purity is extremely high: 99.999%.

In PSA, species other than hydrogen are selectively adsorbed on a solid adsorbent, for example, activated carbon and 5A zeolite (Ruthven et al., 1994), at a relatively high pressure by contacting the gas with the solid in a packed column in order to produce a hydrogen-rich gas stream. The adsorbed species are then desorbed from the solid by lowering the pressure and purging with high-purity product hydrogen, and the PSA waste gas is generated. Continuous flow of product is maintained by using multiple, properly synchronized adsorption beds.

In a pressure swing adsorption (PSA) facility for H_2 purification, the impurities are adsorbed at high pressure, while H_2 passes through the adsorber vessel. When the vessel is full, it is disconnected from the process and the pressure is decreased, thus releasing most of the impurities (Rydén and Lyngfelt, 2006). A small fraction of the produced H_2 is needed for purging and regeneration of the adsorbers, so the H_2 recovery is limited to about 90%. The off-gas from the adsorber vessel consists of CO_2, purge H_2, unreformed CH_4, some CO, and minor fractions of other impurities. PSA is a batch process, but by using multiple adsorbers, it is possible to provide constant flows. The pressure drop for H_2 is usually about 0.5 bar. There is no need for power, heating, or chemicals.

4.4 Environment Protection Processes

When producing hydrogen from coal, pollutants of various forms are released at various conversion stages to the Earth's surface and lower atmosphere. Only gases that form in large quantities, such as H_2S, carbonyl sulfide (COS), and CO_2, and solids, such as ash, tars, and other nonenergy solid wastes, are discussed as part of pollution-control processes. These have to be managed for a sustainable environment without long-term damage to the balance in the ecosystem. This section discusses some of the primary measures in managing these hazardous wastes that result in using coal and other solid fossil fuels. These processes improve the sustainability of hydrogen production systems that use solid fuels.

4.4.1 Acid Gas Removal: Sulfur Removal

Sulfur appears as SO_2 (only in combustion processes), H_2S, and COS in syngas after the energy conversion of sulfur-rich solid fuels (all solid fossil fuels). Sulfur in raw coal is converted to H_2S and COS in the gasifier, and most of the COS is converted to H_2S in the WGS reactors. In systems without the WGS process and CO_2 removal (Kohl, 1997), COS is typically hydrolyzed to H_2S in a catalytic bed at about 200°C.

AGR units (shown in Figure 4.18) are based on chemical absorption and physical absorption for atmospheric and pressurized configurations, respectively (Perna, 2008). AGR removes 99.9% of the H_2S, which is converted to elemental sulfur via Claus and shell claus

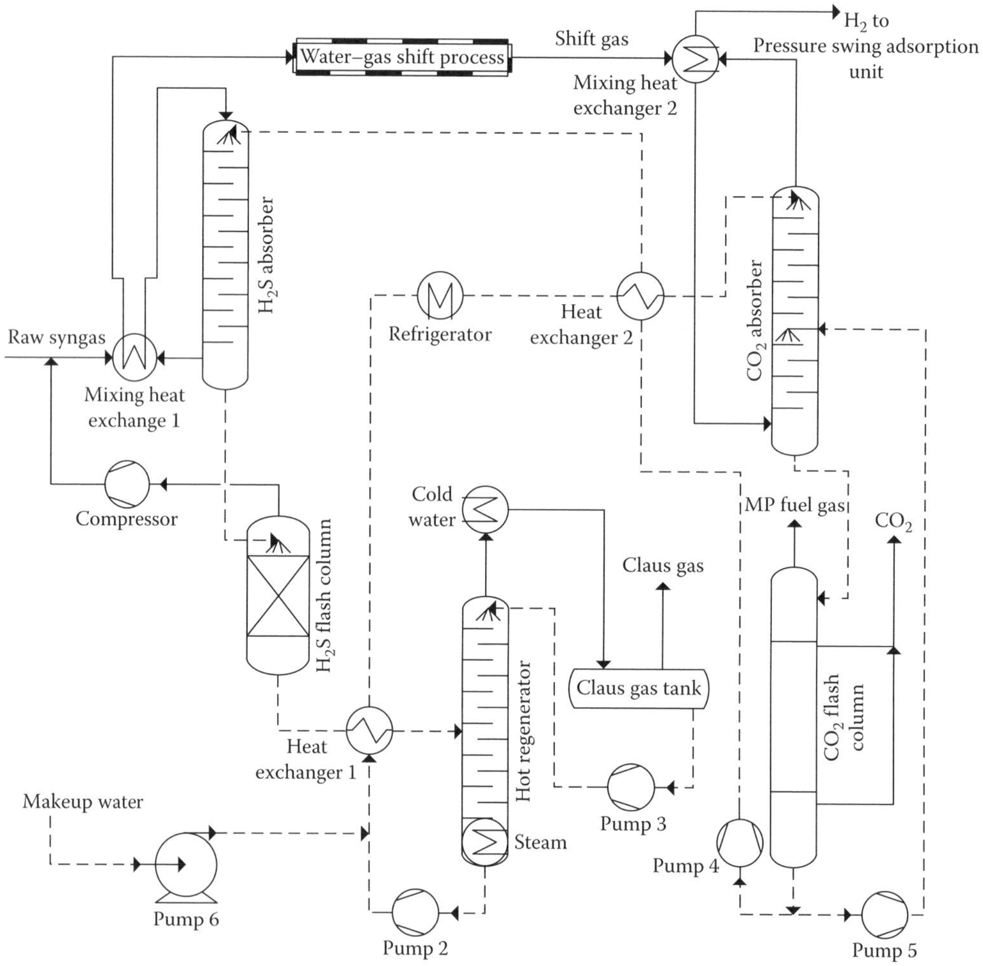

FIGURE 4.18
AGR process. (From Korens, N. et al., Process screening analysis of alternative gas treating and sulfur removal for gasification, Report for US Department of Energy by SFA Pacific, Inc., Engineering and Economic Consultants, 2002. Available at http://www.netl.doe.gov/technologies/coalpower/gasification/pubs/pdf/SFA%20Pacific_Process%20Screening%20Analysis_Dec%202002.pdf, Accessed on December 5, 2013.)

off-gas treating (SCOT) plants. The efficiency of the CO_2 capture unit is assumed equal to 95%. Prior to CO_2 capture, H_2S is removed from the syngas (containing 0.6% H_2S by volume) by physical absorption in CH_3OCH_3 of polyethylene glycol (Selexol). The WGS reactors included in H_2 plants and electricity plants with CO_2 capture greatly elevate (relative to electricity plants with CO_2 venting) the CO_2/H_2S ratio in the syngas and thus promote significant cocapture of CO_2 along with H_2S in the H_2S absorption tower.

Since essentially all of the sulfur in the gasifier feedstock is converted to H_2S, the amount of H_2S produced is totally dependent on the sulfur content of the feedstock. Note that coal has a relatively high sulfur content, while biomass has a relatively low sulfur content. Generally, the AGR processes lower the H_2S content of the syngas to less than 4 ppm, which means that, in essence, all of the H_2S produced in the gasifier must be processed in the sulfur recovery system.

4.4.2 Claus and SCOT Processes: Sulfur Recovery

The type of sulfur recovery system required is dependent on the required sulfur recovery efficiency, the quantity of sulfur to be removed, and the concentration of the H_2S in the acid gas. The required sulfur removal/recovery efficiency will vary depending on location; however, the gasification industry claims that the technology has *near-zero* pollution, so it necessitates the industry to install the best available control technology. Currently, H_2S removal efficiencies of 99.9% can be economically achieved (Nagl, 2005).

The Claus process has been the main sulfur recovery process for applications with large amounts of sulfur (greater than 20 long tons per day), relatively high H_2S concentrations (greater than 15%), and consistent inlet conditions. However, the Claus process is limited by chemical equilibrium to removal efficiencies of approximately 98% if three catalytic reactor stages are employed. To achieve higher removal efficiencies, a tail gas treating unit is required.

For over 30 years, the preferred tail gas treating process has been the SCOT process. In that process, the tail gas from the Claus unit is heated to approximately 300°C in an in-line burner, which serves the dual purpose of heating the gas stream and producing a reducing gas, which is needed in the downstream reactor. The effluent from the burner is then passed over a cobalt–molybdenum catalyst. In the reactor, all of the SO_2, COS, and CS_2 are converted to H_2S by a combination of hydrogenation and hydrolysis reactions. The reactor effluent gas is then cooled and processed through a typical amine unit, which is selective to the absorption of H_2S. The recovered H_2S is then recycled back to the Claus unit, and the remaining gas is sent to an incinerator prior to exhausting to the atmosphere.

Although many of these processes are mature, when integrated with hydrogen production, process optimization becomes essential for improving the overall energy efficiency.

4.3 Tar, Ash, and Soot Removal

The presence of tars in the product gas is a significant challenge in the commercial utilization of coal and biomass product gas (syngas) as a source of sustainable energy. Tar is formed in the gasifier and comprises a wide spectrum of organic compounds, generally consisting of several aromatic rings. Simplified tars can be separated into heavy and light tars (Higman and Burgt, 2008). Heavy tars condense as gas temperature drops and cause notable fouling, efficiency loss, and unscheduled plant shutdowns. The tar dew point, that is, the temperature at which tars start to condense, is a critical factor. Light tars like phenol or naphthalene have limited influence on the tar dew point but are not less problematic.

Light tars like phenol chemically pollute the bleed water of downstream condensers and aqueous scrubbers. Naphthalene is important as it is known to crystallize at the inlet of gas engines increasing needs for maintenance and service.

Conventional gas tar cleaning is based on wet scrubbing and on a wet cleaning with an electrostatic precipitator (ESP). In some new processes (Zwart et al., 2008), the tar removal principle is based on a multiple-stage scrubber in which gas is cleaned by a special scrubbing oil. In the first stage, the gas is cooled with scrubbing oil. Heavy tar particles condense and are collected, after which they are separated from the scrubbing oil and can be recycled to the gasifier, together with a small bleed. In the second stage, lighter gaseous tars are absorbed with scrubbing oil. In the absorber column, the scrubbing oil is saturated by these light tars. This saturated oil is regenerated in a stripper. Hot air or steam is used to strip the tars of the scrubbing oil. All heavy and light tars can be recycled to the gasifier where they are destroyed and contribute to the energy efficiency. Tar waste streams are efficiently recycled this way.

The raw syngas generated in a partial oxidation gasifier also includes carbon soot that is removed and recovered from the syngas by scrubbing with water (as shown in Figure 4.19). The scrubbing water contains one or more high-temperature surfactants that allow greater soot concentrations in the water-scrubbing quench zone of the gasifier. The carbon soot is separated from the scrubbing water with the aid of scrubbing oil. The separation of the carbon soot from the scrubbing water is enhanced with the aid of one or more surfactants that render the soot particles hydrophobic and oleophilic (Jahnke, 2003). The recovered carbon soot is ultimately recycled to the gasifier to recover the energy value of the carbon during the partial oxidation reaction. The overall energy efficiency of the gasification process can be

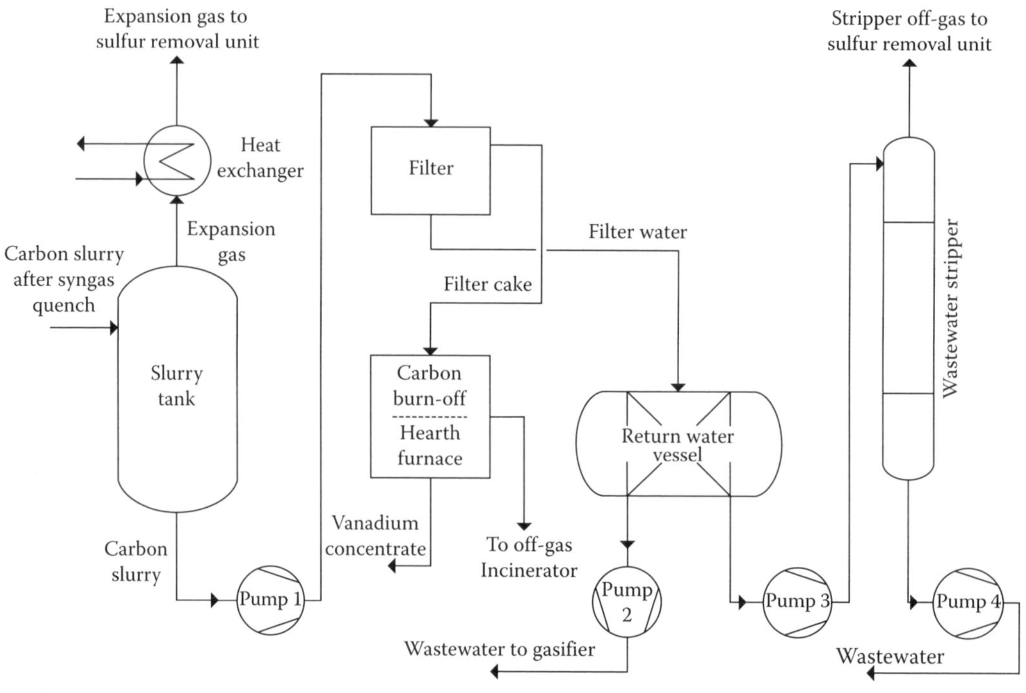

FIGURE 4.19
Syngas carbon soot removal unit. (Adapted from *Gasification*, 2nd edn., Higman, C. and van der Burgt, M., Copyright 2008, with permission from Elsevier.)

increased by removing all or a significant portion of the water from the soot mixture before recycling the soot. Separating the soot from the water allows for independent regulation of the soot and water recycle streams, depending on the reaction conditions in the gasifier.

Other solid particles from mostly syngas are removed at several stages. Char particles are collected in a syngas cooler and cyclones and recycled to the gasifier. Slag from the gasifier is water quenched and removed through a lock hopper (Collot, 2006). Syngas is quenched with recycled cleaned cooled water. Raw syngas is then dedusted in ceramic candle filters. Molten ash is tapped off and quenched with water in the bottom of most types of gasifiers. Ash is commonly removed from the gasifier by a revolving grate and depressurized in a lock hopper.

4.4 CO_2 Capture and Storage

The current challenge for using carbonaceous solid fuels in producing hydrogen is the capture and disposal/storage of CO_2 in an environmentally feasible manner (EcoEnergy, 2008). Based on a performance evaluation of state-of-the-art commercial technology for hydrogen coproduction (Chiesa et al., 2005), a system that converts 57%–58% of the coal lower heating value (LHV) to hydrogen while exporting to the grid electricity amounting to 2%–6% of the coal LHV, in contrast to decarbonizing coal in IGCC electricity generation (which entails a loss of 6%–8% points of electricity conversion when capturing CO_2 as an alternative to venting it), CO_2 capture for hydrogen production incurs a minor energy penalty (2% points of export electricity). Some of these losses are attributable to the number of intermediate processes involved in producing hydrogen from coal, including several for pollution control.

The hydrogen from various gas streams, subsequent to cleaning and particle separation, is accompanied by CO_2, which can be stored (Damen et al., 2007) and used for different applications. After further cooling of syngas from gasifiers, H_2S is removed from the syngas using a physical solvent (Selexol) and CO_2 is removed from the syngas (again using Selexol). After being stripped from the solvent, the CO_2 is dried and compressed to 150 bar for pipeline transport and underground storage (Higman and Burgt, 2008). Three paths for the CO_2 are possible, as shown in Figure 4.1. The commercial route is already applied by several industries for using and storing CO_2 in various forms. The current commercial applications include industrial use of CO_2 in large refrigeration systems making dry ice, enhanced oil recovery, and manufacturing of various chemicals. Also some CO_2 produced within the hydrogen production system may be used for transporting solid fuels into high-pressure reactors.

The second option of storage involves sending the remaining CO_2 for large-scale underground storage (IEA, 2008b). Such processes are being implemented commercially in recent years through a process known as geological sequestration (GS), where the CO_2 is compressed and placed in deep underground aquifers, depleted oil and gas reservoirs, and dried underground coal beds. Some large-scale CO_2 storage projects are already in operation and under construction, while others are the subject of feasibility studies (NETL, 2009).

There is an upcoming and promising third option of disposing CO_2, converting CO_2 into microalgae using sunlight and water, via algae-based artificial photosynthesis (in Figure 4.1). Microalgae are microscopic photosynthetic organisms. They generally produce more of the kinds of natural oils needed for biodiesel extraction (Sheehan et al., 1998). Autotrophic algae enable the photosynthesis process by utilizing light, CO_2, and water to grow the candidate algae (depending on the conditions available for growth). Heterotrophic algae use thermal energy from waste heat applications, CO_2, and nutrients derived from biogas effluents, leachate in landfills, and wastewater from fermenting processes. Some of the advantages of algae include (Sheehan et al., 1998) the following:

- Use of far less water than traditional oilseed crops.
- More efficient conversion of solar energy because of their simple cellular structure.
- Capable of producing 30 times the amount of oil per unit area of land, compared to terrestrial oilseed crops.
- Capable of taking the waste (zero-energy) form of carbon (CO_2) and converting it to a high-density liquid form of energy (natural oil).
- Growth that is directly proportional to the surface area and the area of exposure to sunlight (only for autotrophs).
- Use of optical fiber–based reactor systems (only for autotrophs) developed by Japanese researchers that could dramatically reduce the amount of surface area required for algae production.
- Three main ingredients of microalgae: carbohydrates, protein, and natural oils.
- Fuel production concepts:
  - Production of CH_4 gas through biological or thermal gasification: breakdown of any form or organic carbon into CH_4
  - Production of ethanol through fermentation: most effective for conversion of carbohydrate portion of algae
  - Production of biodiesel: applies exclusively to the natural oil fraction of the microalgae
- Ability to produce up to 60% of their body weight in the form of triacylglycerols (TAGs), making it an alternative source for biodiesel than the oilseeds.
- Transesterification: reacting TAGs with simple alcohols creates a chemical compound known as alkyl ester or, more commonly, biodiesel.
- Not dependent on land area if algae are grown in large existing aquatic environments, although for large-scale production, a sizable portion of the land (nonagricultural land) area is needed.

There is also a fourth option not denoted in Figure 4.1, which is the mineral storage of CO_2, in which CO_2 is reacted with naturally occurring magnesium (Mg)- and calcium (Ca)-containing minerals to form carbonates. This process has several advantages, the most significant of which is the fact that carbonates have a lower energy state than CO_2, which is why mineral carbonation is thermodynamically favorable and occurs naturally (Shackley and Gough, 2006). Thus, the carbonates are stable and are unlikely to convert back to CO_2 under standard conditions. On the same basis, CO_2 recycle or reuse is another option that involves metal oxides such as Fe_2O_3, ZnO, and CaO to split CO_2 into CO and oxygen, for use in various processes (Kodama and Gokon, 2007). The latter option, in which CO_2 is split into CO and oxygen is an artificial photosynthesis process; it is a greenhouse-type concept for controlled feeding of biologically engineered plants that can consume, in a controlled environment, high volumes of CO_2 to store carbon and emit oxygen (Collings and Critchley, 2005).

4.5 Economic Benefits through Process Integration

Several paths exist to obtain hydrogen from water using energy from solid fuels, but the sustainability of these paths remains partly dependent on the economics of the overall system. This section presents an overview of the benefits of appropriate process integration toward economic sustainability of hydrogen production using solid fuels as the energy source.

Biomass is likely to remain a major solid fuel in driving the hydrogen economy. By the year 2050, according to the VTT report estimates (IEA, 2008b), agricultural and forestry residues and wastes will be the most cost-competitive types of biomass, with only around 30% of total bioenergy coming from specialized energy crops. Only current technologies and crop yields are used in this assessment, and population growth is taken to be zero. Larger biomass-based plants can achieve economies of scale, but this can be more than offset by the increased transport distances needed to obtain the required volume of biomass. The costs of delivered biomass vary with country and region due to factors including variations in terrain, labor costs, and crop yields. On average in Europe, the cost of operating a forwarder is USD 67–104/h, chipping costs USD 148–213/h, transport costs USD 91–143/h, and loading/unloading costs USD 40–83/h.

The coal sector is well established from mining to transportation to distribution and utilization, enabling economic development in several countries around the world (IEA, 2008a). Some current and possible future integration benefits are given in the following.

4.5.1 Integration of Energy Conversion and Waste Management Processes

Integrated processes, such as the one shown in Figure 4.20, create opportunities to improve the use of solid fuels, thereby reducing the associated environmental pollution and helping meet growing energy demands (Gnanapragasam et al., 2010). Numerous small-scale integrated projects for end users, such as combined power generation with hydrogen production and CO_2 capture, and their benefits have been investigated (Chiesa et al., 2005;

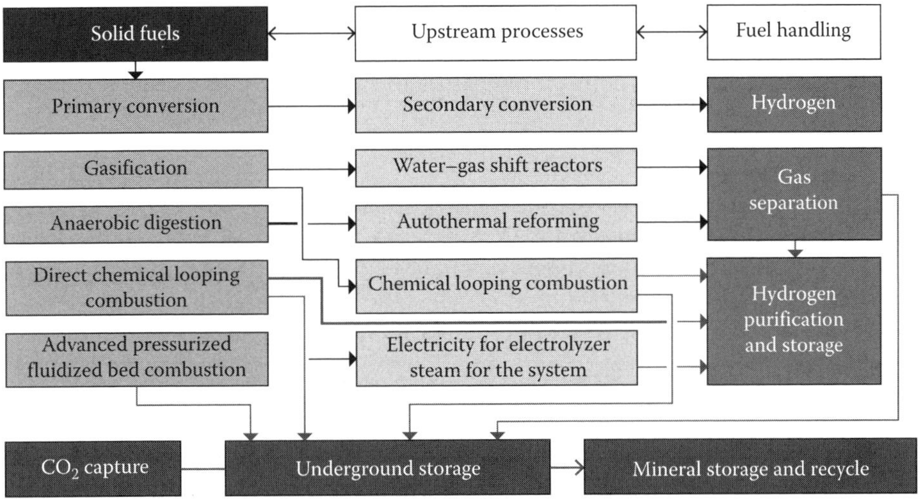

FIGURE 4.20
Example of process integration aimed at centralized hydrogen production for energy, environment, and economic benefits.

Damen et al., 2006; Shoko et al., 2006; Stiegel and Ramezan, 2006; Turner et al., 2008). New energy conversion technologies are also being developed and adopted due to the potential markets for efficient processes. Gasification as shown in Figure 4.7 is a promising conversion process that is expected to achieve significantly higher efficiencies for coal/biomass conversion to electricity and transport fuels (Beer, 2000).

4.5.1.1 Benefits of Process Integration

In large-scale process integration, solid fuels are converted via various industrial-scale integrated processes to convenient energy forms and ultimately to hydrogen. The benefits of such systems include the following:

1. *Multiple uses of feedstock*: Input solid fuels are converted to hydrogen, increasing efficiency and decreasing waste generation.
2. *Demand-based production*: Since each feedstock has a different market value, a proper blend of the cheapest and most efficiently used feedstocks can be used to reduce operating costs as well as to conserve rare fuels.
3. *Process applicability*: Based on the market value and demand for hydrogen, some of the processes are used for producing other products such as chemicals, reducing operating costs while utilizing the full potential of the facility.
4. *Resource availability*: Proven coal reserves and potentially consistent biomass supplies (IEA, 2006, 2007, 2008; Asif and Muneer, 2007), as well as the challenges associated with solid waste disposal (e.g., lack of landfills, special needs depending on the nature of waste), contribute to sustainability by being flexible in terms of allowable feedstocks.
5. *Proven technologies*: The processes considered have been tested or proven in industry so as to ensure high efficiency and low pollution with optimized performance.
6. *Opportunities*: When combining several processes for multiple products, new opportunities are provided for improving and developing these processes.
7. *Centralized pollution control*: Integration provides a major advantage by permitting centralized pollution control (reducing the overall energy consumption), which increases the capture efficiency for many pollutants, for example, SO_2, H_2S, COS, NO_x, and CO_2.

4.5.1.2 Challenges in Large-Scale Process Integration

Integration poses several challenges including potential incompatibility of processes and the need for additional accessory units that increase costs and reduce overall system efficiency. The demand for hydrogen energy is not yet large, in part because infrastructure for hydrogen is not yet widely available. Much of the public and the media view coal as *dirty* based on data from existing power plants, which lack the pollution-control technologies capable of achieving near-zero emissions, confusing information and lack of awareness about cleaner technologies for using solid fuels. The project in Figure 4.6 has various stages of analysis, including thermodynamic, material, economic, environmental, market, and optimization. These stages should detail the benefits of implementing such a system into the existing energy infrastructure. The multiproduct and integrated nature of the system enables efficient use of coal, biomass, and municipal solid waste in producing electricity, hydrogen, and chemicals simultaneously, thereby contributing to making energy systems more sustainable.

4.5.1.3 Economics of Process Integration

Gasification appears to be a more advantageous use of biomass than combustion, but a comprehensive economic analysis of biomass gasification systems suggests that the costs of plant construction are high compared with conventional plants while there are environmental benefits (Dowaki et al., 2005). Hence, an appropriate business model may be available in the future. Gasification has been proven to be the most advantageous conversion process for the initial stages of a hydrogen production facility. However, for gasification to play a major role in the near future, capital and operating cost must be reduced and reliability and performance improved (Collot, 2006).

There are currently several research and industrial development projects worldwide on IGCC and integrated gasification fuel cell (IGFC) systems (Shoko et al., 2006). In such systems, there is a need to integrate complex unit operations including gasifiers, gas separation and cleaning units, WGS reactors, turbines, heat exchangers, steam generators, and fuel cells. The IGFC systems tested in the United States, Europe, and Japan employing gasifiers (types include Texaco, Lurgi, and Eagle) and fuel cells have resulted in energy conversion efficiencies of 47.5% (HHV), which is much higher than the 30%–35% efficiency of conventional coal-fired power generation. IGCC and IGFC are currently not viable economically compared with current coal-utilization technologies, but further efficiency improvements and reductions in gaseous pollutant emissions could render the technologies more competitive. Hydrogen produced from coal-based gasification has recently been shown to be competitive with production from natural gas provided the cost of natural gas remains above $\$4/10^6$ Btu, and the reliability of gasification-based processes can be demonstrated to be high (Stiegel and Ramezan, 2006). These authors suggest that the cost of producing hydrogen from coal could be reduced by 25%–50%, even with the capture and sequestration of CO_2.

Energy conversion efficiencies are greatest for natural gas–based systems due to the fuel characteristics. The costs of hydrogen production for natural gas and coal/biomass are much lower than for electrolysis (which presently has only a 4% market share) due to the volume of production (which is much higher for hydrogen from fossil fuels) and the mature state of the technology. A comparison of efficiencies and costs for various hydrogen production methods (Shoko et al., 2006) shows steam reforming of natural gas to be the most beneficial, with high efficiencies (65%–75% based on LHV) and low production costs (USD 5–8/GJ). Gasification of biomass and coal has an overall efficiency of 42%–47% (LHV) with an average production cost at USD 9–13/GJ, while water electrolysis has the lowest efficiency (35%–42% HHV) and the highest production cost (on average USD 20/GJ).

4.5.2 Cost of Hydrogen Production: Impact of CO_2 Capture on Various Technologies

The costs of hydrogen production with coal-based technologies are compared with those for established technologies in Figure 4.21. With current technologies, large-scale hydrogen production can be attained with SMR and coal gasification with CO_2 capture from the shifted syngas, resulting in a CO_2 capture efficiency of 85%–90% (Damen et al., 2006, 2007). The conversion efficiency, including electricity inputs and outputs in primary terms, is 73% for SMR and 59%–62% for coal gasification. Investment costs are approximately 550 and 840 €/kWh for 1000 MWh SMR and coal gasification plants, respectively. The costs of hydrogen produced by SMR are dominated by fuel and feed costs, which makes coal gasification more favorable, especially if energy prices rise increasingly. Hydrogen production costs for SMR are estimated at 9.5 €/GJ and an optimally designed coal gasification

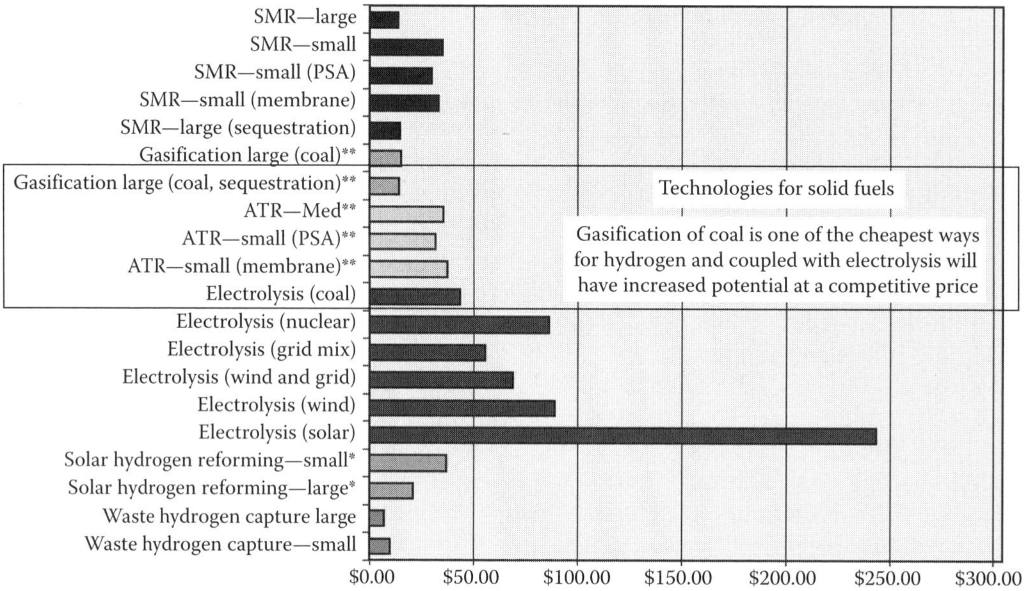

FIGURE 4.21
Total costs in $ per GJ of hydrogen produced with CO_2 valued at $15/ton, for various production processes. (SDTC, Renewable fuel—Hydrogen, SD Business Case, Version 1, Sustainable Development Technology Canada, Report BC_RFH_V7.12.1_EG_061123, 2006. Available at http://www.sdtc.ca/uploads/documents/en/RenewableFuel-Hydrogen.pdf, Accessed on December 5, 2013.)

plant with electricity export may reach 7 €/GJ. CO_2 avoidance costs compared to identical plants without CO_2 capture are 23 and 5 €/ton of CO_2 for SMR and coal gasification, respectively. The penalty for CO_2 capture is compared in Figure 4.22 with other hydrogen production technologies, based on data compiled by SDTC (2006). The rectangular boxes in Figures 4.21 and 4.22 identify processes relating to hydrogen production using solid fuels.

Advanced large-scale ATR and coal gasification systems with ion-transfer membranes enable hydrogen production with 90% CO_2 capture at efficiencies of 73% for ATR and 69% for coal gasification (Damen et al., 2006, 2007). Investment costs for these systems are estimated at nearly 300 and 600 €/kWh, resulting in hydrogen costs of 8.1 and 6.4 €/GJ and CO_2 avoidance costs of 13 and 5 €/ton CO_2 for ATR and coal gasification, respectively.

A membrane reformer enables small-scale hydrogen production with relatively low-cost CO_2 capture. A 2 MWh plant may achieve efficiencies of 65% and investment costs of around 600 €/kWh (including hydrogen compression to 480 bar), resulting in a hydrogen cost of nearly 17 €/GJ, considering gas and electricity prices for small industrial users. Although the desire to reduce CO_2 emissions and capture emitted CO_2 is growing, the investment cost for solid fuel–based hydrogen production will likely remain the main hurdle.

4.5.3 Comparison of Hydrogen Production: Efficiency and Overall Costs

When carbon sequestration is coupled with natural gas steam reforming or coal gasification for hydrogen production, there is 14%–16% expected increase in the production cost of hydrogen (Tzimas and Peteves, 2005). Hydrogen production costs will decrease for coal and biomass once a market is established, likely as shown in Table 4.5, in which current data are compared with data for the year 2020. Natural gas is presently the most

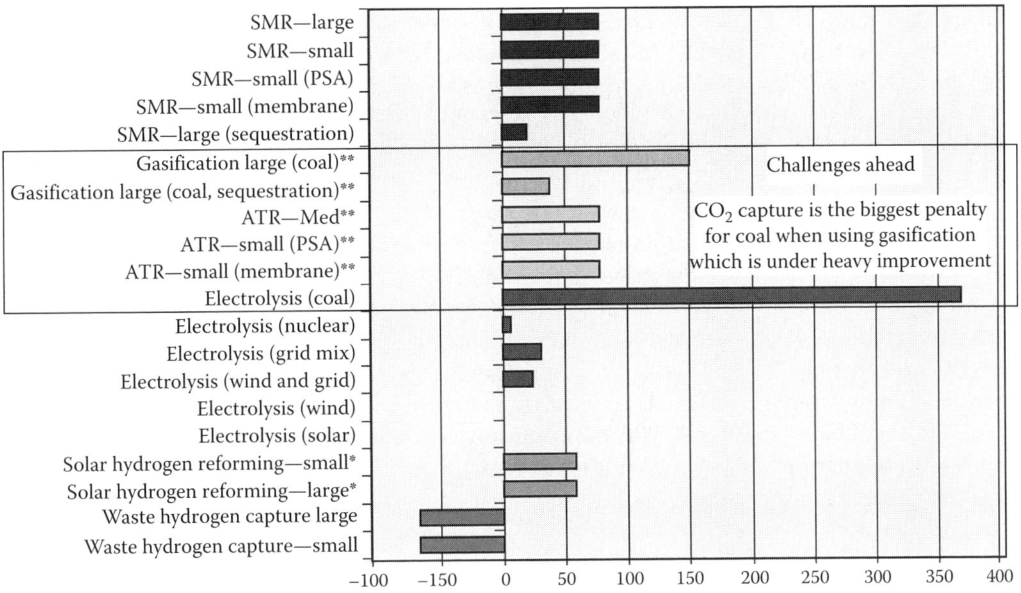

FIGURE 4.22
CO_2 emissions in kg per GJ of hydrogen produced, for various production processes. (SDTC, Renewable fuel—Hydrogen, SD Business Case, Version 1, Sustainable Development Technology Canada, Report BC_RFH_V7.12.1_EG_061123, 2006. Available at http://www.sdtc.ca/uploads/documents/en/RenewableFuel-Hydrogen.pdf, Accessed on December 5, 2013.)

TABLE 4.5

Comparison of Hydrogen Production for Different Feedstocks[a]

Parameter	Natural Gas to H_2	Natural Gas to H_2 with CO_2 Capture	Coal to H_2	Coal to H_2 with CO_2 Capture	Biomass to H_2
Overall efficiency (%)	80	75	55	65	55
Specific capital investment (€/MW_{th})	333,000	380,000	834,000	745,000	932,000
Specific operation and maintenance costs (€/MW_{th})	18,000	43,000	38,000	79,000	39,000
Feedstock costs (€/GJ)	5.4	5.4	2.1	2.1	4.27
H_2 production costs: present (€/GJ)	8.6	10.0	8.6	10.5	12.0
H_2 production costs: 2020 (€/GJ)	10.1	11.1	8.4	8.7	10.2

Source: Data are compiled from Mueller-Langer, F. et al., *Int. J. Hydrogen Energy*, 32(16), 3797, 2007.
[a] Conversion of coal and biomass to hydrogen is via gasification.

advantageous feedstock in terms of efficiency, CO_2 capture, hydrogen production costs, capital investment and specific operation, and maintenance costs (Langer et al., 2007).

Coal is superior to biomass, with lower capital costs, feedstock costs and availability, and hydrogen production costs. But coal has higher operating costs when coupled with CO_2 capture. With CO_2 capture, the investment cost decreases for coal due to the integration of various processes. Hydrogen production costs based on direct conversion concepts have been projected for 2020 (Langer et al., 2007). The comparison is made for the following processes: natural gas steam reforming, conventional small scale (NGSRS) at 22 €/GJ; natural gas steam reforming, conventional large scale with carbon capture (NGSTLCC) at 11 €/GJ; conventional coal gasification, large scale (CCGL) at 10 €/GJ; conventional coal gasification, large scale with carbon capture (CCGLCC) at 11 €/GJ; advanced coal gasification, large scale with carbon capture (ACGLCC) at 9 €/GJ; and biomass gasification, large scale (BGL) at 10 €/GJ.

4.5.4 Cost of Hydrogen in World Markets

Current decentralized hydrogen production costs more than USD 50/GJ, but various centralized production options similar to the process integration shown in Figure 4.20 promise, in the long run, hydrogen prices between USD 10/GJ and USD 15/GJ. While retail hydrogen prices are sensitive to prices of feedstocks (e.g., natural gas and electricity), the cost of hydrogen through natural gas reforming may decrease to less than USD 15/GJ of H_2 by 2030, and through electrolysis, it may decrease to less than USD 20/GJ of H_2. The projected cost of hydrogen from coal gasification in centralized IGCC plants with CCS is even lower—below USD 10/GJ (IEA, 2008b). Long-term costs for high-temperature water splitting could range from USD 10/GJ (using nuclear) to USD 20/GJ (using solar heat). Higher costs are projected for other technologies.

In addition to production facilities, infrastructure needs to be developed to distribute, store, and deliver hydrogen to vehicles. The overall investment cost for this infrastructure, worldwide, is likely to be in the trillions of US dollars (DOE, 2008). Overall, the retail price of hydrogen for transportation users, reflecting all feedstock-related, capital (infrastructure), and operating costs, appears likely to remain well above USD 1.00/L of gasoline equivalent, for the foreseeable future.

4.6 Summary

This review of various stages for obtaining hydrogen using solid fuels, through selected research and development efforts on aspects of hydrogen production from solid fuels, indicates the potential for significant improvements in applications of existing processes in different ways:

- Gasification is an established primary conversion process for producing hydrogen from solid fuels, and research is ongoing in terms of design, modeling, optimization, and market applications.
- Some processes have significant applications in energy markets, including those that use plasma for gasification, SCW for gasification, and USS for gasification, as well as a solar-thermal process for direct dissociation of carbon and hydrogen from solid fuels.
- Several system-integration projects have been proposed and tested based on the applicability of associated processes to increase hydrogen production and improve overall system productivity with additional products.
- Breakthrough conversion technologies or processes that differ fundamentally from those used at present and that can increase efficiency by more than 10%–15% do not exist presently at research or commercial levels. The reason for this observation is that energy research efforts are widespread, with the focus on core technologies driven by market interests and demand.

- The requirements for alternative transportation fuel sources for clean and sustainable energy sources for power generation and less dependence on outside markets will in the near future likely drive increased research and development on new conversion processes. Circumstances at the present time appear adequate for promoting the rapid development of fundamentally different technologies for the conversion of solid fuels to hydrogen. Pressure is increasing on governments, industry, investors, and researchers to find alternatives.

A major shift is expected over the next few years in the way research is performed on the conversion of solid fuels into various energy forms, and hydrogen will likely receive the highest priority. Some research directions for converting solid fuels to hydrogen likely to receive attention include the following:

- Direct conversion of solid fuels to hydrogen and determining ways to reduce the number of steps in the processes
- Decreasing exergy losses for processes and systems and finding ways in process designs to make exergy use more efficient
- Sizing process units to reduce waste heat and entropy creation while maintaining performance
- Improving or optimizing system integration, accounting for convenient installation (closeness to resource location is preferred for municipal solid waste and landfills), low capital costs, low operating and maintenance costs, low emissions, and comparable overall efficiencies (although higher than existing plants is preferred)

The cost of hydrogen from solid fuels is expected to become competitive and may become lower than other sources, based on an adequate hydrogen infrastructure and combined research efforts from the academic and industrial workforce.

The information amalgamated here is intended to provide a useful resource that can aid researchers pursuing advanced methods for hydrogen production using solid fuels.

Acknowledgment

The authors kindly acknowledge the financial support provided by the Natural Sciences and Engineering Research Council of Canada through its Discovery Grant Program.

References

Amelio M, Marrone P, Gallucci F, Basile A. 2007. Integrated gasification gas combined cycle plant with membrane reactors: Technological and economical analysis. *Energy Conversion and Management* 48:2680–2693.

Andrus H. 2009. Chemical looping combustion coal power technology development prototype. *CO_2 Capture Technology Conference*, Pittsburgh, PA, March 24–26, 2009, US DOE/NETL.

Anheden M, Svedberg G. 1998. Exergy analysis of chemical-looping combustion systems. *Energy Conversion and Management* 39(16–18):1967–1980.

Anthony EJ. 2008. Solid looping cycles: A new technology for coal conversion. *Industrial Engineering Chemical Research* 47:1747–1754.

Asif M, Muneer T. 2007. Energy supply, its demand and security issues for developed and emerging economies. *Renewable and Sustainable Energy Reviews* 11:1388–1413.

Aznar MP, Caballero MA, Sancho JA, Frances E. 2006. Plastic waste elimination by co-gasification with coal and biomass in fluidized bed with air in pilot plant. *Fuel Processing Technology* 87:409–420.

Bargigli S, Raugei M, Ulgiati S. 2004. Comparison of thermodynamic and environmental indexes of natural gas, syngas and hydrogen production processes. *Energy* 29:2145–2159.

Basu P. 2006. *Combustion and Gasification in Fluidized Beds*. CRC Press/Taylor & Francis Group, LLC, Boca Raton, FL.

Bates BC, Kundzewicz ZW, Wu S, Palutikof JP (eds.). 2008. Climate change and water. Technical paper of the Intergovernmental Panel on Climate Change. IPCC Secretariat, Geneva, Switzerland.

Beer JM. 2000. Combustion technology developments in power generation in response to environmental challenges. *Progress in Energy and Combustion Science* 26:301–327.

Beer JM. 2007. High efficiency electric power generation: The environmental role. *Progress in Energy and Combustion Science* 33:107–134.

Bredesen R, Jordal K, Bolland O. 2004. High-temperature membranes in power generation with CO_2 capture. *Chemical Engineering and Processing* 43:1129–1158.

Brinkman HW, Van Eijk JPGM, Meinema HA, Terpstra RA. 2000. Processing and characteristics of TNO's hollow fibre ceramic membranes. *International Ceramics* 1:45–47.

British Petroleum. 2003. *BP Statistical Review of World Energy*. British Petroleum, London, U.K.

Brown BB, Yiridoea EK, Gordon R. 2007. Impact of single versus multiple policy options on the economic feasibility of biogas energy production: Swine and dairy operations in Nova Scotia. *Energy Policy* 35:4597–4610.

CanmetENERGY. 2005. Canada's Clean Coal Technology Roadmap. Report, CANMET Energy Technology Center, Natural Resources Canada. Ottawa, Ontario, Canada. Available at: http://canmetenergy.nrcan.gc.ca/clean-fossils-fuels/clean-coal/1806. Accessed on December 5, 2013.

Carabin P, Gagnon J-R. 2007. Plasma gasification and vitrification of ash-conversion of ash into glass-like products and syngas. *2007 World of Coal Ash (WOCA)*, Covington, KY, May 7–10, 2007. Available at: http://www.worldofcoalash.org/2007/ashpdf/148carabin.pdf. Accessed on December 5, 2013.

Chen Y, Xu H, Jin X, Xiong G. 2006. Integration of gasoline prereforming into autothermal reforming for hydrogen production. *Catalysis Today* 116:334–340.

Chiesa P, Consonni S, Kreutz T, Williams R. 2005. Co-production of hydrogen, electricity and CO_2 from coal with commercially ready technology. Part A: Performance and emissions. *International Journal of Hydrogen Energy* 30:747–767.

Chmielniak T, Sciazko M. 2003. Co-gasification of biomass and coal for methanol synthesis. *Applied Energy* 74:393–403.

Collings AF, Critchley C. 2005. *Artificial Photosynthesis: From Basic Biology to Industrial Application*. Wiley, Weinheim, Germany.

Collot AG. 2006. Matching gasification technologies to coal properties. *International Journal of Coal Geology* 65:191–212.

Corella J, Toledo JM, Molina G. 2006. Calculation of the conditions to get less than 2 g tar/m_n^3 in a fluidized bed biomass gasifier. *Fuel Processing Technology* 87:841–846.

Damen K, van Troost M, Faaij A, Turkenberg W. 2006. A comparison of electricity and hydrogen production systems with CO_2 capture and storage. Part A: Review and selection of promising conversion and capture technologies. *Progress in Energy and Combustion Science* 32:215–246.

Damen K, van Troost M, Faaij A, Turkenberg W. 2007. A comparison of electricity and hydrogen production systems with CO_2 capture and storage. Part B: Chain analysis of promising CCS options. *Progress in Energy and Combustion Science* 33:580–609.

Dincer I, Rosen MA. 2007. *Exergy: Energy, Environment and Sustainable Development*. Elsevier, Oxford, U.K.

DOE. 2000. Gasification: Reference shelf—System studies. National Energy Technology Laboratory, U.S. Department of Energy, Washington, DC. Accessible at: http://www.netl.doe.gov/technologies/coalpower/gasification/ref-shelf.html. Accessed on December 5, 2013.

DOE. 2005. International energy outlook. Energy Information Agency, U.S. Department of Energy, Washington, DC. Accessible at: www.eia.doe.gov/oiaf/ieo/index.html. Accessed on December 5, 2013.

DOE. April 2007. Industrial size gasification for syngas, substitute natural gas and power production. Report no. DOE/NETL-401/040607, National Energy Technology Laboratory, U.S. Department of Energy, Washington, DC.

DOE. September 2007. Hydrogen from Coal Program: Research, development and demonstration plan for the period 2007 through 2016. External Report, U.S. Department of Energy, Washington, DC. Available at: http://www.netl.doe.gov/technologies/hydrogen_clean_fuels/refshelf/pubs/External_H2_from_Coal_RDD_Plan_September_13.pdf. Accessed on December 5, 2013.

Dowaki K, Mori S, Fukushima C, Asai N. 2005. A comprehensive economic analysis of biomass gasification systems. *Electrical Engineering in Japan* 153(3):52–63.

Duerr M, Gair S, Cruden A. 2007. Hydrogen and electrical energy from organic waste treatment. *International Journal of Hydrogen Energy* 32:705–709.

EcoEnergy. 2008. Canada's fossil energy future: The way forward on carbon capture and storage. ecoENERGY Carbon Capture and Storage Task Force Report, Government of Alberta, Calgary, Alberta, Canada, January 9, 2008. Available at: http://www.energy.gov.ab.ca/Org/pdfs/Fossil_energy_e.pdf. Accessed on December 5, 2013.

Eldridge M, Sciortino M, Furrey L, Nowak S, Vaidyanathan S, Neubauer M, Kaufman N, Chittum A, Black S. 2009. The 2009 State Energy Efficiency Scorecard. ACEEE; Colin Sheppard, Charles Chamberlin, and Arne Jacobson, Humboldt State University; Yerina Mugica and Dale Bryk, Natural Resources Defense Council (NRDC). Report available at: http://aceee.org/research-report/e097.htm. Accessed on December 5, 2013.

Fan L, Li F, Ramkumar S. 2008. Utilization of chemical looping strategy in coal gasification processes. *Particuology* 6:131–142.

Fargione J, Hill J, Tilman D, Polasky S, Hawthorne P. 2008. Land clearing and the biofuel carbon debt. *Science* 319(5867):235–238.

Fleming GK, Dupuis GE. 1993. Hydrogen membrane recovery estimates. *Hydrocarbon Processing*. Gulf Publishing Company, Houston, TX, pp. 83–86.

Furimsky E. 1998. Gasification of oil sand coke: Review. *Fuel Processing Technology* 56:263–290.

Galvita V, Messerle VE, Ustimenko AB. 2007. Hydrogen production by coal plasma gasification for fuel cell technology. *International Journal of Hydrogen Energy* 32:3899–3906.

Garcia GO, Douglas P, Croiset E, Zheng L. 2006. Technoeconomic evaluation of IGCC power plants for CO_2 avoidance. *Energy Conversion and Management* 47:2250–2259.

Gasafi E, Meyer L, Schebek L. 2007. Exergetic efficiency and options for improving sewage sludge gasification in supercritical water. *International Journal of Energy Research* 31:346–363.

Gerardi MH. 2003. *The Microbiology of Anaerobic Digesters*. John Wiley & Sons, Inc., Hoboken, NJ.

Gnanapragasam NV, Reddy BV, Rosen MA. 2009a. Reducing CO_2 emissions for an IGCC power generation system: Effect of variations in gasifier and system operating conditions. *Energy Conservation and Management* 50:1915–1923.

Gnanapragasam NV, Reddy BV, Rosen MA. 2009b. Hydrogen production from coal using coal direct chemical looping and syngas chemical looping combustion systems: Assessment of system operation and resource requirements. *International Journal of Hydrogen Energy* 34:2606–2615.

Gnanapragasam NV, Reddy BV, Rosen MA. 2010. Feasibility of an energy conversion system in Canada involving large-scale integrated hydrogen production using solid fuels. *International Journal of Hydrogen Energy* 35(10):4788–4807. doi:10.1016/j.ijhydene.2009.10.047.

Goodman BL. 1999. Stabilization: Anaerobic digestion, Chapter 7. In *Environmental Engineer's Handbook*, CRC Press/Taylor & Francis Group, LLC, Boca Raton, FL.

Guo LJ, Lu YJ, Zhang XM, Ji CM, Guan Y, Pei AX. 2007. Hydrogen production by biomass gasification in supercritical water: A systematic experimental and analytical study. *Catalysis Today* 129:275–286.

Gupta P, Velazquez-Vargas LG, Fan L-S. 2007. Syngas redox (SGR) process to produce hydrogen from coal derived syngas. *Energy and Fuels* 21:2900–2908.

Halloran JW. 2008. Extraction of hydrogen from fossil fuels with production of solid carbon materials. *International Journal of Hydrogen Energy* 33(9):2218–2224.

Hamelinck CN, Faaij APC. 2002. Future prospects for production of methanol and hydrogen from biomass. *Journal of Power Sciences* 111:1–22.

Harrabin R. Germany leads 'clean coal' pilot. *BBC News*, September 3, 2008. Available at: http://news.bbc.co.uk/2/hi/science/nature/7584151.stm. Accessed on December 5, 2013.

Higman C, van der Burgt M. 2008. *Gasification*, 2nd edn. Elsevier Science, Oxford, U.K.

Hofbauer H, Rauch R, Bosch K, Koch R, Aichernig C. 2003. Biomass CHP plant Gussing—A success story. In Bridgewater, A.V., ed., *Pyrolysis and Gasification of Biomass and Waste*. CPL Press, Newbury, U.K., pp. 527–536.

Hoffman Z. 2005. Simulation and economic evaluation of coal gasification with SETS reforming process for power production. MS thesis, Louisiana State University, Baton Rouge, LA.

IEA. 2006. World energy outlook 2006. International Energy Agency Report, Paris, France.

IEA. 2008a. World energy outlook 2008. International Energy Agency Report, Paris, France.

IEA. 2008b. Energy technology perspectives 2008: Scenarios and strategies to 2050. International Energy Agency Report, Paris, France.

IEA. 2009. Renewables information 2009 with 2008 data. Available at www.iadb.org/intal/intalcdi/PE/2009/03711.pdf, Accessed on December 5, 2013.

International Energy Agency. 2009. Renewables Information 2009 with 2008 data. Available at www.iadb.org/intal/intalcdi/PE/2009/03711.pdf. Accessed on December 5, 2013.

Ishida M, Jin H. 1995. Chemical-looping combustion power generation plant system. US Patent No. 5,447,024. September 5, 1995.

Jahnke FC. 2003. Removal of soot in gasification system. US Patent No. 6623537.

Jin H, Ishida M. 2004. A new type of coal gas fueled chemical-looping combustion. *Fuel* 83:2411–2417.

de Jong J, Benes NE, Koops GH, Wessling M. 2004. Toward single step production of multilayer inorganic hollow fibres. *Journal of Membrane Science* 239:2656–2669.

Katofsky RE. 1993. The production of fluid fuels from biomass. Center for Energy and Environmental Studies, Princeton University, Princeton, NJ.

Kavidass S, Anderson GL, Norton GS. 2000. Why build a circulating fluidized bed boiler to generate steam and electric power. Report by the Babcock & Wilcox Company, Barberton, OH. Report No. BR-1708.

Kerry FG. 2006. *Industrial Gas Handbook: Gas Separation and Purification*. CRC Press/Taylor & Francis Group, LLC, Boca Raton, FL.

Kodama T, Gokon N. 2007. Thermochemical cycles for high-temperature solar hydrogen production. *Chemistry Review* 107:4048–4077.

Kohl AL. 1997. *Gas Purification*, 5th edn. Gulf Professional Publishing, Houston, TX.

Korens N, Simbeck DR, Wilhelm DJ. 2002. Process screening analysis of alternative gas treating and sulfur removal for gasification. Report for US Department of Energy by SFA Pacific, Inc., Engineering and Economic Consultants. Available at http://www.netl.doe.gov/technologies/coalpower/gasification/pubs/pdf/SFA%20Pacific_Process%20Screening%20Analysis_Dec%202002.pdf, Accessed on December 5, 2013.

LaCava AI, Shirley AI, Ramachandran R. 1998. How to specify pressure-swing adsorption units—Key components of PSA units. *Chemical Engineering* 105(6):110–118.

Leach BE. 1984. *Applied Industrial Catalysis*, vol. 3. Academic Press, New York.

Leung DYC, Wang CL. 2003. Fluidized-bed gasification of waste tire powders. *Fuel Processing Technology* 84(1–3):175–196.

Lewis FM. 2007. Generation of an ultra-superheated steam composition and gasification therewith. US Patent No. US 7,229,483 B2. June 12, 2007.

Lewis FM, Swithenbank J, Hoecke DA, Russell NV, Shabangu SV. September 2002. High temperature steam-only gasification and steam reforming with ultra-superheated steam. *5th International Symposium on High Temperature Air Combustion and Gasification*, Tokyo, Japan.

Li K. 2007. *Ceramic Membranes for Separation and Reaction*. John Wiley & Sons, Ltd., West Sussex, England.
Li X, Grace JR, Watkinson AP, Lim CJ, Èdenler AE. 2001. Equilibrium modeling of gasification: A free energy minimization approach and its application to a circulating fluidized bed coal gasifier. *Fuel* 80:195–207.
Li Y, Guo L, Zhang X, Jin H, Lu Y. 2009. Hydrogen production from coal gasification in supercritical water with a continuous flowing system. *International Journal of Hydrogen Energy* 35(7):3036–3045. doi:10.1016/j.ijhydene.2009.07.023.
Lior N. 2008. Energy resources and use: The present situation and possible paths to the future. *Energy* 33(6):842–857.
Liu G-S, Niksa S. 2004. Coal conversion sub-models for design applications at elevated pressures. Part II. Char gasification. *Progress in Energy and Combustion Science* 30:679–717.
Liu L, Tan X, Liu S. 2006. Yttria stabilized zirconia hollow fibre membranes. *Journal of the American Ceramic Society* 89(3):1156–1159.
Lu YJ, Jin H, Guo LJ, Zhang XM, Cao CQ, Guo X. 2008. Hydrogen production by biomass gasification in supercritical water with a fluidized bed reactor. *International Journal of Hydrogen Energy* 33:6066–6075.
Luyten J, Buekenhoudt A, Adriansens W, Cooymans J, Weyten H, Servaes F, Leysen R. 2000. Preparation of $LaSrCoFeO_{3-x}$ membranes. *Solid State Ionics* 135:637–642.
Mahecha-Botero A, Boyd T, Gulamhusein A, Comyn N, Lima CJ, Gracea JR, Shirasakic Y, Yasudac I. 2008. Pure hydrogen generation in a fluidized-bed membrane reactor: Experimental findings. *Chemical Engineering Science* 63:2752–2762.
Mahecha-Boteroa A, Grace JR, Lim CJ, Elnashaiec SSEH, Boyd T, Gulamhusein A. 2009. Pure hydrogen generation in a fluidized bed membrane reactor: Application of the generalized comprehensive reactor model. *Chemical Engineering Science* 64:3826–3846.
Mahishi MR, Goswami DY. 2007. Thermodynamic optimization of biomass gasifier for hydrogen production. *International Journal of Hydrogen Energy* 32(16):3831–3840.
Manovic V, Anthony EJ, Lu DY. 2008. Sulphation and carbonation properties of hydrated sorbents from a fluidized bed CO_2 looping cycle reactor. *Fuel* 87:2923–2931.
Mattisson T, Lyngfelt A, Leion H. 2009. Chemical-looping with oxygen uncoupling for combustion of solid fuels. *International Journal of Greenhouse Gas Control* 3(1):11–19.
McKendry P. 2002. Energy production from biomass (part 3): Gasification technologies. *Bio-Resource Technology* 83:55–63.
Moghtaderi B. 2007. Effects of controlling parameters on production of hydrogen by catalytic steam gasification of biomass at low temperatures. *Fuel* 86(15):2422–2430.
Minutillo M, Perna A, Di Bona D. 2009. Modelling and performance analysis of an integrated plasma gasification combined cycle (IPGCC) power plant. *Energy Conversion and Management* 50:2837–2842.
Moriarty P, Honnery D. 2007. Intermittent renewable energy: The only future source of hydrogen? *International Journal of Hydrogen Energy* 32(12):1616–1624.
Mountouris A, Voutsas E, Tassios D. 2006. Solid waste plasma gasification: Equilibrium model development and exergy analysis. *Energy Conversion and Management* 47:1723–1737.
Mountouris A, Voutsas E, Tassios E. 2008. Plasma gasification of sewage sludge: Process development and energy optimization. *Energy Conversion and Management* 49:2264–2271.
Moustakas K, Xydis G, Malamis S, Haralambous KJ, Loizidou M. 2008. Analysis of results from the operation of a pilot plasma gasification/vitrification unit for optimizing its performance. *Journal of Hazardous Materials* 151:473–480.
Mozaffarian M, Zwart RWR, Boerrigter H, Deurwaarder EP, Kersten SRA. 2004. "Green gas" as SNG (synthetic natural gas): A renewable fuel with conventional quality. *Proceedings of the International Conference on Science in Thermal and Chemical Biomass Conversion*, August 30–September 2, Victoria, British Columbia, Canada. Paper no. ECN-RX-04-085, 17.
Mueller-Langer F, Tzimas E, Kaltschmitt M, Peteves S. 2007. Techno-economic assessment of hydrogen production processes for the hydrogen economy for the short and medium term. *International Journal of Hydrogen Energy* 32(16):3797–3810.

Mujeebu MA, Abdullah MZ, Bakar MZA, Mohamad AA, Abdullah MK. 2009. Applications of porous media combustion technology—A review. *Applied Energy* 86:1365–1375.

Muradov NZ, Veziroglu TN. 2008. "Green" path from fossil-based to hydrogen economy: An overview of carbon-neutral technologies. *International Journal of Hydrogen Energy* 33:6804–6839.

Nagl GJ. 2005. Cleaning up gasification syngas. Technical article. Available at: http://www.gtp-merichem.com/company/overview/technical-lit/tech-papers/syngas. Accessed on December 5, 2013.

NEDO. 2004. Clean Coal Technologies in Japan, technological innovation in the coal industry. Technical report, New Energy and Industrial Technology Development Organization (NEDO), Kawasaki, Japan. Available at: www.nedo.go.jp/content/100079772.pdf. Accessed on December 5, 2013.

NETL. 2000. British Gas/Lurgi gasifier IGCC base cases. Report, Process Engineering Division, National Energy Technology Laboratory, US Department of Energy, Washington, DC. Report No. PED-IGCC-98-004, June.

NETL. 2009. Worldwide CO_2 capture and geologic storage Projects. Report, National Energy Technology Laboratory, US Department of Energy, Washington, DC. Available at: http://www.netl.doe.gov/technologies/carbon_seq/corerd/storage.html. Accessed on December 5, 2013.

Pandey A. 2004. *Concise Encyclopaedia of Bioresource Technology*. Haworth Press Inc., New York.

Parikha J, Channiwala SA, Ghosal GK. 2005. A correlation for calculating HHV from proximate analysis of solid fuels. *Fuel* 84:487–494.

Patnaik P. 2002. *Handbook of Inorganic Chemicals*. McGraw Hill, New York.

Pei P, Kulkarni M. 2008. Modeling of ultra superheated steam gasification in integrated gasification combined cycle power plant with carbon dioxide capture. *Proceedings of Energy Sustainability*, Jacksonville, FL, August 10–14, 2008. Paper No. 54325, pp. 763–774.

Pennline HW, Luebke DR, Jones KL, Myers CR, Morsi BI, Heintz YJ, Ilconich JB. 2008. Progress in carbon dioxide capture and separation research for gasification-based power generation point sources. *Fuel Processing Technology* 89(9):897–907.

Perna A. 2008. Combined power and hydrogen production from coal: Part A—Analysis of IGHP plants. *International Journal of Hydrogen Energy* 33:2957–2964.

Ptasinski KJ, Prins MJ, Pierik A. 2007. Exergetic evaluation of biomass gasification. *Energy* 32:568–574.

Rezaiyan J, Cheremisinoff NP. 2005. *Gasification Technologies: A Primer for Engineers and Scientists*. CRC Press, Boca Raton, FL.

Rezvani S, Huang Y, Wright DM, Hewitt N, Mondol JD. 2009. Comparative assessment of coal fired IGCC system with CO_2 capture using physical absorption, membrane reactors and chemical looping. *Fuel*. 88(12):2463–2472. doi:10.1016/j.fuel.2009.04.021.

Ross D, Noda R, Horio M, Kosminski A, Ashman P, Mullinger P. 2007. Axial gas profiles in a bubbling fluidised bed biomass gasifier. *Fuel* 86(10–11):1417–1429.

Ruthven DM, Farooq S, Knaebel KS. 1994. *Pressure Swing Adsorption*. John Wiley & Sons, New York.

Rydén M, Lyngfelt A. 2006. Using steam reforming to produce hydrogen with carbon dioxide capture by chemical-looping combustion. *International Journal of Hydrogen Energy* 31:1271–1283.

Ryu C, Nasserzadeh V, Swithenbank J. 2004. Gasification of wood char by ultra-superheated steam in an entrained flow reactor. *Journal of Energy Institute* 77:46–52.

Ryu J-S, Lee K-W, Choi M-J, Yoo H-S. 2004. The synthesis of clean fuels by F-T reaction from CO_2 rich biosyngas. *Studies in Surface Science and Catalysis* 153:47–54.

Saravanakumar A, Haridasan TM, Reed TB, Bai RK. 2007. Experimental investigations of long stick wood gasification in a bottom lit updraft fixed bed gasifier. *Fuel Processing Technology* 88(6):617–622.

SDTC. 2006. Renewable fuel—Hydrogen. SD Business Case, Version 1, Sustainable Development Technology Canada. Report BC_RFH_V7.12.1_EG_061123. Available at http://www.sdtc.ca/uploads/documents/en/RenewableFuel-Hydrogen.pdf. Accessed on December 5, 2013.

Shackley S, Gough C. 2006. *Carbon Capture and Its Storage: An Integrated Assessment*, Illustrated Edition (November). Ashgate Publishing, Aldershot, U.K.

Sheehan J, Dunahay T, Benemann J, Roessler P. 1998. A look back at the U.S. Department of Energy's Aquatic Species Program: Biodiesel from algae. National Energy Technology Laboratory, US Department of Energy, Washington, DC. Report No. NREL/TP-580-24190.

Shoko E, McLellan B, Costa D. 2006. Hydrogen from coal: Production and utilisation technologies. *International Journal of Coal Geology* 65:213–222.

Sieger R, Donovon J. 2002. Biogasification and other conversion technologies. White Paper, Bioenergy Subcommittee, Water Environment Federation, Alexandria, VA. Available at: http://www.wef.org/WorkArea/DownloadAsset.aspx?id=2253. Accessed on December 5, 2013.

Sinag A, Kruse A, Schwarzkopf V. 2003. Key compounds of the hydropyrolysis of glucose in supercritical water in the presence of K_2CO_3. *Industrial Engineering Chemical Research* 42:3516–3521.

Smid J, Avci CG, Gunay V, Terstra RA, van Eijk JPGM. 1996. Preparation and characterization of microporous ceramic hollow fibre membranes. *Journal of Membrane Science* 112:85–90.

de Souza-Santos ML. 2004. *Solid Fuels Combustion and Gasification: Modeling, Simulation, and Equipment Operations*. CRC Press/Taylor & Francis Group, LLC, Boca Raton, FL.

Stiegel G. 1999. Mixed conducting ceramic membranes for gas separation and reaction. *Membrane Technology* 110:5–7.

Stiegel GJ, Ramezan M. 2006. Hydrogen from coal gasification: An economical pathway to a sustainable energy future. *International Journal of Coal Geology* 65:173–190.

Tan X, Liu S, Li K. 2001. Preparation and characterization of inorganic hollow fibre membranes. *Journal of Membrane Science* 188:87–95.

Trapp B. 2005. Coal gasification: When does it make sense? *Process PowerGen Conference*, IGCC Session, Las Vegas, NV. http://www.gasification.org/Docs/Penwell%202005/Eastman.pdf

Turner J, Sverdrup G, Mann MK, Maness P-C, Kroposki B, Ghirardi M, Evans RJ, Blake D. 2008. Renewable hydrogen production. *International Journal of Energy Research* 32:379–407.

Twigg MV. 1996. *Catalyst Handbook*, 2nd edn. Wolfe Publishing, London, U.K., pp. 284–290.

Tzimas E, Peteves SD. 2005. The impact of carbon sequestration on the production cost of electricity and hydrogen from coal and natural-gas technologies in Europe in the medium term. *Energy* 30(14):2672–2689.

USEPA. 2008. Inventory of U.S. Greenhouse Gas Emissions and Sinks: 1990–2011. Report, US Environmental Protection Agency, Washington, DC. Available at: http://www.epa.gov/climatechange/ghgemissions/usinventoryreport.html. Accessed on December 5, 2013.

Van Oost G, Hrabovsky M, Kopecky V, Konrad M, Hlina M, Kavka T. 2009. Pyrolysis/gasification of biomass for synthetic fuel production using a hybrid gas–water stabilized plasma torch. *Vacuum* 83:209–212.

van Rossum G, Potic B, Kersten SRA, van Swaaij WPM. 2009. Catalytic gasification of dry and wet biomass. *Catalysis Today* 145(1–2):10–18.

Wall TF. 2007. Combustion processes for carbon capture. *Process Combustion Institute* 31:31–47.

Wang Z, Zhou J, Wang Q, Fan J, Cen K. 2006. Thermodynamic equilibrium analysis of hydrogen production by coal based on Coal/CaO/H_2O gasification system. *International Journal of Hydrogen Energy* 31(7):945–952.

Ward AJ, Hobbs PJ, Holliman PJ, Jones DL. 2008. Optimisation of the anaerobic digestion of agricultural resources. *Bioresource Technology* 99:7928–7940.

Way JD, Roberts DL. 1992. Hollow fibre inorganic membranes for gas separations. *Separation Science and Technology* 27(1):29–41.

Weinstein RE, Travers RW. 2002. Advanced Circulating Pressurized Fluidized Bed Combustion (APFBC) repowering considerations. National Energy Technology Laboratory, US Department of Energy, Washington, DC. Paper No. 970563.

Xiao G, Ni MJ, Chi Y, Cen KF. 2008. Low-temperature gasification of waste tire in a fluidized bed. *Energy Conversion and Management* 49(8):2078–2082.

Xu X, Yang W, Lui J, Lin L, Stron N, Brunner H. 2004. Synthesis of NaA zeolite membrane on a ceramic hollow fibre. *Journal of Membrane Science* 229:81–85.

Xuan J, Leung MKH, Leung DYC, Ni M. 2009. A review of biomass-derived fuel processors for fuel cell systems. *Renewable and Sustainable Energy Reviews* 13:1301–1313.

Yamaguchi D, Sanderson PJ, Lim S, Aye Lu. 2009. Supercritical water gasification of Victorian brown coal: Experimental characterisation. *International Journal of Hydrogen Energy* 34:3342–3350.

Yoon HC, Erickson PA. 2008. Hydrogen from coal-derived methanol via auto-thermal reforming processes. *International Journal of Hydrogen Energy* 33:57–63.

Yoshida Y, Dowaki K, Matsumura Y, Matsuhashi R, Li D, Ishitani H, Komiyama H. 2003. Comprehensive comparison of efficiency and CO_2 emissions between biomass energy conversion technologies—Position of supercritical water gasification in biomass technologies. *Biomass and Bioenergy* 25:257–272.

Young L, Pian CCP. 2003. High-temperature, air-blown gasification of dairy-farm wastes for energy production. *Energy* 28(7):655–672.

Yu J, Corripio AB, Harrison DP, Copeland RJ. 2003. Analysis of the sorbent energy transfer system (SETS) for power generation and CO_2 capture. *Advances in Environmental Research* 7:335–345.

Yue GX, Yang HR, Lu JF, Zhang H. 2009. Latest development of CFB boilers in China. *20th International Conference on Circulating Fluidized Bed Combustion*. Keynote Speech, May 18–20, Xian City, China.

Yun Y, Yoo YD, Chung SW. 2007. Selection of IGCC candidate coals by pilot-scale gasifier operation. *Fuel Process Technology* 88:107–116.

Zhang R, Cummer K, Suby A, Brown RC. 2005. Biomass-derived hydrogen from an air-blown gasifier. *Fuel Processing Technology* 86:861–874.

Zwart RWR, Bos A, Kuipers J. 2008. Principle of OLGA tar removal system. Report. Available at: https://www.ecn.nl/fileadmin/ecn/units/bio/Leaflets/b-08-022_OLGA_principles.pdf. Accessed on December 5, 2013.

5

Hydrogen Production from Biomass and Fossil Fuels

Madhukar Mahishi
Cummins Power Generation

D. Yogi Goswami
University of South Florida

Gamal Ibrahim
University of Menoufia

Said S.E.H. Elnashaie
University of British Columbia
New Mexico Technical University

CONTENTS

5.1 Introduction

Biomass represents a large potential feedstock for environmentally clean hydrogen production. Biomass resources include herbaceous energy crops, woody energy crops, forest residues, mill wood waste, waste from pulp and paper mills, logging waste, agricultural crop waste, animal waste, industrial waste, municipal solid waste, sewage sludge, and so on.

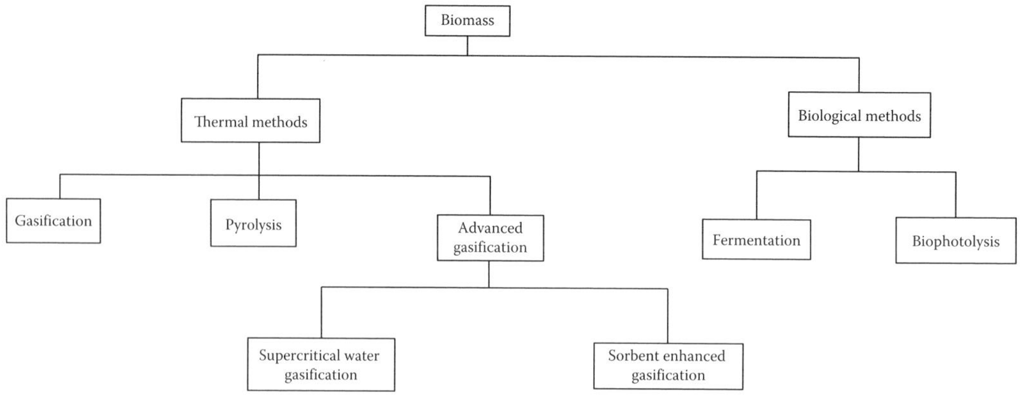

FIGURE 5.1
Biomass conversion to hydrogen.

It lends itself to both biological and thermal conversion processes. In the thermal path, hydrogen can be produced in two ways, namely, direct gasification or pyrolysis. In the biological path, hydrogen can be produced by fermentation or by biophotolysis. In the recent past, some advanced biomass gasification systems for hydrogen production have also been investigated. Figure 5.1 shows the classification of the biomass into hydrogen conversion pathways.

Of all the aforementioned methods, biomass gasification has received considerable attention. Direct gasification of biomass is in many ways similar to coal gasification. The process occurs broadly in three steps: biomass is first gasified (using steam or air) to produce an impure syngas mixture composed of hydrogen, CO, CO_2, CH_4, small amounts of higher hydrocarbons, tars (undesirable by-products of gasification), and water vapor. The gas may also contain particulate matter, which is removed using cyclones and scrubbers. The particulate free gas is compressed and then catalytically steam reformed to eliminate the tars and higher hydrocarbons. This is followed by high and low temperature shift conversion reactions to produce additional hydrogen. Finally, the hydrogen is separated from other products by pressure swing adsorption (PSA) [93] as shown in Figure 5.2. The main reactions taking place in biomass gasification are as follows:

$$\text{Biomass} + \text{heat} + \text{steam} \rightarrow H_2 + CO + CO_2 + CH_4 + \text{light and heavy hydrocarbons} + \text{char} \tag{5.1}$$

The gases produced can be steam reformed to produce H_2, and this process can be further improved by the water gas shift (WGS) reaction as shown next:

$$CO + H_2O \rightarrow CO_2 + H_2 \qquad \Delta H_R = -41.2\ \text{kJ/mol (water gas shift)} \tag{5.2}$$

It is to be noted that unlike pyrolysis (described later), gasification is carried out in the presence of oxygen. Gasification aims to produce gases, whereas pyrolysis aims to produce bio-oil and charcoal. Biomass typically contains about 6% hydrogen by weight. However, in the presence of steam, the hydrogen yield can be considerably improved above the minimum 6% [100]. Gasification temperatures are typically in the range 600°C–850°C, which is lower than many thermochemical water-splitting cycles, thereby making biomass gasification an attractive technology to produce hydrogen. Steam gasification of biomass is endothermic, and the energy required for the process is supplied by

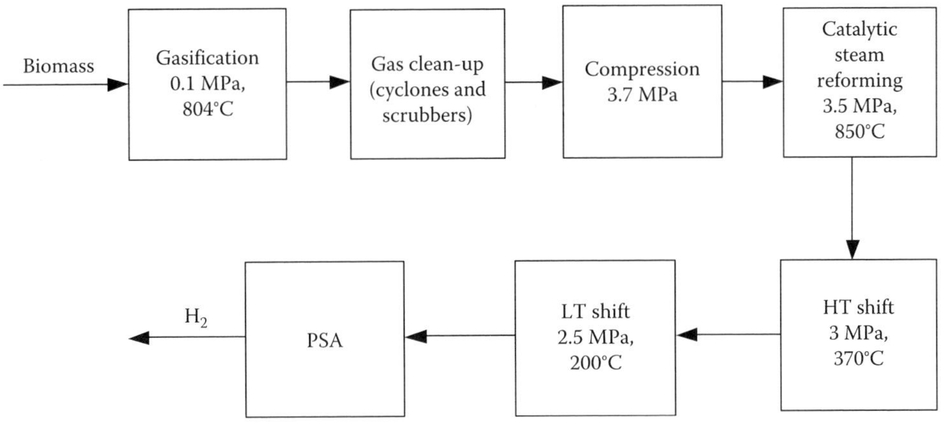

FIGURE 5.2
Gasification followed by steam reforming. (Adapted from Spath, P.L. et al., Update of hydrogen from biomass—Determination of the delivered cost of hydrogen, NREL/MP-510-33112, National Renewable Energy Laboratory Technical Report, Golden, CO, 2000.)

burning part of the biomass feedstock or uncombusted char. One of the major challenges in biomass gasification is to deal with tar formation that occurs during the process. Tars are polyaromatic hydrocarbon by-products of biomass gasification and are undesirable as they clog filters, pipes, and valves and damage downstream equipments such as engines and turbines. The unwanted tar may cause the formation of aerosols and polymerization to a more complex structure, which are not favorable for hydrogen production through steam reforming. Efforts are being made to minimize or reform the tars to produce additional hydrogen [28,30].

Hydrogen can alternately be produced by reforming the biomass to a liquid bio-oil in a process called pyrolysis. Pyrolysis is an endothermic thermal decomposition of biomass carried out in an inert atmosphere typically at 450°C–550°C [61]. The objective of pyrolysis is to produce bio-oil and charcoal. The bio-oil so produced is a liquid composed of 85% oxygenated organics and 15% water. The bio-oil is then steam reformed in the presence of a nickel-based catalyst at 750°C–850°C followed by shift conversion of CO to CO_2 [66] as shown in Figure 5.3. The reactions can be written as

$$\text{Biomass} \rightarrow \text{Bio-oil} + \text{char} + \text{gas} \quad \text{(pyrolysis)} \tag{5.3}$$

$$\text{Bio-oil} + H_2O \rightarrow CO + H_2 \quad \text{(reforming)} \tag{5.4}$$

$$CO + H_2O \rightarrow CO_2 + H_2 \quad \text{(shift)} \tag{5.5}$$

5.2 Research in Biomass Gasification

Biomass gasification has been extensively studied over the last three decades. Different research groups investigated with different objectives like maximizing the hydrogen yield, optimizing syngas yield, product gas cleaning for trouble-free downstream

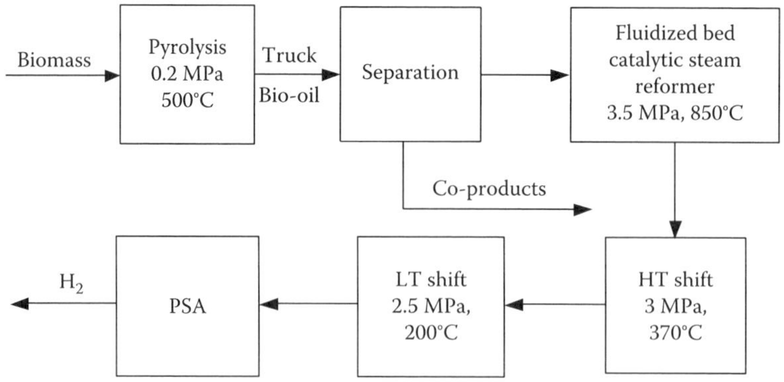

FIGURE 5.3
Pyrolysis followed by steam reforming. (Adapted from Spath, P.L. et al., Update of hydrogen from biomass—Determination of the delivered cost of hydrogen, NREL/MP-510-33112, National Renewable Energy Laboratory Technical Report, Golden, CO, 2000.)

operation, effective waste utilization, and so on. The road blocks in the biomass gasification technology are

- *Gas conditioning*: Product gas needs to be free from particulate matter, tar, entrained char, or catalyst particles; also improved catalysts with high tar-reforming ability and low deactivation must be developed.
- *Feedstock preparation and handling*: Issues related to preparation (drying, storage, and palletizing) and nature of different feeds for high-temperature and -pressure reactors need to be addressed; also since biomass is available seasonally, feedstock-flexible reactors must be developed.
- *Valuable coproduct integration*: Development of technology that can convert by-product streams into useful chemicals or processes.

As products of gasification are mainly gases, the process is more favorable for H_2 production than pyrolysis. In order to optimize the process for H_2 production, efforts have been made to test H_2 production from biomass gasification with various biomass types and at various operating conditions as listed in Table 5.1. Tar reforming is by far one of the major challenges in biomass gasification technology. Currently, three methods are available to minimize tar formation:

1. Catalysis
2. Pretreatment technologies
3. Proper design of gasifier and optimization of design conditions

5.2.1 Catalysis

Biomass thermochemical gasification produces organic impurities in addition to gases (H_2, CO, CO_2, and CH_4) and solids (char and ash). The impurities range from low-molecular aliphatic hydrocarbons to high-molecular-weight polynuclear aromatic hydrocarbons. The high-molecular-weight hydrocarbons are collectively known as *tars*. Tars are undesirable for a number of reasons, for example, tars condense in exit pipes and on particulate filters leading to blockages and clogged filters. Biomass gasification product gas requires

TABLE 5.1
Investigation of Biomass Gasification for Hydrogen Production

	Reactor Type	Catalyst Used	Hydrogen Production (vol.%)	Reference
Sawdust	Not known	Na_2CO_3	48.31 at 700°C	[113]
			55.4 at 800°C	
			59.8 at 900°C	
Sawdust	Circulating Fluidized bed	Not used	10.5 at 810°C	[22]
Wood	Fixed bed	Not used	7.7 at 550°C	[108]
Sawdust	Fluidized bed	Not known	57.4 at 800°C	[100]
Not known	Fluidized bed	Ni	62.1 at 830°C	[80]
Sawdust	Fluidized bed	K_2CO_3	11.27 at 964°C	[46]
		CaO	13.32 at 1008°C	CaO
		Na_2CO_3	14.77 at 1012°C	Na_2CO_3
Pine sawdust	Fluidized bed	Not known	26–42 at 700°C–800°C	[115]
Bagasse			29–38 at 700°C–800°C	
Cotton stem			27–38 at 700°C–800°C	
Eucalyptus globulus			35–37 at 700°C–800°C	
			27–35 at 700°C–800°C	
Sewage sludge	Downdraft	Not known	10–11	[63]
Almond shell	Fluidized bed	La–Ni–Fe	62.8 at 800°C	[81]
		Perovskite	63.7 at 900°C	
Switchgrass	Moving bed	Cu–Zn–Al	27.1	[20]

Source: Ni, M. et al., *Fuel Process. Technol.*, 87, 461, 2006.

substantial cleaning, including tar conversion or removal, before it is used in downstream equipments like fuel cells or turbines.

There are a number of methods to separate or reform tars such as wet scrubbing, thermal cracking, or catalytic cracking. Of these, catalytic cracking has been extensively investigated. Catalytic steam reforming of biomass product gas is an attractive technology as it not only reduces tars but also improves the gas quality and hydrogen yield. Catalytic tar destruction has been studied for several decades, and a number of reviews have been written [28,30,98]. Broadly, three groups of catalyst materials have been used for biomass gasification systems: alkali metals, nonmetallic oxides, and supported metallic oxides. Of these, the nonmetallic oxides and supported metallic oxides have received considerable attention.

5.2.1.1 Nonmetallic Oxide Catalysts

These mainly include calcined dolomites, which are calcium- and magnesium-based salts. These naturally occurring catalysts are relatively inexpensive and disposable. These can be used as primary (in bed) catalysts or secondary catalysts (i.e., in downstream reactors). Several research groups have conducted extensive studies on the tar conversion effectiveness of calcined dolomites and other nonmetallic oxide catalysts. Simell et al. [92] performed a number of studies using model compounds to test the reforming effectiveness of dolomites. The catalysts were calcined at 900°C and showed high toluene conversion efficiencies (>97%); however, catalyst activity was almost completely lost when the CO_2 partial pressure was higher than equilibrium decomposition pressure of dolomite. Simell et al. also reported decomposition of benzene when it was passed over Finnish dolomite at 900°C.

Aznar and Frances [10] constructed a biomass gasification pilot plan where they used air, steam, and a mixture of steam and oxygen as gasifying agents; pinewood was used as the biomass feedstock. It was found that when 20 g of calcined dolomite per kg of biomass was added, the tar content in the product gas decreased by a factor of 4–6, the hydrogen content of the product gas doubled, and CO content reduced by a factor of 2. Several other groups have also studied catalytic tar reforming with dolomites [99,102]. All these studies demonstrate the effectiveness of dolomites as tar-reforming catalysts. High-molecular-weight hydrocarbons are efficiently removed at moderately high temperatures (800°C) with steam-and-oxygen mixtures as the gasifying agent; however, methane concentration is not greatly affected, and benzene and naphthalene are often not completely reformed. A problem with dolomites is the decrease in mechanical strength over time, which leads to catalytic attrition.

5.2.1.2 Commercial Nickel-Reforming Catalyst

A wide variety of Ni-based reforming catalysts are commercially available because of their application in the petrochemical industry. Nickel-based catalysts have also proven to be very effective for hot conditioning of biomass gasification product gases. They have high activity for tar destruction; methane in the gasification product gas is reformed, and they have some WGS activity to adjust the H_2/CO ratio of the product gas. The H_2 and CO contents of the product gas increase, while hydrocarbons and methane are eliminated or substantially reduced for catalyst operating above approximately 740°C. The groups that were active in studying calcined dolomite catalysts have also conducted several studies involving nickel steam reforming catalysts for hot gas conditioning. Aznar et al. [9] conducted several experiments with Ni catalyst at temperatures between 750°C and 850°C and found initial tar conversion efficiency to be greater than 99%. Simell [89] have also investigated commercial Ni steam reforming catalyst for tar conversion using toluene as a model tar compound. They observed complete tar decomposition for catalyst operating at 900°C and 5 MPa. Kinoshita et al. [47] reported results from parametric studies on catalytic reforming of tars produced in a bench-scale gasification system. A commercial Ni catalyst (UCG-90 B) was tested at various temperatures (650°C–800°C), space times (0.6–2.0 s), and steam-to-biomass ratios (SBRs; 0–1.2) in a fluidized bed catalytic reactor. They reported achieving 97% tar conversion; product gas yield was higher in the presence of the catalyst.

Several other groups (Bangala et al. [15] and Wang and Chornet [104]) have reported high effectiveness of Ni catalyst (>90%) in tar reforming. However, there are several factors that still limit the use of Ni catalyst in commercial gasifiers, which need to be addressed. Main limitations include sulfur, chlorine, and alkali metals present in the gasification product gas, which act as catalyst poisons. Coke formation on the catalyst surface can also be substantial when tar levels in the product gas are high. Coke can be removed by regenerating the catalyst; however, repeated high-temperature processing of nickel catalyst can lead to sintering, phase transformations, and volatilization. These catalysts also suffer from frequent deactivation due to poisoning by sulfur, halides, and alkaline impurities.

5.2.1.3 Additional Catalyst Formulations

There are several limitations of Ni reforming catalysts used for tar conversion such as deactivation by coke formation, sulfur and chlorine poisoning, and sintering. Addition of various promoters and support modifiers has been attempted by several groups to improve catalyst activity, lifetime, poison resistance, and resistance to coke formation. Rapagna et al. [80] developed a catalyst with a lanthanum additive (chemical formula $LaNi_{0.3}Fe_{0.7}O_3$) that was prepared by sol–gel process. The prepared catalyst displayed high CH_4 reforming activity at 500°C, resulting in 90% CH_4 conversion. Garcia et al. [34] have prepared a number of Ni-based catalysts with different additives for optimal hydrogen production. They added magnesium and lanthanum as support modifiers, and cobalt and chromium were added to reduce coke formation. The cobalt- and chromium-promoted nickel catalyst on a $MgO–La_2O_3–\alpha-Al_2O_3$ support performed best in terms of yield and lifetime. Sutton et al. [96] studied the effect of different supports using Ni catalyst. The research group impregnated Ni on various supports including Al_2O_3, ZrO_2, TiO_2, SiO_2, and a proprietary tar destruction support. High tar conversion was observed for all of the prepared catalysts.

Drawing a parallel from the auto-industry, Asadullah et al. [7,8] have developed a novel series of catalysts using noble metals on oxide supports. These catalysts were prepared with rhodium, ruthenium, platinum, and palladium and were tested on bench-scale fluidized-bed reactors using cellulose as a model biomass compound. The group found more than 80% tar conversion at temperatures as low as 550°C. Different supports were used such as CeO_2, LiO_2, ZrO_2, Al_2O_3, MgO, and SiO_2. It was found that Rh/CeO_2 gave 100% tar conversion at 550°C. Although these catalysts give 100% tar conversion at relatively low temperatures (500°C–600°C), they are not economically viable due to the high cost of noble metals.

Several catalysts have been investigated for tar reforming of biomass product gases. A critical gap identified for catalytic tar reforming technology in biomass gasification processes is the need for extended lifetime of promising commercial or novel catalysts. Catalytic hot gas conditioning will not become a commercial technology unless adequate catalyst lifetimes can be demonstrated, even for inexpensive, disposable catalysts like calcined dolomite. Frequent disposal of dolomite generates an additional waste stream, and disposal of toxic spent Ni catalysts becomes an environmental burden. Assessment of catalyst lifetimes will allow biomass gasification developers to actually evaluate the cost of operating a biomass gasification plant. The effect of catalyst poisons like sulfur, chlorine, and alkali metals and continued catalyst regeneration must be critically evaluated with long-term catalyst testing. Accurate catalyst cost and lifetime figures will provide important inputs for techno-economic analysis of developing gasification technologies.

5.2.2 Pretreatment Technologies

A challenge in biomass gasification technology is ash formation that leads to deposition, sintering, slagging, fouling, and agglomeration [5,107]. To resolve ash-associated problems, fractionation and leaching of biomass feedstocks have been employed inside the reactor [4–6,101]. Experimental and theoretical studies on different types of biomass have showed that pretreating the biomass feedstock affects the ash and mineral content and augments the volatile (gas and liquid) yield of feedstocks. Pretreatment is carried out by washing the biomass with mild acid or alkali or by impregnating them with salts before gasification. It is hypothesized that during pretreatment, the biomass undergoes de-ashing (removal of mineral matter), which leads to higher gas and hence hydrogen yields. Pretreatment for gasification or pyrolysis also increases the active surface area of biomass.

Das and Ganesh [27] subjected sugarcane bagasse to three different pretreatments (water leaching, mild HCl treatment, and mild HF treatment) and found that the HF treatment reduces ash content of biomass to a negligible amount. The researchers also observed that the char produced in the process had a higher adsorption capacity as compared to untreated biomass. Raveendran et al. [82] impregnated a variety of biomass feedstocks with chloride (KCl and $ZnCl_2$) and carbonate (K_2CO_3 and $ZnCO_3$) salts and found that the gas yield increased substantially. The group later developed a correlation to predict the percentage change in gas yield when any biomass is subjected to potassium and zinc salt pretreatments.

Conesa et al. [23] subjected different almond shell samples to acidic and basic pretreatment followed by $CoCl_2$ (cobalt chloride) impregnation. The samples were then gasified, and the gas composition was determined. The group found that the hydrogen yield of $CoCl_2$-treated almond shells was higher than that of plain almond shells. More recently, gasification of leached olive oil waste in a circulating fluid bed gasifier was reported for gas production that demonstrated the feasibility of leaching as pretreatment for gas production [35]. All the research groups have hypothesized that acid, alkaline, or salt pretreatment alters the mineral matter content of raw biomass. This in turn affects the product yields since the mineral matter generally tends to have a catalytic effect during the gasification process. Biomass pretreatment is a simple and cost-effective way of influencing the product yield of the biomass gasification process. The process generally applies well to feedstocks with large mineral matter (Na, K, Ca, Mg, Fe, and P) content such as switch grass and rice husk.

5.2.3 Proper Design of Gasifier and Optimization of Design Conditions

The operating parameters such as temperature, pressure, gasifying agent, and residence time play an important role in the formation and decomposition of tars and also influence the product gas yield. Gasifier design improvement and experimental studies on biomass gasification have focused on various aspects like parametric analysis, catalytic tar cleaning, cogasification of biomass with coal/plastic, hot gas cleaning, using multiple feedstocks, different gasifier reactor configurations, and so on. In most cases, the end objective was to maximize syngas production. Turn et al. [100] studied the effect of gasifier temperature, SBR, and equivalence ratio (ER) (a measure of air supplied in biomass gasification) on gas yield (mainly H_2, CO, CO_2, and CH_4) in fluidized-bed gasification of sawdust. They found the highest hydrogen yield to occur at a gasifier temperature of 850°C and steam biomass ratio of about 1. The maximum hydrogen yield was found to be 0.128 g/kg dry ash-free biomass. Narvaez et al. [73] have analyzed the effects of temperature, ER, and the addition of dolomite in the air gasification of pine sawdust. The group found that maintaining an ER of 0.3, SBR of 2.2, and gasifier temperature greater than 800°C gave good quality (maximum heating value) gas with minimum tar content. Herguido et al. [44] used different feedstocks (pine sawdust, pinewood chips, cereal straw, and thistles) with steam as the gasification medium and studied the product yield (H_2, CO, CO_2, and CH_4 contents). The group found marked differences in product composition at low gasification temperature, but at temperatures exceeding 780°C, the gas composition was similar for all biomass feedstocks. Gil et al. [36] have studied the effect of different gasification media (air, steam, and steam-and-oxygen mixture) on product gas composition. They observed that using steam in place of air gave a product gas with almost five times more hydrogen. Also, the heating value of product gas in steam gasification (12.2–13.8 MJ/Nm^3) was higher than in air gasification (3.7–8.4 MJ/Nm^3). Pinto et al. [78] have conducted experiments by cogasifying biomass with plastic wastes and observed an increase in the hydrogen yield by about 17% when 40% (wt) of polyethylene was mixed with pinewood. One of the objectives of this research was the effective utilization of plastic waste. It was found that when plain plastic was gasified, it softened and stuck to the walls of the gasifier. Neither cooling nor palletizing of the waste plastic helped solve the problem. Mixing of biomass with plastic avoided the problem of plastic softening and effectively gasified all feedstock. It was typically observed that high gasification temperature (>800°C), presence of steam as a gasifying medium, and cogasification with plastic all led to higher hydrogen yields. Some studies have focused on the basic gasifier design; these studies have shown that process modification by two-stage gasification and secondary air injection reduces the tar yield [45].

5.2.3.1 Direct Solar Gasification

As against conventional biomass gasification where a portion of biomass feed is oxidized to supply the energy to the process, direct solar gasification aims to utilize solar energy for providing the required thermal energy. In 1974, Antal et al. [2] examined the feasibility of using solar process heat for the gasification of organic solid wastes and the production of hydrogen. A detailed review with many references of the technology describes solar gasification of carbonaceous materials to produce a syngas-quality intermediate for production of hydrogen and other fuels. Shahbazov and Usupov [88] have shown good yield of hydrogen from agricultural wastes using a parabolic mirror reflector. The use of a palladium diaphragm in this respect is reported to achieve solar-assisted hydrogen separations from the gases generated by pyrolysis of hazelnut shells at 500°C–700°C [64]. Walcher et al. [103] have mentioned a plan to utilize agricultural wastes in a heliothermic gasifier. Direct solar gasification is a promising technology as it combines two renewable resources (solar and biomass) for hydrogen production.

5.2.3.2 Fast Pyrolysis and Bio-Oil Steam Reforming

Pyrolysis is the heating of biomass at a temperature of 650–800 K at 1–5 bar in the absence of air to convert biomass into liquid oils, solid charcoal, and gaseous compounds. Pyrolysis can be further classified into slow pyrolysis and fast pyrolysis. As the products for slow pyrolysis include mainly charcoal, it is normally not considered for hydrogen production. Fast pyrolysis is a high-temperature process, in which the biomass feedstock is heated rapidly in the absence of air, to form vapor and subsequently condensed to a dark brown mobile bioliquid. The products of fast pyrolysis can be found in all gas, liquid, and solid phases [45] as described next:

1. Gaseous products include H_2, CH_4, CO, CO_2, and other gases depending on the pyrolysis biomass.
2. Liquid products include tar and oils that remain in liquid form at room temperature like acetone, acetic acid, etc.
3. Solid products are mainly composed of char and almost pure carbon plus other inert materials. Although most pyrolysis processes are designed for biofuel production, hydrogen can be produced directly through fast or flash pyrolysis if high temperature and sufficient volatile phase residence time are allowed as follows:

$$\text{Biomass} + \text{heat} \rightarrow H_2 + CO + CH_4 + \text{other products} \tag{5.6}$$

Methane and other hydrocarbon vapors produced can be steam reformed for more hydrogen production:

$$CH_4 + H_2O \rightarrow CO + 3H_2 \tag{5.7}$$

In order to further increase the hydrogen yield, WGS can be applied as follows:

$$CO + H_2O \rightarrow CO_2 + H_2 \tag{5.8}$$

Both gaseous products and oily products can be processed for hydrogen production [33]. The pyrolysis oil can be separated into two fractions based on water solubility.

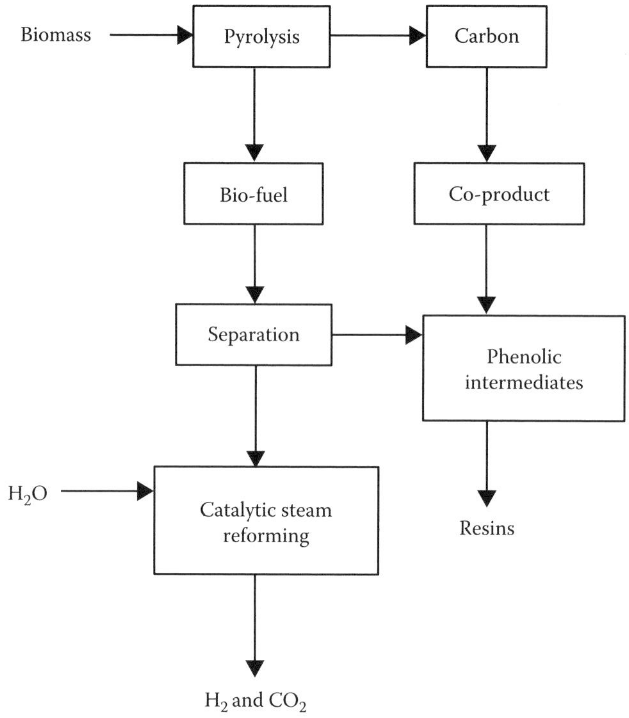

FIGURE 5.4
Hydrogen to biomass pyrolysis with a coproduct strategy. (From Milne, T.A. et al., Biomass gasifier "tars": Their nature, formation, and conversion, NREL/TP_570_25357, ON DE00003726, National Renewable Energy Laboratory Technical Report, Golden, CO, 1998.)

The water-soluble fraction can be used for hydrogen production while the water-insoluble fraction for producing adhesives. The material flow is summarized in Figure 5.4.

Biomass to hydrogen based on pyrolysis with a coproducts strategy has been investigated [33]. Experimental study has shown that when a Ni-based catalyst is used, the maximum yield of hydrogen can reach 90%. With additional steam reforming and water–gas shift reaction, the hydrogen yield can be increased significantly. Temperature, heating rate, residence time, and type of catalyst used are important pyrolysis process control parameters.

To favor gaseous products especially for hydrogen production, high temperature and high heating rate and long volatile phase residence time are required [29]; and they can be regulated by selection among different reactor types and heat transfer modes, such as gas–solid convective heat transfer and solid–solid conductive heat transfer. The heat transfer modes and features of various reactors are listed in Table 5.2, which shows that fluidized bed reactor type exhibits higher heating rates and is thus the promising reactor type for hydrogen production from biomass pyrolysis.

Some inorganic salts, such as chlorides, carbonates, and chromates, exhibit beneficial effect on pyrolysis reaction rate [79]. As tar is difficult to be gasified, extensive studies on the catalytic effect of inexpensive dolomite and CaO on decomposition of hydrocarbon compounds in tar have been conducted [90,91]. The catalytic effects of other catalysts (Ni-based catalysts [72], Y-type zeolite [105], K_2CO_3, Na_2CO_3, and $CaCO_3$ [21]) and various metal oxides (Al_2O_3, SiO_2, ZrO_2, TiO_2 [97], and Cr_2O_3 [21]) have also been investigated. Among the different metal oxides, Al_2O_3 and Cr_2O_3 exhibit better catalytic effect than others. Among the

TABLE 5.2

Pyrolysis Reactor Types, Heat Transfer Modes, and Typical Features

Reactor Type	Mode of Heat Transfer	Typical Features
Ablative	95% conduction	Accepts large-size feedstocks
	4% convection	Very high mechanical char abrasion from biomass
	1% radiation	Compact design
		Heat transfer gas not required
		Particulate transport gas not always required
	Fluidized bed 90% conduction	High heat transfer rates
	9% convection	Heat supply to fluidizing gas or to bed directly
	1% radiation	Limited char abrasion
		Very good solid mixing
		Particle size limit <2 mm in smallest dimension
		Simple reactor configuration
Circulating fluidized bed	80% conduction	High heat transfer rates
	19% convection	High char abrasion from biomass and char erosion
	1% radiation	Leading to high char in product
		Char/solid heat carrier separation required
		Solid recycle required
		Increased complexity of system
		Maximum particle sizes up to 6 mm
		Possible liquids cracking by hot solids
		Possible catalytic activity from hot char
		Greater reactor wear possible
Entrained flow	4% conduction	Low heat transfer rates
	95% convection	Particle size limit <2 mm
	1% Radiation	Limited gas/solid mixing

Source: Bridgwater, A.V., *J. Anal. Appl. Pyrolysis*, 51, 3, 1999.

carbonate catalysts, Na_2CO_3 is better than K_2CO_3 and $CaCO_3$. Although noble metals Ru and Rh are more effective than Ni catalyst and less susceptible to carbon formation, they are not commonly used due to their high cost [34].

To evaluate hydrogen production through pyrolysis of various types of biomass, extensive experimental investigations have been conducted. Agricultural residues [1], peanut shell [33], postconsumer wastes such as plastics, trap grease, mixed biomass and synthetic polymers [26], and rapeseed [75] have been widely tested. And to solve the problem of decreasing reforming performance caused by char and coke deposition on the catalyst surface and in the bed itself, fluidized catalyst beds are usually used to improve hydrogen production from biomass-pyrolysis-derived biofuel [12,13]. Yeboah et al. [112] constructed a demonstration plant for hydrogen production from peanut shell pyrolysis and steam reforming in a fluidized bed reactor, and a production rate of 250 kg H_2/day was achieved.

Pyrolysis is a promising technology for renewable hydrogen production. This technology is being investigated by the US Department of Energy (US DOE) and several national laboratories around the world. Development of low-cost catalysts for bio-oil to hydrogen reforming that can maintain activity for sufficient period of time is one of the critical challenges in the technology.

5.3 Advanced Gasification Technologies

5.3.1 Supercritical Water Gasification

5.3.1.1 Introduction

In recent times, biomass gasification in the presence of supercritical water (SCW) has received increased attention by the renewable energy community. The motivation was the high moisture content in some biomass feedstocks such as sewage sludge, animal manure, and so on. While conventional biomass gasification has low efficiency (about 10%–15% when the moisture content is more than 80%), *s*uper*c*ritical *w*ater biomass *g*asification (SCWG) has efficiencies as high as 70% for the same moisture content [87]. Other advantages include elimination of feedstock drying process, short residence times, and a compact reactor due to high operating pressures. When biomass has high moisture content (typically >35%), it is favorable to gasify biomass in SCW conditions.

SCW possesses unique properties much different from liquid water and subcritical or superheated steam. SCW behaves like an organic solvent, and many organic compounds have very high solubility in it. Moreover, gases are miscible in SCW. Chemical reactions can be conducted in a single supercritical phase, thus eliminating the interphase mass and heat transfer and chemical kinetic resistances [85,86]. Due to lower resistances, SCWG has high efficiency even at temperatures as low as 500°C–600°C.

5.3.1.2 Background and Research Overview

Research in SCWG dates almost 30 years back when the very first experiments to convert carbohydrates to liquid products in SCW were performed by Modell at MIT in late 1970s [69–71]. No char or solid residue was observed in the product gas stream. In the mid-1980s and early 1990s, several investigations were carried out at the Pacific Northwest Laboratories by gasifying waste model compounds and real waste under SCW conditions. The research efforts of the group (Elliott et al. [32]) focused on converting biomass into a methane-rich gas stream. The operating conditions were 350°C and 200 bar at reaction times of 60–120 min. It was shown that aliphatic and aromatic hydrocarbons can be converted into a methane-rich fuel gas in the presence of hydrogenation catalysts (such as Ni or Ru) [31,32]. In the mid-1990s, Antal and coworkers [111] investigated the effects of temperature, pressure, reactant concentration, and the type of catalyst on the conversion of glucose, some real biomass feedstocks, and some waste streams in an SCW atmosphere. Four types of carbon-based catalysts were used for the study, which was conducted in a tubular reactor at 600°C and 34.5 MPa. The researchers observed complete conversion of glucose into a H_2-rich gas stream [106,110,111,114]. The wastes were also completely gasified. The carbon catalysts however got deactivated after about 4 h of operation. No char or tar was observed in the product gas. At around the same time, investigations were being conducted by Minowa et al. [67,68] at the National Institute for Resources and Environment, Japan, on the gasification of a model biomass (cellulose) and real biomass (Japanese oak) in the presence of water at 350°C and 17 MPa using a reduced Ni catalyst and sodium carbonate. The researchers could produce a gas stream rich in hydrogen and having high gas yields (94 wt.% for the model biomass and 55 wt.% for the real biomass). Their work showed that a char and solid-free hydrogen-rich gas stream can be produced from a real biomass (Japanese oak) in a single step using non-Nobel metal catalyst (nickel). In the recent past, Schmieder from Forschungszentrum Karlsrhue, Germany, has been actively involved in studying hydrothermal gasification of biomass and organic wastes [48,49,87]. The group investigated the effect of catalyst additions on the SCW gasification of model and real-world biomass compounds such as glucose and straw wood, respectively. The gasification was carried out in the presence of KOH or K_2CO_3 at an operating temperature of 400°C–600°C and a pressure of 300–450 bar. They observed that the product gas was rich in hydrogen (50%–60%) and CO_2 (30%–35%) with CO, CH_4, and higher hydrocarbons making up the balance. Hydrogen yield was highest at 600°C. It was also observed that the presence of catalyst made a big difference in the product gas composition. The CO yield went up 20-fold when the gasification was carried out in the absence of a catalyst as compared to when a catalyst was present. This showed that the WGS reaction is promoted in the presence of a catalyst. More recently, Hao, Lu, and coworkers [37,42,57,43] at the State Key Laboratory of Multiphase Flow (Xi'an Jiaotong University, China) have been actively involved in researching hydrogen production by SCWG. The researchers developed a thermodynamic model to predict the effect of process parameters (temperature, pressure, water recirculation ratio, and so on) on the product gas yield. Gibbs energy minimization approach was followed. Their model suggested the optimum temperature for SCWG for hydrogen production to be at 600°C–650°C. The researchers also conducted energy and exergy analyses and predicted the efficiencies to be 44% and 42%, respectively, when the process was carried out at 600°C and 250 bar. The researchers have also conducted many experimental studies using glucose and cellulose as the model biomass compounds. They observed that glucose at low concentration can be completely gasified at 650°C and 250 bar with no char/tar in the product gas. They obtained gasification efficiency in excess of 95%.

Table 5.3 summarizes recent attempts to produce H_2 from biomass at SCW conditions. Tubular reactors are widely used in SCWG because of their robust structure to withstand high pressure.

In order to make the SCWG process even more attractive, validation and optimization of the process parameters (pressure, temperature, and molar concentration of biomass in the feed) are necessary. On the practical side, challenges such as heat exchanger fouling, material handling equipment for handling corrosive slurries, and an inexpensive liquid treatment stream need to be addressed. Most work till date has been on lab scale and is still in early stage of development. Several research groups around the world are investigating SCW technology not only for hydrogen production but also for other applications such as waste remediation, hydrocarbon reforming, and power generation [62].

Although SCWG is still at early stage of development, the technology has shown economic competitiveness with other H_2 production methods. Spritzer and Hong have estimated the cost of H_2 produced by SCWG to be about US $3/GJ ($0.35/kg) [94].

5.3.2 Sorbent-Enhanced Biomass Gasification

Recently, a novel gasification method using sorbents has been receiving considerable attention from the research community [51–55]. It is an integration of steam–hydrocarbon reaction, WGS, and CO_2 absorption reactions in a single reactor. As against conventional steam gasification, biomass can be gasified in the presence of a metal oxide sorbent (such as calcium oxide), and this would involve an in situ CO_2 capture as per the following reaction:

$$CaO + CO_2 \rightarrow CaCO_3 \qquad \Delta H = -170.5\ kJ/mol \tag{5.9}$$

If designed properly, the exothermic CO_2 absorption reaction can be coupled with the endothermic biomass gasification reaction. In order to complete the process, the sorbent

TABLE 5.3

Recent Studies on Hydrogen Production by Biomass Gasification in Supercritical Water Conditions

Feedstock	Gasifier Type	Catalyst Used	Temperature and Pressure	Hydrogen Yield	Reference
Glucose	Not known	Not used	600°C, 34.5 MPa	0.56 mol H_2/mol	[109]
Glucose	Not known	Activated carbon	600°C, 34.5 MPa	2.15 mol H_2/mol of feed	[109]
Glucose	Not known	Activated carbon	600°C, 25.5 MPa	1.74 mol H_2/mol	[109]
Glucose	Not known	Activated carbon	550°C, 25.5 MPa	0.46 mol H_2/mol of feed	[109]
Glycerol	Not known	Activated carbon	655°C, 28 MPa	48 vol.%	[3]
Glycerol/methanol	Not known	Activated carbon	720°C, 28 MPa	64 vol.%	[3]
Corn starch	Not known	Activated carbon	650°C, 28 MPa	48 vol.%	[3]
Sawdust/corn starch mixture	Not known	Activated carbon	690°C, 28 MPa	57 vol.%	[3]
Glucose	Tubular reactor	KOH	600°C, 25 MPa	59.7 vol.% (9.1 mol H_2/mol glucose)	[3]
Catechol	Tubular reactor	KOH	600°C, 25 MPa	61.5 vol.% (10.6 mol H_2/mol catechol)	
Sewage	Autoclave	K_2CO_3	450°C, 31.5–35	47 vol.%	
Glucose	Tubular reactor	Not used	600°C, 25 MPa	41.8 vol.%	[43]
Glucose	Tubular reactor	Not used	500°C, 30 MPa	32.9 vol.%	[43]
Glucose	Tubular reactor	Not used	550°C, 30 MPa	33.1 vol.%	[43]
Glucose	Tubular reactor	Not used	650°C, 30 MPa	40.8 vol.%	[43]
Glucose	Tubular reactor	Not used	650°C, 30 MPa	41.2 vol.%	[43]
Sawdust	Tubular reactor	Sodium carboxymethylcellulose (CMC)	650°C, 22.5 MPa	30.5 vol.%	[43]

Source: Ni, M. et al., *Fuel Process. Technol.*, 87, 461, 2006.

needs to be regenerated. The energy requirements associated with the regeneration of the sorbent must be taken into account while determining the overall process efficiency.

$$CaCO_3 \rightarrow CaO + CO_2 \qquad \Delta H = +170.5 \text{ kJ/mol} \tag{5.10}$$

The regeneration (also called calcination) typically takes place at a higher temperature than the CO_2 absorption (also called carbonation) process. Figure 5.5 shows a schematic of the biomass gasification in the presence of a sorbent [58].

The typical reaction assuming char/solid carbon as the feed using CaO as sorbent is as follows:

$$C + 2H_2O + CaO \rightarrow CaO_3 + 2H_2 \qquad \Delta H_{298} = -88 \text{ kJ/mol} \tag{5.11}$$

The capture of CO_2 by reaction with CaO is not a new concept. The idea was first patented in 1867 by DuMotay and Marechal, who used lime to enhance the gasification of carbon

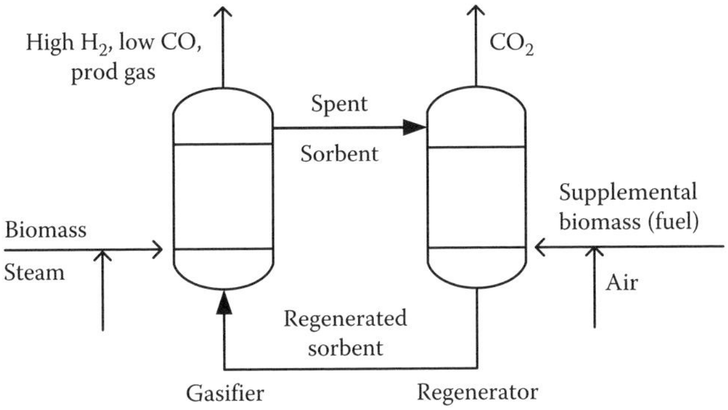

FIGURE 5.5
Schematic of sorbent-enhanced biomass gasification. (Adapted from Mahishi, M.R., Theoretical and experimental investigation of hydrogen production by gasification of biomass, PhD dissertation, 2006.)

with steam [95]. Later Curran and others developed the CO_2 acceptor process utilizing the carbonation reaction for the gasification of coal to H_2-rich gas [25]. Still later, Barker and Bhatia and Perlmutter have independently studied the kinetics of the gas–solid reaction between CaO and CO_2 [16,17].

More recently, the desire to reduce CO_2 emissions and the search for clean energy solutions has renewed interest in the sorbent–CO_2 capture processes. A novel coal gasification process using CaO for CO_2 capture (hydrogen production by reaction-integrated novel gasification (called HyPr-RING)) has recently been proposed by Kuramoto et al. [50]. The CO_2 capture process has also been investigated for producing H_2 by steam methane reforming in the presence of calcium oxide sorbent [14,40]. Sato et al. have investigated H_2 production from oil in the presence of calcium hydroxide [84]. In comparison with conventional gasification, the HyPr-RING process can be carried out in a much simpler manner in a single reactor at a lower temperature as shown in Figure 5.6 [55]. As far as biomass fuel is concerned, only a few but promising investigations have been carried out by some research groups [38,39,41,59,60,74,77]. Mahishi and Goswami studied the effects of adding calcium oxide on the steam reforming of ethanol [60]. The effect of varying the operating parameters on the hydrogen yield was studied (temperature: 500°C–900°C, pressure: 1–25 atm, steam/biomass ratio: 3–8, sorbent/biomass ratio: 0–6) using ASPEN Plus modeler and Gibbs free energy minimization approach. Their study showed that the H_2 yield can be increased by almost 20% in the presence of sorbent. Moreover, the thermodynamic efficiency of ethanol steam reforming in the presence of sorbent (72.1%) was higher than the conventional ethanol reforming (62.9%). Based on the promising results of the model, Mahishi and Goswami conducted experimental studies on solid biomass gasification in the presence of sorbent [59]. Southern pine bark was used as the model biomass with calcium oxide as the model sorbent, and temperature was varied from 500°C to 700°C. The researchers found the H_2 yield, total gas yield, and carbon conversion efficiency to increase by 48.6%, 62.2%, and 83.5%, respectively, in the presence of calcium oxide. They also found the H_2 yield at 500°C in the presence of sorbent to be higher than the yield at 700°C of plain biomass gasification, which shows that there is a substantial potential to reduce the gasifier operating temperature. The reduced tars and hydrocarbons in the product gas coupled with the additional hydrogen showed that calcium oxide played the dual role of sorbent and catalyst.

Owing to the novelty of the process, there are only a few published experimental studies investigating H_2 production via steam gasification of solid biomass in the

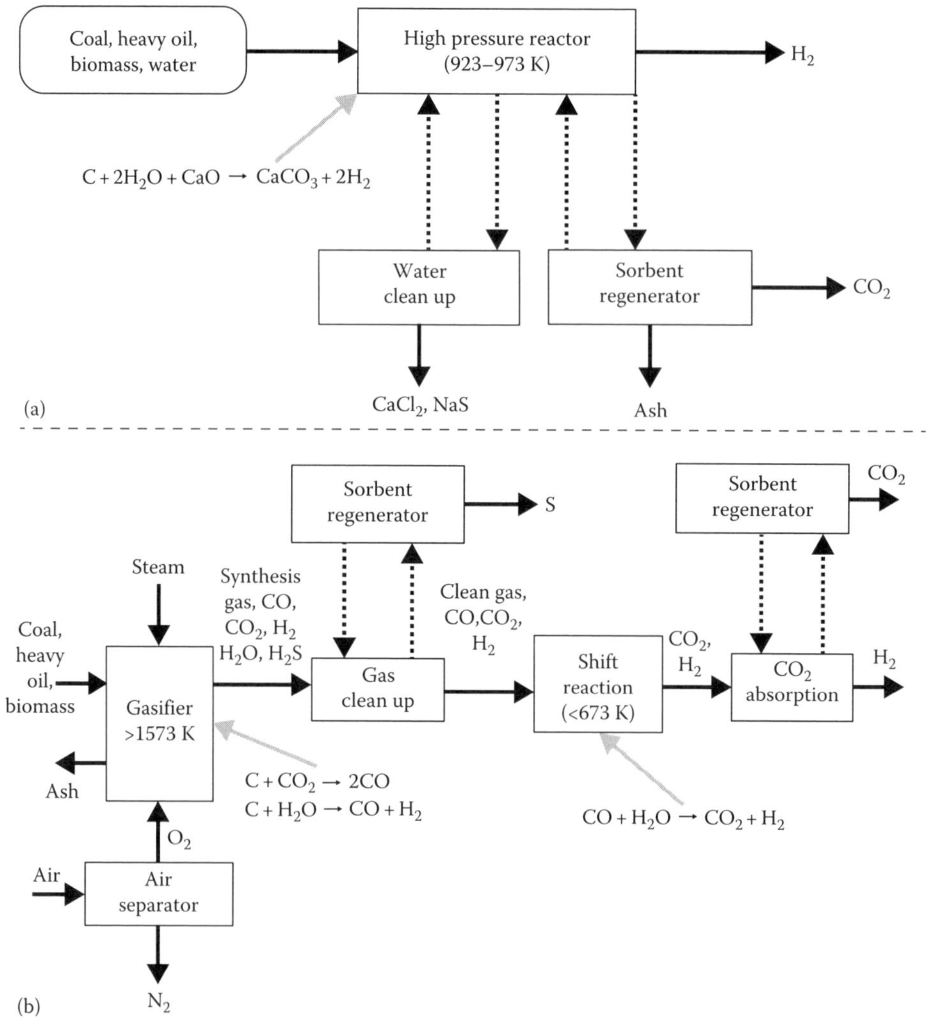

FIGURE 5.6
Comparison of (a) HyPr-RING and (b) conventional gasification hydrogen production process. (From Lin, S.Y. et al., *Energy Convers. Manage.*, 43, 1283, 2002.)

presence of a sorbent. Hatano et al. investigated the steam gasification of Japanese oak in the presence of $Ca(OH)_2$ powder [41]. The group studied the effects of sorbent to biomass ratio (Ca/C molar ratio in the feed) from 0 to 4 at a constant temperature of 650°C and a pressure of 5.9 atm. The group obtained a maximum hydrogen yield of 80.7% (vol.) with very little CH_4 and no CO_2. They also studied the effect of pressure (3.0–82.9 atm) at a fixed temperature of 650°C. They obtained a H_2 yield of 78% (vol.) at a pressure of 12.8 atm. With no sorbents, the highest H_2 yield was 50% (vol.); the significant increase in H_2 yield from 50% (vol.) to almost 80% (vol.) is very promising. Pfeifer et al. steam gasified wood pellets using an atmospheric circulating fluidized bed gasifier (100 kW_{th} pilot plant) with CaO and olivine as the bed material and obtained a comparable H_2-rich product gas (75% vol.) [77]. Steam and wood pellets were fed into the gasifier. The bed material (which included sorbent and unconverted char residue) was circulated to the combustion zone where air was introduced to burn the char and generate heat to drive the endothermic sorbent regeneration reaction. Guoxin et al. experimented wet biomass (pine sawdust) gasification in the presence of $Ca(OH)_2$ sorbent [38]. The group studied the effect of moisture content (varied from 0.09 to 0.9), sorbent-to-biomass molar ratio (0–1), and the reactor temperature (900–1150 K) on the H_2 yield. A moisture content of 0.09 represented relatively dry biomass whereas 0.9 represented wet biomass. The drying of wet biomass created a steam-rich atmosphere in which tar and hydrocarbon reforming and WGS reactions were all strongly favored. For experiments with a moisture content of 0.9, an increase in H_2 yield of 51.5% was obtained. The group found that calcium hydroxide played the dual role of sorbent and catalyst. Based on their experimental studies, the researchers recommended an optimal sorbent-based biomass gasification temperature of 923–973 K (650°C–700°C), which is about 150°C–200°C lower than the conventional biomass gasification temperature.

The analytical and experimental studies show that sorbent-enhanced biomass gasification is a promising technology to produce H_2 by steam gasifying biomass. Several studies have been conducted on using sorbents for coal, oil, and natural gas. However, sorbent-enhanced biomass gasification for hydrogen production is a relatively novel concept with only a few groups having conducted experimental investigations. Although a hydrogen-rich product gas stream is produced, there are some key issues that need to be addressed such as sorbent durability, decay in reactivity due to multiple CO_2 capture/release cycles, and waste sorbent stream disposal.

Hydrogen production from biomass thermochemical processes has the potential to be technically and economically attractive. However, it should be noted that hydrogen gas is normally produced along with other gas constituents. Thus, separation and purification of hydrogen gas from other gases are required. Nowadays, several methods, such as CO_2 absorption, drying/chilling, and membrane separation have been successfully developed for hydrogen gas purification [56,83]. It is expected that biomass thermochemical conversion processes, especially the newly developed gasification types, will be available for large-scale hydrogen production in the near future.

5.4 Hydrogen Production from Fossil Fuels (Coal Gasification)

Hydrogen production by coal gasification involves three steps:

- Treatment of coal feedstock with high-temperature steam (1300°C) to produce syngas
- A catalytic shift conversion of CO to additional H_2
- Purification of the hydrogen product

In the first step, coal is chemically broken down at high temperature (1330°C) by high-pressure steam to produce raw synthesis gas, as per the following reaction:

$$C + H_2O \rightarrow CO + H_2 \qquad \Delta H = +135.8 \text{ kJ/mol} \qquad (5.12)$$

The heat required for this gasification step comes from controlled addition of oxygen, which allows partial oxidation of a small amount of the coal feedstock. Because of this, the reaction is carried out in either an air-blown or an oxygen-blown gasifier. The oxygen-blown gasifier is generally used in order to minimize NO_x formation and make the process more compatible for carbon dioxide sequestration. In the second step, the syngas passes through a shift reactor, converting a portion of the carbon monoxide to carbon dioxide and thereby producing additional hydrogen:

$$CO + H_2O \rightarrow CO_2 + H_2 \qquad \Delta H = -41.2\ \text{kJ/mol} \tag{5.13}$$

In the third step, the hydrogen product is purified. Physical absorption removes 99% of impurities. The majority of H_2 in the shifted syngas is then removed in a PSA unit. In case of CO_2 sequestration, a secondary absorption tower removes CO_2 from the remaining shifted syngas. Coal is an attractive energy source due to its abundance in the United States, Asia, and other parts of the world and low and traditionally stable prices. Coal gasification is an established technology used in hydrogen production today, but additional technical and economic considerations for capture and storage of CO_2 will be necessary in future.

5.5 Present Status and Economics of Biomass Gasification for Hydrogen Production

There are currently no commercial biomass gasification processes for hydrogen production, but there are several demonstration plants of biomass gasifiers for producing electricity or other chemicals [11]. The FERCO SilvaGas process employs low-pressure Battelle Columbus gasification, which consists of two physically separate reactors—a gasification reactor in which biomass is converted into a medium heating value gas and residual char at a temperature of 850°C–1000°C and a combustion reactor that burns residual char to provide heat for gasification. A typical product gas using wood chips consists of about 21% H_2 (using steam/biomass ratio of 0.45). The fast internally circulating fluidized bed (FICFB) process is another example. In FICFB, there is a single reactor with two zones—a gasification zone and a combustion zone; biomass is gasified in the gasification zone with steam at 850°C–950°C followed by circulation to the combustion zone where char is burned to supply thermal energy. A demonstration plant producing 2 MW electrical power was constructed in Europe in 2001 (details in [15]).

Biomass resources have the advantage of being renewable and hence can make an important contribution to the future hydrogen economy. However, an important factor that prohibits commercialization is the difficulty of transporting large amounts of low-energy-density biomass feedstocks over long distances. The cost associated with growing, harvesting, and transporting biomass may be up to 40% of the total biomass plant operating cost (Table 5.4) [24,93].

Economies of scale can be realized if we have large quantities of biomass available. Figure 5.7 [18] depicts the sources for the cost of hydrogen production. It shows the potential to reduce the hydrogen production cost with a plant processing capacity of 2000 tons/day.

For pyrolysis, Padro and Putsche [76] estimated the hydrogen production cost to be in the range of US\$8.86/GJ–US\$15.52/GJ, depending on the facility size and biomass type. For comparison, the costs of hydrogen production by wind-electrolysis systems and PV-electrolysis systems are US\$20.2/GJ and US\$41.8/GJ, respectively. It can be seen that both biomass gasification and biomass pyrolysis are cost-competitive methods for renewable hydrogen production.

TABLE 5.4
Hydrogen Production Cost

Technology	Cost of H_2 ($/GJ)	Cost of H_2 ($/kg)
Direct biomass gasification	12.5–21.6	1.51–2.59
Steam reforming of bio-oil	10.3–19.9	1.24–2.38

Sources: Spath, P.L. et al., Update of hydrogen from biomass—Determination of the delivered cost of hydrogen, NREL/MP-510-33112, National Renewable Energy Laboratory Technical Report, Golden, CO, 2000; Craig, K.R. and Mann, M.K., Cost and performance analysis of Biomass based Integrated Gasification Combined Cycle BIGCC power systems, NREL/TP-430-21657, National Renewable Energy Laboratory, Golden, CO, 1996.

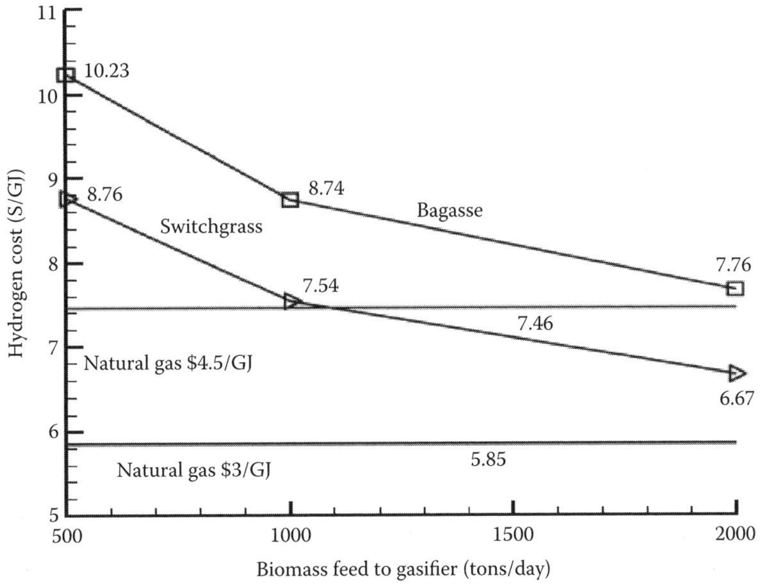

FIGURE 5.7
Costs of hydrogen production by biomass gasification. (From Bowen, D.A. et al., Techno-economic analysis of hydrogen production by gasification of biomass, NREL FY 2003 Progress Report, National Renewable Energy Laboratory, Golden, CO, 2003; Ni, M. et al., *Fuel Process. Technol.*, 87, 461, 470, 2006.)

Additional technical and economic consideration for capture and storage of CO_2 may also be necessary as both gasification and pyrolysis produce carbon dioxide.

The US DOE has been aggressively conducting research on renewable hydrogen production using thermochemical methods. However, in order to make biomass hydrogen production commercially viable, many areas need to be addressed. The following are some of the areas identified by the US DOE:

1. *Feedstock preparation and handling*: Issues related to preparation (drying, storage, and palletizing) and nature of different feeds for high-temperature and -pressure reactors need to be addressed; also since biomass is available seasonally, feedstock-flexible reactors must be developed.
2. *Gas conditioning*: Product gas from biomass gasifiers may be supplied to fuel cells, and hence it needs to be free from particulate matter, tar, entrained char, or

catalyst particles and trace levels of nitrogen, sulfur, and chlorine components; also improved catalysts with high tar-reforming ability and low deactivation must be developed.

3. *System integration*: Successful development of biomass hydrogen technology is based on the optimal integration of appropriate gasification/pyrolysis system with available feedstocks in order to make a product of necessary quality for the intended application.
4. *Valuable coproduct integration*: Development of technology that can convert by-product streams into useful chemicals or processes.

References

1. Abedi, J., Yeboah, Y.D., Realff, M., McGee, D., Howard, J., Bota, K.B. An integrated approach to hydrogen production from agricultural residues for use in urban transportation. *Proceedings of the 2001 DOE Hydrogen Program Review*, Coral Gables, FL, NREL/CP-570-30535, National Renewable Energy Laboratory, Golden, CO, 2001, pp. 1–27.
2. Antal Jr., M.J., Feber, R.C., Tinkle, M.C. Synthetic fuels from solid waste and solar energy. *Proceedings of the 1st World Hydrogen Energy Conference*, Miami Beach, FL, March 1–3, 1976, 1, pp. 3A-69–3A-88.
3. Antal, M.J., Xu, X.D. Hydrogen production from high moisture content biomass in supercritical water. *Proceedings of the 1998 U.S. DOE Hydrogen Program Review*, Coral Gables, FL, NREL/CP-570-25315, National Renewable Energy Laboratory, Golden, CO, 1998, pp. 1–24.
4. Arvelakis, S., Gehrmann, H., Beckmann, M., Koukios, E.G. Effect of leaching on the ash behavior of olive residue during fluidized bed gasification. *Biomass and Bioenergy*, 22, 55–69, 2002.
5. Arvelakis, S., Koukios, E.G. Physicochemical upgrading of agroresidues as feedstocks for energy production via thermochemical conversion methods. *Biomass and Bioenergy*, 22, 331–348, 2002.
6. Arvelakis, S., Vourliotis, P., Kakaras, E., Koukios, E.G. Effect of leaching on the ash behavior of wheat straw and olive residue during fluidized bed combustion. *Biomass and Bioenergy*, 20, 459–470, 2001.
7. Asadullah, M., Fujimoto, K., Tomishige, K. Catalytic performance of Rh/CeO_2 in the gasification of cellulose to synthesis gas at low temperature. *Industrial & Engineering Chemistry Research*, 40, 5894–5900, 2001.
8. Asadullah, M., Tomishige, K., Fujimoto, K. A novel catalytic process for cellulose gasification to synthesis gas. *Catalysis Communications*, 2(2), 63–68, 2001.
9. Aznar, M.P., Caballero, M.A., Gil, J., Martin, J.A., Corella, J. Commercial steam reforming catalyst to improve biomass gasification with steam–oxygen mixtures: Part 2 Catalytic tar removal. *Industrial & Engineering Chemistry Research*, 37(7), 2668–2680, 1998.
10. Aznar, M.P., Frances, E. Testing of downstream catalysts for tar destruction with a guard bed in a fluidized bed biomass gasifier at pilot scale. *VTT Symposium*, 164, 263–268, 1996.
11. Babu, S.P. Biomass gasification for hydrogen production process description and research needs, Gas Technology Institute, des Plaines, IL.
12. Bair, K.A.M., Czernik, S., French, R., Chornet, E. Fluidizable catalysts for hydrogen production from biomass pyrolysis/steam reforming, Hydrogen, Fuel Cells, and Infrastructure Technologies, FY 2003 Progress Report, National Renewable Energy Laboratory, Golden, CO, 2003, pp. 1–4.
13. Bair, K.A.M., Czernik, S., French, R., Parent, Y., Ritland, M., Chornet, E. Fluidizable catalysts for producing hydrogen by steam reforming biomass pyrolysis liquids. *Proceedings of the 2002 U.S. DOE Hydrogen Program Review*, Coral Gables, FL, NREL/CP-610-32405, National Renewable Energy Laboratory, Golden, CO, 2002, pp. 1–11.

14. Balasubramanian, B., Ortiz, A.L., Harrison, D.P. Hydrogen from methane in a single-step process. *Chemical Engineering Science*, 54, 3543–3552, 1999.
15. Bangala, D.N. et al. Steam reforming of naphthalene on Ni-Cr/Al_2O_3 catalysts doped with MgO, TiO_2 and La_2O_3. *AIChE Journal*, 44, 927–936, 1998.
16. Barker, R. The reactivity of CaO towards CO_2 and its use for energy storage. *Journal of Applied Chemistry and Biotechnology*, 24, 221–227, 1974.
17. Bhatia, S.K., Perlmutter, D.D. Effect of the production layer on the kinetics of the CO_2-lime reaction. *AIChE Journal*, 29, 79–86, 1983.
18. Bowen, D.A., Lau, F., Zabransky, R., Remick, R., Slimane, R., Doong, S. Techno-economic analysis of hydrogen production by gasification of biomass, NREL FY 2003 Progress Report, National Renewable Energy Laboratory, Golden, CO, 2003; Ni, M. et al. *Fuel Processing Technology*, 87, 461–472, 470, 2006.
19. Bridgwater, A.V. Principles and practice of biomass fast pyrolysis processes for liquids. *Journal of Analytical and Applied Pyrolysis*, 51, 3–22, 1999.
20. Brown, R.C. Biomass-derived hydrogen from a thermally ballasted gasifier, FY 2003 Progress Report, National Renewable Energy Laboratory, Golden, CO, 2003.
21. Chen, G., Andries, J., Spliethoff, H. Catalytic pyrolysis of biomass for hydrogen rich fuel gas production. *Energy Conversion and Management*, 44, 2289–2296, 2003.
22. Chuangzhi, W., Xiuli, Y., Bingyan, X., Zhengfan, L., Ping, L. The performance study of biomass gasification with oxygen-rich air. *Acta Energiae Solaris Sinica*, 18, 237, 1997 (in Chinese).
23. Conesa, J., Marcilla, A., Cabarello, J.A. Evolution of gases from the pyrolysis of modified almond shells: Effect of impregnation with $CoCl_2$. *Journal of Analytical and Applied Pyrolysis*, 43, 59–69, 1997.
24. Craig, K.R., Mann, M.K. Cost and performance analysis of biomass based integrated gasification combined cycle BIGCC power systems, NREL/TP-430-21657, National Renewable Energy Laboratory, Golden, CO, 1996.
25. Curran, G.P., Fink, C.E., Gorin, E. CO_2 acceptor gasification process: Studies of acceptor properties. In: *Advances in Chemistry Series, Fuel Gasification*, American Chemical Society, Washington, DC, 1967, Vol. 69, pp. 141–165.
26. Czernik, S., French, R., Evans, R., Chornet, E. Hydrogen from postconsumer residues. *U.S. DOE Hydrogen and Fuel Cells Merit Review Meeting*, Berkeley, CA, May 19–23, 2003, pp. 1–17.
27. Das, P., Ganesh, A. Influence of pretreatment of biomass on pyrolysis products, Indian Institute of Technology, Mumbai, India (Internet Resource: http://bioproducts-bioenergy.gov/pdfs/bcota/abstracts/3/z246.pdf).
28. Dayton, D. A review of the literature on catalytic biomass tar destruction, Milestone Completion Report, NREL/TP-510-32815, National Renewable Energy Laboratory, Golden, CO, December 2002.
29. Demirbas, A. Gaseous products from biomass by pyrolysis and gasification: Effects of catalyst on hydrogen yield. *Energy Conversion and Management*, 43, 897–909, 2002.
30. Devi, L. et al. A review of the primary measures for tar elimination in biomass gasification process. *Biomass and Bioenergy*, 24, 125–140, 2003.
31. Elliott, D.C., Phelps, M.R., Sealock, L.J., Baker, E.G.C. Chemical processing in high-pressure aqueous environments 4. Continuous flow reactor process development experiments for organic destruction. *Industrial & Engineering Chemistry Research*, 33, 566–574, 1994.
32. Elliott, D.C., Sealock, L.J., Baker, E.G.C. Chemical processing in high-pressure aqueous environments 3. Batch reactor process development experiments for organic destruction. *Industrial and Engineering Chemistry Research*, 33, 558–565, 1994.
33. Evans, R., Boyd, L., Elam, C., Czernik, S., French, R., Feik, C., Philips, S., Chaornet, E., Parent, Y. Hydrogen from biomass-catalytic reforming of pyrolysis vapors, FY 2003 Progress Report, National Renewable Energy Laboratory, Golden, CO, 2003.
34. Garcia, L., French, R., Czernik, S., Chornet, E. Catalytic steam reforming of bio-oils for the production of hydrogen: Effect of catalyst composition. *Applied Catalysis A: General*, 201, 225–239, 2000.

35. Garcia-Ibanez, P., Cabanillas, A., Sanchez, J.M. Gasification of leached orujillo (olive oil waste) in a pilot plant circulating fluidized bed reactor. Preliminary results. *Biomass and Bioenergy*, 27, 183, 2004.
36. Gil, J., Corella, J., Aznar, M.P., Caballero, M.P. Biomass gasification in atmospheric and bubbling fluidized bed: Effect of the type of gasifying agent on the product distribution. *Biomass and Bioenergy*, 17, 389–403, 1999.
37. Guo, L.J., Lu, Y.J., Zhang, X.M., Ji, C.M., Guan, Y., Pei, A.X. Hydrogen production by biomass gasification in supercritical water: A systematic experimental and analytical study. *Catalysis Today*, 129, 275–286, 2007.
38. Guoxin, H., Hao, H. Hydrogen rich fuel gas production by gasification of wet biomass using a CO_2 sorbent. *Biomass and Bioenergy*, 33, 899–906, 2009.
39. Guoxin, H., Hao, H., Yanhong, L. The gasification of wet biomass using $Ca(OH)_2$ as CO_2 absorbent: The microstructure of char and absorbent. *International Journal of Hydrogen Energy*, 23, 5422–5429, 2008.
40. Han, C., Harrison, D.P. Simultaneous shift reaction and carbon-dioxide separation for the direct production of hydrogen. *Chemical Engineering Science*, 49(24B), 5875–5883, 1994.
41. Hanaoka, H., Yoshida, T., Fujimoto, S., Kamei, K., Harada, M., Suzuki, Y., Hatano, H., Yokoyama, S.Y., Minowa, T. Hydrogen production from woody biomass by steam gasification using a CO_2 sorbent. *Biomass and Bioenergy*, 28, 63–68, 2005.
42. Hao, X., Guo, L., Zhang, X., Guan, Y. Hydrogen production from catalytic gasification of cellulose in supercritical water. *Chemical Engineering Journal*, 110, 57–65, 2005.
43. Hao, X.H., Guo, L.J., Mao, X., Zhang, X.M., Chen, X.J. Hydrogen production from glucose used as a model compound of biomass gasified in supercritical water. *International Journal of Hydrogen Energy*, 28, 55–64, 2003.
44. Herguido, J., Corella, J., Gonzalez-Saiz, J. Steam gasification of lignocellulosic residues in a fluidized bed at a small pilot scale: Effect of the type of feedstock. *Industrial & Engineering Chemistry Research*, 31, 1274–1282, 1992.
45. Jalan, R.K., Srivastava, V.K. Studies on pyrolysis of a single biomass cylindrical pellet-kinetic and heat transfer effects. *Energy Conversion and Management*, 40, 467–494, 1999.
46. Jian-chun, J., Chun, J., Jin-Ping, Z., Hao, Y., Wei-di, D., Yuan-bo, T. Study on industrial applied technology for biomass catalytic gasification. *Chemistry and Industry of Forest Products*, 21, 21, 2001 (in Chinese).
47. Kinoshita, C.M., Wang, Y., Zhou, J. Effect of reformer condition on catalytic reforming of biomass gasification tars. *Industrial & Engineering Chemistry Research*, 34(9), 2949–2954, 1995.
48. Kruse, A., Gawalik, A. Biomass conversion in water at 330–410°C and 30–50 MPa. Identification of key compounds for indicating different chemical reactions pathways. *Industrial & Engineering Chemistry Research*, 42, 267–279, 2003.
49. Kruse, A., Hennignsen, T., Pfeiffer, J., Sinag, A. Biomass gasification in supercritical water; influence of the dry matter content and the formation of phenols. *Industrial & Engineering Chemistry Research*, 42, 3711–3717, 2003.
50. Kuramoto, K., Furuya, T., Suzuki, Y., Hatano, H., Kumabe, K., Yoshiie, R., Moritomi, H., Lin, S.Y. Coal gasification with a subcritical steam in the presence of a CO_2 sorbent: Products and conversion under transient heating. *Fuel Processing Technology*, 82, 61–73, 2003.
51. Lin, S.Y., Harada, M., Suzuki, Y., Hatano, H. Continuous experiment regarding hydrogen production by coal/CaO reaction with steam (I) gas products. *Fuel* 83, 869–874, 2004.
52. Lin, S.Y., Harada, M., Suzuki, Y., Hatano, H. Gasification of organic material/CaO pellets with high-pressure steam. *Energy and Fuels*, 18, 1014–1020, 2004.
53. Lin, S.Y., Harada, M., Suzuki, Y., Hatano, H. Process analysis for hydrogen production by reaction integrated novel gasification (HyPr-RING). *Energy Conversion and Management*, 46, 869–880, 2005.
54. Lin, S.Y., Suzuki, Y., Hatano, H., Harada, M. Hydrogen production from hydrocarbon by integration of water–carbon reaction and carbon dioxide removal (HyPr-RING) method. *Energy and Fuels*, 15, 339–343, 2001.

55. Lin, S.Y., Suzuki, Y., Hatano, H., Harada, M. Development an innovative method, HyPr-RING, to produce hydrogen from hydrocarbons. *Energy Conversion and Management*, 43, 1283–1290, 2002.
56. Lin, Y.S. Microporous and dense inorganic membranes: Current status and prospective. *Separation and Purification Technology*, 25, 39–55, 2001.
57. Lu, Y., Guo, L., Zhang, X., Yan, Q. Thermodynamic modeling and analysis of biomass gasification for hydrogen production in supercritical water. *Chemical Engineering Journal*, 131, 233–244, 2007.
58. Mahishi, M.R. Theoretical and experimental investigation of hydrogen production by gasification of biomass. PhD dissertation, 2006.
59. Mahishi, M.R., Goswami, D.Y. An experimental study of hydrogen production by gasification of biomass in the presence of a CO_2 sorbent. *International Journal of Hydrogen Energy*, 32(14), 2803–2808, 2007.
60. Mahishi, M.R., Sadrameli, S.M., Vijayaraghavan, S., Goswami, D.Y. A novel approach to enhance the hydrogen yield of biomass gasification using CO_2 sorbent. *ASME Journal of Engineering for Gas Turbines and Power*, 130, 011501-1–011501-8, January 2008.
61. Mann, M.K. Technical and economic analysis of hydrogen production via indirectly heated gasification and pyrolysis, *Proceedings of the 1995 USDOE Hydrogen Program Review*, Coral Gables, FL, NREL/CP-430-20036, National Renewable Energy Laboratory, Golden, CO, vol. 1, pp. 205–236, 1995.
62. Matsumura, Y., Minowa, T., Potic, B., Kersten, S.R.A., Prins, W., Swaaij, W.P.M., Beld, B. et al. Biomass gasification in near and super critical water: Status and prospects. *Biomass and Bioenergy*, 29, 269–292, 2005.
63. Midilli, A., Dogru, M., Akay, G., Howarth, C.R. Hydrogen production from sewage sludge via a fixed bed gasifier product gas. *International Journal of Hydrogen Energy*, 27, 1035, 2002.
64. Midilli, A., Rzayev, P., Hayati, O., Teoman, A. Solar hydrogen production from hazelnut shells. *International Journal of Hydrogen Energy*, 25, 723–732, 2000.
65. Milne, T.A., Abatzoglou, N., Evans, R.J. Biomass gasifier "tars": Their nature, formation, and conversion, NREL/TP_570_25357, ON DE00003726, National Renewable Energy Laboratory Technical Report, Golden, CO, 1998.
66. Milne, T.A., Elam, C.C., Evans, R.J. Hydrogen from biomass: State of the art and research challenges, International Energy Agency Technical Report, Task 16, Hydrogen from Carbon Containing Materials, NREL IEA/H_2/TR-02/001, National Renewable Energy Laboratory, Golden, CO, 2001.
67. Minowa, T., Ogi, T., Yokoyama, S.Y. Hydrogen production from wet cellulose by low temperature gasification using a reduced Ni catalyst. *Chemistry Letters*, 10, 937–938, 1995.
68. Minowa, T., Zhen, F., Ogi, T. Cellulose decomposition in hot compressed water with alkali or nickel catalyst. *Journal of Supercritical Fluids*, 21, 105–110, 2001.
69. Modell, M. Processing methods for the oxidation of organics in supercritical water, US Patent 4,338,199, 1982.
70. Modell, M. Gasification and liquefaction of forest products in supercritical water. In: R.P. Overend, T.A. Milne, L.K. Mudge (eds.), *Fundamentals of Thermochemical Biomass Conversion*, Elsevier, London, U.K., 1985, pp. 95–119.
71. Modell, M., Reid, R.C., Amin, S.I. Gasification process, US Patent 4,113,446, 1978.
72. Narvaez, I., Corella, J., Orio, A. Fresh tar (from a biomass gasifier) elimination over a commercial steam-reforming catalyst. Kinetics and effect of different variables of operation. *Industrial & Engineering Chemistry Research*, 36, 317–327, 1997.
73. Narvaez, I., Orio, A., Aznar, M.P., Corella, J. Biomass gasification with air in an atmospheric bubbling fluidized bed: Effect of six operating variables on the quality of the product raw gases. *Industrial & Engineering Chemistry Research*, 35, 2110–2120, 1996.
74. Ni, M., Leung, D.Y.C., Leung, M.K.H., Sumathy, K. An overview of hydrogen production from biomass. *Fuel Processing Technology*, 87, 461–472, 2006.
75. Onay, O., Kockar, O.M. Fixed-bed pyrolysis of rapeseed (*Brassica napus* L.). *Biomass and Bioenergy*, 26, 289–299, 2004.

76. Padro, C.E.G., Putsche, V. Survey of the economics of hydrogen technologies, Technical Report, NREL/TP_570_27079, National Renewable Energy Laboratory, Golden, CO, September 1999, pp. 1–57.
77. Pfeifer, C., Puchner, B., Hofbauer, H. In-situ CO_2 absorption in a dual fluidized bed biomass steam gasifier to produce a hydrogen rich syngas. *International Journal of Chemical Reactor Engineering*, 5, A9, 1542–6580, 2007.
78. Pinto, F., Franco, C., André, R.N., Miranda, M., Gulyurtlu, I., Cabrita, I. Co-gasification study of biomass mixed with plastic wastes. *Fuel*, 81, 291–297, 2002.
79. Rabah, M.A., Eldighidy, S.M. Low cost hydrogen production from waste. *International Journal of Hydrogen Energy*, 14, 221–227, 1989.
80. Rapagna, S., Jand, N., Foscolo, P.U. Catalytic gasification of biomass to produce hydrogen rich gas. *International Journal of Hydrogen Energy*, 23(7), 551–557, 1998.
81. Rapagna, S., Provendier, H., Petit, C., Kiennemann, A., Foscolo, P.U. Development of catalysts suitable for hydrogen or syn-gas production from biomass gasification. *Biomass and Bioenergy*, 22, 377, 2002.
82. Raveendran, K., Ganesh, A., Khilar, K.C. Influence of mineral matter as biomass pyrolysis characteristics. *Fuel*, 74(12), 1812–1822, 1995.
83. Reij, M.W., Keurentjes, J.T.F., Hartmans, S. Membrane bioreactors for waste gas treatment. *Journal of Biotechnology*, 59(3), 155–167, 1998.
84. Sato, S., Lin, S.Y., Suzuki, Y., Hatano, H. Hydrogen production from heavy oil in the presence of calcium hydroxide. *Fuel*, 82, 561–567, 2003.
85. Savage, P.E. Organic chemical reactions in supercritical water. *Chemical Reviews*, 99, 603–621, 1999.
86. Savage, P.E. Heterogeneous catalysis in supercritical water. *Catalysis Today*, 62, 167–173, 2000.
87. Schmieder, H., Abeln, J., Boukis, N., Dinjus, E., Kruse, A., Kluth, M., Petrich, G., Sadri, E., Schacht, M. Hydrothermal gasification of biomass and organic wastes. *Journal of Supercritical Fluids*, 17, 145–153, 2000.
88. Shahbazov, Sh.J., Usupov, I. Technical communication: Non-trading sources of energy for hydrogen I. *International Journal of Hydrogen Energy*, 19, 863–864, 1994.
89. Simell, P., Kurkela, E., Stahlberg, P., Hepola, J. Catalytic hot gas cleaning of gasification gas. *Catalysis Today*, 27(1–2), 55–62, January 1996.
90. Simell, P.A., Hakala, N.A.K., Haario, H.E., Krause, A. Catalytic decomposition of gasification gas tar with benzene as the model compound. *Industrial & Engineering Chemistry Research*, 36, 42–51, 1997.
91. Simell, P.A., Hirvensalo, E.K., Smolander, V.T., Krause, A. Steam reforming of gasification gas tar over dolomite with benzene as a model compound. *Industrial & Engineering Chemistry Research*, 38, 1250–1257, 1999.
92. Simell, P.A., Leppälahti, J.K., Kurkela, E.A. Tar decomposing activity of carbonate rocks under high CO_2 partial pressures. *Fuel*, 74(6), 938–945, 1995.
93. Spath, P.L., Mann, M.K., Amos, W.A. Update of hydrogen from biomass—Determination of the delivered cost of hydrogen. NREL/MP-510-33112, National Renewable Energy Laboratory, Golden, CO, 2000.
94. Spritzer, M.H., Hong, G.T. Supercritical water partial oxidation, FY 2003 Progress Report, National Renewable Energy Laboratory, Golden, CO, pp. 1–5, 2003.
95. Squires, A.M. Cyclic use of calcined dolomite to desulfurize fuels undergoing gasification. *Advanced in Chemistry Series*, 69, 205–229, 1967.
96. Sutton, D., Kelleher, B., Doyle, A., Ross, J.R.H. Investigation of nickel-supported catalysts for the upgrading of brown peat derived gasification products. *Bioresource Technology*, 80, 111–116, 2001.
97. Sutton, D., Kelleher, B., Ross, J. Catalytic conditioning of organic volatile products produced by peat pyrolysis. *Biomass and Bioenergy*, 23, 209–216, 2002.
98. Sutton, D., Kelleher, B., Ross, J.R.H. Review of literature on catalysts for biomass gasification. *Fuel Processing Technology*, 73, 155–173, 2001.

99. Taralas, G., Vassilatos, V., Krister, S., Jesus, D. Thermal and catalytic cracking of n-heptane in presence of calcium oxide, magnesium oxide and calcined dolomites. *Canadian Journal of Chemical Engineering*, 69(6), 1413–1419, 1991.
100. Turn, S.Q. et al. An experimental investigation of hydrogen production from biomass gasification. *International Journal of Hydrogen Energy*, 23(8), 641–648, 1998.
101. Turn, S.Q., Kinoshita, C.M., Ishimura, D. Removal of inorganic constituents of biomass feedstocks by mechanical dewatering and leaching. *Biomass and Bioenergy*, 12, 241–252, 1997.
102. Vassilatos, V., Taralas, G., Sjostrom, K., Bjornbom, E. Catalytic cracking of tar in biomass pyrolysis gas in the presence of calcined dolomites. *Canadian Journal of Chemical Engineering*, 70, 1008–1013, 1992.
103. Walcher, G., Girges, S., Weingartner, S. Hydrogen energy progress XI. *Proceedings of the 11th World Hydrogen Conference*, Stuttgart, Germany, 1996, Vol. 1, pp. 413–418.
104. Wang, D., Chornet, E. Catalytic steam reforming of biomass-derived oxygenates: Acetic acid and hydroxyacetaldehyde. *Applied Catalysis A: General*, 143, 245–270, 1996.
105. Williams, P.T., Brindle, A.J. Catalytic pyrolysis of tyres: Influence of catalyst temperature. *Fuel*, 81, 2425–2434, 2002.
106. Williams, P.T., Onwudili, J. Composition of products from the supercritical water gasification of glucose: A model biomass compound. *Industrial & Engineering Chemistry Research*, 32, 1542–1548, 1993.
107. Wornat, J.M., Hurt, R.H., Yang, N.Y.C., Headley, T.J. Structural and compositional transformations of biomass chars during combustion. *Combustion and Flame*, 100, 131–143, 1995.
108. Xia, Y., Dunsong, W. Study of gasification treatment of biomass in fixed bed gasifier. *Mei Qi Yu Re Li*, 20, 243, 2000 (in Chinese).
109. Xiaodong, X., Yukihiko, M., Jonny, S., Michael, J.A. Carbon-catalyzed gasification of organic feedstocks in supercritical water. *Industrial & Engineering Chemistry Research*, 35, 2522–2528, 1996.
110. Xu, X., Antal, M.J. Gasification of sewage sludge and other biomass for hydrogen production in super critical water. *Environmental Progress*, 17(4), 215–220, 1998.
111. Xu, X., Matsumura, Y., Stenberg, J., Antal, M.J. Carbon-catalyzed gasification of organic feedstocks in supercritical water. *Industrial & Engineering Chemistry Research*, 35, 2522–2530, 1996.
112. Yeboah, Y., Bota, K., Day, D., McGee, D., Realff, M., Evans, R., Chornet, E., Czernik, S., Feik, C., French, R., Philips, S., Patrick, J. Hydrogen from biomass for urban transportation. *Hydrogen, Fuel Cells and Infrastructure Technologies Program Review Meeting*, Berkeley, CA, May 18–22, 2003, pp. 1–5.
113. Yongjie, Y. Exploring energy from biomass—The gasification of residues from hydrolyzed sawdust. *Acta Energiae Solaris Sinica*, 17, 209, 1996 (in Chinese).
114. Yu, D., Aihara, M., Antal, M.J. Hydrogen production by steam reforming glucose in supercritical water. *Energy and Fuels*, 7(5), 574–577, 1993.
115. Zhiwei, W., Songtao, T., Xueyong, S., Zian, L., Congming, C., Dingkai, L. A study on model for biomass pyrolysis and gasification in fluidized bed. *Journal of Fuel Chemistry and Technology*, 30, 342, 2002 (in Chinese).

6

Hydrogen Production by Supercritical Water Gasification

Emhemmed A.E.A. Youssef
University of Western Ontario

George Nakhla
University of Western Ontario

Paul Charpentier
University of Western Ontario

CONTENTS

6.1 Supercritical Fluids

6.1.1 Fundamentals of Supercritical Fluids

A fluid heated to above the critical temperature and compressed to above the critical pressure is known as a supercritical fluid. The phenomena and behavior of supercritical fluids has been the subject of research right from the 1800s (Browne 1987). Two supercritical fluids are of particular interest: carbon dioxide and water. Carbon dioxide has a low critical temperature of 31°C and a moderate critical pressure of 7.3 MPa. It is nonflammable, nontoxic, and environmentally friendly. It is often used to replace toxic Freons and certain organic solvents. Furthermore, it is miscible with a variety of organic solvents and is readily recovered after processing.

6.1.2 Supercritical Water and Pressurized Hot Water

Any fluid is defined as a supercritical fluid when its temperature and pressure are above its critical conditions, that is, 374°C, 22 MPa for water. In Figure 6.1, the upper right quadrant portrays the phase in which water is at its supercritical state. It can be noted that at the critical point there is no distinction between the vapor and liquid phases in the P–T phase diagram. Therefore, if the liquid, which is water here, is heated at constant

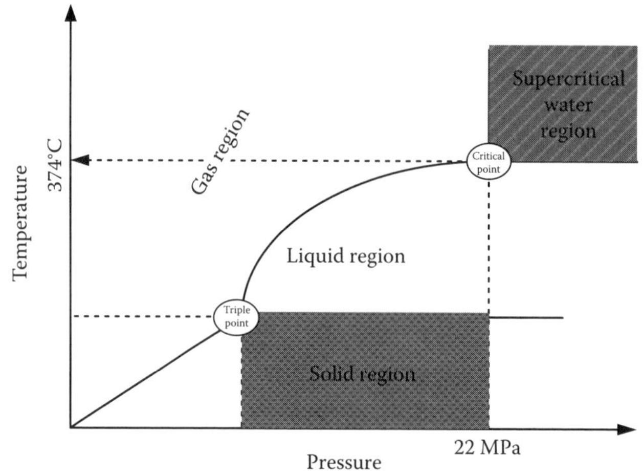

FIGURE 6.1
P–T phase diagram of supercritical water (SCW).

pressure until its temperature exceeds its critical temperature, it expands and becomes in a state that is like vapor but with no phase transition.

6.1.3 Properties of Supercritical Water (Kritzer)

Under normal conditions, water has three states: ice, steam, and liquid. When water is heated and compressed to a high temperature and pressure, above 374°C and 22 MPa, the water enters its supercritical state. Water above its critical point has unique gas-like and liquid-like properties, which are conducive to destruction of organic compounds. The solvation properties of water at 22 MPa as a function of temperature are portrayed in Figure 6.1. The viscosity and diffusivity are more gas-like whereas the density is comparable with liquid water.

The mass transfer resistance in supercritical water (SCW) becomes insignificant because of the high solubility and diffusivity of gases and organic material. Under supercritical conditions, the dielectric constant of water and hydrogen bonding drops sharply, thus providing high solvating power for organic compounds. This results in SCW acting as a single-phase, nonpolar dense gas that has solvation properties similar to low-polarity organics (Modell 1989; Tester et al. 1993; Savage et al. 1995). Moreover, increasing the value of the ionic product increases the concentration of both hydrogen and OH ions, which leads to a significant increase in the power of the hydrolysis reaction. Hence, hydrocarbons and gases such as CO_2, N_2, and O_2 are highly soluble while ionic species, namely, inorganic salts, are practically insoluble in SCW. This enhances the oxidation kinetics of organic species, especially because of the absence of mass transfer limitations. Therefore (although SCW is applicable to a wide range of feed mixtures and is not limited to aqueous organics), organic wastes containing carbon, hydrogen, oxygen, and nitrogen atoms are of particular interest and oxidized to primarily carbon dioxide, water, molecular nitrogen, and other small molecules.

6.2 Development of the SCW Process

6.2.1 Current Status of the SCW Process

The SCW process was first known in the late 1970s, thanks to the pioneering work of Modell (1989) at the Massachusetts Institute of Technology (MIT); SCW was first developed for the treatment of hazardous wastes and materials such as warfare agents. However, SCW has emerged rapidly in the last two decades mainly because of the need to develop an alternative and environmentally friendly fuel source. This attraction was driven by the fact that SCW produces considerable amounts of gaseous fuel such as hydrogen and methane. Although it is still in developmental shape, Yoshida et al. (2003) reported that, from an energy point of view, biomass gasification using SCW is technically feasible compared to other existing technologies such as anaerobic digestion or incineration. This is driven by the fact that SCW converts fully all the feedstock organic contents leaving no residue, which simultaneously achieves the ultimate goal of energy production as well as treatment.

Several laboratory studies have focused on improving the understanding of the SCW process, in order to overcome its commercialization challenges. Preeminent research groups in SCW include Tester and coworkers at the MIT, Savage and coworkers at the University of Michigan, the Advanced Institute of Science and Technology in Korea, the Japanese National Institute for Resources and Environment, the Forschungszentrum Karlsruhe in Germany, University of Twente in the Netherlands, the green energy group at the University of Western Ontario, Canada, and the State Key Laboratory of Multiphase Flow in Power Engineering in China.

Pilot-scale plant operations of SCW are countable and being conducted either by agencies such as General Atomics of the United States, Chematur of United Kingdom, Mitsubishi Heavy Industries of Japan, or by institutions such as the High Pressure Process Group, Department of Chemical Engineering and Environmental Technology, University of Valladolid (Spain), or private companies.

6.2.2 Modeling and Thermophysical Properties of SCW

Modeling of the SCW process has gained more attention recently facilitated by the wealth of experimental results. Properties of SCW significantly differ from those at the ambient conditions as the increase in temperature increases the tendency for ion pairing, which increases the salt concentrations in the mixture. Thus, using numerical techniques for the calculation of the SCW properties is of utmost importance since these properties are usually unknown and difficult to experimentally estimate. The aim of this important step is the scaling up of the SCW process for commercial applications. Therefore, the required essential knowledge of the SCW thermophysical properties such as densities, enthalpies, and heat capacities has motivated several researchers to investigate and overcome the poor prediction capability of the conventional cubic equation of state (EoS) (Bermejo et al. 2005). However, the Peng–Robinson EoS with volume translation correlation as reported by Bermejo et al. (2005) showed a quite accurate reproducibility of both water–air system and the behavior of real supercritical water oxidation (SCWO) reactors. In their high-temperature electrolyte solution comprehensive review Anderko et al. (2002) reported that the development and integration of a universal model using the combined high-temperature ion-pair-based with low-and-moderate temperature approach would be invaluable. Such a model, if computationally sufficient, could cover the wide range of temperatures required for practical applications and, thus, circumvent the impractical thermodynamic property prediction resulting from the current EoS.

6.2.3 Reactions in SCW

The fundamental understanding of reactions in SCW has increased remarkably in the last two decades. According to Savage et al. (1995), the unique physical and transport properties of an SCW are intermediate between those of a liquid and a gas. This makes SCW an attractive medium for chemical reactions. The aforementioned author pointed out that among ambient reactions in SCW are homogeneous and heterogeneous catalysis, polymerization, waste oxidation, and green fuel production such as hydrogen.

Before proceeding into detailed reactions in SCW, and from an organizational point of view, SCW research activities fall into four main categories, that is, pyrolysis, total oxidation, gasification, and partial oxidation. In the literature, supercritical water is the term that stands for the gasification of organic compounds at or above a critical condition that is not well defined. An attempt to classify the cases in which SCW is used to gasify different compounds is presented as follows:

1. *Pyrolysis in SCW*, which stands for the hydrolysis of the organic compounds at SCW conditions in the absence of both a catalyst and an oxidant.
2. *Total oxidation in SCW*, which stands for gasifying the organic compounds in the absence of a catalyst and the presence of excess oxidant at oxygen-to-carbon molar ratios (MRs) of 1.0 or higher than the stoichiometric requirement, that is, $MR \geq 1.0$.

3. *Gasification in SCW*, which stands for gasifying the organic compounds in the absence of an oxidant and the presence of a catalyst, is also known as low-temperature gasification.
4. *Partial oxidation in SCW (SCWPO)*, which stands for oxidizing the organic compounds in the presence of an oxidant at an oxygen-to-carbon MR of less than 1.0 (MR < 1.0) and absence of a catalyst.
5. *Sequential gasification partial oxidation*, which stands for gasifying the organic compounds in the presence of an active catalyst (heterogeneous or homogenous) followed by further gasifying the organic compounds in the presence of an oxidant at an oxygen-to-carbon molar ratio (MR) below the theoretical requirements, that is, an approach using a combination of step 3 followed by step 4.

Sequential gasification and partial oxidation were studied in detail for glucose as a waste biomass model compound and hog manure by Youssef et al. (2010) respectively to determine the product distribution in both gas and liquid phases. Product distribution is important as it facilitates the detailed analysis of reactions and transformation paths that take place in SCW as well as providing insight into the process design and operating conditions, that is, temperature, pressure, oxidant stoichiometric ratio, as well as the selection of the catalyst.

The chemistry of total oxidation in SCW has undoubtedly received the most attention. In this process, organic compounds are converted to carbon dioxide and water at typical operating conditions of about 25 MPa and 400°C–800°C and in the presence of excess oxidant (O_2). The general reaction equation is given as

$$CxHy + \left(x + \frac{y}{4}\right)O_2 \rightarrow xCO_2 + \left(\frac{y}{2}\right)H_2O \tag{6.1}$$

As a waste destruction process, SCWO has several advantages over conventional processes and even some of the relatively modern processes such as wet air oxidation (WAO) and incineration. As a media for chemical reactions, depending on its density, low dielectric constant of SCW promotes dissolution of nonpolar organic compounds whereas its gas-like low viscosity promotes mass transfer, and the solvation is enhanced by the liquid-like density. The pioneering work of Thomason and Modell (Modell 1978) was first to apply SCWO as a powerful technology for the transformation of hazardous and toxic organic wastes and reported destruction efficiency of 99.99% of organic compounds, including polychlorinated biphenyls. Bermejo et al. (2005) reported the use of the SCWO technique for the treatment of municipal sewage sludge and paper mill waste as sewage sludge was converted to clear water and gases.

Goto et al. (1999a,b) reported a study on the destruction of municipal sludge and alcohol distillery wastewater. The composition of the molasses-rich wastewater mainly consisted of saccharide, carbohydrate, and ash (mainly KCl), whereas the sewage sludge composition consisted of lipid (about 10%), protein (about 40%), carbohydrate, lignin (about 17%), and ash. The experiments were conducted at three different temperatures of 400°C, 450°C, and 500°C with hydrogen peroxide as oxidant at 300% of the stoichiometric demand. The authors characterized the liquid effluent by means of total organic carbon (TOC) and used it as a tool for developing a global rate model to express the reduction of components by SCWO of the used wastes since the development of more rigorous kinetic models was more intricate given the complexity of the feed and the intermediates originating from the reactions involved in SCWO.

For partial oxidation in SCW, and in order to generate H_2 from cellulose or glucose as biomass model compounds effectively, water is used as a suitable solvent and reactant in accordance with

$$C_6H_{10}O_5 + 7H_2O \rightarrow 12H_2 + 6CO_2 \tag{6.2}$$

$$C_6H_{12}O_6 + 6H_2O \rightarrow 12H_2 + 6CO_2 \tag{6.3}$$

According to Cortright et al. (2002), the H_2 selectivity is evaluated to know how many hydrogen atoms in an organic compound can be taken out as H_2 as

$$H_2\,\text{selectivity (mol\%)} = \left\{\frac{H_2\,\text{produced (mol)}}{\text{C atom in the gas phase}}\right\} \times \left\{\frac{1}{2}\right\} \times 100 \tag{6.4}$$

The aforementioned authors considered H_2/CO_2 reforming ratio as ½. For glucose, which is represented as $C_6H_{12}O_6$, the maximum hydrogen is 12 mol of H_2 with 6 mol of CO_2. In other words, 6 mol of H_2 produced from the reforming of glucose and 6 mol of CO produced from glucose react with moles of H_2O to produce another 6 mol of H_2 through water–gas shift reaction. The aforementioned equations suggest that the hydrogenation of biomass by employing partial oxidation followed by the water–gas shift reaction could enhance the H_2 production rate. The general equation of this theme could be approximated by the following general reactions:

Gasification

$$C_mH_{n\,\{\text{Catalyst}\}} \rightarrow CO_2 + H_2 + CH_4 + \text{intermediate products} \tag{6.5}$$

Partial oxidation

$$\left[C_mH_n\right]_{\{\%O_2<1\}} + \text{Intermediate products} \rightarrow CO + H_2 + \text{other components} \tag{6.6}$$

Water–gas shift reaction

$$CO + H_2O \rightarrow \text{active hydrogenating species} \rightarrow CO_2 + H_2 \tag{6.7}$$

The water–gas shift reaction represented by Equation 6.7 is of utmost importance for hydrogen production from waste biomass. However, Watanabe et al. (2003) reported that at a pressure below 30 MPa the pressure dependence of kinetics is insignificant. Thus, the acceleration of the water–gas shift reaction by increasing the temperature or employing the appropriate catalyst is unavoidable. Rice et al. (1998), who evaluated the water–gas shift reaction in SCW in the absence of catalyst, reported that the increase in the reaction temperature coupled with the increase in the pressure to up to 60 MPa resulted in a significant acceleration of conversion rates, although the observed water–gas shift reaction conversion rates in SCW were much lower than those of catalytic industrial processes.

6.2.4 SCW Product Distribution

Identifying the SCW intermediate products could help understand the reaction mechanisms and eventually the development of kinetic models that are required for reactor design.

Thus, several studies regarding the analysis of the intermediate products from the SCW process have been reported in the literature (Holgate et al. 1995; Sasaki et al. 1998; Kabyemela et al. 1999; Kruse et al. 2003; Williams and Onwudili 2006; Abdelmoez et al. 2007; Youssef et al. 2010). However, most of these studies have focused on the identification of the intermediate products from model compounds such as glucose, cellulose, and starch with fewer data reported for the real waste biomass such as municipal wastewater sludge and hog manure.

6.2.4.1 Pyrolysis Products

Tester group at MIT reported the SCW liquid effluent characteristics above the water critical temperature from glucose hydrolysis (pyrolysis) and oxidation in a continuous-flow reactor at a pressure of 246 bar and temperature range of 425°C–600°C (Holgate et al. 1995). Since the identification and analysis of all liquid-phase products were challenging, the aforementioned authors employed literature results (Woerner 1976; Antal 1982; Evans and Milne 1987; Antal et al. 1990) for glucose hydrolysis below and near critical water to obtain the list of 40 suspected products. In order to clearly identify each peak, two different wavelengths of 210 and 290 nm, which allow for the discrimination between the peaks by functional groups, were employed and allowed the identification of additional species that are not detectable at a single wavelength. The authors reported that the number of products in the oxidation experiments was substantially reduced compared to the hydrolysis experimental products although many peaks formed during glucose hydrolysis were unidentified. This conclusion was supported by the good carbon balance closure obtained for the oxidation experiments of glucose.

Sasaki et al. (1998) studied cellulose hydrolysis in sub- and supercritical water in a continuous-flow reactor in a temperature range of 290°C–400°C and at a pressure of 25 MPa. The aforementioned authors employed an H-NMR and FAB-MS for the liquid-phase product analysis identification and an HPLC for the quantitative analysis of the identified compounds. At temperatures of 320°C, 350°C, and 400°C, the main components at 100% conversion of cellulose liquid-phase were erythrose, dihydroxyacetone, fructose, glucose, glyceraldehydes, pyruvaldehyde, and oligomers such as cellobiose, cellotriose, cellotetraose, cellopentaose, and cellohexaose. Comparing the results obtained in this study with the previous one for glucose decomposition in sub- and supercritical water (Sasaki et al. 1998), the authors pointed out that glucose had a faster conversion rate than cellulose even with the formed glucose as a hydrolysis product from cellulose. Furthermore, the authors reported that around and above the water critical point, that is, SCW, the hydrolysis products yield was higher than those of the subcritical water. Figure 6.2 portrays the suggested reaction pathways of cellulose decomposition in SCW.

6.2.4.2 Oxidation Products

Williams and Onwudili (2006) investigated the decomposition of cellulose, starch, and glucose as a biomass model compound and real biomass in the form of cassava waste in subcritical and supercritical water. The experiments were conducted in a batch reactor at a temperature range from 330°C to 380°C, a pressure of 25 MPa, and hydrogen peroxide (H_2O_2) as a source of oxygen. The liquid-phase product for each model compound was divided into three main products, that is, water-soluble products (WSPs), char, and oil. The product oil was analyzed using Fourier transform infrared spectrometry, mass of the WSP was determined as total dissolved solids, and the char was removed by filtration. For the oil analysis, the authors reported that using the functional group assignment could help identify the various peaks of

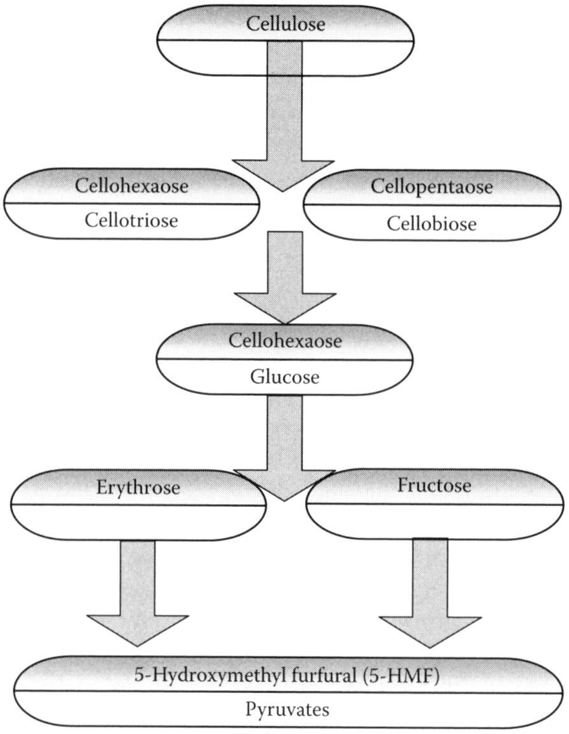

FIGURE 6.2
Cellulose decomposition pathways in supercritical water (SCW).

the spectrogram. Thus, the presence of carboxyl acids was confirmed by the presence of C=O between the wavelengths of 1650 and 1850 cm^{-1} together with the –OH functional groups. Similarly, the authors confirmed the presence of other functional groups such as ketones and aldehydes, primary, secondary, and tertiary alcohols, and alkane and alkene groups.

Wahyudiono et al. (2008) reported in an experimental study on the behavior of catechol as a model compound representing lignin. Experiments were conducted in a batch reactor at temperatures of 370°C–420°C and pressures of 25–40 MPa. The aforementioned authors reported that the variation of temperature resulted in a variation of higher- and lower-molecular-weight products as identified by GC-MS. Phenol was the major component identified with several other components presented in small amounts as summarized in Table 6.1.

The presence of phenol as a major intermediate compound was attributed to the fact that higher temperatures hydrolyzed the dissolved compounds with ether linkages to a single ring phenolic compounds as reported by Wahyudiono et al. (2008) and Watanabe et al. (2003).

6.2.4.3 Gasification Products

Onwudili and Williams (2009) reported the product distribution and characterization of crude glycerol gasification as a model of waste biomass derived from a biodiesel production plant. The experiments were conducted in a batch reactor at a temperature range of 300°C–450°C and pressures between 8.5 and 31 MPa. The liquid-phase analysis was performed by first transferring the entire contents of the reactor to a separating funnel in which the organic fraction was partitioned using liquid–liquid extraction technique. The extracted

TABLE 6.1
Main Intermediate Compounds Produced during SCW of Different Model and Real Waste Compounds

Name of the Feed Compound	Process Description	Main Intermediate Compounds		Reference
Glucose	Plug flow reactor made of Inconel 625 tubing having 6.36 mm OD and 1.7 mm ID with an internal volume of 10.7 cm^3	Acetic acid 5-Hydroxymethylfurfural Propenoic acid Acetaldehyde Acetonylacetone Formic acid Lactic acid Glycolic acid		Holgate et al. (1995)
Cellulose, coconut oil solutions, brewery, and dairy effluents	A batch reactor made of stainless steel 316 coiled tubing having a length of 76.2 cm, OD of 1.27 cm, and 0.21 cm wall thickness A continuous-flow reactor made of stainless steel 316 tubing length of 400 cm, OD of 0.318 cm, and ID of 0.14 cm	Acetic acid Formic acid Lactic acid Glycolic	Major component Major component Minor component Minor component	Calvo and Vallejo (2002)
Crude glycerol	Hastelloy-C batch reactor having 75 mL capacity, design pressure of 40 MPa, and temperature of 600°C	Compound	Concentration (μg/g oil)	Onwudili and Williams (2009)
		Methyl oleate	108,000	
		Methyl stearate	32,800	
		Oleic acid	68,300	
		Palmitic acid	67,500	
		Methyl palmitate	63,700	
		Dimethyl phenol	26,400	
		1-Nonene	39,600	
		Xylene	44,600	
		1-Octene	46,900	
Glucose	Subcritical experiments (stainless steel (SUS 316) reactor with ID of 0.118 cm); supercritical experiments (stainless steel (SUS 316) reactor with ID of 0.077 cm)	Saccharinic acids Fructose Glyceraldehyde Erythrose Pyruvaldehyde 1,6-Anhydroglucose 1,3-Dihydroxypropan-2-one		Kabyemela et al. (1999)
Catechol	A Hastelloy C-276 tube reactor having an internal volume of 5 cm^3	Phenol Cyclopentanone 2-Cyclopentenone 1,2-Benzenedicarboxylic acid Nonylphenol		Goto et al. (2009a,b)

(*continued*)

TABLE 6.1 (continued)

Main Intermediate Compounds Produced during SCW of Different Model and Real Waste Compounds

Name of the Feed Compound	Process Description	Main Intermediate Compounds	Reference
		1,4-Dipropylbenzene Acetophenone 2,6-Di-*tert*-butylnaphthalene 4-Butoxyphenol *o*-Ethoxyphenol 2-Methylterephthalaldehyde *p*-Diethoxybenzene 5-Methoxy-2,3,4-trimethylphenol	
Cellulose	An Inconel 600 tube reactor having a length of 35.56 cm, OD of 0.635 cm, ID of 0.279 cm, and a volume of 2.18 mL	Cellobiose Glucose Fructose Glycolaldehyde D-Fructose 1,3-Dihydroxyacetone Anhydroglucose 5-HMF Furfural	Kumar and Gupta (2008)

oil was analyzed using GC-FID and GC-MS for qualitative and quantitative analyses. The extracted WSPs were analyzed by evaporating the water content, drying and weighing each sample to determine the mass of the WSP. The distribution of oil products present in the liquid phase was examined by varying the reaction temperature and reaction time. Table 6.1 reports some of the results obtained in this study with methyl oleate, methyl stearate, methyl palmitate, palmitic acid, and maleic acid as the major compounds in the liquid phase that exhibit significant resistance to temperature increase. The degree of decomposition of fatty acids was influenced by the temperature increase with the major product at lower temperature, a waxy material containing glycerol, fatty acids, and methyl esters.

6.3 SCW Process Description

Generally, SCW research is conducted in either batch or continuous-flow reactors. Batch reactor systems are easy to construct but require longer heating periods and residence times, and it is usually difficult to conduct batch experiments without avoiding the subcritical reactions that may occur (Guo et al. 2007). On the other hand, the continuous SCW flow process comprises four main sections or zones (Bermejo et al. 2005):

- Feed preparation
- Reactor section
- Heat recovery and depressurization
- Product separation section

There have been several lab-scale and pilot plants for the SCW process built so far in several institutions. In the following section, a brief description of the system built by the green engineering group at the University of Western Ontario, London, Ontario, Canada, is provided. This system consists of a small-scale SCW batch and continuous-flow units.

6.3.1 Batch Unit Setup and Experimental Procedures

The main reactor body was made of Hastelloy C-276 with a capacity of 600 mL and obtained from Autoclave engineers (Erie, PA, the United States). The reactor was rated for a maximum pressure of 41.4 MPa and a maximum temperature of 343°C. The reactor was facilitated to sustain a maximum temperature of 500°C at a lower pressure level of 5100 psi by consulting the pressure–temperature rating of the Hastelloy alloy published on the manufacturer's website. Thus, the reactor was able to operate at the supercritical conditions of water (374°C, 22 MPa). Figure 6.3 portrays the batch unit setup schematic diagram.

The batch experimental procedure started by filling the reactor with 60 mL of double deionized water collected from a compact ultrapure water system (EASY pure LF, Mandel Scientific co, model BDI-D7381). To drive away any dissolved oxygen, the deionized water was heated for about an hour on a hot plate stirrer at a temperature of 60°C. The whole reactor system including all tubing and other parts was purged using helium gas at a constant pressure rate of 50 psi for 15 min to drive away all the air and oxygen that may be presented in the system. After purging with helium, the outlet valve (VO1) was closed and the pressure in the reactor was increased to 100 psi to prevent water evaporation through the heating phase, and both the gas inlet and feed inlet valves (VP2) were closed and the reactor started heating. The pressure in the reactor increased accordingly as the temperature increased. After reaching the desired temperature, an ISCO syringe pump (Model 100 DX, Lincoln, NE) was utilized to inject the feed to the reactor against its pressure. As the desired reaction time reached, valve VO1 is opened to allow the effluent gases to pass through the condenser (Double pipe H/E) where they are cooled and depressurized using a Swagelok piston-sensing high-pressure reducing regulator (KHP series, Solon, OH). The cooled depressurized effluent passes to a gas liquid separator where gases and liquid products are separated. The gaseous products leaving the separator pass through a Swagelok in-line filter (F series, Solon, OH) to remove any moisture before passing through an OMEGA mass flowmeter (FMA 1700/1800 series 0–2 L/min, Laval, Quebec, Canada). The mass flowmeter is equipped with a totalizer that utilizes a K-factor to relate the mass flow rate of an actual gas to nitrogen as a calibrated reference gas. The actual gas flow rate is calculated by determining the average K factor for the produced gas by the mean of the mole fraction of each gas in the stream. After passing through the mass flowmeter, the product gases are collected in a 3 L volume Tedlar gas sampling bag. As soon as the gas bag is filled, the flow of gaseous product is redirected to the fume hood where it is vented.

6.3.2 Continuous Flow System Design and Experimental Procedures

Figure 6.4 shows the schematic of the constructed experimental apparatus that is suitable for continuous operation at temperatures and pressures up to 650°C and 350 bar, respectively. The system consists of four main zones: the pumping section, the reactor section, the cooling-depressurizing section, and the gas liquid separating-GC analysis section. The four zones are described as follows.

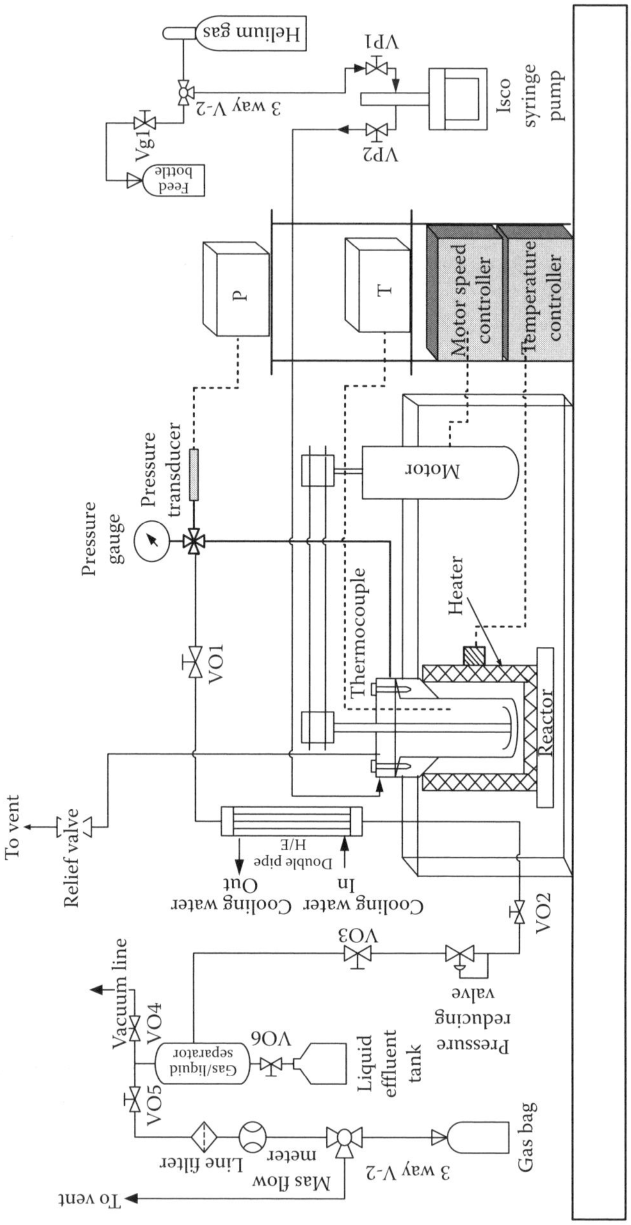

FIGURE 6.3
Schematic diagram of the supercritical water (SCW) batch unit.

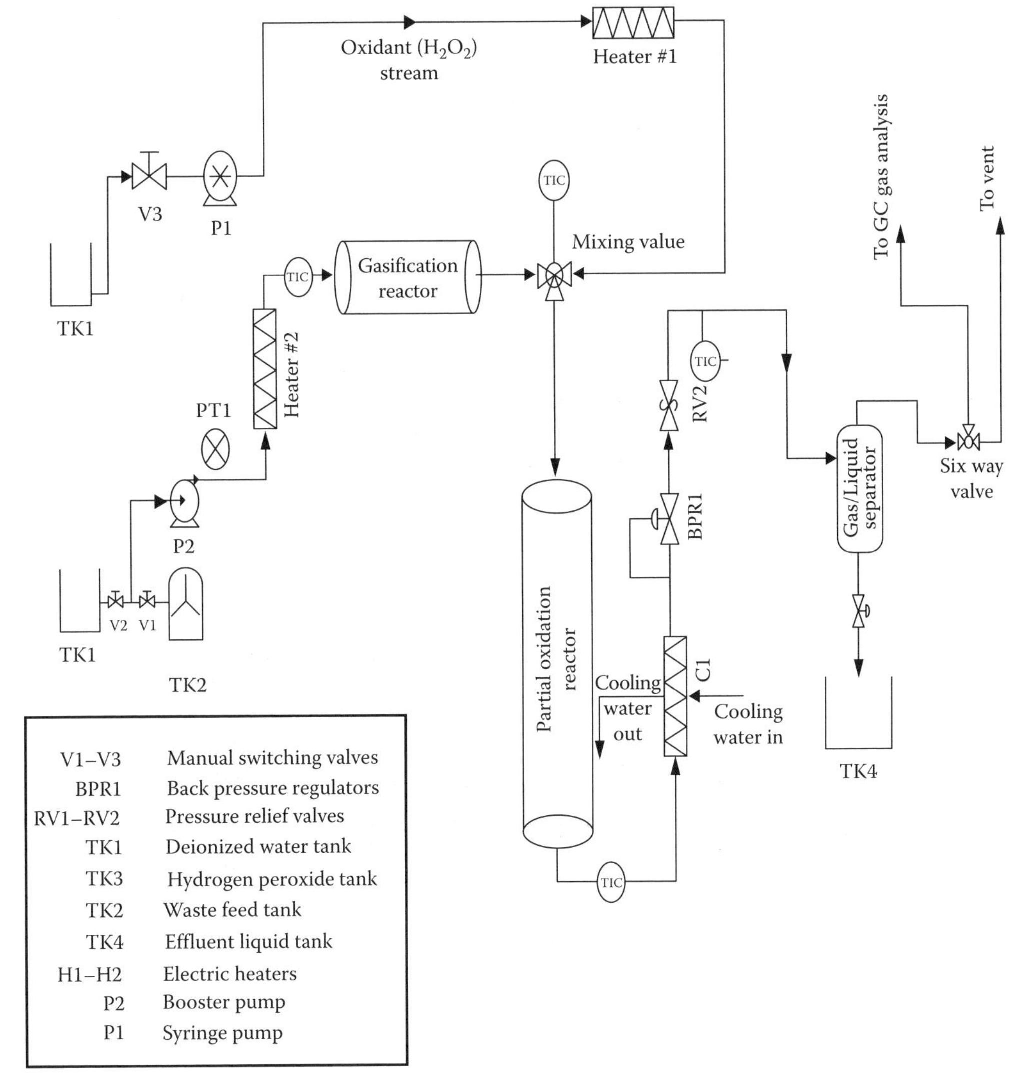

FIGURE 6.4
Schematic PFD diagram of the supercritical water (SCW) continuous-flow unit.

6.3.2.1 Pumping Zone

A high-pressure pump (P1) capable of delivering 10,000 psi (Linc Milton Roy-86 Series Electric Pump, Ivyland, PA) with a nominal flow range between 0.31 and 1.29 L/h is used to feed in the organic solution. This pump is capable of delivering sewage sludge with a mesh size <70 μm to a high-pressure environment at the earlier-mentioned small flow rate that meets the goal to fit the other existing equipment in the lab. A ball-type check valve (High Pressure Equipment Company catalog no 60-41HF9, Erie, PA) was installed in the pump discharge line to ensure flow in one direction only. The pump suction was connected to both the aqueous feed tank and the deionized water tank (TK1 and TK2).

The feed line contains a relief valve (RV1) (High Pressure Equipment Company, Erie, PA) proceeded by a 0–10,000 psi pressure transducer (Noshok, Berea, OH), which was connected to the control and data acquisition system. The relief valve was installed to relieve pressure in case of excessive pressure. Additionally, a check valve (High Pressure Equipment Company, Erie, PA) was installed at the pump discharges to prevent back pressure during shutdown. The oxidant—aqueous H_2O_2 solution—is pumped from the H_2O_2 tank (TK3) to the system by means of an ISCO syringe pump (Model 100 DX, Lincoln, NE).

The oxidant and the feed are pumped to the reactor through two separate lines, with each line preheated by means of two ultra-high-temperature Omega heating tapes (Model Number STH102-080*, Laval, Quebec, Canada). The heating tapes were rated to 1400°F, and the wires were double insulated with braided Samox and knitted into flat tapes for maximum flexibility. The feed and oxidant lines were connected to each other with a high-pressure valve (20-11LF9, High Pressure Equipment Company, Erie, PA) that was used for flushing after each experimental run. After passing through the preheaters in the two separate lines, and just before they enter the reactor, the oxidant and the feed were mixed together in a tee (30-23-HF16 High Pressure Equipment Company, Erie, PA) at an angle of 90° to minimize the mixing effects as per Marrone et al. (2004).

6.3.2.2 Reactor Zone

The reactor section consists of two sections, namely, the gasification section followed by the partial oxidation section. The gasification section is where the preheated feed passed through the reactor and the catalyst. The catalyst was supported by stainless steel frits, fabricated at the University of Western Ontario machine service, and placed at the inlet and outlet of the reactor. Figures 6.5 and 6.6 show a schematic drawing and jpg picture of the catalyst support frits, respectively. The design of gasification and partial oxidation reactors was based on the idea of double pipe tube. In the gasification reactor, the outer pipe was made of 316 SS with dimensions of 0.562 in. ID, 1 in. OD, 16 in. length, and a volume of 60 mL to sustain the process operating pressure. An inner pipe made of titanium (grade 2) was swaged perfectly inside the outer 316 SS pipe.

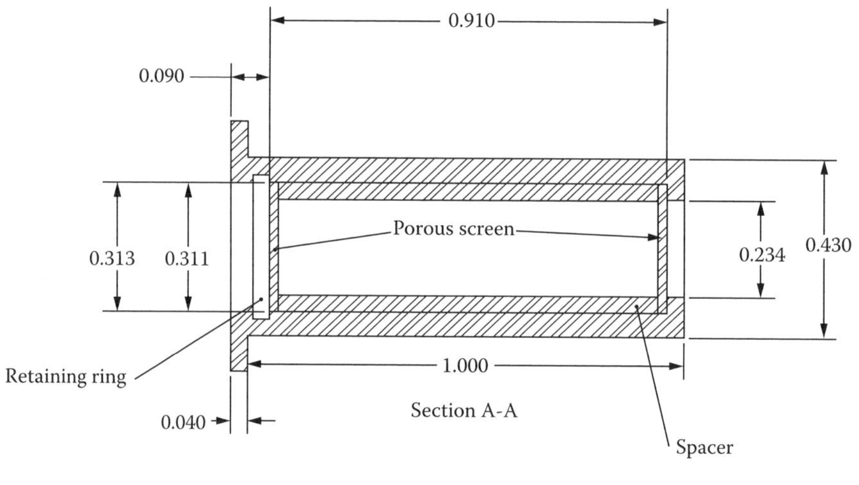

FIGURE 6.5
Schematic diagram of the catalyst support frits.

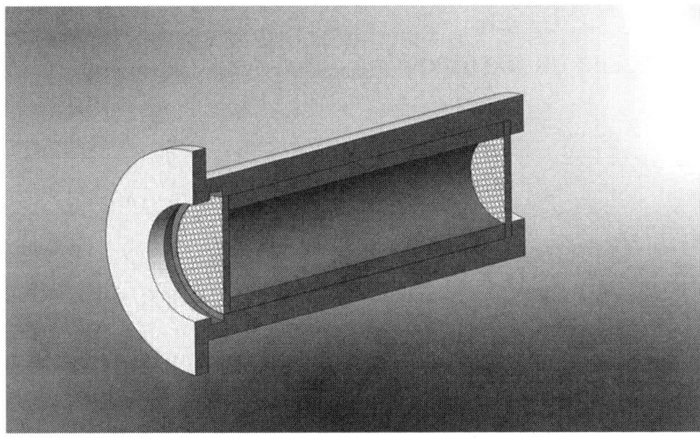

FIGURE 6.6
jpg picture of the catalyst support frits.

The main purpose of using the titanium inner tube was to prevent the fluid from contacting the outer tube and eventually prevent it from being corroded. Another purpose of using a titanium inner tube was to eliminate the interference of the outer reactor wall as a catalyst in the gasification process as it is well known that 316 SS has high percent of nickel that can catalyze the reaction and mislead the gasification process results.

The partial oxidation reactor section was also designed based on the double pipe idea. However, the outer tube was made of 316 SS with dimensions of 0.562 in. ID, 1 in. OD, 16 in. length, and a volume of 127 mL. An inner titanium tube (grade 2) was swaged perfectly through the inner surface of the reactor. Therefore, the actual reactor internal volume was 120 mL. Figure 6.4 shows a simplified schematic for both the reactor and the inner tube. This inner tube was primarily employed to study the corrosion phenomenon in the SCW process. Moreover, for flexibility, the inner tube of the partial oxidation reactor section is easily replaceable. This design flexibility allows for easy replacement and testing of different alloys such as Inconel 625 and Hastelloy other than titanium for detailed corrosion studies and comparison between different alloys as mentioned earlier.

It also should be pointed out that most reactors used in the literature are either from Hastelloy C-276 or Inconel because of their favorable anticorrosion characteristics. Experience has shown that corrosion rates can be rapid when treating wastes containing halogens, such as chlorine. Corrosion-resistant alloys such as Hastelloy C-276 and Inconel 625 do not provide adequate protection against chloride attack under the harsh oxidizing conditions found in SCW systems.

The products leave the second reactor (partial oxidation section) through a high-pressure tee (30-22-HF16 High Pressure Equipment Company, Erie, PA). The reactors were heated by two ultra-high-temperature Omega heating tapes (Model Number STH102-080*, Laval, Quebec, Canada). The heating tapes were rated at 750°C, and the wires were double insulated with braided Samox and knitted into flat tapes for maximum flexibility. The heating tapes were procured to cover all the outer surface of the reactor tube to ensure consistent heating and temperature profiles. The temperature at the inlet and outlet of the reactor was monitored by means of thermocouples (Omega K type, Omega Engineering, Laval, Quebec, Canada) mounted on the inlet and outlet of the tee. The thermocouples were mounted deep enough to be exposed to the fluid to ensure correct measurement of the feed inlet and outlet temperatures.

6.3.2.2.1 Reactor Sizing

The reactor was sized based on the following equation:

$$V_{reactor} = \frac{\{\tau \times \rho_L (F_1 + F_2)\}}{\rho_{SC}} \tag{6.8}$$

where

τ is the reactor residence time

$F = (F_1 + F_2)$ is the total volumetric flow rate of both oxidant and organic feed

ρ_L is the density of the pumped organic feed at the pump conditions

ρ_{SC} is the density of the pumped feed at the reactor conditions (T and P)

6.3.2.3 Cooling-Depressurization Section

The effluent leaves the second reactor through an elbow (60-22HF9, High Pressure Equipment Company, Erie, PA). The elbow is connected to a tee (60-23HF9, High Pressure Equipment Company, Erie, PA) in which the thermocouple is inserted deep until it reaches the outlet of the reactor. The effluents pass through a double pipe heat exchanger that employs service water as a coolant. To ensure efficient cooling, the coolant flow is countercurrent to the effluent. Another thermocouple is installed right after the heat exchanger and is connected to the data acquisition system. A back-pressure regulator (BPR; 26-1700 Series BPR-Tescom model numbers, Emerson Process Management, McKinney, TX) is mounted at the end of the heat exchanger to control the system pressure. The pressure is monitored by the same Noshok pressure transducer just upstream of the back-pressure regulator. Another pressure-reducing valve is mounted after the BPR for safety precautions in case the BPR fails to control the pressure. The pressure-reducing valve is equipped with a pressure gauge (Swagelok, 0–7500 psig). A second relief valve (RV2) (High Pressure Equipment Company, Erie, PA) is connected to the pressure-reducing valve in case of overpressurization.

6.3.2.4 Gas Liquid Separating-GC Analysis Section

After cooling and depressurizing, the reactor effluents are passed through a membrane gas liquid separator (Genie-Filters, Model 120, Gonzales, LA). It is ideal for low-flow applications and can withstand high pressures of 210 bar in the housing. It also provides protection against liquids for online GC analysis. The separator has three ports, inlet, outlet, and bypass. Liquid products are collected from the bypass, and the outlet port is connected to a gas chromatograph equipped with TCD, FID, and FPD detectors for online gas analysis purposes.

6.3.3 Continuous Flow System Procedures

Before starting any experiment, the data acquisition and all control system components are switched on. Each experiment starts with the feed and oxidant preparation in the feed and oxidant tanks (TK1 and TK3). The organic aqueous solution is purged with helium gas from the helium cylinder prior to each experiment to reduce the amount of dissolved oxygen. The feed pump circulates purified water through the heating period, and the BPR1 in

Figure 6.2 regulates the pressure. As the desired temperature is reached, the feed pump is switched to the feed tank by means of switching the three-way valve that connects both the feed and purified water tanks (TK1 and TK2). The high-pressure feed is preheated until it stabilizes to the desired experimental temperature. As the feed pressure and temperature stabilize, the unit is considered at steady state, and samples are taken for the liquid effluent from the gas liquid separator as well as the gaseous stream from the top of the same device. A typical run would last between 2 and 6 h based on the number of gas samples analyzed. At least, three GC injections are measured to ensure the accuracy of the gas stream analysis.

The oxidant (H_2O_2) solution is prepared in a similar manner to the feed stream and placed in the oxidant tank (TK3) and purged with helium gas to drive away any oxygen gas in the tank head space. The oxidant is then pumped to the reactor using an ISCO syringe pump. Thus, there is no need to install any pressure transducer in its line since the ISCO syringe pump is equipped with a control system that provides the flow rate and pressure parameters.

6.4 SCW Process Design Considerations

6.4.1 SCW Reactors (Concepts and Facts)

The discovery of the advantages of the SCWO process generated widespread euphoria in anticipation of a large number of possible applications. The underestimation of the main SCW problems hindered the process from realizing widespread industrial application. To date, a considerable number of SCWO reactor concepts and process designs have been described in the literature. These reactors are either for laboratory use, that is, small-scale applications, or for pilot-scale applications (Guo et al. 2007). The types of these reactors can be classified into five main categories as follows:

- Batch reactors
- Tubular reactors
- Transpiring wall reactors
- Flame reactors
- Other types such as quartz reactors

For the batch reactors, they vary considerably in size from small bomb reactors of volume 2–9 cm^3 to as large as 1000 cm^3. The advantages of the batch reactor system is that the structure is simple, no high-pressure pump transport system is necessary, and can be used for almost all biomass and waste biomass feedstock gasification (Guo et al. 2007). However, there is a finite time required for heat-up and cooldown of the reactor system including the feedstock and reaction products. For biomass gasification in SCW, there are a significant amount of reactions occurring during the heat-up stages of the experiments in the batch reactor. Feedstock transformation appears to become significant before reaching 250°C although little gas formation occurs at these lower temperatures.

For the case of tubular reactors (Bermejo et al. 2005; Brunner 2009; Marrone and Hong 2009), primarily simple tubular reactors consisting of preheat, reaction, and cooldown sections are employed. Although these reactors are prone to corrosion and plugging problems, they are easily accessible, cheap, and practical for lab-scale units, taking into account the tremendous cost of other types of reactors because of the special requirements for the SCW process, that is, high pressure and temperature. Because of the nature of SCW environment, not only high-pressure (mechanical load) and high-temperature (thermal load) conditions but also the very corrosive aqueous environment of the input reactor materials. In addition to reactor plugging, the corrosion of the SCW reactor and components is the other main technical problem pertaining to SCW systems (Marrone et al. 2004; Schubert et al. 2009). As previously mentioned, heteroatoms are converted to the corresponding acids, and the reactive ions lead to corrosion. It must be stressed that chloride is the most important corrosive species in SCWO processes. Furthermore, the pH of the solutions is frequently very low after oxidation. Therefore, the reactor design also needs to account for material fatigue resulting from temperature cycles, loads exerted by thermal shock, and the weakening of material through corrosion.

6.4.2 Corrosion in SCW

The SCW process operates at a high temperature of 400°C–800°C and high pressure of 24–40 MPa. Thus, the SCW process must be able to sustain the resulting mechanical and thermal loads. Moreover, the process must be able to withstand the high-corrosion environment. Several studies pertaining to corrosion in SCW have been reported in the literature. Kritzer and Dinjus (2001) and Kritzer et al. (1997) investigated the corrosion behavior of several Ni-based alloys and stainless steels in oxygen- and chloride-containing aqueous solutions. All materials exhibited a similar corrosion pattern of slight intergranular corrosion below approximately 150°C; pitting between approximately 150°C and 300°C; and shallow pitting and penetration of whole surface at near-critical temperatures. The aforementioned authors classified the types of corrosion that occur in the SCW into four types as follows.

6.4.2.1 Pitting Corrosion

Pitting corrosion is defined as a localized form of corrosion that occurs in the passive state of the metal. This type of corrosion is caused mainly by penetration and local destruction of some aggressive anions to the previously formed metal oxide film as a result of initial oxidation. The pits start small and increase as the oxidation and dissolution of metal components and their following reaction with water proceed. This eventually leads to a strong acidification environment inside the pits (Kritzer and Dinjus 2001). Thus, the corrosion becomes more aggressive especially after the oxide film weakens as the temperature increases.

6.4.2.2 General Corrosion

Taking advantage of the weakness of the metal oxide film, general corrosion attacks the entire surface although its morphology is typically shallow. It is common in high-temperature oxidation environments. General corrosion usually occurs when the alloy is unable to form a protective layer. Bogaerts and Bettendorf (1986) reported that some forms of pitting corrosion transformed to general corrosion at certain temperatures of around 250°C, and this usually occurs with stainless steel alloys.

6.4.2.3 Stress Corrosion Cracking

The stochastic occurrence behavior of stress corrosion cracking makes it the most dangerous form of corrosion as the alloy failure is also stochastic and not predictable. It usually occurs either in the transition ranges between the active such as hydrogen and the passive such as oxygen, or the passive and the transpassive potentials, respectively. Careful and professional design of the SCW process is a must because even low values of stress corrosion cracking can lead to the failure or leakage of the SCW reactor that can cause fires or even explosions by the presence of hydrogen and high temperatures.

6.4.3 Materials of Construction

In the light of the high corrosion potential in SCW, it is obvious that the components and material of construction of the SCW process must be carefully selected. This selection is based on the ability of the alloy to resist corrosion as well as withstand the mechanical and thermal loads of the process. As reported by Kritzer (2004), the corrosion of alkaline solutions is improved by nickel whereas chromium improves the resistance against acidic and oxidizing media and reduces pitting corrosion. Moreover, molybdenum (Kritzer 2004; Marrone and Hong 2009) causes the lowest passivating effect compared to other elements and leads to high corrosion rates as the conventional high chromium-, nickel-, and iron-based alloys tend to lose their protective layers in highly oxidizing acidic solutions at moderately high temperatures. Thus, an alloy that forms a protective layer or oxide film that covers the entire alloy is preferable than other types of alloys. Some elements such as zirconium, titanium, and yttrium form stable oxides in high-temperature oxidizing environments. As suggested by Kritzer (2004), this could be a potential for future research on some alloys based on the previously mentioned elements.

Mitton et al. (2000) investigated the behavior of different alloys by exposing these alloys to a highly chlorinated oxygenated organic feed stream at a temperature of 600°C for different times to a maximum of 65 h. The aforementioned authors reported that the G-30 alloy exhibited the most corrosion resistance whereas the Ni-based alloys C-22, C-276, 625 showed comparable corrosion rates, with the 316 L stainless steel corroding rate 10 times faster than G-30. In conclusion, although there is still no optimum alloy for use in the SCW process, Inconel 625 and Hastelloy C-276 have been recommended in the literature as the most corrosion-resistant alloys available so far (Brunner 2009).

6.5 Applications of SCW for Hydrogen Production

Considerable attention has been focused on the development of alternative energy sources since the energy crisis in the 1970s because the worldwide energy demand has been increasing exponentially. Also, the reserve of fossil fuels has been dwindling, not to mention the negative effects of the fossil fuels on the environment because of the carbon dioxide (CO_2) emission. Hydrogen (H_2) has been gaining more attention as a promising alternative energy form. It is considered as an alternative fuel, and its use is gaining more acceptance as the environmental impacts of hydrocarbon fuels become more evident. Hydrogen is the lightest element, and it is the most abundant element in the universe although it is found in free state in trace amounts. Hydrogen has a high energy yield of 122 MJ/kg, which is 2.75 times greater than an average hydrocarbon fuel. Hydrogen is the cleanest fuel since its combustion produces only water. Nevertheless, hydrogen's viability as a clean fuel is greatly enhanced if it is produced from renewable sources. On the other hand, and in the past two decades, wastewater treatment has gained significant interest due to water shortages (Rulkens 2007). As a result of expanded wastewater treatment, the amount of sewage sludge has also increased in accordance with this development. Compost produced from sewage sludge is a small percent of the total amount produced. Consequently, the valuable energy contained in organic sludge is lost without utilization in the energy cycle since the main use of the obtained product of composting is agricultural use—fertilization of soils and planting of trees and shrubs (Kosobucki et al. 2000). Sewage sludge has a potential to produce hydrogen-rich gaseous fuel. Thus, hydrogen production from sewage sludge may be a solution for cleaner fuel as well as sewage sludge disposal problems. Sewage sludge typically consists of 41% protein by weight, 25% lipid, 14% carbohydrate, and the rest is ash and biodegradable and recalcitrant organic compounds, as well as pathogens and heavy metals (Goto et al. 1999a,b). It is considered as a waste biomass that has a chemical energy content of 9–29 MJ/kg of total suspended solids that can be potentially recovered by various biological or thermochemical processes (Ptasinski et al. 2002). Sludge has a very high water content of greater than 90% on a wet mass basis, which makes it more suitable for the SCWO process than other conventional thermochemical treatment processes. The latter require high energy input for drying the sludge to make them suitable for processes such as combustion and pyrolysis (Furness et al. 2000). Therefore, supercritical water gasification (SCWG), which primarily involves catalytic conversion in water without oxidation, eliminates the need for drying processes since water participates in the steam reforming and water gas shift reactions during the SCWG process. In addition, the organic compounds in sewage sludge are soluble in the SCW, which makes it easier to gasify them into useful gases such as hydrogen and methane (Demirbas 2004; Osada et al. 2006; Guo et al. 2007; Yanik et al. 2007). Due to the complexity of SCW, previous research activities have focused primarily on compounds that model sewage sludge. Employing SCW to produce green energy from waste streams such as sewage sludge has many advantages. Indeed, using SCW as a reaction medium avoids the expensive step of drying. In fact, estimated feedstocks of 30% or higher moisture content are preferable and more economical in SCW (Savage 2009). Yu et al. (1993), Xu et al. (1996), and Antal et al. (2000) research group at the University of Hawaii was the first to publish an extensive work on catalyzed hydrogen production from biomass and sewage sludge. The aforementioned authors realized and pointed out the advantages of SCW compared to conventional methods such as pyrolysis. This is due to SCW supporting ionic chemistry, which could be much more selective and more controllable than the free radical chemistry that occurs in pyrolysis.

6.5.1 Catalytic SCW Gasification

The high temperatures required for some recalcitrant compounds, such as ammonia besides the formation of undesired products in the uncatalyzed SCWO, free-radical reactions have increased the interest in employing catalysts in order to increase the selectivity to complete oxidation products as well as to decrease the temperature required (Ding et al. 1996; Savage 2009). Homogeneous and heterogeneous catalyses in SCW have been systematically examined for few years using various catalysts. Table 6.2 reports a brief review of the investigated catalysts reported in the literature. It is clear that catalysts enhance the efficiency of SCW, that is, substantially higher conversions and higher selectivities to H_2 and CH_4 at lower reaction temperatures, and selectivity toward CO (Yu et al. 1993; Xu et al. 1996; Antal et al. 2000). Catalysts also increase the economic viability of the biomass gasification process (Savage 2009). In SCW, the catalyst selection is extremely important.

TABLE 6.2
SCW Literature-Reported Catalysts

Compound	Catalysts	Main Product Gas	Reference
Ammonia	MnO_2	N/A	Webley and Tester (1988); Ding et al. (1996)
Benzene	V_2O_5, Cr_2O_3, MnO_2	N/A	Ding and Abraham (1995)
MEK	Pt/TiO_2	N/A	Frisch et al. (1994)
Acetic acid	CuO/ZnO, TiO_2, MnO_2	N/A	Frisch et al. (1994)
Alcohols	CuO/ZnO	N/A	Krajnc and Levec (1994)
Pyridine	Pt/TiO_2	N/A	Frisch et al. (1994)
Quinoline	$ZnCl_2$	N/A	
Glucose	Ni/AC	Hydrogen	
Glucose and glycerol	Ru/TiO_2	Hydrogen and methane	
Glucose	ZrO_2	Methane	Watanabe et al. (2003)
Glucose	KOH and Na_2CO_3	Hydrogen and methane	Guo et al. (2007)
Glucose and glycerol	Ru/Al_2O_3	Hydrogen	Byrd et al. (2007, 2008)
Catechol and vanillin	KOH and K_2CO_3	Hydrogen	Schmieder et al. (2000)
Catechol	KOH	Hydrogen	Kruse et al. (2000)
Cellulose, xylan, and lignin mixture	Engelhard 5132P nickel catalyst	Hydrogen and methane	Yoshida et al. (2003)
Lignin	NaOH and ZrO_2	Hydrogen	Watanabe et al. (2003)
Cellulose	K_2CO_3 and $Ca(OH)_2$	Hydrogen and methane	Guan et al. (2008)
Cellulose	Ru/C, Pd/C, CeO_2 particles, nano-CeO_2 and nano-$(CeZr)x\ O_2$	Hydrogen	Guo et al. (2007)
Phenol and glycerol	Nickel wire	Hydrogen	DiLeo et al. (2007)
Phenol	Activated carbon (AC)	N/A	Nunouraa et al. (2002)
Cellulose	Reduced nickel (Ni)	Hydrogen	Minowa et al. (1998)
Glucose	Ni, K_2CO_3	Hydrogen	
Cellulose, softwood, and grass lignin	Ni	Hydrogen	
Organic wastewater	Ni/carbon	Hydrogen and methane	Osada et al. (2006)
Glycerol, glucose, cellobiose, bagasse, and sewage sludge	Charcoal activated carbon	Hydrogen	Xu et al. (1996)
Organosolv-lignin	ZrO_2	Hydrogen	
Vanillin, glycine, straw, sewage, sludge, lignin, pyrocatechol	K_2CO_3 and KOH	Hydrogen	Kruse et al. (2000)
Glycine and glycerol	Na_2CO_3	Hydrogen and methane	
Pyridine	$Pt/\gamma\text{-}Al_2O_3$, $MnO_2/\gamma\text{-}Al_2O_3$, and MnO_2/CeO_2	N/A	Aki et al. (1996)
N/A	Cr_2O_3	N/A	Aki et al. (1996)

The need to produce a tar-free product gas from the gasification of biomass, the removal of tars, and the reduction of the methane has been the main focus of several literature studies. Dunn et al. (2002) examined the effects of reaction time, water density, and the concentrations of catalyst, *p*-xylene, and oxidant on terephthalic acid synthesis in SCW at 380°C using simple tubing-bomb mini batch reactors. Several catalysts have been identified for complete oxidation in SCW. These include heteropolyacids, alkali carbonates, and carbons. However, the criteria for the catalyst are fundamentally the same and may be summarized as follows:

- The catalysts must be effective in the removal of tars.
- The catalysts should be resistant to deactivation as a result of carbon fouling and sintering.
- The catalysts should be easily regenerated, strong, and inexpensive.
- Higher hydrogen yield is a highly desirable criterion in the case of hydrogen production using SCW.

Therefore, the SCW catalytic process is an optimized combination of catalysts (components, manufacturing process, and morphology); reactants; reaction environment; process parameters such as temperature, pressure, and residence time; and reactor configuration. Ding et al. (1996) reported an extensive review of SCW catalysts and kinetics. The aforementioned authors evaluated several catalysts employed for SCWO of selected model compounds and highlighted the concern of the catalyst stability, activity, and mechanical structure within the SCW environment. In their conclusion, Ding et al. (1996) reported that catalysts improve the oxidation rates of both organic compounds and refractory intermediates; however, the authors pointed out that the harsh SCW environment and variability of sludge characteristics possess a technical challenge for SCWO process development.

Several researches demonstrated various catalytic SCWO studies aiming to investigate the effect of the SCW on the catalyst surface, activity, stability, and selectivity. As an example, deactivation of Pt/ZrO_2 and Pt/TiO_2 occurred in a short time (Frisch et al. 1994), partially due to the crystalline growth of platinum particles. Baker et al. (1989) and Elliott et al. (1993) reported the softening and swelling of the Ni/Al_2O_3 catalyst in SCW due to physical strength of the catalyst. Fierro and de la Banda (1986) reported that lattice oxygen may come from vanadium oxide or from oxides of Mn, Fe, and Ni. Thus, oxygen may participate in the SCW reactions as it is adsorbed on the catalyst surface in the case of SCWO and as part of the lattice oxygen present in the metal oxides. Tiltscher and Hofman (1987) and Wu et al. (1991) reported that the mechanism involved in catalytic SCWO may be interpreted similarly to those describing gas-phase or liquid-phase oxidation. However, special factors, such as high concentrations of water and large differences in physicochemical properties between ambient water and SCW, can influence the catalytic oxidation pathways, the product distributions, and the catalyst stability.

An extensive review of the catalysts reported in the literature for SCW revealed that most of the catalysts were employed in WAO processes. It should be pointed out that the typical (WAO) operating pressures range from 20 to 200 bar and temperatures range from 200°C to 330°C, respectively (Mishra et al. 1995). Adschiri et al. (1991), Baptist-Nguyen and Subramaniam (1992), Ding and Abraham (1995), and Frisch et al. (1994) reported that the deactivation of a catalyst is mainly caused by coking, poisoning, and the solid-state transformations of catalysts in the gas-phase oxidation case. However, the aforementioned authors pointed out that the advantage of catalytic reaction in SCW is the prevention of coke formation on catalyst surface. In order to clarify and clearly understand the role of the catalyst in SCW, a simplified approach of tabulating the catalysts that have been employed in the literature is presented in Table 6.2.

6.5.1.1 Model Compounds Gasification in SCW

6.5.1.1.1 Glucose

Since sewage sludge typically consists of 41% protein, 25% lipid, 14% carbohydrate, with the rest ash, biodegradable, and recalcitrant organic compounds, a better understanding of sewage sludge behavior in SCW requires both extensive experimental and theoretical studies. This has motivated several research groups to study model compounds for both biomass and sewage sludges. As an example, glucose gasification in SCW can be considered as a good model for gasification of more complex cellulosic sludges in SCW. Glucose is a refractory intermediate formed during the gasification of biomass for the SCWG process. Amin et al. (1975) obtained hydrogen-rich gas from the catalytic gasification of glucose in water at 374°C and 22.1 MPa, mainly through water gas shift reaction with a low efficiency (20%) of carbon gasification. This means that as much as 20% of the feed carbon was detected in the gaseous decomposition products, which demonstrates the need for higher temperature and active catalysts to improve the gasification efficiency.

As mentioned before, Yu et al. (1993) were the first to investigate hydrogen production from biomass and waste streams using SCW. Yu et al. (1993), who gasified different glucose concentrations in SCW reported that 0.1 M glucose was completely converted to hydrogen-rich synthesis gas whereas higher glucose concentrations were not completely gasified at 600°C, 34.5 MPa, and 34 s reactor residence time. This could be attributed to the reduction in the amount of water present since the reaction is taking place in SCW medium. The gas yield and composition were found to depend on the condition of the reactor wall and the reactant concentration. The observed results motivated the same group in a subsequent study (Xu et al. 1996) to employ activated carbon (AC) as a catalyst to improve glucose gasification efficiency in SCW. Carbon is interesting because it is very stable in SCW, especially when hydrogen gas is present. At a glucose concentration of 0.1 M, the authors reported that the gasification efficiency reached 98% at 600°C and 34 MPa, and decreased sharply to 51% at 500°C while maintaining the same pressure.

Byrd et al. (2007) gasified glucose in SCW in the presence of Ru/Al_2O_3 catalyst. Although the aforementioned authors reported yields as high as 12 mol H_2/mol glucose, they employed relatively high temperatures of 700°C and 800°C. Thus, for enhancing H_2 yield and selectivity from glucose, the desired catalytic property and the optimum conditions must be determined. Wang et al. (2000) reported that nickel catalysts are expected to crack tar and promote water gas shift reaction, methanation, and hydrogenation reaction. From economic and energy efficiency points of view, high gasification efficiency at low temperature with more hydrogen production is favorable.

Elliott et al. (1993) from Pacific Northwest Laboratory conducted crucial studies on the reaction chemistry in a high-pressure aqueous environment. They used nickel and ruthenium catalysts for organic waste gasification in sub- and supercritical water in a batch reactor for 2 h running time, at 350°C and 20 MPa. The authors showed that aromatic and aliphatic hydrocarbons can be transformed into hydrogen- and methane-rich gases, thanks to hydrogenation catalysts. The authors also confirmed these results in a continuous reactor with a residence time of less than 10 s. In order to enhance H_2 production yield, metallic catalysts should be examined. Nickel, by virtue of its high melting point of 1453°C, its availability, and low cost, is a suitable and reasonable choice for SCWG and SCWO. Generally, glucose, as a model constituent of waste biomass, was used in SCW gasification studies. Table 6.3 summarizes some of the literature-reported results for SCW gasification and oxidation of glucose in SCW.

TABLE 6.3

Some Reported Literature Studies of Glucose Gasification in SCW

Feed	Reactor Type and Material Operating Parameters and Catalyst	Gas Yield (mol/mol)							Observations	Reference
Glucose	Continuous-flow system (Hastelloy C-276 tubing reactor having 9.53 mm OD, 6.22 mm ID, and 670 mm functional length. T = 575°C–725°C, P = 28 MPa, LHSV = 6–24 h^{-1}, 16 wt% Ni/AC and AC, Conc. 0.3, 0.6, and 0.9 M.	Catalyst	Ni/AC			AC			1. The 16% wt Ni/AC catalyst gave almost complete gasification efficiency. 2. The H_2 yield increased with increasing temperature from 575°C to 725°C. 3. The hydrogen yield obtained with the 16% wt Ni/AC was about two times higher than that with the AC only or without.	
		Conc. (M)	0.3	0.6	0.9	0.3	0.6	0.9		
		H_2	1.69	0.88	0.69	2.82	2.45	2		
		CO	0.86	1.2	1.26	0.19	0.29	0.42		
		CO_2	2.4	1.94	1.39	3.48	3.24	2.57		
		CH_4	0.99	0.67	0.51	1.17	1.11	0.96		
		COD%	91	81	81	96	93	93		
Glucose	Continuous-flow system (Hastelloy C-276 tubing reactor having 9.53 mm OD, 6.22 mm ID, and 670 mm functional length. T = 480°C–750°C, P = 28 MPa, τ (residence time) = 10–50 s, No catalyst, Conc. 0.6 M.	Temp (°C)	480	600	600	750			1. The hydrogen yield increased sharply by increasing temperature over 660°C. 2. Pseudo-first-order kinetics was obtained for glucose and COD degradations by assuming a plug flow, and it is discussed in the appended paragraph.	
		Flow rate (g/h)	240	120	360	240				
		τ (s)	35	50	16	19				
		H_2	0.08	2.63	0.52	4.78				
		CO	0.47	0.59	1.3	0.27				
		CO_2	0.4	1.72	0.32	3.52				
		CH_4	0.03	0.71	0.21	1.26				
		Liquid effluent characterization								
		Glucose	82.1	99.9	91.5	100				
		COD	38.6	86.7	62.7	99.8				
Glucose	Batch SS 316 stainless steel tube bomb reactor V = 6 cm^3. T = 400°C–440°C, P = 30–35 MPa, τ (residence time) = 10–15 min, ZrO_2 and NaOH catalysts, Conc. 0.5 M.	T = 400 C, τ = 15 min	No catalyst	ZrO_2	NaOH				1. The gasification efficiency with NaOH was the highest. 2. The gasification efficiency did not change with changing residence time. 3. Carbon balance not reported because of lack of analysis for liquid compounds.	Watanabe et al. (2003)
		Yield (mol%)	2	5	21					
		T = 440 C, τ = 10 min	2	5	25					

Several observations can be made from Table 6.3. First, the effect of catalysts on the production of hydrogen gas was clearly pronounced. These results point to the importance of identifying effective catalysts for the production of a hydrogen-rich synthesis gas from waste biomass. Precious metals and AC appear to have a strong influence on the hydrogen production yield even in the cases where glucose concentration in the feed was 17% wt. Second, the operating pressure has no influence on the hydrogen yield whereas the increase in residence time led to only slight increases in the hydrogen yield and the gasification of the glucose. Third, the oxidant MR plays an important role in increasing the hydrogen yield. Although none of the earlier studies reported the optimum MR, it is clearly understood that the hydrogen yield increases with increasing MR to an optimum point expected to be less than unity beyond which hydrogen yield starts to decrease. Fourth, the effect of temperature on the hydrogen yield is clearly reported. As the temperature increases, the hydrogen yield increases, mainly because it promotes the water gas shift reaction in which carbon monoxide reacts with steam to produce hydrogen and carbon dioxide. Furthermore, the increase in temperature increases the degradation of the intermediate compounds, which leads to an increase in the gas yield.

6.5.1.1.2 Proteins

Several reported studies in the literature have dealt with the gasification of lipids, protein, carbohydrates, and lignin as waste biomass model compounds. Thus, the following section pertains to some of these reported studies. Since cellulose and hemicellulose are carbohydrates and lignin includes aromatic rings, several studies have been reported in the literature with model compounds for these categories.

Schmieder et al. (2000) studied glucose as carbohydrate model compound, catechol and vanillin as a lignin model compounds, and glycine as a protein model compound. The experiments were conducted in two batch reactors and continuous-flow reactor. The first batch reactor was an autoclave, designed for a temperature of 700°C and pressure of 1000 bar with a volume of 100 mL whereas the second batch autoclave, designed for 500°C and P of 500 bar, was made from Inconel 625 with volume of 1000 mL. The tubular flow reactor was an Inconel 625 with an ID of 8 mm and a length of 15 m. The aforementioned authors employed alkaline KOH as a homogeneous catalyst. The main gaseous products were CO_2, CO, H_2, and CH_4. For catechol, at an operating temperature of 600°C, residence time of 30 s, pressure of 200–300 bar, and feed concentration of 0.2 M, as much as 10.5 mol H_2 per mole of feed that corresponds to 82% of the theoretical hydrogen formation was achieved considering only carbon dioxide and hydrogen as products, that is, traces amounts of CO and CH_4 observed. As a protein model compound, vanillin gasification was easier than catechol. In fact, more than 99% destruction efficiency was reported even without added KOH, which can be attributed to the low feed concentration and the higher reaction time. The authors did not report the H_2 yield; however, they employed thermodynamic calculations to improve the understanding of the reaction pathways occurring during gasification and to compare the experimental results with the theoretical predicted ones. Results of the comparison showed that the temperature and pressure trends found experimentally were confirmed theoretically. However, the CH_4 yield predicted was higher than the experimental yield.

6.5.1.1.3 Lignin

Lignin is one of the three main components of plant biomass, beside cellulose, and hemicellulose (DiLeo et al. 2007) and possesses the most gasification resistant compound. The abundant presence of aromatic rings in lignin structure provides the baseline selection of its representative model compounds. Nunouraa et al. (2002) reported on the catalytic effect of AC on the SCWO of phenol. By applying AC as a heterogeneous catalyst, the decomposition rate of phenol was enhanced, and the yield of carbon dioxide increased largely as well as the yield of dimeric compounds and tarry materials decreased remarkably with the addition of AC.

Similarly, Kruse et al. (2000) studied the gasification of catechol as lignin model compound in SCW for H_2 production. The authors employed a batch autoclave, made from Inconel 625 designed for 500°C and P of 500 bar with a volume of 1000 mL, tubular flow reactor (Inconel 625, ID 8 mm, and OD 15.4 mm, length 500 mm), and KOH and LiOH as catalysts for their experimental investigation. The effects of two feed concentrations of 0.6 and 1.2 mol/L, temperatures of 600°C–700°C, as well as pressure range of 200–400 bar on the gas product distribution were experimentally studied and discussed. Furthermore, the authors concluded that the increase in temperature from 600°C to 700°C did not influence the catechol gasification since more than 99% of catechol was gasified at 600°C. However, the addition of KOH as catalyst increased the relative yields of hydrogen and carbon dioxide and decreased the relative CO yield. For example, increasing the KOH amount from 0 to 5 wt% enhanced the H_2 yield by 40 vol% at a temperature of 500°C, pressure of 250 bar, and a reaction time of 1 h. However, the CO yield decreased from 40 vol% at 0 wt% of KOH to 0.7 vol% with 5 wt% KOH amount. This can be attributed to the formation of formates by the addition of alkali salts and LiOH, which subsequently degraded to hydrogen and carbon dioxide.

Yoshida et al. (2003) studied cellulose, xylan, and lignin mixture gasification in SCW. They conducted their experiments in a batch reactor, made from SS 316 with an OD of 9.53 mm and an ID of 6.53 mm with commercial Engelhard 5132P nickel as a catalyst at a temperature of 400°C and pressure of 25 MPa. At a cellulose:xylan:lignin mixture of 1:1:4 on gas formation, the experimental results confirmed the pronounced negative influence of lignin relative to other two compounds, that is, cellulose and xylan. This was observed by a decrease of 31% in the H_2 yield to a 1.7 mol/g of reactant. However, the use of equal amounts of cellulose:xylan:lignin mixture of 1:1:1 increased the H_2 yield by 50% to about 2.5 mol/g of reactant. The highest amount of gas of 17 mol of H_2, CO_2, CH_4, and C_2H_6 per gram of reactant was observed in the case at a cellulose:xylan:lignin mixture of 4:1:1.

Savage (2009) reported a study of noncatalytic gasification of lignin in SCW. The experiments were conducted in a mini batch quartz capillary tube with dimensions of 2 mm ID, 6 mm OD, 18.4 cm length, and a volume of 0.58 mL. The operating conditions were residence times from 2.5 to 75 min, temperature range of 365°C–725°C, and a pressure of 31 MPa with three feed concentrations of 5.0, 9.0, and 33 wt%. The aforementioned authors compared their results with cellulose gasification in terms of the product gas yield. As a catalyst, a nickel wire was inserted inside the reactor prior to each experiment. In the absence of catalysts, CH_4 and CO_2 were always the major products from SCWG of lignin. However, cellulose formed 1.8 mmol/g of H_2 after 10 min whereas lignin formed only 0.7 mmol/g. The authors also reported that higher temperatures increased the rate of formation of H_2, CO_2, and CH_4 and the rate of consumption of CO, and the biomass loading was an important parameter to control the CH_4/H_2 ratio for SCWG of both lignin and cellulose.

Watanabe et al. (2003) reported on the gasification of lignin in the presence of NaOH as a homogeneous catalyst and ZrO_2 as a heterogeneous catalyst. A batch 316-SUS stainless steel tube bomb reactor with a volume of 6 cm^3 was employed in all experiments. A temperature of 400°C, pressure of 40 MPa, feed concentration of 0.5 M, and a residence time range of 15–60 min were the operating conditions. The authors reported that without catalyst, the yield of H_2 and CO_2 slightly increased with increasing reaction time, which indicated that the water gas shift reaction proceeded gradually even without catalyst. However, in the presence of ZrO_2 as catalyst, the yield of H_2 and CO_2 increased with increasing reaction time, and a H_2 yield of 4 mol% twice of that without catalyst was observed. On the other hand, adding NaOH significantly enhanced the formation of H_2 and CO_2, whereas the CO yield with NaOH was remarkably low at all the reaction times, which confirms that the addition of alkali promoted the water gas shift reaction. The authors pointed out that the use of base catalysts (ZrO_2 and NaOH) coupled with an oxygen-to-carbon ratio (O/C) of 1.0 enhanced the decomposition rates of aldehydes and ketones, which are assumed to be intermediates of lignin decomposition and eventually increased the H_2 yield.

6.5.1.1.4 Cellulose

Generally, waste biomass such as sludges is known to have a high fiber content, most of which is cellulose. Thus, cellulose can be considered as a model compound for biomass as well as sludges. Minowa et al. (1998) investigated cellulose gasification in near-critical water at 350°C and 16.5 MPa with a reduced nickel catalyst and reported that 70% of the carbon could be gasified. Osada et al. (2004) reported the formation of a high amount of methane in the presence of ruthenium catalyst during the gasification of cellulose and lignin in SCW. Guan et al. (2008) investigated the H_2 production from cellulose as a biomass model compound. The authors conducted their experiments in a batch autoclave reactor made of 316 SS with a capacity of 140 mL designed to withstand the reaction temperature of 650°C and pressure of 35 MPa. The operating conditions at which experiments were performed were a 20 min residence time, temperature of 450°C and 500°C, and pressure of 24–26 MPa with K_2CO_3 and $Ca(OH)_2$ selected as the catalysts. The H_2 and CH_4 yields increased by 70% and 40% as the temperature was raised from 500°C to 550°C and the pressure was kept at 26 MPa, 6.3 mol/kg H_2, and 4.1 mol/kg CH_4. As temperature increases, CH_4 tends to react with water to form H_2 and CO_2, and the net production of CH_4 decreases. However, the amount of CO decreased dramatically by increasing the amount of K_2CO_3 whereas those of H_2, CH_4, and CO_2 increased at the same time. This is because K_2CO_3 enhanced the water gas shift reaction, which resulted in higher production of both H_2 and CO_2. A comparison between the catalyst amounts showed that when no catalyst was added, the H_2 yield was 4.4 mol/kg cellulose, but increased to 9.4 mol/kg cellulose and 8.3 mol/kg cellulose with 0.2 g K_2CO_3 and 1.6 g $Ca(OH)_2$. However, when K_2CO_3 and $Ca(OH)_2$ were present simultaneously, the H_2 yield was 11.958 mol/kg cellulose, which is 2.5 times that without catalyst, and 25% and 45% more than that when K_2CO_3 or $Ca(OH)_2$ were present alone.

In another cellulose study conducted by Hao et al. (2005), a batch autoclave reactor made of 316 SS with a capacity of 140 mL was designed to withstand the reaction temperature of 650°C and pressure of 35 MPa. The operating conditions at which experiments were performed were a 20 min residence time, a temperature of 450°C and 500°C, and a pressure of 24–26 MPa. The authors selected Ru/C, Pd/C, CeO_2 particles, nano-CeO_2, and nano-(CeZr) x O_2 as a different set of catalysts. The authors compared the partial oxidation technique with gasification in terms of H_2 production. The H_2 yield in gasification experiments was 4.4 mmol/g of reactant, which corresponds to 40% higher than the partial oxidation experiments. However, a CO yield of 2.8 mmol/g of reactant in gasification experiments was observed, which corresponds to 25% of the CO yield produced during partial oxidation experiments. This implies that catalytic gasification promotes the water gas shift reaction, which consumes CO and produces H_2. Among all the employed catalysts, Ru/C was the most active catalyst in terms of cellulose gasification, which was nearly completely gasified. In general, and from the earlier literature review, the following conclusions can be drawn:

- For catechol, although homogeneous catalysts such as alkali seem to enhance hydrogen yield, corrosion is a major drawback (Kruse et al. 2000; Schmieder et al. 2000). Also, catechol had undergone complete gasification at a temperature of around 600°C and a pressure of 34 MPa. This result underlines the importance of optimizing the SCW operating conditions for maximum hydrogen yield coupled with complete gasification of sewage sludge model compounds as catechol as a model compound represents lignin, which is gasification resistant (DiLeo et al. 2007). Yoshida et al. (2003) pointed out that oxidation of cellulose, xylan, and lignin mixture showed the clear negative effect of lignin on hydrogen yield in the product. This emphasizes the role lignin and other aromatic ring compounds present in some industrial waste streams, on the hydrogen production. Furthermore, Savage (2009) utilized higher temperatures to test lignin gasification in SCW and compared their results with obtained results from cellulose gasification. The aforementioned authors confirmed the role of lignin as a control compound in the gasification process governing the effect of gasification rate. Moreover, aldehydes and ketones are believed to be intermediates of lignin gasification in SCW (Watanabe et al. 2003), thus pointing to the necessity of studying these intermediate compounds extensively by employing different catalysts and other SCW variables. This can lead to a better understanding and improvement of lignin gasification in SCW.
- Vanillin can be gasified completely at a relatively low SCW temperature of 500°C (Schmieder et al. 2000). Also, by employing thermodynamic calculations to predict the product distribution, the predicted methane yield was higher than the experimental yield, which highlights the need for improvement of the predictive capability of these models.
- In the case of cellulose, Guan et al. (2008) reported that the addition of K_2CO_3 doubled the hydrogen yield with no temperature change. The authors also pointed out that by employing both K_2CO_3 and Ca $(OH)_2$ as catalysts, the hydrogen yield increased by about 2 mol/kg. These results raise an interesting point, which is the synergistic effects of multiple catalysts on the gas yield of the SCW process. Furthermore, no studies coupled the gasification with partial oxidation despite extensive independent evaluation of each. Hao et al. (2005) reported partial oxidation and gasification of cellulose and highlighted that gasification using Ru/C, Pd/C, CeO_2 particles, nano-CeO_2, and nano-$(CeZr)_xO_2$ as catalysts gave higher hydrogen yield. However, the aforementioned authors did not report the oxygen stoichiometric ratio at which partial oxidation experiments were conducted. Also, the authors did not report any experimental results regarding coupled gasification with partial oxidation, that is, SCW partial oxidation.

6.5.1.2 Real Waste Gasification in SCW

Glycerol, which is a by-product from the bioethanol and biodiesel production plants, was also investigated for H_2 production in SCW. This study was conducted by Byrd and coworkers at the University of Auburn, Auburn, AL (Byrd et al. 2008). The aforementioned authors performed their experiments using a continuous-flow system with Inconel 600 tubular reactor having dimensions of 0.5 m length, 0.635 cm OD, and 0.304 cm ID. The operating temperature was in the range of 700°C–800°C, pressure 25 MPa, and residence time 1–6 s. The authors employed 5 wt% Ru/Al_2O_3 as a catalyst and feed concentration of 5–40 wt% glycerol. It is well known that the desired overall reaction of glycerol for hydrogen production was given by

$$C_3H_8O_3 + 3H_2O \rightarrow 7H_2 + 3CO_2 \tag{6.9}$$

Thus, the maximum theoretical H_2 yield that can be obtained is 7 mol. The aforementioned authors reported that the highest H_2 yield of 6.5 mol H_2/mol glycerol was obtained at a feed concentration of 5 wt%, a temperature of 800°C, and reactor residence time of 1 s. The authors also pointed out that the shortest residence time gave high hydrogen yield; however at longer residence times, the hydrogen yield dropped sharply with a decline in CO_2 yields as well. Furthermore, the increase in the feed concentration was coupled with a decrease in the yield of hydrogen and an accompanying increase in the methane yield. As the gasification reaction proceeds in the presence of water, less water was present at higher concentrations, rationalizing the drop in the H_2 yield.

With all the previously mentioned attempts to provide a stable, active, and durable catalyst for the catalytic SCW for H_2 production, none of these catalytic processes are in the stage of commercial application. In fact, this task still has major challenges especially overcoming the problem of deactivation by sulfur compounds that are presented in the waste biomass such as sludges. However, the research in this field has also provided new scientific insights into catalytic chemistry in SCW as well as identifying processes that are technically feasible in SCW.

6.5.2 Catalysts Activity and Stability in SCW

Butt and Peterson (1988) and Lambroua et al. (2005) reported that the deactivation of a catalyst is mainly caused by coking, poisoning, and solid-state transformations of the catalyst for gas-phase oxidation. Coking is the result of carbon deposition on active sites, poisoning is the physical or chemical adsorption of impurities on active sites, and solid-state transformations are caused by the phase transition.

Yu and Savage (2001) compared the activity and stability of three different catalysts, that is, bulk MnO_2, bulk TiO_2, and CuO/Al_2O_3, during SCWO of phenol. Catalyst stability and long-term activity are prerequisites for commercial applications of the SCW process.

The experiments were conducted in a tubular flow reactor at a temperature of 400°C. The authors used the BET surface area, x-ray diffraction, and XPS or electron spectroscopy for chemical analysis techniques to characterize both fresh and spent catalysts. The authors highlighted that on a catalyst mass basis, CuO/Al_2O_3 was the most active whereas MnO_2 was the most active on a surface area basis. On the other hand, CuO/Al_2O_3 was deactivated during the first 12 h on stream and maintained its activity at longer times. However, the observation of both Cu and Al in the reactor effluent limited the usefulness of CuO/Al_2O_3 in the SCW.

6.5.3 Catalyst Preparation for SCW

The chemical stability of a catalyst is governed by its physical stability, which depends on the preparation technique. In SCW, catalyst stability, ability to preserve its characteristics

TABLE 6.4

Catalyst Characterization Techniques

Property	Technique
Surface area and pore volume	Brunauer–Emmett–Teller
Phase transformation	Powder x-ray diffraction (XRD)
Catalyst reducibility	Temperature-programmed reduction of H_2
Bulk metal loading	Inductive couple plasma atomic emission spectroscopy
Acidic and basic site density	Temperature-programmed desorption of NH_3 and CO_2
Surface topography	Scanning electron microscopy

such as surface crystal structure, is of utmost importance due to the harsh SCW environment. Catalysts are prepared by several techniques such as impregnation, coprecipitation, fused metal oxide, supercritical deposition, and fused alloy. The catalyst preparation method (precipitation, impregnation, sol–gel, etc.), type of precursor, calcination, and reduction have an impact on the catalyst activity. The impregnation and coprecipitation techniques are the most popular for metal and metal oxides (Aki et al. 1996; Ding et al. 1996).

Kaddouri and Mazzocchia (2004) prepared Co/SiO_2 and $Co/\gamma\text{-}Al_2O_3$ catalysts using a combined incipient wetness and sol–gel method. The aforementioned authors reported that the catalytic performance toward hydrogen production over a Co/SiO_2 catalyst was enhanced by the preparation method.

Lee and Ihm (2009) prepared a 16 wt% Ni/AC catalyst by an incipient wetness method. The authors compared their experiments with prepared catalyst by a control one with only AC. The aforementioned authors reported that AC provided catalytic activity toward methane production at temperatures above 650°C, whereas the Ni/AC catalyst was relatively stable in SCW. A comparison between the fresh and spent catalysts revealed that unlike AC only, the Ni/AC catalyst was deactivated. Characterization of the spent catalyst showed that the nickel particles were larger than the fresh one, which is due to the crystallite growth of the nickel.

Examining the catalyst activity and to identify catalytic properties such as bulk metal loading, total surface area, crystalline phase and oxidation state, and phase transformation, catalyst characterization is usually carried out. The most used techniques in catalyst characterization are reported in Table 6.4.

6.6 Oxidation Kinetics in SCW

The knowledge of reaction kinetics is of great importance since it is used to determine the required reactor volume V and the residence time τ for the desired conversion X. Considering a reaction of a model compound in SCW that contains only carbon, H_2, and oxygen atoms, the general reaction equation is given by

$$\text{Organic compound (A)} + \text{Oxidant} \rightarrow \text{Carbon dioxide} + \text{Water} \quad (6.10)$$

According to most chemical reaction engineering books, the reaction rate equation of component A is given by

$$-r_A = \left(-\frac{1}{V}\right)\left(\frac{dN_A}{dt}\right) = -\left(\frac{dC_A}{dt}\right) = -\frac{d[A]}{dt} \text{ mol/m}^3\,\text{s} \tag{6.11}$$

where
- r_A is the reaction rate
- V is the reactor volume
- N_A is the number of moles of component A
- t is the residence time
- C_A is the concentration
- A is defined as any component

The reaction rate constant temperature dependence is represented by Arrhenius law, $k = k_o \exp(-E/RT)$, where k_o is called the pre-exponential factor, E is the activation energy, and R is the universal gas constant. Thus, the global reaction rate for SCW is expressed as

$$-r_A = -\frac{d[A]}{dt} = k_o \exp\left(-\frac{E}{RT}\right)[A]^a[O_2]^b[H_2O]^c \tag{6.12}$$

where a, b, and c are the reaction orders of the organic compound, oxygen, and water respectively. The aforementioned equation parameters can be obtained by employing multiple linear regression of experimental data after transforming Equation 6.11 into the following linear form:

$$\ln(-r_A) = \ln k_o - \frac{E}{RT} + a\ln[A] + b\ln[O_2] + c\ln[H_2O] \tag{6.13}$$

The approximation of pseudo-first-order kinetics has dominated most kinetic equations derived previously in the literature (Helling and Tester 1987; Webley and Tester 1988; Lee et al. 1990; Shanableh 1990; Wilmanns 1990; Goto et al. 1999a,b). However, predictions from kinetics models obtained below and above the critical point of water are completely different (Li and Gloyna 1991). Furthermore, predictions from kinetic expressions obtained in the same range of operating conditions vary considerably. Therefore, given the complexity of the real waste biomass streams, both batch and continuous-flow reactors should be studied extensively to compare gasification and partial oxidation kinetics at supercritical conditions.

A great number of kinetic studies have been performed in batch reactors, flow-through reactors, quartz tube reactors, and optical accessible cells. According to process conditions (i.e., pressure, temperature, and oxygen excess), oxidation, hydrolysis, or pyrolysis reactions occur. Different classes of organic compounds have been subjected to SCWO conditions, which include hydrocarbons, nitrogen-containing compounds, sulfur-containing compounds, chlorine- and fluorine-containing compounds, and oxygen-containing compounds (Li and Gloyna 1991). Moreover, considering the fact that many wastes are complex organic mixtures whose constituents vary widely in their susceptibility to SCW oxidation, the reaction rate predictions using Equation 6.13 may be misleading especially because of the critical role of the experimental setup, that is, the Reynolds numbers in plug flow reactors are often so small that laminar effects might falsify the rates. Thus, the need for kinetic comparison for both batch and continuous-flow reactors as well as the need for the development of more complex models using advanced numerical techniques seems to be unavoidable.

Li and Gloyna (1991) reported a generalized kinetic model of organic compounds based on the formation and destruction scheme of rate-controlling intermediate compounds. The aforementioned authors assumed that ethanol, methanol, and acetic acid were the main intermediates as they exhibit higher activation energy. From the reported concentrations of the three previously mentioned compounds in the literature (Taylor and Weygandt 1974; Baillod et al. 1982; Conditt and Sievers 1984; Keen and Baillod 1985; Shanableh 1990), methanol and ethanol concentrations were insignificant compared to acetic acid concentration. Thus, the authors assumed that acetic acid is the rate-controlling intermediate compound in the generalized model. Another factor supporting the authors' assumptions is the fact that the pre-exponential factors for methanol and ethanol are considerably higher than that for acetic acid.

Based on the proposed mathematical model reported by Takamatsu et al. (1970), which describes the reaction pathways for thermal decomposition of activated sludges derived, Li and Gloyna (1991) proposed their generalized kinetic model. The generalized model scheme is comprised of triangular pathways, that is, the organic components feed (*A*) reacts with oxygen to give the final product (C) as well as an intermediate product (*B*), which further decomposes to give the final product C. The authors suggested that the concentrations of the feed (*A*) and intermediate product (*B*) are expressed in forms of either TOC or chemical oxygen demand.

Goto et al. (1999a,b) studied the kinetics of destruction of municipal sewage sludge and alcohol distillery wastewater of molasses in SCW. A batch reactor made of stainless steel tube sealed with Swagelok caps (about 4 mL in volume) was employed to perform the experimental work. Hydrogen peroxide (about 30%) was used as an oxidant. Operating conditions of temperatures 400°C, 450°C, and 500°C; estimated reaction pressure of 30.0 MPa. The amount of hydrogen peroxide used was 300% of the stoichiometric demand, and the water content of municipal sludge and alcohol distillery wastewater was 96% and 66% respectively. Since oxygen and water are present in excess, and because of the existence of O_2 as one phase in SCW, the reaction rate becomes independent of the oxygen concentration. Thus, the authors assumed that the constants b and c in Equation 6.12 are equal to zero, and a first-order reaction with respect to organics only was adopted as it was extensively reported in the literature.

The reported rate constants by the aforementioned authors were compared with the data reported by Foussard et al. (1989) for biological sludge and Shanableh (1990) for activated sludge. Furthermore, the activation energy for the sewage sludge decomposition of 76.3 kJ/mol was somewhat comparable to 67 and 54 kJ/mol obtained by the aforementioned authors, respectively.

Goto et al. (1999a,b) reported another kinetic study pertaining to ammonia decomposition in SCW of sewage sludge using the same experimental technique used in studying the kinetics of destruction of municipal excess sewage sludge and alcohol distillery wastewater of molasses in SCW. However, the amount of hydrogen peroxide (H_2O_2) was 200% of the stoichiometric requirements. The authors adopted the reaction pathway proposed by Li et al. (1993), which is the modified reaction pathway reported by Li and Gloyna (1991). The modified pathway was based on the addition of ammonia as a secondary refractive intermediate. The rate constants for ammonia decomposition determined in the study by Goto et al. (1999a,b) were compared with those reported by Webley and Tester (1988) for ammonia decomposition in SCW when ammonium hydroxide was used as a reactant and oxygen was used as an oxidant although the temperature range in the work reported by Webley and Tester (1988) was higher than those reported by Goto et al. (1999a,b). The activation energy of 139 kJ/mol reported by Webley and Tester (1988) was comparable with the activation energy of 157 kJ/mol evaluated in Goto et al.'s (1999a,b) work. Moreover, the data reported by Webley and Tester (1988) were for the decomposition rate of ammonia, whereas the data reported by Goto et al. (1999a,b) were for the decomposition rate of ammonia that was formed as an intermediate product during the sewage sludge oxidation in SCW.

6.7 Challenges of H_2 Production from Wastes and Potential Future Research

Hydrogen production from waste biomass using SCWG is a promising technology. In fact, the reported research publications pertaining to this research area have been increasing steadily in the past 10 years as the number of publications surpassed 50 papers per year in the last 3 years compared to a few per year in late 1980s (Savage 2009). However, this technology faces a number of challenges that require more research. The challenges can be summarized in the following points:

- Corrosion, as mentioned earlier, is the most difficult challenge to solve for using SCW for hydrogen production from waste biomass (sludge, manure, and industrial wastewater) due to the harsh SCW environment and the corrosive nature of inorganics contained in certain types of waste biomass. Therefore, SCW unit components have to be from expensive corrosion-resistant alloy such as Hastelloy or titanium. The application and use of the previously mentioned alloys should be limited to the components that are more prone to corrosion such as the reactor and heat exchanger. However, the use of such alloys increased the capital cost, which in turn affects the overall process economic viability. Another possible solution for the corrosion problem is to utilize more research efforts for discovering new types of alloys that exhibit a better corrosion resistance in the SCW environment. Furthermore, exploring the use of coating materials that can withstand the corrosion is of most interest, such materials if provided could add a valuable incentive to SCW technology and make it more competitive. In other words, the ability of SCW for gasification of organic compounds as well as support the nanoparticles can be utilized to limit the corrosion problem and provide the required catalyst for gasification. This was clearly demonstrated by the reported work by Gadhe and Gupta (2007) that generated catalytic Cu nanoparticles hydrothermally in situ and used the particles to catalyze methanol SCWG in the reactor.
- The second challenge is pumping, as it is well known that SCW requires high pressures above the water critical pressure. To obtain such high pressures as well a successful pumping of waste biomass that contains a considerable amount of biosolids, a special type of pumps has to be designed and commercially available especially for lab-scale unit since it was a major challenge to find a suitable pump operating on the laboratory scale of few milliliters per hour and can handle the solids (Youssef et al. 2010).
- Catalyst, which is the heart of the SCW process for hydrogen production, and its support need to be stable and should exhibit a reasonable activity and selectivity toward hydrogen generation. Catalyst also needs to be regenerable, which corresponds to a significant impact on the capital cost of SCW plant.

References

Abdelmoez, W., Nakahasi, T., and Yoshida, H. 2007. Amino acid transformation and decomposition in saturated subcritical water conditions. *Ind. Eng. Chem. Res.* 46:5286–5294.

Adschiri, T., Suzuki, T., and Arai, K. 1991. Catalytic reforming of coal tar pitch in supercritical fluid. *Fuel* 70:1483–1484.

Aki, S.N.V.K., Ding, Z.Y., and Abraham, M.A. 1996. Catalytic supercritical water oxidation: Stability of Cr_2O_3 catalyst. *AICHE J.* 42:1995–2004.

Amin, S., Reid, R.C., and Modell, M. 1975. Reforming and decomposition of glucose in an aqueous phase. *The Intersociety Conference on Environmental Systems*, ASME Paper 75-ENAS-21, July 21–24, San Francisco, CA.

Anderko, A., Wang, P., and Rafal, M. 2002. Electrolyte solutions: From thermodynamic and transport property models to the simulation of industrial processes. *Fluid Phase Equilib.* 197:123–142.

Antal, Jr., J.J. 1982. Biomass pyrolysis: A review of the literature. Part 1—Carbohydrate pyrolysis. In: *Advances in Solar Energy*, K.W. Boer and J.A. Duffie, eds. Publisher is American Solar Energy Society, Boulder, CO, vol. 1, pp. 61–111.

Antal, Jr., M.J., Allen, S.G., Schulman, D., Xu, X., and Divilio, R.J. 2000. Biomass gasification in supercritical water. *Ind. Eng. Chem. Res.* 39:4040–4053.

Antal, Jr., M.J., and Mok, W.S. L., and Richards, G.N. 1990. Mechanism of formation of 5-(hydroxymethyl)-2-furaldehyde from D-fructose and sucrose. *Carbohydr. Res.* 199(1):91–109.

Baillod, C.R., Faith, B.M., and Masi, O. 1982. Fate of specific pollutants during wet oxidation and ozonation. *Environ. Prog.* 1(3):217–227.

Baker, E.G., Sealock, Jr., L.J., Butner, R.S., Elliott, D.C., and Neuenschwander, G.G. 1989. Catalytic destruction of hazardous organics in aqueous wastes: Continuous reactor system experiments. *Hazard. Waste Hazard. Mater.* 6:1.

Baptist-Nguyen, S. and Subramaniam, B. 1992. Coking and activity of porous catalysts in supercritical reaction medium. *AICHE J.* 38:1027–1037.

Bermejo, M.D., Fdez-Polanco, F., and Cocero, M.J. 2005. Transpiring wall reactor for the supercritical water oxidation process: Operational results, modeling and scaling. *Proceedings of the 10th European Meeting on Supercritical Fluids: Reactions, Materials and Natural Products Processing*, Colmar, France.

Bogaerts, C. and Bettendorf, A. 1986. Electrochemistry and corrosion of alloys in high-temperature water. EPRI Report NP-4705, Electric Power Research Institute, Palo Alto, CA.

Browne, M. 1987. Neither liquid nor gas, odd substances find uses. *The New York Times*.

Brunner, G. 2009. Near and supercritical water. Part II: Oxidative processes. *J. Supercrit. Fluids* 47:382–390.

Butt, J.B. and Peterson, E.E. 1988. *Activation Deactivation and Poisoning of Catalysts*. Academic Press, San Diego, CA.

Byrd, A.J., Pant, K.K., and Gupta, R.B. 2007. Hydrogen production from glucose using Ru/Al_2O_3 catalyst in supercritical water. *Ind. Eng. Chem. Res.* 46:3574–3579.

Byrd, A.J., Pant, K.K., and Gupta, R.B. 2008. Hydrogen production from glycerol by reforming in supercritical water over Ru/Al_2O_3 catalyst. *Fuel* 87:2956–2960.

Calvo, L. and Valejo, D. 2002. Formation of organic acids during the hydrolysis and oxidation of several wastes in sub- and supercritical water, *Industrial and Engineering Chemistry Research*, Vol. 41, Issue 25, pp. 6503–6509.

Conditt, M.K. and Sievers, E. 1984. Microanalysis of reaction products in sealed tube wet air oxidations by capillary gas chromatography. *Anal. Chem.* 56:2620–2622.

Cortright, R.D., Davada, R.R., and Dumesic, J.A. 2002. Hydrogen from catalytic reforming of biomass-derived hydrocarbons in liquid water, *Nature*, Vol. 418, 29 August, pp. 964–967, doi:10.1038/nature01009.

Demirbas, A. 2004. Hydrogen-rich gas from fruit shells via supercritical water extraction. *Int. J. Hydrogen Energy* 29:1237–1243.

DiLeo, G.J., Neff, M.E., and Savage, P.E. 2007. Gasification of guaiacol and phenol in supercritical water. *Energy Fuels* 21:2340–2345.

Ding, Z.Y. and Abraham, M.A. 1995. Catalytic supercritical water oxidation of aromatic compounds: Pathways, kinetics and modeling. *First International Workshop on Supercritical Water Oxidation*, Jacksonville, FL.

Ding, Z.Y., Frisch, M.A., Li, L., and Gloyna, E.F. 1996. Catalytic oxidation in supercritical water. *Ind. Eng. Chem. Res.* 35(10): 3257–3279.

Dunn, J.B., Urquhart, D.I., and Savage, P.E. 2002. Terephthalic acid synthesis in supercritical water. *Adv. Synth. Catal.* 344:385.

Elliott, D.C., Sealock, Jr., L.J., and Baker, E.G. 1993. Chemical processing in high-pressure aqueous environments. 2. Development of catalysts for gasification. *Ind. Eng. Chem. Res.* 32:1542–1548.

Evans, R.J. and Milne, T.A. 1987. Molecular characterization of the pyrolysis of biomass: 1. Fundamentals. *Energy Fuels* 1(2):123.

Fierro, J.L.G. and de la Band, J.F.G. 1986. Chemisorption of probe molecules on metal oxides, *Catalysis Reviews Science and Engineering*, Vol. 28, p. 265.

Foussard, J.N., Debellefontaine, H., and Besombes-Vailhe, J. 1989. Efficient elimination of organic liquid wastes: Wet air oxidation. *J. Environ. Eng.* 115:367–385.

Frisch, M.A., Li, L., and Gloyna, E.F. 1994. Catalyst evaluation: Supercritical water oxidation process. *Proceedings of 49th Annual Purdue University Industrial Wastewaters Conference*, West Lafayette, IN.

Furness, D.T., Hoggett, L.A., and Judd, S.J. 2000. Thermochemical treatment of sewage sludge. *Water Environ. J.* 14:57–65.

Gadhe, J.B. and Gupta, R.B. 2007. Hydrogen production by methanol reforming in supercritical water: Catalysis by in-situ-generated copper nanoparticles. *Int. J. Hydrogen Energy* 32:2374–2381.

Goto, M., Nada, T., Kodama, A., and Hirose, T. 1999a. Kinetic analysis for destruction of municipal sewage sludge and alcohol distillery wastewater by supercritical water oxidation. *Ind. Eng. Chem. Res.* 38(5):1863–1865.

Goto, M., Shiramizu, D., Kodama, A., and Hirose, T. 1999b. Kinetic analysis for ammonia decomposition in supercritical water oxidation of sewage sludge. *Ind. Eng. Chem. Res.* 38:4500–4503.

Guan, Y., Pei, A., and Guo, L. 2008. Hydrogen production by catalytic gasification of cellulose in supercritical water. *J. Chem. Eng. China* 2(2):176–180.

Guo, L.J., Lu, Y.J., Zhang, X.M., Ji, C.M., Guan, Y., and Pei, A.X. 2007. Hydrogen production by biomass gasification in supercritical water: A systematic experimental and analytical study. *Catal. Today* 129:275–286.

Hao, Y., Guo, L., Zhang, X., and Guan, Y. 2005. Hydrogen production from catalytic gasification of cellulose in supercritical water. *Chem. Eng. J.* 110:57–65.

Helling, R.K. and Tester, J.W. 1987. Oxidation kinetics of carbon monoxide in supercritical water. *Energy Fuel* 1:417–423.

Holgate, H.R., Meyer, J.C., and Tester, J.W. 1995. Glucose hydrolysis and oxidation in supercritical water. *AIChE J.* 41:637–648.

Kabyemela, B.M., Adschiri, T., Malaluan, R.M., and Arai, K. 1999. Glucose and fructose decomposition in subcritical and supercritical water: Detailed reaction pathway, mechanisms, and kinetics. *Ind. Eng. Chem. Res.* 38:2888–2895.

Kaddouri, A. and Mazzocchia, C. 2004. A study of the influence of the synthesis conditions upon the catalytic properties of Co/SiO_2 or Co/Al_2O_3 catalysts used for ethanol steam reforming. *Catal. Commun.* 6:339–345.

Keen, R. and Baillod, C.R. 1985. Toxicity to *Daphnia* of the end products of wet oxidation of phenol and substituted phenols. *Water Res.* 19(6):767–772.

Kosobucki, P., Chmarzyński, A., and Buszewski, B. 2000. Sewage sludge composting. *Pol. J. Environ. Stud.* 9(4):243–248.

Krajnc, M. and Levec, J. 1994. Catalytic oxidation of toxic compounds in supercritical water, *Applied Catalysis B: Environmental*, Vol. 3, Issues 2–3, February, pp. L101–L107.

Kritzer, P. 2004. Corrosion in high-temperature and supercritical water and aqueous solutions: A review. *J. Supercrit. Fluids* 29:1–29.

Kritzer, P., Boukis, N., and Dinjus, E. 1997. Transpassive dissolution of nickel-base alloys and stainless steels in oxygen- and chloride containing high-temperature water materials and corrosion. *Werkstoffe und Korrosion* 48(12):799–805.

Kritzer, P. and Dinjus, E. 2001. An assessment of supercritical water oxidation (SCWO): Existing problems, possible solutions and new reactor concepts. *Chem. Eng. J.* 83:207–214.

Kruse, A., Meier, D., Rimbrecht, P., and Schacht, M. 2000. Gasification of pyrocatechol in supercritical water in the presence of potassium hydroxide. *Ind. Eng. Chem. Res.* 39:4842–4848.

Kruse, A., Sinag, A., and Schwarzkopf, V. 2003. Formation and degradation pathways of intermediate products formed during the hydropyrolysis of glucose as a model substance for wet biomass in a tubular reactor. *Eng. Life Sci.* 3:469–473.

Lambroua, P.S., Christoua, S.Y., Fotopoulosb, A.P., Fotib, F.K., Angelidisb, T.N., and Efstathiou, A.M. 2005. The effects of the use of weak organic acids on the improvement of oxygen storage and release properties of aged commercial three-way catalysts. *Appl. Catal. B: Environ.* 59:1–11.

Lee, D.-S., Li, L., and Gloyna, E.F. 1990. Efficiency of hydrogen peroxide and oxygen in supercritical water oxidation of acetic acid and 2,4-dichlorophenol. *AIChE Meeting*, Orlando, FL.

Lee, I.-G. and Ihm, S.-K. 2009. Catalytic gasification of glucose over Ni/activated charcoal in supercritical water. *Ind. Eng. Chem. Res.* 48(3):1435–1442.

Li, L., Chen, P., and Gloyna, E.F. 1993. Kinetic model for wet oxidation of organic compounds in subcritical and supercritical water. ACS Symposium Series, American Chemical Society, Washington, DC, p. 305.

Li, P.C. and Gloyna, E.F. 1991. Generalized kinetic model for wet oxidation of organic compounds. *AIChE J.* 37(11):1687.

Marrone, P.A., Hodes, M., Smith, K.A., and Tester, J.W. 2004. Salt precipitation and scale control in supercritical water oxidation—Part B: Commercial/full-scale applications. *J. Supercrit. Fluids* 29:289–312.

Marrone, P.A. and Hong, G.T. 2009. Corrosion control methods in supercritical water oxidation and gasification processes. *J. Supercrit. Fluids.* 51(2):83–103.

Modell, M. 1978. Gasification process. U.S. Pat. No. 4,113,446; Pat. No. 4,338,199, 1982.

Modell, M. 1989. Supercritical water oxidation. In: H.M. Freeman, ed., *Standard Handbook of Hazardous Waste Treatment and Disposal.* McGraw Hill, New York. pp. 8.153–8.168.

Minowa, T., Zhen, F., and Ogi, T. 1998. Cellulose decomposition in hot-compressed water with alkali or nickel catalyst. *J. Supercrit. Fluids* 13:253–259.

Mishra, V.S., Mahajani, V.V., and Joshi, J.B. 1995. Wet air oxidation. *Ind. Eng. Chem. Res.* 34(1):2–48.

Mitton, D.B., Yoon, J.H., Cline, J.A., Kim, H.S., Eliaz, N., and Latanision, R.M. 2000. Corrosion behavior of nickel-based alloys in supercritical water oxidation systems. *Ind. Eng. Chem. Res.* 39(12):4689–4696.

Nunouraa, T., Lee, G.H., Matsumurab, Y., and Yamamotob, K. 2002. Modeling of supercritical water oxidation of phenol catalyzed by activated carbon. *Chem. Eng. Sci.* 57:3061–3071.

Onwudili, J.A. and Williams, P.T. 2009. Hydrothermal reforming of bio-diesel plant waste: Products distribution and characterization. *Fuel* 89(2):501–509.

Osada, M., Sato, T., Watanabe, M., Adschiri, T., and Arai, K. 2004. Low temperature catalytic gasification of lignin and cellulose with a ruthenium catalyst in supercritical water. *Energy Fuels* 18:327–333.

Osada, M., Sato, T., Watanabe, M., Shirai, M., and Arai, K. 2006. Catalytic gasification of wood biomass in subcritical and supercritical water. *Combust. Sci. Technol.* 178:537–552.

Ptasinski, K.J., Hamelinck, C., and Kerkhof, P.J.A.M. 2002. Exergy analysis of methanol from the sewage sludge process. *Energy Convers. Manage.* 43:1445–1457.

Rice, S.F., Steeper, R.R., and Aiken, J.D. 1998. Water density effects on homogeneous water–gas shift reaction kinetics. *J. Phys. Chem.* 102:2673–2678.

Rulkens, W. 2007. Sewage sludge as a biomass resource for the production of energy: Overview and assessment of the various options. *Energy Fuels* 22:9–15.

Sasaki, M., Kabyemela, B., Malaluan, R., Hirose, S., Takeda, N., Adschiri, T., and Ara, K. 1998. Cellulose hydrolysis in subcritical and supercritical water. *J. Supercrit. Fluids* 13:261–268.

Savage, P.E. 2009. A perspective on catalysis in sub- and supercritical water. *J. Supercrit. Fluids* 47:407–414.
Savage, P.E., Gopalan, S., and Mizan, T.I. 1995. Reactions at supercritical conditions: Application and fundamentals. *AIChE J.* 41(7):1723–1778.
Schmieder, H., Abeln, J., Boukis, N., Dinjus, E., Kruse, A., Kluth, M., Petrich, G., Sadri, E., and Schacht, M. 2000. Hydrothermal gasification of biomass and organic wastes. *J. Supercrit. Fluids* 17:145–153.
Schubert, M., Regler, J.W., and Vogel, F. 2010. Continuous salt precipitation and separation from supercritical water. Part 2: Type 2 salts and mixtures of two salts. *J. Supercrit. Fluids*. 52(1):113–124.
Shanableh, A.M. 1990. Subcritical and supercritical water oxidation of industrial excess activated sludge. PhD dissertation, Civil Engineering Department, University of Texas, Austin, TX.
Takamatsu, T., Hashimoto, I., and Sioya, S. 1970. Model identification of wet air oxidation process thermal decomposition. *Water Res.* 4:33–59.
Taylor, J.E. and Weygandt, J.C. 1974. A kinetic study of high pressure aqueous oxidations of organic compounds using elemental oxygen. *Can. J. Chem.* 52:1925.
Tester, J.W., Holgate, H.R., Armellini, F.J., Webley, P.A., Killilea, W.R., Hong, G.T., and Barner, H.E. 1993. Supercritical water oxidation technology. In: W.D. Tedder and F.G. Pohland, eds., *Emerging Technologies in Hazardous Waste Management*. Publisher is American Chemical Society, Washington, DC, pp. 35–76.
Tiltscher, H. and Hofman, H. 1987. Trends in high pressure chemical reaction engineering. *Chem. Eng. Sci.* 42:959–977.
Wahyudiono, W., Sasaki, M., and Goto, M. 2008. Recovery of phenolic compounds through the decomposition of lignin in near and supercritical water, *Chemical Engineering and Processing*, Vol. 47, Issues 9–10, September, pp. 1609–1619.
Watanabe, M., Inomata, H., Osada, M., Sato, T., Adschiric, T., and Araia, K. 2003. Catalytic effects of NaOH and ZrO_2 for partial oxidative gasification of n-hexadecane and lignin in supercritical water. *Fuel* 82:545–552.
Webley, P.A. and Tester, J.W. 1988. Fundamental kinetics and mechanistic pathways for oxidation reactants in supercritical water. SAE Technical Paper Series 881039.
Williams, P.T. and Onwudili, J. 2006. Subcritical and supercritical water gasification of cellulose, starch, glucose, and biomass waste. *Energy Fuels* 20:1259–1265.
Wilmanns, E.G. 1990. Supercritical water oxidation of volatile acids. MS thesis, Civil Engineering Department, University of Texas, Austin, TX.
Woerner, G.A. 1976. Thermal decomposition and reforming of glucose and wood at the critical conditions of water. MS thesis, Department of Chemical Engineering, Massachusetts Institute of Technology, Cambridge, MA.
Wu, B.C., Klein, M.T., and Sandler, S.I. 1991. Solvent effects on reactions in supercritical fluids. *Ind. Eng. Chem. Res.* 30:822.
Xu, X., Matsumura, Y., Stenberg, J., and Antal, Jr., M.J. 1996. Carbon-catalyzed gasification of organic feedstocks in supercritical water. *Ind. Eng. Chem. Res.* 35:2522–2530.
Yanik, J., Ebale, S., Kruse, A., Saglam, M., and Yüksel, M. 2007. Biomass gasification in supercritical water: Part 1. Effect of the nature of biomass. *Fuel* 86:2410–2415.
Yoshida, Y., Dowaki, K., Matsumura, Y., Matsuhashi, R., Li, D., Ishitani, Y., and Komiyama, H. 2003. Comprehensive comparison of efficiency and CO_2 emissions between biomass energy conversion technologies-position of supercritical water gasification in biomass technologies. *Biomass Bioenergy* 25(3):257–272.
Youssef, E.A., Chowdhury, M.B., Nakhla, G., and Charpentier, P., 2010, Effect of nickel loading on hydrogen production and chemical oxygen demand (COD) destruction from glucose oxidation and gasification in supercritical water, *Int. J. Hydrogen Energy* 35(10):5034–5042.
Yu, D., Aihara, M., and Antal, Jr., M.J. 1993. Hydrogen production by steam reforming glucose in supercritical water. *Energy Fuels* 7:574–577.
Yu, J. and Savage, P.E. 2001. Catalyst activity, stability, and transformations during oxidation in supercritical water. *Appl. Catal. B: Environ.* 31:123–132.

Section III

Hydrogen Production: Electrolysis

7

Solid Oxide Electrolyzer Cells

Michael K.H. Leung
The University of Hong Kong

Meng Ni
The University of Hong Kong

Dennis Y.C. Leung
The University of Hong Kong

CONTENTS

7.1 Introduction

Production of renewable hydrogen with zero net emission of greenhouse gases plays an important role in the development of clean and sustainable hydrogen economy. Renewable hydrogen can be produced by using solar energy directly in thermochemical water splitting to dissociate water into hydrogen and oxygen. However, a recent study has shown that in terms of potential economic benefit, thermochemical water splitting is not as attractive as water electrolysis (Graf 2008). Photocatalysis is another potential method for production of renewable hydrogen by direct use of solar radiation to reduce water. For the present development of the technology, the photocatalytic efficiency is too low to be economically competitive (Ni et al. 2007a). More research and development are needed to enhance the catalysts and reactor designs.

Water electrolysis is presently considered as the most practical and promising technology for renewable hydrogen production using the renewable energy sources indirectly (Barbir 2005, Ni et al. 2008a). The well-developed solar photovoltaics and wind power technologies can be employed to generate electricity, and the electricity can energize electrolyzers to dissociate water into hydrogen and oxygen as a by-product. There are various types of electrolyzer cells mainly differentiated by the types of electrolytes used. A proton exchange membrane electrolyzer cell (PEMEC) makes use of a proton-conducting polymer electrolyte membrane to separate protons and oxygen ions in the electrolytic hydrogen production process. The operating temperature is low (320–360 K) because water should be fed as a liquid to maintain high ionic conductivity of the PEM electrolyte (Ni et al. 2008b).

Steam-fed high-temperature solid oxide electrolyzer cells (SOECs) can produce hydrogen with a lower electrical energy requirement than the low-temperature PEMEC because the SOEC electrodes are more reactive and the electrolyte is more ion conductive at a high temperature. SOEC has been recognized as a promising technology for large-scale stationary hydrogen production plants. The basic principles, cell configurations, and characteristics of SOEC are discussed in this chapter. It is revealed that the performance of SOEC highly depends on multiple parameters of the cell materials, structural design, and operating condition. For further development of the technology to achieve higher electricity-to-hydrogen efficiency, it is important to understand the details of the mass transport and electrochemical reactions taking place in parallel.

7.2 Fundamentals of SOEC

As illustrated in Figure 7.1, a conventional SOEC consists of an oxygen ion–conducting electrolyte, a cathode, and an anode. When steam is fed to the porous cathode and an electrical potential is applied between the two electrodes, the water molecules diffusing to the reaction sites are dissociated to form hydrogen gas and oxygen ions at the cathode–electrolyte interface. The hydrogen gas produced diffuses to the cathode surface where the hydrogen can be collected. The oxygen ions are transported through the dense electrolyte to the anode. On the anode side, the oxygen ions are oxidized to oxygen gas and the oxygen produced is transported through the porous anode to the surface. The net reaction of SOEC can be written as

$$H_2O \rightarrow H_2 + \frac{1}{2}O_2 \tag{7.1}$$

The total energy demand (ΔH) for SOEC hydrogen production can be expressed as

$$\Delta H = \Delta G + T\Delta S \tag{7.2}$$

where

ΔG is the electrical energy demand (free Gibson energy change)
$T\Delta S$ is the thermal energy demand

Figure 7.2 plots the theoretical energy demands versus operating temperature. The total energy demand is quite independent of the operating temperature. However, the detailed breakdown

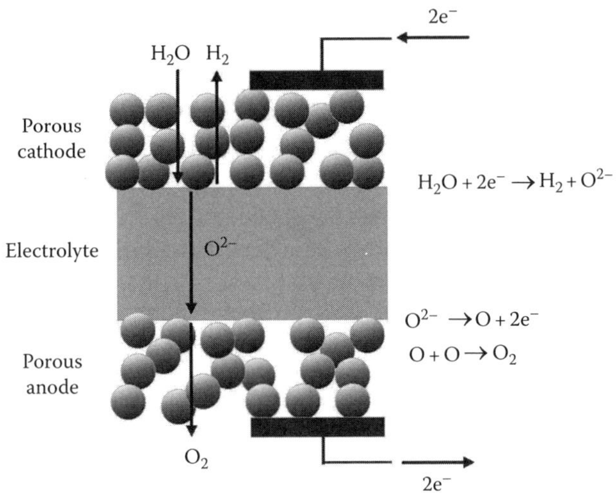

FIGURE 7.1
Schematics of SOSE hydrogen production.

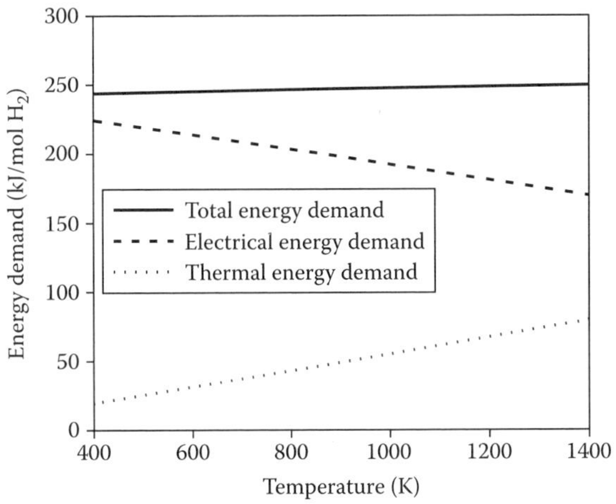

FIGURE 7.2
Calculated of energy demands for electrolytic H_2 with varying temperatures.

shows that at a higher operating temperature, the thermal energy demand increases while the electrical energy demand decreases. Therefore, high-temperature SOEC is advantageous as it provides more opportunities to utilize industrial waste heat for hydrogen production.

Operating at a high temperature, the SOEC components must meet certain requirements for efficient and cost-effective hydrogen production as summarized in the following (Wendt 1990):

- The dense electrolyte should have good ionic conductivity and poor electronic conductivity for effective dissociation of water into hydrogen and oxygen. The electrolyte should be also chemically stable at high temperature.
- The dense electrolyte must be gastight to avoid recombination of hydrogen and oxygen, but it should be as thin as possible to minimize the ohmic overpotential.

- Both electrodes should be chemically stable in the highly reducing/oxidizing environments and have good electronic conductivity.
- Both electrodes should have suitable porosity and pore size: (1) to support gas transportation between the electrode surface and the electrode–electrolyte interface and (2) to provide sufficient electrolyte–electrode–gas triple-phase boundary (TPB) (reaction sites).
- The thermal expansion coefficients of both electrodes should be close to that of the electrolyte to prevent material failure due to exceedingly high mechanical stress induced by thermal expansion mismatch.
- The interconnect materials are needed in large-scale hydrogen production plants. The interconnect materials must be chemically stable in the reducing/oxidizing environments as they are exposed to the steam, hydrogen, and oxygen simultaneously.
- The costs of raw materials and cell fabrication should be low.

7.3 Electrolyte

7.3.1 Oxygen Ion–Conducting Electrolyte

7.3.1.1 Stabilized Zirconia

The electrolyte is the key component of an SOEC cell. The most common electrolyte material used in SOEC is yttria-stabilized zirconia (YSZ), which exhibits high oxygen ion conductivity and good mechanical strength. Zirconia dioxide has a high melting point of around 2973 K but suffers from phase transformation from the monoclinic to the tetragonal form at around 1373 K and to the cubic fluorite form at around 2643 K (Strickler and Carlson 1964), which can lead to large and disruptive volume change. The adverse phase transformation can be prevented by addition of alkaline-earth or rare-earth oxides. Dopant is added on the zirconia lattice site to stabilize the cubic and tetragonal structures and to increase the concentration of oxygen vacancies so as to increase the oxygen ion conductivity. Some experimental data on the ionic conductivity of doped ZrO_2 are summarized in Table 7.1 (Etsell and Flengas 1970, Muccillo and Kleitz 1995, Badwal et al. 2000, Hirano et al. 2000, Ivanov et al. 2006, Brahim et al. 2007, Dahl et al. 2007, Jiang et al. 2007, Prabhakaran et al. 2007). It can be seen that the electrolyte has high oxygen ion conductivity when it is doped with Y_2O_3, Yb_2O_3, and Sc_2O_3. For comparison, the electrolyte doped with MgO, La_2O_3, and CaO has relatively low conductivity.

The Sc_2O_3-doped ZrO_2 exhibits the highest ionic conductivity because the ionic radius of Sc^{3+} is closest to the radius of Zr^{4+}, as shown in Figure 7.3 (Kilner and Brook 1982). According to the work done by Kilner and Brook (1982), the ionic size mismatch is related to the association enthalpy, which influences the oxygen ion conduction. Despite its high ionic conductivity, ScSZ is not widely used as an electrolyte for high-temperature operation mainly because of its high cost (Singhal and Kendall 2003). On the other hand, YSZ is much more economical and offers the best combination of ionic conductivity and stability. Thus, YSZ is a suitable material for SOEC electrolytes.

The ionic conductivity of the electrolyte is influenced by the concentration of the dopants. As shown in Figure 7.4 (Arachi et al. 1999), the maximum oxygen ion conductivity of YSZ can be obtained when the molar fraction of Y_2O_3 is around 8%. The oxygen vacancies are proportional to the concentration of the dopant; thus, at low doping, the conductivity

TABLE 7.1

Reported Ionic Conductivity of Different Doped Zirconia Materials at Typical Temperatures

Material	Conductivity (S/cm)	Temperature (K)	Remarks	References
8YSZ	0.13	1273	Prepared by spay drying of nitrate precursor solution	Prabhakaran et al. (2007)
10.5YSZ	0.034	1073	Thin film prepared by aerosol-assisted metal-organic chemical vapor deposition (AAMOCVD)	Jiang et al. (2007)
10YSZ	4.52×10^{-6}	673	A 300 nm thick film was prepared by atomic laser deposition (ALD)	Brahim et al. (2007)
9.5YSZ	0.057	1173	A 15–25 μm thick film prepared by magnetic pulse compaction of tapes cast of nanopowders	Ivanov et al. (2006)
8YSZ	0.083	1173	Spark plasma sintering at 1573 K and at pressure 70 MPa	Dahl et al. (2007)
$CaO–ZrO_2$ with 12.5 mol% CaO	0.055	1273	Nil	Etsell and Flengas (1970)
$La_2O_3–ZrO_2$ with 5 mol% La_2O_3	0.0044	1273	Nil	Etsell and Flengas (1970)
$MgO–ZrO_2$ with 13.7 mol% MgO	0.098	1273	Prepared by conventional ceramic processing involving wet mixing, pressing, sintering, and machining	Muccillo and Kleitz (1995)
$Sc_2O_3–ZrO_2$ with 9–11 mol% Sc_2O_3	0.28–0.34	1273	Prepared by sintering of coprecipitated powders	Badwal and Ciacchi (2000)
$Sc_2O_3–ZrO_2$ with 6 mol% Sc_2O_3	0.18	1273	The sintered film was treated by hot isostatic pressing (HIP) to improve the mechanical strength	Hirano et al. (2000)
Fe–YSZ with 4 mol% Fe	0.029	973	YSZ power doped with 4 mol% of Fe was sintered at 1200 and 1400 K	Gao et al. (2008)

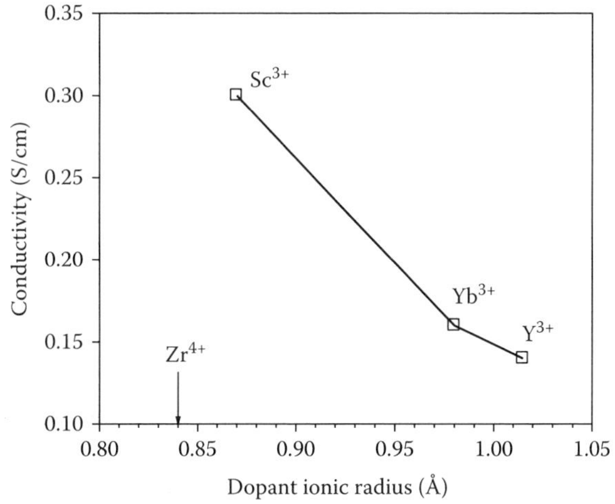

FIGURE 7.3
Relationship between ionic conductivity and dopant ionic radius. (From Kilner, J.A. and Brook, R.J., *Solid State Ionics*, 6(3), 237, 1982.)

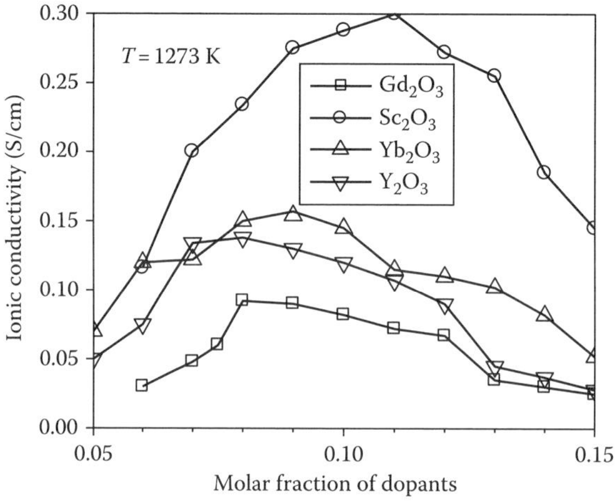

FIGURE 7.4
Dependence of electrolyte ionic conductivity on dopant concentration. (Experimental data from Arachi, Y. et al., *Solid State Ionics*, 121(1–4), 133, 1999.)

increases with increasing molar fraction of Y_2O_3. At high dopant concentration, the oxygen vacancies and the dopant cations will form complex defects, which in turn results in lower conductivity (Kharton et al. 1999).

The ionic conductivity of YSZ electrolytes can be also increased by Fe dopant (Gao et al. 2008). High ionic conductivity was found when YSZ was doped with 4 mol% Fe and sintered at 1673 K. Besides the increase in ionic conductivity of the electrolyte, Fe dopant might improve the electrolyte–electrode interface resulting in better SOEC performance.

Although the YSZ with 8 mol% Y_2O_3 has the optimal ionic conductivity, it suffers from degradation during long-term operation (Gibson et al. 1998). With 7.7 mol% Y_2O_3, the ionic conductivity of YSZ decreases from 16 to 13.7 S/cm after 5000 min. However, this problem can be avoided by increasing slightly the doping concentration (9 mol% or 10 mol% Y_2O_3). Compared with YSZ, ScSZ suffers from more serious degradation in ionic conduction because of phase transformation (Yamamoto et al. 1995). As shown in Figure 7.5 (Haering et al. 2005), it can be seen that the resistivity of 9ScSZ is increased by more than 50% after 60,000 min. Similarly, the aging effect of ScSZ can be reduced by increasing the content of Sc_2O_3 or by codoping with Al_2O_3 or TiO_2, which can suppress the phase transformation of ScSZ. Another benefit of codoping Al_2O_3 is that the mechanical strength can be enhanced because Al impedes the grain growth during sintering. However, the oxygen ion conductivity decreases with increasing content of Al_2O_3 or TiO_2.

Figure 7.6 shows the temperature effect on the oxygen ion conductivities of 9YSZ, 9YbSZ, and 9ScSZ. The temperature–conductivity relationship follows an Arrhenius form as (Strickler and Carlson 1964)

$$\sigma = A\exp\left(-\frac{E}{kT}\right) \tag{7.3}$$

where
- A is a preexponential factor
- E is the activation energy
- k is the Boltzmann constant
- T is the absolute temperature

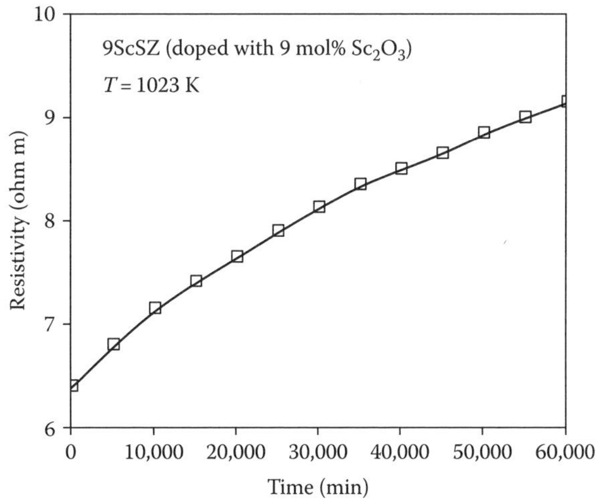

FIGURE 7.5
Variation of 9ScSZ resistivity with time. (Measured data from Haering, C. et al., *Solid State Ionics*, 176(3–4), 261, 2005.)

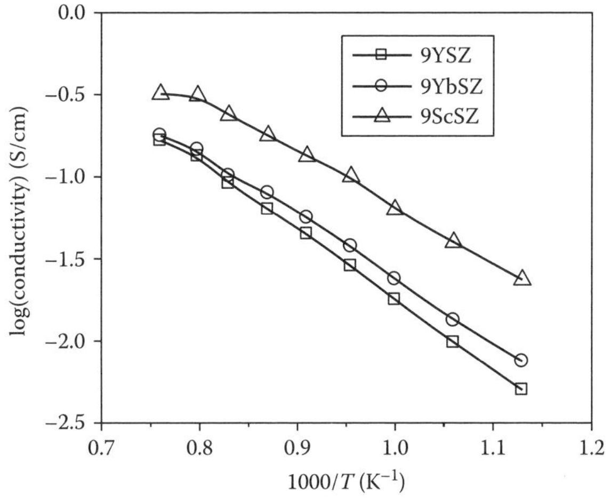

FIGURE 7.6
Temperature effect on oxygen ionic conductivity of 9YSZ, 9YbSZ, and 9ScSZ. (Experimental results from Strickler, D.W. and Carlson, W.G., *J. Am. Ceram. Soc.*, 47(3), 122, 1964.)

It can be seen that in the temperature range of concern, the oxygen ion conductivity decreases in the following order: ScSZ > YbSZ > YSZ. At lower temperature, the difference among them is more obvious, indicating that the ScSZ is more suitable for intermediate-temperature electrolytes, provided that its cost can be considerably reduced.

The grain boundaries of the doped ZrO_2 are also important in determining its oxygen ion conduction (Guo and Waser 2006). For both YSZ and ScSZ, the grain boundary resistance is insignificant at a high temperature. At a low temperature (<1073 K for ScSZ), the grain boundary resistance becomes important and governs the oxygen ion conduction in the electrolyte. This finding indicates that the ionic conductivity of ScSZ at an intermediate temperature can be increased by reducing the grain boundary resistance. Another study conducted by Kosacki et al. (2005a,b) reveals that the electrolyte thickness also influences the oxygen ion conductivity. The conductivity of a YSZ electrolyte decreases with film thickness for its thickness between 60 and 2000 nm. For YSZ thickness less than 60 nm, its conductivity increases with decreasing film thickness. These phenomena indicate the transition from lattice to interface-controlled diffusivity.

The effect of codoping with other oxides has been examined in recent years. Doping of ZnO into 8YSZ promoted densification of the YSZ ceramics considerably. The ionic conductivity of 8YSZ can be increased by doping small amounts of ZnO. With a doping of 0.5 wt.% ZnO, the conductivity of 8YSZ increases from 0.0131 to 0.0289 S/cm at 1073 K (Liu and Lao 2006). Doping of Al_2O_3 into YSZ leads to the creation of space-charge regions, which in turn can enhance the ionic conductivity. On the contrary, the presence of Al_2O_3 also leads to a blocking effect that suppresses conductivity. The combined effects result in minimal effect of Al_2O_3 doping on YSZ conductivity (Kumar et al. 2005). Recently, it was found that addition of Bi_2O_3 could stabilize the cubic phase of ScSZ at a low temperature. At 873 K, a conductivity of 0.18 S/cm of ScSZ was obtained with 2 mol% of Bi_2O_3 (Sarat et al. 2006). The aforementioned studies indicate that the codoping may be effective for improving the mechanical strength or the ionic conductivity. Research works in this direction are expected to be fruitful.

7.3.1.2 Doped $LaGaO_3$

Despite the high reaction rate and ion conduction, the high temperature limits the selection of interconnect materials and poses problems for long-term stability of SOEC components. Therefore, it is demanded to operate the SOEC at an intermediate temperature, between 673 and 1073 K. The doped $LaGaO_3$ material was found to have good ionic conductivity. Early works on $LaGaO_3$-based electrolytes were conducted by Ishihara et al. (1994). They found that doping Sr for the La sites could increase the conductivity of $LaGaO_3$ (Figure 7.7a), while the conductivity could be further enhanced by doping Mg for the Ga sites (Figure 7.7b). The conductivity of $La_{0.9}Sr_{0.1}Ga_{0.8}Mg_{0.2}O_3$ was found higher than conventional YSZ and ScSZ.

Some of the conductivity data from the literature are summarized in Table 7.2 (Stevenson et al. 1998, Huang and Goodenough 2000, Zhang et al. 2000, Cong et al. 2003, Subasri et al. 2003, Polini et al. 2004, Liu et al. 2006). It can be seen that LSGM exhibits high ionic conductivity at high and intermediate temperatures. At 1073 K, the conductivity of LSGM can be around 0.17 S/cm. For comparison, the conductivity of conventional YSZ is only about 0.026 S/cm (Cong et al. 2003). The difference is more pronounced at a lower temperature, that is, 0.03 S/cm for LSGM and 0.00173 S/cm for YSZ at 873 K. These data indicate that LSGM could be a promising electrolyte for intermediate-temperature steam electrolysis. In addition, the conductivity of LSGM varies with different synthesis methods, due to different grain size and grain boundaries.

Similar to YSZ, the conductivity of LSGM depends on the concentration of dopants. Figure 7.8 shows the dependence of electrolyte conductivity on the doping content of Sr and Mg (Gorelov et al. 2001). It can be seen that the conductivity is quite sensitive to the doping content, and the optimal doping content is about 15 mol%. It should be noted that the optimal doping content reported by various groups are different because of different starting materials used and different preparation procedures (Gorelov et al. 2001, Shi et al. 2006). This may cause formation of a secondary phase during the preparation process. Generally, the optimal Sr and Mg doping content is around 15–20 mol%.

Zheng et al. (2004) investigated the secondary phases formed in terms of doping content. It was found that when the doping contents of Sr and Mg were less than 20 mol%, no secondary

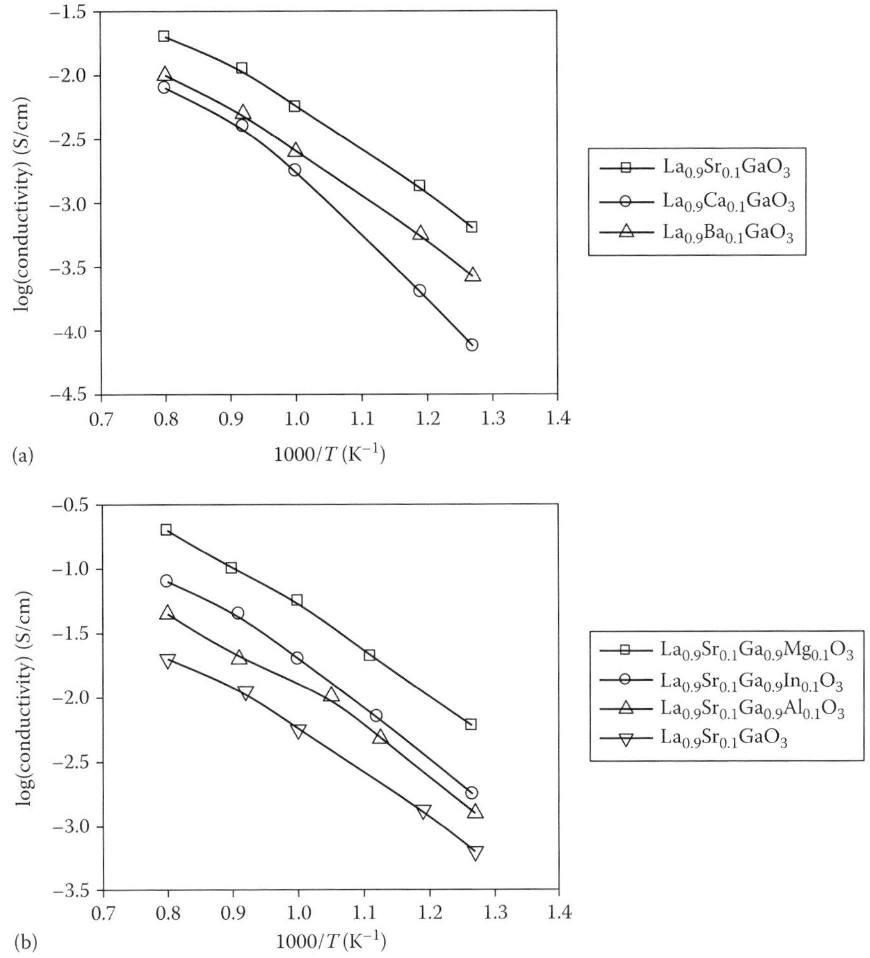

FIGURE 7.7
Oxygen ionic conductivity of doped $LaGaO_3$—(a) Sr-doped $LaGaO_3$ and (b) Sr and Mg–codoped $LaGaO_3$. (Measured data from Ishihara, T. et al., *J. Am. Chem. Soc.*, 16(9), 3801, 1994.)

phase would be formed. However, when the doping contents exceeded 20 mol%, the secondary phase of $SrLaGaO_4$ was formed, decreasing the conductivity of the LSGM electrolyte. In another study conducted by Liu et al. (2006), the secondary phase, $LaSrGa_3O_7$, was formed with a doping content of Sr and Mg of 15 mol%, which caused relatively lower conductivity.

The LSGM exhibits a small amount of electron and hole conductivity. Kharton et al. (2003) studied the effects of sintering time and operating temperature on ionic and electronic conduction of LSGM. It was found that the electronic conductivity highly depended on the ceramic microstructure. Increasing the sintering time from 0.5 to 40 h decreased the electronic conductivity considerably. For comparison, the ionic conductivity was found less sensitive to the material microstructures. Generally, the detrimental electron/hole conductivity is around four orders of magnitude smaller than the oxygen ion conductivity (Baker et al. 1997). Thus, the energy loss due to internal short circuit is negligible.

In addition to doping of Sr and Mg, codoping with other metal ions has been investigated with an aim to further enhance the conductivity of LSGM and its other properties, such as

TABLE 7.2

Reported Ionic Conductivity of Different $LaGaO_3$-Based Materials at Typical Temperatures

Material	Conductivity (S/cm)	Temperature (K)	Remarks	References
$La_{0.8}Sr_{0.2}Ga_{0.8}Mg_{0.2}O_3$	0.45	1273	Prepared by a combustion synthesis technique	Stevenson et al. (1998)
	0.17	1073		
	0.025	873		
$La_{0.8}Sr_{0.2}Ga_{0.83}Mg_{0.17}O_{2.815}$	0.17	1073	Stable over a weeklong test	Huang and Goodenough (2000)
	0.08	973		
	0.03	873		
$La_{0.9}Sr_{0.1}Ga_{0.8}Mg_{0.2}O_{3-\delta}$	0.197	1173	Prepared by solid-state reactions of sintering, ball milling, and calcination	Zhang et al. (2000)
	0.1193	1073		
	0.0532	973		
$La_{0.8}Sr_{0.2}Ga_{0.85}Mg_{0.15}O_{3-\delta}$	0.0782	1123	Synthesized using glycine–nitrate combustion method	Cong et al. (2003)
	0.0606	1073		
	0.0263	973		
	0.00809	873		
$La_{0.8}Sr_{0.2}Ga_{0.83}Mg_{0.17}O_{2.815}$	0.0196	900	Prepared by microwave-assisted processing in a very short time of 10 min	Subasri et al. (2003)
$La_{0.9}Sr_{0.1}Ga_{0.9}Mg_{0.1}O_{2.9}$	0.051	1073	Prepared by citrate solgel method and by subsequent calcination at 1673 K	Polini et al. (2004)
$La_{0.85}Sr_{0.15}Ga_{0.85}Mg_{0.15}O_{2.85}$	0.051	973	Prepared by a novel method based on acrylamide polymerization technique	Liu et al. (2006)
	0.015	873		

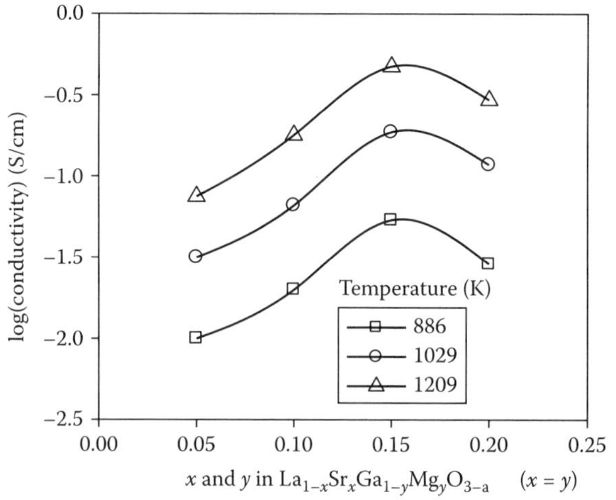

FIGURE 7.8
Dependence of oxygen ionic conductivity of LSGM on doping content of Sr and Mg. (Experimental results from Gorelov, V.P. et al., *J. Eur. Ceram. Soc.*, 21(13), 2311, 2001.)

mechanical strength and thermal expansion characteristics. Stevenson et al. (2000) investigated the effect of doping with Co or Fe in the LSGM electrolyte. It was found that doping a small amount (<10 mol%) of Co or Fe at the Ga sites could enhance the electrical conductivity of LSGM at a low temperature but could not change the electrical conductivity at a high temperature. It should be noted that the higher conductivity at a low temperature was caused by the introduction of electronic charge carriers into the lattice, which brought no beneficial effect for oxygen ion conductivity. Further addition of Co or Fe could cause the electronic conduction to dominate the electrolyte conductivity. Khorkounov et al. (2006) investigated the ionic and electronic conductivities of LSGM electrolytes doped with Co. It was found that doping of Co increased both ionic conductivity and electronic conductivity. However, the increase in electronic conductivity was more significant than the increase in oxygen ion conductivity. Yaremchenko et al. (2003) compared the conductivity of LSGM doped with Co, Fe, and Ni. It was found that the ionic conductivities of these transition metal–doped LSGM electrolytes were considerably lower than their parental LSGM materials, probably because of increased defect association.

The main problem of LSGM is its reactivity with Ni electrodes to form lanthanum nickelates (Huang and Goodenough 2000). Zhang et al. (2000) investigated the reaction between the LSGM electrolyte with NiO and Sm-doped CeO_2 (SDC). It was found that a $LaNiO_3$-based compound was formed in the powder mixture of NiO and LSGM after firing at 1423 K, leading to significant loss of conductivity. For comparison, the reaction between NiO and SDC was not significant. It was found that the reaction of LSGM with Ni could be prevented by using a thin interlayer ($Ce_{0.8}Sm_{0.2}O_{1.9}$) at the electrode–electrolyte interface (Huang and Goodenough 2000, Wang and Tatsumi 2003).

7.3.1.3 Ceria-Based Oxides

Similar to LSGM, the doped ceria is considered as a promising intermediate-temperature electrolyte because of its high ionic conductivity at a temperature between 773 and 1073 K. The conductivity of ceria can be enhanced by proper doping of divalent or trivalent cations. Similar to zirconia, the maximum conductivity of ceria is obtained when the radius mismatch between the dopant cation and the Ce^{4+} is minimized. Gd^{3+}, Sm^{3+}, Y^{3+}, Yb^{3+}, La^{3+}, and Ca^{2+} have been tested as dopant for ceria-based electrolytes (Van Herle et al. 1996, Chinarro et al. 2007, Hui et al. 2007, Im et al. 2007, Thangadurai and Kopp 2007). Generally, Gd^{3+}-doped and Sm^{3+}-doped CeO_2 electrolytes (GDC and SDC) exhibit high ionic conductivity and thus have been investigated extensively. Some conductivity data of ceria-based electrolytes are presented in Table 7.3 (Van Herle et al. 1996, Huang et al. 1997, Dikmen et al. 2002, Seo et al. 2006, Xu et al. 2006, Zhang et al. 2006). It can be seen that the GDC and SDC exhibited good ionic conductivity at an intermediate temperature. For example, the conductivity of 25GDC could be 1.01×10^{-2} S/cm at 873 K. For comparison, the conductivity of the conventional YSZ is about 10^{-4} S/cm at this temperature, two orders of magnitude lower than GDC.

Comparable to YSZ, the conductivity of ceria-based electrolytes depends on the dopant concentration. Generally, their conductivity increases with increasing dopant concentration, and after reaching the maximum, the conductivity decreases with further increase in dopant concentration. In literature, the optimal dopant concentration is in the range of 15–25 mol%. For example, the maximum conductivities of GDC and SDC were observed at about 15 mol% doping of Gd^{3+} or Sm^{3+} (Zha et al. 2003). Seo et al. (2006) reported that the maximum ionic conductivity could be obtained with a Gd doping of 25 mol%. Zhang et al. (2004) obtained the optimal doping content of 20 mol% for GDC. For illustration, the effect of dopant concentration (Gd^{3+}) on ionic conductivity of GDC is shown in Figure 7.9

TABLE 7.3

Reported Ionic Conductivity of Different Ceria-Based Materials at Typical Temperatures

Material	Conductivity (S/cm)	Temperature (K)	Remarks	References
15GDC($Ce_{0.85}Gd_{0.15}O_{2-\delta}$)	4.07×10^{-2}	973	High-purity CeO_2 and Gd_2O_3 powders were used as the starting materials.	Zhang et al. (2006)
25GDC($Ce_{0.75}Gd_{0.25}O_{1.875}$)	1.01×10^{-2}	873	Prepared by the flame spray pyrolysis method.	Seo et al. (2006)
25GDC($Ce_{0.75}Gd_{0.25}O_{1.875}$)	7.5×10^{-3}	873	Prepared by hydrothermal method.	Dikmen et al. (2002)
20GDC	9.0×10^{-2} 4.2×10^{-2}	1073 973	Prepared by oxalate coprecipitation method.	Van Herle et al. (1996)
20SDC	8.8×10^{-2} 4.1×10^{-2}	1073 973	Prepared by oxalate coprecipitation method.	Van Herle et al. (1996)
20YDC	7.7×10^{-2} 3.5×10^{-2}	1073 973	Prepared by oxalate coprecipitation method.	Van Herle et al. (1996)
17SDC($Ce_{0.83}Sm_{0.17}O_{1.95}$)	5.7×10^{-3}	873	Prepared by hydrothermal method.	Huang et al. (1997)
20YDC($Ce_{0.8}Y_{0.2}O_{1.9}$)	3.4×10^{-2}	973	Prepared by citric acid–nitrate low-temperature combustion process.	Xu et al. (2006)

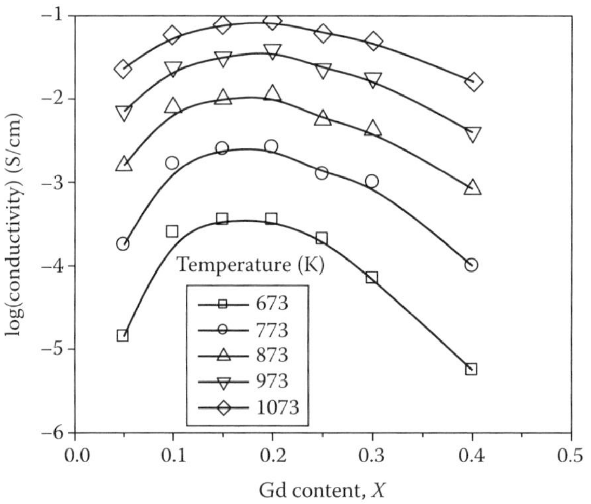

FIGURE 7.9
Oxygen ionic conductivity of GDC. (Experimental data from Zhang, T.S. et al., *Solid State Ionics*, 148(3–4), 567, 2002.)

(Zhang et al. 2002). It can be seen that the electrolyte conductivity is sensitive to the dopant concentration, especially at intermediate and low temperatures.

The difference in conductivity data reported by different research groups in literature may be caused by different preparation methods and preparation conditions, which will result in different microstructures and grain boundaries (Wang and Tatsumi 2003,

Ai et al. 2006). The conductivity consists of two parts: bulk conductivity and grain boundary conductivity, both depending on the microstructures of the electrolyte. Therefore, different microstructure and grain boundary conditions can lead to different conductivities and different optimal dopant concentrations. Perez-Coll et al. (2007) investigated in detail the bulk conductivity and the grain boundary conductivity of GDC. It was found that an increase in the Gd content would yield a decrease in the bulk conductivity and an increase in the grain boundary conductivity.

Codoping could enhance the conductivity of the ceria-based electrolyte. Sha et al. (2007) investigated the La and Y–codoped ceria electrolyte. In the temperature range from 973 to 1123 K, the La and Y–codoped ceria electrolyte ($Ce_{0.8}La_{0.14}Y_{0.06}O_{1.9}$) exhibits higher conductivity than ceria using one dopant ($Ce_{0.8}La_{0.2}O_{1.9}$ and $Ce_{0.8}Y_{0.2}O_{1.9}$). For example, at 1073 K, the conductivity of $Ce_{0.8}La_{0.14}Y_{0.06}O_{1.9}$ is 5.64×10^{-2} S/cm, while the conductivities of $Ce_{0.8}La_{0.2}O_{1.9}$ and $Ce_{0.8}Y_{0.2}O_{1.9}$ are 4.74×10^{-2} and 5.02×10^{-2} S/cm, respectively. However, it should be mentioned that this conductivity is still lower than some ceria electrolytes using one dopant, such as GDC and SDC. Mori et al. (2002) studied the microstructure and conductivity of ceria-based electrolytes codoped by Sm, La, Sr, Ca, and Ba. The highest conductivity was obtained with $(La_{0.75}Sr_{0.2}Ba_{0.05})_{0.175}\ Ce_{0.825}O_{1.891}$. Wang et al. (2005) evaluated the conductivity of Sm and Gd–codoped ceria electrolytes. It was found that the codoped electrolyte showed higher conductivity than the electrolyte using one dopant at a temperature between 673 and 973 K when the total dopant concentration was 15%. These studies indicated that codoping might be useful to enhance the ionic conductivity of the ceria-based electrolyte. However, the mechanisms have not been fully understood. The long-term stability of the codoped electrolyte remained to be demonstrated. More research works are needed in this direction.

The main problem of ceria-based electrolytes is that Ce^{4+} can be partially reduced to Ce^{3+} under a reducing environment. This is detrimental because (1) electronic conduction causes a partial internal electronic short circuit in the electrolyte, reducing the current efficiency, and (2) this can result in lattice expansion, which may lead to mechanical failure (Mogensen et al. 2000a, Kharton et al. 2004). The electronic conductivity of GDC has been investigated in detail by Steele (2000). It was found that the electronic conduction was significant, especially at a high temperature. At an intermediate temperature (773–1073 K), the 10GDC was more resistant to electronic conduction than the 20GDC. The partial reduction of a ceria electrolyte can be partially solved by combining the ceria electrolyte with another electrolyte, such as YSZ or LSGM, to block the reducing environment from the ceria electrolyte (Virkar 1991, Marques and Navarro 1996, 1997, Mehta et al. 1998, Kim et al. 2002, Bi et al. 2004). It has been demonstrated that 1–2 μm thick YSZ was sufficient to block electron conduction (Marques and Navarro 1997, Kim et al. 2002). In addition, the thermal expansion mismatch between the two layers could be acceptable if the cell was operated at an intermediate temperature (below 1073 K).

The problem of reduction of Ce^{4+} to Ce^{3+} can also be solved by adding electrically insulating particles, such as manganese- or cobalt-doped alumina grains, that can trap electrons to reduce the electronic conductivity without any change in the ionic conductivity (Chockalingam 2007). The electron-trapping mechanism works only over a limited layer. Further improvement is needed to develop nanocomposites.

7.3.1.4 Other Oxygen Ion–Conducting Ceramics

The oxygen ion conductivity of bismuth at an intermediate temperature is about 1–2 orders of magnitude higher than conventional YSZ. Therefore, bismuth has been investigated as possible electrolyte materials (Shuk et al. 1996, Sammes et al. 1999).

However, bismuth decomposes at low oxygen partial pressure, limiting its practical applications. This problem can be solved to some extent by combining bismuth with other electrolyte materials. Huang et al. (1996) evaluated the Bi_2O_3–Y_2O_3–CeO_2 solid solution as possible electrolyte material. A high ionic conductivity of 0.1 S/cm was obtained at 873 K without degradation after annealing at 923 K for over 300 h. Sarat et al. (2006) evaluated Bi_2O_3-doped ScSZ for use as an electrolyte. A high ionic conductivity of 0.18 S/cm at 873 K was obtained for 2 mol% Bi_2O_3-doped ScSZ. Nimat et al. (2006) prepared $Bi_2V_{0.9}Cu_{0.1}O_{5.35}$ on a glass substrate by the spray pyrolysis technique. The conductivity was found to be 0.057 S/cm at 698 K. These studies indicated that Bi_2O_3-based materials had high ionic conductivity at an intermediate temperature. Their performance could be enhanced by doping suitable ions or combining with other electrolyte materials. However, the long-term stability and the interaction with other materials remain to be investigated.

7.3.2 Proton-Conducting Electrolyte

In addition to the oxygen ion–conducting ceramics, proton-conducting ceramics can be used in solid oxide steam electrolysis (SOSE). Kreuer (2003) has reviewed the fundamentals and developments of proton-conducting materials for use in electrochemical devices. The cerate-based materials, such as $BaCeO_3$ and $SrCeO_3$, have been studied for a number of years. Paria and Maiti (1984) measured the conductivity of $BaCeO_3$ doped with La, Nd, and Ho. Nd was found most effective to enhance the ionic conductivity of $BaCeO_3$. Bonanos et al. (1989) studied the conductivity of $BaCe_{0.9}Gd_{0.1}O_{1.45}$ at an intermediate temperature. At high temperature and high oxygen partial pressure, $BaCe_{0.9}Gd_{0.1}O_{1.45}$ exhibited both ionic and electronic conduction, while at low oxygen partial pressure, it was a pure ionic conductor. The conductivity of $BaCe_{0.9}Gd_{0.1}O_{1.45}$ was found to be 0.011 and 0.016 S/cm at 873 and 1073 K, respectively. Another study conducted by Taniguchi et al. (1992) reported that the optimal doping content of Gd in $BaCeO_3$ was around 20–25 mol%. Flint et al. (1996) measured the conductivity of Ca-doped $BaCeO_3$ with different doping contents. With a doping content of 10 mol%, the conductivity of Ca-doped $BaCeO_3$ was found to be 1.0×10^{-3} and 2.2×10^{-3} S/cm at 1023 and 1173 K, respectively. Ma et al. (1999) evaluated the conductivity of $BaCeO_3$ with simultaneous doping with La and Y. The codoped $BaCeO_3$ exhibited a mixed oxide ionic and electronic conduction, while at low oxygen partial pressure, its conduction was almost protonic. Peng et al. (2006) evaluated the performance of solid oxide fuel cell (SOFC) with Sm-doped $BaCeO_3$ as electrolyte. The conductivity of $BaCe_{0.8}Sm_{0.2}O_{2.9}$ was found to be 4.16×10^{-3}, 6.62×10^{-3}, and 9.38×10^{-3} S/cm at 773, 873, and 973 K, respectively. Tomita et al. (2004) obtained a conductivity of 8.24×10^{-3} S/cm for 25 mol% Y^{3+}-doped $BaCeO_3$ at 673 K. Su et al. (2006) investigated the electrical properties of the codoped $BaCeO_3$ ceramics. The conductivity of Y and Nd–codoped $BaCeO_3$ was found to be 0.079 S/cm at 1073 K. $BaCe_{0.9}Y_{0.1}O_{3-\delta}$ has a high protonic conductivity (0.018 S/cm at 1100 K), and it is a suitable electrolyte for reversible proton-conducting fuel cells (Stuart et al. 2008, Ni et al. 2008c). The conductivity of proton-conducting electrolytes can be enhanced by controlling the annealing process to change the nanograin size and grain boundary. A high annealing temperature (1500°C) can produce large grains (200 nm) in Y-doped $BaZrO_3$ resulting in high protonic conductivity (0.004 S/cm) (Cervera et al. 2008). Other proton-conducting ceramics, such as doped $SrCeO_3$ and $SrZrO_3$, have also been investigated. However, these materials generally showed lower conductivity than doped $BaCeO_3$ (Muller et al. 1997, Sammes et al. 2004).

7.3.3 Summary on Electrolyte Materials for SOSE

YSZ is presently the most widely used electrolyte for SOSE working at high temperature. Although ScSZ has higher ionic conductivity than YSZ, it is not favorable because of its high cost. At reduced operating temperature, LSGM is a promising candidate for using as an SOSE electrolyte. The problem of reaction between LSGM and Ni electrodes remains to be solved. Alternatively, ceria-based ceramics, such as GDC and SDC, are also promising materials for intermediate-temperature electrolytes. However, under a reducing environment, Ce^{4+} can be partially reduced to Ce^{3+}, which in turn decreases the current efficiency and may result in mechanical failure. This problem can be solved by codoping or by introducing a blocking layer to avoid reduction of ceria electrolytes. Proton-conducting ceramics are alternative materials for use as intermediate-temperature electrolytes. The doped $BaCeO_3$ shows good conductivity and can be applied in SOSE. The selection of suitable electrolyte materials depends on the working temperature, the cost, and the compatibility with other components, that is, electrode materials.

7.4 Cathode

As previously mentioned, the electrode materials must be chemically and physically stable in highly oxidizing/reducing environments and compatible with other cell components. They should also be favorable to gas transport and electrochemically active.

The cathode of an SOEC supports the diffusion of steam and hydrogen gas and provides active sites for steam reduction. As the oxygen partial pressure on the cathode side generally falls into the range of 10^{-12} to 10^{-16} bar, the use of metallic electrode materials is possible (Doenitz et al. 1980). Noble metals, such as Pt, and nonprecious metals, such as Ni and Co, can be used as an SOSE cathode (Iwahara et al. 1987). However, the use of a noble metal electrode is not preferred due to its high cost. In addition, a noble metal electrode has other disadvantages, such as formation of volatile oxides and aging of porous structures at a high temperature. As Ni exhibits high electrochemical reactivity, it is widely used in SOSE and SOFC, which proceeds in the reverse reaction with SOSE. Although Ni can induce hydrogen reduction, it only conducts electrons. As a result, the electrochemical reactions only take place at the TPB of the cathode–electrolyte interface. In order to extend the electrochemical reaction zone, the Ni particles can be mixed with ionic conducting particles, usually the same material as the electrolyte, such as YSZ. This type of electrode is called cermet electrode and is presently widely used (Jiang and Chan 2004).

Conventionally, the cermet Ni–YSZ can be prepared by sintering the NiO and YSZ particles, followed by reduction under a hydrogen atmosphere (Martinez-Frias et al. 2003, Jiang and Chan 2004). Alternatively, the porous electrode can be fabricated using colloidal spray deposition, a low-cost thin-film deposition technique. Recently, it has been reported that the Ni–YSZ cermet could be fabricated by the high-energy ball milling of Ni and YSZ powders (Hong et al. 2005, 2008). However, it should be mentioned that sintering is still needed to improve the electrical conductivity of the ball-milled Ni–YSZ cermet.

The electrochemical performance of Ni–YSZ cermet electrodes has been investigated by several research groups. Eguchi et al. (1996) studied the *J*-*V* and polarization characteristics of SOSE using Ni–YSZ and Pt as the cathode. Measurements were conducted for both SOSE and SOFC modes. It was found that the Ni–YSZ cermet performed well in the SOFC mode but suffered from higher overpotentials in the SOSE mode. It was postulated that the Ni particles were oxidized to form a less active layer, resulting in low activity. As the Pt electrode exhibited lower overpotential than the Ni–YSZ cermet, the authors recommended using Pt as the cathode of SOSE. However, other studies reported that the area-specific resistance of Ni–YSZ cermet electrodes did not differ much between the SOFC mode and the SOSE mode (Momma et al. 2005, O'Brien et al. 2005). The discrepancy among these studies has not been fully understood yet. More detailed experimental investigations are needed to gain in-depth understanding of the electrochemical performance of the Ni–YSZ cathode for operation in SOSE.

In literature, there are some studies on long-term performance of SOSE hydrogen production. Maskalick (1986) tested the hydrogen production characteristics by Westinghouse SOSE up to 500 h. The electronic resistance of the electrolyzer was found constant at a value of 5.5 mΩ for approximately 450 h of operation in the current density range of 1,000–10,000 A/m^2. Other tests also demonstrated that no degradation of SOSE stack performance was observed over long-term operation (Isenberg 1981, Doenitz and Schmidberger 1982). Recently, more detailed information on the durability of a single SOSE cell was reported by Hauch et al. (2006). The electrolyzer internal resistance increased by about 0.1 V during the first 100 h and stabilized or even decreased during the following 600 h. The difference between Hauch et al.'s results and the previous life tests may be caused by different material properties. The increase in cell voltage was ascribed to the buildup of impurities containing silicon at the TPB for Ni–YSZ cermet electrodes. The segregation of impurities formed a thin glass, which could cover partially the TPB surface, resulting in lower electrochemical reactivity (Mogensen et al. 2002, Liu and Jiao 2005, Norrman et al. 2006). The decrease in electrolyzer voltage could be explained by the breakup of the glass. However, this explanation has not been confirmed by experimental observation yet. Related research is in progress at the Riso National Laboratory, Denmark (Hauch et al. 2006).

In addition to the Ni–YSZ cermet, there are only very limited studies on the use of alternative cathode materials. Recently, the use of mixed conducting samaria-doped ceria (SDC) cathodes with highly dispersed Ni catalysts in SOSE was evaluated (Uchida et al. 2004, Osada et al. 2006). The highest performance was obtained at 17 vol.% Ni loading due to the effective enhancement of the reaction rate by increasing the active reaction sites and lowering the electronic resistance.

From the aforementioned studies on SOSE cathodes, it can be seen that the working mechanisms have not been fully understood yet. The relationship between the microstructure and the electrochemical performance as well as its long-term stability has not been established yet. More research works are needed to investigate the electrochemical reaction at the microscale level and to understand the effect of important impurities on cell performance.

7.5 Anode

Only two classes of materials are feasible as SOSE anode materials under highly oxidizing environments: (1) noble metals such as Pt and Au and (2) electronically conducting mixed oxides. Similar to the cathode, the use of noble metals is excluded due to cost consideration. Thus, only some electronically conducting oxides are suitable materials for using as SOSE anodes. So far, the most commonly used anode materials are mixed oxides with perovskite structure, such as the lanthanum strontium manganate (LSM).

Similar to Ni–YSZ cermet cathodes, the thermal expansion coefficient of LSM anodes is close to that of the YSZ electrolyte ($10.6–11.0 \times 10^{-6}$ K^{-1}). However, during operation, MnO_x may diffuse from the LSM into the YSZ, resulting in reaction of La_2O_3 or SrO with YSZ to form poorly conducting $La_2Zr_2O_7$ or $SrZrO_3$ (Ostergard et al. 1995). This problem can be solved by overdoping MnO_x in the LSM by a few percent so that neither free La_2O_3 nor SrO is available for reaction with YSZ (Mogensen et al. 2000b).

Eguchi et al. (1996) measured the *J-V* characteristics of SOSE with zirconia- or ceria-based electrolytes. Because of the reduction of Ce^{4+} to Ce^{3+}, the use of an YSZ–SDC double-layer electrolyte was demonstrated to be effective to lower the electrical resistance of SOSE cells. During electrolysis, the operating potential of the SOSE cell with Ni–YSZ cathodes and $La_{0.6}Sr_{0.4}MnO_3$ (LSM) anodes was found lower than that with Ni–YSZ cathodes and $La_{0.6}Sr_{0.4}CoO_3$ (LSC) anodes. However, when Pt is used as the cathode, the SOSE cell with the LSC anode showed better performance than the cell with the LSM anode.

Recently, the electrochemical properties of $La_{0.8}Sr_{0.2}MnO_3$ (LSM), $La_{0.8}Sr_{0.2}FeO_3$ (LSF), and $La_{0.8}Sr_{0.2}CoO_3$ (LSC) for use in SOSE have been compared (Wang et al. 2006). The potential of SOSE decreased in an order with the following anode materials: LSM–YSZ > LSF–YSZ > LSC–YSZ. It was also found that the LSC–YSZ composite electrode showed a steady decrease in performance over a period of 100 h due to reaction between LSC and YSZ. For comparison, LSF–YSZ showed reasonably good stability in short term at a temperature below 1073 K. From this study, it can be seen that the LSM–YSZ composite may not be the optimal material for SOSE anodes. Since there are limited reports on SOSE electrodes, more research works are needed to gain fundamental understanding of the electrode electrochemical performance as well as to identify ways to improve their long-term stability.

7.6 System Designs

7.6.1 SOSE Cell and Stack Configuration

Single cells are the smallest units of SOSE and can be in either tubular configuration or planar configuration, as shown in Figure 7.10a and b, respectively. Conventional SOSE cells are made in cylindrical shape, such as the *h*igh *o*perating *te*mperature *el*ectro*ly*sis (HOTELLY) cells and the Westinghouse electrolysis cells (Doenitz et al. 1980, Doenitz and Schmidberger 1982, Doenitz and Erdle 1985, Maskalick 1986). In the tubular SOSE cell, steam is fed through the inside of the tube and reduced to hydrogen gas and oxygen ions. The oxygen gas is extracted from the outer layer of the tubular SOSE cell. Compared with planar SOSE cells, the tubular SOSE cells exhibit higher mechanical strength and facilitate sealing. Despite larger sealing length between anode and cathode compartments, the planar cells have received more and more attention in recent years due to their better manufacturability.

The electrochemical characteristics of tubular SOSE cells and planar SOSE cells have been investigated and discussed by Hino et al. (2004). It was found that the planar SOSE cell performed significantly better than its tubular counterpart. It is because the distribution of gas species on the planar SOSE cells is more uniform. Taking into account the aforementioned factor as well as easier mass production of planar cells, the planar SOSE system configuration is advantageous and should be further investigated.

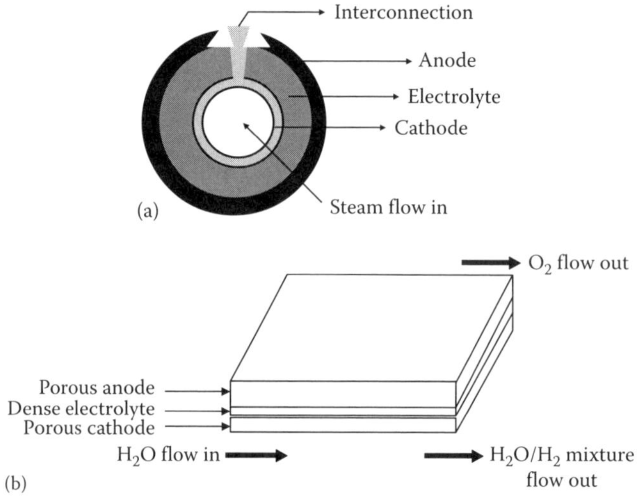

FIGURE 7.10
SOSE cell configurations—(a) tubular SOSE cell (end view) and (b) planar SOSE cell.

In the literature, several novel configurations have been proposed to improve the performance of SOFCs, such as flat-tube high–power density SOFC (HPD-SOFC), monoblock layer built SOFC (MOLB-SOFC), and the thin-walled SOFC (Kim et al. 2003, Hwang et al. 2005, Lu et al. 2005, Ramakrishna et al. 2006). In addition, there are increasingly more interests in single-chamber SOFCs and micro-SOFCs recently (Hao et al. 2006, Liu et al. 2007, Riess 2008). The single chamber concept greatly simplifies the gas supply system, but the selectivity of the electrode materials remains to be improved. The micro-SOFCs are expected to be useful for small-scale applications. Since SOFC uses the same materials but proceeds in the reverse direction with respect to SOSE, the aforementioned novel configurations for SOFC may be applicable to SOSE. More works need to be done to evaluate their applicability in SOSE.

In order to increase the hydrogen production rate, the active area of the electrolyzer should be increased. As it is difficult to achieve this task by simply increasing the single SOSE cell dimension, it is important to connect a large number of single cells to build a desired stack. A large electrolysis stack with tubular SOSE cells in a serial connection has been tested and demonstrated to be feasible (Doenitz and Schmidberger 1982). Stacks with planar SOSE cells and other novel SOSE cells are also possible, but more works are needed to optimize the stack performance.

7.6.2 Waste Heat Utilization

Since a large fraction of the heat added to the feed stream is retained in the product gas stream, it is desirable to recover the waste heat from the product gases. Doenitz et al. (1980) proposed an SOSE plant for hydrogen production, as shown in Figure 7.11. In this plant, considerable waste heat from the product gases can be recovered by preheating the feedwater through a heat exchanger. Subsequently, the feedwater/steam is superheated to reach the temperature of the SOSE cell. A recent energy/exergy analysis showed that more than half of the waste heat from the product gases could be recovered by preheating the feedwater, provided that an efficient counterflow heat exchanger is used (Ni et al. 2007b). However, it should be mentioned that since the waste heat is not sufficient for heating the feedwater to SOSE operating temperature, superheating of the feedwater/steam is

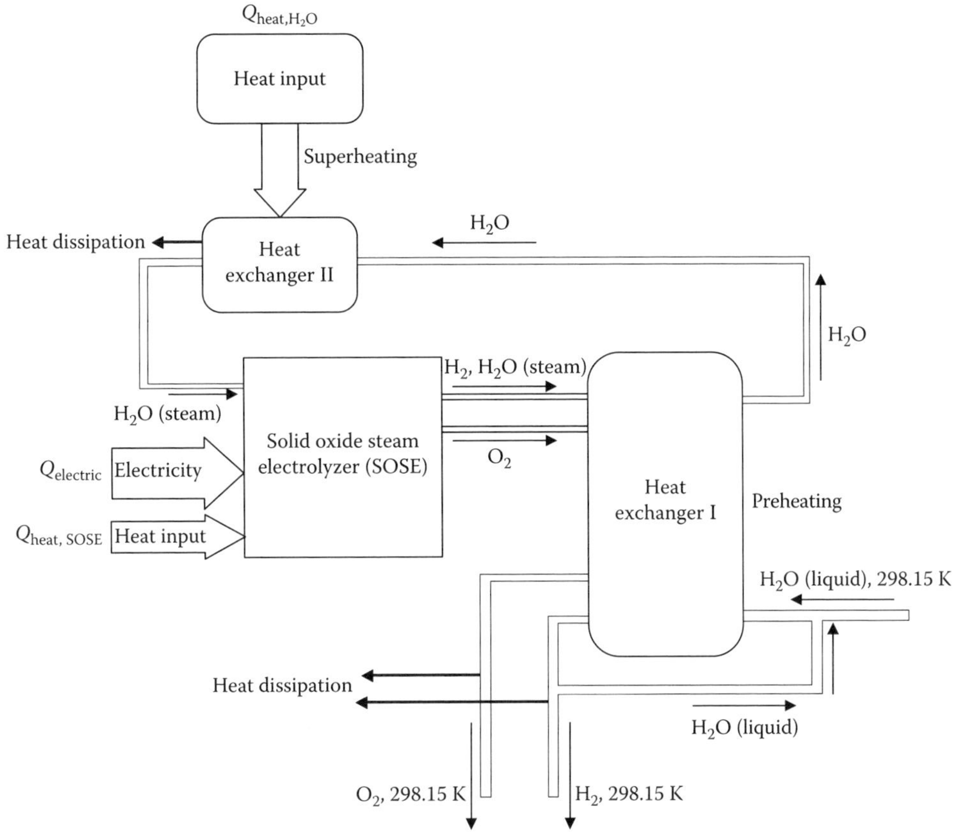

FIGURE 7.11
SOSE plant with waste heat recovery. (From Doenitz, W. et al., *Int. J. Hydrogen Energy*, 5(1), 55, 1980.)

required, and the process consumes a large amount of thermal energy. The overall energy/exergy efficiency depends on how the electricity and heat are generated as well as the type of heat exchangers used. It should be also mentioned that the practical system may be more complex than the one studied. Integration of the SOSE cell and the heat recovery unit need to be carefully studied. The SOSE sealing for practical use also needs to be carefully handled. SOSE sealing should be carefully. Both the efficiency and cost of the overall plant need to be considered for design optimization.

7.6.3 Electrode Depolarization

The SOSE hydrogen production generally requires an electrical voltage around 1.0–1.5 V for practical operation. Since electricity generation from renewable sources is still expensive, there is a need to further reduce the electrical energy demand. In order to achieve this, the anodic depolarization process can be applied. In principle, carbon and hydrocarbons, which can react with oxygen at the anode side, can be used to bring down the chemical potential between the two electrodes of the SOSE (Wendt 1990).

Martinez-Frias et al. (2001, 2003) proposed a natural gas–assisted SOSE for hydrogen production at reduced electrical energy consumption. In the natural gas–assisted SOSE, natural gas reacts with the oxygen produced in the electrolysis, reducing the electrical potential of the SOSE anode. The oxygen produced at the SOSE anode can be consumed by

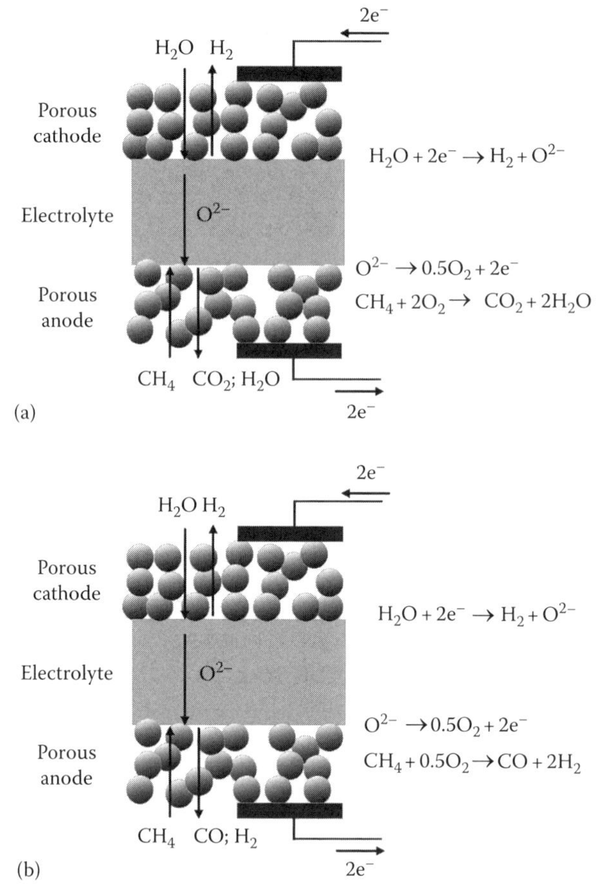

FIGURE 7.12
Schematics of natural-gas-assisted SOSE for hydrogen production—(a) natural gas total oxidation and (b) natural gas partial oxidation. (From Martinez-Frias, J. et al., *Int. J. Hydrogen Energy*, 28(5), 483, 2003.)

either total oxidation or partial oxidation of natural gas, as shown in Figure 7.12. Martinez-Frias et al. chose to study only total oxidation because it was regarded as simpler and did not need any additional water–gas shift or CO cleanup units. Experiments performed on single cells demonstrated that the operating potential of natural gas–assisted SOSE could be as low as about 0.5 V at a current density of 10,000 A/m^2 and a temperature of 973 K. The thermodynamic analysis showed that the system could reach up to 70% efficiency based on primary energy or up to 90% efficiency based on total energy input into the SOSE cell. Martinez-Frias et al.'s studies focused on total oxidation of natural gas. However, it should be mentioned that in the mode of partial oxidation, more hydrogen can be obtained due to water–gas shift reaction. Therefore, thermodynamic analyses on partial oxidation are needed to identify the optimal operating parameters for efficient hydrogen production.

Recently, the concept of a reversible coal-based SOSE was proposed (Wang et al. 2008). Similar to the natural gas–assisted SOSE, the idea is to depolarize the anode by consuming oxygen through reaction with coal, so that the hydrogen can be produced at a lower electrical energy demand. This coal-based SOSE is capable of functioning in two modes: (1) as an electrolyzer to produce hydrogen from steam and coal and (2) as a fuel cell to generate power from hydrogen and air. Theoretical analysis has shown that the hydrogen production efficiency heavily depends on the total area-specific resistance of the cell but is independent of the steam-to-carbon ratio, while the overall efficiency depends on both the cell resistance and the steam-to-carbon ratio. Increasing steam-to-carbon ratio decreases the overall energy efficiency of the process.

These studies showed that hydrogen could be produced by SOSE at a considerably lower electrical energy demand. Heat produced from partial oxidation or total oxidation at the anode can be utilized for SOSE hydrogen production since electrolysis is an endothermic process. In addition, as mentioned in the previous section, the waste heat can be recovered to further enhance the system efficiency. Demonstrated by experimental studies, the electrode depolarization should be a feasible method for electrolytic hydrogen production. However, since fossil fuels (coal or other hydrocarbons) are used for anode depolarization, it is not a clean way for hydrogen production. Considering the present high cost for hydrogen production from renewable sources, the SOSE with anode depolarization could be a good transitional technology to produce hydrogen efficiently and economically. Since the anode reactions become more complex than conventional steam electrolysis, the anodic reaction kinetics and long-term stability need to be investigated. More research works in this direction are expected to be fruitful.

7.7 Mathematical Modeling of SOSE

Most of the SOSE hydrogen production studies discussed in the previous sections are experimental in nature, with a focus on improving the electrolyte conductivity and electrode reactivity. Mathematical modeling studies on SOSE are also available in the literature. The main advantage of mathematical modeling is to facilitate detailed parametric analysis in an efficient and cost-effective manner. The numerical analysis can be conducted to characterize the complex effects of physical mechanisms (gas transport and electron/ion transport) and electrochemical reactions of SOSE. Mathematical modeling can also perform tasks that cannot be accomplished by experiments. For instance, if one is interested in the distribution of gas composition or temperature profile in the microporous electrodes, he needs to carry out a microscale modeling study instead of experimental investigation as such measurements are extremely difficult.

Therefore, mathematical modeling of SOSE is valuable in many aspects, such as microscale analysis of observed phenomena, prediction of SOSE performance, and design optimization. Although there are some SOFC modeling studies available in literature, they are not applicable to SOSE because SOSE and SOFC have different gas transport behaviors (Gopalan et al. 2006).

A macroscale electrochemical model was developed to characterize the SOEC performance (Ni et al. 2006). In this macroscale model, it was assumed that the electrochemical reaction only takes place at the electrode–electrolyte interface. According to the macroscale model, the working potential (V) of the SOEC can be calculated as

$$V = V_R + \eta_{act,a} + \eta_{act,c} + \eta_{con,a} + \eta_{con,c} + \eta_{ohm} \tag{7.4}$$

where

V_R is the reversible cell potential, which can be determined with the Nernst equation

$\eta_{act,a}$, $\eta_{act,c}$, $\eta_{con,a}$, $\eta_{con,c}$, η_{ohm} are the anode activation overpotential, cathode activation overpotential, anode concentration overpotential, cathode concentration overpotential, and ohmic overpotential, respectively

The activation overpotentials are related to the resistance of electrochemical reactions, which can be calculated with the Butler–Volmer equation:

$$J = J_0\left[\exp\left(\frac{\alpha z F\eta_{act}}{RT}\right) - \exp\left(-\frac{(1-\alpha)zF\eta_{act}}{RT}\right)\right] \tag{7.5}$$

where
J is the working current density
J_0 is the exchange current density
α is the symmetric factor
z is the number of electrons involved in the electrochemical reaction
F is the Faraday constant
R is the ideal gas constant
T is the working temperature

The ohmic overpotential is caused by the flow of ions in the dense electrolyte and can be calculated with Ohm's law. The concentration overpotentials represent the resistance of the porous electrodes to the transport of gas species. By applying Fick's law, analytical solution of the mass diffusion in the porous electrodes can be obtained:

$$\eta_{conc,c} = \frac{RT}{2F}\ln\left[\frac{\left(P^0_{H_2} + \dfrac{JRTd_c}{2FD^{eff}_{H_2O}}\right)P^0_{H_2O}}{P^0_{H_2}\left(P^0_{H_2O} - \dfrac{JRTd_c}{2FD^{eff}_{H_2O}}\right)}\right] \tag{7.6}$$

$$\eta_{conc,a} = \frac{RT}{2F}\ln\left[\left(1 + \frac{RT}{D^{eff}_{O_2}P^0_{O_2}}\frac{J}{4F}d_a\right)^{\frac{1}{2}}\right] \tag{7.7}$$

where
d_a and d_c are the thicknesses of the anode and cathode, respectively
$D^{eff}_{H_2O}$ and $D^{eff}_{O_2}$ are the effective diffusion coefficients of H_2O and O_2, respectively
P^0 is the gas partial pressure at the electrode surface

The resulting analytical model showed good agreement with experimental data from literature. Through parametric simulations, the anode-supported structure was identified as the most favorable configuration for SOEC (Figure 7.13) (Ni et al. 2006).

Recently, an electrochemical modeling study on SOEC based on proton-conducting electrolytes (SOEC-H) was conducted (Ni et al. 2008c). Different from the oxygen ion conducting–based SOEC (SOEC-O), steam is presented at the anode in an SOEC-H, which leads to complex multicomponent mass transport in the porous anode.

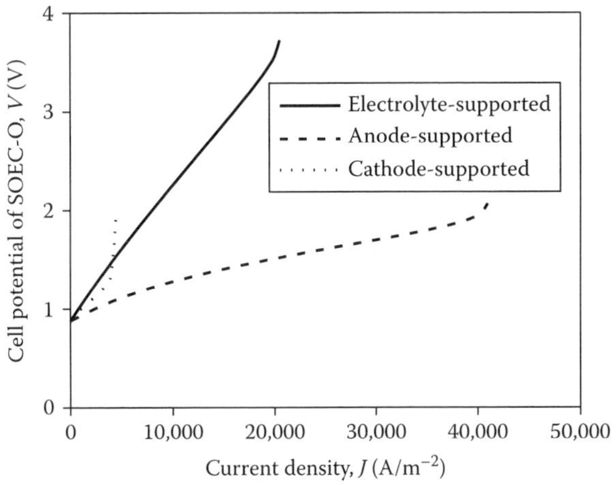

FIGURE 7.13
Theoretical *J-V* characteristics by cathode-supported, electrolyte-supported, and anode-supported SOEC cells based on oxygen-ion-conducting electrolyte (SOEC-O). (From Ni, M. et al., *Chem. Eng. Technol.*, 29(5), 636, 2006.)

As a result, a numerical method is needed to calculate the anode concentration overpotential through the dusty gas model:

$$\frac{N_i}{D_{i,k}^{eff}} + \sum_{j=1,j\neq i}^{n} \frac{y_j N_i - y_i N_j}{D_{ij}^{eff}} = -\frac{1}{RT}\left[P\frac{dy_i}{dx} + y_i \frac{dP}{dx}\left(1 + \frac{B_g P}{D_{i,k}^{eff}\mu_m}\right)\right] \tag{7.8}$$

where

- N_i and y_i are the flux and molar fraction of species i
- μ_m is the viscosity of the gas mixture
- B_g is the permeation coefficient of the porous electrode

As H_2 is the only transporting gas at the cathode of the SOEC-H, its transportation can be modeled with Darcy's law, and the analytical solution can be derived as

$$\eta_{conc,c} = \frac{RT}{2F}\ln\left(\frac{\sqrt{\left(P_{H_2}^0\right)^2 + \frac{JRT\mu_{H_2}d_c}{FB_g}}}{P_{H_2}^0}\right) \tag{7.9}$$

It is found that the use of different electrolyte materials not only causes quite different ion conduction mechanisms of the dense electrolyte but also completely different mass transfer behaviors. As a result, the cathode-supported configuration is identified as the most favorable design for SOEC-H (Figure 7.14).

The weakness of these macroscale models was that the effect of the electrode microstructure on SOEC performance could not be accurately predicted since the electrochemical reactions were assumed independent of the electrode structural properties. In order to solve this problem, a microscale model has been developed (Ni et al. 2007c,d). In this microscale model, it was assumed that the electrochemical reactions could occur inside

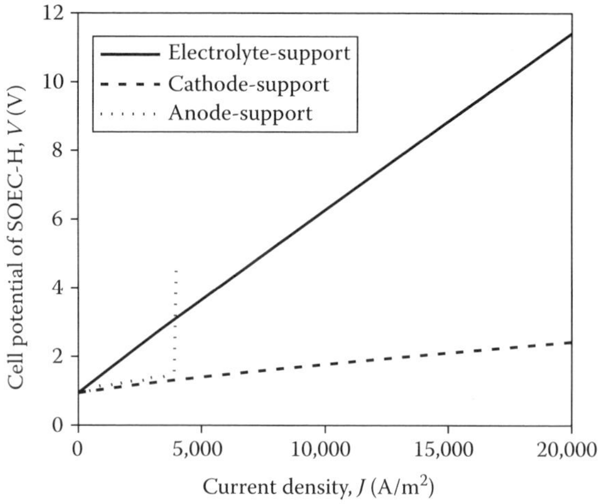

FIGURE 7.14
Theoretical *J-V* characteristics by cathode-supported, electrolyte-supported, and anode-supported SOEC cells based on proton conducting electrolyte (SOEC-H). (From Ni, M. et al., *Int. J. Hydrogen Energy*, 33(15), 4040, 2008d.)

the porous electrode. It was found that the electrochemical reactions mainly occurred in a thin layer near the electrode–electrolyte interface, while most of the electrode layer was only used to provide mechanical strength and to support gas transport (Figure 7.15). Based on this finding, the concept of *particle-size graded electrode* was proposed to improve the electrode performance, as shown in Figure 7.16. The idea was to use small particles at the electrode–electrolyte interface to maximize the reactive surface area and to use larger particles at the electrode outer layer to support gas transport.

These modeling studies focused on the transport and electrochemical reactions inside the porous electrode. Demin et al. (2007) developed a theoretical model to study the performance of SOEC based on solid oxide co-ionic electrolytes. It was found that the counterflow mode (reactants) was preferred to the coflow mode. Recently, the distributions of current density, temperature, and gas composition in a single SOEC and a stack with

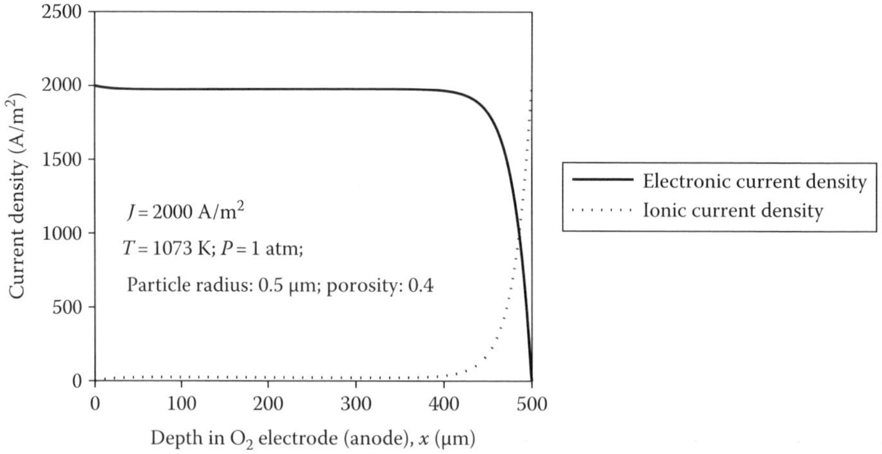

FIGURE 7.15
Distribution of current densities in the porous anode of an SOEC-O. (From Ni, M. et al., *Electrochim. Acta*, 52(24), 6707, 2007c.)

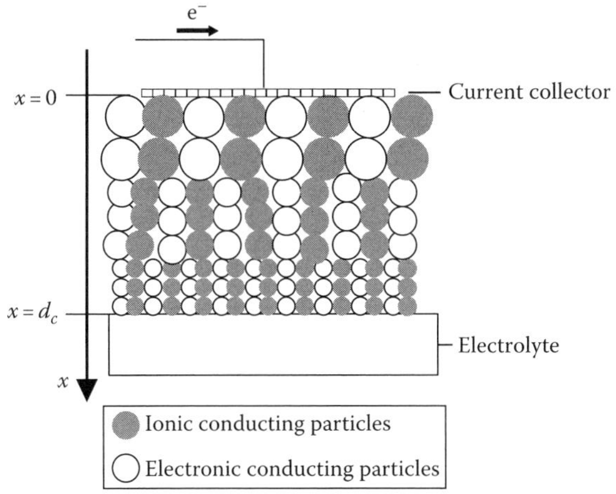

FIGURE 7.16
SOEC electrode with particle-size graded structure. (From Ni, M. et al., *Electrochim. Acta*, 52(24), 6707, 2007c.)

10 planar SOECs were calculated with commercial software *Fluent* (Hawkes et al. 2007, Herring et al. 2007). These studies provided important information on the SOEC and stack performance and could be useful for design optimization. In another study performed by Dutta et al. (1997), the laminar flow and heat transfer in a tubular SOEC was investigated. Pressure and velocity coupling was solved by the semi-implicit method for pressure-linked equations revised (SIMPLER) methodology. It was found that the cooling effect due to endothermic reactions was significant. As a result, heat radiation from the casing surface was found to play an important role in determining the temperatures of the SOEC and the fluids. However, this fluid flow and heat transfer model did not include the heating effect of the overpotentials and, thus, overestimated the cooling effect of the endothermic reactions as well as the radiation effect. Obviously, in order to obtain accurate results, such as the distributions of temperature, current density, and gas composition along the tubular cell, the fluid flow and heat transfer model must be coupled with the electrochemical model. More recently, a thermodynamic analysis of the efficiency of hydrogen production by SOEC integrated with nuclear power plants has been conducted (Liu et al. 2008). It was found that the electrical efficiency of the high-temperature gas-cooled reactor had the most significant effect on the overall efficiency for hydrogen production. However, the operating parameters remain to be optimized for achieving the highest overall efficiency.

7.8 Conclusion

SOSE offers a practical solution for clean hydrogen production from renewable resources. In this work, a comprehensive review of the state-of-the-art SOSE technology for hydrogen production is presented.

The developments of important SOSE components, such as electrolyte and electrode materials, have been reviewed. For SOSE working at high temperature, YSZ appears to be the best option of electrolytes because of its high ionic conductivity and low cost. In addition, YSZ is physically and chemically compatible with the electrode materials.

At an intermediate temperature, LSGM seems to be suitable as the electrolyte due to its high ionic conductivity. However, Ni is excluded as a cathode material due to its reaction with the LSGM electrolyte. SDC and GDC are potentially electrolyte materials for SOSE working at an intermediate temperature. In order to avoid reduction of Ce^{4+} to Ce^{3+}, codoping or using of a blocking layer is needed to ensure reliable operation. The Ni–YSZ and LSM–YSZ cermets are the most widely used cathode and anode materials for SOSE. However, recent studies indicated that other materials, such as LSF, might be better than LSM. More research works are required to study the electrode–electrolyte compatibility and the long-term stability of the SOSE cell. The fundamental mechanisms of electrochemical reactions at the electrodes have not been fully understood yet. The effect of impurity segregation on the electrode long-term stability needs to be further investigated.

The SOSE cells can be in tubular or planar form. Although tubular SOSE cells have higher mechanical strength than the planar SOSE cells, planar cells are preferred due to better manufacturability and higher electrochemical performance. The novel configurations designed for SOFC may not be applicable to SOSE. More works are required to evaluate their feasibility for SOSE applications. In order to achieve high efficiency of the overall SOSE plant, the waste heat from the product gas can be utilized to preheat the feedwater. Anode depolarization by introducing carbon or hydrocarbons at the anode side provides an effective way to reduce the electrical energy demand for SOSE hydrogen production. The overall efficiency of the depolarized SOSE process needs to be optimized since the hydrocarbons contain a large amount of energy. The depolarization process could be a transitional method for renewable hydrogen production, and research works in this direction are expected to be fruitful.

Mathematical modeling of SOSE is an economical and powerful method to predict SOSE performance and to optimize the operating and structural parameters. A macroscale and a microscale model have been recently developed to characterize the SOSE performance. A new design of SOSE electrodes with particle-size graded structure has been proposed to enhance SOSE performance. In addition, a few studies have been performed to investigate the fluid flow and heat transfer in the SOSE flow channels. The coupled heat/mass transfer and electrochemical reaction phenomena remain to be investigated. The literature currently available lacks detailed analyses and optimization of the SOSE stack performance. In addition, mathematical modeling studies considering SOSE anode depolarization will be useful to understand the electrochemical reactions and the depolarization process as well as to enhance the SOSE performance. With the developments in both experimental and mathematical modeling investigations, it is anticipated that SOSE will play an important role in hydrogen production and contribute much to the coming hydrogen economy.

References

Ai, N., Lu, Z., Chen, K.F., Huang, X.Q., Liu, Y.W., Wang, R.F., and Su, W.H. 2006. Preparation of $Sm_{0.2}Ce_{0.8}O_{1.9}$ membranes on porous substrates by a slurry spin coating method and its application in IT-SOFC. *Journal of Membrane Science* 286(1–2):255–259.

Arachi, Y., Sakai, H., Yamamoto, O., Takeda, Y., and Imanishai, N. 1999. Electrical conductivity of the ZrO_2–Ln_2O_3 (Ln = lanthanides) system. *Solid State Ionics* 121(1–4):133–139.

Badwal, S.P.S., Ciacchi, F.T., and Milosevic, D. 2000. Scandia–zirconia electrolytes for intermediate temperature solid oxide fuel cell operation. *Solid State Ionics* 136–137:91–99.

Baker, R.T., Gharbage, B., and Marques, F.M.B. 1997. Ionic and electronic conduction in Fe and Cr doped (La, Sr)$GaO_{3-\delta}$. *Journal of the Electrochemical Society* 144(9):3130–3135.

Barbir, F. 2005. PEM electrolysis for production of hydrogen from renewable energy sources. *Solar Energy* 78:661–669.

Bi, Z.H., Yi, B.L., Wang, Z.W., Dong, Y.L., Wu, H.J., She, Y.C., and Cheng, M.J. 2004. A high-performance anode-supported SOFC with LDC-LSGM bilayer electrolytes. *Electrochemical and Solid State Letters* 7(5):A105–A107.

Bonanos, N., Ellis, B., Knight, K.S., and Mahmood, M.N. 1989. Ionic conductivity of gadolinium-doped barium cerate perovskites. *Solid State Ionics* 35(1–2):179–188.

Brahim, C., Ringuede, A., Cassir, M., Putkonen, M., and Niinisto, L. 2007. Electrical properties of thin yttria-stabilized zirconia overlayers produced by atomic layer deposition for solid oxide fuel cell applications. *Applied Surface Science* 253(8):3962–3968.

Cervera, R.B., Oyama, Y., Miyoshi, S., Kobayashi, K., Yagi, T., and Yamaguchi, S. 2008. Structural study and proton transport of bulk nanograined Y-doped BaZrO3 oxide protonics materials. *Solid State Ionics* 179:236–242.

Chinarro, E., Jurado, J.R., and Colomer, M.T. 2007. Synthesis of ceria-based electrolyte nanometric powders by urea-combustion technique. *Journal of the European Ceramic Society* 27(13–15):3619–3623.

Chockalingam, R., Amarakoon, V.R.W., and Giesche, H. 2007. Alumina/cerium oxide nano-composite electrolyte for solid oxide fuel cell applications. *Journal of the European Ceramic Society* 28(5):959–963.

Cong, L.G., He, T.M., Ji, Y., Guan, P.F., Huang, Y.L., and Su, W.H. 2003. Synthesis and characterization of IT-electrolyte with perovskite structure $La_{0.8}Sr_{0.2}Ga_{0.85}Mg_{0.15}O_{3-\delta}$ by glycine-nitrate combustion method. *Journal of Alloys and Compounds* 348(1–2):325–331.

Dahl, P., Kaus, I., Zhao, Z., Johnsson, M., Nygren, M., Wiik, K., Grande, T., and Einarsrud, M.A. 2007. Densification and properties of zirconia prepared by three different sintering techniques. *Ceramics International* 33(8):1603–1610.

Demin, A., Gorbova, E., and Tsiakaras, P. 2007. High temperature electrolyzer based on solid oxide co-ionic electrolyte: A theoretical model. *Journal of Power Sources* 171(1):205–211.

Dikmen, S., Shuk, P., Greenblatt, M., and Gocmez, H. 2002. Hydrothermal synthesis and properties of $Ce_{1-x}Gd_xO_{2-\delta}$ solid solutions. *Solid State Science* 4(5):585–590.

Doenitz, W. and Erdle, E. 1985. High-temperature electrolysis of water vapor-status of development and perspectives for application. *International Journal of Hydrogen Energy* 10(5):291–295.

Doenitz, W. and Schmidberger, R. 1982. Concepts and design for scaling up high temperature water vapor electrolysis. *International Journal of Hydrogen Energy* 7(4):321–330.

Doenitz, W., Schmidberger, R., Steinheil, E., and Streicher, R. 1980. Hydrogen production by high temperature electrolysis of water vapour. *International Journal of Hydrogen Energy* 5(1):55–63.

Dutta, S., Morehouse, J.H., and Khan, J.A. 1997. Numerical analysis of laminar flow and heat transfer in a high temperature electrolyzer. *International Journal of Hydrogen Energy* 22(9):883–895.

Eguchi, K., Hatagishi, T., and Arai, H. 1996. Power generation and steam electrolysis characteristics of an electrochemical cell with a zirconia- or ceria-based electrolyte. *Solid State Ionics* 86–88:1245–1249.

Etsell, T.H. and Flengas, S.N. 1970. Electrical properties of solid oxide electrolytes. *Chemical Reviews* 70(3):339–376.

Flint, S.D., Hartmanova, M., Jones, J.S., and Slade, R.C.T. 1996. Microstructure of Ca-doped barium cerate electrolytes $BaCe_{1-x}Ca_xO_{3-x}$ (= 0, 0.02, 0.05, 0.10 and 0.15). *Solid State Ionics* 86–88(1):679–683.

Gao, H., Liu, J., Chen, H., Li, S., He, T., Ji, Y., and Zhang, J. 2008. The effect of Fe doping on the properties of SOFC electrolyte YSZ. *Solid State Ionics* 179:1620–1624.

Gibson, I.R., Dransfield, G.P., and Irvine, J.T.S. 1998. Influence of yttria concentration upon electrical properties and susceptibility to ageing of yttria-stabilised zirconias. *Journal of the European Ceramic Society* 18(6):661–667.

Gopalan, S., Ye, G.S., and Pal, U.B. 2006. Regenerative, coal-based solid oxide fuel cell-electrolyzers. *Journal of Power Sources* 162(1):74–80.

Gorelov, V.P., Bronin, D.I., Sokolova, J.V., Nafe, H., and Aldinger, F. 2001. The effect of doping and processing conditions on properties of $La_{1-x}Sr_xGa_{1-y}Mg_yO_{3-a}$. *Journal of the European Ceramic Society* 21(13):2311–2317.

Graf, D., Monnerie, N., Roeb, M., Schmitz, M., and Sattler, C. 2008. Economic comparison of solar hydrogen generation by means of thermochemical cycles and electrolysis. *International Journal of Hydrogen Energy* 33:4511–4519. doi:10.1016/j.ijhydene.2008.05.08.

Guo, X. and Waser, R. 2006. Electrical properties of the grain boundaries of oxygen ion conductors: Acceptor-doped zirconia and ceria. *Progress in Material Sciences* 51(2):151–210.

Haering, C., Roosen, A., Schichl, H., and Schnoller, M. 2005. Degradation of the electrical conductivity in stabilized zirconia system part II: Scandia-stabilized zirconia. *Solid State Ionics* 176(3–4):261–268.

Hao, Y., Shao, Z.P., Mederos, J., Lai, W., Goodwin, D.G., and Haile, S.M. 2006. Recent advances in single-chamber fuel cells: Experiment and modelling. *Solid State Ionics* 177(19–25):2013–2021.

Hauch, A., Jensen, S.H., Ramousse, S., and Mogensen, M. 2006. Performance and durability of solid oxide electrolysis cells. *Journal of the Electrochemical Society* 153(9):A1741–A1747.

Hawkes, G.L., O'Brien, J.E., Stoots, C.M., Herring, J.S., and Shahnam, M. 2007. Computational fluid dynamics model of a planar solid-oxide electrolysis cell for hydrogen production from nuclear energy. *Nuclear Technology* 158(2):132–144.

Herring, J.S., O'Brien, J.E., Stoots, C.M., Hawkes, G.L., Hartvigsen, J.J., and Shahnam, M. 2007. Progress in high-temperature electrolysis for hydrogen production using planar SOFC technology. *International Journal of Hydrogen Energy* 32(4):440–450.

Hino, R., Haga, K., Aita, H., and Sekita, K. 2004. R&D on hydrogen production by high-temperature electrolysis of steam. *Nuclear Engineering and Design* 233:363–375.

Hirano, M., Inagaki, M., Mizutani, Y., Nomura, K., Kawai, M., and Nakamura, Y. 2000. Mechanical and electrical properties of Sc_2O_3-doped zirconia ceramics improved by postsintering with HIP. *Solid State Ionics* 133(1–2):1–9.

Hong, H.S., Chae, U.S., and Choo, S.T. 2008. The effect of ball milling parameters and Ni concentration on a YSZ-coated Ni composite for a high temperature electrolysis cathode. *Journal of Alloys and Compounds* 449:331–334.

Hong, H.S., Chae, U.S., Choo, S.T., and Lee, K.S. 2005. Microstructure and electrical conductivity of Ni/YSZ and NiO/YSZ composites for high-temperature electrolysis prepared by mechanical alloying. *Journal of Power Sources* 149:84–89.

Huang, K.Q., Feng, F., and Goodenough, J.B. 1996. Bi_2O_3-Y_2O_3-CeO_2 solid solution oxide-ion electrolyte. *Solid State Ionics* 89(1–2):17–24.

Huang, K.Q. and Goodenough, J.B. 2000. A solid oxide fuel cell based on Sr- and Mg-doped $LaGaO_3$ electrolyte: The role of a rare-earth oxide buffer. *Journal of Alloys and Compounds* 303–304:454–464.

Huang, W., Shuk, P., and Greenblatt, M. 1997. Hydrothermal synthesis and properties of $Ce_{1-x}Sm_xO_{2-x/2}$ solid solutions. *Chemistry of Materials* 9(10):2240–2245.

Hui, S.Q., Yang, D.F., Wang, Z.W., Yick, S., Deces-Petit, C., Qu, W., Tuck, A., Maric, R., and Ghosh, D. 2007. Metal-supported solid oxide fuel cell operated at 400–600°C. *Journal of Power Sources* 167(2):336–339.

Hwang, J.J., Chen, C.K., and Lai, D.Y. 2005. Computational analysis of species transport and electrochemical characteristics of a MOLB-type SOFC. *Journal of Power Sources* 140(2):235–242.

Im, J.M., You, H.J., Yoon, Y.S., and Shin, D.W. 2007. Synthesis of nano-sized gadolinia doped ceria powder by aerosol flame deposition. *Journal of the European Ceramic Society* 27(13–15):3671–3675.

Isenberg, A.O. 1981. Energy conversion via solid oxide electrolyte electrochemical cells at high temperature. *Solid State Ionics* 3–4:431–437.

Ishihara, T., Matsuda, H., and Takita, Y. 1994. Doped $LaGaO_3$ perovskite type oxide as a new oxide ionic conductor. *Journal of America Chemical Society* 16(9):3801–3803.

Ivanov, V.V., Lipilin, A.S., Kotov, Y.A., Khrustov, V.R., Shkerin, S.N., Pararnin, S.N., Spirin, A.V., and Kaygorodov, A.S. 2006. Formation of a thin-layer electrolyte for SOFC by magnetic pulse compaction of tapes cast of nanopowders. *Journal of Power Sources* 159(1):605–612.

Iwahara, H., Uchida, H., and Yamasaki, I. 1987. High-temperature steam electrolysis using $SrCeO_3$-based proton conductive solid electrolyte. *International Journal of Hydrogen Energy* 12(2):73–77.

Jiang, S.P. and Chan, S.H. 2004. A review of anode materials development in solid oxide fuel cells. *Journal of Materials Science* 39:4405–4439.

Jiang, Y.Z., Gao, J.F., Liu, M.F., Wang, Y.Y., and Meng, G.Y. 2007. Fabrication and characterization of Y_2O_3 stabilized ZrO_2 films deposited with aerosol-assisted MOCVD. *Solid State Ionics* 177(39–40):3405–3410.

Kharton, V.V., Marques, F.M.B., and Atkinson, A. 2004. Transport properties of solid oxide electrolyte ceramics: A brief review. *Solid State Ionics* 174(1–4):135–149.

Kharton, V.V., Naumovich, E.N., and Vecher, A.A. 1999. Research on the electrochemistry of oxygen ion conductors in the former Soviet Union. I. ZrO_2-based ceramic materials. *Journal of Solid State Electrochemistry* 3(2):61–81.

Kharton, V.V., Shaula, A.L., Vyshatko, N.P., and Marques, F.M.B. 2003. Electron-hole transport in $(La_{0.9}Sr_{0.1})_{0.98}Ga_{0.8}Mg_{0.2}O_{3-\delta}$ electrolyte: Effects of ceramic microstructure. *Electrochimica Acta* 48(13):1817–1828.

Khorkounov, B.A., Nafe, H., and Aldinger, F. 2006. Relationship between the ionic and electronic partial conductivities of co-doped LSGM ceramics from oxygen partial pressure dependence of the total conductivity. *Journal of Solid State Electrochemistry* 10(7):479–487.

Kilner, J.A. and Brook, R.J. 1982. A study of oxygen ion conductivity in doped non-stoichiometric oxides. *Solid State Ionics* 6(3):237–252.

Kim, J.H., Song, R.H., Song, K.S., Hyun, S.H., Shin, D.R., and Yokokawa, H. 2003. Fabrication and characteristics of anode-supported flat-tube solid oxide fuel cell. *Journal of Power Sources* 122(2):138–143.

Kim, S.G., Yoon, S.P., Nam, S.W., Hyun, S.H., and Hong, S.A. 2002. Fabrication and characterization of a YSZ/YDC composite electrolyte by a sol-gel coating method. *Journal of Power Sources* 110(1):222–228.

Kosacki, I., Anderson, H.U., Mizutani, Y., and Ukai, K. 2005b. Nonstoichiometry and electrical transport in Sc-doped zirconia. *Solid State Ionics* 152–153:431–438.

Kosacki, I., Rouleau, C.M., Becher, P.F., Bentley, J., and Lowndes, D.H. 2005a. Nanoscale effects on the ionic conductivity in highly textured YSZ thin film. *Solid State Ionics* 176(13–14):1319–1326.

Kreuer, K.D. 2003. Proton-conducting oxides. *Annual Review of Materials Research* 33:333–359.

Kumar, B., Chen, C., Varanasi, C., and Fellner, J.P. 2005. Electrical properties of heterogeneously doped yttria stabilized zirconia. *Journal of Power Sources* 140(1):12–20.

Liu, M.Y., Yu, B., Xu, J.M., and Chen, J. 2008. Thermodynamic analysis of the efficiency of high temperature steam electrolysis for hydrogen production. *Journal of Power Sources* 177(2):493–499.

Liu, N., Yuan, Y.P., Majewski, P., and Aldinger, F. 2006. Synthesis of $La_{0.85}Sr_{0.15}Ga_{0.85}Mg_{0.15}O_{2.85}$ materials for SOFC application by acrylamide polymerisation. *Materials Research Bulletin* 41(3):461–468.

Liu, Y., Hashimoto, S.I., Nishino, H., Takei, K., Mori, M., Suzuki, T., and Funahashi, Y. 2007. Fabrication and characterization of micro-tubular cathode-supported SOFC for intermediate temperature operation. *Journal of Power Sources* 174(1):95–102.

Liu, Y. and Lao, L.E. 2006. Structural and electrical properties of ZnO-doped 8 mol% yttria-stabilized zirconia. *Solid State Ionics* 177(1–2):159–163.

Liu, Y.L. and Jiao, C.G. 2005. Microstructure degradation of an anode/electrode interface in SOFC studied by transmission electron microscopy. *Solid State Ionics* 176(5–6):435–442.

Lu, Y.X., Schaefer, L., and Li, P.W. 2005. Numerical study of flat-tube high power density solid oxide fuel cell: Part I. Heat/mass transfer and fluid flow. *Journal of Power Sources* 140(2):331–339.

Ma, G.L., Shimura, T., and Iwahara, H. 1999. Simultaneous doping with La^{3+} and Y^{3+} for Ba^{2+} and Ce^{4+} sites in $BaCeO_3$ and the ionic conduction. *Solid State Ionics* 120(1–4):51–60.

Marques, F.M.B. and Navarro, L.M. 1996. Performance of double layer electrolyte cells part I: Model behavior. *Solid State Ionics* 90(1–4):183–192.

Marques, F.M.B. and Navarro, L.M. 1997. Performance of double layer electrolyte cells part II: GCO/YSZ, a case study. *Solid State Ionics* 100(1–2):29–38.

Martinez-Frias, J., Pham, A.Q., and Aceves, S.M. 2001. Analysis of a high-efficiency natural gas-assisted steam electrolyzer for hydrogen production. *American Society of Mechanical Engineers, International Mechanical Engineering Congress and Exposition*, New York, November 11–16.

Martinez-Frias, J., Pham, A.Q., and Aceves, S.M. 2003. A natural gas-assisted steam electrolyzer for high-efficiency production of hydrogen. *International Journal of Hydrogen Energy* 28(5):483–490.

Maskalick, N.J. 1986. High temperature electrolysis cell performance characterization. *International Journal of Hydrogen Energy* 11(9):563–570.

Mehta, K., Xu, R., and Virkar, A.V. 1998. Two-layer fuel cell electrolyte structure by sol-gel processing. *Journal of Sol-Gel Science and Technology* 11(2):203–207.

Mogensen, M., Jensen, K.V., Jorgensen, M.J., and Primdahl, S. 2002. Progress in understanding SOFC electrodes. *Solid State Ionics* 150(1–2):123–129.

Mogensen, M., Primdahl, S., Jorgensen, M.J., and Bagger, C. 2000b. Composite electrodes in solid oxide fuel cells and similar solid state devices. *Journal of Electroceramics* 5(2):141–152.

Mogensen, M., Sammes, N.M., and Tompsett, G.A. 2000a. Physical, chemical and electrochemical properties of pure and doped ceria. *Solid State Ionics* 129(1–4):63–94.

Momma, A., Kaga, Y., Takano, K., Nozaki, K., Negishi, A., Kato, K., Kato, T. et al. 2005. Experimental investigation of anodic gaseous concentration of a practical seal-less solid oxide fuel cell. *Journal of Power Sources* 145(2):169–177.

Mori, T., Drennan, J., Lee, J.H., Li, J.G., and Ikegami, T. 2002. Oxide ionic conductivity and microstructures of Sm- or La-doped CeO_2-based systems. *Solid State Ionics* 154–155:461–466.

Muccillo, E.N.S. and Kleitz, M. 1995. Ionic conductivity of fully stabilized ZrO_2: MgO and blocking effects. *Journal of the European Ceramic Society* 15(1):51–55.

Muller, J., Kreuer, K.D., Maier, J., Matsuo, S., and Ishigame, M. 1997. A conductivity and thermal gravimetric analysis of a Y-doped $SrZrO_3$ single crystal. *Solid State Ionics* 97(1–4):421–427.

Ni, M., Leung, M.K.H., and Leung, D.Y.C. 2006. An electrochemical model of solid oxide steam electrolyzer for hydrogen production. *Chemical Engineering & Technology* 29(5):636–642.

Ni, M., Leung, M.K.H., and Leung, D.Y.C. 2007b. Energy and exergy analysis of hydrogen production by solid oxide steam electrolyzer plant. *International Journal of Hydrogen Energy* 32(8):4648–4660.

Ni, M., Leung, M.K.H., and Leung, D.Y.C. 2007c. Mathematical modelling of the coupled transport and electrochemical reactions in solid oxide steam electrolyzer for hydrogen production. *Electrochimica Acta* 52(24):6707–6718.

Ni, M., Leung, M.K.H., and Leung, D.Y.C. 2007d. Micro-scale modelling analyses for advanced design of solid oxide steam electrolyser to enhance hydrogen production. *Proceedings of the 5 International Conference-Hydrogen Economy and Hydrogen Treatment of Materials (HTM-2007)*, May 21–25, 2007, Donetsk, Ukraine, pp. 199–205.

Ni, M., Leung, M.K.H., and Leung, D.Y.C. 2008a. Technological development of hydrogen production by solid oxide electrolyzer cell (SOEC). *International Journal of Hydrogen Energy* 33:2337–2354.

Ni, M., Leung, M.K.H., and Leung, D.Y.C. 2008b. Energy and exergy analysis of hydrogen production by a proton exchange membrane (PEM) electrolyzer plant. *Energy Conversion and Management* 49:2748–2756.

Ni, M., Leung, M.K.H., and Leung, D.Y.C. 2008c. Theoretical analysis of reversible solid oxide fuel cell based on proton-conducting electrolyte. *Journal of Power Sources* 177:369–375.

Ni, M., Leung, M.K.H., and Leung, D.Y.C. 2008d. Electrochemical modelling of hydrogen production by proton-conducting solid oxide steam electrolyzer. *International Journal of Hydrogen Energy* 33(15):4040–4047.

Ni, M., Leung, M.K.H., Leung, D.Y.C., and Sumathy, K. 2007a. A review and recent developments in photocatalytic water-splitting using TiO_2 for hydrogen production. *Renewable and Sustainable Energy Reviews* 11(3):401–425.

Nimat, R.K., Betty, C.A., and Pawar, S.H. 2006. Spray pyrolytic deposition of solid electrolyte $Bi_2V_{0.9}Cu_{0.1}O_{5.35}$ films. *Applied Surface Science* 253(5):2702–2707.

Norrman, K., Hansen, K.V., and Mogensen, M. 2006. Time-of-flight secondary ion mass spectrometry as a tool for studying segregation phenomena at nickel-YSZ interfaces. *Journal of the European Ceramic Society* 26(6):967–980.
O'Brien, J.E., Stoots, C.M., Herring, J.S., Lessing, P.A., Hartvigsen, J.J., and Elangovan, S. 2005. Performance measurements of solid oxide electrolysis cells for hydrogen production. *Journal of Fuel Cell Science and Technology* 2(3):156–163.
Osada, N., Uchida, H., and Watanabe, M. 2006. Polarization behavior of SDC cathode with highly dispersed Ni catalysts for solid oxide electrolysis. *Journal of the Electrochemical Society* 153(5):A816–A820.
Ostergard, M.J.L., Clausen, C., Bagger, C., and Mogensen, M. 1995. Manganite-zirconia composite cathodes for SOFC: Influence of structure and composition. *Electrochimica Acta* 40(12):1971–1981.
Paria, M.K. and Maiti, H.S. 1984. Electrical conduction in barium cerate doped with M_2O_3 (M = La, Nd, Ho). *Solid State Ionics* 13(4):285–292.
Peng, R.R., Wu, Y., Yang, L.Z., and Mao, Z.Q. 2006. Electrochemical properties of intermediate temperature SOFCs based on proton conducting Sm-doped $BaCeO_3$ electrolyte thin film. *Solid State Ionics* 177(3–4):389–393.
Perez-Coll, D., Nunez, P., Ruiz-Morales, J.C., Pena-Martinez, J., and Frade, J.R. 2007. Re-examination of bulk and grain boundary conductivities of $Ce_{1-x}Gd_xO_{2-\delta}$ ceramics. *Electrochimica Acta* 52(5):2001–2008.
Polini, R., Pamio, A., and Traversa, E. 2004. Effect of synthetic route on sintering behaviour, phase purity and conductivity of Sr- and Mg-doped $LaGaO_3$ perovskites. *Journal of the European Ceramic Society* 24(6):1365–1370.
Prabhakaran, K., Beigh, M.O., Lakra, J., Gokhale, N.M., and Sharma, S.C. 2007. Characteristics of 8 mol% yttria stabilized zirconia powder prepared by spray drying process. *Journal of Materials Processing Technology* 189(1–3):178–181.
Ramakrishna, P.A., Yang, S., and Sohn, C.H. 2006. Innovative design to improve the power density of a solid oxide fuel cell. *Journal of Power Sources* 158(1):378–384.
Riess, I. 2008. On the single chamber solid oxide fuel cells. *Journal of Power Sources* 175(1):325–337.
Sammes, N., Phillips, R., and Smirnova, A. 2004. Proton conductivity in stoichiometric and sub-stoichiometric yttrium doped $SrCeO_3$ ceramic electrolytes. *Journal of Power Sources* 134(2):153–159.
Sammes, N.M., Tompsett, G.A., Nafe, H., and Aldinger, F. 1999. Bismuth based oxide electrolytes—Structure and ionic conductivity. *Journal of the European Ceramic Society* 19(10):1801–1826.
Sarat, S., Sammes, N., and Smirnova, A. 2006. Bismuth oxide doped scandia-stabilized zirconia electrolyte for the intermediate temperature solid oxide fuel cells. *Journal of Power Sources* 162(2):892–896.
Seo, D.J., Ryu, K.O., Park, S.B., Kim, K.Y., and Song, R.H. 2006. Synthesis and properties of $Ce_{1-x}Gd_xO_{2-x/2}$ solid solution prepared by flame spray pyrolysis. *Materials Research Bulletin* 41(2):359–366.
Sha, X.Q., Lu, Z., Huang, X.Q., Miao, J.P., Ding, Z.H., Xin, X.S., and Su, W.H. 2007. Study on La and Y co-doped ceria-based electrolyte materials. *Journal of Alloys and Compounds* 428(1–2):59–64.
Shi, M., Liu, N., Xu, Y.D., Yuan, Y.P., Majewski, P., and Aldinger, F. 2006. Synthesis and characterization of Sr- and Mg-doped $LaGaO_3$ by using glycine-nitrate combustion method. *Journal of Alloys and Compounds* 425(1–2):348–352.
Shuk, P., Wiemhofer, H.D., Guth, U., Gopel, W., and Greenblatt, M. 1996. Oxide ion conducting solid electrolytes based on Bi_2O_3. *Solid State Ionics* 89(3–4):179–196.
Singhal, S.C. and Kendall, K. 2003. *High Temperature Solid Oxide Fuel Cells: Fundamentals, Design, and Applications*, Elsevier Science Publishers B.V., New York.
Steele, B.C.H. 2000. Appraisal of $Ce_{1-y}Gd_yO_{2-y/2}$ electrolytes for IT-SOFC operation at 500°C. *Solid State Ionics* 129(1–4):95–110.
Stevenson, J.W., Armstrong, T.R., Pederson, L.R., Li, J., Lewinsohn, C.A., and Baskaran, S. 1998. Effect of A-site cation nonstoichiometry on the properties of doped lanthanum gallate. *Solid State Ionics* 113–115:571–583.

Stevenson, J.W., Hasinska, K., Canfield, N.L., and Armstrong, T.R. 2000. Influence of cobalt and iron additions on the electrical and thermal properties of (La,Sr)(Ga,Mg)O_3-delta. *Journal of the Electrochemical Society* 147(9):3213–3218.

Strickler, D.W. and Carlson, W.G. 1964. Ionic conductivity of cubic solid solutions in the system CaO-Y_2O_3-ZrO_2. *Journal of the American Ceramic Society* 47(3):122–127.

Stuart, P.A., Unno, T., Kilner, J.A., and Skinner, S.J. 2008. Solid oxide proton conducting steam electrolysers. *Solid State Ionics* 179:1120–1124.

Su, X.T., Yan, Q.Z., Ma, X.H., Zhang, W.F., and Ge, C.C. 2006. Effect of co-dopant addition on the properties of yttrium and neodymium doped barium cerate electrolyte. *Solid State Ionics* 177(11–12):1041–1045.

Subasri, R., Mathews, T., and Sreedharan, O.M. 2003. Microwave assisted synthesis and sintering of $La_{0.8}Sr_{0.2}Ga_{0.83}Mg_{0.17}O_{2.815}$. *Materials Letters* 57(12):1792–1797.

Taniguchi, N., Hatoh, K., Niikura, J., Gamo, T., and Iwahara, H. 1992. Proton conductive properties of gadolinium-doped barium cerates at high temperatures. *Solid State Ionics* 53–56:998–1003.

Thangadurai, V. and Kopp, P. 2007. Chemical synthesis of Ca-doped CeO_2-intermediate temperature oxide ion electrolytes. *Journal of Power Sources* 168(1):178–183.

Tomita, A., Hibino, T., Suzuki, M., and Sano, M. 2004. Proton conduction at the surface of Y-doped $BaCeO_3$ and its application to an air/fuel sensor. *Journal of Materials Science* 39(7):2493–2497.

Uchida, H., Osada, N., and Watanabe, M. 2004. High-performance electrode for steam electrolysis mixed conducting ceria-based cathode with highly-dispersed Ni electrocatalysts. *Electrochemical and Solid State Letters* 7(12):A500–A502.

Van Herle, J., Horita, T., Kawada, T., Sakai, N., Yokokawa, H., and Dokiya, M. 1996. Low temperature fabrication of (Y,Gd,Sm)-doped ceria electrolyte. *Solid State Ionics* 86–88(2):1255–1258.

Virkar, A.V. 1991. Theoretical analysis of solid oxide fuel cells with 2-layer, composite electrolytes—Electrolyte stability. *Journal of the Electrochemical Society* 138(5):1481–1487.

Wang, F.Y., Wan, B.Z., and Cheng, S. 2005. Study on Gd^{3+} and Sm^{3+} co-doped ceria-based electrolytes. *Journal of Solid State Electrochemistry* 9(3):168–173.

Wang, S.Z. and Tatsumi, I. 2003. Improvement of the performance of fuel cells anodes with Sm^{3+} doped CeO_2 interlayer. *Acta Physico-Chimica Sinica* 19(9):849–853.

Wang, W.S., Gorte, R.J., and Vohs, J.M. 2008. Analysis of the performance of the electrodes in a natural gas assisted steam electrolysis. *Chemical Engineering Science* 63:765–769.

Wang, W.S., Huang, Y.Y., Jung, S., Vohs, J.M., and Gorte, R.J. 2006. A comparison of LSM, LSF, and LSCo for solid oxide electrolyzer anodes. *Journal of the Electrochemical Society* 153(11):A2066–A2070.

Wang, Y.R., Mori, T., Li, J.G., and Yajima, Y. 2003. Low-temperature fabrication and electrical property of 10 mol% Sm_2O_3-doped CeO_2 ceramics. *Science and Technology of Advanced Materials* 4(3):229–238.

Wendt, H. 1990. *Electrochemical Hydrogen Technologies: Electrochemical Production and Combustion of Hydrogen*, p. 345. Elsevier Science Publishers B.V., New York.

Xu, H.M., Yan, H.G., and Chen, Z.H. 2006. Sintering and electrical properties of $Ce_{0.8}Y_{0.2}O_{1.9}$ powders prepared by citric acid-nitrate low-temperature combustion process. *Journal of Power Sources* 163(1):409–414.

Yamamoto, O., Arati, Y., Takeda, Y., Imanishi, N., Mizutani, Y., Kawai, M., and Nakamura, Y. 1995. Electrical conductivity of stabilized zirconia with ytterbia and scandia. *Solid State Ionics* 79:137–142.

Yaremchenko, A.A., Shaula, A.L., Logvinovich, D.I., Kharton, V.V., Kovalevskyn, A., Naumovich, E.N., Frade, J.R., and Marques, F.M.B. 2003. Oxygen-ionic conductivity of perovskite-type $La_{1-x}Sr_xGa_{1-y}Mg_yM_{0.2}O_3$-delta (M = Fe, Co, Ni). *Materials Chemistry and Physics* 82(3):684–690.

Zha, S.W., Xia, C.R., and Meng, G.Y. 2003. Effect of Gd(Sm) doping on properties of ceria electrolyte for solid oxide fuel cells. *Journal of Power Sources* 115(1):44–48.

Zhang, T.S., Hing, P., Huang, H.T., and Kilner, J. 2002. Ionic conductivity in the CeO_2-Gd_2O_3 system ($0.05 \leq Gd/Ce \leq 0.4$) prepared by oxalate co-precipitation. *Solid State Ionics* 148(3–4):567–573.

Zhang, T.S., Ma, J., Cheng, H., and Chan, S.H. 2006. Ionic conductivity of high-purity Gd-doped ceria solid solutions. *Materials Research Bulletin* 41(3):563–568.

Zhang, T.S., Ma, J., Kong, L.B., Chan, S.H., and Kilner, J.A. 2004. Aging behavior and ionic conductivity of ceria-based ceramics: A comparative study. *Solid State Ionics* 170(3–4):209–217.

Zhang, X.G., Ohara, S., Maric, R., Okawa, H., Fukui, T., Yoshida, H., Inagaki, T., and Miura, K. 2000. Interface reactions in the NiO-SDC-LSGM system. *Solid State Ionics* 133(3–4):153–160.

Zheng, F., Bordia, R.K., and Pederson, L.R. 2004. Phase constitution in Sr and Mg doped $LaGaO_3$ system. *Materials Research Bulletin* 39(1):141–155.

Section IV

Nuclear Hydrogen Production

8

Nuclear Hydrogen Production by Thermochemical Cycles

Greg F. Naterer
Memorial University of Newfoundland

Kamiel S. Gabriel
University of Ontario Institute of Technology

M. Lewis
Argonne National Laboratory

S. Suppiah
Atomic Energy of Canada Limited

CONTENTS

8.1 Introduction

8.1.1 Overview

This chapter focuses on thermochemical cycles of hydrogen production, particularly the copper–chlorine cycle. Over 200 thermochemical cycles have been identified previously (McQuillan et al. 2002). Lewis and Taylor (2006) reported that a survey of the open literature between 2000 and 2005 did not reveal any other cycles. Very few have progressed beyond theoretical calculations to working experimental demonstrations that establish scientific and practical feasibility of the thermochemical processes. After considering factors of availability and abundance of materials, simplicity, chemical viability, thermodynamic feasibility, and safety issues, a number of cycles were identified in a nuclear hydrogen initiative (Lewis and Taylor 2002) as the most promising cycles.

Thermochemical hydrogen production is a method of splitting water by a series of chemical and physical processes. All chemical intermediates are recycled internally, without any emissions to the environment, so that water and heat are the only inputs, while hydrogen and oxygen gas are the only products. Hybrid thermochemical cycles require energy inputs in the form of both heat and electricity, although the predominant majority is typically in the form of heat input. The maximum temperature requirement of most thermochemical cycles is within the temperature range of 530°C–1100°C (Casper 1978). A comprehensive review of thermochemical cycles was presented by Goel et al. (2003) and Mirabel et al. (2004). Thermochemical cycles can be generally described based on the four following basic steps:

1. Water-splitting reaction
2. Hydrogen production
3. Oxygen production
4. Recycling and regeneration of intermediate compounds

This section will provide a brief overview of selected thermochemical cycles, with a sample including the sulfur–iodine (S–I) (Carty et al. 1981; Sakurai et al. 2000; Schultz 2003), hybrid sulfur, Cu–Cl (Sadhankar et al. 2005), University of Tokyo 3 (UT-3) (Sakurai et al., 1996; Lee et al. 2009), Ispra Mark 9, and vanadium–chlorine (V–Cl) (Knoche et al. 1984) cycles.

8.1.2 Sulfur–Iodine (S–I) Cycle

The S–I cycle is a leading example where equipment has been scaled up to a pilot plant level, with active development by General Atomics (United States), Sandia National Laboratory

(United States), Japan Atomic Energy Agency (JAEA), CEA (France), and others (Sakurai et al. 2000; Schultz 2003). JAEA has demonstrated a pilot facility up to 30 L/h of hydrogen with the S–I cycle. It aims to scale up the S–I cycle to much larger production capacities that could eventually support a significant volume of fuel cell vehicles.

The three reactions that produce hydrogen in the S–I cycle are shown as follows:

- $I_2 + SO_2 + 2H_2O \rightarrow 2HI + H_2SO_4$ (120°C)
- $2H_2SO_4 \rightarrow 2SO_2 + 2H_2O + O_2$ (830°C)
- $2HI \rightarrow I_2 + H_2$ (450°C)

In the first step, sometimes called the Bunsen reaction, HI is produced and then separated by distillation. Concentrated H_2SO_4 may react with HI, giving I_2, SO_2, and H_2O (back reaction). In the second step, the water, SO_2, and residual H_2SO_4 must be separated from the oxygen, typically by condensation. Iodine and any accompanying water or SO_2 are separated by condensation in the last step of the cycle, while hydrogen gas is produced as the main product.

The S–I cycle has been extensively developed, as it has a number of key advantages over other thermochemical cycles. For example, all of the working fluids exist in the liquid or gas phase; therefore, there is no solid handling, and the processes are well suited for continuous operation. The S–I cycle is more developed than other cycles. Another advantage is that hydrogen can be generated at high pressure (50 atmospheres), which eliminates the need for compressing the hydrogen for downstream utilization.

However, very high temperatures are required (at least 850°C). Corrosive reagents are used as intermediaries (iodine, sulfur dioxide, hydriodic acid, sulfuric acid); thus, advanced materials of construction are needed. Also, separation of the dense liquid phase from the acid-generating reaction into HI and I_2 is accomplished by extracting water into concentrated phosphoric acid, leading to a large recycle of phosphoric acid through a dehydration system that is capital intensive.

8.1.3 Hybrid Sulfur Cycle (Also Known as Westinghouse, GA-22, or Ispra Mark)

This variation of the S–I cycle reduces the number of steps by incorporating an electrolysis step:

- $H_2SO_4(g) \rightarrow SO_2(g) + H_2O(g) + 1/2O_2(g)$ (850°C)
- $SO_2(aq) + 2H_2O(l) \rightarrow H_2SO_4(aq) + H_2(g)$ (80°C electrolysis)

The cycle is still an all-fluid process, and it has the additional advantage of reducing the number of reactions and cycle complexity. The thermodynamic properties of the chemical compounds are well known, and side reactions are minimal. A demonstration pilot plant was developed and built by the Commission of the European Communities at the Ispra Research Establishment. The first step, called the sulfuric acid decomposition step, was demonstrated using concentrated solar energy from a solar power tower. However, there are some disadvantages, including an electrochemical process that has inherent scale-up problems and limitations due to the surface area of the electrodes. Scaling up beyond the maximum practical electrode area requires additional modules.

8.1.4 University of Tokyo 3 Cycle

This cycle consists of four chemical reaction steps as follows:

- $2Br_2(g) + 2CaO(s) \rightarrow 2CaBr_2(s) + 1/2O_2(g)$ (672°C)
- $3FeBr_2(s) + 4H_2O(g) \rightarrow Fe_3O_4(s) + 6HBr(g) + H_2(g)$ (560°C)
- $CaBr_2(s) + H_2O(g) \rightarrow CaO(s) + 2HBr(g)$ (760°C)
- $Fe_3O_4(s) + 8HBr(g) \rightarrow Br_2(g) + 3FeBr_2(s) + 4H_2O(g)$ (210°C)

This cycle is based on solids; however, the solid materials remain in fixed beds and only gaseous products are transported. The cycle has a reported efficiency of 40%. Disadvantages include an inability to operate in a steady-state mode unless there is movement of solids. The solid materials must be periodically changed from one temperature to another, which in addition occurs near the melting point of the bromides. If melting occurs due to temperature variations, the transport of molten bromides could lead to problematic blockage of the beds.

A substantial amount of research has been done on this cycle since the initial invention and development at the University of Tokyo (Aihara et al. 1990; Sakurai et al. 1996). This cycle has been operated on a large pilot plant scale in Japan (Nakayama et al. 1984). The most recent theoretical and experimental work on this cycle has been done at the University of Florida and the University of South Florida by the Goswami Group (Lee et al. 2006, 2007, 2009). The group has developed an innovative method of immobilizing the solid reactants on woven alumina and yttria fabrics, which provides a much larger surface area for reactions than the porous pellets used by the University of Tokyo group and allows the gaseous reactants to pass through while reacting. This method also provides adequate volumetric space for the reactants as they expand and contract during the cyclic reactions. They evaluated the performance by cyclic bromination and hydrolysis reactions. Based on the experimental results in the cyclic reactions, the calcium oxide on the yttria fabric had continuous higher reactivity in the bromination reaction, and the rate of the hydrolysis reaction was comparable to or faster than that of the calcium oxide pellets in the previous studies (Lee et al. 2009).

8.1.5 Ispra Mark 9 Cycle

This cycle involves three chemical reactions, including processes that involve separation and movement of solids:

- $3FeCl_3(l) \rightarrow 3/2Cl_2(g) + 3FeCl_2(s)$ (420°C)
- $3/2Cl_2(g) + Fe_3O_4 + 6HCl \rightarrow 3FeCl_3 + 3H_2O + 1/2O_2(g)$ (150°C)
- $3FeCl_2(s) + 4H_2O(g) \rightarrow Fe_3O_4(s) + 6HCl(g) + H_2(g)$ (650°C)

It has been shown experimentally that $FeCl_3$ decomposition and hydrolysis of $FeCl_2$ to iron oxides were critical and difficult problems for which there has been no suitable solution found; thus, there has been no further development of this cycle.

8.1.6 Vanadium–Chlorine Cycle

The V–Cl cycle consists of four reaction steps as follows:

- $Cl_2(g) + H_2O(g) \rightarrow 2HCl(g) + 1/2O_2(g)$ (850°C)
- $2HCl(g) + 2VCl_2(s) \rightarrow 2VCl_3(s) + H_2(g)$ (25°C)

- $4VCl_3(s) \rightarrow 2VCl_4(g) + 2VCl_2(s)$ (700°C)
- $2VCl_4(l) \rightarrow Cl_2(g) + 2VCl_3(s)$ (25°C)

All of the process chemistry has been demonstrated by proof-of-principle experiments. One of the disadvantages is that the cycle involves processing and handling of solids. The HCl(g) and $O_2(g)$ in the first step should be separated without the use of water. Any remaining water in the HCl would produce VOCl as a by-product of the second step. A variation of this V–Cl cycle was examined with flow sheets by the University of Aachen, with a calculated efficiency of 42.5%.

8.1.7 Copper–Chlorine Cycle

Most of the thermochemical cycles require process heat over 800°C. Due to lower temperature requirements of about 550°C and lower, the Cu–Cl cycle is a promising alternative that could be eventually linked with next-generation nuclear reactors, such as Canada's supercritical water reactor (SCWR). The Cu–Cl cycle has numerous advantages over the other existing cycles of hydrogen production. It has much lower operating temperatures than other thermochemical cycles, thereby potentially reducing material and maintenance costs. Also, it can effectively utilize low-grade waste heat, thereby improving cycle and power plant efficiencies. Other advantages include lower demands on materials of construction, common chemical agents, and reactions going to completion without side reactions. Disadvantages include solids handling and a high steam-to-copper ratio in the hydrolysis step, although ongoing advances are steadily decreasing this steam requirement.

The Cu–Cl cycle has been identified by Atomic Energy of Canada Limited (AECL) (Chalk River Laboratories [CRL]) as the most promising cycle for thermochemical hydrogen production with SCWR. Current collaboration between the University of Ontario Institute of Technology (UOIT), AECL, and the Argonne National Laboratory is focusing on enabling technologies for the Cu–Cl cycle, through the Generation IV International Forum (GIF). The remainder of this chapter presents the recent advances in the development of the copper–chlorine cycle for thermochemical water decomposition and hydrogen production.

8.2 Overview of the Thermochemical Cu–Cl Cycle

The Cu–Cl thermochemical cycle uses a series of reactions to achieve the overall splitting of water into hydrogen and oxygen as follows: $H_2O(g) \rightarrow H_2 + 1/2O_2$. The Cu–Cl cycle splits water into hydrogen and oxygen through intermediate copper and chlorine compounds. These chemical reactions form a closed internal loop that recycles all chemicals on a continuous basis, without emitting any greenhouse gases. Steps in the Cu–Cl cycle and a schematic realization of the cycle are shown in Table 8.1 and Figure 8.1, respectively. Table 8.2 shows the heat requirements of each individual step within the overall cycle.

Decomposing water into hydrogen and oxygen requires that sufficient energy be provided to break the chemical bond of hydrogen and oxygen atoms. From a thermodynamic perspective, the minimum enthalpy is equal to the formation enthalpy of water, that is, 286 kJ/mol (denoted as ΔH_f), if water, hydrogen, and oxygen are all contained at atmospheric pressure and 25°C. However, a series of intermediate processes is required to split water in the Cu–Cl cycle. As a consequence, the energy required to split water will also be a function of the processes, not solely the thermodynamic state alone. Table 8.2 summarizes the heat requirements for each individual step within the Cu–Cl cycle (Naterer et al. 2011a,b).

TABLE 8.1
Chemical Reaction Steps in the Cu–Cl Cycle

Step	Reaction	Temperature Range (°C)		Feed/Output[a]
1	$2Cu(s) + 2HCl(g) \rightarrow 2CuCl(l) + H_2(g)$	430–475	Feed:	Electrolytic Cu + dry HCl + Q
			Output:	H_2 + CuCl(l) salt
2	$2CuCl(s) \rightarrow 2CuCl(aq) \rightarrow CuCl_2(aq) + Cu(s)$	Ambient (electrolysis)	Feed:	Powder/granular CuCl and HCl + V
			Output:	Electrolytic Cu and slurry containing HCl and $CuCl_2$
3	$CuCl_2(aq) \rightarrow CuCl_2(s)$	<100	Feed:	Slurry containing HCl and $CuCl_2$ + Q
			Output:	Granular $CuCl_2$ + H_2O/HCl vapors
4	$2CuCl_2(s) + H_2O(g) \rightarrow CuO{*}CuCl_2(s) + 2HCl(g)$	400	Feed:	Powder/granular $CuCl_2$ + $H_2O(g)$ + Q
			Output:	Powder/granular $CuO{*}CuCl_2$ + 2HCl(g)
5	$CuO{*}CuCl_2(s) \rightarrow 2CuCl(l) + 1/2O_2(g)$	500	Feed:	Powder/granular $CuO{*}CuCl_2(s)$ + Q
			Output:	Molten CuCl salt + oxygen

Note: Alternative 4-step cycle combines above steps 1 and 2 to produce hydrogen directly as follows: $2CuCl(aq) + 2HCl(aq) \rightarrow H_2(g) + 2CuCl_2(aq)$ (Figure 8.1 illustrates this 4-step version of Cu–Cl cycle).

[a] Q, thermal energy; V, electrical energy.

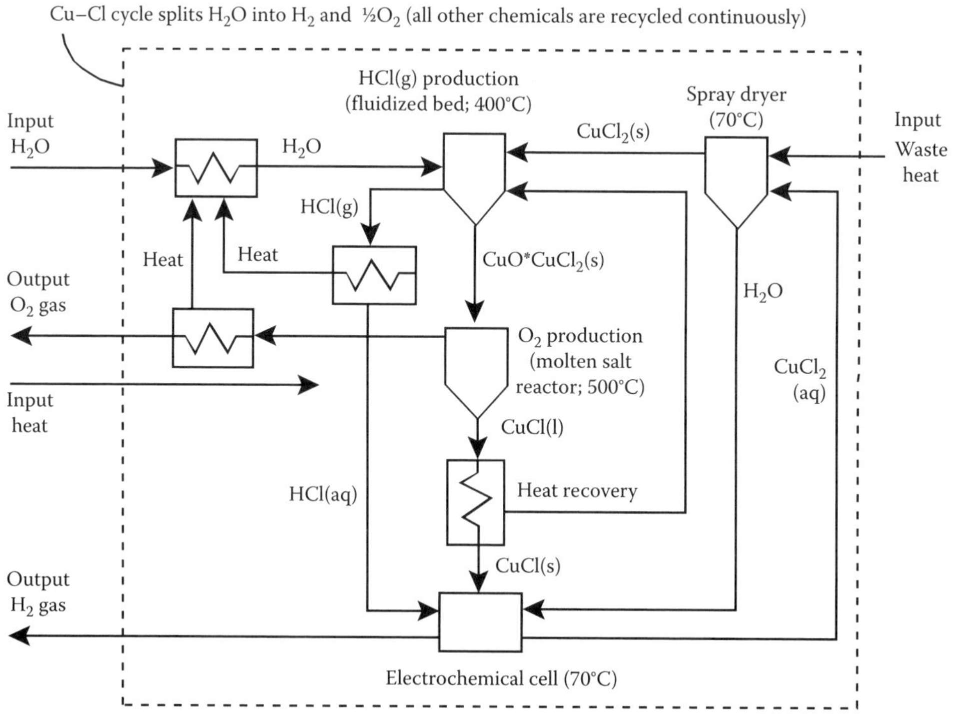

FIGURE 8.1
Schematic of the 4-step copper–chlorine cycle.

TABLE 8.2

Heat Requirements of Individual Steps and the Efficiency of Cu–Cl Cycle

			Heat in Case 3-A Drying		Heat in Case 3-B Drying		
Step	Processes in the Step	T °C	Total kJ/mol H_2	Low grade kJ/mol H_2	Total kJ/mol H_2	Low grade kJ/mol H_2	Heat Output kJ/mol H_2
1	$2Cu(s) + 2HCl = 2CuCl(l) + H_2(g)$ Hydrogen production step						
	Vaporizing moisture form Cu(s)	35 → 70	29.0	29.0	29.0	29.0	
	Heating Cu(s)	25 → 450	23.4	0.3	23.4	0.3	
	Heating HCl(g)	25 → 450	3.0	0.1	3.0	0.1	
	Heat of reaction	450					46.8
	Cooling and solidification of molten CuCl	450 → 25					80.8
	Cooling of $H_2(g)$ product	450 → 25					12.2
2	$4CuCl(aq) = 2CuCl_2(aq) + 2Cu(s)$ Electrolysis step in HCl solution	35–70	Electrical energy 62.6 kJ/mol H_2				
3	$2CuCl_2(aq) \rightarrow 2CuCl_2(s)$ Drying step						
3-A	Vaporizing water from $CuCl_2$ *precipitate* $2.2H_2O(l) \rightarrow 2.2\ H_2O(g)$	35 → 70	122.0	122.0			
3-B	Vaporizing water from $CuCl_2$ *solution* $16.7H_2O(l) \rightarrow 16.7\ H_2O(g)$	35 → 70			931	931	
4	$2CuCl_2(s) + H_2O(g) = CuOCuCl_2(s) + 2HCl(g)$ Hydrolysis step						
	Heating $CuCl_2(s)$	25 → 375	54.2	8.5	54.2	8.5	
	Heat of reaction	375	116.6		116.6		
	Steam production $H_2O(l) = H_2O(g)$	25 → 375	57.1	3.4	57.1	3.4	
5	$CuOCuCl_2(s) \rightarrow 2CuCl(l) + 1/2O_2(g)$ Oxygen production step						
	Heating $CuOCuCl_2(s)$	375 → 530	20.2		20.2		
	Heat of reaction	530	129.2		129.2		
	Cooling and solidification of molten CuCl	530 → 25					84.8
	Cooling of $O_2(g)$ product	530 → 25					7.4
	Sum	kJ/mol H_2	554.7	163.3	1363.7	972.3	232
		kJ/g H_2	277.4	81.7	681.9	486.2	116
	Percentage of low-grade heat			29.5%		71.3%	

Efficiency of Cu–Cl cycle								
Percentage of heat loss to environment (%)	0	0	20	20	20	20	20	20
Percentage of recycled heat internally within the cycle (%)	0	100	0	30	50	70	90	100
Practical efficiency of Cu–Cl cycle (%)	46	74	37	41	44	47	51	53

Low-grade heat: The heat with a temperature that is lower than 70°C.

Thermochemical water decomposition with the Cu–Cl cycle is much more efficient than electrolysis via thermal power plants, because heat is used directly to produce hydrogen, rather than indirectly to first produce electricity, after which hydrogen is generated. A 42% efficiency (electricity) with Canada's next-generation reactor leads to about 30% net efficiency by electrolysis for hydrogen production. In contrast, a 54% heat-to-hydrogen efficiency has been demonstrated from Aspen Plus simulations for the Cu–Cl cycle (Lewis et al. 2005), although 43% is more realistic. This implies a significant margin of superior overall conversion efficiency, with more than one-third improvement over electrolysis, excluding even larger gains if *waste heat* is utilized in the thermochemical cycle.

A unique advantage of the Cu–Cl cycle is an ability to utilize low-grade *waste heat* from power plants or other sources to aid hydrogen production, rather than rejecting that heat to the environment (such as a nearby lake). This could significantly improve the economics of hydrogen production, as well as potentially provide for the generation of electricity. For example, a nuclear plant's efficiency usually refers to electricity output alone, but the *efficiency* or output of an overall system to cogenerate electricity and hydrogen is higher than electricity generation alone, if waste heat contributes to the production of another valuable energy carrier, hydrogen, from the same nuclear heat source. The Cu–Cl cycle's efficiency implies that the heat supply comes at some *cost*. But if low-grade waste heat is utilized as a heat source at minimal or no cost, then the economics of hydrogen production is further improved and more attractive than other high-temperature thermochemical cycles that cannot use such low-grade heat.

Heat exchangers represent an important component within the Cu–Cl cycle. Between each step of the Cu–Cl plant, heat exchangers are needed for heat input/recovery and fluid transport between different steps of the thermochemical cycle. The oxygen production step has the highest temperature requirement (530°C) in the cycle. Since temperature drops are experienced across each heat exchanger, the inflow stream to the oxygen reactor must exceed 530°C. Furthermore, multiple heat exchangers are needed between the heat source and the oxygen reactor (step 5), through an intermediate loop. An intermediate loop with several heat exchangers in series is needed to give progressively higher temperatures. Ongoing research is underway to determine how flow rates, temperatures, and fluid streams should be most effectively partitioned for each step of the Cu–Cl cycle. Also, heat losses must be identified and evaluated throughout the piping network, with new methods developed for heat recovery.

Several types of fluid devices are used throughout the hydrogen plant, including chemical reactors, electrochemical cells, heat exchangers, pumps, valves, pipes, and blowers (for gases). Certain components require substantial modifications of existing technologies with new materials to be economically viable in extreme operating conditions of high-temperature corrosive fluids. For example, common refractory materials for heat exchangers have poor thermal conductivity. Also, high-performance metal alloys cannot withstand corrosive fluids over a long duration. High-temperature alloys with coatings of silicone-based ceramics appear promising, but their thermal behavior and surface interactions in high-temperature multiphase conditions must be further developed. Further data are also needed to better understand the functionality of these materials with the working fluids in the Cu–Cl cycle. This includes thermal behavior, mechanical stresses, fracture toughness, strength, and corrosion resistance over time. Improvements to existing equipment can then be achieved through new materials developed specifically for operating conditions of the Cu–Cl cycle.

8.3 Thermochemical Reaction of Hydrogen Production

8.3.1 Proof-of-Principle Demonstrations

Two possible variations of the Cu–Cl cycle are depicted in Table 8.1: a 5-step and a 4-step (Figure 8.1) cycle. Step 1 in the 5-step Cu–Cl cycle is the H_2 production step, which occurs at 430°C–475°C, characterized by the reaction $2Cu(s) + 2HCl(g) \rightarrow H_2(g) + 2CuCl(l)$. Copper particles enter the reactor vessel and react with HCl gas to generate H_2 gas and molten CuCl at the exit. Scientific practicality of this chemical reaction has been demonstrated experimentally by Serban et al. (2004). The experiments were conducted in a ½ in. by 14 in. quartz vertical reactor, as shown in Figure 8.2. All connecting lines and valves were made of Teflon. Anhydrous hydrochloric acid (99.99%), chlorine (99.99%), and argon (99.996%) were fed to the microreactors. The temperatures were controlled by Omega CN375–type temperature controllers connected to a Lindberg single-zone heated furnace, monitored with K-type thermocouples. The vertical reactor was connected online to a quadrupole mass spectrometer (QMS) 200 to continuously monitor the hydrogen production. The mass spectrometer was calibrated for H_2, O_2, HCl, and Cl_2 using mixtures of pure gases with Ar over a wide range of compositions.

The reaction kinetics were investigated using only reagent-grade reactants of high purity. The experiments were conducted in beds of solid material with a continuous flow of excess gaseous reactants. Constant temperatures in any one experiment were assumed.

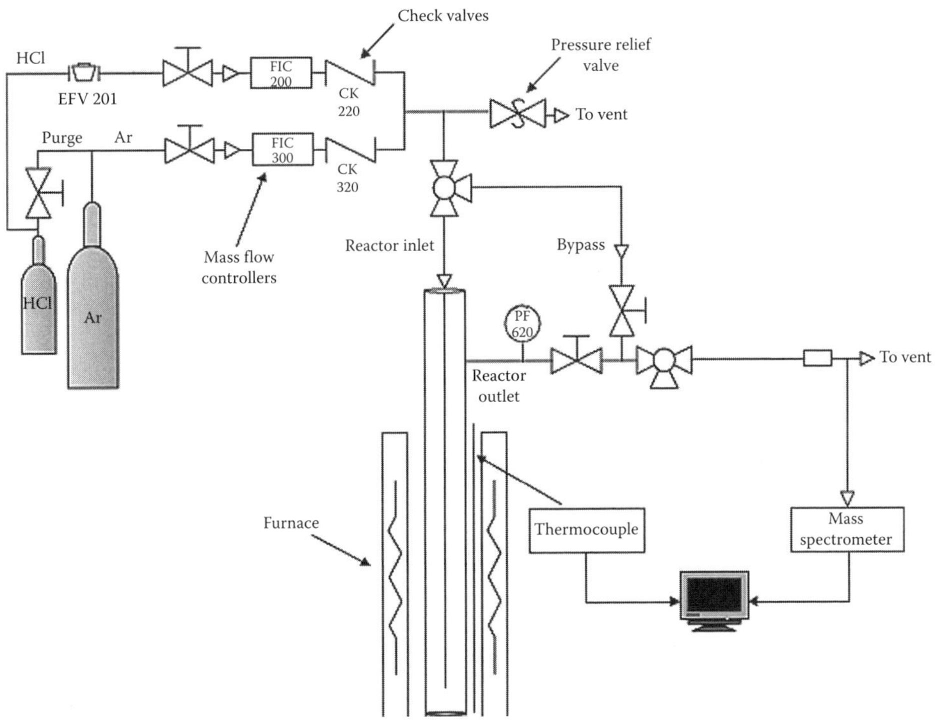

FIGURE 8.2
Lab-scale apparatus for the hydrogen and oxygen production reactions. (Adapted from Serban, M. et al. Kinetic study of the hydrogen and oxygen production reactions in the copper-chloride thermochemical cycle, in *AIChE 2004 Spring National Meeting*, New Orleans, LA, April 25–29, 2004.)

The kinetics analysis provided apparent activation energy values, where the internal mass diffusion limitations were not included (Serban et al. 2004). The reaction between HCl and Cu is a heterogeneous exothermic and reversible reaction. Experiments indicated that at low temperatures, the kinetics of the reaction were very slow. The kinetics of the reaction were accelerated at temperatures above the melting point of CuCl (423°C), facilitating the interaction between HCl and Cu. These results were surprising, since it was reported that the exothermic reaction between Cu and HCl should proceed rapidly at 230°C and 93% of HCl should be decomposed. However, no hydrogen was detected at this temperature, even though the Gibbs free energy change is −5.66 kcal at 230°C. In addition, it is believed that CuCl forms a passivating layer on the surface of the Cu particles. At 430°C–435°C, the CuCl starts to melt, facilitating the interaction between HCl and Cu. Similar experiments performed with Ag and HCl indicate the same behavior.

In an attempt to find the best experimental conditions for complete Cu conversion to CuCl and H_2, experiments were performed with varying sizes and shapes of Cu particles as well as varying HCl flow rates. The conversion of HCl to hydrogen was found to be a function of Cu particle size as shown in Figure 8.3 (Serban et al. 2004). The yields varied between 65% and 100%, with complete conversion in the case of 3 μm Cu particles. Yields were lower in the case of the larger 100 μm Cu particles, which were obtained in an electrolytic cell by disproportionating CuCl to Cu and $CuCl_2$, a step that is an integral part of the overall thermochemical water-splitting cycle. The size and shape of the electrolytic copper particles varied with the operating parameters of the cell, which were not optimized for the subsequent experiments.

All kinetic experiments were conducted using commercial dendritic 3 μm Cu particles. To directly measure the kinetics of the reaction between HCl and Cu, the reaction rates were measured at four different temperatures (400°C, 425°C, 450°C, and 475°C), and the molar ratio of HCl in Ar was varied between 0.33 and 0.67. A rate expression was obtained.

The hydrogen yields were determined by comparing the measured amount of hydrogen produced vs. the stoichiometric amount of hydrogen that would be formed if all of

FIGURE 8.3
Hydrogen yields as a function of Cu particle size—fixed bed reactor, temperature = 450°C, HCl flow rate = 2.5 cm^3/min, 0.5 g of commercial (3 and 10 μm particle size), and electrolytic Cu (100 μm particle size). (Adapted from Serban, M. et al. Kinetic study of the hydrogen and oxygen production reactions in the copper-chloride thermochemical cycle, in *AIChE 2004 Spring National Meeting*, New Orleans, LA, April 25–29, 2004.)

the Cu were oxidized to CuCl (Serban et al. 2004). Hydrogen yields decreased with increasing Cu particle size, indicating that high surface areas of contact between HCl and Cu are necessary for high hydrogen yields (could be due to diffusion limitation). Figure 8.3 shows scanning electron microscope (SEM) images of the spheroidal and dendritic Cu particles used in the hydrogen generation reaction. No gaseous products other than hydrogen and HCl were observed in the effluent stream. X-ray diffraction (XRD) examination of the solid product resulting from the reaction showed patterns only for CuCl and Cu. Apparently, no secondary reactions are favored in the temperature range studied. Thermodynamic computations indicate that when CuCl(s) is heated to very high temperatures, CuCl(g) is formed, which would then spontaneously decompose to Cu and Cl_2. No Cl_2 was detected on the mass spectrometer at any time, suggesting that the extent of CuCl vaporization was negligible.

8.3.2 Design Issues of Hydrogen Reactor Scale-Up

If the net heat output from the exothermic reactor is only absorbed by hydrogen as the product of the reaction, then it can be shown that the temperature of hydrogen will reach about 1244°C. On the other hand, if the net heat is solely consumed by molten CuCl, then its temperature will rise by 180°C up to 639°C. If the net heat is consumed by both products, H_2 and molten CuCl, both temperatures will rise by 153°C up to a product temperature of 603°C. If this product temperature is higher than 550°C, then generation of CuCl vapor will become significant, thereby requiring separation of gaseous products from the exit stream. An operating temperature near 530°C is needed in the oxygen-producing process (step 5 of the Cu–Cl cycle; Table 8.1), where molten CuCl is also produced. As a result, heat generated from the exothermic hydrogen reactor can be supplied to the endothermic oxygen reactor, by controlling the heat output from the exothermic reactor. Assuming the highest temperature allowed in the hydrogen reactor is 530°C, it can be shown that the required heat to be removed from the exothermic reactor per kg of hydrogen is

$$\Delta H_{remove} = \left(c_P^{H_2} + 2^* c_P^{CuCl}\right) * (603 - 530) = 11.4 \text{ kJ} \tag{8.1}$$

where c_P refers to the specific heat. In practice, an excess quantity of HCl reactant gas is needed, which reduces the product temperature. Assuming a lower temperature of 450°C for the products (H_2 gas and molten CuCl), inlet temperature of Cu(s) of 80°C from the upstream cell, and 400°C inlet temperature of HCl gas, the net heat removed from the reactor per kg of hydrogen becomes

$$\Delta H_{remove} = \left(c_P^{H_2} + 2^* c_P^{CuCl}\right) * (603 - 450) = 23.9 \text{ kJ} \tag{8.2}$$

Effective recycling of heat released by the reactor to other steps within the Cu–Cl cycle is critical to improving the heat-to-hydrogen efficiency of the cycle. This thermochemical step provides a valuable high-temperature source of exothermic heat supply to other reactors.

8.4 Electrochemical Process of Copper or Hydrogen Production

8.4.1 Introduction

An electrochemical process is required, either separately from the hydrogen reaction (5-step Cu–Cl cycle) or combined together as $2CuCl + 2HCl \rightarrow 2CuCl_2 + H_2$ (4-step Cu–Cl cycle in Table 8.1 and Figure 8.1). This section investigates the latter process of cuprous chloride/HCl electrolysis, whereby oxidation of cuprous chloride (CuCl) during an electrochemical reaction occurs in the presence of hydrochloric acid (HCl) to generate hydrogen. In the electrolysis process, the cuprous ion is oxidized to cupric chloride at the anode, and the hydrogen ion is reduced at the cathode. Electrolysis has similar characteristics as fuel cells but has the opposite objectives.

The two-phase flow in a direct methanol fuel cell (DFMC) is similar to flow that occurs in an electrolytic cell. Yan and Jen (2008) simulated the latter flow conditions and examined how the performance of the fuel cell was affected by temperature, pressure, and concentration of methanol. Naterer and Tokarz (2006) examined the entropy production in proton exchange membrane fuel cells (PEMFC). This section will present a similar study to characterize irreversible losses in an electrochemical cell of cuprous chloride electrolysis. Electrochemical mass transfer generates entropy, which increases the voltage losses. Mass transfer processes in rotating and large–Lewis number electrochemical cells have been examined by Weng et al. (1998) and Jiang et al. (1996), respectively. Voltage losses in electrochemical systems can be characterized effectively by entropy and the second law (Naterer and Camberos 2008).

Consider an electrochemical process involving electrolysis of cuprous chloride to produce hydrogen in Figure 8.4. The electrical circuit is completed by a resistor, ammeter, and dc voltage source connected by external wiring from one electrode to another. It is assumed that CuCl is supplied continuously to the anode, while HCl is supplied to the cathode. The working electrode is assumed to be platinum (Pt) on a gas diffusion layer (GDL). The electrolyte used on the anode side of the cell is a mixture of HCl and CuCl, while HCl is adopted on the cathode side. The operating parameters replicate conditions of experimental studies conducted by AECL (Stolberg et al. 2007).

Electrochemical reactions differ from chemical redox reactions in the way that reactions occur within the cell. Chemical redox reactions occur at the same location, while electrochemical reactions occur in two different sections of the cell, that is, anode and cathode, respectively. For this reason, the reactions in an electrolytic cell are split up into two half-cell reactions: anodic and cathodic reactions. The anodic half-cell reaction for the oxidation of copper (I) to copper (II) can be represented as

$$2CuCl + 2HCl \rightarrow 2CuCl_2 + 2H^+ + 2e^- \tag{8.3}$$

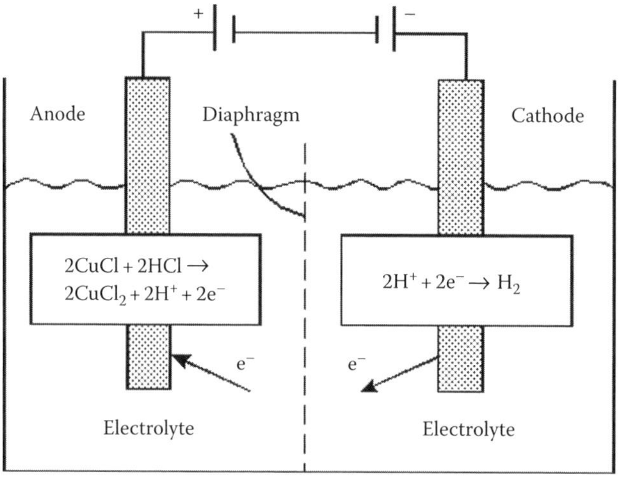

FIGURE 8.4
Schematic of the electrochemical cell.

When CuCl dissolves in 6 M HCl, anionic copper(I) chloride complexes such as $CuCl_2^-$ and $CuCl_3^{2-}$ are formed. Some possible oxidation reactions are given in the following:

$$CuCl_2^-(aq) \rightarrow Cu^{2+}(aq) + 2Cl^-(aq) + e^- \tag{8.4}$$

$$CuCl_2^-(aq) + Cl^-(aq) \rightarrow CuCl_3^-(aq) + e^- \tag{8.5}$$

$$CuCl_3^{2-}(aq) \rightarrow Cu^{2+}(aq) + 3Cl^-(aq) + e^- \tag{8.6}$$

The anodic reactions may involve a number of complexes, many of which have not yet been completely identified. The cathodic half-cell reaction for the reduction of hydrogen protons to hydrogen gas is represented as

$$2H^+ + 2e^- \rightarrow H_2 \tag{8.7}$$

Both the cathode and desired anode reactions can be combined as one reaction, based on the principles of conservation of charge and electron neutrality. As a result, the overall cell reaction can be written as

$$2CuCl + 2HCl \rightarrow 2CuCl_2 + H_2 \tag{8.8}$$

Introducing an electric field across the electrolyte generates a driving force for the ions dissolved in it. While electrons flow through the electrode, as a result of the applied current, ions in the electrolyte migrate as a result of diffusive and convective modes of transport.

8.4.2 Proof-of-Principle Experiments

Scientific viability of the process has been demonstrated experimentally by AECL (Stolberg et al. 2007). Important parameters studied include the chemical kinetics in the electrochemical cell, as a function of temperature, pressure, and compositions. A schematic of the experimental layout is illustrated in Figure 8.5. The copper(I) oxidation reaction is carried out using a 6 M HCl electrolyte. The anodic reaction has been found to require no noble metal catalyst and proceeds quite readily on graphite electrodes. However, the hydrogen production cathodic reaction requires a catalyst. Half-cell studies show that the rate of the CuCl/HCl electrolysis reaction should increase with temperature and CuCl concentration, when the reaction proceeds under mass transfer control. Also, it has been shown that copper crossover from the anolyte to the catholyte causes an increase in the cell voltage during constant current electrolysis experiments. Copper must, therefore, be removed from the catholyte.

In the benchtop experiments at AECL, a variety of electrode materials was tested, and hydrogen production was consistently observed at potentials as low as 0.5 V. However, typically around 0.65 V was required to achieve good current density. The results indicate that the reaction is feasible at reasonably low potentials and with inexpensive electrode materials. Current densities are low and fall from an initial value to a stable plateau. Sample results are shown in Figure 8.6. The current density can be improved considerably by changes to the cell's configuration and improvement of the electrodes.

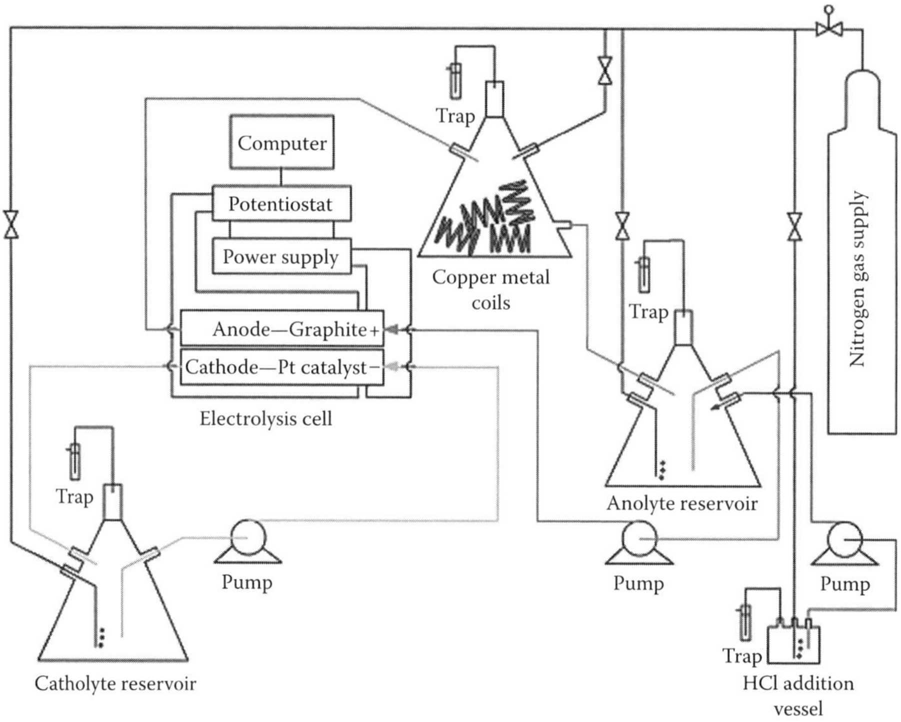

FIGURE 8.5
Experimental configuration of the electrochemical cell. (Adapted from Stolberg, L., et al. Recent advances at AECL in the Cu-Cl cycle for nuclear-produced hydrogen, in *ORF Workshops on Thermochemical Nuclear-Based Hydrogen Production*, Oshawa, ON, December 2007 and Chalk River, ON, October 2008.)

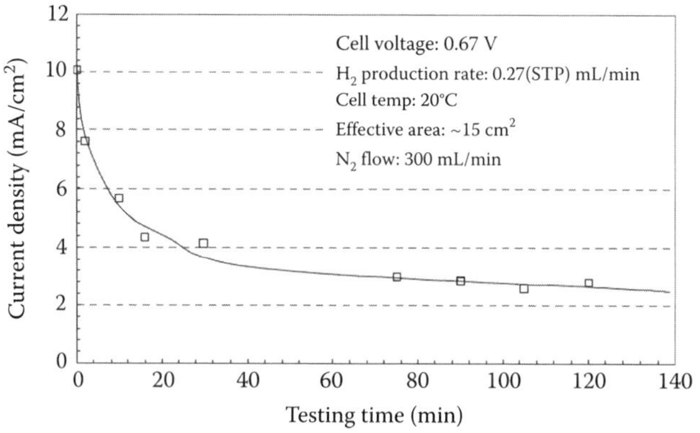

FIGURE 8.6
Typical current density for electrolysis of HCl/CuCl in a continuous system. (Adapted from Stolberg, L., et al. Recent advances at AECL in the Cu-Cl cycle for nuclear-produced hydrogen, in *ORF Workshops on Thermochemical Nuclear-Based Hydrogen Production*, Oshawa, ON, December 2007 and Chalk River, ON, October 2008.)

8.4.3 Predictive Formulation of Hydrogen Electrolysis

The voltage loss in the electrolysis cell includes the surface overpotential, ohmic overpotential, and the concentration overpotential. These losses are known to significantly affect the voltage losses across the cell, and hence they should be included in design considerations for the electrolytic cell. The activation losses can be interpreted as the voltage required in overcoming the open circuit voltage of the cell, when no load is connected in the circuit. Ohmic losses occur as a result of the internal resistance of the cell. Entropy generation in the electrolytic cell is a measure of these irreversibilities and voltage losses within the cell. The entropy generation associated with overpotentials discussed previously can be expressed as (Naterer et al. 2006)

$$\eta = \frac{TP_{act}}{2F} \tag{8.9}$$

where
P_{act} is the entropy generation
T is the operating temperature of the cell

It can be shown that

$$P_{act} = \frac{2F}{T}\left[\mathrm{ROCV} - \left(ir + \frac{RT}{Z\alpha F}\ln\left(\frac{i}{i_o}\right)\right)\right] \tag{8.10}$$

From this equation, it can be observed that entropy production is dependent on temperature, current density, exchange current density, and charge transfer coefficient. In the previous equations, F is Faraday's constant (96,785 C/mol), η is overpotential (mV), Z is number of electrons, i is current density (mA/cm^2), i_o is exchange current density (A/cm^2), α is charge transfer coefficient, and r is electrical resistance (kΩ).

Comparisons between the predicted results and Tafel's model, as well as past experimental data (Stolberg et al. 2007), are shown in Figure 8.7 and Table 8.3, respectively. Close agreement between the results provides a useful validation of the predictive formulation.

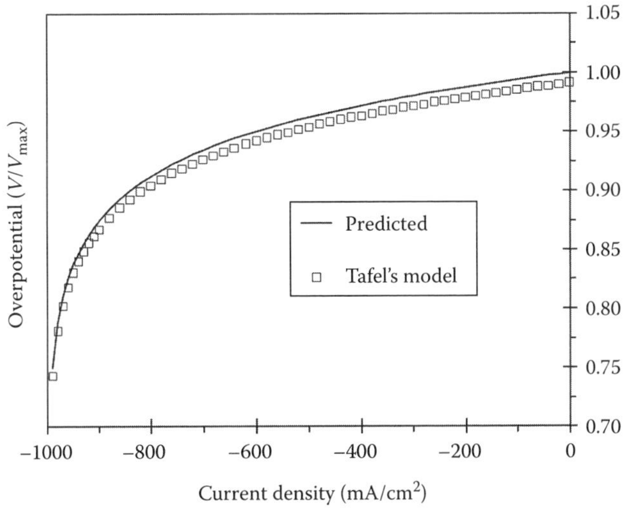

FIGURE 8.7
Comparison between predicted results and Tafel's model for CuCl electrolysis.

TABLE 8.3

Comparison between Predicted and Measured Data

Operating Parameters	Measured	Predicted
Operating temperature of cell (T)	298.15 K	298 K
Activation potential (V_{act})	0.277 mV	0.261 mV
Current density (i)	250 mA/cm^2	250 mA/cm^2
ROCV	0.511 mV	0.515 mV

Source: Adapted from Naterer, G.F. et al. *International Journal of Hydrogen Energy* 36, 15472–15485, 2011a.

Table 8.3 shows that the magnitudes of the predicted overpotential in the formulation show close agreement with experimental half-cell data reported previously for cupric chloride electrolysis. The activation overpotential was measured experimentally (Stolberg et al. 2007) and compared against predicted results in Table 8.3, where it can be observed that the activation potential shows only about a 6% difference between the experimental activation potential and the predicted results. This relatively small difference may have occurred due to the measured uncertainty of reversible open circuit voltage (ROCV), as well as other forms of copper ions or impurities present in the cell.

8.5 Drying of Aqueous Cupric Chloride

8.5.1 Introduction

The drying step of the Cu–Cl cycle is given by $2CuCl_2(aq) \rightarrow 2CuCl_2(s)$. An aqueous $CuCl_2$ stream exiting from the electrochemical cell is supplied to a spray dryer to produce solid $CuCl_2(s)$, which is required for a subsequent hydrolysis step that produces copper oxychloride ($CuO*CuCl_2$) and HCl gas. The apparatus must add sufficient heat to evaporate and remove the water. The process is an energy-intensive step within the Cu–Cl cycle. Although the amount of heat required for the drying step is much higher than the other steps in the cycle, it is of lower temperature (lower quality) and therefore more readily available. Within the drying step, the energy requirement increases from 1 to 5 times for slurry feed to solution, depending on the $CuCl_2$ concentration. The overall cycle efficiency is higher with slurry feed in the drying step than drying aqueous solution. The heat required can be supplied from low-grade waste heat to improve the cycle efficiency. Commercial spray dryers are available for similar processes (used commercially for the production of powders, detergents, food products, and so forth). But further research is needed in the equipment for aqueous cupric chloride, particularly to determine whether aqueous or precipitate slurry from the electrochemical cell should be spray dried. Spray drying is an efficient method of water removal, due to the relatively large surface area available for heat and mass transfer, provided the liquid atomizes into sufficiently small droplets, on the order of several hundred microns.

Many past studies (Walton 2004; Ambike et al. 2005; Chen and Lin 2005) have examined the spray drying of solutions, emulsions, colloidal mixtures, temperature-sensitive materials, and spray conversions. Lin and Chen (2006) developed a droplet evaporation model based on a reaction engineering approach (REA), which uses a minimum energy barrier (called the activation energy) to activate the evaporation process. They used a characteristic drying rate curve (CDRC) approach, in addition to REA, and performed an assessment of accuracy of the different methods. The CDRC approach considers constant and falling-rate periods of drying and relates the specific drying rate to the unhindered drying or constant drying period. The CDRC approach has been widely used in industry, due to its simplicity and ability to handle various chamber parameters.

Drying of aqueous cupric chloride occurs in multiple stages. There is heating of the droplet to the wet-bulb temperature of the surrounding hot air, then droplet evaporation and shrinkage at a constant temperature, after which the concentration of water reaches the critical point and solid crystals precipitate on the surface. Past studies have often assumed a receding interface with a dry core and a constant drying rate period, where fluid transport to the external surface occurs at a rate that matches evaporation. Then, in a falling-rate period, the fluid can no longer reach the surface at the same rate, so the rate of drying falls rapidly and continues until the moisture content falls down to the equilibrium value, then finally drying stops.

Consider a $CuCl_2$ solution with a concentration of 8.2 mol H_2O/mol $CuCl_2$ between 80°C and 100°C, to be dried at a minimum of 300 kPa absolute pressure to 2 mol H_2O/mol of $CuCl_2$ ($CuCl_2 \cdot 2H_2O$ solid). The solubility of $CuCl_2$ in water varies from 9.3 to 7.5 mol H_2O/mol $CuCl_2$ from 30°C to 80°C, and cooling the solution to 30°C leads to precipitate solids and a slurry and solution mixture, from which the solution is eventually recirculated within the Cu–Cl cycle. At these conditions, the slurry with 55% by volume solids contains 3.5 mol H_2O/mol $CuCl_2$. The spray drying occurs when the slurry contacts the drying medium (air) at 80°C. The drying of slurry instead of solution can drastically reduce the energy consumption. Under this approach, air heats the droplets until they reach the air wet-bulb temperature. The water vapor concentration difference in air is a driving factor within the process.

Analysis of the spray drying process often assumes that both external convection and internal conduction affect the drying process. Typically, hot air heats the droplets (see line J-A in Figure 8.8) until the liquid reaches the saturation point at atmospheric pressure. At this point, evaporation starts and continues (line A-E), as long as the heating temperature is higher than the saturation temperature of the liquid at atmospheric pressure, or the partial vapor pressure of the surrounding air is lower than that of the vapor on the droplet surface. The driving potential becomes a small vapor concentration differential between the droplet surface and surrounding air. This approach makes the drying rate substantially slower. Another drying alternative combines flashing and spraying into a chamber at atmospheric pressure, in contact with air preheated at a temperature above the solution temperature. A flashing effect evaporates a small portion of water, instantaneously out of the solution, into a cloud of droplets. The remaining droplets evaporate, subject to the temperature differential between droplets and air.

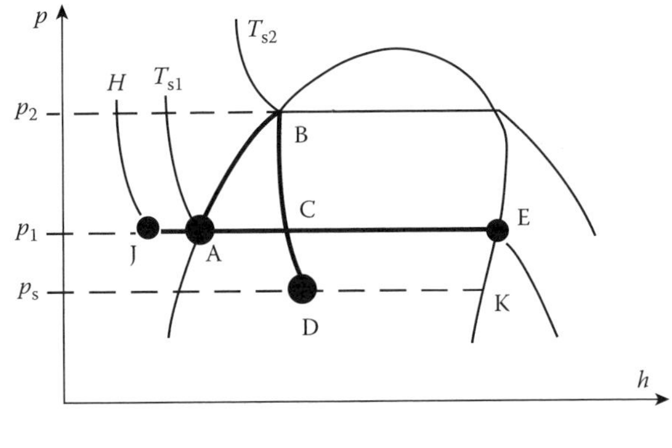

FIGURE 8.8
Pressure–enthalpy diagram for flashing (ABCD) and spraying (JACE) processes.

8.5.2 Predictive Modeling and Scale-Up Design Issues

This section presents a numerical formulation that predicts the spray drying of aqueous cupric chloride droplets. The analysis assumes that the mass and momentum changes of dispersed droplets do not contribute to corresponding changes of the drying medium (air). It also assumes that evaporation results in a continuous shrinkage of the droplet with no solid core formation but rather a continuous growth of $CuCl_2$ crystals, with the average thermal conductivity of the droplet remaining constant.

When the temperature differential between the droplet and air stream is significant, the heat transfer has the following influence on the droplet mass:

$$\frac{dm}{dt} = \frac{hA\left(T_g - T_d\right)}{\lambda} \tag{8.11}$$

where m, t, h, A, T_g, T_d, and λ refer to the droplet mass, time, heat transfer coefficient, droplet surface area, gas temperature, droplet temperature, and evaporation constant, respectively. The Nusselt number (Nu) is defined by the Ranz and Marshall correlation (1952) as follows:

$$\text{Nu} = \frac{hD_d}{k_d} \tag{8.12}$$

$$\text{Nu} = 2 + 0.6\,\text{Re}^{1/2}\,\text{Pr}^{1/3} \tag{8.13}$$

where k_d, D_d, Re, and Pr refer to the droplet conductivity, diameter, Reynolds number, and Prandtl number, respectively. Combining the previous equations and rewriting the result for droplet change in terms of the Spalding number, B, with the correction factor, $(1 + B)^{-0.7}$ (Levi-Hevroni et al. 1995) for the Nusselt and Sherwood numbers, with the Ranz–Marshall correlation, leads to

$$\frac{dD_d}{dt} = -\frac{2k_d\left(1+B\right)^{-0.7}}{\rho_{d,w}D_d\lambda}\left(2 + 0.6\,\text{Re}^{1/2}\,\text{Pr}^{1/3}\right)\left(T_g - T_d\right) \tag{8.14}$$

where the Reynolds number is

$$\text{Re} = \frac{\rho_g\left(V_g - V_d\right)D_d}{\mu_g} \tag{8.15}$$

where V_g, V_d, ρ_g, and μ_g refer to the gas velocity, droplet velocity, density, and viscosity, respectively. The factor $(1 + B)^{-0.7}$ takes into consideration the flow in the boundary layer, and $B = C_{pv}(T_g - T_d)/\lambda$ is the Spalding number. In order to find the droplet shrinkage rate, it can be assumed that the droplet mass decrease is proportional to the droplet diameter shrinkage, thereby leading to the following mass transfer rate:

$$M_v = -0.5\pi\rho_{d,w}D_d^2\frac{dD_d}{dt} \tag{8.16}$$

Numerical integration of these equations over time results in determination of the drying time of the particle under varying conditions of air humidity and temperature.

The following energy balance equation is used for calculating the droplet/particle temperature change with time:

$$mC_{p,d}\frac{dT_d}{dt} = hA(T_b - T_d) + \lambda\frac{dm}{dt} \tag{8.17}$$

During a spray drying operation, the selection of inlet and outlet air temperatures has a key influence on the drying characteristics. The diffusive concentration gradient between the wet surface of the droplet and unsaturated air leads to evaporative mass transfer. To minimize the moisture content of the final product, the inlet temperature of air must be as high as possible, and the temperature difference between inlet and outlet air must be as small as possible. Due to intense heat/mass transfer and loss of humidity, the droplet particle temperature rises to the exiting temperature of the air. The increase of air outlet temperature raises the droplet temperature and subsequently lowers the equilibrium moisture content in the droplet. Increasing the temperature difference between the inlet and outlet air, while holding the inlet air temperature constant, raises the moisture content in the final product. The outlet temperature depends on the mass flow rate of air, solid concentration of feed, and inlet air temperature.

Figure 8.9 illustrates the variation of drying time with air velocity, temperature, and operating pressure. It was observed that the drying time depends strongly on the inlet air humidity, temperature, particle size, and operating pressure. At a low humidity of 0.0025 kg water/kg dry air, the drying time is less than 6 s for droplet sizes less than 200 μm, at 35°C. At a humidity of 0.01 kg water/kg dry air, the drying time is less than 6 s for droplet sizes less than 100 μm, at 35°C, and droplet sizes less than 150 μm, at 70°C and 1 bar operating pressure. At low operating pressures of 0.5 bar, the drying time is less than 8 s for droplet sizes less than 200 μm, at 35°C. The results indicate that evaporative drying is possible down to low temperatures as low as 35°C, although such low-temperature drying may limit the product quality and throughput.

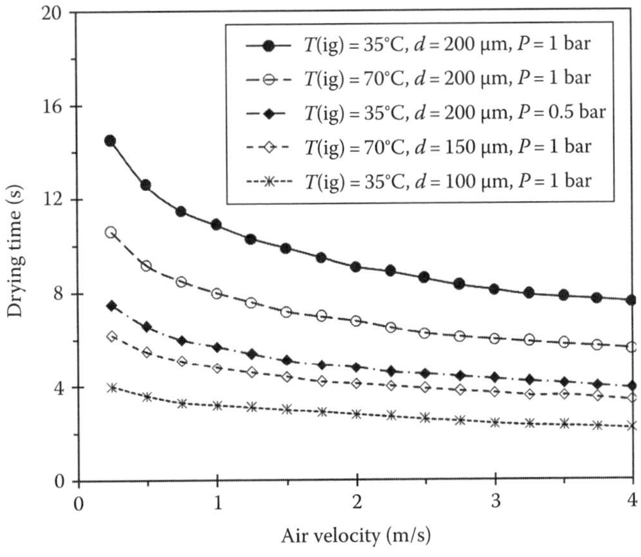

FIGURE 8.9
Predicted drying time at varying air velocities, temperatures, and operating pressures at $H = 0.01$ kg moisture/kg dry air.

These results provide useful data for equipment selection of spray drying in the Cu–Cl cycle. Either an industrial drum dryer or spray dryer could be used to handle liquids or slurries. Drum dryers produce flakes, while spray dryers produce porous, small rounded particles that are often preferable to flakes. A particular advantage of spray dryers is their capability of large evaporation rates. Also, dust control is intrinsic to spray dryer construction, but not drum dryers. Drying proceeds until the moisture content of cupric chloride is reduced sufficiently, and the product is separated from air. Using the previous results such as the drying time data, the operating parameters can be established for the actual spray drying equipment, that is, atomizer design (type, rotational speed, nozzle diameter, and number of nozzles), feed (flow rate, temperature, and concentration), and air supply (flow rate, inlet temperature, outlet temperature, and humidity). Scale-up of the spray drying equipment to larger flow capacities will require advances related to the atomizer, feed supply, air conditions, and flow rates to determine the residence time of droplets.

8.6 Hydrolysis Reaction for Copper Oxychloride Production

8.6.1 Introduction

The following hydrolysis reaction occurs within the Cu–Cl cycle (see Table 8.1): $H_2O(g) + 2CuCl_2(s) \rightarrow Cu_2OCl_2(s) + 2HCl(g)$. The reaction is an endothermic noncatalytic gas–solid reaction that operates between 350°C and 400°C. The solid feed to the hydrolysis reaction is cupric chloride, which comes from step 2 (electrolysis) in the form of an aqueous solution. The aqueous cupric chloride is dried to produce $CuCl_2$ solids, transported to the hydrolysis reactor, and then reacted with steam to produce copper oxychloride and hydrochloric gas. This section examines the transport phenomena of reactive spray drying (a combination of drying and hydrolysis), as well as a noncatalytic gas–solid reaction that separates drying and chemical reaction processes during hydrolysis.

Past experimental studies have examined various types of hydrolysis reactions. The reduction of metal oxides, followed by hydrolysis of the elementary metal with water, was proposed for conversion of solar energy to chemical energy, followed by hydrogen production (Berman and Epstein 2000). Wegner et al. (2006) reported the hydrolysis of zinc using Zn nanoparticles in the gas phase. The chemical conversion of Zn(g) was 83% after a single pass of H_2O with 0.85 s residence time. Hydrolysis is a complex multiphase process, involving transport phenomena of solid–gas interactions and heat transfer (Naterer 2002). This section examines spray flows with $CuCl_2$ during hydrolysis. Ortega-Lopez and Morales-Acevedo (1998) studied mixing and spraying of aqueous $CuCl_2$, in applications of pyrolysis to produce $CuInS_2$ for solar cells.

An alternative to reactive spray drying for the hydrolysis reaction is a conventional fluidized bed. Fluidized beds are widely used in chemical, pharmaceutical, and mineral industries for gas–solid systems. They are suitable for high-temperature gas–solid noncatalytic reactions. Cui et al. (2003) performed hydrodynamic measurements at temperatures up to 420°C in a fluidized bed reactor. Kocakusak et al. (1996) studied the production of anhydrous and crystalline boron oxide by dehydrating boric acid in a fluidized bed reactor with a gradual increase of bed temperature up to 250°C. Konttinen et al. (1997) examined sulfide zinc titanate regeneration in a fluidized bed reactor. Thurnhofer et al. (2005) reported the effects of iron ore reduction on the performance of a fluidized bed.

Depending on the output from an upstream reactor that produces solid $CuCl_2$ particles for hydrolysis in the Cu–Cl cycle, the type of solid feed may be spherical (single or mixed sizes) or flat particles (again single or mixed sizes). Possible reactor types for hydrolysis could be a nebulizer (reactive spray dryer), moving bed, moving feeder, rotary kiln, or a fluidized bed reactor. The three main factors that control the selection of the gas–solid reactor are (1) the reaction kinetics of particles, (2) the size distribution of the solids being handled, and (3) the flow pattern of solids and gas in the reactor. If solids and gas both enter the reactor in plug flow conditions, the composition and temperature change along the length of the reactor, thereby making the process nonisothernal. Examples are a countercurrent type of reactor (moving bed or rotary kiln), cross-flow type (moving belt feeder reactor), or cocurrent flow (rotary kilns). In fluidized beds, the solids move in mixed flow, and the gas flow pattern is more difficult to characterize. Both the solids and gas move in continuous flow, generally under near-isothermal conditions.

8.6.2 Proof-of-Principle Experiments

Ferrandon et al. (2008) have demonstrated experimentally the scientific feasibility of the hydrolysis reaction to produce HCl gas and solid copper oxychloride. Experimental studies were conducted in two types of reactors. The first used a microreactor system consisting of a 0.5 or 0.25 in. OD quartz tube heated by a split temperature-programmed furnace. The typical sample loading was 0.30 g of copper chloride dihydrate (as received $CuCl_2 \cdot 2H_2O$, 99.99%, Sigma Aldrich) packed between two layers of quartz wool. In order to vary the particle sizes, the material was first dried in a dessicator, crushed and sieved to the desired particles sizes, and then rehydrated. Ballmilled materials were obtained by placing dried $CuCl_2$ particles in a shaker with zirconia balls overnight and then rehydrating. The argon (99.999%) flowed between 50 and 250 mL/min through a humidifier with a set temperature varying between 30°C and 95°C. During the hydrolysis experiments, the sample was heated rapidly (within 10 min) in humidified Ar to reaction temperature, between 300°C and 400°C, and then held at the test temperature for a period, between 30 and 90 min. After the end of the experiment, the furnace was opened to allow the sample to cool rapidly in humidified Ar down to 250°C. Dry Ar was then used to cool the sample down to ambient temperature. During the experiments, water and HCl in the gaseous product stream were condensed. Any remaining HCl was trapped using a NaOH solution and Cl_2 was trapped using Fe wool. The solid product samples were stored in closed vials in a dessicator to prevent any air and moisture exposure. The results of these experiments showed relatively high conversion to the desired products when the molar steam to CuCl ratio was high (17:1) and the residence time of the $CuCl_2$ was short, which is desirable to prevent the formation of CuO and CuCl.

Experimental tests were performed at 375°C in the 0.5 in. OD quartz tube using $CuCl_2 \cdot 2H_2O$ as received. The test conditions and the product analyses are listed in Table 8.4. The first group of tests (Tests 1–3 in Table 8.4) was performed keeping the space velocity and steam-to-Cu molar ratio constant. The second group of tests (Tests 4–6 in Table 8.4) was performed keeping the space velocity and the H_2O vapor concentration constant. The test results were promising as they showed high yields of Cu_2OCl_2, up to 89 wt.%. The steam-to-copper molar ratio in these tests varied from 28 to 66, while the CuCl content in the product varied from 8% to 12%. The reason for the higher CuCl yield in the product for Test 1, compared to that in Test 5, cannot be easily determined. For Test 1 compared to Test 5, the steam-to-copper molar ratio is lower, but the test duration is longer. In addition, the small variation in CuCl concentration might be due to experimental error. While every

TABLE 8.4

Experimental Conditions and Elemental Analyses of the Products after the Hydrolysis of $CuCl_2 \cdot 2H_2O$ in a 0.5-in. OD Tube

Test #	Steam/Cu Molar Ratio	Ar (mL/min)	Time (min)	GHSV (h^{-1})	H_2O Vapor (%)	CuCl[a] (wt.%)	$CuCl_2$[b]	$CuCl_2$[c]
1	28.3	200	60	43,327	8	12.1	10.	86.9
2	27	195	43	43,313	10	15.5	1.3	83.2
3	27.1	185	28	43,079	14	14.8	5.2	80.0
4	38	160	20	43,135	26	16.4	10.8	72.8
5	52	160	30	43,135	26	8.0	3.2	88.8
6	66	160	40	43,135	26	11.1	1.5	87.4

[a] Error: ±0.3.
[b] Error: ±0.3–0.7.
[c] Error: ±0.5–0.7.

effort was made to obtain a representative sample, the small sample size precludes multiple analyses. The product in Test 4 had the shortest duration and yet the highest CuCl content. Thus, it was determined that a fixed-bed reactor and solids analysis does not provide sufficient information to determine reaction mechanisms and kinetics.

In order to minimize the decomposition reaction while keeping a low steam-to-Cu ratio, operating parameters, such as the test temperature, test duration, particle size of the starting $CuCl_2$ material, carrier gas flow rate, and steam concentration, were varied. In order to increase the gas velocity in the bed, the reactor tube diameter was reduced to 0.25 in. The amount of CuCl significantly decreased when the reaction temperature was decreased. At 340°C, the amount of CuCl in the sample was only 2.1 wt.%. However, as the temperature was decreased, less Cu_2OCl_2 was produced and more $CuCl_2$ remained unreacted.

8.6.3 Predictive Formulation of Hydrolysis Processes

During the hydrolysis reaction, the particles may grow or shrink. The average conversion of solids depends on the rate of reaction and the residence time of a particle. A uniform-reaction model or shrinking-core model can be used to estimate the reaction rate of the gas–solid reaction. If the diffusion of a gaseous reactant, A, into a particle is much faster than the chemical reaction, the solid reactant B is consumed nearly uniformly throughout the particle (see Figure 8.10a). In this situation, a uniform-reaction model can be used. On the other hand, if the diffusion of a gaseous reactant, A, is much slower and it restricts the reaction zone to a thin layer that advances from the outer surface into the particle (Figure 8.10b), then the shrinking-core model can be adopted.

The rate of conversion of solid, X_B, for a uniform concentration of gas reactant, C_{Ag}, is

$$\frac{dX_B}{dt} = k_r C_{Ag}(1 - X_B) \tag{8.18}$$

which can be integrated over time to yield

$$(1 - X_B) = \exp(-k_r C_{Ag} t) \tag{8.19}$$

where k_r is the reaction rate coefficient, based on a unit volume of solid.

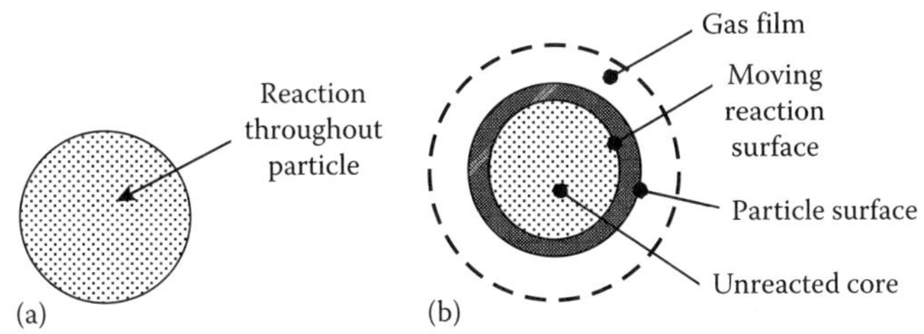

FIGURE 8.10
Schematic of (a) uniform conversion of the particle and (b) shrinking particle core with an ash surface layer and gas film.

In the shrinking-core model, it is assumed that the particles are spherical and the thickness of the reaction zone is smaller than the particle. During the reaction, the reaction zone advances from the outer surface into the particle, completely converting the reactant solid into product. The sequence of steps in the solid conversion process is listed as follows:

1. Diffusion of gaseous reactant, A, through a gas film surrounding the particle to the surface of the solid
2. Penetration and diffusion of A through a layer of ash to the surface of the unreacted core
3. Reaction of gaseous A with solid at the reaction surface
4. Diffusion of gaseous products through the ash back to the exterior surface of the solid
5. Diffusion of gaseous products through the gas film back to the main body of the fluid

In certain conditions, some of these steps may not exist. The thermochemical resistance of each step varies, depending on the flow conditions, and the step of highest resistance will normally control the overall rate of particle conversion.

In the following analysis, the conversion equations will be analyzed for spherical particles, in which steps 1–3 are the rate-controlling steps. For step 1, with diffusion through the gas film, the kinetic equation based on the available surface, S_{ex}, can be written as

$$-\frac{1}{S_{ex}}\frac{dN_B}{dt} = -\frac{\rho_B r_C^2}{R^2}\frac{dr_c}{dt} = bk_g\left(C_{Ag} - C_{As}\right) = bk_g C_{Ag} \tag{8.20}$$

where N_B, ρ_B, r_C, R, k_g, C_{Ag}, and C_{As} refer to the molar density in the solid, density, radius of the unreacted core, radius of the solid particle, mass transfer coefficient, concentration of the gaseous reactant outside of the gas film, and concentration of the gaseous reactant at the particle surface.

The solution yields the following conversion of a single particle with time:

$$\frac{t}{\tau} = 1 - \frac{r_C}{R} = 1 - \left(1 - X_B\right)^{1/3} \tag{8.21}$$

where τ is the time of complete conversion of the solid:

$$\tau = t = \frac{\rho_B R}{bk''C_{Ag}} \tag{8.22}$$

In this result, k'' is the first-order rate constant for the surface reaction. As soon as a product layer forms, the resistance for diffusion through the ash dominates the resistance through the gas film surrounding the particle, so the gas film resistance can be neglected. When the resistances of the chemical reaction and diffusion steps are comparable, the conversion of the particle in terms of the overall resistance is determined by

$$\frac{1}{S_p}\frac{dN_A}{dt} = k''C_A = \frac{C_A}{\left(d_p/(12D_e)\right)_{ash} + \left(1/k''\right)_{reaction}} \tag{8.23}$$

where S_p, d_p, and D_e refer to the surface area of the particle, particle diameter, and diffusion coefficient, respectively. The previous equations represent conversion of a single particle. In plug flow of solids, it is assumed that all solids stay in the reactors for the same length of time. The contact time, or reaction time needed for the solid conversion, can be calculated from the shrinking-core model, depending on the rate-controlling step.

For particles of a single unchanging size, in mixed flow of solids with a uniform gas composition in a fluidized bed, the conversion of reactant in a single particle depends on its length of time in the bed and the reaction-controlling step, or the main resistance mode (film, ash, or reaction). The length of time is not the same for different particles in the reactor. Hence, the conversion can be calculated through the exit age distribution of solids in the bed as follows:

$$\begin{pmatrix}\text{Mean value for}\\ \text{the fraction of}\\ B\text{ unconverted}\end{pmatrix} = \sum_{particles}\begin{pmatrix}\text{Fraction of particles,}\\ \text{unconverted, between}\\ \text{time } t \text{ and } t+dt\end{pmatrix}\begin{pmatrix}\text{Fraction of exit stream}\\ \text{staying in the reactor}\\ \text{between } t \text{ and } t+dt\end{pmatrix} \tag{8.24}$$

$$1-\overline{X_B} = \int_0^{\tau}(1-X_B)E\,dt \tag{8.25}$$

where E is the exit age distribution of solids in the reactor.

The conversion of a mixture of particles in a mixed or plug flow has been predicted previously [35,36] for particles of spherical and nonspherical shapes. Ferrandon et al. (2008) presented data for measurements collected in a microreactor for the hydrolysis of $CuCl_2$ at 375°C, using sieved 150–250 μm particles. The solid conversion data were reported for different steam-to-$CuCl_2$ mole ratios and different time intervals. The vapor fraction for the corresponding steam-to-$CuCl_2$ ratio varied from 0.08 to 0.26 in the gas phase. The authors reported the influence of the reaction time, particle size, and inert gas flow on the thermal decomposition of $CuCl_2$ into CuCl. The following correlations have similar trends as the experimental data:

$$t = k_1\left[1-3(1-X_B)^{2/3}+2(1-X_B)\right] \tag{8.26}$$

$$t = k_2\left[1-(1-X_B)^{1/3}\right] \tag{8.27}$$

where

$k_1 = 0.022\ \text{min}^{-1}$ is the diffusion time constant
$k_2 = 0.02\ \text{min}^{-1}$ is the reaction time constant

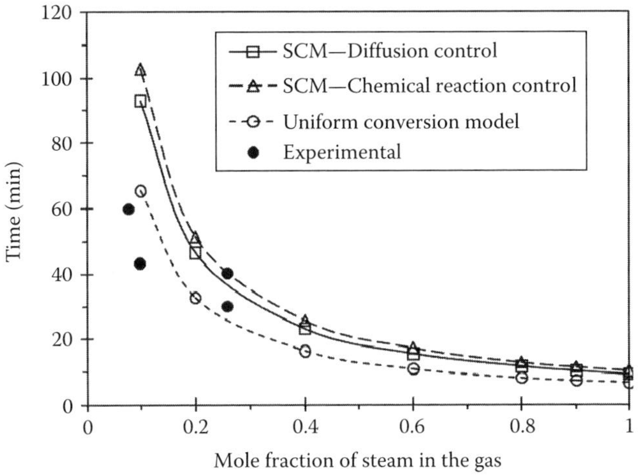

FIGURE 8.11
Comparison of solid conversion time with experimental data. (Adapted from Naterer, G.F. et al. *International Journal of Hydrogen Energy* 36, 15472–15485, 2011a.)

These k_1 and k_2 values will be used to evaluate the reaction rate constant, k'', and D_e, the effective diffusivity for an average steam concentration of $C_{Ag} = 3.8$ mol/m³. The corresponding rate constants are $k'' = 0.00011$ m³ (STP)/m² solids and $D_e = 2.03 \times 10^{-9}$ m²/s.

The variation of solid conversion time with vapor fraction in the gas is illustrated in Figure 8.11. The experimental data of Ferrandon et al. (2008) are also used to predict the solid conversion for uniformly sized particles of 200 μm diameter in plug flow of solids. As the mole fraction of steam in the gas increases, the time of solid conversion decreases. The reaction rate constant for the uniform-reaction model is $k_r = 0.00063$ m³ (STP)/m² for solid conversion of $X_B = 0.99$. Also, at high conversions, all of the models predict essentially the same rate of conversion. The fraction of reacted core volume to the total particle volume increases over time, regardless of the resistance mode for mixed flow of solids.

8.7 Molten Salt Reactor for Oxygen Production

8.7.1 Introduction

The oxygen production step (see Table 8.1 and Figure 8.1) receives solid feed of $CuO*CuCl_2$ and produces O_2 gas and liquid cuprous chloride. The reaction is given by $CuOCuCl_2(s) = 2CuCl$ (molten) + $0.5O_2$ (gas) at 530°C. Gas species leaving the oxygen reactor include oxygen gas and potential impurities of products from side reactions, such as CuCl vapor, chlorine gas, HCl gas (trace amount), and H_2O vapor (trace amount). The substances exiting the reactor are molten CuCl, potentially solid $CuCl_2$ from the upstream reaction ($CuCl_2$ hydrolysis step), due to the incomplete decomposition of $CuCl_2$ at a temperature lower than 750°C, as well as reactant particles entrained by the flow of molten CuCl. In the oxygen reactor, copper oxychloride particles will decompose to molten salt and oxygen. The reactant particles absorb decomposition heat from the surrounding molten bath.

Within the reactor, heat is transferred from liquid (molten CuCl) to solid $CuO*CuCl_2$ (reactant) particles, and the molten bath can be sustained by the reaction product itself.

The oxygen product will leave the reactant particles immediately, due to buoyancy of gas in the molten salt. This fast separation aids the minimization of heat transfer resistance to the reactant particles, which then helps make the overall reaction rate closer to the intrinsic reaction rate itself. The design of the reactor requires a high efficiency of heat exchange and separation of reactants from products, as well as product (oxygen) from product (molten salt).

A suitable method to heat the reactor is needed. A heated inert gas could be introduced, but there are two disadvantages with this approach: poor heat transfer and enhanced entrainment of undecomposed $CuO*CuCl_2$ (reactant) particles. To recycle these entrained $CuO*CuCl_2$ particles using sieves or cyclone/centrifugal separation would require significant power input. Also, the entrainment of fine particles would clog and corrode the equipment, especially components like flanges and screws. Another alternative is radiation, but again this is challenging because the process is more difficult to control, especially if the power input is high.

A more practical and efficient option is to heat the molten salt from the wall with a double-shell structure without using electricity. Flowing in the double shell, a secondary flow of molten salt provides heat to the reactor. The secondary flow of molten salt could be provided from a solar energy storage reservoir or nuclear heat exchanger, which uses molten salt as the heat storage and transferring medium. Since molten CuCl itself can be the heat storage medium used in the solar energy reservoir, the molten CuCl can also be pumped out of the oxygen production reactor to a solar energy reservoir and pumped back into the reactor. If necessary, agitation could be introduced into the bath. This approach would take advantage of both free and forced convection with heterogeneous mixing that facilitates both heat and mass transfer.

A difficulty is bubble spouting and separation of solid and molten salt. To address this issue, techniques used in iron-making or steel-making furnaces could be adopted, which commonly handle the removal of high-temperature molten iron, floating impurities/dross, slag, and the feeding of iron ore. The presence of bubbles in the molten bath has several advantages. For example, they enhance the local rates of heat transfer. The contact area between the $CuO*CuCl_2$ (reactant) particles and molten CuCl (product bath) increases when they are more heterogeneously mixed by ascending bubbles and descending particles. The aggregation of $CuO*CuCl_2$ (reactant) particles will reduce the contact area between particles and liquid, but fortunately the oxygen (product) bubbles can help to break these aggregations. Also, the upflow of oxygen (product) bubbles will resist the descent of solid particles, which will help to increase the residence time of $CuO*CuCl_2$ (reactant) particles in the molten salt; hence, they help decrease the reactor size. Due to the large density difference between oxygen and molten CuCl, the oxygen bubbles formed at the surface of the reactant particles will leave the surface immediately after the bubbles form, which ensures that the reaction is driven toward the direction of oxygen production.

The liquid level within the reactor may abruptly change with a sudden increase of bubbles due to flash decomposition, which may result in the entrainment of bubbles in the exit flow of molten CuCl, upon which the pipe could become choked by these bubbles. The liquid level may also change due to a sudden pressure fluctuation. If there is a sudden decrease of pressure, the bubbles may lift part of the liquid up to the top of the reactor, which could cause flooding or spouting of liquid. This problem can be addressed by a molten salt removal pipe of large diameter that is located on the reactor side, with fewer bubbles than the other side, so the liquid level is kept unchanged.

The reactant $CuO*CuCl_2$ does not melt in molten CuCl because it is first decomposed to oxygen and molten CuCl at 470°C. The reactor to produce oxygen is recommended to operate at about 530°C. In this temperature range, many metals would be oxidized by oxygen. The reactor and oxygen removal equipment must be corrosion resistant to oxygen, molten

CuCl, dry HCl (or even moist HCl), and possibly Cl_2 gas. If $CuCl_2$ is entrained in the reactor with CuO*$CuCl_2$ particles, then chlorine gas may appear and corrode the equipment material at a temperature of 530°C. Oxygen cooling in a heat exchanger downstream of the reactor requires a wall material that is also resistant to O_2 oxidization at 530°C, as well as dry HCl corrosion at such a high temperature.

8.7.2 Proof-of-Principle Experiments

Serban et al. (2004) have demonstrated experimentally the scientific feasibility of the oxygen production reaction. The reaction was conducted in a ½ in. by 14 in. quartz vertical reactor, as shown previously in Figure 8.1. The reaction was performed in a vertical reactor (Figure 8.1) connected to the mass spectrometer in order to monitor the oxygen evolution. Melanothallite was not commercially available. It was synthesized in the laboratory by heating stoichiometric amounts of $CuCl_2$ and CuO at temperatures between 370°C and 470°C. The resulting material was used in the kinetic study. The oxygen evolution was followed using a mass spectrometer, of Serban et al. (2004). The area under the peaks in the mass spectra was combined. Using separately obtained calibration data, it was determined that the oxygen yield at 500°C was 85% of the theoretical amount, and at 530°C, the yield was 100%. The XRD of the solid product indicates that the solid phase is pure CuCl. No chlorine was detected on the mass spectrometer, suggesting that no side reactions occurred. Thermodynamic data indicate that at 500°C, the equilibrium conversion is 60%. However, the reactions are always operated open to the atmosphere, and O_2 is continuously removed from the reaction zone. Under these conditions, complete conversions are possible because the equilibrium is shifted toward the right (Figure 8.12).

Mechanistic studies indicated that the overall oxygen generation reaction proceeds in two steps: (1) the decomposition of $CuCl_2$ to CuCl and Cl_2 and (2) the reaction of CuO

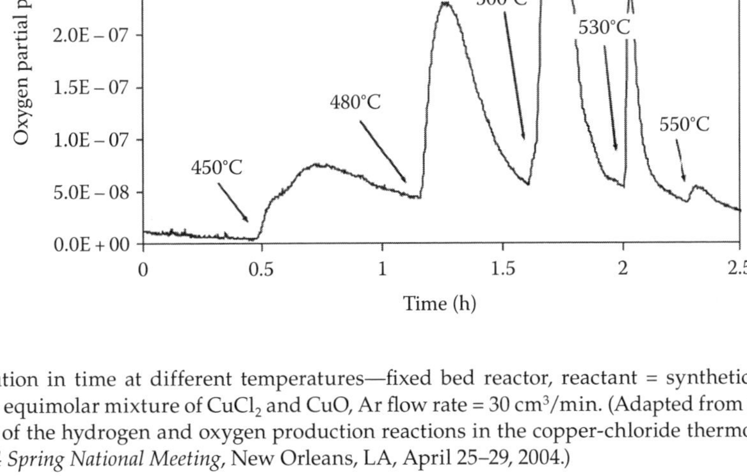

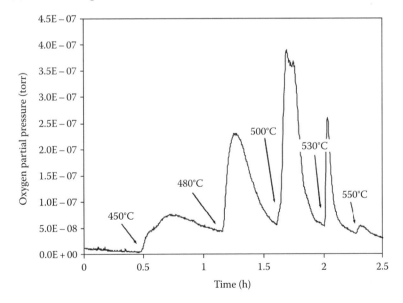

FIGURE 8.12
Oxygen evolution in time at different temperatures—fixed bed reactor, reactant = synthetic melanothallite made from an equimolar mixture of $CuCl_2$ and CuO, Ar flow rate = 30 cm^3/min. (Adapted from Serban, M. et al. Kinetic study of the hydrogen and oxygen production reactions in the copper-chloride thermochemical cycle, in *AIChE 2004 Spring National Meeting*, New Orleans, LA, April 25–29, 2004.)

with Cl_2. In the reaction between CuO and $CuCl_2$, oxygen started to evolve at 450°C, the temperature at which pure $CuCl_2$ starts to decompose and liberate Cl_2, which reacts with the CuO, producing CuCl and free oxygen. The overall oxygen generation reaction is limited by the decomposition of $CuCl_2$. Even though at 530°C only 55% of $CuCl_2$ is decomposed, the presence of CuO seems to facilitate its decomposition such that the products of reaction are O_2 and CuCl generated in stoichiometric amounts. Complete $CuCl_2$ decomposition is reported to be achieved at 630°C. Above 550°C, however, the CuCl vaporization becomes significant, and the experiment was stopped at this temperature.

The Cl_2 from the $CuCl_2$ decomposition is consumed completely in the reaction with CuO, since no Cl_2 was detected on the mass spectrometer. The kinetics of the reaction between CuO and $CuCl_2$ were analyzed in more detail by independently studying the $CuCl_2$ decomposition and the CuO and Cl_2 reactions in a thermal balance and a vertical reactor, respectively. A mixture of CuO and $CuCl_2$ generates O_2 at 500°C. In the experiment, 0.01 mol of CuO (0.8 g) and $CuCl_2$ (1.34 g) were used. Anhydrous $CuCl_2$ was obtained from dehydrated $CuCl_2$ by drying it in air at 175°C. The experiment was run in a flow-through-type quartz vertical microreactor using off-the-shelf reagents and an Ar flow rate of 15 cm^3/min. With the bulk density of 4.853 g/cm^3 for the mixture CuO and $CuCl_2$, the calculated space velocity is 2040 h^{-1}. The reaction was terminated after 45 min, and the oxygen yield was 85%. According to the thermodynamic equilibrium curves, at 500°C, the reaction is incomplete with only 0.3 mol of O_2 (72 cm^3 O_2 from 0.02 mol of solids) being generated. However, since the reaction was run under nonequilibrium conditions, that is, by continuously removing the O_2 from the reaction zone, the amount of oxygen generated surpassed the amount predicted by the equilibrium calculations. The only solid products of reaction were CuCl with traces of $CuCl_2$ as indicated by XRD examination.

8.7.3 Design Issues of Oxygen Reactor Scale-Up

For equipment scale-up beyond small test tubes and microreactors, several design issues must be addressed. This section discusses the broad implications of selected issues in this equipment scale-up. Solid feed of anhydrous solid CuO*$CuCl_2$ from the $CuCl_2$ hydrolysis reaction of the Cu–Cl cycle is provided to the oxygen production reactor. The bulk density of this material depends on the particle size and structure, which influence the volume of the reactor, feeder, and discharge vessels. The structure and shape of solid CuO*$CuCl_2$ are formed in the upstream $CuCl_2$ hydrolysis reactor. Aggregation into blocks of CuO*$CuCl_2$ particles may occur during the process of removing, conveying, and feeding of particles. The aggregation may choke or clog the feeder and cause sudden spouting of particles, so appropriate measures are needed to overcome these problems.

It is advantageous that the reactant CuO*$CuCl_2$ particles do not have a melting point below the decomposition temperature of 450°C. Unfortunately, there may exist embedded particles of $CuCl_2$ from the upstream hydrolysis reactor. The existence of $CuCl_2$ particles in CuO*$CuCl_2$ would lead to undesirable products and side reactions. For example, $CuCl_2$ may decompose to CuCl and Cl_2 gas. The existence of Cl_2 gas is highly corrosive. However, the existence of CuCl generated from $CuCl_2$ is not problematic, because molten CuCl is also a product generated from the decomposition of CuO*$CuCl_2$ particles. The most effective way to remove $CuCl_2$ particles from CuO*$CuCl_2$ is to ensure that the upstream $CuCl_2$ hydrolysis reaction to produce CuO*$CuCl_2$ is fully completed (conversion of $CuCl_2$ to CuO*$CuCl_2$ is 100%). The optimal conditions for full completion of the hydrolysis reaction are still under investigation.

In the oxygen reactor, the operation will be safer if the lowest feeding temperature is higher than the melting point of CuCl, that is, 430°C. If particles enter the reactor at a temperature lower than 430°C, a difficulty with the presence of bubbles in the molten salt may occur. Some CuCl vapor might condense, and molten CuCl might solidify around the $CuO*CuCl_2$ particles, thereby leading to blockage or confinement of bubbles. If an aggregation develops with particles, molten salt, and bubbles, the contact area between a reactant particle and heating medium (molten CuCl) will decrease, and the aggregations may float along the surface of the molten salt. This would deter the decomposition of reactant particles and potentially lead to choking of the reactor (a major safety concern). Fortunately, this issue can be avoided in several ways: hot feeding of the reactant particles, using an agitator, or protective mixing with hot oxygen (one of the decomposition products).

The decomposition temperature of $CuO*CuCl_2$ is higher than the melting point of CuCl. Therefore, the CuCl product will leave the reactor in molten form. The heat carried by molten CuCl is high-grade (high-temperature) heat that can be recovered to supply endothermic processes, such as evaporating water to superheated steam for use in the hydrolysis reaction in step 4 at a temperature of 375°C. Another option is to heat HCl gas from 375°C to 450°C for use in the hydrogen production reaction in step 1. The latter option is less preferable because the temperature difference between HCl gas and molten CuCl is much smaller than that between water and molten CuCl. In addition, the hydrogen generation reaction is exothermic, which implies that HCl gas could be self-heated. Also, to heat HCl gas using molten CuCl requires that the equipment material must be corrosion resistant to HCl gas at a high temperature of 450°C. For example, HCl gas may react with many common metals and their protective oxide film at a temperature of 400°C. In particular, if any moisture is present in the HCl gas, it will be more corrosive. However, this challenge is not as difficult with heating of H_2O.

To recover the heat carried by molten salt, the molten salt can be sprayed downward, while countercurrent motion of gas is flowing upward. For example, due to the very low solubility of CuCl in water (6.2×10^{-5} g/g water, 20°C), the molten salt can be first quenched in the quenching cell, and the vapor generated is then heated to a temperature of 375°C. The molten CuCl effluent from the oxygen production reactor might contain entrained particles of unreacted $CuO*CuCl_2$. In addition, fine oxygen bubbles may appear on the surface of the reactant particles when decomposing. If the bubbles detach slowly from the particles, $CuO*CuCl_2$ particles would not descend properly, thereby enhancing the likelihood of their entrainment in the molten CuCl. Some bubbles may be entrained by the molten CuCl, so the bubbles may plug the pipe where molten salt flows out of the reactor. The entrainment of reactant particles could also introduce safety issues involving the removal equipment. The particles may continue to decompose in the pipe after leaving the reactor, which would generate gas bubbles in the pipe and simultaneously decrease the temperature of molten salt. Since the decomposition requires heat input, this may result in solidification of molten salt. To prevent the entrainment problems, various potential methods are currently under investigation by the authors.

The Molten CuCl product is quenched to solid at a low temperature in the quenching unit. The solubility of CuCl in water is very low, and CuCl will solidify as the molten CuCl descends into liquid water. Simultaneously, part of the liquid water absorbs the heat released from molten CuCl, which then evaporates to steam. Rapid cooling and solidification of molten CuCl is desired in the quenching unit. The heat loss in the spray nozzle of molten CuCl must be small to avoid plugging of solid CuCl in the nozzle.

Another option to quench molten CuCl is that CuCl descends into an aqueous solution of HCl and possibly $CuCl_2$. An advantage of this method is to promote the direct dissolution of CuCl. However, this will introduce other challenges. One challenge is that the mixture of HCl and steam at high temperatures will exist in the gas phase inside the equipment, due to the volatile nature of HCl gas. The mixture is strongly corrosive to many materials. As a result, the use of steam with HCl would be more problematic. Therefore, this quenching method is not recommended in the design.

The removal of oxygen must not be slower than the decomposition rate of the reactant, in order to prevent a sudden rise of the reactor pressure due to possible *flash decomposition* of the reactant particles in the molten salt. A sudden decrease of the reactor pressure could also lead to flooding or spouting of bubbles within the liquid. Prior to steady-state continuous operation, tests should be performed under batch operation of a continuous flow reactor, in order to assess the dependency of reaction kinetics on scale-up, that is, decomposition rate of reactant particles and the removal of oxygen and molten salt. In the processing of oxygen gas, trace amounts of CuCl vapor may be present in the oxygen gas after the oxygen bubbles ascend and leave the molten CuCl bath. Hot oxygen gas can be used to purge lower-temperature $CuO*CuCl_2$ particles. In this way, CuCl would condense and return back to the oxygen production reactor. Trace amounts of CuCl in the oxygen gas product are not problematic if passed through a cyclone first, as any CuCl will deposit on a cooler surface.

Acknowledgments

Support of this research from the following partners is gratefully acknowledged: AECL, Argonne National Laboratory (International Nuclear Energy Research Initiative; US Department of Energy), University Network of Excellence in Nuclear Engineering (notably Ontario Power Generation and Bruce Power), Ontario Research Excellence Fund, Natural Sciences and Engineering Research Council of Canada, and the Canada Research Chairs program.

References

Aihara, M., Umida, H., Tsutsumi, A., Yoshida, K. Kinetic study of UT-3 thermochemical hydrogen production process. *International Journal of Hydrogen Energy* 1990; 15(1): 7–11.

Ambike, A.A., Mahadik, K.R., Paradkar, A. Spray dried amorphous solid dispersions of simvastatin a low Tg drug: In vitro and in vivo evaluations. *Pharmaceutical Research* 2005; 22(6): 990–998.

Berman, A., Epstein, M. The kinetics of hydrogen production in the oxidation of liquid zinc with water vapor. *International Journal of Hydrogen Energy* 2000; 25: 957–967.

Carty, R.H., Mazumder, M., Schreider, J.D., Panborn, J.B. *Thermochemical Hydrogen Production*, vol. 1. Gas Research Institute for the Institute of Gas Technology, Chicago, IL, GRI Report 80-0023, 1981.

Casper, M.S. *Hydrogen Manufacture by Electrolysis, Thermal Decomposition and Unusual Techniques*, Noyes Data Corporation, Park Ridge, NJ, 1978.

Chen, X.D., Lin, S.X.Q. Air drying of milk droplet under constant and time dependent conditions. *AIChE Journal* 2005; 51(6): 1790–1799.

Cui, H., Sauriol, P., Chaouki, J. High temperature fluidized bed reactor: Measurements, hydrodynamics and simulation. *Chemical Engineering Science* 2003; 58(3): 1071–1077.

Ferrandon, M.S., Lewis, M.A., Tatterson, D.F., Nankanic, R.V., Kumarc, M., Wedgewood, L.E., Nitsche, L.C. The hybrid Cu-Cl thermochemical cycle. I. Conceptual process design and H2A cost analysis. II. Limiting the formation of CuCl during hydrolysis, in *NHA Annual Hydrogen Conference*, Sacramento Convention Center, Sacramento, CA, March 30–April 3, 2008.

Goel, N., Mirabel, S.T., Ingley, H., Goswami, D.Y. Solar hydrogen production, SEECL Report 2003-01, Solar Energy & Energy Conversion Lab, University of Florida, Gainesville, FL—An Interim Report submitted to NASA Glenn Research Center, 2003.

Jiang, H.D., Ostrach, S., Kamotani, Y. Thermosolutal transport phenomena in large Lewis number electrochemical systems. *International Journal of Heat and Mass Transfer* 1996; 39: 841–850.

Knoche, K.F., Schuster, P., Ritterbex, T. Thermochemical production of hydrogen by a vanadium/chlorine cycle. II: Experimental investigation of the individual reactions. *International Journal of Hydrogen Energy* 1984; 9: 473–482.

Kocakusak, S., Akcay, K., Ayok, T., Koroglu, H.J., Koral, M., Savasci, O.T., Tolun, R. Production of anhydrous, crystalline boron oxide in fluidized bed reactor. *Chemical Engineering Science* 1996; 35: 311–317.

Konttinen, J.T., Zevenhoven, C.A.P., Hupa, M.M. Modeling of sulfided zinc titanate regeneration in a fluidized bed reactor. 2. Scale-up of the solid conversion model. *Industrial Engineering Chemistry Research* 1997; 36: 5439–5446.

Kunii, D., Levenspiel, O. *Fluidization Engineering*, 2nd edn., Butterworth-Heinemann, Boston, MA, 1991.

Lee, M.S., Goswami, D.Y., Hettinger, B., Vijayaraghavan, S. Preparation and characteristics of calcium oxide pellets for UT-3 thermochemical cycle, in *the Proceedings of the 2006 ASME International Mechanical Engineering Congress and Exposition*, Chicago, IL, November 2006.

Lee, M.S., Goswami, D.Y., Kothurkar, N., Stefanakos, E.K. Fabrication of porous calcium oxide film for UT-3 thermochemical hydrogen production cycle, in *the Proceedings of the ASME Energy Sustainability 2007*, June 27–30, 2007, Long Beach, CA.

Lee, M.S., Goswami, D.Y., Stefanakos, E.K. Immobilization of calcium oxide solid reactant on an yttria fabric and thermodynamic analysis of UT-3 thermochemical hydrogen production cycle. *International Journal of Hydrogen Energy* 2009; 34: 745–752.

Levenspiel, O. *Chemical Reaction Engineering*, 3rd edn., John Wiley & Sons, New York, 1998.

Levi-Hevroni, D., Levy, A., Borde, I. Mathematical modeling of drying of liquid/slurries in a steady state one dimensional flow. *Drying Technology* 1995; 13: 1187–1201.

Lewis, M., Taylor, A. High temperature thermochemical processes, DOE Hydrogen Program, Annual Progress Report, Washington, DC, pp. 182–185, 2006.

Lewis, M.A., Masin, J.G., Vilim, R.B., Serban, M. Development of the low temperature Cu-Cl thermochemical cycle, in *International Congress on Advances in Nuclear Power Plants*, Seoul, Korea, May 15–19, 2005.

Lin, S.X.Q., Chen, X.D. Prediction of air-drying of milk droplet under relatively high humidity using the reaction engineering approach. *Drying Technology* 2005; 23: 1395–1406.

Lin, S.X.Q., Chen, X.D. A model for drying of an aqueous lactose droplet using the reaction engineering approach. *Drying Technology* 2006; 24: 1329–1334.

McQuillan, B.W., Brown, L.C., Besenbruch, G.E., Tolman, R., Cramer, T., Russ, B.E., Vermillion, B.A. et al. High efficiency generation of hydrogen fuels using solar thermochemical splitting of water: Annual Report, GA-A24972, General Atomics, San Diego, CA, 2002.

Mirabel, S.T., Goel, N., Ingley, H.A., Goswami, D.Y. Utilization of domestic fuels for hydrogen production. *International Journal of Power and Energy Systems* 2004; 24(3): 239–245.

Nakayama, T., Yoshioka, H., Furutani, H., Kameyama, H., Yoshida, K. MASCOT—A bench-scale plant for producing hydrogen by the UT-3 thermochemical decomposition cycle. *International Journal of Hydrogen Energy* 1984; 9(3): 187–190.

Naterer, G.F. *Heat Transfer in Single and Multiphase Systems*, CRC Press, Boca Raton, FL, 2002.

Naterer, G.F., Camberos, J.A. *Entropy Based Analysis and Design of Fluids Engineering Systems*, CRC Press, Boca Raton, FL, 2008.

Naterer, G.F., Suppiah, S., Stolberg, L., Lewis, M., Ferrandon, M., Wang, W., Dincer, I. et al., Clean hydrogen production with the Cu-Cl cycle—Progress of international consortium, I: Experimental unit operations. *International Journal of Hydrogen Energy*, 2011a; 36: 15472–15485.

Naterer, G.F., Suppiah, S., Stolberg, L., Lewis, M., Ferrandon, M., Wang, W., Dincer, I. et al. Clean hydrogen production with the Cu-Cl cycle—Progress of international consortium, II: Simulations, thermochemical data and materials. *International Journal of Hydrogen Energy*, 2011b; 36: 15486–15501.

Naterer, G.F., Tokarz, C.D., Avsec, J., Fuel cell entropy production with ohmic heating and diffusive polarization. *International Journal of Heat and Mass Transfer* 2006; 49: 2673–2683.

Ortega-Lopez, M., Morales-Acevedo, A. Characterization of CuInS2 thin films for solar cells prepared by spray pyrolysis. *Thin Solid Films* 1998; 330: 96–101.

Ranz, W.E., Marshall, W.R. Evaporation from droplets. *Chemical Engineering Progress* 1952; 48: 141–146.

Sadhankar, R.R., Li, J., Li, H., Ryland, D., Suppiah, S. Hydrogen generation using high-temperature nuclear reactors, in *55th Canadian Chemical Engineering Conference*, Toronto, Ontario, Canada, October 2005.

Sakurai, M., Bilgen, E., Tsutsumi, A., Yoshida, K. Adiabatic UT-3 thermochemical process for hydrogen production. *International Journal of Hydrogen Energy* 1996; 21(10): 865–870.

Sakurai, M., Nakajima, H., Amir, R., Onuki, K., Shimizu, S. Experimental study on side-reaction occurrence condition in the iodine-sulfur thermochemical hydrogen production process. *International Journal of Hydrogen Energy* 2000; 23: 613–619.

Schultz, K. Thermochemical production of hydrogen from solar and nuclear energy, Technical Report for the Stanford Global Climate and Energy Project, General Atomics, San Diego, CA, 2003.

Serban, M., Lewis, M.A., Basco, J.K. Kinetic study of the hydrogen and oxygen production reactions in the copper-chloride thermochemical cycle, in *AIChE 2004 Spring National Meeting*, New Orleans, LA, April 25–29, 2004.

Stolberg, L., Boniface, H., Deschenes, L., York, S., McMahon, S., Suppiah, S. Recent advances at AECL in the Cu-Cl cycle for nuclear-produced hydrogen, in *ORF Workshops on Thermochemical Nuclear-Based Hydrogen Production*, Oshawa, ON, December 2007 and Chalk River, ON, October 2008.

Thurnhofer, A., Schachinger, M., Winter, F., Mali, H., Schenk, J.L. Iron ore reduction in a laboratory-scale fluidized bed reactor-effect of pre-reduction on final reduction degree. *ISIJ International* 2005; 45(2): 151–158.

Walton, D.E. The evaporation of water droplets: A single droplet drying experiment. *Drying Technology* 2004; 22(3): 431–456.

Wegner, K., Ly, H.C., Weiss, R.J., Pratsnis, S.E., Steinfeld, A. In situ formation and hydrolysis of Zn nanoparticles for H2 production by the 2-step ZnO/Zn water-splitting thermochemical cycle. *International Journal of Hydrogen Energy* 2006; 31: 55–61.

Weng, F.B., Kamotani, Y., Ostrach, S. Mass transfer rate study in rotating shallow electrochemical cells. *International Journal of Heat and Mass Transfer* 1998; 41: 2725–2733.

Yan, T.Z., Jen, T. Two-phase flow modeling of liquid-feed direct methanol fuel cell. *International Journal of Heat and Mass Transfer* 2008; 51: 1192–1204.

Section V

Biological Hydrogen Production

9

Biological Hydrogen Production: Dark Fermentation

Kuo-Shing Lee
Central Taiwan University of Science and Technology

Liang-Ming Whang
National Cheng Kung University

Ganesh D. Saratale
National Cheng Kung University

Shing-Der Chen
National Cheng Kung University

Jo-Shu Chang
National Cheng Kung University

Hisham Hafez
University of Western Ontario
Greenfield Specialty Alcohols

George Nakhla
University of Western Ontario

Hesham El Naggar
University of Western Ontario

CONTENTS

9.1 Introduction

Over the past few decades, the global population and the demand for energy have been growing at an exponential rate. Fossil fuels are the major source of energy (about 80%) in modern society (Demirbas, 2007). On the basis of the current consumption rate, it is estimated that the present known reserves of fossil fuels will last from 41 to 700 years (Goldemberg and Johansson, 2004; Goldemberg, 2007). The limited life and the unfettered use of fossil fuels and concerns about energy security have had a negative impact on the environment because of the emission of greenhouse gases (CO_2, CH_4, and CO), which has resulted in global warming and environmental pollution (Koh and Ghazoul, 2008; Saratale et al., 2008). For these reasons, in the present century, significant efforts are being made globally toward the development of technologies that generate clean, sustainable energy sources that can be a substitute for fossil fuels. One of these is biofuel (Gong et al., 1999; Ragauskas et al., 2006).

Biofuels are part of the CO_2 cycle, and their ecofriendly and sustainable nature makes them an important and promising alternative energy source for fossil fuels to protect the biosphere and prevent more localized forms of pollution (Puppan, 2002; Schubert, 2006). Due to these advantages, the worldwide production of biofuels has increased from 4.4 to 50.1 billion liters in the past few years. However, political and public support for biofuels has been shaky regarding less carbon emission of biofuel after combustion (Licht, 2008). Also, the cultivation of food crops for, and croplands dedicated to, biofuel production has resulted in food shortages and associated increasing costs of staple food crops such as maize and rice (James et al., 2008; Keeney and Hertel, 2008). Nevertheless, some scientists remain optimistic that biological processes such as H_2 and methane production under anaerobic conditions and ethanol fermentation using renewable carbon sources (waste biomass) will minimize the negative environmental and social impacts and that these resources will not compete directly with food production, or with land that may be needed for food production (Slade et al., 2009). Some also assume that biomass energy will play a pivotal role in future energy supply for there are large amounts of waste biomass available in the form of organic residues such as municipal wastes, manure, and forest and agricultural residues (Kapdan and Kargi, 2006; Ragauskas et al., 2006; Field et al., 2008).

Among the processes mentioned earlier, H_2 is considered a clean and renewable energy resource, not contributing to the greenhouse effect because it produces water instead of greenhouse gases after undergoing combustion, with a high energy yield of 122 MJ/kg, which is up to three times greater than that of hydrocarbon fuels (Kapdan and Kargi, 2006; Lo et al., 2008b). Moreover, H_2 has a heating value of 61,100 Btu/lb, almost three times that of methane (23,879 Btu/lb) (Bossel, 2003). Thus, it has been predicted that H_2 will be the main source of energy and a potential substitute for fossil fuels by the year 2100 (Dunn, 2002). Conventional physicochemical methods for H_2 gas production do not accomplish the dual goals of waste reduction and energy production. Furthermore, these methods require such large inputs of energy derived from fossil fuel combustion that they do not qualify as alternative renewable energy sources (Vijayaraghavan and Soom, 2007; Das and Veziroglu, 2008). For global environmental considerations, microbial H_2 production represents an important area of bioenergy production.

Over the past quarter century, many hundreds of publications have focused on microbial H_2 production, but advances toward practical applications have been minimal (Hallenbeck and Benemann, 2002; Show et al., 2008). From an economic point of view, it cannot be expected that biohydrogen production will compete with chemically synthesized H_2 in the next few decades (Akkerman et al., 2002; Kotay and Das, 2008). However, H_2 production from waste materials through dark fermentation by using specific microorganisms is an economically feasible process and is expected to become important in bioenergy production. In the past, the microbial communities in the bioreactor were usually regarded as a *black box*. Many reviews focusing on the fundamentals of biohydrogen production (Hallenbeck and Benemann, 2002; Vijayaraghavan and Soom, 2007; Kotay and Das, 2008; Saratale et al., 2008; Wang and Wan, 2009), metabolism, and bioreactor design for photofermentation (Akkerman et al., 2002; Koku et al., 2002) have been published of late. Nevertheless, there are only few studies on how molecular biotechnology is related to the dark fermentation process. In this chapter, we assess the effect of various factors influencing biohydrogen production, kinetics, and types of reactors for dark fermentation. We also discuss the existing molecular tools for improving microbial H_2 production under dark fermentation.

9.2 Hydrogen Energy: Importance and Production

9.2.1 Features of Hydrogen Energy and Methods of Hydrogen Production

Present-day energy consumption is unsustainable because of equity issues, geopolitical concerns, and the energy paradigm based on fossil fuel dependency that is contributing toward economic and environmental challenges (UNDP, 2000). H_2 is strategically important because of its low emission, representing a cleaner and more sustainable energy system and having many social, economic, and environmental benefits (Bockris, 2002; Maddy et al., 2003). H_2 acts as a versatile energy carrier with the potential for extensive use in power generation and in many other applications. H_2 gas is widely used as a feedstock for the production of chemicals (ammonia and methanol), in oil refineries for removing impurities or for upgrading heavier oil fractions into lighter and more valuable products, for the production of electronic devices, and for processing steel, desulfurization, and the reformulation of gasoline in refineries; it is also used in the world's space programs (1%) (Elam et al., 2003; Vijayaraghavan and Soom, 2007). The concept of H_2 economy has been presented as a clean and efficient replacement for the petroleum-based economy and recognized by many national and international bodies, for example, the US Department of Energy (US-DOE), the International Partnership for the H_2 Economy (IPHE), and the European H_2 Association (EHA) (Kyazze et al., 2008). Based on the National H_2 Program of the United States, the contribution of H_2 to the total energy market is expected to be 8%–10% by 2025 (Armor, 1999). Technical challenges in attaining a H_2 economy are the cost-effective process for H_2 production, storage, delivery, and their practical applications. Conventional physicochemical methods for H_2 production were based on steam reforming of natural gas (methane and other hydrocarbons), partial oxidation of heavier hydrocarbons, coal gasification, electrolysis of water, and pyrolysis or gasification. However, all these methods usually involve complicated procedures and require a source of energy derived from fossil fuels. This makes them economically unfeasible and not always environmentally benign (Das and Veziroglu, 2001; Momirlan and Veziroglu, 2002).

H_2 has attracted worldwide attention as a clean energy source. However, due to its unavailability in nature, there has been focus on the development of cost-effective and efficient H_2 production technologies in recent years (Kapdan and Kargi, 2006; Gustavo et al., 2009). Biological H_2 production has assumed importance as an alternative and renewable bioenergy resource because of certain advantages such as mild reaction conditions, low cost, low energy requirement; also, it is well suited for decentralized energy production in small-scale installations in locations where biomass or wastes are available, thus avoiding energy expenditure and transport costs (Saratale et al., 2008; Wang and Wan, 2009). Biohydrogen production also provides a feasible means for the sustainable supply of H_2 with low energy and high efficiency, and is therefore being considered for producing H_2 (Hallenbeck, 2005). In addition, biohydrogen production is important for energy security reasons, environmental concerns, foreign exchange savings, and socioeconomic issues related to the rural sector of all countries in the world (Hallenbeck, 2009). In the past few decades, research on biological H_2 production has been carried out and a broad spectrum of biological H_2 production processes has been investigated. Methods adopted to produce H_2 from biological methods can be classified into the following groups: (1) biophotolysis of water using algae/cyanobacteria, (2) photodecomposition (photofermentation) of organic compounds using photosynthetic bacteria, (3) dark fermentative H_2 production using anaerobic or facultative anaerobic bacteria, (4) bioelectrohydrogenesis or microbial fuel cell (MFC) (Asada and Miyake, 1999; Ghirardi et al., 2000; Das and Veziroğlu, 2001; Koku et al., 2002; Chen et al., 2008b).

Among the processes mentioned earlier, biohydrogen production by anaerobic dark fermentation has the best potential for practical applications (Levin et al., 2004) (Table 9.1). After the mid-1990s, significant attention has been paid to H_2 production by anaerobic

TABLE 9.1

Comparison of Important Biological Hydrogen Production Processes

Biological Process	Advantages	Disadvantages
Direct and indirect biophotolysis	Can produce H_2 directly from water and sunlight Solar conversion energy increased by 10-fold as compared to trees, crops, can produce H_2 from water, and has the ability to fix N_2 from atmosphere	Natural-borne organisms of these species examined so far show rather low rates of H_2 production due to the complicated reaction systems and inhibition of the hydrogenase enzyme by oxygen. Another drawback encountered is the requirement of a carrier gas to collect the evolved gas from the culture.
Photofermentation	A wide spectral light energy can be used by these bacteria Can use different waste materials like distillery effluents and waste	This process requires high activation energy to drive nitrogenase, which is the enzyme responsible for H_2 production in photosynthetic bacteria. Low solar conversion efficiencies, typically not much higher than that for algal biophotolysis systems. In addition, phototrophic H_2 production with photosynthetic bacteria is extremely doubtful due to ammonia and oxygen contents, making it difficult in practical applications.
Dark fermentation	Some basic advantages relative to other processes include process simplicity on technical grounds, low energy requirements, higher rates of H_2 production, economically feasible or better process economy, and the ability to generate H_2 from a large number of carbohydrates (or other organic materials) frequently obtained as refuse or waste products. It can produce H_2 all day long without light It produces valuable metabolites such as butyric, lactic, and acetic acids as by-products It is an anaerobic process, so there is no O_2 limitation problem	Low H_2 yield compared to physicochemical methods. The major drawback is that these bacteria are unable to overcome the inherent thermodynamic energy barrier to full substrate decomposition due to which low H_2 yields was observed.

dark fermentation (Levin et al., 2004). Compared to other biological processes, dark fermentation has some basic advantages such as the simplicity of the process on technical grounds, low energy requirements, higher rates of stable H_2 production, economically more feasible or better process economy, and the production of H_2 from a large number of carbohydrates frequently obtained as refuse or waste products (Benemann, 1996; Nandi and Sengupta, 1998; Levin et al., 2004; Hallenbeck et al., 2009; Wang and Wan, 2009). Dark fermentation normally achieves a much higher H_2 production rate than photolysis and photofermentation, and this higher rate depends upon various factors such as the fermentation metabolic pathway, hydrogenase activity, substrate types. The theoretical maximum H_2 production yield (12 mol H_2/mol glucose) can be engineered by using suitable organisms, but it seems to be difficult to carry out in living cells (Woodward, 2000). To achieve this yield, most of the research has focused on biohydrogen production from organic wastes by using photosynthetic bacteria over the past two decades (Sasikala et al., 1993). Recent studies show that biohydrogen production using negative-value organic waste streams under dark fermentation is a more promising approach than the use of photosynthetic bacteria (Ueno et al., 1995; Benemann, 1998; Chen et al., 2001; Lo et al., 2009). However, in dark H_2 fermentation, the major drawback is that these bacteria are unable to overcome the inherent thermodynamic energy barrier to full substrate decomposition due to which low H_2 yields are observed (Hallenbeck, 2005; Kotay and Das 2008).

9.2.2 Hydrogen Production Using Dark Fermentation Process

Fermentative conversion of substrates to H_2 is a complex biochemical process manifested by diverse groups of bacteria by a series of biochemical reactions under anoxic conditions (i.e., no oxygen present as an electron acceptor) (Valdez-Vazquez and Poggi-Varaldo, 2009). In case of complex organic polymers, bacteria grow on these organic substrates (heterotrophic growth), which are degraded by oxidation to yield monomer units and metabolic energy for growth. It is also noteworthy that when bacteria degrade organic substrates, electrons needed to be disposed to maintain electrical balance are produced. In aerobic environments, reduction of oxygen yields water as the product, whereas in anaerobic environments, protons (H^+) derived from water are reduced to molecular H_2 (H_2) (Das and Veziroglu, 2001; Levin et al., 2004). In general, hydrogen-producing acidogenic bacteria oxidize fermentation products to acid intermediates and H_2 (e.g., production of H_2 and CO_2 by acetogens and homoacetogens), and finally acetoclastic methanogens convert the acid intermediates into methane and CO_2 (Liu et al., 2008; Hallenbeck et al., 2009) (Figure 9.1). In addition, when sulfates or nitrates are present, sulfate-reducing bacteria (SRB) and nitrate-reducing bacteria (NRB) are capable of using H_2 as an electron donor and generating sulfides and ammonia, respectively (Valdez-Vazquez and Poggi-Varaldo, 2009).

When glucose is used as the model substrate for fermentative H_2 production, it is first converted by hydrogen-producing bacteria to pyruvate, via the Embden–Meyerhof–Parnas or glycolytic pathway, producing the reduced form of nicotinamide adenine dinucleotide (NADH). In most *Clostridial* fermentation, pyruvate is then converted to acetyl-CoA and CO_2 by pyruvate:ferredoxin oxidoreductase. In addition to the formation of acetyl-CoA and CO_2, an iron–sulfur protein called ferredoxin, which acts as electron carrier of low redox potential, is also reduced in this reaction. As described earlier,

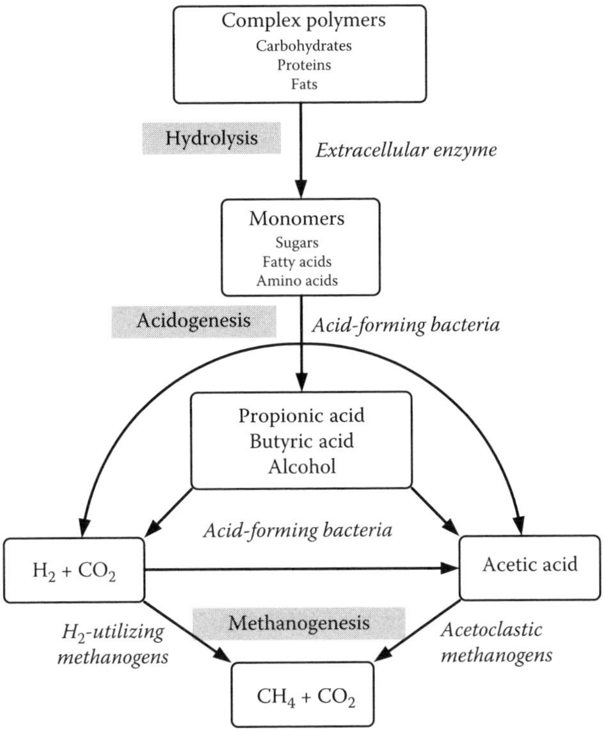

FIGURE 9.1
Schematic description of anaerobic digestion process. (Adapted from Malina, J.F. and Pohland, J.F.G., *Design of Anaerobic Processes for the Treatment of Industrial and Municipal Wastes*, Technomic Pub Co, Lancaster, PA, 1992.)

electrons from the reduced ferredoxin are transferred via hydrogenase to H^+, so that molecular H_2 is formed (Equations 9.1 and 9.2). Pyruvate may also be converted to acetyl-CoA and formate, which may readily be converted to H_2 and carbon dioxide (Nath and Das, 2003; Lee et al., 2004a; Turner et al., 2008). The transfer of electrons from ferredoxin to H^+ is catalyzed by the enzyme called hydrogenase, as illustrated in Figure 9.2 (Brock et al., 1994).

$$\text{Pyruvate} + \text{CoA} + 2\text{Fd (ox)} \rightarrow \text{acetyl-CoA} + 2\text{Fd (red)} + CO_2 \tag{9.1}$$

$$2H^+ + \text{Fd (red)} \rightarrow H_2 + \text{Fd (ox)} \tag{9.2}$$

The fate of acetyl-CoA varies among different *Clostridial* species. It can be converted into some soluble metabolites such as ethanol, acetate, butyrate, and lactate. The amount of H_2 evolved depends on the ratio of the end products formed (Das, 2009). Theoretically, the maximum H_2 yield is 4 mol H_2/mol glucose when glucose is completely metabolized to

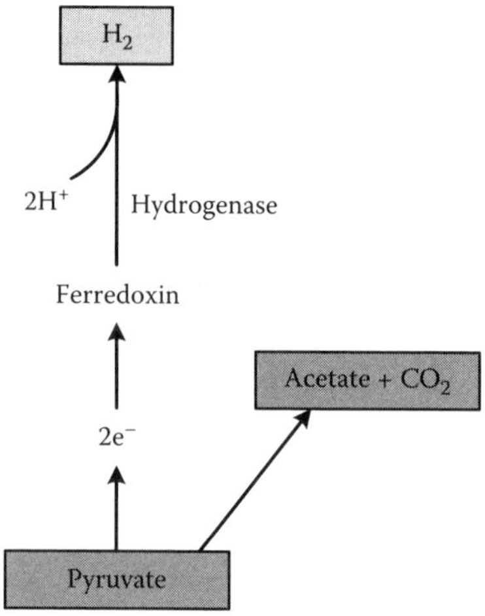

FIGURE 9.2
Production of molecular H_2 from pyruvate. (From Brock, T.D. et al., *Biology of Microorganisms*, 7th edn., Prentice-Hall, Englewood Cliffs, NJ, 1994.)

acetate. The overall stoichiometry is indicated by Equation 9.3 (Ueno et al., 1995; Hawkes et al., 2002; Levin et al., 2004):

$$C_6H_{12}O_6 + 2H_2O \rightarrow 2CH_3COOH + 4H_2 + 2CO_2 \tag{9.3}$$

And according to Equation 9.4, half of this H_2 yield (2 mol H_2/mol glucose) is obtained when butyrate is the fermentation end product:

$$C_6H_{12}O_6 \rightarrow C_3H_7COOH + 2H_2 + 2CO_2 \tag{9.4}$$

The reactions involved in dark fermentative H_2 production (Equations 9.3 and 9.4) are faster and do not rely on the availability of light sources, making them useful for treating a wide spectrum of potentially utilizable substrates, including refuse and waste products, by using appropriate fermenters. Strictly anaerobic bacteria give 4 mol H_2 mole of glucose, whereas facultative anaerobic bacteria give 2 mol of H_2/mole of glucose. Facultative anaerobes were found to be less sensitive to oxygen and were able to recover H_2 production activity after accidental oxygen damage to them by rapidly consuming oxygen present in the broth. Thus, facultative anaerobes are considered to be better than strict anaerobes for the fermentative H_2 production process. However, one of the main constraints in the practical application of anaerobic fermentation is a H_2 yield lower than 4 mol H_2/mol glucose (usually in the range of 0–2 mol H_2/mol glucose). This might be due to the formation of mixed acidogenic (VFA) and alcoholic fermentation end products. Reduced fermentation end products, such as ethanol, butanol, and lactate, contain H_2 that has not been liberated as gas. It was assumed that by modifying the fermentation pathways, a maximum of 4 mol H_2/mol glucose can be expected from ideal acetate fermentation. However, this yield is still low compared to existing chemical or electrochemical processes of H_2 generation (Hallenbeck and Benemann, 2002) and makes biological H_2 production economically unfeasible. Therefore, in this chapter, we will focus on research and development of dark fermentative H_2 production as well as on a critical discussion of the various practical and theoretical approaches toward improvement of overall H_2 production in a dark fermentative process.

9.3 Kinetics of Anaerobic Biohydrogen Production

Reaction kinetic study provides a rational basis for process analysis, control, and design. The kinetic constants obtained are important to develop H_2-producing processes. Most popularly, the modified Gompertz equation (Equation 9.5) has been used to describe the cumulative H_2 production process obtained from a batch experiment (Lay et al., 1999; Van Ginkel et al., 2001; Chen et al., 2002; Khanal et al., 2004; Lin and Lay, 2004b; Lin and Lay, 2005; Fang et al., 2006; Mu et al., 2006a). In this model, the population of bacteria is expressed as a function of the reaction time, and the overall stoichiometry of the model (Equation 9.5) is as follows:

$$H = P \cdot \exp\left\{-\exp\left[\frac{R_m \cdot e}{P}(\lambda - t) + 1\right]\right\} \tag{9.5}$$

where

H (mL) is the total amount of H_2 produced at incubation time t (h)
P (mL) is the potential maximal amount of H_2 produced
R_m (mL/h) is the maximal H_2 production rate
λ (h) is the lag time to exponential H_2 production
$e = 2.71828$

Lin and Lay (2004b) checked the effects of carbonate and phosphate concentrations on biological H_2 production obtained from the Taguchi orthogonal array by fitting the cumulative H_2 production data with the modified Gompertz equation. The correlation coefficient values ranged over 0.996–0.999. Recently, the kinetic parameters of H_2 production from rice slurry with acidophilic hydrogen-producing sludge at feedstock concentrations from 2.7 to 22.1 g carbohydrate/L were determined by using the modified Gompertz equation (Fang et al., 2006). In this study, the highest H_2 yield (346 mL/g carbohydrate) was obtained at a concentration of 5.5 g carbohydrate/L. In addition, the individual and interactive effects of pH, temperature, and substrate concentration on biohydrogen production from sucrose by mixed anaerobic cultures were determined by combining the modified Gompertz equation with response surface methodology (RSM) (Mu et al., 2006b). The results showed that pH, temperature, and sucrose concentration have a significant impact on the specific H_2 production potential. The maximum sucrose yield of 252 mL H_2/g was estimated under the optimum conditions of pH 5.5, temperature 34.8°C, and sucrose concentration 24.8 g/L, while R_{max} of 504 mL H_2/h/L was calculated under the optimum conditions of pH 5.5, temperature 35.5°C, and sucrose concentration 25.4 g/L.

The Monod model is also useful and employed most frequently to describe the kinetics of anaerobic biohydrogen production (Bailey and Ollis, 1986). The overall stoichiometry of the Monod model (Equation 9.6) is as follows:

$$\mu = \frac{\mu_{max} S}{K_s + S} \tag{9.6}$$

where
- μ is the specific growth rate
- μ_{max} is the maximum specific growth rate
- S is the substrate concentration
- K_s is the Monod constant, the concentration of the limiting nutrient at which the specific growth rate is half of its maximum value

Chen et al. (2001) used the Monod model to describe and predict the kinetic properties of H_2 production from sucrose in the anaerobic mixed culture. The μ_{max}, K_s, and yield coefficient for cell growth ($Y_{x/s}$) were estimated as 0.172 h^{-1}, 68 mg COD/L, and 0.1 g/g, respectively. In addition, the Michaelis–Menten equation (Bailey and Ollis, 1986) was also employed to describe the kinetics of many anaerobic biohydrogen productions. The overall stoichiometry (Equation 9.7) is as follows:

$$v = \frac{v_{max} S}{K_m + S} \tag{9.7}$$

where
- v is the H_2 production rate
- v_{max} is the maximum H_2 production rate
- K_m is the half-saturation constant
- S is the concentration of the substrate

The Michaelis–Menten equation was used to determine the growth kinetics of hydrogen-producing bacteria and the H_2 production rate using different substrates and concentrations, namely, sucrose, nonfat dry milk (NFDM), and food waste, under dark fermentation in batch mode. Results showed that the H_2 production potential and H_2 production rate increased with increasing substrate concentration. The maximum H_2 yields from sucrose, NFDM, and food waste were 234, 119, and 101 mL/g COD, respectively (Chen et al., 2006).

Wu et al. (2002) studied the kinetic performance of H_2 fermentation from sucrose using a variety of immobilized-cell systems, activated carbon (AC), polyurethane (PU), and acrylic latex plus silicone (ALSC). The dependence of the specific H_2 production rate (v_s) of the three types of immobilized cells on sucrose concentration was interpreted with Michaelis–Menten kinetics. In this study, the $v_{s,max}$ and K_m values were estimated from the experimental data via numerical simulation. Results showed that CA/AC (addition of activated carbon into alginate gel) cells have considerably higher $v_{s,max}$ and K_m values than those obtained from the other two types of immobilized cells. In the Michaelis–Menten model, the maximal rate (e.g., $v_{s,max}$) is considered a linear function of total active biocatalyst concentration, while the K_m

value is inversely proportional to the affinity between the biocatalyst and the substrate (Bailey and Ollis, 1986). Thus, the higher $v_{s,max}$ value of CA/AC cells represents a higher concentration of *active* biocatalyst than that of PU and ALSC cells. The H_2 production by EVA (ethylene–vinyl acetate copolymer)–immobilized acclimated sewage sludge using sucrose as the sole carbon substrate was reported earlier (Wu et al., 2005). Kinetic studies show that the Michaelis–Menten equation is able to describe the dependence of specific H_2 production rate on sucrose concentration. The EVA-immobilized cells achieved an optimal specific H_2 production rate of 488 mL H_2/g VSS/h (sucrose concentration = 40 g COD/L) and the highest H_2 yield of 1.74 mol H_2/mol (sucrose concentration = 20 g COD/L).

Some scientists developed a new least-squares method to describe the kinetics and to estimate the kinetic parameters (Y_g and k_d) (Equation 9.8) for H_2 production from sucrose in an upflow anaerobic sludge blanket (UASB) (Yu and Mu, 2006). These two parameters could provide useful information for designing an effective granule-based UASB system for H_2 production. The parameters Y_g and k_d could be estimated using the least-squares method by plotting $R_{S/X}$ versus (1/HRT). The Y_g and k_d of the H_2-producing granules were estimated to be 0.334 g VSS/g COD and 0.004 h^{-1}, respectively. In some reports, the sludge yield in CSTR (ranging from 0.116 to 0.240 g VSS/g COD) was higher than that of the H_2-producing granule using sucrose as substrate (Liu and Fang, 2002). The kinetic parameters of biological H_2 production in the literature are summarized in Table 9.2.

$$R_{S/X} = \frac{1}{Y_g} \times \frac{1}{\text{HRT}} + \frac{k_d}{Y_g} \tag{9.8}$$

where

$R_{S/X}$ (g COD/g VSS/h) is the specific substrate utilization rate

Y_g (g VSS/g COD) is the sludge yield, expressing the cell mass produced per unit substrate consumed

k_d (h^{-1}) is the endogenous decay coefficient

9.4 Effects of Physicochemical Factors on Anaerobic Biohydrogen Production

Biohydrogen production is a complex process and is greatly influenced by many factors. These include microorganism seeding, substrate specificity, organic loading rate (OLR), hydraulic retention time (HRT), pH, temperature, metal ion, oxidation–reduction potential (ORP), nutritional requirements, H_2 partial pressure in the reactor, and reactor configuration. The effects of these factors on fermentative H_2 production have been reported by a number of studies worldwide in the past few years (Hawkes et al., 2002, 2007; Nishio and Nakashimada, 2004; Kapdan and Kargi, 2006; Kraemer and Bagley, 2008; Li and Fang 2007a; Das and Veziroglu, 2008; Wang and Wan, 2009). The literature reviews regarding the effects of these factors on biohydrogen production are summarized in Tables 9.3 through 9.8.

TABLE 9.2

Comparison of Kinetic Parameters for H_2 Fermentation in Different H_2-Producing Systems

Hydrogen Producer	Culture Type	Substrate	Temperature (°C)	μ_{max}[a] (h^{-1})	v_{max}[b] (mL/h/L)	$v_{s,max}$[b] (mL/h/g VSS)	K_s[a]/K_m[b] (g COD/L)	$Y_{X/S}$ (g Biomass/g Substrate)	References
Sewage sludge	CSTR	Sucrose	35	0.172	NA		0.068	0.1	Chen et al. (2001)
Anaerobically digested sludge	Batch	Sucrose	36	ND	92.7		1.446	ND	Chen et al. (2006)
	Batch	Nonfat dry milk	36	ND	170.7		6.616	ND	
	Batch	Food waste	36	ND	199.3		8.692	ND	
Anaerobically digested sludge	CSTR	Glucose	37	0.33	NA		NA	0.26	Horiuchi et al. (2002)
Sewage sludge (immobilized cells)	Batch								Wu et al. (2002)
	CA/AC	Sucrose	35			8033	8.962		
	PU	Sucrose	35			1225	4.279		
	ALSC	Sucrose	35			2640	9.46		
Sewage sludge (EVA-immobilized cells)	Batch/without acclimation	Sucrose	40			278	20.0		Wu et al. (2005)
	Batch/with acclimation	Sucrose	40			468	10.0		

Note: VSS, volatile suspended solid; COD, chemical oxygen demand; ND, not determined.

[a] For Monod equation: $\mu = \frac{\mu_{max} S}{K_s + S}$.

[b] For Michaelis–Menten equation: $v = \frac{v_{max} S}{K_m + S}$, $v_s = \frac{v_{s,max} S}{K_m + S}$.

TABLE 9.3

Hydrogen Production Rates and Yield Coefficients from Pure and Complex Substrates Using Pure and Mixed Culture under Batch Operation

Substrate (Concentration, g/L)	H_2 Producer	Temperature (°C)	pH	H_2 Yield (mol H_2/mol Hexose)	H_2 Production Rate (mol H_2/L/day)	References
Glucose (4)	*Escherichia coli* strains	37	7.0	2	NA	Bisaillon et al. (2006)
Raw sludge (11.5 g COD/L)	Raw sludge	36	11 (initial)	0.45 mmol H_2/g COD	NA	Cai et al. (2004)
Alkaline-treated sludge (11.5 g COD/L)	Alkaline-treated sludge	36	11 (initial)	0.83 mmol H_2/g COD		
Sucrose (4.5 g COD/L)	Anaerobic digester	36	5.5 (initial)	9.55 mmol H_2/g COD	NA	Chen et al. (2006)
Nonfat dry milk (4.0 g COD/L)	Sludge	36	5.5 (initial)	4.86 mmol H_2/g COD		
Food waste (4.6 g COD/L)		36	5.5 (initial)	4.12 mmol H_2/g COD		
Sucrose (17.8)	*Clostridium butyricum*	37	5.5	1.39	0.160	Chen et al. (2005)
	CGS5		6.0	0.73	0.205	
Hydrolyzed starch (26.7 g COD/L)	*Clostridium butyricum* CGS2	37	7.5 (initial)	1.23	0.162	Chen et al. (2008b)
	Clostridium butyricum CH5			1.25	0.171	
Glucose (21.3)	Anaerobic sludge (acid treatment)	35	5.7	0.98	0.214	Cheong and Hansen (2006)
Glucose (20)	*Clostridium acetobutylicum*	37	6.0	2.0	NA	Chin et al. (2003)
Hydrolyzed corn stover biomass (glucose 138 mM)	Dried mixed sludge	35	5.5	2.86	0.567	Datar et al. (2007)
N-acetyl-v-glucosamine (GlcNAc) (10)	*Clostridium paraputrificum* M-21	45	6.0	2.2 mol/mol GlcNAc	NA	Evvyernie et al. (2001)
Starch hydrolysate (20)	*Enterobacter aerogenes*	40	6.5	1.09	0.417	Fabiano et al. (2002)
HCl pretreated wheat straw (25)	Microflora from a cow dung compost (heat treatment)	36	7.0 (initial)	2.78 mmol H_2/g TVS/h	NA	Fan et al. (2006b)

(continued)

TABLE 9.3 (continued)

Hydrogen Production Rates and Yield Coefficients from Pure and Complex Substrates Using Pure and Mixed Culture under Batch Operation

Substrate (Concentration, g/L)	H_2 Producer	Temperature (°C)	pH	H_2 Yield (mol H_2/mol Hexose)	H_2 Production Rate (mol H_2/L/day)	References
Rice slurry (5.5 g-carbohydrate/L)	Anaerobic digester sludge	37	4.5	346 mL/g carbohydrate	NA	Fang et al. (2006)
			6.5	223 mL/g carbohydrate		
Xylose (3.2)	*Thermoanaerobacter finnii*	60	6.4	0.49 mol/mol xylose	0.14 mL/h/L	Fardeau et al. (1996)
Crude cheese whey (ca. 41.4 g lactose/L)	*Clostridium saccharoperbutylacetonicu* ATCC 27021	30	6.0 (initial)	2.7 mol H_2/mol lactose	0.139	Ferchichi et al. (2005)
Sterilized waste sludge (16 g TOCD/L)	*Pseudomonas* sp. GZ1	35	7.38	15.02 mL/g TCOD	NA	Guo et al. (2008)
Glucose (118 mM)	*Enterobacter aerogenes*	38	6.13	NA	425.8 mL H_2/g dry cell/h	Jo et al. (2008)
Glucose (10)	*Caldicellulosiruptor saccharolyticus*	70	6.4	2.53 mol/mol glucose	0.257	Kadar et al. (2004)
Xylose (10)				2.24 mol/mol xylose	0.273	
Glucose (20)	Aerobic and anaerobic sludges, soil and lake sediment (acid and heat conditioned)	35	6.0	1.4	NA	Kawagoshi et al. (2005)
Sucrose (10)	Compost material	37	4.5	214 mL H_2/g COD	NA	Khanal et al. (2004)
Starch (10)			4.5	125 mL H_2/g COD		
Glucose (10)	Defined consortium (1:1:1, and separately tested): *Enterobacter cloacae* IIT-BT 08, *Citrobacter freundii* IIT-BT L139, *Bacillus coagulans* IIT-BT S1	37	6.0	41.23 mL H_2/g $COD_{removed}$	NA	Kotay and Das (2006)

Cellobiose (10)	*Enterobacter cloacae*	36	6.0	2.7 mol H_2/mol hexose	0.637	Kumar and Das (2000)
L-Arabinose (10)	IIT-BT 08			1.5 mol H_2/mol arabinose	0.353	
Fructose (10)				1.6 mol H_2/mol hexose	0.431	
Maltose (10)				0.7 mol H_2/mol hexose	0.176	
Potato starch (10)				NA	0.196	
C.M. cellulose (10)				NA	0.086	
D-Xylose (10)				0.95 mol H_2/mol xylose	0.341	
Dextrose (10)				2.2 mol H_2/mol hexose	0.438	
Sucrose (10)				3.0 mol H_2/mol hexose	0.647	
Fodder maize	Heat-treated anaerobically digested sludge	35	5.2–5.3	62.4 mL/g dry matter of fodder maize	NA	Kyazze et al. (2008)
Chicory Fructooligosaccharides				218 mL/g chicory fructooligosaccharides		
Perennial ryegrass (*Lolium perenne*)				75.6 mL H_2/g dry matter of wilted perennial ryegrass		
				21.8 mL H_2/g dry matter of fresh perennial ryegrass		
Microcrystalline cellulose (12.5)	Anaerobic digested sludge	37	7.0 (initial)	2.18 mmol H_2/g cellulose	NA	Lay (2001)
Microcrystalline cellulose (25)				1.60 mmol H_2/g cellulose		
Sucrose (10)	Mixed microbial flora	37	9	1.77	NA	Lee et al. (2002)
Cassava starch (20.3)	Municipal sewage sludge	37	6.0 (initial)	9.47 mmol H_2/g starch	1.10	Lee et al. (2008)
Cassava starch (20.3)				11.25 mmol H_2/g starch	1.05	
α-Cellulose (0.1)	*Clostridium thermocellum* 27405	60	7.0	1.9	NA	Levin et al. (2006)
Filter paper (0.1)				1.7		
Delignified wood (0.1)				2.3		
Cellobiose (0.1)				1.6		

(continued)

TABLE 9.3 (continued)

Hydrogen Production Rates and Yield Coefficients from Pure and Complex Substrates Using Pure and Mixed Culture under Batch Operation

Substrate (Concentration, g/L)	H_2 Producer	Temperature (°C)	pH	H_2 Yield (mol H_2/mol Hexose)	H_2 Production Rate (mol H_2/L/day)	References
Peptone (0.1)	Anaerobic sludge	35	7.0 (initial)	1.7 mmol H_2/g peptone	NA	Liang (2003)
Peptone (0.3)				2.2 mmol H_2/g peptone		
Peptone (0.6)				2.1 mmol H_2/g peptone		
Xylose (18.75)	Sludge	35	6.5	2.25	0.086	Lin and Cheng (2006)
Cellulose (3)	Anaerobic cow dung microflora	55	7.0	2.8 mmol H_2/g cellulose	NA	Lin and Hung (2008)
Sucrose (8.9) (C/N ratio = 47)	Sewage sludge	35	6.8	2.4	NA	Lin and Lay (2004a)
Sucrose (8.9)	Sewage sludge	35	6.8	1.72	NA	Lin and Lay (2005)
Starch (20 g COD/L)	Anaerobic sludge	35	5.5	1.1	0.250	Lin et al. (2008a)
Microcrystalline cellulose (5)	*Clostridium thermocellum* JN4 + *Thermoanaerobacterium thermosaccharolyticum* GD17	60	7.0	1.8	NA	Liu et al. (2008)
Xylose (18.75)	*Clostridium butyricum* CGS5	37	7.5 (initial)	0.73 mol/mol xylose	0.208	Lo et al. (2008b)
Xylose (37.5)	*Clostridium butyricum* CGS5	37	7.5 (initial)	0.68 mol/mol xylose	0.264	
Sucrose (35.6)	*Clostridium pasteurianum* CH_4	37	7.5 (initial)	1.04 mol/mol hexose	0.557	
Hydrolyzed carboxymethyl cellulose (10)	*Clostridium pasteurianum*	35	7.0	1.09 mmol H_2/g cellulose	NA	Lo et al. (2008a)
Rice husk hydrolysate (10)	*Clostridium butyricum* CGS5	35	7.0	17.24 mmol H_2/g rice husk hydrolysate	NA	Lo et al. (2009)
Glucose (4 g COD/L)	Heat-shocked mixed cultures	26	6.0	125 mL H_2/g COD	NA	Logan et al. (2002)
Sucrose (4 g COD/L)				131 mL H_2/g COD		
Molasses (4 g COD/L)				134 mL H_2/g COD		

Potato starch (4 g COD/L)				90 mL H_2/g COD		
Sodium lactate (4 g COD/L)				2.3 mL H_2/g COD		
Cellulose (4 g COD/L)				0.6 mL H_2/g COD		
Xylose (1.26)	*Clostridium uliginosum* sp. nov.	20	6.5	26.7 mol/mol xylose	2.59 mL/h/L	Matthies et al. (2001)
Sucrose (30)	Anaerobic sludge	35	5.5	1.01	0.20	Mu et al. (2006a)
Sucrose (24.8)	Anaerobic sludge	34.8	5.5	1.76	NA	Mu et al. (2006b)
Sucrose (25.4)		35.5	5.5	NA	0.49	
Glucose (10)	Sewage sludge	41	5.5	1.67	0.42	Mu et al. (2006c)
Potassium gluconate (10)	*Enterobacter aerogene* HU-101	37	7.0	1.44 mmol/g substrate	NA	Nakashimada et al. (2002)
Glucose (10)				1.97 mmol/g substrate		
Fructose (10)				2.17 mmol/g substrate		
Galactose (10)				1.90 mmol/g substrate		
Sorbitol (10)				4.96 mmol/g substrate		
Mannitol (10)				5.20 mmol/g substrate		
Glycerol (10)				6.69 mmol/g substrate		
Cellulose (5)	*Thermotoga maritima* (DSM 3109)	80	6.5	0.96 mmol/g cellulose	NA	Nguyen et al. (2008)
CMC (5)				3.29 mmol/g cellulose		
Cellulose (5)	*Thermotoga neapolitana* (DSM 4359)	75	7.0	1.07 mmol/g cellulose	NA	Nguyen et al. (2008)
CMC (5)				3.37 mmol/g cellulose		
Glucose (5)	*Ruminococcus albus*	37	NR	2.76	NA	Ntaikou et al. (2008)
Sorghum extract (3)		37	6.4–6.5	2.61		
Sorghum stalks (3)		37	6.4–6.5	3.15		
Sorghum residues (3)		37	6.4–6.5	2.59		
Glucose (23.75)	*Clostridium* sp. Fanp2	36	6.47	2.53	0.306	Pan et al. (2008)
Glucose (2)	Aerobic sludge (heat conditioned)	30	6.2	2.0	NA	Park et al. (2005)

(continued)

TABLE 9.3 (continued)

Hydrogen Production Rates and Yield Coefficients from Pure and Complex Substrates Using Pure and Mixed Culture under Batch Operation

Substrate (Concentration, g/L)	H_2 Producer	Temperature (°C)	pH	H_2 Yield (mol H_2/mol Hexose)	H_2 Production Rate (mol H_2/L/day)	References
Glucose (3.76)	Anaerobic sludge (heat treated)	30	6.2	1.9	NA	Salerno et al. (2006)
Glucose (3)	*Clostridium* sp. strain No. 2	36	6.8	14.6 mmol H_2/g substrate	NA	Taguchi et al. (1995)
Xylose (3)				16.1 mmol H_2/g substrate		
Avicel hydrolysate (3)				19.6 mmol H_2/g substrate		
Xylan hydrolysate (3)				18.6 mmol H_2/g substrate		
Cellulose (10)	Sludge compost	60	NA	2.4	NA	Ueno et al. (1995)
Cellulose (10)	Anaerobic digestion sludge			0.9		
Glucose (10)	*Thermoanaerobacterium thermosaccharolyticum* KU001	60	6.6	2.4	NA	Ueno et al. (2001a)
Glucose (10)	Thermophilic anaerobic microflora enriched from sludge compost	60	NA	2	NA	Ueno et al. (2001b)
Sucrose (6.68)	Mixed bacteria	37	5.5 (initial)	2.05	0.073	Van Ginkel et al. (2001)

Sucrose (10)	*Caldicellulosiruptor saccharolyticus*	70	7.0	3.3	0.202	van Niel et al. (2002a)
Glucose (10)	*Thermotoga elfiii*	65	7.4	3.3	0.065–0.108	
Glucose (10)	Hydrogen-producing bacterial B49	35	6.0 (initial)	1.73	NA	Wang et al. (2007)
Microcrystalline cellulose (10)	*Clostridium acetobutylicum* X_9 and *Ethanoigenens harbinense* B49	37	5.0	8.1 mmol H_2/g cellulose	NA	Wang et al. (2008)
Sucrose (17.8)	Immobilized sewage sludge	35		1.98	NA	Wu et al. (2002)
Sucrose (17.8)	Immobilized sewage sludge	40	6.7	0.87	NA	Wu et al. (2005)
Sucrose (35.6)		40	6.7	0.72		
Sucrose (25)	Mixed bacteria	35	8.0 (initial)	1.30	NA	Zhang and Shen (2006)
Starch (4.6)	Hydrogen-producing sludge	37	7.0	47 mL H_2/g starch	NA	Zhang et al. (2003)
Starch (4.6)		55	7.0	78 mL H_2/g starch		
Starch (4.6)		55	6.0	92 mL H_2/g starch		
Starch (9.2)		55	6.0	67 mL H_2/g starch		
Sucrose (44)	Mixed bacteria	35	7.5 (initial)	1.37	NA	Zhang et al. (2005)
Glucose (10)	Anaerobic sludge (heat conditioned)	37	6.0	1.75	NA	Zheng and Yu (2005)
Xylose (28.0)	*Clostridium tyrobutyricu* ATCC 25755	37	6.3	187 mol/mol xylose	0.77 mL/h/L	Zhu and Yang (2004)

TABLE 9.4

Hydrogen Production Rates and Yield Coefficients from Pure and Complex Substrates Using Pure and Mixed Culture under Semicontinuous and Continuous Operation

Substrate (Concentration, g/L)	H_2 Producer	Culture Type	HRT (h)	Temp. (°C)	pH	H_2 Yield (mol/mol Hexose)	H_2 Production Rate (mol/L/day)	References
Starch (20 g COD/L)	Anaerobic digester sludge	CSTR	12	35	5.3	0.92	0.175	Arooj et al. (2008)
Palm oil mill effluent (POME)	Mixed microflora	Fed batch	60	60	5.5	2382 mL H_2/L POME	0.307	Atif et al. (2005)
			20			1597 mL H_2/L POME	0.427	
Glucose (2)	Windrow yard waste compost	Fed batch	76	55	5.4	1.75	0.179	Calli et al. (2006)
Starch (10) and xylose (1:1 w/w)	Mixed culture	CSTR and UASB	32.9	35	7	NA	0.108	Camilli and Pedroni (2005)
			6.7	35	7		0.114	
			20.5	35	7		0.061	
Sucrose (17.8)	Mixed culture	UASB	8	35	6.7	0.92	0.274	Chang and Lin (2004)
Sucrose (17.8)	Sewage sludge	Packed-bed	1	35	6.7	1.19	1.29	Chang et al. (2002)
Glucose (18.7)	Sewage sludge	CSTR	13.3	35	6.7	1.63	NA	Chen and Lin (2001)
Sucrose (17.8)			13.3	35	6.7	2.225		
Sucrose (17.8)	Municipal sewage sludge	CSTR	8	35	7.0	2.26	0.649	Chen and Lin (2003)
			3	35	6.9	1.44	1.096	
Sucrose (17.8)	Sewage sludge	CSTR	8	35	6.7	1.735	0.630	Chen et al. (2001)
Sucrose (17.8)	Municipal sewage sludge	CSTR	4	35	5.5	2.35	0.874	Chen et al. (2008a)
Hydrolyzed starch (29.1 g COD/L)	*Clostridium butyricum* CGS2.	CSTR	12	37	NA	2.03	ca.0.5	Chen et al. (2008b)
			2	37		1.50	1.47	

Broken kitchen wastes (10 kg COD/m^3-day) and corn starch (10 kg COD/m^3-day)	Mixed culture	CSTR	96	35	5.3–5.6	NA	0.041	Cheng et al. (2006)
Lactose (10)	*Clostridium thermolacticum*	CSTR	17.9	58	7.0	2.7 mol/mol lactose	0.060	Collet et al. (2004)
Brewery waste (50 g COD/L)	Cattle dung compost	CSTR	18	37	5.5	1.76 mmol/g COD	0.124	Fan et al. (2006a)
Glucose (7)	Sludge	CSTR	6	36	5.5	2.1	NA	Fang and Liu (2002)
Sucrose (12.15)	Sludge	CSTR (granular sludge)	6	26	5.5	1.96	0.529	Fang et al. (2002)
Glucose (10)	Anaerobic sludge	CSTR	2	35	6.0	1.26	0.185	Gavala et al. (2006)
Glucose (10)		CSTR	2	55	6.4	1.16	0.163	
Glucose (10)		UASB	2	35	4.4	0.79	0.418	
Glucose (8)	Sludge	CSTR	5.4	37	6.0	NA	0.169	Horiuchi et al. (2002)
Wheat starch (10)	Sewage sludge	CSTR						Hussy et al. (2003)
Without N_2 sparging			18	30	5.2	1.3	0.071	
With N_2 sparging			15	30	5.2	1.9	0.120	
Sucrose (without N_2 sparging)	Mixed microflora	CSTR	14–15	32	5.2	1.0	NA	Hussy et al. (2005)
Sucrose (with N_2 sparging)						1.9		
Sugarbeet juice (without N_2 sparging)						0.9		
Sugarbeet juice (with N_2 sparging)						1.7		
(OLR = 16 kg total sugar/m^3/day)								

(*continued*)

TABLE 9.4 (continued)

Hydrogen Production Rates and Yield Coefficients from Pure and Complex Substrates Using Pure and Mixed Culture under Semicontinuous and Continuous Operation

Substrate (Concentration, g/L)	H_2 Producer	Culture Type	HRT (h)	Temp. (°C)	pH	H_2 Yield (mol/mol Hexose)	H_2 Production Rate (mol/L/day)	References
Glucose (20)	Sewage digester sludge	Biofilm system	20	37	5.0	NA	0.049	Kim et al. (2005)
Glucose (20)		Granular sludge system (stirred reactor)	20	37	5.0		0.118	
Glucose (10)	Mixed culture	MBR	0.79	37	5.5	1.1	1.71	Kim et al. (2006)
Glucose (10)	Mixed culture	UASB	26.7	70	4.8–5.5	2.47	0.051	Kotsopoulos et al. (2006)
Glucose (10)	*Enterobacter cloace* IIT-BT 08	Packed bed (lignocellulosic agroresidue)	1.1	37	NA	2.04	1.81	Kumar and Das (2001)
Soluble starch (4.25)	Anaerobic digested sludge	CSTR	17	37	5.2	1.29 L H_2/g starch-COD	0.066	Lay (2000)
Sucrose (17.8)	Municipal sewage sludge	Packed bed	0.5	35	6.7	1.44	7.20	Lee et al. (2003)
Sucrose (17.8)	Municipal sewage sludge	CIGSB	0.5	35	6.7	1.52	7.18	Lee et al. (2004a)
Sucrose (17.8)	Municipal sewage sludge	CIGSB	1	35	6.7	2.01	4.97	Lee et al. (2006a)
			0.5			1.96	9.12	
Sucrose (17.8)	Municipal sewage sludge	CIGSB	1	40	6.7	1.94	4.65	Lee et al. (2006b)
			0.5			1.58	7.50	
Glucose (18.75)	Municipal sewage sludge	CSTR	8	35	6.7	1.23	0.362	Lee et al. (2007)
Glucose (18.75)		MBR	1			1.27	1.450	
Sucrose (17.8)		CSTR	4			1.49	0.686	

Sucrose (17.8)		MBR	1			1.39	2.028	
Fructose (18.75)		CSTR	4			1.59	0.823	
Fructose (18.75)		MBR	1			1.36	2.694	
Glucose (17.8)	Sewage sludge	CSTR	6	35	5.7	1.71	0.711	Lin and Chang (1999)
				35	6.4	1.66	0.574	
Glucose (17.8)	Sewage sludge	CSTR	6	30–34	6.2	1.15	0.359	Lin and Chang (2004)
Xylose (18.75)	Sludge	CSTR	12	35	6–7	0.7 mol H_2/mol xylose	0.247	Lin and Cheng (2006)
Sucrose (17.8)	Sewage sludge	CSTR	4	35	7.1	1.75	0.408	Lin et al. (2006)
Starch (20 g COD/L)	Anaerobic sludge	CSTR	4	35	5.5	1.5	0.450	Lin et al. (2008a)
Sucrose (14.3)	Sludge	CSTR (granular sludge)	13.7	26	5.5	1.89	0.274	Liu and Fang (2002)
Glucose (10)	Anaerobic microflora (predominantly *Clostridium* sp.)	CSTR (N_2 sparging)	5	35	6	1.43	0.196	Mizuno et al. (2000)
Sucrose (4.75)	Mixed culture	UASB	18	38	4.5	1.44	0.049	Mu and Yu (2006)
Glucose (8.9)	Heat-treated soil	CSTR	3.3	—	5.5	0.88	0.294	Oh et al. (2004)
Glucose (8.9)		MBR	3.3	—	5.5	1.00	0.372	
Molasses (3 g COD/L)	Sewage sludge	CSTR (pilot)	3.9	35	4–5	26.13 mmol/g COD	0.225	Ren et al. (2006)
Food waste (6 g VSS/L)	Thermophilic acidogenesis	Semicontinuous operation	5	55	5.6	1.80	NA	Shin et al. (2004)
Molasses (20)	*Enterobacter aerogenes* E 82005	Continuous fermentor (packing urethane foam)	5	NA	6	1.75	0.313	Tanisho and Ishiwata (1994)

(continued)

TABLE 9.4 (continued)

Hydrogen Production Rates and Yield Coefficients from Pure and Complex Substrates Using Pure and Mixed Culture under Semicontinuous and Continuous Operation

Substrate (Concentration, g/L)	H_2 Producer	Culture Type	HRT (h)	Temperature (°C)	pH	H_2 Yield (mol/mol Hexose)	H_2 Production Rate (mol/L/day)	References
Sugary wastewater (carbohydrates 9.847 g/L; COD 31.85 g/L)	Sludge compost	CSTR	12	60	NA	2.52	0.028	Ueno et al. (1996)
Cellulose (10) (peptone 5 g/L)	Thermophilic anaerobic microflora enriched from sludge compost	CSTR	72	60	6.4	2	0.029	Ueno et al. (2001)
Municipal solid waste (Fruit+vegetable+office paper, TS 35%, VS 21%)	Mixed culture	Fed batch	21 day	37	5.5	6.73 mmol/g $VS_{removed}$	NA	Valdez-Vazquez et al. (2005)
				55	6.4	14.7 mmol/g $VS_{removed}$		
Glucose (10)	Agricultural soil	CSTR	1	30	5.5	1.7	2.126	Van Ginkel and Logan (2005)
Glucose (2.5)			10	30	5.5	2.8	0.079	
Palm oil mill effluent (20 g COD/L)	Cow dung	UACF	7 day	NA	5	9.77 mmol/g $COD_{destroyed}$	NA	Vijayaraghavan and Ahmad (2006)
Jackfruit peel (33 g VS/L)	Cow dung	UACF	12 day	NA	5.1	16.2 mmol/g $VS_{destroyed}$	0.024	Vijayaraghavan et al. (2006)
Mixed fruit peelwaste (46–84 g VS/L)	Cow dung	UACF	>5 day	NA	5.4–5.7	18.8 mmol H_2/g $VS_{destroyed}$	NA	Vijayaraghavan et al. (2007)
Organic wastewater from a beet sugar refinery (4 g COD/L)	Mixed culture	CSTR	12	30	4.4	NA	0.120	Wang et al. (2006)

Sucrose (17.8)	Immobilized sewage sludge	Fluidized bed	2	35	5.8–6.8	1.34	0.911	Wu et al. (2003)
Glucose (18.75)	Municipal anaerobic sludge	CSTR	10	35	6.5	1.63	0.411	Wu et al. (2008b)
		CSTR	6	35	6.6	0.75	0.588	
		IC-CSABR	4	35	6.3	0.87	0.520	
		AGSB	4	35	6.6	1.57	0.950	
Xylose (18.75)	Municipal sewage sludge	CSTR	12	50	6.5	0.95 mol/mol xylose	0.183	Wu et al. (2008c)
		IC-CSABR	2	40	5.8	0.80 mol/mol xylose	1.060	
		AGSB	2	40	6.5	0.61 mol/mol xylose	0.852	
Sucrose (17.8)	Municipal sewage sludge	IC-CSABR	0.5	40	6.6	1.05	4.79	Wu et al. (2006)
Sucrose (26.7)			0.5	40	6.6	1.93	14.2	
Sucrose (35.6)			0.5	40	6.6	1.59	14.8	
Citric acid wastewater (18 kg COD/L)	Mixed culture	UASB	12	35–38	6.8–7.2	0.84	0.029	Yang et al. (2006)
Sweet potato starch residue (5–20 g/L) + 0.1% Polypeptone	Two-step cultures (by a mixed culture of *C. butyricum* and *E. aerogenes* and by *Rhodobacter* sp. M-19)	Repeated batch		37 (dark), 35 (photo)	5.25 (dark), 7.5 (photo)	7.0 (2.4 + 4.6)		Yokoi et al. (2001b)
Sweet potato starch residue (10 g/L) + 0.5% corn steep liquor as nitrogen source	Mixed culture of *C. butyricum* and *E. aerogenes*	Repeated batch		37	5.25	2.7		Yokoi et al. (2002)
	Two-step cultures (by a mixed culture of *C. butyricum* and *E. aerogenes* and by *Rhodobacter* sp. M-19)	Repeated batch		37 (dark), 35 (photo)	5.25 (dark), 7.5 (photo)	7.2		

(*continued*)

TABLE 9.4 (continued)

Hydrogen Production Rates and Yield Coefficients from Pure and Complex Substrates Using Pure and Mixed Culture under Semicontinuous and Continuous Operation

Substrate (Concentration, g/L)	H_2 Producer	Culture Type	HRT (h)	Temperature (°C)	pH	H_2 Yield (mol/mol Hexose)	H_2 Production Rate (mol/L/day)	References
Rice winery wastewater (34)	Mixed bacterial flora	Upflow anaerobic reactor	24	35	5.5	2.14		Yu et al. (2002)
Rice winery wastewater (34)			2	35		1.74		
Rice winery wastewater (13)			2	35		1.89		
Rice winery wastewater (36)			2	35		1.79	0.156	
Rice winery wastewater (34)			2	55		1.37–2.14		
Glucose (10)	Anaerobically digested sludge	CSTR	10	37	5.5	1.95	0.249	Zhang et al. (2006)
			6	37	5.5	1.88	0.317	
Glucose (10)	Municipal sewage sludge	CSTR (granular sludge)	0.5	37	5.5	1.81	3.135	Zhang et al. (2007)
Glucose	Anaerobically digested sludge	Anaerobic biofilm fluidized bed reactors	0.25	37	5.5	1.7	7.45	Zhang et al. (2008)
		Anaerobic granule fluidized bed reactors				1.6	6.47	

CIGSB, carrier induced granular sludge bed; IC-CSABR, immobilized cell continuously stirred anaerobic bioreactor; CSTR, continuous flow stirred tank reactor; MBR, membrane bioreactor; UASB, upflow anaerobic sludge blanket; UACF, upflow anaerobic contact filter; AGSB, agitated granular sludge bed.

TABLE 9.5

Comparison of Optimal pH for H_2 Fermentation in Different H_2-Producing Systems

Authors	Microorganism	Culture Type	Substrate	Operation Temperature (°C)	Examined pH	Optimal pH for Maximum HY	Optimal pH for Maximum HPR	Optimal pH for Maximum SHPR
Chen et al. (2005)	*Clostridium butyricum* CGS5	Batch	Sucrose	37	5.0–6.5	5.5	6.0	
Cheong and Hansen (2006)	Anaerobic digester sludge	Batch	Glucose	35	4.6, 5.7, 6.8	5.7		
Fan et al. (2006a)	Cattle dung compost	CSTR	Brewery waste	37	5.0–6.5	5.5	5.5	
Fan et al. (2006b)	Cow dung compost	Batch	HCl pretreated wheat straw	36	4.0–9.0	6.0–8.0		
Fang and Liu (2002)	Sludge	CSTR	Glucose	36	4.0–7.0	5.5		
Fang et al. (2006)	Anaerobic digester sludge	Batch	Rice slurry	37	4.0–7.0	4.5		6.5
Horiuchi et al. (2002)	Anaerobic digester sludge	CSTR	Glucose	37	5.0–7.0		6.0	
Khanal et al. (2004)	Compost material	Batch	Sucrose/starch	37	4.5–6.5	4.5		5.5
Lay (2000)	Sludge	CSTR	Soluble starch	37	4.0–7.0		5.2	
Lee et al. (2002)	Mixed microbial flora	Batch	Sucrose	37	5.5–10.0	9.0	9.0	
Lin and Cheng (2006)	Sewage sludge	Batch	Xylose	35	5.0–9.5	6.5	6.5	

(continued)

TABLE 9.5 (continued)

Comparison of Optimal pH for H_2 Fermentation in Different H_2-Producing Systems

Authors	Microorganism	Culture Type	Substrate	Operation Temperature (°C)	Examined pH	Optimal pH for Maximum HY	Optimal pH for Maximum HPR	Optimal pH for Maximum SHPR
Lin and Hung (2008)	Cow dung sludge	Batch	Cellulose	55	5.5–9.0	7.0		
Lin et al. (2008a)	Anaerobic sludge	Batch	Starch	35	5.0–7.0	5.0	5.5	
Liu and Shen (2004)	Cracked cereals	Batch	Starch	35	4.0–9.0	8.0	8.0	
Mu et al. (2006b)	Anaerobic sludge	Batch (RSM)	Sucrose	35	4.5–6.5	5.5	5.5	
O-Thong et al. (2008b)	*Thermoanaerobacterium thermosaccharolyticum* PSU-2	Batch	Sucrose	60	4.0–8.5	6.25	6.25	
Ren et al. (1997)	Sludge	CSTR	Molasses	30	3.0–5.5		4.5	
Wu and Lin (2004)	Waste-activated sludge	Batch	Food wastewater	35	4.0–8.0	6.0	6.0	
Yu et al. (2002)	Mixed bacterial flora	Upflow reactor	Rice winery wastewater	35	4.5–6.0	5.5	5.5	
Zhang et al. (2003)	Sludge	Batch	Starch	55	4.0–9.0	6.0		7.0

HY, H_2 yield; HPR, H_2 production rate; SHPR, specific H_2 production rate; RSM, response surface methodology.

TABLE 9.6

Comparison of Optimal Temperature for H_2 Fermentation in Different H_2-Producing Systems

Authors	Microorganism	Culture Type	Substrate	Examined Temperature (°C)	Optimal H_2-Producing Temperature (°C)
Chin et al. (2003)	*Clostridium acetobutylicum* ATCC824	Fedbatch	Glucose	30, 33, 37, 40	37
Evvyernie et al. (2000)	*Clostridium paraputrificum* M-21	Batch	*N*-Acetyl-D-glucosamine (GlcNAc)	30, 35, 40, 45, 50	45
Fang et al. (2006)	Anaerobic digester sludge	Batch	Rice slurry	37, 55	37
Gavala et al. (2006)	Anaerobic sludge	CSTR	Glucose	35, 55	35
Lee et al. (2006b)	Municipal sewage sludge	CIGSB	Sucrose	30, 35, 40, 45	40
Lee et al. (2008)	Municipal sewage sludge	Batch	Starch	37–55	55
Lin et al. (2008a)	Municipal sewage sludge	Continuous	Xylose	30–55	50
Mu et al. (2006b)	Anaerobic sludge	Batch (RSM)	Sucrose	25, 30, 35, 40, 45	35.5
Mu et al. (2006c)	Sewage sludge	Batch	Glucose	33, 35, 37, 39, 41	41
O-Thong et al. (2008b)	*Thermoanaerobacterium thermosaccharolyticum* PSU-2	Batch	Sucrose	40–80	60
Tanisho et al. (1994)	*Enterobacter aerogenes* E. 82005	Batch	Glucose	32, 35, 37.5, 40.5, 44	40.5
Valdez-Vazquez et al. (2005)	Anaerobic digester sludge	Semicontinuous	Organic waste	37–55	55

(continued)

TABLE 9.6 (continued)

Comparison of Optimal Temperature for H_2 Fermentation in Different H_2-Producing Systems

Authors	Microorganism	Culture Type	Substrate	Examined Temperature (°C)	Optimal H_2-Producing Temperature (°C)
Wang and Wan (2008)	Anaerobic sludge	Batch	Glucose	25–55	40
Wang et al. (2005)	Anaerobic sludge	Batch	Sucrose	25–45	35.1
Wu et al. (2005)	Immobilized sewage sludge	Batch	Sucrose	30, 35, 40, 45	40
Xing et al. (2008)	*Ethanoligenens harbinense* YUAN-3	Batch	Glucose	20–44	37
Yokoi et al. (1995)	*Enterobacter aerogenes* HO-39	Batch	Glucose	30, 33, 35, 38, 40, 43	38
Yokoi et al. (2001a)	*Enterobacter* sp. BY-29	Batch	Glucose	30, 35, 40, 45	37
Yokoyama et al. (2007b)	Cow waste slurry	Batch	Cow waste slurry	37–85	60
Yu et al. (2002)	Mixed bacterial flora	Upflow anaerobic reactor	Rice winery wastewater	20, 30, 35, 55	55
Zhang and Shen (2006)	Mixed bacteria	Batch	Sucrose	25, 35, 40, 45	35
Zhang et al. (2003)	Hydrogen-producing sludge	Batch	Starch	37, 55	55

CIGSB, carrier induced granular sludge bed; RSM, response surface methodology.

TABLE 9.7

Effect of Different Nitrogen Concentrations on Fermentative Hydrogen Production

H_2 Producer	Substrates	Culture Type	Nitrogen Source	Optimum Nitrogen Concentration	Hydrogen Yield (mol H_2/mol Glucose)	References
Escherichia coli	Glucose	Batch	NH_4Cl	0.01 g N/L	1.7	Bisaillon et al. (2006)
Grass compost	Food wastes	Batch	NH_4HCO_3	0.4 g N/L	77 mL/g TVS	Lay et al. (2005)
Cracked cereals	Starch	Batch	NH_4HCO_3	1 g N/L	146 mL/g starch	Liu and Shen (2004)
Compost	Glucose	Batch	Yeast extract	4% yeast extract	2.1	Morimoto et al. (2004)
Dewatered and thickened sludge	Glucose	Batch	NH_4Cl	7 g N/L	1.17	Salerno et al. (2006)
Enterobacter aerogenes HO-39	Glucose	Batch	Polypeptone	2% Polypeptone	1.0	Yokoi et al. (1995)

9.4.1 Substrate and Microorganism Species

9.4.1.1 Pure Culture

The literature survey suggests that dark fermentative H_2 production is undertaken by a diverse group of bacteria including facultative anaerobes, facultative aerobes, and obligate anaerobes. Mainly the obligate anaerobes and spore-forming organisms such as *Clostridium buytricum* (sweet potato starch) (Yokoi et al., 2001), *C. thermolacticum* (lactose) (Collet et al., 2004), *C. pasteurianum* (starch) (Liu and Shen, 2004) and *C. paraputrificum* M-21 (chitinous waste) (Evvyernie et al., 2001), *C. butyricum* CGS5 (rice husk hydrolysate) (Lo et al., 2009), and *C. bifermentants* (wastewater sludge) (Wang et al., 2003b) demonstrate maximum H_2 production during the exponential growth phase. *Clostridium* sp. can form endospores in response to unfavorable environmental conditions such as lack of nutrients or elevation of temperature and can be activated when required environmental conditions are provided for H_2 production (Mallette et al., 1974; Lay, 2000; Wu et al., 2006). A study on the microbial community of mesophilic hydrogen-producing sludge showed the presence of *Clostridia* species up to 64.6%, indicating that these species are dominant microbes for H_2 production (Fang et al., 2002). H_2 production by *Thermotogales* sp. and *Bacillus* sp. were detected in mesophilic acidogenic cultures (Shin et al., 2004). In anaerobic granular sludge, some anaerobic cultures like *Actinomyces* sp. and *Porphyromonos* sp. have been detected along with *Clostridium* sp., and these strains showed H_2 yields between 1 and 1.2 mmol H_2/mol glucose when cultivated under anaerobic conditions (Oh et al., 2003) (Tables 9.3 and 9.4). Facultative anaerobes such as *Enterobacter* species have the ability to metabolize carbohydrates and produce some valuable products like gaseous H_2 and CO_2 and a mixture of acids, ethanol, and 2,3-butanediol. The capacity of H_2 production associated with *Enterobacter aerogenes* using different substrates has been widely studied (Tanisho and Ishiwata, 1994; Yokoi et al., 1997; Rachman et al., 1998; Palazzi et al., 2000; Fabiano and Perego, 2002). Enhancement of H_2 production (2.2 mol H_2/mol glucose) by

TABLE 9.8

Effect of Metal Ion Concentrations on Fermentative Hydrogen Production

Inoculum	Substrates	Reactor Type	Metal Ion	Concentration (mg/L)		Hydrogen Yield	References
				Examined Concentration	Optimal Concentration	(mol H_2/mol Hexose)	
Digested sludge	Sucrose	Continuous	Ca^{2+}	0–300	150	1.8	Chang and Lin (2006)
Grass compost	Food wastes	Batch	Fe^{2+}	0–250	132	77 mL/g TVS	Lay et al. (2005)
Anaerobic sludge	Sucrose	Batch	Fe^{2+}	0–1763.8	352.8	1.84	Lee et al. (2001)
Municipal sewage sludge	Sucrose	Continuous	Ca^{2+}	0–27.2	27.2	1.10	Lee et al. (2004)
Anaerobic sludge	Palm oil mill effluent	Batch	Fe^{2+}	2–400	257	6.33 L/L substrate	Thong et al. (2008)
Digested sludge	Glucose	Batch	Fe^{2+}	0–1500	350	2.29	Wang and Wan (2008)
Digested sludge	Glucose	Batch	Ni^{2+}	0–50	0.1	2.18	Wang and wan (2008)
Hydrogen-producing bacterial B49	Glucose	Batch	Mg^{2+}	1.2–23.6	23.6	2360.5 mL/L culture	Wang et al. (2007)
Anaerobic sludge	Starch	Batch	Fe^{2+}	0–1473.7	55.3	296.2 mL/g starch	Yang and Shen (2006)
Anaerobic sludge	Glucose	Batch	Cu^{2+}	0–400	400	1.74	Zheng and Yu (2004)
Anaerobic sludge	Glucose	Batch	Zn^{2+}	0–500	250	1.73	Zheng and Yu (2004)

using *Enterobacter cloacae* ITT-BY 08 has been reported earlier (Kumar and Das, 2000). H_2 production from glucose by *E. coli* and *Hafnia alvei* was studied by Podestá et al. (1997), and a trace amount of H_2 yield was detected. Some thermophilic anaerobic organisms belonging to the genus *Thermoanaerobacterium*, viz. *T. thermosaccharolyticum* and *Desulfotomaculum geothermicum*, have the ability to produce H_2 gas in a thermophilic acidogenic culture (Shin et al., 2004). The *Thermococcus kodakaraensis* KOD1 and *Clostridium thermolacticum* strains can produce H_2 at 85°C and 58°C (Kanai et al., 2005), respectively, whereas *Klebisalle oxytoca* HP1 shows maximal H_2 production rate at 35°C reported earlier (Minnan et al., 2005). In a recent study, isolated *Klebsiella* sp. HE1 demonstrated the ability to produce 2,3-butanediol, ethanol, and H_2 using sucrose as substrate under the dark fermentation process (Wu et al., 2008a). Some studies show that aerobic cultures such as *Aeromonos* spp., *Pseudomonos* spp., and *Vibrio* spp. can produce H_2. *Thermotogales* species and *Bacillus* sp. can produce H_2 under mesophilic acidogenic cultures according to a recent report (Shin et al., 2004).

9.4.1.2 Mixed Culture

Increasing industrialization and urbanization of modern society have resulted in massive amounts of organic wastes being generated, and it would be desirable to turn these wastes into a useful product, such as a biohydrogen. The major criteria for the selection of waste materials for biohydrogen production are the availability, cost, carbohydrate content, and biodegradability (Kapdan and Kargi, 2006). These waste materials can be ideal inexpensive feedstocks for biological H_2 production. Literature survey suggests that a single microorganism has limited hydrolytic potential in degrading complex wastes for biohydrogen production. The degradation of organic matter in anaerobic environments by microbial consortia involves the cooperation of different species existing in the system and synergistic metabolic activity, which generate a stable, self-regulating fermentation. For this purpose, worldwide efforts are under way to design a suitable mixed microbial consortium capable of decomposing various organic waste streams and to improve the H_2 production yields and rates by using these cultures. The mixed microbial consortia presenting in anaerobic digester sludge, sludge compost, soil, animal feces, and so on may be useful for the direct utilization of starch- and cellulose-containing agricultural wastes, food industry wastes, carbohydrate-rich industrial wastewaters, and organic wastewater for biohydrogen production (details are given in the next section) (Kapdan and Kargi, 2006; Datar et al., 2007). Before preparation of the inoculum of various sludges, different pretreatment methods such as incubation at high temperatures or acidic conditions or combination of these are necessary to remove the methanogens (Valdez-Vazquez and Poggi-Varaldo, 2009). These methods favor the survival of spore-forming *Clostridia* and are found to be effective in enriching for H_2 producers (Tables 9.3 and 9.4). In general, anaerobic activated sludge is used for the production of H_2 from cellulose and biomass. A recent study about H_2 production from chemical wastewater, using anaerobic mixed consortia in a biofilm-configured reactor operated in periodic discontinuous batch mode, suggests that reactor configuration and mode of operation also affect the composition of the microbial consortia (Mohan et al., 2007) (Tables 9.3 and 9.4).

9.4.1.3 Potential of Simple and Complex Substrates for Biohydrogen Production

Many studies have investigated the potential of various substrates, from simple sugars (glucose) to more complex substrates such as biomass, for dark fermentation H_2 production. A brief summary of the H_2 production yields and rates (batch and continuous) with various substrates is given in Tables 9.3 and 9.4. Glucose and sucrose are the most common pure substrates used in both batch and continuous processes because of their relatively simple structures, ease of biodegradability, and presence in several industrial effluents (Kapdan and Kargi, 2006) and different agricultural and biomass wastes in polymeric form. Theoretically, bioconversion of 1 mol of glucose yields 12 mol of H_2, but the highest H_2 yield from glucose in recent studies is only around 2.0–2.4 mol H_2/mol glucose (Ueno et al., 2001b; Morimoto et al., 2004). This lower H_2 yield may be the result of the utilization of glucose as an energy source for bacterial growth. The highest H_2 yield (2.76 mol H_2/mol glucose) from glucose was observed in *Ruminococcus albus* (Ntaikou et al., 2008). In the presence of sucrose, a higher yield was obtained (4.52 mol H_2/mol sucrose) by employing CSTR operation at 8 h HRT (Chen and Lin, 2003). Collet et al. (2004) reported a maximum H_2 yield of 3 mol H_2/mol lactose from lactose, although the theoretical yield is 8 mol H_2/mol lactose. The results indicate that sucrose gives a higher yield compared to other simple sugars. However, the yield per mole of hexose remains almost the same. Starch is abundant in nature and has great potential to be used as a carbohydrate source for H_2 production. Studies for both batch and continuous operations have been done in recent years. According to the reaction stoichiometry, 1 g of starch yields 553 mL H_2 gas with acetate as the by-product (Zhang et al., 2003). However, the actual yield is lower than the theoretical value because of the utilization of substrate for cell synthesis. The maximum specific H_2 production rate was observed in *C. pasteurianum* (237 mL H_2/g VSS/day from 24 g/L edible corn starch) (Liu and Shen, 2004) and *Thermoanaerobacterium* (365 mL H_2/g VSS/day at 55°C) (Zhang et al., 2003). Mixed culture of *C. butyricum* and *E. aerogenes*, obtained in long-term repeated batch operations with starch residue (2.0%) containing wastewater as substrate, gives better H_2 yield (2.4 mol H_2/mol glucose) (Yokoi et al., 2001). Lay (2000) used anaerobic digested sludge to ferment substrate containing 4.25 g/L soluble starch, attaining a rate of 66.7 mL H_2/h/L. Lin and Cheng (2006) demonstrated H_2 production from xylose (20 g COD/L) by mesophilic sewage sludge in a chemostat anaerobic bioreactor at an HRT of 12 h. Each mole of xylose yields 0.7 mol of H_2 and each gram of biomass produces 0.038 mol H_2 day. Some researchers have studied H_2 fermentation using protein as substrate by anaerobic sludge (Bai et al., 2001; Liang et al., 2001; Liang, 2003). Liang (2003) showed a H_2 yield of 2.2 mmol H_2/g peptone with anaerobic sludge in batch culture and noted that H_2 fermentation of peptone has a lower H_2 yield but a higher growth yield than glucose based on 1 g of substrate. In H_2 fermentation of peptone using mixed culture, H_2 consumption was always observed after H_2 production reached a peak value (Bai et al., 2001; Liang, 2003).

The literature survey suggests that some nontoxic carbohydrate-rich industrial effluents such as food industry, dairy industry, olive mill, baker's yeast, and brewery wastewaters can be used as raw material for biohydrogen production. In addition, biologically derived organic materials and residues currently constitute a large source of waste biomass (Giallo et al., 1985). Bioenergy derived from water containing biomass (sewage sludge, agricultural and livestock effluents, as well as animal excreta) was mainly produced by microbial fermentation. However, only limited data on H_2 yield from wastewater sludge have been reported (Wang et al., 2003b; Chang and Yang, 2006).

Lignocellulosic products such as agricultural crops and their waste by-products, effluents from the paper industry, wood and wood waste, food, aquatic plants, and algae are the major biomass resources. Use of these biomass-rich resources for bioenergy and related bioproducts can contribute to the displacement of fossil fuels as our primary energy source and reduce greenhouse gas emissions. Thus, biomass acts as an energy source and is characterized in the form of both flow and stock. Production of biohydrogen from renewable resources such as lignocellulosic wastes is expected to become an attractive and major source of energy in the future (Kapdan and Kargi, 2006; Saratale et al., 2008). Bioconversion of biomass utilizing anaerobic fermentation to produce H_2 has been demonstrated in many studies (Lin and Chang, 1999; Lay, 2000; Ren et al., 2006; Wu et al., 2006; Lo et al., 2008a). The main source of H_2 production during a biological fermentative process is carbohydrates, which are very common in plant tissues, in the form of either oligosaccharides or polymers (e.g., cellulose, hemicellulose, and starch). Among the polymeric forms, cellulose is the predominant constituent and is widely available in agricultural wastes and industrial effluents such as pulp/paper and food industries (Nowak et al., 2005; Lo et al., 2009); it is considered as a very promising feedstock for biohydrogen production. Significant amounts of H_2 may be produced from cellulosic feedstocks (straw, woodchips, grass residue, paper waste, sawdust, etc.) using conventional anaerobic dark fermentation technology and natural mixed microflora under conditions that favor hydrogen-producing acetogenic bacteria and inhibit methanogens (Sparling et al., 1997; Valdez-Vazquez et al., 2005). However, depending on the metabolic shift used by the organisms within the consortium, H_2 yields may be variable (Benemann, 1996; Hallenbeck and Benemann, 2002). Some recent studies reported effective H_2 production from agricultural waste using *Ruminococcus albus* (sorghum extract) (Ntaikou et al., 2008), heat-treated anaerobically digested sludge (fodder maize, chicory fructooligosaccharides, perennial ryegrass) (Kyazze et al., 2008), *Clostridium butyricum* CGS5 (rice husk hydrolysate) (Lo et al., 2009), and *Clostridium paraputrificum* M-2; (chitinous wastes) (Evvyernie et al., 2001). To get a higher H_2 yield directly from cellulose materials using dark fermentation, pretreatment processes for delignification and hydrolysis of cellulose are required. Cellulolytic microorganisms or cellulase complexes are usually used to accomplish this (Taguchi et al., 1995, 1996; DeVrije et al., 2002). Recently, Fan et al. (2006b) conducted some batch tests to compare the efficiency of H_2 production from raw wheat straw and HCl-pretreated wheat straw, using cow dung compost as seeding. Results showed that acid pretreatment of the substrate played a key role in efficient conversion of the wheat straw wastes into H_2 gas. The maximum cumulative H_2 yield of 68.1 mL H_2/g TVS was observed from the pretreated wheat straw. This value is about 136 times higher than that from raw wheat straw wastes. Ueno et al. (1996) obtained a yield of 2.52 mol H_2/mol glucose with a laboratory-scale continuous mixed-culture flora chemostat reactor. In addition, effective H_2 production using mixed cultures under CSTR operation has been reported. The combinations of mixed-culture type versus the substrate used are as follows: broken kitchen wastes versus corn starch (Cheng et al., 2006), mixed culture versus organic wastewater (Wang et al., 2003b), sewage sludge versus molasses (Ren et al., 2006). Moreover, Yu et al. (2002) constructed a mixed-culture upflow anaerobic reactor, using high-strength rice winery wastewater as substrate, to investigate individual effects of HRT, chemical oxygen demand (COD) concentration, pH, and temperature on biohydrogen production. The optimal H_2 production rate was achieved at an HRT of 2 h, a COD of 34 g/L, a pH of 5.5, and a temperature of 55°C. The H_2 yield was in the range of 1.37–2.14 mol/mol hexose. Some studies also used solid waste containing jackfruit peel as substrate of H_2 with microflora isolated from cow dung. This sucrose-rich wastewater under UASB operation gave a H_2 yield of 16.2 mmol H_2/g VS destroyed and 1.61 mol H_2/mol glucose, respectively (Mu and Yu, 2006; Vijayaraghavan et al., 2006).

On the other hand, several studies have focused on H_2 production from sludge by anaerobic fermentation, since the sludge contains large quantities of polysaccharides and proteins (Wang et al., 2003a,b; Cai et al., 2004). Wang et al. (2003b) examined the effects of five pretreatment processes (namely, ultrasonication, acidification, sterilization, freezing/thawing, and adding a methanogenic inhibitor) on the production of H_2 from wastewater sludge by *Clostridium bifermentans*. Results showed that freezing/thawing and sterilization increased the H_2 yield from 0.6, for the original sludge, to 1.5–2.1 mmol H_2/g COD. However, fast consumption of H_2 and interference of methanogens still hamper H_2 production from sludge (Liang, 2003; Wang et al., 2003a,b; Cai et al., 2004). This makes it difficult to maintain high and stable H_2 production and limits the application of biohydrogen production from sludge. Cai et al. (2004) showed that biohydrogen production from sludge could be enhanced and stably maintained by the combination of high initial pH and alkaline pretreatment (Tables 9.3 and 9.4).

9.4.2 Substrate/Product Concentration

Some studies suggested that concentration of carbon substrates (or organic loading) in wastewater had a slightly negative effect on the H_2 yield (Yu et al., 2002). It was also observed that the specific H_2 production rate substantially increased with increasing substrate concentration (Lo et al., 2008b). However, in some cases, high substrate concentration may be unfavorable for H_2 production due to substrate inhibition (Lay, 2001; van Ginkel et al., 2001). Lay (2001) performed batch studies on converting microcrystalline cellulose into H_2 at 12.5 and 25 g cellulose/L and a H_2 yield of 2.18 mmol H_2/g cellulose and 1.60 mmol H_2/g cellulose, respectively. This decrease in H_2 production at higher substrate concentration might be due to substrate inhibition. It was also reported that higher substrate concentrations may quickly become inhibitory through pH depletion, acid production, or increased H_2 partial pressures (van Ginkel et al., 2001). Hence, the removal of these inhibitory mechanisms may be necessary to achieve high H_2 conversion efficiencies and production rates at higher substrate concentrations.

During dark fermentation, anaerobic bacteria have the ability to produce H_2 while converting organic substrates into soluble metabolites such as volatile fatty acids (e.g., acetic acid, butyric acid) and alcohols. The unstable and low H_2 production is possibly attributed to the different metabolic shift of hydrogen-producing bacteria under dark fermentation. Recently, Fang et al. (2006) investigated the effect of feedstock concentration on H_2 production from rice slurry with acidophilic hydrogen-producing sludge. The results showed that the concentration of 5.5 g carbohydrate/L gave the highest H_2 yield of 346 mL/g carbohydrate, and above this concentration the H_2 yield decreased steadily. The authors mentioned that this may be due to the increased concentration of VFA and alcohols. Similarly, Cheng et al. (2002) also pointed out that high VFA content in the system sped up the consumption of H_2 gas. To study the effect of VFA, some scientists determined the inhibitory effects of butyrate addition on H_2 production from glucose by using anaerobic mixed cultures in batch experiments (Zheng and Yu, 2005). Experimental results showed that addition of butyrate at 25.08 g/L had a strong inhibitory effect on substrate degradation and H_2 production. They employed a noncompetitive and nonlinear inhibition model to describe the inhibition of butyrate addition on H_2 production. The $C_{I,50}$ values (the butyrate concentration at which bioactivity is reduced by 50%) for H_2 production rate and yield were estimated as 19.39 and 20.78 g/L of butyrate added, respectively. Ren et al. (2006) undertook a study on biohydrogen production from molasses in a pilot-scale CSTR (with an available volume of 1.48 m^3) for over 200 days. They reported that the H_2 yield was affected by the presence of ethanol and acetate in the liquid phase, and the maximum H_2 production rate occurred while the ratio of ethanol to acetate was close to 1, due to the adjustment of $NAD^+/(NADH+H^+)$ through fermentation pathways. The H_2 yield reached 26.13 mol/kg $COD_{removed}$ within an OLR range of 35–55 kg COD/m^3 reactor/day. (The value of 26.13 mol H_2/kg COD does not seem to be correct as it is equivalent to 5.02 mol H_2/mol hexose, higher than the theoretical 4 mol H_2/mol hexose.)

9.4.3 Hydraulic Retention Time

Hydraulic retention time (HRT), also known as the inverse of the dilution rate (*D*), is defined as the volume of the reactor/volumetric flow. Continuous-flow stirred-tank reactor (CSTR) is used to select microbial populations whose growth rates are able to catch up to the dilution caused by continuous volumetric flow. The effects of HRT on H_2 production are summarized in Table 9.4. In continuous cultures, microbial populations with growth rates higher than the dilution rate can remain in the reactor ($\mu_{max} > D$). Methanogenesis can be limited by operating at a short HRT (high dilution rate) to cause the washout of slowly growing methanogens and/or operating under acidic environment to inhibit the pH-sensitive methanogens (Lin and Chang, 1999; Chen et al., 2001). The specific growth rates of methanogens are much shorter than those of H_2-producing bacteria (0.0167 and 0.083 h^{-1}, respectively) (Wang and Wan, 2009). Chen et al. (2001) operated a CSTR seeding with nonpretreated sewage sludge to convert sucrose to H_2 with a gradual decrease of HRT. They observed that the H_2 production rate was dramatically enhanced by a shift of HRT from 5 to 13.3 h and that the H_2 production rate continued to increase when HRT was shifted down further to 8 h. It is likely that an HRT of 13.3 h was short enough to cause the complete washout of methane-producing bacteria, while the H_2-producing population remained in the culture. Fan et al. (2006a) investigated the HRT (from 48 to 8 h) effect on H_2 production from brewery waste by cattle dung compost in a CSTR system at pH 5.5 in which maximum H_2 production rate and H_2 yield were obtained at an HRT of 18 h. In these studies, they expressed that the lower yield at high HRT was probably due to the interference of methanogens and inhibition via major intermediates of acidogenic and alcoholic products. Lin et al. (2006) used a base-enriched anaerobic mixed microflora to perform H_2 fermentation from sucrose in a CSTR bioreactor operating the HRT from 12 to 2 h at 35°C. The results showed that H_2 yield and H_2 production rate were HRT-dependent, and their values ranged from 0.9 to 3.5 mol H_2/mol sucrose and from 263 to 408 mmol H_2/L/day, respectively, with an HRT of 4 h having peak H_2 production. The biomass activity was also HRT-dependent, with each gram of biomass producing 65–145 mmol H_2/day. Further analysis by using denaturing gradient gel electrophoresis (DGGE) showed that microbial species shifted during the HRT reduction operation, but *Clostridium ramosum* was dominant throughout the experiments. One important hydrogenic organism, namely, *C. pasteurinum*, disappeared at an HRT of 2 h. It was observed that the H_2 fermentation pattern may shift to methanogenic fermentation if the HRT increased (Wang and Wan, 2009). Moreover, to understand the effect of glucose loading rate on H_2 production, van Ginkel and Logan (2005) performed the experiments with a combination of different glucose concentrations and HRTs in chemostat reactors seeding a heat-shocked agricultural soil. The results showed that the H_2 production rate increased with increasing glucose loading rate (high HRT or low glucose concentration), but the H_2 yield decreased with increasing glucose loading rate. They observed that the decrease in yield resulted from inhibition of H_2 supersaturation in the liquid phase due to a high H_2 production rate at a higher feeding rate.

Liu and Fang (2002) investigated the influence of HRT and sucrose concentration on H_2 production by the acidogenic granular sludge in a CSTR reactor at a constant loading rate of 25 g sucrose/L/day. The results show that the H_2 yield was highly dependent on HRT and sucrose concentration. The H_2 yield ranged from 0.19 to 0.27 L H_2/g sucrose (2.65–3.77 mol H_2/mol sucrose), with the maximum occurring at a higher HRT of 13.7 h and with a sucrose concentration of 14.3 g/L in the wastewater. Similarly, Yu et al. (2002) operated an upflow reactor for H_2 production from rice winery wastewater under different HRTs (2–24 h) in which the specific H_2 production rate increased as HRT decreased, but the H_2 yield decreased with decreasing HRT. The results suggest that more carbohydrates in the wastewater were converted into H_2 at longer HRT. The findings obtained from these studies demonstrate that a decrease in HRT can achieve a methanogen washout, although this method can be applied only when noncomplex or soluble substrates are used.

9.4.4 pH

One of the major factors affecting dark fermentative H_2 production is pH. Most investigators reported that the maximum H_2 yield or specific H_2 production rate was obtained at pH between 5.0 and 6.0 (Lay et al., 1999; Chen et al., 2001; Lay, 2001; Fang and Liu, 2002), whereas some reported a pH range between 6.8 and 8.0 (Lay, 2001; Fabiano and Perego, 2002; Collet et al., 2004; Liu and Shen, 2004; Lin and Cheng, 2006) (Table 9.5). During dark fermentative H_2 production, formation of organic acids depletes the buffering capacity of the medium and results in low final pH. This lower pH inhibits H_2 production since pH affects the activity of the iron-containing hydrogenase enzyme and metabolic pathway (Dabrock et al., 1992). Horiuchi et al. (2002) showed that the product spectrum in the CSTR reactor strongly depended on the culture pH. Under acidic and neutral conditions, the main product was butyric acid, while acetic and propionic acids were the main products under basic conditions. This phenomenon was reversible between the acidic and basic conditions and was caused by the change in the dominant microbial populations as a pH shift led to transition of the dominant population in the reactor from butyric acid–producing bacteria to propionic acid–producing bacteria. Medium pH also affects H_2 production yield, biogas content, types of organic acids produced, and the specific H_2 production rate (Khanal et al., 2004). Therefore, control of pH at the optimum level may be useful for getting better yield. Since pH affects the growth rate of microorganisms, pH changes may cause drastic shifts in the relative numbers of different species in a heterogeneous population present in the hydrogen-producing reactors (Horiuchi et al., 1999). Many aspects of microbial metabolism are greatly influenced by pH variations over the range within which the microorganisms can grow. These aspects include utilization of carbon and energy sources, efficiency of substrate degradation, synthesis of proteins and various types of storage materials, and release of metabolic products from cells (Baily and Ollis, 1986). Thus the foregoing results suggest that, in an appropriate range, increasing pH could increase the ability of hydrogen-producing bacteria to produce H_2 during fermentative H_2 production, but pH at much higher levels could reduce production. A number of studies were conducted in batch mode without pH control; only the effect of initial pH on fermentative H_2 production was investigated in these studies. Table 9.5 summarizes several studies investigating the effect of initial pH on fermentative H_2 production in batch mode. In batch systems, the final pH is around 4–5 regardless of the initial pH. This is due to the production of organic acids, which decreases the buffering capacity of the medium, resulting in a low final pH.

Table 9.5 also shows the reported optimal pH range for fermentation of carbohydrates by mixed-culture microorganisms. Zhang et al. (2003) examined the effect of pH (from 4.0 to 9.0) on starch conversion into H_2 with sludge at 55°C. The maximum H_2 yield of 92 mL/g of starch added (17% of the theoretical value) was found at pH 6.0, and the maximum specific H_2 production rate of 365 mL/day/g VSS occurred at pH 7.0. Zoetemeyer et al. (1982b) obtained an optimum H_2 production rate at a pH of 5.7 and HRT of 4 h with glucose as substrate. Acetate and butyrate with negligible propionate was produced in this batch. Fang and Liu (2002) systematically investigated the optimum pH (from 4.0 to 7.0) for H_2 production from glucose by a mixed culture at an HRT of 6 h and found the optimum yield at pH 5.5. Lay (2000) considered the pH effect on H_2 production from soluble starch by anaerobic digested sludge in a chemostat reactor. Experimental results indicated that the maximum H_2 production rate was obtained at pH = 5.2 and HRT = 17 h, while the culture favored alcohol production when the pH was lower than 4.1. This implies that the metabolism shifted from hydrogen/VFA production to alcohol production at pH < 4.1. Similarly, Mu et al. (2006c) showed that volatile fatty acids (VFA) and ethanol formation was pH-dependent. Ethanol production decreased when pH was decreased from 4.2 to 3.4. In contrast, butyrate concentration decreased when pH was increased from 4.2 to 6.3. It has been reported that high VHPR is associated with butyrate and acetate production and that inhibition of H_2 production is associated with propionic acid formation (Oh et al., 2004; Wang et al., 2006).

Yu et al. (2002) suggested that the optimum pH level was 5.5 for continuous H_2 production from rice winery wastewater by a mixed bacterial flora using an upflow reactor. However, Ren et al. (1997) indicated that the operating pH must be maintained at about 4.5 in an anaerobic sludge system, a CSTR reactor receiving molasses as the feed, to avoid onset of propionic fermentation and to gain a maximum fermentation rate. Cheong and Hansen (2006) studied H_2 production from glucose by mixed anaerobic bacteria in batch culture with pH controlled at 4.6, 5.7, and 6.8. The results showed that higher H_2 yield was significantly related to butyric acid formation. The reactor operated at a pH controlled at 5.7 showed higher H_2 yield than those operated at other pH values, where butyric acid (as 50.8%–75.8% of total acidogenic liquid products) was dominant. The presence of propionic acid and ethanol indicated a decrease in H_2 production. Fan et al. (2006a) investigated the pH effect on H_2 production from brewery waste by cattle dung compost in a CSTR system at HRT = 18 h. The results indicated that maximum H_2 production (47% H_2 concentration, 43 mL H_2/g COD_{added}, and 3.1 L H_2/L reactor/day) was achieved at pH = 5.5.

There is still some disagreement on the optimal initial pH for fermentative H_2 production. For example, batch operation with a mixed culture (using sucrose and starch as substrate) and hydrogen-producing sludge (using rice slurry as substrate) obtained the highest H_2 production yield at a pH of 4.5 (Khanal et al., 2004; Fang et al., 2006). However, Lee et al. (2002) reported an unusual result in which the optimal pH was 9.0 for the batch biohydrogen fermentation of sucrose. The possible reason for this disagreement was the difference among these studies in terms of inoculum, substrate, and initial pH range studied. The foregoing results suggest that a proper control of pH appears to be essential for successful operation of biohydrogen-producing reactors.

9.4.5 Temperature

Temperature is an important environmental factor that influences the growth rate and metabolic activities of hydrogen-producing bacteria and fermentative H_2 production (Bailey and Ollis, 1986). Fermentation reactions can be operated at mesophilic (25°C–40°C), thermophilic (40°C–65°C), extreme thermophilic (65°C–80°C), or hyperthermophilic (>80°C) temperatures. Most of the biohydrogen-producing studies were performed under mesophilic conditions and some under thermophilic conditions at a constant temperature. In comparable studies reported in the literature, the optimal temperature for H_2 production via dark fermentation varied widely, mainly depending on the type of H_2 producers and carbon substrates (Table 9.6). Nevertheless, the optimal temperatures were in the range of 37°C–45°C for pure culture *Clostridium* or *Enterobacter* species, whereas the mixed bacterial flora gave distinct optimal temperatures and were more effective at thermophilic conditions (Table 9.6).

Recently, semicontinuous H_2 production at mesophilic and thermophilic conditions was studied by Valdez-Vazquez et al. (2005). They found that volumetric H_2 production rate (VHPR) was 60% greater at thermophilic than at mesophilic conditions. The result suggests that this behavior might be due to the optimal temperature for the enzyme hydrogenase (50°C and 70°C) present in thermophilic *Clostridia* sp. Yu et al. (2002) studied the effect of temperature (20°C–55°C) on H_2 production from rice winery wastewater by mixed anaerobic cultures in an upflow anaerobic reactor. The results showed that the partial pressure of H_2, H_2 yield, and specific H_2 production rate increased with temperature. Wu et al. (2005) utilized ethylene–vinyl acetate copolymer (EVA) to immobilize acclimated sewage sludge for investigating the effect of temperature on H_2 production from sucrose in a batch test. The results showed that operation at 40°C attained the best specific H_2 production rate, with a H_2 yield of 389 mL/g VSS/h and 1.41 mol H_2/mol sucrose, respectively. Thermophilic continuous operation on glucose had been reported by Zoetemeyer et al. (1982a) with an optimum temperature of 52°C. It was also reported that shorter retention times may be used under thermophilic conditions. However, the process could become less stable and more sensitive to small changes in HRT or organic load, highlighting the need for sound online monitoring and control (Hawkes et al., 2002). It was observed that fermentation at temperatures above 37°C may be useful in inhibiting the activity of H_2 consumers (Lay et al., 1999) and suppressing lactate-forming bacteria (Oh et al., 2004), resulting in higher yield in both cases. Moreover, Lee et al. (2006b) studied the effect of temperature on H_2 production in a carrier-induced granular sludge bed bioreactor (CIGSB) and found that increasing the temperature from 35°C to 45°C inhibits cell growth or granular sludge formation due primarily to denaturation of essential enzymes to paralyze normal metabolic functions or to the decrease in production of extracellular polymeric substances. Another potential disadvantage of thermophilic processes is the increased energy costs.

Zhang et al. (2003) compared the H_2 production from starch at 37°C and 55°C by mesophilic sludge with batch experiments. The results showed that more starch was converted into H_2 at 55°C, although the sludge required a longer lag time for H_2 production. On the other hand, Fang et al. (2006) compared H_2 production from rice slurry at 37°C and 55°C by anaerobic digester sludge with batch experiments. The results showed that the mesophilic sludge was more effective than the thermophilic sludge in treating rice slurry. The maximum specific H_2 production rate and H_2 yield at 37°C were calculated as 2.1 L/g VSS/day and 346 mL/g carbohydrate. Also, the biohydrogen production from glucose in the CSTR at the mesophilic and thermophilic temperature ranges was studied and compared (Gavala et al., 2006). In this study, Gavala et al. found that the thermophilic conditions resulted in low volumetric H_2 production rate with poor microbial mass production, whereas they gave a higher specific H_2 production rate compared to mesophilic conditions at 6–12 h HRT.

Mu et al. (2006c) investigated H_2 production from glucose by mixed-culture anaerobes at various temperatures in the mesophilic range (33°C–41°C). Results showed that glucose degradation rate and efficiency, H_2 yield, and growth rate of H_2-producing bacteria all increased as the temperature increased from 33°C to 41°C. However, the specific H_2 production rate increased with increasing temperature from 33°C to 39°C, then decreased as the temperature was further increased to 41°C (Mu et al., 2006c). Through thermodynamic analysis, the activation energies for H_2 production and microbial growth were estimated as 107.66 and 204.77 kJ/mol, respectively. It has been demonstrated that within an appropriate range, increasing temperature could enhance the ability of hydrogen-producing bacteria to produce H_2 during fermentative H_2 production. However, at a higher temperature, decrease in H_2 yield was observed. The literature survey for the effect of temperature on fermentative H_2 production indicates that glucose and sucrose are the most widely used substrates and that most of the reviewed studies were conducted in batch mode (Table 9.6). Thus, investigations on the effect of temperature on fermentative H_2 production using organic wastes as substrate and conducted in continuous mode are recommended.

9.4.6 Oxidation–Reduction Potential

The oxidation–reduction potential is a good indicator to determine the anaerobic conditions useful for higher H_2 production. The value of ORP depends upon the strain as well as the substrate used during the study. Cohen et al. (1984) showed a strong linear inverse correlation between ORP and butyrate production over the range of −300 mV (maximum butyrate) and −100 mV (zero butyrate) using activated sludge inoculum and continuous operation on a glucose–mineral salts medium. This was likely due to the selection of propionate-producing species as the ORP rose. It is thus important to maintain the ORP near −300 mV to attain maximal H_2 yield for the culture. Ren et al. (2001) studied the effects of pH and ORP on fermentation types in continuous flow acidogenic reactors using molasses as substrate. They indicated that higher ORP (> −100 mV) would lead to propionic-type fermentation because propionogens are facultative anaerobic bacteria. When ORP came down to −200 mV, typical butyric-type fermentation occurred.

9.4.7 Inorganic Nutrients

The carbon-to-nitrogen (C/N) ratio is important for biological processes since nitrogen is an important component in proteins, nucleic acids, and enzymes and one of the most essential nutrients needed for the growth of hydrogen-producing bacteria. Thus, providing an appropriate level of nitrogen is beneficial for the growth of hydrogen-producing bacteria and for fermentative H_2 production (Bisaillon et al., 2006). The optimum C/N ratio at 47 provided efficient conversion of sucrose to H_2 gas by using acclimated anaerobic sewage sludge with a yield of 4.8 mol H_2/mol sucrose in batch experiments reported earlier (Lin and Lay, 2004a). Experimental results indicated that the H_2 production ability depended on the influent C/N ratio. Under optimum C/N ratio, H_2 production increased by 500% and 80%, respectively, when compared with that obtained from the blank experiments. In addition, Cheng et al. (2002) demonstrated that the use of a proteinaceous substrate, that is, peptone, could avoid the abrupt pH drops in the H_2 fermentation system due to the production of ammonia from fermenting peptone. Thus, with the addition of peptone in the feed, their H_2 fermentation system can be stably maintained without additional pH control. They also investigated the concentration effects of carbonate and phosphate on biological H_2 production using a Taguchi orthogonal array (Lin and Lay, 2004b). Experimental results indicated that the H_2 production ability of the anaerobic microflora in sewage sludge was affected by NH_4HCO_3, Na_2HPO_4, and Na_2CO_3, with Na_2HPO_4 being the most significant supplement. Optimal H_2 production was observed with the addition of 600 mg/L Na_2HPO_4. When proper carbonate and phosphate concentration formulation was used, the H_2 production rate was enhanced 1.9 times compared with an acidogenic nutrient formulation. Ueno et al. (2001b) used a continuous reactor for H_2 production from cellulose powder with thermophilic anaerobic microflora enriched from sludge compost. The H_2 production yield was 2.0 mol/mol hexose when the medium contained both NH_4Cl as an inorganic nitrogen source and peptone as an organic nitrogen source. However, the medium without peptone demonstrated a lower H_2 production yield of 1.0 mol/mol hexose, while a lesser amount of butyrate was formed. This implies the need for an appropriate organic nitrogen source in media rich in starch, cellulose, or hemicellulose products. Table 9.7 summarizes several studies investigating the effect of nitrogen concentration on fermentative H_2 production. The results show some disagreement on the optimal ammonia nitrogen concentration for fermentative H_2 production. For example, Bisaillon et al. (2006) and Salerno et al. (2006) reported the optimal ammonia nitrogen concentration for fermentative H_2 production to be 0.01 and 7.0 g N/L, respectively. The possible reason for this variation may be due to the differences among these studies in terms of the inoculum and ammonia nitrogen concentration range studied. However, this type of study is most widely carried out using glucose as substrate and ammonia as nitrogen source in batch mode. Further investigation should be done using more complex organic waste and different nitrogen sources under chemostat operation.

Phosphate is also needed for H_2 production due to its nutritional value as well as buffering capacity. Within an appropriate range, increasing phosphate concentration could increase the ability of hydrogen-producing bacteria to produce H_2 during fermentative H_2 production. But decrease in H_2 yield was observed at higher levels (Lay et al., 2005; Bisaillon et al., 2006). Thus an appropriate C/N and C/P ratio is essential for better fermentative H_2 production. The optimal C/N and C/P for fermentative H_2 production reported by Argun et al. (2008) were 200 and 1000, respectively. On the other hand, O-Thong et al. (2008a) reported the values to be 74 and 559, respectively. The possible reason for this disagreement was the difference among these studies in terms of the substrate, C/N range, and C/P range studied. The results suggest that the C/N and C/P ratios have a significant effect on fermentative H_2 production.

9.4.8 Trace Metal Elements

Biohydrogen production requires certain essential trace metal elements (micronutrients) for bacterial metabolism and growth. Many different micronutrient formulations have been used in studies on biohydrogen, and they exhibited variations in H_2 production effectiveness. Fe^{2+} was the most widely investigated metal ion for fermentative H_2 production, as iron is a cofactor of hydrogenase, which catalyzes the reaction to generate H_2. Limitation in iron in the culture medium lowers hydrogenase activity (Wang and Wan, 2008). Lee et al. (2001) reported that low iron concentrations favored ethanol and butanol production from batch studies with mixed cultures on sucrose, whilst maximum H_2 yields were observed when 800 mg $FeCl_2$/L was added to the growth medium. Lin and Lay (2005) explored the effect of nutrient formulation for biological H_2 production by anaerobic microflora of sewage sludge in a batch experiment using Taguchi orthogonal arrays. Experimental results indicated that the sewage sludge enriched with the proposed nutrients had a H_2 productivity of 3.43 mol H_2/mol sucrose, which was about 30% higher than those of a control and an acidogenic nutrient formulation. Magnesium, sodium, zinc, and iron were found to be important trace metals affecting H_2 production, with magnesium being the most significant. Zhang et al. (2005) investigated the effect of iron concentration (0–5000 mg $FeSO_4$/L) on H_2 yield in batch tests by heat-shocked mixed cultures growing on a sucrose mineral medium at 35°C. The maximum H_2 production (2.73 mol of H_2/mol sucrose) was obtained at an iron concentration of 1600 mg $FeSO_4$/L. The H_2 production yield and the butyric acid and acetic acid ratio followed a similar trend, suggesting that the formation of butyrate favors H_2 production. Yang and Shen (2006) investigated the role of ferrous ion concentration ($FeSO_4$) on anaerobic mixed bacteria for conversion of soluble starch to H_2 in both batch experiments at initial pH 7.0 and 8.0, respectively. At initial pH = 8.0, the H_2 yield significantly increased from 106.4 to 274.0 mL/g starch with increasing iron concentration from 0 to 200 mg $FeSO_4$/L. When iron concentration continued to increase from 200 to 4000 mg $FeSO_4$/L, iron inhibition did not happen. Furthermore, Zhang and Shen (2006) explored the effects of temperature and iron concentration on H_2 production from sucrose with mixed bacteria dominated by *Clostridium pasteurianum* in batch experiments. Experimental results showed that the optimum iron concentrations for H_2 production decreased with increasing culture temperature. For 25°C, 35°C, and 40°C, the maximum H_2 yield was obtained at iron concentrations of 800, 200, and 25 mg $FeSO_4$/L, respectively. In addition, several studies were also conducted to investigate the toxicity of heavy metals to fermentative H_2 production.

Recently, Li and Fang (2007b) reported that the relative toxicity of six heavy metals to fermentative H_2 production was in the following order: Cu > Ni–Zn > Cr > Cd > Pb. In contrast, Lin and Shei (2008) reported that the relative toxicity of three heavy metals to fermentative H_2 production was in the following order: Zn > Cu > Cr. Moreover, Zheng and Yu (2004) investigated the influence of copper and zinc on H_2 production from glucose by enriched anaerobic culture in batch experiments. Results showed that the specific H_2 production rate was enhanced by the dosage of Cu at 50–100 mg/L or the dosage of Zn at 100–250 mg/L, but was inhibited by Cu over 200 mg/L or Zn at 500 mg/L. The H_2 production yield was enhanced by 5–400 mg/L of Cu or 5–500 mg/L of Zn. In terms of the $C_{I,50}$ values, Cu was observed to be more toxic than Zn. Thus at a higher concentration, metal ions may inhibit the activity of hydrogen-producing bacteria, and a trace level of metal ion might be required for fermentative H_2 production (Li and Fang, 2007b). Table 9.8 summarizes several studies investigating the effect of metal ion concentration on fermentative H_2 production.

9.4.9 Hydrogen Partial Pressure (P_{H_2})

Generally, in biological production of H_2, all of the observed H_2 can be attributed to electrons derived from a single reaction: oxidative decarboxylation of pyruvate by pyruvate:ferredoxin oxidoreductase (Figure 9.3). Hexoses can be metabolized to pyruvate through several pathways, often involving the Embden–Meyerhof–Parnas (i.e., glycolysis) or the Entner–Doudoroff pathways. Both of these pathways produce 2 mol of pyruvate and 2 mol of NADH for every mole of hexose that is transformed. Therefore, hexose metabolism by bacteria that contain pyruvate:ferredoxin oxidoreductase can result in the formation of 2 mol of H_2 per mole of hexose. If the H_2 partial pressure is sufficiently low (<60 Pa), the NADH produced may also be used to generate H_2 (at best, an additional 2 mol

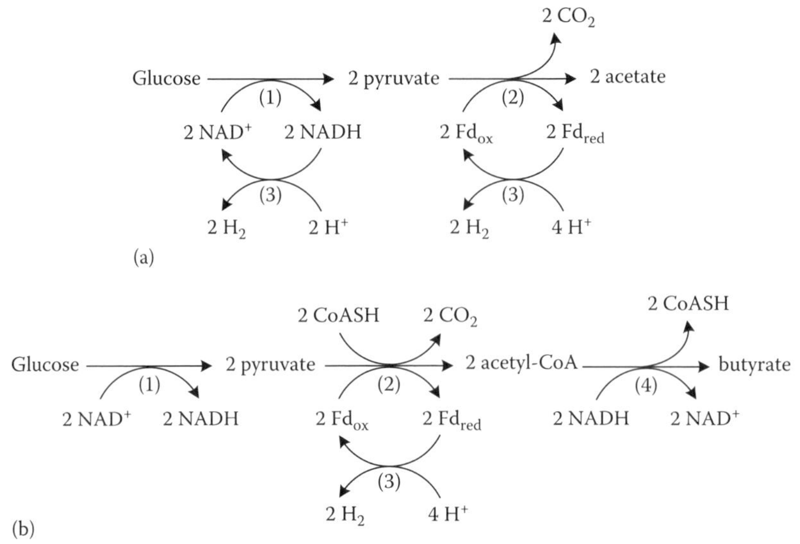

FIGURE 9.3
Effect of H_2 partial pressure on biological production of hydrogen: (a) Oxidation of NADH by production of H_2 is thermodynamically favorable only when the H_2 partial pressure is less than 60 Pa, otherwise, (b) other fermentation products must be formed. Reactions [(a) $P_{H_2} < 60$ Pa and (b) $P_{H_2} > 60$ Pa]: 1, glucose metabolism through glycolysis or the Entner–Doudoroff pathway; 2, oxidative decarboxylation of pyruvate by pyruvate:ferredoxin oxidoreductase; 3, formation of H_2 by hydrogenase; and 4, butyrate fermentation.

of H_2 mol of glucose). However, most of the NADH is probably oxidized through other fermentation pathways, such as butyrate fermentation (Figure 9.3, step 4). Some fermentation products (e.g., ethanol and lactate) represent the operation of alternative pathways for pyruvate metabolism that compete with pyruvate:ferredoxin oxidoreductase. As such, they are usually associated with systems that produce less than 2 mol of H_2 mol of glucose (Angenent et al., 2004).

The partial pressure of H_2 (P_{H_2}) is an extremely important factor for continuous H_2 generation. The increase in the partial pressure of H_2 (P_{H_2}) to a certain level in the reactor headspace might cause a diversion of metabolic shift toward alcohol production and thus lead to less H_2 production. H_2 synthesis pathways are sensitive to H_2 concentrations and subject to end-product inhibition. As H_2 concentration increases, H_2 synthesis decreases and metabolic pathways shift to production of more reduced substrates such as lactate, ethanol, acetone, butanol, or alanine (Levin et al., 2004). As the temperature increases, H_2 concentration affects H_2 synthesis more significantly. For continuous H_2 synthesis, P_{H_2} of <50 kPa at 60°C (Lee and Zinder, 1988), <20 kPa at 70°C (van Niel et al., 2002b), and <2 kPa at 98°C (Adams, 1990) were reported.

Several strategies—mainly sparging with inert gas and using a silicone rubber membrane to avoid the negative effect of the accumulation of H_2 in the gaseous space—have been developed. Mizuno et al. (2000) have demonstrated that lowering H_2 partial pressure by sparging with N_2 gave a 68% increase in H_2 yield from a CSTR reactor containing an enriched mixed microflora with 10 g/L glucose at pH 6.0 and HRT 5 h. Liang et al. (2002) used a silicone rubber membrane to separate biogas from the liquid medium in the H_2 fermentation reactor. Their results showed that the silicone rubber effectively reduced the biogas partial pressure in the H_2 fermentation reactor and improved the H_2 evolution rate by 10% and the H_2 yield by 15%. Hussy et al. (2005) investigated the effect of sparging on continuous H_2 production from sucrose and sugarbeet juice using mixed microflora. With N_2 sparging, H_2 yields were 1.9 and 1.7 mol H_2/mol hexose for sucrose and sugarbeet juice, respectively, at 14–15 h HRT, which was about 90% higher than those without sparging.

9.5 Bioreactor Configuration

Many studies were conducted in batch mode to examine the characteristics of H_2-producing bacteria and to optimize culture-operating conditions. The worse microbial cultures led to lower H_2 production rates. Enhancing H_2 production efficiency, stability, and sustainability is thus a major challenge in batch H_2 systems. Recently, Valdez-Vazquez et al. (2005a) examined H_2 production from municipal solid wastes in another semicontinuous pattern. Reactors that were fed with substrate twice a week in a fill-and-draw mode in an anaerobic chamber and operated continuously at 35°C and 55°C for 40 days demonstrated steady H_2 production. The high-rate biodegradation processes are based on the concept of retaining high-concentration viable biomass by some mode of bacterial sludge immobilization. These are achieved by one of the following methods (Hulshoff and Lettinga et al., 1981; Rajeshwari et al., 2000):

1. Entrapment of sludge aggregates between packing materials supplied to the reactor, for example, anaerobic filter (packed bed).
2. Bacterial attachment to high-density particulate carrier materials, for example, fluidized bed reactors and anaerobic expanded bed reactors.
3. Formation of highly settleable sludge aggregates combined with gas separation and sludge settling, for example, upflow anaerobic sludge blanket reactor and anaerobic baffled reactor.

9.5.1 Continuous-Flow Stirred-Tank Reactor

The literature survey suggests that most of the studies on fermentative H_2 production were conducted in batch mode because of its simplicity and easy operation and control. However, for large-scale H_2 production, continuous processes are required because of practical engineering reasons. Table 9.4 summarizes a number of studies using continuous reactors for fermentative H_2 production. Continuous-flow stirred-tank reactors (CSTRs) are the most frequently used reactors for continuous production of H_2 from organic matter (Das and Veziroglu, 2001; Hawkes et al., 2002; Chen and Lin, 2003; Levin et al., 2004; Gavala et al., 2006; Zhang et al., 2006, 2007; Lee et al., 2007; Chen et al., 2008b; Wu et al., 2008b; Wang and Wan, 2009). As shown in Table 9.4, glucose and sucrose were the most widely used substrates for continuous H_2 production under CSTR operation. The average yield obtained was between 1.5 and 2.3 mol H_2/mol glucose. A pilot-scale CSTR (with an available volume of 1.48 m^3) for biohydrogen production was operated for over 200 days by Ren et al. (2006). The system was operated under the organic loading rate (OLR) of 3.11–85.57 kg COD/m^3/day with molasses as the substrate. A maximum H_2 production rate of 5.57 L/L/day was obtained at an OLR of 68.21 kg COD/m^3/day (HRT = 3.9 h).

In CSTR operation, the biomass is well suspended in the mixed liquor, which has the same composition as the effluent. It was observed that the biomass has the same retention time as the HRT and that washout of biomass may occur at shorter HRT. Recently, the effect of organic loading rate on H_2 production from glucose in chemostat reactors using a heat-shocked agricultural soil as inoculum was studied. In this study, at influent glucose concentrations of 5–10 g COD/L, substantial flocculation was observed particularly as the feeding rate increased due to a reduction in the HRT from 10 to 2.5 h. At an HRT of 2.5 h, the biomass concentration reached values as high as 25 g/L. The flocculant nature of the biomass allowed reactor operation at low HRTs with steady H_2 production and >90% glucose removal. However, when the HRT was reduced to 1 h at a glucose feed concentration of 2.5 g COD/L, there was little flocculation evident resulting in washout of the culture (van Ginkel and Logan, 2005). It has been demonstrated that in an appropriate range, increasing HRT could increase the ability of hydrogen-producing bacteria to produce H_2 during fermentative H_2 production. But marginal reduction in yield was observed at a much higher HRT level (Chen et al., 2008b). Furthermore, there exists certain disagreement on the optimal HRT for chemostat reactors, even for the same type of reactor. The optimal HRT for a CSTR was 96 h as reported by Cheng et al. (2002) and 12 h as reported by Arooj et al. (2008), whereas the optimal HRT for a CSTR was 0.5 h as reported by Zhang et al. (2007) (Table 9.4). The possible reason for the variation in optimal HRT might be due to the difference in terms of inoculum, substrate, and HRT range studied. Fang et al. (2002) demonstrated that hydrogen-producing acidogenic sludge agglutinated into granules in a well-mixed reactor treating a synthetic sucrose-containing wastewater at HRT = 6 h. They obtained a H_2 production rate of 13.0 L/L/day and a yield of 0.28 L/g sucrose (3.91 mol H_2/mol sucrose). However, CSTRs do have certain limitations. First, the CSTR system is very sensitive to environmental conditions such as changes in pH and HRT. Additionally, operation at a high dilution rate (or short residence time) can lead to washout of biomass, limiting its hydrogen-producing rate (Show et al., 2008). Therefore, several researchers attempt to retain high biomass by various bioreactor designs and/or operation strategies.

9.5.2 Anaerobic Sequencing Batch Reactor

A high-rate anaerobic sequencing batch reactor (ASBR) has been used to evaluate H_2 production from sucrose with acclimated sewage sludge by Lin and Jo (2003). Results show that the H_2 production depended on hydraulic retention time (HRT) and reaction/settling period (R/S) ratio. A short equivalent HRT, even up to 4 h, resulted in good H_2 productivity and high H_2 production rate (HPR). For each equivalent HRT, R/S ratio control also increased the H_2 productivity and HPR (Table 9.4). Reactor operation at an intimate control of HRT and R/S ratio was preferable for H_2 production. At an HRT of 8 h, an R/S ratio of 5.6, and an organic loading rate of 0.23 mol sucrose/L/day, the H_2 yield of the mesophilic hydrogenic reactor was 2.6 mol H_2/mol sucrose and the specific H_2 production rate was 0.069 mol H_2/day/g biomass. H_2 productivity by using an acid-enriched sewage sludge with sucrose (20 g COD/L) at 35°C was studied recently by using ASBR. The HRT was initially maintained at 12–120 h and thereafter at 4–12 h, with the R/S ratio maintained at 1.7. Hydrogenic activity of sludge microflora was found to be HRT-dependent, and proper pH control was necessary for a stable operation of the bioreactor. Higher H_2 production was noted at an HRT of 8 h and an OLR of 80 kg COD/m^3/day, giving the highest yield of 2.8 mol H_2/mol sucrose and a specific production rate of 39 mmo H_2/g biomass/day. Extremely low HRT might deteriorate H_2 productivity. Concentration ratios of acidogenic and other soluble microbial products can be used as monitoring indicators for a hydrogenic bioreactor (Lin and Chou, 2004).

9.5.3 Membrane Bioreactor

Membrane bioreactors (MBRs) have recently emerged as an effective means for increasing biomass retention in wastewater treatment (Wen et al., 1999). Although MBRs allow high biomass concentration in the reactor without suffering mass transfer limitations, there is little information regarding the use of MBRs for fermentative H_2 production. Kim et al. (2005) used an MBR system with glucose as a substrate, which gave a yield of 1.1 mol H_2/mol glucose with a maximum VHPR of about 71.4 mmol/L/h. Liang et al. (2003) used a microfiltration membrane to recover hydrogen-producing bacteria, allowing accumulation of biomass from 500 mg/L to 5200 mg VSS/L. The fermentor employed mixed-culture consortia to ferment multiple substrates including peptone and glucose at 55°C. The maximum H_2-producing rate was 28 mmol H_2/L h and the maximum specific H_2-producing rate was 4.6 mmol H_2/g VSS/h when the fermentor was operated at more than 80 g COD/L day. The H_2 yield was 8.41 mmol H_2/g COD. Oh et al. (2004) reported that the H_2 production rate from glucose with mixed culture was enhanced 27% with MBR at a short HRT of 3.3 h, while the biomass concentration increased 164%. This demonstrates the possibility of using MBR to improve the performance of biohydrogen production (Table 9.4).

Lee et al. (2007) used an MBR system fabricated by connecting a hollow-fiber microfiltration membrane module with a CSTR to enhance H_2 production through high–dilution rate operations. They demonstrated that the MBR system was effective in retaining biomass within the reactor as the system can be stably operated at a low HRT of 1 h with an optimal steady-state HPR of 1.48, 2.07, and 2.75 L/h/L, respectively, with glucose, sucrose, and fructose as the sole carbon source (Table 9.4). Nevertheless, the use of MBR systems has been limited to the laboratory scale. This may be due to the high operating cost as well as membrane caking and fouling problems. However, it has been shown that using MBRs in wastewater treatment could be more cost-effective when compared to conventional wastewater treatment plants with secondary clarification (Gander et al., 2000). Moreover, revolutionary advances in membrane technology have reduced the cost of membranes and also created methods to solve the problem of membrane fouling, such as the addition of coagulants, crossflushing, imposing a pulsed electric field, backwashing, and rapid backpulsing (Ma et al., 2000). Therefore, the MBR is still a promising process for H_2 production from organic substrates.

9.5.4 Packed-Bed Reactor

Fixed- or packed-bed reactors offer the advantages of simplicity of construction, elimination of mechanical mixing, better stability at higher loading rates, and the capability to withstand large toxic shock loads and organic shock loads. The reactors can recover very quickly after a period of starvation. In stationary packed-bed (fixed-bed) reactors, the reactor has a biofilm support structure (media) such as activated carbon, PVC (polyvinyl chloride) supports, hard rock particles, or ceramic rings for biomass immobilization. The substrate (wastewater) is evenly distributed from above or below the media. The main limitation of this design is that the reactor volume is relatively high compared to other high-rate processes due to the volume occupied by the media. Another constraint is clogging of the reactor due to the increase in biofilm thickness and/or high suspended solids concentration in the wastewater (Kennedy and Droste, 1985; Van den Berg et al., 1985; Rajeshwari et al., 2000) (Table 9.4).

Most of the studies on H_2 fermentation by employing packed-bed processes focused on pure cultures (Yokoi et al., 1997; Rachman et al., 1998; Palazzi et al., 2000; Kumar and Das, 2001). Support materials also have important effects on biomass retention and consequently H_2 production in fixed-bed reactors. Chang et al. (2002) conducted biohydrogen production by mixed culture with packed-bed bioreactors in batch and continuous modes. Three porous materials, namely, loofah sponge (LS), expanded clay (EC), and activated carbon (AC), were assessed because of their effectiveness in biofilm formation. It was found that LS was inefficient for biomass retention, while EC and AC exhibited better biomass yields. The packed-bed reactors with EC or AC were thus used for continuous H_2 fermentation at an HRT of 0.5–5 h. PBR with cylindrical activated carbon exhibited a better H_2 production rate of 1.32 L/L/h at an HRT of 1 h. However, the packed-bed reactor was of limited use in reducing the effective volume in the reactor due to the presence of solid supports, which occupied a significant portion of the working volume. Thus, the effect of the void fraction of the bed (ε_b, 70–90%) on H_2 fermentation in packed-bed systems was examined (Lee et al., 2003). The results showed that higher ε_b favored H_2 production, and the optimal H_2 production rate (7.35 L/L/h) was obtained with ε_b = 90% at HRT = 0.5 h with 20 g COD/L of sucrose in the feed. During the investigation of void fraction effect, flocculation of cells to form granular sludge was accidentally observed, which was more abundant when the void space increased.

Kumar and Das (2001) observed that both H_2 production and substrate conversion rate of a packed-bed reactor increased with recycling ratio and the maximum H_2 production rate (1.69 L/L/h) was observed at a recirculation ratio of 6.4. Moreover, Rachman et al. (1998) observed that high H_2 molar yield could not be maintained consistently in a packed-bed reactor, although the pH in the effluent was controlled at >6.0. This is because pH gradient distribution along the reactor column resulted in a heterogeneous distribution of microbial activity. In order to overcome the mass transfer resistance and pH heterogeneous distribution, a fluidized-bed or an expanded-bed reactor system with recirculation flow was recommended to be more appropriate in further enhancing the H_2 production rate and yield. Increasing slurry recycle ratio can thus alleviate mass transfer resistance in a packed-bed reactor.

9.5.5 Fluidized-Bed Reactor

Cell immobilization techniques have been applied to improve cell retention and were shown to be suitable for continuous H_2 production (Wu et al., 2003). Most H_2 fermentation associated with immobilized cells was conducted in packed-bed reactors (Yokoi et al., 1997; Rachman et al., 1998; Palazzi et al., 2000; Kumar and Das, 2001), which often suffer from inefficient mass transfer even though they are relatively inexpensive and easy to operate. However, for immobilized cells created by entrapment methods, mass transfer efficiency is often a limiting factor. In addition, when using immobilized systems, it is important to consider that the excessive amount of gases (H_2 and CO_2) in the reactor could lead to inhibition of hydrogen-producing microorganisms. This may be due to changes in pH (Table 9.4).

Generation of gaseous products from H_2 fermentation further highlights the importance of mass transfer efficiency for the immobilized-cell system. As a result, fluidized-bed reactors may be preferable to the packed-bed reactors in biohydrogen production. However, reports regarding the use of fluidized beds for H_2 fermentation have been quite rare. Wu et al. (2003) used three-phase fluidized beds to produce H_2 from sucrose with immobilized hydrogen-producing sludge. The fluidized bed can be stably carried out at a high loading rate (HRT as low as 2 h), and the maximal H_2 production rate and yield were 0.93 L/L/h and 2.67 mol H_2/mol sucrose, respectively. Recently, fluidized-bed reactor (FBR) and draft tube fluidized-bed reactor (DTFBR) systems with immobilized cells were studied for the production of H_2 using sucrose (Wu et al., 2006). The results showed that the VHPR obtained with DTFBR (95.23 mmol/L/h) was higher than that obtained with FBR (50.27 mmol/L/h) (Table 9.4).

9.5.6 Granular Sludge Reactors

Upflow anaerobic sludge blanket (UASB) technology is being used extensively for effluents from different sources such as distilleries, food processing units, tanneries, and municipal wastewater. The active biomass in the form of sludge granules is retained in the reactor by direct settling for achieving high cell retention time, thereby achieving highly cost-effective designs. A major advantage of this technology is the comparatively less investment costs when compared to a packed-bed or a fluidized-bed system. Moreover, a UASB reactor essentially consists of a gas–solid separator (to retain the anaerobic sludge within the reactor), an influent distribution system, and effluent draw-off facilities. Effluent recycling (to fluidize the sludge bed) is not necessary as sufficient contact between wastewater and sludge is guaranteed even at low organic loads with the influent distribution system (Rajeshwari et al., 2000). However, a long start-up period and the requirement for a sufficient amount of granular seed sludge for faster start-up are among notable disadvantages.

Vijayaraghavan et al. (2006) demonstrated H_2 production from solid waste consisting of jackfruit peel with microflora isolated from cow dung in an upflow anaerobic contact filter packed rigid with circular porous plastic balls of 40 mm diameter. The effect of HRT on the destruction efficiency of volatile solids was investigated for an influent volatile solids content of 33 g/L at an HRT of 7 and 12 days. The results showed a volatile solids destruction efficiency of 22% and 50%, respectively. In this study H_2 production rate and H_2 yield were 23.6 mL/L/h and 16.2 mmol/g VS destroyed, respectively, at an HRT of 12 days (Table 9.4).

Liu et al. (2003) reviewed the existing mechanisms and models for anaerobic granulation in the UASB reactor, and proposed a general three-step model for anaerobic granulation as follows:

Step 1: Physical movement to initiate bacterium-to-bacterium contact or bacterial attachment onto nuclei. The forces involved in this step are hydrodynamic force, diffusion force, gravity force, thermodynamic forces like Brownian movement, and cell mobility.

Step 2: Initial attractive forces to keep stable multicellular contacts. Those attractive forces are physical, chemical, and biochemical forces:

Physical forces: Van der Waals forces, opposite charge attraction, thermodynamic forces including free energy of surface; surface tension, hydrophobicity, and filamentous bacteria that can serve as bridges to link or grasp individual cells together.

Chemical forces: H_2 liaison, formation of ionic pairs and triplets, and interparticulate bridge.

Biochemical forces: Cellular surface dehydration, cellular membrane fusion, signaling, and collective action in bacterial community.

Step 3: Microbial forces making cell aggregation mature. The process is as follows: Extracellular polymers such as exopolysaccharides are produced by bacteria. Then growth of cellular clusters is observed, and finally metabolic change and genetic competence are induced by the environment, which facilitate cell–cell interaction and result in a highly organized microbial structure. In the last stage, the steady-state three-dimensional structure of microbial aggregates is shaped by hydrodynamic shear forces. The microbial aggregates are finally shaped by hydrodynamic shear force to form a certain structured community. The outer shape and size of microbial aggregates are determined by the interactive pattern between aggregates and by the hydrodynamic shear force, microbial species, and substrate loading rate.

Yu et al. (2002) investigated continuous H_2 production from a high-strength rice winery wastewater by a mixed bacterial flora, using an upflow reactor. The H_2 yield was in the range of 1.37–2.14 mol/mol hexose. Chang and Lin (2004) examined the H_2 production from sucrose with a UASB reactor and demonstrated the feasibility of using the UASB system for H_2 production. Gavala et al. (2006) examined and compared the biological H_2 production from glucose in a CSTR and a UASB at various HRTs (2–12 h) under mesophilic conditions. The results showed that the H_2 production rate in the UASB reactor was significantly higher than that of the CSTR at low retention times (19.05 and 8.42 mmol H_2/h/L, respectively, at 2 h HRT), while H_2 yield was higher in the CSTR reactor at all HRTs tested due to low glucose utilization. Yu and Mu (2006) evaluated the performance of a UASB for H_2 production from sucrose-rich synthetic wastewater at various substrate concentrations (5.33–28.07 g COD/L) and HRTs (3–30 h) for over 3 years. Experimental results showed that the H_2 production rate increased with both increasing substrate concentration and decreasing HRT. The H_2 yield was in the range of 0.49–1.44 mol H_2/mol glucose. The physicochemical characteristics of the granules were also evaluated (Mu and Yu, 2006). The mature granules had a diameter ranging from 1.0 to 3.5 mm and an average density of 1.036 g/mL, having good settling ability and a high settling velocity of 32–75 m/h (Table 9.4). The low ratio of proteins/carbohydrates of the extracellular polymeric substances (EPS) in the granules suggests that carbohydrates rather than proteins might play an important role in the formation of H_2-producing granules. The contact angle of the mature granules was larger than that of the seeding sludge, indicating that the microbial cells in the H_2-producing granules had higher hydrophobicity. Results also suggested that molecular diffusion plays an important role in mass transfer through the H_2-producing granules.

It was reported that the presence of nuclei or biocarriers for microbial attachment is one of the contributing factors to the development of granules from suspended sludge (Imai et al., 1997; Teo et al., 2000). The attachment of cells to these particles has been proposed as the initiating step for granulation. The second step was the formation of a dense and thick biofilm on the cluster of inert carriers and could be considered as biofilm formation. In other words, once the initial aggregates are formed, subsequent granulation could be regarded as an increase in biofilm thickness. Hence, the sludge granulation process in UASB reactors with added inert particles might be interpreted as a biofilm-forming phenomenon (Yu et al., 1999). Lettinga et al. (1981) claim that clay and other inorganic particles are harmful for the formation of granular sludge. Several investigators have studied the effect of inert particles on granulation. According to Morgan et al. (1991), the addition of supplements to a nongranular inoculum during the start-up of UASB reactors is beneficial. A granular activated carbon (GAC) supplement offers two advantages: (1) sheltered ecological niches that enhance biological attachment and thus initiate granule formation and, possibly, (2) a capacity for the adsorption of pollutants, which can then be degraded in the immobilized state. The activated carbon particles also enhance the development of an attached biofilm and thus act as a nucleus for granule formation. Yu et al. (1999) studied the effects of powdered activated carbon (PAC) and GAC on sludge granulation during the start-up of UASB reactors. Results showed that the addition of PAC or GAC clearly enhanced the sludge granulation process and accelerated the start-up process. Sludge granulation, defined as the time by which 10% of the granules are larger than 2.0 mm, took approximately 95 days to achieve in the reactor without addition of inert materials, while the time required for sludge granulation reduced by 25 and 35 days in the PAC- and GAC-added reactors, respectively. Besides, the addition of GAC and PAC provoked higher biomass concentrations throughout the experiment and earlier observation of visible granules. It also improved the volumetric COD removal capacity. Moreover, the addition of GAC showed slightly more beneficial effects during the start-up of UASB reactors than PAC. The enhanced granulation by the addition of PAC or GAC was attributed to a better attachment of the filamentous bacteria on the activated carbon. Imai et al. (1997) studied the effects of adding water-absorbing polymer (WAP) particles into the inoculated sludge. WAP is a resin that is mainly composed of acrylic compounds and shows a complex network structure with a high specific surface for microbial attachment. Moreover, WAP shows a low density (wet density of 1.0 g/mL), which means that the contact between the particles and biomass is improved when compared to sand and other materials. Although the average granule size was not affected, the addition of WAP clearly enhanced the granulation in the lab-scale and pilot-scale UASB reactors using glucose or VFA as substrate by serving as a biocarrier to allow more biomass attachment. Yu et al. (1999) proposed the following guidelines for the choice of inert materials to enhance sludge granulation:

- High specific surface area
- Specific gravity similar to anaerobic sludge
- Good hydrophobicity
- Spherical shape

Since bacteria have negatively charged surfaces under normal pH conditions, a basic idea to expedite anaerobic granulation processes is to reduce electrostatic repulsion between negatively charged bacteria by introducing multivalence positive ions, such as calcium, ferric, aluminum, or magnesium ions, into the seeding sludge (Figure 9.4). Reduced electrostatic repulsion between bacterial particles could promote anaerobic sludge granulation (Liu et al., 2003). Kim et al. (2005) employed two immobilization methods, biofilm formation on polyvinyl alcohol (PVA) and granulation of the sludge with cationic and anionic polymers, to immobilize sewage digester sludge and conducted continuous H_2 production tests in a stirred reactor simultaneously. The result showed that the granular sludge system produced much more H_2 gas than the biofilm system (ca. 120 mL/L/h vs. 20 mL/L/h at 20 h HRT). The reason was that granular sludge harbored more microorganisms (approximately

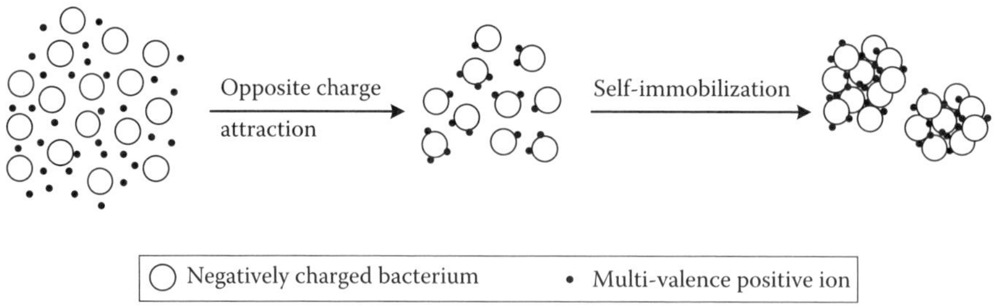

FIGURE 9.4
Schematic representation of the multivalence positive ion-bonding model. (From Liu, Y. et al., *Water Res.*, 37, 661, 2003.)

7 g/L vs. 4 g/L) and its density was higher than that of the biofilm on a PVA medium. Therefore, granulation was considered to be a better method for continuous H_2 production.

Lee et al. (2004a) used the inert carriers with biofilm to induce sludge granulation for enhancing H_2 production, and thereby a novel carrier-induced granular sludge bed (CIGSB) bioreactor was developed (Figure 9.5). A variety of carrier matrices were examined for their effectiveness in stimulating sludge granulation. Among the carriers examined, spherical activated carbon (SAC) and cylindrical activated carbon (CAC) were the more effective inducers for granular sludge formation. The SAC–CIGSB bioreactor achieved an optimal volumetric H_2 production rate of 7.33 L/L/h and a maximal H_2 yield of 3.03 mol H_2/mol sucrose when operated at an HRT of 0.5 h with an influent sucrose concentration of 20 g COD/L. In their next study, Lee et al. (2004b) stated that supplementation of calcium ion enhanced the mechanical strength of the granular sludge. Addition of 5.4–27.2 mg/L of Ca^{2+} led to an over threefold increase in biomass concentration and a nearly fivefold increase in the H_2 production rate when the CIGSB bioreactor caused washout of biomass at an HRT of 0.5 h. Lee et al. (2004b) utilized two reflux strategies, both liquid and gas reflux, to enhance the mass transfer efficiency when the dead zones were found in the CIGSB due to poor mixing efficiency of the reactor. The liquid reflux (LR) strategy enhanced the H_2 production rate 2.2-fold at an optimal liquid upflow velocity of 1.09 m/h, which also gave a maximal biomass concentration of ca. 22 g VSS/L. A similar optimal H_2 production rate of 1.0–1.49 m/h was also obtained with the gas reflux (GR) strategy, whereas the biomass concentration decreased to 2–7 g VSS/L, and thereby the specific H_2 production rate was higher than that with LR. Lee et al. (2006a) also designed

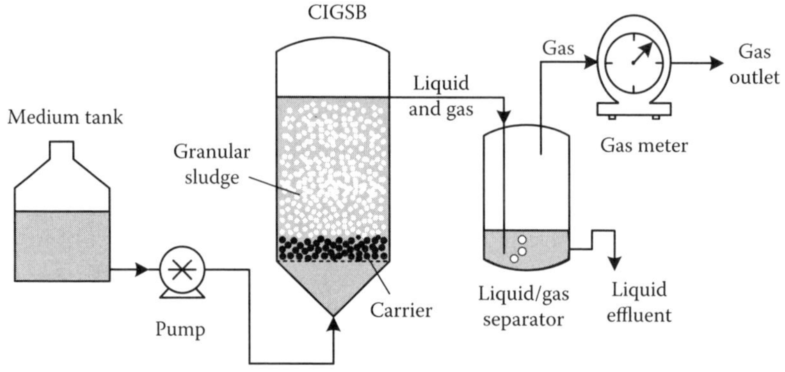

FIGURE 9.5
Schematic description of carrier-induced granular sludge bed (CIGSB) reactor.

experiments to adjust the height-to-diameter (H/D) ratios of the CIGSB bioreactor to improve the mixing efficiency for better biomass–substrate contact. The results show that the liquid upflow velocity (v_{up}) is a key factor to influence the H_2 production efficiency with a different H/D ratio. Increasing the H/D ratio gives higher v_{up}, allowing better hydraulic mixing to enhance the biomass–substrate contact, but an excessively large H/D ratio may also cause sludge floatation to diminish the sludge retention. Supply of additional mechanical agitation for the CIGSB reactor (H/D = 12) alleviated the phenomenon of sludge piston floatation, leading to further increases in the H_2 production rate and H_2 yield to 9.31 L/L/h and 4.02 mol H_2/mol sucrose, respectively.

In a traditional UASB reactor, channeling of the wastewater through the bed is frequently encountered, resulting in poor substrate–biomass contact (Jeison and Chamy, 1999). In light of this, the expanded granular sludge bed (EGSB) reactor was designed to cope with the aforementioned drawbacks of the UASB reactors. The EGSB reactors are operated at high liquid upflow velocities by increasing the height/diameter (H/D) ratio and by recirculation of effluent liquid, thereby providing better mixing of the reactor contents and allowing more efficient contact between substrate and biomass (Hwu et al., 1998). Francese et al. (1998) also showed that upflow velocity affects physical characteristics and the specific activity of granules.

Culture temperature is often an important environmental factor for the performance of a bioreactor, because it may strongly affect the growth rate and metabolic activity of bacteria in the reactor. In addition, Schmidt and Ahring (1996) observed that the concentration of extracellular polymeric substance (EPS), which plays a crucial role in cell adhesion, was lower in granules at high temperatures. Lee et al. (2006b) explored the temperature effects on biohydrogen production in the CIGSB system. This implies that the culture temperature may also affect the efficiency of sludge granulation.

Recently, Wu et al. (2006) developed a novel continuously stirred anaerobic bioreactor (CSABR) seeded with silicone-immobilized sludge for high-rate fermentative H_2 production. The CSABR system was operated at an HRT of 0.5–6 h and an influent sucrose concentration of 10–40 g COD/L. Formation of self-flocculated granular sludge occurred during operation at a short HRT. A high biomass concentration of up to 35.4 g VSS/L was achieved even though the reactor was operated at an extremely low HRT (i.e., 0.5 h). With a high feeding sucrose concentration (i.e., 30–40 g COD/L) and a short HRT (0.5 h), the CSABR reactor produced H_2 more efficiently with the highest volumetric rate of 15 L/L/h and an optimal yield of ca. 3.5 mol H_2/mol sucrose. By or after comparing the different configurations of reactors to draw a conclusion with regard to what configuration is better even under specific set conditions since many factors in particular H_2 yield and H_2 production rate depends significantly upon experimental conditions such as temperature, pH, substrate concentration, type of substrate, metal ion and HRT as well as long-term stability of the reactor and scale-up performance which directly influence on the economics of fermentative H_2 production.

9.6 Molecular Techniques for Microbial Community Characterization

For years, typically, microbial species were isolated and subsequently characterized based on their physiological and biochemical properties and culture-based studies to maintain and evaluate process conditions. However, these methods are limited by the lack of knowledge on preparation of suitable cultivation media. Most microorganisms, presumably more than 99% of naturally occurring microbes in a particular environment, cannot be readily isolated or cultured in laboratories with conventional enrichment techniques (Amann et al., 1995). This lack of ability in isolating unculturable microbes from a particular environment has been an obstacle in microbial community analyses. In addition, these cultivation methods are time-consuming, labor-intensive, and susceptible to bias toward nonpredominant culturable microorganisms (Vazquez and Poggi-Varaldo, 2009). As such, modern molecular techniques have revolutionized and expanded the scope of microbial systematics and physiology. The development and application of molecular techniques for microbial diversity analysis has proven superior to cultivation methods since the culture-independent molecular phylotype approaches, based on comparative analysis of gene sequences within the genome, can also infer evolutionary relationships among organisms studied (Hugenholtz and Pace, 1996). Because of its highly conserved nature, the bacterial 16S rRNA gene has emerged as one of the premier phylogenetic markers, allowing for bacterial diversity studies of isolated strains and uncultured populations present in complex microbial communities. In addition, further characterization was carried out using indirect methods (metabolite distribution, enrichment methods and microscopic examination, etc.) and some advanced molecular biological techniques (DGGE, DNA cloning analysis, dot-blot hybridization, terminal restriction fragment length polymorphism) in order to determine microbial composition in hydrogenogenic processes. The strategies and some of the methods/tools used for characterizing microbial communities of environmental samples without cultivation are outlined in Figure 9.6, and the detailed discussion can be found in Hugenholtz and Pace (1996).

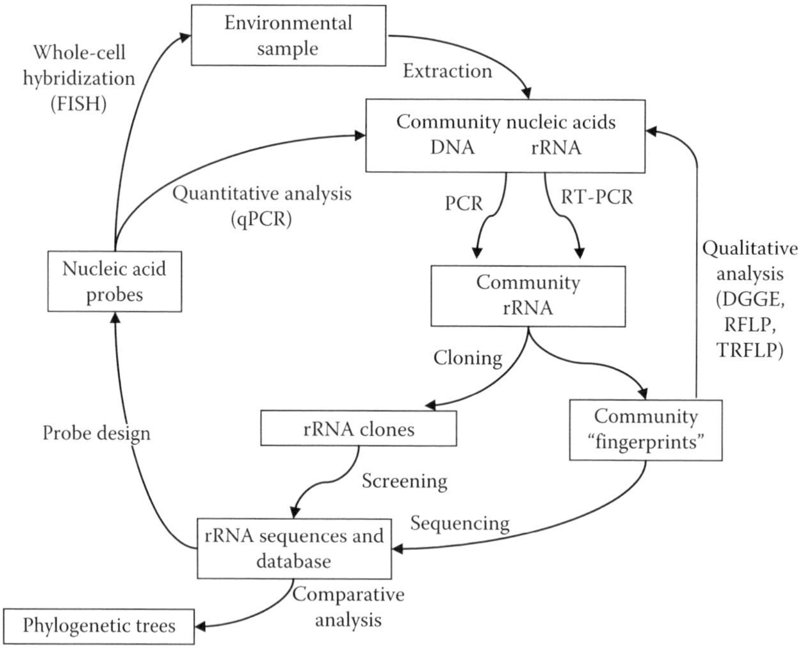

FIGURE 9.6
Strategies for characterizing microbial communities of environmental samples without cultivation. Abbreviations: DGGE, denaturing gradient gel electrophoresis; PCR, polymerase chain reaction; RFLP, restriction fragment length polymorphism; rRNA, ribosomal RNA; RT–PCR, reverse transcription PCR, qPCR, quantitative PCR, FISH, fluorescent in situ hybridization. (Modified from Hugenholtz, P. and Pace, N.R., *Trends Biotechnol.*, 14, 190, 1996.)

9.6.1 Phylogenetic Analysis of Microbial Community in Biohydrogen Reactors

Researchers have attempted, using culture-independent molecular phylotype methods, to characterize the complex microbial community present in bioreactors producing H_2 from pure substrates such as glucose, sucrose, and cellulose as well as organic waste. It was observed that H_2 production using anaerobic consortia has many advantages, the main one being that organic waste or wastewater can be used without sterilization, which makes the process economically beneficial. H_2 production using microbial consortia have some major factors that need to be taken into account: a developed microbial consortium is likely to be quite complex, with not every organism present directly producing H_2, and the species composition of the consortium might vary with substrate, but this has not been examined in a systematic way. To enhance the process performance and maintain attractive H_2 production, it is prudent to gain insight on the community structure using techniques such as DGGE and phylogenetic assignments made through analysis of 16S RNA. In addition, one way to determine which organisms that are present could potentially be responsible for the observed H_2 production is by probing the population for metabolism-specific genes.

Ueno et al. (2001), using 16S rDNA phylogenetic analyses, studied the microbial community of hydrogen-producing thermophilic anaerobic microflora enriched from sludge compost treating cellulose with or without peptone addition. In the chemostat reactor with peptone addition, most of the dominant isolates belonged to the cluster of the *Clostridium/Bacillus/Thermoanaerobacterium* subphylum of low G+C Gram-positive bacteria, including *Clostridium thermocellum, Clostridium cellulosi, Clostridium* sp., *Bacillus sporothermodurans,* and *Thermoanaerobacterium thermosaccharolyticum.* Without peptone addition, *Thermoanaerobacterium thermosaccharolyticum* and *Thermobacteroides acetoethylicus* were dominant but with a 50% reduced H_2 yield performance. Also, under thermophilic conditions but using a trickling biofilter fed with glucose, Ahn et al. (2005), based on 16S rDNA phylogenetic analyses, showed that *Thermoanaerobacterium thermosaccharolyticum* and *Mitsuokella jalaludinii* were the predominant microbes with a H_2 yield of 1.1 mol of H_2/mol glucose. For mesophilic conditions, Fang et al. (2002a) analyzed the microbial diversity of a mesophilic hydrogen-producing sludge with glucose as the substrate, using 16S rDNA–based techniques involving PCR amplification followed by DGGE screening, cloning, and sequencing. Phylogenetic analysis indicated that 64.6% of the major clones were closely related to *Clostridium* species including *C. cellulosi, C. acetobutylicum,* and *C. tyrobutyricum,* and 19% were closely related to *Citrobacter* spp. A subsequent microbial community characterization of hydrogen-producing granular sludge treating synthetic sucrose-containing wastewater detected 69% of cloned 16S rDNA sequences closely related to *Clostridium* species including *C. pasterurianum, C. tyrobutyricum,* and *C. acidisoli,* and 14% closely related to the *Bacillus and Staphylococcus* group including *Sporolactobacillus racemicus* (Fang et al., 2002b). Iyer et al. (2004) operated a CSTR, inoculated with heat-treated soil, to evaluate hydrogen-producing bacterial communities at different temperatures and hydraulic retention times (HRTs). Using PCR-based detection of bacterial populations by ribosomal intergenic spacer analysis (RISA), *Bacillus myxolacticus, Bacillus racemilacticus, Clostridium acidisoli,* and *Klebsiella ornithinolytica* were found to be dominant at an HRT of 30 h with a H_2 yield of 0.9 mol of H_2/mol glucose, while *Clostridium* sp., *C. acidisoli, C. acetobutylicum, Citrobacter braakii,* and *Enterobacter cloacae,* but mostly *Clostridium* species, were dominant at an HRT of 10 h with a yield of 1.8 mol H_2/mol glucose. Also, with a mesophilic CSTR inoculated with heat-treated sludge but fed with dual substrates, that is, glucose and peptone, Wang et al. (2003b), using 16S rDNA phylogenetic analyses for microbial diversity, found that 62% of total obtained clones were affiliated with the genus *Clostridium*, including *C. sporogenes*, *C. celerecrescens*, *C. butyricum*, *C. sartagoforme*, and *C. chauvoei*, 6% were affiliated with *Dendrosporobacter quercicolus* (formally *Clostridium quercicolum*), 3% were affiliated with the genus *Enterobacter*, and 2% were affiliated with the genus *Klebsiella*. Among the cloned *Clostridium* species, *C. sporogenes*, *C. celerecrescens*, and *C. butyricum*, representing 52% of relative abundance, were capable of producing H_2 from glucose fermentation (Hippe et al., 1992). Furthermore, *C. sporogenes* was revealed to metabolize various amino acids through the Stickland reaction (Stickland, 1934, 1935), and *C. celerecrescens* was able to hydrolyze certain proteins such as gelatin (Hippe et al., 1992), presumably because they were the predominant hydrogen-producing bacteria (HPB) in the CSTR fed with glucose and peptone.

In addition to pure substrates, several studies have investigated microbial communities of hydrogen-producing reactors that were capable of converting organic wastes to H_2. In a thermophilic hydrogen-producing CSTR fed with food waste (Shin and Youn, 2005), *Thermoanaerobacterium thermosaccharolyticum*, based on 16S rDNA phylogenetic analyses, was found to be the ubiquitously predominant HPB with a H_2 yield of 1–2.2 mol/mol hexose at different pH (5, 5.5, 6) studied. Chin et al. (2003) conducted an investigation on microbial diversity of HPB, using both culture-dependent and culture-independent approaches, and obtained 174 isolates from composts and hydrogen-producing reactors fed with different substrates including glucose, sucrose, cellulose, rice husk, wheat husk, and wasted yeast powder. Among these isolates, seven possessed superior hydrogen-producing ability, and, therefore, were further identified based on 16S rDNA phylogenetic analyses. Four out of seven superior hydrogen-producing isolates were confirmed as members of clostridial clusters I and II (Collins et al., 1994) including *C. tyrobutyricum*, *C. butyricum*, *C. pasteurinasum*, and *C. tertium*, two were closely related to *Klebsiella pneumoniae*, and the remaining one was closely related to *Megasphaera elsdenii*.

9.6.2 Fingerprint Methods

The microbial community characterization studies mentioned earlier, based on phylogenetic analyses, mostly relied on a one-time sampling during reactor operation. Since microbial populations change over time, monitoring temporal changes is critical to understand the performance of a reactor. In most studies, reactor operating parameters are varied to increase H_2 production. It is vital to identify the shift in microbial populations due to changes in operating conditions in order to fully understand a reactor system and to optimize reactor performance.

The accumulation of genetic information and the development of polymerization chain reaction (PCR)-based techniques allow us to obtain phylotypes from the environmental sample without cultivation or isolation and make it possible to better understand the microbial ecology of that specific sample. The separation or detection of subtle differences in PCR-amplified specific sequences may provide important information about the microbial community structure. Several different polymorphism-based procedures have been developed as DNA fingerprint procedures and applied to study microbial ecology systems (Rosenbaum and Riesner, 1987; Hayashi, 1992; Muyzer et al., 1993; Hallenbeck, 2009). With the development of DNA fingerprinting techniques, the analysis of complex microbial communities has advanced rapidly during recent years. Among them, denaturing gradient gel electrophoresis (DGGE) and terminal restriction fragment length polymorphism (T-RFLP) are some DNA fingerprinting techniques that have found popularity in the analysis of microbial diversity in anaerobic hydrogen-producing reactors (Lo et al., 2008b).

9.6.2.1 Denaturing Gradient Gel Electrophoresis

DGGE is one of the commonly used techniques to address questions related to diversity and dynamics of microbial communities. In this method, PCR-amplified 16S rRNA gene fragments from a bacterial community can be separated into discrete bands during electrophoresis in a polyacrylamide gel containing a linearly increasing gradient of DNA denaturant (mixture of urea and formamide) (Muyzer et al., 1993). The separation is based on the decreasing electrophoretic mobility of unwinding double-stranded DNA molecules, partially denatured according to their sequence (Figure 9.6). Partial denaturation causes unique bands to form in the gel. The attachment of a GC-rich sequence (GC clamp) to DNA fragments optimizes the detection of sequence variants, by generating only a single band for each denaturing fragment. The diversity of a microbial community can be visualized in terms of banding patterns in the electrophoresis gel. These individual bands can be excised, reamplified, and sequenced or hybridized with oligonucleotide probes to determine the composition of the bacterial community. Many researchers have applied the PCR–DGGE fingerprinting approach to study microbial communities in various environments such as ocean mats (Muyzer et al., 1993), activated sludge (Nielsen et al., 1999), and biofilms (Zhang and Fang, 2000). Recently, this technique was applied to characterize the microbial communities in hydrogen-producing reactors (Ueno et al., 2001; Fang et al., 2002a; Shin and Youn, 2004; Ahn et al., 2006; Wu et al., 2006). Based on PCR–DGGE analysis, Ueno et al. (2001) observed different predominant HPB communities present in hydrogen-producing reactors with and without peptone addition, before performing nucleotide sequences and phylogenetic analyses. By applying the PCR–DGGE fingerprinting approach to the samples obtained from thermophilic hydrogen-producing reactors fed with food wastes at different pH conditions, Shin and Youn (2005) demonstrated that the ubiquitously predominant HPB, *Thermoanaerobacterium thermosaccharolyticum*, was present at all tested pHs, but different *T. thermosaccharolyticum* strains were sensitive to the tested pHs (Figure 9.7). Ahn et al. (2006) performed the PCR–DGGE technique for the HPB samples collected from different heights of a thermophilic trickling biofilter, and the results suggested similar microbial populations present at these levels. In a high-rate hydrogen-producing bioreactor seeded with silicone-immobilized sludge and operated at different HRTs (0.5–6 h), Wu et al. (2006) observed a significant increase in H_2 production when the HRT was reduced to 0.5 h, presumably due to a corresponding HPB community shift, which was revealed with the PCR–DGGE procedure.

9.6.2.2 Terminal Restriction Fragment Length Polymorphism

The analysis of terminal restriction fragment length polymorphism (T-RFLP) is another more recently developed fingerprinting technique to assess the diversity of complex microbial communities. It allows rapid comparisons of community structure and diversity of different ecosystems (Liu et al., 1997). In this method, PCR is used to amplify a selected region of the 16S rRNA gene. One or both of the PCR primers are labeled with a fluorochrome. The resulting fluorescently labeled PCR products are digested with restriction enzymes, and an automated DNA sequencer determines the length and intensity of

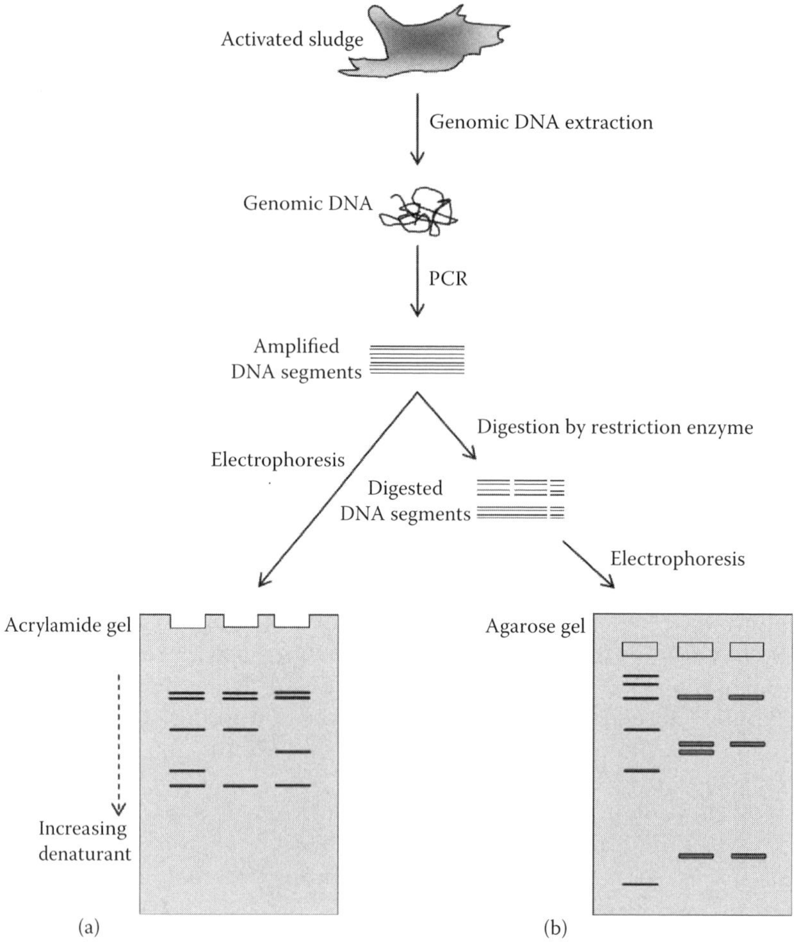

FIGURE 9.7
Schematic description of (a) DGGE and (b) RFLP analyses.

the resulting fluorescently labeled terminal fragments (Figure 9.7). Direct reference can be made to sequence databases to determine the composition of the microbial community (Marsh et al., 2000). The T-RFLP technique is gaining usage in microbial community characterization because of its rapidity and high resolving power. However, it has all the pitfalls associated with other molecular techniques in that it depends on efficient extraction of community DNA and on unbiased PCR amplifications (Figures 9.7 and 9.8). Raghavendra and Padmasree (2003) presented a T-RFLP fingerprinting methodology based on the clostridial 16S rDNA cluster classification defined by Collins et al. (1994) and Mitchell (1998) to monitor the dynamics and abundance of *Clostridium* species in hydrogen-producing reactors. By using the T-RFLP methodology to investigate microbial population dynamics of a continuous flow reactor, Duangmanee et al. (2003) found that *Clostridium* cluster I and II organisms were abundant in the reactor when the H_2 production was at its maximum and confirmed that *Clostridium* clusters I and II contained hydrogen-producing microorganisms capable of elevated levels of H_2 production. Moreover, a decrease in the levels of *Clostridium* clusters I and II coincided with an increase in the levels of *Clostridium* clusters III and XI, corresponding to a decrease in H_2 production. Recently, Sung et al. (2002) used

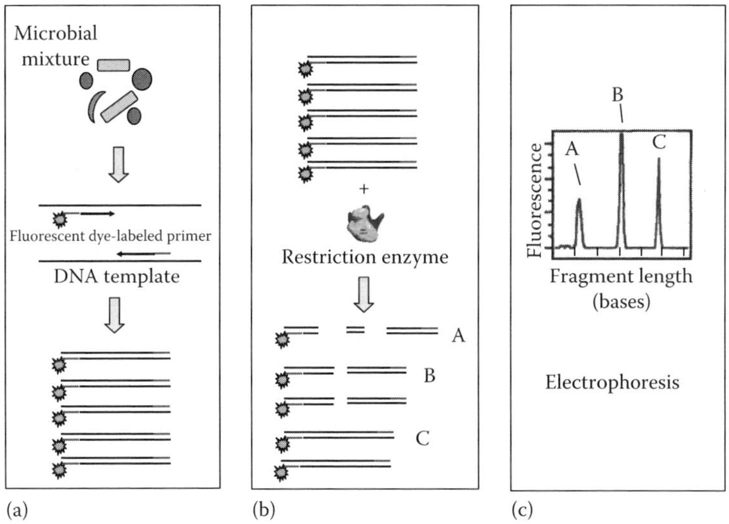

FIGURE 9.8
Schematic description of T-RFLP analyses: (a) PCR amplify and label target gene(s), (b) digest labeled PCR products, and (c) detect labeled terminal fragments.

terminal restriction fragment length polymorphism (T-RFLP) to identify mixed communities of H_2 producers in mesophilic continuous flow reactors using sucrose and heat-treated inocula. The reactor was equipped with an activation chamber that exposed a fraction of the settled sludge to a temperature of 90°C for 20 min. Their results indicated that two major groups of *Clostridium* species were dominant during the first 15 days of operation. The first dominant *Clostridium* group was composed of the following species: *C. beijerinckii, C. botulinum, C. putrificum,* and *C. sporogenes.* The second dominant population was identified to be *C. butyricum.* The authors found that a decrease in H_2 production was accompanied with a decrease in the total of *Clostridium* species and vice versa in the first 15 days of operation. However, after day 18, the two *Clostridium* species groups ceased to grow and the *Bacillus* species became dominant after day 22. They concluded that it was necessary to apply repeated heat treatments to maintain H_2 production in a continuous flow reactor.

9.7 Conclusions

The world population and consequently the energy demands are growing at an exponential rate; however, the impending shortage of energy resources and the negative impact on the environment due to the unsustainable use of fossil fuels are leading many scientists to search for alternative energy sources that are imperative and more advantageous. Among these energy resources, biohydrogen seems promising. Biohydrogen production has been established as a prospective alternative and an integral part of green sustainable energy. Biological systems have significant potential as an environmentally friendly means of producing H_2, a widely touted possible fuel of the future. However, a number of obstacles must be overcome if this potential is to be realized on a practical scale. One attractive route that was discussed in detail in this chapter is the use of dark fermentation. Several factors influencing dark fermentative H_2 production were summarized and analyzed in this review, and the effect of each factor on fermentative H_2 production and the advances in research on better fermentative H_2 production were briefly introduced and discussed. We have also reviewed various approaches to this problem using either pure cultures or microbial consortia in a variety of reactor configurations with different substrates and the tools of molecular biotechnology. Optimization of the bioreactor design and operation can be useful in maintaining H_2 production at a high and stable level. It was assumed that biohydrogen is the key to accelerate the coming of the *H_2 economy* and solve the problem of *global warming* and *shortage of fossil fuel*. Pure cultures and defined substrates have proven useful for probing limiting factors presented by different microbial physiologies whereas more engineering-type *black box* studies can highlight areas that need to be targeted in bioprocess development. In addition, it was observed that during dark H_2 fermentation, anaerobic bacteria could produce H_2 while converting organic substrates into volatile fatty acids and alcohols. To achieve better energy yield and lower chemical oxygen demand (COD) level in the effluent, these soluble metabolites (organic acids and alcohols) can be further utilized via photofermentation using photosynthetic bacteria, such as purple nonsulfur bacteria, resulting in more H_2 production as well as higher COD removal. Thus, an integration process combining dark and photo H_2 fermentation could be an effective and efficient energy process to increase H_2 production capacity, and enhancing energy recovery is thus a future prospective.

References

Adams MWW. 1990. The metabolism of hydrogen by extremely thermophilic sulphur-dependent bacteria. *FEMS Microbiol Rev* 75:219–238.

Ahn JM, Kim BC, Gu MB. 2006. Characterization of gltA: LuxCDABE fusion in *Escherichia coli* as a toxicity biosensor. *Biotechnol Bioprocess Eng* 11:513–521.

Akkerman I, Janssen M, Rocha J, Wijffels RH. 2002. Photobiological hydrogen production: Photochemical efficiency and bioreactor design. *Int J Hydrogen Energy* 27:1195–1208.

Amann RI, Ludwig W, Schleifer KH. 1995. Phylogenetic identification and in situ detection of individual microbial cells without cultivation. *Microbiol Rev* 59:143–169.

Angenent LT, Karim K, Al-Dahhan MH, Wrenn BA, Domíguez-Espinosa R. 2004. Production of bioenergy and biochemicals from industrial and agricultural wastewater. *Trends Biotechnol* 22:477–485.

Argun H, Kargi F, Kapdan IK, Oztekin R. 2008. Biohydrogen production by dark fermentation of wheat powder solution: Effects of C/N and C/P ratio on hydrogen yield and formation rate. *Int J Hydrogen Energy* 33:1813–1819.

Armor JN. 1999. The multiple roles for catalysis in the production of H_2. *Appl Catal A Gen* 176:159–176.

Arooj MF, Han SK, Kim SH, Kim DH, Shin HS. 2008. Continuous biohydrogen production in a CSTR using starch as a substrate. *Int J Hydrogen Energy* 33:3289–3294.

Asada Y, Miyake J. 1999. Photobiological hydrogen production. *J Biosci Biotechnol* 88:1–6.

Atif AAY, Fakhru'l-Razi A, Ngan MA, Morimoto M, Iyuke SE, Veziroglu NT. 2005. Fed batch production of hydrogen from palm oil mill effluent using anaerobic microflora. *Int J Hydrogen Energy* 30(13–14):1393–1397.

Bai MD, Cheng SS, Tseng IC. 2001. Biohydrogen produced due to peptone degradation by pretreated seed sludge. *The IWA Asia-Pacific Regional Conference (WaterQual 2001)*, Fukuoka, Japan, pp. 315–320.

Bailey JE, Ollis DF. 1986. *Biochemical Engineering Fundamentals*, 2nd edn. McGraw-Hill, New York.

Benemann J. 1996. Hydrogen biotechnology: Progress and prospects. *Nat Biotechnol* 14:1101–1103.

Bisaillon A, Turcot J, Hallenbeck PC. 2006. The effect of nutrient limitation on hydrogen production by batch cultures of *Escherichia coli. Int J Hydrogen Energy* 31:1504–1508.

Brock TD, Madigan MT, Martinko JM, Parker J. 1994. *Biology of Microorganisms*, 7th edn. Prentice-Hall, Englewood Cliffs, NJ.

Bockris J. 2002. The origin of ideas on a hydrogen economy and its solution to the decay of the environment. *Int J Hydrogen Energy* 27:731–740.

Bossel U. 2003. Well-to-wheel studies, heating values, and the energy conservation principle. European Fuel Cell Forum, Oberrohrdorf, Switzerland.

Cai ML, Liu JX, Wei YS. 2004. Enhanced biohydrogen production from sewage sludge with alkaline pretreatment. *Environ Sci Technol* 38:3195–3202.

Calli B, Boënne W, Vanbroekhoven K. 2006. Bio-hydrogen potential of easily biodegradable substrate through dark fermentation. In: *Proceedings of the 16th World Hydrogen Energy Conference*, Lyon, France, pp. 215–216.

Camilli M, Pedroni PM. 2005. Comparison of the performance of three different reactors for biohydrogen production via dark anaerobic fermentations. In: *Proceedings of the International Hydrogen Energy Congress and Exhibition*, BHP 250, CD-ROM, Istanbul, Turkey.

Chang FY, Lin CY. 2004. Biohydrogen production using an up-flow anaerobic sludge blanket reactor. *Int J Hydrogen Energy* 29:33–39.

Chang JS, Lee KS, Lin PJ. 2002. Biohydrogen production with fixed-bed bioreactors. *Int J Hydrogen Energy* 27:1167–1174.

Chang JS, Yang SM. 2006. Application of artificial neural networks coupled with sequential pseudo-uniform design to optimization of membrane reactors for hydrogen production. *J Chin Inst Chem Eng* 37:395–400.

Chen CC, Chen HP, Wu JH, Lin CY. 2008a. Fermentative hydrogen production at high sulfate concentration. *Int J Hydrogen Energy* 33:573–1578.

Chen CC, Lin CY. 2003. Using sucrose as a substrate in an anaerobic hydrogen producing reactor. *Adv Environ Res* 7:695–699.

Chen CC, Lin CY, Chang JS. 2001. Kinetics of hydrogen production with continuous anaerobic cultures utilizing sucrose as the limiting substrate. *Appl Microbiol Biotechnol* 57:56–64.

Chen CC, Lin CY, Lin MC. 2002. Acid-base enrichment enhances anaerobic hydrogen production process. *Appl Microbiol Biotechnol* 58:224–228.

Chen SD, Lee KS, Lo YC, Chen WM, Wu JF, Lin CY, Chang JS. 2008b. Batch and continuous biohydrogen production from starch hydrolysate by *Clostridium* species. *Int J Hydrogen Energy* 33(7):1803–1812.

Chen WH, Chen SY, Khanal SK, Sung S. 2006. Kinetic study of biological hydrogen production by anaerobic fermentation. *Int J Hydrogen Energy* 31(15):2170–2178.

Chen WM, Tseng ZJ, Lee KS, Chang JS. 2005. Fermentative hydrogen production with *Clostridium butyricum* CGS5 isolated from anaerobic sewage sludge. *Int J Hydrogen Energy* 30:1063–1070.

Cheng SS, Chang SM, Chen ST. 2002. Effects of volatile fatty acids to a thermophilic anaerobic hydrogen fermentation process degrading peptone. *Water Sci Technol* 46:209–214.

Cheong DY, Hansen CL. 2006. Acidogenesis characteristics of natural, mixed anaerobes converting carbohydrate-rich synthetic wastewater to hydrogen. *Process Biochem* 41(8):1736–1745.

Chin HL, Chen ZS, Chou CP. 2003. Fedbatch operation using *Clostridium acetobutylicum* suspension culture as biocatalyst for enhancing hydrogen production. *Biotechnol Prog* 19:383–388.

Cohen A, van Gemert JM, Zoetemeyer RJ, Breure AM. 1984. Main characteristics and stoichiometric aspects of acidogenesis of soluble carbohydrate containing wastewaters. *Proc Biochem* 19:228–232.

Collet C, Adler N, Schwitzguebel JP, Peringer P. 2004. Hydrogen production by *Clostridium thermolacticum* during continuous fermentation of lactose. *Int J Hydrogen Energy* 29(14):1479–1485.
Collins MD, Lawson PA, Willems A, Cordoba JJ, Fernandez-Garayzabal J, Garcia P, Cai J, Hippe H, Farrow JAE. 1994. The phylogeny of the genus *Clostridium*: Proposal of five new genera and eleven new species combinations. *Int J Syst Bacteriol* 44:812–826.
Dabrock B, Bahl H, Gottschalk G. 1992. Parameters affecting solvent production by *Clostridium pasteurium. Appl Environ Microbiol* 58:1233–1239.
Das D, Veziroglu TN. 2001. Hydrogen production by biological processes: A survey of literature. *Int J Hydrogen Energy* 26:13–28.
Das D, Veziroglu TN. 2008. Advances in biological hydrogen production processes. *Int J Hydrogen Energy* 33:6046–6057.
Das D. 2009. Advances in biohydrogen production processes: An approach towards commercialization. *Int J Hydrogen Energy* 34:7349–7357.
Datar R, Huang J, Maness PC, Mohagheghi A, Czernik S, Chornet E. 2007. Hydrogen production from the fermentation of corn stover biomass pretreated with a steam-explosion process. *Int J Hydrogen Energy* 32:932–939.
Demirbas A. 2007. Progress and recent trends in biofuels. *Prog Energy Combust Sci* 33:1–18.
DeVrije T, deHaas GG, Tan GB, Keijsers ERP, Claassen PAM. 2002. Pretreatment of *Miscanthus* for hydrogen production by *Thermotoga elfii. Int J Hydrogen Energy* 27:1381–1390.
Duangmanee T, Padmasiri S, Simmons JJ, Raskin L, Sung S. 2002. Hydrogen production by anaerobic microbial communities exposed to repeated heat treatment. In: *WEFTEC 75th Annual Conference and Exhibition of the Water Environment Federation,* Chicago, IL; *Water Environment Federation,* Alexandria, VA.
Dunn S. 2002. Perspectives towards a hydrogen future. *Cogen On Site Power Product* 3:55–60.
Elam CC, Gregoire Padro CE, Sandrock G, Luzzi A, Lindblad P, Hagen EF. 2003. Realizing the hydrogen future: The international energy agency's efforts to advance hydrogen energy technologies. *Int J Hydrogen Energy* 28:601–607.
Evvyernie D, Morimoto K, Karita S, Kimura T, Sakka K, Ohmiya K. 2001. Conversion of chitinous waste to hydrogen gas by *Clostridium paraputrificum* M-21. *J Biosci Bioeng* 91:339–343.
Fabiano B, Perego P. 2002. Thermodynamic study and optimization of hydrogen production by *Enterobacter aerogenes. Int J Hydrogen Energy* 27:149–156.
Fan KS, Kan NR, Lay JJ. 2006a. Effect of hydraulic retention time on anaerobic hydrogenesis in CSTR. *Bioresour Technol* 97(1):84–89.
Fan YT, Zhang YH, Zhang SF, Hou HW, Ren BZ. 2006b. Efficient conversion of wheat straw wastes into biohydrogen gas by cow dung compost. *Bioresour Technol* 97(3):500–505.
Fang HHP, Li CL, Zhang T. 2006. Acidophilic biohydrogen production from rice slurry. *Int J Hydrogen Energy* 31:683–692.
Fang HHP, Liu H. 2002. Effect of pH on hydrogen production from glucose by a mixed culture. *Bioresour Technol* 82(2):87–93.
Fang HHP, Liu H, Zhang T. 2002. Characterization of a hydrogen producing granular sludge. *Biotechnol Bioeng* 78(1):44–52.
Fang HHP, Liu H, Zhang T. 2002a. Characterization of a hydrogen-producing granular sludge. *Biotechnol Bioeng* 78:44–52.
Fang HHP, Zhang T, Liu H. 2002. Microbial diversity of mesophilic hydrogen producing sludge. *Appl Microbiol Biotechnol* 58:112–118.
Fardeau ML, Faudon C, Cayol JL, Magot M, Patel BKC, Ollivier B. 1996. Effect of thiosulphate as electron acceptor on glucose and xylose oxidation by *Thermoanaerobacterfinnii* and a *Thermoanaerobacter* sp. isolated from oil field water, *Res Microbiol* 147(3):159–165.
Ferchichi M, Crabbe E, Gil GH, Hintz W, Almadidy A. 2005. Influence of initial pH on hydrogen production from cheese whey. *J Biotechnol* 120(4):402–409.
Field CB, Campbell JE, Lobell DB. 2008. Biomass energy: The scale of the potential resource. *Trends Ecol Evol* 23:65–72.

Francese A, Cordoba P, Duran J, Sineriz F. 1998. High upflow velocity and organic loading rate improves granulation in upflow anaerobic sludge blanket reactors. *World J Microbiol Biotechnol* 14(3):337–341.

Gander M, Jefferson B, Judd S. 2000. Aerobic, MBRs for domestic wastewater treatment: A review with cost considerations. *Sep Purif Technol* 18:119–130.

Gavala HN, Skiadas IV, Ahring BK. 2006. Biological hydrogen production in suspended and attached growth anaerobic reactor systems. *Int J Hydrogen Energy* 31:1164–1175.

Gatze L, de Zeeuw W, Ouborg E. 1981. Anaerobic treatment of wastes containing methanol and higher alcohols. *Water Res* 15:171–182.

Ghirardi ML, Zhang L, Lee JW, Flynn T, Seibert M, Greenbaum E, Melis. 2000. Microalgae: A green source of renewable H_2. *Trend Biotechnol* 18:506–511.

Giallo J, Gaudin C, Belaich JP. 1985. Metabolism and solubilization of cellulose by *Clostridium cellulolyticum* H10. *Appl Environ Microbiol* 49:1216–1221.

Goldemberg J. 2007. Ethanol for a sustainable energy future. *Science* 315:808–810.

Goldemberg J, Johansson TB. 2004. World energy assessment overview: 2004 update. United Nations Development Programme, New York, http://www.undp.org/energy/weaover2004.htm.

Gong CS, Cao NU, Du J, Tsao GT. 1999. Ethanol production by renewable sources. *Adv Biochem Eng Biotechnol* 65:207–241.

Guo L, Li XM, Bo X, Yang Q, Zeng GM, Liao DX, Liu JJ. 2008. Impacts of sterilization, microwave and ultrasonication pretreatment on hydrogen producing using waste sludge. *Bioresour Technol* 99:3651–3658.

Gustavo DV, Sonia A, Felipe AM, Antonio de LR, Luis MRC, Elías RF. 2009. Fermentative biohydrogen production: Trends and perspectives. *Rev Environ Sci BioTechnol* (doi 10.1007/s11157-007-9122-7).

Hallenbeck PC. 2005. Fundamentals of the fermentative production of hydrogen. *Water Sci Technol* 52:21–29.

Hallenbeck PC. 2009. Fermentative hydrogen production: Principles, progress, and prognosis. *Int J Hydrogen Energy* 34:7379–7389.

Hallenbeck PC, Benemann JR. 2002. Biological hydrogen production; fundamentals and limiting processes. *Int J Hydrogen Energy* 27:1185–1193.

Hallenbeck, PC, Ghosh D, Skonieczny MT, Yargeau V. 2009. Microbiological and engineering aspects of biohydrogen production. *Indian J Microbiol* 49:48–59.

Hawkes FR, Dinsdale R, Hawkes DL, Hussy I. 2002. Sustainable fermentative hydrogen production: Challenges for process optimization. *Int J Hydrogen Energy* 27:1339–1347.

Hawkes FR, Hussy I, Kyazze G, Dinsdale R, Hawkes DL. 2007. Continuous dark fermentative hydrogen production by mesophilic microflora: Principles and progress. *Int J Hydrogen Energy* 32:172–184.

Hayashi K. 1992. Genetic analysis: Review PCR-SSCP: A method for detection of mutations. *Biomol Eng* 9(3):73–79.

Hippe H, Andreesen JR, Gottschalk G. 1992. The genus *Clostridium*—Nonmedical. In *The Prokaryotes*, 2nd edn., A. Balows, H. C. Trüper, M. Dworkin, W. Harder, and K.-H. Schleifer, eds. Springer, New York.

Horiuchi J, Shimizu T, Kanno T, Kobayashi M. 1999. Dynamic behavior in response to pH shift during anaerobic acidogenesis with a chemstat culture. *Biotechnol Tech* 13:155–157.

Horiuchi JI, Shimizu T, Tada K, Kanno T, Kobayashi M. 2002. Selective production of organic acids in anaerobic acid reactor by pH control. *Bioresour Technol* 82:209–213.

Hugenholtz P, Pace NR. 1996. Identifying microbial diversity in the natural environment: A molecular phylogenetic approach. *Trends Biotechnol* 14:190–197.

Hulshoff Pol LW, Lettinga G. 1986. New technologies for anaerobic wastewater treatment. *Water Sci Technol* 18(12):41–53.

Hussy I, Hawkes FR, Dinsdale R, Hawkes DL. 2003. Continuous fermentative hydrogen production from a wheat starch co-product by mixed microflora. *Biotechnol Bioeng* 84(6):619–626.

Hussy I, Hawkes FR, Dinsdale R, Hawkes DL. 2005. Continuous fermentative hydrogen production from sucrose and sugarbeet. *Int J Hydrogen Energy* 30(5):471–483.
Hwu CS, van Lier JB, Lettinga G. 1998. Physicochemical and biological performance of expanded granular sludge bed reactors treating long-chain fatty acids. *Process Biochem* 33(1):75–81.
Imai T, Ukita M, Liu J, Sekine M, Nakanishi H, Fukagawa M. 1997. Advanced start up of UASB reactors by adding of water absorbing polymer. *Water Sci Technol* 36:399–406.
Iyer P, Bruns MA, Zhang H, van Ginkel S, Logan BE. 2004. H_2 producing microbial communities from a heat treated soil inoculum. *Appl Microbiol Biotechnol* 66:166–173.
James WE, Jha S, Sumulong L, Son HH, Hasan R, Khan ME, Sugiyarto G, Zhai F. 2008. *Food Prices and Inflation in Developing Asia: Is Poverty Reduction Coming to an End?* Asian Development Bank, Manila, Philippines. http://www.adb.org/reports/food-prices-inflation/food-prices-inflation.pdf.
Jeison D, Chamy R. 1999. Comparison of the behaviour of expanded granular sludge bed (EGSB) and upflow anaerobic sludge blanket (UASB) reactors in dilute and concentrated wastewater treatment. *Water Sci Technol* 40:91–97.
Jo JH, Lee DS, Park D, Choe WS, Park JM., 2008. Optimization of key process variables for enhanced hydrogen production by *Enterobacter aerogenes* using statistical methods. *Bioresour Technol* 99:2061–2066.
Kadar Z, Vrije TD, van Noorden GE, Budde MAW, Szengyel Z, Reczey K, Claassen PAM. 2004. Yields from glucose, xylose, and paper sludge hydrolysate during hydrogen production by the extreme thermophile *Caldicellulosiruptor saccharolyticus. Appl Biochem Biotechnol* 113–116(12):497–508.
Kanai T, Imanaka H, Nakajima A, Uwamori K, Omori Y, Fukui T. Atomi H, Imanaka T. 2005. Continuous hydrogen production by the hyperthermophilic archaeon, *Thermococcus kodakaraensis* KOD1. *J Biotechnol* 116:271–282.
Kapdan IK, Kargi F. 2006. Biohydrogen production from waste materials. *Enzyme Microbial Technol* 38:569–582.
Kawagoshi Y, Hino N, Fujimoto A, Nakao M, Fujita Y, Sugimura S, Furukawa K. 2005. Effect of inoculum conditioning on hydrogen fermentation and pH effect on bacterial community relevant to hydrogen production. *J Biosci Bioeng* 100:524–530.
Keeney R, Hertel TW. 2008. *The Indirect Land Use Impacts of US Biofuel Policies: The Importance of Acreage, Yield, and Bilateral Trade Responses*. Center for Global Trade Analysis, Purdue University, West Lafayette, IN. http://www.gtap.agecon.purdue.edu/resources/download/3904.pdf.
Kennedy JL, Droste RL. 1985. Startup of anaerobic downflow stationary fixed film (DSFF) reactors. *Biotechnol Bioeng* 27:1152–1165.
Khanal SK, Chen WH, Li L, Sung S. 2004. Biological hydrogen production: Effects of pH and intermediate products. *Int J Hydrogen Energy* 29(11):1123–1131.
Kim JO, Kim YH, Ryu JY, Song BK, Kim IH, Yeom SH. 2005. Immobilization methods for continuous hydrogen gas production biofilm formation versus granulation. *Process Biochem* 40:1331–1337.
Kim MS, Oh YK, Yun YS, Lee DY. 2006. Fermentative hydrogen production from anaerobic bacteria using a membrane bioreactor. In: *Proceedings of the 16th World Hydrogen Energy Conference*. Lyon, France, p. 50.
Koh LP, Ghazoul J. 2008. Biofuels, biodiversity, and people: Understanding the conflicts and finding opportunities. *Biolog Conserv* 141:2450–2460.
Koku H, Eroğlu İ, Gündüz U, Yücel M, Türker L. 2002. Aspects of the metabolism of hydrogen production by *Rhodobacter sphaeroides. Int J Hydrogen Energy* 27:1315–1329.
Kotay SM, Das D. 2006. Feasibility of biohydrogen production from sewage sludge using defined microbial consortium. In: *Proceedings of the 16th World Hydrogen Energy Conference*. Lyon, France, pp. 209–210.
Kotay SM, Das D. 2008. Biohydrogen as a renewable energy resource—Prospects and potentials. *Int J Hydrogen Energy* 33:258–263.
Kotsopoulos TA, Zeng RJ, Angelidaki I. 2006. Biohydrogen production in granular up-flow anaerobic sludge blanket (UASB) reactors with mixed cultures under hyper-thermophilic temperature (70°C). *Biotechnol Bioeng* 94(2):296–302.

Kraemer JT, Bagley DM. 2008. Optimisation and design of nitrogen-sparged fermentative hydrogen production bioreactors. *Int J Hydrogen Energy* 33:6558–6565.

Kumar N, Das D. 2000. Enhancement of hydrogen production by *Enterobacter cloacae* IIT-BT 08. *Process Biochem* 35:589–593.

Kumar N, Das D. 2001. Continuous hydrogen production by immobilized *Enterobacter cloacae* IIT-BT 08 using lignocellulosic materials as solid matrices. *Enzyme Microb Technol* 29:280–287.

Kyazze G, Dinsdale R, Hawkes FR, Guwy AJ, Premier GC, Donnison IS. 2008. Direct fermentation of fodder maize, chicory fructans and perennial ryegrass to hydrogen using mixed microflora. *Bioresour Technol* 99(18):8833–8839.

Lay JJ. 2000. Modeling and optimization of anaerobic digested sludge converting starch to hydrogen. *Biotechnol Bioeng* 68:269–278.

Lay JJ. 2001. Biohydrogen generation by mesophilic anaerobic fermentation of microcrystalline cellulose. *Biotechnol Bioeng* 74:280–287.

Lay JJ, Fan KS, Hwang J, Chang JI, Hsu PC. 2005. Factors affecting hydrogen production from food wastes by *Clostridium*-rich composts. *J Environ Eng* 131:595–602.

Lay JJ, Lee YJ, Noike T. 1999. Feasibility of biological hydrogen production from organic fraction of municipal solid waste. *Water Res* 33(11):2579–2586.

Lee KS, Hsu YF, Lo YC, Lin PJ, Lin CY, Chang JS. 2008. Exploring optimal environmental factors for fermentative hydrogen production from starch using mixed anaerobic microflora. *Int J Hydrogen Energy* 33:1565–1572.

Lee KS, Lin PJ, Chang JS. 2006b. Temperature effects on biohydrogen production in a granular sludge bed induced by activated carbon carriers. *Int J Hydrogen Energy* 31(4):465–472.

Lee KS, Lin PJ, Fang K, Chang JS. 2007. Continuous hydrogen production by anaerobic mixed microflora using a hollow-fiber microfiltration membrane bioreactor. *Int J Hydrogen Energy* 32:950–957.

Lee KS, Lo YC, Lin PJ, Chang JS. 2006a. Improving biohydrogen production in a carrier-induced granular sludge bed by altering physical configuration and agitation pattern of the bioreactor. *Int J Hydrogen Energy* 31(12):1648–1657.

Lee KS, Lo YS, Lo YC, Lin PJ, Chang JS. 2003. H_2 production with anaerobic sludge using activated-carbon supported packed-bed bioreactors. *Biotechnol Lett* 25:133–138.

Lee KS, Lo YS, Lo YC, Lin PJ, Chang JS. 2004b. Operation strategies for biohydrogen production with a high-rate anaerobic granular sludge bed bioreactor. *Enzyme Microb Technol* 35(6–7):605–612.

Lee KS, Wu JF, Lo YS, Lo YC, Lin PJ, Chang JS. 2004a. Anaerobic hydrogen production with an efficient carrier-induced granular sludge bed bioreactor. *Biotechnol Bioeng* 87(5):648–657.

Lee MJ, Zinder SH. 1988. Hydrogen partial pressures in a thermophilic acetate-oxidizing methanogenic co-culture. *Appl Environ Microbiol* 54:1457–1461.

Lee YJ, Miyahara T, Noike T. 2001. Effect of iron concentration on hydrogen fermentation. *Bioresour Technol* 80:227–231.

Lee YJ, Miyahara T, Noike T. 2002. Effect of pH on microbial hydrogen fermentation. *J Chem Technol Biotechnol* 77:694–698.

Levin DB, Pitt L, Love M. 2004. Biohydrogen production: Prospects and limitations to practical application. *Int J Hydrogen Energy* 29:173–185.

Li CL, Fang HHP. 2007a. Fermentative hydrogen production from wastewater and solid wastes by mixed cultures. *Crit Rev Env Sci Technol* 37:1–39.

Li CL, Fang HHP. 2007b. Inhibition of heavy metals on fermentative hydrogen production by granular sludge. *Chemosphere* 67:668–673.

Liang TM. 2003. Application of membrane separation on anaerobic hydrogen-producing processes. Doctor dissertation. National Cheng Kung University, ROC.

Liang TM, Cheng SS, Wu KL. 2002. Behavioral study on hydrogen fermentation reactor installed with silicone rubber membrane. *Int J Hydrogen Energy* 27:1157–1165.

Liang TM, Wu KL, Cheng SS. 2001. Hydrogen production of chloroform inhibited granular sludge. *The IWA Asia-Pacific Regional Conference (WaterQual 2001)*, Fukuoka, Japan, pp. 863–868.

Licht FO. 2008. World Ethanol & Biofuels Report. Agra Informa Ltd., Kent, U.K. http://www.agra-net.com/portal/puboptions.jsp?Option=menu&pubId=sag072.

Lin CY, Cheng CH. 2006. Fermentative hydrogen production from xylose using anaerobic mixed microflora. *Int J Hydrogen Energy* 31:832–840.

Lin CY, Chang CC, Hung CH. 2008a. Fermentative hydrogen production from starch using natural mixed cultures. *Int J Hydrogen Energy* 33(10):2445–2453.

Lin CY, Chang RC. 1999. Hydrogen production during the anaerobic acidogenic conversion of glucose. *J Chem Technol Biotechnol* 74(6):498–500.

Lin CY, Chang RC. 2004. Fermentative hydrogen production at ambient temperature. *Int J Hydrogen Energy* 29(7):715–720.

Lin CY, Cheng CH. 2006. Fermentative hydrogen production from xylose using anaerobic mixed microflora. *Int J Hydrogen Energy* 31(7):832–840.

Lin CY, Chou CH. 2004. Anaerobic hydrogen production from sucrose using an acid-enriched sewage sludge microflora. *Eng Life Sci* 4:66–70.

Lin CY, Hung WC. 2008. Enhancement of fermentative hydrogen/ethanol production from cellulose using mixed anaerobic cultures. *Int J Hydrogen Energy* 33(14):3660–3667.

Lin CY, Jo CH. 2003. Hydrogen production from sucrose using an anaerobic sequencing batch reactor process. *J Chem Technol Biotechnol* 78:678–684.

Lin CY, Lay CH. 2004a. Carbon/nitrogen-ratio effect on fermentative hydrogen production by mixed microflora. *Int J Hydrogen Energy* 29:41–45.

Lin CY, Lay CH. 2004b. Effects of carbonate and phosphate concentrations on hydrogen production using anaerobic sewage sludge microflora. *Int J Hydrogen Energy* 29:275–281.

Lin CY, Lay CH. 2005. A nutrient formulation for fermentative hydrogen production using anaerobic sewage sludge microflora. *Int J Hydrog Energy* 30(3):285–292.

Lin CY, Lee CY, Tseng IC, Shiao IZ. 2006. Biohydrogen production from sucrose using base-enriched anaerobic mixed microflora. *Process Biochem* 41(4):915–919.

Lin CY, Shei SH. 2008. Heavy metal effects on fermentative hydrogen production using natural mixed microflora. *Int J Hydrogen Energy* 33:587–593.

Lin CY, Wu CC, Hung CH. 2008b. Temperature effects on fermentative hydrogen production from xylose using mixed anaerobic cultures. *Int J Hydrogen Energy* 33:43–50.

Liu GZ, Shen JQ. 2004. Effects of culture and medium conditions on hydrogen production from starch using anaerobic bacteria. *J Biosci Bioeng* 98:251–256.

Liu H, Fang HHP. 2002. Hydrogen production from wastewater by acidogenic granular sludge. *Water Sci Technol* 47:153–158.

Liu J, Chen JM, Cihlar J, Park WM. 1997. A process-based boreal ecosystem productivity simulator using remote sensing inputs. *Remote Sens Environ* 62(2):158–175.

Liu Y, Xu HL, Yang SF, Tay JH. 2003. Mechanisms and models for anaerobic granulation in upflow anaerobic sludge blanket reactor. *Water Res* 37:661–673.

Liu Y, Yu P, Song X, Qu Y. 2008. Hydrogen production from cellulose by co-culture of *Clostridium thermocellum* JN4 and *Thermoanaerobacterium thermosaccharolyticum* GD17. *Int J Hydrogen Energy* 33:2927–2933.

Lo YC, Bai MD, Chen WM, Chang JS. 2008a. Cellulosic hydrogen production with a sequencing bacterial hydrolysis and dark fermentation strategy. *Bioresour Technol.* 99:8299–8303.

Lo YC, Chen WM, Hung CH, Chen SD, Chang JS. 2008b. Dark H_2 fermentation from sucrose and xylose using H_2-producing indigenous bacteria: Feasibility and kinetic studies. *Water Res* 42:827–842.

Lo YC, Saratale GD., Chen WM, Bai MD, Chang JS. 2009. Isolation of cellulose-hydrolytic bacteria and applications of the cellulolytic enzymes for cellulosic biohydrogen production. *Enzyme Microb Technol* 44(6–7):417–425.

Logan BE. 2004. Extracting hydrogen and electricity from renewable resources. *Environ Sci Technol* 38(9):160A–167A.

Logan BE, Oh SE, Kim IS, Van Ginkel S. 2002. Biological hydrogen production measured in batch anaerobic respirometers. *Environ Sci Technol* 36:2530–2535.

Ma H, Brwman CN, Davis RH. 2000. Membrane fouling reduction by backpulsing and surface modification. *J Membr Sci* 173:191–200.
Maddy J, Cherryman S, Hawkes FR, Hawkes DL, Dinsdale RM, Guwy AJ, Premier GC, Cole S. 2003. *Hydrogen.* University of Glamorgan, Pontypridd, U.K.
Malina JF, Pohland JFG. 1992. *Design of Anaerobic Processes for the Treatment of Industrial and Municipal Wastes.* Technomic Pub Co, Lancaster, PA.
Mallette MF, Reece P, Dawes EA. 1974. Culture of *Clostridium pasteurianum* in defined medium and growth as a function of sulfate concentration. *Appl Microbiol* 28:999–1003.
Marsh TL, Saxman P, Cole J, Tiedje J. 2000. Terminal restriction fragment length polymorphism analysis program, a web-based research tool for microbial community analysis. *Appl Environ Microbiol* 66:3616–3620.
Matthies C, Kuhner CH, Acker G, Drake HL. 2001. *Clostridium uliginosum* sp. nov., a novel acid-tolerant, anaerobic bacterium with connecting filaments. *Int J Syst Evol Microbiol* 51(3):1119–1125.
Minnan L, Jinli H, Xiaobin W, Huijuan X, Jinzao C, Chuannan L, Fengzhang Z, Liangshu X. 2005. Isolation and characterization of a high H_2-producing strain *Klebsialle oxytoca* HP1 from a hot spring. *Res Microbiol* 156:76–81.
Mitchell WJ. 1998. Physiology of carbohydrate to solvent conversion by clostridia. *Adv Microbial Physiol* 39:31–130.
Mizuno O, Dinsdale R, Hawkes FR, Kawkas DL, Noike T. 2000. Enhancement of hydrogen production from glucose by nitrogen gas sparging. *Biotechnol Bioeng* 73(1):59–65.
Mohan SV, Bhaskar YV, Sharma PN. 2007. Biohydrogen production from chemical wastewater treatment in biofilm configured reactor operated in periodic discontinuous batch mode by selectively enriched anaerobic mixed consortia. *Water Res* 41:2652–1664.
Momirlan M, Veziroglu T. 2002. Current status of hydrogen energy. *Renew Sustain Energy Rev* 6:141–179.
Morgan JW, Evison LM, Forster CF. 1991. Changes to the microbial ecology in anaerobic digesters treating ice cream wastewater during start-up. *Water Res* 25:639–653.
Morimoto M, Atsuko M, Atif AAY, Ngan MA, Fakhru'l-Razi A, Iyuke SE, Bakir AM. 2004. Biological production of hydrogen from glucose by natural anaerobic microflora. *Int J Hydrogen Energy* 29:709–713.
Mu Y, Wang G, Yu HQ. 2006a. Kinetic modeling of batch hydrogen production process by mixed anaerobic cultures. *Bioresour Technol* 97(11):1302–1307.
Mu Y, Wang G, Yu HQ. 2006b. Response surface methodological analysis on biohydrogen production by enriched anaerobic cultures. *Enzyme Microb Technol* 38(7):905–913.
Mu Y, Yu HQ. 2006. Biological hydrogen production in a UASB reactor with granules. I: Physicochemical characteristics of hydrogen-producing granules. *Biotechnol Bioeng* 94(5):980–987.
Mu Y, Zheng XJ, Yu HQ, Zhu RF. 2006c. Biological hydrogen production by anaerobic sludge at various temperatures. *Int J Hydrogen Energy* 31(6):780–785.
Muyzer G, Waal CDE, Uitierlinden AG. 1993. Profiling of complex microbial populations by denaturing gradient gel electrophoresis analysis of polymerase chain reaction-amplified genes coding for 16S rRNA. *Appl Environ Micrbiol* 59:695–700.
Nandi R, Sengupta S. 1998. Microbial production of hydrogen: An overview. *Crit Rev Microbiol* 24:61–84.
Nath K, Das D. 2003. Hydrogen from biomass. *Curr Sci* 85:265–271.
Nath K, Das D. 2004. Improvement of fermentative hydrogen production—Various approach. *Appl Microbiol Biotechnol* 65:520–529.
Nishio N, Nakashimada Y. 2004. High rate production of hydrogen/methane from various substrates and wastes. *Adv Biochem Eng Biotechnol* 90:63–87.
Nguyen TAD, Kim JP, Kim MS, Oh YK, Sim SJ. 2008. Optimization of hydrogen production by hyperthermophilic eubacteria, *Thermotoga maritima* and *Thermotoga neapolitana* in batch fermentation. *Int J Hydrogen Energy* 33(5):1483–1488.

UNDP. 2000. *Energy and the Challenge of Sustainability*, World Energy Assessment, United Nations Development Programme, New York.
Valdez-Vazquez I, Elvira RL, Esparza-García F, Cecchi F, Poggi-Varaldo HM. 2005. Semi-continuous solid substrate anaerobic reactors for H_2 production from organic waste: Mesophilic versus thermophilic regime. *Int J Hydrogen Energy* 30:1383–1391.
van Niel EWJ, Budde MAW, de Haas GG, van der Wal FJ, Claassen PAM, Stams AJM. 2002a. Distinctive properties of high hydrogen producing extreme thermophiles, *Caldicellulosiruptor saccharolyticus* and *Thermotoga elfiii. Int J Hydrog Energy* 27:1391–1398.
van Niel EWJ, Claassen PAM, Stams AJM. 2002b. Substrate and product inhibition of hydrogen production by the extreme thermophile *Caldicellulosiruptor saccharolyticus. Biotechnol Bioeng* 81:255–262.
Nielsen AT, Liu WT, Filipe C, Grady L Jr, Molin S, Stahl DA. 1999. Identification of a novel group of bacteria in sludge from a deteriorated biological phosphorus removal reactor. *Appl Environ Microbiol* 65:1251–1258.
Nowak J, Florek M, Kwiatek W, Lekki J, Chevallier P, Zieba E. 2005. Composite structure of wood cells in petrified wood. *Mater Sci Eng* 25:119–30.
Ntaikou I, Gavala HN, Kornaros M, Lyberatos G. 2008. Hydrogen production from sugars and sweet sorghum biomass using *Ruminococcus albus. Int J Hydrogen Energy* 33:1153–1163.
Oh SE, Iyer P, Bruns MA, Logan BE. 2004. Biological hydrogen production using a membrane bioreactor. *Biotechnol Bioeng* 87(1):119–127.
Oh YK, Park MS, Seol EH, Lee SJ, Park S. 2003. Isolation of hydrogen-producing bacteria from granular sludge of an upflow anaerobic sludge blanket reactor. *Biotechnol Bioprocess Eng* 8:54–57.
O-Thong S, Prasertsan P, Intrasungkha N, Dhamwichukorn S, Birkeland NK. 2008a. Optimization of simultaneous thermophilic fermentative hydrogen production and COD reduction from palm oil mill effluent by *Thermoanaerobacterium*-rich sludge. *Int J Hydrogen Energy* 33(4):1221–1231.
O-Thong S, Prasertsan P, Karakashev D, Angelidaki I. 2008b. Thermophilic fermentative hydrogen production by the newly isolated *Thermoanaerobacterium thermosaccharolyticum* PSU-2. *Int J Hydrogen Energy* 33(4):1204–1214.
Palazzi E, Fabino B, Perego P. 2000. Process development of continuous hydrogen production by *Enterobacter aerogenes* in a packed column reactor. *Bioprocess Eng* 22:205–213.
Pan CM, Fan YT, Xing Y, Hou HW, Zhang ML. 2008. Statistical optimization of process parameters on biohydrogen production from glucose by *Clostridium* sp. Fanp2. *Bioresour Technol* 99:3146–3154.
Park W, Hyun SH, Oh SE, Logan BE, Kim IS. 2005. Removal of headspace CO_2 increases biological hydrogen production. *Environ Sci Technol* 39(12):4416–4420.
Podestá JJ, Navarro AMG, Estrella CN, Esteso MA. 1997. Electrochemical measurements of trace concentrations of biological hydrogen produced by *Enterobacteriaceae. Inst Pasteur* 148:87–93.
Puppan D. 2002. Environmental evaluation of biofuels. *Period Polytech Ser Soc Man Sci* 10:95–116.
Rachman MA, Nakashimada Y, Kakizono T, Nishio N. 1998. Hydrogen production with high yield and high evolution rate by self-flocculated cells of *Enterobacter aerogenes* in a packed-bed reactor. *Appl Microbiol Biotechnol* 49:450–454.
Ragauskas AJ, Williams CK, Davison BH, Britovsek G, Cairney J, Eckert CA, Frederick WJJR, Hallett JP, Leak DJ, Liotta CL. 2006. The path forward for biofuels and biomaterials. *Science* 311:484–489.
Raghavendra AS, Padmasree K. 2003. Beneficial interactions of mitochondrial metabolism with photosynthetic carbon assimilation. *Trend Plant Sci* 8:546–553.
Rajeshwari KV, Balakrishnan M, Kansal A, Lata K, Kishore VVN. 2000. State-of-the-art of anaerobic digestion technology for industrial wastewater treatment. *Renew Sust Energ Rev* 4(2):135–156.
Ren N, Li J, Li B, Wang Y, Liu S. 2006. Biohydrogen production from molasses by anaerobic fermentation with a pilot-scale bioreactor system. *Int J Hydrogen Energy* 31(15):2147–2157.
Ren NQ, Chen XL, Zhao D. 2001. Control of fermentation types in continuous-flow acidogenic reactors: Effects of pH and redox potential. *J Harbin Inst Technol (New Ser)* 8(2):116–119.
Ren NQ, Wang BZ, Huang JC. 1997. Ethanol-type fermentation from carbohydrate in high rate acidogenic reactor. *Biotechnol Bioeng* 54(5):428–433.

Rosenbaum V, Riesner D. 1987. Temperature-gradient gel electrophoresis; thermodynamic analysis of nucleic acids and proteins in purified form and in cellular extracts. *Biophys Chem* 26:235–246.
Salerno MB, Park W, Zuo Y, Logan BE. 2006. Inhibition of biohydrogen production by ammonia. *Water Res* 40:1167–1172.
Saratale GD, Chen SD, Lo YC, Saratale RG, Chang JS. 2008. Outlook of biohydrogen production from lignocellulosic feedstock using dark fermentation—A review. *J Sci Ind Res* 67:962–979.
Sasikala K, Ramana CV, Rao PR, Kovacs KL. 1993. Anoxygenic phototrophic bacteria: Physiology and advances in hydrogen technology. *Adv Appl Microbiol* 38:211–295.
Schmidt JE, Ahring BK. 1996. Granular sludge formation in upflow anaerobic sludge blanket (UASB) reactors. *Biotechnol Bioeng* 49(3):229–246.
Schubert C. 2006. Can biofuels finally take center stage? *Nat Biotechnol* 24:777–784.
Shin HS, Youn JH. 2005. Conversion of food waste into hydrogen by thermophilic acidogenesis. *Biodegradation* 16:33–44.
Shin HS, Youn JH, Kim SH. 2004. Hydrogen production from food waste in anaerobic mesophilic and thermophilic acidogenesis. *Int J Hydrogen Energy* 29:1355–1363.
Show KY, Zhang ZP, Lee DJ. 2008. Design of bioreactors for biohydrogen production. *J Sci Ind Res* 67:941–949.
Stickland LH. 1934. The chemical reactions by which *Cl. sporogenes* obtains its energy. *Biochem J* 28(5):1746–1759.
Stickland LH. 1935. Studies in the metabolism of the strict anaerobes (Genus: *Clostridium*): The oxidation of alanine by *Cl. sporogenes*. IV. The reduction of glycine by *Cl. sporogenes*. *Biochem J* 29(4):889–898.
Slade R, Bauen A, Shah N. 2009. The commercial performance of cellulosic ethanol supply-chains in Europe. *Biotechnol Biofuels* (doi:10.1186/1754–6834-2-3).
Sparling R, Risbey D, Poggi-Varaldo HM. 1997. Hydrogen production from inhibited anaerobic composters. *Int J Hydrogen Energy* 22:563–566.
Sung S, Raskin L, Duangmanee T, Padmasiri S, Simmons JJ. 2002. Hydrogen production by anaerobic microbial communities exposed to repeated heat treatment. *Proceedings of the 2002 U.S.* DOE Hydrogen Program Review NREL/CP-610-32405.
Taguchi F, Mizukami N, Yamada K, Hasegawa K, Saito TT. 1995. Direct conversion of cellulosic materials to hydrogen by *Clostridium* sp. strain No. 2. *Enzyme Microbiol Technol* 17:147–150.
Taguchi F, Yamada K, Hasegawa K, Takisaito T, Hara K. 1996. Continuous hydrogen production by *Clostridium* sp. Strain No. 2 from cellulose hydrolysate in aqueous two phase system. *J Ferment Bioeng* 82:80–83.
Tanisho S, Ishiwata Y. 1994. Continuous hydrogen production from molasses by bacterium *Enterobacter aerogenes. Int J Hydrogen Energy* 19:807–812.
Teo KC, Xu HL, Tay JH. 2000. Molecular mechanism of granulation. II: Proton translocating activity. *J Environ Eng* 126:411–418.
Turner J, Sverdrup G, Mann MK, Maness PC, Kroposki B, Ghirardi M, Evans RJ, Blake D. 2008. Renewable hydrogen production. *Int J Energy Res* 32:379–407.
Ueno Y, Haruta S, Ishii M, Igarashi Y. 2001a. Characterization of a microorganism isolated from the effluent of hydrogen fermentation by microflora. *J Biosci Bioeng* 92:397–400.
Ueno Y, Haruta S, Ishii M, Igarashi Y. 2001b. Microbial community in anaerobic hydrogen-producing microflora enriched from sludge compost. *Appl Microbiol Biotechnol* 57(11):555–562.
Ueno Y, Kawai T, Sato S, Otsuka S, Morimoto M. 1995. Biological production of hydrogen from cellulose by natural anaerobic microflora. *J Fermen Bioeng* 79:395–397.
Ueno Y, Otsuka S, Morimoto M. 1996. Hydrogen production from industrial wastewater by anaerobic microflora in chemostate culture. *J Ferment Bioeng* 82(2):194–197.
Valdez-Vazquez I, Poggi-Varaldo HM. 2009. Hydrogen production by fermentative consortia. *Renew Sustain Energy Rev* 13:1000–1013.
Valdez-Vazquez I, Rios-Leal E, Carmona-Martinez A, Munoz-Paez KM, Poggi-Varaldo HM. 2006. Improvement of biohydrogen production from solid wastes by intermittent venting and gas flushing of batch reactors headspace. *Environ Sci Technol* 40(10):3409–3415.

Valdez-Vazquez I, Rios-Leal E, Esparza-Garcia F, Cecchi F, Poggi-Varaldo HA. 2005a. Semi-continuous solid substrate anaerobic reactors for H_2 production from organic waste: Mesophilic versus thermophilic regime. *Int J Hydrogen Energy* 30:1383–1391.

Van den Berg L, Kennedy KJ, Samson R. 1985. Anaerobic downflow stationary fixed film reactor: Performance under steady-state and non-steady conditions. *Water Sci Technol* 17(1):89–102.

Van Ginkel S, Sung S, Lay JJ. 2001. Biohydrogen production as a function of pH and substrate concentration. *Environ Sci Technol* 35:4726–4730.

Van Ginkel SW, Logan B. 2005. Increased biological hydrogen production with reduced organic loading. *Water Res* 39:3819–3826.

Vazquez IV, Sparling R, Risbey D, Rinderknecht SN, Poggi-Varaldo HM. 2005. Hydrogen generation via anaerobic fermentation of paper mill wastes. *Bioresour Technol* 96:1907–1913.

Vijayaraghavan K, Ahmad D. 2006. Biohydrogen generation from palm oil mill effluent using anaerobic contact filter. *Int J Hydrogen Energy* 31(10):1284–1291.

Vijayaraghavan K, Ahmad D, Ibrahim MKB. 2006. Biohydrogen generation from jackfruit peel using anaerobic contact filter. *Int J Hydrogen Energy* 31(5):569–579.

Vijayaraghavan K, Ahmad D, Soning C. 2007. Bio-hydrogen generation from mixed fruit peelwaste using anaerobic contact filter. *Int J Hydrogen Energy* 32(18):4754–4760.

Vijayaraghavan K, Soom MAM. 2007. Trends in biological hydrogen production—A review. *Int J Hydrogen Energy*.

Wang A, Ren N, Shi Y, Lee DJ. 2008. Bioaugmented hydrogen production from microcrystalline cellulose using co-culture-*Clostridium acetobutylicum* X9 and *Ethanoigenens harbinense* B49. *Int J Hydrogen Energy* 33(2):912–917.

Wang CC, Chang CW, Chu CP, Lee DJ, Chang BV, Liao CS. 2003b. Producing hydrogen from wastewater sludge by *Clostridum bifermentans. J Biotechnol* 102:83–92.

Wang CC, Chang CW, Chu CP, Lee DJ, Chang BV, Liao CS, Tay JH. 2003a. Using filtrate of waste biosolids to effectively produce bio-hydrogen by anaerobic fermentation. *Water Res* 37(11):2789–2793.

Wang JL, Wan W. 2008. Effect of Fe^{2+} concentrations on fermentative hydrogen production by mixed cultures. *Int J Hydrogen Energy* 33:1215–1220.

Wang JL, Wan W. 2009. Factors influencing fermentative hydrogen production: A review. *Int J Hydrogen Energy* 34:799–811.

Wang L, Zhou Q, Li FT. 2006. Avoiding propionic acid accumulation in the anaerobic process for biohydrogen production. *Biomass Bioenergy* 30(2):177–182.

Wang XJ, Ren NQ, Xiang WS, Guo WQ. 2007. Influence of gaseous end-products inhibition and nutrient limitations on the growth and hydrogen production by hydrogen-producing fermentative bacterial B49. *Int J Hydrogen Energy* 32:748–754.

Wen C, Huang X, Qian Y. 1999. Domestic wastewater treatment using an anaerobic bioreactor coupled with membrane filtration. *Process Biochem* 35:335–340.

Woodward J, Orr M, Cordray K, Greenbaum E. 2000. Enzymatic production of biohydrogen. *Nature* 405:1014–1015.

Wu JH, Lin CY. 2004. Biohydrogen production by mesophilic fermentation of food wastewater. *Water Sci Technol* 49:223–228.

Wu KJ, Saratale GD, Lo YC, Chen SD, Chen WM, Tseng ZJ, Chang JS. 2008a. Fermentative production of 2,3 butanediol, ethanol and hydrogen with *Klebsiella* sp. isolated from sewage sludge. *Bioresour Technol* 99:7966–7970.

Wu SY, Hung CH, Lin CN, Chen HW, Lee AS, Chang JS. 2006. Fermentative hydrogen production and bacterial community structure in high-rate anaerobic bioreactors containing silicone-immobilized and self-flocculated sludge. *Biotechnol Bioeng* 93(5):934–946.

Wu SY, Hung CH, Lin CY, Lin PJ, Lee KS, Lin CN, Chang FY, Chang JS. 2008b. HRT-dependent hydrogen production and bacterial community structure of mixed anaerobic microflora in suspended, granular and immobilized sludge systems using glucose as the carbon substrate. *Int J Hydrogen Energy* 33:1542–1549.

Wu SY, Lin CN, Chang JS. 2003. Hydrogen production with immobilized sewage sludge in three-phase fluidized-bed bioreactors. *Biotechnol Prog* 19:828–832.
Wu SY, Lin CN, Chang JS, Chang JS. 2005. Biohydrogen production with anaerobic sludge immobilized by ethylene-vinyl acetate copolymer. *Int J Hydrogen Energy* 30(13–14):1375–1381.
Wu SY, Lin CN, Chang JS, Lee KS, Lin PJ. 2002. Microbial hydrogen production with immobilized sewage sludge. *Biotechnol Prog* 18:921–926.
Wu SY, Lin CY, Lee KS, Hung CH, Chang JS, Lin PJ, Chang FY. 2008c. Dark fermentative hydrogen production from xylose in different bioreactors using sewage sludge microflora. *Energy Fuels* 22:113–119.
Xing DF, Ren NQ, Wang AJ, Li QB, Feng YJ, Ma F. 2008. Continuous hydrogen production of auto-aggregative *Ethanoligenens harbinense* YUAN-3 under non-sterile condition. *Int J Hydrogen Energy* 33:1489–1495.
Yang H, Shao P, Lu T, Shen J, Wang D, Xu Z, Yuan X. 2006. Continuous bio-hydrogen production from citric acid wastewater via facultative anaerobic bacteria. *Int J Hydrogen Energy* 31(10):1306–1313.
Yang H, Shen J. 2006. Effect of ferrous iron concentration on anaerobic bio-hydrogen production from soluble starch. *Int J Hydrogen Energy* 31:2137–2146.
Yeonghee A, Park E-J, Oh Y-K, Park S, Webster G, Weightman AJ. 2005. Biofilm microbial community of a thermophilic trickling biofilter used for continuous biohydrogen production. *FEMS Microbiol Lett* 249:31–38.
Yoda M, Kitagawa M, Miyaji Y. 1989. Granular sludge formation in the anaerobic expanded micro carrier process. *Water Sci Technol* 21:109–122.
Yokoi H, Maki R, Hirose J, Hayashi S. 2002. Microbial production of hydrogen from starch-manufacturing wastes. *Biomass Bioenerg* 22:389–395.
Yokoi H, Ohkawara T, Hirose J, Hayashi S, Takasaki Y. 1995. Characteristics of hydrogen production by aciduric *Enterobacter aerogenes* strain HO-39. *J Ferment Bioeng* 80:571–574.
Yokoi H, Saitsu AS, Uchida H, Hirose J, Hayashi S, Takasaki Y. 2001. Microbial hydrogen production from sweet potato starch residue. *J Biosci Bioeng* 91:58–63.
Yokoi H, Tokushige T, Hirose J, Hayashi S, Takasaki Y. 1997. Hydrogen production by immobilized cells of aciduric *Enterobacter aerogenes* strain HO-39. *J Ferment Bioeng* 83:481–484.
Yokoyama H, Waki M, Moriya N, Yasuda T, Tanaka Y, Haga K. 2007b. Effect of fermentation temperature on hydrogen production from cow waste slurry by using anaerobic microflora within the slurry. *Appl Microbiol Biotechnol* 74:474–483.
Yokoyama H, Waki M, Ogino A, Ohmori H, Tanaka Y. 2007a. Hydrogen fermentation properties of undiluted cow dung. *J Biosci Bioeng* 104:82–85.
Yu HQ, Mu Y. 2006. Biological hydrogen production in a UASB reactor with granules. II: Reactor performance in 3-year operation. *Biotechnol Bioeng* 94(5):988–995.
Yu HQ, Tay JH, Fang HHP. 1999. Effects of added powdered and granular activated carbons on start-up performance of UASB reactors. *Environ Technol* 20:1095–1101.
Yu HQ, Zhu ZH, Hu WR, Zhang HS. 2002. Hydrogen production from rice winery wastewater in an upflow anaerobic reactor by using mixed anaerobic cultures. *Int J Hydrogen Energy* 27:1359–1365.
Zhang T, Fang HHP. 2000. Digitization of DGGE (denaturing gradient gel electrophoresis) profile and cluster analysis of microbial communities. *Biotechnol Lett* 22:399–405.
Zhang T, Liu H, Fang HHP. 2003. Biohydrogen production from starch in wastewater under thermophilic condition. *J Environ Manage* 69:149–156.
Zhang Y, Liu G, Shen J. 2005. Hydrogen production in batch culture of mixed bacteria with sucrose under different iron concentrations. *Int J Hydrogen Energy* 30:855–860.
Zhang Y, Shen J. 2006. Effect of temperature and iron concentration on the growth and hydrogen production of mixed bacteria. *Int J Hydrogen Energy* 31(4):441–446.
Zhang ZP, Show KY, Tay JH, Liang DT, Lee DJ. 2008. Biohydrogen production with anaerobic fluidized bed reactors—A comparison of biofilm-based and granule-based systems. *Int J Hydrogen Energy* 33:1559–1564.
Zhang ZP, Show KY, Tay JH, Liang DT, Lee DJ, Jiang WJ. 2006. Effect of hydraulic retention time on biohydrogen production and anaerobic microbial community. *Process Biochem* 41:2118–2123.

Zhang ZP, Show KY, Tay JH, Liang DT, Lee DJ, Jiang WJ. 2007. Rapid formation of hydrogen-producing granules in an anaerobic continuous stirred tank reactor induced by acid incubation. *Biotechnol Bioeng* 96:1040–1050.

Zheng XJ, Yu HQ. 2004. Biological hydrogen production by enriched anaerobic cultures in the presence of copper and zinc. *J Environ Sci Health Part A Toxic/Hazard Subst Environ Eng* 39(1):89–101.

Zheng XJ, Yu HQ. 2005. Inhibitory effects of butyrate on biological hydrogen production with mixed anaerobic cultures. *J Environ Manage* 74(1):65–70.

Zhu Y, Yang ST. 2004. Effect of pH on metabolic pathway shift in fermentation of xylose by *Clostridium tyrobutyricum*. *J Biotechnol* 110(2):143–157.

Zoetemeyer RJ, Arnoldy P, Cohen A, Boelhouwer C. 1982a. Infuence of temperature on the anaerobic acidification of glucose in a mixed culture forming part of a two-stage digestion process. *Water Res* 16(3):313–321.

Zoetemeyer RJ, Van den Heuvel JC, Cohen A. 1982b. pH influence on acidogenic dissimilation of glucose in an anaerobic digester. *Water Res* 16:303–311.

10

Biological Hydrogen Production: Light-Driven Processes

Hisham Hafez
University of Western Ontario
Greenfield Specialty Alcohols

George Nakhla
University of Western Ontario

Hesham El Naggar
University of Western Ontario

Gamal Ibrahim
University of Menoufia

Said S.E.H. Elnashaie
University of British Columbia
New Mexico Technical University

CONTENTS

10.1 Introduction

The conventional hydrogen production methods involve fossil fuel reforming. This process usually generates large amounts of CO_2. Alternative methods of hydrogen generation include electrolysis of water, biophotolysis, and biological production from organic waste materials. In biological hydrogen production, bacteria or microalgae are used to process the organic material through metabolic action and produce hydrogen. Biological hydrogen production has two main advantages over the conventional methods: it generates less greenhouse gases and treats wastes rich in organics (e.g., biomass residues and domestic or food industry wastewaters), thus reducing their negative impact on the environment.

Although some observations of hydrogen production by microalgae and bacteria extend back over 100 years, basic research in this field started only in the late 1920s with bacterial H_2 production and in the 1940s with microalgal hydrogen production. Applied research and development (R&D) started only in the early 1970s. Most applied R&D has emphasized photosynthetic processes. Over the past quarter century, numerous studies have been focused on microbial H_2 production, but advances toward practical applications have been minimal. This chapter reviews the various types of biohydrogen processes and discusses some of their potential practical limitations.

10.2 Biological Hydrogen Production

Biological hydrogen production from renewable sources (biomass, water, and organic wastes) has the potential to meet the growing demand for energy. It offers a feasible means for sustainable supply of H_2 with low pollution and high efficiency, thereby considered a promising ecofriendly energy source. Biological hydrogen production can be achieved by anaerobic and photosynthetic microorganisms using carbohydrate-rich and nontoxic raw materials. Under anaerobic conditions, hydrogen is produced as a by-product during conversion of organic wastes into organic acids, which can subsequently be used for methane generation. The acidogenic phase of anaerobic digestion of wastes can be manipulated to improve hydrogen production.

The main processes for biohydrogen production include the following:

1. Light-driven processes
2. Dark fermentation
3. Hybrid (two-stage integrated) systems

Anaerobic (or dark) fermentation and photosynthetic degradation are the two most widely studied biohydrogen production techniques. Anaerobic fermentation is advantageous due to its rapid hydrogen evolution rate, while in the photosynthetic process, complete conversion of organic matter is theoretically possible. Either process, however, suffers from relatively low yields when used independently. For biohydrogen production to be economically feasible, hydrogen yields on dissolved organic material should be approximately 60%–80%. Combining the two methods, either directly or in a two-stage series-type configuration, has then the potential to increase the hydrogen yield to achieve this conversion target. The rest of this chapter is focused on light-driven processes and hybrid processes.

10.2.1 Light-Driven Processes

10.2.1.1 General Considerations

Biological hydrogen production using light-driven processes can be classified into four different groups (Levin et al. 2004):

1. Direct biophotolysis
2. Indirect biophotolysis
3. Photofermentation

It is clear that the first two bioprocesses are non-biowaste-based processes, where the direct photolysis depends on the splitting of water molecules to hydrogen and oxygen and the indirect photolysis includes two main steps: the first step is utilizing CO_2 to build up biomolecules like glucose and the second step is the decomposing of these molecules to hydrogen. The third bioprocess is a biowaste-based process.

In light-driven biological hydrogen production, microorganisms are utilized to convert solar energy to hydrogen. The diffuse nature of solar energy and the consequent low energy density places severe economic restrictions on the efficiency of this process. Hallenbeck and Benemann (2002) estimated that \$10 worth of H_2/m^2/year would utilize an average solar radiation of 5 kWh/m^2/day, which corresponds to 6.6 GJ/year and represents a conversion efficiency of 10%. Even with this overly optimistic estimate, the US Department of Energy (US DOE) puts the price for H_2 at \$15/GJ. This renders the light-driven process uneconomical in view of the required expenditures for infrastructure (construction of facilities) and operation (considering energy input for mixing, gas exchange and compression, and cooling, etc.). Obviously, lower efficiencies translate to increased bioreactor surface areas and consequently cost for the production per unit output of hydrogen fuel. In spite of the importance of economic feasibility, only relatively few studies on photobiological hydrogen production reported conversion efficiencies, which typically fall well below 1% (i.e., <\$1 H_2/m^2/year).

In theory, photosynthesis in general and microalgal cultures in particular could achieve as much as a 10% total light energy conversion into a primary product, such as CO_2 fixed into biomass or even H_2. However, such expectations are based on data obtained under low-light conditions and do not account for the rather low efficiencies typically observed at the full sunlight intensity to which algal cultures would be exposed to in practice. Note that the dark reactions are the rate-limiting factor (i.e., the rate of transfer of electrons between the two photosystems [photosystem I, PSI, and photosystem II, PSII]), which is roughly 10 times lower than the rate of light captured by photosynthetic pigments (e.g., chlorophyll). This results in up to 90% of the photons captured by the photosynthetic apparatus under full sunlight not being used in photosynthesis but rather dissipated as heat or fluorescence. Similar considerations apply to photosynthetic bacteria whose pigment antenna systems absorb light at high rates for high light intensities, which cannot be productively used due to slow dark reactions. This energy waste is only somewhat ameliorated in dense cultures where most cells are not close to the culture surface and thus are not subjected to maximum solar intensity. This so-called light saturation effect is a major reason that algal productivity is not nearly as high as those projected from extrapolations of laboratory data at low light intensity.

The light saturation effect can be addressed by employing different techniques such as rapid mixing, dilution of light incident on the surface of the algal cultures, and using algal mutants with reduced chlorophyll contents that would absorb fewer photons. Rapid mixing may create a *flashing light effect* in eddies of turbulence surrounding the algal cells, wherein millisecond flashes of high-intensity light, followed by an approximately five- to tenfold longer dark period, maximize overall photon use efficiency. This is explained as follows.

During a brief light flash, each photosynthetic unit (containing several hundred chlorophyll molecules) captures only one photon. The subsequent longer dark phase allows for a slower transfer of electrons between photosystems. Achieving this flashing light effect with algal mass cultures, however, requires very high mixing power inputs, rendering the process impractical. Nonetheless, some mixing is required in algal mass cultures to supply nutrients (in particular, CO_2) in order to prevent settling and to maintain uniformity of the culture. From a practical perspective, mixing velocities of 25–35 cm/s should not be exceeded, in particular for fuel production processes, as the power inputs required for mixing increase as a cubic function of velocity. The effect of such gentler mixing regimes on algal productivity is somewhat controversial. Weissman (1988) observed no enhancement of productivity over a wide range of mixing velocities (5–60 cm/s), while controlling other potentially confounding factors (pH, O_2 concentrations). However, other researchers noted that mixing (and the associated periodic light regime experienced by individual algal cells in mass cultures) has a profound effect on productivity.

Another approach to overcome the light saturation effect is using light attenuation devices to transfer sunlight into the depths of a dense algal culture. The simplest approach is to arrange photobioreactors in vertical arrays to reduce direct sunlight. However, this configuration increases the area of required photobioreactors proportionally, which is the limiting economic factor in any photobiological fuel production process. Another alternative is the use of optical fiber photobioreactors that collect light energy by employing large concentrating mirrors and piping light energy into small photobioreactors through optical fibers. More recently, this concept became the centerpiece of significant R&D efforts carried out in Japan on microalgae biofixation of CO_2 and greenhouse gas mitigation, including H_2 production, notwithstanding its technical and economic challenges.

Using mutants of algal cells with reduced pigment content (smaller amounts of *antenna* chlorophyll or other *light-harvesting* pigments such as phycobili proteins in cyanobacteria) offers a practical alternative to rapid mixing. A photosynthetic apparatus with less light-harvesting chlorophyll absorbs fewer photons at high light intensity, and thus wastes fewer photons, suggesting that such algal strains do not readily exist in nature, as they are intermittently exposed to both high light intensity (at or near the surface) and low light intensity (deeper in the culture). This results in an evolutionary selection for strains with more light-harvesting pigments per photosynthetic reaction center. Generally, these cell types can capture more photons per cell than strains with a reduced ratio of antenna chlorophyll per reaction center. However, such cells waste most of the captured photons near the culture surface, thus decreasing the overall culture productivity compared to cultures of cells containing a reduced antenna to reaction center pigment ratio. Even though similar observations were made over 50 years ago, the concept of increasing productivity by reducing antenna size has remained rather dormant.

This approach has recently been revived in Japan and the United States. In the Japanese project, laboratory experiments on cultures operating at high light intensity showed that microalgal mutants with reduced antenna sizes exhibited a 50% increase in productivity, compared to the wild type. This increase demonstrated the potential of this approach. Another study adopted a physiological approach using cultures of the green alga *Dunaliella salina*. Under stress conditions, these algae exhibited both damaged PSII centers and reduced antenna size. However, PSII centers were repaired before their antenna size increased when the stress was relieved. This allowed for a brief period of relatively high rates of photosynthesis under high light intensity. More recently, antenna mutants have been isolated and studied in the United States, with limited success in enhancing efficiency at high light intensity. The use of algal mutants believed to be deficient in the PSI complex but reportedly still able to produce H_2 and fix CO_2 is an alternative approach. This suggested the possibility of a *PSII-only* photosynthesis, which could double the overall solar conversion efficiency by requiring only one photon per electron transported from water to CO_2 or H_2. However, that work proved to be irreproducible.

Currently, photosynthetic efficiency (PE) of about 3% of solar energy converted into algal biomass can be achieved considering microalgae biomass production in outdoor cultures. The ultimate goal of photosynthetic H_2 production is therefore to increase solar conversion efficiency by at least threefold. This goal is crucial to the realization of practical microalgae processes for H_2 production.

10.2.1.2 Direct Biophotolysis

The potential use of microorganisms for biological production of hydrogen as an energy resource places hydrogen metabolism at the forefront of research on biological hydrogen production. The concept of *direct biophotolysis* encompasses light-driven simultaneous O_2 evolution on the oxidizing side of PSII and H_2 production on the reducing side of PSI, with a maximum H_2–O_2 (mol–mol) ratio of 2:1. Such a reaction with green algae could serve to provide a clean, renewable, and economically viable H_2 fuel. This process of photosynthetic hydrogen production does not entail CO_2 fixation or energy storage into cellular metabolites. Hydrogenase is the name given to the family of enzymes that catalyze the reversible oxidation of hydrogen into its elementary particle constituents, two protons (H^+) and two electrons.

These enzymes are divided into three classes depending on the metal content at the active site: Fe hydrogenase, NiFe hydrogenase, and nitrogenase. Fe hydrogenase enzyme is used in the biophotolysis processes whereas photofermentation processes utilize nitrogenase. The Fe hydrogenases are enzymes that exhibit high sensitivity to O_2 and light. This may pose an additional problem for their biotechnological use in photosynthetic organisms such as algae.

Biological hydrogen can be generated from plants by biophotolysis of water using microalgae (green algae and cyanobacteria), fermentation of organic compounds, and photodecomposition of organic compounds by photosynthetic bacteria. Photosynthetic production of hydrogen from water is a biological process that can convert sunlight into useful, stored chemical energy by the following general reaction (Equation 10.1):

$$2H_2O \rightarrow 2H_2 + O_2 \quad (10.1)$$

This is an attractive process since solar energy is used to convert a readily available substrate, water, to oxygen and hydrogen. This reaction was first demonstrated with a cell-free chloroplast–ferredoxin (Fd)–hydrogenase system, but the existence of such a reaction in green algae had been suggested earlier. Indeed, due to the inherent liability of the various components in a cell-free oxic environment, more reliable systems using whole cells have been studied since then by many researchers. The hydrogen is generated in the case of the unicellular algae via the hydrogenase enzyme. In the case of cyanobacteria, on the other hand, the water-splitting process involves two enzymes, hydrogenase and nitrogenase, both of which catalyze the hydrogen generation process. The light energy is absorbed by the pigments at PSI, PSII, or both, which raises the energy level of electrons from water oxidation when they are transferred from PSI via PSII to Fd. The hydrogenase accepts the electrons from Fd to produce hydrogen as shown in Figure 10.1. Since hydrogenase is sensitive to oxygen, it is necessary to maintain the oxygen content at a level lower than 0.1% so that hydrogen production can be sustained. This condition can be achieved by utilizing green algae *Chlamydomonas reinhardtii* that can deplete oxygen during oxidative respiration.

In fact, photon energy conversion efficiency of some 22% of visible light energy into H_2 by direct biophotolysis corresponds to solar conversion efficiency of some 10%. This was demonstrated in vivo with the green microalga *C. reinhardtii* in laboratory experiments under low light intensities and very low partial pressures of O_2. This avoided the limitation of the light saturation effect and inhibition by photosynthetically produced O_2. In addition, when the culture of green microalga *C. reinhardtii* is deprived of inorganic S, the rates of O_2 synthesis and CO_2 fixation decline significantly within 24 h (in the light). The reason for this loss of activity is due to the need for frequent replacement of the H_2O-oxidizing protein D1 in the PSII reaction center.

The essential anaerobiosis in cells under light is achieved by depriving sulfur from the medium (<0.45 mM) to reduce the oxygenic PSII activity. Also, it is equally important to remove the generated oxygen by keeping the cell's respiration rate on exogenous acetate and/or endogenous carbohydrates. Acetate is a good carbon substrate for *C. reinhardtii* cells

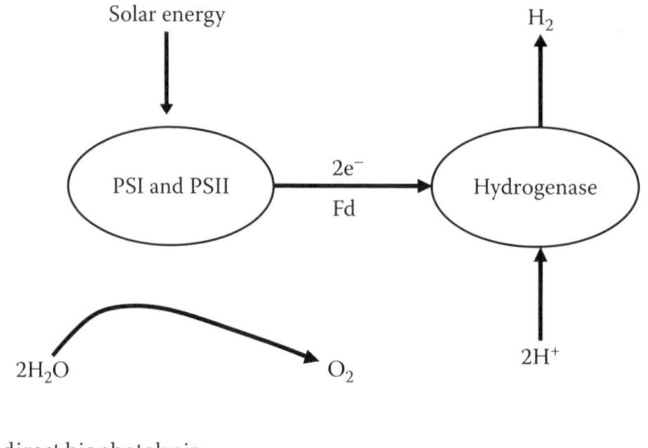

FIGURE 10.1
Schematic diagram for direct biophotolysis.

to maintain a high respiration rate. The primary processes of photosynthesis (i.e., light-triggered charge separation) occur with an unsurpassed efficiency of >95% but all subsequent processes up to the fixation of CO_2 in the so-called dark reactions dissipate a considerable amount of energy. Although the conversion efficiency under ideal conditions (low irradiation, red light) is about 27% (21% for photosynthetic active radiation), the typical conversion efficiency to biomass is about 5% or can be even lower. Photosynthetic microorganisms, such as green algae and cyanobacteria (blue-green algae), have the potential to store efficiently the energy of incident sunlight as high-energy H_2 molecules, with a maximum theoretical efficiency of approximately 13%.

Simultaneous hydrogen and oxygen production with this system has only been demonstrated under very high inert gas sparging leading to very low concentrations of hydrogen and oxygen. It is necessary to use special conditions since Fe hydrogenase activity is extremely oxygen sensitive. Although not specifically studied, it appears that in order for simultaneous production of H_2 and O_2 to occur, the O_2 partial pressures must be below 0.001 atm. It would not be possible to maintain such low partial pressure in any practical direct biophotolysis process due to the large amount of diluent gas and power input required for gas transfer. A direct biophotolysis process must therefore operate at a partial pressure of near one atmosphere of O_2, which is a thousand times the maximum likely to be tolerated. Thus, the O_2 sensitivity of the hydrogenase enzyme reaction and the supporting reductant-generating pathway remains the key problem, as it has been for the past 30 years.

The O_2 absorbers can be used to overcome the O_2 sensitivity. The O_2 absorbers are of two types: irreversible, such as glucose/glucose oxidase and dithionite, and reversible such as hemoglobin. However, using irreversible or regeneration of reversible O_2 absorbers would not be practical in any scaled-up process as they reduce its net efficiency. Endogenous respiration by microalgae can reduce O_2 levels, as observed in a recent study. It should be noted that the results of this study were mistakenly attributed to an indirect biophotolysis process. However, respiratory O_2 uptake decreases by half the potential H_2 evolution, since for each H_2 mole produced, an equivalent amount of substrate is respired. This is an unacceptable loss of efficiency.

Uptake hydrogenases have been made more O_2 tolerant through protein engineering, but these enzymes operate at a much higher redox potential than reversible (H_2 evolving) hydrogenases. In an attempt to achieve oxygen-tolerant hydrogenase activity through classical mutagenesis Ghirardi (1997) developed an experimental technique for the selection of O_2-tolerant, H_2-producing variants of *C. reinhardtii* based on the ability of wild-type cells to survive a short (20 min) exposure to metronidazole in the presence of controlled concentrations of O_2. In this approach, the number of survivors depends on the metronidazole concentration, light intensity, preinduction of the hydrogenase, and the presence or absence of O_2. The authors demonstrated that some of the selected survivors in fact exhibit H_2 production capacity that is less sensitive to O_2 than the original wild-type population.

One approach that can resolve this problem is the use of heterocystous cyanobacteria. The heterocyst is the site of nitrogen fixation, with the vegetative cells carrying out normal photosynthesis (e.g., CO_2 fixation and O_2 evolution) and providing the heterocyst with reduced carbohydrates. A heavy cell wall that reduces diffusion of gases in combination with high respiration rates allows the heterocysts to maintain an essentially anaerobic internal environment. Under an inert gas, such as argon, such cyanobacteria simultaneously evolve H_2 and O_2. Subsequent work demonstrated the long-term sustainability of such a reaction, even in an outdoor demonstration of H_2 production using solar energy. However, the efficiencies of light energy conversion into H_2 were disappointing, at most 1%–2% in laboratory experiments at low light intensities and less than 0.3% outdoors with sunlight. Extensive work from many laboratories has expanded on this research without resolving the fundamental problem of such systems: the very low efficiency of solar energy conversion into H_2. Although in part this also reflects limitations of photosynthesis, discussed in the succeeding text, the main reason for the low efficiencies was the very high metabolic cost of nitrogenase-based, and in particular heterocyst-based, H_2 production. Heterocysts biosynthesis and maintenance account for about half of the energy metabolism of these cyanobacteria, without even considering the actual nitrogenase reaction, which requires at least 4 ATP per H_2 evolved. This metabolic energy must be provided by light through PSI-mediated cyclic phosphorylation or by respiration, in either case reducing H_2 production and achievable solar conversion efficiencies to well below half those expected from simpler, hydrogenase-based, indirect biophotolysis processes. Furthermore, the simultaneous production of H_2 and O_2 by heterocystous cyanobacteria requires gas separation, at a significant economic cost, as would also be the case for direct biophotolysis processes. For all these reasons, heterocystous cyanobacteria specifically, and nitrogen-fixing cyanobacteria in general, can no longer be recommended as subjects for applied microalgal H_2 production R&D.

Practical direct biophotolysis processes must be highly resistant to O_2 inhibition, but such a process is still lacking. Other challenges facing direct biophotolysis processes include the requirement of large photobioreactors and the need for separation of H_2 and O_2, both of which make such a process impractical. Direct biophotolysis requires that the entire production area be enclosed in photobioreactors in order to produce and capture H_2 and O_2. Few cost estimates are available for large-scale closed photobioreactors. One conceptual study of photobiological H_2 production using an inclined tubular photobioreactor with internal gas exchange arrived at an optimistic capital cost of only \$50/m^2. The study also assumed minimal operating costs, considering extrapolation of as little as \$15/GJ H_2 produced for a direct (or other single-stage) biophotolysis process operating at 10% solar conversion efficiency. The cost analysis, however, did not include costs associated with contingencies, engineering, gas separation, and handling. A more realistic cost for such photobioreactors would be in the range of \$100/m^2, without gas separation or even contingencies and engineering costs. This cost would render any single-stage process (e.g., direct biophotolysis or heterocystous cyanobacterial processes) uneconomical. Indeed, any handling of H_2/O_2 mixtures in large volumes and over larger areas would likely be impractical.

In conclusion, the solar conversion efficiency of a direct biophotolysis process is approximately 10% requiring a large bioreactor surface area and thus high capital costs. Furthermore, the overall rate of hydrogen evolution is slow as the algal hydrogenase is inhibited by oxygen that is generated during the process. Efforts have been made to genetically modify the microorganisms to obtain a system that is less sensitive to oxygen, but to date, there are no known significant successes in this field. Therefore, direct biophotolytic processes, though fundamentally attractive from an environmental point of view, are economically unfeasible due to the insurmountable barriers of oxygen sensitivity, intrinsic limitations of poor light conversion efficiencies, problems with gas capture and separation, and slow hydrogen production rate.

10.2.1.3 Indirect Biophotolysis

Reversible hydrogenase-based indirect biophotolysis processes have, at least conceptually, major advantages over the nitrogenase-based systems: The specific H_2 evolution activities of reversible hydrogenases are almost a thousandfold higher than those of nitrogenase and, most importantly, require no ATP. On the other hand, there are also some drawbacks. Reversible hydrogenases in microalgae are generally expressed at relatively low activities, resulting in H_2 production rates typically much lower than those observed with nitrogenase-based processes. Also, H_2 production by reversible hydrogenases is typically

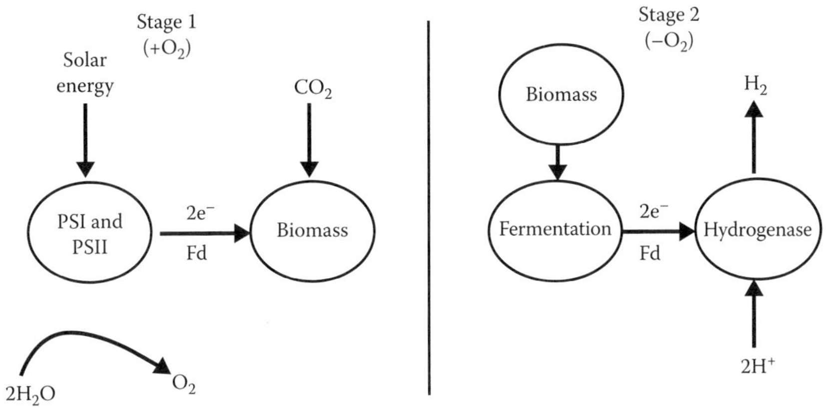

FIGURE 10.2
Schematic diagram for indirect biophotolysis.

inhibited by relatively low (<10%) partial pressures of H_2. Moreover, problems of sensitivity of the hydrogen evolution process are potentially circumvented by separating temporally and/or spatially oxygen evolution and hydrogen evolution. Thus, indirect biophotolysis processes involve separation of the H_2 and O_2 evolution reactions into separate stages, coupled through CO_2 fixation/evolution (see Figure 10.2).

One elaboration of this concept involved four distinct steps:

1. Production in open ponds at 10% solar efficiency of biomass high in storage carbohydrates. After carbohydrate accumulation, the cells would be removed from the growth pond or photobioreactor.
2. Concentrated if required.
3. Through respiratory O_2 uptake, allowed to become anaerobic and activate or induce the hydrogenase enzyme. H_2 evolution would then commence, first in the dark through endogenous fermentations.
4. In a light-driven (PSI-mediated) electron transport to convert remaining stored carbohydrates and fermentation products (e.g., acetate) to H_2. The depleted cells would be reused for additional cycles of CO_2 fixation–H_2 production.

Much information is available on each of these steps, but significant gaps remain. At present, the light-driven step of H_2 production has been demonstrated only in green algae. For unknown reasons, cyanobacteria that have reversible hydrogenases only seem to produce H_2 fermentatively in the dark, not in a light-driven or stimulated reaction. For green algae, hydrogenase levels are often not the limiting factor and it is possible to obtain high initial rates for short periods (seconds to minutes). However, sustained high rates have not been achieved, possibly due to competition with other reactions. Progress in the genetics of the hydrogenases, both reversible and uptake, in cyanobacteria and green algae has been rapid in recent years. Cyanobacteria have the unique characteristics of using CO_2 in the air as a carbon source and solar energy as an energy source (Equation 10.2). The cells take up CO_2 first to produce cellular substances, which are subsequently used for hydrogen production (Equation 10.3):

$$12H_2O + 6CO_2 \rightarrow C_6H_{12}O_6 + 6O_2 \quad (10.2)$$

$$C_6H_{12}O_6 + 12H_2O \rightarrow 12H_2 + 6CO_2 \quad (10.3)$$

A cost estimate of \$7/m^2 was developed for the highly conceptual process of indirect biophotolysis based on prior work on the economics of large-scale microalgae mass cultures in open ponds. The photobioreactors, requiring only one-tenth the area of the ponds, were not further specified and merely assumed to cost \$100/m^2, which with contingencies, engineering, and other overheads gave a total capital cost of \$135/m^2. With these assumptions, overall operational costs of such a process were estimated at about \$10/GJ of H_2 (the cost of the photobioreactors was over half of the total costs). This was based on the assumption that less than one photon per H_2 evolved would be sufficient to drive H_2 evolution from the acetate. This assumption may not be correct with some experts believing that a minimum of two photons per H_2 is more realistic. Accordingly, the photobioreactor area required in such a process would increase from 10% to over 25% of the area of the pond cultures, that is, more than doubling the overall cost of H_2 produced by such a process. Another key issue is design, performance, and cost of the photobioreactors. Subsequently, Tredici (1998) devised that a simple, slightly inclined internal gas exchange tubular photobioreactor, consisting of forty 50 m long, 4.2 cm diameter tubes (glass or Teflon®), ganged together with a bottom manifold and top degasser, positioned on a slightly inclined (about 10% slope) earthen platform, and mixed by airlift. Despite simplicity, the Tredici-type reactor, suggested for biological H_2 production, was deemed to be too costly for such applications. In conclusion, indirect biophotolysis processes are still at the conceptual stage. More research has to be done, and its economical feasibility needs to be more qualified.

10.2.1.4 Photofermentation

Photosynthetic bacteria have long been studied for their capacity to produce hydrogen through the action of their nitrogenase system (Figure 10.3). Calculations of the efficiency of conversion of light energy into hydrogen often show values approaching 10%, but these estimates generally ignore the energy content of the organic substrate. True PE is much lower (less than 3%) under ideal (low) light conditions. Although little data are available, PE must be even lower under high-light (full sunlight) conditions since the

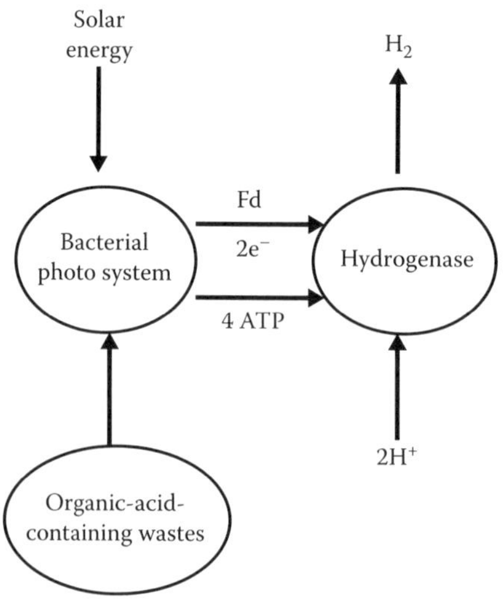

FIGURE 10.3
Hydrogen production via the action of the nitrogenase system using photosynthetic bacteria.

photosystem of the photosynthetic bacteria is, like that of microalgae and cyanobacteria (discussed earlier), optimized for low-light conditions.

Photosynthetic bacteria evolve molecular hydrogen catalyzed by nitrogenase under nitrogen-deficient conditions using light energy and reduced compounds.

Photoheterotrophs utilize dissolved organic compounds, as well as sunlight, and produce hydrogen under anaerobic/anoxic conditions. Many studies have demonstrated that photofermentation by photosynthetic bacteria can convert low–molecular weight fatty acids into H_2 and CO_2 with high efficiency.

Much work has been devoted to the development of systems that utilize photoheterotrophic bacteria for simultaneous hydrogen generation and waste disposal. Photosynthetic bacteria, such as *Rhodobacter sphaeroides*, generate hydrogen mainly through the action of the nitrogenase enzyme. This activity is inhibited by the presence of oxygen and high nitrogen/carbon ratios >1.4 N/C. In fact, high nitrogen concentrations have been linked to a metabolic shift where organic substrate is utilized mainly for biomass generation instead of hydrogen production. High biomass concentration is not desirable due to the reduction of light diffusion into the bioreactor. As with all biological systems, careful consideration must be given to operating conditions in order to maximize the desired parameters, in this case, hydrogen yield and rate of production. Depending on the temperature, pH, bacterial strain, irradiation, and substrate used, *R. sphaeroides* has exhibited variable results. The optimum temperature and pH were in the range of 30°C–35°C and 7.0–8.0, respectively, and irradiation was approximately 200 W/m^2. Though hydrogen production has been achieved with *R. sphaeroides* growing on various substrates, the bacteria seem to prefer organic acids such as acetic, butyric, malic, and lactic acid.

The utilization of simple sugars such as glucose and sucrose for hydrogen production is possible but usually results in lower substrate conversion and rate of H_2 evolution and is secondary if an organic acid cosubstrate is present. Some success in using industrial wastewater such as sugar refinery and tofu has been shown, but due to either the toxic nature of the effluent or color/opaqueness, pretreatment is needed prior to photosynthetic biohydrogen gas production.

Despite the success of hydrogen generation via photosynthetic degradation of organic compounds, much work is still needed to create a large-scale, economically attractive process. Even though the conversion of substrate is generally high, H_2 production rate is slow and the yield of hydrogen is far from the theoretical maximum. There are several other drawbacks to this process: (1) the use of the nitrogenase enzyme with its inherent high energy demand; (2) the low solar energy conversion efficiencies; and (3), as noted earlier, the requirement for elaborate anaerobic photobioreactors covering large areas. As with other light-based H_2 production processes, light diffusion and intensity play a key role in maximizing hydrogen yield. Increasing light intensity from 50 to 1000 W/m^2 decreased the light conversion efficiency from approximately 8% to around 2%. Expensive equipment and the need for a large reactor surface area remain a drawback. Albeit cyclic light process operation (i.e., light/dark cycles) has been shown to increase the amount of hydrogen evolved when compared to continuous illumination, many questions remain unanswered regarding stability of overall light conversion efficiency to warrant large-scale systems. It is apparent that photosynthetic hydrogen production might have to be coupled with another process in order to make it an economically viable option of biogas production.

It is interesting to note that after extensive R&D on this system in Japan under the Research on Innovative Technologies for the Earth (RITE) program, it was dropped from further study. In conclusion, the rate and efficiency of hydrogen production by systems directly involving photosynthesis H_2 production (photobiological hydrogen production) fall far short of economic feasibility. Since organic substrates are the ultimate source of hydrogen in photofermentations or indirect biophotolysis processes, it can be argued that it should be simpler and more efficient to extract the hydrogen from such substrates using a dark fermentation process. This approach is discussed thoroughly in the following.

10.2.2 Anaerobic (Dark) Fermentation

Anaerobic fermentation is similar to the photosynthetic process in that it makes use of microorganisms to produce hydrogen from organic matter. Anaerobic systems have an advantage over their photosynthetic counterparts in that they are simpler and less expensive and produce hydrogen at a much faster rate. The drawback, however, is that anaerobes are unable to utilize light energy and thus lack the ability to overcome the inherited thermodynamic energy barrier to fully decompose a substrate. Anaerobic fermentation presents an interesting yet largely unexplored avenue for the biological production of hydrogen. Much is presently known about the molecular biology and biochemistry of the hydrogen-producing enzymes, reductant-generating systems, and physiology of many hydrogen-producing fermentative microorganisms. Very little is presently known about the currently attainable yields or the possibilities for improving these yields in the future.

10.2.3 Hybrid Systems

Since hydrogen production from waste can only achieve partial conversion efficiency, as low as 33%, this means that the majority of the COD of waste streams remains untreated. A second-stage process downstream of the hydrogen reactor is a promising approach that renders the overall process more effective and economic. A variety of methods have been tested as a second stage in hydrogen production systems.

10.2.3.1 Second Stage: Methane Fermentation

Organic pollutants are anaerobically converted to methane in two distinct stages: acidification and methanogenesis. Acidification produces hydrogen as a by-product, which in turn is used as an electron donor by many methanogens at the second stage of the process. Separation of the two stages is feasible for hydrogen collection from the first stage. The second stage is further used for the treatment of the remaining acidification products, mainly volatile fatty acids (VFAs).

A number of recent studies have reported a successful operation of such two-stage systems. The disadvantage here is that two different gas streams are created and in reality, since the majority of the substrate is converted to methane, this is a methane fermentation in which one is pulling off a little hydrogen. However, at least with some substrates (organic solid wastes), there may be advantages over a traditional methane fermentation, since the first stage acts effectively in solubilization. In addition, combining the two gas streams would create a hydrogen–methane mixture (20%–30% H_2 [after removal of CO_2], 80%–70% CH_4) shown to burn cleaner than methane alone.

Ueno (2007) using artificial organic solid waste (COD = 37,100 mg/L, VSS = 23,700 mg/L) in laboratory scale, showed that not only relatively high hydrogen yields coupled with methane yields twofold higher than a comparable single-stage process could be sustained but also overall chemical oxygen demand (COD) removal efficiency and volatile suspended solids (VSSs) decomposition were 80% and 96%, respectively. Recently, Hafez et al. (2009) used a hybrid system that comprised both a novel biohydrogen reactor with a gravity settler followed by a second-stage conventional anaerobic digester for the production of methane gas. The authors tested their system with a synthetic wastewater/leachate solution and was operated at 37°C for 45 days. The maximum hydrogen yield was 400 mL H_2/g glucose with an average of 345 mL H_2/g glucose, and the methane yield for the second stage approached a maximum value of 426 mL CH_4/gCOD removed, with an overall COD removal efficiency of 94%.

10.2.3.2 Second Stage: Photofermentation

In another proposed process to recover additional hydrogen from the products of a dark hydrogen-generating fermentation, nonsulfur purple photosynthetic bacteria can be used to capture light and quantitatively convert organic acids to hydrogen in a reaction that has been known for nearly 60 years and has been named photofermentation. A great deal of research has gone into this area, which has been recently reviewed. In photofermentation, hydrogen evolution is driven by nitrogenase, and since this enzyme is ATP dependent (the requisite ATP is formed via the photosynthetic capture of light energy), hydrogen evolution is essentially irreversible. Thus, in principle, photofermentation is capable of complete conversion of organic acids to hydrogen.

There have been a few studies where this capacity has been used to demonstrate the conversion of organic acids, principally acetate, produced during a first stage dark fermentation of sugars (glucose, sucrose) to hydrogen, thus increasing the overall hydrogen yield. The most severe constraint, as discussed previously, is the fact that photosynthetic efficiencies, with either solar radiation or tungsten lamps, are very low. At even moderate light intensities, the majority (>80%) of captured light is dissipated as heat. Obviously, this increases greatly the surface area required for the production of a given quantity of hydrogen. This, combined with the potentially high cost of transparent, hydrogen-impermeable photobioreactors, presents very significant hurdles for the development of a practical process based on photofermentation.

10.2.3.3 Second Stage: Electrohydrogenesis

In this recent (2005) approach, an electrical input provided to a modified microbial fuel cell (MFC) provides the energy necessary to convert organic acids, such as acetate, to hydrogen. In fact, these types of cells are rather versatile and have been shown to be able to generate hydrogen from a variety of substrates, including some wastewaters. Moreover, it is not necessary to start with a pure, defined bacterial culture with the desired characteristics, since the appropriate bacteria can be selected during operation. As originally constituted, these bioelectrochemical cells consisted of an anode chamber where, like in MFCs, electrogenic bacteria catabolize substrate using the electrode as an electron acceptor for their metabolism, thereby generating an electrical potential. Both the anode and cathode were operated anaerobically and connected by a proton (cation) exchange membrane. The potential generated from a substrate such as acetate (300 mV) isn't sufficient to drive hydrogen evolution at the cathode, so supplementary voltage input is required, around 110 mV in theory, but because of electrode resistance, this is actually >200 mV.

Although admittedly elegant in its demonstration of the thermodynamic principles of the manipulation of microbial metabolism for use in the biotechnological production of fuels, hydrogen evolution rates and efficiencies remain quite low. A number of problems can be pointed out, and some of these have been at least partially dealt with in more recent publications. First, as is the case with MFCs, current densities obtained at the anode (graphite) are quite low. Some improvement has been attained by ammonia treatment of carbon cloth anodes, but further gains are desirable. Secondly, the membrane separating the two chambers allows passage of cations, Na^+ and K^+, which are present at much higher concentrations than protons at the near-neutral pHs required. The resulting pH rise at the cathode limits hydrogen production, while acidification at the anode inhibits microbial activity. Thus, replacement with an anion-selective membrane (protons are transferred via negatively charged phosphate ions) leads to some improvement, and in fact, it has been demonstrated that it may be possible to do away with the membrane altogether.

A further obstacle is that the employed cathodes usually contain platinum, too expensive for the obtained current densities. In a novel approach to overcoming this problem, it has been proposed to replace the usual cathode with a biocathode where bacteria attached to an appropriate graphite surface catalyze hydrogen evolution. However, it is not clear that a sufficiently high catalytic density can be achieved using this approach. Thus, electrohydrogenesis is a novel approach, but it remains to be seen if sufficiently high volumetric rates can be achieved with nominal electrical inputs. Currently, high volumetric rates (around 3 m^3 H_2/m^3 reactor/day), which are however still almost two orders of magnitude lower than those seen with dark fermentations, are only observed at high (around 800 mV) applied voltages.

10.2.4 Summary and Conclusion

Photosynthetic hydrogen production is a theoretically perfect process for transformation of the *free* solar energy into hydrogen by photosynthetic bacteria; practical applications are cost prohibitive due to the low utilization efficiency of light, large footprint, low volumetric hydrogen production rate, and difficulties in designing the reactors for hydrogen production. On the other hand, comparison of the rates of H_2 production by various biohydrogen systems coupled with the associated complexity level of operation suggests that dark fermentation systems offer an excellent potential for practical application. Although fermentative biohydrogen technologies are still in their infancy, developing more efficient technologies offer potential for practical application. Therefore, fermentative hydrogen production has been receiving increasing attention in recent years.

10.3 Photobioreactors

A reactor for photobiological hydrogen production has to meet several conditions. Since the hydrogen gas has to be collected, a prerequisite of the batch photobioreactor is for it to be an enclosed system. It has to be possible to maintain a monoculture for an extended time (it must be practical to sterilize the reactor). Preferably, sunlight is the energy source. The productivity of photobioreactors is light limited, and a high surface-to-volume ratio is a prerequisite for a photobioreactor. The photochemical efficiencies are low (theoretically a maximum of 10% and in laboratory experiments ranges from 3% to 10% [Miron et al. 2000]) and tend to decrease at higher light intensities (the effect of light saturation; photons cannot all be used for reaction energy but are dissipated as heat energy). This means that in order to create an efficient biological process, it is important to either dilute the light or distribute it as much as possible over the reactor volume and/or mix the culture at a high rate, so that cells are light exposed only for a short period.

Janssen (2002) reviewed three types of photobioreactors: the vertical-column reactors (airlift loop reactor and bubble column), a panel reactor, and tubular reactors. He looked at the light gradients and the mixing-induced light/dark cycles PE and scalability. Depending on the reactor type and the way it is operated, cells are exposed for a certain period of time at the irradiated surface or in a dark part of the reactor. The (mixing-induced) light/dark cycles, when they are in the range of micro- or milliseconds, can enhance PE, approaching the theoretical PEs at low light intensities. When the cycles are from several seconds to tens of seconds, there is no improvement and even a decrease in PE has been reported. The depth (and volume) of the photic zone depends on the dimensions and operations of the reactor, the algal concentration, and the specific absorption coefficient of the algae (and the wavelength of the incoming light). On the basis of model calculations and/or empirical data, several reactor types were compared. The results (for biomass production of photoautotrophs) are shown in Table 10.1. Flat panel reactors show a high photochemical efficiency or biomass yield on light energy, while biomass density is also high. Tubular reactors in theory should show better efficiencies because of the shorter average light/dark cycles.

The analysis of typical examples of microalgal cultivations in enclosed (outdoor) photobioreactors showed that the PE and productivity are determined by the light regime inside the reactors. In addition, only oxygen accumulation and shear stress limit productivity in certain designs.

TABLE 10.1

Enclosed Photobioreactors, Photosynthetic Efficiency (PE), and Biomass Yield on Light Energy

Photobioreactor Type	PE or Ydw; E a;b (% or g dw/mol)	Reference (Microorganism)
Bubble-column and airlift-column reactors	0:84 (Ydw;E)[c]	Camacho et al. (1999) (*Phaeodactylum tricornutum*)
Internal draught tube		
Split cylinder		
Airlift column	0:82 (Ydw;E)	Hu and Richmond (1996) (*Phaeodactylum tricornutum*)
Flat panel		
(a) Vertical	(a) 1:48 (Ydw;E) ≈ 16 (PE)	(a) Hu et al. (1998)
(b) Tilted	(b) 10–20 (PE)	(b) Torzillo et al. (1996) (*Spirulina platensis*)
Tubular reactor	0:60 (Ydw;E); 6.5 (PE)[d]	Thimijan and Heins (1983) (*Spirulina platensis*)
Tubular reactor		
Diameter 2:5 cm	(a) 0.480–0:63 (Ydw;E)[e]	Fontoynont et al. (1998) (*Phaeodactylum tricornutum*)
Diameter 5:3 cm	(b) 0.680–0:95 (Ydw;E)[e]	

Source: Janssen, M., Cultivation of microalgae: Effect of light/dark cycles on biomass yield. Thesis, Wageningen University, Wageningen, the Netherlands, 2002.

Ydw;E, the biomass yield in dry weight basses.

[a] Daily irradiance values in $MJ/m^2/day$ were divided by 12 × 3600 s, assuming a day length of 12 h; multiplied with 0.429, the fraction PAR in the solar spectrum [193]; and multiplied with 4.57, mol photons/MJ (Thimijan and Heins 1983).

[b] Monthly averages of daily solar irradiance on a vertical cylindrical surface in the same period of the year were obtained from the European Database of Daylight and Solar Radiation, www.satel-light.com (Fontoynont et al. 1998).

[c] Based on a linear growth phase with a productivity of about 0:49 g/L/day observed in an outdoor batch culture in three different reactor types.

[d] Before calculating PE and Ydw;E, irradiance data were corrected for transmittivity tubes (Thimijan and Heins 1983).

[e] Calculated by Janssen (2002).

The comparison of the bioreactors described earlier referred to processes with microalgae, diatoms, or cyanobacteria. Though photoheterotrophic bacteria differ, for instance, in photochemical efficiency, absorption coefficient, and size, the relative difference in performance of the reactor types might be extrapolated to the case of photoheterotrophics. The light regime, including mixing-induced light/dark cycles, is assumed to be much more determining than biological factors.

Considering the findings that panel reactors and tubular reactors (at least in theory) show highest efficiencies, it is worthwhile to look further into all types of reactors and their possibilities to be scaled up for practical purposes.

Ugwu et al. (2008) developed a recent review of the photobioreactors for mass cultivation of algae; this critical review includes some aspects of hydrodynamics and mass transfer characteristics of these photobioreactors. Algal culture systems can be illuminated by artificial light, solar light, or by both. Naturally illuminated algal culture systems with large illumination surface areas include open ponds (Hase et al. 2000), flat-plate (Hu 1996), horizontal/serpentine tubular airlift (Rubio et al. 1999), and inclined tubular photobioreactors (Ugwu et al. 2002). Generally, laboratory-scale photobioreactors are artificially illuminated (either internally or externally) using fluorescent lamps or other light distributors. Some of these photobioreactors include bubble-column (Degen et al. 2001), airlift-column (Kaewpintong et al. 2007), stirred-tank (Ogbonna et al. 1999), helical tubular, conical (Watanabe and Saiki 1997), torus (Pruvost et al. 2006), and seaweed-type photobioreactors (Hallenback and Benemann 2002). Furthermore, some photobioreactors can be easily tempered. Tempering could simply be achieved by placing a photobioreactor in a constant temperature room. This approach is limited to compact photobioreactors. Large-scale outdoor systems such as tubular photobioreactors cannot be easily tempered without high technical efforts. However, several commercially available photobioreactors, for example, BIOSTAT photobioreactors (developed by Sartorius BBI Systems Inc.) (Pohl et al. 1988) can be readily tempered. Also, some efforts were undertaken to design temperature-controlled photobioreactors, such as double-walled internally illuminated photobioreactors with a heating and cooling water circuit (Pohl et al. 1988).

10.3.1 Open Ponds

Cultivation of algae in open ponds has been extensively studied in the past few years (Boussiba et al. 1988; Tredici and Materassi 1992; Uguwu 2002). Open ponds can be categorized into natural waters (lakes, lagoons, ponds) and artificial ponds or containers. The most commonly used systems include shallow big ponds, tanks, circular ponds, and raceway ponds. One of the major advantages of open ponds is that they are easier to construct and operate than most closed systems. However, major limitations in open ponds include poor light utilization by the cells, evaporative losses, diffusion of CO_2 to the atmosphere, and requirement of large areas of land. Furthermore, contamination by predators and other fast-growing heterotrophs have restricted the commercial production of algae in open-culture systems to only those organisms that can grow under extreme conditions. Also, due to inefficient stirring mechanisms in open cultivation systems, their mass transfer rates are very poor, resulting in low biomass productivity (Tredici and Materassi 1992).

In order to overcome the problems with open ponds, much attention is now focused on the development of suitable closed systems such as flat-plate, tubular, vertical-column, and internally illuminated photobioreactors.

10.3.2 Flat-Plate Photobioreactors

Flat-plate photobioreactors have received much attention for cultivation of photosynthetic microorganisms due to their large illumination surface area. The work presented by Milner (1953) paved way to the use of flat culture vessels for cultivation of algae. Following this work, Samson and Leduy (1985) developed a flat reactor equipped with fluorescence lamps. A year later, Ramos de Ortega and Roux (1986) developed an outdoor flat panel reactor by using thick transparent PVC materials. As time went on, extensive works on various designs of vertical alveolar panels and flat-plate reactors for mass cultivation of different algae were reported (Tredici and Materassi 1992; Hu and Richmond, 1996; Hoekema et al. 2002; Zhang et al. 2002). Generally, flat-plate photobioreactors are made of transparent materials for maximum utilization of solar light energy. Accumulation of *d*issolved *o*xygen (*DO*) concentrations in flat-plate photobioreactors is relatively low compared to horizontal tubular photobioreactors. It has been reported that with flat-plate photobioreactors, high photosynthetic efficiencies can be achieved (Hu and Richmond 1996; Richmond 2000). Flat-plate photobioreactors are very suitable for mass cultures of algae. However, they also have some limitations as indicated in Table 10.2.

10.3.3 Tubular Photobioreactors

Among the proposed photobioreactors, tubular photobioreactor is one of the most suitable types for outdoor mass cultures. Most outdoor tubular photobioreactors are usually constructed with either glass or plastic tube, and their cultures are recirculated either with pump or preferably with airlift system. They can be in the form of horizontal/serpentine (Chaumont et al. 1988; Molina et al. 2001), vertical (Pirt et al. 1983), near horizontal

TABLE 10.2

Prospects and Limitations of Various Culture Systems for Algae

Culture Systems	Prospects	Limitations
Open ponds	Relatively economical, easy to clean up after cultivation, good for mass cultivation of algae	Little control of culture conditions, difficulty in growing algal cultures for long periods, poor productivity, are easily contaminated, occupy small illumination surface area, their construction require sophisticated materials, shear stress to algal cultures, decrease of illumination surface area upon scale-up
Vertical-column photobioreactors	High mass transfer, good mixing with low shear stress, low energy consumption, high potentials for scalability, easy to sterilize, readily tempered, good for immobilization of algae, reduced photoinhibition and photooxidation	Require sophisticated materials, shear stress to algal cultures, decrease of illumination surface area upon scale-up
Flat-plate photobioreactors	Large illumination surface area, suitable for outdoor cultures, good for immobilization of algae, good light path, good biomass productivities, relatively cheap, easy to clean up, readily tempered, low oxygen buildup	Scale-up require many compartments and support materials, difficulty in controlling culture temperature, some degree of wall growth, possibility of hydrodynamic stress to some algal strains
Tubular photobioreactors	Large illumination surface area, suitable for outdoor cultures, fairly good biomass productivities, relatively cheap	Gradients of pH, dissolved oxygen and CO_2 along the tubes, fouling some degree of wall growth, requires large land space

(Tredici et al. 1998), conical (Watanabe and Saiki 1997), or inclined (Lee and Low 1991; Ugwu et al. 2002) photobioreactor. Aeration and mixing of the cultures in tubular photobioreactors are usually done by air-pump or airlift systems. Advantages and limitations of tubular photobioreactors are shown in Table 10.2. Table 10.2 summarizes the advantages and limitations of open ponds.

Tubular photobioreactor are very suitable for outdoor mass cultures of algae since they have large illumination surface area. On the other hand, one of the major limitations of tubular photobioreactor is poor mass transfer. It should be noted that mass transfer (oxygen buildup) becomes a problem when tubular photobioreactors are scaled up. For instance, some studies have shown that very high *DO* levels are easily reached in tubular photobioreactors (Torzillo et al. 1986; Richmond et al. 1993; Molina et al. 2001). Also, photoinhibition is very common in outdoor tubular photobioreactors (Vonshak and Torzillo 2004). When a tubular photobioreactor is scaled up by increasing the diameter of tubes, the illumination surface-to-volume ratio would decrease. On the other hand, the length of the tube can be kept as short as possible while a tubular photobioreactor is scaled up by increasing the diameter of the tubes. In this case, the cells at the lower part of the tube will not receive enough light for cell growth (due to light shading effect) unless there is a good mixing system. In any case, efficient light distribution to the cells can be achieved by improving the mixing system in the tubes (Ugwu et al. 2003, 2005a). Also, it is difficult to control culture temperatures in most tubular photobioreactors. Although they can be equipped with thermostat to maintain the desired culture temperature, this could be very expensive and difficult to implement. It should also be noted that adherence of the cells of the walls of the tubes is common in tubular photobioreactors.

Furthermore, long tubular photobioreactors are characterized by gradients of oxygen and CO_2 transfer along the tubes (Camacho et al. 1999; Ugwu et al. 2003). The increase in pH of the cultures would also lead to frequent recarbonation of the cultures, which would consequently increase the cost of algal production.

10.3.4 Vertical-Column Photobioreactors

Various designs and scales of vertical-column photobioreactors have been tested for cultivation of algae (Choi et al. 2003; Lopez et al. 2006; Kaewpintong et al. 2007). Vertical-column photobioreactors are compact, low cost, and easy to operate monoseptically (Miron et al. 2002). Furthermore, they are very promising for large-scale cultivation of algae. It was reported that bubble-column and airlift photobioreactors (up to 0.19 m in diameter) can attain a final biomass concentration and specific growth rate that are comparable to values typically reported for narrow tubular photobioreactors (Miron et al. 2002). Some bubble-column photobioreactors are equipped with either draft tubes or constructed as split cylinders. In the case of draft tube photobioreactors, intermixing occurs between the riser and the downcomer zones of the photobioreactor through the walls of the draft tube. A summary of the prospects and limitations of vertical-column photobioreactors is shown in Table 10.2.

10.3.5 Internally Illuminated Photobioreactors

As mentioned earlier, some photobioreactors can be internally illuminated with fluorescent lamps. Figure 10.4 shows a typical internally illuminated photobioreactor. This photobioreactor is equipped with impellers for agitation of the algal cultures. Air and CO_2 are supplied to the cultures through the spargers.

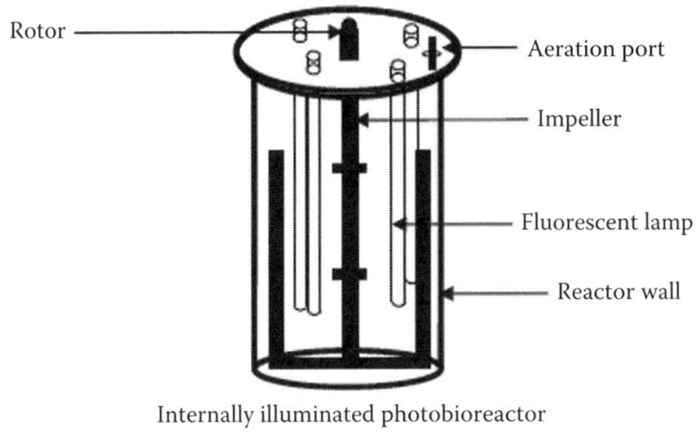

FIGURE 10.4
Schematic diagram of internally illuminated photobioreactor.

This type of photobioreactor can also be modified in such a way that it can utilize both solar and artificial light system (Ogbonna et al. 1999). In that case, the artificial light source is switched on whenever the solar light intensity decreases below a set value (during cloudy weather or at night). There are also some reports on the use of optical fibers to collect and distribute solar light in cylindrical photobioreactors (Mori 1985; Matsunaga et al. 1991). One of the major advantages of an internally illuminated photobioreactor is that it can be heat sterilized under pressure, and thus, contamination can be minimized. Furthermore, supply of light to the photobioreactor can be maintained continuously (both day and night) by integrating artificial and solar light devices. However, outdoor mass cultivation of algae in this type of photobioreactor would require some technical efforts (Ugwu et al. 2008).

10.3.6 Hydrodynamics and Mass Transfer Characteristics of Photobioreactors

Although relationship between hydrodynamics and mass transfer has been extensively investigated and correlated in bioreactors for heterotrophic cultures, only a few studies are available in phototrophic cultures that are applicable in photobioreactors. This includes the overall mass transfer coefficient (k_{La}), mixing, liquid velocity, gas bubble velocity, and gas holdup. k_{La} is the most commonly used parameter for assessing the performance of photobioreactors. The following equation governs mass transfer from the gas phase to the liquid phase:

$$N = k_L a * (C^*_{,l} - C_{,l}) \tag{10.4}$$

where

- N is the mass transfer rate
- k_L is the mass transfer rate coefficient
- a is the interface area
- $C^*_{,l}$ is the equilibrium gas concentration at the interface of the gas and liquid
- $C_{,l}$ is the gas concentration in the liquid
- k_{La} is generally used to describe the overall volumetric mass transfer coefficient in photobioreactors

The volumetric mass transfer coefficient of photobioreactors is dependent on various factors such as agitation rate, the type of sparger, surfactants/antifoam agents, and temperature. Mixing time can be defined as the time taken to achieve a homogenous mixture after injection of tracer solution (Ugwu et al. 2008). Lee and Bazin (1990) define mixing time as the time taken for a small volume of dye solution added to the liquid to transverse the reactor. Generally, mixing time is determined in photobioreactors using tracer substances such as dyes. However, mixing time can also be measured by signal–response method using tracer and pH electrode (Camacho et al. 1999; Ugwu et al. 2005a; Pruvost et al. 2006). Furthermore, *c*omputational *f*luid *d*ynamic (*CFD*) models were used to evaluate global mixing in torus photobioreactor (Pruvost et al. 2006; Sato et al. 2006). Mixing time is a very important parameter in designing photobioreactors for various biological processes. Good mixing would ensure high cell concentration, keep algal cells in suspension, eliminate thermal stratification, help nutrient distribution, improve gas exchange as well as reduce the degree of mutual shading, and lower the probability of photoinhibition (Janvanmardian and Palsson 1991). It was also reported that when the nutritional requirements are sufficient and the environmental conditions are optimized, mixing aimed at inducing turbulent flow would result in high yield of algal biomass (Hu and Richmond 1996). Bosca et al. (1991) demonstrated that the productivity of alga is higher in mixed culture than in an unmixed one under the same condition. Various mixing systems are currently used in algal cultures depending on the type of photobioreactors. In open-pond systems, paddle wheels were used to induce turbulent flow (Boussiba et al. 1988; Hase et al. 2000). In stirred-tank photobioreactors, impellers were used in mixing algal cultures (Ogbonna and Tanaka 2001; Mazzuca et al. 2006). In tubular photobioreactors, mixing can be done by bubbling air directly or indirectly via airlift systems (Tredici et al. 1998; Ogbonna et al. 1999) or by installing static mixers inside the tubes (Ugwu et al. 2002). Mixing systems that utilized baffles in bubble-column photobioreactors were also demonstrated in algal cultures (Merchuk et al. 2000; Degen et al. 2001). In bubble-column and large-diameter tubular photobioreactors, demarcation exists between the light-illuminated and dark surfaces. Thus, mixing strategies should be introduced in cultures to circulate algal cells between the light-illuminated and dark regions of the photobioreactors (Molina et al. 1999; Ugwu et al. 2005b; Mazzuca et al. 2006). Increase in aeration rate would improve mixing, liquid circulation, and mass transfer between gas and liquid phases in algal cultures. However, high aeration could cause shear stress to algal cells (Mazzuca et al. 2006; Kaewpintong et al. 2007). Gas bubble velocity is a measure of culture flow rates in tubular photobioreactors (plug flow regime) since algal cultures are circulated along with gas bubbles. When fine spargers are used to increase gas dispersion inside horizontal tubular photobioreactors, relatively large bubbles are produced. However, the bubbles coalesce during flow to form interface between the liquid broth, gas, and the walls of the tube. The contact area between the liquid and the gas is reduced, thereby, resulting in poor mass transfer rates. Gas bubble velocity and size of the bubbles are dependent on the liquid flow rate. By increasing the gas flow rate, the bubble diameter increases, which consequently, would increase the gas bubble velocity. The rate of gas circulation may be interrupted when baffles or static mixers are installed inside the reactors to increase gas dispersion. However, the mixer elements would help to break down the large bubbles into fine ones, thereby improving the mass transfer rates. Some studies have indicated that relationship exists between superficial gas velocity, bubble velocity, and the overall mass transfer coefficient in bioreactors (Lu et al. 1995; Wongsuchoto et al. 2003; Couvert et al. 2004). In some photobioreactors, the relationship between superficial gas velocity and the overall mass transfer coefficient (k_{La}) studied in various algal cultures can be evaluated (Table 10.3).

In a concentric tube airlift photobioreactor (which was used for *Phaeodactylum tricornutum* cultures), it was reported that at a superficial gas velocity of 0.055 m/s, the k_{La} of about 0.02 s^{-1} was obtained (Contreras et al. 1998). This k_{La} value was about the same as the one

TABLE 10.3

Relationship between the Superficial Velocities and Overall Mass Transfer Coefficient (k_{La}) in Various Cultures Systems

Photobioreactor	Volume (L)	Superficial Velocity (m/s)	k_{La} (s^{-1})	Strain	References
Concentric tube airlift	12	0.055	0.020	*Phaeodactylum tricornutum*	Contreras et al. (1998)
Internally illuminated	3	0.009	0.020	*Chlorella pyrenoidosa*	Ogbonna et al. (1998)
Airlift tubular horizontal	200	0.160	0.014	*Porphyridium cruentum*	Camacho et al. (1999)
Bubble column	13	$5.4–82 \times 10^{-4}$	$1.7–4.7 \times 10^{-3}$	*Porphyridium* sp.	Merchuk et al. (2000)
External-loop airlift tubular	200	0.250	0.006	*Phaeodactylum tricornutum*	Fernandez et al. (2001)
Inclined tubular	6	0.020	0.003	*Chlorella sorokiniana*	Ugwu et al. (2002)
Flat-plate	3	0.009	0.002	*Synechocystis aquatilis*	Boussiba et al. (1988)
Split-cylinder internal-loop airlift	2	0.024	0.009	*Haematococcus pluvialis*	Estrada et al. (2005)

Source: Ugwu, C.U. et al., *Bioresour. Technol.*, 99, 4021, 2008.

reported by Ogbonna et al. (1998) (with 3 L internally illuminated photobioreactor) for *Chlorella pyrenoidosa* cultures at a superficial gas velocity of 0.009 m/s. In 6 L inclined tubular, which was used for cultivation of *Chlorella sorokiniana*, k_{La} of about 0.003 s^{-1} (at a superficial gas velocity of 0.02 m/s) was obtained [198]. Merchuk et al. (2000) reported that by varying the superficial gas velocity ($5.4–82 \times 10^{-4}$) in 13 L bubble-column photobioreactor (which was tested for cultivation of *Porphyridium* sp.), the k_{La} obtained was in the range of 1.7 and 4.7×10^{-3} s^{-1}. With 3 L flat-plate photobioreactor (which was used for cultivation of *Synechocystis aquatilis* cultures), the k_{La} obtained was 0.002 s^{-1} at a superficial gas velocity of 0.009 m/s (Zhang et al. 2002). At a superficial gas velocity of 0.024 m/s, the k_{La} reported in 2 L split-cylinder internal-loop airlift photobioreactors (which was used for cultivation of *Haematococcus pluvialis* cultures) was 0.009 s^{-1} (Vega et al. 2005). By using 200 L airlift tubular horizontal photobioreactors (at a superficial gas velocity of 0.16 m/s), the k_{La} reported was about 0.014 s^{-1} (Camacho et al. 1999). In the case of 200 L external-loop airlift tubular, which was tested for outdoor cultures of *Phaeodactylum tricornutum*, k_{La} of 0.006 s^{-1} was obtained at a superficial gas velocity of 0.25 m/s (Fernandez et al. 2001). It should be noted that comparison of the k_{La} based on only superficial gas velocity could be misleading, considering the differences in photobioreactor scales (volume), geometry, algal strains, and cultures as well as the methods used for such studies. Furthermore, liquid velocity is a measure of liquid flow and degree of turbulence in photobioreactors that is required to ensure that all the cells are frequently exposed to light (Carlozzi 2003; Provost et al. 2006). Also, solid velocity would give an indication of how algal cells can be uniformly transported along the tube length as the cultures are aerated. Solid velocity is also a very important parameter for the determination of hydrodynamics and mass transfer characteristics of bioreactors. Couvert et al. (2004) reported that the nature (i.e., shape, size, and porosity) and quantity of solids have an influence on the mass transfer in bioreactors. In intense algal cultures, cells can aggregate to form some clumps inside photobioreactors. In narrow-bore tubes, these clumps may settle such that they cannot be recirculated uniformly along the tubes.

Another important aspect of hydrodynamics that has been used in characterizing photobioreactor design is gas holdup. Gas holdup is described as the fraction of the reactor volume taken by the gas. This can be estimated as the volume of the liquid displaced by the gas (expansion of liquid volume) due to aeration. Gas holdup is very important in photobioreactor design as it determines the circulation rate, the gas residence time, as well as the overall mass transfer rate (k_{La}). Some studies have demonstrated that a relationship exists between gas holdup, bubble size, gas–liquid interfacial surface area, and the overall mass transfer coefficient, k_{La} (Chisti 1998; Vandu et al. 2005) as shown in Table 10.3.

10.3.7 Mass Cultivation of Algae

A good number of photobioreactors can be used in production of various algal products. Apparently, while many photobioreactors are easily operated at laboratory scale, only few of them can be successfully scaled up to pilot scale. Scale-up of photobioreactors can be done by increasing the length, diameter, height, or the number of compartments of the culture systems (depending on the type of photobioreactor). These scale-up strategies are very challenging, mainly due to difficulty in maintaining optimum light, temperature, mixing, and mass transfer in photobioreactors. Nevertheless, few large-scale photobioreactors with relatively good biomass productivities have been developed.

Table 10.4 shows the algal biomass productivities reported with different types and scales of outdoor photobioreactors. In 200 L airlift tubular photobioreactor (which was used for outdoor cultivation of *Phaeodactylum tricornutum*), biomass productivity of 1.20–1.50 g/L/day was obtained (Camacho et al. 1999; Fernandez et al. 2001).

Furthermore, with 200 L airlift-driven-tubular photobioreactor tested for outdoor cultivation of *Phaeodactylum tricornutum*, biomass productivity of 1.90 g/L/day was attained (Molina et al. 2001). In 11 L undular row tubular photobioreactor (for *Arthrospira platensis*), the productivity reported was about 2.7 g/L/day (Carlozzi 2003). Furthermore, about 0.27 g/L/day was obtained in 440 L outdoor flat-plate photobioreactor, which was used for cultivation of *Nannochloropsis* (Wu et al. 2001). In 55 L bubble-column photobioreactor (for outdoor cultivation of *H. pluvialis*), the biomass productivity obtained was 0.06 g/L/day (Lopez et al. 2006). Also, in 25,000 L outdoor photobioreactor (developed for commercial production of astaxanthin from *H. pluvialis*), about 0.05 g/L/day was obtained (Olaizola 2000). It should be noted that aside from volumetric productivity (productivity per unit of reactor volume per unit of time), algal biomass productivity can be evaluated in photobioreactors based on areal productivity (productivity per unit of occupied-land area per unit of time), PE, or biomass yield (g-biomass per unit of solar radiation) (Ugwu et al. 2008).

TABLE 10.4

Productivity of Algal Strains Reported in Some Outdoor Photobioreactors

Photobioreactors	Volume (L)	Photosynthetic Strain	Productivity (g/L/day)	References
Airlift tubular	200	*Porphyridium cruentum*	1.50	Camacho et al. (1999)
Airlift tubular	200	*Phaeodactylum tricornutum*	1.20	Fernandez et al. (2001)
Airlift tubular	200	*Phaeodactylum tricornutum*	1.90	Molina et al. (2001)
Inclined tubular	6.0	*Chlorella sorokiniana*	1.47	Ugwu et al. (2002)
Undular row tubular	11	*Arthrospira platensis*	2.70	Carlozzi (2003)
Parallel tubular (AGM)	25,000	*Haematococcus pluvialis*	0.05	Lopez et al. (2006)
Flat-plate	440	*Nannochloropsis* sp.	0.27	Wu et al. (2001)

Source: Ugwu, C.U. et al., *Bioresour. Technol.*, 99, 4021, 2008.

10.3.8 Conclusion and Perspectives of Photobioreactors

Despite the fact that a great deal of work has been done to develop photobioreactors for algal cultures, more efforts are still required to improve photobioreactor technologies and know-how of algal cultures. Photobioreactor development is perhaps one of the major steps that should be undertaken for efficient mass cultivation of algae. The major issue in the design of efficient photobioreactors should be their capacity to maximize the outdoor solar radiation. Large-scale outdoor photobioreactors should have large volume and occupy less land space. In addition, they should have transparent surfaces, high illumination surfaces, and high mass transfer rates and should as well be able to give high biomass yields. Furthermore, design and construction of any photobioreactor should depend on the type of strain, the target product, geographical location, as well as the overall cost of production. It should be noted that for mass cultivation of algae, vast areas of land are required. This is actually a very serious setback of algal cultivation in many developed countries. Thus, the increasing population and, consequently, the exorbitant cost of land have attracted the attention of many scientists to look for alternative cultivation sites. In order to reduce the cost of producing algal biomass and products, intensive efforts should be made to increase the algal biomass productivity. Also, high-value metabolites should be produced to compromise the technical costs involved in algal production. Given that outdoor photobioreactors are usually naturally illuminated using solar light, biomass productivities (in such systems) would depend on the prevailing weather conditions in a particular locality. Although commercial cultivation of algae is done in developed countries, there are seasonal variations in temperatures and solar light energy throughout the year in most of these regions. Due to these problems, it is difficult to carry out outdoor mass cultivation of algae all year round in such regions. However, in most tropical developing countries, outdoor cultures of algae can be maintained for relatively long periods of time in a year because there is neither winter nor cold seasons in those regions. Thus, tropical developing countries might be potential cultivation sites for commercial production of algal products (Ugwu et al. 2008).

10.4 Hydrogen from Lignocellulosic Biomass

This section is based on the review article "Bioconversion of lignocellulosic biomass to hydrogen: Potential and challenges" by Ren et al. (2009): Hydrogen is a promising alternative to conventional fossil fuels since it does not have the same environmental impact associated with fossil fuels. A major doubt on hydrogen as a clean energy alternative is how it is produced. Hydrogen production by microbial biochemical reactions has attracted worldwide attention, due to its potential as an inexhaustible, low-cost, and renewable source of clean energy. Studies on biohydrogen production have been focused on biophotolysis of water using algae and cyanobacteria, photodecomposition of organic compounds by photosynthetic bacteria, and dark fermentation from organic compounds with anaerobes. Among these biological processes, anaerobic hydrogen fermentation seems to be favorable, since hydrogen can be yielded at higher rates and at lower costs in degrading various organic wastes enriched with carbohydrates (Kumar and Das 2000; Chen et al. 2001; Ginkel et al. 2001; Fang et al. 2003; Xing et al. 2006; Lo et al. 2008a; Wang et al. 2008a). Lignocellulosic biomass in nature is by far the most abundant raw material from hardwood, softwood, grasses, and agricultural residues as well as municipal wastes (Table 10.5).

TABLE 10.5
Hydrogen Yields and Production Rates by Different Microorganisms

	Conditions						
Microorganism	**Culture**	**D (h^{-1})**	**pH**	**T (°C)**	**Substrate**	**H_2 Yield mol/mol Monosaccharide**	**H_2 Production Rate (Maximal) mmol/h/L**
Strict anaerobes							
Clostridia							
Clostridium sp. no 2	Batch		6.0	36	Glucose	2.0	23.9
	Batch		6.0	36	Xylose	2.1	21.7
C. paraputrificum M-21	Batch		Uncontrolled	37	GlcNAc[a]	2.5	31.0
C. butyricum LMG1213tl	Continuous	0.222	5.8	36	Glucose	1.5	21.7
Clostridium sp. no 2	Continuous	0.18	6.0	36	Glucose	2.4	7.1
	Continuous	1.16	6.0	36	Glucose	1.4	20.4
	Continuous	0.96	6.0	36	Xylose	1.7	21.0
Thermophiles							
Thermotoga maritime	Batch		Uncontrolled	80	Glucose	4.0	10
Thermotoga elfii	Batch		7.4	65	Glucose	3.3	2.7
Caldicellulosiruptor Saccharolyticus	Batch		7.0	70	Sucrose	3.3	8.4
Facultative anaerobes							
Enterobacter							
E. aerogenes E 82005	Batch		6.0	38	Glucose	1.0	21
E. cloacae IIT-BT 80 wt	Batch		Uncontrolled	36	Glucose	2.2	
	Batch		Uncontrolled	36	Sucrose	3.0	35
E. cloacae IIT-BT 08 m DM_{11}	Batch		Uncontrolled	36	Glucose	3.4	
E. aerogenes E 82005	Continuous	0.32	6.0	38	Molasses	0.7	20
E. aerogenes HU-101 wt	Continuous	0.67	Uncontrolled	37	Glucose	0.6	31
E. aerogenes HU-101 m AY-2	Continuous	0.67	Uncontrolled	37	Glucose	1.1	58

Coculture							
C. butyricum IFO13949+ *E. aerogenes* HO-39	Continuous	1.0	5.2	36	Starch	2.6	33
		HRT, h					
Mixed cultures from:							
Sludge compost	Continuous	12	6.8	60	Waste water Sugar factory	2.5	8.3
Sewage sludge	Continuous	6	5.7	35	Glucose	1.7	29.6
	Continuous	8	6.7	35	Sucrose	1.7	26.2
Fermented soybean meal	Continuous	8.5	6.0	35	Glucose	1.4	8

Source: Reith, J.H. et al., *Dutch Biol. Hydrogen Found.*, 2003.

[a] GlcNAc = *N*-Acetyl-D-glucosamine.

The annual yields of lignocellulosic biomass residues worldwide were estimated to exceed 220 billion tons, equivalent to 60–80 billion tons of crude oil. The lignocellulosic biomass hence presents an attractive, low-cost feedstock for hydrogen production. Its direct conversion to hydrogen needs pretreatment to hydrolyze the incorporated heterogeneous and crystalline structure. This section presents a critical review on the bioconversion of lignocellulosic biomass to hydrogen in terms of prehydrolysis and hydrolysis technology, bioconversion technology, and fermentation tactics for enhancing hydrogen production. Future perspectives of bioconversion of lignocellulosic biomass to hydrogen are discussed.

10.4.1 Pretreatment and Hydrolysis of Lignocellulosic Biomass

Lignocellulosic biomass is composed of carbohydrate polymers (cellulose and hemicellulose) and lignin. The agricultural residues, that is, wheat straw, corn stover, and rice straw, are comprised of 32%–47% cellulose, 19%–27% hemicellulose, and 5%–24% lignin. The composition of different softwoods varies widely. The cellulose and hemicellulose, which typically comprise two-thirds of the dry lignocellulosic materials, are polysaccharides. Both hemicellulose and lignin provide a protective sheath around the cellulose, which must be hydrolyzed prior to efficient utilization of the embedded polysaccharides. The hydrolysis generally includes prehydrolysis and cellulose hydrolysis: prehydrolysis of lignocellulosic materials is used to remove lignin and partly hydrolyze hemicellulose, while the cellulose hydrolysis is used to ferment reducing sugars. Figure 10.5 shows the general scheme for converting lignocellulose to biohydrogen.

10.4.1.1 Prehydrolysis (Lignocelluloses Fractionation)

Prehydrolysis, often referred to as pretreatment, is required to alter the structure of lignocellulosic biomass to make cellulose more accessible to the enzymes that convert the carbohydrate polymers to fermentable sugars (Wyman 1994). The prehydrolysis process can be performed by physical (mechanical comminution and hydrothermolysis), chemical (ozonolysis, acid hydrolysis, alkaline hydrolysis, oxidative delignification, and organosolv process), combined (acid-catalyzed steam explosion, ammonia fiber explosion, and CO_2 explosion), and biological (white or fungi) techniques (Fan et al. 1982; Wyman 1994; Fang et al. 1999; Ballesteros et al. 2002; Kim et al. 2002; Sun and Cheng 2002; Mosier et al. 2005; Pan et al. 2005; Yang and Wyman 2004; Yang and Wyman 2008). The selection of a pretreatment method affects the cost and performance in the subsequent hydrolysis and fermentation stages. An ideal prehydrolysis process should achieve high yields of fermentable reducing sugars, avoid degradation or loss of yielded sugars and the formation of inhibitors to the subsequent fermentation, and improve the later cellulose hydrolysis stage in terms of minimal energy, chemicals, and capital equipment use (Hamelinck et al. 2005).

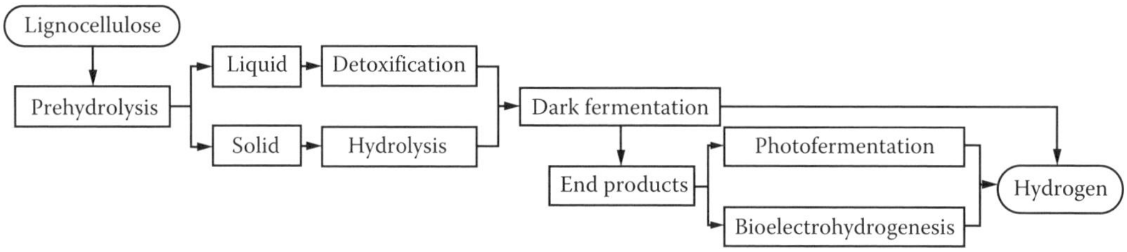

FIGURE 10.5
Flowchart of biohydrogen production from lignocellulosic biomass.

Of all pretreatments, the combined and the biological technologies are the most promising. The combined method, that is, sulfur-catalyzed steam explosion, is preferred for industrial applications, although it has some disadvantages such as destruction of a portion of the xylan fraction, removal/neutralization of the acid before fermentation, and generation of compounds that may be inhibitory to enzymes and microorganisms in the downstream processes (Nguyen et al. 1998; Larsson et al. 1999; Palmquist et al. 2000a,b). This pretreatment retains reasonably high sugar yields from hemicellulose and significantly increases the accessibility of enzymes to cellulose. Biological methods, although uncontrollable and insufficient, offer some unique advantages such as low energy requirement, low quantity of inhibitors produced, and operability at room temperature. White fungi, by secreting lignin-degrading enzymes including lignin peroxidases, manganese-dependent peroxidases, lignin peroxidases, and laccase during secondary metabolism (Sun and Cheng 2002), are widely adopted for lignin removal and lignocelluloses degradation although part of the cellulose would be consumed (Yang and Wyman 2008).

Through pretreatment, lignin and all or part of the hemicellulose is dissolved in the prehydrolysates. The free hemicellulose polymer is hydrolyzed to a mixture of monomeric and oligomeric sugars, including xylose, xylan, mannose, arabinose, galactose, and glucose that comes from cellulose hydrolysis. The residual cellulose and lignin can be filtered and washed. Following this way, cellulose could be collected in a separate process for hydrolysis—H_2 fermentation or direct H_2 fermentation (Hamelinck et al. 2005).

10.4.1.2 Cellulose Hydrolysis

The hydrolysis step is usually performed by enzymes or chemicals. The cellulose hydrolysates comprise of reduced saccharides with mainly glucose (Hamelinck et al. 2005). Although chemical processes are technically more mature, significant technical progress for enzymatic processes was made to make comparably low costs in terms of investments and operations. Enzymatic hydrolysis has some important advantages such as high yield of reduced sugars acquired under mild process conditions and no corrosion problems (Duff and Murray 1996; Hamelinck et al. 2005). Enzymatic hydrolysis of cellulose is carried out by cellulases that are from bacteria and fungi. These microorganisms could be aerobic or anaerobic, mesophilic or thermophilic. Among them, the Trichoderma is intensively studied for cellulase production (Sternberg 1976; Ghose and Bisaria 1979; Mononmani and Sreekantiah 1987; Beldman et al. 1988; Duff and Murray 1996). Simultaneous saccharification and fermentation (SSF) is adopted owing to reduced inhibition of end products from glucose and cellobiose. Using genetically engineered microorganisms harbored in the cellulase coding sequences provides new cellulase production system with improvement in enzyme production and activity (Sun and Cheng 2002; Hamelinck et al. 2005).

10.4.2 Bioconversion of Lignocellulosic Hydrolysate to Hydrogen

10.4.2.1 Functional Microorganisms

Functional microorganisms hydrolyzed the lignocellulosic biomass to fermentable reducing sugars, mainly composed of hexose and pentose (Sun and Cheng 2002; Mosier et al. 2005; Chen et al. 2008a,b). Identification of fermentative microorganisms that can effectively degrade hexose and/or pentose is essential to develop practical biohydrogen production process from lignocellulose.

10.4.2.1.1 Pentose Fermentation Microorganisms

The pentose fraction in hemicellulose consists mainly of xylose. Table 10.6 lists the functional microorganisms and mixed cultures reported for hydrogen production from xylose. Compared to hexose-fermenting microorganisms, only a few pentose-fermenting microorganisms have been identified. Among the known pentose-fermenting microorganisms, the mesophilic *Clostridium* sp. no. 2 (Taguchi et al. 1995) revealed the highest H_2 yield (2.36 mol H_2/mol xylose), even higher than the thermophiles of *Caldicellulosiruptor saccharolyticus* with a yield of 2.24 mol H_2/mol xylose (Kadar et al. 2004) and that of *Thermoanaerobacterium thermosaccharolyticum* W16 with a yield of 2.19 mol H_2/mol xylose (Ren et al. 2008a,b). Further isolation of the pentose-fermenting microorganisms is needed to explore the lignocellulosic hydrolysate utilization.

10.4.2.1.2 Simultaneous C5/C6 Fermentation Microorganisms

Lignocellulosic prehydrolysates contain a mixture of hexose and pentose. Typical hydrogen fermentation microorganisms, that is, *Ethanoligenens harbinense* B49 (Wang et al. 2007), could not metabolize pentoses. van Niel et al. (2003) and Kadar et al. (2004) demonstrated that their extreme thermophile *C. saccharolyticus* can produce H_2 from both mono- and disaccharides. Datar et al. (2007) reported that sewage sludge following heat pretreatment can simultaneously utilize mixed sugars to generate hydrogen from corn stover treated by a steam explosion process. The molar H_2 yields of 2.84 and 3.0 mol were obtained from the hydrolysates containing mixed sugars. Ren et al. (2008a,b) reported that a thermophilic strain of *T. thermosaccharolyticum* W16 could simultaneously ferment the mixture of glucose

TABLE 10.6

H2 Yields and Production Rates Reported in Xylose Fermentation Studies

Microorganism	Substrate	Conditions	*T* (°C)	pH	Q_{H_2} [a]	Y_{H_2} [b]
Pure cultures						
Clostridium sp. no. 2	Xylose	Continuous	25	6.0	21.0	2.36
Clostridium sp. HR-1	Xylose	Batch	35	6.5		2.14
C. tyrobutyricum ATCC 25755	Xylose	Batch	37	6.3	8.35	0.77
C. butyricum CGS5	Xylose	Batch	37	7.5	9.64	0.73
Enterobacter cloacae IIT-BT 08	Xylose	Batch	36	6.0	15.5	0.95
Caldicellulosiruptor saccharolyticus	Xylose	Batch	70	7.2	11.3	2.24
	Mix	Batch	70	7.2	9.2	2.32
	Glucose	Batch	70	7.2	10.7	2.5
T. thermosaccharolyticum W16	Xylose	Batch	60	6.5	10.7	2.19
	Mix	Batch	60	6.5	11.2–12.7	2.23–2.37
	Glucose	Batch	60	6.5	12.9	2.42
Mix cultures						
Sewage sludge	Xylose	Batch	35	6–7	5.94–8.93	1.92–2.25
	Xylose	Continuous	35	7.1	4.15	0.7
Sewage sludge	Xylose	Continuous	40	6.5	25.4	0.8
Microflora	Xylose	Batch	75	7.3	—	0.54
Sewage sludge	Xylose	Continuous	50	7.0	5.76	1.4
Compost	Xylose	Fed batch	55	5.0	0.27	1.7

Source: Ren, N. et al., *Biotechnol. Adv.*, 2009.

[a] Q_{H_2} is the maximal hydrogen production rate (mmol H_2/L/h).

[b] Y_{H_2} is the hydrogen yield (mol H_2/mol (substrate).

and xylose with a hydrogen yield of up to 2.37 mol H_2/mol substrate. Also, the W16 strain has a high tolerance to common prehydrolysate inhibitors, that is, acetate and furfural.

10.4.2.1.3 Microorganisms Involved in Cellulose Hydrolysates Fermentation

Hexose is the predominant component in the cellulose hydrolysates. Table 10.7 summarizes the reported microorganisms that can yield hydrogen from cellulose hydrolysate. The highest H_2 yield of approximately 83% of the theoretical value (theoretical is 4 mol/mol hexose) has been reported using thermophilic anaerobic bacteria (van Niel 2003).

The strict anaerobic *E. harbinense*, a new genus identified by Xing et al. (2006), also revealed high H_2 yield. The model strain, *E. harbinense* YUAN-3T, showed a special autoaggregation ability during shake cultivation. Hydrogen yield and hydrogen production rate of the strain *E. harbinense* YUAN-3T were 2.81 mol H_2/mol glucose and 27.6 mmol H_2/g dry cell/h, respectively. The autoaggregative property of the strain *E. harbinense* YUAN-3 renders this strain the potential to form cell granules for the development of compact hydrogen bioreactors for industrial use. The strict anaerobic Clostridia and facultative anaerobes produce hydrogen with comparable yields (approximately 2 mol H_2/mol glucose) (Heyndrickx et al. 1986; Tanisho et al. 1987; Taguchi et al. 1994; Kumar and Das 2000). The H_2 yield <2 mol H_2/mol glucose was reported for mixed cultures (Lin and Chang 1999; Chen et al. 2001). Wang et al. (2008) suggested that bioaugmentation with the addition of certain functional H_2-producing strains might be promising for further improving H_2 yield of mixed culture.

TABLE 10.7

Hydrogen Yields and Production Rates by Microorganisms as Reported in the Literature

Microorganism	Conditions	pH	T (°C)	Substrate	Y_{H_2}[a]	Q_{H_2}[b]
Strict anaerobes						
Clostridia						
C. sp. no 2	Batch	6.0	36	Glucose	2.0	23.9
C. butyricum LM G1213tl	Continuous	5.8	36	Glucose	1.5	21.7
C. beijerinckii AM21B	Batch	6.5	36	Glucose	2.0	
Thermophiles						
Thermotoga elfii	Batch	7.4	65	Glucose	3.3	2.7
Caldicellulosiruptor saccharolyticus	Batch	7.0	70	Sucrose	5.9	8.4
Ethanoligenens harbinense						
E. harbinense B49	Batch	3.9–4.2	35	Glucose	16–1.8	
E. harbinense Yuan-3	Batch		35	Glucose	2.81	
Facultative anaerobes						
Enterobacter						
E. aerogenes E.82005	Batch	6.0	38	Glucose	1.0	21
E. cloacae IIT-BT 08	Batch		36	Glucose	2.2	
E. aerogenes HO-39	Continuous	6–7	38	Glucose	1.0	
Mixed cultures						
Sludge compost	Continuous	6.8	60	Waste water		8.3
Sewage sludge	Continuous	5.7	35	Glucose	1.7	29.6
Sewage sludge	Continuous	6.7	35	Sucrose	1.7	26.2

Source: Ren, N. et al., *Biotechnol. Adv.*, 2009.

[a] Y_{H_2} is the hydrogen yield (mol H_2/mol) (substrate).

[b] Q_{H_2} is the maximal hydrogen production rate (mmol H_2/L/h).

10.4.2.2 Key Factors Affecting Bioconversion of Prehydrolysates/Hydrolysates to H_2

10.4.2.2.1 Inhibitors

Besides monomeric sugars, various toxic compounds were derived from lignocellulosic prehydrolysates such as fermentation inhibitors (Palmquist et al. 2000a,b; Mussatto et al. 2004). The fermentation inhibitors in the lignocellulosic prehydrolysates are composed of the following three groups:

1. The substances released during degradation of hemicellulose, including acetic acid and terpenes, alcohols, and aromatic compounds, whose toxicity was interpreted as penetration of the undissociated acid into the cell and dissociation of acid at higher intracellular pH (Pampulha et al. 1989). The fermentability of lignocellulosic hydrolysate can hence be improved by raising the pH.
2. The second group includes furfural, 5-hydroxymethyl furfural, levulinic acid, formic acid, and humic substances, which are the by-products for sugar degradation. Direct inhibition of both glycolytic (Banerjee et al. 1981) and nonglycolytic (Taherzadeh et al. 2000) enzymes has been suggested as a potential mechanism for furfural toxicity.
3. The third group of inhibitors is the lignin degradation products. This group of inhibitors includes a wide range of aromatic and polyaromatic compounds (Lo et al. 2008a; Wang et al. 2008).

10.4.2.2.2 pH and Temperature

These are two principal process factors that affect biohydrogen production performance. The reported optimal pH range for hydrogen production ranged 5.0–7.5 (Chen et al. 2001; Fang et al. 2002; Khanal et al. 2004; Calli et al. 2008). Ren et al. (1997) stated that the optimal pH for hydrogen production can be less than 5.0. Decline in pH value is noted in hydrogen production process owing to the production of organic acids that depletes the buffering capacity of the medium (Khanal et al. 2004). Since pH affects the activity of iron-containing hydrogenase enzyme, decreases in pH will inhibit hydrogen production (Dabrock et al. 1992).

Hydrogen production was tested under mesophilic conditions using glucose, sucrose, and wastewater as substrates (Mononmani et al. 1987; Ueno et al. 1996; Yokoi et al. 1998). Mesophilic biohydrogen production is preferred for preventing the need of external heating. However, hydrogen fermentation at high temperatures has high hydrogen yield, owing to the suppression of H_2-consuming bacteria and the capability to utilize numerous substrates (Talabardon et al. 2000; Fang et al. 2003). Lin et al. (2008) demonstrated a two-time hydrogen yield at thermophilic (50°C–55°C) compared with mesophilic (30°C–40°C) fermentation. Moreover, in terms of future application, it is possible to couple the H_2 fermentation with bioethanol production from lignocellulosic biomass under thermophilic condition.

10.4.2.2.3 Nutrient Supplementation

The medium components added to the hydrolysate affect the fermentation performance. Medium components should fulfill the carbon and nitrogen need of the involved microorganisms. Also, trace metals, growth factors, and vitamins are also essential for cell growth. Studies on hexose fermentation or lignocellulosic hydrolysates mainly conducted experiments using laboratory media with yeast extract and peptone. The strain *Thermotoga elfii*, which is capable of producing high yields of hydrogen from glucose, grows on a medium enriched with yeast extract (van Niel et al. 2002). The strain *T. thermosaccharolyticum* W16 can simultaneously intake glucose and xylose for hydrogen production without extract or tryptone. However, with these, the strain W16 can grow even better when extract and tryptone are present (Ren 2008a,b). The strain *Clostridium butyricum* AS1.209 exhibited a C/N-ratio-dependent growth characteristic, and a proper C/N ratio enhances biohydrogen production in SSF from steam-exploded corn straw (Li and Chen 2007).

10.4.2.3 Fermentation Tactics for Lignocellulosic Prehydrolysates to Hydrogen

10.4.2.3.1 Detoxification

Removal of inhibitors from lignocellulosic hydrolysates enhances the fermentation efficiencies (de Vrije et al. 2002). Effects of adding activated charcoal, extracting with organic solvents, ion exchange, ion exclusion, or molecular sieves have been investigated (Palmquist et al. 2000a,b; Mussatto et al. 2004). Different detoxification methods removed partially on the hydrolysate toxicity in different magnitude. The choice of detoxification method depends on the hydrolysate origin and the involved fermentation microorganisms.

10.4.2.3.2 Bioaugmentation for Higher H_2 Production

Increasing cell density has been shown in several cases to be a suitable way of increasing the volumetric productivity of biohydrogen (Zhu et al. 1999; Jo et al. 2008). Moreover, high cell density is beneficial to overcome toxicity of lignocellulosic hydrolysate (Shreenath and Batch 1987). Cell recycling to reactor or cell immobilization enrich cell density. Cheap attachment material is developed (Kumar et al. 2001). Cells can be immobilized by adhesion to a surface (electrostatic or covalent), entrapment in polymeric matrices, or retention by membranes for hydrogen production. The strain *Enterobacter cloacae* IIT-BT 08 was immobilized in an environmentally friendly solid matrix via adsorption technique. The maximum hydrogen production rate was 75.6 mmol/L/h at a dilution rate of 0.93 h^{-1}. The substrate conversion efficiency was increased by 15% at immobilized conditions (Kumar et al. 2001).

Single culture may not be capable of degrading a few monosaccharides in the lignocellulosic prehydrolysates. We believe that cocultured microorganisms should be a better configuration than single culture for most efficient hydrogen production from the lignocellulosic hydrolysates.

10.4.3 Direct Microbial Conversion of Cellulose to Hydrogen

10.4.3.1 Microorganisms Involved in Converting Cellulose to Hydrogen

10.4.3.1.1 Pure Culture

A few pure cultures can directly degrade cellulose to produce hydrogen. Wang et al. (2008) reported that *Clostridium acetobutylicum* X9 generated the maximum hydrogen production and cellulose hydrolysis rate of 6.4 mmol H_2/h/g dry cell and 68.3%, respectively, using microcrystalline cellulose as the substrate. These authors also showed that X9 can degrade pretreated acetic steam-explored corn stalks to hydrogen with specific hydrogen production rate of 3.4 mmol/g steam-exploded corn stalks. Levin et al. (2006) demonstrated that the hydrogen yield of strain *Clostridium thermocellum* 27405 could reach 1.6 mol H_2/mol glucose using *de*lignified *w*oods (*DLWs*).

The pure culture system is attractive and is preferred for mechanistic research and genetic reconstruction for improving cellulose hydrolysis rate and hydrogen yield. However, strain isolation technique is complicated and time-consuming. In addition, only a small fraction of microbes can be cultivated.

TABLE 10.8

Biohydrogen Production Performance from Cellulosic Biomass with Mixed Culture

Microorganism	Cellulosic Substrate	Experimental Conditions	Y_{H_2} [a]	Specific Q_{H_2} Rate
Heat-shock digested sludge	Microcrystalline cellulose	37°C, 12 days	2.18 mmol H_2/g Avice	18 mmol H_2/g VSS/day
Microflora	Cellulose	55°C, 200 h	4.55 mmol H_2/g cellulose	12.8 mmol H_2/g VSS/day
Pig-dung compost	Rice straw	37°C, 17 days	53.5 mL H_2/g TS	
Anaerobic digested sludge	Corn stover	36°C, 300 h	5.66 mmol H_2/g TVS	
	Beer lees		2.42 mmol H_2/g TVS	
	Wheat bran		4.55 mmol H_2/g TVS	
Cow-dung compost	Wheat straw	126.5 h	3.04 mmol H_2/g TVS	10.14 mL H_2/g TVS/h
Cow-dung microflora	α-Cellulose and saccharification products		2.8 mmol H_2/g cellulose	

Source: Ren, N. et al., *Biotechnol. Adv.*, 2009.

[a] The listed data were calculated based on the original data.

10.4.3.1.2 Mixed Cultures

Recently, the research on biological production of cellulose hydrogen has been focused on mixed-culture systems (Lay et al. 1999; Ginkel et al. 2001; Fang et al. 2002) for its greater cellulose conversion rates and broader carbon sources. In addition, the mixed culture, such as natural microflora, anaerobic digested sludge, or compost, contains large amounts of organisms that could serve as the ideal microorganism resources for cellulose hydrolysis. Table 10.8 lists the hydrogen yields from cellulosic materials with mixed cultures. Among them, the highest hydrogen yield reported to be 4.55 mmol H_2/g cellulose by microflora (Liu et al. 2003).

10.4.3.2 Bioaugmentation for Cellulose Degradation to Hydrogen

Bioaugmentation via coculture or community system to further enhance H_2 production was studied for a few years. Rarely, bioaugmentation is applicable to practical cellulose degradation to hydrogen process. Complementarily, functions between the augmented strains and the indigenous strains overcome the often encountered repression problem in cellulose hydrolysis process, such as the characteristic instability, structure, and mechanical pathway complication. Wang et al. (2008) reported dark fermentation of microcrystalline cellulose to produce biohydrogen using coculture of *C. acetobutylicum* X9 and *E. harbinense* B49. The maximum hydrogen yield in the bioaugmented cocultured system was 16.2 mg H_2/g cellulose. B49 can rapidly remove reduced sugar produced by X9 and hence improved cellulose hydrolysis and subsequent hydrogen production rates. *C. thermocellum* JN4 and its counterpart bacterium, a hydrogen-producing *T. thermosaccharolyticum* GD17, were investigated in terms of the mechanisms of interactive cooperation in cellulose degradation and hydrogen production. The data showed that when *C. thermocellum* JN4 was cocultured with *T. thermosaccharolyticum* GD17, hydrogen production increased about twofold and H_2 yield increased to a high level of 1.8 mol H_2/mol glucose (Liu et al. 2008a,b).

Liu et al. (2008a) studied the cellulose hydrolysis activity of two mixed bacterial consortia (NS and QS). Using the cellulosic hydrolysate containing 0.8 g/L reducing sugar,

Clostridium pasteurianum CH_7 attained the best H_2 production and yield of approximately 23.8 mL/L and 1.21 mmol H_2/g reducing sugar, respectively.

Intensive research on microbial communities for cellulose degradation systems was conducted in terms of their stable functional characteristic, microbial community composition and structure, reaction pathways, and mechanism to tolerant harsh environments. Haruta et al. (2002) developed a microbial community capable of degrading N60% rice straw within 4 days at 50°C. The stability of the community was demonstrated using multiple subcultures for several times in medium with/without cellulosic material, being heated to 95°C and being frozen at −80°C. Ren et al. (2009) proposed a functional community (rumen cellulose degradation bacteria consortia) that could simultaneously degrade cellulose and produce hydrogen and establish the isolation strategy.

10.4.3.3 Key Factors Affecting Cellulose Degradation and Hydrogen Production

Factors affecting the cellulose degradation process mainly lie in environmental factors, such as initial and end fermentation pH, the addition quantity and type of nitrogen resource, feedback inhibition of cellulose hydrolysis products–soluble saccharide, as well as the inhibition of the end-products accumulation, such as acetate, lactate, and butyrate. In addition, substrate type and concentration also have strong effect on cellulose degradation. *Bacteroides cellulosolvens* efficiently hydrolyzed a variety of cellulosic substrates. This strain can grow up to degradation of 20 g cellulose/L. Further, cellulose hydrolysis and sugar accumulation were accomplished by cellulose enzymes. This growth cessation was not due to low pH, to nutrient depletion, or to toxic accumulation of any of the major end products (CO_2, ethanol, acetic acid, lactic acid, cellobiose, glucose, or xylose) (Murray 1986; Narahiro and Sekiguchi 2007).

Initial pH may influence the lag phase in batch cellulose-to-hydrogen tests. Some studies reported that the low initial pH of 4.0–4.5 causes long lag periods of about 20 h (Khanal et al. 2004; Liu and Shen 2004). Alkaline pH decreases lag time; however, it may cause low yield of hydrogen (Zhang et al. 2003). Gradual decreases in pH during fermentation inhibit cellulose degradation since pH affects the activity of iron-containing cellulase enzyme (Dabrock et al. 1992). The pH could also affect the mechanisms of fermentation, the specific hydrogen production rate, and the hydrogen content in the gas phase. The optimal pH range for maximum hydrogen yield or specific hydrogen production rate is 5–6 (Lay 1999; Chen et al. 2001; Fang et al. 2003; Khanal et al. 2004), with some others declaring 6.8–8 (Lay 2001; Zhang et al. 2003; Collet et al. 2004; Kanai et al. 2005) or around 4.5 for the thermophilic culture (Shin et al. 2004).

Most studies revealed that the final pH in cellulose hydrolysis to hydrogen ranges 4.0–4.8 regardless of initial pH (Evvyernie et al. 2001; Lay 2001; Liu et al. 2003; Zhang et al. 2003; Liu and Shen 2004). The intense drop in pH level is due to the production of organic acids that weaken the buffering capacity of the medium (Khanal et al. 2004). The pH value should be properly controlled in biohydrogen production. The principal end products for anaerobic wastewater pretreatment from hard-to-decompose carbohydrates (cellulose) are propionate, acetate, CO_2, and butyrate (Karapinar and Fikret 2006). However, the production of propionate could have adverse effect on hydrogen production. Methane was not detected in most of the hydrogen production studies since heat treatment removes methane producers in sludge (Yu et al. 2002; Lin and Lay 2004; Liu and Shen 2004). Han and Shin (2004) probed methane by mesophilic cultures at long retention times of reactors.

Nitrogen is an essential nutrient for hydrogen production by dark fermentation under anaerobic conditions. Yokoi et al. (2002) reported that the highest level of hydrogen

(2.4 mol H_2/mol glucose) could be obtained from starch in the presence of 0.1% polypeptone. But no hydrogen production was observed when inorganic nitrogen salts were used as nitrogen source. Corn-steep liquor is a waste of corn starch manufacturing process that can be used as a nitrogen source (Yokoi et al. 2002). Lin and Lay (2004) reported that the C/N ratio affects hydrogen productivity more significantly than the specific hydrogen production rate.

The inhibition induced by accumulated saccharides can affect the performance of cellulose-to-hydrogen process. Ramos et al. (1986) observed lower hydrolysis rates at higher sugar concentrations. The removal of soluble sugars liberated during hydrolysis should enhance the efficiency of hydrolysis of the residual substrate. Levin et al. (2006) conducted batch tests with *C. thermocellum* 27,405 to produce biohydrogen from cornstalk wastes. Results showed that initial substrate type, substrate concentration, substrate moisture content, supplemental nutrient concentration, and duration of cultivation time could markedly affect the product quality. Zhang et al. (2007) converted cellulose in various wastewaters to H_2 by a mixed culture in batch tests at 55°C, pH 5.5–8.5, and cellulose concentrations of 10–40 g/L. The maximum cumulative H_2 yield of 149.69 mL H_2/g TVS was obtained at initial pH 7.0 and substrate concentration 15 g/L.

10.4.3.4 Integrated Process System for Multi-Output Solutions

Cellulose is the most abundant biopolymer in the world [384]. The 1.4 billion dry tons of lignocellulosic biomass produced annually in China can yield up to 100 billion kg H_2 (70% of the amount of total cellulose and hemicellulose was used for this preliminary calculation). Here, producing 1.0 kg of H_2 (roughly equivalent to the energy content of a gallon of gasoline) would take 10.3 kg cellulose based on 73% of H_2 recovery reported by Cheng and Logan (2007a). This amount of H_2 is equivalent to 500 billion kg standard coal, which is around 22.4% of the total coal consumption (2.2 billion tons) amount in China in 2006. Hence, bioconversion of cellulose provides a viable approach to produce renewable hydrogen from organic matter, provided a high efficient combinatory process system is adopted.

The combined dark fermentation coupling with photofermentation or dark fermentation coupling with bioelectrohydrogenesis is a promising hydrogen production process from lignocellulosic biomass if the technological barriers can be overcome. First, the development of a well-integrated balanced system is the key. Because hydrogen production rates were much lower from cellulose (0.11 m^3/m^3/day) than from glucose (1.23 m^3/m^3/day), and the photofermentation rates were 100 times lower than the dark processes, the rates of hydrolysis and fermentation by electrohydrogenesis in microbial electrolysis cell (MEC) system or in photofermentation are different. The fermentation and electrohydrogenesis rates need to be regulated using feeding strategies or a two-stage process, using reactors especially designed to handle particulate substrates. Increasing the H_2 production efficiency by the functional microorganisms is another essential issue. Aiming at this, the metabolic modification at both molecular and physioecological levels is necessary. At the molecular level, as indicated in US DOE Genomics: GTL roadmap systems biology for energy and environment, studies are needed in modifying the key enzymes and functional pathways, controlling the organic matter transportation, and even regulating or reconstructing the entire metabolic network in order to increase the hydrogen yield and lignocelluloses degradation. At the microbial community level, establishing the linkage between process system function and microbial diversity is even more challenging. This problem is partially due to insufficient experimental information on community-wide spatial and temporal dynamics of microbial community structure, function, and activity for rigorous examination. Overall, to develop a mature hydrogen production technology, bioconversion performance from lignocellulosic biomass will need to be further improved in production rates, cost-effectiveness, and system scale-up. Rapid advancements in the development of multiprocesses coupled with additional funding and mechanistic research into microbial processes should make rapid commercialization of this new biohydrogen technology possible (Ren et al. 2009).

10.4.3.5 Coupling Process of Hydrogen–Methane Fermentation

Biological hydrogen production process does not significantly reduce the organic content in terms of the COD of the feedstock. Usually, COD removal is below 20% during hydrogen production (Antonopoulou et al. 2008). The residual VFAs such as acetic acid exist in the fermenting waste liquor, which can be utilized if a methane fermentation stage is followed. The combined hydrogen fermentation and methane fermentation system will benefit the maximal energy recovery from feed biomass (Han and Shin 2004; Liu et al. 2006; Cooney et al. 2007; Ting and Lee 2007; Ueno et al. 2007; Zhu et al. 2008). For instance, Zhu et al. reported that, via coupling hydrogen and methane fermentation from potato waste, the total COD0 in the feedstock was removed by 64% and the hydrogen and methane yields were totally 30 L/kg TS (total solids) (with a maximum of 68 L/kg) or 183 L/kg TS (with a maximum of 225 L/kg). Cooney et al. reported that methane production in the two-phase anaerobic digestion process is more stable and effective than the one-phase process. The need for clean production of renewable energy from lignocellulosic biomass will thus hasten the development of combination of hydrogen and methane production (Ren et al. 2009).

10.4.3.6 Integrated Process System for Multioutput Solutions

Based on the aforementioned review, it is clear that the bioconversion of lignocelluloses to H_2 shows a very feasible solution to produce hydrogen via biotechnology. Concern remains on how to establish an applicable and affordable lignocellulose-to-H_2 process. A considerable pathway is to develop an integrated process for multioutputs from raw wastes including lignocellulosic H_2–CH_4, lignocellulosic ethanol, and for high- and low-value-added bioproducts (i.e., biobutanol, bioflocculants). Through this approach, a multichoice scheme might be performed according to the needs for the future bioenergy and bioproducts market. Aside from the scientific assessment, a careful and detailed economic programming is needed in order to boost this entire blueprint into practice.

References

Acién Fernández, F.G., Fernández Sevilla, J.M., Sánchez Peréz, J.A., Molina Grima, E., and Chisti, Y. Airlift-driven external loop tubular photobioreactors for outdoor production of microalgae: Assessment of design and performance. *Chem. Eng. Sci.* 56 (2001): 2721–2732.

Amos, W.A. Biological water-gas shift conversion of carbon monoxide to hydrogen: Milestone completion report NREL/MP-560-35592 (January 2004).

Antonopoulou, G., Gavala, H.N., Skiadas, I.V., Angelopoulos, K., and Lyberatos, G. Biofuels generation from sweet sorghum: Fermentative hydrogen production and anaerobic digestion of the remaining biomass. *Bioresour. Technol.* 99 (2008): 110–119.

Aoyama, K., Uemura, I., Miyake, J., and Asada, Y. Fermentative metabolism to produce hydrogen gas and organic compounds in a cyanobacterium *Spirulina Platensis. J. Ferment. Bioeng.* 83 (1997): 17–20.
Artero, V. and Fontecave, M. Some general principles for designing electrocatalysts with hydrogenase activity. *Coor. Chem. Rev.* 249 (2005): 1518–1535.
Axelsson, R. and Lindblad, P. Transcriptional regulation of *Nostoc* hydrogenases: Effects of oxygen, hydrogen, and nickel. *Appl. Environ. Microbiol.* 68(1) (2002): 444–447.
Ballesteros, I., Oliva, J.M., Negro, M.J., Manzanares, P., and Ballesteros M. Enzymic hydrolysis of steam exploded herbaceous agricultural waste (*Brassica carinata*) at different particle sizes. *Process Biochem.* 38 (2002): 187–192.
Banerjee, N., Bhatnagar, R., and Viswanathan, L. Inhibition of glycolysis by furfural in *Saccharomyces cerevisiae. Eur. J. Appl. Microbiol. Biotechnol.* 11 (1981): 224–228.
Basak, N. and Das, D. The prospect of purple non-sulfur [Pns] photosynthetic bacteria for hydrogen production: The present state of the art. *World J. Microbiol. Biotechnol.* 23(1) (2007): 31–42.
Bauer, C.G. and Forest, T.W. Effect of hydrogen addition on the performance of methane-fueled vehicles. Part I. Effect on S.I. engine performance. *Int. J. Hydrogen Energy* 26 (2001): 55–70.
Beldman, G., Voragen, A.G.J., Rombouts, F.M., and Pilnik, W. Synergism in cellulose hydrolysis by endoglucanases and exoglucanases purified from *Trichoderma viride. Biotechnol. Bioeng.* 31 (1988): 173–178.
Benemann, J. Hydrogen biotechnology: Progress and prospects. *Nat. Biotechnol.* 14(9) (1996): 1101–1103.
Benemann, J.R. Hydrogen production by microalgae. *J. Appl. Phycol.* 12 (2000): 291–300.
Benemann, J.R., Berenson, J.A., Kaplan, N.O., and Kamen, M.D. Hydrogen evolution by chloroplast-ferredoxin-hydrogenase system. *Proc. Natl. Acad. Sci. USA* 70 (1973): 2317–2320.
Benemann, J.R. and Weare, N.M. Hydrogen evolution by nitrogen-fixing *Anabaena cylindrica* cultures. *Science* 184 (1974): 174–175.
Bisaillon, A., Turcot, J., and Hallenbeck, P.C. The effect of nutrient limitation on hydrogen production by batch cultures of *Escherichia coli. Int. J. Hydrogen Energy* 31 (2006): 1504–1508.
Bolton, J.R. Solar photoproduction of hydrogen, IEA Agreement on the Production and Utilization of Hydrogen. (1996).
Bosca, C., Dauta, A., and Marvalin, O. Intensive outdoor algal cultures. How mixing enhances the photosynthetic production rate. *Bioresour. Technol.* 38 (1991): 185–188.
Boussiba, S., Sandbank, E., Shelef, G., Cohen, Z., Vonshak, A., Ben-Amotz, A., Arad, S., and Richmond, A. Outdoor cultivation of the marine microalga *Isochrysis galbana* in open reactors. *Aquaculture* 72 (1988): 247–253.
Cai, M., Liu, J., and Wei, Y. Enhanced biohydrogen production from sewage sludge with alkaline pretreatment. *Environ. Sci. Technol.* 38(11) (2004): 3195–3202.
Call, D. and Logan, B.E. Hydrogen production in a single chamber microbial electrolysis cell lacking a membrane. *Environ. Sci. Technol.* 42(9) (2008): 3401–3406.
Calli, B., Schoenmaekers, K., Vanbroekhoven, K., and Diels, L. Dark fermentative H_2 production from xylose and lactose—Effects of on-line pH control. *Int. J. Hydrogen Energy* 33 (2008): 522–530.
Camacho, F.G., Gomez, A.C., Fernandez, F.G.A., Sevilla, J.M.F., and Grima, E.M. Use of concentric-tube airlift photobioreactors for microalgal outdoor mass cultures. *Enzyme Microb. Technol.* 24 (1999): 161–172.
Camacho Rubio, F., Acién Fernández, F.G., Sánchez Pérez, J.A., García Camacho, F., and Molina Grima, E. Prediction of dissolved oxygen and carbon dioxide concentration profiles in tubular photobioreactors for microalgal culture. *Biotechnol. Bioeng.* 62 (1999): 71–86.
Carlozzi, P. Dilution of solar radiation through culture lamination in photobioreactor rows facing south–north: A way to improve the efficiency of light utilization by cyanobacteria (*Arthrospira platensis*). *Biotechnol. Bioeng.* 81 (2003): 305–315.
Chaumont, D., Thepenier, C., and Gudin, C. Scaling up a tubular photobioreactor for continuous culture of *Porphyridium cruentum*—From laboratory to pilot plant. In: Stadler, T., Morillon, J., Verdus, M.S., Karamanos, W., Morvan, H., Christiaen, D. (Eds.), *Algal Biotechnology*. Elsevier Applied Science, London, U.K. (1988): pp. 199–208.

Chen, C.C., Lin, C.Y., and Chang, J.S. Kinetics of hydrogen production with continuous anaerobic cultures utilizing sucrose as the limiting substrate. *Appl. Microbiol. Biotechnol.* 57 (2001): 56–64.

Chen, C.Y., Lu, W.B., Liu, C.H., and Chang, J.S. Improved phototrophic H_2 production with Rhodopseudomonas palustris WP3-5 using acetate and butyrate as dual carbon substrates. *Bioresour. Technol.* 99 (2008a): 3609–3616.

Chen, C.Y., Lu, W.B., Wu, J.F., and Chang, J.S. Enhancing phototrophic hydrogen production of *Rhodopseudomonas palustris* via statistical experimental design. *Int. J. Hydrogen Energy* 32(8) (2007): 940–949.

Chen, C.Y., Yang, M.H., Yeh, K.L., Liu, C.H., and Chang, J.S. Biohydrogen production using sequential two-stage dark and photo fermentation processes. *Int. J. Hydrogen Energy* 33 (2008): 4755–4762.

Chen, M., Zhao, J., and Xia, L.M. Enzymatic hydrolysis of maize straw polysaccharides for the production of reducing sugars. *Carbohydr Polymers* 71 (2008b): 411–415.

Chen, S.D., Sheu, D.S., Chen, W.M., Lo, Y.C., Huang, T.I., Lin, C.Y. et al. Dark hydrogen fermentation from hydrolyzed starch treated with recombinant amylase originating from *Caldimonas taiwanensis* On1. *Biotechnol. Prog.* 23 (2007): 1312–1320.

Cheng, S. and Logan, B.E. Ammonia treatment of carbon cloth anodes to enhance power generation of microbial fuel cells. *Electrochem. Commun.* 9 (2007a): 492–496.

Cheng, S. and Logan, B.E. Sustainable and efficient biohydrogen production via electrohydrogenesis. *Proc. Natl. Acad. Sci. USA* 104(47) (2007b): 18871–18873.

Cheng, S. and Logan, B.E. Sustainable and efficient biohydrogen production via electro-hydrogenesis. *PNAS* 104 (2007c): 18871–18873.

Cheng-Wu, Z., Zmora, O., Kopel, R., and Richmond, A. An industrial size flat glass reactor for mass production of *Nannochloropsis* sp. (Eustigmatophyceae). *Aquaculture* 195 (2001): 35–49.

Cheong, D.Y. and Hansen, C.L. Bacterial stress enrichment enhances anaerobic hydrogen production in cattle manure sludge. *Appl. Microbiol. Biotechnol.* 72 (2006): 635–643.

Chin, H.L., Chen, Z.S., and Chou, C.P. Fedbatch operation using *Clostridium acetobutylicum* suspension culture as biocatalyst for enhancing hydrogen production. *Biotechnol. Prog.* 19 (2003): 383–388.

Chisti, Y. Pneumatically agitated bioreactors in industrial and environmental bioprocessing: hydrodynamics, hydraulic, and transport phenomena. *Appl. Mech. Rev.* 51 (1998): 33–112.

Choi, S.L., Suh, I.S., and Lee, C.G. Lumostatic operation of bubble column photobioreactors for *Haematococcus pluvialis* cultures using a specific light uptake rate as a control parameter. *Enzyme Microb. Technol.* 33 (2003): 403–409.

Collet, C., Adler, N., Schwitzguebel, J.P., and Peringer, P. Hydrogen production by clostridium thermolacticum during continuous fermentation of lactose. *Int. J. Hydrogen Energy.* 29 (2004): 1479–1485.

Contreras, A., García, F., Molina Grima, E., and Merchuk, J.C. Interaction between CO_2-mass transfer, light availability and hydrodynamic stress in the growth of *Phaeodactylum tricornutum* in a concentric tube airlift photobioreactor. *Biotechnol. Bioeng.* 60 (1998): 318–325.

Cooney, M., Maynard, N., Cannizzaro, C., and Benemann, J. Two-phase anaerobic digestion for production of hydrogen–methane mixtures. *Bioresour. Technol.* 98 (2007): 2641–2651.

Couvert, A., Bastoul, D., Roustan, M., and Chatellier, P. Hydrodynamic and mass transfer study in a rectangular three-phase airlift loop reactor. *Chem. Eng. Proc.* 43 (2004): 1381–1387.

Dabrock, B., Bahl, H., and Gottschalk, G. Parameters affecting solvent production by *Clostridium pasteurium. Appl. Environ. Microbiol.* 58 (1992): 1233–1239.

Das, D., Dutta, T., Nath, K., Kotay, S.M., Das, A.K., and Veziroglu, T.N. Role of Fe-hydrogenase in biological hydrogen production. *Curr. Sci.* 90 (2006): 1627–1637.

Datar, R., Huang, J., Maness, P.C., Mohagheghi, A., Czernik, S., and Chornet, E. Hydrogen production from the fermentation of corn stover biomass pretreated with a steam-explosion process. *Int. J. Hydrogen Energy* 32 (2007): 932–939.

Degen, J., Uebele, A., Retze, A., Schmidt-Staigar, U., and Trosch, W.A. A novel airlift photobioreactor with baffles for improved light utilization through flashing light effect. *J. Biotechnol.* 92 (2001): 89–94.

Demirbas, A. Biodiesel: A realistic fuel alternative for diesel engines. *Chem. Mater. Sci.* (2008).

Ditzig, J., Liu, H., and Logan, B.E. Production of hydrogen from domestic wastewater using a bioelectrochemically assisted microbial reactor [BEAMR]. *Int. J. Hydrogen Energy* 32(13) (2007): 2296–2304.

Duff, S.J.B. and Murray, W.D. Bioconversion of forest products industry waste cellulosics to fuel ethanol: A review. *Bioresour. Technol.* 55 (1996): 1–33.

Ensign, S.A. and Ludden, P.W. Characterization of the CO oxidation/H_2 evolution system of Rhodospirillum rubrum. Role of a 22-kDa iron–sulfur protein in mediating electron transfer between carbon monoxide dehydrogenase and hydrogenase. *J. Biol. Chem.* 266 (1991): 18395–18403.

Eroglu, I., Aslan, K., Gunduz, U., Yucel, M., and Turker, L. Substrate consumption rates for hydrogen production by *Rhodobacter sphaeroides* in a column photobioreactor. *J. Biotechnol.* 70(1–3) (1999): 103–113.

Esper, B., Badura, A., and Rögner, M. Photosynthesis as a power supply for (bio-)hydrogen production. *Trends Plant Sci.* 11 (2006): 543–549.

Evyernie, D., Morimoto, K., Karita, S., Kimura, T., Sakka, K., and Ohmiya, K. Conversion of chitinous wastes to hydrogen gas by *Clostridium paraputrificum* M-21. *J. Biosci. Bioeng.* 91 (2001): 339–343.

Fabiano, B. and Perego, P. Thermodynamic study and optimization of hydrogen production by Enterobacter aerogenes. *Int. J. Hydrogen Energy* 27 (2002): 149–156.

Fan, L.T., Lee, Y.H., and Gharpuray, M.M. The nature of lignocellulosics and their pretreatments for enzymatic hydrolysis. *Adv. Biochem. Eng. Biotechnol.* 23(1982):158–187.

Fang, H.H.P., Li, C.L., and Zhang, T. Acidophilic biohydrogen production from rice slurry. *Int. J. Hydrogen Energy* 31 (2006): 683–692.

Fang, H.H.P. and Liu, H. Effect of pH on hydrogen production from glucose by a mixed culture." *Bioresour. Technol.* 82 (2002): 87–93.

Fang, H.H.P., Liu, H., and Zhang, T. Phototrophic hydrogen production from acetate and butyrate in wastewater. *Int. J. Hydrogen Energy* 30(7) (2005): 785–793.

Fang, H.H.P., Zhang, T., and Liu, H. Biohydrogen production from starch in wastewater under thermophilic condition. *J. Environ. Manag.* 69 (2003): 149–156.

Fang, H.H.P., Zhu, H., and Zhang, T. Phototrophic hydrogen production from glucose by pure and co-cultures of *Clostridium butyricum* and *Rhodobacter sphaeroides. Int. J. Hydrogen Energy* 31(15) (2006): 2223–2230.

Fang, J.M., Sun, R.C., Salisbury, D., Fowler, P., and Tomkinson, J. Comparative study of hemicelluloses from wheat straw by alkali and hydrogen peroxide extractions. *Polym. Degrad. Stab.* 66 (1999): 423–432.

Fedorov, A.S., Tsygankov, A.A., Rao, K.K., and Hall, D.O. Hydrogen photoproduction by *Rhodobacter sphaeroides* immobilised on polyurethane foam. *Biotechnol. Lett.* 20 (1998): 1007–1009.

Fontoynont, M., Dumortier, D., Heinemann, D., Hammer, A., Olseth, J., Skarveit, A., Ineichen, P., Reise, Ch.P.J., Roche, L., Beyer, H.G., and Wald, L. Satellight: A www server which provides high quality daylight and solar radiation data for western and central Europe. *Proceedings of the Ninth Conference on Satellite Meteorology and Oceanography*, Paris, France (1998).

Gaffron, H. and Rubin, J. Fermentative and photochemical production of hydrogen in algae. *J. Gen. Physiol.* (1942) 219–240.

García Malea López, M.C., Del Río Sánchez, E., Casas López, J.L., Acién Fernández, F.G., Fernández Sevilla, J.M., Rivas, J., Guerrero, M.G., and Molina Grima, E. Comparative analysis of the outdoor culture of *Haematococcus pluvialis* in tubular and bubble column photobioreactors. *J. Biotechnol.* 123 (2006): 329–342.

Gest, H. and M.D. Kamen. Photoproduction of molecular hydrogen by *Rhodospirillum rubrum. Science* 109 (1949): 558–559.

Ghirardi, M.L., King, P.W., Posewitz, M.C., Maness, P.C., Fedorov, A., Kim, K., Cohen, J., Schulten, K., and Seibert, M. Approaches to developing biological H_2-photoproducing organisms and processes. *Biochem. Soc. Trans.* 33 (2005): 70–72.

Ghirardi, M.L., Togasaki, R.K., and Seibert, M. Oxygen sensitivity of algal H_2-production." *Appl. Biochem. Biotechnol.* 63 (1997): 141–151.

Ghose, T.K. and Bisaria, V.S. Studies on mechanism of enzymatic hydrolysis of cellulosic substances. *Biotechnol. Bioeng.* 21(1979): 131–146.

Ginkel, S.V., Sung, S.W., and Lay, J.J. Biohydrogen production as a function of pH and substrate concentration. *Environ. Sci. Technol.* 35 (2001): 4726–4730.

Gonzalez, J.M. and Robb, F.T. Genetic analysis of *Caboxydothemus hydrogenoformans* carbon monoxide dehydrogenase genes *CooF* and *CooS*. *FEMS Microbiol. Lett.* 191 (2000): 243–247.

Greenbaum, E. Energetic efficiency of hydrogen photoevolution by algal water splitting. *Biophys. J.* 54 (1988): 365–368.

Guo, L., Li, X.M., Bo, X., Yang, Q., Zeng, G.M., Liao, D.X. et al. Impacts of sterilization, microwave and ultrasonication pretreatment on hydrogen producing using waste sludge. *Bioresour. Technol.* 99 (2008): 3651–3658.

Gyoo, Y.J., Jung, R.K., Ji-Young, P. and Sunghoon, P. Hydrogen production by a new chemoheterotrophic bacterium *Citrobacter sp.* Y19. *Int. J. Hydrogen Energy* 27(6) (2002): 601–610.

Hafez, H., Nakhla, G., and El Naggar, H. An integrated system for hydrogen and methane production during landfill leachate treatment. *Int. J. Hydrogen Energy* in Press, Corrected Proof (2009).

Hallenbeck, P.C. Integration of hydrogen evolving systems with cellular metabolism: The molecular biology and biochemistry of electron transport factors and associated reductases. *Biohydrogen* II (2001): 171–184.

Hallenbeck, P.C. Fermentative hydrogen production: Principles, progress, and prognosis. *Int. J. Hydrogen Energy* 34 (2009): 7379–7389.

Hallenbeck, P.C. and Benemann, J.R. Biological hydrogen production; fundamentals and limiting processes. *Int. J. Hydrogen Energy* 27 (2002): 1185–1193.

Hallenbeck, P.C., Kochian, L.V., Weissman, J.C., and Benemann, J.R. Solar energy conversion with hydrogen producing cultures of the blue-green alga, *Anabaena cylindrica. Biotechnol. Bioeng. Symp.* 8 (1978): 283–297.

Hamelinck, C.N., van Hooijdonk, G., and Faaij, A.P.C. Ethanol from lignocellulosic biomass: Techno-economic performance in short-, middle- and long-term. *Biomass Bioenergy* 28 (2005): 384–410.

Han, S.K. and Shin, H.S. Performance of an innovative two-stage process converting food waste to hydrogen and methane. *J. Air. Waste Manag.* 54 (2004): 242–249.

Haruta, S., Cui, Z., Huang, Z., Li, M., Ishii, M., and Igarashi, Y. Construction of a stable microbial community with high cellulose-degradation ability. *Appl. Microbiol. Biotechnol.* 59 (2002): 529–534.

Hase, R., Oikawa, H., Sasao, C., Morita, M., and Watanabe, Y. Photosynthetic production of microalgal biomass in a raceway system under greenhouse conditions in Sendai City. *J. Biosci. Bioeng.* 89 (2000): 157–163.

Heyndrickx, M., Vansteenbeeck, A., Vos, P., and deLey, L. Hydrogen gas production from continuous fermentation of glucose in a minimal medium with *Clostridium butyricum* LMG 1213tl. *Syst. Appl. Microbiol.* 8 (1986): 239–244.

Hillmer, P. and Gest, H. H_2 metabolism in photosynthetic bacterium Rhodopseudomonas capsulata: H_2 production by growing cultures. *J. Bacteriol.* 129(2) (1977): 724–731.

Hoekema, S., Bijmans, M., Janssen, M., Tramper, J., and Wijffels, R.H. A pneumatically agitated flat-panel photobioreactor with gas recirculation: Anaerobic photoheterotrophic cultivation of a purple nonsulfur bacterium. *Int. J. Hydrogen Energy* 27(2002): 1331–1338.

Hoekema, S., Douma, R.D., Janssen, M., Tramper, J., and Wijffels, R.H. Controlling light-use by *Rhodobacter capsulatus* continuous cultures in a flat-panel photobioreactor. *Biotech. Bioeng.* 95(4) (2006): 613–626.

Hu, Q., Faiman, D., and Richmond, A. Optimal tilt angles of enclosed reactors for growing photoautotrophic microorganisms outdoors. *J. Ferment. Bioeng.* 8 (1998): 230–236.

Hu, Q. and Richmond, A. Productivity and photosynthetic efficiency of *Spirulina platensis* as a1ected by light intensity, algal density and rate of mixing in a flat plate photobioreactor. *J. Appl. Phycol.* 8 (1996): 139–145.

Hussy, I., Hawkes, F.R., Dinsdale, R., and Hawkes, D.L. Continuous fermentative hydrogen production from sucrose and sugarbeet. *Int. J. Hydrogen Energy* 30 (2005): 471–483.

Ishikawa, M., Yamamura, S., Takamura, Y., Sode, K., Tamiya, E., and Tomiyama, M. Development of a compact high-density microbial hydrogen reactor for portable bio-fuel cell system. *Int. J. Hydrogen Energy* 31 (2006): 1484–1489.

Jackson, D.D. and Ellms, J.W. On odors and tastes of surface waters with special reference to anabaena, a microscopial organism found in certain water supplies of Massachusetts. *Rep. Mass. State Board Health* (1896): 410–420.

Janssen, M. Cultivation of microalgae: Effect of light/dark cycles on biomass yield. Thesis, Wageningen University, Wageningen, the Netherlands (2002).

Janvanmardian, M. and Palsson, B.O. High density photoautotrophic algal cultures: Design, construction and operation of a novel photobioreactor system. *Biotechnol. Bioeng.* 38 (1991): 1182–1189.

Jo, J.H., Lee, D.S., Park, D., and, Park, J.M. Biological hydrogen production by immobilized cells of Clostridium tyrobutyricum JM1 isolated from a food waste treatment process. *Bioresour. Technol.* 99 (2008): 6666–6672.

Jung, G.Y., Kim, J.R., Park, J.Y., and Park, S. Hydrogen production by a new chemoheterotrophic bacterium *Citrobacter* sp. Y19. *Int. J. Hydrogen Energy* 27 (2002): 601.

Kadar, Z., Vrije, T.D., van Noorden, G.E., Budde, M.A.W., Szengyel, Z., Reczey, K. et al. Yields from glucose, xylose, and paper sludge hydrolysate during hydrogen production by the extreme thermophile *Caldicellulosiruptor saccharolyticus. Appl. Biochem. Biotechnol.* 113–116 (2004): 497–508.

Kaewpintong, K., Shotipruk, A., Powtongsook, S., and Pavasant, P. Photoautotrophic high-density cultivation of vegetative cells of *Haematococcus pluvialis* in airlift bioreactor. *Bioresour. Technol.* 98 (2007): 288–295.

Kanai, T., Imanaka, H., Nakajima, A., Uwamori, K., Omori, Y., Fukui, T. et al. Continuous 860 hydrogen production by the hyperthermophilic archaeon, *Thermococcus kodakaraensis* KOD1. *J. Biotechnol.* 116 (2005): 271–282.

Karapinar, K.I. and Fikret, K. Bio-hydrogen production from waste materials. *Enzyme Microb. Technol.* 38 (2006): 569–582.

Kellum, R. and Drake, H.L. Effects of cultivation gas phase on hydrogenase of the acetogen *Clostridium thermoaceticum. J. Bacteriol.* 160(1) (1984): 466–469.

Kerby, R.L. Hong, S.S., Ensign, S.A., Coppoc, L.J., Ludden, P.W., and Roberts, G.P. Genetic and physiological characterization of the *Rhodospirillum rubrum* carbon monoxide dehydrogenase system. *J. Bacteriol.* 174(16) (1992): 5284–5294.

Kerby, R.L., Ludden, P.W., and Roberts, G.P. Carbon monoxide-dependent growth of *Rhodospirillum rubrum. J. Bacteriol.* 177 (1995): 2241.

Khanal, S.K., Chen, W.H., Li, L., and Sung, S. Biological hydrogen production: Effects of pH and intermediate products. *Int. J. Hydrogen Energy* 29 (2004): 1123–1131.

Kim, B.S. and Lee, Y.Y. Diffusion of sulfuric acid within lignocellulosic biomass particles and its impact on dilute-acid pretreatment. *Bioresour. Technol.* 8 (2002): 165–171.

Kim, M.S. An integrated system for the biological hydrogen production from organic wastes and waste-waters. Paper presented at the *International Symposium on Hydrogen and Methane Fermentation of Organic Waste*, Tokyo, Japan (2002).

Kim, M.S., Baek, J.S., and Lee, J.K. Comparison of H_2 accumulation by *Rhodobacter sphaeroides* Kd131 and its uptake hydrogenase and Phb synthase deficient mutant. *Int. J. Hydrogen Energy* 31(1) (2006a): 121–127.

Kim, M.S., Baek, J.-S., Yun, Y.-S., Jun Sim, S. et al. Hydrogen production from *Chlamydomonas reinhardtii* biomass using a two-step conversion process: Anaerobic conversion and photosynthetic fermentation. *Int. J. Hydrogen Energy* 31(6) (2006b): 812–816.

Klasson, K.T., Lundback, K.M.O., Clausen, E.C., and Gaddy, J.L. Kinetics of light limited growth and biological hydrogen production from carbon monoxide and water by *Rhodospirillum rubrum. J. Biotechnol.* 29 (1993): 177–188.

Kok, B. Algal culture: From laboratory to pilot plant. In: Burlew, J.S. (Ed.). Carnegie Institute of Washington, Washington, DC (1953): pp. 235–272.

Kok, B. Photosynthesis. *Proceedings of the Workshop on Bio Solar Hydrogen Conversion*, September 5–6 (1973).
Koku, H., Eroglu, I., Gunduz, U., Yucel, M., and Turker, L. Aspects of the metabolism of hydrogen production by *Rhodobacter sphaeroides*. *Int. J. Hydrogen Energy* 27 (2002): 1315–1329.
Koku, H., Eroglu, I., Gunduz, U., Yucel, M., and Turker, L. Kinetics of biological hydrogen production by the photosynthetic bacterium *Rhodobacter sphaeroides* O.U. 001. *Int. J. Hydrogen Energy* 28(4) (2003): 381–388.
Kraemer, J.T. and Bagley, D.M. Continuous fermentative hydrogen production using a two-phase reactor system with recycle. *Environ. Sci. Technol.* 39(10) (2005): 3819–3825.
Kumar, N. and Das, D. Enhancement of hydrogen production by *Enterobacter cloacae* IIT-BT 08. *Process Biochem.* 35 (2000): 589–593.
Kumar, N., Ghosh, A. and Das, D. Redirection of biochemical pathways for the enhancement of H_2 production by *Enterobacter cloacae*. *Biotechnol. Lett.* 23 (2001): 537–541.
Lambert, G.R. and Smith, G.D. The hydrogen metabolism of cyanobacteria (blue-green algae). *Biol. Rev.* 56 (1981): 589–660.
Larsson, S., Palmqvist, E., Hahn-Hägerdal, B., Tengborg, C., Stenberg, K., Zacchi, G. et al. The generation of fermentation inhibitors during dilute acid hydrolysis of softwood. *Enzyme Microb. Technol.* 24 (1999): 151–159.
Lay, J. Biohydrogen generation by mesophilic anaerobic fermentation of microcrystalline cellulose. *Biotechnol. Bioeng.* 74 (2001): 280–287.
Lay, J.J., Lee, Y.J., and Noike, T. Feasibility of biological hydrogen production from organic fraction of municipal solid waste. *Water Res.* 33 (1999): 2579–2586.
Lee, E.T.Y. and Bazin, M.J. A laboratory scale airlift photobioreactor to increase biomass output rate of photosynthetic algal cultures. *New Phytol.* 116 (1990): 331–335.
Lee, J.Z., Klaus, D.M., Maness, P.C., and Spear, J.R. The effect of butyrate concentration on hydrogen production via photofermentation for use in a Martian habitat resource recovery process. *Int. J. Hydrogen Energy* 32 (2007): 3301–3307.
Lee, K.S., Hsu, Y.F., Lo, Y.C. Lin, P.J., Lin, C.Y., and Chang, J.S. Exploring optimal environmental factors for fermentative hydrogen production from starch using mixed anaerobic microflora. *Int. J. Hydrogen Energy* 33 (2008): 1565–1572.
Lee, K.S., Lin, P.J., and Chang, J.S. Temperature effects on biohydrogen production in a granular sludge bed induced by activated carbon carriers. *Int. J. Hydrogen Energy* 31 (2006): 465–472.
Levin, D.B., Islam, R., Cicek, N., and Sparling, R. Hydrogen production by *Clostridium thermocellum* 27405 from cellulosic biomass substrates. *Int. J. Hydrogen Energy* 31 (2006): 1496–1503.
Levin, D.B., Pitt, L., and Love, M. Biohydrogen production: Prospects and limitations to practical application. *Int. J. Hydrogen Energy* 29 (2004): 173–185.
Li, D.M. and Chen, H.Z. Biological hydrogen production from steam-exploded straw by simultaneous saccharification and fermentation. *Int. J. Hydrogen Energy* 32 (2007): 1742–1748.
Lin, C.-Y. and Chang, R.-C. Hydrogen production during the anaerobic acidogenic conversion of glucose. *J. Chem. Technol. Biotechnol.* 74 (1999): 498–500.
Lin, C.Y. and Lay, C.H. Carbon/nitrogen ratio effect on fermentative hydrogen production by mixed microflora. *Int. J. Hydrogen Energy* 29 (2004): 41–45.
Lin, C.Y., Wu, C.C., and Hung, C.H. Temperature effects on fermentative hydrogen production from xylose using mixed anaerobic cultures. *Int. J. Hydrogen Energy* 33 (2008): 43–50.
Liu, D.W., Liu, D.P., Zeng, R.J., and Angelidaki, I. Hydrogen and methane production from household solid waste in the two-stage fermentation process. *Water Res.* 40 (2006): 2230–2236.
Liu, G. and Shen, J. Effects of culture medium and medium conditions on hydrogen production from starch using anaerobic bacteria. *J. Biosci. Bioeng.* 98 (2004): 251–256.
Liu, H., Grot, S., and Logan, B.E. Electrochemically assisted microbial production of hydrogen from acetate. *Environ. Sci. Technol.* 39 (2005): 4317–4320.
Liu, H., Zhang, T., and Fang, H.H.P. Thermophilic H_2 production from a cellulose-containing wastewater. *Biotechnol. Lett.* 25 (2003): 365–369.
Liu, W.Z., Wang, A.J., Ren, N.Q., Zhao, X.Y., Liu, L.H., Yu, Z.G. et al. Electrochemically assisted biohydrogen production from acetate. *Energy Fuels* 22 (2008a): 159–163.

Liu, Y., Yu, P., Song, X., and Qu, Y.B. Hydrogen production from cellulose by co-culture of Clostridium thermocellum JN4 and *Thermoanaerobacterium thermosaccharolyticum* GD17. *Int. J. Hydrogen Energy* 33 (2008b): 2927–2933.
Lo, Y.C., Bai, M.D., Chen, W.M., and Chang, J.S. Cellulosic hydrogen production with a sequencing bacterial hydrolysis and dark fermentation strategy. *Bioresour. Technol.* 99(17) (2008a): 8299–8303.
Lo, Y.C., Chen, W.M., Hung, C.H., Chen, S.D., and Chang, J.S. Dark H_2 fermentation from sucrose and xylose using H_2-producing indigenous bacteria: Feasibility and kinetic studies. *Water Res.* 42 (2008): 827–842.
Logan, B., Oh, S.E., Kim, I.K., and Van Ginkel, S.W. Biological hydrogen production measured in batch anaerobic respirometers. *Environ. Sci. Technol.* 36(11) (2002): 2530–2535.
Lu, W.J., Hwang, S.J., and Chang, C.M. Liquid velocity and gas hold up in three-phase internal loop airlift reactors with low density particles. *Chem. Eng. Sci.* 50 (1995): 1301–1310.
Manish, S. and Banerjee, R. Comparison of biohydrogen production processes. *Int. J. Hydrogen Energy* 33(1) (2008): 279–286.
Markov, S.A., Bazin, M., and Hall, D.O. The potential of using cyanobacteria in photobioreactors for hydrogen production. *Adv. Biochem. Eng.* (52) (1995): 59–86.
Matsunaga, T., Takeyama, H., Sudo, H., Oyama, N., Ariura, S., Takano, H., Hirano, M., Burgess, J.G., Sode, K., and Nakamura, N. Glutamate production from CO_2 by marine cyanobacterium *Synechococcus* sp. using a novel biosolar reactor employing light diffusing optical fibers. *Appl. Biochem. Biotechnol.* 28/29 (1991): 157–167.
Mazzuca Sobczuk, T., García Camacho, F., Molina Grima, E., and Chisti, Y. Effects of agitation on the microalgae *Phaeodactylum tricornutum* and *Porphyridium cruentum*. *Bioproc. Biosyt. Eng.* 28 (2006): 243–250.
McTavish, H., Sayavedra-Soto, L.A., and Arp, D.J. Substitution of *Azotobacter vinelandii* hydrogenase small subunit cysteines by serines can create insensitivity to inhibition by O_2 and preferentially damages H_2 oxidation over H_2 evolution. *J. Bacteriol.* 177(14) (1995): 3960–3964.
Melis, A. and Happe, T. Trails of green alga hydrogen research—From Hans Gaffron to new frontiers. *Photosyn. Res.* 80 (2004): 401–409.
Melis, A., Neidhardt, J., and Benemann, J.R. *Dunaliella salina* (Chlorophyta) with small chlorophyll antenna sizes exhibit higher photosynthetic productivities and photon use efficiencies than normally pigmented cells. *J. Appl. Phycol.* 10 (1999): 515–525.
Melis, A., Zhang, L., Foestier, M., Ghirardi, M.L., and Seibert, M. Sustained photobiological hydrogen gas production upon reversible inactivation of oxygen evolution in the green alga *Chlamydomonas reinhardtii*. *Plant Physiol.* 122 (2000): 127–136.
Merchuk, J.C., Gluz, M., and Mukmenev, I. Comparison of photobioreactors for cultivation of the microalga *Porphyridium* sp. *J. Chem. Technol. Biotechnol.* 75 (2000): 1119–1126.
Milner, H.W. Rocking tray. In: Burlew, J.S. (Ed.) *Algal Culture from Laboratory to Pilot Plant*. Carnegie Institution, Washington, DC, No. 600 (1953): p. 108.
Miron, A.S., Camacho, F.G., Gomez, A.C., Grima, E.M., and Chisti, M.Y. Bubble-column and airlift photobioreactors for algal culture. *AIChE J.* 46(9) (2000): 1872–1887.
Miyake, J. The science of biohydrogen. In: *BioHydrogen*. Plenum Press, New York, 1998.
Miyake, J., Mao, X.Y., and Kawamura, S. Efficiency of light energy conversion to hydrogen by the photosynthetic bacterium *Rhodobacter sphaeroides*. *Int. J. Hydrogen Energy* 12 (1987): 1147–1149.
Molina, E., Fernández, J., Acién, F.G., and Chisti, Y. Tubular photobioreactor design for algal cultures. *J. Biotechnol.* 92 (2001): 113–131.
Molina Grima, E., Acién Fernández, F.G., García Camacho, F., and Chisti, Y. Photobioreactors: Light regime, mass transfer, and scale up. *J. Biotechnol.* 70 (1999): 231–247.
Mononmani, H.K. and Sreekantiah, K.R. Saccharification of sugarcane bagasse with enzymes from *Aspergillus ustus* and *Trichoderma viride*. *Enzyme Microb. Technol.* 9 (1987): 484–488.
Mori, K. Photoautotrophic bioreactor using visible solar rays condensed by Fresnel lenses and transmitted through optical fibers. *Biotechnol. Bioeng. Symp.* 15 (1985): 331–345.
Mosier, N., Wyman, C., Dale, B., Elander, R., Lee, Y.Y., Holtzapple, M. et al. Features of promising technologies for pretreatment of lignocellulosic biomass. *Bioresour. Technol.* 96 (2005): 673–686.

Mu, Y., Wang, G., and Yu, H.Q. Response surface methodological analysis on biohydrogen production by enriched anaerobic cultures. *Enzyme Microb. Technol.* 38 (2006): 905–913.
Murray, W.D. Symbiotic relationship of *Bacteroides cellulosolvens* and *Clostridium saccharolyticum* in cellulose fermentation. *Appl. Environ. Microbiol.* 51 (1986): 710–714.
Mussatto, S.I. and Roberto, I.C. Alternatives for detoxification of diluted-acid lignocellulosic hydrolyzates for use in fermentative processes: A review. *Bioresour. Technol.* 93 (2004): 1–10.
Myers, J. Algal culture. In: *Encyclopedia of Chemical Technology*. New York (1957): pp. 649–680.
Najafpour, G., Younesi, H., and Mohamed, A.R. Effect of organic substrate on hydrogen production from synthesis gas using *Rhodospirillum rubrum* in batch culture. *Biochem. Eng. J.* 21 (2004): 123–130.
Nakajima, Y., Tsuzuki, M., and Ueda, R. Improved productivity by reduction of the content of light harvesting pigment in *Chlamydomonas perigranulata*. *J. Appl. Phycol.* 13 (2001): 95–101.
Nakajima, Y. and Ueda, R. Improvement of microalgal photosynthetic productivity by reducing the content of light harvesting pigment. *J. Appl. Phycol.* 11 (1999): 195–201.
Nakajima, Y. and Ueda, R. The effect of reducing light-harvesting pigment on marine microalgal productivity. *J. Appl. Phycol.* 12 (2000): 285–290.
Nakashimada, Y., Rachman, M.A., Kakizono, T., and Nishio, N. Hydrogen production of *Enterobacter aerogenes* altered by extracellular and intracellular redox states. *Int. J. Hydrogen Energy* 27 (2002): 1399–1405.
Narihiro, T. and Sekiguchi, Y. Microbial communities in anaerobic digestion processes for waste and wastewater treatment: A microbiological update. *Curr. Opin Biotechnol.* 18 (2007): 273–278.
Nath, K., Kumar, A., and Das, D. Hydrogen production by *Rhodobacter sphaeroides* strain O.U.001 using spent media of Enterobacter cloacae strain Dm11. *Appl. Microbiol. Biotechnol.* 68 (2005): 533–541.
Neidhardt, J., Benemann, J.R., Baroli, I., and Melis, A. Maximizing photosynthetic productivity and light utilization in microalgae by minimizing the light-harvesting chlorophyll antenna size of the photosystems. *Photosyn. Res.* 56 (1998): 175–184.
Nguyen, Q., Tucker, M., Boynton, B., Keller, F., and Schell, D. Dilute acid pretreatment of softwoods. *Appl. Biochem. Biotechnol.* (1998): 77–87.
van Niel, E.W.J., Budde, M.A.W., de Haas, G.G., van der Wal, F.J., Claassen, P.A.M., and Stams, A.J.M. Distinctive properties of high hydrogen producing extreme thermophiles, *Caldicellulosiruptor saccharolyticus* and *Thermotoga elfii*. *Int. J. Hydrogen Energy* 27 (2002): 1391–1398.
van Niel, E.W.J., Claassenb, P.A.M., and Stamsa, A.J.M. Substrate and product inhibition of hydrogen production by the extreme thermophile, *Caldicellulosiruptor saccharolyticus*. *Biotechnol. Bioeng.* 81 (2003): 255–262.
Noike, T. Biological hydrogen production of organic wastes-development of the two-phase hydrogen production process. Paper presented at the *International Symposium on Hydrogen and Methane Fermentation of Organic Waste*, Tokyo, Japan (March 2002).
Ntaikou, I., Gavala, H.N., Kornaros, M., and Lyberatos, G. Hydrogen production from sugars and sweet sorghum biomass using *Ruminococcus albus*. *Int. J. Hydrogen Energy* 33 (2008): 1153–1163.
Ogbonna, J.C., Soejima, T., and Tanaka, H. Development of efficient large scale photobioreactors. In: Zaborosky, O.R. (Ed.), *Biohydrogen*. Plenum Press, New York (1998): pp. 329–343.
Ogbonna, J.C., Soejima, T., and Tanaka, H. An integrated solar and artificial light system for internal illumination of photobioreactors. *J. Biotechnol.* 70 (1999): 289–297.
Ogbonna, J.C. and Tanaka, H. Photobioreactor design for photobiological production of hydrogen. In: Miyake, J., Matsunaga, T., and San Pietro, A. (Eds.), *Biohygrogen II—An Approach to Environmentally Acceptable Technology*. Pergamon Press, London, U.K. (2001): pp. 245–261.
Oh, Y.K., Kim, H.J., Park, S., Kim, M.S., and Ryu, D.D.Y. Metabolic-flux analysis of hydrogen production pathway in *Citrobacter amalonaticus* Y19. *Int. J. Hydrogen Energy* 33 (2008): 1471–1482.
Olaizola, M. Commercial production of astaxanthin from *Haematococcus pluvialis* using 25,000-liter outdoor photobioreactors. *J. Appl. Phycol.* 12(2000): 499–506.
Palmqvist, E. and Hahn-Hägerdal, B. Fermentation of lignocellulosic hydrolysates. I: Inhibition and detoxification. *Bioresour. Technol.* 74 (2000a): 17–24.

Palmqvist, E. and Hahn-Hägerdal, B. Fermentation of lignocellulosic hydrolysates. II: Inhibitors and mechanisms of inhibition. *Bioresour. Technol.* 74 (2000b): 25–33.

Pampulha, M.E. and Lourero-Dias, M.C. Combined effect of acetic acid, pH, and ethanol on intracellular pH of fermenting yeast. *Appl. Microb. Biotechnol.* 31 (1989): 547–550.

Pan, X.J., Arato, C., Gilkes, N., Gregg, D., Mabee, W., Pye, K. et al. Biorefining of softwoods using ethanol organosolv pulping: Preliminary evaluation of process streams for manufacture of fuel-grade ethanol and co-products. *Biotechnol. Bioeng.* 90 (2005): 473–481.

Pezacha, E. and Wood, H.G. The synthesis of acetyl-coA by *Clostridium thermoaceticum* from carbon dioxide, hydrogen, coenzyme A, and methyltetrahydrofolate. *Arch. Microbiol.* 137 (1984): 63–69.

Philips, J.R., Clausen, E.C., and Gaddy, J.L. Synthesis gas as substrate for the biological production of fuels and chemicals. *Appl. Biochem. Biotechnol.* 45–46 (1994): 145–156.

Pin-Ching, M. and Weaver, P.F. Biological H_2 from fuel gasses and from H_2O. *Proceedings of the 2000 DOE Hydrogen Production Review*, NREL/CP-570-28890 (2000).

Pirt, S.J., Lee, Y.K., Walach, M.R., Pirt, M.W., Balyuzi, H.H.M., and Bazin, M.J. A tubular photobioreactor for photosynthetic production of biomass from carbon dioxide: Design and performance. *J. Chem. Tech. Biotechnol.* 33B (1983): 35–38.

Planchard, A., Mignot, L., Jouenne, T., and Junter, G.A. Photoproduction of molecular hydrogen by *Rhodospirillum rubrum* immobilized in composite agar layer/microporous membrane structure. *Appl. Microbiol. Biotechnol.* 31 (1989): 49–54.

Pohl, P., Kohlhase, M., and Martin, M. Photobioreactors for the axenic mass cultivation of microalgae. In: Stadler, T., Mollion, J., Verdus, M.C., Karamanos, Y., Morvan, H., and Christiaen, D. (Eds.), *Algal Biotechnology*. Elsevier Applied Science, New York (1988): pp. 209–217.

Pruvost, J., Pottier, L., and Legrand, J. Numerical investigation of hydrodynamic and mixing conditions in a torus photobioreactor. *Chem. Eng. Sci.* 61 (2006): 4476–4489.

Ramos de Ortega, A. and Roux, J.C. Production of Chlorella biomass in different types of flat bioreactors in temperate zones. *Biomass* 10 (1986): 141–156.

Redding, K., Cournac, L., Vassiliev, I., Golbeck, J., Peltier, G., and Rochaix, J.D. Photosystem I is required for photoautotrophic growth, Co_2 fixation, and H_2 photoevolution in *Chlamydomonas reinhardtii. J. Biol. Chem.* 274(104) (1999): 66–73.

Reith, J.H., Wijffels, R.H., and Barten, H. Bio-methane & bio-hydrogen: Status and perspectives of biological methane and hydrogen production. *Dutch Biol. Hydrogen Found.* (2003): ISBN: 90-9017165-7.

Ren, N. et al. Bioconversion of lignocellulosic biomass to hydrogen: Potential and challenges. *Biotechnol. Adv.* (2009): doi:10.1016/j.biotechadv.2009.05.007.

Ren, N.Q., Wang, B.Z., and Huang, J.C. Ethanol type fermentation from carbohydrate in high rate acidogenic reactor. *Biotechnol. Bioeng.* 54 (1997): 428–433.

Richmond, A. Efficient utilization of high irradiance for production of photoautotrophic cell mass: A survey. *J. Appl. Phycol.* 8 (1996): 381–387.

Richmond, A. Microalgal biotechnology at the turn of the millennium: A personal view. *J. Appl. Phycol.* 12 (2000): 441–451.

Richmond, A., Boussiba, S., Vonshak, A., and Kopel, R. A new tubular reactor for mass production of microalgae outdoors. *J. Appl. Phycol.* 5 (1993): 327–332.

Rosenkranz, A. and Krasna, A.J. Stimulation of hydrogen photoproduction in algae by removal of oxygen by reagents that combine reversibly with oxygen. *Biotechnol. Bioeng.* 26(11) (1984): 1334–1342.

Rozendal, R.A., Hamelers, H.V.M., Euverink, G.J.W., Metz, S.J., and Buisman, C.J.N. Principle and perspectives of hydrogen production through biocatalyzed electrolysis. *Int. J. Hydrogen Energy* 31 (2006): 1632–1640.

Rozendal, R.A., Hamelersm, H.V.M., Molenkamp, R.J., and Buisman, C.J.N. Performance of single chamber biocatalyzed electrolysis with different types of ion exchange membranes. *Water Res.* 41 (2007): 1984–1994.

Rozendal, R.A., Jeremiasse, A.W., Hamelers, H.V.M., and Buisman, C.J.N. Hydrogen production with a microbial biocathode. *Environ. Sci. Technol.* 42(2) (2008): 629–634.

Samson, R. and Leduy, A. Multistage continuous cultivation of blue-green alga *Spirulina maxima* in the flat tank photobioreactors. *Can. J. Chem. Eng.* 63 (1985): 105–112.
Sánchez Miron, A., Cerón García, M.C., García Camacho, F., Molina Grima, E., and Chisti, Y. Growth and characterization of microalgal biomass produced in bubble column and airlift photobioreactors: Studies in fed-batch culture. *Enzyme Microb. Technol.* 31 (2002): 1015–1023.
Sato, T., Usui, S., Tsuchiya, Y., and Kondo, Y. Invention of outdoor closed type photobioreactor for microalgae. *Energy Convers. Manage.* 47 (2006): 791–799.
Schulz, R. Hydrogenases and hydrogen production in eukaryotic organisms and cyanobacteria. *J. Mar. Biotechnol.* 4 (1996): 15–22.
Shin, H.S., Youn, J.H., and Kim, S.H. Hydrogen production from food waste in anaerobic mesophilic and thermophilic acidogenesis. *Int. J. Hydrogen Energy* 29 (2004): 1355–1363.
Soboh, B., Linder, D., and Hedderich, R. Purification and catalytic properties of a CO-oxidizing: H_2-evolving enzyme complex from *Carboxydothermus hydrogenoformans. Eur. J. Biochem.* 269 (2002): 5712.
Spruit, C.J.P. Simultaneous photoproduction of hydrogen and oxygen by *Chlorella. Meded Landbouwhogesch Wageningen* 58 (1958): 1–17.
Sreenath, H.K. and Jeffries, T.W. Batch and membrane-assisted cell recycling in ethanol production by *Candida shehatae. Biotechnol. Lett.* 9 (1987): 293–298.
Sternberg, D. Production of cellulase by trichoderma. *Biotechnol. Bioeng. Symp.* (1976): 35–53.
Strickland, L.H. The bacterial decomposition of formate. *Biochem. J.* 23 (1929): 1187.
Sun, Y. and Cheng, J. Hydrolysis of lignocellulosic materials for ethanol production: A review. *Bioresour. Technol.* 83 (2002): 1–11.
Taguchi, F., Mizukami, N., Hasegawa, K, and Saito-Taki, T. Microbial conversion of arabinose and xylose to hydrogen by a newly isolated *Clostridium* sp. No. 2. *Can. J. Microbiol.* 40 (1994): 228–233.
Taguchi, F., Mizukami, N., Saito-Taki, T., and Hasegawa, K. Hydrogen production from continuous fermentation of xylose during growth of *Clostridium* sp. strain no. 2. *Can. J. Microbiol.* 41 (1995): 536–540.
Taguchi, F., Mizukami, N., and Yamada, K., Hasegawa, K., and Saito-Taki, T. Direct conversion of cellulosic materials to hydrogen by *Clostridium* sp. strain no. 2. *Enzyme Microb. Technol.* 17 (1993): 147–150.
Taguchi, F., Yamada, K., Hasegawa, K., Taki-Saito, T., and Hara, K. Continuous hydrogen production by *Clostridium* sp. Strain No. 2 from cellulose hydrolysate in an aqueous two-phase system. *J. Ferment. Bioeng.* 82 (1996): 80–83.
Taherzadeh, M.J., Gustafsson, L., Niklasson, C., and Lidén, G. Inhibition effects of furfural on aerobic batch cultivation of *Saccharomyces cerevisiae* growing on ethanol and/or acetic acid. *J. Biosci. Bioeng.* 90 (2000): 374–380.
Talabardon, M., Schwitzguebel, J.P., and Peringer, P. Anaerobic thermophilic fermentation for acetic acid production from milk permeate. *J. Biotechnol.* 76 (2000): 83–92.
Tanisho, S. and Ishiwata, Y. Continuous hydrogen production from molasses by fermentation using urethane foam as a support of flocks. *Int. J. Hydrogen Energy* 20 (1995): 541–545.
Tanisho, S., Kuromoto, M., and Kadokura, N. Effect of CO2 removal on hydrogen production by fermentation. *Int. J. Hydrogen Energy* 23(7) (1998): 559–563.
Tanisho, S., Suzuki, Y., and Wakao, N. Fermentative hydrogen evolution by *Enterobacter aerogenes* strain E.82005. *Int. J. Hydrogen Energy* 12 (1987): 623–627.
Tao, Y., Chen, Y., Wu, Y., He, Y., and Zhou, Z. High hydrogen yield from a two-step process of dark- and photo-fermentation of sucrose. *Int. J. Hydrogen Energy* 32 (2007): 200–206.
Thimijan, R.W. and Heins, R.D. Photometric, radiometric and quantum light units of measure: A review of procedures for interconversion. *Hort. Sci.* 18(6) (1983): 818–822.
Ting, C.H. and Lee, D.J. Production of hydrogen and methane from wastewater sludge using anaerobic fermentation. *Int. J. Hydrogen Energy* 32 (2007): 677–682.
Ting, C.H., Lin, K.R., Lee, D.J., and Tay, J.H. Production of hydrogen and methane from wastewater sludge using anaerobic fermentation. *Water Sci. Technol.* 50 (2004): 223–228.

Torzillo, G., Accolla, P., Pinzani, E., and Masojidek, J. In situ monitoring of chlorophyll fluorescence to assess the synergistic of low temperature and high irradiance stresses in Spirulina cultures grown outdoors in photobioreactors. *J. Appl. Phycol.* 8 (1996): 283–291.

Torzillo, G., Pushparaj, B., Bocci, F., Balloni, W., Materassi, R., and Florenzano, G. Production of Spirulina biomass in closed photobioreactors. *Biomass* 11 (1986): 61–74.

Tredici, M.R. and Chini Zittelli, G. Efficiency of sunlight utilization: Tubular versus flat photobioreactors. *Biotechnol. Bioeng.* 57 (1998): 187–197.

Tredici, M.R. and Materassi, R. From open ponds to vertical alveolar panels: The Italian experience in the development of reactors for the mass cultivation of photoautotrophic microorganisms. *J. Appl. Phycol.* 4 (1992): 221–231.

Tredici, M.R., Zittelli, G.C., and Benemann, J.R. A tubular internal gas exchange photobioreactor for biological hydrogen production. *BioHydrogen* (1998): 391–402.

Turcot, J., Bisaillon, A., and Hallenbeck, P.C. Hydrogen production by continuous cultures of *Escherichia coli* under different nutrient regimes. *Int. J. Hydrogen Energy* 33 (2008): 1465–1470.

Turpin, D.H., Layzell, D.B., and Elrifi, I.R. Modeling the carbon economy of *Anabaena flos-aquae. Plant Physiol.* 78(4) (1985): 746–752.

Ueno, Y., Fukui, H., and Goto, M. Operation of a two-stage fermentation process producing hydrogen and methane from organic waste. *Environ. Sci. Technol.* 41(4) (2007): 1413–1419.

Ueno, Y., Morimoto, M., Otsuka, S., Kawai, T., and Sato, S. Process for the production of hydrogen by microorganisms. US Patent 5 (1995): 464–539.

Ueno, Y., Otsuka, S., and Morimoto, M. Hydrogen production from industrial wastewater by anaerobic microflora in chemostat culture. *J. Ferment. Bioeng.* 82 (1996): 194–197.

Ueno, Y., Tatara, M., Fukui, H., Makiuchi, T., Goto, M., and Sode, K. Production of hydrogen and methane from organic solid wastes by phase-separation of anaerobic process. *Bioresour. Technol.* 98(9) (2007): 1861–1815.

Ugwu, C.U., Aoyagi, H., and Uchiyama, H. Review: Photobioreactors for mass cultivation of algae. *Bioresour. Technol.* 99 (2008): 4021–4028.

Ugwu, C.U., Ogbonna, J.C., and Tanaka, H. Improvement of mass transfer characteristics and productivities of inclined tubular photobioreactors by installation of internal static mixers. *Appl. Microbiol. Biotechnol.* 58 (2002): 600–607.

Ugwu, C.U., Ogbonna, J.C., and Tanaka, H. Design of static mixers for inclined tubular photobioreactors. *J. Appl. Phycol.* 15 (2003): 217–223.

Ugwu, C.U., Ogbonna, J.C., and Tanaka, H. Light/dark cyclic movement of algal cells in inclined tubular photobioreactors with internal static mixers for efficient production of biomass. *Biotechnol. Lett.* 27 (2005a): 75–78.

Ugwu, C.U., Ogbonna, J.C., and Tanaka, H. Characterization of light utilization and biomass yields of *Chlorella sorokiniana* in inclined outdoor tubular photobioreactors equipped with static mixers. *Proc. Biochem.* 40 (2005b): 3406–3411.

Vandu, C.O., Liu, H., and Krishna, R. Mass transfer from Taylor bubbles rising in single capillaries. *Chem. Eng. Sci.* 60 (1993): 6430–6437.

Van Ginkel, S.W., Oh, S.E., and Logan, B.E. Biohydrogen gas production from food processing and domestic wastewaters. *Int. J. Hydrogen Energy* 30(15) (2005): 1535–1542.

Vega Estrada, J., Montes Horcasitas, M.C., Dominígues Bocanegra, A.R., and Cañizares Villanueva, R.O. *Haematococcus pluvialis* cultivation in split-cylinder internal-loop airlift photobioreactor under aeration conditions avoiding cell damage. *Appl. Microbiol. Biotechnol.* 68 (2005): 31–35.

Vignais, P.M. and Colbeau, A. Molecular biology of microbial hydrogenases. *Curr. Issues Mol. Biol.* 6 (2004): 159–188.

Vonshak, A. and Torzillo, G. Environmental stress physiology. In Richmond, A. (Ed.), *Handbook of Microalgal Culture*. Blackwell Publishers, Oxford, U.K. (2004): pp. 57–82.

de Vrije, T., de Haas, G.G., Tan, G.B., Keijsers, E.R.P., and Claassen, P.A.M. Pretreatment of Miscanthus for hydrogen production by *Thermotoga elfii. Int. J. Hydrogen Energy* 27 (2002): 1381–1390.

Wakayama, T., Toriyama, A., Kawasugi, T., Arai, T., Asada, Y., and Miyake, J. Photohydrogen production using photosynthetic bacterium *Rhodobacter sphaeroides* Rv: Simulation of the light cycle of natural sunlight using an artificial source. *Biohydrogen* (1998).
Wang, A.J., Ren, N.Q., Shi, Y.J., and Lee, D.J. Bioaugmented hydrogen production from microcrystalline cellulose using co-culture-*Clostridium acetobutylicum* X9 and *Ethanoigenens harbinense* B49. *Int. J. Hydrogen Energy* 33 (2008): 912–917.
Wang, C.C., Chang, C.W., Chu, C.P., Lee, D.J., Chang, B.V., and Liao, C.S. Producing hydrogen from wastewater sludge by *Clostridium bifermentans. J. Biotechnol.* 102 (2003a): 83–92.
Wang, C.C., Chang, C.W., Chu, C.P., Lee, D.J., Chang, B.V., Liao, C.S., and Tay, J.H. Using filtrate of waste biosolids to effectively produce bio-hydrogen by anaerobic fermentation. *Water Res.* 37(11) (2003b): 2789–2793.
Wang, G., Mu, Y., and Yu, H.Q. Response surface analysis to evaluate the influence of pH, temperature and substrate concentration on the acidogenesis of sucrose-rich wastewater. *Biochem. Eng. J.* 23 (2005): 175–184.
Wang, J.L. and Wan, W. Effect of temperature on fermentative hydrogen production by mixed cultures. *Int. J. Hydrogen Energy* 33 (2008): 5392–5397.
Wang, X.J., Ren, N.Q., Xiang, W.S., and Guo, W.Q. Influence of gaseous end-products inhibition and nutrient limitations on the growth and hydrogen production by hydrogen-producing fermentative bacterial B49. *Int. J. Hydrogen Energy* 32 (2007): 748–754.
Watanabe, Y. and Saiki, H. Development of photobioreactor incorporating *Chlorella* sp. for removal of CO_2 in stack gas. *Energy Convers. Manage.* 38 (1997): 499–503.
Weare, N.M. and Benemann, J.R. Nitrogen fixation by *Anabaena cylindrica*. I. Localization of nitrogen fixation in the heterocysts. *Arch. Mikrobiol.* 90 (1973): 323–332.
Weissman, J.C. and Benemann, J.R. Hydrogen production by nitrogen-fixing cultures of *Anabaena cylindrica. Appl. Environ. Microbiol.* 33 (1977): 123–131.
Weissman, J.C., Goebel, R.P., and Benemann, J.R. Photobioreactor design: Comparison of open ponds and tubular reactors. *Biotechnol. Bioeng.* 31 (1988): 336–344.
Wilde, E.W. and Benemann, J.R. Bioremoval of heavy metals by the use microalgae. *Biotechnol. Adv.* 11 (2005): 781–812.
Wolfrum, E.J. and Maness, P. Biological water gas shift. *U.S. DOE Hydrogen, Fuel Cell and Infrastructure Technologies Program Review;* May 19–22, 2003. Berkeley, CA (2003).
Wolfrum, E.J. and Watt, A.S. Bioreactor design studies for a novel hydrogen-producing bacterium. *Proceedings of the 2001 U.S. DOE Hydrogen Program Review. NREL/CP-570-30535.* National Renewable Energy Laboratory, Golden, CO (2001).
Wolfrum, E.J., Watt, A.S., and Huang, J. Bioreactor development for biological hydrogen production. *Proceedings of the 2002 U.S. DOE Hydrogen Program Review. NREL/CP-610-32405.* National Renewable Energy Laboratory, Golden, CO (2002).
Wongsuchoto, P., Charinpanitkul, T., and Pavasant, P. Bubble size distribution and gas–liquid mass transfer in airlift contactors. *Chem. Eng. J.* 92 (2003): 81–90.
Wu, K.J. and Chang, J.S. Batch and continuous fermentative production of hydrogen with anaerobic sludge entrapped in a composite polymeric matrix. *Proc. Biochem.* 42 (2007): 279–284.
Wyman, C.E. Ethanol from lignocellulosic biomass: Technology, economies, and opportunities. *Bioresour. Technol.* 50 (1994): 3–15.
Xing, D.F., Ren, N.Q., Li, Q.B., Lin, M., and Wang, A. *Ethanoligenens harbinense* gen nov., sp. nov., isolated from molasses wastewater. *Int. J. Syst. Evol. Microbiol.* 56 (2006): 755–760.
Yang, B. and Wyman, C.E. Effect of xylan and lignin removal by batch and flow through pretreatment on the enzymatic digestibility of corn stover cellulose. *Biotechnol. Bioeng.* 86 (2004): 88–95.
Yang, B. and Wyman, C.E. Pretreatment: The key to unlocking low-cost cellulosic ethanol. *Biofuels Bioprod. Biorefining* 2 (2008): 26–40.
Yetis, M., Gunduz, U., Eroglu, I., Yucel, M., and Turker, L. Photoproduction of hydrogen from sugar refinery wastewater by *Rhodobacter sphaeroides* O.U. 001. *Int. J. Hydrogen Energy* 25(11) (2000): 1035–1041.

Yokoi, H., Maki, R., Hirose, J., and Hayashi, S. Microbial production of hydrogen from starch manufacturing wastes. *Biomass Bioenergy* 22 (2002): 389–395.

Yokoi, H., Ohkawara, T., Hirose, J., Hayashi, S., and Takasaki, Y. Characteristics of hydrogen production by aciduric *Enterobacter aerogenes* strain Ho-39. *J. Ferment. Bioeng.* 80 (1995): 571–574.

Yokoi, H., Tokushige, T., Hirose, J., Hayashi, S., and Takasaki, Y., H_2 production from starch by a mixed culture of *Clostridium butyricum* and *Enterobacter aerogenes. Biotechnol. Lett.* 20 (1998): 143–147.

Yokoyama, H., Waki, M., Moriya, N., Yasuda, T., Tanaka, Y. and Haga, K. Effect of fermentation temperature on hydrogen production from cow waste slurry by using anaerobic microflora within the slurry. *Appl. Microbiol. Biotechnol.* 74 (2007): 474–483.

Yu, H.Q., Zhu, Z.H., Hu, W.R., and Zhang, H.S. Hydrogen production from rice wastewater in an upflow anaerobic reactor by using mixed anaerobic cultures. *Int. J. Hydrogen Energy* 27 (2002): 1359–1365.

Yu, J. and Takahashi, P. Biophotolysis-based hydrogen production by cyanobacteria and green microalgae. *Commun. Curr. Res. Educ. Top. Trends Appl. Microbiol.* (2007): 79–89.

Zhang, H., Bruns, M.A., and Logan, B.E. Biological hydrogen production by *Clostridium acetobutylicum* in an unsaturated flow reactor. *Water Res.* 40 (2006): 728–734.

Zhang, K., Kurano, N., and Miyachi, S. Optimized aeration by carbon dioxide gas for microalgal production and mass transfer characterization in a vertical flat-plate photobioreactor. *Bioproc. Biosys. Bioeng.* 25 (2002): 97–101.

Zhang, M.L., Fan, Y.T., Xing, Y., Pan, C.M., Zhang, G.S., and Lay, J.J. Enhanced biohydrogen production from cornstalk wastes with acidification pretreatment by mixed anaerobic cultures. *Biomass Bioenergy* 31 (2007): 250–254.

Zhang, T., Liu, H., and Fang, H.H.P. Biohydrogen production from starch in wastewater under thermophilic conditions. *J. Environ. Manag.* 69 (2003): 149–156.

Zhu, H., Ueda, S., Asada, Y., and Miyake, J. Hydrogen production as a novel process of wastewater treatment—Studies on Tofu wastewater with entrapped *R. sphaeroides* and mutagenesis. *Int. J. Hydrogen Energy* 27 (11–12) (2002): 1349–1157.

Zhu, H.G., Stadnyk, A., Béland, M., and Seto, P. Co-production of hydrogen and methane from potato waste using a two-stage anaerobic digestion process. *Bioresour. Technol.* 99 (2008): 5078–5084.

Zhu, H.G., Suzuki, T., Tsygankov, A.A., Asada, Y., and Miyake, J. Hydrogen production from tofu wastewater by *Rhodobacter sphaeroides* immobilized in agar gels. *Int. J. Hydrogen Energy* 24 (1999): 305.

11

Photobiological Hydrogen Production

Laurent Pilon
University of California, Los Angeles

Halil Berberoğlu
University of Texas at Austin

CONTENTS

11.1 Introduction

11.1.1 Motivation

Industrial and developing nations are facing an unprecedented combination of economic, environmental, and political challenges. First, they face the formidable challenge to meet ever-expanding energy needs without further impacting the climate and the environment. Second, the continued population growth in developing countries and the emergence of a global economy are creating unprecedented stress on the resources of the Earth. Emerging countries are claiming access to the same standard of living as industrial nations, resulting in large needs for energy sources, fast and reliable transportation systems, and industrial equipment. From the standpoint of international security, energy issues include the potential for conflict over access to remaining supplies of inexpensive fossil fuels, which are often concentrated in politically unstable regions.

Currently, fossil fuels supply more than 81% of the world's energy needs estimated at about 137 PWh/year (1 PW = 1^{15} W) or 493 EJ/year (1 EJ = 10^{18} J) [1]. Oil meets more than 92% of the world transportation energy needs [1]. However, its production is expected to peak between 2000 and 2050 after which its production will enter a terminal decline [2–5]. Simultaneously, the world energy consumption is expected to grow by 50% between 2005 and 2030 [5]. Thus, the end of easily accessible and inexpensive oil is approaching.

Moreover, intensive use of fossil fuels increases concentrations of carbon dioxide (CO_2) in the atmosphere, and their contribution to world climate changes is a topic of worldwide concerns [6]. For example, 71.4% of the electricity consumed in the United States is generated from fossil fuel, especially coal [7], making the United States responsible for about 21% of the world CO_2 emission in 2005 [8]. It is predicted that atmospheric CO_2 levels above 450 ppm will have severe impacts on sea levels, global climate patterns, and survival of many species and organisms [6].

Consequently, the growing energy needs will necessitate much greater reliance on a combination of fossil fuel–free energy sources and on new technologies for capturing and converting CO_2. Hydrogen offers a valuable alternative as an energy carrier for stationary and mobile power generation. It has much larger gravimetric energy content than fossil fuels [9]. In addition, its combustion with oxygen does not produce CO_2 but simply water vapor.

11.1.2 Current Hydrogen Production and Usage

Hydrogen is not a fuel but an energy carrier; as such, it is as clean as the production method. Worldwide, 48% of hydrogen is currently produced by steam reforming, partial oxidation, or autothermal reforming of natural gas, 30% from petroleum refining, and 18% from coal gasification [10]. However, all these thermochemical processes require fossil fuel and produce CO_2. The remaining 4% of hydrogen is produced via water electrolysis [11]. This technology used to be the most common process for hydrogen production, but it now represents a small fraction of the world's production. It is used mainly for producing high-purity hydrogen. Thus, current H_2 production fails to address outstanding issues related to depleting oil reserves, energy security, and global warming.

In 2005, 45% of the US hydrogen production was used in oil refineries and 38% in the ammonia industry [10]. It is also used in rocket propulsion applications [12]. In the future, hydrogen could be used in different energy conversion systems such as (1) internal combustion engines for surface transportation [13], (2) high-pressure H_2/O_2 steam generators for power generation [14–16], and (3) proton exchange membrane (PEM) fuel cells [17]. The demand for hydrogen is expected to increase significantly in the next decades as these technologies become more affordable and reliable.

11.1.3 Sustainable Hydrogen Production Technologies

Several technologies offer the advantage of producing hydrogen in a sustainable manner without either relying on fossil fuels or producing carbon dioxide. They can be listed as follows:

1. *Water electrolysis* can be performed using electricity generated in a sustainable manner, by photovoltaic solar cells, for example. Both photovoltaics and electrolyzers are very expensive and cost remains the major challenge of this technology. Typical efficiency of such a system is less than 8% [10]. Alternatively, wind electrolysis uses electricity generated from wind energy to carry out water electrolysis.
2. *Photoelectrochemical* hydrogen production uses catalysts that absorb solar radiation and generate large current densities on the order of 10–30 mA/cm^2 [18]. This enables the water-splitting reactions to take place at a significantly lower voltage than conventional electrolysis. Research results for the development of photoelectrochemical water-splitting systems have shown a solar-to-hydrogen efficiency of 12.4% for the lower heating value (LHV) of hydrogen using concentrated sunlight [10]. Catalyst stability and large band gap are the current challenges to be overcome in this technology [10].

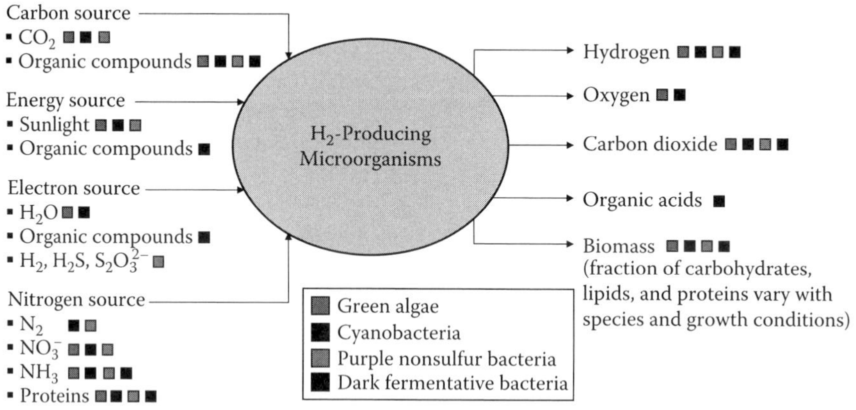

FIGURE 11.1
Schematic of the microalgae/cyanobacteria consuming CO_2 and producing H_2.

3. *Solar-driven thermochemical hydrogen production* uses solar collectors to concentrate thermal radiation from the sun to produce H_2 from various high-temperature thermochemical cycles. This process is an active area of research [19].
4. *Biological hydrogen production* by cultivation of microorganisms offers a clean and sustainable alternative to thermochemical or electrolytic hydrogen production technologies with the possible advantage of CO_2 capture. Under certain physiological conditions, some microorganisms can produce H_2. Biological hydrogen production offers several advantages over currently used technologies: (a) it occurs under mild temperatures and pressures; (b) the reaction specificity is typically higher than that of inorganic catalysts; (c) it is tolerant to sulfur gases, thus reducing the cleanup cost of the gas prior to use; and (d) a diverse array of raw materials can serve as feedstock. The major drawbacks of this technology lie in its currently low efficiency and the fact that it requires large surface area and amounts of water.

Like all living organisms, algae or bacteria need (1) an electron source, (2) an energy source, (3) a carbon source, and (4) a nitrogen source to produce biomass. The latter can further be used as a value-added by-product or as animal feed, fertilizer, and raw material for biofuel production [20]. There are four different hydrogen-producing microorganisms, namely, (1) green algae, (2) cyanobacteria, (3) purple nonsulfur bacteria, and (4) dark fermentative bacteria. Figure 11.1 schematically shows these microorganisms as *black boxes* with the different combinations of input and output parameters. More details about photosynthesis and hydrogen production pathways are provided in the next sections.

11.1.4 Solar Radiation

Solar radiation is the most abundant and renewable energy source on Earth. Through photosynthesis, it has provided human beings with food, fuel, heat, and even fossil fuels generated as a result of geologically deposited biomass chemical transformation of over billions of years under extreme pressures and temperatures [9].

Seen from the Earth, the sun is approximately a disk of radius 6.96×10^8 m at an average distance of 1.496×10^{11} m and viewed with a solid angle of 6.8×10^{-5} sr. The sun is often approximated as a blackbody at 5800 K emitting according to Planck's law [21].

The solar constant is defined as the total energy incident per unit surface area at the outer surface of Earth's atmosphere and oriented perpendicular to the sun's rays. It is estimated to be 1367 W/m^2 [21,22].

As the solar radiation travels through the Earth's atmosphere, it is (1) absorbed by atmospheric gases (e.g., CO_2, H_2O) and (2) scattered by gas molecules and larger aerosol particles, ice crystals, or water droplets. Once it reaches the Earth's surface, most of the ultraviolet (UV) component has been absorbed by oxygen and ozone molecules. Attenuation in the visible is mainly due to Rayleigh scattering by small gas molecules such as oxygen (O_2) and water vapor (H_2O). In the near-infrared (NIR) part of the spectrum, the main absorber is water vapor with contributions from carbon dioxide. Other minor absorbers include nitrous oxide (N_2O), carbon monoxide (CO), and methane (CH_4) [21].

The solar radiation reaching the Earth's atmosphere consists of 6.4% of UV radiation ($\lambda < 380$ nm), 48% of visible light ($380 \leq \lambda \leq 780$ nm), and 45.6% of infrared radiation ($\lambda > 780$ nm) [22]. Overall, the sun delivers 1.73×10^{17} W or 6.38×10^{19} Wh/year on the surface of the atmosphere [22]. This should be compared with the 2006 world energy consumption rate of 1.56×10^{13} W or an annual total energy of 1.37×10^{17} Wh/year [1].

The American Society for Testing and Materials (ASTM) G173-03 standard [23] provides reference terrestrial solar spectral irradiance distributions for wavelength from 280 to 4000 nm averaged over 1 year and over the 48 contiguous states of the continental United States under atmospheric conditions corresponding to the US standard atmosphere [24]. Figure 11.2 shows

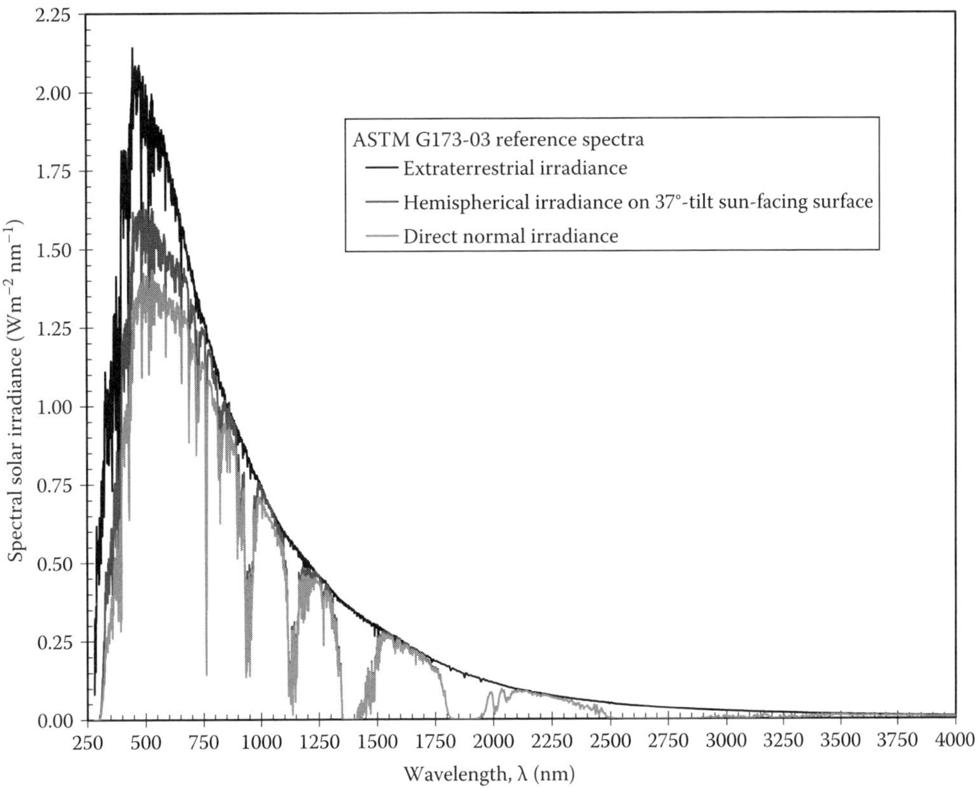

FIGURE 11.2
Averaged daily extraterrestrial solar irradiance and ASTM G173-03 (direct and hemispherical, 37° sun-facing tilted) sea level irradiance in W/m^2 nm. (From Gueymard, C. et al., *Sol. Energy*, 73(6), 443, 2002.)

(1) the extraterrestrial spectral irradiance [23], (2) the direct normal spectral irradiance at sea level with an air mass of 1.5, and (3) the hemispherical (or global) spectral irradiance on an inclined plane at sea level, tilted at 37° toward the equator and facing the sun. The data were produced using the Simple Model for Atmospheric Transmission of Sunshine (SMARTS2 version 2.9.2) [25]. Absorption due to atmospheric O_3, O_2, CO_2, and H_2O is apparent in the direct normal irradiance.

Moreover, Figure 11.3 shows the amount of daily solar irradiance in hours incident on an optimally tilted surface during the worst month of the year based on worldwide solar insolation data [26]. The most promising regions for harvesting solar energy are the southwest United States and northern Mexico, the Andes, northern and southern Africa and the Middle East, as well as Australia. Other regions with favorable conditions include southern Europe, southern China, Southeast Asia, Brazil, and most of Africa. Note that many of these regions have limited freshwater resources, and microorganisms should be selected accordingly. Selection criteria to minimize water use include tolerance to wastewater or seawater and to high microorganism concentrations.

11.1.5 Scope of This Chapter

This chapter focuses on photobiological hydrogen production by green algae, cyanobacteria, and purple nonsulfur bacteria. During photobiological hydrogen production, these microorganisms are cultivated in enclosures known as photobioreactors [27]. Due to the multidisciplinary nature of photobiological hydrogen production, this chapter provides the reader with the background on (1) the fundamentals of photosynthesis and photobiological hydrogen production, (2) photobioreactor technologies, and (3) the associated challenges. Finally, economic and environmental considerations along with prospects for this technology are discussed.

11.2 Photosynthesis

Photosynthesis is a series of biochemical reactions converting sunlight into chemical energy [28]. Fixation of CO_2 into organic matter, such as carbohydrates, lipids, and proteins, through photosynthesis also provides food for all living creatures [28]. In other words, photosynthesis is the process to convert solar energy (energy source) and CO_2 (carbon source) into organic material essential for life on Earth.

Photosynthesis involves two types of reactions, namely, (1) light and (2) dark reactions. During light reactions, photons are absorbed by the microorganisms and are used to produce (1) adenosine triphosphate (ATP), the principal energy-carrying molecule in cells, and (2) the electron carrier nicotinamide adenine dinucleotide phosphate (NADPH). These products of the light reaction are then used in the subsequent dark reactions such as CO_2 fixation [28] and H_2 production [29]. The electrons that drive these reactions usually come (1) from reduced sulfur sources such as hydrogen sulfide (H_2S), sulfur (S^0), or thiosulfate ($S_2O_3^{2-}$) in photosynthetic bacteria and (2) from water (H_2O) in plants, algae, and cyanobacteria [30]. When water is used as the electron source, O_2 is produced as a by-product. These processes are known as oxygenic photosynthesis. Those that do not produce O_2 are known as anoxygenic photosynthesis [30]. The reader is referred to Section 11.2.2 for detailed discussion of anoxygenic and oxygenic photosynthesis.

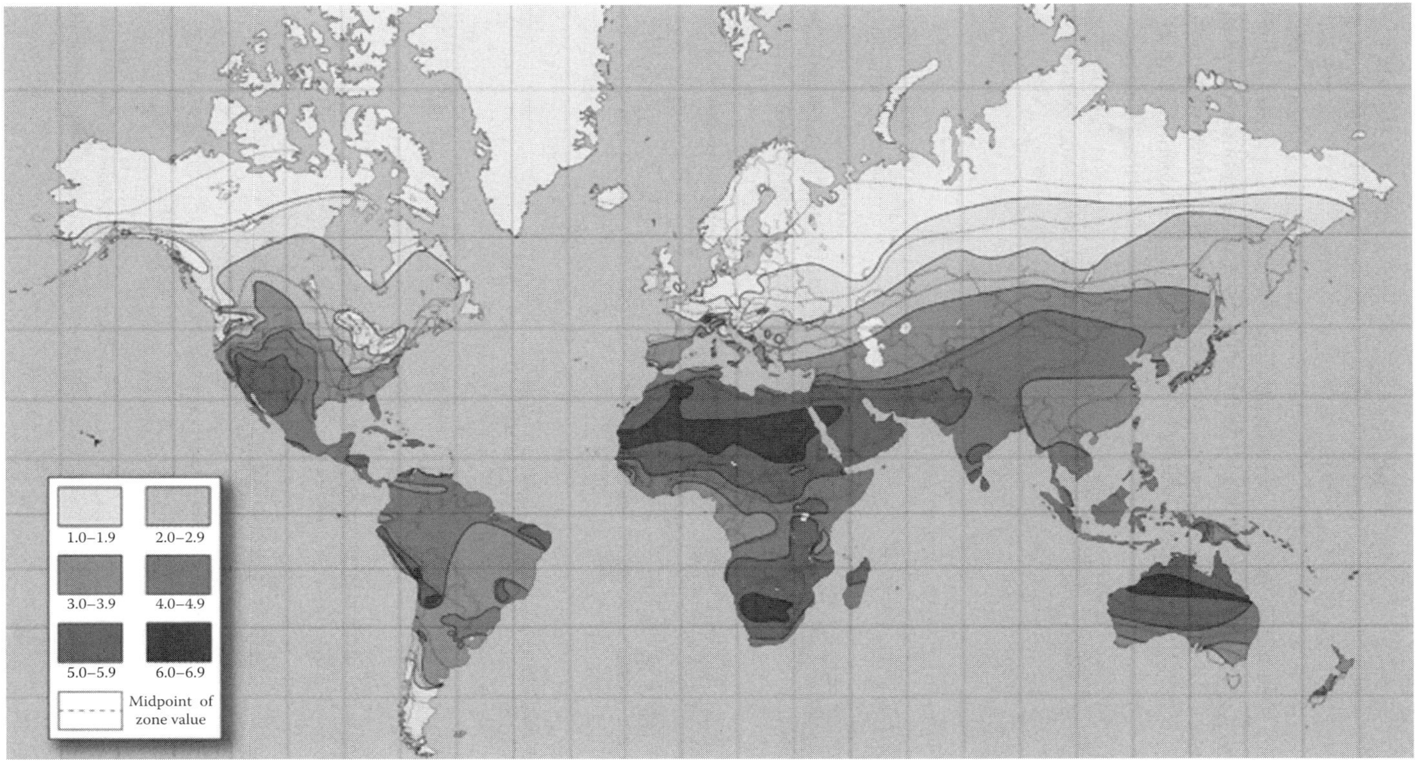

FIGURE 11.3
Average daily local solar irradiance on an optimally tilted surface during the worst month of the year (units are in $kWh/m^2/day$). (Courtesy of SunWize Technologies, Kingston, New York, World insolation map, http//www.sunwize.com/info_center/solar-insolation-map.php, 2008, Used by permission. All rights reserved. [copyright] 2009 SunWize Technologies.)

11.2.1 Photosynthetic Apparatus and Light-Harvesting Pigments

Photosynthesis begins with the absorption of photons by the photosynthetic apparatus. The latter consists of three major parts: (1) the reaction center, (3) the core antenna, and (3) the peripheral antenna. Photochemical charge separation and electron transport take place in the reaction center [28]. The core antenna contains the minimum number of pigments, consisting only of chlorophylls or bacteriochlorophylls, which are necessary for photosynthesis. It is surrounded by the peripheral antenna, which is an assembly of chlorophylls, bacteriochlorophylls, and other accessory pigments such as carotenoids and phycobiliproteins. The peripheral antenna is particularly important in channeling additional photon energy to the reaction center at small light intensities. In algae and cyanobacteria, the photosynthetic apparatus is located on the photosynthetic membrane called thylakoid as shown in transmission electron microscope (TEM) micrographs in Figures 11.8 [31] and 11.9 [32]. In purple bacteria, it is located on vesicular photosynthetic membranes as shown in Figure 11.10 [30,33]. Each of the pigments used in the photosynthesis process is described in detail in the next sections.

11.2.1.1 Chlorophylls and Bacteriochlorophylls

The main pigments necessary for oxygenic photosynthesis are called chlorophylls and those responsible for anoxygenic photosynthesis are called bacteriochlorophylls [30]. Both are molecules containing a magnesium atom at their center. Figure 11.4a shows the structural formula of chlorophyll and bacteriochlorophyll molecules where R1 through R7 are organic chains [30]. The nature of the substituents present in the positions R1 through R7 defines different chlorophylls and bacteriochlorophylls. For example, Figure 11.4b and c shows the structure of chlorophyll *a* and bacteriochlorophyll *a*, respectively.

Moreover, the absorption peak wavelengths of common chlorophyll and bacteriochlorophyll pigments are summarized in Table 11.1. It shows that chlorophylls *a* and *b* have two absorption peaks, one in the blue and one in the red part of the visible spectrum [28]. Chlorophyll *a* absorbs around 430 and 680 nm, while chlorophyll *b* absorbs around 450 and 660 nm. Since they do not absorb green light ($\lambda \approx 520$–570 nm), they appear green to the human eye. These pigments are also responsible for the green color of plants. On the other

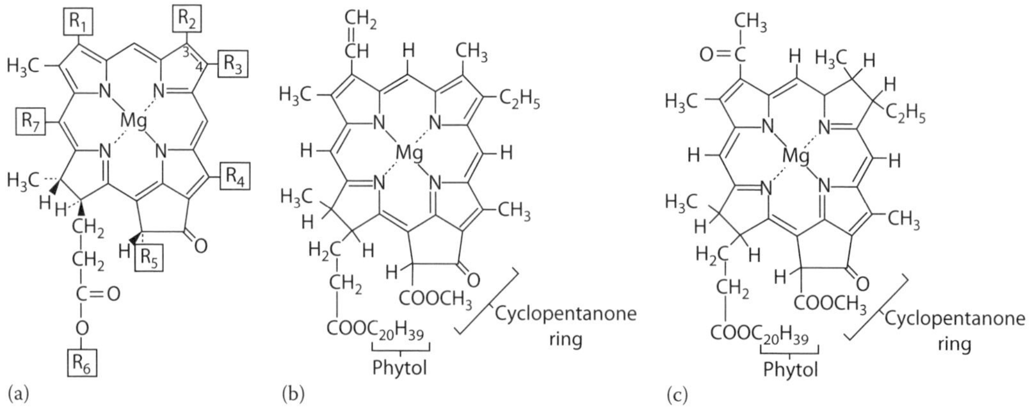

FIGURE 11.4
Structural formulae of (a) general chlorophyll molecule, (b) chlorophyll *a*, and (c) bacteriochlorophyll *a*. (From Madigan, M.T. and Martinko, J.M., *Biology of Microorganisms*, Pearson Prentice Hall, Upper Saddle River, NJ, 2006.)

TABLE 11.1

Common Photosynthetic Pigments in Photosynthetic Microorganisms

Pigment Group	Pigment Name	Absorption Maxima (nm)	Microorganism Type
Chlorophylls	Chl *a*	430, 680	Cyanobacteria, green algae
	Chl *b*	450, 660	Green algae
Bacteriochlorophylls	Bchl *a*	805, 830–890	Purple bacteria
	Bchl *b*	835–850, 1020–1040	Purple bacteria
	Bchl *c*	745–755	Green sulfur bacteria
	Bchl c_s	740	Green nonsulfur bacteria
	Bchl *d*	705–740	Green sulfur bacteria
	Bchl *e*	719–726	Green sulfur bacteria
	Bchl *g*	670–788	Heliobacteria
Carotenoids	*B*-carotene	425, 448, 475	All photosynthetic microbes
	Lutein	421, 445, 474	All photosynthetic microbes
	Violaxanthin	418, 442, 466	All photosynthetic microbes
	Neoxanthin	418, 442, 467	All photosynthetic microbes
	Spheroidene	429, 455, 486	All photosynthetic microbes
Phycobilins	Phycocyanin	620	Cyanobacteria
	Phycoerythrin	550	Cyanobacteria

Sources: Ke, B., *Photosynthesis, Photobiochemistry and Photobiophysics*, Kluwer Academic Publishers, Dordrecht, the Netherlands, 2001; Madigan, M.T. and Martinko, J.M., *Biology of Microorganisms*, Pearson Prentice Hall, Upper Saddle River, NJ, 2006.

hand, bacteriochlorophylls absorb light mainly in the far-infrared to NIR part of the electromagnetic spectrum ($700 \leq \lambda \leq 1000$ nm) [28].

11.2.1.2 Carotenoids

Carotenoids are accessory pigments found in all photosynthetic microorganisms. They absorb mainly the blue part of the spectrum ($400 \leq \lambda \leq 550$ nm) and are responsible for the yellow color of leaves in autumn and orange color of carrots [28]. Carotenoids serve two major functions: (1) shielding the photosynthetic apparatus from photooxidation under large light intensities and (2) increasing the solar light utilization efficiency by expanding the absorption spectrum of the microorganism. They are hydrophobic pigments composed of long hydrocarbon chains and are embedded in the photosynthetic membrane. There are numerous carotenoids [28]. The most common ones are listed in Table 11.1 along with their absorption peak wavelength.

11.2.1.3 Phycobiliproteins

Phycobiliproteins are also accessory pigments that play a role in light harvesting and transferring this energy to the reaction centers. They are found in cyanobacteria and red algae [30]. Two major ones are phycoerythrin absorbing mainly around 550 nm and phycocyanin absorbing strongly at 620 nm [30]. They are essential to the survival of these microorganisms at low light intensities.

Different pigment molecules absorb at different spectral bands of the solar spectrum enabling more efficient utilization of solar energy. They also allow for the coexistence of different photosynthetic microorganisms by sharing different bands of the solar spectrum. Figure 11.5 shows the absorption spectra of chlorophylls *a* and *b*, β-carotenoid,

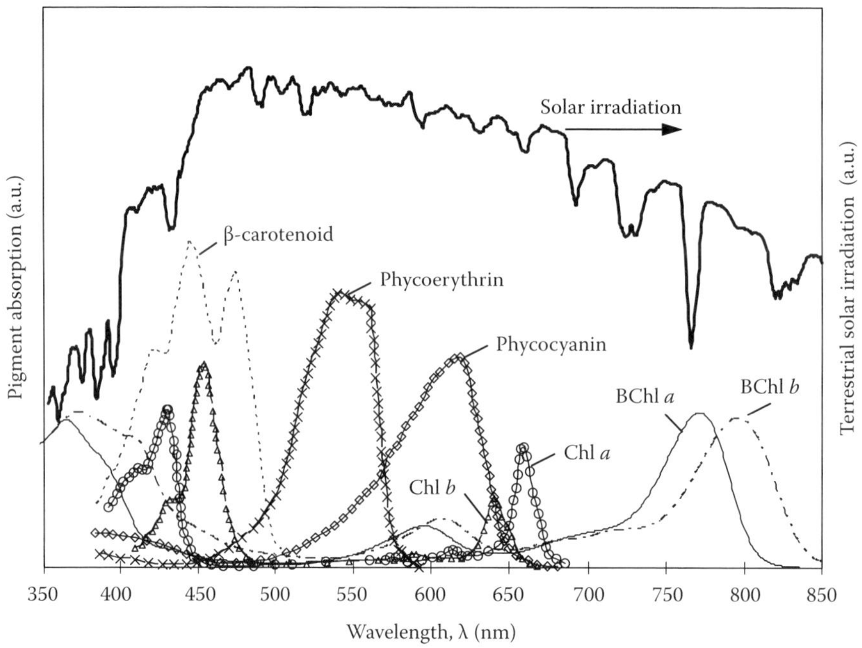

FIGURE 11.5
Absorption spectra of β-carotenoid, phycoerythrin, phycocyanin, chlorophyll *a* (Chl *a*), chlorophyll *b* (Chl *b*), bacteriochlorophyll *a* (BChl *a*), and bacteriochlorophyll *b* (BChl *b*) in relation to the terrestrial solar irradiation spectrum. The figure is in arbitrary units and the spectra of different molecules have been scaled arbitrarily for the sake of clarity. (From Ke, B., *Photosynthesis, Photobiochemistry and Photobiophysics*, Kluwer Academic Publishers, Dordrecht, the Netherlands, 2001.)

phycoerythrin, and phycocyanin over the spectral region from 400 to 700 nm, known as the photosynthetically active radiation (PAR) [28]. It also shows the profile of solar radiation spectrum (in arbitrary units) indicating that these pigments have evolved to absorb at wavelengths where the solar energy is most abundant.

11.2.2 Anoxygenic and Oxygenic Photosynthesis

Two types of photosynthetic processes exist depending on whether molecular oxygen is evolved as a by-product [30]. Anoxygenic photosynthesis is mainly conducted by purple and green sulfur and nonsulfur bacteria, whereas oxygenic photosynthesis is conducted by green algae, cyanobacteria, and plants [28,30]. The source of electrons in anoxygenic photosynthesis can be molecular hydrogen, sulfide, or organic acids. However, in oxygenic photosynthesis, the source of electrons is always water [28]. Details of the electron transport in both types of photosynthesis are described in the following sections.

11.2.2.1 Electron Transport in Anoxygenic Photosynthesis

Figure 11.6 shows the electron flow in anoxygenic photosynthesis, conducted by purple bacteria, for example, with respect to the reduction potential E_o' of the molecules expressed in volts [30]. Anoxygenic photosynthesis begins when a photon with wavelength 870 nm is absorbed by the antenna and transferred to the reaction center. The reaction center, known as P870, is a strong electron donor P870* with very low reduction potential. The electrons

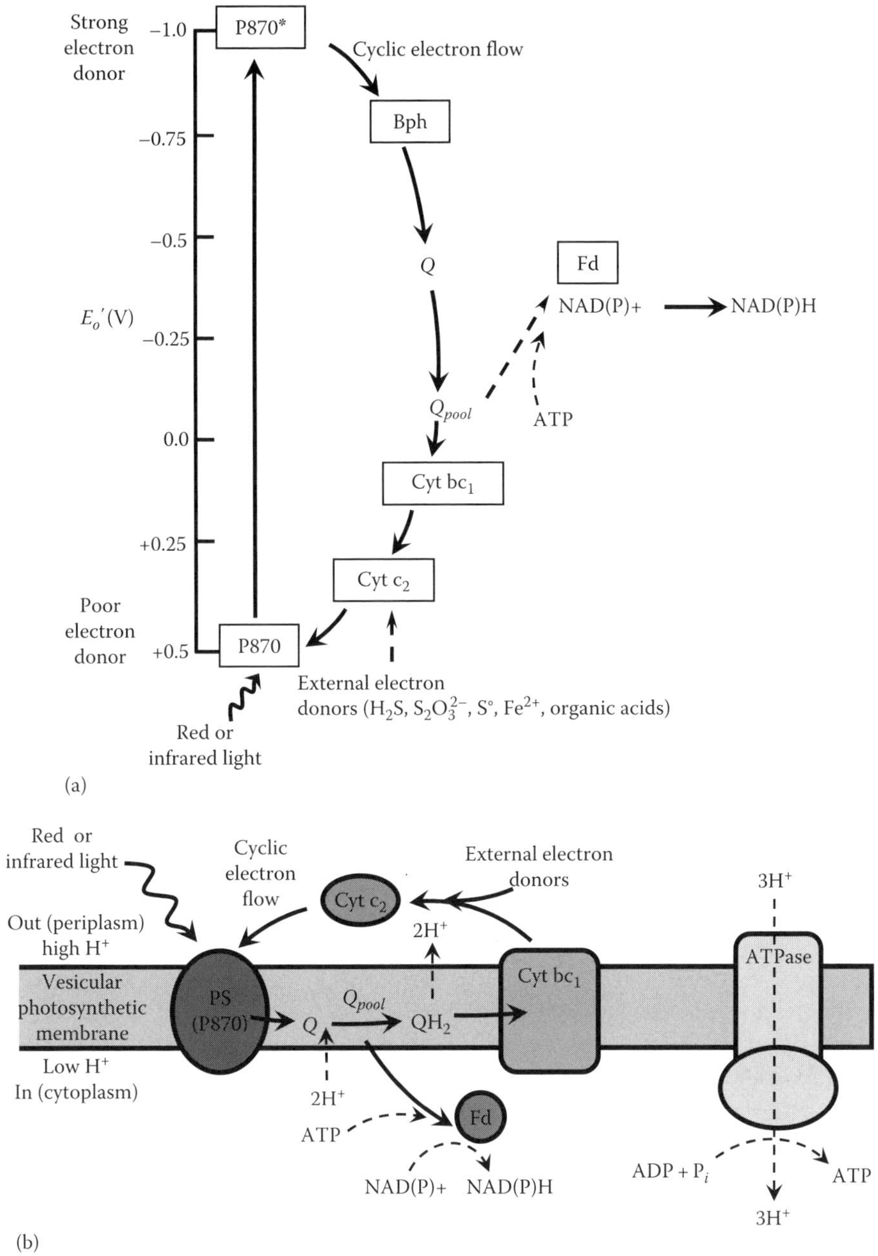

FIGURE 11.6
Electron flow in anoxygenic photosynthesis in purple bacteria illustrated as (a) redox potentials and (b) membrane schematic. (From Madigan, M.T. and Martinko, J.M., *Biology of Microorganisms*, Pearson Prentice Hall, Upper Saddle River, NJ, 2006.)

from P870* are donated very quickly to bacteriopheophytin *a* (Bph) within the reaction center to prevent electron recombination. These electrons are transported from the reaction center through a series of quinone molecules within the photosynthetic membrane denoted by *Q* (Figure 11.6). The electron flow in the photosynthetic membrane is also shown schematically in Figure 11.6b. The transport of electrons induces a proton gradient across the membrane known as the proton motive force that drives the synthesis of ATP through a reaction known as phosphorylation [30]. With additional energy in the form of ATP, electrons are transferred from quinone pool of the photosynthetic membrane denoted by Q_{pool} either (1) to ferredoxin, which carries electrons to be used in nitrogen fixation and/or hydrogen production, or (2) to $NAD(P)^+$ to be converted to NAD(P)H, which carries electrons to biosynthetic reactions such as CO_2 fixation in the Calvin cycle [30,34]. Unused electrons return back to the reaction center via cytochromes Cyt bc_1 and Cyt c_2, thus forming an electron cycle [30]. The electrons lost during the electron cycle are replaced during photofermentation by the cytochrome Cyt c_2 that oxidizes organic acids such as acetate or reduced compounds such as H_2S [30,34].

11.2.2.2 Electron Transport in Oxygenic Photosynthesis

In contrast, during oxygenic photosynthesis conducted by algae and cyanobacteria, electron transport is not cyclic but follows the *Z-scheme* shown in Figure 11.7 [30].

In this scheme, two distinct but interconnected photochemical reactions known as photosystem I (PSI) and photosystem II (PSII) function cooperatively. Oxygenic photosynthesis begins when photons with wavelength around 680 nm are absorbed and transferred to the reaction center known as P680 located in PSII. This converts P680 to a strong reductant that can oxidize water to liberate electrons and protons and evolve molecular O_2 according to $2H_2O \rightarrow 4e^- + 4H^+ + O_2$. Electrons from the reduced P680* are quickly transferred to pheophytin (Ph) within the reaction center to prevent electron recombination. Subsequent electron transfer in the photosynthetic membrane from the reaction center drives the proton motive force responsible for the generation of ATP. The electrons reaching the cytochrome Cyt *bf* are transported to P700 of PSI with plastocyanin (PC). Absorbing light energy at about 700 nm, P700 is reduced to P700*, which has a very low reduction potential. The electrons are quickly donated to a special chlorophyll *a* molecule (Chl a_0) within the reaction center. These electrons are then donated to $NAD(P)^+$ to synthesize NAD(P) H through a cascade of quinone molecules (Q), nonheme iron–sulfur protein (FeS), ferredoxin (Fd), and flavoprotein (Fp) as shown in Figure 11.7. Since the electrons generated from water splitting are not returned back to P680, this form of ATP generation is known as noncyclic phosphorylation. However, if sufficient reducing power is present in the cells, a cyclic phosphorylation can also take place around PSI as shown in Figure 11.7 [35,36]. In both oxygenic and anoxygenic photosynthesis, ATP and NAD(P)H produced are used by the microorganisms as their energy and electron carriers in order to fix CO_2.

11.3 Microbiology of Photobiological Hydrogen Production

11.3.1 Hydrogen-Producing Microorganisms

There are various methods for biological hydrogen production depending on the type of microorganism used in the process. Thus, it is necessary to classify the different hydrogen-producing microorganisms and understand their metabolism. On the most basic premise, microorganisms can be divided into two major groups known as prokaryotes and eukaryotes. Unlike prokaryotes, eukaryotes have a nucleus where the genetic material is stored and other membrane-enclosed organelles [30]. Members of *bacteria* such

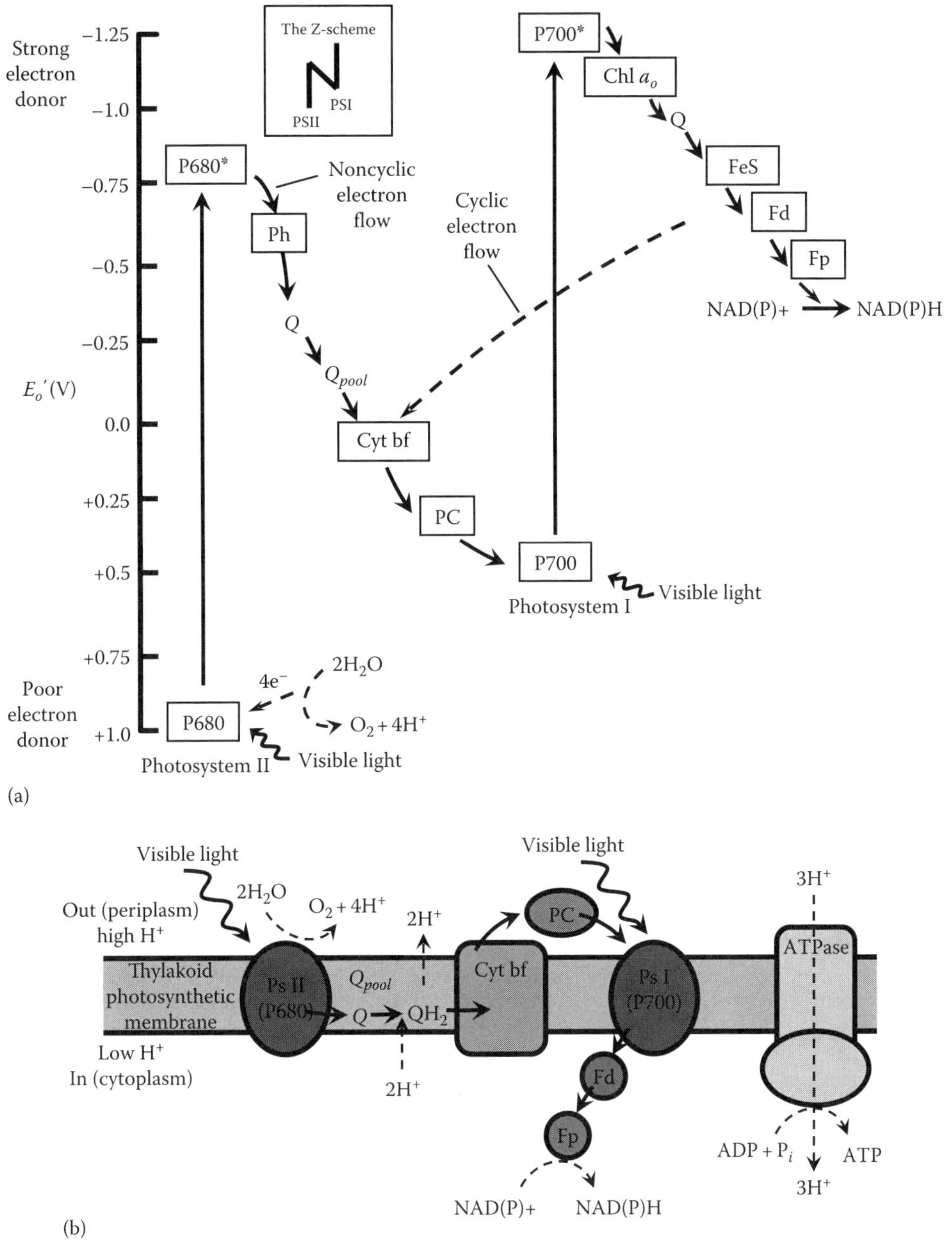

FIGURE 11.7
Electron flow in oxygenic photosynthesis in cyanobacteria and green algae illustrated as (a) redox potentials and (b) membrane schematic. (From Madigan, M.T. and Martinko, J.M., *Biology of Microorganisms*, Pearson Prentice Hall, Upper Saddle River, NJ, 2006.)

as cyanobacteria, purple nonsulfur bacteria, and fermentative bacteria are prokaryotes. Algae, on the other hand, are eukaryotes.

Microorganisms that can use solar radiation as their energy source and CO_2 as their sole carbon source are known as photoautotrophs. Cyanobacteria, algae, and purple nonsulfur bacteria are capable of a photoautotrophic life style. Among these, cyanobacteria and algae use water as their electron source and conduct oxygenic photosynthesis (see Section 11.2.2.2). Purple nonsulfur bacteria use molecular hydrogen, sulfide, or organic acids as their electron source and conduct anoxygenic photosynthesis (see Section 11.2.2.1). Some of these photoautotrophs can also live as photoheterotrophs, that is, they can utilize light as their energy source and organic compounds as both their carbon and electron sources. Organic compounds include (1) organic acids (e.g., acetic acid, amino acids, lactic acid, citric acid, butyric acids) and (2) carbohydrates (sugars) such as monosaccharides (e.g., glucose, sucrose, fructose, $(CH_2O)_n$) and polysaccharides (e.g., starch, cellulose, $(C_6H_{10}O_5)_n$).

On the other hand, some microorganisms use chemical compounds as their energy source and are known as chemolithotrophs. Among these, those using organic compounds as their energy, carbon, and electron sources are known as chemoorganotrophs. This is the case of dark fermentative bacteria. Finally, prokaryotes, like most cyanobacteria and all purple nonsulfur bacteria, are capable of using molecular nitrogen as their nitrogen source. They achieve this through nitrogen fixation that uses special enzymes called nitrogenase and requires energy in the form of ATP. On the other hand, eukaryotes cannot fix molecular nitrogen and require sources of nitrogen in the form of ammonia, nitrates, or proteins (e.g., albumin, glutamate, yeast extract) [30,37].

11.3.1.1 Green Algae

Green algae are eukaryotic organisms that contain chlorophylls and conduct oxygenic photosynthesis [38,39]. They live in freshwater and most of them have cellulose cell walls. They can produce hydrogen through direct and indirect biophotolysis as well as photofermentation (Section 11.3.2.2). All these processes require anaerobic conditions, that is, the absence of oxygen from the algae environment. Examples of green algae capable of photobiological hydrogen production include (1) freshwater species such as *Chlamydomonas reinhardtii* [40], *Chlamydomonas moewusii* [41], and *Scenedesmus obliquus* [42] as well as (2) saltwater species such as *Chlorococcum littorale* [43], *Scenedesmus obliquus* [44], and *Chlorella fusca* [45]. Figure 11.8 depicts the TEM micrograph of the green algae *Chlamydomonas reinhardtii* [31]. It shows the location of the nucleus; the chloroplast, where photosynthetic pigments are located; and the pyrenoid, where CO_2 fixation takes place.

11.3.1.2 Cyanobacteria

Cyanobacteria, also known as blue-green algae, are photoautotrophic prokaryotes that are capable of conducting oxygenic photosynthesis [30]. These microorganisms are the first organisms that could evolve oxygen and are responsible for converting Earth's atmosphere from anoxic (oxygen lacking) to oxic (oxygen containing) [28]. There exist unicellular and filamentous forms and their size can range from 0.5 to 40 μm in diameter depending on the strain [30]. Most species are capable of fixing atmospheric nitrogen using the nitrogenase enzyme and play an important role in the global nitrogen cycle [30]. Some filamentous forms have evolved to contain the nitrogenase enzyme in special cells called heterocysts. Heterocysts protect nitrogenase from oxygen inhibition.

Just like green algae, cyanobacteria can produce hydrogen through direct and indirect biophotolysis as well as photofermentation (Section 11.3.2.2). In addition to anaerobic conditions, nitrogen-fixing cyanobacteria also require the absence of nitrogen sources (N_2, NO_3^-, or NH_4) in order to produce H_2. Examples of cyanobacteria capable of photobiological

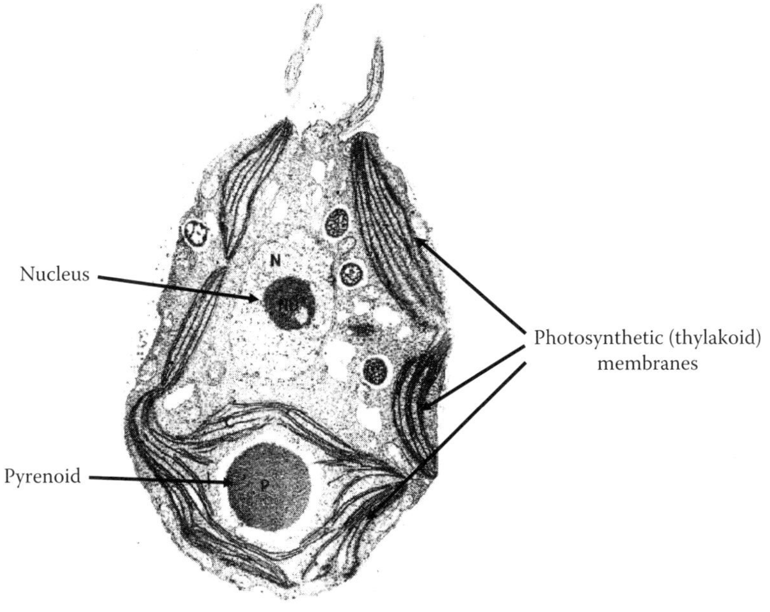

FIGURE 11.8
TEM micrograph of *Chlamydomonas reinhardtii*. A typical cell is ellipsoidal with major and minor diameters equal to about 9 and 8 μm, respectively. (From Harris, E.H., *The Chlamydomonas Sourcebook*, Vol. 1, Academic Press, San Diego, CA, 1989. With permission; Berberoğlu, H. et al., *Int. J. Hydrogen Energy*, 33, 6467, 2008. With permission.)

hydrogen production include (1) freshwater species such as *Anabaena variabilis* [46], *Anabaena azollae* [46], and *Nostoc punctiforme* [47] as well as (2) saltwater species such as *Oscillatoria* Miami BG7 [48] and *Cyanothece* 7822 [49]. Figure 11.9 presents the TEM micrograph of the filamentous cyanobacterium *Anabaena variabilis* ATCC 29413 [32]. It shows the location of the thylakoid membrane where the photosynthetic apparatus is located. Note the absence of nucleus and organelles.

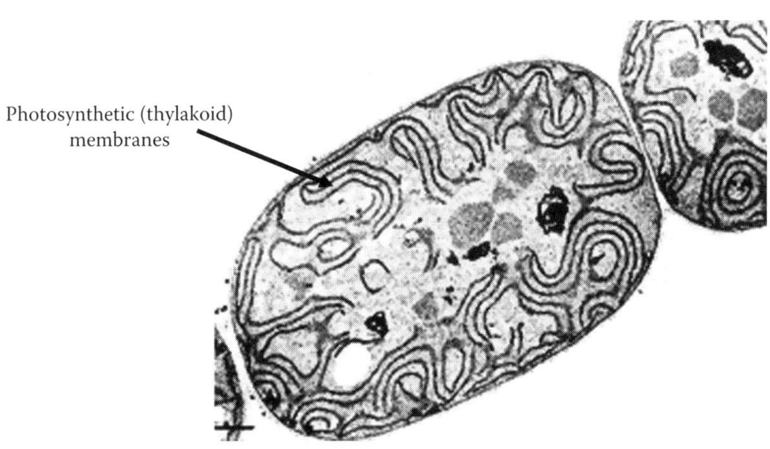

FIGURE 11.9
TEM micrograph of *Anabaena variabilis* ATCC 29413. Typical cell is 5 μm wide. (From Lang, N.J. et al., *J. Bacteriol.*, 169(2), 920, 1987. With permission.)

11.3.1.3 Purple Nonsulfur Bacteria

Purple nonsulfur bacteria are prokaryotes that conduct anoxygenic photosynthesis, that is, they do not produce oxygen. In general, purple nonsulfur bacteria survive by photoheterotrophy using light as their energy source and organic compounds as their carbon source. Organic compounds include fatty, organic, or amino acids; sugars; alcohols; and aromatic compounds [30]. Purple nonsulfur bacteria contain bacteriochlorophylls and carotenoids and have a brown/dark red color, hence their name. The photosynthetic apparatus of purple nonsulfur bacteria is located on an intracytoplasmic photosynthetic membrane [30]. Some species can also grow in the dark conducting fermentation and anaerobic respiration, while others can grow photoautotrophically fixing CO_2 using H_2 or H_2S as their electron source [30]. All species have the nitrogenase enzyme and can fix molecular nitrogen.

Purple nonsulfur bacteria produce hydrogen by photofermentation, which requires removal of both oxygen and nitrogen from the environment (Section 11.3.2.2). Examples of purple nonsulfur bacteria capable of producing hydrogen include *Rhodobacter sphaeroides* [37,41] and *Rhodospirillum rubrum* [50]. Figure 11.10 illustrates the TEM micrograph of the purple nonsulfur bacterium *Rhodobacter sphaeroides* during cell division [33]. It shows the location of the intracytoplasmic membranes where the photosynthetic apparatus is located.

11.3.1.4 Dark Fermentative Bacteria

Dark fermentative bacteria are chemoorganotrophs deriving their energy, carbon, and electrons from the degradation of organic compounds including carbohydrates, amino acids, cellulose, purines, and alcohols [30]. Their size is typical of bacteria, that is, 0.5–1.5 μm. They live in soil and organic nutrient–rich waters. Some grow best at temperatures ranging from 25°C to 40°C and are known as mesophiles, and others grow best at even higher temperatures ranging from 40°C to 80°C and are known as thermophiles. Examples of hydrogen-producing mesophiles and thermophiles are *Enterobacter cloacae* IIT BT-08 [51] and *Clostridium butyricum* [52], respectively.

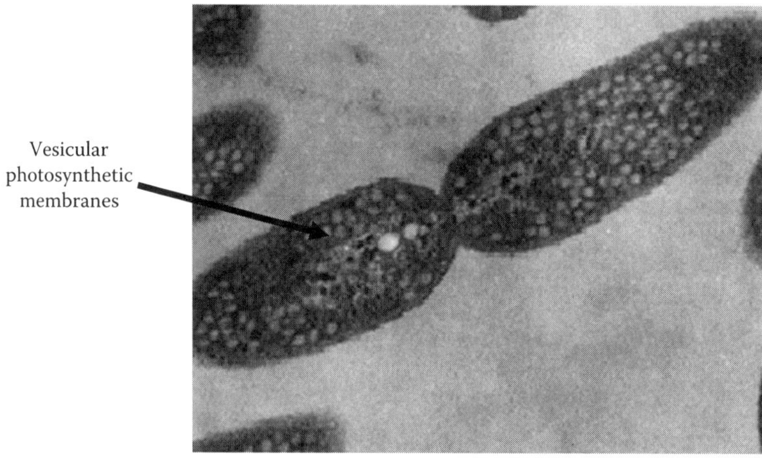

FIGURE 11.10
TEM micrograph of *Rhodobacter sphaeroides*. A typical cell is about 1.5 μm long. (From Department of Energy, *Rhodobacter sphaeroides*, Joint Genome Institute, Walnut Creek, CA, http://genome.jgi-psf.org/finished-microbes/rhosp/rhosp.home.html, accessed on April 19, 2008.)

TABLE 11.2

Energy, Carbon, Electron, and Nitrogen Sources of H_2-Producing Microorganisms

Microorganism	Green Algae	Cyanobacteria	Purple Nonsulfur Bacteria	Dark Fermentative Bacteria
Energy source	Light (oxygenic)	Light (oxygenic)	Light (anoxygenic)	Organic matter
Carbon source	CO_2 or organic matter	CO_2	Organic matter	Organic matter
Electron source	H_2O	H_2O	Organic matter	Organic matter
Nitrogen source	Ammonia or NO_3^-	NH_4, NO_3^-, N_2, or proteins	NH_4, NO_3^-, N_2, or proteins	NH_4, NO_3^-, or proteins
Photosynthetic pigments	Chl *a*, Chl *b*, carotenoids	Chl *a*, carotenoids, phycobilins	Bchl *a*, Bchl *b*, carotenoids	None
H_2-producing enzyme	[Fe-Fe]-hydrogenase	[NiFe]-hydrogenase and/or nitrogenase	Nitrogenase	[NiFe]-hydrogenase
Products	H_2, O_2, carbohydrates	H_2, O_2, CO_2, carbohydrates	H_2, organic acids	H_2, organic acids

Table 11.2 summarizes the energy, carbon, electron, and nitrogen sources for each of the microorganisms described earlier along with the enzymes responsible for H_2 production and the by-products of this process in these microorganisms. The table also lists the pigments present in these microorganisms that are responsible for absorbing and utilizing solar radiation. These enzymes and pigments are presented in detail in the next section.

11.3.2 Enzymatic Pathways

Photobiological hydrogen production takes place when the electrons generated during (1) the light reactions, (2) the degradation of carbohydrates, or (3) respiration are directed to specific enzymes. There are two types of enzymes that catalyze the production of hydrogen in microorganisms, namely, nitrogenase and hydrogenase [53]. Table 11.3 summarizes the advantages and disadvantages of each enzyme group.

11.3.2.1 Enzyme Systems

11.3.2.1.1 Nitrogenase

Nitrogenase is found in prokaryotes such as most photosynthetic bacteria and some cyanobacteria [30]. It is not present in eukaryotes such as green algae [53]. The main role of nitrogenase is to reduce molecular nitrogen to ammonia during fixation of nitrogen dissolved

TABLE 11.3

Advantages and Disadvantages of Nitrogenase and Hydrogenase Enzymes in Producing Hydrogen

Enzyme	Microorganisms	Advantages	Disadvantages
Nitrogenase	Most photosynthetic bacteria and some cyanobacteria	Robust H_2 production Able to generate H_2 under large H_2 partial pressure	Low efficiency (16%) Small turnover rate Requires 2 ATP/electron Sensitive to O_2
Hydrogenase	Cyanobacteria, green algae, and purple nonsulfur bacteria	Does not require ATP High efficiency (41%) Large turnover rate (very active)	Unable to generate H_2 under large H_2 partial pressure Very sensitive to O_2

Source: Prince, R.C. and Kheshgi, H.S., *Crit. Rev. Microbiol.*, 31(1), 19, 2005.

in the liquid phase [41] that provides the nitrogen source needed by the microorganism to produce biomass. This primary reaction catalyzed by nitrogenase is given by [53]

$$N_2 + 8e^- + 8H^+ \rightarrow 2NH_3 + H_2 \tag{11.1}$$

In this reaction, H_2 is produced at low rates as a by-product of nitrogen fixation. In the absence of N_2, nitrogenase catalyzes the irreversible production of H_2 provided that reductants, that is, electrons, and ATP are present via

$$2H^+ + 2e^- + 4ATP \rightarrow H_2 + 4ADP + 4P_i \tag{11.2}$$

where ADP and P_i are adenosine diphosphate and inorganic phosphate, respectively. Since the cell energy carrier ATP is used by nitrogenase, this hydrogen production route is energy intensive. The electrons for nitrogenase are donated either by ferredoxin (Fd) or by flavoprotein (Fp) flavodoxin from the photosynthetic electron flow shown in Figures 11.6 and 11.7 [53]. This enables nitrogenase to evolve H_2 even at a partial pressure of H_2 larger than 50 atm making the process robust [53]. However, nitrogenase-based H_2 production suffers from (1) a small turnover rate of less than 10 s^{-1}, that is, nitrogenase can catalyze less than 10 reactions per second, and (2) low quantum efficiency, defined as the ratio of the number of moles of H_2 produced to the number of photons absorbed by the photosystems. Thus, both the rate of H_2 production and the solar-to-H_2 energy conversion efficiency are low.

Nitrogenase enzymes have an organometallic reaction center. The efficiency of H_2 production by the nitrogenase enzyme varies depending on the type of transition metal located at the reaction center [53]. The most common type of nitrogenase enzyme uses molybdenum at its reaction center, but vanadium and iron can also be found [53].

11.3.2.1.2 Hydrogenase

There are two types of bidirectional (or reversible) hydrogenase enzymes, namely, (1) [Fe-Fe]-hydrogenase and (2) [FeNi]-hydrogenase, also called uptake hydrogenase. The iron [Fe-Fe]-hydrogenase is present in green algae [53]. It is a very active bidirectional enzyme with a large turnover rate of 10^6 s^{-1} [53]. It receives electrons from ferredoxin (Fd in Figures 11.6 and 11.7) and does not require energy (ATP) to produce H_2. Thus, [Fe-Fe]-hydrogenase has better quantum efficiency than nitrogenase. The bidirectional [Fe-Fe]-hydrogenase catalyzes both the production and consumption of hydrogen through the reaction

$$2H^+ + 2e^- \rightleftharpoons H_2 \tag{11.3}$$

The rate at which [Fe-Fe]-hydrogenase can catalyze the production of hydrogen decreases significantly with increasing partial pressure of H_2.

Finally, [NiFe]-hydrogenase is the commonly known uptake hydrogenase that is found in nitrogen-fixing microorganisms [54]. It is present only in nitrogen-fixing microorganisms such as cyanobacteria and purple nonsulfur bacteria [29]. It catalyzes both H_2 evolution and uptake. In purified form, it has been shown to evolve hydrogen at a low turnover rate of 98 s^{-1}. It also enables microorganisms to consume back H_2 produced as a by-product of nitrogen fixation (Equation 11.1) and, thus, recover some energy.

Both [Ni-Fe]- and [Fe-Fe]-hydrogenases are very sensitive to the presence of O_2. In particular, the [Fe-Fe]-hydrogenase is irreversibly inhibited by O_2, whereas the [NiFe]-hydrogenase is reversibly affected (see Section 11.6.1.2) [53].

11.3.2.2 Pathways for Biological Hydrogen Production

Biological processes resulting in hydrogen production can be grouped in four categories, namely, (1) direct and (2) indirect biophotolysis, (3) photofermentation, and (4) dark fermentation.

11.3.2.2.1 Direct Biophotolysis

In this mechanism, H_2 is produced by diverting the electrons generated from water splitting from the Calvin cycle to the bidirectional hydrogenase enzyme according to Equation 11.3 [41]. The energy source is the sunlight in the spectral range from 400 to 700 nm. This mechanism is theoretically the most energy efficient for H_2 production with a theoretical maximum of 40.1% [53]. However, the oxygen produced during water splitting irreversibly inhibits the functioning of the [Fe-Fe]-hydrogenase and makes the process impractical for industrial applications [54]. Green algae such as *Chlamydomonas reinhardtii*, *Chlamydomonas moewusii*, *Scenedesmus obliquus*, and *Chlorococcum littorale* are capable of producing H_2 via direct biophotolysis [41].

11.3.2.2.2 Indirect Biophotolysis

The source of electrons in indirect biophotolysis is also water. However, in this mechanism, the electrons are first used to reduce CO_2 into organic compounds during photosynthesis where O_2 is generated. Then, the electrons are recovered from the degradation of the organic compounds and used in generating H_2 through the action of nitrogenase [54]. Thus, no O_2 is generated during H_2 production. The maximum possible light-to-H_2 energy conversion efficiency of indirect biophotolysis is only 16.3% [53] due to the facts that (1) multiple steps are involved in converting solar energy to H_2 and (2) the use of nitrogenase enzyme requires ATP. Cyanobacteria such as *Anabaena variabilis*, *Anabaena azollae*, *Nostoc muscorum* IAM M-14, and *Oscillatoria limosa* are capable of indirect biophotolysis [49]. The nitrogenase enzyme also gets inhibited by O_2; however, cyanobacteria have evolved in many ways to circumvent this problem [55]. For example, *A. variabilis* has evolved to contain the nitrogenase enzyme in special O_2 protective cells called heterocysts as illustrated in Figure 11.11.

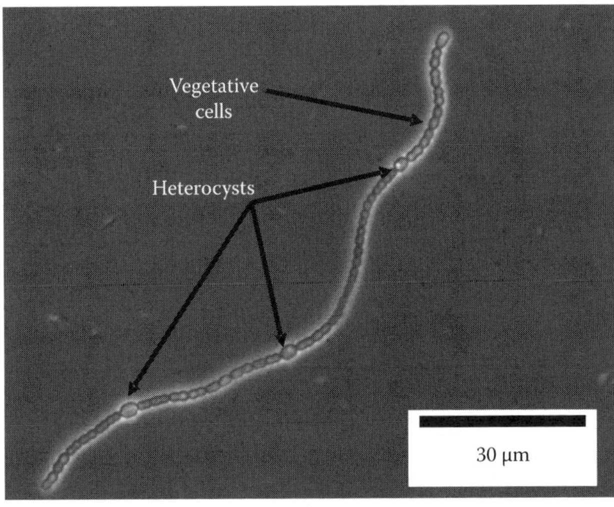

FIGURE 11.11
Micrograph of *Anabaena variabilis* ATCC 29413-U. (From Berberoğlu, H. and Pilon, L., *Int. J. Hydrogen Energy*, 32(18), 4772, 2007.)

11.3.2.2.3 Photofermentation

This mechanism is similar to indirect biophotolysis with the distinction that the organic compounds used are produced outside the cells via the photosynthesis of other organisms, for example, plants. These extracellular organic materials, such as organic acids, carbohydrates, starch, and cellulose [37], are used as the electron source, and sunlight is used as energy source to produce H_2 by nitrogenase enzyme [41]. Due to the fact that the cells do not need to carry out photosynthesis, no O_2 is generated and all the solar energy can be used to produce H_2. Thus, this mechanism is viewed as the most promising microbial system to produce H_2 [41]. The major advantages of this route are (1) the absence of O_2 evolution that inhibits the H_2-producing enzymes and (2) the ability to consume a wide variety of organic substrates found in wastewaters. Due to their ability to harvest a wider spectrum of light, from 300 to 1000 nm, purple nonsulfur bacteria such as *Rhodobacter sphaeroides*, pictured in Figure 11.12, *Rhodospirillum rubrum*, and *Rhodopseudomonas sphaeroides* hold promise as photofermentative H_2 producers.

11.3.2.2.4 Dark Fermentation

In dark fermentation, anaerobic bacteria use the organic substances (e.g., glucose, hexose monophosphate, and pyruvate [41]) as both their energy and electron sources. These bacteria mainly use the [NiFe]-hydrogenase enzyme to produce H_2. Due to the absence of O_2 in the environment and the use of hydrogenase, they can produce H_2 at a higher rate without inhibition. Moreover, H_2 production is continuous throughout the day and night because these microorganisms do not depend on sunlight as their energy source. The hydrogen production is accompanied by CO_2 production as well. Examples of fermentative hydrogen producers include *Enterobacter cloacae* IIT BT-08, *Enterobacter aerogenes, Clostridium butyricum*, and *Clostridium acetobutylicum* [41]. This process falls outside the scope of this chapter and will not be discussed further.

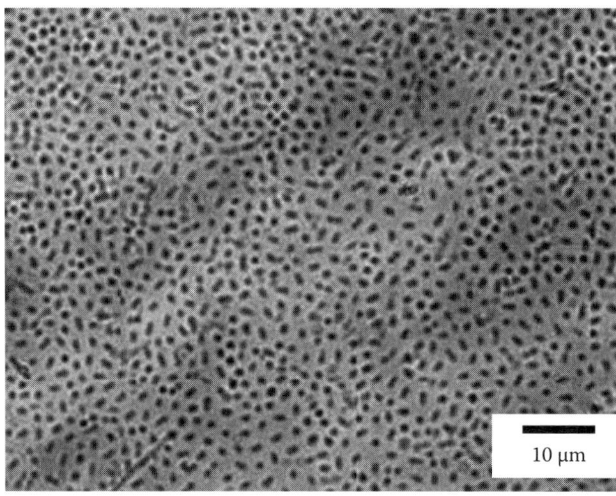

FIGURE 11.12
Micrograph of *Rhodobacter sphaeroides* ATCC 494119.

11.3.2.2.5 Quantum Efficiency of Different Photobiological H_2 Production Pathways

In order to compare the different photobiological pathways, Prince and Kheshgi [53] defined the maximum theoretical quantum efficiency of H_2 production based on monochromatic illumination at 680 nm as

$$\eta_{th} = \frac{n_{H_2} \Delta H_{H_2}}{n_{p,680} hc/\lambda} \tag{11.4}$$

where

n_{H_2} is the number of moles of produced H_2

ΔH_{H_2} is the enthalpy of formation of hydrogen equal to 285.83 kJ/mol at standard state (25°C and 0.1 MPa) [56]

Light input energy is computed as the product of the number of moles of photons needed at 680 nm denoted by $n_{p,680}$ and the energy of photons at 680 nm that is equal to hc/λ = 176 kJ/mol. In their definition, photons at 680 nm are used as this wavelength corresponds to the absorption peak of the reaction center in PSII. Note that the absorption peak of PSI is at 700 nm and that of purple bacteria is at 790 and 850 nm. The minimum number of photons needed is evaluated by considering the electron and ATP requirement of the reactions and assuming that two photons are needed per electron and two ATPs are needed per electron [53]. Finally, note that this definition of efficiency does not apply to process efficiency of a photobioreactor.

11.3.3 Performance Assessment and Units

Photobiological hydrogen production and/or carbon dioxide mitigation are very interdisciplinary topics involving researchers from many different disciplines including microbiologists and plant biologists as well as engineers. Owing to its diversity, it lacks standards in units to report experimental conditions and results and to enable direct comparisons between studies and microorganisms. This section aims to inform the reader on the different units found in the literature.

11.3.3.1 Microorganism Concentration

The microorganism concentration is usually reported in kilogram of dry cell per cubic meter of the liquid medium denoted by kg dry cell/m^3 or simply kg/m^3. This requires sampling a known volume of the cell suspension, drying it using an oven overnight, and weighing it using a high-precision analytical balance. Another method is to report the number of cells per cubic meter of liquid medium. This also requires sampling a known volume of cell suspension and counting the cells in that volume as observed under an optical microscope. This technique may be challenging if the microorganisms are (1) very small (<1 μm) such as purple nonsulfur bacteria, (2) in high concentration, and/or (3) having a morphology that is complex such as filamentous cyanobacteria (see Figure 11.11). Finally, some researchers perform chlorophyll *a* extraction and report the chlorophyll *a* concentration, denoted by mg Chl *a*/m^3, as a measure of the microorganism concentration. Chlorophyll *a* is chosen as it is present in all plants, algae, and cyanobacteria that photosynthesize. The results reported in mg Chl *a*/m^3 make comparisons difficult with other results as the chlorophyll *a* content per microorganism depends on the species and growth conditions. Moreover, different microorganisms can have different pigment concentrations or different pigments altogether. Then, reporting the chlorophyll *a* concentration can make results difficult to compare.

Thus, it is recommended that the microorganism concentration be reported in kg dry cell/m^3 or in number density (#/m^3). Rapid measurements of microorganism concentrations can be performed by measuring the optical density (OD) of microorganism suspension at one or more wavelengths using a UV–visible spectrophotometer. Then, convenient calibration curves can be developed to relate the OD to the dry cell weight or the number density.

11.3.3.2 Hydrogen Production Rate

The hydrogen production rate is reported either as the total mass or volume of hydrogen produced per unit time by the photobioreactor, expressed in kg/h, mmol/h, or m^3/h, or as the specific rate per kg dry cell or per milligram of Chl *a*, denoted by kg/kg dry cell/h or kg/mg Chl *a*/h. When reported in volumetric units, the pressure and temperature of the sample must be specified. Unfortunately, many volumetric production rates reported in the literature lack this detail making the results impossible to compare with other studies.

It is recommended that hydrogen production be reported in kg of H_2/h as this rate does not depend on the temperature and pressure of the measurement conditions. Similarly, the specific hydrogen production rates can be reported in kg of H_2/h per kg dry cell or in kg of H_2/h per unit volume of the photobioreactor.

11.3.3.3 Illumination

Light irradiance G_{in} is reported (1) in total luminous flux expressed in lux (1 lux = 1 cd·sr/m^2), (2) in photon flux expressed in μmol/m^2/s, or (3) in energy flux expressed in W/m^2. The total luminous flux, also known as illuminance, is a photometric unit that measures light accounting for the human eye sensitivity. The energy emitted by the source is wavelength weighted by the luminosity function that describes the average sensitivity of the human eye to light at different wavelengths between 400 and 700 nm. Different light sources with different emission spectra could have the same illuminance. Thus, illuminance is not recommended for reporting illumination.

On the other hand, photon flux refers to the number of moles of photons (6.02×10^{23} photons/mol) incident on a unit surface area per unit time in the PAR, that is, from 400 to 700 nm. This is a more appropriate unit for reporting the incident energy in photosynthetic systems. It is measured with a quantum sensor that is calibrated to measure the photon flux in the PAR. However, both illuminance and photon flux are valid only in the spectral range from 400 to 700 nm and cannot be used to quantify energy in the NIR part of the spectrum. Thus, in experiments using microorganisms that absorb in NIR such as purple nonsulfur bacteria, illumination reported in illuminance or photon flux cannot be used. Instead, the energy flux should be recorded with a pyranometer having the sensitivity from about 300 to 2800 nm and reported in W/m^2 μm. Moreover, it is recommended that the spectral sensitivity of the detector and the emission spectrum of the light source be also reported for clarity and reproducibility of the experiments. For benchtop experiments using artificial light, it is useful to report not only the total but also the spectral irradiance.

11.3.3.4 Light-to-Hydrogen Energy Conversion Efficiency

Light-to-hydrogen energy conversion efficiency of photobioreactors is defined as the ratio of (1) the energy that would be released from the reaction of the produced hydrogen with oxygen to produce water and (2) the energy input to the system as light, that is [57],

$$\eta_{H_2} = \frac{[\Delta G_o + RT\ln(P_{H_2}/P_o)]R_{H_2}}{G_{in}A_s} \quad (11.5)$$

where

R_{H_2} is the rate of production of H_2 in mol/s

A_s is the illumination area in m^2

ΔG_o is the standard-state free energy of the formation of H_2 from the water-splitting reaction, equal to 236,337 J/mol at 303 K and 1 atm

The term $RT\ln(P_{H_2}/P_o)$ is the correction factor for ΔG_o when H_2 production takes place at H_2 partial pressure P_{H_2} instead of the standard pressure P_o of 1 atm. The term G_{in} is the power input to the system as light, that is, irradiance in W/m^2. In reporting the light energy conversion efficiency, it is important to report the spectral range over which G_{in} is measured. Indeed, the efficiency computed using G_{in} defined over the PAR is about 2.22 times larger than that obtained with G_{in} computed over the entire solar spectrum.

11.4 Photobioreactor Systems

Photobiological hydrogen production by direct and indirect biophotolysis and by photofermentation typically consists of a first stage when microorganisms are grown by photosynthesis in the presence of air and CO_2. It is followed by a second stage when hydrogen is produced at constant microorganism concentration in the absence of CO_2, O_2, and N_2. During the growth phase, cyanobacteria fix CO_2 and nitrogen from the atmosphere to grow and produce photosynthates. In the H_2 production phase, they utilize the photosynthates to produce H_2. Alternatively, green algae *C. reinhardtii* are grown in a medium containing acetate. Then, the microorganisms are transferred into a sulfur-deprived medium where anoxia is induced by algae respiration resulting in H_2 production under relative high light irradiance, as first proposed by Melis et al. [40].

For economic reasons, the growth phase should be performed in open ponds [58]. In the hydrogen production phase, open systems will not be appropriate as the method of collection of hydrogen will pose serious difficulties. Similarly, closed indoor systems are not economically feasible since using artificial lighting defeats the purpose of solar energy utilization. Figure 11.13 schematically illustrates the typical process flow envisioned for photobiological hydrogen production at industrial scale [59]. This section presents the different types of photobioreactors used for the hydrogen production phase with emphasis on closed outdoor photobioreactors. It also discusses performances and modeling of mass and light transfer in photobioreactors.

11.4.1 Photobioreactor Types

Photobioreactors have been used for a wide range of applications including the production of pharmaceutics, food additives for humans, feed for animals, and cosmetic chemicals

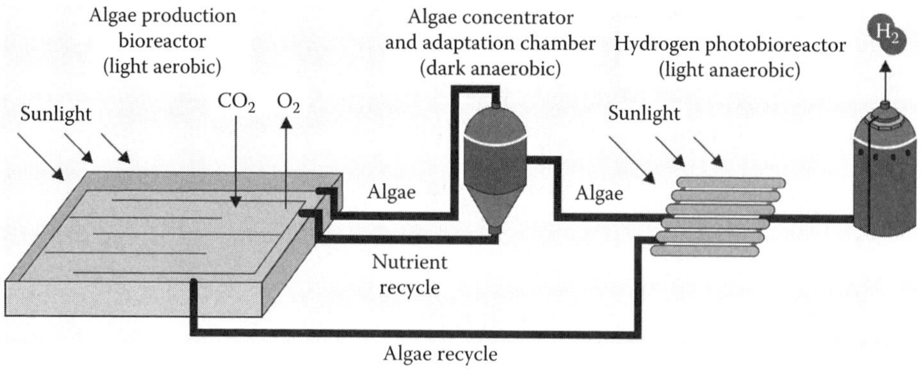

FIGURE 11.13
Typical process flow for industrial-scale photobiological hydrogen production. (From Riis, T. et al., Hydrogen production and storage—R&D priorities and gaps, International Energy Agency—Hydrogen Co-Ordination Group—Hydrogen Implementing Agreement, 2006, www.iea.org/Textbase/papers/2006/hydrogen.pdf; Courtesy of the International Energy Agency, Paris, France.)

using photosynthetic microorganisms [60]. They also found applications in environmental engineering such as wastewater treatment, heavy metal removal, and CO_2 mitigation [60]. More recently, they have been considered for hydrogen production [61–65].

On the most basic premises, photobioreactors can be grouped into three main categories, namely, (1) open cultivation systems, (2) closed outdoor systems, and (3) closed indoor systems [66]. The open systems are constructed in natural or artificial ponds and utilize sunlight. Closed outdoor photobioreactors consist of specially designed light transparent containers usually in the form of tubes or flat plates and also utilize sunlight [66]. Closed indoor systems, on the other hand, use artificial lighting such as fluorescent lights and light-emitting diodes (LEDs). Their construction is usually a light transparent adaptation of the conventional fermenter systems including stirred-tank bioreactors and vertical cylindrical columns. The advantages and disadvantages of closed and open systems are summarized in Table 11.4.

TABLE 11.4
Comparison of Photobioreactors

Type	Advantages	Disadvantages
Open Systems	Inexpensive to build and operate Uses sunlight	Small cell densities (0.1–0.2 g/L) Large space requirements Difficult to maintain monoculture Large water and CO_2 losses
Closed Outdoor	Improved control Limited water losses Uses sunlight Large cell densities (2–8 g/L) Easy to maintain monoculture	Relatively high installation costs Susceptible to ambient temperature variations Difficult to scale up Thermal management challenges
Closed Indoor	Large cell densities (2–8 g/L) Total control over physiological conditions	Relatively high installation and operation costs Inefficient

Sources: Pulz, O., *Appl. Microbiol. Biotechnol.*, 57(3), 287, 2001; Suh, I.S. and Lee, C.G., *Biotechnol. Bioprocess Eng.*, 8(6), 313, 2003.

11.4.2 Closed Outdoor Photobioreactor Designs and Performances

Most common types of closed photobioreactors are vertical or horizontal tubular, helical, and inclined or horizontal thin panel types [66]. Figure 11.14 shows some of these photobioreactor types that can be listed as follows:

- *Stirred-tank photobioreactors* are mechanically stirred photobioreactors to enhance mass transfer as shown in Figure 11.14a [67]. They are mostly suited for reactor volumes between 0.2 and 3 L. Examples include torus photobioreactors [68]. These reactors are often used for research purposes as they enable the control and uniformity of the growth conditions including hydrodynamics conditions, light exposure, concentrations, and pH. This permits full analysis of the system based on mass and energy conservation principles [69–71]. The drawback of these reactors is that very high stirring rates (>600 rpm) might be required to avoid the reactor to become mass transfer limited. Thus, continuously stirring the photobioreactor increases the operating costs, makes scale-up difficult, and decreases the reliability of the system.
- *Sparged- and stirred-tank photobioreactors*. To effectively stir the photobioreactor, gas can be sparged into the photobioreactor as shown in Figure 11.14b [67]. Sparging consists of injecting gas into the liquid phase through a porous medium called sparger or diffuser. This creates bubbles that could be further broken up by mechanical stirring to increase the interfacial area between the liquid and gas and enhances mass transfer. These photobioreactors require lower stirring rates than nonsparged ones and can accommodate liquid volumes greater than 500,000 L [67].
- *Bubble-column photobioreactors* use only sparging for agitation and aeration purposes as shown in Figure 11.14c [67]. They have a high liquid height to base width ratio to increase the bubble residence time and consequently the interfacial area available for mass transfer. Compared with stirred-tank photobioreactors, bubble columns provide less shear on the microorganisms and thus are more suitable for cultivation of plant cells [67].
- *Airlift photobioreactors* are very similar to bubble-column reactors except that they house a draft tube to regulate the flow of bubbles in the reactor as shown in Figure 11.14d [67]. This draft tube provides better heat and mass transfer efficiencies as well as more uniform shear levels. Excessive foaming and cell damage due to bubble bursting are among the drawbacks of airlift photobioreactors. Bubble-column and airlift designs include cylindrical or flat-plate types that can be oriented vertically upright or tilted at an angle [34,72].
- *Packed-bed photobioreactors*. The volume of the reactor is packed with small particles that provide a high surface area substrate on which microorganisms can grow as shown in Figure 11.14e [67]. The packed bed is completely filled with nutrient medium that is constantly circulated. Packed-bed photobioreactors suffer from clogging that inhibits effective mass transfer and limited light transfer to the microorganisms.
- *Trickle-bed photobioreactors* are very similar to packed-bed photobioreactors. However, the reactor liquid does not completely submerge the packing where the microorganisms are immobilized but, instead, trickles on the particles' surface as illustrated in Figure 11.14f [67]. This offers the advantage of minimizing water use. Usually, a gas flow counter to the liquid flow is also provided in trickle-bed photobioreactors for enhanced aeration. One drawback of these reactors is that

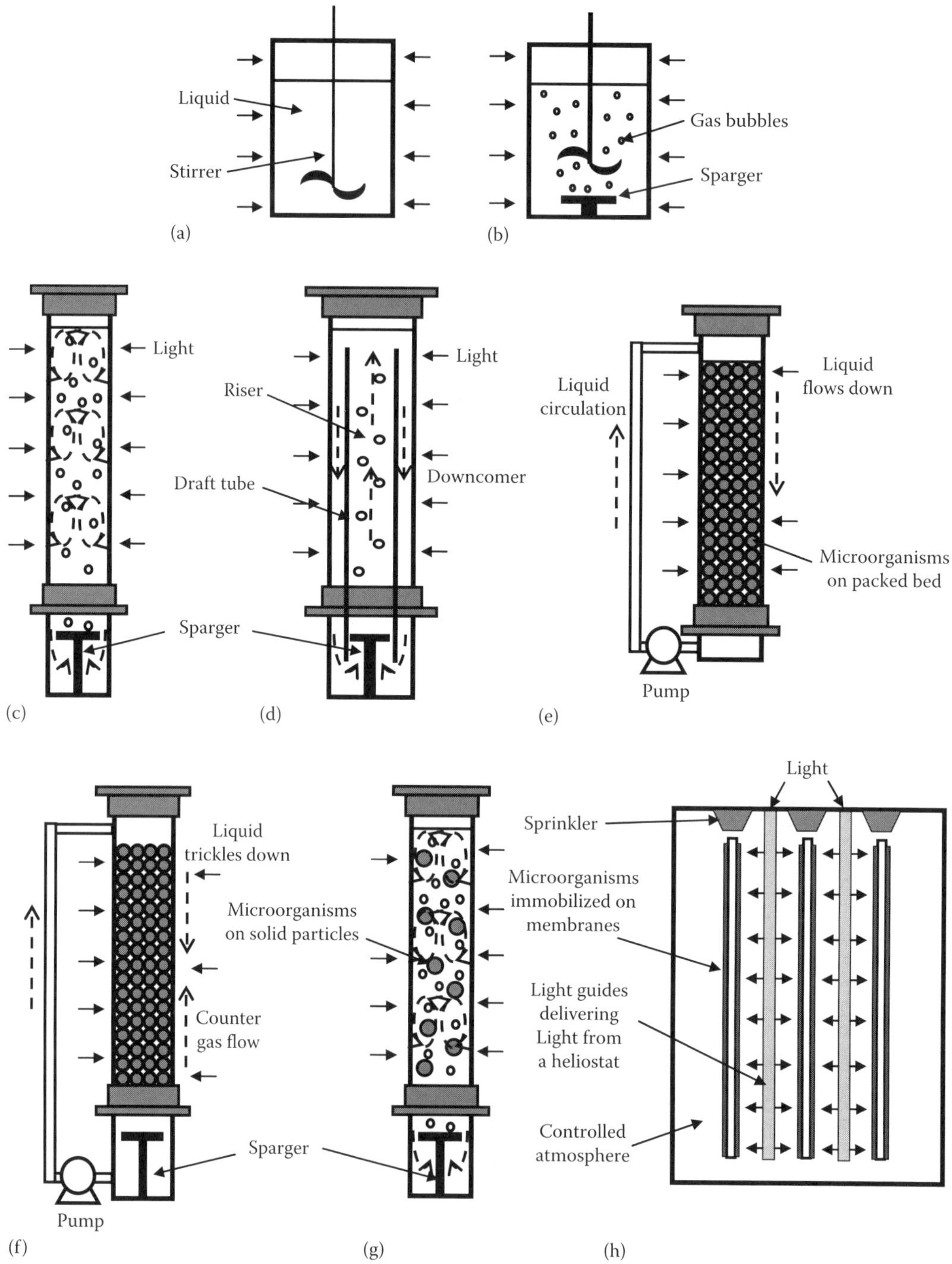

FIGURE 11.14
Typical photobioreactor designs: (a) stirred-tank, (b) sparged- and stirred-tank, (c) bubble-column, (d) airlift, (e) packed-bed, (f) trickle-bed, (g) fluidized-bed, and (h) membrane photobioreactors. (After Bayless, D.J. et al., *J. Environ. Eng. Manage.*, 16(4), 209, 2006.)

they have relatively low gas transfer per unit volume compared with sparged systems and have limited light transfer.

- *Fluidized-bed photobioreactors.* Microorganisms are immobilized within and/or on lightweight solid particles that are free to move with the fluid and are circulated within the photobioreactor as depicted in Figure 11.14g [67]. This can achieve large mass transfer rates at high cell concentrations.
- *Membrane photobioreactors.* Microorganisms are immobilized on membranes that are constantly being wet by a drip or a sprinkler system as shown in Figure 11.14h [73,74]. Light is collected by a heliostat unit and delivered to the microorganism via lightguides. These photobioreactors minimize the use of water, deliver controlled irradiance to the microorganisms, and can achieve large growth rates. However, they are expensive to build and operate.

Although most of these photobioreactor types were designed for fermenters, their adaptations to photobiological technologies have been discussed in the literature and need not be repeated here [66,75,76]. Despite recent advances, the performances of photobioreactors remain far from theoretical maxima even for benchtop systems [76].

11.4.3 Microorganism and Photobioreactor Performances

Table 11.5 provides a sample of selected studies conducted in laboratories reporting hydrogen production rates by a wide variety of microorganisms. The performance of various microorganisms and growth conditions and photobioreactors has been compared and discussed in Refs. [34,77,78].

Instead of providing an exhaustive review of past studies in this rapidly evolving field, the reader is referred to Ref. [79] for consulting and contributing to the latest experimental results. This wikipage makes use of the cyberinfrastructure to develop a virtual community focusing on photobiological CO_2 fixation and H_2 production. This resource should benefit this community in several ways. First, it provides a platform to share recent advances, experimental tips and data, database of bacterial properties, medium, as well as teaching material. Unlike textbooks, this resource can be regularly updated to reflect new advances so its content is less likely to become outdated. The content of the wiki is not dictated by a few experts but is entirely editable by the readers. Consequently, peer review is built into the publishing process and remains an active and continuous component throughout the life of the repository. Finally, it will bridge currently distinct communities in microbiology and plant biology on the one hand and engineering on the other.

11.4.4 Simulating Photobioreactors

In order to design, scale up, optimize, and compare the various photobioreactor designs, it is essential to develop experimentally validated simulation tools that account for light transfer, hydrodynamic conditions, and microorganism growth or H_2 production, along with mass conservation for nutrients and gas species. This section briefly reviews efforts in this area. For the sake of brevity, only selected studies are discussed.

11.4.4.1 Simulating Light Transfer

As light penetrates in the photobioreactor, it is absorbed by the microorganisms or by the medium and scattered by microorganisms and, possibly, by gas bubbles. These scatterers are much larger than visible wavelengths and therefore scattering is strongly

TABLE 11.5
Maximum Reported Hydrogen Production Rates for Various Types of Microorganisms

	Strain Name	Maximum Reported H_2 Production Rate	References
Freshwater Green Algae	*Chlamydomonas moewusii*	460 mmol/g chl *a*/h	[136]
	Chlamydomonas reinhardtii	200 mmol/g chl *a*/h	[136]
	Lobochlamys segnis	96 mmol/g chl *a*/h	[136]
	Chlamydomonas noctigama	31 mmol/g chl *a*/h	[136]
Marine Green Algae	*Scenedesmus vacuolatus*	155 mmol/g chl *a*/h	[136]
	Scenedesmus obliquus	150 mmol/g chl *a*/h	[44]
	Chlorococcum littorale	52 mmol/g chl *a*/h	[43]
Freshwater Cyanobacteria	*Anabaena variabilis* PK 84	167.6 mmol/g chl *a*/h	[46]
	Anabaena variabilis PK 17R	59.18 mmol/g chl *a*/h	[46]
	Anabaena variabilis ATCC 29413	45.16 mmol/g chl *a*/h	[46]
	Anabaena azollae	38.5 mmol/g chl *a*/h	[46]
	Anabaena CA	2.14 mol/kg/h	[49]
	Gloeobacter PCC 7421	1.38 mmol/g chl *a*/h	[77]
	Anabaena cylindrica	1.3 mol/kg/h	[77]
	Nostoc muscorum IAM M-14	0.6 mmol/g chl *a*/h	[77]
	Synechococcus PCC 602	0.66 mmol/g chl *a*/h	[77]
Marine Cyanobacteria	*Cyanothece* 7822	0.92 mmol/g chl *a*/h	[137]
	Oscillatoria limosa	0.83 mmol/g chl *a*/h	[49]
	Oscillatoria Miami BG7	0.3 mol/kg/h	[49]
Purple Nonsulfur Bacteria	*Rhodopseudomonas sphaeroides*	133 mol/kg/h	[41]
	Rhodobacter sphaeroides	5.9 mol/kg/h	[138]
	Rhodospirillum rubrum	2.5 mol/kg/h	[139,140]
Dark Fermentative Bacteria	*Enterobacter cloacae* IIT BT-08	211.63 mol/kg/h	[51]
	Enterobacter aerogenes	17 mol/kg/h	[41]
	Citrobacter intermedius	11.5 mol/kg/h	[141]
	Clostridium butyricum	7.3 mol/kg/h	[142]

forward [80,81]. Light transfer through the photobioreactor is governed by the radiative transfer equation (RTE) that expresses an energy balance in a unit solid angle $d\Omega$, about the direction $\hat{s}$ at location $\hat{r}$. The steady-state RTE in a well-mixed photobioreactor containing microorganisms and bubbles is expressed as [82]

$$\hat{s}\cdot\nabla I_\lambda(\hat{r},\hat{s}) = -\kappa_{eff,\lambda} I_\lambda(\hat{r},\hat{s}) - \sigma_{eff,\lambda} I_\lambda(\hat{r},\hat{s}) + \frac{\sigma_{X,\lambda}}{4\pi}\int_{4\pi} I_\lambda(\hat{r},\hat{s}_i)\Phi_{X,\lambda}(\hat{s}_i,\hat{s})d\Omega_i + \frac{\sigma_{B,\lambda}}{4\pi}\int_{4\pi} I_\lambda(\hat{r},\hat{s}_i)\Phi_{B,\lambda}(\hat{s}_i,\hat{s})d\Omega_i \quad (11.6)$$

where

$I_\lambda(\hat{r}, \hat{s})$ is the radiation intensity at wavelength λ in direction $\hat{s}$ at location $\hat{r}$

$\kappa_{eff,\lambda}$ and $\sigma_{eff,\lambda}$ are the effective spectral absorption and scattering coefficients, respectively

The coefficients $\sigma_{X,\lambda}$ and $\sigma_{B,\lambda}$ are the spectral scattering coefficients of the microorganisms and the bubbles, respectively

The scattering phase functions of microorganisms and bubbles are denoted by $\Phi_{X,\lambda}(\hat{s}_i, \hat{s})$ and $\Phi_{B,\lambda}(\hat{s}_i, \hat{s})$, respectively

They describe the probability that radiation traveling in the solid angle $d\Omega_i$ around the direction $\hat{s}_i$ will be scattered into the solid angle $d\Omega$ around the direction $\hat{s}$. They equal unity when scattering is isotropic. The effective absorption coefficient $\kappa_{eff,\lambda}$ accounts for the absorption by the liquid phase and by the microorganisms at wavelength λ. It can be written in terms of the bubble void fraction f_B and of the microorganism concentration X (in kg/m^3),

$$\kappa_{eff,\lambda} = \kappa_{L,\lambda}(1 - f_B - Xv_X) + A_{abs,\lambda}X \tag{11.7}$$

where v_X is the specific volume of microorganisms. The absorption coefficient of the liquid phase $\kappa_{L,\lambda}$ is expressed in m^{-1}, and the mass absorption cross section of microorganisms $A_{abs,\lambda}$ is expressed in m^2/kg. The term $\kappa_{X,\lambda} = A_{abs,\lambda}X$ corresponds to the absorption coefficient of microorganisms. Finally, the term Xv_X represents the volume fraction of the photobioreactor occupied by microorganisms. Assuming independent scattering, the effective scattering coefficient of the composite medium $\sigma_{eff,\lambda}$ can be expressed as the sum of the scattering coefficients of the microorganisms $\sigma_{X,\lambda}$ and of the bubbles $\sigma_{B,\lambda}$ as

$$\sigma_{eff,\lambda} = \sigma_{X,\lambda} + \sigma_{B,\lambda} = S_{sca,\lambda}X + \frac{A_i}{4}Q_{sca,B}(a,\lambda) \tag{11.8}$$

where
- $S_{sca,\lambda}$ is the mass scattering cross section of microorganisms expressed in m^2/kg
- $Q_{sca,B}(a,\lambda)$ is the scattering efficiency factor of monodisperse bubbles of radius a at wavelength λ obtained from the Mie theory [83]

The interfacial area concentration A_i is defined as the total surface area of bubbles per unit volume and expressed as $A_i = 3f_B/a$. Note that a similar approach can be used to model (1) mixed cultures, (2) scattering by beads in packed beds, and/or (3) polydispersed bubbles [84], for example.

Beer–Lambert's law provides the solution of the 1D steady-state RTE accounting for both absorption and out-scattering but ignoring in-scattering. It physically corresponds to cases when photons experience at most one scattering event as they travel through the reactor, that is, single scattering prevails. It gives the local spectral irradiance $G_\lambda(z)$ within the photobioreactor as

$$G_\lambda(z) = \int_0^{4\pi} I_\lambda(z,\hat{s})\,d\Omega = G_{\lambda,in}\exp(-\beta_{eff,\lambda}z) \tag{11.9}$$

where
- $G_{\lambda,in}$ is the spectral irradiance incident on the photobioreactor
- z is the distance from the front surface
- $\beta_{eff,\lambda}$ is the effective extinction coefficient of the suspension at wavelength λ defined as $\beta_{eff,\lambda} = \kappa_{eff,\lambda} + \sigma_{eff,\lambda}$

Beer–Lambert's law has been used extensively to predict the local irradiance within photobioreactors [85,86].

Moreover, Cornet et al. [87–89] solved the RTE using the Schuster–Schwarzschild two-flux approximation to model light transfer in filamentous cyanobacterium *Spirulina platensis* cultures. This approach consists of solving a pair of coupled ordinary differential equations obtained by integrating the RTE over two complementary hemispheres. It can account for in-scattering terms as well as anisotropic scattering [84]. It can also provide an analytical solution for $G_\lambda(z)$ albeit more complex than Beer–Lambert's law [68,69,90]. Finally, most of the aforementioned studies did not account for the spectral dependency of the radiation characteristics and/or for the presence of bubbles. More recently, Berberoğlu and Pilon [82] simulated light transfer in a bubble sparged photobioreactor accounting for absorption and anisotropic scattering by bubbles and microorganisms. Spectral variations of radiation characteristics over the spectral range from 400 to 700 nm were accounted for using the box model [82]. Genetically engineered microorganisms with reduced pigment content were also considered. The authors established that (1) Beer–Lambert's law cannot be applied to predict the irradiance inside the photobioreactor, that is, multiple scattering must be accounted for, (2) isotropic scattering can be assumed for wild-strain microorganisms for all practical purposes in the absence of bubbles, (3) anisotropic scattering by the bubbles must be accounted for particularly as the interfacial area concentration increases, (4) for microorganisms with reduced pigment concentration, their anisotropic scattering should be considered.

In order to simulate light transfer in photobioreactors and use any of the aforementioned light transfer models, the spectral radiative characteristics, namely, κ_λ, σ_λ, and $\Phi_\lambda(\hat{s}_i, \hat{s})$ of the microorganisms and/or the bubbles are required. They can be determined either through experimental measurements [80,81] or theoretically by using the Mie theory [69]. Theoretical predictions often assume that the scatterers have relatively simple shapes (e.g., spherical) and ignore their heterogeneous nature by assuming that the complex index of refraction is uniform. Pottier et al. [69] acknowledged that for complex microorganism shapes (e.g., cylinders and spheroids), advanced numerical tools are required to predict radiative characteristics. Alternatively, experimental measurements account for the actual shape and morphology and size distribution of the microorganisms. A comprehensive review of the experimental techniques for measuring the radiation characteristics has been reported by Agrawal and Mengüç [91] and need not be repeated. Pilon and Berberoğlu [80,81] experimentally measured the radiation characteristics of H_2-producing microorganisms, namely, (1) purple nonsulfur bacteria *R. sphaeroides* [80], (2) cyanobacteria *A. variabilis* [80], and (3) green algae *Chlamydomonas reinhardtii* strain CC125 and its truncated chlorophyll antenna transformants *tla*1, *tlaX*, and *tla*1-CW$^+$ [81]. The absorption and scattering cross sections of all strains studied were obtained over the spectral range from 300 to 1300 nm along with their scattering phase function at 632.8 nm. The latter can be assumed to be independent of wavelength in the PAR [92,93]. It was established that *R. sphaeroides* absorbs mainly in two distinct spectral regions from 300 to 600 nm and from 750 to 900 nm. The major absorption peaks can be observed around 370, 480, 790, and 850 nm and can be attributed to the presence of bacteriochlorophyll *b* and carotenoids in the antenna complexes B850 and the reaction center complex [30,94]. Moreover, *A. variabilis* and the wild strain *C. reinhardtii* CC125 absorb mainly in the spectral region from 300 to 700 nm with absorption peaks at 435 and 676 nm corresponding to in vivo absorption peaks of chlorophyll *a*. *A. variabilis* also absorbs at 621 nm corresponding to absorption by the pigment phycocyanin [30], while *C. reinhardtii* has additional absorption peaks at 475 and 650 nm corresponding to absorption by chlorophyll *b*. The genetically engineered strains of *C. reinhardtii* were shown to have less chlorophyll pigments than the wild strain and thus smaller absorption cross sections as illustrated in Figure 11.15. In particular, the mutant *tlaX* features a significant reduction in chlorophyll *b* concentration. For all mutants, however, the reduction in the absorption

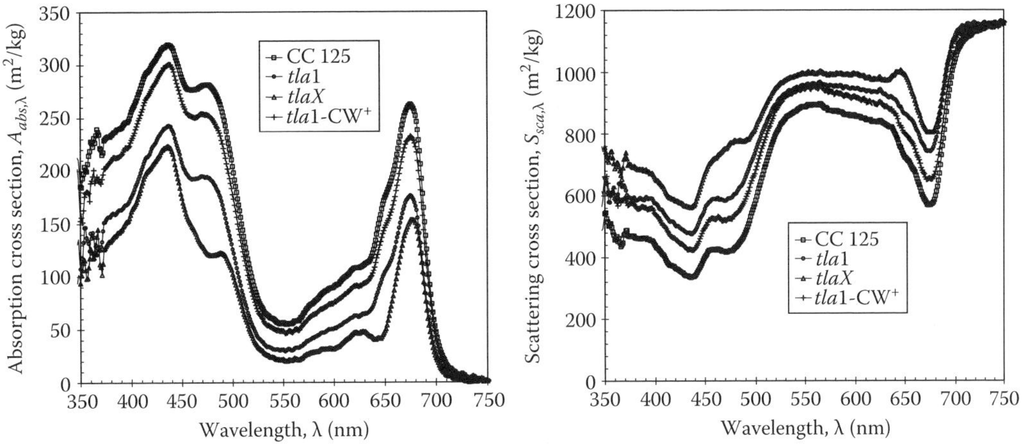

FIGURE 11.15
Absorption and scattering cross sections of the green algae *C. reinhardtii* CC 125 and its truncated chlorophyll antenna transformants *tla*1, *tla*X, and *tla*1-CW+. (From Berberoğlu, H. et al., *Int. J. Hydrogen Energy*, 33(22), 6467, 2008.)

cross section is accompanied by an increase in scattering cross section [81] (see Figure 11.15). Although scattering becomes the dominant phenomenon contributing to the overall extinction of light, it is mainly in the forward direction so light penetrates within the reactor.

11.4.4.2 Simulating Hydrodynamic Conditions

Hydrodynamic conditions in a photobioreactor affect the shear stress and the amount of light to which individual microorganisms are subjected. Both affect the system's productivity [90]. Simulations of hydrodynamic conditions consist of, first, solving mass and momentum (or Navier–Stokes) conservation equations for a specific reactor geometry to find the local fluid velocity within the photobioreactors. The Lagrangian approach is used to identify the trajectory of individual microorganisms as they are carried by the medium [85]. The instantaneous light flux received by a microorganism can, then, be determined as a function of time [85], and the frequency of light–dark cycles can be determined. These cycles are known to have a strong effect on the microorganism growth, and the cycle frequency should optimally range between 1 Hz and 1 kHz, which is difficult to achieve at industrial scale [76]. Finally, the average light energy received by a microorganism can be computed and used to estimate their growth or hydrogen production rate based on kinetic models.

11.4.4.2.1 Photosynthetic Growth Kinetics

During the growth phase, the time rate of change of microorganism concentration X can be modeled as [95]

$$\frac{dX}{dt} = \mu X \tag{11.10}$$

where μ is the specific growth rate expressed in s^{-1} and function of the average available irradiance denoted by G_{av}. The specific growth rate has been modeled using the modified Monod model taking into account light saturation and inhibition as [67]

$$\mu = \mu_{max}\left(\frac{G_{av}}{K_{s,G} + G_{av} + (G_{av}^2/K_{i,G})}\right) \tag{11.11}$$

where μ_{max} is the maximum specific growth rate while the coefficients $K_{s,G}$ and $K_{i,G}$ are the light half-saturation and inhibition constants, respectively. Similar models can be formulated to account for saturation and inhibition due to limited or excessive carbon dioxide concentrations or excessive microorganism concentrations, for example. The average available irradiance G_{av} can be estimated by averaging the local irradiance over the depth of the culture L as

$$G_{av} = \frac{1}{L}\int_0^L G(z)dz = \frac{1}{L}\int_0^L\int_0^\infty G_\lambda(z)d\lambda dz \tag{11.12}$$

where $G_\lambda(z)$ is estimated by (1) solving the RTE, (2) using approximate solutions such as Beer–Lambert's law (Equation 11.9), or (3) averaging the light energy received by microorganisms as predicted by hydrodynamics simulations. Fouchard et al. [71] identified μ_{max}, $K_{s,G}$, and $K_{i,G}$ for *C. reinhardtii* in TAP medium without acetate to be 0.2274 h^{-1}, 81.38 μmol photon/m^2/s, and 2500 μmol photon/m^2/s, respectively.

11.4.4.2.2 Photobiological H_2 Evolution Kinetics

Similarly, the specific production rate π_{H_2} has been modeled with a modified Michaelis–Menten-type equation given by [95]

$$\pi_{H_2}(z) = \pi_{H_2,max}\frac{G_{av}(z)}{K_{s,H_2} + G_{av}(z) + \left(G_{av}^2(z)/K_{i,H_2}\right)} \tag{11.13}$$

where $\pi_{H_2,max}$ is the maximum specific hydrogen production rate expressed in kg H_2/kg dry cell/h. The parameters K_{s,H_2} and K_{i,H_2} account for the saturation and the inhibition of hydrogen production due to excessive irradiation or limited light irradiance.

Nogi et al. [96] measured the specific hydrogen production rate π_{H_2} of the purple non-sulfur bacteria *Rhodopseudomonas rutila* as a function of incident irradiance. The authors reported the absorption spectrum, the hydrogen production rate as a function of spectral incident radiation, and the specific hydrogen production rate as a function of usable radiation. The parameters $\pi_{H_2,max}$, K_{s,H_2}, and K_{i,H_2} were estimated by least-squares fitting of the experimental data reported over the usable incident radiation range from 0 to 80 W/m^2 [96]. The values of $\pi_{H_2,max}$, K_{s,H_2}, and K_{i,H_2} were found to be 1.3×10^{-3} kg H_2/kg dry cell/h, 25 W/m^2, and 120 W/m^2, respectively. Figure 11.16 compares the prediction of Equation 11.13 for π_{H_2} with data reported by Nogi et al. [96].

11.4.4.2.3 Mass Conservation Equations

Mass conservation principles should be satisfied by all gas and nutrient species such as dissolved oxygen, hydrogen, starch, and/or sulfur for green algae. For a well-mixed photobioreactor, the concentrations are assumed to be uniform throughout the reactor and their time rate of change is expressed as

$$\frac{dC_i}{dt} = r_i - k_L a(C_i - C_{i,eq}) \tag{11.14}$$

where

r_i is the net production rate of species "i"
$k_L a$ is the specific gas–liquid mass transfer coefficient
$C_{i,eq}$ is the equilibrium concentration between the gas and the liquid phases

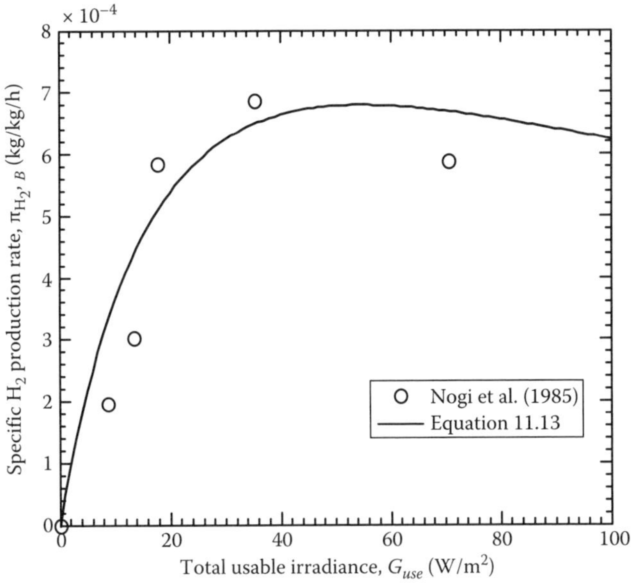

FIGURE 11.16
Experimental data and the modified Michaelis–Menten model (Equation 11.13) for the specific hydrogen production rate of *R. rutila* as a function of the usable incident radiation G_{use} with parameters $\pi_{H_2,max} = 1.3 \times 10^{-3}$ kg H_2/kg dry cell/h, $K_G = 25$ W/m^2, and $K_I = 120$ W/m^2. (From Nogi, Y. et al., *Agric. Biol. Chem.*, 49, 3538, 1985.)

For oxygen, for example, r_{O_2} accounts for O_2 generated during photosynthesis and consumed by respiration. It is often assumed to be proportional to the microorganism growth rate, that is, $r_{O_2} = Y_{O_2}\mu X$, where Y_{O_2} is the yield coefficient of O_2 conversion. The reader is referred to Ref. [71] for an illustration of modeling of photobiological H_2 production by *C. reinhardtii* accounting for light transfer, algal growth, and extracellular and intracellular sulfur, starch, and oxygen consumption and/or production along with an estimate of the associated model parameters.

11.5 Technical Challenges and Limitations

Current photobiological hydrogen production suffers from low solar-to-hydrogen energy conversion efficiency that is typically less than 1% [97] under outdoor conditions. In addition, scale-up and economic viability remain major challenges mainly due to issues related to (1) light transfer limitation, (2) mass transfer and hydrodynamics limitations, (3) thermal management, (4) contamination and maintenance of monoculture, and (5) the photobioreactor cost.

11.5.1 Light Transfer

Light transfer in photobioreactors is one of the main barriers to the technology [98,99]. Indeed, photosynthetic microorganisms require an optimum irradiance to achieve the most efficient photosynthesis and H_2 production. This optimum depends on the particular microorganism but, in general, is about 100 W/m^2 for naturally occurring strains [100]. Thus, photobioreactors can suffer from the following:

- *Light inhibition*. Excessive irradiance inhibits the microbial photosynthesis and H_2 production through a process called photooxidative damage [101,102]. In outdoor systems, where this technology is meant to be deployed, solar irradiance can reach as high as 1000 W/m^2 [103]. For more efficient use of sunlight, it has to be redistributed uniformly throughout the photobioreactor.
- *Limited light penetration*. Due to light absorption by the microorganisms and the medium and scattering by both the microorganisms and gas bubbles, the local irradiance within the photobioreactor may decrease below the required levels for photosynthesis and/or H_2 production [87–89,97]. This, in turn, limits the productivity and scale-up of the system.

Ways to address these light transfer challenges are discussed in Section 11.6.

11.5.2 Mass Transfer and Hydrodynamics

Hydrogen and possibly oxygen produced by the microorganisms have inhibitory effects on photobiological hydrogen production [55,103]. Therefore, these gas species must be efficiently removed from the photobioreactor for sustained H_2 production. The issues related to mass transfer and hydrodynamics are the following:

- *By-product buildup*. Algae and cyanobacteria produce O_2 as a by-product of photosynthesis. Excessive O_2 concentrations result in inhibition of nitrogenase and hydrogenase [97,101]. Similarly, excessive buildup of H_2 decreases the production rate in hydrogenase-based systems [101,104]. Methods such as applying partial vacuum [61] and sparging with an inert gas [105] have been suggested to remove H_2 and O_2. However, these techniques have been considered economically unfeasible at industrial scale [106]. Currently, this challenge is being addressed by microbiologists where O_2-tolerant enzymes are being isolated/developed and expressed in selected microorganisms [106]. Alternatively, the inhibition of O_2 production by sulfur deprivation of green algae *C. reinhardtii* has been demonstrated [40].
- *Low interfacial area concentration*. Mass transfer from gas to liquid phase requires large gas–liquid interfacial area. As the reactor is scaled up, the surface area available for gas transfer per unit volume of the reactor decreases in nonsparged reactors. This makes CO_2 transfer a limiting factor, during the growth phase, for achieving large cell densities in scaled-up systems having volumes larger than 50 L [55,66]. It also limits the removal of O_2 and H_2 during the H_2 production phase and may result in by-product buildup and reduce the overall efficiency.
- *Sedimentation of microorganisms and nutrients*. This causes limitations on the availability of nutrients and light to the settled microorganisms [27,66]. The photobioreactor can be stirred or sparged with bubbles to keep microorganisms in suspension in addition to alleviating the mass transfer limitations. However, unfavorably high shear can be detrimental to microorganisms during liquid flow and/or bubble collapse [66,107]. In addition, stirring is prohibitively expensive for large-scale systems.

11.5.3 Nutrient Composition

The concentration of both macro- and micronutrients has significant effect on the CO_2 consumption and H_2 production by the microorganisms [55,101,105]. However, the effects of different nutrient concentrations and different media on photobiological hydrogen production and the associated pathways are not known precisely. The optimization of nutrient composition could contribute to increasing the efficiency of the photobiological system. For example, Berberoğlu et al. [108] reported a factor 5.5 increase in hydrogen production rate by *Anabaena variabilis* ATCC 29413 using Allen–Arnon medium compared with BG-11 and BG-11_0 media under otherwise identical conditions (light, concentration, pH, temperature). Moreover, the heterocyst frequency was 5%, 4%, and 9% for *A. variabilis* grown in BG-11, BG-11_0, and Allen–Arnon media, respectively. The authors also reported larger specific hydrogen production rates, efficiencies, and cyanobacteria concentrations achieved using Allen–Arnon medium. This was attributed to higher concentrations of magnesium, calcium, sodium, and potassium in the medium. Finally, the presence of vanadium in Allen–Arnon medium could have induced the transcription of vanadium-based nitrogenase that is capable of evolving more hydrogen than the molybdenum-based one. Further research is needed in this area.

11.5.4 Thermal Management

Economic and practical difficulties are faced in maintaining an optimum reactor temperature over night and day and over the season cycle [58,66]. Active temperature control has been considered in pilot systems but this adds cost and decreases the economic viability of the technology [58]. A practical solution for this problem can be the choice of favorable geographic locations or marine systems for which the ocean can act as a temperature bath whose temperature varies slightly over the course of the year. Alternatively, infrared solar radiation that would otherwise heat up the photobioreactor could be filtered before delivering only usable light to the photobioreactor.

11.5.5 Sterility and Monoculture

It is essential that the photobioreactors do not get contaminated by other microorganisms that could compete for light and nutrients, thus, adversely affecting the performance of the system [53]. This becomes a major challenge for large-scale systems. Alternatively, in order to overcome contamination, commercial algae growers have been using strains that survive in harsh environments such as high salinity and/or low pH media where most other microorganisms cannot live [58].

11.5.6 Freshwater Consumption

Many of the algal and cyanobacterial strains considered for CO_2 mitigation and H_2 production grow in freshwater. Once scaled up, such a system may require large quantities of freshwater competing with the resources used for domestic and agricultural needs. Several approaches for overcoming the need for large quantities of freshwater are (1) using trickle-bed, fluidized-bed, and membrane photobioreactors (Figure 11.14) minimizing the water usage [73,74], (2) using saltwater species where water from the oceans can be utilized [109], and (3) using wastewater where more value can also be added to the process through wastewater treatment [37].

11.6 Prospects

To date, the aforementioned limitations still remain challenges to the commercial realization of the technology. Several strategies are being pursued to increase the efficiency of photobiological hydrogen production. First, with the advent of genetic engineering, microorganisms can be engineered to have the desired pigment concentrations, optimum enzymatic pathways and electron transport, as well as reduced O_2 inhibitions. Second, processes and photobioreactors can be designed to increase efficiency by achieving optimum light delivery, maximizing sunlight utilization, and providing ideal conditions for growth and H_2 production. Third, the construction, operation, and maintenance costs of the photobioreactor systems should also be reduced.

11.6.1 Bioengineering of Microorganisms

As previously discussed, there are several intrinsic limitations to hydrogen production by the enzymes. These issues are being addressed by microbiologists and genetic engineers. Efforts include (1) truncating the light-harvesting antenna, (2) developing O_2-tolerant enzymes, (3) eliminating the expression of uptake hydrogenase, and (4) inserting proton channels in thylakoid membranes. Each of these approaches is briefly described in detail in the following sections.

11.6.1.1 Truncating the Light-Harvesting Antenna

Microorganisms that are found in nature are not always subjected to optimum illumination. Therefore, as a survival mechanism, they have adapted to produce relatively large amounts of pigments. This maximizes the probability of capturing and utilizing photons at low light intensities. However, when these microorganisms are mass cultured in photobioreactors, they absorb more photons than they can utilize and waste the light energy as heat and fluorescence [102]. In addition, light does not penetrate deep into the photobioreactor. Thus, the quantum efficiency of photobiological hydrogen production decreases. Moreover, high intensities can catalyze the formation of harmful oxides that can damage the photosynthetic apparatus, a process known as photooxidation [28]. Therefore, it is desirable to decrease the chlorophyll antenna size down to the size of the core antenna [100].

Melis et al. [102,110] physiologically reduced the pigment content of the green algae *Dunaliella salina* from 1×10^9 chlorophyll molecules per cell (Chl/cell) to 0.15×10^9 Chl/cell. More recently, Polle et al. [100] genetically engineered the green algae *Chlamydomonas reinhardtii* to have a truncated light-harvesting chlorophyll antenna size. The authors reported that the microorganisms with less pigments had higher quantum yield, photosynthesis rate, and light saturation irradiance [100].

Figure 11.17 shows the in vivo differential interference contrast (DIC) and chlorophyll fluorescence micrographs of green algae *C. reinhardtii* CC125 and its truncated chlorophyll antenna transformants *tla*1, *tla*X, and *tla*1-CW$^+$ [81]. The images were obtained using a Zeiss 510 confocal scanning laser microscope in the transmission and epifluorescence mode simultaneously as reported by Chen and Melis [111]. The excitation was provided by a helium–neon laser at 543 nm, while the chlorophyll fluorescence emission was detected in the red region with a longpass filter with a cutoff wavelength of 560 nm placed in front of the detector. It illustrates the size and shape of each strain as well as the location of the

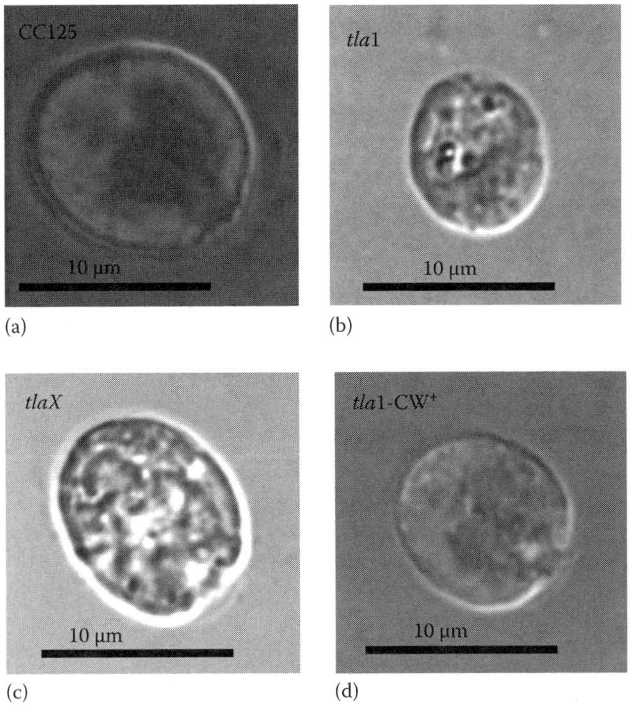

FIGURE 11.17
DIC and fluorescence micrographs of (a) CC125, (b) *tla*1, (c) *tlaX*, and (d) *tla*1-CW^+. The scale bars correspond to 10 μm. (From Berberoğlu, H. et al., *Int. J. Hydrogen Energy*, 33(22), 6467, 2008.)

chlorophyll pigments that fluoresce in red [31]. The strong red fluorescence observed in the wild strain CC125 qualitatively shows that it has the largest concentration of chlorophyll while *tlaX* has the least.

11.6.1.2 Oxygen-Tolerant Enzymes

As previously discussed, there have been numerous approaches in overcoming the O_2 inhibition of hydrogen production such as applying partial vacuum [61], sparging with an inert gas [105], and sulfur deprivation [40]. Developing O_2-tolerant enzymes concerns mainly green algae and cyanobacteria as they produce O_2 as a result of oxidation of water during oxygenic photosynthesis. Moreover, it depends on the type of enzyme used by these microorganisms. Green algae use bidirectional [Fe-Fe]-hydrogenase, whereas cyanobacteria use either bidirectional [NiFe]-hydrogenase or nitrogenase.

[Fe-Fe]-hydrogenase in green algae: Hydrogen production by green algae is due to bidirectional [Fe-Fe]-hydrogenase [53]. However, the functioning of this enzyme is irreversibly inhibited by micromolar concentrations of O_2 [106]. Inhibition takes place when O_2 molecules bind to the catalytic site of the enzyme. Based on the structural modeling of the algal hydrogenase, Forestier et al. [112] suggested that inhibition takes place due to the presence of a large gas channel leading to the catalytic center. Figure 11.18 shows the structural model of the *C. reinhardtii* [Fe-Fe]-hydrogenase [112]. The channel enables the formed H_2 molecule to escape. However, due to its large size, it can also let O_2 diffuse to the catalytic center and

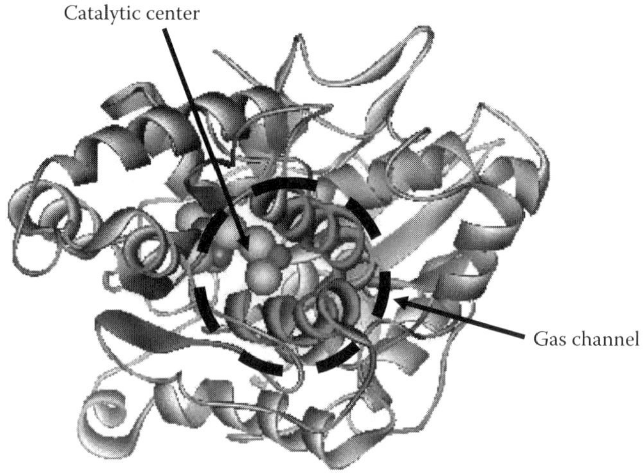

FIGURE 11.18
The structural model of the *C. reinhardtii's* [Fe-Fe]-hydrogenase enzyme. (From Forestier, M. et al., *Eur. J. Biochem.*, 270(13), 2750, 2003.)

inactivate the enzyme. Strategies to engineer a novel hydrogenase with a steric restriction to O_2 diffusion and express it in algae are being pursued [112]. This will allow the photobiological hydrogen production to take place in air making the process more efficient and less expensive.

[NiFe]-hydrogenase in Cyanobacteria: Another approach for overcoming O_2 inhibition is to identify naturally occurring oxygen-tolerant [NiFe]-hydrogenases in photosynthetic bacteria (e.g., *Rubrivivax gelatinosus* and *Thiocapsa roseopersicinas*) and expressing them in selected cyanobacteria (e.g., *Synechocystis* sp. PCC6803) [113,114].

Nitrogenase in Cyanobacteria: The thermophilic filamentous actinobacteria *Streptomyces thermoautotrophicus* synthesize a unique nitrogenase that is both structurally and functionally different from the classic [Mo]-nitrogenase. This unusual nitrogenase is reported to be completely insensitive to O_2 and consumes half the amount of ATP consumed by the classic [Mo]-nitrogenase [30]. However, hydrogen production by this enzyme has not been demonstrated. The expression of such a nitrogenase in cyanobacteria through genetic engineering can prove beneficial for more robust and cost-effective photobiological H_2 production.

11.6.1.3 Eliminating the Expression of Uptake Hydrogenase

Uptake hydrogenase is found in cyanobacteria that produce hydrogen using nitrogenase. It catalyzes the consumption of H_2 in the presence of O_2 to recover energy and produce water [53]. This decreases the net H_2 production rate by cyanobacteria. To address this issue, several researchers have proposed to improve the cyanobacterial H_2 production by eliminating the expression of uptake hydrogenase [46,115,116]. Sveshnikov et al. [46] reported that *Anabaena variabilis* mutant PK84 lacking the uptake hydrogenase had 3–4.3 times larger hydrogen production rates compared with the wild strain. In an independent study, Tsygankov et al. [117] reported that at a dissolved O_2 concentration of 315 μM in the medium, the net hydrogen production rate of the wild strain *A. variabilis* was only 7%

of the control experiment with no dissolved oxygen in the medium. On the other hand, under the same conditions, the mutant PK84 showed 75% of the hydrogen production rate of the same control experiment. Note that the equilibrium concentration of dissolved O_2 in water exposed to air at 1 atm and 25°C is about 250 μM.

11.6.1.4 Inserting Proton Channels in Photosynthetic (Thylakoid) Membranes

The rate of photobiological H_2 production from water is slowed down by inefficient electron coupling to ferredoxin due to the large proton gradient across the algal photosynthetic (thylakoid) membrane [113,118]. Lee [118] proposed that genetic insertion of proton channels in the photosynthetic membrane can decrease the proton gradient and overcome this limitation. Moreover, this will aid in preventing the electrons from participating in the Calvin cycle and in diverting the electron flow to hydrogenases, thus improving the rate of H_2 production.

11.6.2 Process Optimization

Several process optimizations have been considered to further develop photobiological hydrogen production technologies including (1) symbiotic or mixed cultures of different types of microorganisms, (2) controlled and optimum light delivery, and (3) cell immobilization.

11.6.2.1 Symbiotic and Mixed Cultures

To date, the majority of research efforts have concentrated on cultivating single species of microorganisms for photobiological hydrogen production. Among these, cyanobacteria and green algae that utilize solar energy in the spectral range from 400 to 700 nm to produce hydrogen have been studied extensively [119]. On the other hand, purple nonsulfur bacteria have also been identified as potential hydrogen producers that mainly use solar energy in the NIR part of the spectrum from 700 to 900 nm [96]. Note that only about 45% of the total solar radiation is emitted between 400 and 700 nm and an additional 20% is emitted between 700 and 900 nm [50].

Thus, mixed cultivation of green algae and purple bacteria has the potential to achieve higher solar-to-hydrogen energy conversion efficiencies than single cultures by using solar radiation in the spectral range from 400 to 900 nm where 65% of the total solar radiation is concentrated. Such a mixed culture has been demonstrated by Melis and Melnicki [50] where the green algae *C. reinhardtii* were cocultured with the purple bacteria *Rhodospirillum rubrum*. The authors suggested that once the photosynthesis to respiration (P/R) ratio of the green algae is reduced to 1, such a coculture could be used for more efficient photobiological hydrogen production. Currently, the wild-strain algae have a P/R of about 4 [50]. Unfortunately, the purple bacteria also absorb light in the visible part of the spectrum due to the presence of bacteriochlorophyll *b* and carotenoids [80]. Consequently, the two species may compete for light during both the growth and the hydrogen production phases.

Recently, Berberoğlu and Pilon [120] reported a numerical study aiming to maximize the solar-to-hydrogen energy conversion efficiency of a mixed culture containing the green algae *Chlamydomonas reinhardtii* and the purple nonsulfur bacteria *Rhodobacter sphaeroides*. The authors used the radiation characteristics measured experimentally [80,81] as input parameters for calculating the local spectral incident radiation within a flat-panel photobioreactor. Their results show that for monocultures, the solar-to-H_2 energy conversion efficiency depends only on the optical thickness of the system. The maximum solar energy conversion efficiency of monocultures of *C. reinhardtii* and *R. sphaeroides*, considering the entire solar spectrum, was found to be 0.061% and 0.054%, respectively, corresponding to optical thicknesses of 200 and 16, respectively. Using mixed cultures, a total conversion efficiency of about 0.075% could be achieved corresponding to an increase of about 23% with respect to that of a monoculture of *C. reinhardtii*. The choice of microorganism concentrations for maximum solar energy conversion efficiency in mixed cultures was nontrivial and requires careful radiation transfer analysis coupled with H_2 production kinetics.

Another strategy is to grow symbiotic cultures such as combining purple nonsulfur bacteria and anaerobic fermentative bacteria. For example, Miyake et al. [121] used symbiotic cultures of the anaerobic fermentative bacteria *Clostridium butyricum* and the purple nonsulfur bacteria *R. sphaeroides* to produce H_2. In this symbiotic culture, the anaerobic bacteria converted sugars to H_2 and organic acids, whereas the purple nonsulfur bacteria converted the organic acids to H_2. Overall, their symbiotic system produced 7 mol of H_2 per mole of glucose.

11.6.2.2 Advanced Light Delivery Systems

The saturation irradiance of a photosynthetic apparatus is on the order or 5000–6000 lux [108]. This corresponds to about one-tenth of the total solar irradiance where the rest of the energy is wasted as heat and fluorescence. Thus, light can be delivered to a 10 times larger surface area using solar collectors and lightguides to enhance the solar energy utilization efficiency. To do so, cost-effective light delivery technologies need to be developed and integrated into the design of future photobioreactors.

System engineers are designing novel photobioreactors that collect and deliver solar light in a controlled manner within the photobioreactor [34,73,109,122–124]. These systems usually involve a heliostat comprised of either Fresnel lenses [109] or reflective dishes [34,73] that concentrate the solar radiation to be distributed via fiber optics or lightguides. The lightguides are made of glass or acrylic and can deliver sunlight deep into the photobioreactors by total internal reflection. At desired locations, the lightguides have rough surfaces and light *leaks* out providing the desired irradiance. In some elaborate designs, LEDs are also incorporated into the lightguide delivery system to provide artificial light to the microorganisms at night [74]. Lightguides and fiber optics have been used to increase the light irradiance in the center of photobioreactors where it is typically the smallest [124]. However, this technology is judged too costly to be adopted at industrial scale [98].

Alternatively, Kondo et al. [98] proposed the simultaneous culture of the purple nonsulfur bacteria *Rhodobacter sphaeroides* RV and its reduced pigment mutant MTP4 in two separate but stacked plate-type photobioreactors. MTP4 produces H_2 more efficiently under large irradiance, while *R. sphaeroides* RV is more efficient under low irradiance. The authors showed that two stacked flat-plate photobioreactors with MTP4 in the front reactor facing the light source and *R. sphaeroides* RV in the rear reactor produced more H_2 than any other configuration. The front reactor acted as an absorption filter to the second.

11.6.2.3 Immobilized Cell Photobioreactors

In order to achieve high H_2 production rates, Markov et al. [61] immobilized *A. variabilis* on hollow fibers. They operated the photobioreactor in two stages alternating between (1) growth and (2) H_2 production phases. The authors reported a CO_2 consumption rate of 7000 mmol/kg dry cell/h and an H_2 production rate of 830 mmol H_2/kg dry cell/h.

Moreover, Bagai and Madamwar [125] immobilized a mixed culture of the nonheterocystous cyanobacteria *Phormidium valderianum*, the halophilic bacteria *Halobacterium halobium*, and the hydrogenase containing *Escherichia coli* in polyvinyl alcohol (PVA) alginate beads for prolonged production of H_2. The authors demonstrated H_2 production by the mixed culture for over 4 months.

More recently, Laurinavichene et al. [126] immobilized *C. reinhardtii* on glass mesh cloth in an enclosed photobioreactor of a total volume of 160 mL. The immobilized system enabled easy switch between sulfur-containing and sulfur-depleted media during growth and H_2 production stages, respectively. The authors reported a maximum hydrogen yield of 380 mL over 23 days with a maximum H_2 production rate of 45 mL/day. The immobilized cell system prolonged the H_2 production up to 4 weeks compared with suspended cell systems investigated by Kosourov et al. [127].

11.7 Economic and Environmental Considerations

Photobiological hydrogen production aims to produce hydrogen in an environmentally friendly and sustainable manner. Thus, its environmental impacts must be discussed in terms of toxicity, water usage, and lifecycle analysis. Economic considerations are also essential to assess the feasibility of the technology and ensure it provides a viable and competitive alternative to fossil fuel or other H_2 production technologies.

11.7.1 Economic Analysis

The US Department of Energy (DOE) set a hydrogen cost goal of \$2.00–\$3.00 (2005 US dollars) per kilogram of delivered and untaxed H_2 by 2015 [128]. For comparison with gasoline cost, note that the energy contained in 1 kg of H_2 is equivalent to that contained in 1 gal of gasoline. The average 2002 price for compressed hydrogen gas produced from fossil fuel and delivered in tube trailers was \$11.0/kg of H_2. The price of pipeline-delivered compressed-H_2 merchant hydrogen ranged from \$0.8 to \$3.4/kg of H_2 in 2003 [129]. Prices for commercial hydrogen have risen steadily in recent years primarily due to growing demand in the refinery sector and increase in oil and natural gas prices. Similarly, the price of gasoline has increased significantly and may make hydrogen more competitive if produced from renewable energy. For example, the cost of H_2 produced by wind electrolysis was \$5.90/kg of H_2 in 2006 [10].

Economic analysis of photobiological hydrogen production considers (1) the construction and maintenance costs of the photobioreactor; (2) the operating cost including labor, power, and water supplies for mixing, periodic cleaning, and powering compressors, for example; (3) the purification of hydrogen gas and its compression for transportation or storage; (4) land purchase; and (5) daily solar irradiance of the site. It was estimated that to achieve a 10% return on investment, the photobioreactor cost should be less than \$165/m^2 of footprint, for a system having 10% light-to-H_2 energy conversion [57]. Moreover, to be economically viable, the system should achieve conversion efficiencies larger than 10% [57]. Note that theoretically, H_2-producing microorganisms have a maximum light-to-H_2 efficiency ranging between 16% and 41% depending on the metabolic pathway used [53].

Photobioreactor cost is a major contributing factor to the cost of photobiological H_2 production and the most important parameter to the economic feasibility of the technology [57,130]. For example, it was estimated that the price of a photobioreactor could be in excess of \$100/m^2 if glass or plexiglass were used as transparent windows. It could be reduced to about \$1/m^2 if low-density polyethylene (LDPE) films were used [130]. Amos [130] estimated the selling price of H_2 at \$ 2.04/kg for hydrogen delivered via a pipeline and produced by *C. reinhardtii* at a concentration of 0.2 g/L in a 10 cm deep photobioreactor and accounting for a 15% return on investment. However, the price can rise up to \$13.53/kg for purified and compressed H_2 at 2 MPa for a photobioreactor cost of \$10/m^2. In this case, the compression and storage cost alone contributed \$2.25/kg. The author acknowledged that the algal hydrogen production rate was the maximum rate biologically possible that is much larger than that achieved in practice even in benchtop photobioreactors. Finally, cost associated with energy consumption for mixing and gas injection was stated not to exceed 2 W/m^2 [76].

The annual average solar irradiance (over 24 h and over 365 days) at favorable locations such as the southwestern United States is about 210 W/m^2 [131]. In addition, fuel cells running on hydrogen have fuel efficiency of about 30%–50%, whereas internal combustion engines using gasoline are only about 30% efficient. Thus, a hydrogen refueling station providing similar service to public will have to supply about one-third the fuel supplied by a gasoline station. A light-to-hydrogen conversion efficiency of 10% would represent a production rate of 5.45 kg of H_2/m^2/year. To put this in perspective, it would require a surface area of 54,000 m^2 or 14 ac to supply the equivalent of 800,000 gal of gasoline sold every year by one of the 168,000 gasoline service stations in the United States [53]. Further improving the efficiency of this technology through photobioreactor design and genetic engineering would reduce the footprint requirement.

Moreover, to achieve sustainable energy production, the total energy used to build the system should be much lesser than that produced by the system during its entire lifetime. Thus, the service lifetime and the energy cost of materials used for building the photobioreactors should be considered in addition to their financial cost. Burgess and Fernandez-Velasco [132] defined the so-called net energy ratio (NER) as the ratio of the higher heating value of the produced hydrogen to the total primary energy input into the construction of the system. The authors reported that for tubular photobioreactors, LDPE film and glass have significantly higher NER than rigid polymers such as polymethyl methacrylate (acrylic) [132]. Similar lifecycle analysis should be performed for other novel photobioreactor designs to assess their sustainability.

In brief, photobiological hydrogen production is at a very early stage of development. It currently does not constitute an economically viable hydrogen production method and needs additional basic and applied research to approach practical efficiency and production rates. Thorough economic analysis has to be performed for various organisms and photobioreactor design in order to assess the viability in the short run and the sustainability in the long run of photobiological hydrogen production. More realistic economic analysis will require operation, maintenance, and field data from pilot photobioreactor systems taking into account the seasonal performance variations.

11.7.2 Environmental Impacts

Some strains of cyanobacteria are known to produce toxins that are harmful to human and animal health such as anatoxins, microcystins, or saxitoxins [133]. Most algae species are harmless to animals and humans. These toxins can cause acute or chronic illnesses such as gastroenteritis, adverse respiratory effects, skin irritations, allergic responses, and liver damage [134]. In addition to producing toxins, freshwater cyanobacteria can cause nuisance effects such as excessive accumulations of foams and scums and discoloration of water [135].

Moreover, current photobiological hydrogen production would require large amounts of water if the benchtop processes were scaled up for industrial production. This would create large demand for freshwater that would compete with domestic and agriculture uses both of which are scarce in regions with high solar irradiance. One may argue, however, that the system would be closed and water would simply need to be supplied to compensate for evaporation and separation needs. Thus, water consumption for H_2 production is likely to be much lesser than that of agricultural crops. In addition, photobioreactors designed to be constructed on land are likely to compete with land use for humans, forests, or agricultural use.

Alternatively, wastewater or seawater in combination with cyanobacteria and marine algae could be used to reduce demand on freshwater and land space. Marine green algae *Chlorococcum littorale* and marine cyanobacteria *Oscillatoria* sp. as well as *Miami BG7*, *Anabaena cylindrical* B-629, and *Calothrix scopulorum* 1410/5 are known hydrogen producers [77]. Marine-based systems benefit from (1) thermal regulation of the ocean preventing the overheating of the photobioreactors under direct sunlight and (2) agitation of the reactor fluid that could be achieved by the ocean waves. Some of the major issues concerning the development of marine photobioreactors include (1) contingency engineering for possible microorganism leakage into the ocean, (2) durability of the materials used in marine environment, and (3) contamination and damage to the photobioreactors by the marine animals and microorganisms. Thus, research should be directed to address these issues for developing cost-effective marine photobioreactors.

Finally, public perception and potential fear associated with the use and possible release of genetically modified microorganisms in the environment will need to be addressed. Public reaction is expected to vary from one region of the world to another. However, it may constitute a major obstacle in some countries. Past experiences with genetically modified crops (e.g., corn or soybean) can constitute a valuable reference and provide useful lessons.

11.8 Conclusion

This chapter presented the current state of knowledge in photobiological H_2 production as well as CO_2 fixation. It provided the reader with a basic background in the microbiology of photosynthesis and photobiological H_2 production. Then, photobioreactor designs, operations, performances, and simulation tools were reviewed. The challenges associated with the technology were discussed followed by strategies to overcome the biological barriers and to optimize the process. Finally, economic analysis and potential environmental impacts were presented. In brief, photobiological hydrogen production is at a very early stage of development and requires additional basic and applied research efforts. However, progresses from genetic engineering to innovative photobioreactor designs with advanced light delivery and reduced water consumption are promising. If successful, this technology can offer a long-term solution for sustainable hydrogen production. It can also alleviate concerns over energy security with the advantage of capturing CO_2.

References

1. International Energy Agency, Key world energy statistics 2013, http://www.iea.org/publications/, accessed on December 6, 2013.
2. M.K. Hubbert, Nuclear energy and the fossil fuels, Presented before the *Spring Meeting of the Southern District*, American Petroleum Institute, San Antonio, TX, 1956.
3. B. Holmes and N. Jones, Brace yourself for the end of cheap oil, *New Scientist*, 179(2406), 9–11, 2003.
4. K.S. Deffyes, *Hubbert's Peak: The Impending World Oil Shortage*, Princeton University Press, Princeton, NJ, 2001.
5. Energy Information Administration, International energy outlook 2008—Highlights, Report DOE/EIA-0484, June 2008.
6. IPPC, Climate change 2007: Impacts, adaptation and vulnerability. Contribution of working group ii to the fourth assessment report of the intergovernmental panel on climate change, Cambridge, U.K., 2007, Cambridge University Press, Cambridge, U.K.
7. U.S. Central Intelligence Agency, *The World Factbook 2013*, https://www.cia.gov/cia/publications/factbook/index.html, accessed on December 6, 2013.
8. Energy Information Administration, Emissions of greenhouse gases in the United States 2005, Report DOE/EIA-0573, 2005.
9. P. Kruger, *Alternative Energy Resources: The Quest for Sustainable Energy*, John Wiley & Sons, Hoboken, NJ, 2006.
10. U.S. DOE, Hydrogen, fuel cells, and infrastructure technologies program, http://www1.eere.energy.gov/hydrogenandfuelcells/, 2006, accessed on December 6, 2013.
11. U.S. Department of Energy, *Hydrogen Data Book*, Hydrogen Analysis Resource Center, Pacific Northwest National Laboratory, Richland, WA, http://hydrogen.pnl.gov/cocoon/morf/hydrogen/article/103, 2008, accessed on December 6, 2013.
12. G.P. Sutton and O. Biblarz, *Rocket Propulsion Elements*, 7th edn., John Wiley & Sons, New York, 2000.
13. J.M. Norbeck, J.W. Heffel, T.D. Durbin, B. Tabbara, J.M. Bowden, and M.C. Montano, *Hydrogen Fuel for Surface Transportation*, SAE International, Warrendale, PA, 1996.
14. H.J. Sternfeld and P. Heinrich, A demonstration plant for the hydrogen/oxygen-spinning reserve, *International Journal of Hydrogen Energy*, 14(10), 703–716, 1989.
15. H.J. Sternfeld, Capacity control of power stations by O_2/H_2 rocket combustor technology, *Acta Astronautica*, 37, 11–19, 1995.
16. S.P. Malyshenko, A.N. Gryaznov, and N.I. Filatov, High-pressure H_2/O_2-steam generators and their possible applications, *International Journal of Hydrogen Energy*, 29(6), 589–596, 2004.
17. G. Hoogers, *Fuel Cell Technology Handbook*, CRC Press, Boca Raton, FL, 2002.
18. P.J. Sebastian, Photoelectrochemical water splitting for hydrogen generation, *International Journal of Hydrogen Energy*, 26(2), 115, 2001.
19. J.E. Funk, Thermochemical hydrogen production: Past and present, *International Journal of Hydrogen Energy*, 26(3), 185–190, 2001.
20. D.O. Hall, S.A. Markov, Y. Watanabe, and K.K. Rao, The potential applications of cyanobacterial photosynthesis for clean technologies, *Photosynthesis Research*, 46(1–2), 159–167, 1995.
21. K.N. Liou, *An Introduction to Atmospheric Radiation*, 2nd edn., Academic Press, San Diego, CA, 2002.
22. J.A. Duffie and W.A. Beckman, *Solar Engineering of Thermal Processes*, 3rd edn., John Wiley & Sons, Hoboken, NJ, 2006.
23. C. Gueymard, D. Myers, and K. Emery, Proposed reference irradiance spectra for solar energy systems testing, *Solar Energy*, 73(6), 443–467, 2002.
24. The National Oceanic and Atmospheric Administration (NOAA), U.S. standard atmosphere, 1976, in NOAA/NASA/USAF-NOAA-S/T76-1562. Washington, DC, 1976.
25. C. Gueymard, Smarts code, version 2.9.2 users direct beam spectral irradiance data for photovoltaic cell manual, Solar Consulting Services, Golden, CO, http://www.nrel.gov/rredc/smarts/, 2002, accessed on December 6, 2013.

26. SunWize Technologies, World insolation map, Kingston, New York, 2008.
27. O. Pulz, Photobioreactors: Production systems for phototrophic microorganisms, *Applied Microbiology and Biotechnology*, 57(3), 287–293, 2001.
28. B. Ke, *Photosynthesis, Photobiochemistry and Photobiophysics*, Kluwer Academic Publishers, Dordrecht, the Netherlands, 2001.
29. P.F. Weaver, S. Lien, and M. Seibert, Photobiological production of hydrogen, *Solar Energy*, 24(1), 3–45, 1980.
30. M.T. Madigan and J.M. Martinko, *Biology of Microorganisms*, Pearson Prentice Hall, Upper Saddle River, NJ, 2006.
31. E.H. Harris, *The Chlamydomonas Sourcebook*, Vol. 1, Academic Press, San Diego, CA, 1989.
32. N.J. Lang, J.M. Krupp, and A.L. Koller, Morphological and ultrastructural-changes in vegetative and heterocysts of *Anabaena variabilis* grown with fructose, *Journal of Bacteriology*, 169(2), 920–923, 1987.
33. Department of Energy, *Rhodobacter sphaeroides*, Joint Genome Institute, Walnut Creek, CA. http://genome.jgi-psf.org/finished-microbes/rhosp/rhosp.home.html, accessed on April 19, 2008.
34. I. Akkerman, M. Jansen, J. Rocha, and R.H. Wijffels, Photobiological hydrogen production: Photochemical efficiency and bioreactor design, *International Journal of Hydrogen Energy*, 27, 1195–1208, 2002.
35. P. Joliot and A. Joliot, Cyclic electron transport in plant leaf, *Proceedings of the National Academy of Sciences USA*, 99(15), 10209–10214, 2002.
36. Y. Munekaga, M. Hashimoto, C. Miyake, K.I. Tomizawa, T. Endo, M. Tasaka, and T. Shikanai, Cyclic electron flow around photosystem I is essential for photosynthesis, *Nature*, 429(6991), 579–582, 2004.
37. I.K. Kapdan and F. Kargi, Bio-hydrogen production from waste materials, *Enzyme and Microbial Technology*, 38(5), 569–582, 2006.
38. A. Melis, Bioengineering of green algae to enhance photosynthesis and hydrogen production, in *Artificial Photosynthesis: From Basic Biology to Industrial Applications*, A.F. Collings and C. Critchley, eds., pp. 229–240. Wiley-VCH Verlag GmbH, Weinheim, Germany.
39. M.L. Ghirardi, A. Dubini, J. Yu, and P.-C. Maness, Photobiological hydrogen-producing systems, *Chemical Society Review*, 38, 52–61, 2009.
40. A. Melis, L. Zhang, M. Forestier, M.L. Ghirardi, and M. Seibert, Sustained photobiological hydrogen gas production upon reversible inactivation of oxygen evolution in the green alga *Chlamydomonas reinhardtii*, *Plant Physiology*, 117(1), 129–139, 2000.
41. D. Das and T.N. Veziroğlu, Hydrogen production by biological processes: A survey of literature, *International Journal of Hydrogen Energy*, 26(1), 13–28, 2001.
42. Y. Ueno, N. Kurano, and S. Miyachi, Purification and characterization of hydrogenase from the marine green alga *Chlorococcum littorale*, *FEBS Letters*, 443(2), 144–148, 1999.
43. L. Florin, A. Tsokoglou, and T. Happe, A novel type of iron hydrogenase in the green alga *Scenedesmus obliquus* is linked to the photosynthetic electron transport chain, *Journal of Biological Chemistry*, 276(9), 6125–6132, 2001.
44. M. Pavetic, H. Tausch, G. Stehlik, and M. Roehr, Influence of ruthenium, rhodium and vanadium ions on photoproduction of hydrogen by *Chlorella fusca*, *Photochemistry and Photobiology*, 40(1), 15–21, 1984.
45. D.A. Sveshnikov, N.V. Sveshnikova, K.K. Rao, and D.O. Hall, Hydrogen metabolism of mutant forms of *Anabaena variabilis* in continuous cultures and under nutritional stress, *FEMS Microbiology Letters*, 147, 297–301, 1997.
46. P. Lindberg, K. Schutz, T. Happe, and P. Lindblad, A hydrogen-producing, hydrogenase-free mutant strain of *Nostoc punctiforme* ATCC29133, *International Journal of Hydrogen Energy*, 27(11–12), 1291–1296, 2002.
47. S. Kumazawa and A. Mitsui, Characterization and optimization of hydrogen photoproduction by a saltwater blue green alga, *Oscillatoria* sp. miami bg7. enhancement through limiting the supply of nitrogen nutrients, *International Journal of Hydrogen Energy*, 6(4), 339–348, 1981.

48. F.A.L. Pinto, O. Troshina, and P. Lindblad, A brief look at three decades of research on cyanobacterial hydrogen evolution, *International Journal of Hydrogen Energy*, 27(11–12), 1209–1215, 2002.
49. A. Melis and M. Melnicki, Integrated biological hydrogen production, *International Journal of Hydrogen Energy*, 31(11), 1563–1573, 2006.
50. N. Kumar, P.S. Monga, A.K. Biswas, and D. Das, Modeling and simulation of clean fuel production by *Enterobacter cloacae* IIT-BT 08, *International Journal of Hydrogen Energy*, 25(10), 945–952, 2000.
51. H. Yokoi, Y. Maeda, J. Hirose, S. Hayashi, and Y. Takasaki, H_2 production by immobilized cells of *Clostridium butyricum* on porous glass beads, *Biotechnology Techniques*, 11(6), 431–433, 1997.
52. R.C. Prince and H.S. Kheshgi, The photobiological production of hydrogen: Potential efficiency and effectiveness as a renewable fuel, *Critical Reviews in Microbiology*, 31(1), 19–31, 2005.
53. P.C. Hallenbeck and J.R. Benemann, Biological hydrogen production: Fundamentals and limiting processes, *International Journal of Hydrogen Energy*, 27, 1185–1193, 2002.
54. D. Madamwar, N. Garg, and V. Shah, Cyanobacterial hydrogen production, *World Journal of Microbiology and Biotechnology*, 16(8–9), 757–767, 2000.
55. G.M. Bodner and H.L. Pardue, *Chemistry: An Experimental Science*, John Wiley & Sons, New York, 1995.
56. J.R. Bolton, Solar photoproduction of hydrogen, IEA Technical Report, International Energy Agency, Paris, France, IEA/H2/TR-96, 1996.
57. J. Sheehan, T. Dunahay, J. Benemann, and P. Roessler, A look back at the U.S. Department of Energy's aquatic species program—Biodiesel from algae, National Renewable Energy Laboratory, Golden, CO, 1998.
58. T. Riis, E.F. Hagen, P.J.S. Vie, and Ø. Ulleberg, Hydrogen production and storage—R&D priorities and gaps, International Energy Agency—Hydrogen Co-Ordination Group—Hydrogen Implementing Agreement, 2006, http://www.iea.org/publications/freepublications/publication/hydrogen.pdf, accessed on December 6, 2013.
59. N.J. Kim, I.S. Suh, B.K. Hur, and C.G. Lee, Simple monodimensional model for linear growth rate of photosynthetic microorganisms in flat-plate photobioreactors, *Journal of Microbiology and Biotechnology*, 12(6), 962–971, 2002.
60. S.A. Markov, R. Lichtl, K.K. Rao, and D.O. Hall, A hollow fibre photobioreactor for continuous production of hydrogen by immobilized cyanobacteria under partial vacuum, *International Journal of Hydrogen Energy*, 18(11), 901–906, 1993.
61. S.A. Markov, M.J. Bazin, and D.O. Hall, Hydrogen photoproduction and carbon dioxide uptake by immobilized *Anabaena variabilis* in a hollow-fiber photobioreactor, *Enzyme and Microbial Technology*, 17(4), 306–310, 1995.
62. S.A. Markov, A.D. Thomas, M.J. Bazin, and D.O. Hall, Photoproduction of hydrogen by cyanobacteria under partial vacuum in batch culture or in a photobioreactor, *International Journal of Hydrogen Energy*, 22, 521–524, 1997.
63. A.A. Tsygankov, A.S. Fedorov, S.N. Kosourov, and K.K. Rao, Hydrogen production by cyanobacteria in an automated outdoor photobioreactor under aerobic conditions, *Biotechnology and Bioengineering*, 80(7), 777–783, 2002.
64. J.H. Yoon, J.H. Shin, M.S. Kim, S.J. Sim, and T.H. Park, Evaluation of conversion efficiency of light to hydrogen energy by *Anabaena variabilis*, *International Journal of Hydrogen Energy*, 31(6), 721–727, 2006.
65. I.S. Suh and C.G. Lee, Photobioreactor engineering: Design and performance, *Biotechnology and Bioprocess Engineering*, 8(6), 313–321, 2003.
66. J.A. Asenjo and J.C. Merchuk, *Bioreactor System Design*, Marcel Dekker, New York, 1995.
67. S. Fouchard, J. Pruvost, B. Degrenne, and J. Legrand, Investigation of H_2 production using the green microalga *Chlamydomonas reinhardtii* in a fully controlled photobioreactor fitted with online gas analysis, *International Journal of Hydrogen Energy*, 33(13), 3302–3310, 2008.

68. L. Pottier, J. Pruvost, J. Deremetz, J.F. Cornet, J. Legrand, and C.G. Dussap, A fully predictive model for one-dimensional light attenuation by *Chlamydomonas reinhardtii* in a torus photobioreactor, *Biotechnology and Bioengineering*, 91(5), 569–582, 2005.
69. J. Pruvost, L. Pottier, and J. Legrand, Numerical investigation of hydrodynamic and mixing conditions in a torus photobioreactor, *Chemical Engineering Science*, 61(14), 4476–4489, 2006.
70. S. Fouchard, J. Pruvost, B. Degrenne, M. Titica, and J. Legrand, Kinetic modeling of light limitation and sulfur deprivation effects in the induction of hydrogen production with *Chlamydomonas reinhardtii*: Part I. Model development and parameter identification, *Biotechnology and Bioengineering*, 102(1), 232–245, 2009.
71. J.C. Merchuk, Y. Rosenblat, and I. Berzin, Fluid flow and mass transfer in a counter current gas liquid inclined tubes photo-bioreactor, *Chemical Engineering Science*, 62(24), 7414–7425, 2007.
72. D.J. Bayless, G. Kremer, M. Vis, B. Stuart, L. Shi, J. Cuello, and E. Ono, Photosynthetic CO_2 mitigation using a novel membrane-based photobioreactor, *Journal of Environmental Engineering and Management*, 16(4), 209–215, 2006.
73. E. Ono and J.L. Cuello, Design parameters of solar concentrating systems for CO_2 mitigating algal photobioreactors, *Energy*, 29(9–10), 1651–1657, 2004.
74. E.J. Wolfrum, A.S. Watt, and J. Huang, Bioreactor development for biological hydrogen production, *Proceedings of the 2002 U.S. DOE Hydrogen Program Review (NERL CP-610-32405)*, Golden, CO, 6–10 May 2002, pp. 10–19; NREL Report No. CP-510-32407.
75. C. Posten, Design principles of photo-bioreactors for cultivation of microalgae, *Engineering in Life Sciences*, 9(3), 165–177, 2009.
76. D. Dutta, D. De, S. Chaudhari, and S.K. Bhattacharya, Hydrogen production by cyanobacteria, *Microbial Cell Factories*, 4(36), 1–11, 2005.
77. D. Das and T.N. Veziroğlu, Advances in biological hydrogen production processes, *International Journal of Hydrogen Energy*, 33, 6046–6057, 2008.
78. H. Berberoğlu and L. Pilon, Experimental measurement of the radiation characteristics of *Anabaena variabilis* ATCC 29413-U and *Rhodobacter sphaeroides* ATCC 49419, *International Journal of Hydrogen Energy*, 32(18), 4772–4785, 2007.
79. H. Berberoğlu, A. Melis, and L. Pilon, Radiation characteristics of *Chlamydomonas reinhardtii* CC125 and its truncated chlorophyll antenna transformants *tla1*, *tlaX*, and *tla1*-CW$^+$, *International Journal of Hydrogen Energy*, 33(22), 6467–6483, 2008.
80. H. Berberoğlu, J. Yin, and L. Pilon, Simulating light transfer in a bubble sparged photobioreactor for simultaneous hydrogen fuel production and CO_2 mitigation, *International Journal of Hydrogen Energy*, 32(13), 2273–2285, 2007.
81. G. Mie, Beiträge zur Optik trüber Medien, speziell kolloidaler Metallsungen, *Annalen der Physik*, 25(3), 377–445, 1908.
82. M.F. Modest, *Radiative Heat Transfer*, Academic Press, San Diego, CA, 2003.
83. J. Pruvost, J. Legrand, P. Legentilhomme, and A. Muller-Feuga, Simulation of microalgae growth in limiting light conditions: Flow effect, *AIChE Journal*, 48(5), 1109–1120, 2002.
84. F.G. Acién Fernández, F. García Camacho, J.A. Sánchez Pérez, J.M. Fernández Sevilla, and E. Molina Grima, A model for light distribution and average solar irradiance inside outdoor tubular photobioreactors for the microalgal mass culture, *Biotechnology and Bioengineering*, 55(5), 701–714, 1997.
85. J.F. Cornet, C.G. Dussap, and G. Dubertret, A structured model for simulation of cultures of the cyanobacterium spirulina platensis in photobioreactors: I. Coupling between light transfer and growth kinetics, *Biotechnology and Bioengineering*, 40(7), 817–825, 1992.
86. J.F. Cornet, C.G. Dussap, P. Cluzel, and G. Dubertret, A structured model for simulation of cultures of the cyanobacterium spirulina platensis in photobioreactors: II. Identification of kinetic parameters under light and mineral limitations, *Biotechnology and Bioengineering*, 40(7), 826–834, 1992.
87. J.F. Cornet, C.G. Dussap, J.B. Gross, C. Binois, and C. Lasseur, A simplified monodimensional approach for modeling coupling between radiant light transfer and growth kinetics in photobioreactors, *Chemical Engineering Science*, 50(9), 1489–1500, 1995.

88. J. Pruvost, J.-F. Cornet, and J. Legrand, Hydrodynamics influence on light conversion in photobioreactors: An energetically consistent analysis, *Chemical Engineering Science*, 63(14), 3679–3694, 2008.
89. B.M. Agrawal and M.P. Mengüç, Forward and inverse analysis of single and multiple scattering of collimated radiation in an axisymmetric system, *International Journal of Heat and Mass Transfer*, 34, 633–647, 1991.
90. M.N. Merzlyak and K.R. Naqvi, On recording the true absorption spectrum and scattering spectrum of a turbid sample: Application to cell suspensions of cyanobacterium *Anabaena variabilis*, *Journal of Photochemistry and Photobiology B*, 58, 123–129, 2000.
91. D. Stramski and C.D. Mobley, Effect of microbial particles on oceanic optics: A database of single-particle optical properties, *Limnology and Oceanography*, 42, 538–549, 1997.
92. R.M. Broglie, C.N. Hunter, P. Delepelaire, R.A. Niederman, N.H. Chua, and R.K. Clayton, Isolation and characterization of the pigment–protein complexes of *Rhodopseudomonas sphaeroides* by lithium dodecylsulfate/polyacrylamide gel electrophoresis, *Proceedings of the National Academy of Sciences of the United States of America*, 77(1), 87–91, 1980.
93. I.J. Dunn, E. Heinzle, J. Ingham, and J.E. Prenosil, *Biological Reaction Engineering; Dynamic Modelling Fundamentals with Simulation Examples*, 2nd edn., Wiley-VCH Verlag GmbH, Weinheim, Germany, 2003.
94. Y. Nogi, T. Akiba, and K. Horikoshi, Wavelength dependence of photoproduction of hydrogen by *Rhodopseudomonas rutila*, *Agricultural and Biological Chemistry*, 49, 3538, 1985.
95. J.R. Benemann, Hydrogen production by microalgae, *Journal of Applied Phycology*, 12(3–5), 291–300, 2000.
96. T. Kondo, M. Arakawa, T. Wakayama, and J. Miyake, Hydrogen production by combining two types of photosynthetic bacteria with different characteristics, *International Journal of Hydrogen Energy*, 27(11–12), 1303–1308, 2002.
97. J.C. Ogbonna, T. Soejima, and H. Tanaka, An integrated solar and artificial light system for internal illumination of photobioreactors, *Journal of Biotechnology*, 70(1–3), 289–297, 1999.
98. J.E. Polle, S.D. Kanakagiri, and A. Melis, *tla1*, a DNA insertional transformant of the green alga *Chlamydomonas reinhardtii* with a truncated light-harvesting chlorophyll antenna size, *Planta*, 217(1), 49–59, 2003.
99. J.H. Yoon, S.J. Sim, M.S. Kim, and T.H. Park, High cell density culture of *Anabaena variabilis* using repeated injections of carbon dioxide for the production of hydrogen, *International Journal of Hydrogen Energy*, 27(11–12), 1265–1270, 2002.
100. A. Melis, J. Neidhardt, and J.R. Benemann, *Dunaliella salina (Chlorophyta)* with small chlorophyll antenna sizes exhibit higher photosynthetic productivities and photon use efficiencies than normally pigmented cells, *Journal of Applied Phycology*, 10(6), 515–525, 1999.
101. A. Melis, Green alga hydrogen production: Process, challenges and prospects, *International Journal of Hydrogen Energy*, 27(11–12), 1217–1228, 2002.
102. J.W. Van Groenestijn, J.H.O. Hazewinkel, M. Nienoord, and P.J.T. Bussmann, Energy aspects of biological hydrogen production in high rate bioreactors operated in the thermophilic temperature range, *International Journal of Hydrogen Energy*, 27(11–12), 1141–1147, 2002.
103. V.B. Borodin, A.A. Tsygankov, K.K. Rao, and D.O. Hall, Hydrogen production by *Anabaena variabilis* PK84 under simulated outdoor conditions, *Biotechnology and Bioengineering*, 69(5), 478–485, 2000.
104. M.L. Girardi, P. King, S. Kosourov, M. Forestier, L. Zhang, and M. Seibert, Development of algal systems for hydrogen photoproduction: Addressing the hydrogenase oxygen sensitivity problem, in *Artificial Photosynthesis: From Basic Biology to Industrial Applications*, A.F. Collings and C. Critchley, eds., pp. 211–228. Wiley-VCH Verlag GmbH, Weinheim, Germany.
105. K. van't Riet and J. Tramper, *Basic Bioreactor Design*, Marcel Dekker, New York, 1991.
106. H. Berberoğlu, J. Jay, and L. Pilon, Effect of nutrient media on photobiological hydrogen production by *Anabaena variabilis* ATCC 29413, *International Journal of Hydrogen Energy*, 33(3), 1172–1184, 2008.
107. S. Hirata, M. Hayashitani, M. Taya, and S. Tone, Carbon dioxide fixation in batch culture of *Chlorella* sp. using a photobioreactor with a sunlight-collection device, *Journal of Fermentation and Bioengineering*, 81(5), 470–472, 1996.

108. A. Melis and T. Happe, Trails of green alga hydrogen research—From Hans Gaffron to new frontiers, *Photosynthesis Research*, 80(1–3), 401–409, 2004.
109. H.C. Chen and A. Melis, Localization and function of SulP, a nuclear-encoded chloroplast sulfate permease in *Chlamydomonas reinhardtii*, *Planta*, 220, 198–210, 2004.
110. M. Forestier, P. King, L. Zhang, M. Posewitz, S. Schwarzer, T. Happe, M.L. Girardi, and M. Seibert, Expression of two Fe-hydrogenases in *Chlamydomonas reinhardtii* under anaerobic conditions, *European Journal of Biochemistry*, 270(13), 2750–2758, 2003.
111. U.S. Department of Energy, Prospectus on biological hydrogen production, hydrogen program, http://www1.eere.energy.gov/hydrogenandfuelcells/production/photobiological.html, 2008, accessed on December 6, 2013.
112. Q. Xu, P.C. Maness, and R. Gerald, Hydrogen from water in a novel recombinant oxygen-tolerant cyanobacterial system, U.S. Department of Energy, Hydrogen Program, 2007 Annual Progress Report, pp. 175–177, 2007.
113. T. Happe, K. Schutz, and H. Bohme, Transcriptional and mutational analysis of the uptake hydrogenase of the filamentous cyanobacterium *Anabaena variabilis* ATCC 29413, *Journal of Biotechnology*, 182(6), 1624–1631, 2000.
114. L.E. Mikheeva, O. Schmitz, and S. Shestakov, Mutants of the cyanobacterium *Anabaena variabilis* altered in hydrogenase activities, *Zeitschrift Fur Naturforschung C—A Journal of Biosciences*, 50(7–8), 505–510, 1995.
115. A.A. Tsygankov, L.T. Serebryakova, K.K. Rao, and D.O. Hall, Acetylene reduction and hydrogen photoproduction by wild-type and mutant strains of *Anabaena* at different CO_2 and O_2 concentrations, *FEMS Microbiology Letters*, 167(1), 13–17, 1998.
116. J.W. Lee, Photobiological hydrogen production systems: Creation of designer alga for efficient and robust production of H_2 from water, U.S. Department of Energy, Hydrogen Program, 2006 Annual Progress Report, pp. 125–127, 2006.
117. A. Melis and T. Happe, Hydrogen production: Green algae as a source of energy, *Plant Physiology*, 127(3), 740–748, 2001.
118. H. Berberoğlu and L. Pilon, Maximizing solar to H_2 energy conversion efficiency of outdoor photobioreactors using mixed cultures, *International Journal of Hydrogen Energy*, 35, 500–510, 2010.
119. J. Miyake, X.Y. Mao, and S. Kawamura, Photoproduction of hydrogen from glucose by a co-culture of a photosynthetic bacterium and *Clostridium butyricum*, *Journal of Fermentation Technology*, 62(6), 531–535, 1984.
120. R.M.A. El-Shishtawy, S. Kawasaki, and M. Morimoto, Biological H_2 production using a novel light-induced and diffused photobioreactor, *Biotechnology Techniques*, 11(6), 403–407, 1997.
121. C. Stewart and M.A. Hessami, A study of methods of carbon dioxide capture and sequestration—The sustainability of a photosynthetic bioreactor approach, *Energy Conversion and Management*, 46(3), 403–420, 2005.
122. C.Y. Chen, C.M. Lee, and J.S. Chang, Hydrogen production by indigenous photosynthetic bacterium *Rhodopseudomonas palustris* WP3-5 using optical fiber-illuminating photobioreactors, *Biotechnology Engineering Journal*, 32(1), 33–42, 2006.
123. R. Bagai and D. Madamwar, Long-term photo-evolution of hydrogen in packed bed reactor containing a combination of *Phormidium valderianum*, *Halobacterium halobium*, and *Escherichia coli* immobilized in polyvinyl alcohol, *International Journal of Hydrogen Energy*, 24(4), 311–317, 1999.
124. T.V. Laurinavichene, A.S. Fedorov, M.L. Ghirardi, M. Seibert, and A.A. Tsygankov, Demonstration of sustained hydrogen production by immobilized, sulfur-deprived *Chlamydomonas reinhardtii* cells, *International Journal of Hydrogen Energy*, 31(5), 659–667, 2006.
125. S. Kosourov, M. Seibert, and M.L. Ghirardi, Effects of extracellular ph on the metabolic pathways in sulfur-deprived H_2 producing *Chlamydomonas reinhardtii* cultures, *Plant Cell Physiology*, 44(2), 146–155, 2003.
126. U.S. Department of Energy Hydrogen, Targets for on-board hydrogen storage systems, http://www.eere.energy.gov/hydrogenandfuelcells/pdfs/freedomcar_targets_explanations.pdf, accessed on December 20, 2008.

127. M.K. Mann and J.S. Ivy, Renewable hydrogen: Can we afford it? *Solar Today*, May/June 2004, pp. 28–31; NREL Report No. JA-560-36056 (2004) http://www.americanhydrogenassociation.org/H2today23-1.pdf, 2004, accessed on December 6, 2013.
128. W.A. Amos, Updated cost analysis of photobiological hydrogen production from *Chlamydomonas reinhardtii* green algae, Milestone Completion Report, National Renewable Energy Laboratory, Golden, CO, NREL/MP-560-35593, January 2004.
129. B.G. Liepert, Observed reductions of surface solar radiation at sites in the United States and worldwide from 1961 to 1990, *Geophysical Research Letters*, 29(10), 1421–1426, 2002.
130. G. Burgess and J.G. Fernandez-Velasco, Materials, operational energy inputs, and net energy ratio for photobiological hydrogen production, *International Journal of Hydrogen Energy*, 32(9), 1225–1234, 2007.
131. Purdue University, A webserver for cyanobacterial research, http://www-cyanosite.bio.purdue.edu, 2004, accessed on December 6, 2013.
132. Department of Health and Human Services, Harmful algal blooms, Center for Disease Control and Prevention, http://www.cdc.gov/nceh/hsb/hab/default.html, 2008, accessed on December 6, 2013.
133. D. Westwood, *The Microbiology of Drinking Water—Part 1—Water Quality and Public Health*. U.K. Environmental Agency, Bristol, U.K., 2002.
134. M. Winkler, A. Hemschemeier, C. Gotor, A. Melis, and T. Happe, [Fe]-hydrogenases in green algae: Photo-fermentation and hydrogen evolution under sulfur deprivation, *International Journal of Hydrogen Energy*, 27(11–12), 1431–1439, 2002.
135. J. Van der Oost, B.A. Bulthuis, S. Feitz, K. Krab, and R. Kraayenhof, Fermentation metabolism of the unicellular cyanobacterium *Cyanothece* PCC-7822, *Archives of Microbiology*, 152(5), 415–419, 1989.
136. K. Sasikala, C.V. Ramana, and P.R. Rao, Environmental regulation for optimal biomass yield and photoproduction of hydrogen by *Rhodobacter sphaeroides* O.U. 001, *International Journal of Hydrogen Energy*, 16(9), 597–601, 1991.
137. H. Zürrer and R. Bachofen, Hydrogen production by the photosynthetic bacterium *Rhodospirillum rubrum*, *Applied and Environmental Microbiology*, 37, 789–793, 1979.
138. H. Zürrer and R. Bachofen, Aspects of growth and hydrogen production of the photosynthetic bacterium *Rhodospirillum rubrum* in continuous culture, *Biomass*, 2(3), 165–174, 1982.
139. J.D. Brosseau and J.E. Zajic, Continual microbial production of hydrogen gas, *International Journal of Hydrogen Energy*, 7(8), 623–628, 1982.
140. I. Karube, T. Matsunaga, S. Tsuru, and S. Suzuki, Continuous hydrogen production by immobilized whole cells of *Clostridium butyricum*, *Biochimica et Biophysica Acta*, 444(2), 338–343, 1976.

Section VI

Solar Hydrogen Production

12

Solar Thermochemical Production of Hydrogen

Aldo Steinfeld
ETH Zurich
and
Paul Scherrer Institute

CONTENTS

This chapter reviews the underlying science and describes the technological advances in the field of solar thermochemical cycles and processes for producing hydrogen that use concentrated solar radiation as the energy source of high-temperature process heat.

12.1 Thermodynamics of Solar Thermochemical Processes

A comprehensive thermodynamic analysis of solar thermochemical processes is described by Fletcher (2001) and by Steinfeld and Palumbo (2001). The principal concepts are summarized herein. Solar thermochemical processes are based on the use of concentrated solar radiation as the energy source of high-temperature process heat for driving an endothermic chemical transformation. Three main optical configurations based on parabolic-shaped reflectors are at present commercially available for large-scale concentration of solar energy: the trough system, the tower system, and the dish system. These three systems are schematically shown in Figure 12.1.

Trough systems use linear, 2D, parabolic mirrors to focus sunlight onto a solar tubular receiver positioned along their focal line. Tower systems use a field of heliostats (two-axis

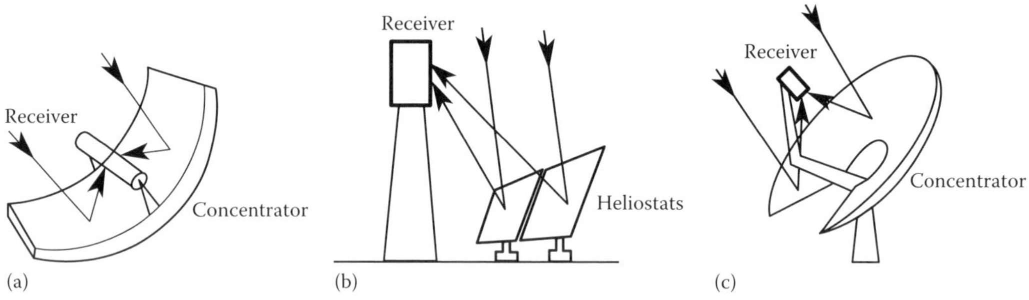

FIGURE 12.1
Schematic of the three main optical configurations for large-scale collection and concentration of solar energy: (a) the trough system, (b) the tower system, and (c) the dish system.

tracking parabolic mirrors) that focus the sun rays onto a solar receiver mounted on top of a centrally located tower. Dish systems use paraboloidal mirrors to focus sunlight onto a solar receiver positioned at their focus. The total amount of power collected by any of these systems is proportional to the projected area of the mirrors. Cassegrain optical configurations for the tower system make use of a hyperboloidal reflector at the top of the tower to redirect sunlight onto a receiver located on the ground level (Yogev et al., 1998). The capability of these collection systems to concentrate solar energy is described in terms of their mean flux concentration ratio $\tilde{C}$ over a targeted area A at the focal plane, normalized with respect to the direct normal solar irradiation (DNI):

$$\tilde{C} = \frac{Q_{solar}}{I \cdot A} \tag{12.1}$$

where Q_{solar} is the solar power intercepted by the target. $\tilde{C}$ is often expressed in units of *suns* when normalized to $I = 1$ kW/m^2. The solar flux concentration ratio typically obtained is at the level of 100, 1,000, and 10,000 suns for trough, tower, and dish systems, respectively. Higher concentration ratios imply lower heat losses from smaller areas and, consequently, higher attainable temperatures at the receiver. To some extent, the flux concentration can be further augmented with the help of nonimaging secondary concentrators, for example, compound parabolic concentrators (CPC), when positioned in tandem with the primary parabolic concentrating systems (Welford and Winston, 1989). The aforementioned solar concentrating systems have been proven to be technically feasible in large-scale (MW) pilot and commercial plants aimed at the production of electricity in which a working fluid (typically air, water, synthetic oil, helium, sodium, or molten salt) is solar-heated and further used in traditional Rankine, Brayton, and Stirling cycles. Solar thermochemical applications, although not as far developed as solar thermal electricity generation, employ the same solar concentrating technologies.

Solar reactors for highly concentrated solar systems usually feature the use of a cavity-receiver-type configuration, that is, a well-insulated enclosure with a small opening—the *aperture*—to let in concentrated solar radiation. Because of multiple internal reflections, the fraction of the incoming energy absorbed by the cavity greatly exceeds the surface absorptance of the inner walls. As the ratio of the cavity's characteristic length to the aperture diameter increases, the cavity-receiver approaches a blackbody absorber. The solar energy absorption efficiency of a solar reactor, $\eta_{absorption}$, is defined as the net rate at which energy is being absorbed divided by the solar radiative power coming from the solar concentrator.

For a perfectly insulated cavity-receiver (no convection or conduction heat losses), it is given by (Fletcher and Moen, 1977)

$$\eta_{absorption} = \frac{\alpha_{eff} Q_{aperture} - \varepsilon_{eff} A_{aperture} \sigma T^4}{Q_{solar}} \tag{12.2}$$

where

Q_{solar} is the solar power coming from the solar concentrator

$Q_{aperture}$ is the amount intercepted by the aperture of area $A_{aperture}$

α_{eff} and ε_{eff} are the effective absorptance and emittance of the solar cavity-receiver, respectively

T is the nominal reactor temperature

σ is the Stefan–Boltzmann constant

The numerator denotes the difference between the power absorbed and reradiated, which should match the enthalpy change of the chemical reaction. The incoming solar power is determined by the normal beam insolation I, by the collector area, and by taking into account for the optical imperfections of the collection system (e.g., reflectivity, specularity, tracking imperfections). For simplification, we assume an aperture size that captures all incoming solar power so that $Q_{aperture} = Q_{solar}$. With this assumption, and for a perfectly insulated isothermal blackbody cavity-receiver ($\alpha_{eff} = \varepsilon_{eff} = 1$), Equations 12.1 and 12.2 are combined to yield

$$\eta_{absorption} = 1 - \left(\frac{\sigma T^4}{I\tilde{C}} \right) \tag{12.3}$$

The absorbed concentrated solar radiation drives an endothermic chemical reaction. The measure of how well solar energy is converted into chemical energy for a given process is the solar-to-fuel energy conversion efficiency, $\eta_{solar\text{-}to\text{-}fuel}$, defined as

$$\eta_{solar\text{-}to\text{-}fuel} = \frac{-\dot{n}\,\Delta G|_{298\,\mathrm{K}}}{Q_{solar}} \tag{12.4}$$

where ΔG is the maximum possible amount of work that may be extracted from the products as they are transformed back to reactants at 298 K. The Second Law is now applied to calculate the maximum $\eta_{solar\text{-}to\text{-}fuel}$ for an ideal cyclic process, limited by both the solar absorption and Carnot efficiencies:

$$\eta_{solar\text{-}to\text{-}fuel,ideal} = \eta_{absorption} \cdot \eta_{Carnot} = \left[1 - \left(\frac{\sigma T_H^4}{IC} \right) \right] \times \left[1 - \left(\frac{T_L}{T_H} \right) \right] \tag{12.5}$$

where T_H and T_L are the upper and lower operating temperatures of the equivalent Carnot heat engine. $\eta_{solar\text{-}to\text{-}fuel,ideal}$ is plotted in Figure 12.2 as a function of T_H for T_L = 298 K, and $I = 1$ kW/m^2, and for various solar flux concentrations. Because of the Carnot limitation, one should try to operate thermochemical processes at the highest upper temperature possible; however, from a heat transfer perspective, higher T_H implies higher reradiation losses. The highest temperature an ideal solar cavity-receiver is capable of achieving, defined as the stagnation temperature $T_{stagnation}$, is calculated by setting Equation 12.5 equal to zero, to yield

$$T_{stagnation} = \left(\frac{I\tilde{C}}{\sigma} \right)^{0.25} \tag{12.6}$$

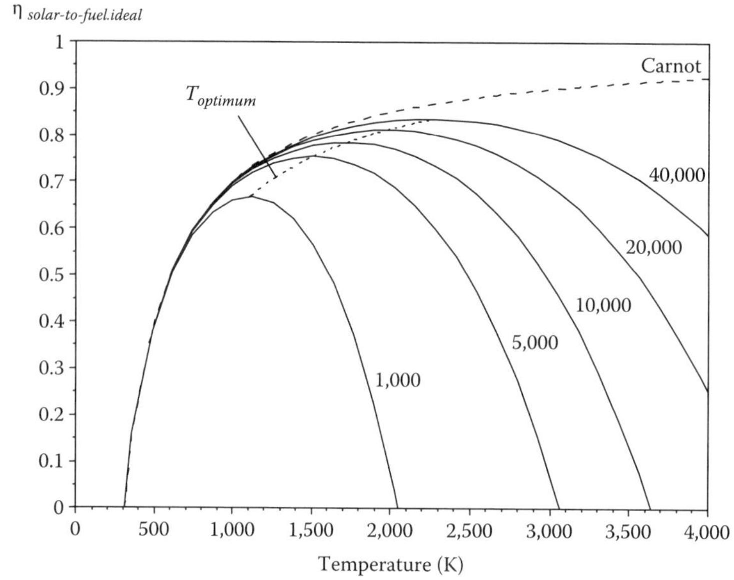

FIGURE 12.2
Variation of the ideal solar-to-fuel efficiency as a function of the operating temperature T_H, for a blackbody cavity-receiver converting concentrated solar energy into chemical energy. The mean solar flux concentration is the parameter: 1,000, … 40,000. Also plotted are the Carnot efficiency and the locus of the optimum cavity temperature $T_{optimum}$, Equation 12.7. (From Steinfeld, A. and Palumbo, R., *Encyclopedia of Physical Science and Technology*, 15, 237–256, 2001.)

At this temperature, $\eta_{solar\text{-}to\text{-}fuel,ideal} = 0$ because energy is being reradiated as fast as it is absorbed. Stagnation temperatures exceeding 3000 K are attainable with solar concentration ratios above 5000. However, an energy-efficient process must run at temperatures substantially below $T_{stagnation}$. There is an optimum temperature $T_{optimum}$ for maximum efficiency, obtained by setting $\partial\eta_{solar\text{-}to\text{-}fuel,ideal}/\partial T = 0$. Assuming uniform power-flux distribution, it yields

$$T_{optimum}^5 - (0.75T_L)T_{optimum}^4 - \left(\frac{\alpha_{eff} T_L I \tilde{C}}{4\varepsilon_{eff}\sigma}\right) = 0 \tag{12.7}$$

The locus of $T_{optimum}$ is shown in Figure 12.2 and varies between 1,100 and 1,800 K for uniform power-flux distributions with concentrations between 1,000 and 13,000 (Steinfeld and Schubnell, 1993). For example, for $\tilde{C} = 5000$, the maximum $\eta_{solar\text{-}to\text{-}fuel,ideal}$ of 75% is achieved at $T_{optimum}$ = 1500 K. For a Gaussian incident power-flux distribution having peak concentration ratios between 1,000 and 12,000 suns, the optimal temperature varies from 800 to 1,300 K. In practice, when considering convection and conduction losses in addition to radiation losses, the efficiency will peak at a somewhat lower temperature.

In order to illustrate the use of these equations, we consider as an example a two-step solar thermochemical process for splitting H_2O using ZnO/Zn redox reaction, comprising (1) the solar endothermal dissociation of ZnO(s) into its elements and (2) the nonsolar exothermal steam hydrolysis of Zn into H_2 and ZnO(s) and represented by

$$\text{First step (solar ZnO decomposition)}: \text{ZnO} \rightarrow \text{Zn} + 0.5\text{O}_2 \tag{12.8}$$

$$\text{Second step (nonsolar Zn hydrolysis)}: \text{Zn} + \text{H}_2\text{O} \rightarrow \text{ZnO} + \text{H}_2 \tag{12.9}$$

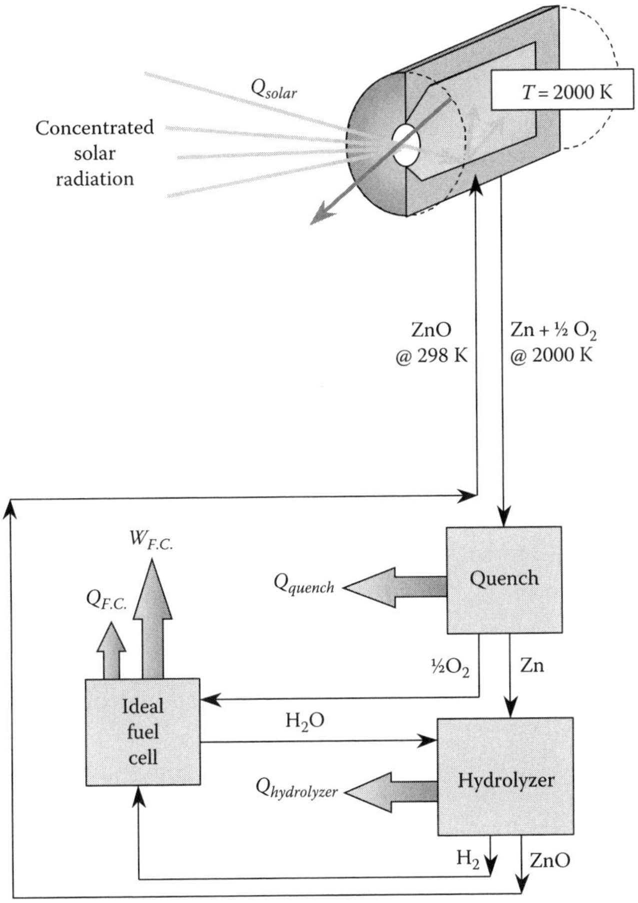

FIGURE 12.3
Schematic of an ideal cyclic process for calculating the maximum solar-to-fuel energy conversion efficiency of the two-step water-splitting cycle using ZnO/Zn redox reactions.

A model flow diagram for the proposed two-step solar thermochemical cycle is shown schematically in Figure 12.3. It uses a solar reactor, a quenching device, a hydrolyzer reactor, and a H_2/O_2 fuel cell. All materials are recycled. The complete process is carried out at a constant pressure of 1 bar. In practice, pressure drops will occur throughout the system and pumping work will be required. The solar reactor is assumed to be a cavity-receiver operating at 2000 K. The molar feed rate of ZnO to the reactor, $\dot{n}$, is set to 1 mol/s and is equal to that of H_2O fed to the hydrolyzer. Chemical equilibrium is assumed inside the solar reactor. The net power absorbed in the solar reactor should match the enthalpy change per unit time of the reaction:

$$Q_{reactor,net} = \dot{n}\Delta H\big|_{ZnO(s)@298\,K \rightarrow Zn(g)+0.5O_2\,@2000\,K} = 557\ \text{kJ/mol} \tag{12.10}$$

For $\tilde{C} = 5000$, $\eta_{absorption} = 82\%$ and $Q_{solar} = 680$ kJ/mol. Products Zn(g) and O_2 exit the solar reactor at 2000 K and are cooled rapidly to 298 K. It is assumed that the chemical composition of the products remains unchanged upon cooling in the quencher. Since the quench step is required for avoiding recombination of products, no heat exchanger is used for recovering their sensible and latent heat. Thus, the amount of power lost during quenching is

$$Q_{quench} = -\dot{n}\Delta H\big|_{Zn(g)+0.5O_2 @ 2000\,K \rightarrow Zn(s)+0.5O_2 @ 298\,K} = -209 \text{ kJ/mol} \quad (12.11)$$

After quenching, the products separate naturally (without expending work) into gaseous O_2 and condensed phase zinc. Zinc is sent to the hydrolyzer to react exothermally with water and form hydrogen, according to reaction (12.9). The heat liberated is assumed lost to the surroundings, as given by

$$Q_{hydrolyzer} = -\dot{n}\Delta H\big|_{Zn+H_2O @ 298\,K \rightarrow ZnO+H_2 @ 298\,K} = -62 \text{ kJ/mol} \quad (12.12)$$

The cycle is closed by introducing an ideal H_2/O_2 fuel cell, in which the products recombine to form H_2O and thereby generate electrical power. $W_{F.C.}$ and $Q_{F.C.}$ are the work output and heat rejected, respectively given by

$$W_{F.C.} = -\dot{n}\Delta G\big|_{H_2+0.5O_2 @ 298\,K \rightarrow H_2O @ 298\,K} = 237 \text{ kJ/mol} \quad (12.13)$$

$$Q_{F.C.} = -T_L \times \dot{n}\Delta S\big|_{H_2+0.5O_2 @ 298\,K \rightarrow H_2O @ 298\,K} = -49 \text{ kJ/mol} \quad (12.14)$$

Finally, assuming no heat recovery during quenching and hydrolysis,

$$\eta_{solar\text{-}to\text{-}fuel} = \frac{W_{F.C.}}{Q_{solar}} = 35\% \quad (12.15)$$

The major sources of irreversibility are associated with the reradiation losses from the solar reactor and the heat lost during quenching and hydrolysis. To some extent, the sensible heat of the hot products exiting the reactor may be recovered to preheat the reactants, increasing the efficiency up to 50%. $\eta_{solar\text{-}to\text{-}fuel}$ can be further increased with higher $\tilde{C}$, for example, by incorporating a CPC at the aperture, which results in a smaller aperture to intercept the same amount of solar power, and, consequently, lower reradiation losses. Note that, for a given $\tilde{C}$, smaller apertures intercept a reduced fraction of the incoming solar power. Thus, the optimum aperture size of the solar cavity-receiver becomes a compromise between maximizing solar radiation capture and minimizing reradiation losses. Reradiation losses can also be diminished by implementing selective windows with high transmissivity in the solar spectrum around 0.5 μm where the solar irradiation peaks and high reflectivity in the infrared range around 1.45 μm where the Planck's spectral emissive power for a 2000 K blackbody peaks.

This kind of process modeling establishes a base for evaluating and comparing different solar thermochemical processes for ideal, closed cyclic systems that recycle all materials. For open materials cycles, in which fuels are the reactants being solar-upgraded (see next section: cracking, reforming, gasification), the solar-to-fuel energy conversion efficiency is calculated as

$$\eta_{solar\text{-}to\text{-}fuel} = \frac{W_{F.C.}}{Q_{solar} + HHV_{reactants}} \quad (12.16)$$

where $HHV_{reactants}$ is the high-heating value of the fuel being processed, for example, about 890 kJ/mol for natural gas (NG) and 35,700 kJ/kg for anthracite coal. The higher $\eta_{solar\text{-}to\text{-}fuel}$, the lower is the required solar collection area for producing a given amount of solar fuel and, consequently, the lower are the costs incurred for the solar concentrating system, which usually correspond to half of the total investments for the entire solar chemical plant. Thus, high $\eta_{solar\text{-}to\text{-}fuel}$ implies favorable competitiveness.

12.2 Solar Thermochemical Processes and Reactors

Five thermochemical routes for solar hydrogen production are depicted in Figure 12.4. Indicated is the chemical feedstock: H_2O and/or carbonaceous feedstock (e.g., NG, oil, coal, biomass). All of these routes are highly endothermic processes that proceed at high temperatures and make use of concentrated solar radiation as the energy source of process heat (Steinfeld, 2005).

12.2.1 H_2 from H_2O by Solar Thermolysis

The single-step thermal dissociation of water is known as water thermolysis:

$$H_2O \rightarrow H_2 + 0.5O_2 \tag{12.17}$$

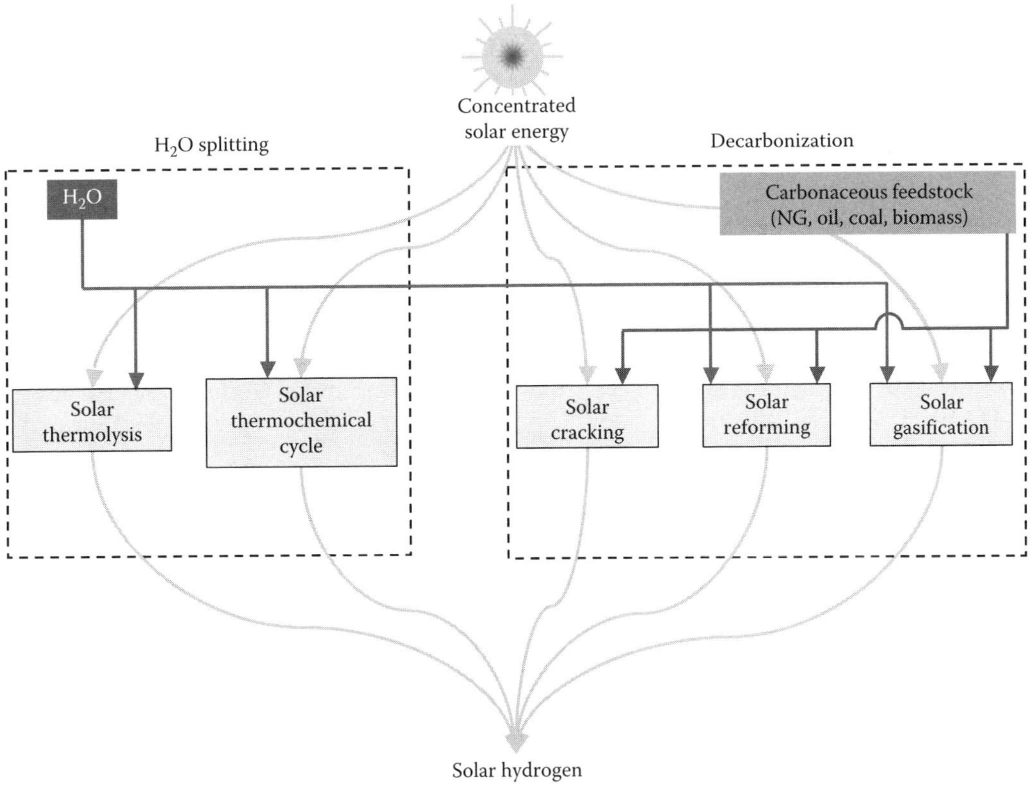

FIGURE 12.4
Thermochemical routes for solar hydrogen production using concentrated solar radiation as the energy source of high-temperature process heat. (From Steinfeld, A., *Sol. Energy*, 78, 603, 2005.)

Although conceptually simple, reaction (12.17) has been impeded by the need for a high-temperature heat source at above 2500 K for achieving a reasonable degree of dissociation and by the need for an effective technique for separating H_2 and O_2 to avoid ending up with an explosive mixture. Among the ideas proposed for separating H_2 from the products are effusion separation (Fletcher and Moen, 1977; Bilgen, 1984; Kogan, 1998) and electrolytic separation (Ihara, 1980; Fletcher, 1999). Semipermeable membranes based on ZrO_2 and other high-temperature materials have been tested at up to 2500 K (Diver et al., 1983; Kogan, 1998), but these ceramics usually fail to withstand the severe thermal shocks that often occur when working under high-flux solar irradiation. Rapid quench by injecting a cold gas (Lédé et al., 1987) is simple and workable, but the quench introduces a significant drop in the efficiency and produces an explosive gas mixture. Furthermore, the very high temperatures demanded by the thermodynamics of the process (e.g., 3000 K for 64% dissociation at 1 bar) pose severe material problems and can lead to significant reradiation from the reactor, thereby lowering the absorption efficiency (Equation 12.2).

12.2.2 H_2 from H_2O by Solar Thermochemical Cycles

Water-splitting thermochemical cycles bypass the H_2/O_2 separation problem and further allow operating at relatively moderate upper temperatures. Previous studies performed on H_2O-splitting thermochemical cycles were mostly characterized by the use of process heat at temperatures below about 1200 K, available from nuclear and other thermal sources. These cycles required multiple steps (more than two) and were suffering from inherent inefficiencies associated with heat transfer and product separation at each step. Status reviews on multistep cycles are given by Serpone et al. (1992) and by Funk (2001) and include the leading three-step sulfur–iodine cycle based on the thermal decomposition of H_2SO_4 at 1140 K and the four-step UT3 cycle based on the hydrolysis of $CaBr_2$ at 1020 K. In recent years, significant progress has been accomplished in the development of optical systems for large-scale collection and concentration of solar energy capable of achieving solar concentration ratios of 5000 suns and higher. Such high solar radiation fluxes allow the conversion of solar energy to thermal reservoirs at 1500 K and above, which are needed for the more efficient two-step thermochemical cycles using metal oxide redox reactions. The cycle is depicted in Figure 12.5.

The first endothermic step is the solar thermal dissociation of the metal oxide to the metal or the lower-valence metal oxide. The second, nonsolar, exothermic step is the hydrolysis of the metal to form H_2 and the corresponding metal oxide. The net reaction is $H_2O = H_2 + 0.5O_2$, but since H_2 and O_2 are formed in different steps, the need for high-temperature gas separation is thereby eliminated. The second hydrolysis step can be accomplished on demand at the H_2 consumer site, as it is decoupled from the availability of solar energy. Alternatively, CO_2 can be co-fed with H_2O to react with the metal and produce syngas, which can be further processed to liquid fuels.

This cycle was originally proposed for the redox pair Fe_3O_4/FeO (Nakamura, 1977). The solar step, that is, the thermal dissociation of magnetite to wustite at above 2300 K, has been thermodynamically examined (Steinfeld et al., 1999) and experimentally studied in a solar furnace (Sibieude et al., 1982). It was found necessary to quench the products in order to avoid re-oxidation, but quenching introduced an energy penalty of up to 80% of the solar energy input. Other redox pairs, such as Mn_3O_4/MnO and Co_3O_4/CoO have also been considered, but the yield of H_2 has been too low to be of any practical interest (Sibieude et al., 1982). H_2 may be produced instead by reacting MnO with NaOH at above 900 K in a three-step cycle (Sturzenegger and Nüesch, 1999). One promising redox system is ZnO/Zn; see Equations 12.8 and 12.9 (Bilgen et al., 1977; Palumbo et al., 1998; Steinfeld et al., 1998; Lédé et al., 2001;

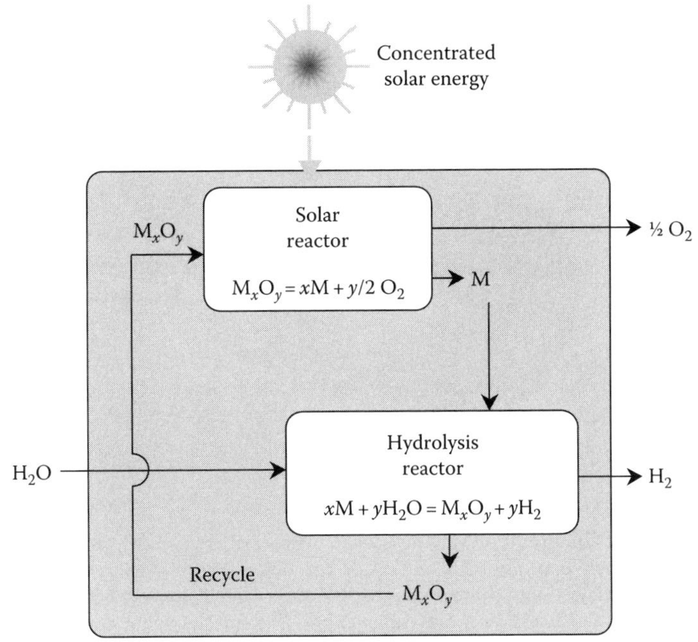

FIGURE 12.5
Scheme of a two-step solar thermochemical cycle based on metal oxide redox reactions. Here, M_xO_y denotes a metal oxide, and M the corresponding metal or lower-valence metal oxide. In the first, endothermic, solar step, M_xO_y is thermally dissociated into the metal or lower-valence metal oxide M and oxygen. Concentrated solar radiation is the energy source for the required high-temperature process heat. In the second, exothermic, nonsolar step, M reacts with water to produce hydrogen. The resulting metal oxide is then recycled back to the first step. (From Steinfeld, A. and Palumbo, R., *Encyclopedia of Physical Science and Technology*, 15, 237–256, 2001.)

Steinfeld, 2002; Perkins and Weimer, 2004; Loutzenhiser et al., 2010). Figure 12.6 shows a solar chemical reactor configuration for performing the thermal dissociation of ZnO (Schunk et al., 2008; Schunk et al., 2009). It has been shown that the process is strongly dependent on the efficient separation of gaseous products, Zn(g) and O_2, to avoid recombination upon cooling (Weidenkaff et al., 2000). Research has focused on diluting and rapid quenching of gaseous products below the Zn saturation and solidification points (Gstoehl et al., 2008). Alternatively, electrolytic methods have been proposed for in situ separation of Zn(g) and O_2 at high temperatures and experimentally demonstrated to work in small-scale reactors (Fletcher, 1999). As for the second step of the cycle, the Zn hydrolysis was performed in aerosol flow reactors, designed for the formation of Zn nanoparticles followed by their in situ reaction with H_2O (Wegner et al., 2006; Abu Hamed et al., 2009; Melchior et al., 2009b). These aerosol flow configurations offered high Zn-to-ZnO conversions over short residence times due to augmented reaction kinetics and heat/mass transfer.

Spinel ferrites of the form $M_xFe_{3-x}O_4$, where M generally represents Ni, Zn, Co, Mn, or other transition metals, have been applied for the two-step redox cycle of Figure 12.5 (Ehrensberger et al., 1995; Tamaura et al., 1995; Roeb et al., 2006; Charvin et al., 2007; Allendorf et al., 2008; Ishihara et al., 2008; Miller et al., 2008; Fernando et al., 2009; Gokon et al. 2009). These mixed oxides may be reducible at lower temperatures than those required for the reduction of Fe_3O_4, while the reduced phase remains capable of splitting water. However, the extent of reduction is thermodynamically limited to low conversion and metal oxide solutions play a critical role in the overall process (Allendorf et al., 2008). Cerium-oxide-based materials

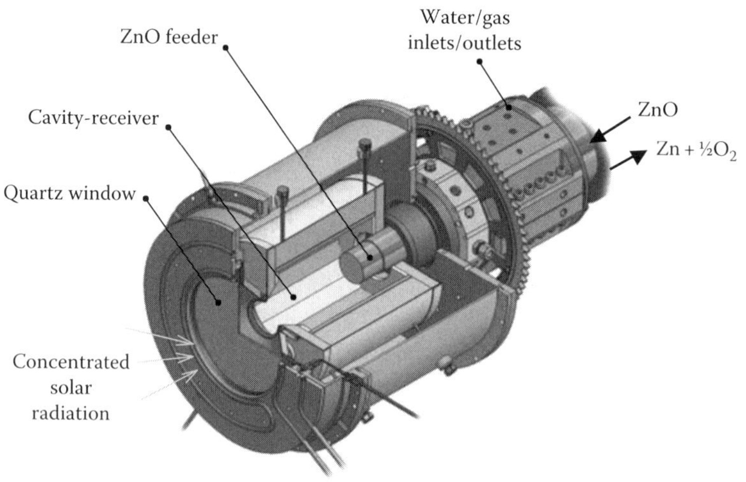

FIGURE 12.6
Scheme of the solar reactor configuration for the thermal dissociation of ZnO, as part of a two-step water-splitting thermochemical cycle based on ZnO/Zn redox reactions. It consists of a windowed rotating cavity-receiver lined with ZnO particles. With this arrangement, ZnO is directly exposed to high-flux solar irradiation and serves simultaneously the functions of radiant absorber, thermal insulator, and chemical reactant. (From Loutzenhiser et al., *Materials*, 3, 4922, 2010.)

have recently emerged as attractive candidates due to relatively rapid kinetics, good stability, and high selectivity (Abanades and Flamant, 2006; Kaneko et al., 2007; Chueh and Haile, 2009; Singh and Hegde, 2009; Abanades et al., 2010). The nonstoichiometric ceria fluorite-type structure and phase are maintained as oxygen vacancies are created, making cyclability possible. Figure 12.7 shows the configuration of a solar reactor designed for performing both steps of thermochemical cycle with the ceria redox reactions (Chueh et al., 2010).

12.2.3 H_2 from Gaseous Carbonaceous Feedstock by Solar Cracking

The solar cracking route refers to the thermal decomposition of NG, oil, and other hydrocarbons and can be represented by the simplified net reaction:

$$C_xH_y = xC(gr) + \frac{y}{2}H_2 \tag{12.18}$$

Other compounds may also be formed, depending on the presence of impurities in the raw materials. The thermal decomposition yields a carbon-rich condensed phase and a hydrogen-rich gas phase. The carbonaceous solid product can either be sequestered without CO_2 release or used as material commodity under less severe CO_2 restraints. It can also be applied as reducing agent in metallurgical processes. The hydrogen-rich gas mixture can be further processed to high-purity hydrogen that is not contaminated with oxides of carbon and, thus, can be used in proton exchange membrane (PEM) fuel cells without inhibiting platinum-made electrodes. From the point of view of carbon sequestration, it is easier to separate, handle, transport, and store solid carbon than gaseous CO_2. Assuming carbon sequestration, Equation 12.16 yields $\eta_{solar\text{-}to\text{-}fuel}$ = 0.55 (von Zedtwitz et al., 2006). Reaction (12.18) has been effected using solar process heat with CH_4 and C_4H_{10} at 823 K for the catalytic production of filamentous carbon (Steinfeld et al., 1997; Meier et al., 1999). Figure 12.8 shows a scheme of a vortex-type solar reactor for performing the solar cracking

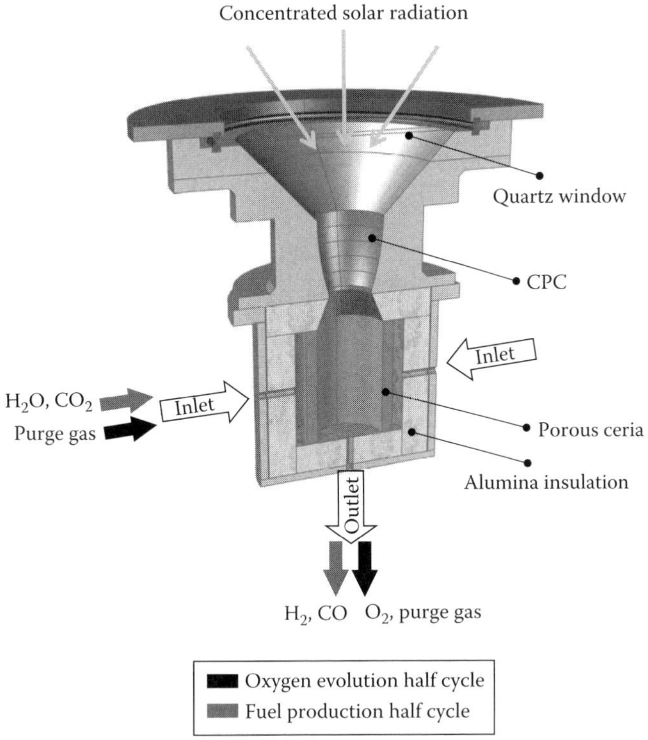

FIGURE 12.7
Schematic of the solar reactor configuration for the two-step solar-driven thermochemical production of fuels. It consists of a cavity-receiver containing a porous monolithic ceria cylinder. Concentrated solar radiation enters through a windowed aperture and impinges on the ceria inner walls. Reacting gases flow radially across the porous ceria, while product gases exit the cavity through an axial outlet port. Black arrow indicates ceria reduction (oxygen evolution); gray arrow indicates oxidation (fuel production). (From Chueh, W.C. et al., *Science*, 330, 1797, 2010.)

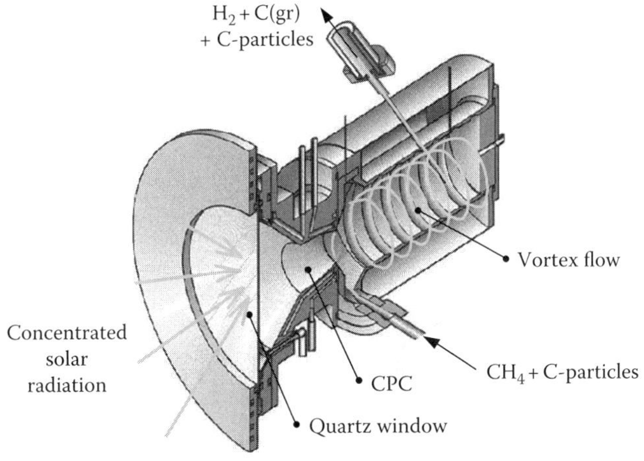

FIGURE 12.8
Scheme of the solar chemical reactor for the coproduction C and H_2 by thermal decomposition of CH_4. It consists of a continuous flow of CH_4 laden with µm-sized carbon black particles, confined to a cavity-receiver and directly exposed to concentrated solar irradiation. The carbon particles fed serve the functions of radiant absorbers and nucleation sites for the heterogeneous reaction. (From Maag et al., *Int J. Hydrogen Energy*, 34, 7676, 2009.)

of methane (Hirsch and Steinfeld, 2004; Maag et al., 2009). Other reactor concepts have been demonstrated in solar furnaces (Dahl et al., 2004a and b; Rodat et al., 2009).

12.2.4 H_2 from Gaseous Carbonaceous Feedstock by Solar Reforming

The steam reforming of gaseous carbonaceous feedstock, for example, NG, oil, and other hydrocarbons, can be represented by the simplified net reaction:

$$C_xH_yO_z + (x-z)H_2O = \left(\frac{y}{2} + x - z\right)\cdot H_2 + xCO \qquad (12.19)$$

Other compounds may also be formed (e.g., H_2S), depending on the reaction rate and on the impurities contained in the raw materials. The principal product is high-quality synthesis gas (syngas), which can be further processed via water–gas shift reaction and CO_2 separation to a pure stream of hydrogen.

The solar reforming of NG, using either steam or CO_2 as partial oxidant, has been extensively studied in solar concentrating facilities with small-scale solar reactor prototypes using Rh-based catalyst (Levy et al., 1989; Hogan et al., 1990; Buck et al., 1991; Levy et al., 1992; Buck et al., 1994; Muir et al., 1994; Wörner and Tamme, 1998), in molten salt using other metallic catalysts (Kodama et al., 2001; Gokon et al., 2002), and in the absence of catalysts (Dahl et al., 2004). The solar reforming process has been scaled up to power levels of 300–500 kW and tested at 1100 K and 8–10 bar in a solar tower using two solar reforming reactor concepts: an indirect-irradiation tubular reactor (Epstein and Spiewak, 1996) and a direct-irradiation ceramic foam reactor (Tamme et al., 2001; Moeller et al., 2002). The latter, also referred to as *volumetric* reactor, is shown in Figure 12.9. High operating pressures require shaped windows, for example, conical (Kribus et al., 2001) and hemispherical (Moeller et al., 2006).

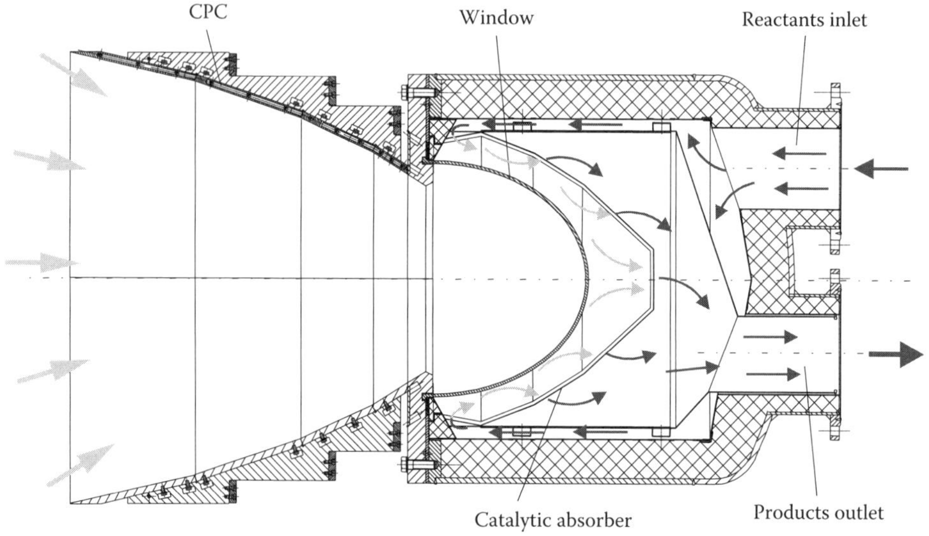

FIGURE 12.9
Scheme of the *volumetric* solar reactor concept for the reforming of NG. The main component is the porous ceramic absorber, made of SiC and coated with Rh catalyst, which is directly exposed to the concentrated solar radiation. A concave quartz window, mounted at the aperture, minimizes reflection losses and permits operation at elevated pressures. A CPC is implemented at the aperture. (Courtesy of DLR, Cologne, Germany.)

12.2.5 H_2 from Solid Carbonaceous Feedstock by Solar Gasification

Thermochemical gasification transforms solid carbonaceous feedstocks into widely applicable, clean, and energy-rich synthesis gas (syngas)—mainly H_2 and CO—which can be further processed via water–gas shift reaction and CO_2 separation to a pure stream of hydrogen. Conventional autothermal gasification requires a significant portion of the injected feedstock mass to be combusted internally with pure O_2 to supply high-temperature process heat for the endothermic reactions, which inherently decreases coal utilization and contaminates the product gases with combustion products. In contrast, solar-driven steam gasification is free of combustion by-products and yields higher syngas output per unit of feedstock. Based on the stoichiometric gasification of pure carbon, the product syngas may have up to 33% lower CO_2 intensity because its calorific value is solar-upgraded over that of the original carbon feedstock by an amount equal to the enthalpy change of the endothermic reaction, while retaining the same amount of carbon. A second-law (exergy) analysis indicated that combined Brayton–Rankine power cycles running on solar-made syngas can double the specific electric output per unit mass of coal and, consequently, achieve specific CO_2 intensities of 0.49–0.56 kg CO_2/kWh_e—approximately half that of conventional coal-fired power plants (von Zedtwitz and Steinfeld, 2003). If biomass is used as a feedstock, the syngas produced may be considered CO_2-neutral. Ultimately, solar-driven gasification is a means of storing intermittent solar energy in a transportable and dispatchable chemical form (Piatkowski et al., 2011).

The steam gasification of carbonaceous feedstocks involves principally two chemical processes: pyrolysis and char gasification, both of which are critical in the handling of carbonaceous fuels over a wide temperature range. The overall net reaction for stoichiometric water delivery can be represented by

$$C_1H_xO_yS_uN_v + (1-y)H_2O = \left[\frac{x}{2}+1-y-u\right]H_2 + CO + uH_2S + vN_2 \quad (12.20)$$

where x, y, u, and v are the elemental molar ratios of H/C, O/C, S/C, and N/C in the feedstock. Mineral matter, intrinsic water content, and other impurities contained in the feedstock are omitted from consideration in Equation 12.20. Their presence may have an effect on the kinetics and final product composition, but their exclusion does not affect the main conclusions of this analysis. Pyrolysis, occurring typically in the temperature range 450–900 K, involves the thermal decomposition of bonds within the carbonaceous chain and release of hydrogen-rich gaseous (CH_4, C_2H_6 etc.) and condensable (tars) compounds, while the fixed carbon portion of the feedstock forms char—effectively pure solid carbon. The net pyrolysis reaction can be represented by

$$C_1H_xO_yS_uN_v \xrightarrow{\text{heat}} C(s) + CO + CO_2 + H_2 + C_xH_y + \text{tars} \quad (12.21)$$

Solar-driven pyrolysis was investigated in early studies on biomass, coal, and shale oil (Gregg et al., 1980; Fletcher and Berber, 1988; Lede, 1999). Subsequent to pyrolysis, char serves as the reactant for the highly endothermic carbon-steam gasification reaction:

$$C(s) + H_2O = CO + H_2 \quad \Delta H^\circ_{298\,K} = 131\ \text{kJ/mol} \quad (12.22)$$

Favorable conditions for this reaction are temperatures above 1100 K, where the reaction kinetics is fast and equilibrium is entirely on the side of the products. Equation 12.3 summarizes the overall reaction, but a number of intermediate competing reactions need to be considered. The Boudouard reaction, $C(s) + CO_2 = 2CO$, becomes important for CO_2-based gasification at above 1000 K, while CH_4 cracking, Equation 12.18, and CH_4 reforming, Equation 12.19, proceed catalytically at above 900 K. All these reactions depend strongly on temperature, pressure, and residence times in a solar reactor, and their combination yields the final product gas composition. As an example, the equilibrium composition of reacting beech charcoal ($C_1H_{0.47}O_{0.055}S_{0.022}N_{0.004}$) with stoichiometric water per Equation 12.20 is shown in Figure 12.10 at 1 bar (Piatkowski and Steinfeld, 2008). Species with mole fractions lower than 10^{-3} (e.g., H_2S, HCN) are omitted. At temperatures below about 800 K, the formation of CH_4 and CO_2 is thermodynamically favored but unlikely to proceed due to kinetic limitations. The reaction goes to completion at above 1200 K, producing a syngas mixture of 53% H_2 and 47% CO. For higher, industrially preferred pressures, the equilibrium is shifted toward the reactants according to Le Châtelier's principle, as shown in Figure 12.10 for beech charcoal gasification as a function of pressure at 1000 K (Figure 12.11a) and as a function of temperature at 10 bar (Figure 12.11b).

Solar gasification reactors may be classified as: (1) directly irradiated reactors, where the solid carbonaceous reactants are directly exposed to the concentrated solar irradiation; and (2) indirectly irradiated reactors, where heat is transferred to the reaction site through an opaque wall. While directly irradiated reactors provide efficient heat transfer directly to the reaction site, they require a transparent window for the access of concentrated solar radiation, which becomes a critical and troublesome component under high pressures, severe gas environments, and particularly at large scales. On the other hand, the implementation of spectrally selective windows, such as fused quartz, with high transmissivity in the solar spectrum around 0.5 μm where the solar irradiation peaks and high reflectivity in the infrared range around 1.93 μm where the Planck's spectral emissive power for a 1500 K blackbody peaks, can diminish significantly reradiation losses and, consequently, augment $\eta_{absorption}$. Indirectly irradiated reactors eliminate the need for a window at the expense of having less efficient heat transfer—by conduction—through the walls of an opaque absorber. Thus, the disadvantages are linked to the limitations imposed by the materials of the absorber, with regard to maximum operating temperature, inertness to the chemical reaction, thermal conductivity, radiative absorptance, and resistance to thermal shocks.

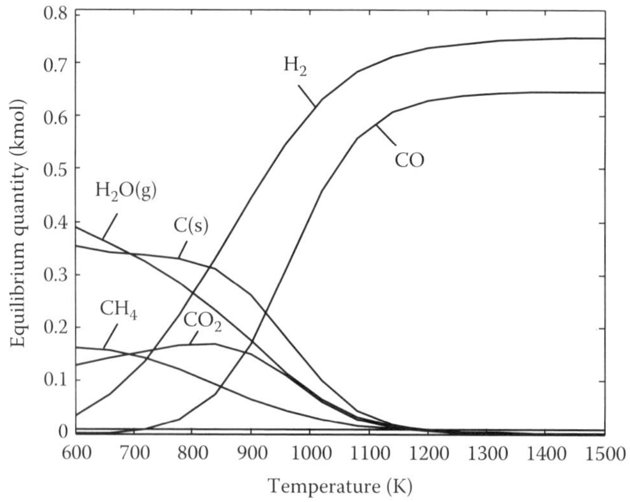

FIGURE 12.10
Equilibrium composition as a function of temperature of the stoichiometric system of Equation 12.1 for beech charcoal $C_1H_{0.47}O_{0.055}S_{0.022}N_{0.004}$ at 1 bar. Species with mole fractions less than 10^{-3} have been omitted.

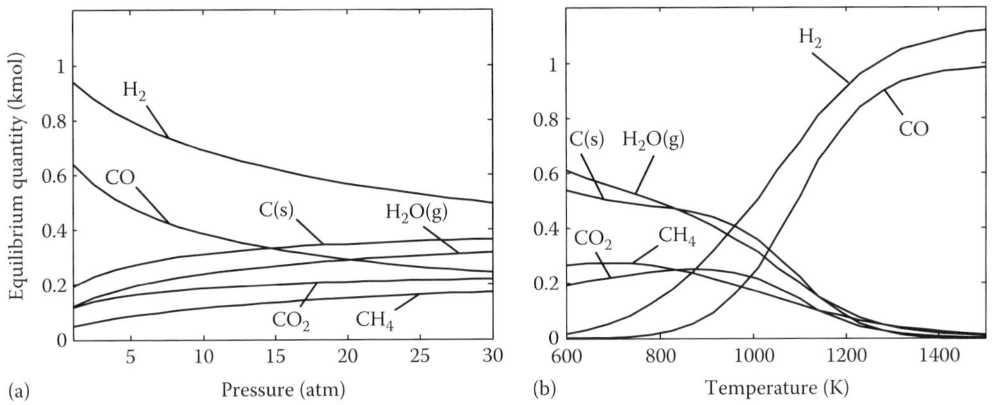

FIGURE 12.11
Equilibrium composition of the stoichiometric system of Equation 12.1 for beech charcoal $C_1H_{0.47}O_{0.055}S_{0.022}N_{0.004}$: (a) as a function of pressure at 1000 K and (b) as a function of temperature at 10 bar. Species with mole fractions less than 10^{-3} have been omitted.

Early studies on solar gasification used packed-bed reactors in which the feedstock charge moves in a counterflow to the reacting gas, progressively drying, pyrolyzing, gasifying, and finally slagging ashes for removal (Gregg et al., 1980; Beattie et al., 1983). Packed beds were further applied for kinetics studies as functions of particle size, temperature, and moisture content (Flechsenhar and Sasse, 1995). Packed-bed reactors are generally simple and robust, can accommodate a wide range of feedstock sizes and forms, and do not require excess steam flows, making them correspondingly cheap. Their main drawback is associated with the rate of heat/mass transfer through the porous packed bed, which limits the reaction rate and throughput. Additionally, ash buildup decreases the surface absorptivity of the irradiated packed bed and may lead to slagging and sintering, particularly for high ash content feedstocks. In contrast, fluidized-bed reactors achieve efficient mass and heat transfer at the expense of requiring feedstock preparation with small particles—typically less than 5 mm—and narrow particle size distributions (Ingel et al., 1992; Murray and Fletcher, 1994; Kodama et al., 2002; Müller et al., 2003; von Zedtwitz and Steinfeld, 2005). Additionally, excess (over-stoichiometric) steam or inert gas is often necessary to achieve fluidization, which translates into an energy penalty while displacing possible syngas production with a given allowable bed pressure drop. Recently, innovative reactor designs based on packed beds and entrained flows have been proposed and experimentally demonstrated for the combined pyrolysis and gasification of petcoke, biomass charcoal, coal, and carbonaceous waste feedstocks (Z'Graggen et al., 2006; Melchior et al., 2009a; Piatkowski et al., 2009a; Lichty et al., 2010). Three examples are shown in Figures 12.12 through 12.14.

Of the three solar reactor concepts, the one based on the packed bed is the most flexible in handling and processing heterogeneous feedstocks with varying compositions and particle sizes (typically 0.1–10 mm). Because of long residence times, it can also tolerate lower-reactivity feedstocks. However, its energy conversion efficiency is constrained by the rate of heat and mass transfer (Piatkowski and Steinfeld, 2008). In contrast, the solar reactors based on the entrained flow exhibit more efficient transport properties, but at the expense of being sensitive to particle sizes, typically <10 μm (Z'Graggen and Steinfeld, 2009). Because of the short residence times, they are most suited to high-reactivity feedstocks. The directly irradiated concept bypasses the limitations imposed by conductive heat transfer through ceramic walls and, consequently, promises high-energy conversion

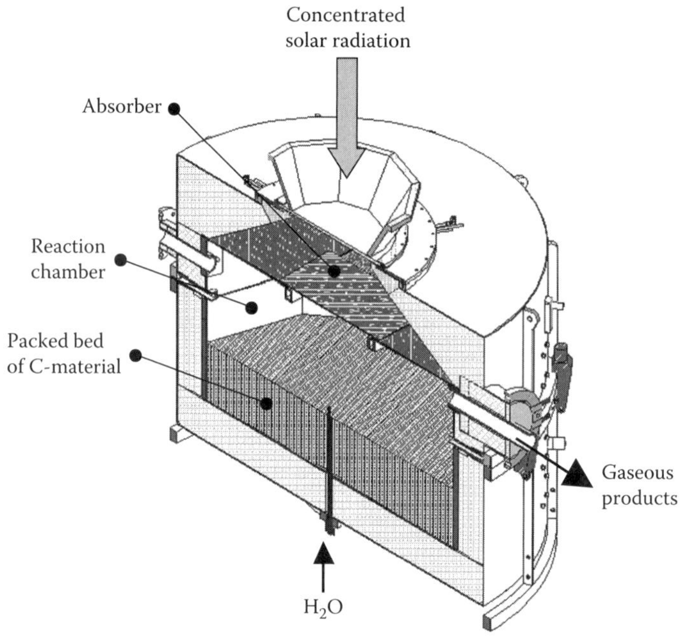

FIGURE 12.12
Scheme of the indirectly irradiated packed-bed solar reactor configuration, featuring two cavities separated by an emitter plate, with the upper one serving as the radiative absorber and the lower one containing the reacting packed bed that shrinks as the reaction progresses. (From Piatkowski, N. et al., *Energy Environ. Sci.*, 4, 73, 2011.)

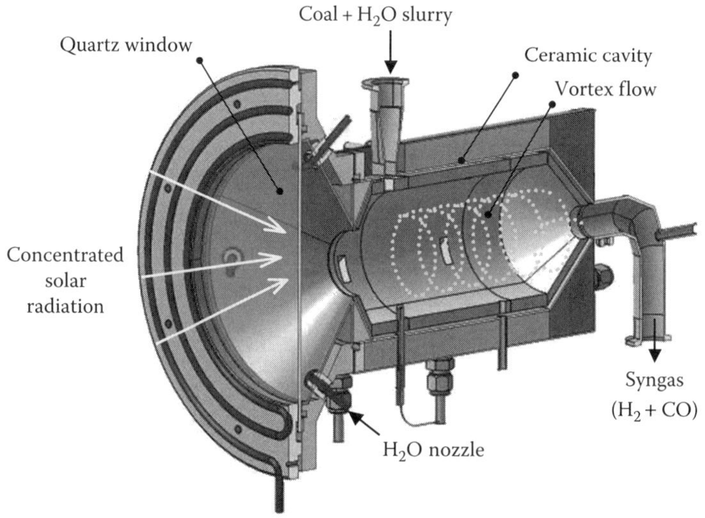

FIGURE 12.13
Scheme of the directly irradiated vortex-flow solar reactor configuration, featuring a helical flow of carbonaceous particles and steam confined to a cavity-receiver and directly exposed to concentrated solar radiation. (From Piatkowski, N. et al., *Energy Environ. Sci.*, 4, 73, 2011.)

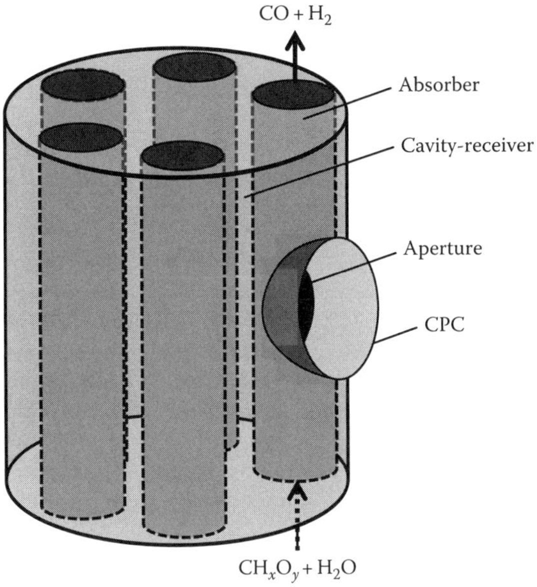

FIGURE 12.14
Scheme of the indirectly irradiated entrained-flow solar reactor configuration, featuring a cylindrical cavity-receiver containing an array of tubular absorbers through which a continuous flow of water vapor laden with carbonaceous particles reacts to form syngas. (From Piatkowski, N. et al., *Energy Environ. Sci.*, 4, 73, 2011.)

efficiencies; however, it introduces a critical quartz window. The indirectly irradiated concept is a more technically feasible approach but with an associated heat transfer penalty.

12.2.6 H_2 from H_2S by Solar Thermolysis

An optional source of H_2 is H_2S, a toxic industrial by-product derived from the NG, petroleum, and coal processing. Current industrial practice uses the Claus process to recover sulfur from H_2S, but H_2 is wasted by oxidizing it to H_2O. Alternatively, H_2S can be thermally decomposed at 1800 K to coproduce H_2 and sulfur, which after quenching have a natural phase separation:

$$H_2S = H_2 + 0.5S_2 \tag{12.23}$$

In contrast to H_2O thermolysis, solar experimental studies on H_2S thermolysis indicate that high degree of chemical conversion is attainable and that the reverse reaction during quench is negligible (Noring and Fletcher, 1982; Kappauf et al., 1985; Kappauf and Fletcher, 1989). A study delineating the chemical kinetics gives a quantitative rate expression for H_2S decomposing in an Al_2O_3 reactor (Harvey et al., 1998).

12.3 Economical Assessments

The economics of solar hydrogen production have been assessed for H_2 produced via thermochemical cycles (Kromer et al., 2011), solar reforming (Spiewak et al., 1992), solar cracking (Spath and Amos, 2003), and H_2S thermolysis (Diver and Fletcher, 1985; Villasmil and Steinfeld, 2010). These assessments indicate that the solar thermochemical production of hydrogen can be competitive vis-à-vis the electrolysis of water using solar-generated electricity and, under certain conditions, might become competitive with conventional fossil-fuel-based processes, provided credit is applied for CO_2 mitigation and pollution avoidance. The weaknesses of these economic evaluations are related primarily to the uncertainties in the viable efficiencies and investment costs of the various components due to their early stage of development and their economy of scale. Further development and large-scale demonstration are warranted.

Nomenclature

$A_{aperture}$	Area of aperture
C	Solar flux concentration ratio
I	Normal beam insolation
$\dot{n}$	Molar flow rate
HHV	High-heating value
$Q_{aperture}$	Incoming solar energy intercepted by the aperture
Q_{FC}	Heat rejected to the surroundings by an ideal fuel cell
Q_{quench}	Heat rejected to the surroundings by the quenching process
$Q_{reactor,net}$	Net energy absorbed by the solar reactor
Q_{solar}	Total solar energy coming from the solar concentrator
T	Nominal solar reactor temperature
$T_{stagnation}$	Maximum temperature of a blackbody absorber
$T_{optimum}$	Optimal temperature of the solar reactor for maximum $\eta_{solar\text{-}to\text{-}fuel}$
W_{FC}	Work output by an ideal fuel
α_{eff}	Effective absorptance of the solar cavity-receiver
ε_{eff}	Effective emittance of the solar cavity-receiver
ΔG	Gibbs free energy change
ΔH	Enthalpy change
$\eta_{absorption}$	Solar energy absorption efficiency
η_{Carnot}	Efficiency of a Carnot heat engine operating between T_H and T_L
$\eta_{solar\text{-}to\text{-}fuel}$	Solar-to-fuel energy conversion efficiency
σ	Stefan–Boltzmann constant (5.6705×10^{-8} W/m^2 K^4)

References

Abanades, S. and Flamant, G. (2006). Thermochemical hydrogen production from a two-step solar-driven water-splitting cycle based on cerium oxides. *Sol. Energy* **80**, 1611–1623.

Abanades, S., Legal, A., Cordier, A., Peraudeau, G., Flamant, G., and Julbe, A. (2010). Investigation of reactive cerium-based oxides for H_2 production by thermochemical two-step water-splitting. *J. Mater. Sci.* **45**, 4163–4173.

Abu Hamed, T., Venstrom, L., Alshare, A., Brülhart, M., and Davidson, J.H. (2006). Study of a quench device for simultaneous synthesis and hydrolysis of Zn nanoparticles: Modeling and experiments. *J. Sol. Energy Eng.* **131**, 031018-1–031018-9.

Allendorf, M.D., Diver, R.B., Siegel, N.P., and Miller, J.E. (2008). Two-step water splitting using mixed-metal ferrites: Thermodynamic analysis and characterization of synthesized materials. *Energy Fuels* **22**, 4115–4124.

Beattie, W.H., Berjoan, R., and Coutures, J.P. (1983). High-temperature solar pyrolysis of coal. *Sol. Energy* **31**(2), 137–143.

Bilgen, E. (1984). Solar hydrogen production by direct water decomposition process: A preliminary engineering assessment. *Int. J. Hydrogen Energy* **9**, 53–58.

Bilgen, E., Ducarroir, M., Foex, M., Sibieude, F., and Trombe, F. (1977). Use of solar energy for direct and two-step water decomposition cycles. *Int. J. Hydrogen Energy* **2**, 251–257.

Buck, R., Abele, M., Bauer, H., Seitz, A., and Tamme, R. (1994). Development of a volumetric receiver-reactor for solar methane reforming. *ASME—J. Sol. Energy Eng.* **116**, 449.

Buck, R., Muir, R.E., Hogan, E.H., and Skocypec, R.D. (1991). Carbon dioxide reforming of methane in a solar volumetric receiver/reactor: The CAESAR Project. *Sol. Energy Mater.* **24**, 449–463.

Charvin, P., Abanades, S., Flamant, G., and Lemort, F. (2007). Two-step water splitting thermochemical cycle based on iron oxide redox pair for solar hydrogen production. *Energy* **32**, 1124–1133.

Chueh, W.C., Falter, C., Abbott, M., Scipio, D., Furler, P., Haile, S.M., and Steinfeld, A. (2010). High-flux solar-driven thermochemical dissociation of CO_2 and H_2O using nonstoichiometric ceria. *Science* **330**, 1797–1801.

Chueh, W.C. and Haile, S.M. (2009). Ceria as a thermochemical reaction medium for selectively generating syngas or methane from H_2O and CO_2. *ChemSusChem* **2**, 735–739.

Dahl, J., Buechler, K., Weimer, A.W., Lewandowski, A., and Bingham, C. (2004a). Solar-thermal dissociation of methane in a fluid-wall aerosol flow reactor. *Int. J. Hydrogen Energy* **29**, 725–736.

Dahl, J.K., Weimer, A.W., Lewandowski, A., Bingham, C., Bruetsch, F., and Steinfeld, A. (2004b). Dry reforming of methane using a solar-thermal aerosol flow reactor. *Ind. Eng. Chem. Res.* **43**, 5489–5495.

Diver, R.B. and Fletcher, E.A. (1985). Hydrogen and sulfur from H_2S—III. The economics of a quench process. *Energy* **10**, 831–842.

Diver, R.B., Pederson, S., Kappauf, T., and Fletcher, E.A. (1983). Hydrogen and oxygen from water—VI. Quenching the effluent from a solar furnace. *Energy* **12**, 947–955.

Ehrensberger, K., Frei, A., Kuhn, P., Oswald, H.R., and Hug, P. (1995). Comparative experimental investigations on the water-splitting reaction with iron oxide $Fe_{1-y}O$ and iron manganese oxides $(Fe_{1-x}Mn_x)_{1-y}O$. *Solid State Ionics* **78**, 151–160.

Epstein, M. and Spiewak, I. (1996). Solar experiments with a tubular reformer. In *Proceedings of the 8th International Symposium on Solar Thermal Concentrating Technologies*, Cologne, Germany. Müller Verlag, Heidelberg, Germany, pp. 1209–1229.

Fernando, F., Fernandez-Saavedra, R., Gomez-Mancebo, M., Vidal, A., Sanchez, M., Rucandio, M., Quejido, A., and Romero, M. (2009). Solar hydrogen production by two-step thermochemical cycles: Evaluation of the activity of commercial ferrites. *Int. J. Hydrogen Energy* **34**, 2918–2924.

Flechsenhar, M. and Sasse C. (1995). Solar gasification of biomass using oil shale and coal as candidate materials. *Energy* **20**(8), 803–810.

Fletcher, E.A. (1999). Solarthermal and solar quasi-electrolytic processing and separations: Zinc from zinc oxide as an example. *Ind. Eng. Chem. Res.* **38**, 2275–2282.

Fletcher, E.A. (2001). Solarthermal processing: A review. *J. Sol. Energy Eng.* **123**, 63–74.

Fletcher, E.A. and Berber, R. (1988). Extracting oil from shale using solar energy. *Energy* **13**, 13–23.

Fletcher, E.A. and Moen, R.L. (1977). Hydrogen and oxygen from water. *Science* **197**, 1050–1056.

Funk, J. (2001). Thermochemical hydrogen production: Past and present. *Int. J. Hydrogen Energy* **26**, 185–190.

Gregg, D.W., Taylor, D.W., Campbell, J.H., Taylor, J.R., and Cotton, A. (1980). Solar gasification of coal, activated carbon, coke, and coal and biomass mixtures. *Sol. Energy* **25**, 353–364.

Gokon, N., Murayama, H., Nagasaki, A., and Kodama, T. (2009). Thermochemical two-step water splitting cycles by monoclinic ZrO_2-supported $NiFe_2O_4$ and Fe_3O_4 powders and ceramic foam devices. *Sol. Energy* **83**, 527–537.

Gokon, N., Oku, Y., Kaneko, H., and Tamaura, Y. (2002). Methane reforming with CO_2 in molten salt using FeO catalyst. *Sol. Energy* **72**, 243–250.

Gstoehl, D., Brambilla, A., Schunk, L.O., and Steinfeld, A. (2008). A quenching apparatus for the gaseous products of the solar thermal dissociation of ZnO. *J. Mater. Sci.* **43**, 4729–4736.
Harvey, S., Davidson, J.H., and Fletcher, E.A. (1998). Thermolysis of hydrogen sulfide in the temperature range 1350 to 1600 K. *Ind. Eng. Chem. Res.* **37**, 2323–2332.
Hirsch, D. and Steinfeld, A. (2004). Solar hydrogen production by thermal decomposition of natural gas using a vortex-flow reactor. *Int. J. Hydrogen Energy* **29**, 47–55.
Hogan Jr., R.E., Skocypec, R.D., Diver, R.B., Fish, J.D., Garrait, M., and Richardson, J.T. (1990). A direct absorber reactor/receiver for solar thermal applications. *Chem. Eng. Sci.* **45**, 2751–2758.
Ihara, S. (1980). On the study of hydrogen production from water using solar thermal energy. *Int. J. Hydrogen Energy* **5**, 527–534.
Ingel, G., Levy, M., and Gordon, J. (1992). Oil shale gasification by concentrated sunlight: An open-loop solar chemical heat pipe. *Energy* **17**, 1189–1197.
Ishihara, H., Kaneko, H., Hasegawa, N., and Tamaura, Y. (2008). Two-step water-splitting at 1273–1623 K using yttria-stabilized zirconia-iron oxide solid solution via co-precipitation and solid-state reaction. *Energy* **33**, 1788–1793.
Kaneko, H., Miura, T., Ishihara, H., Taku, S., Yokoyama, T., Nakajima, H., and Tamaura, Y. (2007). Reactive ceramics of CeO_2–MO_x (M = Mn, Fe, Ni, Cu) for H_2 generation by two-step water splitting using concentrated solar thermal energy. *Energy* **32**, 656–663.
Kappauf, T. and Fletcher, E.A. (1989). Hydrogen and sulfur from hydrogen sulfide—VI. Solar thermolysis. *Energy* **14**, 443–449.
Kappauf, T., Murray, J.P., Palumbo, R., Diver, R.B., and Fletcher, E.A. (1985). Hydrogen and sulfur from hydrogen sulfide—IV. Quenching the effluent from a solar furnace. *Energy* **10**, 1119–1137.
Kodama, T., Kondoh, Y., and Tamagawa, T. (2002). Fluidized bed coal gasification with CO_2 under direct irradiation with concentrated visible light. *Energy Fuels* **16**(5), 1264–1270.
Kodama, T., Koyanagi, T., Shimizu, T., and Kitayama, Y. (2001). CO_2 reforming of methane in a molten carbonate salt for use in solar thermochemical processes. *Energy Fuels* **15**, 60–65.
Kogan, A. (1998). Direct solar thermal splitting of water and on-site separation of the products. II. Experimental feasibility study. *Int. J. Hydrogen Energy* **23**, 89–98.
Kribus, A., Doron, P., Rubin, R., Reuven, R., Taragan, E., Duchan, S., and Karni, J. (2001). Performance of the directly-irradiated annular pressurized receiver (DIAPR) operating at 20 bar and 1200°C. *J. Sol. Energ.-T. ASME* **123**, 10–17.
Kromer, M., Roth, K., Takata, R., and Chin, P. (2011). Support for cost analyses on solar-driven high temperature thermochemical water-splitting cycles. DOE Final Report DE-DT0000951.
Lédé, J. (1999). Solar thermochemical conversion of biomass. *Solar Energy* **65**, 3–13.
Lédé, J., Boutin, O., Elorza-Ricart, E., and Ferrer, M. (2001). Solar thermal splitting of zinc oxide: A review of some of the rate controlling factors. *J. Sol. Energy Eng.* **123**, 91–97.
Lédé, J., Villermaux, J., Ouzane, R., Hossain, M.A., and Ouahes, R. (1987). Production of hydrogen by simple impingement of a turbulent jet of steam upon a high temperature zirconia surface. *Int. J. Hydrogen Energy* **12**, 3–11.
Levy, M., Rosin, H., and Levitan, R. (1989). Chemical reactions in a solar furnace by direct irradiation of the catalyst. *ASME—J. Sol. Energy Eng.* **111**, 96–97.
Levy, M., Rubin, R., Rosin, H., and Levitan, R. (1992). Methane reforming by direct solar irradiation of the catalyst. *Sol. Energy* **17**, 749–756.
Lichty, P., Perkins, C., Woodruff, B., Bingham, C., and Weimer, A.W. (2010). Rapid high temperature solar thermal biomass gasification in a prototype cavity reactor. *J. Sol. Energy Eng.* **132**, 011012-1.
Loutzenhiser, P., Meier, A., and Steinfeld, A. (2010). Review of the two-step H_2O/CO_2-splitting solar thermochemical cycle based on Zn/ZnO redox reactions. *Materials* **3**, 4922–4938.
Maag, G., Zanganeh, G., and Steinfeld, A. (2009). Solar thermal cracking of methane in a particle-flow reactor for the co-production of hydrogen and carbon. *Int. J. Hydrogen Energy* **34**, 7676–7685.

Meier, A., Kirillov, V., Kuvshinov, G., Mogilnykh, Y., Reller, A., Steinfeld, A., and Weidenkaff, A. (1999). Solar thermal decomposition of hydrocarbons and carbon monoxide for the production of catalytic filamentous carbon. *Chem. Eng. Sci.* **54**, 3341–3348.

Melchior, T., Perkins, C., Lichty, P., Weimer, A.W., and Steinfeld, A. (2009a). Solar-driven biochar gasification in a particle-flow reactor. *Chem. Eng. Process.* **48**, 1279–1287.

Melchior, T., Piatkowski, N., and Steinfeld, A. (2009b). H_2 production by steam-quenching of Zn vapor in a hot-wall aerosol flow reactor. *Chem. Eng. Sci.* **64**, 1095–1101.

Miller, J.E., Allendorf, M.D., Diver, R.B., Evans, L.R., Siegel, N.P., and Stuecker, J.N. (2008). Metal oxide composites and structures for ultra-high temperature solar thermochemical cycles. *J. Mater. Sci.* **43**, 4714–4728.

Moeller, S., Buck, R., Tamme, R., Epstein, M., Liebermann, D., Moshe, M., Fisher, U., Rotstein, A., and Sugarmen, C. (2002). Solar production of syngas for electricity generation: SOLASYS project test-phase. In *Proceedings of the 11th SolarPACES International Symposium on Concentrated Solar Power and Chemical Energy Technologies*, Steinfeld, A. (ed.), Zurich, Switzerland, pp. 231–237.

Moeller, S., Kaucic, D., and Sattler C. (2006). Hydrogen production by solar reforming of natural gas: A comparison study of two possible process configurations. *J. Sol. Energ.-T. ASME* **128**, 16–23.

Muir, J., Hogan, R., Skocypec, R., and Buck, R. (1994). Solar reforming of methane in a direct absorption catalytic reactor on a parabolic dish: I. Test and analysis. *Sol. Energy* **52**, 467–477.

Müller, R., von Zedtwitz, P., Wokaun, A., and Steinfeld, A. (2003). Kinetic investigation on steam gasification of charcoal under direct high flux irradiation. *Chem. Eng. Sci.* **58**, 5111–5119.

Murray, J.P. and Fletcher, E.A. (1984). Reaction of steam with cellulose in a fluidized bed using concentrated sunlight. *Energy* **19**, 1083–1098.

Nakamura, T. (1977). Hydrogen production from water utilizing solar heat at high temperatures. *Sol. Energy* **19**, 467–475.

Noring, J.E. and Fletcher, E.A. (1982). High temperature solar thermochemical processing—Hydrogen and sulfur from hydrogen sulfide. *Energy* **7**, 651–666.

Palumbo, R., Lédé, J., Boutin, O., Elorza Ricart, E., Steinfeld, A., Moeller, S., Weidenkaff, A., Fletcher, E.A., and Bielicki, J. (1998). The production of Zn from ZnO in a single step high temperature solar decomposition process. *Chem. Eng. Sci.* **53**, 2503–2518.

Perkins, C. and Weimer, A.W. (2004). Likely near-term solar-thermal water splitting technologies. *Int. J. Hydrogen Energy* **29**, 1587–1599.

Piatkowski, N. and Steinfeld, A. (2008). Solar-driven coal gasification in a thermally irradiated packed-bed reactor. *Energy Fuels* **22**, 2043–2052.

Piatkowski, N., Wieckert, C., and Steinfeld, A. (2009). Experimental investigation of a packed-bed solar reactor for the steam-gasification of carbonaceous feedstocks. *Fuel Process. Technol.* **90**, 360–366.

Piatkowski, N., Wieckert, C., Weimer, A.W., and Steinfeld, A. (2011). Solar-driven gasification of carbonaceous feedstock—A review. *Energy Environ. Sci.* **4**, 73–82.

Rodat, S., Abanades, S., Sans, J.-L., and Flamant, G. (2009). Hydrogen production from solar thermal dissociation of natural gas: Development of a 10 kW solar chemical reactor prototype. *Sol. Energy* **83**, 1599–1610.

Roeb, M., Sattler, C., Klueser, R., Monnerie, N., De Oliveira, L., Konstandopoulos, A.G., Agrafiotis, C. et al. (2006). Solar hydrogen production by a two-step cycle based on mixed iron oxides. *J. Sol. Energy Eng.* **128**, 125–133.

Schunk, L., Haeberling, P., Wepf, S., Wuillemin, D., Meier, A., and Steinfeld, A. (2008). A solar receiver-reactor for the thermal dissociation of zinc oxide. *ASME J. Sol. Energy Eng.* **130**, 021009.

Schunk, L., Lipinski, W., and Steinfeld, A. (2009). Heat transfer model of a solar receiver-reactor for the thermal dissociation of ZnO—Experimental validation at 10 kW and scale-up to 1 MW. *Chem. Eng. J.* **150**, 502–508.

Serpone, N., Lawless, D., and Terzian, R. (1992). Solar fuels: Status and perspectives. *Sol. Energy* **49**, 221–234.
Sibieude, F., Ducarroir, M., Tofighi, A., and Ambriz, J. (1982). High-temperature experiments with a solar furnace: The decomposition of Fe_3O_4, Mn_3O_4, CdO. *Int. J. Hydrogen Energy* **7**, 79–88.
Singh, P. and Hegde, M.S. (2009). $Ce_{0.67}Cr_{0.33}O_{2.11}$: A new low-temperature O_2 evolution material and H_2 generation catalyst by thermochemical splitting of water. *Chem. Mater.* **22**, 762–768.
Spath, P. and Amos, W.A. (2003). Using a concentrating solar reactor to produce hydrogen and carbon black via thermal decomposition of natural gas: Feasibility and economics. *J. Sol. Energy Eng.* **125**, 159–164.
Spiewak, I., Tyner, C.E., and Langnickel, U. (1992) Solar reforming applications study summary. In *Proceedings of the 6th International Symposium on Solar Thermal Concentrating Technologies*, Mojacar, Spain, September 28–October 2, pp. 955–968.
Steinfeld, A. (2002). Solar hydrogen production via a 2-step water-splitting thermochemical cycle based on Zn/ZnO redox reactions. *Int. J. Hydrogen Energy* **27**, 611–619.
Steinfeld, A. (2005). Solar thermochemical production of hydrogen—A review. *Sol. Energy* **78**, 603–615.
Steinfeld, A., Kirillov, V., Kuvshinov, G., Mogilnykh, Y., and Reller, A. (1997). Production of filamentous carbon and hydrogen by solar thermal catalytic cracking of methane. *Chem. Eng. Sci.* **52**, 3599–3603.
Steinfeld, A., Kuhn, P., Reller, A., Palumbo, R., Murray, J.P., and Tamaura, Y. (1998). Solar-processed metals as clean energy carriers and water-splitters. *Int. J. Hydrogen Energy* **23**, 767–774.
Steinfeld, A. and Palumbo, R. (2001). Solar thermochemical process technology. In *Encyclopedia of Physical Science and Technology*, Vol. 15, Meyers, R.A. (ed.), Academic Press, San Diego, CA, pp. 237–256.
Steinfeld, A., Sanders, S., and Palumbo, R. (1999). Design aspects of solar thermochemical engineering. *Sol. Energy* **65**, 43–53.
Steinfeld, A. and Schubnell, M. (1993). Optimum aperture size and operating temperature of a solar cavity-receiver. *Sol. Energy* **50**, 19–25.
Sturzenegger, M. and Nüesch, P. (1999). Efficiency analysis for a manganese-oxide-based thermochemical cycle. *Energy* **24**, 959–970.
Tamaura, Y., Steinfeld, A., Kuhn, P., and Ehrensberger, K. (1995). Production of solar hydrogen by a novel, 2-step, water-splitting thermochemical cycle. *Energy* **20**, 325–330.
Tamme, R., Buck, R., Epstein, M., Fisher, U., and Sugarmen, C. (2001). Solar upgrading of fuels for generation of electricity. *J. Sol. Energy Eng.* **123**, 160–163.
Villasmil, W. and Steinfeld, A. (2010). Hydrogen production by hydrogen sulfide splitting using concentrated solar energy—Thermodynamics and economic evaluation. *Energy Convers. Manage.* **51**, 2353–2361.
von Zedtwitz, P., Petrasch, J., Trommer, D., and Steinfeld, A. (2006). Solar hydrogen production via the solar thermal decarbonization of fossil fuels. *Sol. Energy* **80**, 1333–1337.
von Zedtwitz, P. and Steinfeld, A. (2003). The solar thermal gasification of coal—Energy conversion efficiency and CO_2 mitigation potential. *Energy—Int. J.* **28**(5), 441–456.
von Zedtwitz, P. and Steinfeld, A. (2005). Steam-gasification of coal in a fluidized-bed/packed-bed reactor exposed to concentrated thermal radiation—Modeling and experimental validation. *Ind. Eng. Chem. Res.* **44**, 3852–3861.
Wegner, K., Ly, H.C., Weiss, R.J., Pratsinis, S.E., and Steinfeld, A. (2006). In situ formation and hydrolysis of Zn nanoparticles for H_2 production by the two-step ZnO/Zn water-splitting thermochemical cycle. *Int. J. Hydrogen Energy* **31**, 55–61.
Weidenkaff, A., Reller, A., Sibieude, F., Wokaun, A., and Steinfeld, A. (2000). Experimental investigations on the crystallization of zinc by direct irradiation of zinc oxide in a solar furnace. *Chem. Mater.* **12**, 2175–2181.
Welford, W.T. and Winston, R. (1989). *High Collection Nonimaging Optics*. Academic Press, San Diego, CA.
Wörner, A. and Tamme, R. (1998). CO_2 reforming of methane in a solar driven volumetric receiver-reactor. *Catal. Today* **46**, 165–174.

Yogev, A., Kribus, A., Epstein, M., and Kogan, A. (1998). Solar tower reflector systems: A new approach for high-temperature solar plants. *Int. J. Hydrogen Energy* **23**, 239–245.

Z'Graggen, A., Haueter, P., Trommer, D., Romero, M., Jesus, J.C., and Steinfeld, A. (2006). Hydrogen production by steam-gasification of petroleum coke using concentrated solar power—II. Reactor design, testing, and modeling. *Int. J. Hydrogen Energy* **31**, 797–811.

Z'Graggen, A. and Steinfeld, A. (2009). Heat and mass transfer analysis of a suspension of reacting particles subjected to concentrated solar radiation—Application to the steam-gasification of carbonaceous materials. *Int. J. Heat Mass Transfer* **52**, 385–395.

13

Solar Photoelectrochemical Production of Hydrogen

Janusz Nowotny
University of Western Sydney

Tadeusz Bak
University of Western Sydney

Wenxian Li
University of Western Sydney

CONTENTS

13.1 Introduction

This chapter is focused on photoelectrochemical water splitting using titanium dioxide, TiO_2, as the photoelectrode for photoelectrochemical cells (PECs). It is shown that commercial TiO_2 may be used as a raw material for processing of well-defined TiO_2-based photosensitive oxide semiconductors (POSs) [1,2].

The use of TiO_2 for water photolysis has been reported for the first time by Fujishima and Honda [3]. Because of the importance of TiO_2 in photoelectrochemical energy conversion, as well as in other environmentally friendly technologies, this chapter provides an extensive outline on its performance-related properties, including electronic structure, charge transport, surface and near-surface properties, and photoreactivity. It is shown that these functional properties are closely related to the disorder of point defects [2]. This relationship applies for all nonstoichiometric compounds. Therefore, defect chemistry may be used as a framework for the formation of oxide semiconductors with enhanced performance in a range of applications. Consequently, the properties of nonstoichiometric oxides must be considered in terms of all lattice species, including the basic lattice elements and lattice imperfections, such as intrinsic and extrinsic defects. This chapter outlines the basic concepts of defect chemistry for nonstoichiometric oxides in general and titanium oxide in particular. It is shown that the performance of oxide semiconductors in energy conversion is closely related to their defect disorder in the bulk phase and at the surface.

The key PEC performance indicator is the energy conversion efficiency (ECE), which is the ratio of the generated chemical energy (e.g., in the form of hydrogen) to the total amount of incoming light energy striking the photoelectrode. It is shown that the performance of PECs should be considered in terms of the multifactorious approach involving all performance-related properties, which are interrelated [4]. Since all these properties are related to defect disorder, this chapter also considers defect chemistry for TiO_2 as an example representing POSs.

13.2 Photosensitive Oxide Semiconductors

13.2.1 Development Strategy for Photoelectrochemical Hydrogen Generation

The efforts to develop the high-performance PEC for solar hydrogen production are focused on processing novel materials that are based on binary, ternary, and quaternary oxides as well as their solid solutions, which are needed for photoelectrodes. The most promising candidate for this application is TiO_2 for the following reasons:

- TiO_2 exhibits an outstanding chemical stability in aqueous environments [5]. While TiO_2 is reactive with water, TiO_2 itself remains intact and exhibits stable properties over a prolonged period of time.
- The properties of TiO_2 (rutile), including its functional properties, may substantially be modified within the stability of the same crystalline structure by changes of its oxygen content and the concentration of foreign ions (anions and cations) introduced into the TiO_2 lattice.

One of the key performance-related properties of photoelectrode is electronic structure, which impacts on the amount of the sunlight energy that can be absorbed. The bandgap for TiO_2, rutile, is 3.05 eV (Table 13.1 [6–15]). However, the optimal bandgap desired for water splitting is between 1.8 and 2.2 eV.

The reports of Hoffmann et al. [16] and Wang et al. [17] indicated that below a certain critical grain size (10 nm), the bandgap has a tendency to increase as the particle size decreases.

TABLE 13.1

Bandgap Energy for the Rutile Phase Reported in the Literature

Authors	Bandgap (eV)	Method	Specimen (Temperature Range)
Cronemeyer [6]	3.05	Electrical conductivity	Single crystal (773–1223 K)
	3.05	Electrical conductivity	Single crystal (623–1123 K)
	3.03–3.06	Optical method	Single crystal (room temperature)
Rudolph [7]	3.12	Electrical conductivity	Ceramic specimen (1125–1300 K)
Frova et al. [8]	3.0	Optical method	Single crystal (room temperature)
Vos and Krusemeyer [9]	3.026	Optical method	Parallel to *c* axis (room temperature)
	3.059		Perpendicular to *c* axis (room temperature)
Pascual et al. [10]	3.031	Optical method	Parallel to *c* axis (1.6 K)
	3.031		Perpendicular to *c* axis (1.6 K)
Daude et al. [11]	2.91	Theoretical calculation	Parallel to *c* axis
	3.05		Perpendicular to *c* axis
Vos [12]	3.03	Theoretical calculation	Parallel to *c* axis (1.6 K)
	3.07		Perpendicular to *c* axis (1.6 K)
Gupta and Ravindra [13]	3.0329	Optical method	Single crystal (both perpendicular and parallel to *c* axis)
Khan et al. [14]	3.06	Theoretical calculation	Single crystal (both perpendicular and parallel to *c* axis)
Nowotny [15]	3.16	Electrical conductivity	High-purity single crystal (1073–1323 K)
Average	3.05		

It has been shown, however, that the bandgap is not the only property that controls the photocatalytic performance. Karakitsou and Verykios [18] have shown that the anatase form of TiO_2 exhibits the hydrogen production rate that is higher than that of rutile by the factor of 7, despite that its bandgap is larger (3.2 eV). It has been shown that the performance of photoelectrodes depends on several properties, including charge transport, the chemical potential of electrons, as well as surface and near-surface properties, in addition to electronic structure.

13.2.2 System Selection

The search for high-performance photoelectrodes for solar hydrogen generation includes a wide range of compounds of different compositions, structure, microstructure, and electronic structure. While the highest conversion efficiency has been achieved for valence semiconductors, such as GaAs, GaInP, and AlGaAs [19,20], these compounds are not promising for practical application because of high costs and poor chemical stability in water. On the other hand, metal oxides (MOs) exhibit much better stability in water and are less expensive than valence semiconductors.

A wide range of oxide materials have been studied for photoelectrochemical properties, including binary oxides (Fe_2O_3, ZnO, WO_3, Cu_2O), ternary oxides ($SrTiO_3$, $BaTiO_3$, $CaTiO_3$), as well as quaternary oxides. The most promising compound is TiO_2, which exhibits high reactivity with both light and water and is inexpensive. The additional advantage of TiO_2 is its high nonstoichiometry and complex defect disorder, which can be used for manipulation with defect-related properties.

The most common strategy in the modification of TiO_2 properties includes the following procedures:

- Annealing at different temperatures in air. The resulting changes of properties were considered in terms of either surface area [21] or crystalline structure [22].
- Doping with cations and anions. However, most of the reported systems (discussed in the following) are not compatible because the applied processing procedures are not well defined.
- Formation of nanosize systems. While such systems exhibit outstanding properties [16], most of the reported data are not reproducible and cannot be compared.

The studies of the authors on electrical properties of TiO_2 have been recently overviewed [1]. These studies show that properties of TiO_2 are closely related to lattice imperfections, such as point defects. Therefore, defect chemistry may be used as a framework in the development of novel TiO_2-based oxide semiconductors, which may be modified in a controlled manner using defect engineering [1].

The following sections consider defect chemistry of TiO_2 and defect-related properties, including the following matters:

- The reactivity of the TiO_2 lattice with oxygen resulting in the formation or removal of point defects (oxidation or reduction)
- Application of defect engineering in the formation of TiO_2-based semiconductors with controlled chemical potential of electrons

13.2.3 Point Defects and Defect-Related Properties

It has been documented that properties of nonstoichiometric compounds, such as TiO_2, are controlled by point defects and the related defect disorder [23,24]. Therefore, defect chemistry may be used for the conversion of any TiO_2, which is not well defined, into a semiconductor with controlled semiconducting properties. The formation of TiO_2, which is well defined, requires knowledge of basic concepts of defect chemistry, which are formulated in the following discussion [2].

13.2.4 Summary

The production of solar hydrogen by photoelectrochemical water splitting requires development of a new generation of solar materials, which are free from corrosion and photocorrosion in aqueous environments. The TiO_2-based POSs are the most promising candidates for solar water splitting. Their performance-related properties, which are closely related to defect disorder, may be optimized by defect engineering [2]. The concept of defect engineering may be applied for other oxide materials.

13.3 Defect Chemistry for TiO_2

13.3.1 Basic Properties

Titanium dioxide exists in three different structures: rutile, anatase, and brookite. The rutile structure, which is the only thermodynamically stable structure, is shown in Figure 13.1. The commercial specimen of Degussa (P25), which is frequently used as a reference TiO_2 specimen, contains 20% of the rutile phase and 80% of the anatase phase. The Millennium specimens (PC-10, PC-50, PC-500) exhibit the anatase structure [25].

Wu et al. [22], who studied nanocrystalline TiO_2 prepared by solgel, observed that heating of the anatase form of TiO_2 in air leads to its transition into the rutile form at approximately 600 K (Figure 13.2). Annealing of TiO_2 in extremely reduced conditions results in its transition into a wide range of lower titanium oxides, including $Ti_{20}O_{39}$ [26] (Figure 13.3).

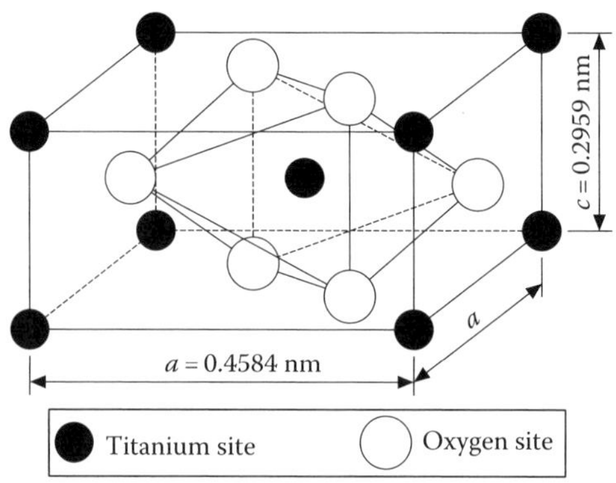

FIGURE 13.1
Structure of the rutile phase.

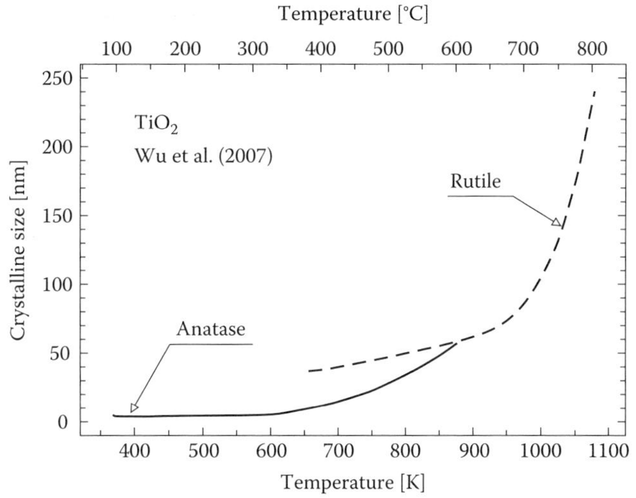

FIGURE 13.2
Crystalline structure versus temperature showing the transition between the anatase and rutile structures. (Reprinted from *Mater. Chem. Phys.*, 102, Wu, Q., Li, D., Hou, Y., Wu, L., Fu, X., and Wang, X., Study of relationship between surface transient photoconductivity and liquid-phase photocatalytic activity of titanium dioxide, 53–59, Copyright 2007, with permission from Elsevier.)

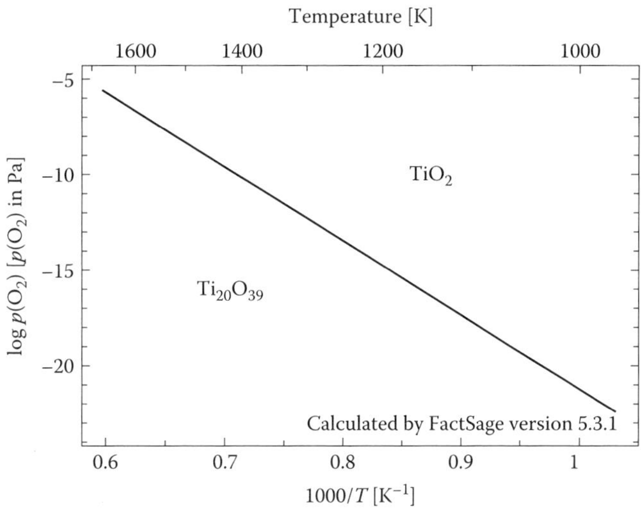

FIGURE 13.3
The stability range of the TiO_2 rutile phase in extremely reduced conditions.

Undefected (ideal) TiO_2 lattice does not exist. The real TiO_2 lattice includes a range of defects, such as point defects, linear, planar, and space defects. This chapter is focused on point defects and their impact on properties. The latter can be explained by defect chemistry [23,24].

TiO_2 has been commonly considered as an oxygen-deficient compound of the formula TiO_{2-x}, where x is the effective deviation from stoichiometry [2,24,27–33]. The extent of oxygen deficit can be determined by thermogravimetry at elevated temperatures when TiO_2 is in equilibrium with the gas phase [27,29–33]. As seen in Figure 13.4,

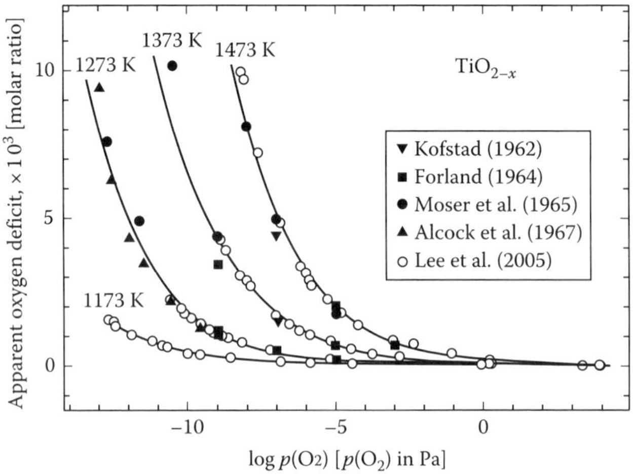

FIGURE 13.4
The effect of oxygen partial pressure on the deviation from stoichiometry, x, for TiO_{2-x} determined by using thermogravimetry according to Kofstad [28], Forland [29], Moser et al. [30], Alcock et al. [32], and Lee et al. [33]. (Reprinted with permission from Bak, T., Nowotny, J., and Nowotny, M.K., Defect disorder of titanium dioxide, *J. Phys. Chem. B*, 110, 21560–21567. Copyright 2006, American Chemical Society.)

there is a good agreement between different reports on the effect of oxygen activity on the apparent oxygen deficit.

TiO_2 has been commonly considered as an n-type semiconductor [23]. Its valence band is formed of filled 2*p* orbitals of doubly valent oxygen ions, and the conduction band is formed of empty 3*d* states of four valent Ti ions. The difference between the energies of the top of the valence band, E_V, and the bottom of the conduction band, E_C, forming the forbidden energy gap, is 3.05 eV. The n-type conduction in TiO_2 is associated with the transport of electrons, which are formed as a result of ionization of oxygen vacancies, which are the predominant defects in TiO_2 [23]. TiO_2 also involves titanium interstitials, which are minor-type defects. TiO_2 may be reduced or oxidized within a single phase leading to the formation, or removal, of point defects [23].

Reduction results in the formation of donor-type defects: oxygen vacancies and titanium interstitials [2,23]. Their ionization results in the formation of quasi-free electrons, which are the predominant electronic defects in n-type TiO_2. Recent studies have shown that prolonged oxidation of pure TiO_2 leads to p-type semiconductivity, which is associated with the presence of titanium vacancies [1]. Ionization of these defects results in the formation of electron holes. While the properties of oxides, including TiO_2, are closely related to defect disorder, their reactivity and the related charge transfer are determined by the chemical potential of electrons, μ_n, which is defined as

$$\mu_n = \mu_n^0 + kT \ln a_n \tag{13.1}$$

where

a_n denotes the activity of electrons
μ_n^0 is the standard term
n in subscript denotes the terms associated with electrons

Therefore, knowledge of the concentration of the electronic charge carriers is essential in the assessment of the reactivity of TiO_2, including the reactivity with water. Defect chemistry may be used in the determination of this quantity.

TiO_2 exhibits several interesting properties, such as:

- TiO_2 is reactive with both light and water. This reactivity can be attributed to the ease with which the Ti ions (Ti^{3+} and Ti^{4+}) alter their valence.
- TiO_2 has excellent chemical stability in aqueous environments [5].
- The properties of TiO_{2-x} can be altered by varying the defect chemistry and related electronic structure through alteration of nonstoichiometry [2,23].
- TiO_2 is substantially less expensive than other photosensitive materials and, therefore, may be a candidate for a new generation of solar materials.
- TiO_2 has several spin-off applications, which are environmentally friendly [34].

The purpose of the following section is to outline the basic concepts of defect chemistry for TiO_2 and the impact of point defects on semiconducting properties, reactivity, and photoreactivity. It is shown that defect chemistry may be used as a framework for the processing of TiO_2 with controlled properties, including photoreactivity with water and oxygen. Defect disorder models considered in this chapter are based on the most recent studies for pure TiO_2 and its solid solutions [2,4,35–60]. The studies aimed to understand the relationship between defect disorder of TiO_2 and its performance as photoelectrode for solar hydrogen PEC.

13.3.2 Nonstoichiometry and Point Defects of Rutile

The properties of TiO_2 can be explained assuming the presence of the following point defects [2,27]:

1. *Oxygen vacancies*: oxygen ions are missing from their lattice sites.
2. *Titanium vacancies*: titanium ions are missing in their lattice sites.
3. *Titanium interstitials*: titanium ions are located in interstitial sites.
4. *Electrons*: these defects are located on Ti^{3+} ions in their lattice sites.
5. *Electron holes*: these defects are located on O^- ions in their lattice sites.

Ionization of ionic defects (1)–(3) leads to the formation of electronic defects (electrons and electron holes), which are mainly responsible for the charge transport.

All point defects in MOs, including TiO_2, have specific functional properties, which impact on the performance of TiO_2-based photocatalysts and photoelectrodes:

- Oxygen vacancies
  - Form adsorption sites for oxygen and water.
  - Their ionization leads to the formation of quasi-free electrons.
- Titanium vacancies
  - Form adsorption sites for water, leading to the formation of an active complex [1].
  - Their ionization leads to the formation of quasi-free electron holes.
- Titanium interstitials
  - Their ionization leads to the formation of quasi-free electrons.
  - Form adsorption sites for acceptor-type molecules, such as oxygen.

- Electrons

  These defects are responsible for charge transfer within the conduction band and midgap bands.
- Electron holes

  These defects are responsible for charge transfer within the valence band and midgap bands.

The effect of defect disorder on electronic structure is shown schematically in Figure 13.5 for a binary MO (where *M* is a bivalent metal, such as Ni). The left side of Figure 13.5 represents different types of point defects, including metal vacancies (5b), oxygen vacancies (5c), and metal interstitials (5d). The right side of Figure 13.5 represents ionization of the ionic defects leading to the formation of electronic defects.

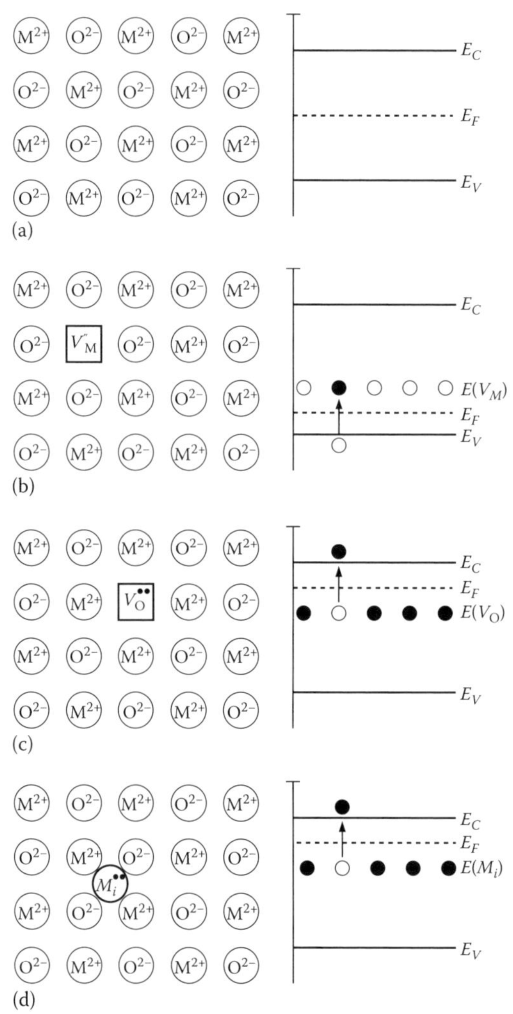

FIGURE 13.5
Schematic representation of point defects in a binary MO and the related band model: (a) undefected structure, (b) doubly ionized cation vacancy, (c) doubly ionized oxygen vacancy, and (d) doubly ionized interstitial cation. (Reproduced with permission from Bak, T., *Oxide Semicond. Res. Rep.*, Copyright 2010.)

The point defects in Figure 13.5 are represented using the Kröger–Vink notation [24] (this notation allows to assess easily the lattice charge neutrality condition). It has been shown that defect disorder is closely related to reactivity, photoreactivity, and the related charge transfer. Therefore, knowledge of defect disorder is essential to assess the reactivity and photoreactivity. TiO_2 is not an exception. Accordingly, defect engineering may be used in processing TiO_2 with enhanced photoreactivity with water, oxygen, hydrogen, and alternative species, such as microbial cells.

In equilibrium, the concentration of thermodynamically reversible defects in oxides is a function of temperature and oxygen activity [23]. Quantitative assessment of the effect of defect disorder on properties, such as semiconducting properties, requires knowledge of the related equilibrium constants. These constants may be used for derivation of defect disorder diagrams representing the effect of oxygen activity on the concentration of defects [42].

The following sections report the recent progress in defect chemistry of TiO_2. It is shown that photocatalytic properties of nonstoichiometric compounds, such as TiO_2, are dependent on defect disorder rather than other commonly studied properties, such as crystal structure, microstructure, and surface area.

Defect chemistry considers point defects in crystalline solids as a solid solution. In the case of dilute solutions, when the concentration of defects is very small, the mass action law may be applied to calculate defect concentrations. At larger concentrations, application of the mass action law requires to use activities instead of concentrations.

The formation of defects in crystals is governed by the general law of thermodynamics, which requires that spontaneous chemical reactions at constant temperature, T, and pressure, p, result in a decrease of free enthalpy:

$$\left(\Delta G = \Delta H - T\Delta S\right)_{T,p} < 0 \tag{13.2}$$

where ΔG, ΔH, and ΔS denote the change of free enthalpy, the enthalpy, and the related change of entropy, respectively. When defects are formed in perfect crystals, the change of entropy should be considered in terms of both configuration entropy change and the vibrational entropy change. Therefore, the change of free energy assumes the following form:

$$\Delta G = [d]\left(\Delta H_f - T\Delta S_v\right) - T\Delta S_{con} \tag{13.3}$$

where

$[d]$ is the concentration of defects
H_f denote the partial enthalpy of defect formation
S_v is the partial vibrational entropy
S_{con} is configuration entropy

While both the ΔH_f and ΔS_{con} terms are positive, the ΔS_v term may assume either positive or negative values, depending on defect disorder. Therefore, the equilibrium concentration of point defects is the result of competition between the thermodynamic terms of ΔS_{con}, ΔS_v, and ΔH_f. Consequently, the following conclusions could be made:

1. Ideal (undefected) crystals are thermodynamically unstable (at any temperature above absolute zero).
2. The equilibrium concentration of thermodynamically reversible defects in oxide crystals depends on equilibrium conditions described by the temperature and oxygen activity.

It is shown in the following sections that photosensitivity of TiO_2 and its performance as photoelectrode are closely related to the concentration and the valence of point defects.

13.3.3 Defect Disorder for Undoped TiO_2

TiO_2 is a nonstoichiometric compound. It has been generally considered that TiO_2 is an oxygen-deficient compound [23]. This picture, which has been supported by gravimetric studies [27–33], suggests that the predominant defects in TiO_2 are oxygen vacancies and/or interstitial titanium ions. Recent studies show, however, that strong oxidation of TiO_2 (at elevated temperatures) leads to the formation of a metal-deficient oxide [39,42]. In this case, TiO_2 may be represented by the formula $Ti_{1-x}O_{2-y}$, where $x > y/2$. The properties of the metal-deficient TiO_2 are determined by titanium vacancies, which are formed during a prolonged oxidation leading, in consequence, to p-type properties [39,52].

Figure 13.6 shows a schematic representation of the periodic lattice for defected TiO_2. The titanium and oxygen ions in their normal lattice sites are represented by the traditional notation using their chemical symbols, and the exponents represent their electrical charge. However, point defects in Figure 13.5, including ion vacancies and the ions located in interstitial sites, are represented according to the Kröger–Vink notation [24]. Introduction of this notation allows representing defect reactions taking into account only their relative electrical charge (compared to the lattice) and ignoring their absolute charge (the classical notation vs. the Kröger–Vink notation is shown in Table 13.2).

13.3.3.1 Definition of Basic Relationships

Description of defect chemistry requires definition of basic relationships, such as defect equilibria, and charge neutralities. These relationships, which have been reported before [1], are briefly outlined in the following.

Using the Kröger–Vink notation [24], the formation of defects at elevated temperatures may be described by the following equilibria [2]:

$$O_O^{\times} \Leftrightarrow V_O^{\bullet\bullet} + 2e' + \frac{1}{2}O_2 \tag{13.4}$$

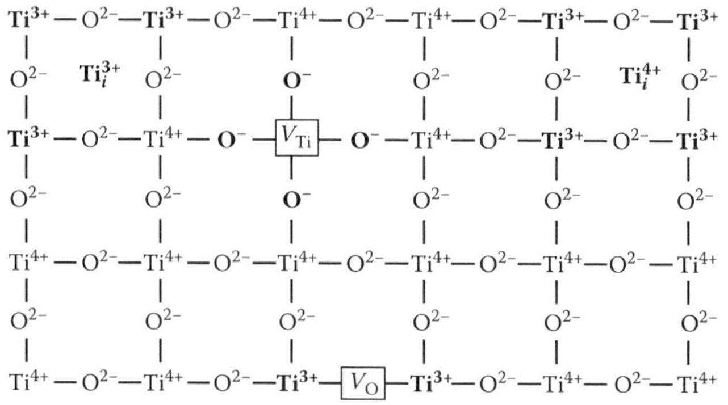

FIGURE 13.6
Schematic representation of thermodynamically reversible point defects in the undoped TiO_2 lattice according to the standard notation showing absolute charges of ions (Table 13.2) [2]. (Reprinted with permission from Nowotny, M.K., Sheppard, L.R., Bak, T., and Nowotny, J., Defect chemistry of titanium dioxide: Application of defect engineering in processing of TiO_2-based photocatalysts, *J. Phys. Chem. C*, 112, 5275–5300. Copyright 2008, American Chemical Society.)

TABLE 13.2

Notations of TiO_2 Lattice Species according to the Kröger–Vink Notation along the Absolute Valency Notation

Notation	Meaning	Kröger–Vink Notation	Index
Ti_{Ti}^{4+}	Ti^{4+} ion in the titanium lattice site	$Ti_{Ti}^{\times}$	*a*
Ti_{Ti}^{3+}	Ti^{3+} ion in the titanium lattice site (quasi-free electron)	e'	*b*
M_{Ti}^{5+}	M^{5+} cation in the titanium lattice site	M_{Ti}^{+}	*c*
M_{Ti}^{3+}	M^{3+} cation in the titanium lattice site	M_{Ti}'	*d*
V_{Ti}	Titanium vacancy	V_{Ti}''''	*f*
Ti_i^{3+}	Ti^{3+} ion in the interstitial site	Ti_i^{+++}	*g*
Ti_i^{4+}	Ti^{4+} ion in the interstitial site	Ti_i^{++++}	*h*
M_i^{+}	M^{+} cation in the interstitial site	M_i^{+}	*j*
O_O^{2-}	O^{2-} ion in the oxygen lattice site	$O_O^{\times}$	*k*
V_O	Oxygen vacancy	V_O^{++}	*l*
O_O^{-}	O^{-} ion in the oxygen lattice site (quasi-free electron hole)	h^{+}	*m*
A_O^{-}	A^{-} anion in the oxygen lattice site	A_O^{+}	*o*
A_O^{3-}	A^{3-} anion in the oxygen lattice site	A_O'	*r*

$$2O_O^{\times} + Ti_{Ti}^{\times} \Leftrightarrow Ti_i^{\bullet\bullet\bullet} + 3e' + O_2 \tag{13.5}$$

$$2O_O^{\times} + Ti_{Ti}^{\times} \Leftrightarrow Ti_i^{\bullet\bullet\bullet\bullet} + 4e' + O_2 \tag{13.6}$$

$$O_2 \Leftrightarrow O_O^{\times} + V_{Ti}'''' + 4h^{\bullet} \tag{13.7}$$

$$\text{Nil} \Leftrightarrow e' + h^{\bullet} \tag{13.8}$$

where e' and $h^{\bullet}$ denote electron and electron hole, respectively.

Any defect disorder must satisfy the charge neutrality condition, which requires that the crystal is electrically neutral. Consequently, the concentration of all charged defects must satisfy the following condition:

$$2\left[V_O^{\bullet\bullet}\right] + 3\left[Ti_i^{\bullet\bullet\bullet}\right] + 4\left[Ti_i^{\bullet\bullet\bullet\bullet}\right] + \left[D^{\bullet}\right] + p = n + 4\left[V_{Ti}''''\right] + \left[A'\right] \tag{13.9}$$

where

n and p denote the concentrations of electrons and electron holes, respectively

$[D^{\bullet}]$ and $[A']$ denote the concentrations of singly ionized donor- and acceptor-type foreign ions, respectively

The condition expressed by Equation 13.9 involves both thermodynamically reversible defects (oxygen vacancies, titanium interstitials, and electronic defects) and also those defects that are thermodynamically irreversible (foreign ions). TiO_2 also includes titanium vacancies, which are thermodynamically reversible (theoretically). These defects, however, are relatively immobile and, therefore, may be considered as quenched in the experimental conditions commonly applied in the determination of defect-related properties. Consequently, these defects may be considered as acceptor-type impurities. Consequently, the titanium vacancies and the impurities (dopants) may be considered as an *effective concentration of acceptors*:

$$A = 4\left[V_{Ti}''''\right] + \left[A'\right] - \left[D^{\bullet}\right] \tag{13.10}$$

For pure TiO_2, the quantity A may be directly related to the concentration of titanium vacancies:

$$A = 4\left[V_{Ti}^{''''}\right] \tag{13.11}$$

According to equilibria (13.4) through (13.7), the concentrations of intrinsic defects depend on oxygen activity. While equilibria (13.4) through (13.6) may be established relatively fast, the formation and the transport of titanium vacancies, represented by equilibrium (13.7), are extremely slow [39,52]. Therefore, the titanium vacancies in Equations 13.9 and 13.10 may be assumed as acceptor-type dopants, that is, their concentration remains practically independent of oxygen activity. This is the reason why these defects are treated differently than other types of ionic defects. Therefore, the concentration of defects, may also be manipulated by the kinetic factor (equilibration time).

The equilibrium constants for equilibria (13.4) through (13.8) are as follows:

$$K_1 = \left[V_O^{\bullet\bullet}\right] n^2 p(O_2)^{1/2} \tag{13.12}$$

$$K_2 = \left[Ti_i^{\bullet\bullet\bullet}\right] n^3 p(O_2) \tag{13.13}$$

$$K_3 = \left[Ti_i^{\bullet\bullet\bullet\bullet}\right] n^4 p(O_2) \tag{13.14}$$

$$K_4 = \left[V_{Ti}^{''''}\right] p^4 p(O_2)^{-1} \tag{13.15}$$

$$K_i = np \tag{13.16}$$

where square brackets represent the concentration of ionic defects (molar fractions) and $p(O_2)$ is the oxygen activity.

Therefore, the concentrations of both electronic and ionic defects may be expressed as follows:

$$\left[V_O^{\bullet\bullet}\right] = K_1 n^{-2} p(O_2)^{-1/2} \tag{13.17}$$

$$\left[Ti_i^{\bullet\bullet\bullet}\right] = K_2 n^{-3} p(O)_2)^{-1} \tag{13.18}$$

$$\left[Ti_i^{\bullet\bullet\bullet\bullet}\right] = K_3 n^{-4} p(O)_2)^{-1} \tag{13.19}$$

$$p = K_i n^{-1} \tag{13.20}$$

Knowledge of the equilibrium constants enables the determination of the concentrations of defects. These equilibrium constants have been recently determined using three independently measured defect-related properties (electrical conductivity, thermoelectric

TABLE 13.3
Equilibrium Constants of Defect Reactions for TiO_2

Equilibrium Constant	ΔH^0 (kJ/mol)	ΔS^0 J/(mol K)	Specimen	Methods	Author
K_1	493.1	106.5	Undoped TiO_2	Electrical conductivity Thermoelectric power	Bak et al. [42]
	334.9	49.9	Nb-doped TiO_2	Electrical conductivity	Bak et al. [42]
K_2	879.2	190.8	Undoped TiO_2	Thermoelectric power	Kofstad [27]
K_3	1025.8	238.3	Undoped TiO_2	Thermoelectric power	Kofstad [27]
K_4	354.5	−202.1	Undoped TiO_2	Electrical conductivity	Bak et al. [42]
	394.5	−378.7	Nb-doped TiO_2	Electrical conductivity	Bak et al. [42]
K_i	222.1	44.6	Undoped TiO_2	Electrical conductivity Thermoelectric power Thermogravimetry	Bak et al. [42]

K_1, K_2, K_3, K_4, and K_i are defined in text.

power, and thermogravimetry) [42]. These constants can be related to the standard-state thermodynamic quantities—entropy ΔS^0 and enthalpy ΔH^0:

$$\ln K = \frac{\Delta S^0}{R} - \frac{\Delta H^0}{RT} \qquad (13.21)$$

Both of these thermodynamic quantities and the associated equilibrium constants represent materials data, which have been reported elsewhere [42,43]. The equilibrium constants and the related data of ΔS^0 and ΔH^0 are presented in Table 13.3.

13.3.3.2 Defect Disorder Diagram

Using the combination of Equations 13.10, 13.17 through 13.20 and the electroneutrality condition (13.9), the concentration of electronic charge carriers may be described by the relationship, involving $p(O_2)$, equilibrium constants, and the effective concentration of acceptors, A [42]:

$$n^5 + An^4 - K_i n^3 - 2K_1 p(O_2)^{-1/2} n^2 - 3K_2 p(O_2)^{-1} n - 4K_3 p(O_2)^{-1} = 0 \qquad (13.22)$$

As seen from Equation 13.22, the effect of $p(O_2)$ on the concentration of electronic charge carriers depends on a combination of all defects. Equation 13.22 may be used for derivation of a defect disorder diagram in the form of the plot of concentrations of reversible defects as a function of $p(O_2)$.

Figure 13.7 shows the defect diagram for TiO_2 in terms of the concentration of defects as a function of oxygen activity at 1273 K [42]. As seen, the concentration of electronic charge carriers and the related electrical properties of TiO_2 are closely related to oxygen activity, which may be used for the imposition of either n- or p-type properties or mixed conduction.

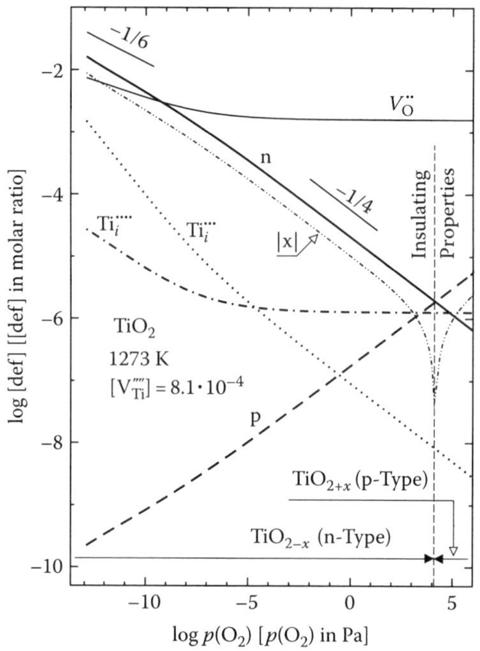

FIGURE 13.7
Defect disorder for undoped TiO_2 showing the effect of oxygen activity on the concentration of both electronic and ionic defects at 1273 K. (Reprinted with permission from Nowotny, M.K., Sheppard, L.R., Bak, T., and Nowotny, J., Defect chemistry of titanium dioxide: Application of defect engineering in processing of TiO_2-based photocatalysts, *J. Phys. Chem. C*, 112, 5275–5300. Copyright 2008, American Chemical Society.)

In equilibrium, the concentration data represented in the diagram shown in Figure 13.7 are well defined by the equilibrium conditions (temperature and oxygen activity). The changes in the concentrations of defects during cooling from the temperature of processing to room temperature depend on the applied cooling procedure, such as rate of cooling and the associated gas-phase composition.

As seen in Figure 13.7, in the vicinity of the n–p transition regime, the effect of $p(O_2)$ on the concentration of electronic charge carriers in the n- and p-type regimes may be expressed as follows [36]:

$$n = n_0 p(O_2)^{-1/4} \tag{13.23}$$

$$p = p_0 p(O_2)^{1/4} \tag{13.24}$$

where n_0 and p_0 denote the concentration of electrons and holes in standard conditions. In strongly reducing conditions, when $p(O_2) < 10^{-5}$ Pa, the concentration of electrons is the following function of $p(O_2)$ [36]:

$$n = (2K_1)^{1/3} p(O_2)^{-1/6} \tag{13.25}$$

The ionic defects form donor and acceptor levels in the electronic structure of TiO_2. Both oxygen vacancies and titanium interstitials form donor levels and titanium vacancies form acceptors [6,61–63]. The effect of these defects on the concentrations of electronic charge carriers depends on their ionization degree. At this stage, the following points can be made:

1. TiO_2 may exhibit both n-type and p-type properties. Its defect disorder and defect-related properties are closely related to oxygen activity in the TiO_2 lattice.
2. The predominant ionic defects in TiO_2 at high oxygen activity are oxygen vacancies and titanium vacancies [1]. Recent studies show that nanosize TiO_2 exhibits much larger concentrations of these defects than that in the bulk phase [64].
3. Defect disorder diagrams may be used for the processing of TiO_2 with controlled properties that are desired for specific performance.

The defect diagrams, such as the diagram in Figure 13.7, may be used to predict the processing conditions in the formation of TiO_2 with controlled semiconducting properties, which can be tailored to suit specific applications.

13.3.4 Extended Defects

In strongly reducing conditions, when nonstoichiometry achieves very large values, oxygen vacancies in TiO_2 have a tendency to interact leading, in consequence, to the formation of larger defect aggregates and superstructures [23]. When the concentration of point defects surpasses a certain critical value, the crystal structure is stabilized by elimination of isolated oxygen vacancies leading to the formation of extended planar defects, which are known as shear planes. The ratio of titanium to oxygen ions within the shear plane, which is charged positively, is higher than in stoichiometric TiO_2. According to Matzke [65], oxygen vacancies aggregate into shear planes when the deviation from stoichiometry exceeds 2×10^{-3}.

There has been an accumulation of the experimental evidence indicating that point defects are present in equilibrium with shear planes. The planar defects, which have been observed by electron microscopy, lead to the formation of a homologous series of so-called Magneli-type phases. These phases may be represented by the chemical formula Ti_nO_{2n-1}, where n vary between 38 ($TiO_{1.97}$) and 4 ($TiO_{1.75}$) [66–68]. So far, little is known about the effect of shear planes on photocatalytic properties of TiO_2.

13.3.5 Defect Disorder for Donor-Doped TiO_2

The properties of TiO_2 (rutile) may be modified by the incorporation of foreign ions with a valency that is different from that of the host lattice ions. Incorporation of such ions leads to the formation of donors or acceptors [23]. The defected structure of TiO_2, including both donor- and acceptor-type extrinsic defects, is shown in Figure 13.8 (the meanings of symbols are in Table 13.2). This section considers the effect of donor-type ions, such as niobium, on the defect disorder of TiO_2. Niobium has been most commonly applied as a donor-type dopant for the modification of semiconducting properties of TiO_2 [2,40,43,44,58,69,70]. The studies of electrical properties for Nb-doped TiO_2 indicate that at low oxygen activity, niobium incorporation leads to the formation of electrons. This reaction may be represented by the following equilibrium:

$$Nb_2O_5 \Leftrightarrow 2Nb^{\bullet}_{Ti} + 4O^{\times}_{O} + 2e' + \frac{1}{2}O_2 \qquad (13.26)$$

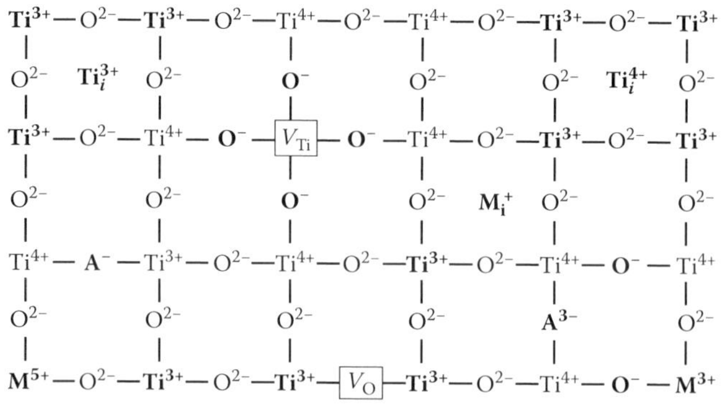

FIGURE 13.8
Schematic representation of both thermodynamically reversible point defects and extrinsic defects in the TiO_2 lattice according to the standard notation showing absolute charges of ions listed in Table 13.2. (Reprinted with permission from Nowotny, M.K., Sheppard, L.R., Bak, T., and Nowotny, J., Defect chemistry of titanium dioxide: Application of defect engineering in processing of TiO_2-based photocatalysts, *J. Phys. Chem. C*, 112, 5275–5300. Copyright 2008, American Chemical Society.)

The defect disorder in this regime is governed by the following (electronic) charge compensation:

$$n = \left[\mathrm{Nb}_{\mathrm{Ti}}^{\bullet}\right] \tag{13.27}$$

As seen, the concentration of electrons in this regime is determined by the content of niobium and is practically independent of $p(O_2)$. It was shown that TiO_2 in this regime exhibits a quasi-metallic conduction [40].

The mechanism of niobium incorporation into TiO_2 at high oxygen activity is entirely different. In this case, niobium incorporation leads to the formation of ionic defects (titanium vacancies) [2,43]:

$$2\mathrm{Nb_2O_5} \Leftrightarrow 4\mathrm{Nb}_{\mathrm{Ti}}^{\bullet} + V_{\mathrm{Ti}}^{''''} + 10\mathrm{O}_{\mathrm{O}}^{\times} \tag{13.28}$$

In this regime, the defect disorder is governed by ionic charge compensation:

$$4\left[V_{\mathrm{Ti}}^{''''}\right] = \left[\mathrm{Nb}_{\mathrm{Ti}}^{\bullet}\right] \tag{13.29}$$

Therefore, the concentration of electrons for Nb-doped TiO_2 may be represented by the following function of oxygen activity:

$$n = K_i \left(\frac{\left[\mathrm{Nb}_{\mathrm{Ti}}^{\bullet}\right]}{4K_4}\right)^{1/4} p(\mathrm{O_2})^{-1/4} \tag{13.30}$$

As seen, the concentration of electrons in this regime is represented by the slope of log n versus log $p(O_2)$ that is equal to −1/4. In conclusion, the effect of niobium on electrical properties of TiO_2 is closely related to oxygen activity in the lattice. It was shown that niobium incorporation into the TiO_2 lattice in extremely reduced conditions leads to a high charge transport [50].

13.3.6 Defect Disorder for Acceptor-Doped TiO_2

Chromium has been commonly used as an acceptor-type dopant of TiO_2 [59,71]. Its incorporation into the TiO_2 lattice at low oxygen activity may be represented by the following equilibrium:

$$Cr_2O_3 \Leftrightarrow 2Cr'_{Ti} + 3O_O^{\times} + V_O^{\bullet\bullet} \quad (13.31)$$

where the charge neutrality requires that chromium ions are compensated by oxygen vacancies:

$$\left[Cr'_{Ti}\right] = 2\left[V_O^{\bullet\bullet}\right] \quad (13.32)$$

Therefore, the concentration of electron holes for Cr-doped TiO_2 may be represented as the following function of $p(O_2)$:

$$p = K_i \left(\frac{\left[Cr'_{Ti}\right]}{2K_1} \right)^{1/2} p(O_2)^{1/4} \quad (13.33)$$

The previous relationship has been verified experimentally by Carpentier et al. [71] by using measurements of electrical conductivity.

Acceptor-type doping may be used for the imposition of p-type properties of TiO_2, which are required for application of TiO_2 as photocathodes.

13.3.7 Real Chemical Formula for Nonstoichiometric TiO_2

The defect disorder models derived previously indicate that the TiO_2 lattice involves a number of point defects. These defects may be grouped according to their location in the TiO_2 lattice, including [2] the following:

1. The titanium sublattice, $\mathcal{A}_{Ti}$
2. Interstitial sites, $\mathcal{B}_i$
3. The oxygen sublattice, $\mathcal{C}_O$

Accordingly, the TiO_2 lattice may be represented by the following general formula:

$$\mathcal{A}_{Ti}\mathcal{B}_i\mathcal{C}_O \quad (13.34)$$

where $\mathcal{A}_{Ti}$, $\mathcal{B}_i$, and $\mathcal{C}_O$ can be considered as modules of the TiO_2 lattice. These modules are expressed by the following specific formulas:

$$\mathcal{A}_{Ti} = \left(Ti_{Ti}^{4+}\right)_a \left(Ti_{Ti}^{3+}\right)_b \left(M_{Ti}^{5+}\right)_c \left(M_{Ti}^{3+}\right)_d \left(V_{Ti}\right)_f \quad (13.35)$$

$$\mathcal{B}_i = \left(Ti_i^{3+}\right)_g \left(Ti_i^{4+}\right)_h \left(M_i^{+}\right)_j \quad (13.36)$$

$$\mathcal{C}_O = \left(O_O^{2-}\right)_k \left(V_O\right)_l \left(O_O^{-}\right)_m \left(A_O^{-}\right)_o \left(A_O^{3-}\right)_r \quad (13.37)$$

where the indexes a, b, c, d, f, g, h, j, k, l, m, o, and r correspond to the amount of the related lattice species (in molar ratio), which are outlined in Table 13.2; M denote foreign cations in the cation sublattice; and A are foreign anions in the oxygen sublattice. The concentrations of these species are interdependent. A wide range of their combinations may be imposed in a controlled manner by appropriate processing conditions at elevated temperatures. The resulting properties are then determined by (i) oxygen activity in the gas phase and (ii) the content of foreign ions forming donors and acceptors. Each combination, however, requires that the following conditions are satisfied:

- The sum of concentrations of all species in the titanium sublattice must be equal to unity:

$$a + b + c + d + f = 1 \tag{13.38}$$

- The sum of concentrations of all species in the oxygen sublattice must be equal to two.

Consequently,

$$k + l + m + o + r = 2 \tag{13.39}$$

The lattice charge neutrality condition can be easily derived when the defects and their electrical charge are expressed in the Kröger–Vink notation [24]. Therefore, Equations 13.35 through 13.37 may be expressed in the following forms:

$$\mathcal{A}_{\mathrm{Ti}} = \left(\mathrm{Ti}_{\mathrm{Ti}}^{\times}\right)_a \left(\mathrm{Ti}_{\mathrm{Ti}}'\right)_b \left(M_{\mathrm{Ti}}^{\bullet}\right)_c \left(M_{\mathrm{Ti}}'\right)_d \left(V_{\mathrm{Ti}}''''\right)_f \tag{13.40}$$

$$\mathcal{B}_i = \left(\mathrm{Ti}_i^{\bullet\bullet\bullet}\right)_g \left(\mathrm{Ti}_i^{\bullet\bullet\bullet\bullet}\right)_h \left(M_i^{\bullet}\right)_j \tag{13.41}$$

$$\mathcal{C}_{\mathrm{O}} = \left(\mathrm{O}_{\mathrm{O}}^{\times}\right)_k \left(V_{\mathrm{O}}^{\bullet\bullet}\right)_l \left(\mathrm{O}_{\mathrm{O}}^{\bullet}\right)_m \left(A_{\mathrm{O}}^{\bullet}\right)_o \left(A_{\mathrm{O}}'\right)_r \tag{13.42}$$

The charge neutrality condition requires that the charges associated with all lattice species are fully compensated electrically. Consequently,

$$c + 3g + 4h + j + 2l + m + o = b + d + 4f + r \tag{13.43}$$

The defects indexed by a and k are electrically neutral with respect to the lattice and, therefore, are not taken into account in the charge neutrality condition.

The diagram in Figure 13.7 shows the effective deviation from stoichiometry, x, which can be expressed as follows:

$$x = \frac{2(g + h - f) + l}{1 + g + h - f} \tag{13.44}$$

The concept of defect engineering is based on the imposition of desired properties by controlled combination of the concentration of the lattice species, which are a function of the following variables:

1. Temperature
2. Oxygen activity
3. Concentration of aliovalent ions

The reactivity and photoreactivity of TiO_2-based oxide semiconductors and the related photocatalytic properties are closely related to their electroactivity, which is determined by defect disorder. The defect disorder diagrams (see Figure 13.7) may be used for assessment of the electroactivity and prediction of optimized processing conditions.

The incorporation of aliovalent ions into the TiO_2 lattice (deliberately and incidentally) has a substantial impact on the semiconducting properties. Correct understanding of the effects of these ions imposes the following requirements:

1. Basic characterization of TiO_2 specimens should include the determination of the concentration of incidentally introduced foreign defects (impurities), which have a substantial impact on properties already at the level of parts per million.
2. The formation of TiO_2-based solid solutions leads to well-defined systems only when the doping procedure is well defined in terms of (1) oxygen activity, (2) time of processing, and (3) cooling procedure.

In conclusion, the formula TiO_2 is not reflective of the complex composition of this nonstoichiometric compound that involves a wide range of ionic and electronic point defects. The real chemical formula of TiO_2, which is reflective of specific properties, is more complex.

13.3.8 Summary

TiO_2 of unknown defect disorder may be used as a raw material for the processing of well-defined POSs, which are required for photoelectrodes of PECs. Therefore, the promising research strategy in the development of TiO_2-based semiconductors for high-performance photoelectrodes is the processing at elevated temperatures in controlled oxygen activity and the subsequent cooling procedure.

13.4 Electrical Properties

13.4.1 General

The electrical properties are defect sensitive if the defects are electrically charged [23]. Therefore, the electrical properties have been most commonly applied in the verification of defect disorder models. The most commonly studied electrical properties are as follows:

1. Electrical conductivity
2. Thermoelectric power (Seebeck effect)
3. Work function (WF)

The electrical properties may easily be determined experimentally at room temperature (during the performance) and also at elevated temperatures (during processing in the gas phase of controlled oxygen activity). Finally, the electrical properties may be used for monitoring the chemical reactions associated with charge transfer at gas/solid and liquid/solid interfaces.

13.4.2 Electrical Conductivity

13.4.2.1 Definition

The measurements of electrical conductivity can be used to assess the charge transport involving the components related to both electrons and electron holes [23,36]:

$$\sigma = en\mu_n + ep\mu_p \tag{13.45}$$

where
- e is the elementary charge
- μ is the mobility
- n and p denote the concentration of electrons and holes
- the subscripts n and p correspond to specific charge carriers

Usually one type of charge carriers is predominant. However, for amphoteric oxides, such as TiO_2, the electrical conductivity in the n–p transition regime involves the components related to both charge carriers. Taking into account Equations 13.23 and 13.24, Equation 13.45 in equilibrium assumes the following form:

$$\sigma = \sigma_n^0 p(O_2)^{-1/4} + \sigma_p^0 p(O_2)^{1/4} \tag{13.46}$$

where σ_n^0 and σ_p^0 are the parameters related to the conductivities of electrons and holes in standard conditions. However, in strongly reducing conditions, the electrical conductivity exhibits the dependence, which is consistent with Equation 13.25 [36]:

$$\sigma = \sigma^0 p(O_2)^{-1/6} \tag{13.47}$$

The defect disorder models, represented by Equations 13.46 and 13.47, have been verified against well-defined experimental data of electrical conductivity for high-purity TiO_2 single crystal (TiO_2-SC), which are shown in Figure 13.9 [36], and also for high-purity

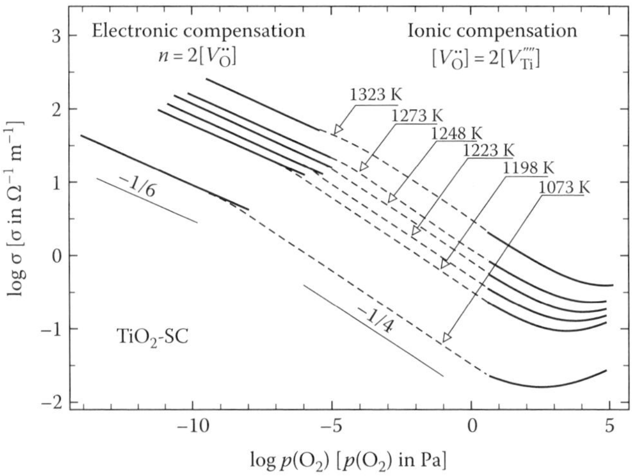

FIGURE 13.9
Effect of oxygen activity on electrical conductivity for undoped TiO_2-SC in the 1073–1323 K range within both extremely reducing conditions and the n–p transition regime. (Reproduced with permission from Nowotny, M.K., Bak, T., and Nowotny, J., Electrical properties and defect chemistry of TiO_2 single crystal. I. Electrical conductivity, *J. Phys. Chem. B*, 110, 16270–16282. Copyright 2006, American Chemical Society.)

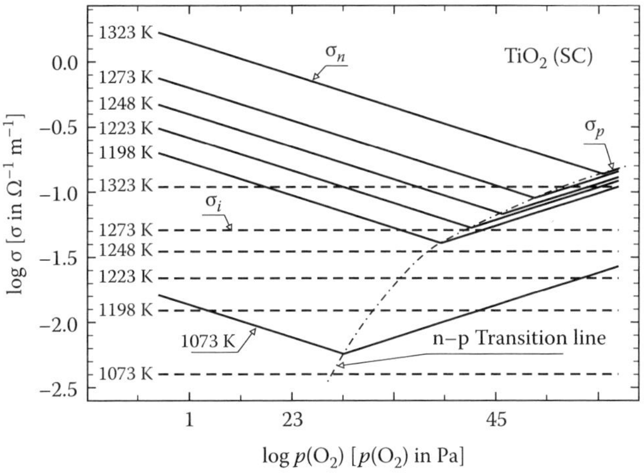

FIGURE 13.10
Effect of oxygen activity on the electrical conductivity components related to different charge carriers for undoped TiO_2-SC in the 1073–1323 K range within the n–p transition. (Reproduced with permission from Nowotny, M.K., Bak, T., and Nowotny, J., Electrical properties and defect chemistry of TiO_2 single crystal. I. Electrical conductivity, *J. Phys. Chem. B*, 110, 16270–16282. Copyright 2006, American Chemical Society.)

polycrystalline TiO_2 (TiO_2-PC) [51]. As seen, the electrical conductivity of TiO_2, determined as a function of oxygen activity in oxidizing and reducing conditions, is consistent with Equations 13.46 and 13.47, respectively. The difference between the electrical conductivity data for TiO_2-SC and TiO_2-PC may be used to assess the local semiconducting properties of grain boundaries [55].

The electrical conductivity at elevated temperatures includes a substantial contribution of ions to the charge transport. Then the ionic conductivity component, σ_i, cannot be ignored, especially in the n–p transition regime. Therefore,

$$\sigma = \sigma_n^0 p(O_2)^{-1/4} + \sigma_p^0 p(O_2)^{1/4} + \sigma_i \tag{13.48}$$

The individual electrical conductivity components outlined in Equation 13.48, which were determined from the experimental data shown in Figure 13.9, are shown in Figure 13.10. As seen, these data allow the determination of the n–p transition point.

The defect disorder models for Nb–TiO_2 and Cr–TiO_2, outlined by Equations 13.30 and 13.33, have been verified using the measurements of the electrical conductivity versus oxygen activity. These data indicate the following:

1. Doping with niobium results in a substantial increase of the electrical conductivity of TiO_2. The effect of niobium on the electrical conductivity depends on oxygen activity and the content of niobium.
2. At low concentrations, chromium doping results in a decrease of electrical conductivity within the n-type regime. However, at higher concentrations, chromium results in a transition of semiconducting properties from n- to p-type.

13.4.2.2 Mobility of Electronic Charge Carriers

Equation 13.45 involves both concentration and mobility terms. The latter term can be determined using the concentration terms derived from defect disorder diagrams and the

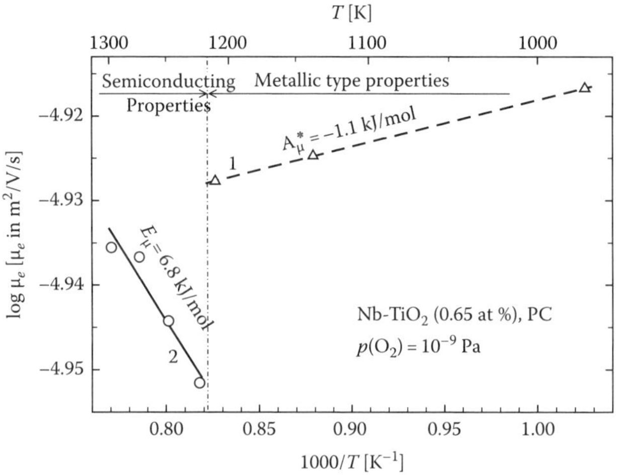

FIGURE 13.11
Arrhenius plot of the mobility terms for electrons in Nb-doped TiO_2 showing metallic-type charge transport in the range 950–1220 K. (Adapted from Sheppard, L.R., Bak, T., and Nowotny, J., Electrical properties of niobium-doped titanium dioxide. 1. Defect disorder, *J. Phys. Chem. B*, 110, 22447–22454. Copyright 2006, American Chemical Society.)

experimental data of electrical conductivity. For high-purity TiO_2-SC, the mobility terms for electrons and holes assume the following respective forms [60]:

$$\mu_n(\mathrm{TiO_2\text{-}SC}) = (6.7 \pm 0.3)\times 10^{-6}\ (\mathrm{m^2/V\,s})$$
$$\mu_p(\mathrm{TiO_2\text{-}SC}) = (1.5 \pm 0.7)\times 10^{-1} \exp\left(-\frac{94 \pm 4\ (\mathrm{kJ/mol})}{RT}\right)(\mathrm{m^2/V\,s}) \tag{13.49}$$

The observed independence of the mobility of temperature for electrons suggests that their transport can be described by the band model and mobility of holes, which is thermally activated, is described by the hopping model. The transport in TiO_2-PC is similar. However, absolute values of the mobility terms are different, indicating that grain boundaries have an effect on the charge transport. Figure 13.11 shows the effect of temperature on the mobility of electrons for Nb-doped TiO_2 in reducing conditions. As seen, a metallic-type charge transport is observed in the range 900–1230 K [40].

13.4.2.3 Bandgap

The bandgap, E_g, may be expressed as the following function of temperature [72,73]:

$$E_g = E_g^0 - \beta T \tag{13.50}$$

where β is the temperature coefficient. According to Becker and Frederikse [73], the component E_g^0 may be determined from the temperature dependence of the minimum value of the electrical conductivity versus $p(O_2)$, σ_{min}, which corresponds to the n–p transition:

$$\sigma_{min} = 2e\sqrt{\mu_n \mu_p N_n N_p}\exp\left(\frac{\beta}{2k}\right)\exp\left(-\frac{E_g^0}{2kT}\right) \tag{13.51}$$

where

N_n and N_p are the densities of states for electrons and electron holes, respectively
k is the Boltzmann constant

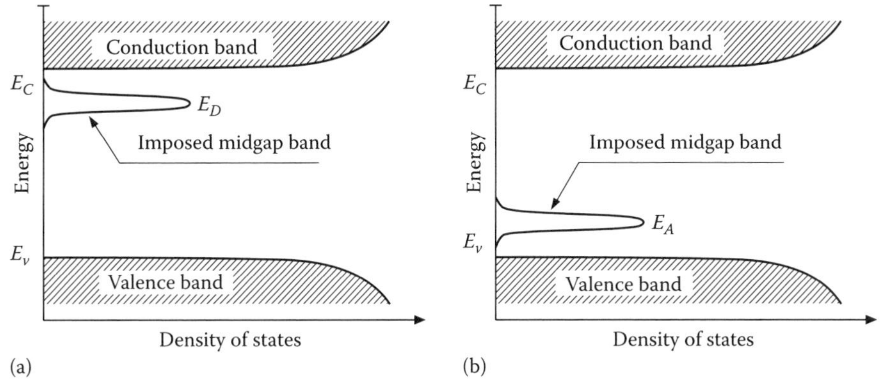

FIGURE 13.12
Schematic representation of midgap bands imposed by donor (a) and acceptor (b) centers, respectively. (Adapted from *Int. J. Hydrogen Energy*, 30, Nowotny, J., Sorrell, C.C., Sheppard, L.R., and Bak, T., Solar hydrogen: Environmentally safe fuel for the future, 521–544, Copyright 2005, International Association for Hydrogen Energy published by Elsevier Ltd.)

Both β and E_g may also be determined from the Jonker analysis that is based on the isothermal plot of S versus log σ [74]. The bandgap determined from the plot of log σ_{min} as a function of $1/T$ for undoped TiO_2-SC at elevated temperatures results in the following value:

$$E_g = 3.16 \pm 0.02 \text{ eV} \tag{13.52}$$

Doping with aliovalent ions may lead to the reduction of the effective bandgap required for ionization. This is the case when doping results in imposition of midgap levels as it is schematically represented in Figure 13.12 [75].

13.4.2.4 Effect of Cooling

The processing procedures of nonstoichiometric oxides, which aim at imposition of controlled oxygen activity, take place at elevated temperatures at which the gas/solid kinetics is relatively fast. On the other hand, the performance of TiO_2 as photoelectrode and photocatalyst takes place at room temperature. The changes of TiO_2 properties during cooling may be considered in terms of the following effects [23]:

1. The changes in the concentration of ionic defects, which are electrically charged. These changes are related to the formation term for defects, ΔH_f.
2. The changes of the mobility term, ΔH_m.
3. The changes of the ionization degree of ionic defects.

The effect of cooling on the electrical conductivity of oxide semiconductors is schematically represented in Figure 13.13. As seen, the activation energy of the electrical conductivity at higher temperatures (above the T_c point) involves both ΔH_m and ΔH_f terms. Then

$$\sigma = \text{const} \cdot \exp\left(-\frac{(2/m_\sigma)\Delta H_f - \Delta H_m}{RT} \right) \tag{13.53}$$

where m_σ is the slope of the following dependence:

$$\frac{1}{m_\sigma} = \frac{\partial \log \sigma}{\partial \log p(O_2)} \tag{13.54}$$

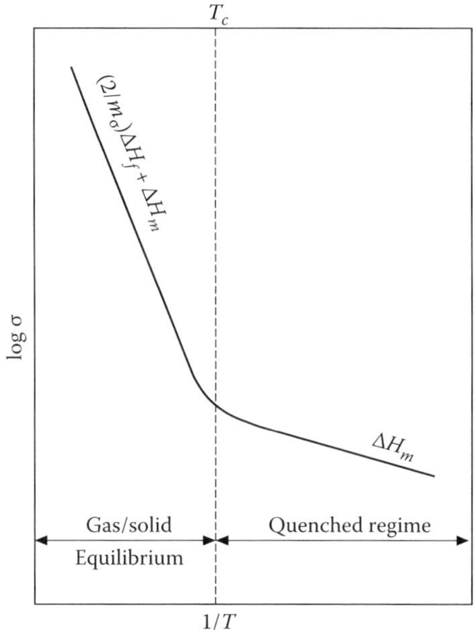

FIGURE 13.13
Temperature dependence of the electrical conductivity for a nonstoichiometric compound showing the regime corresponding to thermodynamic equilibrium, above T_c (the slope can be related to both the formation and mobility terms), and the quenched regime, below T_c (the slope is determined by the mobility term).

Below a certain critical temperature, T_c, the ionic transport in the lattice is quenched and the activation energy is determined by the ΔH_m term:

$$\sigma = \text{const} \cdot \exp\left(-\frac{\Delta H_m}{RT}\right) \tag{13.55}$$

13.4.2.5 Measurements

The electrical conductivity is relatively easy to measure in laboratory conditions. Therefore, this property is frequently reported in studies of defect disorder of MOs.

There is a wide range of approaches for the determination of electrical conductivity at elevated temperatures. The basic principle is shown in Figure 13.14. The external (current)

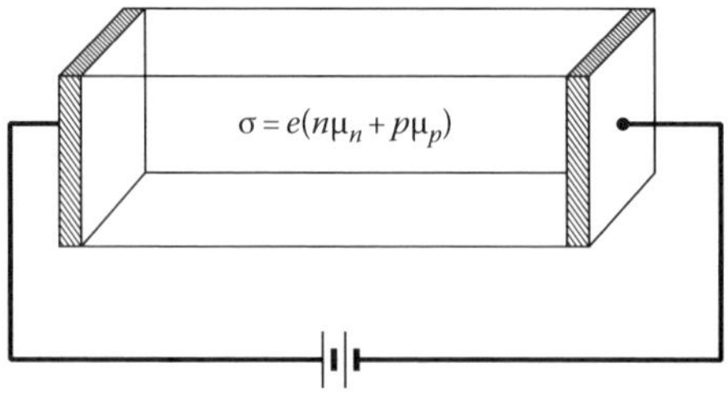

FIGURE 13.14
The concept of the electrical conductivity measurements by the two-probe method. (Reproduced with permission from Bak, T., *Oxide Semicond. Res. Rep.*, Copyright 2010.)

probes, consisting commonly of platinum plates, are attached to both sides of rectangular-shaped specimen. A spring mechanism, located outside the high-temperature zone, is used to maintain effective galvanic contact between the electrodes and the specimen. In the case of four probe method [36], the voltage electrodes formed of two platinum wires are wrapped around the specimen and welded to the platinum wires connected to a voltmeter. The sample holder, placed in an alumina tube, is connected to a gas-flow system that aims to impose the gas phases of controlled oxygen activity in the reaction chamber [76].

The required oxygen activity, $p(O_2)$, in the reaction chamber can be imposed using gas mixtures of appropriate compositions. The $p(O_2)$ can be imposed using mixtures of hydrogen and water vapor or argon/oxygen mixtures to achieve lower and higher oxygen activities, respectively. The oxygen activities can be determined using a zirconia-based electrochemical oxygen sensor. Measurements of electrical conductivity, σ, can be used for monitoring semiconducting properties during both oxidation and reduction. Details of the experimental procedures used to determine the electrical conductivity have been reported elsewhere [36,76].

Figure 13.15 shows a standard sheet for monitoring the equilibration kinetics during oxidation of $CaTiO_3$ at 1223 K. Figure 13.15a shows that imposition of a new gas phase results in a very rapid increase in the $p(O_2)$ to the level of ~95% of its final value within seconds, followed by adoption of the final equilibrium value within 10 min. Figure 13.15b shows that the temperature during the experiment remains constant within ±0.3 K. The observed fluctuations in temperature have a negligible effect on the measured electrical resistance data. Figure 13.15c shows that constant resistance is reached within ~1 h and then remains constant for the following 20 h. These kinetics data can be used to determine the chemical diffusion coefficient [38,39].

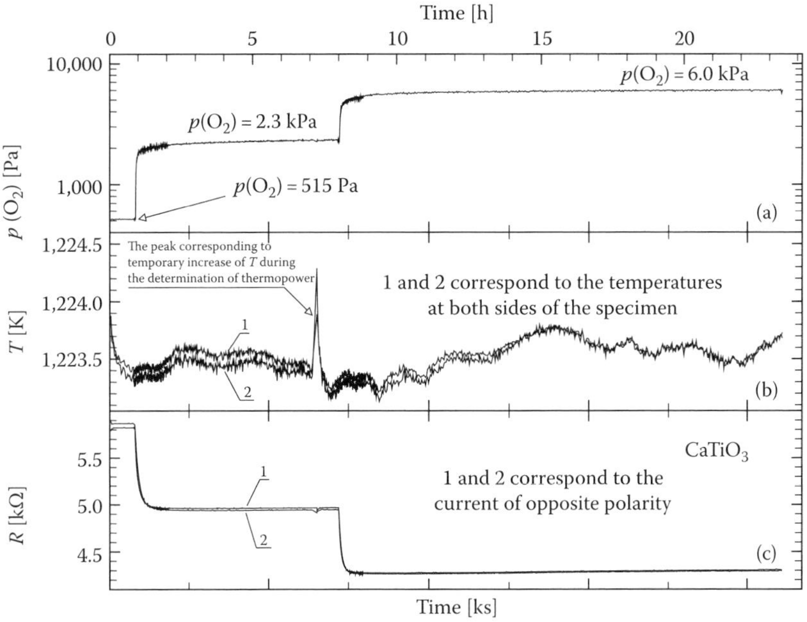

FIGURE 13.15
Typical record for the measurements of electrical resistivity for $CaTiO_3$ during two cycles of isothermal oxidation at 1223 K, showing the time dependence of (a) oxygen activity, (b) temperature at both sides of the specimen, and (c) resistance corresponding to the dc current passing in both directions. (Reproduced with permission from Springer Science+Business Media: *J. Mater. Sci.: Mater. Electron.*, Charge transport in $CaTiO_3$: Electrical conductivity, 15, 2004, 635, Bak, T., Nowotny, J., Sorrell, C.C., Zhou, M.F., and Vance, E.R.)

13.4.3 Thermoelectric Power

The thermoelectric power, also termed the *Seebeck coefficient* or *thermopower*, is an electrical property that may be used to characterize semiconducting properties at elevated temperatures. Specifically, the thermoelectric power may be used to assess the concentrations of electronic charge carriers. Since the electrical conductivity is the product of the concentration and mobility terms, the combination of the electrical conductivity and thermoelectric power data can be used to determine both terms [60].

The principles of the determination of thermoelectric power are given elsewhere [76]. The imposition of a temperature gradient (ΔT) across a specimen results in the generation of a potential difference ($\Delta\Psi$), which is termed the Seebeck voltage or thermovoltage. Knowledge of both $\Delta\Psi$ and ΔT is required to determine the thermoelectric power (S), which can be expressed as follows:

$$S = \lim_{\Delta T \to 0} \frac{\Delta\Psi}{\Delta T} = \frac{d\Psi}{dT} \tag{13.56}$$

For non-degenerated semiconductors thermoelectric power can be related to the concentration of electronic charge carriers according to the following expressions for n- and p-type regimes, respectively [78]:

$$S_n = -\frac{k}{e}\left(\ln\frac{N_n}{n} + A_n\right) \tag{13.57}$$

$$S_p = \frac{k}{e}\left(\ln\frac{N_p}{p} + A_p\right) \tag{13.58}$$

where

- e is the elementary charge
- k is the Boltzmann constant
- N_n and N_p denote the densities of states for electrons and electron holes, respectively
- n and p denote their respective concentrations
- A_n and A_p are the kinetic constants associated with the scattering of electrons and electron holes, respectively

The common way to verify the defect disorder models of MOs is based on the dependence of S as a function of $p(O_2)$:

$$\frac{1}{m_S} = \frac{k}{e}\frac{\partial S}{\partial \log p(O_2)} \tag{13.59}$$

where

- m_S is a parameter related to the specific defect disorder
- the subscript S refers to the case when the parameter is obtained using thermoelectric power data

The equivalent equation allows the assessment of the effect of oxygen activity on the electrical conductivity, which is expressed by Equation 13.54. Both sets of data may be used for the confirmation of the effect of oxygen activity on the concentration of electronic charge carriers (when the mobility term remains independent of oxygen nonstoichiometry).

It is well known that in the case of symmetrical semiconductors thermoelectric power achieves a critical value at the n–p transition, namely, $S = 0$. Figure 13.16 shows a schematic representation of the effect of the $p(O_2)$ on both the electrical conductivity (σ) and the thermoelectric power (S) for an amphoteric oxide semiconductor, which exhibits both n- and p-type regimes. It can be seen that the slope of the logarithm of electrical conductivity versus log $p(O_2)$ dependence in the n- and p-type regimes adopts negative and positive values, respectively, and that the electrical conductivity at the n–p transition point reaches a minimum. The n- to p-type transition point occurs at

$$\sigma_n = \sigma_p \tag{13.60}$$

It also can be seen in Figure 13.16 that the thermoelectric power versus log $p(O_2)$ dependencies in the n- and p-type regimes are linear. In these two regimes, the formalism is to present the slopes ($1/m_S$) as positive values. The parameter $1/m_S$ is well defined when thermoelectric power corresponds to pure n- or p-type regime. Then the thermoelectric power data reflect the effects of the majority charge carriers (in this case the effects of the minority charge carriers are negligible). However, in the n–p transition regime, where two charge carriers are present in comparable concentrations, the meaning of $1/m_S$ is more complex.

The principle of the measurement of thermoelectric power according to Equation 13.56 is shown in Figure 13.17. Figure 13.18 represents the circuit for simultaneous determination

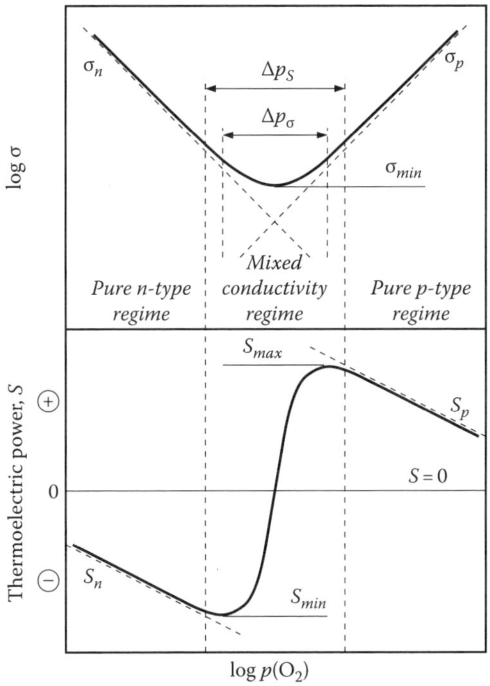

FIGURE 13.16
Schematic representation of the effect of oxygen activity on both electrical conductivity (σ) and thermoelectric power (S) within p-type, n–p transition and p-type regimes (the subscripts n and p are related to n-type and p-type regimes, respectively). (Reproduced with permission from Nowotny, M.K., Bak, T., and Nowotny, J., Electrical properties and defect chemistry of TiO_2 single crystal. II. Thermoelectric power, *J. Phys. Chem. B*, 110, 16283–16291. Copyright 2006, American Chemical Society.)

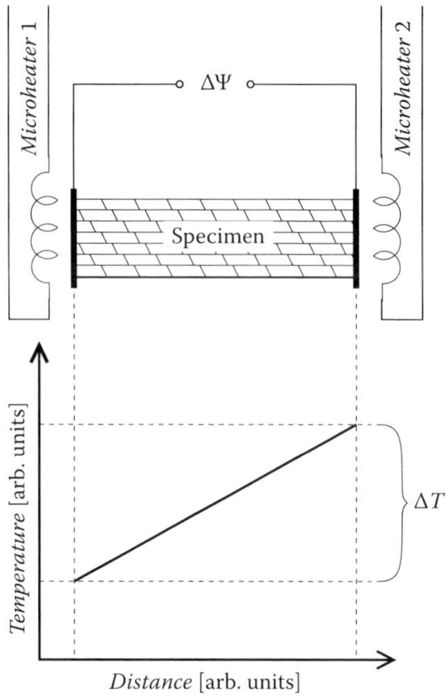

FIGURE 13.17
The principle of measurements of thermovoltage along the temperature gradient imposed by microheaters on both sides of the specimen. (Reproduced with permission from Nowotny, M.K., Bak, T., and Nowotny, J., Electrical properties and defect chemistry of TiO_2 single crystal. II. Thermoelectric power, *J. Phys. Chem. B*, 110, 16283–16291. Copyright 2006, American Chemical Society.)

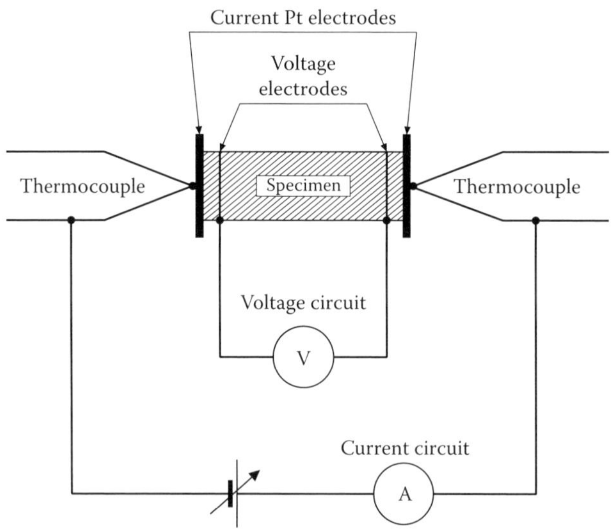

FIGURE 13.18
The circuit for simultaneous determination of electrical conductivity (four probes) and thermoelectric power. (Reproduced with permission from Nowotny, M.K., Bak, T., and Nowotny, J., Electrical properties and defect chemistry of TiO_2 single crystal. I. Electrical conductivity, *J. Phys. Chem. B*, 110, 16270–16282. Copyright 2006, American Chemical Society.)

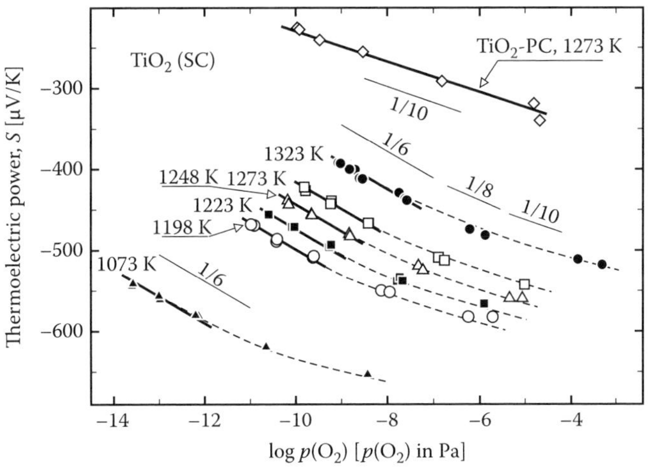

FIGURE 13.19
Thermoelectric power versus oxygen activity for TiO_2-SC in the 1073–1323 K range, along with the data of TiO_2-PC at 1273 K, showing the slope that may be used for the verification of defect disorder models. (Reproduced with permission from Nowotny, M.K., Bak, T., and Nowotny, J., Electrical properties and defect chemistry of TiO_2 single crystal. II. Thermoelectric power, *J. Phys. Chem. B*, 110, 16283–16291. Copyright 2006, American Chemical Society.)

of electrical conductivity and thermoelectric power. The temperature gradient required for the determination of thermoelectric power is imposed by microheaters located close to external (current) probes. The internal voltage electrodes wrapped around the specimen and welded to the platinum wires served for the measurements of the electrical conductivity. The sample holder is placed in tube furnace that is connected to a gas-flow system that imposes the gas phase of controlled oxygen activity.

The experimental data of thermoelectric power for TiO_2-SC in reducing conditions, over the temperature range 1073–1323 K, are represented in Figure 13.19. As seen, the data exhibit a continuous change of the S versus log $p(O_2)$ slope from 1/6 at extremely reduced conditions to lower $1/m_S$ values. The gradually decreasing slope indicates increasing influence of the minority charge carriers. This figure also includes the thermoelectric power data for high-purity TiO_2-PC determined at 1273 K using the same equipment and following the same experimental procedure.

As seen in Figure 13.19, the slope of the thermoelectric power versus log $p(O_2)$ dependence for TiO_2-PC at 1273 K is ~1/10 in the entire low $p(O_2)$ regime (similar slope was determined at other temperatures as well). Since in both cases of TiO_2-PC and TiO_2-SC the specimens were of high purity, the only difference between the two is the presence of grain boundaries for TiO_2-PC. Therefore, the difference between the S versus log $p(O_2)$ slope for these two specimens is reflective of the local properties of grain boundaries of the polycrystalline material.

13.4.4 Work Function

WF is the electrical property that is selectively sensitive to the outermost surface layer and is defined as the work required for removing an electron from its Fermi level (at the surface) to the energy level outside the surface [76,79]. Consequently, the WF measurements may be used for in situ monitoring of the charge transfers during chemical reactions at the gas/solid interface, such as chemisorption of gases.

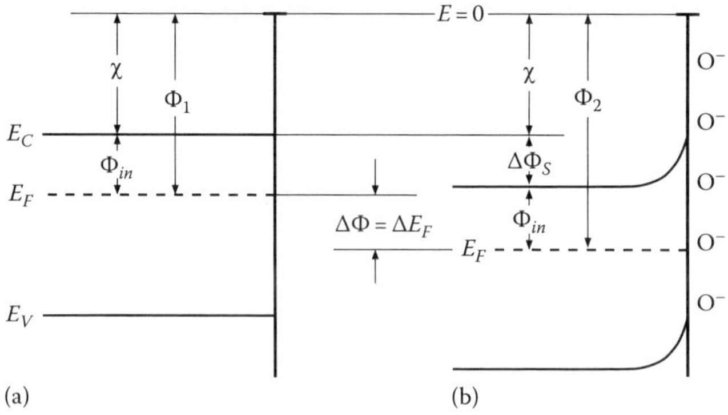

FIGURE 13.20
Band model of n-type semiconductor without surface charge (a) and at the presence of oxygen chemisorption-induced surface charge (b).

The WF of oxide semiconductors involves several components, including the internal WF, Φ_{in}; the WF component related to band bending, Φ_s; and the external WF, χ [80] (Figure 13.20):

$$\Phi = \Phi_{in} + \Phi_s + \chi \tag{13.61}$$

Oxidation and reduction of oxide semiconductors within a single-phase regime result in WF changes that are determined by the components Φ_{in} and Φ_s, while the external WF, χ, remains constant. Therefore,

$$\Delta\Phi = \Delta\Phi_{in} + \Delta\Phi_s \tag{13.62}$$

Equation 13.62 represents the case when oxidation of an MO leads to both oxygen chemisorption and oxygen incorporation.

Oxidation of oxides at elevated temperatures results in a change of the component $\Delta\Phi_{in}$ while the component $\Delta\Phi_s$ is negligibly small and, therefore, may be ignored. Conversely, oxidation of oxides at room temperature results mainly in the WF changes due to the component $\Delta\Phi_s$, while the component $\Delta\Phi_{in}$ may be ignored since the bulk phase is quenched. Then the WF changes may be considered in terms of chemisorptions equilibria.

The WF changes may also be used in studies of segregation-induced effects, where segregation refers to adsorption of species derived from the solid phase. Therefore, it is important to recognize that WF changes at elevated temperatures may also include a component related to segregation. However, the changes of WF at room temperature are determined mainly by the component $\Delta\Phi_s$ because the lattice transport is quenched.

The external WF, χ, is determined by the surface structure [76,79]. Consequently, during chemisorption, leading to changes of $\Delta\Phi_s$, the component χ remains constant. Also during oxidation, leading to oxygen incorporation without structural changes, the WF component χ remains constant. In certain cases, however, a change in the concentration of defects may lead to structural changes, which are induced by strong interactions between the defects [55].

The effect of oxygen on WF of titanium dioxide has been reported by Figurovskaya et al. [81] and Bourasseau et al. [82–87] at room temperature and by Odier et al. [88,89] at elevated temperatures. These WF data allow assessing the reactivity of oxygen with the surface of TiO_2.

The effect of oxygen activity on WF of oxide semiconductors may be expressed by the following equation [79]:

$$\frac{1}{m_\Phi} = \frac{1}{kT}\frac{\partial \Delta\Phi}{\partial \ln p(O_2)} \tag{13.63}$$

where m_Φ is the exponent of the $p(O_2)$ dependence that is related to the WF changes, which are determined by changes of the Fermi level of the outermost surface layer:

$$\Delta\Phi = -\Delta E_F \tag{13.64}$$

The relationship between the surface coverage by oxygen-chemisorbed species and the WF changes may be expressed by the following expression:

$$\frac{1}{m_\Phi} = \frac{1}{z}\left[1 - \frac{\alpha}{x_{\alpha,z}}\frac{\partial x_{\alpha,z}}{\partial \ln p(O_2)}\right] \tag{13.65}$$

where the component x_{α}, z, related to the activity of the specific chemisorbed species, is given by

$$x_{\alpha,z} = \frac{\Theta_{\alpha,z}}{1 - \Theta_1 - \Theta_2} \tag{13.66}$$

where
- α is the number of chemisorbed species formed from a single oxygen molecule
- z is the number of electrons involved in chemisorption
- Θ is the degree of surface coverage with chemisorbed oxygen species

WF may be determined by the dynamic condenser method, which has been developed by Kelvin [90]. The high-temperature Kelvin probe (HTKP), which allows WF measurements at elevated temperatures in controlled gas-phase composition, may be used for in situ monitoring of surface reactions at the gas/solid interface and the related charge transfer [76,91].

The main part of the Kelvin probe is the vibrating capacitor, which is formed of a lower electrode (involving the studied specimen) and an upper reference platinum electrode (Figure 13.21). The HTKP allows determining the WF changes with an accuracy of 0.5 meV, if the external noise level is minimized.

The WF changes can be determined by the measurements of the contact potential difference (CPD), which is equal to the difference between the WF values of the studied specimen and of the reference electrode, Φ_R:

$$\text{CPD} = \frac{1}{e}(\Phi - \Phi_R) \tag{13.67}$$

The formation of the CPD according to Wagner is shown in Figure 13.22 [93]. Platinum can be applied as the reference electrode. It was shown that oxidation of platinum leads to the formation of a PtO_2 layer on surface [76]. Therefore, the WF changes of platinum should be considered in terms of the electrical properties of the PtO_2 surface layer, which can be expressed by Equation 13.63.

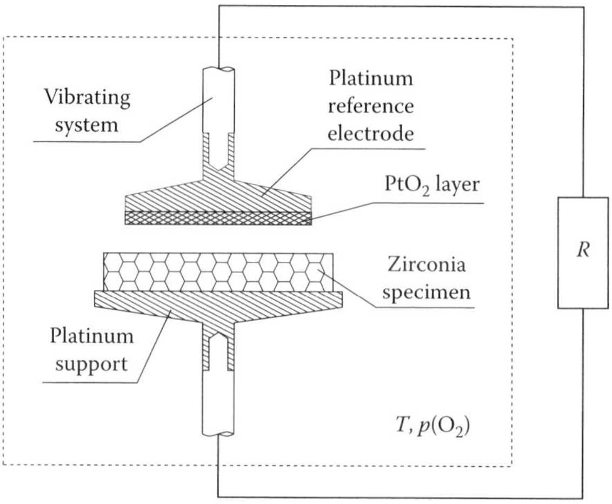

FIGURE 13.21
The entrance circuit for WF measurements at elevated temperatures and the gas phase of controlled oxygen activity, the sample holder, and the vibrating Pt electrode. (Reprinted from Nowotny, J. et al., *Adv. App. Ceram.*, 104, 188, 2005. Copyright 2005, Maney Publishing.)

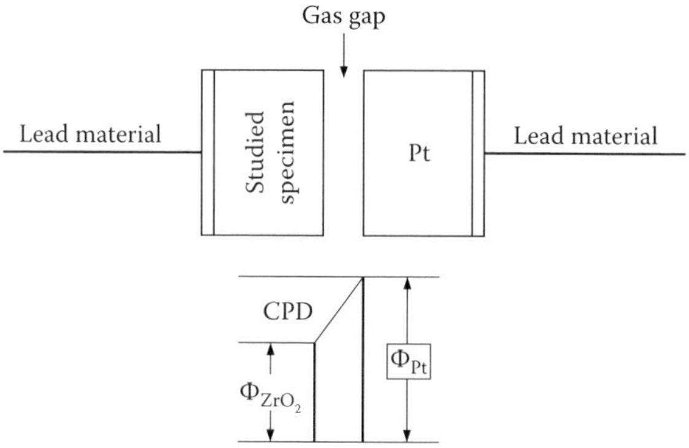

FIGURE 13.22
The principle of the CPD measurements according to Carl Wagner. (Reproduced with permission from Nowotny, J. et al., *Adv. Appl. Ceram.*, 104, 181, 2005. Copyright 2005, Maney Publishing.)

According to Equation 13.67, the WF changes of the studied specimen may be determined from the CPD data only when the WF changes of the reference electrode are known:

$$\Delta\Phi = e\Delta\text{CPD} + \Delta\Phi_R \tag{13.68}$$

Figure 13.23 represents the WF data during oxidation for high-purity TiO_2, which was equilibrated in different conditions, including annealing at 1173 K at $p(O_2) = 10$ Pa (Figure 13.23a) and reduction at 1173 K at $p(O_2) = 10^{-10}$ Pa (Figure 13.23b) [79]. In both cases, the subsequent oxidation experiments were performed at $p(O_2) = 75$ kPa after cooling the specimen to room temperature. As seen, oxidation of the specimen initially reduced in moderate conditions (Figure 13.23a) leads to WF increase within different stages, including oxygen chemisorption (rapid WF changes within 5 h by 0.38 eV) and subsequent slow changes related to oxygen incorporation over the period of 130 h (by ~0.22 eV). These data represent the reactivity of TiO_2 reduced in moderate conditions with oxygen. As seen in Figure 13.23b, oxidation of strongly reduced TiO_2 results in a rapid WF decrease by 0.78 eV. The sign of the

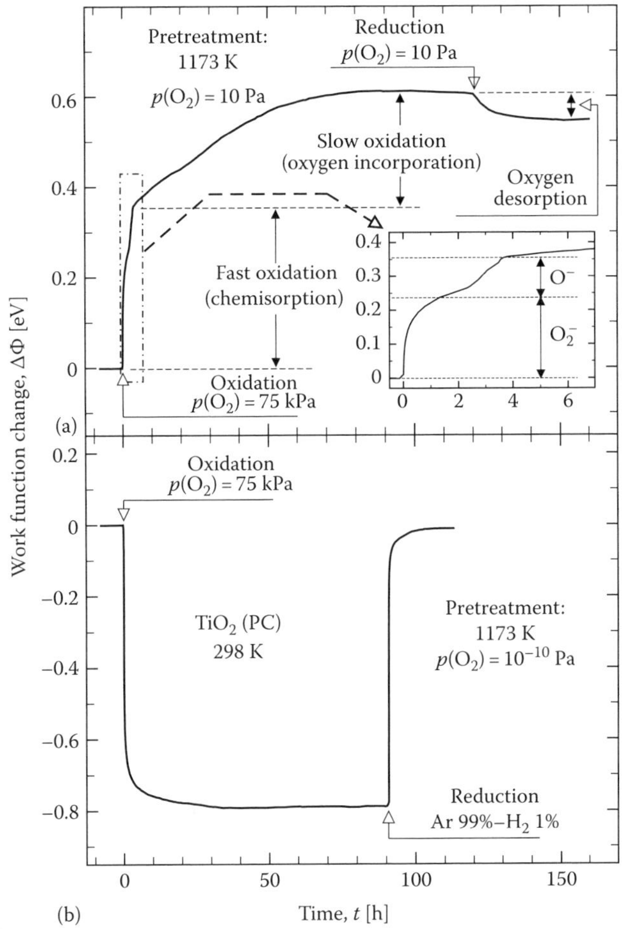

FIGURE 13.23
WF changes of undoped TiO_2 during oxidation and subsequent reduction at 298 K for two specimens: (a) oxidized at 1173 K and $p(O_2) = 10$ Pa then cooled to 298 K and (b) reduced at 1173 K and $p(O_2) = 10^{-10}$ Pa. (Reprinted with permission from Nowotny, J., Bak, T., Sheppard, L.R., and Nowotny, M.K., Reactivity of titanium dioxide with oxygen at room temperature and the related charge transfer, *J. Am. Chem. Soc.*, 130, 9984–9993. Copyright 2008, American Chemical Society.)

WF changes during oxidation of this specimen, which is opposite to that for the specimen reduced in moderate conditions, indicates that oxidation in this case results in the formation or removal of a low-dimensional surface structure of outstanding properties, which is formed when the specimen was exposed to the gas phase involving hydrogen.

13.4.5 Surface Photovoltage Spectroscopy

The concept of surface photovoltage spectroscopy (SPS) is based on the measurements of WF of a photosensitive semiconductor (using the Kelvin probe [76,80]) versus incident photon energy. In this method, the reference electrode is made of a mesh (platinum or gold) that allows light access to the surface of the studied specimen. The SPS provides information on the effect of light on surface semiconducting properties [93,95]. Figure 13.24 represents the plot of the surface photovoltage signal versus photon energy in the range 0.4–4.5 eV for TiO_2 specimens annealed at 1273 K in oxidizing conditions, $p(O_2) = 21$ kPa

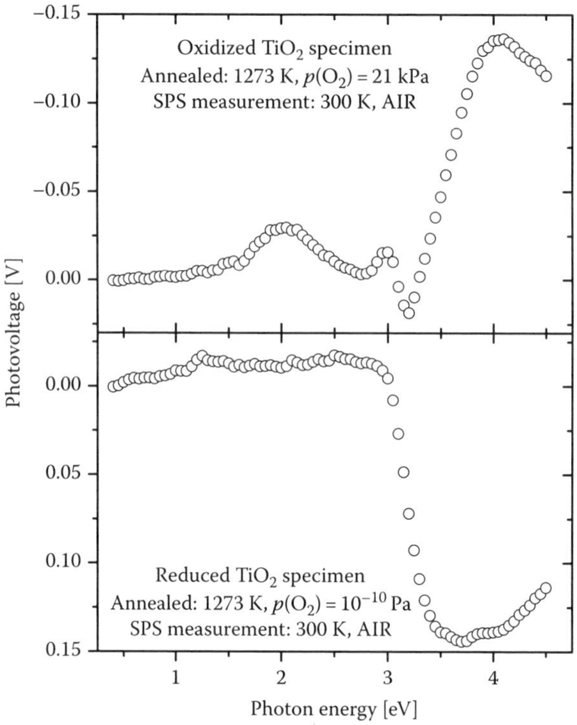

FIGURE 13.24
Surface photovoltage spectra at 300 K in air for TiO_2 both oxidized and reduced at 1273 K. (Reprinted with permission from Tributsch, H. et al., *Energy Mater.*, 3, 158. Copyright 2009, Institute of Materials, Minerals & Mining and Maney Publishing.)

(upper part), and in reducing conditions at 1273 K, $p(O_2) = 10^{-10}$ Pa (lower part) [95]. As seen, at the bandgap illumination ($h\nu > 3$ eV), the photovoltage signals are positive and negative for oxidized and reduced TiO_2, respectively, indicating that the photoinduced electrons are transferred toward the surface and the bulk, respectively. Therefore, the SPS signal informs of the effect of light on the charge transfer and the related photoreactivity.

13.4.6 Effect of Impurities

The effect of dopants on properties, which usually concerns high concentrations of donors or acceptors, has been widely reported. It has been generally assumed that the effect of impurities on properties can be ignored when their concentration is low. The recent studies indicate that this is not the case [57]. Defect disorder diagrams may be used to predict the effect of aliovalent ions (donors and acceptors) in titanium dioxide on the concentration of electronic charge carriers (electrons and electron holes) and the related electrical properties. Such effect of both donors and acceptors, considered in terms of the effective concentration of acceptors [57], is shown in Figure 13.25. These data indicate the following:

- The effect of aliovalent ions on properties may be ignored below certain critical values. The effect of temperature and oxygen activity on the critical value A_c, below which the concentration of electronic charge carriers is independent of A, is shown in Figure 13.26. As seen, the effect of aliovalent ions on the electrical conductivity above this value becomes substantial and cannot be ignored.

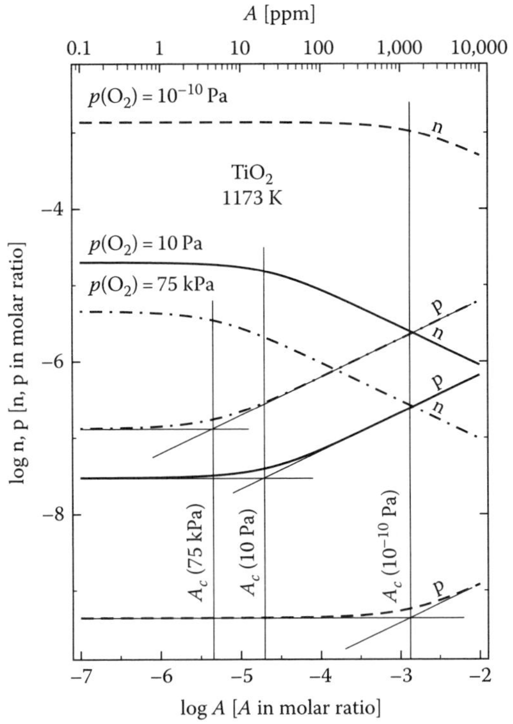

FIGURE 13.25
Effect of acceptors, A, at the ppm level on the concentration of electronic charge carriers at 1173 K for both reduced and oxidized TiO_2. (Reprinted with permission from Nowotny, J., Bak, T., Nowotny, M.K., and Sheppard, L.R., Defect chemistry and electrical properties of titanium dioxide. 2. Effect of aliovalent ions, *J. Phys. Chem. C*, 112, 602–610. Copyright 2008, American Chemical Society.)

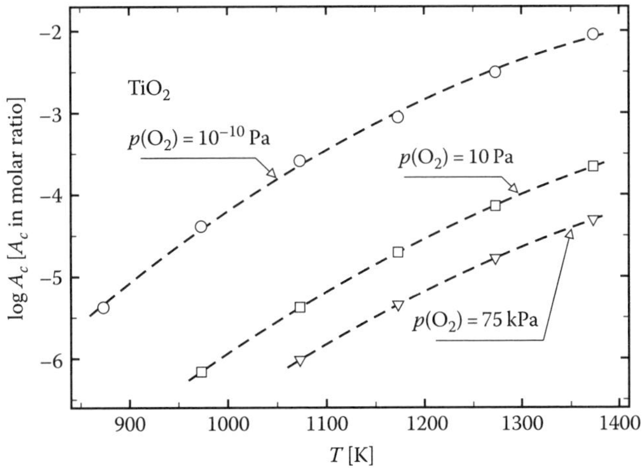

FIGURE 13.26
Isobaric plots of the critical values of the effective concentration of acceptors, above which the effect on electrical conductivity becomes substantial, as a function of temperature for TiO_2. (Reprinted with permission from Nowotny, J., Bak, T., Nowotny, M.K., and Sheppard, L.R., Defect chemistry and electrical properties of titanium dioxide. 2. Effect of aliovalent ions, *J. Phys. Chem. C*, 112, 602–610. Copyright 2008, American Chemical Society.)

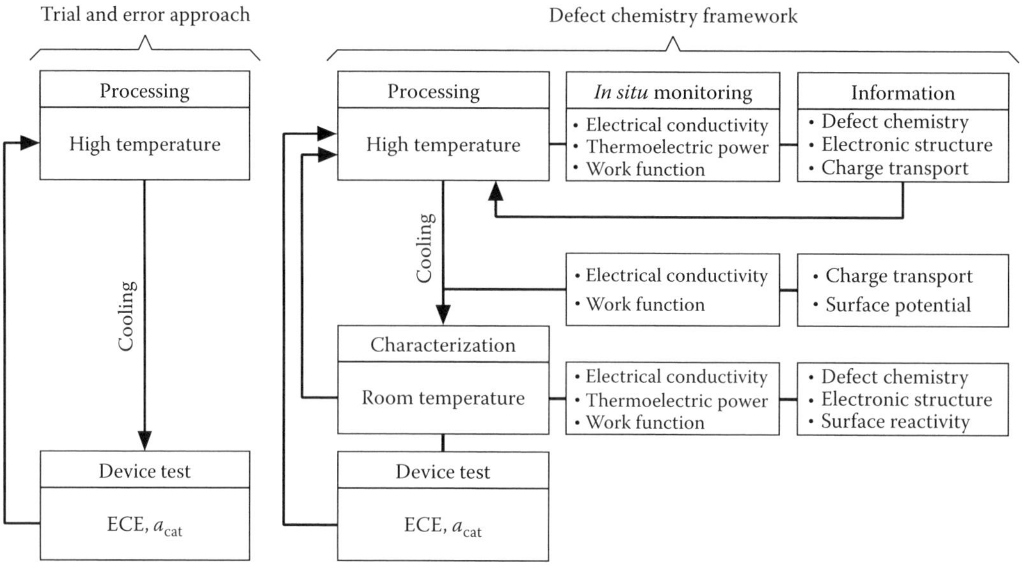

FIGURE 13.27
The *trial and error* versus *in situ monitoring* procedures in processing of oxide semiconductors. (From Nowotny, J., Titanium dioxide-based semiconductors for solar-driven environmentally friendly applications: Impact of point defects on performance, *Energy Environ. Sci.*, 1, 565–572. Reproduced by permission of The Royal Society of Chemistry.)

- The effect of aliovalent ions, added intentionally (dopants) or unintentionally (impurities), on the concentration of electronic charge carriers and related electrical properties depends on temperature and oxygen activity.
- The observed effect of aliovalent ions increases with the decrease of temperature.
- Aliovalent ions may have a substantial effect on properties already at the level of several parts per million and, therefore, cannot be ignored.

13.4.7 Summary

The electrical techniques are sensitive to defect disorder of MOs. Simultaneous measurements of electrical conductivity and thermoelectric power may be used for the determination of several semiconducting quantities, such as the mobility and concentration terms. The measurements of both electrical conductivity and WF provide information about surface versus bulk semiconducting properties. The light-induced WF data may be used for the determination of the effect of light on reactivity. The measurements of the electrical properties may also be used for in situ monitoring of processing at elevated temperatures. The experimental approach for in situ monitoring is shown in Figure 13.27.

13.5 Collective and Local Factors in Reactivity and Photoreactivity

13.5.1 TiO_2/H_2O Interface

The prerequisite of the reactions between TiO_2 and water (and its solutes) is adsorption of the reacting species on the TiO_2 surface and the subsequent charge transfer. The reactivity is determined by the ability of TiO_2 to donate or accept electrons and the chemical affinity or ionization potential of the adsorbed species. There has been a general perception that the reactivity of TiO_2 (with water and organic solutes in water) is closely related to collective properties of TiO_2. So far, little is known about the effect of local surface properties, which are closely related to the presence of point defects, on reactivity.

13.5.2 Collective Factor

The collective factor is related to collective properties in a macroscale, which are reflective of the entire bulk phase, or its surface layer, as a continuum. An important collective factor controlling the reactivity of oxide semiconductors and their ability to charge transfer is the chemical potential of electrons, which is related to the Fermi level. There is a perception that the charge transfer at the TiO_2/liquid interface is determined by collective properties, such as the chemical potential of electrons and the flat-band potential (FBP). According to the electronic theory of chemisorption and catalysis [97], the charge transfer between the surface of a semiconductor and the adsorbed molecule is determined by the Fermi level at the surface and the ionization potential of the molecule.

The ability of the semiconductor to donate or accept electrons, and the related chemical potential of electrons, may be modified by the incorporation of either donors or acceptors. The schematic representation of the effect of donor versus acceptor doping on the Fermi level is shown in Figure 13.28. The effect of acceptors and donors on the concentration of electrons in TiO_2 at 1073 K is shown in Figure 13.29 [57]. The doping procedure may be used for shifting up or down the chemical potential of electrons in TiO_2, compared to the energy levels of the electrochemical couples H^+/H_2 and O_2/H_2O, in order to allow spontaneous charge transfer. Therefore, the collective properties have an essential effect on reactivity of TiO_2.

13.5.3 Local Factor

While the collective factor is the driving force of the charge transfer within the PEC, it has been shown that photoreactivity at the TiO_2 surface, and the related charge transfer, must be considered in terms of both the collective factor and a local factor [35].

The local factor is related to local interactions in an atomic scale between the adsorbed species and specific surface-active sites, which are formed by individual lattice species (ions and defects) at the surface. These defects, which are directly involved in the reactivity

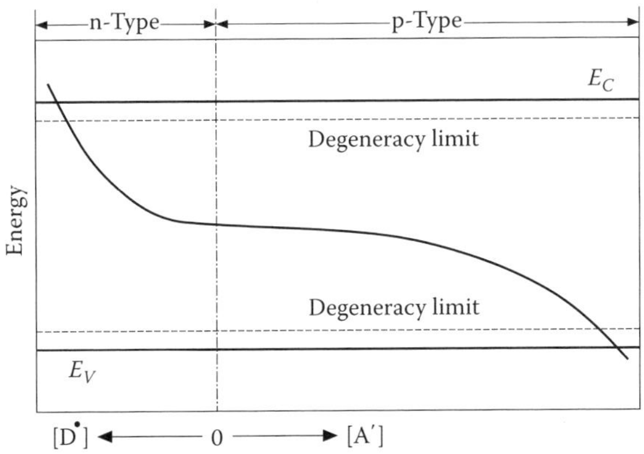

FIGURE 13.28
Schematic representation of the effect of donors and acceptors on the Fermi level for a semiconductor. (Reproduced with permission from Bak, T., *Oxide Semicond. Res. Rep.*, Copyright 2010.)

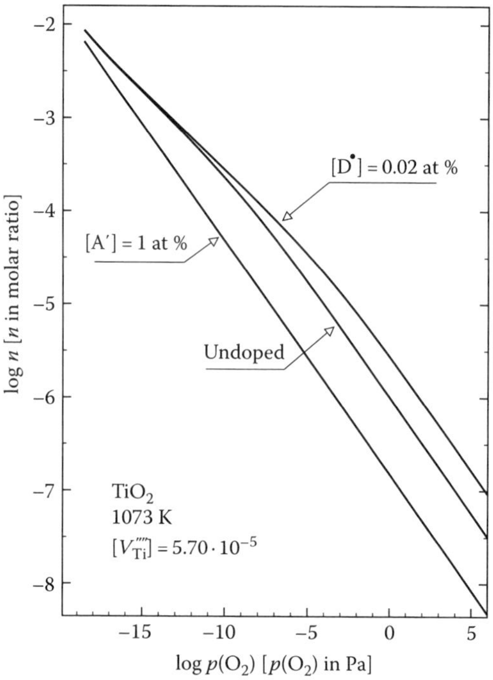

FIGURE 13.29
Effect of oxygen activity on the concentration of electrons for undoped, donor-doped, and acceptor-doped TiO_2 at 1073 K. (Reproduced with permission from Bak, T., *Oxide Semicond. Res. Rep.*, Copyright 2010.)

between H_2O and TiO_2, play an essential role in the charge transfer, such as the transfer of electron holes from the TiO_2 surface to the adsorbed H_2O molecules. Different sites have different ability for the charge transfer. Recent studies have shown that titanium vacancies at the outermost surface layer are the favorable active surface sites that allow effective charge transfer between the H_2O molecule and the TiO_2 surface. The proposed reactivity model, involving the reaction between the H_2O molecule and the TiO_2 surface site, leading to the formation of an active complex, is shown schematically in Figure 13.30 [35].

The photoreactivity between the TiO_2 surface and water may be considered in terms of the following reactions:

- Adsorption of water molecule on the active site, such as titanium vacancy (V_{Ti})

$$2H_2O + V_{Ti} \rightarrow (2H_2O - V_{Ti}) \tag{13.69}$$

- The charge transfer between the TiO_2 lattice and adsorbed water species resulting in the formation of a photocatalytically active complex

$$(2H_2O - V_{Ti}) \rightarrow (2H_2O^{2+} - V_{Ti}^{4-})^* \tag{13.70}$$

- Decomposition of the activated complex into oxygen, proton, and titanium vacancies

$$(2H_2O^{2+} - V_{Ti}^{4-})^* \rightarrow O_2 + 4H^+ + V_{Ti}^{4-} \tag{13.71}$$

where $(2H_2O^{2+} - V_{Ti}^{4-})^*$ is a metastable surface-active complex.

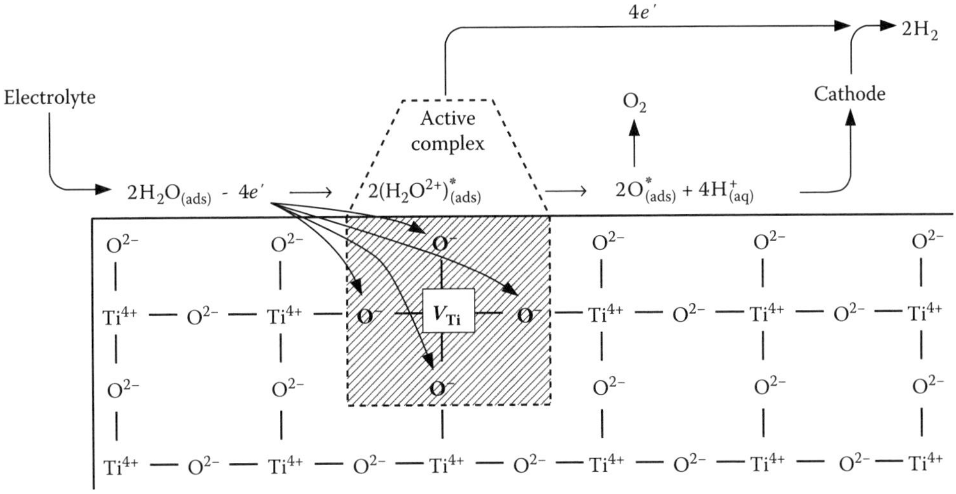

FIGURE 13.30
Theoretical model of water splitting at the TiO_2 surface and the related multielectron charge transfer. (Reprinted with permission from Nowotny, J., Bak, T., Nowotny, M.K., and Sheppard, L.R., TiO_2 surface active sites for water splitting, *J. Phys. Chem. B*, 110, 18492–18495. Copyright 2006, American Chemical Society.)

- Light-induced ionization over the bandgap leading to the formation of an electron–hole pair

$$O_O^{2-} + Ti_{Ti}^{4+} \xrightarrow{hv} O_O^{-} + Ti_{Ti}^{3+} \tag{13.72}$$

- Reactivation of the surface sites that is associated with the following charge transfer

$$4O_O^{-} + V_{Ti}^{4-} \rightarrow 4O_O^{2-} + V_{Ti} \tag{13.73}$$

In analogy, oxygen vacancies and the associated trivalent Ti ions may be considered as local active sites for the formation of chemisorbed oxygen species, which are important in photocatalytic water purification. The related reactivity model is represented in Figure 13.31.

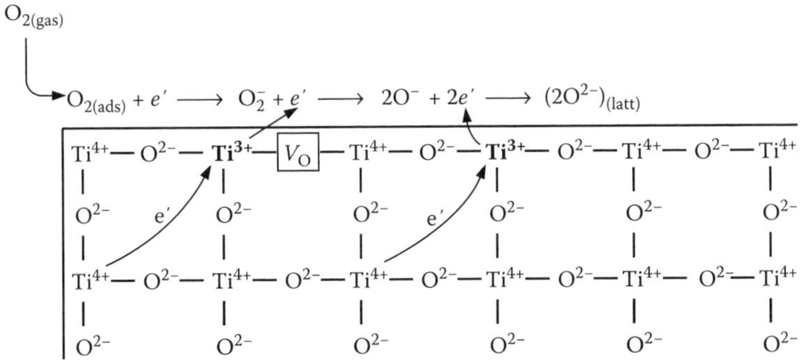

FIGURE 13.31
Theoretical model representing the reactivity of TiO_2 surface with oxygen. (Reprinted with permission from Nowotny, J., Bak, T., Nowotny, M.K., and Sheppard, L.R., TiO_2 surface active sites for water splitting, *J. Phys. Chem. B*, 110, 18492–18495. Copyright 2006, American Chemical Society.)

The development of high-performance photoelectrodes for water splitting and photocatalysts for water purification requires that both collective and local factors are optimized to the level required for maximum performance.

13.5.4 Summary

The photoreactivity of photoelectrode with water should be considered in terms of both the collective and the local reactivity factors. The typical collective properties include electrical conductivity and Fermi level. The local component is related to local structure in the atomic scale.

13.6 Reactions at the Water/TiO_2 Interface

The reactivity of water with TiO_2 should be considered in terms of all species present in water, including oxygen (dissolved in water), hydrogen (in the form of protons), and the dissolved ions. The reactivity includes adsorption and chemisorption of species derived from the liquid phase as well as diffusion and segregation in the solid phase. The latter process leads to the imposition of surface composition that is different from the bulk phase.

13.6.1 Effect of Oxygen

Since oxygen is a part of the oxide lattice, the properties of oxides, such as TiO_2, are closely related to oxygen activity in the lattice and that in the surrounding gas phase. The effect of the gas phase on properties depends critically on temperature.

At lower temperatures, when the lattice transport is quenched, the changes in gas-phase composition lead to changes in chemisorption equilibria. In other words, oxidation of oxides at lower temperatures is limited to the adsorption layer. Then oxygen activity in the oxide lattice is independent of the gas-phase composition.

As temperature increases, oxidation leads to the imposition of strong concentration gradients within the gas/solid interface. Then changes of oxygen activity in the gas phase result in the propagation of the newly imposed oxygen activity into the bulk phase leading, ultimately, to the imposition of new gas/solid equilibrium.

The rate of the gas/solid equilibration reactions is determined by the chemical diffusion coefficient, which can be considered as the rate constant of the diffusion of defects under chemical potential gradient [23,41]. Therefore, knowledge of the diffusion data is essential for understanding the reactivity between oxygen in the gas phase and the oxide lattice. The following section considers the following phenomena related to the O_2/TiO_2 interface:

1. Oxygen chemisorption leading to the formation of oxygen-chemisorbed species
2. Oxygen lattice diffusion leading to imposition of controlled oxygen activity within the entire O_2/TiO_2 system

13.6.2 Oxygen Chemisorption

Oxygen chemisorption on oxide semiconductors may be considered in terms of the following equilibria [79]:

$$O_2 + e' \Leftrightarrow O_2^- \qquad (13.74)$$

$$O_2^- + e' \Leftrightarrow 2O^- \tag{13.75}$$

$$O^- + e' \Leftrightarrow O^{2-} \tag{13.76}$$

Reaction (13.74) represents the formation of singly ionized molecular species, which is considered a weak type of oxygen chemisorption. In dark conditions, these species have the tendency to be transformed into singly ionized atomic species, represented by reaction (13.75), which is considered a strong form of oxygen chemisorption. Imposition of light leads to the transition of these species back to the singly ionized molecular species, which are formed as a result of leftward shift of equilibrium (13.75) [83–86]. According to Henrich and Cox [98], oxygen may also be chemisorbed in the form of doubly ionized atomic species, which are formed according to reaction (13.76). This claim, however, is in conflict with the energy-related data indicating that these species may only be stable within the crystal field, that is, after the incorporation into the oxide lattice [95].

The mechanism of oxygen chemisorption on the surface of TiO_2, leading to the formation of several oxygen species, is represented in Figure 13.31. The active sites for oxygen chemisorption are the donor sites formed by oxygen vacancies or interstitial titanium ions.

The charge transfer related to the reactivity of TiO_2 with oxygen was studied using WF measurements by Bourasseau et al. [82–87] and Nowotny et al. [79].

13.6.3 Oxygen Propagation (Equilibration)

The reaction of TiO_2 with oxygen at elevated temperatures, involving oxidation and reduction, results in shifts of defect equilibria that are represented by Equations 13.4 through 13.7. Then defects are formed or removed at the gas/solid interface. The newly imposed defects diffuse into the bulk phase leading to the imposition of new equilibrium state. The change of the concentration of defects versus distance from the surface (1) before isothermal oxidation and reduction of the oxide lattice, (2) during equilibration, and (3) after equilibration is shown schematically in Figure 13.32. At this point, it is important to note that knowledge of the diffusion data is essential to predict the time required for uniform imposition of oxygen activity over the specimen.

In most cases, the mass transport may be considered in terms of a single diffusion coefficient when the diffusion involves only type of species. However, when the mass transport proceeds via two types of species with different diffusion rates, then the equilibration may be considered in terms of two kinetic regimes [39,52]. This is the case of TiO_2. It has been shown that mass transport in TiO_2 involves both oxygen vacancies and titanium interstitials, which are very fast, and titanium vacancies, which are extremely slow. In this case, the following two kinetic regimes have been identified [49]:

1. *Kinetic Regime I.* This regime corresponds to the transport of fast defects (oxygen vacancies and titanium interstitials) that exhibit high diffusion rates.
2. *Kinetic Regime II.* The kinetics in this regime is determined by the diffusion rate of titanium vacancies, which is exceptionally slow (it takes 3–4 months at 1323 K to impose an equilibrium concentration of titanium vacancies in a TiO_2 disk that is 1 mm thick).

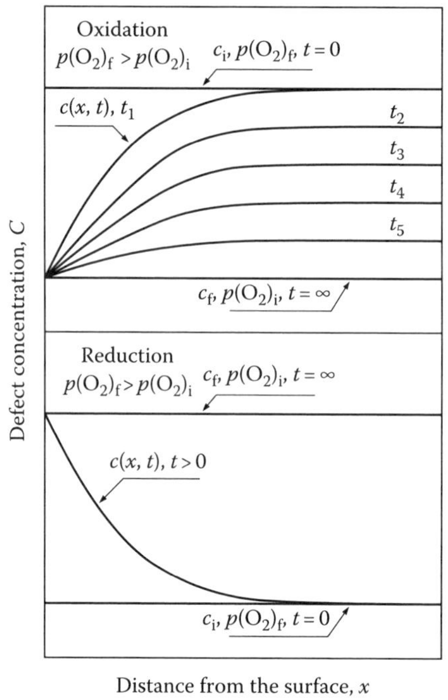

FIGURE 13.32
Schematic representation of defect concentration during isothermal oxidation and reduction (the propagation of defects at different stages of oxidation is represented by isoconcentration curves). (Reproduced with permission from Bak, T., *Oxide Semicond. Res. Rep.*, Copyright 2010.)

The effects of isothermal oxidation of TiO_2 at 1123 and 1323 K on electrical conductivity during the Kinetic Regime I (the left-hand side) and the Kinetic Regime II (the right-hand side) are shown in Figure 13.33 [49]. This figure also includes the change in the concentration of titanium vacancies during both kinetic regimes. These data indicate that prolonged oxidation results in a measurable change of the concentration of titanium vacancies (the Kinetic Regime II).

13.6.4 Reactivity of TiO_2 with Water

The reactivity of TiO_2 with water leads to the incorporation of hydrogen into the TiO_2 lattice even at room temperature. The most recent studies indicate that hydrogen may result in substantial changes of semiconducting properties of TiO_2 [100].

The reactivity of TiO_2 with water may be considered in terms of the defect reaction resulting in removal of oxygen vacancies and the formation of protons (Figure 13.34):

$$H_2O + V_O^{\bullet\bullet} \Leftrightarrow 2H^{\bullet} + O_O^{\times} \quad (13.77)$$

Alternatively, the incorporation of hydrogen may lead to the formation of titanium vacancies (Figure 13.35):

$$2H_2O \Leftrightarrow 4H^{\bullet} + V_{Ti}'''' + 2O_O^{\times} \quad (13.78)$$

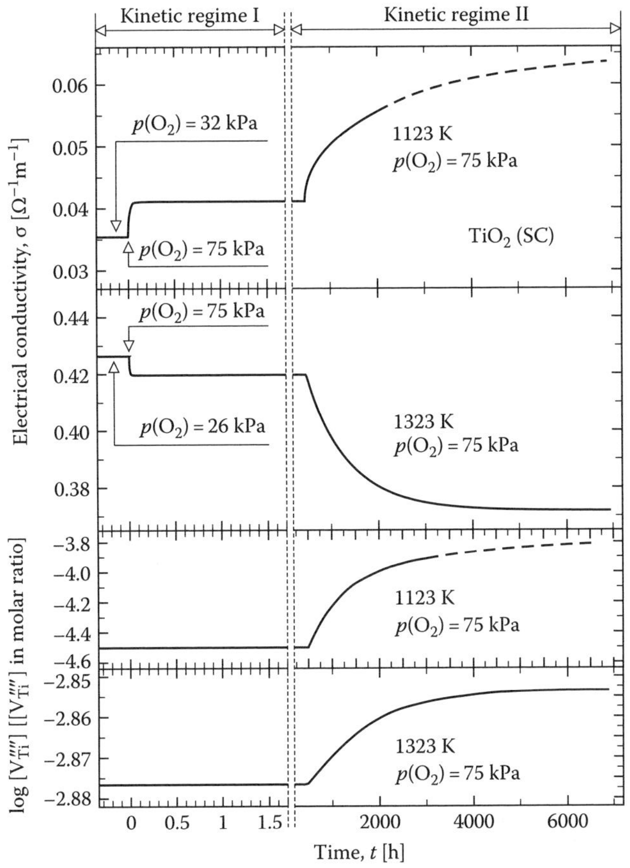

FIGURE 13.33
Equilibration kinetics for TiO_2-SCs during oxidation, represented by the time dependence of the electrical conductivity and the concentration of Ti vacancies at 1123 and 1323 K, within the short- and long-time regimes (Kinetic Regimes I and II, respectively). (Adapted from Bak, T., Nowotny, M.K., Sheppard, L.R., and Nowotny, J., Effect of prolonged oxidation on semiconducting properties of titanium dioxide, *J. Phys. Chem. C*, 112, 13248–13257. Copyright 2008, American Chemical Society.)

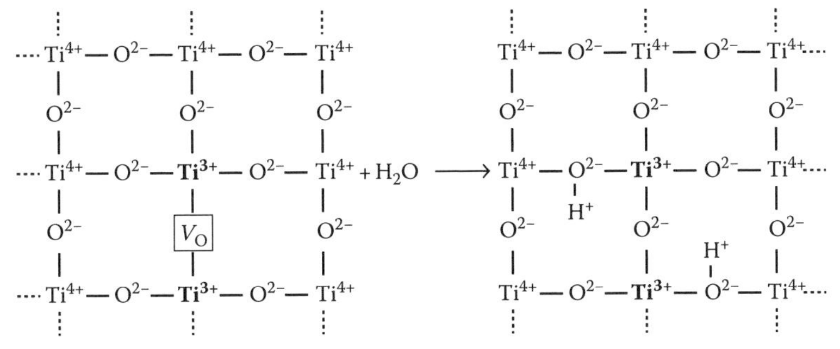

FIGURE 13.34
Reaction of proton incorporation resulting in the removal of oxygen vacancies. (Reprinted with permission from Nowotny, J., Norby, T., and Bak, T., Reactivity between titanium dioxide and water at elevated temperatures, *J. Phys. Chem. C*, 114, 18215–18221. Copyright 2010, American Chemical Society.)

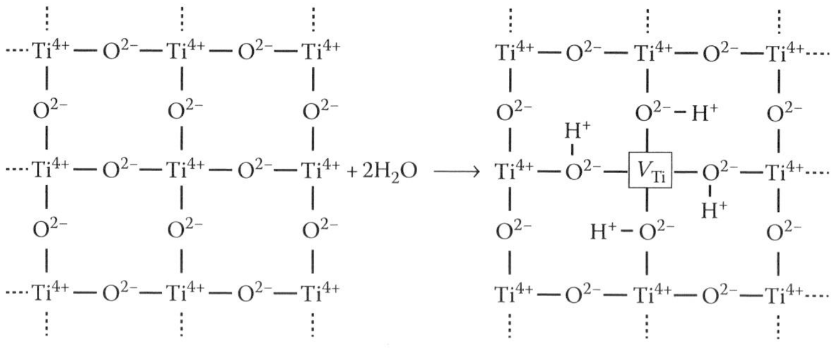

FIGURE 13.35
Reaction of proton incorporation resulting in the formation of titanium vacancies. (Reprinted with permission from Nowotny, J., Norby, T., and Bak, T., Reactivity between titanium dioxide and water at elevated temperatures, *J. Phys. Chem. C*, 114, 18215–18221. Copyright 2010, American Chemical Society.)

where the charge neutrality requires that

$$\left[H^{\bullet}\right]+2\left[V_{O}^{\bullet\bullet}\right]=n+4\left[V_{Ti}^{''''}\right] \tag{13.79}$$

Consequently, hydrogenation of the TiO_2 lattice favors the incorporation of protons. Exposure of n-type TiO_2 to hydrogen at room temperature may lead to the formation of a low-dimensional surface layer of the H_4TiO_4 structure [100].

13.6.5 Summary

The performance of TiO_2-based photoelectrode for PEC is closely related to the reactivity of TiO_2 with water, oxygen, and hydrogen.

13.7 Segregation

13.7.1 Segregation-Induced Effects

The chemical composition and the related properties of materials interfaces are different from those of the bulk phase as a result of segregation [101,102]. Therefore, knowledge of the effect of segregation on surface properties is crucial in correct interpretation of catalytic and photocatalytic reactions.

The driving force for segregation is the excess of interfacial energy. Such segregation is termed *equilibrium segregation* or, equivalently, *thermodynamic segregation*. Details of the physical meaning of segregation and the main driving forces involved are outlined elsewhere [101,102].

Segregation is a diffusional process and so may take place at elevated temperatures at which the mobilities of lattice elements are sufficiently high. However, it is important to recognize that sufficient time also is required to reach the segregation equilibria.

It is clear that segregation-induced concentration gradients at interfaces, such as external surfaces and grain boundaries, have a substantial impact on the functional properties of photoelectrodes. This impact may be beneficial or detrimental. The imposition of segregation-induced concentration gradients in a controlled manner may be used to

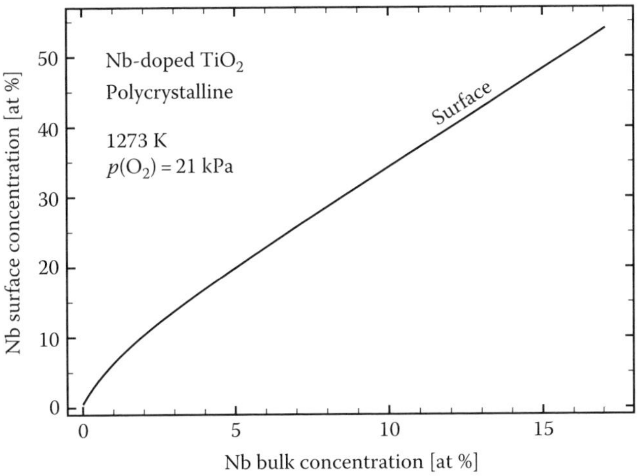

FIGURE 13.36
Effect of segregation of Nb on surface versus bulk concentration in Nb-doped TiO_2 at 1273 K. (Reprinted with permission from Sheppard, L.R. et al., *Proc. SPIE*, 6340, 634015, 2006. Copyright 2006, SPIE.)

engineer interfaces in order to achieve specific properties desired for defined applications. Thus, it is clear that there is a need to understand both the impact of segregation on the functional properties of materials and the theoretical underpinning of this phenomenon. At present, the level of knowledge on segregation in ionic-covalent compounds is limited, although there is a large body of literature on segregation in metals [103]. This limitation is due largely to experimental difficulties in the determination of well-defined data on segregation.

The solid remains in continuous interaction with the gas phase, and so the effect of the gas phase on segregation must be taken into account, particularly when comparing different segregation data.

The study of nonstoichiometric compounds, such as binary MOs, requires consideration of several issues that impact the generation of well-defined data on segregation, including (1) impurities, (2) low-dimensional surface structures, and (3) nonstoichiometry.

Segregation results in the formation of both chemical and electrical potential barrier layers within the near-surface layer. Segregation impacts on the surface and grain boundary composition of both undoped TiO_2 and its solid solutions. It has been shown that grain boundaries of undoped TiO_2 are enriched with donor-type defects, such as oxygen vacancies and titanium interstitials [55]. Surface versus bulk analysis of Nb-doped TiO_2 indicates that the surface is enriched with niobium as it is shown in Figure 13.36 [104].

The phenomenon of segregation may be used as a technology for the imposition of controlled surface composition and chemically induced electric field, *F*. The latter may be used for charge separation. Figure 13.37 represents the WF component related to the surface charge and the related electric field.

13.7.2 Summary

Segregation results in a change of the local properties at interfaces, including chemical composition and the associated semiconducting properties. These properties have a substantial effect on reactivity and photoreactivity of oxide semiconductors.

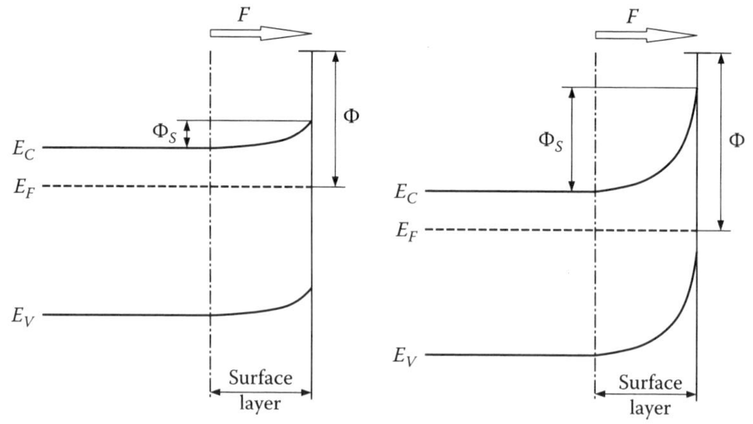

FIGURE 13.37
Effect of surface charge on band bending, WF, and the related electric field.

13.8 Defect Engineering of TiO_2

13.8.1 Concept of Defect Engineering

Defect engineering allows the imposition of controlled defect-related property through the modification of defect disorder. Therefore, defect engineering allows imposition of controlled reactivity of the surface with adsorbed species, such as oxygen, hydrogen, water, and its solutes. Consequently, defect engineering offers an innovative approach for the processing of high-performance TiO_2-based photoelectrodes and photocatalysts with desired properties through the imposition of controlled defect disorder.

It is essential to note at this point that the performance of photoelectrodes and photocatalysts concerns room temperature. On the other hand, most of the processes leading to the modifications of defect disorder require annealing at elevated temperatures, which allow the diffusion transport of defects from the surface into the bulk or vice versa. Consequently, the studies on the development of oxide semiconductors with controlled properties may include the following procedures:

- The formation of oxides with desired oxygen activity. This may be achieved by the imposition of controlled oxygen activity at elevated temperatures.
- The formation of solid solutions by the incorporation of aliovalent ions (cations and anions) at elevated temperatures.
- Surface processing leading to imposition of surface versus bulk controlled concentration gradients of oxygen and aliovalent ions.
- The formation of polycrystalline specimens of controlled grain size.

Defect chemistry may be used in the development of high-performance photoelectrodes. While the concept of defect engineering is described in this chapter for titanium dioxide as an example, its principal approach is valid for all nonstoichiometric oxides.

13.8.2 Summary

Defect engineering may be used for the processing of TiO_2-based photosensitive semiconductors with desired functional properties, which are defect-related.

13.9 Effects of Light

13.9.1 Solar Energy Spectrum

The spectrum of electromagnetic waves ranges between kilometers (radio frequency) and fractions of picometers (gamma radiation), including the UV range (100–400 nm), the visible range (400–700 nm range), and the infrared range (>700 nm), as seen in Figure 13.38. The energy of photons is related to their wavelength by the following expression:

$$E = \frac{hc}{\lambda} \tag{13.80}$$

where
- c is the speed of light
- h is the Planck constant
- λ is the wavelength

The energy provided by the Sun on Earth is substantial and exceeds the present global energy needs by the factor of 3×10^4. This supply of energy has been the support of life. The resulting photosynthesis over billions of years led to the accumulation of the resources of fossil fuels. In order to reverse the effects of climate change, which are already apparent, there is now a need to use solar energy in the formation of fuel that is

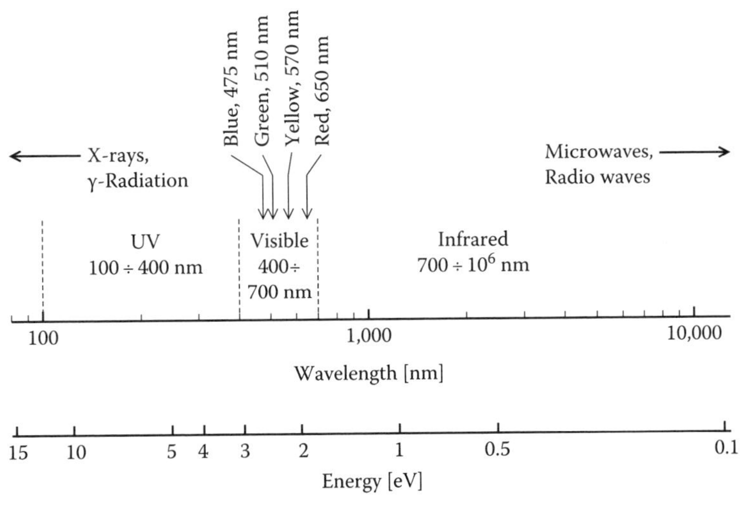

FIGURE 13.38
Electromagnetic wave spectrum in terms of energy and wavelength. (Reproduced with permission from Bak, T., *Oxide Semicond. Res. Rep.*, Copyright 2010.)

environmentally clean. The promising clean fuel is the hydrogen fuel, which can be generated by photoelectrochemical water splitting.

The most critical functional element in the development of solar hydrogen fuel is the photoelectrode. The key functional property of the photoelectrode is its ability to efficiently absorb sunlight.

The solar energy spectrum is frequently considered in terms of radiation energy versus wavelength, as shown in Figure 13.39. The area under this spectrum is the incidence of solar irradiance, I_r.

$$I_r = \int_0^{\lambda_i} E(\lambda)\,d\lambda \tag{13.81}$$

The bandgap is the key functional property of a photocatalyst as it has a critical impact on the ECE [2,4,105]. Since only the photons of energy equal to and larger than the bandgap may be absorbed and used for conversion, there is a need to select the semiconductors with an optimized bandgap that allow to achieve maximized conversion.

Taking into account the amount of solar energy that can be absorbed and the energy losses, the solar energy spectrum can be subdivided into several segments, which are represented in Figure 13.40 [106,107]. Consequently, for standard TiO_2, with the bandgap, E_g, equal to 3.05 eV [6–15], only the most energetic segment is available for conversion. As seen, this is a very small part of the entire solar energy spectrum. Therefore, there have been efforts to reduce the bandgap of TiO_2 from 3 to 2 eV in order to increase the amount of the absorbed energy. This may be achieved, for example, through the imposition of midgap bands [2]. Asahi et al. [108] reported that the bandgap reduction may also be achieved by

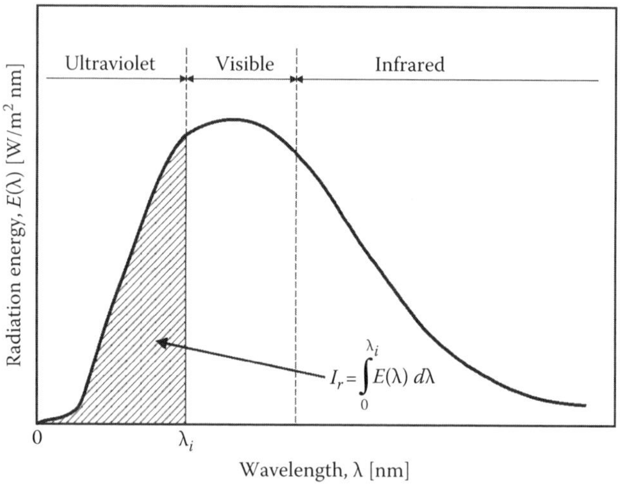

FIGURE 13.39
Schematic illustration of solar spectrum (radiation energy vs. wavelength), showing incidence of solar irradiance (I_r) available for conversion. (Reproduced from *Int. J. Hydrogen Energy*, 27, Bak, T., Nowotny, J., Rekas, M., and Sorrell, C.C., Photoelectrochemical hydrogen generation from water using solar energy. Materials-related aspects, 991–1022, Copyright 2002, with permission from Elsevier.)

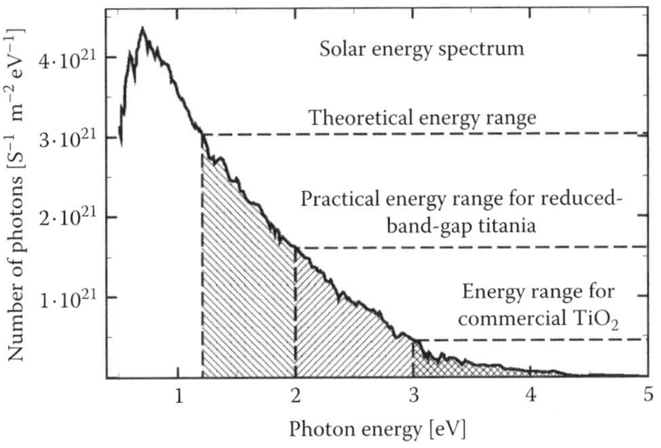

FIGURE 13.40
Solar energy spectrum (*AM* of 1.5) in terms of number of photons versus photon energy, showing different photon flux regimes corresponding to specific properties of photoelectrodes. (Reproduced from *Int. J. Hydrogen Energy*, 27, Bak, T., Nowotny, J., Rekas, M., and Sorrell, C.C., Photoelectrochemical hydrogen generation from water using solar energy. Materials-related aspects, 991–1022, Copyright 2002, with permission from Elsevier.)

lifting the E_V energy level through mixing 2*p* states of oxygen and *s* states of dopant, such as nitrogen. The same effect was reported by Kudo et al. [109].

The effect of the Earth's atmosphere on solar radiation is considered in terms of the so-called air mass (AM) which can be expressed in a simplified form as:

$$\mathrm{AM} = \frac{1}{\cos\alpha} \tag{13.82}$$

where α is the angle between the overhead and actual positions of the Sun. At the Earth's surface, the AM assumes values between unity ($\alpha = 0$) and infinity ($\alpha = 90°$). The AM characterizes the effect of the atmosphere on solar radiation, which depends on geographical position, local time, and date. By definition, outside the Earth's atmosphere, the AM is zero. The radiation standard assumes an AM of 1.5, which corresponds to $\alpha = 0.841$ rad or 48°. Of course, the solar energy available for conversion depends also on local atmospheric conditions, such as cloudiness, air pollution, airborne dust particles, and relative humidity.

The solar energy spectrum for the common case of an AM 1.5 is shown in Figure 13.41 in terms of radiation energy versus the wavelength. Since the energy required for splitting the water molecule is 1.23 eV, the solar radiation with the wavelength greater than 1 μm is not available for conversion.

13.9.2 Light Source

The ECE data reported in the literature have been determined for different light sources, which exhibit a wide range of spectral distributions, usually different from that of the

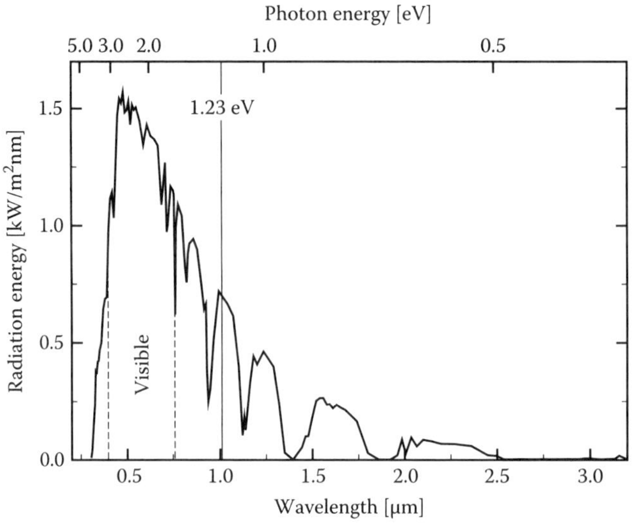

FIGURE 13.41
Solar energy spectrum (AM of 1.5) in terms of radiation energy versus wavelength. (Reproduced from *Int. J. Hydrogen Energy*, 27, Bak, T., Nowotny, J., Rekas, M., and Sorrell, C.C., Photoelectrochemical hydrogen generation from water using solar energy. Materials-related aspects, 991–1022, Copyright 2002, with permission from Elsevier.)

sunlight [106,107]. Therefore, the related ECE data may differ substantially from those corresponding to sunlight. Consequently, the ECE data for artificial light sources should be considered only as indicative.

13.9.3 Summary

Most of light-related data of oxide semiconductors are reported for artificial light sources, which are not well defined in terms of their energy spectrum. On the other hand, the key criterion for the evaluation of the performance is their response to sunlight. Therefore, while artificial light sources may be used to compare sample to sample within the same laboratory, the final test should be performed under sunlight.

13.10 Photoelectrochemical Water Splitting

13.10.1 Photoelectrochemical Cell

The concept of photoelectrochemical water splitting is represented schematically in Figure 13.42. Fujishima and Honda [3] reported their pioneering experiment with the PEC, which was formed of a TiO_{2-x} single crystal as a photoanode and platinum as a cathode. The PEC performance includes water oxidation at photoanode, leading to the formation of gaseous oxygen, hydrogen ions (protons), and electrons. The electrons and protons are transported to cathode (via the external circuit and the electrolyte, respectively) where protons are reduced to hydrogen gas.

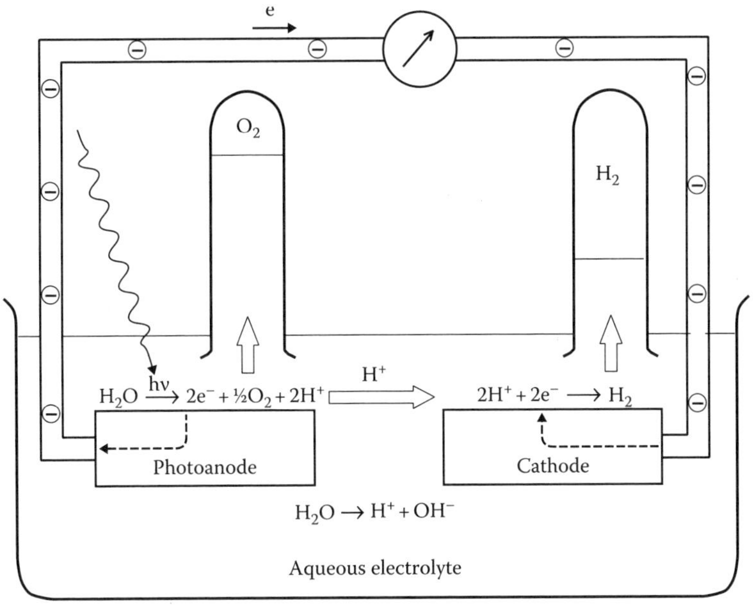

FIGURE 13.42
PEC and the related reactions. (From Nowotny, J., Titanium dioxide-based semiconductors for solar-driven environmentally friendly applications: Impact of point defects on performance, *Energy Environ. Sci.*, 1, 565–572. Reproduced by permission of The Royal Society of Chemistry.)

Water is a very stable compound. Therefore, high temperature is required to split water into oxygen and hydrogen gas:

$$2H_2O \Leftrightarrow O_2 + 2H_2 \tag{13.83}$$

At 3773 K, 30% of water is decomposed. However, both hydrogen and oxygen recombine during cooling.

The key reaction associated with water splitting, which requires a substantial amount of energy, is the removal of electrons from water molecules [2,4,35]:

$$H_2O \Leftrightarrow \frac{1}{2}O_2 + 2H^+ + 2e' \quad E = -1.23\text{ eV} \tag{13.84}$$

Light-induced water splitting at room temperature by a PEC equipped with a single photoelectrode (photoanode), which is an n-type semiconductor, and a metallic cathode involves the following reactions [4]:

1. *Adsorption of water molecules at the photoanode.* This primary reaction leads to the formation of an active complex consisting of the water molecule and the adsorption site, such as surface lattice ion or a defect. In the dark, the reactivity of oxide semiconductors with water is limited to physical adsorption. The formed adsorbed species have a weak tendency to exchange charge.

2. *Absorption of light by photoanode.* Absorption of a photon of energy, which is equal to or larger than the band energy, leads to electron excitation over the bandgap resulting in the formation of electron holes in the valence band and electrons in the conduction band. These light-induced electronic charge carriers are very reactive but usually quickly recombine.
3. *Charge separation.* The light-induced charge carriers have the tendency to recombine what leads to energy loses. The light-induced ionization and subsequent recombination of electrons and holes are shown in Figure 13.43a. The recombination-related energy losses may be reduced when the light-induced electronic charge carriers are separated in an electric field that is formed by the surface charge, resulting in the formation of a space charge within the surface layer (Figure 13.43b). The charge separation at the photoanode leads to the transport of electrons and holes toward the bulk and the surface, respectively.
4. *Photoreactivity of photoanode.* The newly imposed chemical potential of electrons and holes leads to enhanced surface reactivity of photoanode promoting the charge transfer between the adsorbed water molecules and the surface. The resulting multielectron charge transfer leads to water splitting into oxygen gas, hydrogen ions, and electrons.
5. *Reduction of hydrogen ions.* The hydrogen ions are transported to cathode via electrolyte, and the electrons removed from water molecules are transferred to cathode via the external circuit. Both hydrogen ions and electrons combine into hydrogen molecules at the cathode.

The most important aspect of effective water splitting concerns the photocatalytic material, which must exhibit the following key functional properties:

1. Availability of appropriate surface-active sites for adsorption of water molecules
2. Ability to absorb sunlight
3. Ability for multielectron charge transfer

The studies on the development of high-performance PECs are focused on TiO_2-based oxide semiconductors, which are the most promising candidates for photoelectrochemical

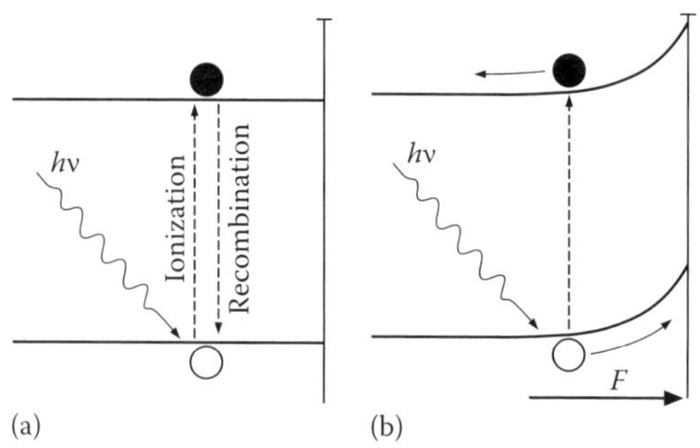

FIGURE 13.43
Schematic representation of recombination (a) and the effect of band bending on charge separation (b).

water splitting. It will be shown later that commercially available TiO_2 may be used as a raw material for the processing of well-defined TiO_2, which exhibits the desired performance as photoelectrode for solar hydrogen cells [2].

13.10.1.1 Anodic Reactions (Oxidation)

The light-induced electron holes diffuse to the surface where they oxidize water at the photoanode/electrolyte interface:

$$2h^{\bullet} + H_2O_{(liquid)} \rightarrow 2H^{+} + \frac{1}{2}O_{2(gas)} \qquad (13.85)$$

Gaseous oxygen evolves at the photoanode and the hydrogen ions migrate to the cathode through the internal circuit (electrolyte), while electrons travel to the cathode through the external circuit where they are available for reduction reaction.

The charge transfer at the semiconductor/electrolyte interface is influenced by the structure of the solid/liquid interface and the related potential distributions within the layers forming this interface, including the Gouy layer and the Helmholtz layer on the liquid side, as well as the space charge layer on the solid side (Figure 13.44).

13.10.1.2 Cathodic Reactions (Reduction)

The cathodic reaction between protons and electrons results in the formation of hydrogen gas, which evolves at the cathode:

$$2H^{+} + 2e' \rightarrow H_{2(gas)} \qquad (13.86)$$

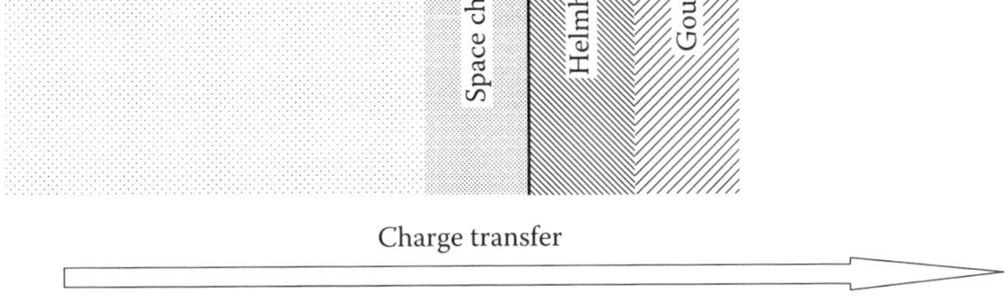

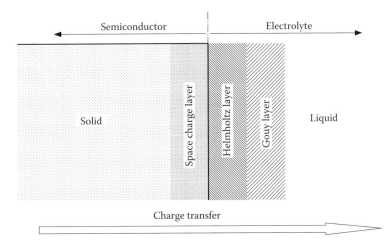

FIGURE 13.44
Schematic representation of the solid/liquid interface structure.

Taking into account the reactions (13.85) and (13.86), the overall reaction of the PEC may be expressed as follows:

$$2h\nu + H_2O_{(liquid)} \rightarrow \frac{1}{2}O_{2(gas)} + H_{2(gas)} \tag{13.87}$$

where
h is the Planck constant
ν is the light frequency

Accordingly, the overall reaction 13.87 takes place when the energy of the photons absorbed by the photoanode is equal to or larger than energy E_t that is needed to split water molecule:

$$E_t = \frac{\Delta G^0}{N_A} = 1.23\ \text{eV} \tag{13.88}$$

where
ΔG^0 is the standard free enthalpy per mole of reaction (13.87)
N_A is the Avogadro number

The economic feasibility of the photoelectrochemical hydrogen generation technology depends on the following criteria:

- The ECE and the related rate of hydrogen formation are above the level required for commercial viability. According to the DOE, such level in 2002 was 10% [111].
- The photoelectrode is chemically stable in an aqueous environment that is used in the PEC as electrolyte.

13.10.2 PEC Circuit

A typical cell involves a photoanode and cathode immersed in an aqueous solution of a salt (electrolyte). The cell reaction results in oxygen and hydrogen evolution at the photoanode and cathode, respectively. The band energy models of both electrodes, including the photoanode formed from an n-type semiconductor, such as TiO_2, and metallic cathode, are shown schematically in Figures 13.45 through 13.48 at different stages of performance. The related performance is represented by several energy-related quantities, including WF, band levels of the electrodes, and band bending.

13.10.2.1 Open Circuit

Figure 13.45 shows the open circuit and the flat-band model, with the two electrodes. The ability to charge transfer between the electrodes is determined by their WFs. The WF

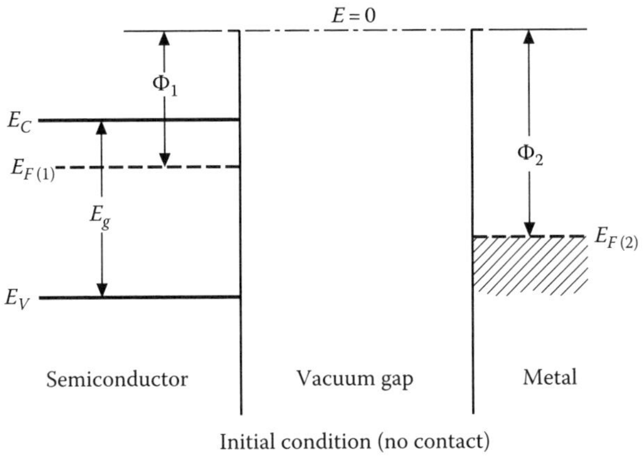

FIGURE 13.45
Energy diagram of PEC components: semiconducting photoanode, vacuum gap, and metallic cathode before galvanic contact. (Reproduced from *Int. J. Hydrogen Energy*, 27, Bak, T., Nowotny, J., Rekas, M., and Sorrell, C.C., Photoelectrochemical hydrogen generation from water using solar energy. Materials-related aspects, 991–1022, Copyright 2002, with permission from Elsevier.)

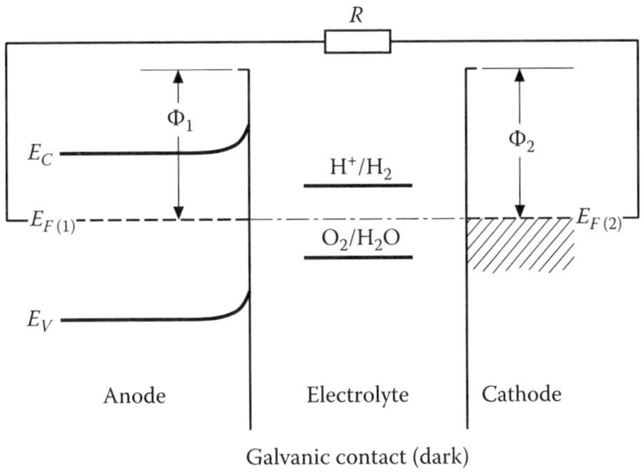

FIGURE 13.46
Energy diagram of PEC components after galvanic contact between the photoanode and cathode immersed in aqueous electrolyte in dark conditions. (Reproduced from *Int. J. Hydrogen Energy*, 27, Bak, T., Nowotny, J., Rekas, M., and Sorrell, C.C., Photoelectrochemical hydrogen generation from water using solar energy. Materials-related aspects, 991–1022, Copyright 2002, with permission from Elsevier.)

values for TiO_2 and platinum are in the range 2.9–3.9 eV and 5.12–5.93 eV, respectively [112–114]:

$$\Phi_{Pt} > \Phi_{TiO_2} \tag{13.89}$$

This difference indicates that electrons have a tendency to be transferred from the TiO_2 semiconductor (higher E_F) to platinum (lower Fermi level).

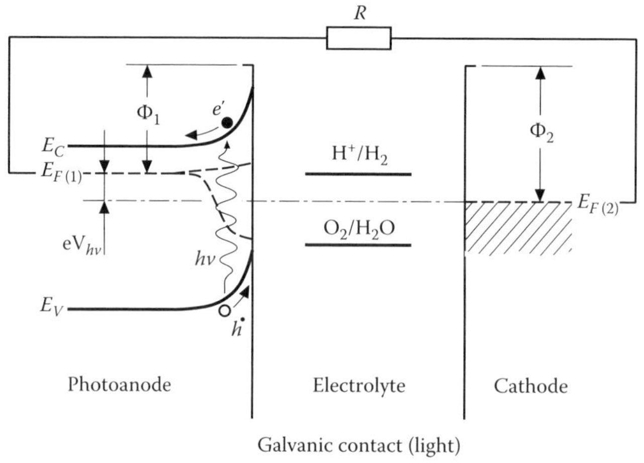

FIGURE 13.47
Effect of light on the band structure of the PEC components under light. (Reproduced from *Int. J. Hydrogen Energy*, 27, Bak, T., Nowotny, J., Rekas, M., and Sorrell, C.C., Photoelectrochemical hydrogen generation from water using solar energy. Materials-related aspects, 991–1022, Copyright 2002, with permission from Elsevier.)

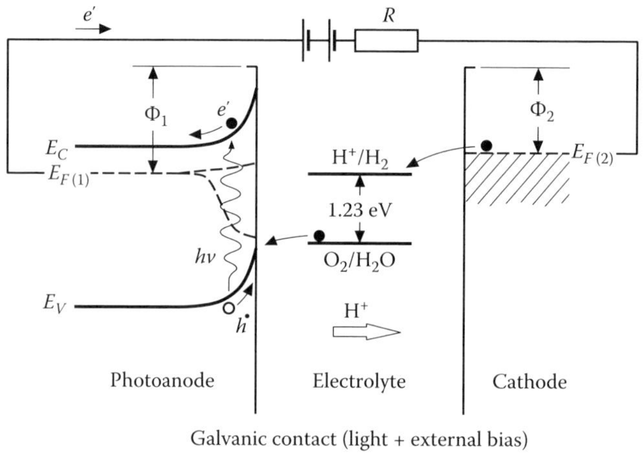

FIGURE 13.48
Effect of light on the band structure of the PEC under light with externally applied bias. (Reproduced from *Int. J. Hydrogen Energy*, 27, Bak, T., Nowotny, J., Rekas, M., and Sorrell, C.C., Photoelectrochemical hydrogen generation from water using solar energy. Materials-related aspects, 991–1022, Copyright 2002, with permission from Elsevier.)

13.10.2.2 Closed Circuit

Figure 13.46 represents the PEC circuit, involving the electrodes that are connected both internally (via electrolyte) and externally (via the external circuit). As seen, the electrons are transferred from the phase of the higher E_F (TiO_2) to the phase of lower E_F (Pt), leading to an upward band bending of the TiO_2 semiconductor. The charge transfer results in the formation of the CPD (CPD = V_s), which prevents further charge transfer:

$$\text{CPD} = \frac{1}{e}\left(\Phi_{\text{Pt}} - \Phi_{\text{TiO}_2}\right) \qquad (13.90)$$

13.10.2.3 Effect of Light

The effect of light on the PEC's band model is shown in Figure 13.47. As seen, illumination results in the following effects:

1. Light-induced ionization over the bandgap
2. Split of the Fermi level leading to the formation of quasi-Fermi levels related to electrons and holes
3. Charge separation in the electric field within the space charge layer

As seen in Figure 13.47, the energy level of the electrochemical couple H^+/H_2 is above the Fermi level of the cathode ($E_{F(2)}$). The difference is responsible for the formation of an electrical potential barrier preventing spontaneous charge transfer at the Pt/electrolyte system.

13.10.2.4 Effect of Light and Electrical Bias

The retarding barrier shown in Figure 13.47 may be removed by the imposition of an electrical bias resulting in lifting the ($E_{F(2)}$) level above that for the electrochemical couple H^+/H_2. Then potential distribution within the electrochemical chain becomes favorable for spontaneous charge flow within the PEC circuit as it is represented in Figure 13.48.

13.10.3 Light-Induced Reactions

13.10.3.1 Light-Induced Ionization over the Bandgap

An essential part of the PEC is the semiconducting photoelectrode. The light-induced ionization over the bandgap results in the formation of an electron–hole pair:

$$2hv \rightarrow 2e' + 2h^{\bullet} \tag{13.91}$$

The effect of light on the photoactivity may be considered in terms of splitting the Fermi level, E_F, into two quasi-Fermi levels related to electrons, $(E_F)_n^*$, and holes, $(E_F)_p^*$, as it is schematically represented in Figure 13.49 [115]. As seen, the effect of light on changes of E_F related to electrons for n-type semiconductors, such as n-type TiO_2, is very small. However, the effect

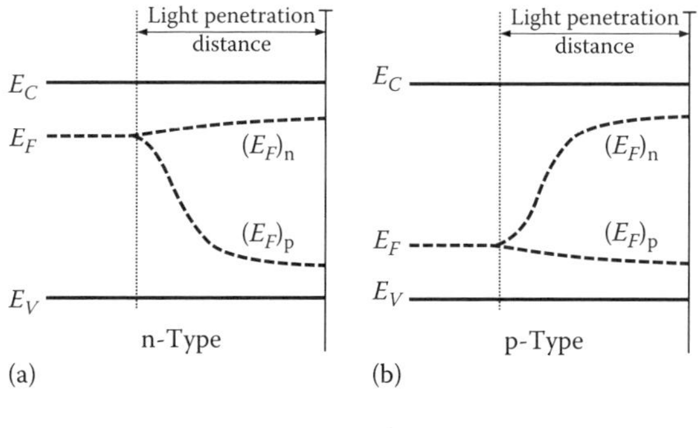

FIGURE 13.49
Light-induced split of the Fermi energy for (a) n-type and (b) p-type semiconductors. (Reproduced with permission from Bak, T., *Oxide Semiconductors Res. Rep.*, Copyright 2010.)

on the minority charge carriers (holes) is substantial. Therefore, the photon-induced ionization for n-type semiconductors results in a large increase of the oxidation potential induced by electron holes. The picture for the p-type semiconductor is similar but the final result is an increase in the reduction potential.

The effect of light on the quasi-Fermi levels of electrons and holes can be considered in terms of the associated increase of the concentrations of electrons (Δn) and electron holes (Δp), which can be expressed by the following dependencies, respectively [115]:

$$(E_F)_n^* = E_C + kT \ln \frac{n_0 + \Delta n}{N_n} \quad (13.92)$$

$$(E_F)_p^* = E_V - kT \ln \frac{p_0 + \Delta p}{N_p} \quad (13.93)$$

where

$(E_F)_n^*$ and $(E_F)_p^*$ are the ionization-induced quasi-Fermi levels related to electrons and electron holes, respectively
E_C and E_V are the energies of the bottom of the conduction band and the top of the valence band, respectively
k is the Boltzmann constant
T is the absolute temperature
n_0 and p_0 denote the concentrations of electrons and electron holes, respectively, before irradiation
Δn and Δp denote the changes in the concentrations of electrons and electron holes, respectively, after irradiation
N_n and N_p are the densities of states in the conduction band and valence band, respectively

13.10.4 Structures of Photoelectrochemical Cells

The PECs frequently require the imposition of an external electrical bias in order to perform. This is not required when PEC consists of two photoelectrodes [116,117]. In that configuration a substantial increase of the ECE may be achieved. The advantage of such a system is that the photovoltages are generated on both electrodes, resulting, in consequence, in the formation of an overall photovoltage that is sufficient for water decomposition without the application of a bias. In this case, light energy is absorbed by two photoelectrodes, including photoanode and photocathode, which are formed of n- and p-type semiconductors, respectively. The performance concept of the PEC equipped with two photoelectrodes, its photoelectrochemical chain, and the related band model are shown in Figures 13.50 through 13.52, respectively.

The efficient operation of PEC equipped with two photoelectrodes requires that both the collective and the local factors are optimized. The collective factor is related to the n- and p-type photoanode and photocathode, respectively. The local factor is related to specific point defects, which form surface-active sites. Figure 13.53 shows the surface defect disorder models for the photoanode and the photocathode, involving acceptor- and donor-type defects, respectively:

1. Titanium vacancies act as acceptor sites at the surface of photoanode. These sites form an active complex with adsorbed water molecule.
2. Oxygen vacancies act as donor sites at the surface of photocathode. These sites provide electrons to protons and reduce them to hydrogen gas.

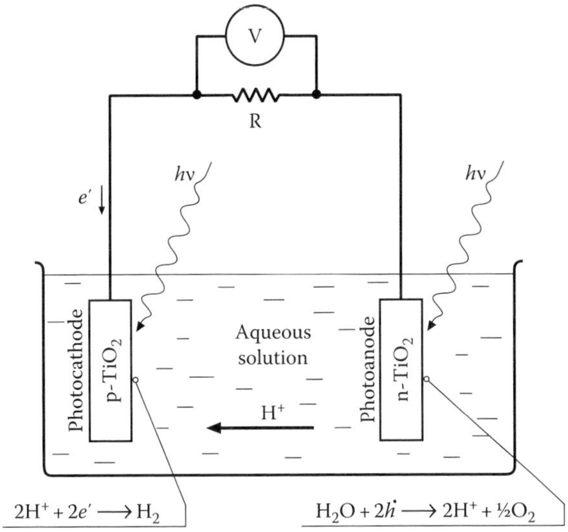

FIGURE 13.50
PEC equipped with two photoelectrodes. (Reproduced with permission from *Int. J. Hydrogen Energy*, 32, Nowotny, J., Bak, T., Nowotny, M.K., and Sheppard, L.R., Titanium dioxide for solar hydrogen 1. Functional properties, 2609–2629. Copyright 2007, International Association for Hydrogen Energy. Published by Elsevier Ltd.)

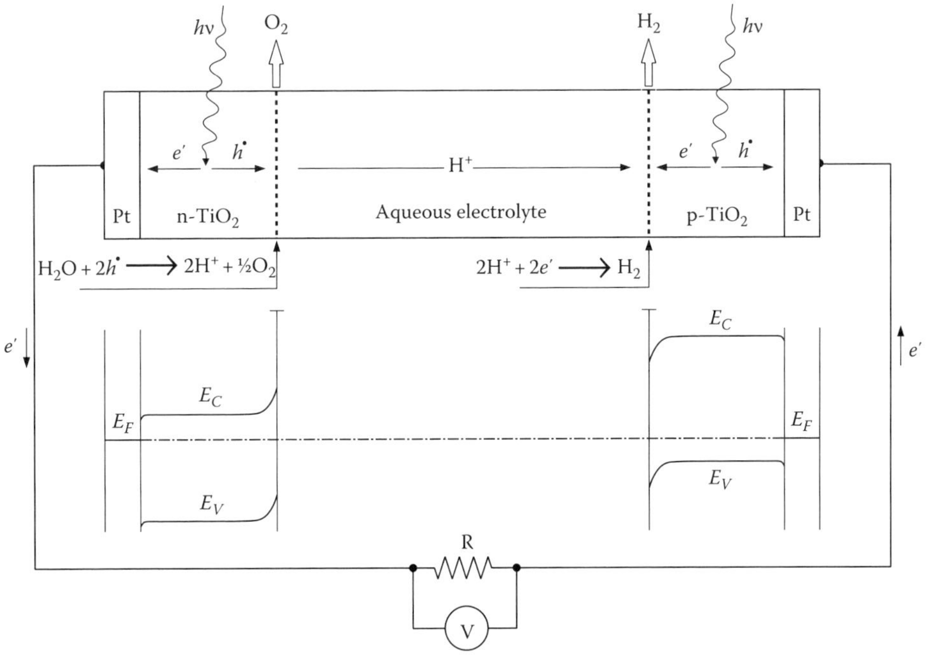

FIGURE 13.51
Electrochemical chain for the TiO_2-based PEC involving two photoelectrodes. (Reproduced with permission from *Int. J. Hydrogen Energy*, 32, Nowotny, J., Bak, T., Nowotny, M.K., and Sheppard, L.R., Titanium dioxide for solar-hydrogen 1. Functional properties, 2609–2629. Copyright 2007, International Association for Hydrogen Energy. Published by Elsevier Ltd.)

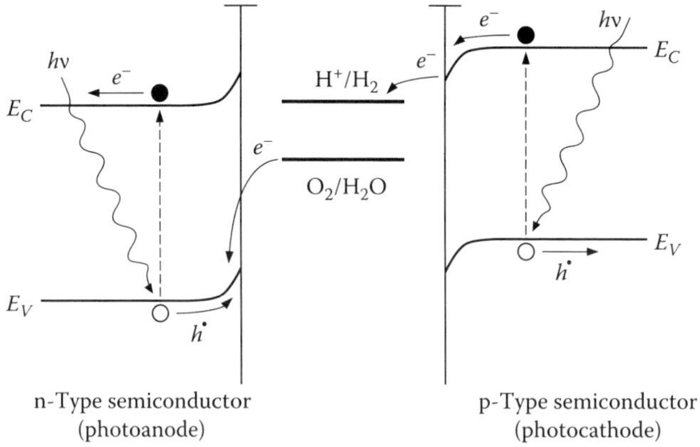

FIGURE 13.52
Band model of the PEC involving two photoelectrodes. (Reproduced with permission from Bak, T., *Oxide Semicond. Res. Rep.*, Copyright 2010.)

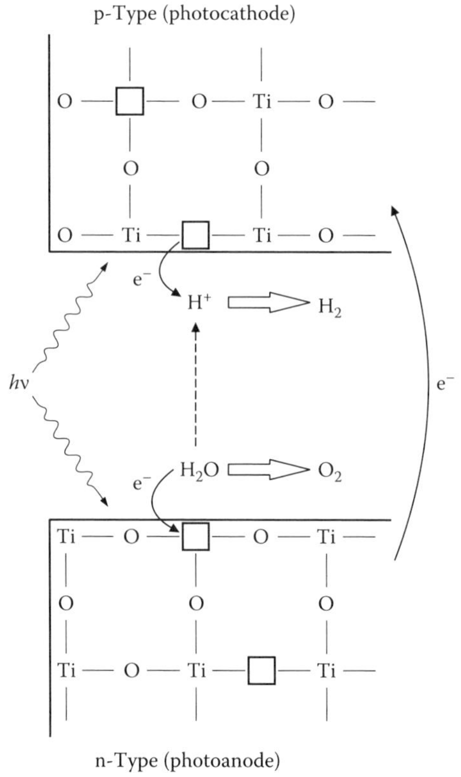

FIGURE 13.53
Schematic representation of the reactivity of two TiO_2-based photoelectrodes with water and the related charge transfer.

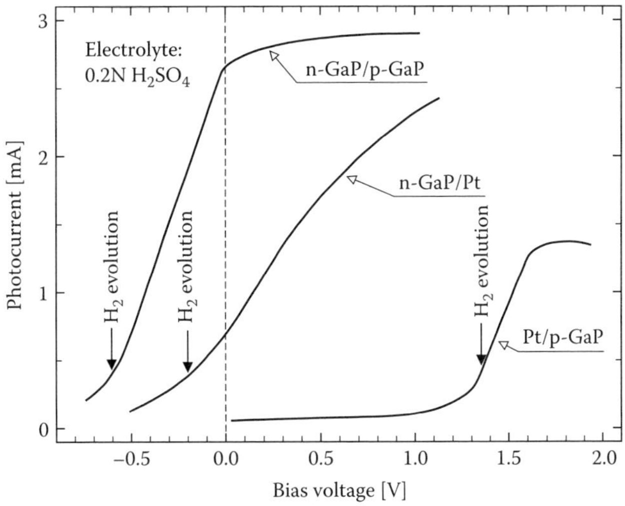

FIGURE 13.54
Photocurrent versus bias voltage for PEC involving two photoelectrodes along with PECs equipped with single photoelectrode, according to Nozik. (Reprinted with permission from Nozik, A.J., p–n photoelectrolysis cells, *Appl. Phys. Lett.*, 29, 150–153. Copyright 1976, American Institute of Physics.)

According to Nozik [116], application of two photoelectrodes (n–p PEC) leads to a substantial increase of the light-induced cell current. As seen in Figure 13.54, the performance of the PEC equipped with two photoelectrodes is substantially enhanced compared to that of single photoelectrode PECs since the energy required for water splitting is derived from two photoelectrodes that are exposed to light, instead of one. According to Nozik [116], the theoretical efficiency of such cells is 45%. However, the p-type photoelectrodes reported so far, such as p-GaP and p-InP, exhibit fast photocorrosion in aqueous environments leading to a substantial deterioration of the PEC performance [112]. This will not be the case when the PEC is equipped with both photoanode and photocathode made of TiO_2. The recent discovery of a p-type semiconductor made of pure TiO_2 [39,42] paves the way for high-performance solar cells involving two TiO_2 photoelectrodes: n-type TiO_2 photoanode and p-type TiO_2 photocathode. The platinum-free high-performance solar hydrogen cells, based entirely on TiO_2, are expected to pave the way for commercialization of solar hydrogen.

Better performance may also be achieved by integrating a photovoltaic system into the PEC. The ECE for the system based on the GaAs/$GaInP_2$ system is 12.4% [19]. The efficiency reported for the system including the tandem cell GaInP and GaInAs and the polymer electrolyte membrane (PEM) electrolyzer is 18% (Figure 13.55) [118]. While these systems exhibit a high ECE level, their relatively high costs are the main concerns.

Morisaki et al. [119] reported a PEC involving a hybrid photoelectrode (HPE), which is formed of a silicon cell and TiO_2 layer on the top. The advantage of this structure is that only the TiO_2 layer is exposed to the aqueous environment, while the silicon solar cell, forming a sublayer, is not in contact with the electrolyte. The purpose of the silicon solar cell is to generate photovoltage that provides an internal electrical bias. This type of solar cell exhibits spontaneous performance in the absence of an external bias. The HPE cell allows very efficient use of solar energy. As the external layer of TiO_2 absorbs only the

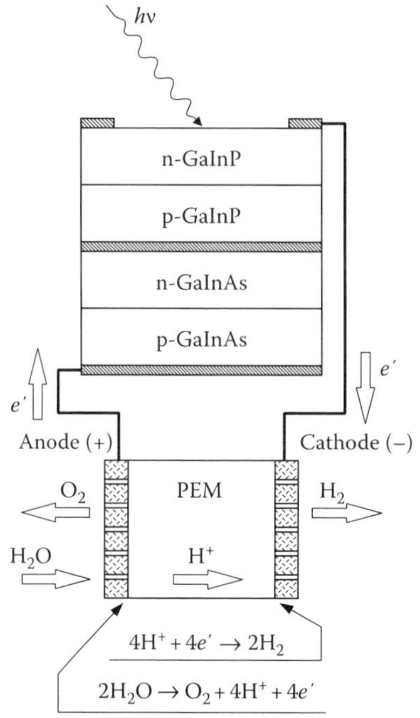

FIGURE 13.55
PEC including the tandem cell, GaInP and GaInAs, and the PEM electrolyzer according to Peharz et al. (Reproduced with permission from Bak, T., *Oxide Semiconductors Res. Rep.*, Copyright 2010.)

photons of energy greater than 3 eV, the remaining part of the solar spectrum is transmitted to the silicon solar cell (beneath the TiO_2 layer), which has E_g = 1.2 eV. Consequently, the silicon cell absorbs the low-energy part of the solar spectrum, involving the photons of energy between 3 and 1.2 eV. Figure 13.56 shows the electrochemical chain of the HPE invented by Morisaki et al. [119].

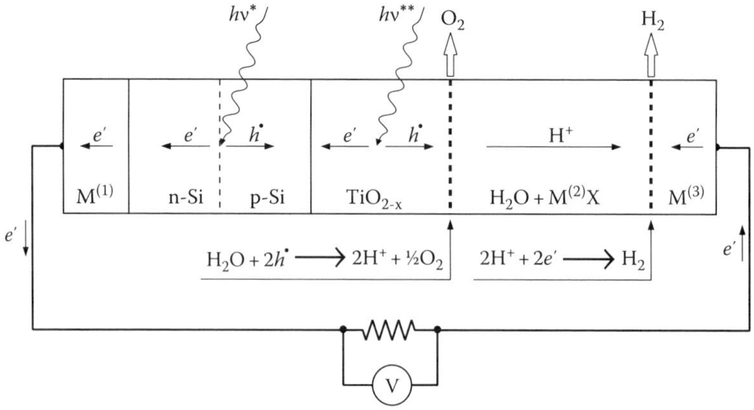

FIGURE 13.56
Electrochemical chain for the hybrid PEC involving inner silicon cell and outer TiO_2 layer according to Morisaki et al. (Reproduced with permission from *Int. J. Hydrogen Energy*, 32, Nowotny, J., Bak, T., Nowotny, M.K., and Sheppard, L.R., Titanium dioxide for solar-hydrogen 1. Functional properties, 2609–2629. Copyright 2007, International Association for Hydrogen Energy. Published by Elsevier Ltd.)

A wide range of approaches have been applied to enhance the performance of photoelectrodes, including sensitization by incorporation of foreign ions; formation of HPEs involving the components that exhibit different functions [119]; incorporation of noble metals in micronized particles, such as Ag and Pt [120]; and dye deposition [121].

Another kind of cells uses dye sensitization [121]. The photosensitizer, which is an organic dye, is attached to the surface of the photoelectrode. Light absorption by the dye leads to excitation of the dye molecules, which may be represented by the following reaction:

$$\text{Dye} + h\nu \rightarrow \text{Dye}^* \tag{13.94}$$

The excited dye state (Dye*) may be oxidized releasing electrons:

$$\text{Dye}^* \rightarrow \text{Dye}^+ + e' \tag{13.95}$$

The reaction between the oxidized dye molecule and I^- ions in the electrolyte results in the formation of I_3^- ions at the photoanode:

$$2\text{Dye}^+ + 3\text{I}^- \rightarrow 2\text{Dye} + \text{I}_3^- \tag{13.96}$$

The I_3^- ions are transported to the cathode where they are reduced:

$$\text{I}_3^- + 2e' \rightarrow 3\text{I}^- \tag{13.97}$$

The dye-sensitized semiconducting photoelectrode exhibits two functions: (1) absorption of light by the dye and (2) charge transport by the semiconductor. Such dye-sensitized cells allow conversion of light into electricity.

13.10.5 Summary

The performance of TiO_2-based PECs for water splitting is relatively well defined in terms of light-induced electrode reactions and the related charge transfer. A wide range of approaches have been reported in the development of high-performance PECs with reduced energy losses. The most promising approach includes the development of PECs equipped with two photoelectrodes.

13.11 Functional Properties

The research strategy on the development of a TiO_2-based PEC with high performance involves maximization of light absorption and minimization of all energy losses. In order to achieve high efficiency, there is a need to optimize the key performance-related properties, such as electronic structure, FBP, charge transport, concentration of surface-active sites, and charge separation. These may be achieved through the imposition of bulk versus interface properties in a controlled manner.

13.11.1 Electronic Structure

The electronic structure for metals, semiconductors, and insulators is schematically represented in Figure 13.57. The most critical quantity of electronic structure of semiconductors

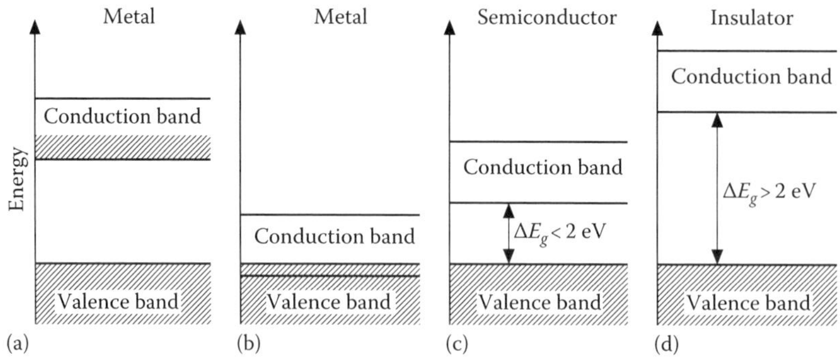

FIGURE 13.57
Band models for (a and b) metal, (c) semiconductor, and (d) insulator.

is the bandgap, E_g. The light is effectively absorbed by semiconductors when the photon energy is equal to or higher than the width of forbidden bandgap. Then light absorption leads to ionization i.e. generation of electron-electron hole pairs. Since the energy required for water splitting is 1.23 eV, the optimum value of the bandgap is the sum of 1.23 eV and the amount of energy losses (approximately 0.5–1 eV). Therefore, the optimal bandgap for water splitting is between 1.8 and 2 eV, depending on the extent of energy losses.

The bandgap of TiO_2, rutile, is 3.05 eV (Table 13.1 [6–15]). Therefore, intensive research aims to process TiO_2 with reduced bandgap. The electronic structure of oxides, including TiO_2, is closely related to defect disorder, which may be modified by varying oxygen content as well as through doping with aliovalent ions [2].

The main research strategy in reducing the bandgap of TiO_2 includes the following approaches:

1. Elevation of the edge of valency band
2. Imposition of midgap bands so that the effective bandgap required for ionization is reduced

Figure 13.58 is an estimated representation of the effect of E_g on the ECE for a single photoelectrode PEC, including both conservative and optimistic scenarios.

Wilke and Breuer [122] reported that incorporation of Cr^{3+} and Mo^{5+} results in reduction of the bandgap to 2 and 2.8 eV, respectively. According to Khan et al. [123], doping TiO_2 with carbon results in bandgap reduction to 2.3 eV. This effect, however, was not confirmed by Barnes et al. [124].

There is no agreement on the reported effect of vanadium doping on electrochemical properties of TiO_2. Phillips et al. [125] reported that addition of 30 mol% V to TiO_2-SC results in bandgap reduction to 1.99 eV; however, the formation of $Ti_{0.7}V_{0.3}O_2$ has detrimental effects on photoactivity. On the other hand, Zhao et al. [126,127] observed that increased amount of vanadium in TiO_2 thin films results in an increase in the ECE.

There have been some studies on the reduction of the bandgap by manipulation the grain size of TiO_2. Hoffmann et al. [16] reported that below a certain critical grain size (~10 nm), the bandgap increases. This effect has been confirmed by Wang et al. [17] who observed that the bandgap of the 2.72 nm grain size TiO_2 is 3.32 eV, while the bandgap for TiO_2-SC is 3.05 eV (Table 13.1 [6,8–15]). On the other hand, there are several experimental and theoretical evidences indicating that TiO_2 nanotubes exhibit reduced bandgap [128–132]. These studies indicate a relationship between the surface shape and bandgap. Namely, the concave curvatures at surfaces result in a decrease of bandgap, while convex curvatures lead to increase of bandgap.

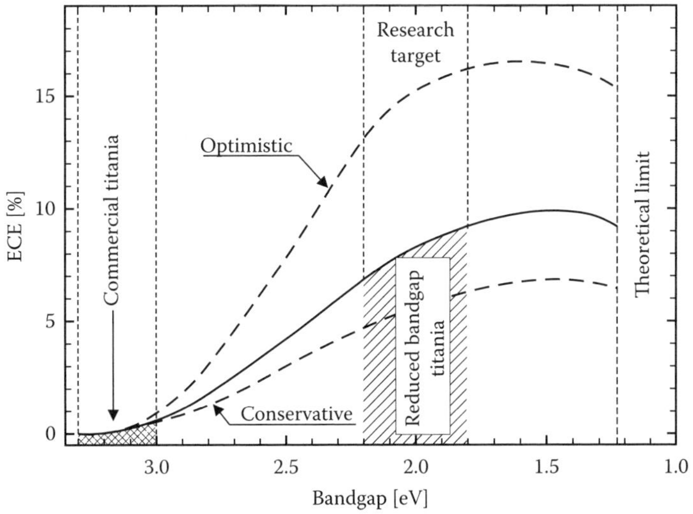

FIGURE 13.58
Estimated effect of bandgap reduction of TiO_2 on the ECE for a PEC equipped with single photoelectrode.

The bandgap is not always the key property, controlling the photocatalytic performance. Karakitsou and Verykios [18] observed that the hydrogen production rate by anatase form of TiO_2 is higher than that of rutile by the factor of 7.

There are several experimental approaches to assess the width of the bandgap, including the optical reflection spectra and the temperature dependence of the electrical conductivity corresponding to the n–p transition point [2].

13.11.2 Flat-Band Potential

When a semiconducting photoelectrode is immersed in an aqueous electrolyte, the charge transfer at the photoelectrode/electrolyte interface results in the formation of an electrical potential barrier, which causes the band bending. This barrier, which helps with the charge separation, is an important property of photoelectrode. The voltage needed to straighten the bands is termed the FBP. It may be determined experimentally by the imposition of an external potential compensating the surface charge and flattening the band bending.

Figure 13.59 shows the FBP relative to the vacuum level and the normal hydrogen electrode (NHE) level, and the bandgap values for several oxide materials [33]. The FBP may be determined from the Mott–Schottky equation:

$$\frac{1}{C^2} = \frac{2}{e\varepsilon_S\varepsilon_0 N_{A,D}}\Delta V_{SC} \tag{13.98}$$

where

C is the measured interfacial capacitance

ε_S and ε_0 denote the dielectric constant of the specimen and dielectric permittivity of vacuum, respectively

$N_{A,D}$ is the concentration of donors or acceptors

ΔV_{SC} is the applied external potential

The FBP is the intercept on the voltage axis (Figure 13.60) of the extrapolated linear dependence between $1/C^2$ and the voltage, V.

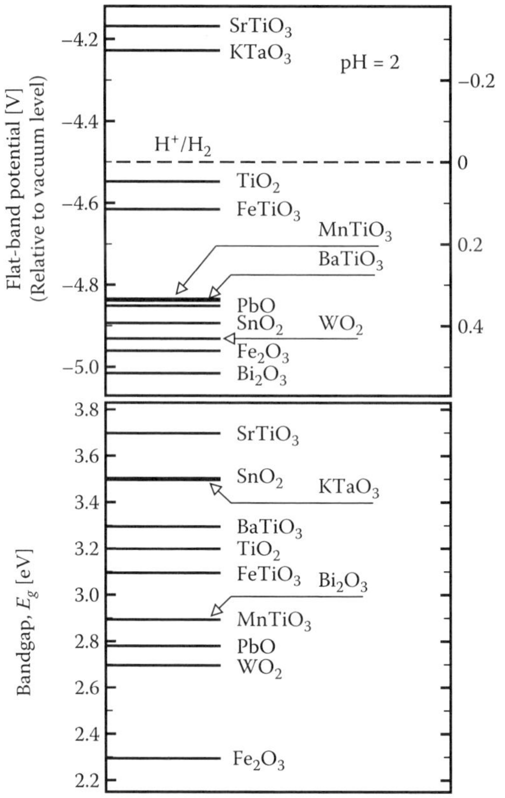

FIGURE 13.59
FBP of different oxide semiconductors versus energy gap and normal hydrogen energy level in electrolyte of pH = 2. (Reproduced with permission from Bak, T., *Oxide Semicond. Res. Rep.*, Copyright 2010.)

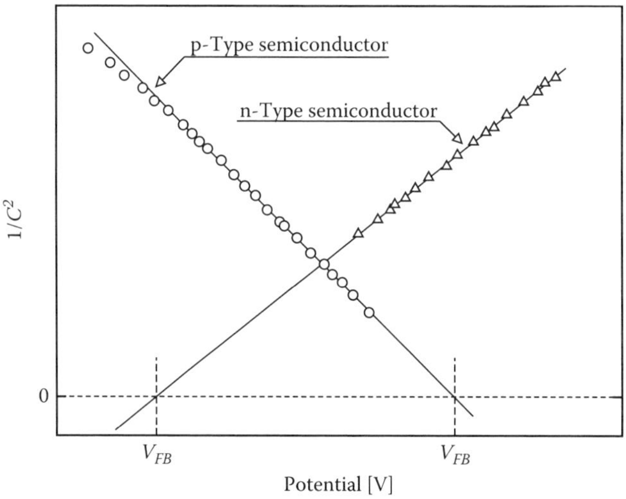

FIGURE 13.60
Graphical determination of the FBP from the Mott–Schottky relationship (plot of $1/C^2$ vs. potential). (Reproduced with permission from Bak, T., *Oxide Semicond. Res. Rep.*, Copyright 2010.)

13.11.3 Charge Transport

The energy losses related to charge transport may be substantial. These may be minimized when the charge transport is maximized and the ohmic resistance is minimized. The charge transport is closely related to the concentration of charge carriers and their mobility. Consequently, the charge transport may be enhanced by the modification of defect disorder in order to enhance the concentration of electronic charge carriers with high mobility.

The amount of hydrogen generated within the PEC is related to the charge transported through the PEC over time. Under the influence of electric field, *F*, randomly moving quasi-free electrons would have acceleration in the direction opposite to the field. The mobility of electrons is

$$\mu_n = \frac{\upsilon}{F} \tag{13.99}$$

where υ is the drift velocity:

$$\upsilon = \frac{et_r}{m_n^*}F \tag{13.100}$$

where
- t_r is the relaxation time
- m_n^* is the effective mass of electron
- F is the electric field
- e is the elementary charge

The corresponding electrical current densities for electrons and holes may be expressed by the following respective equations:

$$\begin{aligned} J_n &= e\mu_n nF \\ J_p &= e\mu_p pF \end{aligned} \tag{13.101}$$

The conductivity is then the sum of both conductivity components for electrons and holes:

$$\sigma = \frac{J}{F} = e\mu_n n + e\mu_p p \tag{13.102}$$

However, the physical meaning of F is more complex in the case when a PEC is equipped with a TiO_2 photoelectrode that exhibits a segregation-induced electric field, F_s. Then F involves two components: one is related to F_s and the second is the cell component related to the electric field imposed by the electromotive force (EMF) induced by light, F_c. The effective electric field will then be a superimposition of these two electric fields that are localized in the surface layer of the photoelectrode. Consequently, the charge transport within a PEC may be enhanced when the direction of F_s is the same as that of F_c. In analogy, the field F_s has a retarding effect on the charge transport when its direction is opposite to that of F_c.

An important electrical property is electrical conductivity, which must be maximized. *In situ* monitoring of the electrical conductivity, thermoelectric power, and work function during processing provides a mean to achieve optimal properties [78,134].

The electrical resistance of the TiO_2 photoelectrode may be reduced by reduction of TiO_2 at high temperatures in a hydrogen/argon mixture. In nonstoichiometric TiO_{2-x}, the higher the x value, the lower the resistance [36].

An alternative method of reducing the resistance is through minimization of the thickness of the photoelectrode by fabricating it in the form of a thin film. This method has the advantage that the substrate can be made of titanium metal, which imposes a strong reduction potential, thereby possibly obviating the need for postreduction.

13.11.4 Surface Properties

Undefected surface of oxides, including TiO_2, is not reactive [98]. In other words, efficient photoreactivity between the adsorbed water molecules and TiO_2 requires the presence of defects. Lo et al. [130] reported that oxygen vacancies are the active sites for oxygen and water adsorption. These active sites may be considered in terms of either point defects in the outermost surface layer or traces of another phase deposited on the TiO_2 surface, such as platinum. High performance requires optimal population of the surface-active sites, able to form photocatalytically active complexes with water, which ultimately leads to its splitting.

An important surface property is the Schottky barrier, which is formed as a result of concentration gradients, surface states, and adsorption states. Such barrier plays an important role in preventing recombination of the charge formed as a result of photoionization. An electrical potential barrier across the surface layer can be formed as a result of structural deformations within the near-surface layer due to an excess of surface energy and segregation-induced chemical potential gradients of aliovalent ions across the surface layer imposed during processing [101,135]. Accordingly, the formation of these gradients may be used for the modification of the Schottky barrier in a controlled manner. The use of this procedure requires in situ monitoring of the surface versus bulk electrochemical properties during the processing of the electrode materials [78,134].

Surface reactivity of TiO_2, including the reactivity with water, is closely related to the concentration and the charge of point defects in the outermost surface layer. Recent studies show that effective water splitting on the surface of TiO_2, which requires multielectron charge transfer, takes place at titanium vacancies, which are strong acceptor sites able to remove electrons from water molecules [35]. Consequently, high reactivity of TiO_2 with water requires an optimal surface population of these active sites.

The concentration of titanium vacancies may be estimated from full defect disorder diagram, which may be derived from defect-related data for TiO_2 exposed to prolonged oxidation [39]. An alternative way to determine their concentration is by using the spectroscopy of soft positrons [136].

13.11.5 Corrosion Resistance

Photoelectrodes may exhibit stable performance when resistant to corrosion and photocorrosion in aqueous environment [4]. Any form of reactivity results in a change in the chemical composition and the related properties. Therefore, this property is critical for the selection of materials for photoelectrodes. Certain oxide materials, such as TiO_2 and its solid solutions, are particularly resistant to these reactivity types [5]. Therefore, they are suitable candidates for photoelectrodes for electrochemical water decomposition.

A large group of valence semiconductors, which exhibit suitable semiconducting properties for solar energy conversion (width of bandgap and direct transition within the gap), are not resistant to these types of reactivity. Consequently, their exposure to aqueous environments during the photoelectrochemical process results in the deterioration of their performance.

13.11.6 Property Limitations

Figure 13.61 shows the positions of band edges for several oxide materials, which are the candidates for photoelectrodes, compared to the energy levels of the electrochemical couples H^+/H_2 and O_2/H_2O [135]. Unfortunately, the most promising materials from the viewpoint of the bandgap width, such as GaP (E_g = 2.23 eV [128]) and GaAs (E_g = 1.4 eV [135]), are not stable in aqueous environments and so suffer from a significant corrosion. Therefore, these materials are not suitable as photoelectrodes in solar cells for water decomposition. The most promising oxide materials, which are corrosion resistant, include TiO_2 and $SrTiO_3$ [137–139].

13.11.7 Summary

The key performance-related properties for PECs include electronic structure (bandgap), FBP, charge transport, and surface properties. These properties are closely related to defect disorder and, therefore, may be modified in a controlled manner by defect engineering.

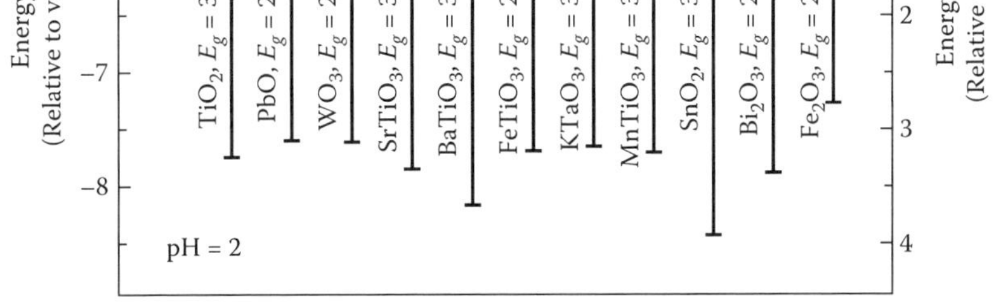

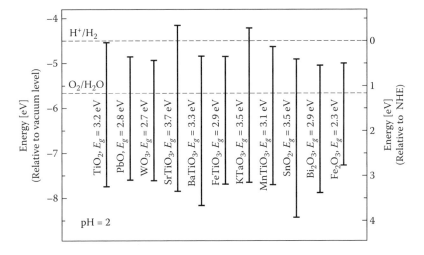

FIGURE 13.61
Diagram showing bandgap width of different oxide materials and relative positions of band edges with respect to vacuum level and NHE level in electrolyte of pH = 2. (Reproduced with permission from Chandra, S., *Photoelectrochemical Solar Cells* (Electrocomponent Science Monographs), Routledge, New York, 1985.)

13.12 Solar Energy Conversion Efficiency

13.12.1 Basic Relationships

The overall efficiency of a PEC unit, which is known as the *solar conversion efficiency* η_c, can be defined according to the following formula [136]:

$$\eta_c = \frac{\Delta G^0_{H_2O} R_{H_2} - V_{bias} I}{I_r A} \quad (13.103)$$

where
$\Delta G^0_{H_2O}$ is the Gibbs free energy of formation for 1 mol of liquid H_2O = 237.141 (kJ/mol)
R_{H_2} is the rate of hydrogen generation (mol/s)
V_{bias} is the bias voltage applied to the cell (V)
I is the current within the cell (A)
I_r denotes the incidence of solar irradiance, which depends on geographical location, time, and weather conditions (W/m^2)
A is the irradiated area (m^2)

Assuming that $R_{H_2} = I/2F$, Equation 13.103 assumes the following form [140]:

$$\eta_c = \frac{I(1.23 - V_{bias})}{I_r A} \quad (13.104)$$

The overall efficiency η_c is the following function of the property-related components [136]:

$$\eta_c = \eta_g \eta_{ch} \eta_{QE} \quad (13.105)$$

where
η_g denotes the solar irradiance efficiency
η_{ch} is the chemical efficiency
η_{QE} is the quantum efficiency

The η_g is defined as the fraction of the incident solar irradiance with photoenergy $\geq E_g$ and may be expressed as

$$\eta_g = \frac{J_g E_g}{E_S} \quad (13.106)$$

where
J_g is the flux density of absorbed photons
E_S is the incident solar irradiance (W/m^2)

The chemical efficiency is defined as the fraction of the excited state energy effectively converted to chemical energy and may be expressed as

$$\eta_{ch} = \frac{E_g - E_{loss}}{E_g} \quad (13.107)$$

where E_{loss} is the energy loss per molecule in the overall conversion process. For ideal systems, E_{loss} is defined as the difference between the internal energy and Gibbs free energy of the excited states. For real systems, E_{loss} assumes considerably larger values.

The quantum efficiency is defined as the following ratio:

$$\eta_{QE} = \frac{N_{eff}}{N_{tot}} \tag{13.108}$$

where

N_{eff} is the number of effective incidents leading to the generation of photoelectron/photohole pairs

N_{tot} is the total number of absorbed photons

13.12.2 Energy Losses

The key performance indicator of PECs is the ECE, which may be defined as the ratio of the energy output, E_{out}, to the energy input, E_{in}:

$$\text{ECE} = \eta_c = \frac{E_{out}}{E_{in}} \tag{13.109}$$

where E_{out} is the difference between the E_{in} and all kinds of energy losses:

$$E_{out} = E_{in} - E_{loss} \tag{13.110}$$

These losses are related to a range of properties/phenomena, including (1) optical reflection, E_{OPT}; (2) recombination, E_{REC}; (3) electrical resistance, E_R; (4) charge transfer, E_{CT}; and (5) heat, E_H. Therefore,

$$E_{loss} = E_{OPT} + E_{REC} + E_R + E_{CT} + E_H \tag{13.111}$$

Figure 13.62 shows the optical processes within PEC associated with different types of reflection and absorption. The E_{REC} component may be reduced by the imposition of an electric field leading to enhanced charge separation. Recent reports indicate that the electric field may be imposed in a controlled manner by surface and near-surface engineering, leading to the formation of concentration gradients and the related potential barriers. The electrical resistance-related losses, E_R, may be decreased by the increase of the concentration of charge carriers and/or their mobility. The energy losses related to charge transfer at the surface, E_{CT}, may be reduced by appropriate engineering of the outermost surface layer, where the charge transfer between the solid and the adsorbed molecules takes place. The heat related energy losses, E_H, are caused by light absorption of the energy lower and also larger than the width of bandgap.

13.12.3 Interdependence of Functional Properties

It has been a general perception that the width of the forbidden gap is the most important property of the photoelectrode. Indeed, it has been shown previously that the amount of the light energy being absorbed is determined by the bandgap. It has been shown, however, that the amount of the energy output is substantially lower due to energy losses.

The key performance-related properties are interdependent. Therefore, the modification of one property also results in a change of other properties. For example, while a

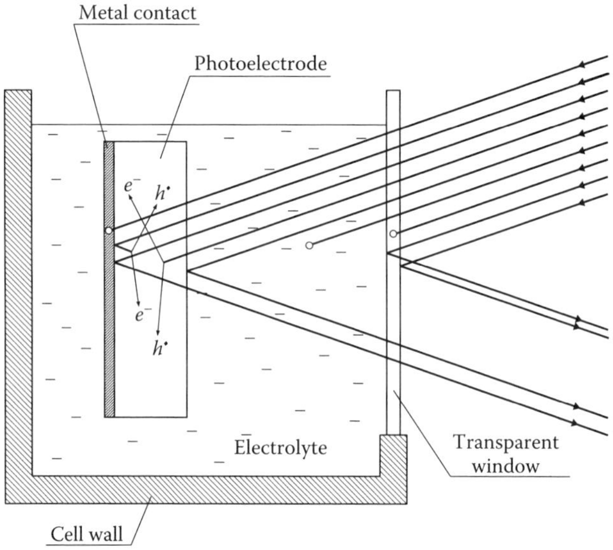

FIGURE 13.62
Illustration of optical processes within a PEC, including (1) reflection from window external surface, (2) reflection from window internal surface, (3) absorption by window, (4) absorption by electrolyte, (5) reflection from the surface of photoanode, (6) photon absorbed by photoanode and efficiently used for the generation of an electron–hole pair, (7) reflection from the surface of metal contact, (8) reflection from the surface of metal contact efficiently used for the generation of an electron–hole pair, and (9) absorption by metal contact. (Reproduced from *Int. J. Hydrogen Energy*, 27, Bak, T., Nowotny, J., Rekas, M., and Sorrell, C.C., Photoelectrochemical hydrogen generation from water using solar energy. Materials-related aspects, 991–1022, Copyright 2002, with permission from Elsevier.)

particular dopant ion introduced into the lattice results in reduction of the band gap, the same dopant may lead to a decrease of charge transport. When the earlier or the latter term predominates, the ECE will increase or decrease, respectively. Therefore, ECE should be maximized using a multivariant approach of all performance-related properties, which are closely related to defect disorder. Therefore, the performance of oxide semiconductors may be modified by using defect chemistry as a framework to enhance the performance. This approach is based on the fact that all functional properties, and the associated energy losses, are related to defect disorder. However, since the effects of defect disorder on these losses are interdependent, the system should be considered as multivariant. Consequently, each variable leading to the modification of the system is expected to have an effect on all properties.

It is difficult to make a graphical representation of the effect of all functional properties, and the related energy losses, on ECE. An attempt to make the 3D representation of the effect of the bandgap, along with the effect of the electric field, on the ECE is shown in Figure 13.63c.

In summary, the research strategy in the development of high-performance oxide semiconductors should lead to minimization of the energy losses by a multifactorious approach:

$$\nabla E_{loss}\,(x_1, x_2, \ldots, x_n) = 0 \tag{13.112}$$

where x_1, x_2, and x_n are independent variables, such as dopant concentration and oxygen activity.

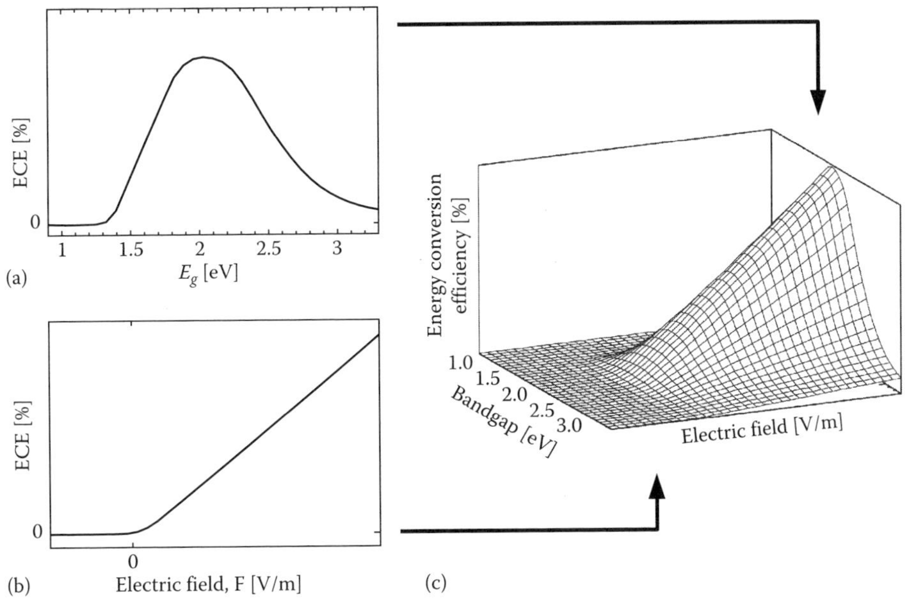

FIGURE 13.63
A 3D representation of the dependence of ECE on both bandgap and the FBP (represented by electric field), the latter in arbitrary units.

13.12.4 Overview of Progress

The increasing recognition of the impact of greenhouse gas emissions on climate change is expected to intensify the development of the technology of hydrogen production from water using solar and other renewable energy sources. Awareness is growing that hydrogen generated by water splitting using solar energy (solar hydrogen) is the most promising clean fuel of the future.

At present, the most common way of solar hydrogen generation is by water electrolysis using photovoltaic electricity. Tani et al. [141] reported that efficiency of the experimental PV-based solar hydrogen systems is 4.47%. The conversion efficiency of the most recently reported home fueling solar system, involving high-efficiency PV modules (16%) and high-pressure (44.8 MPa) electrolyzer, is 9.3% [142]. The two-device approach, however, requires application of two different devices: the silicon-based photovoltaic panel to harness solar energy for the production of electricity and the water electrolyzer, which converts the photovoltaic electricity into hydrogen.

There have been efforts to develop a PEC, which absorbs solar energy and splits water within a single device [2–4,78,80,105,113,116,117,119–121,126,127,137–139,143–179]. This technology has a substantial advantage over the PV technology due to the following reasons:

- The key component of the PEC is expected to be an oxide semiconductor, which is much less expensive than silicon and, first of all, exhibits stable performance when immersed in water.
- The PEC allows generating hydrogen within a single step.

While the advantage of the PEC technology is clear, its performance is still not satisfactory because so far its ECE is below the level that is economically feasible.

The simplest approach for photoelectrochemical hydrogen generation is using a photocatalyst, such as TiO_2, dispersed in aqueous solution. Then water splitting leads to the formation of the gas mixture involving both hydrogen and oxygen. This approach, however, requires energy for gas separation.

For a long time, the ECE for photoelectrochemical water splitting by using TiO_2 has remained below the level required for commercialization [4]. However, the recent progress in solid-state science and materials engineering is expected to allow the development of high-performance photoelectrodes for the production of hydrogen fuel with high efficiency. The most recent progress in defect chemistry for TiO_2 indicates that it is possible to process TiO_2 with controlled properties that are desired for specific applications.

Owing to the promising properties of TiO_2, this compound has been investigated in many laboratories. The research project initiated by the National Space Development Agency (NASDA), Japan, and the Institute for Laser Technology (ILT) aims to generate solar hydrogen using a space-based solar unit harvesting solar energy and transferring this energy (by laser) to a TiO_2-based electrochemical device located on Earth [180].

The future production scale of hydrogen using different technologies will be determined by their production costs. Solar hydrogen can be expected to be a long-distance winner as the ultimate fuel for the following reasons:

- Large parts of continents have an abundance of solar energy.
- Solar energy may be captured by MOs, which are relatively inexpensive.
- Solar hydrogen production technology could be adapted easily to the needs of individual households. The technology of domestic PECs may provide a driving force for mass production of small units.

The focal points of the research on photosensitive compounds for photo-assisted water splitting include the determination of the effects of composition on their performance in water splitting. There has been an accumulation of data indicating that the incorporation of foreign ions may lead to the reduction of the bandgap [122,123,140,181]. However, there are substantial hurdles, which must be overcome. For example, in the case of chromium incorporation, the observed reduction of the bandgap of TiO_2 even to 2 eV [122] leads to a decrease of ECE [182]. The latter data have been considered in terms of the effect of Cr on an (1) increase of the recombination-related energy losses due to reduced lifetime of light-induced electron–hole pairs from 90 μs for undoped TiO_2 to 30 μs for Cr-doped TiO_2 [122] and an (2) increase of the ohmic-resistance-related energy losses.

The majority of performance-related data on photoelectrochemical water splitting are reported in terms of arbitrary units that are related to specific experimental conditions. These data cannot be compared.

Mavroides et al. [138] reported quantum energy conversion efficiencies for different TiO_2 specimens, including single crystals, polycrystals, thin films, and thin layers formed on metallic titanium by oxidation. These data indicate that the TiO_2 layers formed by oxidation on metallic titanium exhibit the best performance.

The use of hybrid PECs, involving inner photovoltaic tandem systems that are covered with a thin layer of corrosion resistance TiO_2, represents a promising research strategy in the development of solar hydrogen technology [118,119].

Photosensitizers made of organic compounds may be used to increase the energy conversion efficiencies up to 10% [121]. The key issue in the development of the dye-sensitized systems is their durability.

Very high total energy conversion efficiencies (in the range 12%–18%) have been reported for photoelectrodes made of GaAs and Al-doped GaAs [19,118]. However, their stable performance in aqueous environments is limited.

The key issues, which must be addressed in the development of a commercial solar hydrogen PEC, include maximization of the ECE and the related hydrogen generation rate and maximization of the lifetime of photoelectrode. The latter impacts on the maintenance cost. Taking this into account, the thick line in Figure 13.64 seems to represent the expected commercial viability line.

While the technology of solar hydrogen has not been commercialized so far, there have been efforts toward installation of pilot plants. A US company, *Nanoptek*, Maynard, MA, claims to have achieved success in the processing of titania photocatalyst to be photoactive well into the visible blue and so is 6× more efficient in sunlight than native titania [183]. The company disclosure, however, neither provides the definition of the *native titania* nor specifies its ECE. Therefore, it is difficult to compare the performance of their photocatalysts with other systems reported in the literature. J. Guerra, the company's CEO, claims that this has been achieved by coating titania on domelike plastic nanostructure surface resulting in *the pulling of atoms apart* [184].

The solar hydrogen technology has not been commercialized so far. One may expect that the approach to develop a commercial unit will include (1) the solar cell exposed to sunlight and (2) two water circulation units including gas collection cylinders and water pump enforcing the circulation in the unit.

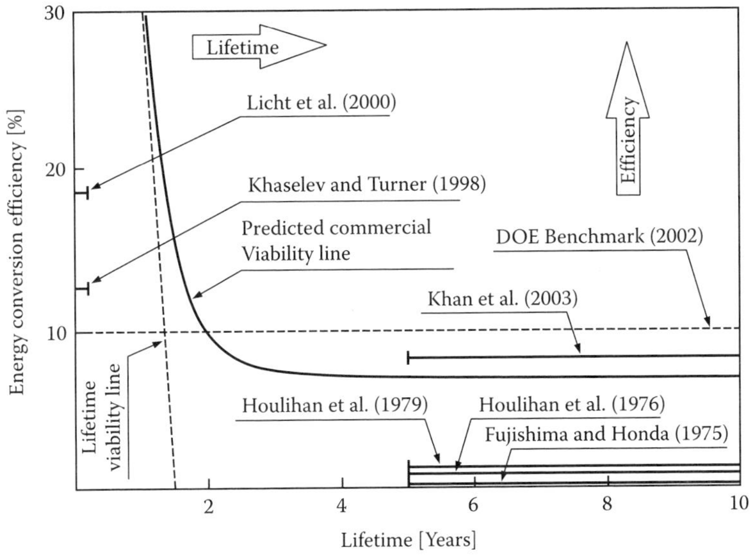

FIGURE 13.64
Schematic representation of critical requirements for the solar hydrogen commercial viability in terms of the ECE and the lifetime of photoelectrode along with selected ECE data reported in the literature. (Reproduced with permission from *Int. J. Hydrogen Energy*, 32, Nowotny, J., Bak, T., Nowotny, M.K., and Sheppard, L.R., Titanium dioxide for solar-hydrogen 1. Functional properties, 2609–2629. Copyright 2007, International Association for Hydrogen Energy. Published by Elsevier Ltd.)

The key function of the solar cell is to absorb solar energy. Therefore, the cell should be flat in order to achieve maximum surface area for the photoelectrode. As shown in the cross section in Figure 13.66, the cell has a layered structure including (from the top) (1) a sunlight transparent window, (2) photoanode deposited on a support, (3) water permeable membrane that allows rapid proton transport between the electrodes, and (4) cathode.

The aqueous electrolyte circulation system aims to remove the gases from both PEC compartments as well as collection of gases. The system includes both oxygen and hydrogen circuits connected to photoanode and cathode spaces, respectively. These gases are initially collected in the gas collection cylinders under atmospheric pressure and are subsequently pumped into storage tanks.

TiO_2 is also a promising candidate for photocatalytic water purification. The performance model of TiO_2-based photocatalyst is shown in Figure 13.65. Such a photocatalyst may be considered as a micro-PEC involving anodic and cathodic sites, which are contained to a single TiO_2 grain. The primary anodic reaction product is hydroxyl radical, OH*, formed according to the following reaction:

$$H_2O + h^{\bullet} \rightarrow H^+ + OH^* \quad (13.113)$$

The most important cathodic reaction is the formation of superoxide species [185,186]:

$$O_2 + e' \rightarrow O_2^- \quad (13.114)$$

Alternatively, cathodic reduction may be represented by the following reaction:

$$O_2 + 2e' + 2H^+ \rightarrow H_2O_2 \quad (13.115)$$

The species OH*, O_2^-, and H_2O_2 then react with toxic organic compounds and bacteria leading ultimately to their oxidation and the formation of stable molecules. An efficient photocatalytic process requires that both cathodic and anodic reactions take place with the same rate, leading to efficient removal of the excess of both charge carriers at both anodic and cathodic sites. In the photocatalytic process, both oxidation and reduction

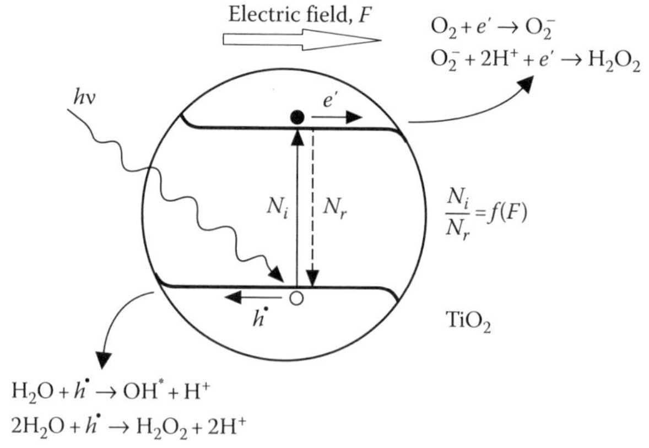

FIGURE 13.65
Theoretical model of TiO_2-based photocatalyst, showing light-induced electron ionization over the bandgap (N_i) and subsequent recombination (N_r), charge separation in the electric field (F), and the reactivity of both anodic and cathodic sites with water. (From Nowotny, J., Titanium dioxide-based semiconductors for solar-driven environmentally friendly applications: Impact of point defects on performance, *Energy Environ. Sci.*, 1, 565–572. Reproduced by permission of The Royal Society of Chemistry.)

occur on the surface of the photocatalyst, which exhibits the functions of both anode and cathode as it is shown in Figure 13.65.

13.12.5 Summary

The solar-to-chemical ECE of PECs is determined by several performance-related properties, which are interrelated. Therefore, the modification of the ECE requires applying a multivariant approach. The most promising is the PEC equipped with two photoelectrodes, photoanode and photocathode, which are made of oxide semiconductors that exhibit stable performance in water.

13.13 Summary

This chapter provided a comprehensive overview of several aspects of hydrogen generation using solar energy. The focus is on several aspects of hydrogen generation by photoelectrochemical water splitting using TiO_2-based POSs. It is shown that these materials are the promising candidates for photoelectrochemical solar cells due to an outstanding corrosion and photocorrosion resistance.

The key performance-related properties of TiO_2-based oxide semiconductors have been considered, including the following:

- *Electronic structure*. This property, and specifically the bandgap, is responsible for the absorption of solar energy. The optimal bandgap of TiO_2 that is required for maximized absorption of sunlight is in the range of 1.8–2.2 eV.
- *Charge transport*. Minimization of the energy losses related to change transport requires minimizing the ohmic resistance.
- *FBP*. An optimal value of the FBP is required for effective charge separation and reduction of recombination-related energy losses in photoelectrodes.
- *Surface defect disorder*. The ECE critically depends on the presence of surface-active sites for water splitting. These sites should be identified and their surface population should be optimized.

All these performance-related properties are closely related to defect disorder. Therefore, the performance of TiO_2 may be tailored using defect engineering.

13.14 Conclusions

There is an increasingly urgent need to develop renewable energy-related technologies. Since solar energy is available in abundance, this energy is expected to be the most attractive option in the development of the modern energy system, including photovoltaic electricity and solar fuel. Therefore, there have been efforts to harness solar energy for a wide range of applications. A spectacular achievement in the race to increase the solar ECE is a triple-junction solar PV panel with ECE at the new record level of 40.8% [187].

There is a general consensus that fossil fuels will be replaced by hydrogen as the fuel in the near future. However, the economical consequences of climate change dictate the need

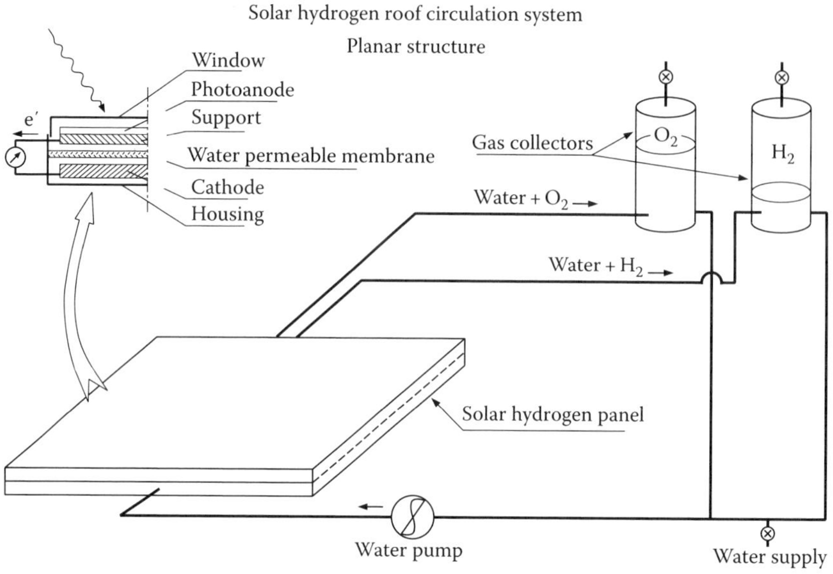

FIGURE 13.66
Concept of solar hydrogen production unit, including planar PEC and water circulation and gas collection systems. (Reproduced with permission from Bak, T., *Oxide Semicond. Res. Rep.*, Copyright 2010.)

to abandon the present steam reforming technology and develop the technology of hydrogen generation using renewable energy, such as solar hydrogen. Therefore, the future of hydrogen economy must be considered in terms of the development of hydrogen generation technologies using renewable energy. While solar power is universally available, equally attractive alternatives are wind, tide, hydroelectric, hydrothermal, and geothermal power that are more regionally based.

Hydrogen generated from solar energy does not contribute to the emission of greenhouse gases and climate change. Awareness is growing that hydrogen generated from water as a raw material using solar energy as the renewable energy is the most promising option for generation of hydrogen (solar hydrogen).

While at present solar hydrogen is generated by water electrolysis using photovoltaic energy, one of the most promising methods of hydrogen generation, which is environmentally friendly, is photoelectrochemical water splitting using PEC. This method allows hydrogen generation in a single step. The concept of a PEC—equipped with one photoelectrode—for solar hydrogen, involving a planar solar unit, the water flow system, and the gas collection system, is represented schematically in Figure 13.66. There are several advantages of solar hydrogen, including the following:

- Solar hydrogen will assist in reducing the levels of greenhouse and pollutant gases.
- Solar hydrogen encompasses both the production and utilization of a fuel that is 100% environmentally clean.
- Solar hydrogen will reduce the reliance on fossil fuels.
- Solar hydrogen will allow its producers to export solar energy.
- When this technology matures, it will allow developing countries to have access to cheap energy.

13.15 Historical Outline

1839: The Becquerel effect. The discovery of the photovoltaic effect by Becquerel [188] was the first observation of the chemical reaction induced by radiation.

1954: Theory of photoelectrochemistry. Brattain and Garret [189] first explained the charge transport during the photoelectrochemical effect in terms of the band model. Subsequent studies of Gerischer [190], Memming [191], and Morrison [192] led to a better understanding of the impact of the semiconducting properties of photoelectrodes on the photoelectrochemical effect.

1972: Photoelectrochemical hydrogen generation. Fujishima and Honda [3] first reported that sunlight results in water splitting into hydrogen and oxygen by using PEC formed of TiO_2 as photoanode and Pt as cathode.

1975: Bi-photoelectrode PEC. Yoneyama et al. [117] first reported PEC equipped with two semiconducting photoelectrodes, involving a photoanode and a photocathode made of n-type and p-type semiconductors. The performance of the bi-PEC was subsequently explained by Nozik [116].

1976: Hybrid PEC. Morisaki et al. [119] first reported an HPE consisting of the inner Si-based photovoltaic cell and external TiO_2-based photoanode. This concept subsequently led to the construction of high-efficiency cells reported by Khaselev and Turner [19] and Peharz et al. [118].

1978: Modification of TiO_2. Ghosh and Maruska [193] reported the effect of doping on semiconducting and photoelectrochemical properties of TiO_2 (the effect of doping was then confirmed by Houlihan et al. [181] and others).

1993–2008: Bioinspired effects. Derivation of theoretical models for biosystems [194–197].

2006: Derivation of defect disorder for 2006: Diagram for TiO_2. First derived defect diagram that allows to determine the effect of oxygen activity, temperature, and the concentration of aliovalent ions on the chemical potential of electrons [42]. This diagram may be used to predict the reactivity and photoreactivity of TiO_2 with water.

Nomenclature

a	Activity
A	Effective concentration of acceptors (atomic ratio)
c	Chemical concentration (atomic ratio)
CPD	Contact potential difference (V)
d	Thickness of the depleted layer (m)
D_{chem}	Chemical diffusion coefficient (m^2/s)
e	Elementary charge (1.602×10^{-19} C)
e'	Quasi-free electron
E_C	Energy of the bottom of the conduction band (eV)
E_F	Fermi level (eV)
ΔE_F	Change in Fermi level after irradiation (eV)
$(E_F)_n^*$	Light-induced quasi-Fermi level associated with electrons (eV)

$(E_F)_p^*$	Light-induced quasi-Fermi level associated with electron holes (eV)
E_g	Bandgap (eV)
E_V	Energy of the top of the valence band (eV)
$E(H^+/H_2)$	Energy level of the redox couple H^+/H_2 (eV)
$E(O_2/H_2O)$	Energy level of the redox couple O_2/H_2O (eV)
E_σ	Activation energy of electrical conductivity (kJ/mol)
F	Electric field (V/m)
h	Planck constant (6.626×10^{-34} J s)
$h^\bullet$	Quasi-free electron hole
ΔH_f	Activation enthalpy of defect formation (kJ/mol)
ΔH_m	Activation enthalpy of defect motion (kJ/mol)
j	Current density (A/m^2)
J	Light flux (lm)
k	Boltzmann constant (1.3807×10^{-23} J/K)
K	Equilibrium constant
L_D	Debye length (m)
m_e	Mass of electron (kg)
m_σ	Parameter related to defect disorder
n	Concentration of electrons (m^{-3})
$N(E)$	Distribution of photons with respect to energy (1/s m^2 eV)
N_n	Density of states in the conducting band (m^{-3})
N_p	Density of states in the valence band (m^{-3})
p	Concentration of electron holes (m^{-3})
p_0	Concentration of electron holes before irradiation (m^{-3})
$p(O_2)$	Oxygen activity (Pa)
PEC	Photoelectrochemical cell
R	Universal gas constant (8.3144 J/mol K)
S	Thermoelectric power (V/K)
SPS	Surface photoelectron spectroscopy
T	Absolute temperature (K)
TiO_2-PC	Polycrystalline titanium dioxide
TiO_2-SC	Single-crystal titanium dioxide
WF	Work function (eV)
x	Distance (m)
z	Valence
β	Temperature coefficient of the bandgap (eV/K)
ε	Dielectric constant
η	Electrochemical potential (eV)
Θ	Surface coverage (ratio)
μ	Chemical potential (eV)
σ	Electrical conductivity (1/Ωm)
ν	Frequency of light (Hz)
$\rho(x)$	Charge distribution (C/m)
Ψ	Electrical potential (V)
χ	External work function (eV)
Φ	Work function (eV)
Φ_s	Work function component related to surface charge (eV)
Φ_{in}	Internal work function component (eV)

References

1. T. Bak, J. Nowotny, N. Sucher and E. Wachsman, Effect of crystal imperfections on reactivity and photoreactivity of TiO_2 (rutile) with oxygen, water and bacteria, *J. Phys. Chem. C*, **115** (2011) 15711–15738.
2. M. K. Nowotny, L. R. Sheppard, T. Bak and J. Nowotny, Defect chemistry of titanium dioxide. Application of defect engineering in processing of TiO_2-based photocatalysts, *J. Phys. Chem. C*, **112** (2008) 5275–5300.
3. A. Fujishima and K. Honda, Electrochemical photolysis of water at a semiconductor electrode, *Nature*, **238** (1972) 37–38.
4. Nowotny, T. Bak, M. K. Nowotny and L. R. Sheppard, Titanium dioxide for solar hydrogen I. Functional properties, *Int. J. Hydrogen Energy*, **32** (2007) 2609–2629.
5. P. Reiche, *A Survey of Weathering Processes and Products*, University of New Mexico Press, Albuquerque, 1950, p. 54.
6. D. C. Cronemeyer, Infrared absorption of reduced rutile TiO_2 single crystals, *Phys. Rev.*, **113** (1959) 1222–1226.
7. J. Rudolph, Mechanism of conduction in oxide semiconductors at high temperatures, *Z. Naturforsch. A: Phys. Sci.*, **14** (1959) 727–737.
8. A. Frova, P. J. Boddy and Y. S. Chen, Electromodulation of the optical constants of rutile in the UV, *Phys. Rev.*, **157** (1967) 700–708.
9. K. Vos and H. J. Krusemeyer, Low temperature electroreflectance of TiO_2, *Solid State Comm.*, **15** (1974) 949–952.
10. J. Pascual, J. Camassel and H. Mathieu, Fine structure in the intrinsic absorption edge of TiO_2, *Phys. Rev. B*, **18** (1978) 5606–5614.
11. N. Daude, C. Gout and C. Jouanin, Electronic band structure of titanium dioxide, *Phys. Rev. B*, **15** (1977) 3229–3235.
12. K. Vos, Reflectance and electroreflectance of TiO_2 single crystals. II. Assignment to electronic energy levels, *J. Phys. C: Solid State Phys.*, **10** (1977) 3917–3939.
13. V. P. Gupta and N. M. Ravindra, Optoelectronic properties of rutile (TiO_2), *J. Phys. Chem. Solids*, **41** (1980) 591–594.
14. M. A. Khan, A. Kotani and J. C. Parlebas, Electronic structure and core level photoemission spectra in TiO2 compounds, *J. Phys.: Condens. Matter*, **3** (1991) 1763–1772.
15. M. K. Nowotny, *Defect Disorder, Semiconducting Properties and Chemical Diffusion of Titanium Dioxide Single Crystal*, Ph.D. thesis, Faculty of Science, University of New South Wales, Sydney, 2006.
16. M. R. Hoffmann, S. T. Martin, W. Choi and D. W. Bahnemann, Environmental applications of semiconductor photocatalysis, *Chem. Rev.*, **95** (1995) 69–96.
17. C.-Y. Wang, D. W. Bahnemann and J. K. Dohrmann, A novel preparation of irondoped TiO_2 nanoparticles with enhanced photocatalytic activity, *Chem. Commun.* **16**, (2000) 1539–1540.
18. K. E. Karakitsou and X. E. Verykios, Effects of altervalent cation doping of titania on its performance as a photocatalyst for water cleavage, *J. Phys. Chem.*, **97** (1993) 1184–1189.
19. O. Khaselev and J. A. Turner, A monolithic photovoltaic-photoelectrochemical device for hydrogen production via water splitting, *Science*, **280** (1998) 425–427.
20. S. Licht, B. Wang and S. Mukerji, Efficient solar water splitting, exemplified by RuO_2 catalyzed AlGaAs/Si photoelectrolysis, *J. Phys. Chem. B*, **104** (2000) 8920–8924.
21. A. G. Agrios and P. Pichat, Recombination rate of photogenerated charges versus surface area: Opposing effects of TiO_2 sintering temperature on photocatalytic removal of phenol, anisole, and pyridine in water, *J. Photochem. Photobiol. A: Chem.*, **180** (2006) 130–135.
22. Q. Wu, D. Li, Y. Hou, L. Wu, X. Fu and X. Wang, Study of relationship between surface transient photoconductivity and liquid-phase photocatalytic activity of titanium dioxide, *Mater. Chem. Phys.*, **102** (2007) 53–59.

23. P. Kofstad, *Nonstoichiometry, Diffusion and Electrical Conductivity in Binary Metal Oxides*, Wiley–Interscience, New York, 1972.
24. F. A. Kröger, *The Chemistry of Imperfect Crystals*, volume 3, North Holland, Amsterdam, 1974.
25 C. Guillard, E. Puzenat, H. Lachheb, A. Houas and J.-M. Herrmann, Why inorganic salts decrease the TiO_2 photocatalytic efficiency, *Int. J. Photoenergy*, **7** (2005) 1–9.
26 Data calculated by the program FactSage version 5.3.1.
27 P. Kofstad, Note on the defect structure of rutile (TiO_2), *J. Less Common Metals*, **13** (1967) 635–638.
28 P. Kofstad, Thermogravimetric studies of the defect structure of rutile (TiO_2), *J. Phys. Chem. Solids*, **23** (1962) 1579–1586.
29 K. S. Førland, The defect structure of rutile, *Acta Chem. Scand.*, **18** (1964) 1267–1275.
30 J. B. Moser, R. N. Blumenthal and D. H. Whitmore, Thermodynamic study of nonstoichiometric rutile (TiO_2-x), *J. Am. Ceram. Soc.*, **48** (1965) 384–384.
31 L. M. Atlas and G. J. Schlehman, Katharometric and resistivity studies of PuO_2-x equilibria, Technical Report ANL-6960, Aragonne National Laboratory, 1965, reported by J. B. Moser *et al.*, *J. Am. Ceram. Soc.*, **48** (1965) 384.
32 C. B. Alcock, S. Zador and B. C. H. Steele, A thermodynamic study of dilute solutions of defects in the rutile structure TiO_{2-x}, NbO_{2-x}, and Ti0.5Nb0.5$O_{2\pm x}$, *Proc. Brit. Ceram. Soc.*, **8** (1967) 231–245.
33 D.-K. Lee, J.-I. Jeon, M.-H. Kim, W. Choi and H.-I. Yoo, Oxygen nonstoichiometry (δ) of $TiO_{2-\delta}$—Revisited, *J. Solid State Chem.*, **178** (2005) 185–193.
34 A. Fujishima, K. Hashimoto and T. Watanabe, *TiO_2 Photocatalysis. Fundamentals and Applications*, BKC Inc., Tokyo, 1999.
35 J. Nowotny, T. Bak, M. K. Nowotny and L. R. Sheppard, TiO_2 surface active sites for water splitting, *J. Phys. Chem. B*, **110** (2006) 18492–18495.
36 M. K. Nowotny, T. Bak and J. Nowotny, Electrical properties and defect chemistry of TiO_2 single crystal. I. Electrical conductivity, *J. Phys. Chem. B*, **110** (2006) 16270–16282.
37 M. K. Nowotny, T. Bak and J. Nowotny, Electrical properties of TiO_2 single crystal. II. Thermoelectric power, *J. Phys. Chem. B*, **110** (2006) 16283–16291.
38 M. K. Nowotny, T. Bak and J. Nowotny, Electrical properties of TiO_2 single crystal. III. Equilibration kinetics and chemical diffusion, *J. Phys. Chem. B*, **110** (2006) 16292–16301.
39 M. K. Nowotny, T. Bak and J. Nowotny, Electrical properties of TiO_2 single crystal. IV. Prolonged oxidation kinetics and chemical diffusion, *J. Phys. Chem. B*, **110** (2006) 16302–16308.
40 L. R. Sheppard, J. Nowotny and T. Bak, Metallic TiO_2, *Phys. Stat. Solidi (A)*, **203** (2006) R85–R87.
41 J. Nowotny, T. Bak, M. K. Nowotny and L. R. Sheppard, Chemical diffusion in metal oxides. example of TiO_2, *Ionics*, **12** (2006) 227–243.
42 T. Bak, J. Nowotny and M. K. Nowotny, Defect disorder of titanium dioxide, *J. Phys. Chem. B*, **110** (2006) 21560–21567.
43 L. R. Sheppard, T. Bak and J. Nowotny, Electrical properties of niobium-doped titanium dioxide. 1. Defect disorder, *J. Phys. Chem. B*, **110** (2006) 22447–22454.
44 L. R. Sheppard, T. Bak and J. Nowotny, Electrical properties of niobium-doped titanium dioxide. 2. Equilibration kinetics, *J. Phys. Chem. B*, **110** (2006) 22455–22461.
45 T. Bak, J. Nowotny, M. K. Nowotny and L. R. Sheppard, Reactivity at the oxygen/titania interface and the related charge transfer, *Ionics*, **12** (2006) 247–251.
46 J. Nowotny, T. Bak and T. Burg, Electrical properties of polycrystalline TiO_2 at elevated temperatures. Electrical conductivity, *Phys. Stat. Solidi (B)*, **244** (2007) 2037–2054.
47 J. Nowotny, T. Bak, M. K. Nowotny and L. R. Sheppard, Titanium dioxide for solar-hydrogen II. Defect chemistry, *Int. J. Hydrogen Energy*, **32** (2007) 2630–2643.
48 J. Nowotny, T. Bak, M. K. Nowotny and L. R. Sheppard, Titanium dioxide for solar-hydrogen III. Kinetic effects at elevated temperatures, *Int. J. Hydrogen Energy*, **32** (2007) 2644–2650.
49 J. Nowotny, T. Bak, M. K. Nowotny and L. R. Sheppard, Titanium dioxide for solar-hydrogen IV. Collective and local factors in photoreactivity, *Int. J. Hydrogen Energy*, **32** (2007) 2651–2659.
50 L. R. Sheppard, T. Bak, J. Nowotny and M. K. Nowotny, Titanium dioxide for solar hydrogen V. Metallic-type conduction of Nb-doped TiO_2, *Int. J. Hydrogen Energy*, **32** (2007) 2660–2663.

51 J. Nowotny, T. Bak and T. Burg, Electrical properties of polycrystalline TiO_2: Equilibration kinetics, *Ionics*, **13** (2007) 71–78.
52 J. Nowotny, T. Bak and T. Burg, Electrical properties of polycrystalline TiO_2: Prolonged oxidation kinetics, *Ionics*, **13** (2007) 79–82.
53 J. Nowotny, T. Bak and T. Burg, Electrical properties of polycrystalline TiO_2: Thermoelectric power, *Ionics*, **13** (2007) 155–162.
54 L. R. Sheppard, A. J. Atanacio, T. Bak, J. Nowotny and K. E. Prince, Bulk diffusion of niobium in single-crystal titanium dioxide, *J. Phys. Chem. B*, **111** (2007) 8126–8130.
55 J. Nowotny, T. Bak, T. Burg, M. K. Nowotny and L. R. Sheppard, Effect of grain boundaries on semiconducting properties of TiO_2 at elevated temperatures, *J. Phys. Chem. C*, **111** (2007) 9769–9778.
56 J. Nowotny, T. Bak, M. K. Nowotny and L. R. Sheppard, Defect chemistry and electrical properties of titanium dioxide. 1. Defect diagrams, *J. Phys. Chem. C*, **112** (2008) 590–601.
57 J. Nowotny, T. Bak, M. K. Nowotny and L. R. Sheppard, Defect chemistry and electrical properties of titanium dioxide. 2. Effect of aliovalent ions, *J. Phys. Chem. C*, **112** (2008) 602–610.
58. L. R. Sheppard, T. Bak and J. Nowotny, Electrical properties of niobium-doped titanium dioxide. 3. Thermoelectric power, *J. Phys. Chem. C*, **112** (2008) 611–617.
59. T. Bak, M. K. Nowotny, L. R. Sheppard and J. Nowotny, Charge transport in Cr-doped titanium dioxide, *J. Phys. Chem. C*, **112** (2008) 7255–7262.
60. T. Bak, M. K. Nowotny, L. R. Sheppard and J. Nowotny, Mobility of electronic charge carriers in titanium dioxide, *J. Phys. Chem. C*, **112** (2008) 12981–12987.
61. J. He, R. K. Behera, M. W. Finnis, X. Li, E. C. Dickey, S. R. Phillpot and S. B. Sinnott, Prediction of high-temperature point defect formation in TiO_2 from combined *ab initio* and thermodynamic calculations, *Acta Materialia*, **55** (2007) 4325–4337.
62. A. K. Ghosh, F. G. Wakim and R. R. Addiss Jr., Photoelectronic processes in rutile, *Phys. Rev.*, **184** (1969) 979–988.
63. F. M. Hossain, G. E. Murch, L. Sheppard and J. Nowotny, The effect of defect disorder on the electronic structure of rutile TiO_{2-x}, *Defect Diffusion Forum*, **251–252** (2006) 1–12, (Defects and Diffusion Ceramics Abstracts).
64. I. E. Grey and N. C. Wilson, Titanium vacancy defects in solgel prepared anatase, *J. Solid State Chem.*, **180** (2007) 670–678.
65. H. Matzke, Diffusion in nonstoichiometric oxides, in O. T. Sørensen (Editor) *Nonstoichiometric Oxides*, Academic Press, New York, 1981, pp. 155–230.
66. S. Andersson, B. Collén, U. Kuylenstierna and A. Magnéli, Phase analysis studies on the titanium-oxygen system, *Acta Chem. Scand.*, **11** (1957) 1641–1652.
67. L. A. Bursill, B. G. Hyde, O. Terasaki and D. Watanabe, New family of titanium oxide and the nature of slightly-reduced rutile, *Phil. Mag.*, **20** (1969) 347–359.
68. B. G. Hyde and L. A. Bursil, in L. Eyring and M. O'Keeffe (Editors) *The Chemistry of Extended Defects in Non-Metallic Solids*, North Holland, Amsterdam, 1970, pp. 347–373.
69. N. G. Eror, Self-compensation in niobium-doped TiO_2, *J. Solid State Chem.*, **38** (1981) 281–287.
70. J. Gautron, J. F. Marucco and P. Lemasson, Reduction and doping of semiconducting rutile (TiO_2), *Mater. Res. Bull.*, **16** (1981) 575–580.
71. J.-L. Carpentier, A. Lebrun and F. Perdu, Point defects and charge transport in pure and chromium-doped rutile at 1273 K, *J. Phys. Chem. Solids*, **50** (1989) 145–151.
72. H. P. R. Fredrikse, Recent studies on rutile (TiO_2), *J. Appl. Phys.*, **32** (1961) 2211–2215.
73. J. H. Becker and H. P. R. Frederikse, Electrical properties of nonstoichiometric semiconductors, *J. Appl. Phys.*, **33** (1962) 447–453.
74. G. H. Jonker, The application of combined conductivity and Seebeck-effect plots for the analysis of semiconductor properties (conductivity vs Seebeck coefficient plots for analyzing n-type, p-type and mixed conduction semiconductors transport properties), *Philips Res. Rep.*, **23** (1968) 131–138.
75. J. Nowotny, C. C. Sorrell, L. R. Sheppard and T. Bak, Solar-hydrogen: Environmentally safe fuel for the future, *Int J Hydrogen Energy*, **30** (2005) 521–544.

76. J. Nowotny, Interface electrical phenomena in ionic solids, in P. J. Gellings and H. J. M. Bouwmeester (Editors) *Handbook of Solid State Electrochemistry*, CRC Press, Boca Raton, 1997, pp. 121–160.
77. T. Bak, J. Nowotny, C. C. Sorrell, M. F. Zhou and E. R. Vance, Charge transport in $CaTiO_3$: Electrical conductivity, *J Mater Sci: Materials for Electronics*, **15** (2004) 635–644.
78. M. Russ, *Cost-Effective Strategies for an Optimised Allocation of Carbon Dioxide Emission Reduction Measures*, Unwelttechnik, Verlag-Shaker, Aachen, 1994.
79. J. Nowotny, T. Bak, L. R. Sheppard and M. K. Nowotny, Reactivity of titanium dioxide with oxygen at room temperature and the related charge transfer, *J. Am. Chem. Soc.*, **130** (2008) 9984–9993.
80. B. O. Seraphin, Spectrally selective surfaces and their impact on photothermal solar energy conversion, in B. O. Seraphin (Editor) *Solar Energy Conversion: Solid State Physics Aspects*, Springer-Verlag, Berlin, volume 31 of *Topics in Applied Physics*, 1979, pp. 5–56.
81. E. N. Figurovskaya, V. F. Kiselev and F. F. Volkenstein, Influence of oxygen chemisorption on work function and electrical conductivity of titanium dioxide, *Doklady Akad. Nauk USSR*, **161** (1965) 1142–1145, in Russian.
82. S. Bourasseau, J. R. Martin, F. Juillet and S. J. Teichner, Variation of thermoelectronic extraction function of powdered semiconductors under electromagnetic irradiation. I. Construction of an experimental device, *J. Chim. Phys. Phys.-Chim. Biol.*, **70** (1973) 1467–1471.
83. S. Bourasseau, J. R. Martin, F. Juillet and S. J. Teichner, Variation of the thermoelectronic extraction function of powdered semiconductors under electromagnetic radiation. II. Photodesorption of oxygen from titanium dioxide (anatase), *J. Chim. Phys. Phys.-Chim. Biol.*, **70** (1973) 1472–1477.
84. S. Bourasseau, J. R. Martin, F. Juillet and S. J. Teichner, Variation in the thermoelectronic work function of powdered semiconductors submitted to electromagnetic radiation. III. Adsorption of oxygen by titania (anatase), *J. Chim. Phys. Phys.-Chim. Biol.*, **71** (1974) 22–26.
85. S. Bourasseau, J. R. Martin, F. Juillet and S. J. Teichner, Variation of the thermoelectronic work function of powdered semiconductors submitted to the action of electromagnetic radiation. IV. Photoadsorption and photodesorption of oxygen on titanium dioxide, *J. Chim. Phys. Phys.-Chim. Biol.*, **71** (1974) 1017–1024.
86. S. Bourasseau, J. R. Martin, F. Juillet and S. J. Teichner, Variation of the thermoelectronic work function of powdered semiconductors submitted to the action of electromagnetic radiation. V. Case of titanium dioxide in the presence of the mechanism of photooxidation of paraffins, *J. Chim. Phys. Phys.-Chim. Biol.*, **71** (1974) 1025–1027.
87. S. Bourasseau, *Variation du Travail d'Extraction Thermoelectrique du Dioxyde de Titane Pulveruent, Soumis a un Rayonnement Ultraviolet*, Ph.D. thesis, L'Universite Claude Bernard, Lyon, 1973.
88. P. Odier and J. P. Loup, Thermal emission of electrons: A sensitive probe for investigation of non-stoichiometry of oxides at high temperature, in J. Nowotny (Editor) *Transport in Non-Stoichiometric Compounds: Proceedings of the First International Conference on Transport in Non-Stoichiometric Compounds Held at Mogilany Near Cracow*, August 27–30, 1980, Elsevier Scientific, Amsterdam, number 15 in Materials Science Monographs, pp. 393–408.
89. J. C. Riflet, P. Odier, M. Anthony and J. P. Loup, Use of thermoelectronic emission for studying point defects in refractory oxides, *J. Am. Ceram. Soc.*, **58** (1975) 493–497.
90. Lord Kelvin, Contact electricity of metals, *Phil. Mag.*, **46** (1898) 82–120.
91. J. Nowotny, Surface re-equilibration kinetics of nonstoichiometric oxides, *J. Mater. Sci.*, **12** (1977) 1143–1160.
92. J. Nowotny, T. Bak and C. C. Sorrell, Charge transfer at oxygen/zirconia interface at elevated temperatures: Part 6 Work function measurements, *Adv App Ceramics*, 104 (2005) 188–194.
93. C. Wagner, 2 June 1977, private communication to J. Nowotny.
94. L. Kronik and Y. Shapira, Surface photovoltage phenomena: theory, experiment, and applications, *Surf. Sci. Rep.*, **37** (1999) 1–206.
95. H. Tributsch, J. Nowotny, M. K. Nowotny and L. R. Sheppard, Photoreactivity models for titanium dioxide with water, *Energy Mater.*, 3 (2008) 158–168.

96. J. Nowotny, Titanium dioxide-based semiconductors for solar-driven environmentally friendly applications: impact of point defects on performance, *Energy Environ Sci*, 1 (2008) 565–572.
97. F. F. Volkenstein, *The Electronic Theory of Catalysis on Semiconductors*, Pergamon Press, New York, 1964.
98. V. E. Henrich and P. A. Cox, *The Surface Science of Metal Oxides*, Cambridge University Press, Cambridge, 1994, pp. 319–321.
99. A. Bielański and J. Haber, Oxygen in catalysis on transition metal oxides, *Catal. Rev. Sci. Eng.*, **19** (1979) 1–41.
100. J. Nowotny, T. Norby and T. Bak, Reactivity between titanium dioxide and water at elevated temperatures, *J. Phys. Chem. C*, **114** (2010) 18215–18221.
101. J. Nowotny, Interface defect chemistry and its impact on properties of oxide ceramic materials, in J. Nowotny (Editor) *Science of Ceramic Interfaces*, Elsevier Science, Amsterdam, volume 75 of Materials Science Monographs, 1991, pp. 79–204.
102. J. Nowotny, Work function of oxide ceramic materials, in J. Nowotny and L.-C. Dufour (Editors) *Surface and Near-Surface Chemistry of Oxide Materials*, Elsevier Scientific, Amsterdam, Materials Science Monographs, 1988, pp. 281–343.
103. J. Cabane and F. Cabane, Equilibrium segregation in interfaces, in J. Nowotny (Editor) *Interface Segregation and Related Processes in Materials*, Trans Tech Publications, Zurich, 1991, pp. 1–159.
104. L. R. Sheppard, A. Atanacio, T. Bak, J. Nowotny and K. E. Prince, Effect of niobium segregation on surface properties of titanium dioxide, *Proc. SPIE*, **6340** (2006) 634015.
105. H. P. Maruska and A. K. Ghosh, Transition-metal dopants for extending the response of titanate photoelectrolysis anodes, *Solar Energy Mats.*, **1** (1979) 237–247.
106. NREL, Renewable Resource Data Center, Reference solar spectral irradiance: Air mass 1.5, http://rredc.nrel.gov/solar/spectra/am1.5/, accessed: February 2009.
107. Newport Oriel Instruments, Light sources—technical information, http://www.newport.com/Technical-Information/381840/1033/catalog.aspx, accessed February 2009.
108. R. Asahi, T. Morikawa, T. Ohwaki, K. Aoki and Y. Taga, Visible-light photocatalysis in nitrogen-doped titanium oxides, *Science*, **293** (2001) 269–271.
109. A. Kudo, K. Omori and H. Kato, A novel aqueous process for preparation of crystal form-controlled and highly crystalline $BiVO_4$ powder from layered vanadates at room temperature and its photocatalytic and photophysical properties, *J. Am. Chem. Soc.*, **121** (1999) 11459–11467.
110. T. Bak, J. Nowotny, M. Rekas and C. C. Sorrell, Photoelectrochemical hydrogen generation from water using solar energy. Materials-related aspects, *Int J Hydrogen Energy*, 27 (2002) 991–1022.
111. R. F. Service, Catalyst boosts hopes for hydrogen bonanza, *Science*, **297** (2002) 2189–2190.
112. V. S. Fomenko, *Emission Properties of Elements and Chemical Compounds*, Naukova Dumka, Kiev, 1964, in Russian.
113. R. Memming, Solar energy conversion by photoelectrochemical processes, *Electrochim. Acta*, **25** (1980) 77–88.
114. P. A. Kohl, S. N. Frank and A. J. Bard, Semiconductor electrodes, *J. Electrochem.Soc.*, **124** (1977) 225–229.
115. H. Gerischer, Solar photoelectrolysis with semiconductor electrodes, in B. O. Seraphin (Editor) *Solar Energy Conversion: Solid-State Physics Aspects*, Springer, Berlin, volume 31 of Topics in Applied Physics, 1979, pp. 115–172.
116. A. J. Nozik, p-n photoelectrolysis cells, *Appl. Phys. Lett.*, **29** (1976) 150–153.
117. H. Yoneyama, H. Sakamoto and H. Tamura, A photo-electochemical cell with production of hydrogen and oxygen by a cell reaction, *Electrochim. Acta*, **20** (1975) 341–345.
118. G. Peharz, F. Dimroth and U. Wittstadt, Solar hydrogen production by water splitting with a conversion efficiency of 18%, *Int. J. Hydrogen Energy*, **32** (2007) 3248–3252.
119. H. Morisaki, T. Watanabe, M. Iwase and K. Yazawa, Photoelectrolysis of water with TiO_2-covered solar-cell electrodes, *Appl. Phys. Lett.*, **29** (1976) 338–340.

120. G. Zhao, H. Kozuka and T. Yoko, Effects of the incorporation of silver and gold nanoparticles on the photoanodic properties of rose bengal sensitized TiO_2 film electrodes prepared by sol-gel method, *Solar Energy Mater. Solar Cells*, **46** (1997) 219–231.
121. M. Gräatzel, Mesoscopic solar cells for electricity and hydrogen production from sunlight, *Chem. Lett.*, **34** (2005) 8–13.
122. K. Wilke and H. D. Breuer, The influence of transition metal doping on the physical and photocatalytic properties of titania, *J. Photochem. Photobiol. A: Chem.*, **121** (1999) 49–53.
123. S. U. M. Khan, M. Al-Shahry and W. B. Ingler Jr, Efficient photochemical water splitting by a chemically modified n-TiO_2, *Science*, **297** (2002) 2243–2245.
124. P. R. F. Barnes, L. K. Randeniya, A. B. Murphy, P. B. Gwan, I. C. Plumb, I. E. Grey and C. Li, TiO_2 photo-electrodes for water splitting prepared by flame pyrolysis, Intern. Conf. on Materials for Hydrogen Energy, 27 August 2004, Sydney, Australia.
125. T. E. Phillips, K. Moorjani, J. C. Murphy and T. O. Poehler, TiO_2–VO_2 alloysreduced bandgap effects in the photoelectrolysis of water, *J. Electrochem. Soc.*, **129** (1982) 1210–1215.
126. G. Zhao, H. Kozuka, H. Lin and T. Yoko, Sol-gel preparation of $Ti_{1x}V_xO_2$ solid solution film electrodes with conspicuous photoresponse in the visible region, *Thin Solid Films*, **339** (1999) 123–128.
127. G. Zhao, H. Kozuka, H. Lin, M. Takahashi and T. Yoko, Preparation and photoelectrochemical properties of $Ti_{1-x}V_xO_2$ solid solution thin film photoelectrodes with gradient bandgap, *Thin Solid Films*, **340** (1999) 125–131.
128. T. Kasuga, M. Hiramatsu, A. Hoson, T. Sekino and K. Niihara, Titania nanotubes prepared by chemical processing, *Adv. Mater.*, **11** (1999) 1307–1311.
129. T. Dittrich, Porous TiO_2: Electron transport and application to dye sensitized injection solar cells, *Phys. Stat. Sol. (A)*, **182** (2000) 447–455.
130. S. K. Mohapatra, M. Misra, V. K. Mahajan and K. S. Raja, A novel method for the synthesis of titania nanotubes using sonoelectrochemical method and its application for photoelectrochemical splitting of water, *J. Catal.*, **246** (2007) 362–369.
131. Z. Liu, Q. Zhang and L.-C. Qin, Reduction in the electronic band gap of titanium oxide nanotubes, *Solid State Comm.*, **141** (2007) 168–171.
132. W. Wunderlich, N. T. Hue and S. Tanemura, Fabrication of nano-structured titania—thin-films on carbon-coated nickel sheets, *AZo J. Mater. Online*, **2** (2006), DOI: 10.2240/azojomo0208.
133. S. Chandra, *Photoelectrochemical Solar Cells*, Electrocomponent Science Monographs, Routledge, New York, 1985.
134. J. Nowotny, Work function in studying mechanism and kinetics of heterogeneous reactions in the system oxygen-nonstoichiometric oxides, *J. Chim. Phys. Phys.-Chim. Biol.*, **75** (1978) 689–702.
135. W. J. Lo, Y. W. Chung and G. A. Somorjai, Electron spectroscopy studies of the chemisorption of O_2, H_2 and H_2O on the TiO_2(100) surfaces with varied stoichiometry: Evidence for the photogeneration of Ti_{+3} and for its importance in chemisorption, *Surf. Sci.*, **71** (1978) 199–219.
136. B. Parkinson, On the efficiency and stability of photoelectrochemical devices, *Acc. Chem. Res.*, **17** (1984) 431–437.
137. T. Watanabe, A. Fujishima and K. Honda, Photoelectrochemical reactions at $SrTiO_3$ single crystal electrode, *Bull. Chem. Soc. Jpn.*, **49** (1976) 355–358.
138. J. G. Mavroides, D. I. Tchernev, J. A. Kafalas and D. F. Kolesar, Photoelectrolysis of water in cells with titanium oxide anodes, *Mat. Res. Bull.*, **10** (1975) 1023–1030.
139. M. Y. El Zayat, A. O. Saed and M. S. El-Dessouki, Photoelectrochemical properties of dye sensitized Zr-doped $SrTiO_3$ electrodes, *Int. J. Hydrogen Energy*, **23** (1998) 259–266.
140. J. F. Houlihan, D. B. Armitage, T. Hoovler, D. Bonaquist, D. P. Madacsi and L. N. Mulay, Doped polycrystalline TiO_2 electrodes for the photo-assisted electrolysis of water, *Mater. Res. Bull.*, **13** (1978) 1205–1212.
141. T. Tani, N. Sekiguchi, M. Sakai and D. Ohta, Optimization of solar hydrogen systems based on hydrogen production cost, *Solar Energy*, **68** (2000) 143–149.
142. N. A. Kelly, T. L. Gibson and D. B. Ouwerkerk, A solar-powered, high-efficiency hydrogen fueling system using high-pressure electrolysis of water: Design and initial results, *Int. J. Hydrogen Energy*, **33** (2008) 2747–2764.

143. D. Morgan and F. Sissine, Congressional Research Service, report for Congress, Technical Report D.C. 20006–1401, The Committee for the National Institute for the Environment, Washington, April 1995.
144. T. N. Veziroglu, Dawn of the hydrogen age, *Int. J. Hydrogen Energy*, **23** (1998) 1077–1078.
145. U.S. Department of Energy, Hydrogen the fuel for the future, Technical Report DOE/GO-1-95-099, National Renewable Energy Laboratory, March 1995.
146. C. E. Thomas, B. D. James and F. D. Lomax Jr, Market penetration scenarios for fuel cell vehicles, *Int. J. Hydrogen Energy*, **23** (1998) 949–966.
147. T. N. Veziroglu, More smelling land, *Int. J. Hydrogen Energy*, **25** (2000) 601–601.
148. T. N. Veziroglu, Quarter century of hydrogen movement 1974–2000, *Int. J. Hydrogen Energy*, **25** (2000) 1143–1150.
149. United Nations Environmental Program, Full list of vital climate graphics, http://www.grida.no/db/maps/collection/climate6/austral.htm, 12th March 2006, GRID, Arendal, Norway.
150. A. Fujishima, K. Kohayakawa and K. Honda, Hydrogen production under sunlight with an electrochemical photocell, *J. Electrochem. Soc.*, **122** (1975) 1487–1489.
151. A. J. Nozik, Photoelectrolysis of water using semiconducting TiO_2 crystals, *Nature*, **257** (1975) 383–386.
152. T. Watanabe, A. Fujishima and K. Honda, Photoelectrochemical hydrogen production, in T. Ohta (Editor) *Solar-Hydrogen Energy Systems*, Pergamon Press, Oxford, 1979, pp. 137–169.
153. T. Ohnishi, Y. Nakato and H. Tsubomura, Quantum yield of photolysis of water on titanium dioxide electrodes, *Ber. Bunsen-Ges.*, **79** (1975) 523–525.
154. G. Hodes, D. Cahen and J. Manassen, Tungsten trioxide as a photoanode for a photoelectrochemical cell (PEC), *Nature*, **260** (1976) 312–313.
155. M. S. Wrighton, A. B. Ellis, P. T. Wolczanski, D. L. Morse, H. B. Abrahamsona and D. S. Ginley, Strontium titanate photoelectrodes. Efficient photoassisted electrolysis of water at zero applied potential, *J. Am. Chem. Soc.*, **98** (1976) 2774–2779.
156. M. A. Butler, R. D. Nasby and R. K. Quinn, Tungsten trioxide as an electrode for photoelectrolysis of water, *Solid State Comm.*, **19** (1976) 1011–1014.
157. J. M. Bolts and M. S. Wrighton, Correlation of photocurrent-voltage curves with flatband potential for stable photoelectrodes for the photoelectrolysis of water, *J. Phys. Chem.*, **80** (1976) 2641–2645.
158. J. H. Carey and B. G. Oliver, Intensity effects in the electrochemical photolysis of water at the TiO_2 electrode, *Nature*, **259** (1976) 554–556.
159. E. C. Dutoit, F. Cardon and W. P. Gomes, Electrochemical reactions involving holes at the illuminated TiO_2 (rutile) single crystal electrode, *Ber. Bunsen-Ges.*, **80** (1976) 1285–1288.
160. J. G. Mavroides, J. A. Kafalas and D. F. Kolesar, Photoelectrolysis of water in cells with $SrTiO_3$ anodes, *Appl. Phys. Lett.*, **28** (1976) 241–243.
161. P. D. Fleischauer and J. K. Allen, Photochemical hydrogen formation by the use of titanium dioxide thin-film electrodes with visible-light excitation, *J. Phys. Chem.*, **82** (1978) 432–438.
162. M. A. Butler, M. Abramovich, F. Decker and J. F. Juliao, Subband gap response of TiO_2 and $SrTiO_3$ photoelectrodes, *J. Electrochem. Soc.*, **128** (1981) 200–204.
163. A. A. Soliman and H. J. J. Seguin, Reactively sputtered TiO_2 electrodes from metallic targets for water electrolysis using solar energy, *Solar Energy Mater.*, **5** (1981) 95–102.
164. J. Kiwi and M. Grätzel, Heterogeneous photocatalysis: enhanced dihydrogen production in titanium dioxide dispersions under irradiation. The effect of magnesium promoter at the semiconductor interface, *J. Phys. Chem.*, **90** (1986) 637–640.
165. G. Prasad, N. N. Rao and O. N. Srivastava, On the photoelectrodes TiO_2 and WSe_2 for hydrogen production through photoelectrolysis, *Int. J. Hydrogen Energy*, **13** (1988) 399–405.
166. R. N. Pandey, K. S. Chandra Babu, D. Singh and O. N. Srivastava, Studies on n-CdSe/Ti semiconductor septum based photoelectrochemical solar cell in regard to the influence of structural and compositional characteristics of the semiconductor electrode, *Bull. Chem. Soc. Japan*, **65** (1992) 1072–1077.
167. J. Augustynski, The role of the surface intermediates in the photoelectrochemical behaviour of anatase and rutile TiO_2, *Electrochim. Acta*, **38** (1993) 43–46.

168. J. P. H. Sukamto, C. S. McMillan and W. Smyrl, Photoelectrochemical investigations of thin metal-oxide films: TiO_2, Al_2O_3, and HfO_2 on the parent metals, *Electrochim. Acta*, **38** (1993) 15–27.
169. L. D. Danny Harvey, Solar-hydrogen electricity generation and global CO_2 emission reduction, *Int. J. Hydrogen Energy*, **21** (1996) 583–595.
170. J. R. Bolton, Solar photoproduction of hydrogen: A review, *Solar Energy*, **57** (1996) 37–50.
171. R. N. Pandey, K. S. Chandra Babu and O. N. Srivastava, High conversion efficiency photoelectrochemical solar cells, *Prog. Surf. Sci.*, **52** (1996) 125–192.
172. G. Zhao, S. Utsumi, H. Kozuka and T. Yoko, Photoelectrochemical properties of solgel-derived anatase and rutile TiO_2 films, *J. Mater. Sci.*, **33** (1998) 3655–3659.
173. Y. Liu, A. Hagfeldt, X.-R. Xiao and S.-E. Lindquist, Investigation of influence of redox species on the interfacial energetics of a dye-sensitized nanoporous TiO_2 solar cell, *Solar Energy Mater. Solar Cells*, **55** (1998) 267–281.
174. A. Stanley, B. Verity and D. Matthews, Minimizing the dark current at the dyesensitized TiO_2 electrode, *Solar Energy Mater. Solar Cells*, **52** (1998) 141–154.
175. R. N. Pandey, M. Misra and O. N. Srivastava, Solar hydrogen production using semiconductor septum (n-CdSe/Ti and n-TiO_2/Ti) electrode based photoelectrochemical solar cells, *Int. J. Hydrogen Energy*, **23** (1998) 861–865.
176. M. Gómez, J. Rodríguez, S. Tingry, A. Hagfeldt, S.-E. Lindquist and C. G. Granqvist, Photoelectrochemical effect in dye sensitized, sputter deposited Ti oxide films: The role of thickness-dependent roughness and porosity, *Solar Energy Mater. Solar Cells*, **59** (1999) 277–287.
177. Y. Hao, M. Yang, C. Yu, S. Cai, M. Liu, L. Fan and Y. Li, Photoelectrochemical studies on acid-doped polyaniline as sensitizer for TiO_2 nanoporous film, *Solar Energy Mater. Solar Cells*, **56** (1998) 75–84.
178. H. Kozuka, Y. Takahashi, G. Zhao and T. Yoko, Preparation and photoelectrochemical properties of porous thin films composed of submicron TiO_2 particles, *Thin Solid Films*, **358** (2000) 172–179.
179. V. Aranyos, H. Grennberg, S. Tingry, S.-E. Lindquist and A. Hagfeldt, Electrochemical and photoelectrochemical investigation of new carboxylatobipyridine (bisbipyridine) ruthenium (ii) complexes for dye-sensitized TiO_2 electrodes, *Solar Energy Mater. Solar Cells*, **64** (2000) 97–114.
180. Yomiuri Shimbun, Satellite system would generate clean fuel, *The Daily Yomiuri*, 18 August 2001.
181. J. F. Houlihan, J. R. Hamilton and D. P. Madacsi, Improved solar efficiencies for doped polycrystalline TiO_2 photoanodes, *Mater. Res. Bull.*, **14** (1979) 915–920.
182. T. Bak, J. Nowotny, M. Rekas and C. C. Sorrell, Photo-electrochemical properties of the TiO_2–Pt system in aqueous solutions, *Int. J. Hydrogen Energy*, **27** (2001) 19–26.
183. Nanoptek Inc., Technology, http://www.nanoptek.com/technology.html, accessed: February 2009.
184. K. Bullis, Cheap hydrogen, http://www.technologyreview.com/energy/20134/, January 2008, accessed: February 2009.
185. D. Chatterjee and S. Dasgupta, Visible light induced photocatalytic degradation of organic pollutants, *J. Photochem. Photobiol. C: Photochem. Rev.*, **6** (2005) 186–205.
186. O. Carp, C. L. Huisman and A. Reller, Photoinduced reactivity of titanium dioxide, *Progr. Solid State Chem.*, **32** (2004) 33–177.
187. National Renewable Energy Laboratory, News release NR-2708: NREL solar cell sets world efficiency record at 40.8 percent, http://www.nrel.gov/news/press/2008/625.html, 13 August 2008, accessed: February 2009.
188. E. Becquerel, Memoire sur les effets electriques produits sous l'influence des rayons solaires, *Compt. Rend.*, **9** (1839) 561–567.
189. W. H. Brattain and C. G. B. Garrett, Experiments on the interface between germanium and an electrolyte, *Bell System Tech. J.*, **34** (1955) 129–177.
190. H. Gerischer, Kinetics of oxidation-reduction reactions on metals and semiconductors. I. General remarks on the electron transition between a solid body and a reductionoxidation electrolyte, *Z. Phys. Chem.*, **26** (1960) 223–247.

191. R. Memming, Mechanism of the electrochemical reduction of persulfates and hydrogen peroxide, *J. Electrochem. Soc.*, **116** (1969) 785–790.
192. S. R. Morrison, *Electrochemistry at Semiconductor and Oxidized Metal Electrodes*, Plenum Press, New York, 1980.
193. A. K. Ghosh and H. P. Maruska, Photoelectrolysis of water in sunlight with sensitized semiconductor electrodes, *J. Electrochem. Soc.*, **124** (1977) 1516–1522.
194. G. Renger, Water cleavage by solar radiation—An inspiring challenge of photosynthesis research, *Photosynth. Res.*, **38** (1993) 229–247.
195. A. W. Rutherford and A. Boussac, Water photolysis in biology, *Science*, **303** (2004) 1782–1784.
196. W. Lubitz, E. J. Reijerse and J. Messinger, Solar water splitting into H_2 and O_2: Design principles of photosystem II and hydrogenases, *Energy Environ. Sci.*, **1** (2008) 15–31.
197. J. J. Concepcion, J. W. Jurss, J. L. Templeton and T. J. Meyer, One site is enough. Catalytic water oxidation by $[Ru(tpy)(bpm)(OH_2)]^{2+}$ and $[Ru(tpy)(bpz)(OH_2)]^{2+}$, *J. Am. Chem. Soc.*, **130** (2008) 16462–16463.

Recommended Readings

Bockris JO'M, *Energy. The Solar-Hydrogen Alternative*, Australia & New Zealand Book Company, Sydney, New South Wales, Australia, 1975.
Bockris JO'M and Khan SUM, *Surface Electrochemistry*, Plenum Press, New York, 1993.
Bockris JO'M, Veziroglu TN, and Smith D, *Solar Hydrogen Energy*, MacDonald & Co, London, U.K., 1991.
Chandra S, *Photoelectrochemical Solar Cells*, Gordon & Breach Science Publishers, New York, 1985.
Gellings PJ and Bouwmeester HJM, eds., *Solid State Electrochemistry*, CRC Press, Boca Raton, FL, 1997.
Green MA, *Solar Cells*, University of New South Wales, Sydney, New South Wales, Australia, 1986.
Hanjalic K. ed., *Sustainable Energy Technologies*, Springer, New York, 2008.
Johansson TB, Kelly H, Reddy AKN, and Williams RH, *Renewable Energy*, Island Press, Washington, DC, 1993.
Kofstad P, *Nonstoichiometry, Electrical Conductivity and Diffusion in Binary Metal Oxides*, Wiley, New York, 1972.
Nowotny J, ed., *Science of Ceramic Interfaces*, Elsevier, Amsterdam, the Netherlands, 1991.
Nowotny J, ed., *Science of Ceramic Interfaces II*, Elsevier, Amsterdam, the Netherlands, 1994.
Nowotny J and Dufour L-C, eds., *Surface and Near-Surface Chemistry of Metal Oxides*, Elsevier, Amsterdam, the Netherlands, 1988.
Rajeshwar K, McConnel R, and Licht S, eds., *Solar Hydrogen Generation*, Springer, New York, 2008.
Seraphin BO, ed., *Solar Energy Conversion*, Springer-Verlag, Berlin, Germany, 1979.
West AR, *Basic Solid State Chemistry*, Wiley, Chichester, U.K., 1988.

14

Life Cycle Analysis and Economic Assessment of Solar Hydrogen

Anton Meier
Paul Scherrer Institute

CONTENTS

14.1 Introduction

H_2 and electricity will likely become the main energy carriers in future low-carbon energy systems. Long-term energy scenarios such as the Energy Technology Perspectives of the International Energy Agency (IEA) [1] and the World Energy Technology Outlook 2050 (WETO-H_2) of the European Commission (EC) [2] predict considerable market penetration of emerging H_2 production and end use technologies [3]. H_2 is expected to be particularly advantageous as transportation fuel due to its versatility, pollutant-free end use, and storage capability. H_2 produced from nonfossil fuels (e.g., renewable energy sources like solar energy) and used in fuel cells (FCs) can provide sustainable energy and thus reduce the adverse energy-related environmental effects on climate change, such as greenhouse gas (GHG) emissions. Of special interest are solar thermochemical H_2 production processes and their prospects in terms of technological and economic potential as well as their role in future H_2 supply [3].

14.1.1 Objectives and Scope

This chapter is based on a comprehensive well-to-wheel (WTW) analysis and an economic assessment for solar-produced H_2 and its utilization in fuel cell vehicles (FCVs) for passenger transportation [4]. Results from the life cycle analysis (LCA) include environmental impacts of GHG emissions and cumulative energy demand (CED), as well as damage caused to human health (HH), ecosystem quality (EQ), and natural resources [5]. The LCA encompasses all environmental interactions like extraction of resources, fuel processing steps, supply, use, and final disposal. The economic assessment provides H_2 production and supply costs for selected scenarios [6]. The eco-efficiency analysis method combines ecological and economic aspects in condensed form [6].

Conventional technologies used by industry to produce H_2 include steam reforming of natural gas (NG) (or steam methane reforming [SMR]), coal gasification (CGA), and water electrolysis (ELE). At the start of a growing hydrogen economy, it is anticipated that H_2 will be produced by advanced, process-optimized SMR [7] as well as CGA followed by CO_2 capture and sequestration [8]. In the long term, strong H_2 markets and a growing infrastructure will create opportunities for renewable H_2 production systems [7]. Of special interest are recent technological advances in the field of solar thermochemical production of H_2 using concentrated solar radiation as the energy source of high-temperature process heat [9]. Among the most promising H_2O-splitting thermochemical processes is the two-step Zn/ZnO cycle [10–12]. The intermediate energy carrier Zn is formed in an endothermic step either by solar thermal dissociation (STD) of ZnO at above 2000 K [13–15] or, alternatively, by solar carbothermic reduction (SCR) of ZnO at about 1500 K [16]. In both cases, H_2 is generated from H_2O in a nonsolar exothermic step by hydrolysis of Zn [17], the benchmark being solar H_2 production by state-of-the-art alkaline electrolysis of H_2O using electricity from solar thermal power plants (solar thermal electricity [STE]).

The WTW analysis includes solar H_2 production, transport, and usage in future passenger car transportation systems. The H_2 is assumed to be produced in a concentrating solar power (CSP) plant located in Southern Spain (ES) and transported to Central Europe exemplified by Switzerland (CH). Solar H_2 production methods (STD, SCR, and STE) are compared with selected conventional production technologies (SMR, CGA, ELE). Energy transport concepts include (1) Zn transport, (2) on-site Zn hydrolysis followed by H_2 pipeline transport, and (3) high-voltage direct current (HVDC) electricity transport from a solar thermal power plant. Utilization of H_2 in an FCV is compared with advanced power trains for the combustion of oil-based energy carriers [18].

14.1.2 System Boundary Definition

The system boundaries are determined by the complete process chain—from the exploitation of natural resources to the functional unit of one passenger kilometer (assuming average load of 1.59 passengers per car [19]). Each process step accounts for relevant energy and resource consumption, land use, and emissions. Data source for conventional and ancillary processes is the Swiss LCA database *ecoinvent v1.2* [20] describing standard technology in Western Europe around the year 2000. This comprehensive dataset is a consistent background for all energy systems compared and is used here as an approximation for the reference year 2025, although alterations of involved processes may be expected. The average European electricity mix (UCTE'25) used for all background processes is predicted based on an EC business-as-usual scenario [21] that utilizes an increased share of NG in more efficient combined cycle plants. Uncertainties associated with the time frame of this study are being addressed with sensitivity calculations to further substantiate the results obtained.

14.2 Methodology

The LCA and WTW methodologies employed for the production of H_2 and its use in mobile applications show similarities but also differences [22]. WTW studies are specifically aimed at transport applications focusing on the production and distribution of different fuels and on the emissions of vehicles during use. They include GHG emissions (e.g., contributions from CO_2, N_2O, and CH_4) and an energy (efficiency) indicator. In contrast, LCA is a general methodology that can be applied to any kind of system or product. LCA studies focus on full life cycles of products or product systems. Applied to transportation, they typically include the three phases of a vehicle (production, use, and end of life) as well as the production and distribution processes of the fuels consumed. LCA studies usually include more impact categories than WTW studies, such as acidification, eutrophication, ozone layer depletion, and carcinogens.

14.2.1 Ecological Assessment

LCA aims at quantifying cumulative environmental interactions in terms of resource consumption and emissions to air, water, and soil. Based on this life cycle inventory (LCI), a life cycle impact assessment (LCIA) can be conducted by rating these interactions using various methodologies. Their application follows a procedure described in ISO norms [23,24]. The basic idea is to use the cumulative inventories and multiply the resulting single elementary flows with substance-specific factors. The following methodologies are employed for the present assessment:

- *GHG* emissions are taken into account based on the GHG species' global warming potential (GWP) using infrared forcing values relative to CO_2 according to the Intergovernmental Panel on Climate Change (IPCC) [25].
- *CED* describes primary energy resource consumption [26].
- The comprehensive *Eco-indicator '99 Hierarchist* (*EI'99-H*) method [27] provides quantification of aggregated environmental impacts of all emitted or used substances in three damage categories: HH, EQ, and both fossil and mineral resources. The damage categories are normalized and weighted according to the *Hierarchist* perspective considered closest to the scientists' point of view.
- The *Eco-scarcity '97* (Umweltbelastungspunkt [*UBP*]) method [28] also yields an aggregated impact score. Conceptually, it is a distance-to-target model that

compares actual flows to the environment with critical flows derived from scientifically sound political targets such as emission limits. Multiplying each mass flow with its corresponding weighting factor yields a specific environmental impact value (so-called UBP). In contrast to *EI'99*, no damage categories are implemented in *Eco-scarcity*.

- *HH* and *EQ* both represent environmental damage categories described in the *Impact 2002+* methodology [29]. *HH* mainly considers respiratory damages and carcinogenic effects; *EQ* includes land use and ecotoxic emissions.

14.2.2 Economic Assessment

The economic competitiveness of the renewable processes considered in this study is judged according to different cost calculations:

- *Fuel and transportation costs* encompass fuel production costs per energy unit and passenger car transportation costs per passenger kilometer (pkm), respectively.
- *External costs* represent a measure of environmental damage normally not included in the direct costs of a technology. Internalization of external costs provides an economic indicator for cost-benefit analyses of alternative options.

14.2.3 Eco-Efficiency

The eco-efficiency methodology [30,31] links both ecological and economic aspects of a specific technology under consideration. With this approach, an environmental indicator—such as GHG emissions—is combined with an economic indicator—such as production or service costs. Both indicators are normalized with a typical reference value, for example, the state-of-the-art technology, the specific technology to be substituted, or the average value of the technology alternatives. A weighting factor describes the importance of costs versus environmental impacts.

14.3 Fuels and Power Trains

14.3.1 Solar H_2 Pathways

H_2 from concentrated solar power (*CSP*). Figure 14.1 shows pathways for the provision of solar H_2 that encompass (1) the production of H_2 or an intermediate energy carrier (Zn or electricity) using concentrated solar energy; (2) the transport of H_2, Zn, or electricity from Southern Spain to Central Europe (Switzerland); and (3) further processing and, finally, use of H_2 in an FCV. The fuel production pathways and power trains considered are listed in Table 14.1.

Note that some important technologies have been excluded from the present study: compressed natural gas (CNG), although a viable option as transition fuel, does not fit into the long-term solar H_2 scenario; biofuels, represented by synthetic natural gas (SNG) from wood, are being addressed in a different work [4]; H_2-fueled internal combustion engines (ICEs) offer distinctly lower efficiencies than FCs [18]; and liquid H_2 is energetically inefficient compared to gaseous H_2 [32].

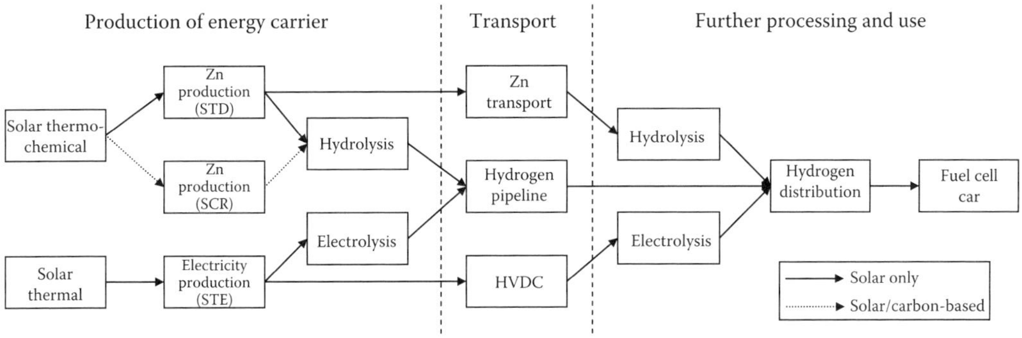

FIGURE 14.1
Pathways for (1) solar production of H_2 at a CSP plant located in Southern Spain; (2) transport of H_2 or an intermediate energy carrier such as Zn or electricity to Central Europe (Switzerland); (3) distribution and utilization of H_2 in an FCV. Acronyms are explained in Table 14.1. (From Felder, R. and Meier, A., *J. Sol. Energy Eng.*, 130(1), 011017-1/10, 2008.)

14.3.2 Specifications and Assumptions

For consistency, all solar technologies considered in this study use the same CSP tower configuration based on the PS10 plant near Seville, Spain [33,34], as specified in Table 14.2. A field of sun-tracking heliostats (624 glass–metal heliostats of 121 m^2 reflecting surface area each) concentrates direct solar radiation onto the top of a tower (100 m height). At the design point (insolation of 860 W/m^2 and optical efficiency of 77%), the reference CSP plant features 50 MW thermal power input into a receiver (for STE) [35] or a chemical reactor (for STD) mounted on top of the tower; for SCR, a tower reflector (TR) redirects the sunrays to a chemical reactor placed on the ground [36]. The average annual optical efficiency of the heliostat field (HF) is $\eta_1 = Q_{HF}/Q_{solar} = 64\%$. For typical sites in Southern Spain, the direct normal irradiance (DNI) is about 2000 $kWh/m^2/a$. A plant lifetime of 30 years is assumed [37]. Infrastructure data are taken from a detailed analysis of CSP tower plants generating STE [38].

It is conceivable that the HF will look different for the various solar applications compared: On one hand, a large receiver (for STE) operating at relatively low temperature and a TR (for SCR) both can accept sunlight from a wide angle; on the other hand, a chemical reactor (for STD) operating at high temperatures and solar concentrations near 5000 suns (1 sun = 1 kW/m^2) requires a compound parabolic concentrator (CPC) with a much smaller acceptance angle [39]. This leads to a different—usually elliptical—shape of the HF with a significant number of heliostats at larger distance from the tower, thus resulting in additional efficiency losses through atmospheric attenuation. It is anticipated that these losses may be compensated by advanced heliostat designs (e.g., with higher mirror quality), however, at the penalty of higher costs for STD. The validation of this assumption requires a detailed analysis of the various HF configurations, which lies outside the scope of this study.

STE. A benchmark for the thermochemical processes (STD, SCR) is STE generation followed by H_2 production via H_2O electrolysis (STE), as shown in Figure 14.2. Table 14.3 summarizes the assumptions for the STE scenario. The CSP tower plants currently in operation, under construction, or in planning use either the saturated steam receiver [34] or the molten salt receiver [40] concept. In both cases, the heat transfer medium absorbs concentrated solar energy and delivers heat via a heat exchanger to a steam turbine to drive an electric generator. Average thermal

TABLE 14.1

Summary of Fuel Production Pathways and Power Trains

Acronym	Fuel Production Pathways and Power Trains	Production Site	Transport
Compressed Gaseous Hydrogen			
Solar			
STD pipe	Solar thermal dissociation of ZnO + Zn hydrolysis	On-site ES	CG H_2 pipeline
STD zinc	Solar thermal dissociation of ZnO + Zn hydrolysis	CH	Zn, ship and rail
STE pipe	Solar thermal electricity + water electrolysis	On-site ES	CG H_2 pipeline
STE hvdc	Solar thermal electricity + water electrolysis	CH	HVDC
Fossil			
SMR	Steam methane reforming	Northern Europe	CG H_2 pipeline
CGA	Coal gasification (advanced)	NE Europe	CG H_2 pipeline
Electrolysis			
ELE CH mix	Electrolysis using predicted Swiss mix 2030	CH	—
ELE nuclear	Electrolysis using nuclear power	CH	—
ELE hydro	Electrolysis using hydro power	CH	—
ELE wind	Electrolysis using wind power	CH/Import	HVDC
ELE PV	Electrolysis using photovoltaics	CH	—
Power Train			
FCV	PEM fuel cell vehicle. 0.94 MJ_{LHV}/km		
Electricity			
Electricity Source			
STE hvdc	Solar thermal electricity	ES	HVDC
CH mix	Predicted Swiss consumer mix 2030	CH	—
Hydro	Hydropower	CH	—
Power Train			
BEV	Battery electric vehicle; 0.53 MJ/km		
CNG			
Wood			
SNG	Synthetic natural gas from wood gasification	CH	—
Fossil			
CNG	Compressed natural gas	EU	CNG pipeline
Power Train			
ICE CNG	Internal combustion engine; 1.93 MJ/km		
Liquid Hydrocarbons			
Fossil			
Gasoline	Gasoline	EU	Ship, truck
Diesel	Diesel	EU	Ship, truck
Power Trains			
ICE G	Internal combustion engine, gasoline; 1.90 MJ/km		
ICE D	Internal combustion engine, diesel; 1.80 MJ/km		

Indicated are acronyms, production sites of end energy carrier, and long-distance transport options for energy carriers.

Note: CH, Switzerland; ES, Spain; EU, European Union; CG, compressed gas; LHV, lower heating value.

TABLE 14.2
Solar Plant Specification (Baseline Case)

	Parameter	Unit	Value	Remarks
Baseline CSP plant	Solar thermal power input	MW	50	Design point
	Insolation	W/m^2	860	Design point
	DNI	$kWh/m^2/a$	2,000	Site dependent
	Optical efficiency:	—	0.77	PS10 [33]; design point
	$\eta_1 = Q_{HF}/Q_{solar}$		0.64	PS10 [33]; mean annual value
	Tower height	m	100	Derived from [33]
	Number of heliostats	—	624	Glass–metal [38]
	Single heliostat area	m^2	121	PS10 [33]
	HF area	m^2	75,504	
	Annual solar energy to receiver	GJ/a	347,922	Based on DNI
	Land use factor	—	0.2	[37]
	Land area	m^2	377,520	Total plant area
	Plant lifetime	a	30	[37]

Source: Felder, R. and Meier, A., *J. Sol. Energy Eng.*, 130(1), 011017-1/10, 2008.

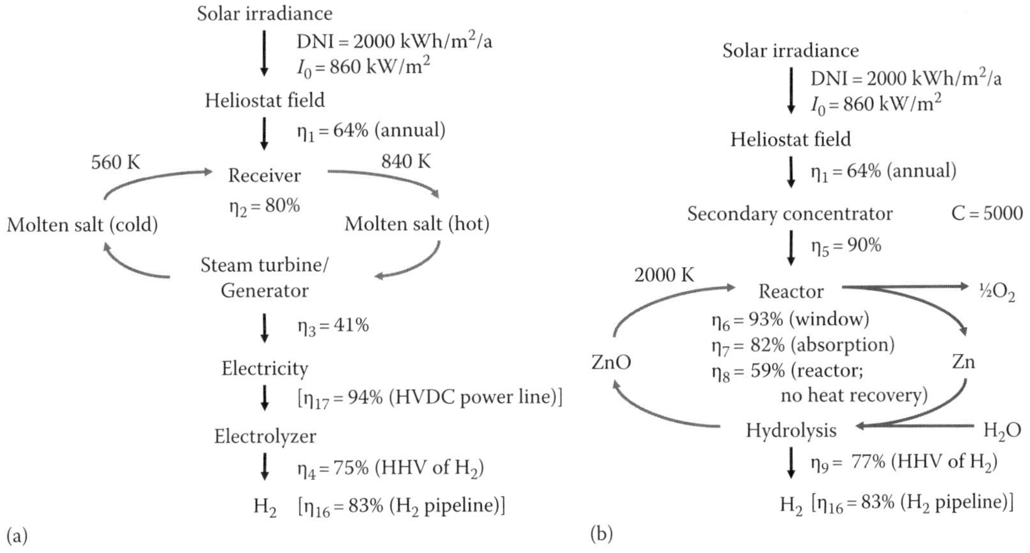

FIGURE 14.2
(a) Flow chart of H_2 production via STE generation followed by H_2O electrolysis. (b) Flow chart of H_2 production via solar thermal ZnO dissociation (STD) and Zn hydrolysis. For efficiencies, see Tables 14.3, 14.4, and 14.6, respectively. Higher heating value (HHV) of H_2 used.

energy conversion efficiencies $\eta_2 = Q_{receiver}/Q_{HF}$ of 80% for a molten salt receiver and $\eta_3 = Q_{el}/Q_{receiver}$ of 41% for the electricity generation cycle are taken from a CSP study [37]. H_2 is produced by H_2O electrolysis based on bipolar alkaline low-temperature technology with a thermal efficiency $\eta_4 = Q_{H_2}/Q_{el}$ of 75% (using HHV of H_2) [41].

STD of ZnO. Solar thermochemical cycles for fuel production offer the potential of high-energy conversion efficiencies [12]. Figure 14.3 presents the model flow diagram of the H_2O-splitting solar thermochemical Zn/ZnO cycle. Table 14.4

TABLE 14.3
STE Generation and H_2O Electrolysis

	Parameter	Unit	Value	Remarks
CSP plant ($16\ MW_{el}$)	Molten salt receiver efficiency: $\eta_2 = Q_{receiver}/Q_{HF}$	—	0.80	[37] (2020); adapted to actual plant size
	Power block efficiency: $\eta_3 = Q_{el}/Q_{receiver}$	—	0.41	[37] (2020); incl. parasitic losses
	Storage capacity (3h peak power)	MWh	180	Based on Solar Two Tower Plant [42]
	Annual plant availability	—	0.94	CSP tower plant [37]
H_2O electrolysis ($12\ MW_{H_2,HHV}$)	H_2O + electricity → $H_2 + ½O_2$			Bipolar alkaline low-temperature technology
	H_2 production capacity at 30 bar	kg/h	290	Based on design point
	Electrolyzer efficiency (HHV): $\eta_4 = Q_{H_2}/Q_{el}$	—	0.75	[41]

Source: Felder, R. and Meier, A., *J. Sol. Energy Eng.*, 130(1), 011017-1/10, 2008.

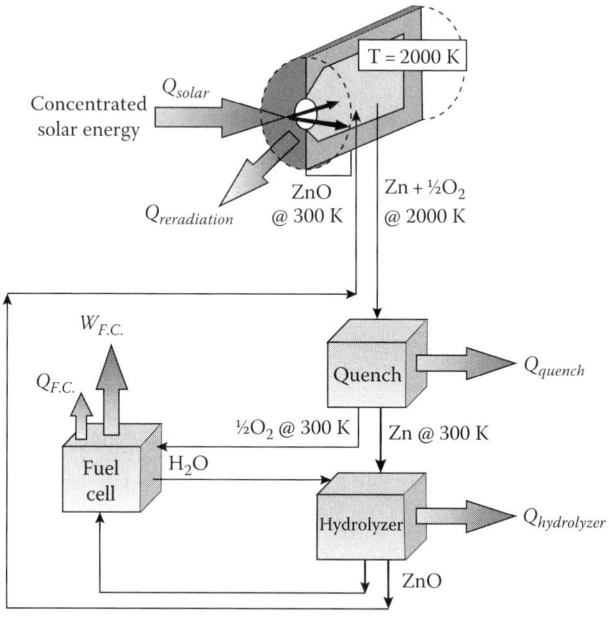

FIGURE 14.3
Model flow diagram of the H_2O-splitting solar thermochemical Zn/ZnO cycle. (Based on Steinfeld, A., *Int. J. Hydrogen Energy*, 27(6), 611, 2002.)

summarizes the assumptions used for the STD scenario. To achieve process temperatures exceeding 2000 K, the solar radiation reflected from the HF has to be further concentrated using a nonimaging CPC. Thus, the optical efficiency $\eta_1 = Q_{HF}/Q_{solar}$ of 64% is decreased by absorption and spillage losses at the CPC (10%) and, in addition, by absorption and reflection losses at the quartz

TABLE 14.4

STD of ZnO and Zn Hydrolysis

	Parameter	Unit	Value	Remarks
CSP plant	Solar reactor:			
(20 MW_{Zn})	$ZnO + heat \rightarrow Zn + 0.5O_2$			
	Zn production capacity	kg/h	10,700	Design point (860 W/m²)
	CPC: $\eta_5 = Q_{CPC}/Q_{HF}$	—	0.90	Own assumption
	Window: $\eta_6 = Q_{window}/Q_{CPC}$		0.93	Own assumption
	Absorption: $\eta_7 = Q_{reactor}/Q_{window/CPC}$		0.82	T = 2000 K; C = 5000 suns [44]
	Solar reactor: $\eta_8 = Q_{Zn}/Q_{reactor}$		0.59	Incl. thermal and quenching losses; min. insolation level
	Annual plant availability	—	0.94	CSP tower plant [37]
H_2 plant	Hydrolyzer:			
(16 $MW_{H_2,HHV}$)	$Zn + H_2O \rightarrow ZnO + H_2$			
Zn hydrolysis	H_2 production capacity at 1 bar	kg/h	300	Design point (860 W/m²)
(16 $MW_{H_2,HHV}$)	Hydrolyzer efficiency (HHV):	—	0.77	95% molar conversion
	$\eta_9 = Q_{H_2,ES}/Q_{Zn}$			

Source: Felder, R. and Meier, A., *J. Sol. Energy Eng.*, 130(1), 011017-1/10, 2008.

window mounted in front of the solar chemical reactor (7%). Reradiation losses through the reactor aperture account for 18%, assuming a solar flux concentration of 5000 suns and a cavity temperature of 2000 K. Such high temperatures require a minimum insolation level for reactor operation [43], thus reducing the annual reactor efficiency $\eta_8 = Q_{Zn}/Q_{reactor}$, which is further decreased by convection and conduction heat losses from the well-insulated reactor (up to 10%) and losses by quenching the highly reactive gaseous mixture of Zn and O_2 to ambient temperature (up to 26% [12]). Quenching losses can somewhat be alleviated if the quench takes effect from 2000 to 700 K (instead of 298 K) and if part (up to 50%) of the sensible energy in the products can be recovered. Finally, H_2 production by Zn hydrolysis [17] is assumed proceeding at a thermal conversion efficiency $\eta_9 = Q_{H_2,ES}/Q_{Zn}$ of 77% (HHV of H_2) with 95% molar conversion.

SCR of ZnO. The technical feasibility of the SCR of ZnO has been successfully demonstrated in a 300 kW pilot plant implemented at a solar facility with beam-down optical configuration [45]. Table 14.5 summarizes the assumptions used for the SCR scenario. The optical efficiency η_1 of 64% is reduced by absorption and spillage losses at the TR (10%) and the CPC (10%). Reradiation losses through the reactor aperture account for 19%, assuming solar a flux concentration of 2000 suns and a cavity temperature of 1600 K. For large-scale industrial plants, maximum reactor efficiency $Q_{Zn}/Q_{reactor}$ of 69% is anticipated [46], resulting in overall reactor efficiency $\eta_{13} = (Q_{Zn} + Q_{CO})/(Q_{reactor} + Q_C)$ of 64% if the energy contents of carbon (Q_C) and CO (Q_{CO}) are included [47] and 60% if, in addition, a minimum solar insolation level for operating the reactor is taken into account [43]. The source of carbon can be different kinds of coal powder. The so-called SOLZINC process made use of beech charcoal, which showed the best overall reaction properties [48], apart from being a renewable resource. However, for

TABLE 14.5

SCR of ZnO and Zn Hydrolysis

	Parameter	Unit	Value	Remarks
CSP plant	Solar reactor:			
(23 MW_{Zn} and 11 MW_{CO})	$ZnO + C + heat \rightarrow Zn + CO$			
	Zn production capacity	kg/h	15,100	Design point (860 W/m^2)
	TR: $\eta_{10} = Q_{TR}/Q_{HF}$	—	0.90	Own assumption
	CPC: $\eta_{11} = Q_{CPC}/Q_{TR}$		0.90	Own assumption
	Absorption: $\eta_{12} = Q_{reactor}/Q_{CPC}$		0.81	T = 1600 K; C = 2000 suns
	Solar reactor:			
	$\eta_{13} = (Q_{Zn} + Q_{CO})/(Q_{reactor} + Q_C)$		0.60	Based on [46]; incl. thermal losses; min. insolation level
	Efficiency (total):			
	$\eta_{tot} = (Q_{Zn} + Q_{CO})/(Q_{solar} + Q_C)$		0.31	
	Annual plant availability	—	0.94	CSP tower plant [37]
H_2 plant	Hydrolyzer: $Zn + H_2O \rightarrow ZnO + H_2$			
(27 $MW_{H_2,HHV}$)	CO shift: $CO + H_2O \rightarrow CO_2 + H_2$			
Zn Hydrolysis (18 $MW_{H_2,HHV}$)	$\eta_9 = Q_{H_2,ES}/Q_{Zn}$ H_2 production capacity at 1 bar	kg/h	700	Design point (860 W/m^2)
+ CO–H_2O shift	Hydrolyzer efficiency (HHV):	—	0.77	95% molar conversion
(9 $MW_{H_2,HHV}$)	CO–H_2O shift efficiency (HHV): $\eta_{14} = Q_{H_2,ES}/Q_{CO}$	—	1.00	Long term: 100% conversion [51]
	Combined efficiency (HHV): $\eta_{15} = Q_{H_2,ES}/(Q_{Zn} + Q_{CO})$	—	0.84	Process efficiency considering Zn and CO mass flows

Source: Felder, R. and Meier, A., *J. Sol. Energy Eng.*, 130(1), 011017-1/10, 2008.

each solar plant, site-specific carbonaceous feedstock needs to be identified. Charcoal is produced using the Reichert procedure [49,50], an advanced technology including treatment and utilization of the wood gases in order to minimize external energy demand and emissions during production; in particular, release of CH_4 is eliminated. H_2 is produced by Zn hydrolysis [17] and also from CO via water–gas shift reaction [51] with a combined thermal efficiency η_{15} of 84% (HHV of H_2).

Transport options. H_2 can be produced at the CSP plant with subsequent pipeline delivery (STD pipe and STE pipe) over a distance of 1650 km assumed between Southern Spain (ES) and Central Europe, exemplified by Switzerland (CH); alternatively, H_2 can be produced at the consumer site after transport of either Zn powder (STD zinc) or electricity (STE hvdc). The assumptions for the transport options are summarized in Table 14.6. H_2 produced on-site uses transport infrastructure based on today's CNG pipelines and storage tanks [52]. Losses for H_2 pipeline transport are quantified with 7.2% per 1000 km, plus additional loss of 5.1% for initial compression to pipeline pressure of 120 bar [52,53]. Electric transmission losses amount to 3.3% per 1000 km in 600 kV HVDC power lines [54] and to additional 1% in the Swiss medium-voltage alternating current (MVAC) grid [55]. For Zn transport by freight ship (1250 km) and rail (400 km), material losses are neglected, but energy consumption is taken into account.

TABLE 14.6

Summary of Transport Options

	Parameter	Unit	Value	Remarks
H_2 pipeline	Transport distance	km	1650	Southern Spain (ES) to Switzerland (CH)
	H_2 operating pressure	bar	120	In 2025 [54]
	Losses	%/1000 km	7.2	+5.1% losses for initial H_2 compression to 120 bar
	Efficiency: $\eta_{16} = Q_{H_2,CH}/Q_{H_2,ES}$	—	0.83	
Electricity transmission lines (HVDC and MVAC)	Transport distance (overland)	km	1650	Southern Spain (ES) to Switzerland (CH)
	Transmission loss (HVDC)	%/1000 km	3.3	600 kV [54]
	Transmission loss (MVAC)	%/100 km	1	20 kV [55]; CH grid
	Efficiency: $\eta_{17} = Q_{el,CH}/Q_{el,ES}$	—	0.94	
Zn/ZnO transport	Transport distance (freight ship)	km	1250	Southern Spain (ES) to Genoa, Italy
	Transport distance (rail)	km	400	Genoa, Italy to Switzerland (CH)
	Efficiency: $\eta_{18} = Q_{Zn,CH}/Q_{Zn,ES}$	—	1	Own assumption

Source: Felder, R. and Meier, A., *J. Sol. Energy Eng.*, 130(1), 011017-1/10, 2008.

Summary of solar H_2 production technologies. Table 14.7 summarizes energy conversion efficiencies for all considered solar H_2 pathways. The average annual solar-to-H_2 production efficiencies at the solar plant vary from 14.7% (STE) to 18.9% (STD) and 29.4% (SCR). If transport losses from Southern Spain to Switzerland are included, the annual solar-to-H_2 conversion efficiency ranges for STE from 12.2% (H_2 pipeline transport) to 13.8% (HVDC transport), for STD from 15.7% (H_2 pipeline transport) to 18.9% (Zn transport), while for SCR it is close to 25%.

Conventional H_2 production technologies and fuel options. The solar scenarios (STD, SCR, and STE) are compared with various pathways for alternative fuel production. Today's cheapest and most mature H_2 production technology is SMR (Table 14.8) with fuel conversion efficiency η_{19} of 89% (HHV of H_2) [41,56]. H_2 production from CGA (Table 14.9) based on autothermal gasification proceeds with thermal conversion efficiency η_{20} of 86% (HHV of H_2) [41,54,57]. H_2 can also be generated via H_2O electrolysis (ELE, Table 14.10) with an efficiency η_{21} of 75% (HHV of H_2) [41]. In this case, electricity is supplied either by the Swiss electricity mix (ELE, CH mix) [55] of locally produced CO_2-low hydro and nuclear power as predicted for 2030 [21] or—as a *worst-case scenario*—by the continental European electricity mix as forecast for the year 2025 (UCTE'25 [21]). Supply of nuclear, hydro, and wind power is documented in *ecoinvent* [58–60]. Photovoltaic (PV) data are taken from a Swiss study on renewable energies [61]. Vehicle data are based on a 55 kW Volkswagen Golf 4 [19]. H_2 powered FCV [62] is compared with advanced gasoline and diesel power trains [18,63]. LCI data for gasoline, diesel, and NG vehicles include car driving emissions [19,64,65] as well as construction, maintenance, and disposal of car and road infrastructure [19,66,67]. Vehicle efficiencies are taken from a study comparing future perspectives of different power trains (FCV, BEV, and ICE) [18].

TABLE 14.7

Average Annual Energy Conversion Efficiencies for Solar H_2 Pathways Depicted in Figure 14.1

Energy Conversion Efficiency (%)			STD pipe	STD zinc	SCR	STE pipe	STE hvdc
HF	Q_{HF}/Q_{solar}	η_1	64	64	64	64	64
TR	Q_{TR}/Q_{HF}	η_{10}	—	—	90	—	—
CPC	$Q_{CPC}/Q_{HF/TR}$	$\eta_{5,11}$	90	90	90	—	—
Window	Q_{window}/Q_{CPC}	η_6	93	93	—	—	—
Absorption	$Q_{reactor}/Q_{window/CPC}$	$\eta_{7,12}$	82	82	81	—	—
	STE: $Q_{receiver}/Q_{HF}$	η_2	—	—	—	80	80
Reactor	STD: $Q_{Zn}/Q_{reactor}$	η_8	59	59	—	—	—
	SCR: $(Q_{Zn} + Q_{CO})/(Q_{reactor} + Q_C)$	η_{13}	—	—	60[a]		
Electricity generation	$Q_{el}/Q_{receiver}$	η_3	—	—	—	41	41
Zn hydrolysis	Zn: $Q_{H_2,ES}/Q_{Zn}$	η_9	77	77	77	—	—
CO shift	CO: $Q_{H_2,ES}/Q_{CO}$	η_{14}	—	—	100		
	Zn + CO: $Q_{H_2,ES}/(Q_{Zn} + Q_{CO})$	η_{15}			84[a]		
Electrolysis	$Q_{H_2,ES}/Q_{el}$	η_4	—	—	—	75	75
Annual plant availability		η_{plant}	94	94	94	94	94
Annual solar-to-H_2 production efficiency	$Q_{H_2,ES}/(Q_{solar} + Q_C)$	η_{prod}	18.9	—	29.4	14.7	—
Transport	$Q_{H_2,CH}/Q_{H_2,ES}$	$\eta_{16\text{–}18}$	83	100	83	83	94
Total efficiency	$Q_{H_2,CH}/(Q_{solar} + Q_C)$	η_{tot}	15.7	18.9	24.4	12.2	13.8

Source: Felder, R. and Meier, A., *J. Sol. Energy Eng.*, 130(1), 011017-1/10, 2008.
Higher heating value (HHV) of H_2 used; acronyms explained in Table 14.1.
[a] SCR also includes energy stored in carbon source (C) and product (CO).

TABLE 14.8

SMR

	Parameter	Unit	Value	Remarks
SMR plant ($155\ MW_{H_2,HHV}$)	Catalytic steam reforming of NG:			Low-temperature shift reactor
	$CH_4 + H_2O \rightarrow CO + 3H_2$			
	$CO + H_2O \rightarrow CO_2 + H_2$			
	H_2 production capacity at 30 bar	Nm^3/h	69,400	Based on [56]
	Fuel conversion efficiency (HHV):	—	0.89	[41]
	$\eta_{19} = Q_{H_2,CH}/(Q_{NG,feed} + Q_{NG,fuel})$			

Source: Felder, R. and Meier, A., *J. Sol. Energy Eng.*, 130(1), 011017-1/10, 2008.

TABLE 14.9
CGA

	Parameter	Unit	Value	Remarks
CGA plant ($275\ MW_{H_2,HHV}$)	Autothermal gasification:			
	$C + H_2O \rightarrow CO + H_2$			
	$CO + H_2O \rightarrow CO_2 + H_2$			
	H_2 production capacity at 50 bar	Nm^3/h	100,000	Based on [54]
	Fuel conversion efficiency (HHV):	—	0.86	[41]
	$\eta_{20} = Q_{H_2,CH}/Q_{hardcoal}$			

Source: Felder, R. and Meier, A., *J. Sol. Energy Eng.*, 130(1), 011017-1/10, 2008.

TABLE 14.10
ELE at Local Gas Station

	Parameter	Unit	Value	Remarks
ELE gas station ($1.3\ MW_{H_2,HHV}$)	ELE:			Bipolar alkaline
	H_2O + electricity $\rightarrow H_2 + ½\ O_2$			low-temperature technology
	H_2 production capacity at 30 bar	kg/h	40	Based on [68]
	Efficiency (HHV): $\eta_{21} = Q_{H_2}/Q_{el}$	—	0.75	[41]
	Electricity demand	kWh/Nm^3	4.30	

Source: Felder, R. and Meier, A., *J. Sol. Energy Eng.*, 130(1), 011017-1/10, 2008.

14.4 Production and Supply Chains

14.4.1 Emissions and Cumulated Energy Demand

14.4.1.1 GHG Emissions and CED of Solar H_2 Production

Figure 14.4 shows GHG emissions for the H_2 production and utilization pathways considered in this study. In general, GHG emissions are distinctly lower for the scenarios based on solar energy than for those relying on fossil resources; the best solar technologies emit GHG up to 6 times less than SMR and more than 10 times less than CGA. Solar thermochemical production of Zn with on-site hydrolysis and subsequent H_2 pipeline transport (STD pipe) performs best. In contrast, transporting Zn (STD zinc) almost doubles GHG emissions because of associated transport emissions. Solar carbothermic production of Zn with charcoal (SCR cc) involves only slightly higher GHG emissions, as long as CH_4-free charcoal production technologies are used. However, it is important to limit the wood/charcoal transport distance to the CSP plant, since total WTW-GHG emissions increase by about 10% (or 1 g CO_2-eq. per MJ H_2) for 1000 km of railway transport or 100 km of lorry transport. Solar carbothermic production of Zn with hard coal (SCR hc) leads to markedly higher GHG emissions, though lower than conventional CGA. Solar thermal electrolysis also emits low amounts of GHG, with an advantage for H_2 pipeline transport (STE pipe) compared to electricity transport (STE hvdc) because of N_2O emissions in the corona of HVDC transmission lines (GWP of N_2O is 296 times

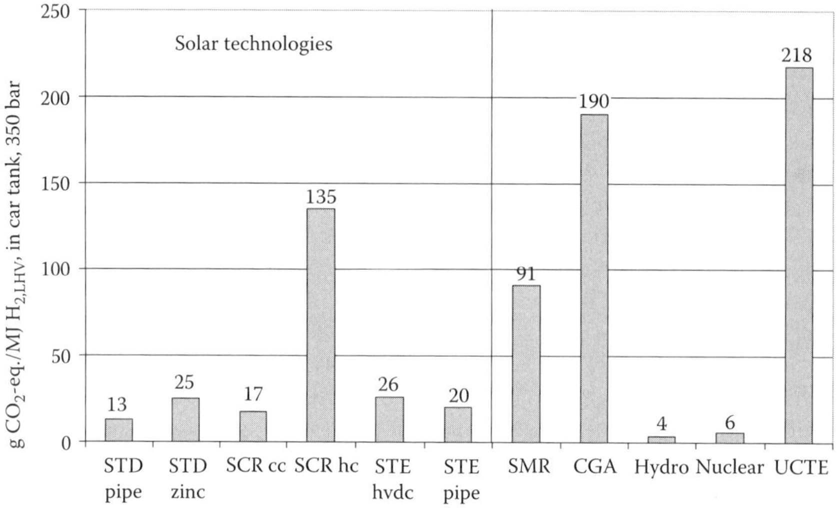

FIGURE 14.4
GHG emissions in g CO_2-eq. per MJ H_2 (LHV) in car tank at 350 bar. (From Felder, R. and Meier, A., *J. Sol. Energy Eng.*, 130(1), 011017-1/10, 2008.)

higher than that of CO_2 [25]). The life cycle GHG budget is lower for H_2O electrolysis using CO_2-low electricity from local Swiss nuclear power or hydropower, because the low GHG emissions at the CSP plant in Spain (7 g CO_2-eq. per MJ H_2) are increased by transport emissions and losses. However, producing H_2 from solar energy instead of using the future European electricity mix (UCTE'25) reduces GHG emissions by a factor of more than 10.

Figure 14.5 presents the CED for supplying H_2 to a car tank at 350 bar. The primary energy demand of solar-only pathways is very low apart from the considerable amount of solar energy needed due to low whole chain conversion efficiency. Fossil resource

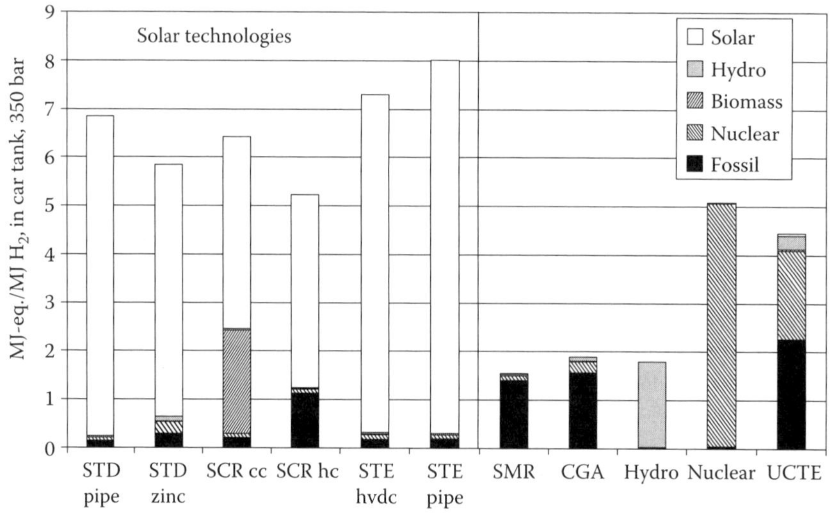

FIGURE 14.5
CED of primary energy sources, in MJ-eq. per MJ H_2 (LHV) in car tank at 350 bar. (From Felder, R. and Meier, A., *J. Sol. Energy Eng.*, 130(1), 011017-1/10, 2008.)

consumption for the best solar option (STD pipe) is only slightly higher than for hydropower but distinctly lower than for all pathways directly using fossil energy. In conclusion, solar technologies offer sustainable H_2 production by saving fossil energy resources by a factor of about 10.

14.4.1.2 GHG Emissions and CED of Passenger Car Technologies

Figure 14.6 presents WTW-GHG emissions for all fuel/power train combinations assessed in this study. Contributions of life cycle stages include construction, maintenance, and disposal of car and road infrastructure as well as fuel chain and combustion emissions. Cars powered by renewable fuels emit about 44 ± 8 g CO_2-eq. per passenger kilometer (pkm). Assuming lifetime mileage of 150,000 km, up to 75% of the emissions are associated with material-intensive car construction. Compared to the gasoline/diesel car, GHG emissions from car infrastructure increase by 28% for the FCV with carbon-fiber aluminum pressurized gas tank and by 19% for the BEV. Differences in total GHG emission mainly result from fuel supply and use. In general, the BEV shows the lowest GHG emissions because of its higher energy conversion efficiency compared to the FCV. The best performer is the BEV using local Swiss hydro, nuclear, and wind power. Even with slightly higher fuel-associated GHG emissions, the ICE car using biofuel (SNG) is among the lowest-emitting options, profiting from simpler and less material-intensive car construction.

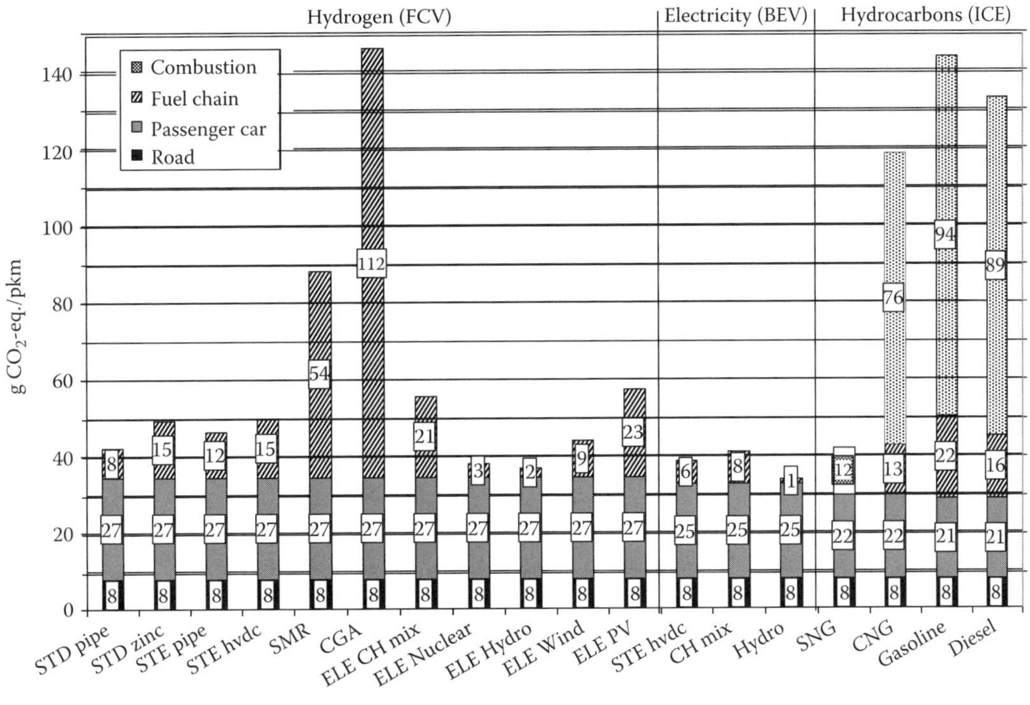

FIGURE 14.6
GHG emissions in g CO_2-eq. per pkm showing contributions of life cycle stages such as construction, maintenance, and disposal of passenger car and road infrastructure; fuel chain (production and supply); and direct combustion. Average load of 1.59 passengers per car is used [19]. For SNG, fuel chain and combustion emissions are summed up. For acronyms, see Table 14.1. (Details in Felder, R., Well-to-wheel analysis of renewable transport fuels: Synthetic natural gas from wood gasification and hydrogen from concentrated solar energy, PhD thesis, ETH No. 17437, ETH Zurich, Zurich, Switzerland, 2007.)

CSP technologies (STD, STE) reduce GHG emissions for the production of H_2 compared to conventional technologies by a factor of 7 (SMR) and 14 (CGA), which results in 50%–70% CO_2 savings in the FCV. CSP technologies also perform better than H_2O electrolysis using electricity from a future Swiss electricity mix (CH mix) or PV. However, compared to electrolysis using locally produced hydro, nuclear, or wind power, GHG emissions from CSP pathways are markedly higher due to steel-intensive infrastructure for the HF and energy transport losses from Southern Spain to Switzerland.

14.4.2 Life Cycle Impact Assessment

14.4.2.1 Eco-Indicator'99 Methodology

Figure 14.7 presents weighted LCIA results using the *Eco-indicator'99* (*EI'99*) methodology with the *Hierarchist* perspective. The dominant environmental impacts of solar H_2 production technologies are associated with the construction of the steel-intensive infrastructure for solar energy collection due to mineral and fossil resource consumption. In particular, today's steel production technology is responsible for GHG and other harmful emissions such as particulates causing respiratory damages. Land consumption and impacts on local flora and wildlife during the build-up of the CSP plant are often quoted as an issue [3]. However, *EI'99* possibly overrates the impact of land use, since CSP installations are preferably located at species-poor arid sites [5]. The damage category *other impacts* encompasses minor effects such as respiratory damages from organics, carcinogens, ecotoxicity, acidification, eutrophication, and ozone layer depletion. Operational impacts from CSP plants are negligible. Ancillary transports have only marginal effects except for Zn transport by freight ship and rail with associated fossil resource consumption and emission of CO_2, NO_x, SO_2, and particulates.

Environmental impacts from the carbothermic Zn/ZnO process (SCR) depend—apart from the construction of CSP plant infrastructure—on the reducing agent used: (1) For the production of charcoal, the main impact factors are land use for growing wood, possible CH_4 release by conventional production procedures, and emission of particulates

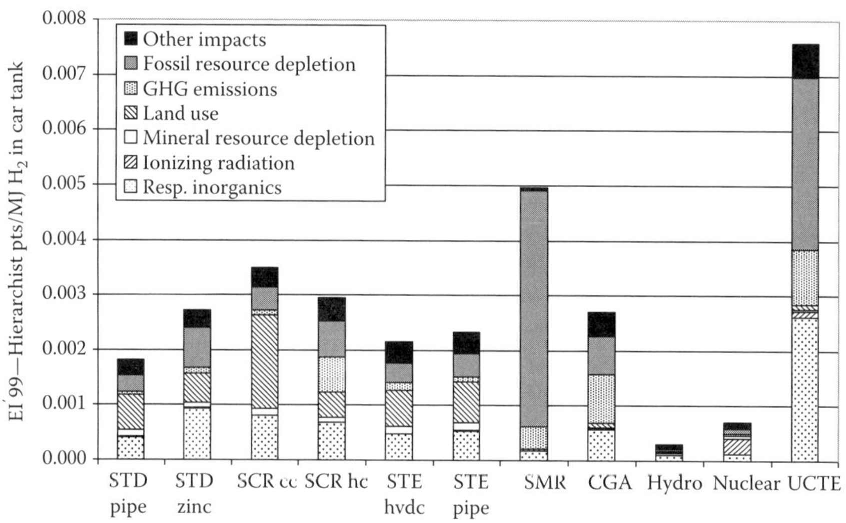

FIGURE 14.7
Eco-indicator'99—Hierarchist impact values per MJ H_2 (LHV) in car tank at 350 bar, presented for the most important impact categories. For details, see text. (From Felder, R. and Meier, A., *J. Sol. Energy Eng.*, 130(1), 011017-1/10, 2008.)

during forestry; (2) using hard coal, negative impacts result from fossil resource consumption and GHG emissions. STE and STD of ZnO (STD pipe) show about the same LCIA impact scores because of the similar energy conversion efficiency. Overall, electricity transport via HVDC power lines is slightly advantageous compared to H_2 pipeline transport due to lower energy losses. Conventional H_2 production technologies are penalized by fossil resource consumption and net GHG emissions. Using local hydropower to produce electricity for H_2O electrolysis shows the lowest environmental impact. Aspects of nuclear power production (ionizing radiation) are rated low in *EI'99*.

14.4.2.2 Eco-Scarcity'97 Methodology

In contrast to *EI'99*, the *Eco-scarcity'97* (*UBP*) methodology strongly penalizes nuclear waste disposal (Figure 14.8). Emissions of GHG and other harmful gases and particulates as well as hazardous heavy metals receive a stronger weight since fossil and mineral resource depletion and land use are neglected. Compared to the best-rated hydropower pathway, the *UBP* score is elevated for all CSP technologies due to the accumulation of various environmental impacts associated with the construction of plant infrastructure, as discussed previously. In general, environmental impacts of the solar H_2 production pathways are similar as long as efficient energy transport systems (H_2 pipelines and HVDC transmission lines) and no continuous fossil resource input (hard coal) are used.

14.4.2.3 Life Cycle Impact Assessment Example

Figure 14.9 depicts LCIA results for the STD scenario with pipeline transport of H_2 and its use in an FCV. Compared to the reference case—advanced gasoline vehicle—GHG emissions can be reduced by 70% to 44 g CO_2-eq. per pkm. Due to the relatively low conversion efficiency of concentrated solar energy, total CED increases by a factor of about

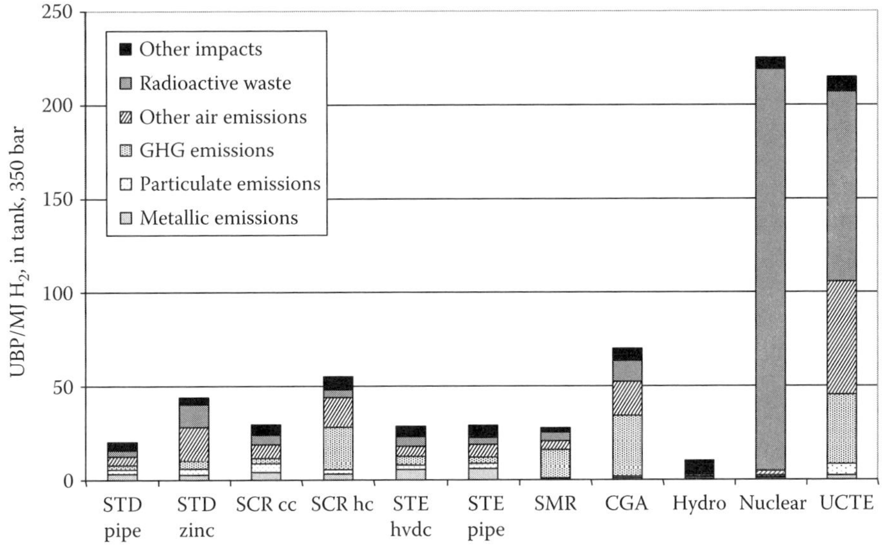

FIGURE 14.8
Eco-scarcity'97 impact values (UBP) per MJ H_2 (LHV) in car tank at 350 bar, presented for the most important impact categories. For details, see text. (From Felder, R. and Meier, A., *J. Sol. Energy Eng.*, 130(1), 011017-1/10, 2008.)

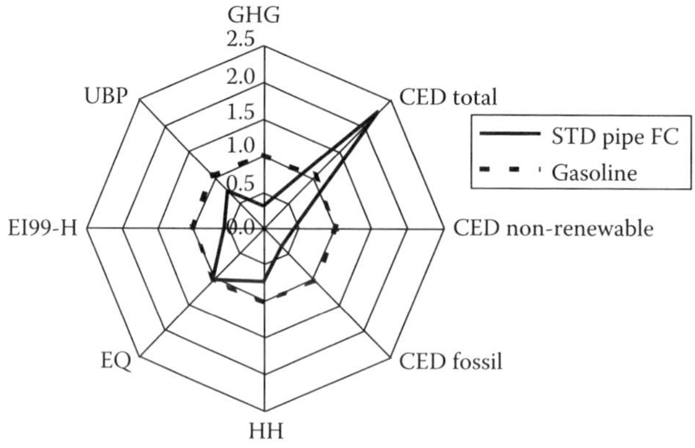

FIGURE 14.9
LCIA results for the STD scenario with pipeline transport of H_2 and its use in an FCV compared to an advanced gasoline vehicle. (For details, see text and Felder, R., Well-to-wheel analysis of renewable transport fuels: Synthetic natural gas from wood gasification and hydrogen from concentrated solar energy, PhD thesis, ETH No. 17437, ETH Zurich, Zurich, Switzerland, 2007.)

2.3 to about 7 MJ-eq./MJ H_2 (LHV) in the car tank at 350 bar. However, CED is reduced by 50% if only nonrenewable energy sources are considered (CED nonrenewable) and by 70% if only fossil energy sources are taken into account (CED fossil). *HH* impacts are reduced by 25%, mainly caused by the absence of emissions such as NO_x and SO_2 in the gasoline fuel chain, aromatic hydrocarbons at the gas station, and combustion. *EQ* does not improve, since dominant ecotoxic emissions from tire abrasion (Zn) are fuel-independent. The aggregated environmental impact is reduced by 50% when using the *EI'99-H* methodology—which strongly weights the consumption of fossil resources—and by 25% when using the *UBP* methodology. Here, the effect of lower GHG emissions and fossil resource consumption is partly counterbalanced by particulate emissions from steel production for the HF, land use, and consumption of mineral resources. The combined LCIA shows better overall environmental performance of the FCV with STD pipe scenario compared to an advanced gasoline vehicle.

14.5 Fuel and Vehicle Costs

Table 14.11 presents an overview of cost assumptions and results for solar H_2 production technologies (STD, STE) compared to conventional H_2 production (SMR, CGA) and H_2O electrolysis (ELE) based on electricity from different sources (CH mix, wind, PV). Cost of solar H_2 is calculated from various literature sources on CSP technologies [35] and H_2 production methods [41,69]. Passenger car transportation costs are taken from two comprehensive references [18,70]. Electricity costs used are 0.07 US$/kWh for conventional industrial plants [41], 0.09 US$/kWh for wind (average midterm scenario for 2020; exchange rate 1 USD = 1.2 CHF) [71], and 0.28 US$/kWh for PV (assumed for large Swiss plant in 2020; exchange rate 1 USD = 1.2 CHF) [71]. Costs for NG [69] of 0.22 US$/Nm^3 and for coal [41] of 30 US$/ton are anticipated for 2025. Assumptions for vehicle transportation calculations encompass capital costs [72,73] and annual costs

TABLE 14.11

Cost Assumptions and Results for Solar H_2 Production (STD, STE) Compared to Conventional H_2 Production (SMR, CGA) and H_2O Electrolysis (ELE) Based on Electricity from Various Sources (CH Mix, Wind, and PV)

	Unit	STD pipe	STD zinc	STE pipe	STE hvdc	SMR	CGA	ELE CH mix	ELE wind	ELE PV
Solar plant (input into receiver)	MW_{th}	50	50	50	50					
H_2 production (LHV)[a]	MWh/a	24,138	28,398	18,951	18,951	1.6E06	1.6E06	4005	4005	4005
H_2 production (LHV)[a]	MW	13.3	15.7	10.4	10.4	177	177	0.55	0.55	0.55
Annual thermal eff. (LHV)	—	0.17	0.20	0.13	0.13	0.76	0.73	0.63	0.63	0.63
Capital cost	*mill. $*									
HF		10.6	10.6	10.6	10.6					
Tower		2.2	2.2	2.2	2.2					
CPC		0.9	0.9							
Receiver–reactor, quencher/periphery		8.2	8.2	8.2	8.2					
Hydrolyzer		2.5								
Molten salt storage system				2.0	2.0					
Power block				16.0	16.0					
Electrolyzer				12.2						
Compressor (1–30 bar)		5.5		4.7						
Zn			3.0							
Land costs		1.0	1.0	1.0	1.0					
Balance of plant,		6.0	6.0	6.0	6.0					
Indirect costs, contingency										
Total capital cost		37.3	31.9	62.8	46.0	79.1	258.5	4.15	4.15	4.15
Annual cost	*mill. $*									
Weighted lifetime	a	22.8	25.7	22.9	28.0	30	30	30	30	30
Capital charge rate	%	16.6	16.4	12.3	10.1	18.0	18.0	18.0	18.0	18.0
Capital costs		6.20	5.23	7.71	4.62	14.24	46.53	0.75	0.75	0.75
O&M		0.75	0.64	1.26	0.92	4.75	15.51	0.25	0.25	0.25
Compression (1–30 bar)		0.12		0.09						

(continued)

TABLE 14.11 (continued)

Cost Assumptions and Results for Solar H_2 Production (STD, STE) Compared to Conventional H_2 Production (SMR, CGA) and H_2O Electrolysis (ELE) Based on Electricity from Various Sources (CH Mix, Wind, and PV)

	Unit	STD pipe	STD zinc	STE pipe	STE hvdc	SMR	CGA	ELE CH mix	ELE wind	ELE PV
Fuel						84.37	10.20			
Electricity		0.07	0.05	0.00	0.00	5.21	18.15	0.46	0.61	1.83
Other inputs		0.12	0.19	0.06	0.06					
Total annual cost, at plant		7.26	6.11	9.12	5.60	108.56	90.39	1.46	1.61	2.83
Specific cost per unit										
Solar Zn, at plant	$/kg	0.25	0.25							
Electricity, at plant	$/kWh			0.19	0.19			0.07	0.09	0.28
H_2 30 bar, at plant (LHV)	$/kg	10.0		16.0		2.2	1.7	12.0	13.0	23.4
Zn transport	$/kg		0.01							
Solar Zn, at gas station	$/kg		0.26							
Electricity transport	$/kWh				0.01					
Solar electricity, at gas station	$/kWh				0.21					
H_2 pipeline transport (LHV)	$/kg	3.7		4.7		0.8	1.3			
H_2 30 bar, at gas station (LHV)	$/kg	13.7	10.0	20.7	20.0	3.0	3.0			
H_2 fueling 30–350 bar (LHV)	$/kg	1.0	0.3	1.0	0.3	1.0	1.0	0.3	0.3	0.3
H_2 350 bar, in car tank (LHV)	$/kg	14.7	10.3	21.7	20.3	4.0	4.0	12.3	13.3	23.7

[a]*Note:* Each scenario has different H_2 production capacities, which has an impact on H_2 costs.

(depreciation, taxes, O&M) [70] for a variety of vehicles with different fuels and power trains (FCV, BEV, CNG, gasoline, and diesel).

14.5.1 Passenger Car Transportation Costs

Figure 14.10 depicts passenger car transportation costs for selected scenarios showing contributions of road infrastructure, operation and maintenance (O&M), fuel production and supply, and external costs. Main cost component is O&M, comprising total annual costs for the consumer (consisting to a large extent of car depreciation)

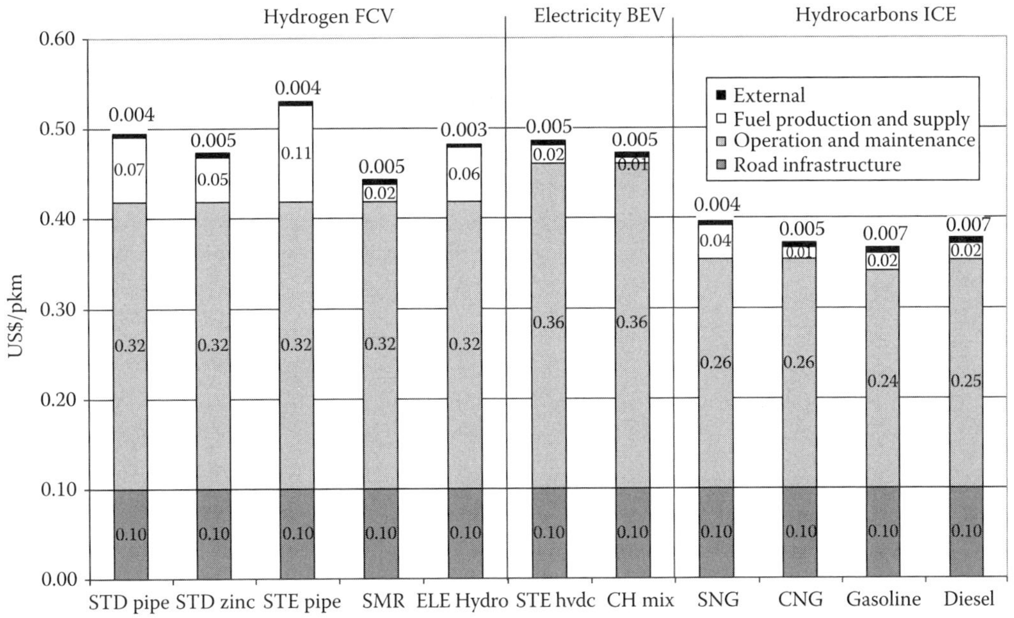

FIGURE 14.10
Transportation costs per pkm for selected scenarios with contribution of road infrastructure, O&M (total annual costs for consumer not including fuel), fuel production and supply, and external costs. Average load of 1.59 passengers per car is used [19]. For acronyms, see Table 14.1. (Details in Felder, R., Well-to-wheel analysis of renewable transport fuels: Synthetic natural gas from wood gasification and hydrogen from concentrated solar energy, PhD thesis, ETH No. 17437, ETH Zurich, Zurich, Switzerland, 2007.)

excluding fuel costs. Assuming equal maintenance and repair costs for all power trains, total costs are higher by about 0.07–0.15 US$/pkm for alternative vehicle concepts (FCV and BEV) compared to conventional hydrocarbon technologies. Among the ICE cars, SNG is relatively cost-competitive, with an increase of total costs by only 0.03 US$/pkm compared to gasoline. Although BEVs have rather low fuel costs, total costs are about 0.10 US$/pkm higher than for conventional vehicles due to their expensive car infrastructure. The most economic alternative option is the FCV using H_2 produced by SMR, costing 0.07 US$/pkm more than a gasoline vehicle while fuel costs are similar (about 0.02 US$/pkm, corresponding to 4 US$/kg H_2 in car tank). Total transportation costs may increase by 0.12–0.16 US$/pkm or 0.20–0.25 US$/km for an FCV powered by solar H_2 compared to an ICE vehicle using gasoline, depending on fossil fuel price development.

H_2 production and supply costs for CSP scenarios are 2.5–5.5 times higher than for SMR (4 US$/kg H_2 in car tank), depending on the chosen scenario. STD with H_2 pipeline transport (STD pipe) costs about 0.07 US$/pkm, corresponding to 14.7 US$/kg H_2 in car tank and 10.1 US$/kg H_2 at plant. STE with H_2 pipeline transport (STE pipe) costs about 0.11 US$/pkm (21.9 US$/kg H_2 in car tank and 16.1 US$/kg H_2 at plant). For comparison, the most economic CSP option using Zn transport (STD zinc) costs around 0.05 US$/pkm or 10.2 US$/kg H_2 in car tank, less than H_2 fuel costs using hydro or wind electricity for H_2O electrolysis (about 0.06 US$/pkm). External costs have marginal impact on total costs. Only for conventional low-cost fuels, external costs increase fuel costs by about 40%. For most of the alternative fuels, internalization of external costs virtually has no effect, since fuel costs are much higher.

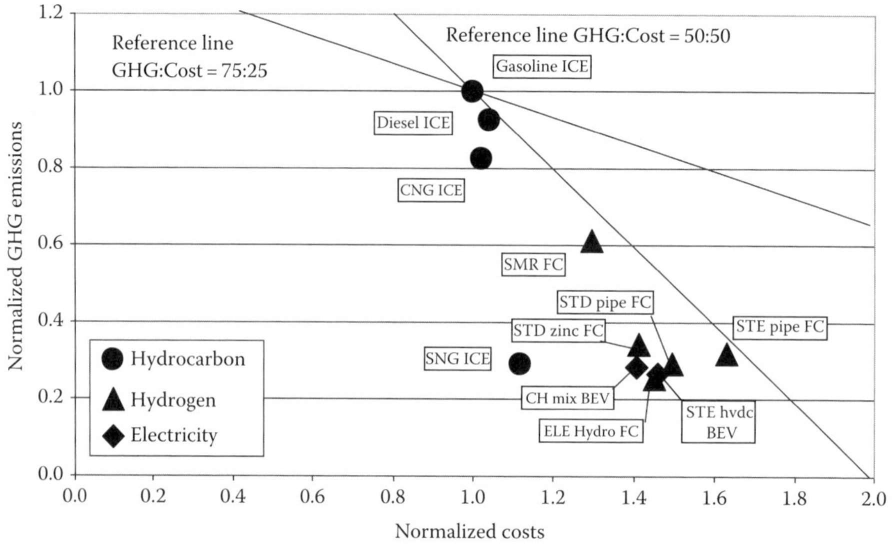

FIGURE 14.11
Eco-efficiency for selected fuel/power train combinations showing normalized GHG emissions versus cost per pkm; reference is the gasoline ICE vehicle. Two reference lines are shown: (1) weighting of 50:50 for GHG emissions versus costs (GHG emissions = 2 – costs) and (2) weighting of 75:25 for GHG emissions versus costs (GHG emissions = 4/3 – 1/3 * costs). For acronyms, see Table 14.1. (Details in Felder, R., Well-to-wheel analysis of renewable transport fuels: Synthetic natural gas from wood gasification and hydrogen from concentrated solar energy, PhD thesis, ETH No. 17437, ETH Zurich, Zurich, Switzerland, 2007.)

14.5.2 Eco-Efficiency

Figure 14.11 presents eco-efficiency results in terms of normalized GHG emissions versus normalized costs per pkm, referenced to the gasoline ICE vehicle. Data points below the reference line indicate higher eco-efficiency than the reference case; moreover, the larger their perpendicular distance to the reference line, the better their relative eco-efficiency performance. Assuming a weighting ratio of 50:50 for GHG emissions versus cost, the SNG ICE vehicle shows highest eco-efficiency. The other technologies are at roughly the same level with slight advantage for FCV using ELE hydro and BEV using CH mix. All CSP pathways exhibit a better eco-efficiency than the gasoline ICE vehicle. Assuming a more environmentally oriented stakeholder perspective by setting a weighting ratio of 75:25 for GHG emissions versus costs favors renewable fuels over fossil-based fuels.

14.6 Conclusions

A WTW analysis and an economic assessment have been conducted for the production, transport, and usage of H_2 as future transportation fuel. Solar H_2, produced by the thermochemical Zn/ZnO cycle in Southern Spain and transported to Central Europe, was compared with SMR, advanced CGA, and ELE based on renewable electricity generated in solar thermal, PV, wind, and hydropower plants. Usage of H_2 in an FCV was anticipated for 2025. It was found that cars powered by renewable fuels emit about

42 ± 8 g CO_2-eq. per passenger kilometer. In general, the battery electric vehicle (BEV) shows the lowest GHG emissions because of its higher energy conversion efficiency compared to the FCV. In the specific case of Switzerland, the best performers are BEVs using local hydro, nuclear, and wind power. CSP technologies reduce GHG emissions for the production and supply of H_2 by a factor of 7 (SMR) and 14 (CGA), resulting in 50%–70% CO_2 savings in an FCV. Total environmental impact of passenger car transportation can be reduced by replacing advanced gasoline vehicles with FCVs powered by H_2 from CSP.

High capital investments for HF infrastructure and long-distance transport result in 2.5–5.5 times higher H_2 production and supply costs for solar scenarios vis-à-vis conventional SMR (4 US$/kg H_2 in car tank). Total transportation costs may increase by 0.20–0.25 US$/km for an FCV powered by solar H_2 compared to an ICE vehicle using gasoline. Assuming long-term market acceptance associated with return on investment being reduced from 15% for new technologies to 5% for established technologies, costs of the most economic CSP technology (STD with Zn transport) would decrease to 6 US$/kg H_2 and become competitive to costs of SMR and CGA, which are expected to rise due to fossil resource shortage. External costs have only marginal impact on total costs of alternative fuels, where fuel costs are dominant. Eco-efficiency is low for all solar H_2 production technologies, since their relatively high costs are not outweighed by their environmental benefit, unless ecological aspects are stronger weighted than economics.

Acknowledgments

The financial support by the Swiss Federal Office of Energy (SFOE) and the Baugarten Foundation is gratefully acknowledged. This work is based on the PhD thesis of R. Felder, who tragically passed away shortly after his doctoral exam. The author would like to thank A. Wokaun, A. Steinfeld, and C. Wieckert for fruitful discussions.

References

1. IEA, 2008, Energy technology perspectives 2008, International Energy Agency, Paris, France, 2008.
2. EC, 2006, World Energy Technology Outlook—2050 (WETO-H2), European Commission, Directorate-General for Research, Brussels, Belgium, 2006, ISBN 92-79-01636-9.
3. Pregger, T., Graf, D., Krewitt, W., Sattler, C., Roeb, M., Möller, S., 2009, Prospects of solar thermal hydrogen production processes, *Int. J. Hydrogen Energy*, **34**, 4256–4267.
4. Felder, R., 2007, Well-to-wheel analysis of renewable transport fuels: Synthetic natural gas from wood gasification and hydrogen from concentrated solar energy, PhD thesis, ETH No. 17437, ETH, Zurich, Switzerland.
5. Felder, R., Meier, A., 2008, Well-to-wheel analysis of solar hydrogen production and utilization for passenger car transportation, *J. Sol. Energy Eng.*, **130**(1), 011017-1/10.

6. Felder, R., Meier, A., Wokaun, A., 2008, Solar hydrogen as future transportation fuel—Well-to-wheel analysis and economic assessment, *Proceedings of the 14th Biennial Concentrating Solar Power SolarPACES Symposium*, Las Vegas, NV, March 4–7, 2008.
7. Midilli, A., Ay, M., Dincer, I., Rosen, M.A., 2005, On hydrogen and hydrogen energy strategies: I: Current status and needs, *Renewable Sustainable Energy Rev.*, **9**(3), 255–271.
8. Winter, C.-J., 2005, Into the hydrogen energy economy—Milestones, *Int. J. Hydrogen Energy*, **30**(7), 681–685.
9. Steinfeld, A., 2005, Solar thermochemical production of hydrogen—A review, *Sol. Energy*, **78**(5), 603–615.
10. Abanades, S., Charvin, P., Flamant, G., Neveu, P., 2006, Screening of water-splitting thermochemical cycles potentially attractive for hydrogen production by concentrated solar energy, *Energy*, **31**(14), 2805–2822.
11. Perkins, C., Weimer, A.W., 2004, Likely near-term solar-thermal water splitting technologies, *Int. J. Hydrogen Energy*, **29**(15), 1587–1599.
12. Steinfeld, A., 2002, Solar hydrogen production via a two-step water-splitting thermochemical cycle based on Zn/ZnO redox reactions, *Int. J. Hydrogen Energy*, **27**(6), 611–619.
13. Abanades, S., Charvin, P., Flamant, G., 2007, Design and simulation of a solar chemical reactor for the thermal reduction of metal oxides: Case study of zinc oxide dissociation, *Chem. Eng. Sci.*, **62**, 6323–6333.
14. Schunk, L.O., Haeberling, P., Wepf, S., Wuillemin, D., Meier, A., Steinfeld A., 2008, A receiver-reactor for the solar thermal dissociation of zinc oxide, *J. Sol. Energy Eng.*, **130**(2), 021009-1/6.
15. Schunk, L.O., Lipiński, W., Steinfeld A., 2009, Heat transfer analysis of a solar receiver-reactor for the thermal dissociation of ZnO—Experimental validation at 10 kW and scale-up to 1 MW, *Chem. Eng. J.*, **150**, 502–508.
16. Wieckert, H.-C., 2005, Design studies for a solar reactor based on a simple radiative heat exchange model, *J. Sol. Energy Eng.*, **127**(3), 425–429.
17. Ernst, F., Tricoli, A., Steinfeld, A., Pratsinis, S.E., 2006, Co-synthesis of H_2 and ZnO by in-situ Zn aerosol formation and hydrolysis, *AIChE J.*, **52**, 3297–3303.
18. Edwards, R., Griesemann, J.-C., Larivé, J.-F., Mahieur, V., 2006, Well-to-wheels analysis of future automotive fuels and powertrains in the European context. Well-to-Wheels report. Version 2b, EUCAR, CONCAWE, JRC/IES.
19. Spielmann, M., Kägi, T., Stadler, P. et al., 2003, Life cycle inventories of transport services. Ecoinvent report No. 14, Swiss Centre for Life Cycle Inventories, Dübendorf, Switzerland.
20. Frischknecht, R., Jungbluth, N., Althaus, H.-J. et al., 2004, Overview and methodology. Ecoinvent report No. 1, Swiss Centre for Life Cycle Inventories, Dübendorf, Switzerland.
21. Bauer, C., Dones, R., 2006, Axpo-Stromperspektiven 2030. Ongoing projects from GaBe project at PSI, Villigen PSI, Switzerland, Personal Communication, 5.4.2006.
22. Geerken, T., Lassaux, S., Renzoni, R., Timmermans, V., 2004, Review of hydrogen LCA's for the Hysociety project, Final Report, EU FP5 Project, Contract No. 11673.
23. ISO, 1998, Environmental management—Life cycle assessment—Goal and scope definition and inventory analysis. International Organization for Standardization, ISO 14041.
24. ISO, 2000, Environmental management—Life cycle assessment—Life cycle impact assessment. International Organization for Standardization, ISO 14042.
25. IPCC, 2001, *Climate Change 2001: The Scientific Basis*. Contribution of Working Group I to the Third Assessment Report of the Intergovernmental Panel on Climate Change, J.T. Houghton, Y. Ding, D.J. Griggs, M. Noguer, P.J. van der Linden, X. Dai, K. Maskell, and C.A. Johnson, eds., Cambridge University Press, Cambridge, U.K.
26. Frischknecht, R., Jungbluth, N., Althaus, H.-J. et al., 2004, Implementation of life cycle impact assessment methods. Ecoinvent report No. 3, Swiss Centre for Life Cycle Inventories, Dübendorf, Switzerland.
27. Goedkoop, M., Effting, S., Collignon, M., 2001, The eco-indicator 99. A damage oriented method for Life Cycle Impact Assessment, PRé Consultants B.V., Amersfoort, the Netherlands.

28. Brand, G., Scheidegger, A., Schwank, O. et al., 1998, Weighting in ecobalances with the ecoscarcity method, Environment Series No. 297, Swiss Agency for the Environment, Forests and Landscape, Berne, Switzerland.
29. Jolliet, O., Margni, M., Charles, R. et al., 2003, Impact 2002+: A new life cycle impact assessment methodology, *Int. J. LCA*, **8**(6), 324–330.
30. Saling, P., Kicherer, A., Dittrich-Krämer, B. et al., 2002, Eco-efficiency analysis by BASF: The method, *Int. J. LCA*, **7**(4), 203–218.
31. Suh, S., Lee, K.M., Ha, S., 2005, Eco-efficiency for pollution prevention in small to medium-sized enterprises, *J. Ind. Ecol.*, **9**(4), 223–240.
32. Colella, W.G., Jacobson, M.Z., Golden, D.M., 2005, Switching to a U.S. hydrogen fuel cell vehicle fleet: The resultant change in emissions, energy use, and greenhouse gases, *J. Power Sources*, **150**, 150–181.
33. Osuna, R., Fernandez, V., Romero, S., 2004, PS10: A 11.0-MWe solar tower power plant with saturated steam receiver [CD-ROM], *Proceedings of the 12th SolarPACES International Symposium*, Oaxaca, Mexico, ISBN 968-6114-18-1.
34. PS10, 2006, 10 MW solar thermal power plant for southern Spain, Final Technical Project Report, EU FP5 Contract No. NNE5-1999-356.
35. ECOSTAR, 2005, ECOSTAR—European Concentrated Solar Thermal Road-Mapping, Final Report, EU FP6 Contract No. SES6-CT-2003-502578, R. Pitz-Paal, J. Dersch, B. Milow, eds., DLR, Cologne, Germany.
36. Segal, A., Epstein, M., 1999, Comparative performances of tower-top and tower-reflector central solar receivers, *Sol. Energy*, **65**, 206–226.
37. Sargent & Lundy LLC Consulting Group, 2003, Assessment of parabolic trough and power tower solar technology cost and performance forecasts, NREL/SR-550-34440, Prepared for the US Department of Energy and National Renewable Energy Laboratory, Golden, CO.
38. Weinrebe, G., 2000, Technische, ökologische und ökonomische Analyse von solarthermischen Turmkraftwerken [Technical, ecological and economic analysis of solar thermal tower power plants], PhD thesis in Energietechnik, University of Stuttgart, Stuttgart, Germany.
39. Pitz-Paal, R., Merz, T., Bayer Botero, N., Steinfeld, A., 2009, Heliostat field layout optimization for high temperature solar thermochemical processing, *Proceedings of the 15th SolarPACES Conference*, Berlin, Germany, September 15–18, 2009.
40. Romero, M., Buck, R., Pacheco, J.E., 2002, An update on solar central receiver systems, projects, and technologies, *J. Sol. Energy Eng.*, **124**(2), 98–108.
41. Simbeck, D., Chang, E., 2002, Hydrogen supply: Cost estimate for hydrogen pathways—Scoping analysis, NREL/SR-540-32525, National Renewable Energy Laboratory, Golden, CO.
42. Reilly, H.E., Kolb, G.J., 2001, An evaluation of molten-salt power towers including results of the Solar Two Project, SAND 2001-3674, Sandia National Laboratories, Albuquerque, NM.
43. Meier, A., Gremaud, N., Steinfeld, A., 2005, Economic evaluation of the industrial solar production of lime, *Energy Convers. Manage.*, **46**, 905–926.
44. Müller, R., Steinfeld, A., 2007, Band-approximated radiative heat transfer analysis of a solar chemical reactor for the thermal dissociation of zinc oxide, *Sol. Energy*, **81**(10), 1285–1294.
45. Wieckert, C., Frommherz, U., Kräupl, S., Guillot, E., Olalde, G., Epstein, M., Santen, S., Osinga, T., Steinfeld, A., 2007, A 300 kW solar chemical pilot plant for the carbothermic production of zinc, *J. Sol. Energy Eng.*, **129**, 190–196.
46. Wieckert, H.-C., 2005, Solzinc reactor efficiency, Villigen PSI, Switzerland, Personal Communication, September 2005.
47. Epstein, M., Olalde, G., Santén, S., Steinfeld, A., Wieckert, C., 2008, Towards the industrial solar carbothermal production of zinc, *J. Sol. Energy Eng.*, **130**, 014505-1/4.
48. Kräupl, S., Frommherz, U., Wieckert, H.-C., 2006, Solar carbothermic reduction of ZnO in a two-cavity reactor: Laboratory experiments for a reactor scale-up, *J. Sol. Energy Eng.*, **128**(1), 8–15.

49. FAO, 1985, Industrial charcoal making. FAO Forestry Paper 63, Chapter 3.3, Food and Agriculture Organization of the United Nations (FAO), Rome, Italy, http://www.fao.org/docrep/X5555E/X5555E00.htm, accessed on 03.12.2013.
50. Winnacker, K., Küchler, L., 1971, Organische Technologie I [Organic technology I], in *Chemische Technologie* [*Chemical Technology*], 3rd edn., Carl Hanser Verlag, Munich, Germany.
51. Criscuoli, A., Basile, A., Drioli, E., 2000, An analysis of the performance of membrane reactors for the water-gas shift reaction using gas feed mixtures, *Catal. Today*, **56**(1–3), 53–64.
52. Faist Emmenegger, M., Jungbluth, N., Heck, T., 2003, Erdgas [Natural Gas]. Ecoinvent report No. 6-V, Swiss Centre for Life Cycle Inventories, Dübendorf, Switzerland.
53. Bossel, U., Eliasson, B., Taylor, G., 2005, The future of the hydrogen economy: Bright or bleak? in European Fuel Cell Forum, Lucerne, Switzerland.
54. Angloher, J., Dreier, T., 2000, Techniken und Systeme zur Wasserstoffbereitstellung. Perspektiven einer Wasserstoff-Energiewirtschaft (Teil 1). [Techniques and systems for hydrogen supply. Perspectives of a hydrogen energy economy (Part 1)], in *Perspektiven einer Wasserstoff-Energiewirtschaft*, U. Wagner, ed., wiba—Koordinationsstelle der Wasserstoff-Initiative Bayern, Munich, Germany.
55. Frischknecht, R., Faist Emmenegger, M., 2003, Strommix und Stromnetz [Electricity mix and network]. Ecoinvent report No. 6, Swiss Centre for Life Cycle Inventories, Dübendorf, Switzerland.
56. Spath, P.L., Mann, M.K., 2001, Life cycle assessment of hydrogen production via natural gas steam reforming, NREL/TP-570-27637, National Renewable Energy Laboratory, Golden, CO.
57. Stiegel, G.J., Ramezan, M., 2006, Hydrogen from coal gasification: An economical pathway to a sustainable energy future, *Int. J. Coal Geol.*, **65**(3–4), 173–190.
58. Dones, R., 2003, Kernenergie [Nuclear energy]. Final report ecoinvent 2000 No. 6-VII, R. Dones, ed., Swiss Centre for Life Cycle Inventories, Dübendorf, Switzerland.
59. Bolliger, R., Bauer, C., 2004, Wasserkraft [Hydro power]. Final report ecoinvent 2000 No. 6-VIII, Swiss Centre for Life Cycle Inventories, Dübendorf, Switzerland.
60. Burger, B., Bauer, C., 2004, Windkraft [Wind power]. Final report ecoinvent 2000 No. 6-XIII, Swiss Centre for Life Cycle Inventories, Dübendorf, Switzerland.
61. Durisch, W., Bauer, C., 2005, Photovoltaik [Photovoltaics], in *Erneuerbare Energien und neue Nuklearanlagen* [*Renewable Energies and New Nuclear Power Plants*], Paul Scherrer Institute, Villigen, Switzerland, pp. 124–172.
62. Pehnt, M., 2003, Life-cycle analysis of fuel cell system components, in *Handbook of Fuel Cells—Fundamentals, Technology and Applications*, W. Vielstich, A. Lamm, and H.A. Gasteiger, eds., John Wiley & Sons, Ltd., Chichester, U.K., pp. 1293–1317.
63. Jungbluth, N., 2004, Erdöl [Crude Oil]. Final report ecoinvent 2000 No. 6-IV, in *Sachbilanzen von Energiesystemen: Grundlagen für den ökologischen Vergleich von Energiesystemen und den Einbezug von Energiesystemen in Ökobilanzen für die Schweiz*, R. Dones, ed., Swiss Centre for Life Cycle Inventories, Dübendorf, Switzerland.
64. Dones, R., Heck, T., Bauer, C. et al., 2005, New Energy Technologies—Final Report on Work Package 6—Release 2, July 2005. ExternE-Pol Project 'Externalities of Energy: Extension of Accounting Framework and Policy Applications', European Commission, Brussels, Belgium.
65. INFRAS, 1998, Ökoprofile von Treibstoffen [Ecological profiles of fuels], Umwelt-Materialien Nr. 104, Swiss Agency for the Environment, Forests and Landscape SAEFL, Berne, Switzerland.
66. Röder, A., 2001, Integration of life-cycle assessment and energy planning models for the evaluation of car powertrains and fuels, PhD Thesis, ETH, Nr. 14291, Zurich, Switzerland.
67. Rade, I., Andersson, B.A., 2001, Requirement for metals of electric vehicle batteries, *J. Power Sources*, **93**, 55–71.
68. Ivy, J., 2004, Summary of electrolytic hydrogen production. Milestone completion report, NREL/MP-560-35948, National Renewable Energy Laboratory, Golden, CO.
69. Hydrogen Analysis Resource Center, US DOE, http://www.hydrogen.energy.gov/resource_center.html, accessed on November 15, 2006.

70. Zwyssig, M., 2005, *Vergleich der Flottenkosten von Erdgas- und konventionell angetriebenen Fahrzeugen* [*Comparison of Fleet Costs of Natural Gas and Conventional Cars*], Departement Wirtschaft und Management, Zentrum für Accounting & Controlling, Winterthur, Switzerland.
71. Hirschberg, S., Bauer, C., Burgherr, P. et al., 2005, *Erneuerbare Energien und neue Nuklearanlagen: Potenziale und Kosten* [*Renewable Energies and New Nuclear Power Plants: Potential and Cost*], Paul Scherrer Institute, Villigen, Switzerland. ISSN 1019-0643.
72. Edwards, R., Griesemann, J-C., Larivé, J.-F. et al., 2006, Well-to-wheels analysis of future automotive fuels and powertrains in the European context. Tank-to-Wheels Report, Version 2b, Appendix 1. EUCAR, CONCAWE, JRC/IES.
73. Weiss, M.A., Heywood, J.B., Drake, E.M. et al., 2000, *On the Road in 2020: A Life-Cycle Analysis of New Automobile Technologies*, Energy Laboratory Report # MIT EL 00-003, Massachusetts Institute of Technology, Cambridge, MA.

Section VII

Hydrogen Storage, Transportation, Handling, and Distribution

15

Overview of Hydrogen Storage, Transportation, Handling, and Distribution

Ned T. Stetson
U.S. Department of Energy

Robert C. Bowman Jr.
Oak Ridge National Laboratory

Gregory L. Olson
HRL Laboratories, LLC

CONTENTS

15.1 Background

As hydrogen finds increasing use in emerging applications, the need for improved storage methods is becoming more important. Hydrogen has a normal boiling temperature of about 20 K and a critical temperature of approximately 33 K, above which a liquid cannot be formed through the application of pressure. It is therefore a gas at essentially all normal use and storage temperatures. Hydrogen is the lightest of all elements, and since it behaves as an ideal gas close to ambient temperatures (~300 K) and pressure conditions, it has a very low normal density of 0.09 g/L (or 11 L/g) at 288 K and 0.1 MPa. To put this density in perspective, with a lower heating value of about 120 kJ/g, the normal energy density of hydrogen is 10 kJ/L. While gasoline has a specific energy of only about 42 kJ/g, its energy density is 32,000 kJ/L [1] and thus has an energy density about 3,200 times greater than normal hydrogen gas.

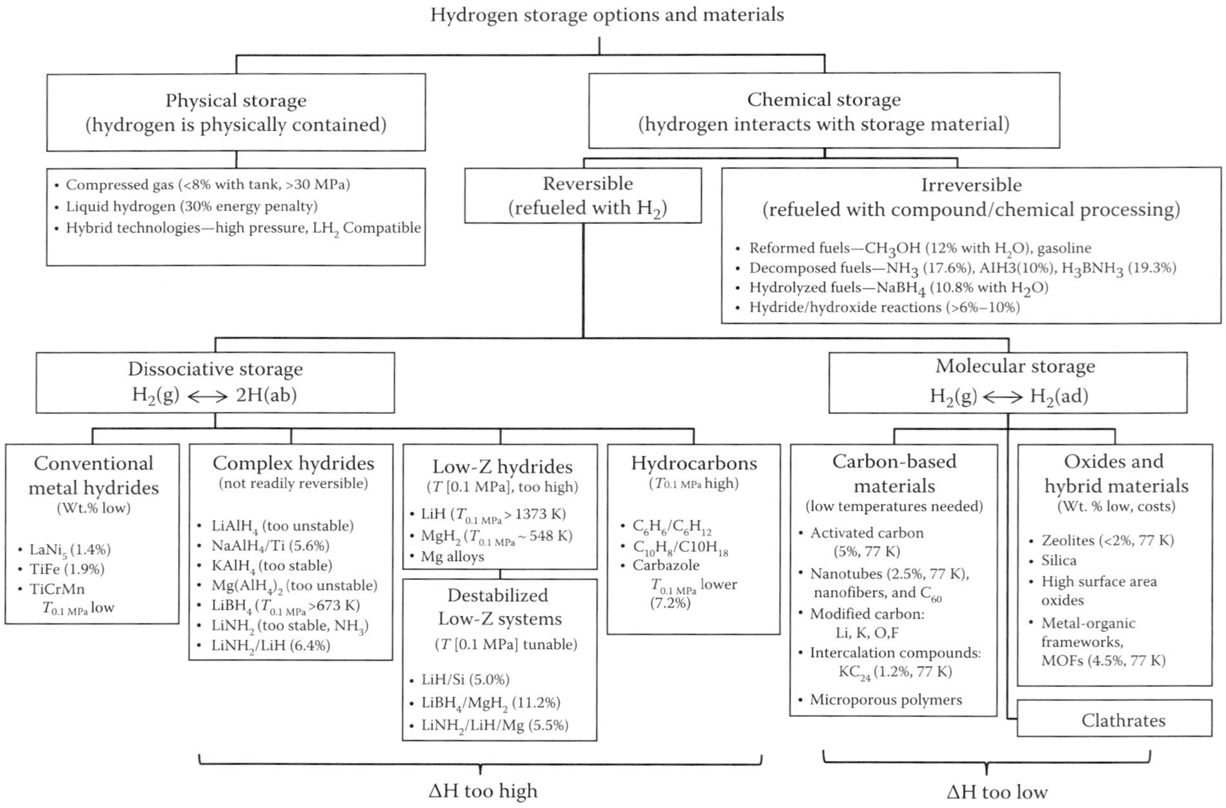

FIGURE 15.1
Various hydrogen storage technologies along with some of their advantages and disadvantages. (Courtesy of Dr. John Vajo.)

To overcome the low density of normal hydrogen gas, it has typically been either produced *over the fence* from the point of use (e.g., near petroleum refineries), liquefied at cryogenic temperatures (i.e., 20 K), or compressed to high pressures (typical merchant hydrogen is delivered at 15–25 MPa). For many new applications, such as a fuel for fuel cells in stationary, mobile, and portable applications, the traditional hydrogen storage options have issues with, for example, cost, weight, volume, dormancy, and/or overall practicality. For instance, in stationary applications for backup power, dormancy (i.e., the length of time the hydrogen can be stored without loss), fast start, ease of refill, and cost may be critical issues, while weight and volume may not, whereas for portable power, weight, volume, robustness, and safety may be vital, while cost and dormancy may be of less importance. Automotive applications, which are likely the most challenging, have very severe requirements with respect to weight, volume, speed of refill, transient operation, and cost. To meet the need of these demanding consumer applications, there have been increasing efforts at developing advanced storage options, such as the use of solid or liquid materials in which the hydrogen may be either weakly physisorbed as the dihydrogen molecule or more strongly chemically bound as monatomic hydrogen. Figure 15.1 shows a number of hydrogen storage options that have been proposed, along with some of their identified strengths and weaknesses. This overview gives a brief introduction to many of these options with more in-depth discussions in the following chapters. Two recent monographs [2,3] also provide comprehensive assessments of hydrogen storage systems and materials.

15.2 Physical Storage as Compressed Gas or Cryogenic Liquid

Hydrogen has been stored and transported as a compressed gas in metal cylinders since at least the 1880s when the British used wrought-iron metal cylinders for transporting hydrogen to inflate war balloons during expeditions across Asia and Africa [4]. While hydrogen behaves almost as an ideal gas with its density increasing linearly with pressure near ambient conditions, as shown in Figure 15.2, significant deviation from ideal behavior occurs at elevated pressure, and the gas density increases much more slowly than the pressure. Storing gas at high pressures requires robust pressure vessels that may incur significant weight and cost penalties. The design, manufacture, transport, and use of gas cylinders and pressure vessels, and their appurtenances, are typically regulated

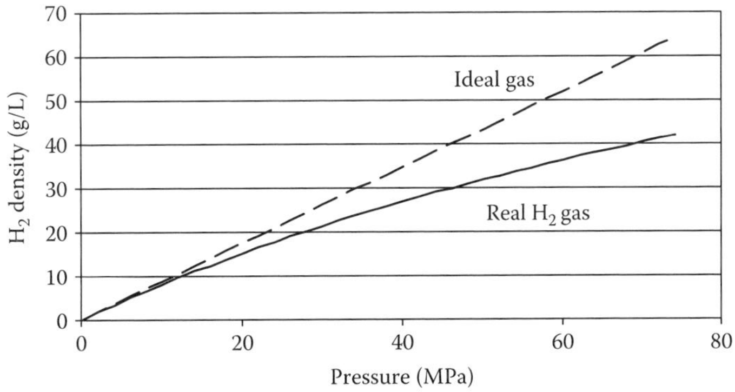

FIGURE 15.2
Comparison of real hydrogen gas density to ideal gas as a function of pressure.

by government agencies. The regulations often require compliance with specific design, manufacture, and/or use codes and standards developed by organizations such as the International Standards Organization (ISO), the Compressed Gas Association (CGA), and the American Society of Mechanical Engineers (ASME).

There are four standard types of cylinders used for the transport and storage of hydrogen: Type I, all metal cylinders; Type II, hoop-wrapped composite cylinders; Type III, fully wrapped composite cylinders with load-bearing metallic liners; and Type IV, fully wrapped composite cylinders with non-load-bearing nonmetallic liners. Type I steel or aluminum cylinders are the most common type found in use for merchant hydrogen delivery and storage. The cylinders are designed for a maximum working pressure, with the minimum wall thickness determined by a formula relating the pressure to the metals' yield and tensile strength. Figure 15.3 shows the minimum wall thickness for 6061-T6 aluminum* and 4140 steel† cylinders with nominal outer diameter of 300 mm calculated using Equation 15.1 in standards ISO 7866:1999(E) [5] and ISO 9806–1:1999(E) [6], respectively. Also shown in Figure 15.3 is the ratio of the internal volume to the total cylindrical volume (i.e., the internal volume divided by the sum of internal plus wall volumes, neglecting the volume of the ends). This shows that at higher pressures, the useable volume, and thus the effective hydrogen density, decreases significantly, especially for lower-strength metals such as aluminum. Additionally, the mass of the metal is substantial such that the mass of hydrogen stored is typically only 1% or 2% of the cylinder mass, dropping to less than 1% at pressures of about 35 MPa and higher. For automotive applications, weight and volume constraints make Type I cylinders impractical; consequently, significant efforts have been devoted to developing Type III and IV cylinders. In these cylinders, a thin, lightweight metal or nonmetallic liner is wrapped by a fiber/epoxy matrix. The fiber wrapping supplies the strength to contain the high pressure, while the liner primarily acts as a gas permeation barrier. Fiberglass and carbon fiber are typical materials used to provide the strength, with high-tensile-strength carbon fiber being the material of choice for 35 and 70 MPa cylinders. Analyses indicate that hydrogen system capacities of 5.9% and 4.7% by weight can be obtained for 35 and 70 MPa

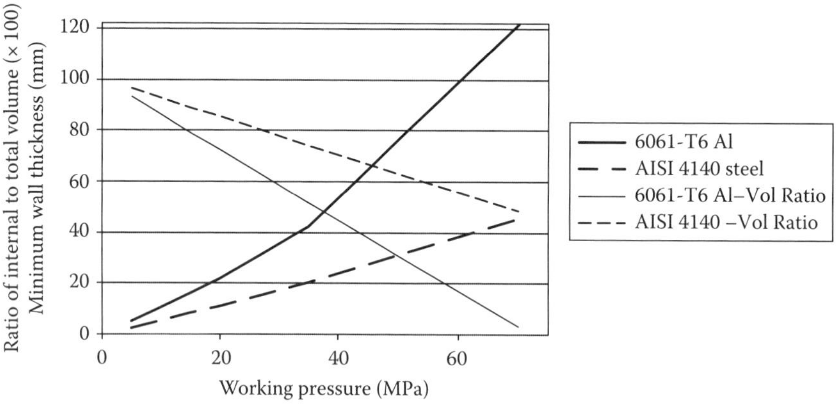

FIGURE 15.3
Minimum wall thickness and the ratio of internal volume to total volume as a function of working pressure for aluminum and steel cylinders.

* 6061-T6 aluminum properties used in calculations: Density = 2.6 g/cc; tensile strength = 290 MPa; yield strength = 241 MPa.

† 4140 steel properties used in calculations: Density = 7.8 g/cc; tensile strength = 655 MPa; yield strength = 417 MPa.

Type IV cylinders [7]. While the fiber wrapping can lower the weight of the cylinders dramatically, they add significant costs, currently projected to be close to three-quarters of the cylinder costs even with high volume manufacture [8].

Lower storage temperatures can also be used to increase hydrogen density. At its normal boiling point of 20 K, liquid hydrogen has a density of about 71 g/L, which is about 1.8 times the density of 70 MPa hydrogen at 288 K. A drawback with liquid hydrogen is that the liquefaction process is energy intensive, consuming an energy equivalence of 30% or more of the energy content of the stored hydrogen [9,10]. One factor that makes it so energy intensive is the need to convert the hydrogen molecules into the para state. The hydrogen molecule is composed of two hydrogen atoms, each with one proton and one electron. In the molecule, each atom can have a nuclear spin of either +½ or −½, leading to a total nuclear spin (I) for the molecule of either $I = 0$ (antiparallel spin or para state) or $I = 1$ (parallel spin or ortho state). The para state is lower in energy than the ortho state; however, normal hydrogen gas at ambient temperature is about 75% ortho and 25% para. At the normal boiling point, the equilibrium concentration of ortho-hydrogen to para-hydrogen is 0.2%, but the noncatalyzed conversion rate from ortho to para is very slow. Thus, if hydrogen is liquefied without forcing an ortho–para conversion, the ortho state will be much higher than its equilibrium concentration and it will slowly self-convert. At 20 K, the ortho–para conversion energy is 523 kJ/kg, which is greater than the latent heat of vaporization, 452 kJ/kg, and thus, the ortho–para self-conversion would cause significant evaporation [10]. Modern commercial hydrogen liquefaction processes therefore include catalytic processes to convert ortho-hydrogen to para-hydrogen.

To store liquid hydrogen, it is critical to minimize heat leakage into the interior of the vessel to prevent boil-off. Boil-off occurs when heat transfer to the stored hydrogen causes sufficient evaporation that the pressure within the inner vessel builds to a critical point where the vessel vents hydrogen, typically at a pressure of 1 MPa or less. Cryogenic liquid hydrogen tanks are typically double-walled vessels with the annular space between the walls evacuated and containing multiple layers of thin metal foil separated by a thermal insulator such as glass wool (known as multilayer vacuum superinsulation [MLVSI]), to prevent both convective and radiative heat transfer. All connections between the outer and inner vessels are specially designed and constructed to minimize thermal conduction to the inner tank. Even with the use of designs to minimize radiative, convective, and conductive thermal transport, some heat leakage to the inner vessel occurs, and eventual boil-off can be expected. Linde has developed a novel approach for further delaying boil-off called *CooLH$_2$* [11]. When cold liquid hydrogen is drawn off the tank, it is passed through a heat exchanger to liquefy dried air as the hydrogen is warmed. The liquefied air at 82 K is circulated through a cooling jacket around the inner vessel, further shielding it. The CooLH$_2$ system can reportedly extend the dormancy period to greater than 12 days versus 3 days or so for tanks without the CooLH$_2$ system [11].

Figure 15.4 is a plot that relates hydrogen density to temperature and pressure. At temperatures slightly above its critical temperature of 33 K [12], the gas density increases rapidly with application of pressure. The use of low, cryogenic temperatures in conjunction with pressure is therefore another potential method to increase the density of hydrogen. The temperature and pressure of the hydrogen fill and storage as well as the target dormancy period determine the type of cryogenic pressure vessel needed. For vessels designed for only moderately cold hydrogen (e.g., 233 K) and short dormancy periods, single-walled insulated cylinders may be sufficient. For applications such as onboard vehicles where it is desired to achieve significant increases in density with dormancy periods of days, lower temperatures, such as 77 K, have been proposed. At these temperatures, the storage vessel typically uses double-walled vessels with MLVSI

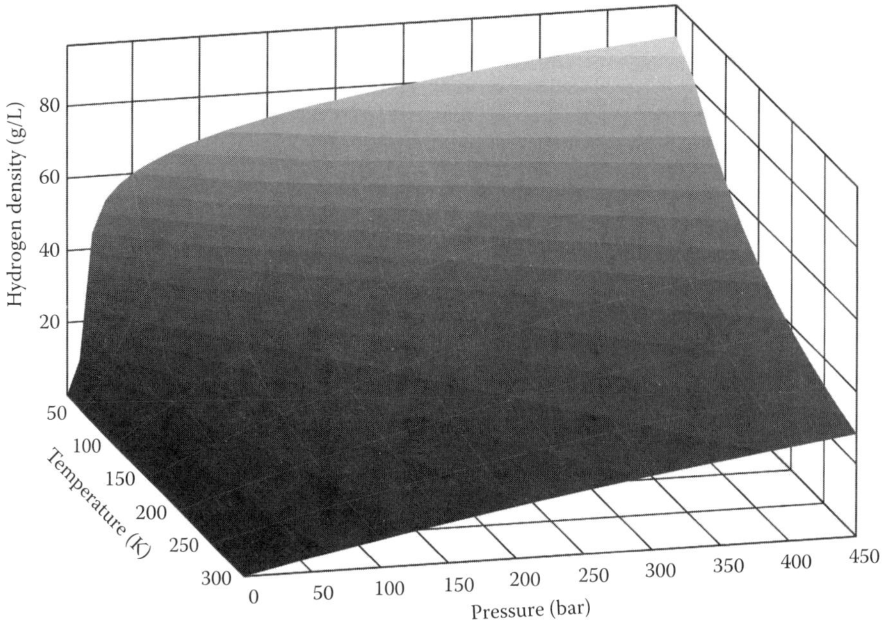

FIGURE 15.4
Plot of hydrogen density as a function of temperature and pressure.

similar to liquid hydrogen; however, the inner vessel is a Type I, II, or III pressure vessel. Prototype cryogenic-capable pressure vessels have demonstrated two to three times higher capacities than ambient pressure vessels and with low evaporative losses [13].

Summary: Hydrogen has typically been stored and transported as either a cryogenic liquid or a compressed gas. All Type I metal cylinders have very low gravimetric capacities at the high pressures required to increase the hydrogen density to meet the needs of the new emerging applications. Type III and IV tanks consisting of a lightweight liner overwrapped with a fiber/epoxy matrix have been developed to have higher gravimetric densities at significantly higher pressures (35–70 MPa) than typically found for merchant hydrogen. Even greater densities are obtained at cryogenic temperatures with either liquid or cryo-compressed hydrogen. These low temperatures require specialty double-walled vessels with MLVSI to minimize heat leakage and extend dormancy times. Physical hydrogen storage technologies are discussed in more detail in Chapters 16 through 19, 25 and 26.

15.3 Transition Metal Hydrides

Metal hydrides (MH_x) are a broad class of materials formed by the usually reversible reactions of metals and alloys with hydrogen as represented by the generic reaction:

$$M + \left(\frac{x}{2}\right) H_2 = MH_x + \Delta Q \qquad (15.1)$$

In Equation 15.1, ΔQ represents the heat either released (i.e., an exothermic reaction) or absorbed (i.e., endothermic reaction) during formation of the MH_x phase. Although some metals (e.g., Pt, Ni, Fe) with endothermic absorption of hydrogen show catalytic activity, only those metals that form hydrides via exothermic reactions are viable candidates for hydrogen storage. The book edited by Mueller et al. [14] is a comprehensive description of the synthesis, properties, and applications of hydrides as studied through the middle of the 1960s and is still an excellent resource for the modern researcher. The book edited by Manchester is another comprehensive reference on the phase diagrams, crystal structures, thermodynamics, as well as other properties of the binary hydrides [15]. While a few binary hydrides (i.e., PdH_x, VH_2, UH_3) have been used for decades to store and purify hydrogen isotopes in laboratory research and niche applications, it was the series of discoveries of hydrogen reactions with intermetallic compounds and alloys (starting during the late 1960s) that simulated extensive development of metal hydrides for various energy storage and conversion applications as described in numerous reviews [9,16–25]. The ideal reversible reaction of an intermetallic alloy A_nB_m where metal A forms stable binary hydride phases and metal B does not form a stable hydride is

$$A_nB_m + xH_2 = A_nB_mH_{x/2} + \Delta Q \tag{15.2}$$

While literally hundreds of intermetallic alloys have been prepared and their hydrogenation potentials assessed [18,26], relatively few have the right combination of properties that permit their use for hydrogen storage or other applications [21,22]. The most viable candidates include alloys with the following compositions: A_2B (e.g., Mg_2Ni), AB (e.g., TiFe), AB_2 (e.g., $ZrMn_2$, $TiCr_{1.9}$), AB_5 (e.g., $LaNi_5$), and the body-centered-cubic (bcc) $A_{1-y}B_y$ alloys (e.g., $V_{0.85}Ti_{0.1}Fe_{0.05}$). A key advantage of many intermetallic alloys is the ability to alter their hydrogen sorption behavior by substitution for either or both the A and B metals that often improve their performance in various applications.

Based upon their crystal structures, the volumetric densities of most metal hydrides can be very high (i.e., >100 g H_2/L) that significantly exceed the density of cryogenic liquid hydrogen [26]. Unfortunately, hydrides are nearly always brittle materials that fracture into fine powders (the notable exception is Pd metal and many of its alloys) during hydrogen absorption and desorption cycling that decreases their filling capacities in storage beds, which reduces effective packing densities by at least a factor of two. The poor thermal conductivity with powders usually requires internal components (i.e., foams, fins) within the hydride beds in order to manage [27] the heat associated with reactions (15.1) and (15.2). Hence, the volumetric capacities of practical hydride storage systems are further impacted.

Nearly all of the readily reversible hydrides are composed of transition or rare earth metals, which results in intrinsic gravimetric capacities of 1–4 wt.%. It is unusual that hydrides with gravimetric capacities below 1 wt.% are considered for any energy storage systems including stationary applications [20–22]. However, metal hydrides are suitable for numerous systems [21]. The greatest current limitation of metal hydrides is for onboard hydrogen storage in passenger vehicles where the system densities exceeding 6 wt.% are desired [25,27], which imposed requirements greater than ~9 wt.% on the hydride materials. None of the hydrides based on transition metals can meet these gravimetric densities.

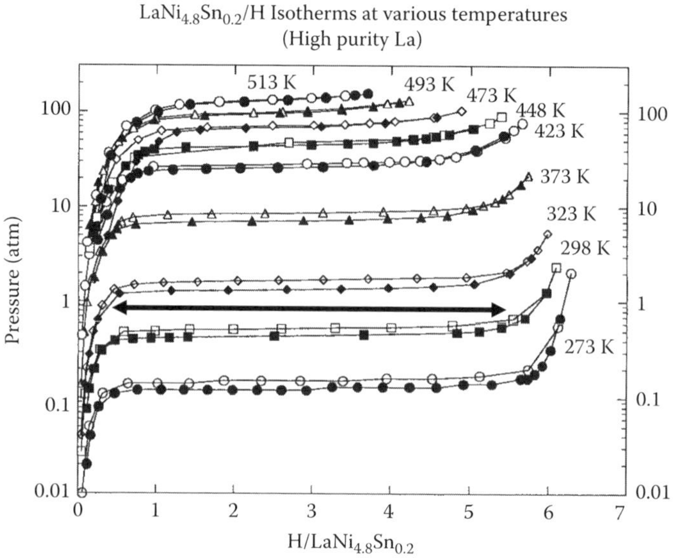

FIGURE 15.5
Hydrogen absorption and desorption isotherms experimentally determined on the alloy $LaNi_{4.8}Sn_{0.2}$. The arrow indicates the reversible storage capacity for this system.

The primary property for the selection of a hydrogen storage material is its ability to provide sufficient amounts of hydrogen gas at the appropriate pressures over the operating temperature of the planned application [9,17–27]. These characteristics of metal–hydrogen systems are usually established from their equilibrium pressure–composition–temperature (PCT) isotherms. Figure 15.5 illustrates the PCT isotherms that had been determined for the $LaNi_{4.8}Sn_{0.2}$ alloy by Luo et al. [28]. As indicated in Figure 15.5, the reversible capacity of a metal hydride is usually defined by the plateau length in its PCT isotherms, which is always somewhat less than the total hydrogen content. For a $LaNi_{4.8}Sn_{0.2}H_x$ system, these values are 1.24 and 1.40 wt.%, respectively. This hydride also has relatively little hysteresis between the absorption and desorption isotherms, which is often an advantage in many applications such as heat pumps and compressors [21]. As described in previous reviews [17–20], the PCT plateau pressures provide fundamental thermodynamic parameters using the relationship

$$\ln P = \frac{\Delta S}{R} - \frac{\Delta H}{RT} \tag{15.3}$$

where
R is the universal gas constant
ΔS is the entropy
ΔH is the enthalpy for the reaction between hydrogen and the metal/alloy

The pressures shown in Figure 15.5 lead to $\Delta H = -32.8$ kJ/mol-H_2 and $\Delta S = 105$ J/K-mol H_2 for the $LaNi_{4.8}Sn_{0.2}H_x$ system. The ideal range of enthalpies for storing hydrogen to use with proton exchange membrane (PEM) fuel cells is often suggested to be 20–50 kJ/mol-H_2 depending on entropy change [25].

While the gravimetric/volumetric capacities and thermodynamics are often the first considerations in choosing a storage candidate, numerous other properties must also be simultaneously satisfied by the hydride material for any given application [21]. Some examples include ease of activation or initial hydriding, sensitivity to air or impurities, volume expansion during hydrogen absorption, decrepitation into fine powder during cycling, pyrophoricity, toxicity, and costs of the raw materials and any processing necessary to produce the storage alloys. In many cases though, either the reaction kinetics or durability of the hydride phases to retain reversibility during extended cycling determines whether a candidate material can be used in a given application. The apparent rates for hydride reactions are sometimes limited by the intrinsic kinetics of the materials via either diffusion within the solids or the chemical reactions on particle surfaces; but in hydrides of the AB, AB_2, and AB_5 alloys, it is usually the heat transfer within the beds of hydride powders [19,21] that dominates. The selective oxidation and segregation of metal A on the surface of the particles leave the catalytic B metal to promote hydrogen molecules to adsorb and dissociate while forming the hydride phases. On the other hand, when A forms a very stable binary hydride phase AH_x, the $A_nB_mH_{x/2}$ phase is metastable and often disproportionates irreversibly into AH_x and B-enriched compounds or alloys [21,26]. This process can severely impact the ideal reversible reaction of Equation 15.2 and has often eliminated candidates from various storage applications.

Summary: The traditional metal hydrides have efficiently and successfully stored hydrogen in numerous laboratory and niche applications where their reversibility and rapid kinetics offset disadvantages from their lower range of gravimetric capacity. Various hydrides remain viable storage candidates for stationary power systems as well as specialty and utility vehicles (e.g., forklifts, carts) since their H_2 delivery pressures can be met using the waste heat from operating PEM fuel cells. Additional information on the properties and current R&D status of hydrides for storage can be found in Chapter 20.

15.4 Complex Hydrides

A particularly interesting class of compounds for hydrogen storage applications include group I or II metal cations (e.g., Li, Na, Mg, Ca) bonded to complex anions comprising light central atoms covalently bound to two or more hydrogen atoms (e.g., $[NH_2]^-$ (amides), $[BH_4]^-$ (borohydrides), and $[AlH_4]^-$ (alanates)). Some noteworthy examples include $LiNH_2$, $LiBH_4$, and $NaAlH_4$. These *complex hydrides* are especially compelling for vehicular hydrogen storage due to their high gravimetric capacity for hydrogen arising from the low atomic weights of the alkali or alkaline earth cations and the central atom in the anion, as well as the existence of multiple hydrogen atoms in compound. For example, $LiBH_4$ has a total gravimetric capacity of approximately 18.5 wt.% hydrogen. For comparison, transition metal hydrides have a capacity that is typically 1–4 wt.%. In addition, these materials have moderate densities that translate into high volumetric capacities for hydrogen, another important feature that makes them desirable for onboard transportation applications. The structure, properties, and hydrogen storage characteristics of alanates, borohydrides, and amides are described in comprehensive reviews [9,10,25,29–36]. In addition, recent research on binary and ternary metal–N–H systems and multinary systems involving metal amides and other complex hydrides (e.g., $NaNH_2$–$LiAlH_4$) is reviewed by Chen et al. [37]. Complex hydrides are also discussed in considerable detail later in this chapter.

Although complex hydrides have hydrogen capacity characteristics that would seem to make them ideal candidates for onboard storage, problems with high thermodynamic stability and prohibitively slow hydrogen adsorption/desorption kinetics have limited their use as reversible hydrogen storage media. Research on these materials over the past 15 years has been devoted almost exclusively to overcoming those obstacles. The thermodynamics and kinetics issues and general approaches that have been used to address the issues are summarized in the following texts.

15.4.1 Thermodynamics

As discussed earlier, transition metal hydrides are characterized by metallic bonding that stabilizes the molecular structure at ambient temperatures and pressures. However, due to the comparatively weak delocalized bonding in the structure, the bonds can be broken and hydrogen released at modest temperatures (≤350 K). In contrast, complex hydrides contain much stronger covalent and ionic/covalent bonds resulting in higher thermodynamic stability and correspondingly lower equilibrium hydrogen pressures. Tables 3–4 in Ref. [29] summarize the hydrogen storage properties of alanates, amides, borohydrides, and related materials. The enthalpy for dehydrogenation is of particular importance for hydrogen storage applications because it provides a quantitative measure of the amount of heat that is required to release hydrogen from the compound. As a rule of thumb, to operate at temperatures consistent with PEM fuel cells, the reaction enthalpy should lie in the range of 20–50 kJ/mol. Of course, it is understood that even though an enthalpy value in this range is necessary for adequate system performance, it is *not sufficient* because, as discussed in the following, kinetics considerations can seriously impact complex hydride sorption behavior.

Two primary approaches for overcoming unfavorable thermodynamics have been explored. The first focuses on new single-phase materials whose thermodynamic properties are altered by atomic substitution or alloying [37–43]. In that approach, an intermetallic alloy with an enthalpy lower than the pure starting material is formed. Although that is usually accomplished using cation substitution, more complex combinations can also favorably alter the reaction enthalpy. For example, it is now well known that reactions between amides and hydrides can occur and that those reactions can facilitate the formation of new alloys with concomitant liberation of hydrogen at reduced temperatures (e.g., see Refs. [37,42]).

Although related, the second approach to relaxing thermodynamic constraints in metal hydrides employs additives that form new compounds or alloys *during dehydrogenation*. The proper selection of an additive can reduce the enthalpy for dehydrogenation, thereby increasing the equilibrium hydrogen pressure and effectively destabilizing the hydride(s) [44–46]. A schematic illustration of destabilization through alloy formation in the dehydrogenated state is shown in Figure 15.6. In the absence of the alloying agent, the hydride undergoes the dehydrogenation reaction with a relatively high enthalpy (ΔH). However, if the original hydride MH_x is combined with additive, X, then the alloy MX can form during hydrogen release. Since the formation of MX is exothermic, the enthalpy of the dehydrogenated state ($\Delta H'$) is reduced. When the additive is present, the system can now cycle between the hydride, MH_x, and the alloy, MX (instead of the nonalloyed material, M). By forcing a change in the reaction pathway through alloy formation in the dehydrogenated state, the hydride is thereby *destabilized*. We point out that the addition of the alloying species does reduce the net gravimetric capacity for hydrogen that is achieved in the reaction system. Therefore, it is desirable to select low-Z destabilization agents.

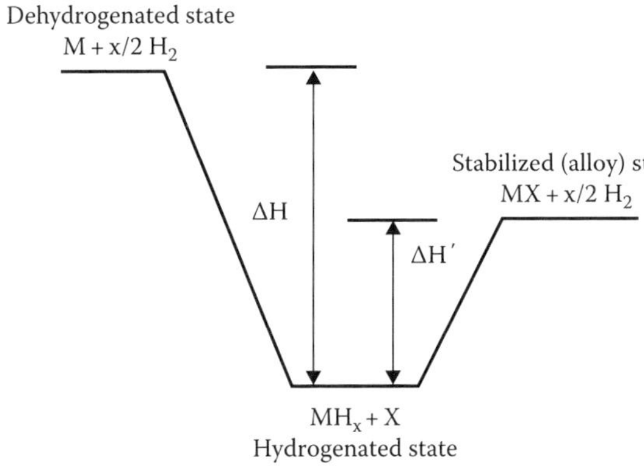

FIGURE 15.6
Generalized enthalpy diagram showing transition between hydrogenated and dehydrogenated states and how enthalpy change is reduced when system is destabilized by alloy formation upon dehydrogenation.

This hydride destabilization process was first introduced by Reilly and Wiswall over 40 years ago in studies of $Mg_2Cu + MgH_2$ [44]. It was found that the temperature to achieve an equilibrium pressure of 0.1 MPa was approximately 49 K lower in the mixed system than in pure MgH_2, suggesting that MgH_2 is destabilized by $MgCu_2$ ($MgCu_2$ forms upon dehydrogenation). This concept was recently extended to a wide range of low-Z binary and complex hydrides [45–48]. For example, upon heating, $LiBH_4$ dissociates by the following reaction:

$$2LiBH_4 \leftrightarrow 2LiH + B + 3H_2 \tag{15.4}$$

The enthalpy of the $LiBH_4$ dehydrogenation reaction is ~67 kJ/mol, resulting in a temperature of 683 K to achieve an equilibrium hydrogen pressure of 0.1 MPa. However, if the destabilizing agent MgH_2 is added, the dehydrogenation reaction proceeds as follows [46]:

$$2LiBH_4 + MgH_2 \leftrightarrow 2LiH + MgB_2 + 4H_2 \tag{15.5}$$

In this case, the formation of MgB_2 in the dehydrogenated state reduces the reaction enthalpy from ~67 to ~45 kJ/mol, which translates to a temperature of 441 K to achieve an equilibrium hydrogen pressure of 0.1 MPa—more than 240° lower than for the simple $LiBH_4$ dissociation reaction. Although the destabilization approach has been employed successfully to modify the thermodynamics in selected complex hydrides, serious kinetics challenges that limit the overall reaction rate remain to be solved. A review of recent work on destabilization of complex hydrides is given in Ref. [47]. Following the initial experimental studies on thermodynamic tuning by destabilization of complex hydrides, a wide range of new systems were explored in a comprehensive series of theory and simulation investigations [49–56]. Many of the new reaction systems discovered in that work are currently being tested experimentally.

Although the two approaches described previously for altering reaction thermodynamics have received the most research attention in recent years, a third approach, based upon thermodynamic changes that accompany reduction in reactant particle size, has also been considered. As will be discussed in the following, a reduction in particle size to nanoscale dimensions can have large effects on reaction kinetics (e.g., reduction in diffusion distances results in a faster net reaction rate). However, it has been shown theoretically that thermodynamics can be affected as well. For example, using ab initio Hartree–Fock method and density functional theory (DFT), Wagemans et al. [57] calculated that when particles of MgH_2 are reduced to a grain size less than 1.3 nm, the hydrogen desorption temperature decreases significantly. For instance, they showed that the desorption temperature for 0.9 nm particles is approximately 100° less than desorption in the bulk material. On the other hand, a more recent theoretical study using a different approach showed a much more modest effect of particle size on desorption temperature [58]. Although changes in thermodynamic properties with particle size are certainly intriguing from a theoretical standpoint, it remains a serious challenge to experimentally produce nanoscale reactants and then maintain the small reactant size during multiple hydrogen sorption cycles. Methods for encapsulation or confinement of the nanoscale particles seem to be the most viable way to maintain small particle sizes in multiple cycled reaction system (for instance, see Ref. [59]).

15.4.2 Kinetics

The approaches summarized earlier are being successfully employed to alter the thermodynamics of hydrogen sorption reactions in complex hydrides so that reversible dehydrogenation/hydrogenation reactions can be potentially achieved at temperatures compatible with PEM fuel cell operation. However, in most complex hydride systems, the hydrogen sorption reactions often proceed via multiple reaction steps, and in many cases, they can involve nucleation and growth of different material phases. The kinetic barriers that exist for these steps can greatly impede the overall reaction rate, thereby producing desorption behavior that is markedly different than what might be expected from thermodynamic considerations alone. The two main approaches for addressing these kinetics limitations generally involve the use of catalysts to enhance the reaction rate and the use of nanoengineering methods to reduce reactant particle size, thereby limiting diffusion distances and increasing the surface reaction area. There are several recent reviews of nanoengineering approaches that have been employed to favorably alter reaction kinetics [31,60–62].

Arguably, the most important development to date in the improvement of reaction kinetics in complex hydrides was made in 1997 by Bogdanović and Schwickardi [63] who demonstrated that $NaAlH_4$ could be induced to reversibly hydrogenate and dehydrogenate at modest temperatures and pressures by the addition of a titanium catalyst. Although the thermodynamic properties and gravimetric storage capacity of $NaAlH_4$ were recognized as being favorable for mobile hydrogen storage/fuel cell applications, it had previously been impossible to dehydrogenate and rehydrogenate the material under practical operating conditions.

The dehydrogenation of $NaAlH_4$ occurs in two consecutive steps involving alanate phases with different crystal structures. Although the total hydrogen release is relatively high (5.6 wt.%), the elevated decomposition temperatures 493 and 523 K preclude the use of the undoped material for practical storage applications. However, Bogdanović and Schwickardi showed that the addition of a few mole percent titanium reduced the dehydrogenation temperatures to 373 and 458 K (equilibrium H_2 pressure ~1 MPa H_2), respectively. Those results were particularly noteworthy because they revealed for the first time that dramatic modifications of reaction kinetics could be achieved in an operationally practical complex hydride material and it stimulated a major resurgence of interest in complex hydrides for hydrogen storage. The precise role of the titanium and other transition metals in catalyzing the sorption reaction(s) has been widely studied, and the reader is referred to several excellent reviews [9,10,25,29–36] for additional detail.

Catalysts are now routinely used to enhance the rate of hydrogen sorption reactions in complex hydrides. In addition to the catalytic behavior due to the presence of metal additives, a novel *self-catalyzing* set of reactions has also been reported [64]. In some systems (e.g., $NaAlH_4$), the catalytic effects are sufficient to promote reactions at temperatures and pressures compatible with PEM fuel cell operation. However, even though beneficial catalytic effects by dopants or other additives are observed in many other complex hydrides, the magnitude of the effects is generally insufficient to increase sorption rates to values needed for hydrogen delivery to PEM fuel cells under standard operating conditions. Consequently, other approaches for improving reaction kinetics are also being considered. The most widely studied approach involves *nanoengineering* to reduce the particle size of the reactants [31,60–62]. Reduction in reactant particle size to the nanoscale does not necessarily alter the intrinsic hydrogenation/dehydrogenation kinetics, but rather, sorption rates are improved simply because the diffusion distances become much shorter and the reactant surface area is increased in the nanoscale material.

Traditionally, energetic ball milling has been used to produce small reactant particles. Even though the particle size remains comparatively large (>1 μm), much smaller crystallite sizes within the ball-milled particle can promote enhanced mass transport and increased interface reaction rates compared to the bulk [62]. Although this approach is now widely used for solid-state reactant synthesis, it is often preferable to reduce the actual size of the particles themselves to <100 nm in order to overcome the notoriously slow kinetics in complex hydrides. This can lead to an increase in the rate of hydrogen diffusion across the particle by at least two orders of magnitude compared to the bulk. A wide variety of methods exist for producing nanoparticles of primary reactants and catalysts [31,60–62]. However, as pointed out earlier, in reversible hydrogen storage systems employing complex hydrides, multiple cycles of hydrogenation and dehydrogenation generally act to induce sintering or particle agglomeration, which produces unwanted particle growth and reduced reaction rates. It is therefore not particularly desirable to employ synthesis methods that produce a large number of free particles in contact with each other (e.g., *nanomarbles in a bowl*).

To overcome the problem of particle growth upon cycling, approaches based on particle confinement in nanoporous host materials are being explored. For example, compared to the bulk, greatly improved kinetics were observed when ammonia borane (a nonreversible chemical hydrogen storage material) was encapsulated in a mesoporous silica network [65]. That work was later extended to include $NaAlH_4$ [66] and $LiBH_4$ [67] reactants encapsulated in carbon aerogels. Carbon was selected for the latter work because of its comparatively lightweight, low reactivity, and ease of forming meso- and nanoporous scaffold frameworks. The aerogel serves both as a structure-directing agent to restrict reactant particle size and as a host to confine the reactant particle before and during the sorption reaction. When $LiBH_4$ was incorporated into a C-aerogel with pore sizes ranging from 10 to 30 nm, the dehydrogenation temperature was reduced by approximately 100° [67] and the desorption rate increased by up to 60 times [68] compared to the bulk material. This result, together with those published previously, suggests that the scaffold facilitates reduced diffusion distances and promotes an increase in the reactant surface area, which, in turn, improves the overall desorption kinetics.

There are clearly advantages to employing nanoscale reactants and catalysts to improve reaction kinetics in complex hydride materials. However, in nanoporous scaffold systems, the framework material imposes a gravimetric penalty that must be accounted for.

Likewise, the nanoscale scaffold must be robust to multiple sorption cycles. Also, in nanoscale systems, there is a greater potential for the formation of unwanted (inert) phases that can impede sorption reactions or terminate them altogether. Those problems notwithstanding, it is likely that development of nanoscale engineering approaches will continue to be an active research topic as further attempts are made to improve the hydrogen sorption kinetics in complex hydride materials.

15.4.3 Summary

The high gravimetric and volumetric capacities for hydrogen in complex hydrides have made these materials attractive candidates for onboard hydrogen storage applications. Significant progress has been made by the international research community over the past 15 years to improve the thermal characteristics, sorption kinetics, and reversibility, obstacles that have historically limited the use of these materials in highly demanding transportation applications. Although they are emerging as potentially viable media for hydrogen storage and delivery to PEM cell systems, there is no single material that meets the stringent onboard storage requirements. Several candidate systems that meet the thermodynamic requirements have been identified, but kinetic barriers and, in most cases, cycling problems still limit their incorporation in practical fuel cell systems.

Current research is focused primarily on five areas: (1) elucidating the detailed reaction mechanisms and identifying the rate-limiting step(s) that reduce the sorption reaction rates, (2) developing a detailed understanding of how metal catalysts influence elementary reaction steps, (3) exploring new ways to alter reaction pathways to eliminate product *sinks* that reduce the cycling efficiency, (4) discovering new nanoengineering approaches that create a reaction environment that improves diffusion rates and limits formation of unwanted material phases, and (5) exploring (theoretically and experimentally) entirely new material combinations that possess improved thermodynamic and kinetic properties.

This is clearly a fertile area for research and development in materials science, and based upon the impressive advances that have been made thus far, it is anticipated that progress will be made in future work to overcome the limitations of current complex hydride materials for practical hydrogen storage applications. The current status of R&D in complex hydrides and plans for addressing existing barriers and obstacles are described in greater detail in Chapter 21. In addition, the discovery and development of new materials, including complex hydrides, were conducted a focused, 5-year, technical effort in the US Department of Energy (DOE) Metal Hydride Center of Excellence (MHCoE), an applied R&D consortium comprising industry, university, and government laboratories. The reader is referred to final report from the MHCoE [69].

15.5 Chemical Hydrogen Storage Materials

Chemical hydrogen storage materials are a class of storage compounds containing large quantities of chemically bound hydrogen that can be released either by heating or by reactions with other materials (e.g., hydrolysis reactions). Unlike metal hydrides or physisorption materials that store and desorb hydrogen reversibly, the chemical hydrogen storage material must be regenerated using separate (off-board) processing steps following release of the hydrogen. Due to their high gravimetric storage capacities (>10 wt.% H_2) and fast desorption kinetics at near ambient temperatures, these systems are ideal for *single-use* applications where disposal of the spent fuel is acceptable or in situations where efficient fuel regeneration schemes exist. A recent review [70] summarizes status and progress in the thermochemistry and engineering of chemical hydrogen storage systems; the topic is also discussed in greater detail later in Chapter 23.

Common chemical hydrogen storage materials can be either liquids or solids, and the thermodynamics and mechanisms of hydrogen release can vary markedly in specific systems. For example, some materials such as cyclohexane [71], decalin [72], and pi-conjugated materials [73] release hydrogen catalytically in endothermic reactions. On the other hand, hydrogen release can also occur exothermically from, for example, hydrolysis reactions involving hydrogen-rich materials such as sodium borohydride ($NaBH_4$) [74] or from catalytically activated release of hydrogen from ammonia borane [70,75].

Hydrolysis reactions have been actively investigated because they can potentially release large amounts of hydrogen in a process that can be readily integrated with PEM fuel cell systems. In the case of sodium borohydride, simple thermal decomposition of the compound occurs at ~400°C, a high temperature that is incompatible with fuel cell operation. However, a metal-catalyzed $NaBH_4$–H_2O hydrolysis reaction proceeds at ambient temperatures to produce hydrogen and sodium metaborate ($NaBO_2$) [74]:

$$NaBH_4 + 2H_2O \rightarrow 4H_2 + NaBO_2. \quad (15.6)$$

In this simplified view of the overall hydrolysis reaction, the participation of water in the reaction effectively doubles the amount of hydrogen that would be obtained by simple thermal decomposition. Unfortunately, $NaBO_2$ has only about half the solubility of $NaBH_4$ at 293 K. Consequently, the $NaBH_4$ concentration must be kept lower than the solubility of the $NaBO_2$ to preclude precipitation of the metaborate [76]. This greatly reduces the overall storage capacity of the system. For example, 20% $NaBH_4$ solution (with a 1% NaOH stabilization additive) was used in a commercial system (Millennium Cell). The low $NaBH_4$ concentration reduces the overall storage capacity to only 4% [9]. In addition to the reduced storage capacity in this system, regeneration of the hydride from the metaborate solution is energy intensive, making the system useful only for special storage applications.

Recent work has focused on hydrogen release from ammonia borane [70,77–80]. This compound has received considerable attention because of its high storage capacity for hydrogen (>19 wt.% H_2 and 150 g H_2/L) and the potential to controllably release hydrogen at temperatures <100°C in a metal-catalyzed reaction system [70]. Although some problems with hydrogen kinetics remain to be solved before ammonia borane can be utilized in a practical hydrogen storage application, they appear to be tractable. On the other hand, the development of an efficient and scalable regeneration method remains a serious challenge. The reader is referred to several recent reviews [70,77–80] that summarize the status of ammonia borane as a practical hydrogen storage medium and address the problems that remain to be solved for applications that require efficient recycling of the starting material. Extensions to other novel amidoborane compounds are discussed in Ref. [81].

A major research and development effort was carried out within the US DOE Chemical Hydrogen Storage Center of Excellence (CHSCoE) to optimize hydrogen release characteristics of ammonia borane and to develop improved regeneration methods. In addition, investigators within the CHSCoE are exploring new systems that have even better performance. The reader is referred to annual progress reports given in Ref. [82] for information concerning the center activities and progress.

Aluminum hydride (AlH_3—alane) is another chemical hydrogen storage material that is attracting increasing attention in the hydrogen storage R&D community. Alane has gravimetric and volumetric capacities of 10.1 wt.% H_2 and 148 g-H_2/L, respectively. These characteristics make alane attractive for onboard storage applications. Although all phases of the compound are thermodynamically unstable under ambient temperature, the existence of a passivating Al_2O_3 surface layer inhibits decomposition at room temperature. Graetz and Reilly [83] recently demonstrated that the α-, β-, and γ-phases of AlH_3 decompose readily at 373 K at rates commensurate with onboard fuel cell operation. Likewise, Sandrock et al. [84] showed that the hydrogen desorption kinetics could be improved at practical fuel cell temperatures using particle size control and doping with alkali metal hydrides. Although the dehydrogenation properties of alane make it a potentially promising candidate for onboard storage applications, direct hydrogenation of aluminum to AlH_3 requires an unacceptably high H_2 pressure (>10^4 MPa at 300 K), which historically has rendered the alane system impractical as a hydrogen storage medium.

Recent work has shown that problems with hydriding of aluminum can be overcome using alternate approaches involving low-temperature and low-pressure organometallic reactions [85] and electrochemical regeneration [86]. In the former process, a slurry of titanium-doped aluminum powder and triethylenediamine (TEDA) in tetrahydrofuran in a H_2 gas atmosphere forms an AlH_3–TEDA adduct at room temperature and at pressures <3.5 MPa. The addition of the Ti catalyst allows the adduct formation reaction to proceed at pressures an order of magnitude lower than had been previously reported [87]. The adduct can be recovered by precipitation or recycled to generate H_2 gas, thereby providing a pathway to achieving efficient dissociation/regeneration cycle performance. A detailed analysis of an onboard storage hydrogen storage system that uses an alane slurry as the hydrogen carrier is given in Ref. [88].

The second method that has been studied recently for alane regeneration employs an electrochemical cycle to avoid high hydriding pressures and formation of unwanted stable by-products [86]. The cycle utilizes electrochemical potential to generate alane and alkali hydride from an ionic alanate salt. The feasibility of a recyclable process has been demonstrated, and work is in progress to improve the efficiency and scale up the reaction system. Both this method and the adduct formation process are producing encouraging results that should significantly impact the use of aluminum hydride in a wide range of hydrogen storage applications. These processes were investigated in the DOE MHCoE; additional information is provided in the MHCoE final reports [69].

The trade-offs between liquid and solid systems must also be carefully considered in the selection of optimum systems for chemical hydrogen storage. Powders and slurries have been investigated for both alane and ammonia borane systems. In general, liquids have advantages for distribution both within a fuel delivery network and within a vehicle. Crabtree has discussed these advantages, specifically with respect to use of liquid organic heterocycle compounds [89]. Likewise, the system implications of storage and delivery strategies in solid and liquid systems are discussed in detail by Aardahl and Rassat [70].

Summary: The high gravimetric and volumetric storage capacities for hydrogen together with favorable hydrogen release kinetics at operationally relevant temperatures and pressures make chemical hydrogen storage materials potentially important candidates for a range of storage applications. However, since these materials require off-board regeneration, the energy efficiency and scalability of the regeneration processes are critical to successful deployment of the materials in situations that require reuse of the materials. These issues are being addressed in laboratories worldwide and good progress is being made. The thermochemistry and kinetics of hydrogen release from chemical hydrogen storage materials as well as processes for efficient regeneration of spent fuel will be discussed in detail in Chapter 23.

15.6 Hydrogen Sorbents

Van der Waals interactions between a hydrogen molecule and the surface of a solid can result in adsorption of hydrogen on localized surface sites. Since these dispersive interactions result in weak bonding (1–10 kJ/mol), significant physisorption occurs only at low temperatures (generally much less than 273 K). Since molecular adsorption occurs predominantly in the first monolayer, the use of a high surface area material is desirable in order to achieve hydrogen capacities that are useful for practical hydrogen storage applications in instances where cooling of the adsorbent is acceptable. In this case, simple monolayer saturation of the adsorption sites can be described by a Langmuir isotherm [10]. A significant benefit of the physisorption mechanism is that the adsorption process is not activated, implying that the adsorption kinetics are fast. Moreover, the low binding energy translates into a low barrier for desorption, leading to good reversibility in system applications. Hydrogen storage systems based upon molecular physisorption are being explored at R&D organizations worldwide, several excellent reviews are available [9,10,25,31,36,60,61,90–96], and the topic is discussed in detail later in this chapter.

As expected for coverage limited to a monolayer, the amount of adsorbed hydrogen is proportional to the specific surface area of the adsorbent, which leads to approximately 2.5 wt.% adsorbent per 1000 m^2 surface area at cryogenic temperatures [97]. This dependence of adsorption capacity on the surface area has motivated the search for new adsorbents with the highest possible number of adsorption sites, including metal oxide frameworks (MOFs) [98–103], covalent organic frameworks (COFs) [104–106], zeolites [107], and nanostructured carbon (e.g., nanotubes, templated carbons, fullerenes, aerogels, substituted polymers, and intercalated graphite) [31,60,61,91–96]. Due to their extremely high porosity, MOFs have become especially intriguing candidates for physisorption applications [98–103]. MOFs are highly porous, 3D polymeric structures comprising inorganic building units connected by organic linkers. In the case of MOF-177 (zinc acetate building units linked by 1,3,5-benzenetribenzoate), the Langmuir surface area is extremely large (~4500 m^2/g), giving rise to a storage capacity of 7.5 wt.% at 7 MPa and 77 K [99,108].

The structure and properties of specific physisorption systems will not be discussed in this introduction. Although many systems have been explored, the hydrogen adsorption capacities at ambient temperatures and modest pressures (10 MPa) are less than 0.8 wt.% H_2; that is, as mentioned earlier, practical gravimetric storage capacities are only achievable at low temperatures (typically 77 K) in these materials. (The general trend of hydrogen uptake as a function of specific surface area at cryogenic temperatures is shown for a wide variety of porous adsorbents in Ref. [96].) Consequently, considerable research attention has been directed toward finding ways to modify the adsorbent surface to increase both the adsorption potential and the number of available sorption sites.

A particularly interesting approach that has been proposed to achieve enhanced storage capacity at ambient temperatures utilizes hydrogen *spillover* from substituted metal atoms on adsorbent surfaces [108]. In this process, the surface of the adsorbent is doped with a catalytic additive, typically a metal, which serves to promote dissociation of molecular hydrogen and facilitate transport of atomic hydrogen to the adsorbent surface sites. The molecular dissociation is believed to occur rapidly, and the rate-determining step for adsorption involves diffusion of the hydrogen atoms to/on the receptor surface. Since binding energy for atomic hydrogen is greater than for H_2, the storage capacity is expected to be enhanced at elevated temperatures. For example, it has been reported that in IRMOF-8 (zinc acetate units with naphthalenedicarboxylate linkers), the heat of adsorption for atomic hydrogen is ~20 kJ/mol [109] compared to ~6 kJ/mol for molecular adsorption [110]. Yang and coworkers reported that the H_2 uptake was increased from 0.5 wt.% in pure IRMOF-8 at 298 K and ~10 MPa to 4 wt.% on bridged IRMOF-8 (bridge building facilitates enhanced secondary spillover between two dissimilar materials) [109].

Hydrogen spillover in a large number of adsorbent systems, including carbon nanotubes, zeolites, MOFs, COFs, and other nanostructured materials, has been investigated, and significant enhancements in hydrogen uptake at room temperature have been reported [109]. However, diminished surface diffusion rates that may arise from site filling near the active metal center remain a serious concern. For example, recent DFT calculations have shown that diffusion of atomic hydrogen from the metal nanoparticle to a graphitic substrate is both kinetically and thermodynamically inefficient and limits the spillover rate [111]. Lack of reproducible adsorption capacity and slow diffusion kinetics in some spillover systems are outstanding issues that must be addressed before the process can be used in practical applications. The detailed aspects of the spillover mechanism and the limits of hydrogen storage capacity in different material systems remain topics of active research.

Another approach that is being explored to enhance adsorption energies employs the so-called Kubas interaction between hydrogen and a metal complex [112–114]. In this process, the hydrogen molecule bonds *side-on* to the metal through donation of its two σ orbitals to a vacant metal orbital, forming a two-electron, three-center bond [114]. Since a stable H_2–metal complex is created in this interaction, the binding energy is greater than what would be expected from weak van der Waals interactions between the H_2 and the adsorbent. The increase in the binding potential via Kubas interactions is being explored in a large number of nanostructured systems; a review of these interactions and some perspectives on utilizing them for enhanced hydrogen storage in different adsorbent systems is given in Refs. [113,114].

In addition to extensive experimental studies on hydrogen storage via physisorption in highly porous media, considerable theoretical work has been performed to explore new materials and surface modification approaches. The current state of the art in molecular and quantum mechanical modeling of physisorption and diffusive transport in MOFS is discussed in a recent review [115]. Also, simulations that predict enhanced hydrogen storage in 3D COFs are described in Ref. [106]. Hydrogen adsorption in doped fullerenes and related carbon nanostructures has also been studied theoretically. For example, first-principles DFT and quantum Monte Carlo calculations have suggested enhanced binding of molecular hydrogen on substitutional B and Be sites in fullerenes [116]. Likewise, a first-principles study showed that a single titanium atom on a titanium-decorated single-walled carbon nanotube can bind up to four hydrogen molecules [117]. In most cases, the theoretical predictions have not been experimentally validated, largely because of the difficulty in preparing the tailored adsorbent and competing reactions of the isolated metal atom or complex with foreign species. A broadly based theoretical effort on enhanced physisorption in highly porous systems was conducted within the US DOE Hydrogen Sorption Center of Excellence. The reader is referred to the final report of the DOE's Hydrogen Sorption Center of Excellence [118]. Additional information concerning theory and modeling of hydrogen adsorption on tailored surface sites is summarized in several review articles [9,10,25,31,36,60,61,90–96].

Summary: A large variety of materials with high surface area and enhanced surface adsorption capacities are being investigated for hydrogen storage applications. Modification of surface properties to increase the binding energy with hydrogen and thereby enhance the sorption capacity at elevated temperatures is an especially active area of study. Compared to other solid-state hydrogen storage approaches that involve the formation and rupture of strong covalent or ionic bonds and complex phase changes during hydrogenation and dehydrogenation cycles, the physisorption process is more conceptually straightforward. However, the development of reproducible approaches for achieving adsorption at elevated temperatures in extremely porous materials remains a significant technical challenge. Likewise, as pointed out by Felderhoff et al. [9], physisorption of hydrogen under cryogenic conditions in a practical engineering system requires a large amount of excess heat to be removed. This places serious demands on the heat management system and can result in prohibitively large amounts of liquid nitrogen or other cryogens to be consumed during charging. The materials and engineering challenges that impact the implementation of the physisorption approach for hydrogen storage in a practical system are described in greater detail in Chapter 24.

15.7 Engineering Aspects of Hydrogen Storage Systems

While the attributes and limitations of each storage medium are the first considerations for any given application, numerous engineering factors impact the design and performance of the integrated hydrogen storage system. Reliable and safe containment of hydrogen is required over the entire range of pressure and temperature during the projected operational lifetime with adequate margins of safety. Hence, the materials used to fabricate these vessels and all hydrogen-exposed components must have appropriate mechanical strengths and not be susceptible to hydrogen embrittlement or other detrimental chemical reactions such as corrosion. Materials with these characteristics are often relatively expensive that contribute to making hydrogen storage systems much more costly when compared to other fuel storage options such as gasoline or natural gas.

Beyond the demands for mechanical integrity, significant heat transfer issues must also be addressed for all hydrogen storage methods [119]. The designs of vessels and refueling stations for H_2 gas compressed to high pressures (i.e., above ~10 MPa) should provide for accommodating heat generated during both compression and the inverse Joule–Thomson effect under ambient conditions in order to achieve efficient filling of the vessel. The cryogenic storage of hydrogen as either liquid or cryo-compressed gas [13] requires designs that provide efficient thermal insulation that can minimize loss of hydrogen by boil-off or excessive vaporization from extraneous heat leaks to extend the dormancy life described previously in this chapter. For hydrogen storage using adsorption on activated carbon or other high surface materials, the thermal conductivity within the sorbent bed also needs to be enhanced by additional components (e.g., metallic foams, fins, or mesh network) to manage the heat of reaction during adsorption or desorption. Ahluwalia and Peng [120] present an engineering analysis on the behavior of such a system based upon the activated carbon AX-21 for potential automotive storage. The much greater heats of reaction of ~20–100+ kJ/mol for hydrogen absorption by the metal and complex hydrides significantly exacerbate the need for improved thermal conductivity within the storage beds [17,19,21,120]. Due to the added complexity of reactions associated with chemical hydrogen systems where both initial fuel material and the products formed during decomposition must be processed and/or separated as hydrogen is released, extensive physical and thermal management engineering is needed for various components and auxiliary subsystems are necessary [70].

Because metal hydrides have been considered for numerous energy storage and conversion applications for over 30 years and these materials are commonly known to have very poor thermal conductivities in powder form [17,19,21], extensive efforts have been made to develop hydride bed designs with various configurations that enhance the effective conductivity within the bed without seriously impacting filling capacity, bed mass, and internal gas flow. Most of these approaches included the use of foams and compacts as was reviewed by Zhang et al. [120]. Numerous analytical models have been developed by various workers that attempt to represent the complex relationship of heat exchange and thermal management during the absorption and desorption processes within hydride beds in order to either predict performance or compare with experimental testing of prototypes. A recent example of these analyses is the work by Hardy [121,122] who also cites a number of earlier studies conducted by others. The primary storage system considered by Hardy is based upon a prototype storage bed design using titanium-catalyzed $NaAlH_4$ as the storage media as described by Mosher et al. [123]. Ahluwalia has performed independent analyses of the kinetics and thermodynamics for $NaAlH_4$ in the same prototype that indicated these properties are insufficient to meet performance targets for storage onboard fuel cell–powered passenger cars [123].

Summary: In order to achieve the best performance from any hydrogen storage material in a given application, careful and thorough system engineering must be considered with respect to the vessel design as well as the chemical and physical properties of the media and the components that used to enhance heat and mass transfers during both the filling and discharge operations. As the performance targets with respect to operating conditions such as temperature and time become more demanding, increasing penalties are incurred with respect to mass, volume, and costs for the resulting storage system.

15.8 Summary

Traditional methods of hydrogen storage do not provide suitable performance for many emerging applications for hydrogen use as a fuel, especially with PEM fuel cells. Therefore, there are considerable efforts directed at developing advanced hydrogen storage methods. These efforts include lightweight Type III and IV high-pressure composite cylinders, cryogenic-capable high-pressure vessels, reversible metal and complex hydrides, and regenerable chemical hydrogen storage materials and hydrogen sorbents. While significant progress has been made in the development of these technologies, none currently meets all of the stringent performance requirements for the most demanding applications, such as onboard vehicle storage. The materials-based technologies require further development of materials with improved sorption kinetics, less extreme operating temperature and pressures, and higher volumetric and gravimetric capacities. Additionally, there are remaining engineering challenges for the materials-based and physical storage methods, not the least of which is to reduce the costs. This chapter provided a brief overview of these technologies and their current status. The following chapters provide a much more in-depth review of the various storage technologies. For physical storage methods, Chapter 16 discusses gaseous storage, Chapters 17 and liquid storage, Chapter 19 cryo-compressed storage, and Chapter 25 storage in hollow microspheres and Chapter 26 slush hydrogen storage. For materials-based storage, Chapter 20 covers metal hydrides, Chapter 21 complex hydrides, Chapter 22 nanomaterials, Chapter 23 chemical hydrogen storage, and Chapter 24 sorbents. Additionally, Chapter 29 hydrogen distribution, and Chapters 30 and 31 safety, codes, and standard efforts.

References

1. Davis, S.C., S.W. Diegel, R.G. Boundy. 2009. *Transportation Energy Data Book*, 28th edn. Table B.4. Oak Ridge National Laboratory, Oak Ridge, TN. http://cta.ornl.gov/data/download28.shtml
2. Broom, D.P. 2011. *Hydrogen Storage Materials: The Characterization of the Storage Properties*. London: Springer.
3. Klebanoff, L. 2012. *Hydrogen Storage Technology: Materials and Applications*. Boca Raton, FL. CRC Press.
4. Irani, R.S. 2002. Hydrogen storage: High-pressure gas containment. *MRS Bull.* 27(9):680–682.
5. ISO 7866:1999(E), Gas cylinders—Refillable seamless aluminum alloy gas cylinders—Design, construction and testing, International Standards Organization, http://www.iso.org/iso/iso_catalogue/catalogue_tc/catalogue_detail.htm?csnumber=14795
6. ISO 9809-1:1999(E), Gas cylinders—Refillable seamless steel gas cylinders—Design, construction and testing—Part 1: Quenched and tempered steel cylinders with tensile strength less than 1 100 MPa, International Standards Organization, http://www.iso.org/iso/iso_catalogue/catalogue_tc/catalogue_detail.htm?csnumber=17683
7. Ahluwalia, R.K., T.Q. Hua, J.-K. Peng, R. Kumar. 2009. *System Level Analysis of Hydrogen Storage Options*, DOE Hydrogen Program Review, Arlington, VA. http://www.hydrogen.energy.gov/pdfs/review09/st_13_ahluwalia.pdf
8. Hua, T.Q., R.K. Ahluwalia, J.-K Peng, M. Kromer, S. Lasher, K. McKenney, K. Law, J. Sinha. 2011. Technical assessment of compressed hydrogen storage tank systems for automotive applications. *Int. J. Hydrogen Energy* 36:3037–3049.
9. Felderhoff, M., C. Weidenthaler, R. von Helmolt, U. Eberle. 2007. Hydrogen storage: The remaining scientific and technological challenges. *Phys. Chem. Chem. Phys.* 9:2643–2653.
10. Züttel, A., M. Hirscher, K. Yvon et al. 2008. Hydrogen storage, chapter 6. In *Hydrogen as a Future Energy Carrier*, A. Züttel, A. Borgschulte, L. Schlapbach, eds., pp. 165–264. Weinheim, Germany: Wiley-VCH Verlag GmBH & Co.
11. Wolf, J. 2002. Liquid-hydrogen technology for vehicles. *MRS Bull.* 27(9):684–687.
12. The plotted temperature/pressure/density values were calculated using the National Institute of Standards and Technology's on-line Chemistry WebBook for the isothermal fluid properties for hydrogen. Lemmon, E.W., M.O. McLinden, D.G. Friend. Thermophysical properties of fluid systems. In *NIST Chemistry WebBook*, P.J. Linstrom, W.G. Mallard, eds., NIST Standard Reference Database Number 69. Gaithersburg, MD: National Institute of Standards and Technology. http://webbook.nist.gov (retrieved February 8, 2010).
13. Aceves, S.M., F. Espinosa-Loa, E. Ledesma-Orozco, T.O. Ross, A.H. Weisberg, T.C. Brunner, O. Kircher. 2010. High-density automotive hydrogen storage with cryogenic capable pressure vessels. *Int. J. Hydrogen Energy* 35(3):1219–1226. DOI: 10.1016/j.ijhydene.2009.11.069.
14. Mueller, W.M., J.P. Blackledge, G.G. Libowitz. 1968. *Metal Hydrides*. New York: Academic Press.
15. Manchester, F.E., ed. 2000. *Phase Diagrams of Binary Hydrogen Alloys*. Materials Park, OH: ASM International.
16. Reilly, J.J., G. Sandrock. 1980. Hydrogen storage in metal hydrides. *Sci. Am.* 242(2):118–129.
17. Sandrock, G., S. Suda, L. Schlapbach. 1992. Applications. In *Hydrogen in Intermetallic Compounds II*, L. Schlapbach, ed., Topics in Applied Physics, vol. 67, pp. 197–258. Berlin, Germany: Springer-Verlag.
18. Sandrock, G. 1995. Applications of hydrides. In *Hydrogen Energy System—2 Production and Utilization of Hydrogen and Future Aspects*, Y. Yurum, ed., NATO ASI Series E 295, p. 253. Amsterdam, the Netherlands: Kluwer.
19. Dantzer, P. 1997. Metal-hydride technology: A critical review. In *Hydrogen in Metals III*, H. Wipf, ed., pp. 279–340. Berlin, Germany: Springer.
20. Sandrock, G. 2003. Hydride storage. In *Handbook of Fuel Cells—Fundamentals, Technology and Applications*, W. Vielstich, H.A. Gasteiger, A. Lamm, eds., vol. 3(2), pp. 101–112. Chichester, U.K.: Wiley.

21. Bowman, Jr. R.C., B. Fultz. 2002. Metal hydrides i: Hydrogen storage and other gas-phase applications. *MRS Bull.* 27(9):688–693.
22. Sandrock, G., R.C. Bowman, Jr. 2003. Gas-based hydride applications: Recent progress and future needs. *J. Alloys Compd.* 356/357:794–799.
23. Sakintuna, B., F. Lamari-Darkrim, M. Hirscher. 2007. Metal hydride materials for solid hydrogen storage: A review. *Int. J. Hydrogen Energy* 32:1121–1140.
24. Sandrock, G., G. Thomas. Metal hydride applications list, in the Hydride Information Center (Hydpark) section of the DOE on-line hydrogen storage materials database. http://hydrogenmaterialssearch.govtools.us/ (accessed February 8, 2010).
25. Yang, J., A. Sudik, C. Wolverton, D.J. Siegel. 2010. High capacity hydrogen storage materials: Attributes for automotive applications and techniques for materials discovery. *Chem. Soc. Rev.* 39:656–675. DOI: 10.1039/b802882f.
26. Sandrock, G. 1999. A panoramic overview of hydrogen storage alloys from a gas reaction point of view. *J. Alloys Compd.* 293–295:877–888.
27. Satyapal, S., J. Petrovic, C. Read, G. Thomas, G. Ordaz. 2007. The U.S. Department of Energy's National Hydrogen Storage Project: Progress towards meeting hydrogen-powered vehicle requirements. *Catal. Today* 120:246–256.
28. Luo, S., W. Luo, J.D. Clewley, T.B. Flanagan, R.C. Bowman, Jr. 1995. Thermodynamic and degradation studies of $LaNi_{4.8}Sn_{0.2}$-H using isotherms and calorimetry. *J. Alloys Compd.* 231:473–478.
29. Orimo, S.-I., Y. Nakamori, J.R. Eliseo, A. Züttel, C.M. Jensen. 2007. Complex hydrides for hydrogen storage. *Chem. Rev.* 107:4111–4132.
30. Schüth, F., B. Bogdanović, M. Felderhoff. 2004. Light metal hydrides and complex hydrides for hydrogen storage. *Chem. Commun.* 2249–2258. DOI: 10.1039/b406522k.
31. Varin, R.A., T. Czujko, Z.S. Wronski. 2009. *Nanomaterials for Solid State Hydrogen Storage.* New York: Springer. ISBN 978-0-387-77711-5.
32. Bogdanović, B., U. Eberle, M. Felderhoff, F. Schüth. 2007. Complex aluminum hydrides. *Scr. Mater.* 56:813–816.
33. Züttel, A., A. Borgschulte, S.-I. Orimo. 2007. Tetrahydroborates as new hydrogen storage materials. *Scr. Mater.* 56:823–828.
34. Hauback, B.C. 2008. Structures of aluminium-based light weight hydrides. *Z. Kristallogr.* 223:636–648.
35. Grochala, W., P.P. Edwards. 2004. Thermal decomposition of the non-interstitial hydrides for the storage and production of hydrogen. *Chem. Rev.* 104:1283–1315.
36. Ross, D.K. 2006. Hydrogen storage: The major technological barrier to the development of hydrogen fuel cell cars. *Vacuum* 80:1084–1089.
37. Chen, P., Z. Xiong, G. Wu, Y. Liu, J. Hua, W. Luoc. 2007. Metal–N–H systems for the hydrogen storage. *Scr. Mater.* 56:817–822.
38. Graetz, J., Y. Lee, J.J. Reilly, S. Park, T. Vogt. 2005. Structures and thermodynamics of the mixed alkali alanates. *Phys. Rev. B* 71:184115.
39. Rönnebro, E., E. Majzoub. 2006. Crystal structure, Raman spectroscopy and ab-initio calculations of a new bialkali alanate K_2LiAlH_6. *J. Phys. Chem. B* 110:25686–25691.
40. Nakamoria, Y., G. Kitaharaa, K. Miwab et al. 2005. Hydrogen storage properties of Li–Mg–N–H systems. *J. Alloys Compd.* 404–406:396–398.
41. Nakamori, Y., S.I. Orimo. 2004. Destabilization of Li-based complex hydrides. *J. Alloys Compd.* 370:271–275.
42. Pinkerton, F.E., G.P. Meisner, M.S. Meyer, M.P. Balogh, M.D. Kundrat. 2005. Hydrogen desorption exceeding ten weight percent from the new quaternary hydride $Li_3BN_2H_8$. *J. Phys. Chem. B* 109:6–8.
43. Nakamori, T., K. Miwa, A. Ninomiya et al. 2006. Correlation between thermodynamical stabilities of metal borohydrides and cation electronegativites: First-principles calculations and experiments. *Phys. Rev. B* 74:045126–045135.
44. Reilly, J.J., R.H. Wiswall. 1967. The reaction of hydrogen with alloys of magnesium and copper. *Inorg. Chem.* 6:2220–2223.

45. Vajo, J.J., F. Mertens, C.C. Ahn, R.C. Bowman, Jr., B. Fultz. 2004. Altering hydrogen storage properties by hydride destabilization through alloy formation: LiH and MgH_2 destabilized with Si. *J. Phys. Chem. B* 108:13977–13983.
46. Vajo, J.J., S.L. Skeith, F. Mertens. 2005. Reversible storage of hydrogen in destabilized $LiBH_4$. *J. Phys. Chem. B* 109:3719–3722.
47. Vajo, J.J., G.L. Olson. 2007. Hydrogen storage in destabilized chemical systems. *Scr. Mater.* 56:829–834.
48. Yang, J., A. Sudik, C. Wolverton. 2007. Destabilizing $LiBH_4$ with a metal (M = Mg, Al, Ti, V, Cr, or Sc) or metal hydride ($MH_2 = MgH_2$, TiH_2, or CaH_2). *J. Phys. Chem. C* 111:19134–19140.
49. Alapati, S.V., J.K. Johnson, D.S. Sholl. 2006. Identification of destabilized metal hydrides for hydrogen storage using first principles calculations. *J. Phys. Chem. B* 110:8769–8776.
50. Siegel, D., C. Wolverton, V. Ozolins. 2007. New hydrogen storage reactions based on destabilized $LiBH_4$. *Phys. Rev. B: Condens. Matter Mater. Phys.* 76:134102.
51. Alapati, S.V., J.K. Johnson, D.S. Sholl. 2007. Using first principles calculations to identify new destabilized metal hydride reactions for reversible hydrogen storage. *Phys. Chem. Chem. Phys.* 9:1438–1452.
52. Alapati, S.V., J.K. Johnson, D.S. Sholl. 2007. Stability analysis of doped materials for reversible hydrogen storage in destabilized metal hydrides. *Phys. Rev. B* 76:104108.
53. Alapati, S.V., J.K. Johnson, D.S. Sholl. 2007. Predicting reaction equilibria for destabilized metal hydride decomposition reactions for reversible hydrogen storage. *J. Phys. Chem. C* 111:1584–1591.
54. Alapati, S.V., J.K. Johnson, D.S. Sholl. 2008. Large-scale screening of metal hydride mixtures for high-capacity hydrogen storage from first-principles calculations. *J. Phys. Chem. C* 112:5258–5262.
55. Ozolins, V., E.H. Majzoub, C. Wolverton. 2009. First-principles prediction of thermodynamically reversible hydrogen storage reactions in the Li-Mg-Ca-B-H system. *J. Am. Chem. Soc.* 131:230–237.
56. Ozolins, V., A.R. Akbarzadeh, H. Gunaydin, K. Michel, C. Wolverton, E.H. Majzoub. 2009. First-principles computational discovery of materials for hydrogen storage. *J. Phys.: Conf. Ser.* 180:012076.
57. Wagemans, R.W.P., J.H. van Lenthe, P.E. de Jongh, A. Jos van Dillen, K.P. de Jong. 2005. Hydrogen storage in magnesium clusters: Quantum chemical study. *J. Am. Chem. Soc.* 127(47):16675–16680.
58. Kim, K.C., B. Dai, J.K. Johnson, D.S. Sholl. 2009. Assessing nanoparticle size effects on metal hydride thermodynamics using the Wulff construction. *Nanotechnology* 20:204001.
59. Wu, H., W. Zhou, K. Wang et al. 2009. Size effects on the hydrogen storage properties on nanoscaffolded $Li_3BN_2H_8$. *Nanotechnology* 20:204002–204006.
60. For recent reviews of nanoengineered materials for hydrogen storage, see articles in: Viewpoint set no. 42—Fichtner, M., ed. 2007. Nanoscale materials for hydrogen storage. *Scr. Mater.* 56:801–858.
61. Thermodynamics, kinetics and structure of hydrogen storage materials at the nanoscale are given in: Vajo, J., Pinkerton, F., Stetson, N.T., eds. 2009. Special issue on nanoscale phenomena in hydrogen storage. *Nanotechnology* 20:200201–204030.
62. Zaluska, A., L. Zaluski, J.O. Ström-Olsen. 2001. Structure, catalysis and atomic reactions on the nano-scale: A systematic approach to metal hydrides for hydrogen storage. *Appl. Phys. A* 72:157–165.
63. Bogdanović, B., M. Schwickardi. 1997. Ti-doped alkali metal aluminum hydrides as potential novel reversible hydrogen storage materials. *J Alloys Compd.* 253/254:1–9.
64. Yang, J., A. Sudik, D.J. Siegel et al. 2007. Hydrogen storage properties of $2LiNH_2 + LiBH_4 + MgH_2$. *J. Alloys Compd.* 446/447:345–349.
65. Gutowska, A., L. Li, Y. Shin et al. 2005. Nanoscaffold mediates hydrogen release and the reactivity of ammonia borane. *Angew. Chem., Int. Ed.* 44:3578–3582.
66. Schüth, F., B. Bogdanović, A. Taguchi. Materials encapsulated in porous matrices for the reversible storage of hydrogen. Feb. 17, 2005. WIPO Patent WO 2005/014469 A1.

67. Gross, A.F., J.J. Vajo, S.L. Van Atta, G.L. Olson. 2008. Enhanced hydrogen storage kinetics of $LiBH_4$ in nanoporous carbon scaffolds. *J. Phys. Chem. C* 112:5651–5657.
68. Liu, P., J.J. Vajo. 2008. Thermodynamically tuned nanophase materials for reversible hydrogen storage. In the 2008 Annual Progress Report of the DOE Hydrogen Program. Report IV.A.1b. pp. 449–453. http://www.hydrogen.energy.gov/pdfs/progress08/iv_a_1b_liu.pdf
69. Klebanoff, L., J. Keller. 2012. Final report for the DOE Metal Hydride Center of Excellence. Sandia Report: SAND2012-0786. http://www1.eere.energy.gov/hydrogenandfuelcells/pdfs/metal_hydride_coe_final_report.pdf.
70. Aardahl, C.L., S.D. Rassat. 2009. Overview of systems considerations for on-board chemical hydrogen storage. *Int. J. Hydrogen Energy* 34:6676–6683.
71. Kariya, N.A., M. Ichikawa. 2002. Efficient evolution of hydrogen from liquid cycloalkanes over Pt-containing catalysts supported on active carbons under 'wet-dry multiphase conditions. *Appl. Catal., A* 233:91–102.
72. Hodoshima, S., H. Arai, Y. Saito. 2003. Liquid-film-type catalytic decalin dehydrogeno-aromatization for long-term storage and long-distance transportation of hydrogen. *Int. J. Hydrogen Energy* 28:197–204.
73. Pez, G.P., A.R. Scott, A.C. Cooper, H. Cheng. 2006. Hydrogen storage by reversible hydrogenation of pi-conjugated substrates. U.S. Patent No. 7101530.
74. Amendola, S.C., S.L. Sharp-Goldman, M.S. Janjua, N.C. Spencer, M.T. Kelly., P.J. Petillo, M. Binder. 2000. A safe, portable hydrogen gas generator using aqueous borohydride solution and Ru catalyst. *Int. J. Hydrogen Energy* 25:969–975.
75. Ramachandran, P.V., P.D. Gagare. 2007. Preparation of ammonia borane in high yield and purity, methanolysis, and regeneration. *Inorg. Chem.* 46:7810–7817.
76. Shang, Y., R. Chen. 2006. Hydrogen storage via the hydrolysis of $NaBH_4$ basic solution: Optimization of $NaBH_4$ concentration. *Energy Fuels* 20:2142–2148.
77. Hamilton, C.W., R.T. Baker, A. Staubitz, I. Manners. 2009. B–N compounds for chemical hydrogen storage. *Chem. Soc. Rev.* 38:279–293.
78. Karkamkar, A., C. Aardahl, T. Autrey. 2007. Recent developments on hydrogen release from ammonia borane. *Mater. Matters* 2:6–9.
79. Marder, T.B. 2007. Will we soon be fueling our automobiles with ammonia–borane? *Angew. Chem. Int. Ed.* 46:8116–8118.
80. Stephens, F.H., V. Pons, R.T. Baker. 2007. Ammonia–borane: The hydrogen source par excellence? *Dalton Trans.* 25:2613–2626.
81. Xiong, Z., C.K. Yong, G. Wu. 2008. High-capacity hydrogen storage in lithium and sodium amidoboranes. *Nat. Mater.* 7:138–141.
82. Ott, K.C. 2012. Final report for the DOE Chemical Hydrogen Storage Center of Excellence. *Los Alamos Report: LA-UR-20074.* http://www1/.eere.energy.gov/hydrogenandfuelcells/pdfs/chemical_hydrogen_storage_coe_final_report.pdf
83. Graetz, J., J.J. Reilly. 2005. Decomposition kinetics of the AlH_3 polymorphs. *J. Phys. Chem. B* 109:22181.
84. Sandrock, G., J. Reilly, J. Graetz, W. Zhou, J. Johnson, J. Wegrzyn, 2004. Accelerated thermal decomposition of AlH_3 for hydrogen-fueled vehicles. *Appl. Phys. A* 80:687–690.
85. Graetz, J., S. Chaudhuri, J. Wegrzyn, Y. Celebi, J.R. Johnson, W. Zhou, J.J. Reilly. 2007. Direct and reversible synthesis of AlH_3 triethylenediamine from Al and H_2. *J. Phys. Chem. C* 111:19148–19152.
86. Zidan, R., B.L. Garcia-Diaz, C.S. Fewox, A.C. Stowe, J.R. Gray, A.G. Harter. 2009. Aluminium hydride: A reversible material for hydrogen storage. *Chem. Commun.* 25:3717–3719.
87. Ashby, E.C. 1964. The direct synthesis of amine alanes. *J. Am. Chem. Soc.* 86:882–883.
88. Ahluwalia, R.K., T.Q. Hua, J.K. Peng. 2009. Automotive storage of hydrogen in alane. *Int. J. Hydrogen Energy* 34:7731–7740.
89. Crabtree, R.H. 2008. Hydrogen storage in liquid organic heterocycles. *Energy Environ. Sci.* 1:134–138.
90. Züttel, A. 2004. Hydrogen storage methods. *Naturwissenschaften* 91:157–172.

91. Bénard, P., R. Chahine. 2007. Storage of hydrogen by physisorption on carbon and nanostructured materials. *Scr. Mater.* 56:803–808.
92. Ströbel, R., J. Garche, P.T. Moseley, L. Jörissen, G. Wolf. 2006. Hydrogen storage by carbon materials. *J. Power Sources* 159:781–801.
93. Nijkamp, M.G., J.E.M.J. Raaymakers, A.J. van Dillen, K.P. de Jong. 2001. Hydrogen storage using physisorption. *Appl. Phys. A* 72:619–623.
94. Hirscher, M., B. Panella. 2005. Nanostructures with high surface area for hydrogen storage. *J. Alloys Compd.* 404/406:399–401.
95. Shenderova, O.A., V.V. Ahimov, D.W. Brenner. 2004. Carbon nanostructures. *Crit. Rev. Solid State Mater. Sci.* 27:227–356.
96. Thomas, K.M. 2007. Hydrogen adsorption and storage on porous materials. *Catal. Today* 120:389–398.
97. Ansón, A., M. Benham, J. Jagiello et al. 2004. Hydrogen adsorption on a single-walled carbon nanotube material: A comparative study of three different adsorption techniques. *Nanotechnology* 15:1503–1508.
98. Rosi, N.L., J. Eckert, M. Eddaoudi et al. 2003. Hydrogen storage in microporous metal-organic frameworks. *Science* 300:1127–1129.
99. Wong-Foy, A.G., A.J. Matzger, O.M. Yaghi. 2006. Exceptional H_2 saturation uptake in microporous metal-organic frameworks. *J. Am. Chem. Soc.* 128:3494–3495.
100. Hirscher, M., B. Panella. 2007. Hydrogen storage in metal–organic frameworks. *Scr. Mater.* 56:809–812.
101. Collins, D.J., H.-C. Zhou. 2007. Hydrogen storage in metal–organic frameworks. *J. Mater. Chem.* 17:3154–3160.
102. Long, J.R., O.M. Yaghi. 2009. The pervasive chemistry of metal-organic frameworks. *Chem. Soc. Rev.* 38:1213–1214.
103. Thomas, K.M. 2009. Adsorption and desorption of hydrogen on metal–organic framework materials for storage applications: Comparison with other nanoporous materials. *Dalton Trans.* (9):1487–1505. DOI: 10.1039/b815583f.
104. Han, S.S., H. Furukawa, O.M. Yaghi, W.A. Goddard. 2008. Covalent organic frameworks as exceptional hydrogen storage materials. *J. Am. Chem. Soc.* 130:11580–11581.
105. Furukawa, H., O.M. Yaghi. 2009. Storage of hydrogen, methane, and carbon dioxide in highly porous covalent organic frameworks for clean energy applications. *J. Am. Chem. Soc.* 25:8876–8883.
106. Klontzas, E., E. Tylianakis, G.E. Froudakis. 2010. Designing 3D COFs with enhanced hydrogen storage capacity. *Nano Lett.* 10:452–454. DOI: 10.1021/nl903068a.
107. See, for example: Yang, Z., Y. Xia, R. Mokaya. 2007. Enhanced hydrogen storage capacity of high surface area zeolite-like carbon materials. *J. Am. Chem. Soc.* 129:1673–1679.
108. Furukawa, H., M.A. Miller, O.M. Yaghi. 2007. Independent verification of the saturation hydrogen uptake in MOF-177 and establishment of a benchmark for hydrogen adsorption in metal–organic frameworks. *J. Mater. Chem.* 17:3197.
109. Wang, L., R.T. Yang. 2008. New sorbents for hydrogen storage by hydrogen spillover—A review. *Energy Environ. Sci.* 1:268–279 and references therein.
110. Dailly, A., J.J. Vajo, C.C. Ahn. 2006. Saturation of hydrogen sorption in Zn benzenedicarboxylate and Zn naphthalenedicarboxylate. *J. Phys. Chem. B* 110:1099–1101.
111. Psofogiannakis, G.M., G.E. Froudakis. 2009. DFT study of enhanced hydrogen storage by spillover on graphite with oxygen surface groups. *J. Phys. Chem. C* 113:14908–14915.
112. Kubas, G.J., R.R. Ryan, B.I. Swanson, P.J. Vergamini, H.J. Wasserman. 1984. Crystalline dihydrogen complexes. Intramolecular and intermolecular interactions and dynamic behavior. *J. Am. Chem. Soc.* 106:451–452.
113. Hoang, T.K.A., D.M. Antonelli. 2009. Exploiting the Kubas interaction in the design of hydrogen storage materials. *Adv. Mater.* 21:1787–1800.
114. Kubas, G.J. 2007. Dihydrogen complexes as prototypes for the coordination chemistry of saturated molecules. *Proc. Nat. Acad. Sci.* 104:6901–6907.

115. Keskin, S., J. Liu, R.B. Rankin, J.K. Johnson, D.S. Sholl. 2009. Progress, opportunities, and challenges for applying atomically detailed modeling to molecular adsorption and transport in metal-organic framework materials. *Ind. Eng. Chem. Res.* 48:2355–2371.
116. Kim, Y.-H., Y. Zhao, A. Williamson, M.J. Heben, S.B. Zhang. 2006. Nondissociative adsorption of H_2 molecules in light-element-doped fullerenes. *Phys. Rev. Lett.* 96:016102–016105.
117. Yildirim, T., S. Ciraci. 2005. Titanium-decorated carbon nanotubes as a potential high-capacity hydrogen storage medium. *Phys. Rev. Lett.* 94:175501–175504.
118. Simpson, L. 2010. Hydrogen Sorption Center of Excellence (HSCoE) final report. http://www1.eere.energy.gov/hydrogenandfuelcells/pdfs/hydrogen_sorption_coe_final_report.pdf
119. Zhang, J., T.S. Fisher, P.V. Ramachandran, J.P. Gore, I. Mudawar. 2005. A review of heat transfer issues in hydrogen storage technologies. *J. Heat Transfer* 127:1391–1399.
120. Ahluwalia, R.K., J.K. Peng. 2009 Automotive hydrogen storage system using cryo-adsorption on activated carbon. *Int. J. Hydrogen Energy* 34:5476–5487.
121. Hardy, B.J. 2007. Geometry, heat removal and kinetics scoping models for hydrogen storage systems. Savannah River National Laboratory Report WSRC-TR-2007-00439. http://www1.eere.energy.gov/hydrogenandfuelcells/pdfs/bruce_hardy_srnl-2007–0043_part1.pdf
122. Hardy, B.J. 2007. Integrated Hydrogen Storage System. Savannah River National Laboratory Report WSRC-TR-2007-00440. http://www1.eere.energy.gov/hydrogenandfuelcells/pdfs/bruce_hardy_srnl-2007–0043_part2.pdf
123. Mosher, D.A., S. Arsenault, X. Tang, D.L. Anton. 2007. Design, fabrication and testing of $NaAlH_4$ based hydrogen storage systems. *J. Alloys Compd.* 446–447:707–712.
124. Ahluwalia, R.K. 2007. Sodium alanate hydrogen storage system for automotive fuel cells. *Int. J. Hydrogen Energy* 32:1251–1261.

16

Gaseous Hydrogen Storage

S.A. Sherif
University of Florida

Frano Barbir
University of Split

CONTENTS

Depending on storage size and application, several types of hydrogen storage may be employed as follows:

1. *Stationary large storage systems*: These are typically storage devices at the production site or at the start or end of pipelines and other transportation pathways.
2. *Stationary small storage systems*: These are typically at the distribution or final user level, for example, a storage system to meet the demand of an industrial plant.
3. *Mobile storage systems for transport and distribution*: These include both large-capacity devices, such as a liquid hydrogen tanker–bulk carrier, and small systems, such as a gaseous or liquid hydrogen truck trailer.
4. *Vehicle tanks*: These are typically used to store hydrogen used as fuel for road vehicles.

Because of hydrogen's low density, its storage always requires relatively large volumes and is associated with either high pressures (thus requiring heavy vessels) or extremely low temperatures and/or a combination with other materials (much heavier than hydrogen itself). Table 16.1 shows achievable storage densities with different types of hydrogen storage methods. Some novel hydrogen storage method may achieve even higher storage densities but have yet to be proven in terms of practicality, cost, and safety.

16.1 Large Underground Hydrogen Storage

Future hydrogen supply systems will have a structure similar to today's natural gas supply systems. Underground storage of hydrogen in caverns, aquifers, depleted petroleum and natural gas fields, and man-made caverns resulting from mining and other activities is likely to be technologically and economically feasible [1–4]. Hydrogen storage systems

TABLE 16.1

Hydrogen Storage Types and Densities

	kg H_2/kg	kg H_2/m^3
Large volume storage (10^2–10^4 m^3 geometric volume)		
Underground storage		5–10
Pressurized gas storage (aboveground)	0.01–0.014	2–16
Metal hydride	0.013–0.015	50–55
Liquid hydrogen	~1	65–69
Stationary small storage (1–100 m^3 geometric volume)		
Pressurized gas cylinder	0.012	~15
Metal hydride	0.012–0.014	50–53
Liquid hydrogen tank	0.15–0.50	~65
Vehicle tanks (0.1–0.5 m^3 geometric volume)		
Pressurized gas cylinder	0.05	15
Metal hydride	0.02	55
Liquid hydrogen tank	0.09–0.13	50–60

of the same type and the same energy content will be more expensive (by approximately a factor of 3) than natural gas storage systems, due to hydrogen's lower volumetric heating value. Technical problems, specifically for the underground storage of hydrogen other than expected losses of the working gas in the amount of 1%–3% per year, are not anticipated. The city of Kiel's public utility has been storing town gas with a hydrogen content of 60%–65% in a gas cavern with a geometric volume of about 32,000 m^3 and a pressure of 80–160 bar at a depth of 1,330 m since 1971 [2–4]. Gaz de France (the French national gas company) has stored hydrogen-rich refinery by-product gases in an aquifer structure near Beynes, France. Imperial Chemical Industries of Great Britain stores hydrogen in the salt mine caverns near Teesside, United Kingdom [3,4].

16.2 Aboveground Pressurized Gas Storage Systems

Pressurized gas storage systems are used today in natural gas business in various sizes and pressure ranges from standard pressure cylinders (50 L, 200 bar) to stationary high-pressure containers (over 200 bar) or low-pressure spherical containers (>30,000 m^3, 12–16 bar). This application range will be similar for hydrogen storage.

16.3 Vehicular Pressurized Hydrogen Tanks

The development of ultralight but strong new composite materials has enabled storage of hydrogen in automobiles. Pressure vessels that allow hydrogen storage at pressures greater than 200 bar have been developed and used in automobiles (such as Daimler-Benz NECAR II). A storage density higher than 0.05 kg of hydrogen per 1 kg of total weight is easily achievable [4].

References

1. Taylor, J.B., Alderson, J.E.A., Kalyanam, K.M., Lyle, A.B., and Phillips, L.A., Technical and economic assessment of methods for the storage of large quantities of hydrogen, *Int. J. Hydrogen Energy*, 11(1), 5–22, 1986.
2. Carpetis, C., Storage, transport and distribution of hydrogen, in C.-J. Winter and J. Nitsch (eds.), *Hydrogen as an Energy Carrier*, Springer-Verlag, Berlin, Germany, pp. 249–289, 1988.
3. Pottier, J.D. and Blondin, E., Mass storage of hydrogen, in Y. Yurum (ed.), *Hydrogen Energy System: Production and Utilization of Hydrogen and Future Aspects*, NATO ASI Series E-295, Kluwer Academic Publishers, Dordrecht, the Netherlands, pp. 167–180, 1995.
4. Mitlitsky, F., Development of an advanced, composite, lightweight, high pressure storage tank for on-board storage of compressed hydrogen, *Proceedings of Fuel Cells for Transportation TOPTEC: Addressing the Fuel Infrastructure Issue*, Alexandria, VA. SAE, Warrendale, PA, 1996.

17

Cryogenic Refrigeration and Liquid Hydrogen Storage

Gary G. Ihas
University of Florida

CONTENTS

17.1 Introduction

The use of cryogenics and access to low temperatures, once confined to the laboratories searching for the liquefaction of the *permanent gases* [1], is now almost worldwide. All low-temperature apparatus must be contained in a *cryostat*, a word derived from Greek meaning literally *frost apparatus*. Cryogenic storage and refrigeration are becoming common in our everyday lives. Liquid nitrogen is used for food and living tissue preservation, liquid oxygen for medical patient respiration, and liquid hydrogen for power generation, transportation, and propulsion. Large superconducting magnets are used for magnetic resonance imaging (MRI). All these applications use either vacuum-insulated containers (dewars) or foam-insulated vessels to contain the cryogens. This is because direct exposure to environmental radiation supplies more than enough heat to evaporate the cryogen in a short time. In the future, the use of cryogenic equipment will rapidly increase with new demands, for example, in computers and communication systems (cold and superconducting electronics), medicine (surgery and diagnostics), materials fabrication, transportation (motor vehicles, ships, and magnetically levitated trains), and space travel (propulsion, cooling, and water generation), particularly in the hydrogen economy. This last item is important because large amounts of hydrogen will be used, stored, and transported. There is no better way to store or transport hydrogen than as a liquid cryogen, which is more safe, less expensive, and at higher density than any other form. Scientifically, cryogenics are used in virtually every area of research, with the widest uses in propulsion (combustion and fuel cells), condensed matter and particle physics, physical chemistry, geology, biology, and medical sciences.

In this short chapter, we offer an overview of the state of the art of cryogenic storage and small-scale refrigeration, with references that provide more detailed information. This chapter is divided into two major parts: Section 17.2 covers dewars and storage techniques, and refrigeration apparatus and techniques are discussed in Section 17.3.

17.2 Storage Vessels, Dewars, and Storage Techniques

17.2.1 Non-Vacuum-Insulated Containers

Cryogens, defined as gases that liquefy below 123 K, including nitrogen, oxygen, neon, hydrogen, and helium, are stored, transported, and used in double-walled, vacuum-insulated dewars (large, improved versions of the thermos bottle) or foam-insulated containers. The purpose of this special container is to protect the cryogen from the heat of its surroundings, thus preserving the cold liquid and reducing evaporation (boil-off). This heat leak cannot be reduced to zero, regardless of cleverness of design, unless an active refrigerator, using external power, is employed (see Sections 17.2.3 and 17.3.4). The damaging heat leak to the interior of the vessel, as usual, may have three components: conduction, convection, and radiation.

Conduction heat is minimized by suspending the vessel inside an outer housing, which is otherwise filled with powder, foam, or other porous media to inhibit gas convection heat leaks. The inner vessel is suspended by relatively thin supports made of poorly conducting metal (stainless steel) or polymer. Because the cryogen-containing vessel may be quite massive, care must be taken to balance forces in the suspension, accounting for thermal contractions in the vessel and suspension. Often, the vessel is hung by a neck tube, through which the cryogen is introduced and removed (see Figure 17.1). For larger vessels, a centering snub at the bottom or snubbing pieces on the sides keep the vessel from swinging during motion. These do not make contact between the inner vessel and outer container during normal use.

The insulation between the inner vessel and outer container must be maintained as a poor thermal conductor. In humid environments, therefore, the space it occupies must be sealed, which is often difficult and/or expensive to accomplish. Because this insulation may require frequent replacement, inexpensive materials such as vermiculite, perlite, glass bubbles, Styrofoam, or polyurethane foam are usually used. All these materials have thermal conductances about four orders of magnitude less than bulk copper. Besides being poor bulk conductors, like cork, their insulating property is enhanced by another order of magnitude since they consist of poorly contacting, small particles.

Under certain circumstances, for example, to save cost or weight, the outer shell of these containers may be eliminated. The most famous example of this is the external tank on the US space shuttle vehicle, which contains two cryogens, liquid hydrogen and liquid oxygen, in two separate vessels, surrounded by ridged insulating foam. Since this container is used in a very humid environment (Florida), there is a large amount of ice buildup on this foam during use (just prior to launch). This contributes to the launch weight, but, more important, it produces extreme loads on the foam insulation during the acceleration

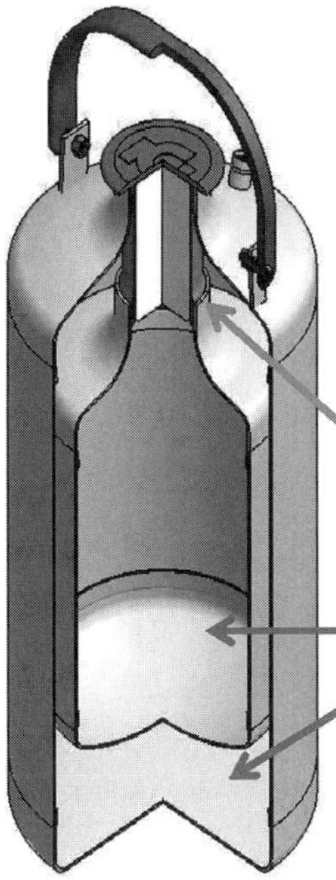

FIGURE 17.1
Basic cryogenic container. The inner vessel containing the cryogen is suspended by a polymer neck with poor thermal conductivity. The space between the inner vessel and outer container must be filled with a poorly conducting powder or foam to stop gas convection or be evacuated. (Courtesy of International Cryogenics, Inc., Indianapolis, IN.)

phase of achieving orbit. It is known that pieces of this foam have caused damage to the shuttle vehicle, particularly its wings and heat shield tiles. This caused the catastrophic SS Columbia accident.

17.2.2 Dewars

If the space between the inner vessel and outer container is evacuated, the whole apparatus is called a dewar, after Sir James Dewar (see next section), and was originally constructed of blown glass. The inner surfaces of the vacuum spaces in this construction are silvered. The vacuum space minimizes conduction heating, and the silvering minimizes radiant heating, but this vacuum is sometimes inconvenient or impossible, requiring stronger container walls, more safety apparatus, and construction of higher integrity.

Vacuum-insulated dewars are the preponderance of cryogenic storage and research containers. Although dewars used in research are physically quite different in appearance from storage and transport dewars, they have similar requirements to reduce heat leaks into the cryogen. Special features of storage and transport dewars necessary for these uses, such as more robust construction, are included in standard designs of manufacturers. Here, we will also discuss research dewars, likely to be of more concern to experimenters who may have to design or write specifications for a dewar.

In most research at low temperatures, liquid cryogen is transferred from a storage dewar (Figure 17.2) to a research dewar (Figure 17.3), designed for specific uses. The dewar in Figure 17.3, constructed of aluminum with a fiberglass neck, is designed for a large

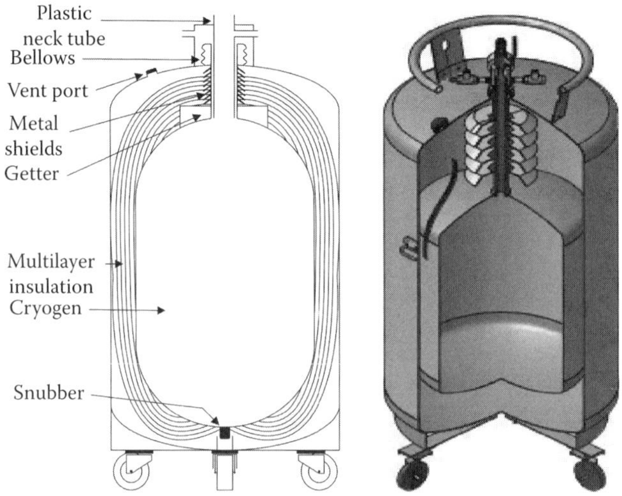

FIGURE 17.2
Cross-sectional view of standard transport (storage dewar), with cutaway drawing of the same. Cross-sectional view, courtesy of the author. (Courtesy of International Cryogenics, Inc., Indianapolis, IN.)

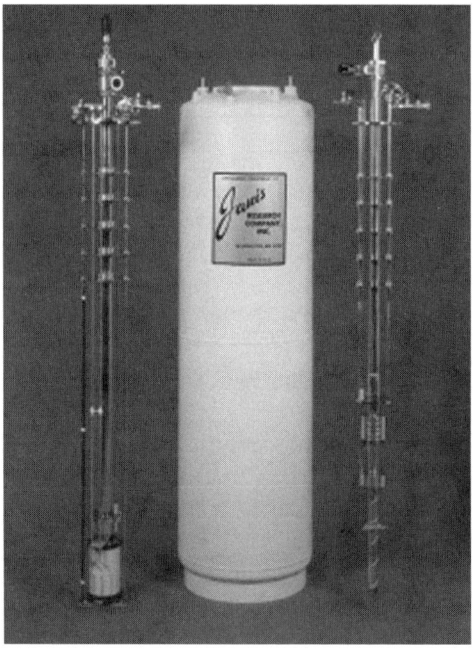

FIGURE 17.3
Research dewar with two inserts: Left is a 3 T magnet with a ^{4}He evaporative cooling stage. Right is a transport measuring probe on a continuous ^{3}He evaporative cooling stage. The dewar has an aluminum outer jacket and fiberglass G10 inner neck and aluminum bucket, with multilayer insulation in the vacuum space between. (Courtesy of Janis Research, Inc., Wilmington, MA.)

superconducting magnet. It has features to reduce helium consumption in common with most other dewars used in research and with storage and transport dewars.

Radiation from room temperature to the inner wall containing the cryogen is reduced dramatically by vapor-cooled heat shields and multilayer (super)insulation (usually aluminized Mylar film separated by thin, loose fiberglass layers) in the vacuum space. Heat transfer by conduction down the low-thermal-conductivity fiberglass neck is reduced further by these vapor-cooled shields attached to the neck at appropriate intervals, which are cooled by the enthalpy of the vapor boiling off the liquid cryogen. The large neck, necessary for inserting the superconducting magnet, also has baffles inside to reduce radiation from the top flange. A thin metal barrier is incorporated inside the fiberglass to prevent diffusion of helium and/or hydrogen into the vacuum space, which occurs if residual cryogen is left in the dewar on warming.

These features of the dewars shown in Figures 17.2 and 17.3 provide a quite low boil-off rate: about 1% of capacity for the storage dewar and less than 2 L/h for the research dewar, even in this relatively large dewar, which is about 3.7 m long with a neck diameter of 46 cm.

The absence of liquid-nitrogen shielding for these dewars has a number of advantages including simpler construction, lighter weight, with lower initial and operating costs, and also bothersome refilling of the nitrogen jacket is avoided. Furthermore, eliminating boiling nitrogen can reduce vibrations that may be critical in some uses.

Some small hand dewars and research dewars are constructed of Pyrex glass, in which case liquid-nitrogen shielding using a separate liquid-nitrogen dewar, surrounding the lower-temperature hydrogen or helium dewar, is common, since incorporation of superinsulation is impractical. In using Pyrex dewars, it is essential to avoid leaving helium and/or hydrogen gas in the dewar when it is warmed in order to prevent diffusion of gas into the vacuum space, which can create an insurmountable heat leak. The simplest way to achieve this is to terminate the vacuum space of the helium dewar with a single-wall neck at a point below the nitrogen level in the outer dewar. This construction also facilitates precooling of the lower dewar with liquid nitrogen.

Research dewars may also be of all-stainless-steel construction, with or without nitrogen shielding. Frequently, storage and transport dewars are also of all-stainless-steel construction in order to provide the necessary ruggedness. Additionally, the nonmagnetic stainless steel allows the dewar to be moved safely near a large magnet.

In cooling a dewar from room temperature, much less cryogen is used if the dewar is precooled with liquid nitrogen. In case a dewar has no nitrogen shielding, nitrogen is transferred directly into the cryogen space. After the equipment inside has been cooled to nitrogen temperature (and the shields, in the case of a vapor-shielded dewar), the nitrogen is back-transferred. This is accomplished by applying a small overpressure above the nitrogen surface to force it out through a tube extending through the top flange of the dewar to the inside bottom. Once liquid ceases to flow, the container space should be pumped, being careful to keep the pressure above the triple point of N_2, until there is very little flow of gas. This assures no liquid nitrogen remains in the dewar at the time cryogen is transferred (which would require a large quantity of cryogen to cool the nitrogen to the boiling point of the cryogen).

If a separate nitrogen dewar is used, then cooling from room temperature to liquid-nitrogen temperature can be accomplished simply by filling the nitrogen dewar and waiting for the inner dewar and the apparatus inside it to be cooled to near 77 K. In this case, it is essential that there be an exchange gas of air or N_2 at a pressure of a few mm Hg in the vacuum jacket of the inner dewar to provide heat transfer. Once the cryogen transfer begins, this exchange gas is frozen on the wall of the dewar, producing the high vacuum necessary for holding the cryogen.

The research dewar is usually closed at the top by a flange that constitutes the top of the cryostat insert used for the particular experiment. The insert has suitable tubulation and electrical leads for the measurements to be made and for operation of the refrigerator. Cryogen is transferred from a storage dewar into the research dewar using a double-walled, vacuum-insulated transfer tube that extends to the bottom of the storage dewar. In the initial part of the transfer, it is important that the cryogen be introduced at the bottom of the research dewar in order to make full use of the enthalpy of the gas in cooling the dewar and equipment inside it. In subsequent topping-off of the cryogen, it is best that the liquid be introduced above the level of cryogen remaining in the dewar.

A port allows recovery of the cryogen's gas, or it is vented into the atmosphere if not recovered (not recommended for hydrogen). In a situation where large amounts of cryogen are transferred causing moisture condensation, it is useful for the port to be through a collar below the top flange, thus avoiding freezing of the top flange. Also, one or more relief ports may be needed to allow gas to escape quickly in cases of rapid boil-off such as when a superconducting magnet quenches. Relief valves are required on transport dewars for safety, but full dewars of liquid helium may be flown on commercial aircraft.

17.2.3 Zero Boil-Off Storage

In some situations, it is desired to have no loss of cryogen from the storage container. These applications include such common installations as MRI machines or special situations as onboard spacecraft or satellites. Since, as mentioned previously, it is impossible to completely avoid heat leaks to the cryogenic storage vessel, this heat leak must be intercepted by some refrigeration technique to achieve zero boil-off (ZBO).

Terrestrial applications, such as MRI machines in hospitals and clinics, research magnets, or particle accelerators, are straightforward to achieve ZBO. Electrical power is readily available, and gravity ensures that the cryogen is well behaved, that is, remains in the bottom of the dewar. Typically, a cryocooler, such as that described in Section 17.3.4, is employed. The cryocooler is mounted on top of the dewar, and a thermal connection, such as a heat pipe, draws in evaporated cryogen and recondenses it. Such a system is shown diagrammatically in Figure 17.4, which uses the latest innovation in cryocoolers, the pulse tube refrigerator.

Space applications, particularly to liquid hydrogen storage, are more challenging. A test of such a ZBO system for nonterrestrial use is described in Ref. [2].

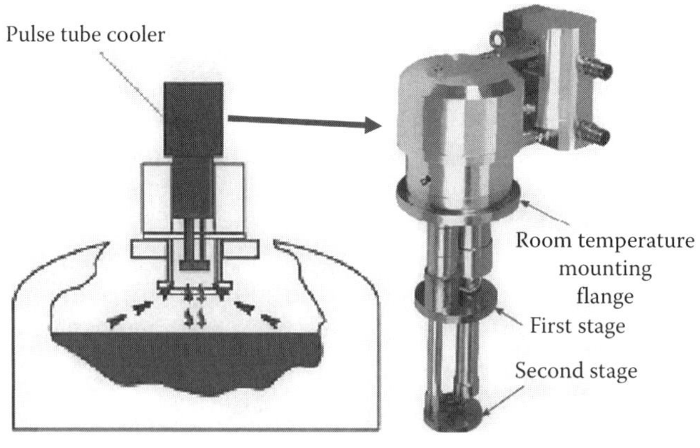

FIGURE 17.4
Dewar incorporating a pulse tube refrigerator (Courtesy of Janis Research, Inc., Wilmington, MA.), reducing the cryogen to ZBO. Detail of pulse tube refrigerator (Courtesy of CryoMech, Inc., New York.) shown to right.

17.3 Small-Scale Refrigeration

17.3.1 ^{4}He Evaporation Refrigeration

Temperatures from the normal boiling point of ^{4}He, 4.2 K, to as low as about 1.2 K (limited by the pumping speed of the system) can be achieved by pumping the vapor above liquid helium [2–4]. The simplest means of doing this, in terms of the equipment inside the dewar, is to pump the entire bath of helium. However, this method has several disadvantages. A large fraction of the helium is used in cooling the remaining helium to the minimum temperature of about 1.2 K. If the temperature is lowered below the lambda point, 2.18 K, a film of the superfluid helium creeps up the wall of the dewar, contributing significantly to the boil-off rate, resulting in higher consumption of helium [4].

The disadvantages just described can be overcome by keeping the main helium bath at 4.2 K and placing the experiment and a small pot of helium to be pumped to a lower temperature inside a vacuum space immersed in the helium bath, as shown in Figure 17.5. The pumped pot of helium can be refilled periodically from the bath using a valve at the top of the vacuum space, operated from outside the cryostat. Instead of filling periodically, a small needle valve may be used, which can be adjusted so that the flow of liquid into the pot is just adequate for the heat load. The arrangement shown requires no valve and allows continuous filling of the pot by using an impedance of appropriate size between the bath and the pot [5]. Even under varying heat loads, the temperature remains almost constant up to a critical power that equals the cooling by the incoming liquid. As shown, the inlet may be below the top of the vacuum space to give longer times between helium transfers into the dewar. The impedance must be in the vacuum space, and a filter is necessary to prevent clogging of the small line.

In experiments using a pumped pot of ^{4}He, care must be taken to ensure adequate thermal contact between the helium and the sample or equipment that must be cooled. This

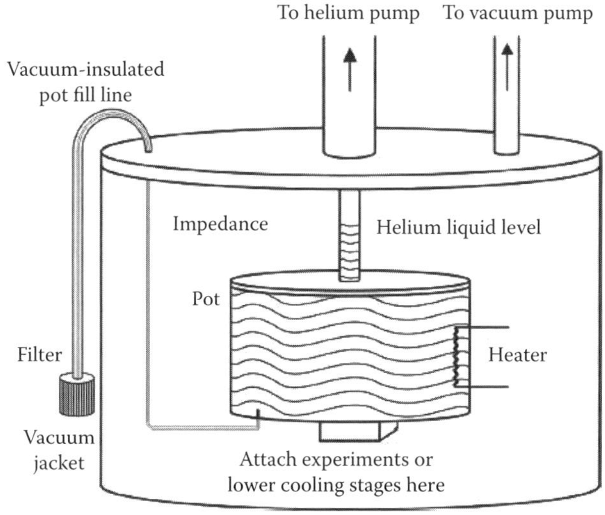

FIGURE 17.5
Schematic diagram of a continuously operating ^{4}He evaporation refrigerator. See Ref. 3 for details.

can be done by immersing the sample in the helium within the pot, with access from the top through the pumping line. Another approach is to mount the sample on the outside of the pot using a screw contact if the sample can be electrically grounded, or for a sample that must be insulated, contact may be made with varnish, epoxy, or grease. Care must be taken to thermally ground leads to the sample and thermometers, otherwise there may be a heat leak into the sample that will keep it at an elevated temperature. These precautions become ever more important as the temperature of the sample is lowered, such as by the techniques discussed in the following [3].

In cases where the experimental parts to be cooled can be made small, miniature cryostats with a pumped inner pot of helium have been constructed that may be inserted directly into a storage dewar with a wide neck (up to 60 mm diameter) [6]. These have the advantages that no transfer of helium is required, helium consumption is low, and the time required to reach the desired temperature is short.

Another arrangement used frequently to lower the temperature of superconducting magnets below 4.2 K is the "lambda-plate" refrigerator (see Figure 17.11 and Section 17.3.3). The lambda plate, situated in the lower portion of the dewar but above the magnet, has a needle valve for admission of liquid ^{4}He into a coil heat exchanger connected through a tube to the top of the cryostat where a pump is connected. The normal ^{4}He above the lambda plate is a very poor thermal conductor and remains at 4.2 K, while the helium in the lower portion of the dewar, where the magnet is located, is pumped to a reduced temperature, allowing the magnet to operate at a higher field.

17.3.2 ^{3}He Evaporation Refrigeration

The vapor pressure of ^{3}He is higher than that of ^{4}He at the same temperature, a consequence of the higher zero-point motion of the lighter isotope [2]. Furthermore, ^{3}He has no creeping film in the range of interest. Consequently, temperatures of 0.3 K or slightly lower can be obtained by pumping to lower the vapor pressure of ^{3}He.

Most ^{3}He is obtained from the decay of ^{3}H; therefore, it is quite expensive. Consequently, ^{3}He is used in relatively small quantities of a few liters of gas STP in a closed cycle, with sealed pumps (including the exhaust) to recover the helium. A ^{3}He refrigerator is usually run in a continuous mode, with the helium gas from the pump returned to the evaporator as a liquid, after it has been brought into thermal contact with a ^{4}He pot to remove the heat of vaporization. Also, the pressure of the returning ^{3}He gas must be brought above its saturated vapor pressure at the temperature of the ^{4}He pot by including sufficient impedance in the return line below the ^{4}He pot. A ^{3}He refrigerator with gas-handling system is shown in Figure 17.6. A typical ^{3}He refrigerator, which may be directly inserted into a ^{4}He transport dewar, is shown in Figure 17.7. The advantage of this design is the avoidance of transferring the liquid ^{4}He for the bath, saving time and money.

Recently, heat exchangers have been developed for closed cycle refrigerators that allow operation of a ^{3}He refrigerator or a dilution refrigerator (see Section 17.3.3) without using pumped ^{4}He [7]. An example is shown in Figure 17.8.

One of the limiting factors in the temperature that may be reached, or the refrigeration available by pumping ^{3}He, is the pumping speed of the system, including the pumping line and the external pumps. The pumping speed can be increased substantially by incorporating a small adsorption pump (usually with charcoal as the adsorbate) within the vacuum container. Shown in Figure 17.9 is such an adsorption pump

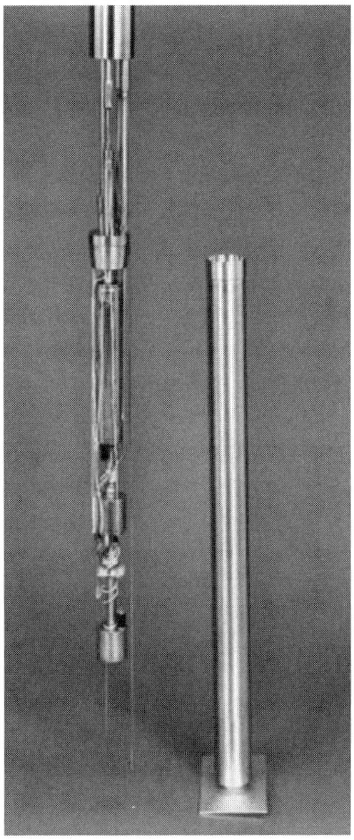

FIGURE 17.6
Continuously operating ^3He refrigerator with gas-handling system (^4He bath dewar not shown). (Courtesy of Janis Research Company LLC, Woburn, MA.)

in a miniature dilution refrigerator (see Section 17.3.3) designed for insertion into a storage dewar [8].

A feature of these miniature cryostats worth pointing out is the compact cone seal with vacuum grease that works well at helium temperatures (see Figure 17.8). Earlier, similar seals were made with Wood's metal, requiring undesirable heating to make the seal, or with bulkier bolted flanges using soft metal (e.g., indium or lead) O-rings.

Adsorption pumps may also be used to advantage in larger systems to increase the pumping speed and to reduce the expense, since large external pumps are not required. One disadvantage is that continuous operation would require a more complicated arrangement with two adsorbers and cold valves. However, this disadvantage can be partially overcome by including a larger quantity of ^3He in the system. A system capable of holding a few cm^3 of liquid ^3He can maintain the desired temperature for a few days if the heat input in the experiment is not great.

Because of the rapid increase of the thermal boundary resistance between materials as the temperature is decreased, the evaporator of a ^3He refrigerator should have several cm^2 of surface area in contact with the ^3He. This can easily be provided by some means such as milling slots in the body or by brazing a spiral sheet of copper to the bottom.

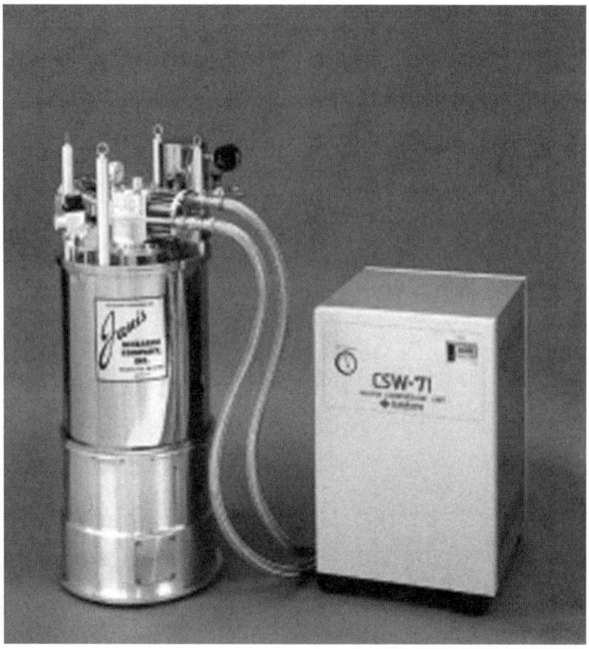

FIGURE 17.7
^3He refrigerator for insertion in storage dewar with vacuum jacket (on right). Vacuum jacket grease seal flange can be seen near top. By not requiring a separate research dewar, cooldown and turnaround time and liquid helium are saved at the expense of experimental volume. (Courtesy of Janis Research Company LLC, Woburn, MA.)

17.3.3 ^3He–^4He Dilution Refrigeration

Refrigeration by pumping ^3He is impractical below about 0.25 K because of the vanishingly small vapor pressure at lower temperatures. The most common means of reaching temperatures as low as a few millikelvin is by ^3He–^4He dilution refrigeration [3]. The physical process itself is quite easy to understand: cooling is provided by the heat of mixing when the two isotopes are mixed. This is possible at ultralow temperatures because the two isotopes remain liquids all the way to absolute zero and a finite amount, up to 6.5%, of ^3He will dissolve in ^4He at the lowest temperature [2,9].

A schematic drawing, depicting the essential components of a dilution refrigerator, is shown in Figure 17.10. This consists of a mixing chamber in which two separated phases exist, a concentrated phase of almost pure ^3He floating on top and a dilute phase of 6.5% ^3He in ^4He at the bottom; heat exchangers for transferring heat from the warm incoming stream of concentrated ^3He to the outgoing cold dilute phase; and a still for extracting the ^3He from the dilute phase, allowing the process to be continuous.

The refrigerator is immersed in a helium dewar to provide a 4.2 K environment. Typically, there is a heat shield (not shown) attached to the still to reduce the heat reaching the mixing chamber through 4.2 K radiation and thermal conduction through spacers. Also, a mixing-chamber heat shield may be used to reduce heat input to a sample or colder stage below the mixing chamber.

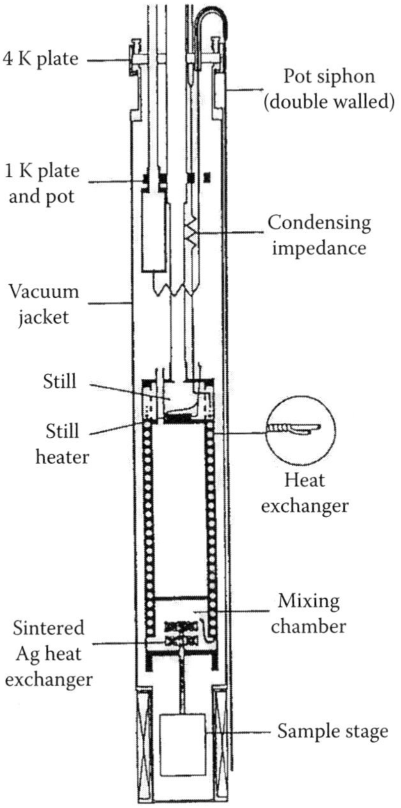

FIGURE 17.8
Completely self-contained cryogen-free ^3He refrigerator system. A pulse tube refrigerator, seen protruding from top, supplies the cooling to condense the circulating ^3He. Insert sample, plug into mains, and cool down. (Courtesy of Janis Research Company LLC, Woburn, MA.)

The very high vapor pressure of ^3He relative to ^4He provides a vapor phase in the still of almost pure ^3He above the liquid, which is less than 6.5% ^3He. Thus, almost pure ^3He is extracted from the still for recirculation, with most of the ^4He remaining in place to serve as a medium in which to dissolve the ^3He. Maintaining sufficiently high vapor pressure to allow the desired circulation rate requires that the still be kept at about 0.7 K by supplying the heat of vaporization with an electric heater.

The returning ^3He gas must be liquefied by removing the latent heat either with a pumped pot of ^4He as described in Section 17.3.1 or with a heat exchanger between the outgoing and incoming streams [3]. Just as with the ^3He refrigerator discussed in Section 17.3.2, the pressure of the incoming stream of gas must be brought above the saturated vapor pressure. Again, this requires a flow impedance below the ^4He pot or heat exchanger.

Perhaps the most critical parts of a dilution refrigerator are the heat exchangers that are necessary to cool the incoming stream from the still temperature of about 0.7 K to near that of the mixing chamber, which is much colder. With a simple counterflow capillary heat exchanger, minimum mixing-chamber temperatures of about 50 mK can be achieved.

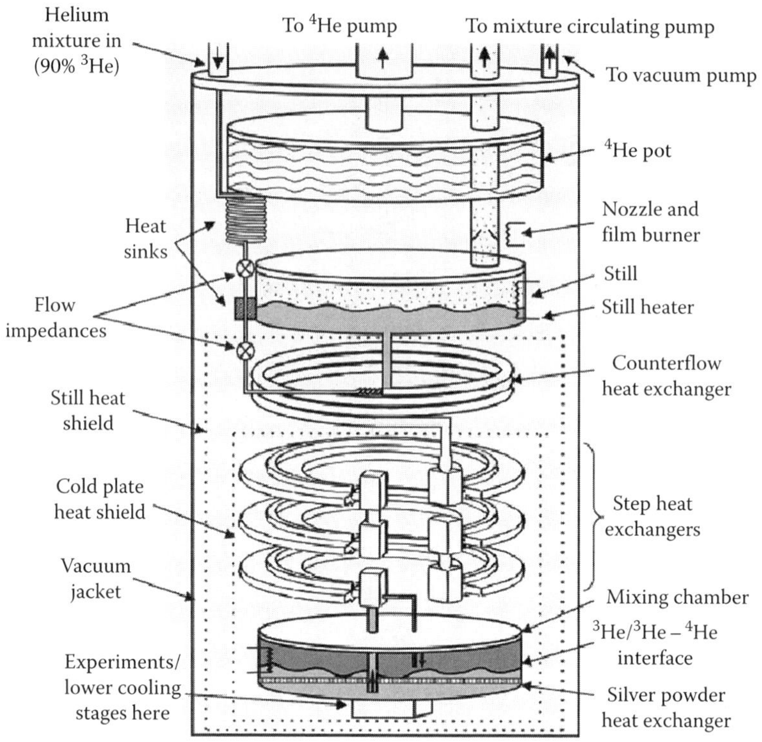

FIGURE 17.9
Schematic diagram of a miniature charcoal-pumped ^3He–^4He dilution refrigerator that may be used in a helium storage dewar. (Courtesy of Nanoway; From Uhlig, K., *Cryogenics*, 27, 454, 1987.)

Typically, a refrigerator designed to achieve the lowest temperatures has a counterflow exchanger followed by several *step exchangers*. These are also counterflow exchangers that have a large inside surface area of very fine sintered metal powder (silver is commonly used) for heat transfer between the two streams and are bulkier in construction than capillary exchangers [10].

One of the main requirements for the mixing chamber is that it has adequate surface area, usually provided with sintered metal powder, to minimize the temperature difference between the chamber body and the cold dilute phase. Thermal contact to the material to be cooled can then be made using bolts screwed tightly to the mixing chamber. Another requirement is that the entrance tube for the dilute phase actually extends into this phase (see Figure 17.11).

An alternative to this "bottom-loading" refrigerator is to insert samples directly into the liquid in the mixing chamber, usually by loading them in through a vacuum port in the pumping line at the top of the cryostat. This "top-loading" refrigerator has the advantage that samples may be changed relatively quickly without warming the refrigerator to room temperature. However, care must be taken to insert the sample slowly and to adequately heat-sink the sample, ensuring it does not remain at a significantly higher temperature than the liquid in the mixing chamber.

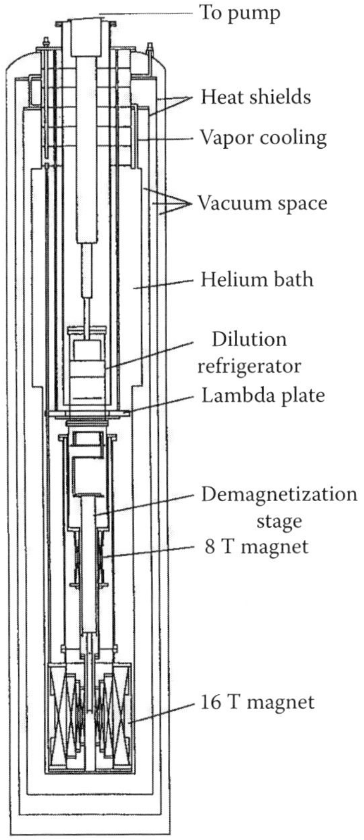

FIGURE 17.10
Block diagram of the essential components of a ^3He–^4He dilution refrigerator.

With a very-well-designed refrigerator, minimum temperatures of about 2.5 mK have been achieved, although commercial refrigerators have typical minimum temperatures of 5–10 mK. The maximum cooling power of a dilution refrigerator, $\dot{Q}(W)$, is given by

$$\dot{Q} = 84\dot{n}T_m^2 \tag{17.1}$$

where

$\dot{n}$ is the number of mol/s of ^3He being circulated

T_m is the mixing-chamber temperature (K)

The coefficient 84 is determined by the entropies of the dilute and concentrated phases [3,11]

Frequently, the cooling power of a dilution refrigerator is specified at 100 mK, which by Equation 17.1 is 840 μW/mmol/s of ^3He in circulation. Near the minimum temperature of the dilution refrigerator, the cooling power drops significantly below the value given by Equation 17.1 [11].

Two examples of commercially available dilution refrigeration units are shown in Figure 17.12. These may be purchased with varying cooling capacities up to 1 mW at

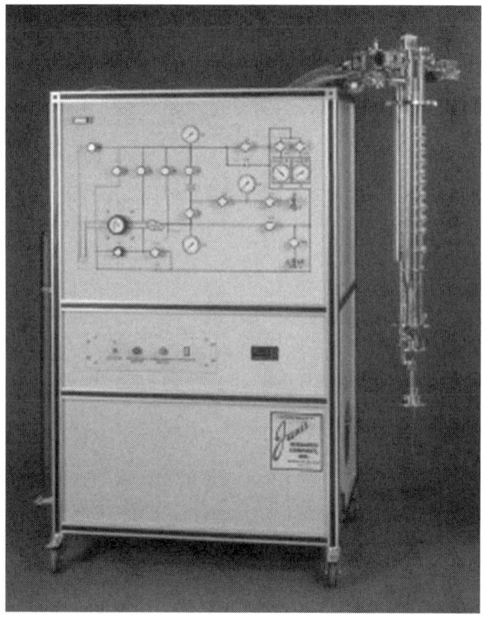

FIGURE 17.11
Schematic diagram of a ^3He–^4He dilution refrigerator in an aluminum and fiberglass research dewar (as in Figure 17.2b) containing a lambda-plate refrigerator and two high field magnets. The lambda plate allows the helium below it to be pumped to about 2.1 K, allowing the 16 T magnet to achieve full field. Pumps and external gas-handling system are not shown. (Courtesy of J.S. Xia, Microkelvin Research Laboratory, University of Florida, Gainesville, FL.)

FIGURE 17.12
Two examples of the business end of dilution refrigeration, with still at the top, mixing chamber at the bottom, and continuous and block heat exchangers in between. (Courtesy of Janis Research Company LLC, Woburn, MA.)

T = 0.1 K and then attached to the users' pumped ^{4}He cryostats. A sealed pumping system must be provided to continuously evaporate the ^{3}He from the still. Also, some care must be taken to isolate the cryostat from electrical oscillations (from low to very high frequency) and mechanical vibration, both of which will overwhelm the cooling power of the dilution unit if not eliminated.

Miniature dilution refrigerators employing adsorption pumping, similar in design to the ^{3}He refrigerator shown in Figure 17.8, for use in storage dewars, have been reported [8,12].

Dilution refrigerators provide the lowest temperature available for continuous cooling. The most common technique for achieving still lower temperatures is magnetic demagnetization using the nuclei of various metals such as copper. This technique, by which temperatures of nuclei of less than 1 nK have been reached, is quite specialized, and interested readers are referred to the literature [3].

17.3.4 Refrigeration above 4.2 K

There is more work done above the temperature of boiling liquid helium than below. Applications include almost anything also done at room temperature: light scattering, microscopy, x-ray Raman scattering, Mossbauer and Hall effect, nuclear magnetic and electron spin resonance (NMR and ESR), magnetization and susceptibility, matrix isolation, resistivity, and radio astronomy, to name a few. Cold traps and cryopumping are also broadly applied. Systems that cool from room temperature down to 4.2 K fall into two classes: open and closed systems. Open systems use cryogenic fluids that eventually boil and are released into the environment. Closed systems use pressurized gases, such as helium, nitrogen, neon, or hydrogen, in some type of expansion process.

Most open systems use helium or nitrogen because of the explosive nature of hydrogen/air mixtures or the high cost of neon. This method of cooling simply transfers the cryogen from a storage dewar, through a triaxial (vacuum-insulated) transfer tube to a cold head. Here, the cryogen evaporates, cooling the apparatus, with the evolved gas returning through heat exchangers, heat shields, and the transfer tube middle-annular space, maximizing efficient use of the cryogen. A heater at the cold head is used in an electric feedback circuit to hold the temperature at the desired value. If a Joule–Thomson expansion valve [13] is added to the cryogenic liquid exit port, temperatures as low as 2 K may be reached using helium.

Open systems are simple, have relatively high cooling capacity, and may be constructed to cool in confined spaces, such as in a magnet or with optical instrumentation. They may also be made quite vibration-free. Their obvious disadvantage is that cryogenic fluid must be continually supplied. Hence, expertise in handling the fluid and a source of the fluid must be maintained. In addition, as the system is adjusted to temperatures farther away from the boiling point, control becomes more and more difficult, with some intermediate temperatures becoming impossible to regulate. New types of microprocessor-based regulators (see Section 17.3.1) are reducing this problem.

Often, temperatures below 4 K and high cooling power are not necessary. In this case, closed cycle systems are often used. These systems require only electricity and a thermal bath (ambient air or cooling water) and run continuously for long periods of time. They may also be made very small and relatively vibration-free. There are three types of closed cycle refrigerators, all involving the expansion of a compressed gas: Joule–Thomson, Gifford–McMahon (GM) [14], pulse tube refrigerator [15], or a linear combination of these three types of expansion.

The Joule–Thomson expansion of a high-pressure gas extracts heat by doing work against the internal forces between the molecules of a nonideal gas. The expansion must begin in the region of the pressure–temperature plane where the isenthalps have positive slopes (the inversion region). For nitrogen, this includes room temperature, so that nitrogen expanded in this way from a gas cylinder can be made to reach 70 K with no moving parts (except the gas molecules). This process is used in *microminiature* refrigerators [16], which are centimeters in size, have essentially no electrical or mechanical vibrations, and require no maintenance. Larger versions can produce watts of cooling power below 100 K. The expanded cold gas must be used to precool the incoming pressurized gas for lowest temperatures to be achieved. As usual, efficient heat exchangers, often of counterflow construction, are necessary.

The GM process uses a displacer moving in a cylinder driven by pressurized gas. In addition, it uses a regenerator, which is essentially a single-chamber heat exchanger with the incoming and outgoing fluids sharing it at different times in the cycle. This regenerator must be extremely efficient (>98%). An advantage of the GM cycle is that stages may be easily put in series and driven on the same shaft to obtain lower temperatures and higher cooling capacities. Cooling powers of 2 W at 10 K and 100 W at 77 K have been achieved by commercial units [17]. A low temperature of less than 5 K may be reached if helium is the working medium.

Most units operate from a simple, small air-conditioning compressor using either helium or nitrogen gas. The compressor, ambient heat exchanger, and ancillary equipment are housed in a small (0.1 m^3) portable unit, connected to the cold head using standard flexible metal hoses. The cold head can be quite small and configured to meet almost any experimental need. It contains the moving displacer and valve disk/motor, which controls the gas cycle [18]. It may be operated in any orientation relative to gravity, or in zero gravity, since bulk cryogenic liquid is not part of the cycle. Experiments or samples may be thermally heat sunk to the cold head using mechanical or gas (heat pipe) contacts.

GM units are commonly used to cool the neck of a large cryostat, thus greatly reducing helium consumption. One dilution refrigerator (see Section 17.3.3) has been constructed using a GM machine to thermally shield the dilution unit and provide the cooling power to condense the circulating ^3He. No liquid cryogens are required, allowing operation anywhere that electrical power is available.

By using a GM machine to precool pressurized helium gas, the inversion region of helium may be reached (<45 K). This is also a closed system, as the helium gas is reused and pressurized in the same compressor that runs the GM cycle. This adds very little to the size of the cold head while achieving a low temperature of 3.6 K with a cooling power of 1 W at 4.2 K.

A pulse tube refrigerator is very similar to a GM refrigerator (also developed by Gifford) but has the advantage of no moving cold parts (except the gas working medium). It has the disadvantage of being gravitation field dependent. This is because there is no displacer per se, such as the piston in the GM machine, so the moving gas is susceptible to convection. Therefore, buoyancy effects are important. However, the pulse tube refrigerator can function in very low or zero gravity.

Recently, the performance of pulse tube refrigerators has been greatly improved. They can reach almost 1 K, have useful capacity, are compact and simple to operate, and produce relatively low vibration levels. The working medium, often helium gas, is self-contained and therefore does not require circulating, cleaning, or replenishing. This device has ushered in the age of the *cryogen-free* refrigerator that can reach temperatures as low as 6 mK and the *cryogen-free* superconducting magnets.

Acknowledgments

The author thanks Professor Dwight Adams for assisting in the assembling of the information presented here. Roman Ciapurin assisted in the preparation of several figures. Lydia Munday and Logan Bessll assisted in assembling references.

References

1. Mendelssohn, K., 1966. *The Quest for Absolute Zero*, McGraw-Hill, New York.
2. Hastings, L.J., Bryant, C.B., Flachbart, R.H., Holt, K.A., Johnson, E., Hedayat, A., Hipp, B., Plachta, D.W., 2010. Large-scale demonstration of liquid hydrogen storage with zero boiloff for in-space applications, NASA/TP-2010-216453, M-1302.
3. Wilks, J., 1967. *The Properties of Liquid and Solid Helium*, Clarendon Press, Oxford, U.K.
4. A full description of the cooling techniques described here can be found in: Lounasmaa, O.V., 1974. *Experimental Principles and Methods Below 1 K*, Academic Press, London, U.K.; Pobell, F., 1992. *Matter and Methods at Low Temperatures*, Springer-Verlag, Berlin, Germany.
5. For a description of general low-temperature techniques, see White, G.K., 1987. *Experimental Techniques in Low Temperature Physics*, Oxford University Press, Oxford, U.K.
6. De Long, L.E., Symco, O.G., and Wheatley, J.C., 1971. Continuously operating 4He evaporation refrigerator. *Rev. Sci. Instrum.* **42**, 147, http://dx.doi.sig/10.1063/1.1684846.
7. Engel, B.N., Ihas, G.G., Adams, E.D., and Fombarlet, C., 1984. Insert for rapidly producing temperatures between 300 and 1 K in a helium storage dewar. *Rev. Sci. Instrum.* **55**, 1489.
8. Uhlig, K., 1987. ^{3}He/^{4}He dilution refrigerator without a pumped ^{4}He stage. *Cryogenics*, **27**, 454.
9. Swartz, E.T., 1986. Efficient He-4 cryostats for storage dewars. *Rev. Sci. Instrum.* **57**, 2848 and references therein.
10. Edwards, D.O., Brewer, D.F., Seligman, P., Skertic, M., and Yaqub, M., 1965. Solubility of He^3 in liquid He^4 at 0°K. *Phys. Rev. Lett.* **15**, 773.
11. Frossati, G., 1978. Obtaining ultralow temperatures by dilution of ^{3}He into ^{4}He. *J. Phys.* **39**(C6), 1578.
12. Takano, Y., 1994. Cooling power of the dilution refrigerator with a perfect continuous counterflow heat exchanger. *Rev. Sci. Instrum.* **65**, 1667.
13. Pobell, F., 1967. Chapters 3 and 4. In *The Properties of Liquid and Solid Helium*, Wilks, J., ed., Clarendon Press, Oxford, U.K.
14. Barron, R.F., 1985. *Cryogenic Systems*, 2nd edn., Oxford, New York, p. 64.
15. Barron, R.F., 1985. *Cryogenic Systems*, 2nd edn., Oxford, New York, p. 270.
16. Radebaugh, R., 2000. Development of the pulse tube refrigerator as an efficient and reliable cryo-cooler. *Proceedings of the Institute of Refrigeration of London*, London, U.K., vol. 96, pp. 11–29.
17. Little, W.A., 1984. Microminiature refrigeration. *Rev. Sci. Instrum.* **55**, 661.
18. CryoMech, Inc., New York; SHI-APD Cryogenics, Inc., Allentown, PA; and CTI-Cryogenics, Chelmsford, MA.

18

Magnetic Liquefaction of Hydrogen

Tom Burdyny
University of Victoria

Andrew Rowe
University of Victoria

CONTENTS

18.1 Introduction

The ability to induce a reversible temperature change in a magnetic material by changing the magnetization is called the magnetocaloric effect (MCE) and has been known for over 100 years. Magnetic cooling is a process that uses the MCE to reduce the temperature of another substance and was originally employed to produce temperatures lower than those possible with liquid helium (LHe). Since the early 1970s, advancements in materials (magnetic refrigerants, superconductors, permanent magnets) and our understanding of the thermodynamics of magnetic cycles have led to increased interest in using magnetic cycles for a variety of applications. One area of particular interest is the liquefaction of hydrogen. Although no commercial liquefiers using the MCE currently exist, the potential for high efficiencies at smaller scales and lower cost than conventional gas expansion cycles is a driver of research and development activities. In addition, future energy systems in which hydrogen is used as an energy carrier require logistical chains for transmission, distribution, and storage. For this to happen, the need for hydrogen liquefaction will grow substantially from current levels.

This chapter describes some of the activities reported in the literature regarding magnetic liquefaction of hydrogen. A brief history of magnetic refrigeration is provided and is followed by a summary of the basic thermodynamic relationships quantifying the performance of magnetocaloric materials and cycles. Following this, there is a review of proposed devices and their configurations. Some of the refrigerants that have been specifically suggested for use in hydrogen liquefaction will be discussed, but, given the substantial research currently happening on material synthesis and characterization, no attempt will be made to cover this area rigorously. Instead, the reader will be directed to reviews where further information can be found. Likewise, much of the work on device development has been in national laboratories and companies through government contract research. Many of the details for these projects are not easily accessible; thus, this review will focus on information available in the open literature.

18.1.1 Overview of Magnetic Refrigeration

Magnetic refrigeration is based upon a phenomenon known as the MCE. The MCE is sometimes described as the isothermal entropy change due to a change in the applied magnetic field. In order to produce useful cooling or heat with a magnetic material, there must be a decrease or increase in temperature due to a change in the applied field—this is the adiabatic temperature change and is another way of reporting the MCE. Unlike eddy-current heating, the MCE is identified by the *reversible* change in temperature when adiabatically subjected to a magnetic field. For a given material, the magnitude of this temperature change is proportional to the strength of the applied magnetic field. Different materials also display this effect to varying degrees depending upon properties such as the Curie temperature, magnetic entropy, and lattice entropy.

Using the MCE, it is possible to form a cycle that parallels a vapor-compression process whereby a volume can be continuously refrigerated through a cyclical process. The compression and expansion stages of a gas-refrigeration system are replaced by magnetizing and demagnetizing a magnetic material. Because the working substance is a solid with limited heat transfer capabilities, a heat transfer fluid is used to thermally couple the refrigerant to hot and cold heat sinks. An example of a particular process for a magnetic refrigeration cycle is summarized in the following paragraph with reference to Figure 18.1.

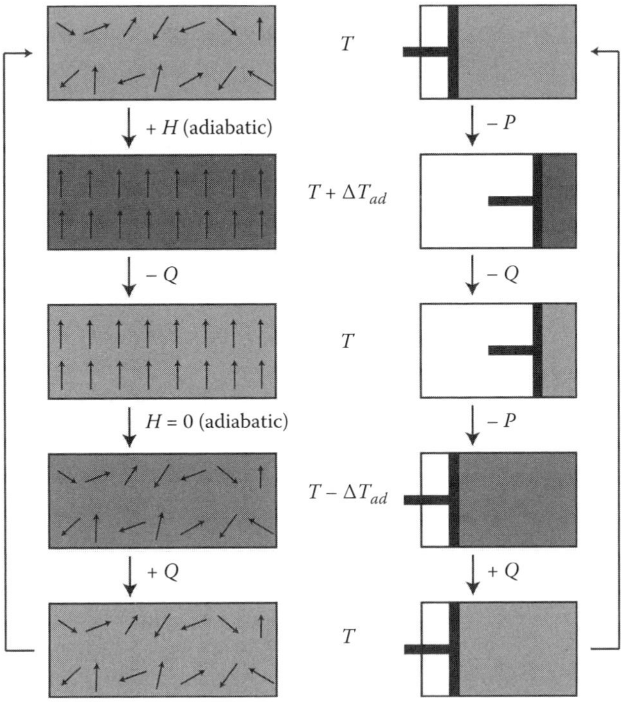

FIGURE 18.1
The general process of magnetic refrigeration as compared to gas systems.

When a conventional magnetocaloric material is reversibly subjected to a magnetic field change, the magnetic dipoles become more ordered. In practice, this can be observed by measuring the force on a substance in a nonuniform magnetic field. If the total entropy is to remain constant, adiabatic magnetization causes magnetic entropy to decrease, which results in an increase in temperature of the lattice. With the magnetic field still applied, the adiabatic condition is relaxed so that the material rejects heat and the temperature decreases. Next, an adiabatic condition is imposed, the magnetic field is then removed so that the magnetic order decreases, and the temperature decreases below its original value. Once again, the material is thermally coupled to absorb heat and the material returns to the original state, completing the cycle.

While all magnetic refrigeration cycles utilize the same basic steps, the actual cycle depends on the application (temperature span and cooling power) and available refrigerants. In addition, device geometries and heat transfer flow configurations can be quite different.

Although the MCE was hypothesized to exist by Kelvin [1] and measured by Weiss and Piccard [2] in 1918, magnetic cooling was not demonstrated until 1933 [3,4]. The idea of using the MCE for hydrogen liquefaction did not appear until the late 1970s.

The first applications of magnetic cooling were to produce sub-Kelvin temperatures with a *single-shot* process where the entire mass of refrigerant was brought to a uniform temperature and demagnetized. Larger amounts of material could cool more mass and maintain low temperatures for longer periods. For paramagnetic materials at low temperatures, the total entropy of the refrigerant is dominated by the magnetic entropy and, therefore, significant temperature changes are easy to produce. At higher temperatures, the magnetic entropy change in a paramagnetic material becomes insignificant relative to the total entropy, making the temperature change small. The primary deterrent to high-temperature cooling was thought to be a lack of materials, which showed a significant temperature change when a moderate magnetic field (0–7 T) was applied.

In the mid-1970s, two important developments for magnetic refrigeration were reported by Brown: (1) a ferromagnetic material near its Curie temperature was used to produce a similar effect to that of the paramagnetic salts and (2) the limitations of the MCE were overcome using recuperation. Shortly after this work, the use of ferromagnetic materials for magnetic refrigeration in the range of 10 K to room temperature was described [5].

Brown demonstrated a magnetic refrigerator producing a temperature span of 47°C using gadolinium [6], and it was the first time that magnetic refrigeration was demonstrated at temperatures above 20 K. Brown's experimental testing was discussed by Steyert [7] who analyzed the reversibility of magnetic refrigeration. He concluded that continued progress should allow for magnetic refrigeration to have higher efficiencies than existing gas-refrigeration technologies. Perhaps the most significant development for magnetic refrigeration at temperatures above 20 K was the idea of using the magnetocaloric material in the form of a regenerator. This idea was called an *active magnetic regenerator* (AMR) and was patented by Barclay and Steyert [8]. In addition to using regeneration, the AMR concept overcomes the limited range of a single magnetocaloric material by using a number of different materials in a layered structure. Similar to a passive regenerator, the solid matrix periodically exchanges heat with a fluid oscillating through the pores. Unlike a passive regenerator, however, the solid refrigerant is the working material that creates the temperature distribution from the hot end to the cold end. Today, nearly all MR development above 20 K is making use of the AMR concept.

These initial papers subsequently led to several studies regarding magnetic materials, device design, and analytic studies of performance [9–12].

18.1.2 History of Magnetic Liquefaction of Hydrogen

The idea of using magnetic cycles for hydrogen liquefaction really started at the beginning of the 1980s. During 1982–1986, a number of related studies were performed by a group at Los Alamos National Laboratory [13]. The work was primarily funded by NASA and various offices in the US DOE. Hydrogen reliquefaction using a magnetic refrigerator was one application, and the development of a device operating between liquid nitrogen (LN_2) and liquid hydrogen was another. The latter was identified as having good commercialization potential due to the significant fraction of capital expenditures required for the 77–20 K range with conventional liquefiers [14]. A number of devices were designed, analyzed, and tested; details of those designed to work below LN_2 will be discussed later.

In 1991, Barclay summarized the status of magnetic refrigeration technology and discussed the future prospects for liquefaction of hydrogen [15]. This summary explained the technical aspects of magnetic refrigeration in addition to describing the thermodynamic advantages they have over conventional systems. The regenerator was identified as being the main design challenge to create a magnetic device spanning 300–20 K. This chapter concluded with a cost estimate of two different magnetic liquefiers operating from 77 to 20 K, indicating the potential for significantly lower cost than a gas-cycle device. In 1992, Janda et al. [16] reported the design of a 0.1 ton/day AMR refrigerator for liquefying hydrogen; this device was later analyzed by Zimm et al. [17] and Degregoria [18]. This study of this design was continued by Zhang et al. in 2000 in the form of an optimization study [19]. The most recent design and experimental unit was created by Kamiya et al. [20]

and Matsumoto et al. [21], which takes into account the actual condensation process of hydrogen using a Carnot magnetic refrigerator (CMR) stage.

The potential for magnetic refrigeration to be used in hydrogen liquefaction has made the 77–20 K range a focus for magnetic refrigeration. Multiple papers have now been dedicated to theoretical studies [22,23], numerical models [24–32], test devices [33–37], and review articles [38,39] in an effort to improve the process of hydrogen liquefaction and make magnetic liquefaction a competitive alternative to traditional gas expansion cycles.

The thermodynamics of the MCE is introduced in the following section and leads to a discussion of magnetic cycles.

18.2 Thermodynamics

This section discusses the fundamental theory and processes behind magnetic refrigeration. This includes the functional form for quantifying magnetic work, entropy, and specific heat. The basic cycles of magnetic refrigeration such as the Carnot, Stirling, Ericsson, and Brayton cycles are then reviewed. The AMR cycle is also described followed by a brief summary of some of the losses that limit magnetic cycle performance.

18.2.1 Magnetic Systems

The following sections review the fundamental thermodynamic quantities of a material with a reversible magnetic work mode. For simplicity, it is assumed that the material has no significant volume change, is homogenous and hysteresis-free, and has continuous specific heat as a function of field and temperature. Materials that fit these assumptions well are some of the rare-earth elements and alloys and materials displaying a second-order magnetic phase change. A number of first-order materials have good potential as magnetic refrigerants; however, their behavior tends to be more complicated than second-order materials. For those interested readers, more details on magnetocaloric materials can be found in the literature [40,41].

A challenge one encounters when dealing with magnetism, magnetic fields, and magnetocaloric materials is the wide range of units and potentials one can work with. Here we will work with the following symbols and magnetic quantities: B_0 is the flux density *applied* to a material and is considered to be due to the magnetic field, H_0, generated by a coil in free space. The magnetism of a material, M, is assumed to be a single-value function of temperature, T, and H. A significant amount of confusion (and error) can arise when discussing the magnetic field because the field *in* a body is generally not the same as the applied field and depends upon the shape of the material. The impact of shape on magnetization is often accounted for by a *demagnetizing* field, H_d. The local field determines the local magnetization, and the *local* internal field in a magnetic material is $H_i = H_0 + H_d$. Fields, fluxes, and magnetizations are vectors, but, for thermodynamic considerations, we will assume that all values are 1D.

18.2.1.1 Work

By strategically applying and removing a magnetic field to a magnetocaloric material, it is possible to develop a process that extracts heat from a cold environment and rejects it to a hot environment. The act of magnetizing and demagnetizing a material parallels that of compressing and expanding a gas and, similarly, is achieved by a work interaction.

The first law of thermodynamics relates the energy change in a magnetic material to heat transfer and magnetic work:

$$du = dq + dw. \tag{18.1}$$

In the previous texts, the work and heat transfer are assumed to be *in* to the material system. A clear derivation of the appropriate generalized force and displacement for a magnetic system can be found in Appendix B of [42]. One can differentiate between work done by a power supply to generate a magnetic field and the work that changes the internal energy of the material. The latter is the magnetic work and the incremental value per unit volume of material is

$$dw = B_0 dM = \mu_0 H_0 dM, \tag{18.2}$$

where μ_0 is the permeability of free space. The applied field is analogous to the generalized force, and the magnetization (total magnetic moment per unit volume) is the generalized displacement. One can now see how the work done on magnetic material compares to that done in gas refrigeration by observing the fundamental relation of a simple compressible system:

$$du = dq - pdv. \tag{18.3}$$

Other potentials for magnetic systems can be derived using Maxwell's relationships [43,44].

18.2.1.2 Entropy

The entropy of a magnetic material can be introduced into Equation 18.1 by replacing the differential heat transfer with the entropy equivalent. The entropy of a simple magnetic material is a function of temperature and field strength, $s = s(T, H)$, and the differential change is

$$\frac{dq}{T} = ds = \left(\frac{\partial s}{\partial T}\right)_H dT + \left(\frac{\partial s}{\partial H}\right)_T dH_0. \tag{18.4}$$

The first term in Equation 18.4 can be replaced using the definition of specific heat at constant field, c_H,

$$c_H \equiv \left(\frac{\partial s}{\partial T}\right)_H T. \tag{18.5}$$

The last term in Equation 18.4 can be written in terms of magnetization and temperature using Maxwell's relations,

$$\left(\frac{\partial M}{\partial T}\right)_H = \left(\frac{\partial s}{\mu_0 \partial H}\right)_T. \tag{18.6}$$

So, the variation in entropy due to temperature and field change is

$$ds = \frac{c_H}{T}dT + \left(\frac{\partial M}{\partial T}\right)_H \mu_0 dH_0. \tag{18.7}$$

Magnetization can be estimated by model calculations (i.e., molecular field theory) or by direct measurement.

The total entropy of the material can also be broken down into a summation of the lattice entropy (s_g), magnetic entropy (s_m), and entropy from conducting electrons (s_E):

$$s_{tot} = s_m + s_E + s_g. \tag{18.8}$$

Standard expressions can be used to quantify the lattice and electronic contributions to entropy (such as the Debye approximation for the lattice and quantum theory for a free-electron gas [45]). The isothermal change in magnetic entropy as a result of the applied magnetic field is determined by integrating Equation 18.7:

$$\Delta s_m(T,H) = \int_0^H \left(\frac{\partial M}{\partial T}\right)_H \mu_0 dH. \tag{18.9}$$

The isothermal magnetic entropy change is often called the MCE, but this is a derived quantity, whereas the adiabatic temperature change can be directly measured.

A reversible adiabatic process implies constant entropy; therefore, using Equation 18.7, the adiabatic temperature change can be calculated from specific heat and magnetization data as

$$\Delta T = -\int_0^H \frac{T}{c_H}\left(\frac{\partial M}{\partial T}\right)_H \mu_0 dH. \tag{18.10}$$

Thus, the dependence of adiabatic temperature change on magnetic entropy change is determined by the variation in specific heat with temperature.

The magnitude of the change in entropy, and the temperature change, is dependent upon the material and the operating temperature. At a certain point known as the Curie temperature, T_C, the MCE is a maximum. As a material is cooled below the Curie point, the material spontaneously magnetizes (orders) and displays properties similar to ferromagnetic materials. As the temperature rises above the Curie point, the dipoles become randomized again and the response to a magnetic field is more similar to a paramagnet. Every material has a distinct Curie temperature based upon its internal magnetic and crystal structure.

The relationship between adiabatic temperature change and isothermal magnetic entropy change for gadolinium is shown in Figure 18.2a. Gadolinium is a good prototype refrigerant ordering near room temperature. Figure 18.2b shows that the MCE decreases quickly as the temperature deviates from the Curie point. As will be discussed later, this characteristic limits the use of magnetic cooling using traditional cycles.

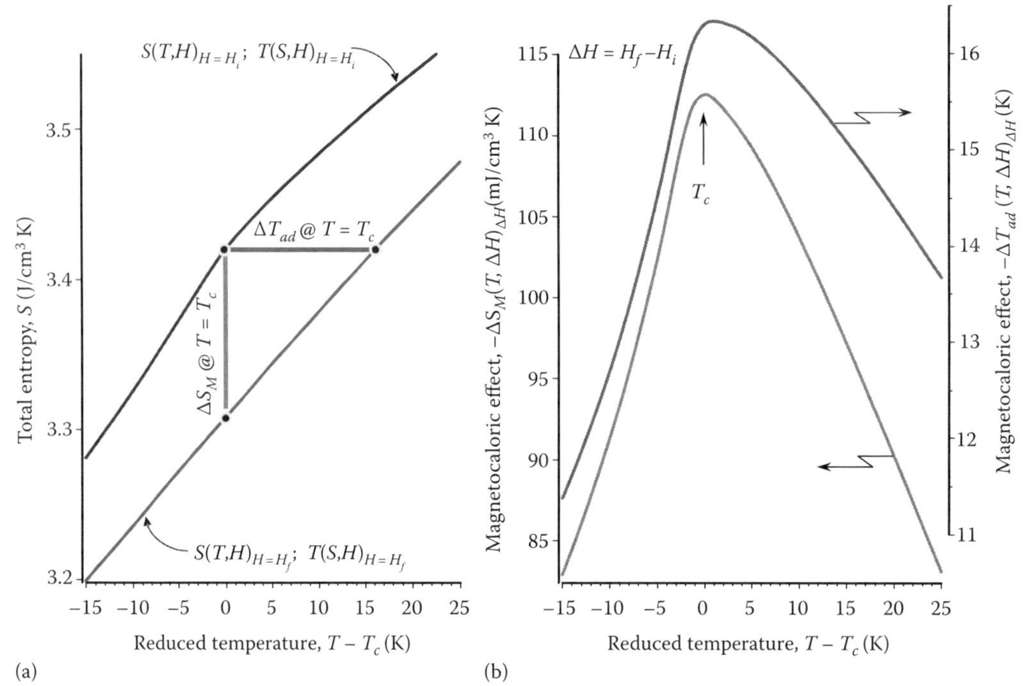

FIGURE 18.2
The total entropy for the initial and final magnetic fields (a) and the MCE (b) for gadolinium near the Curie temperature. (From Gschneidner, K.A., Jr., Pecharsky, V.K., and Tsokol, A.O., Recent developments in magnetocaloric materials, *Rep. Prog. Phys.*, 68, 1479–1539, 2005. With permission from IOP Publishing Ltd.)

The impact of magnetization on specific heat can be shown using Equation 18.7. Noting that this is the total differential of entropy, the following result can be derived due to equivalence of the second partial derivatives [39]:

$$c_H(T,H) = c_0(T,0) + T\int_0^H \left(\frac{\partial^2 M}{\partial T^2}\right)\mu_0 dH. \qquad (18.11)$$

It can be seen that the heat capacity of the magnetic material is dependent upon the variation in magnetization with field and temperature. Near the Curie temperature, specific heat can be a strong function of field and temperature. This behavior complicates modeling and optimal operating strategies.

An overview of magnetic cycles is presented in the following section. The cycle currently favored for magnetic refrigeration is an AMR. As no single material has the ideal properties for this type of process, there is a substantial research effort in the materials science community aimed at developing materials. Magnetic refrigerants for hydrogen liquefaction will be discussed following the overview of cycles.

18.2.2 Magnetic Refrigeration Cycles

Although all magnetic refrigeration processes make use of the same basic physical response, idealized processes can be defined depending upon how and when heat transfer and field changes occur. This section will introduce the current cycles used in magnetic refrigeration processes and discuss the fundamental differences between them.

The primary cycles used with fluids are the classical thermodynamic cycles, that is, Brayton, Ericsson, and Stirling. Analogous cycles can be defined for magnetic solids where magnetization and field replace volume and pressure, respectively. The Carnot cycle quantifies the maximum performance of a cycle using isothermal heat rejection and absorption. An important consideration is that a Carnot cycle is not the most efficient process for gas liquefaction where sensible heat must be removed in addition to latent heat. This will become evident later when discussing proposed processes for hydrogen liquefaction.

18.2.2.1 Carnot Cycle

A magnetic Carnot cycle is composed of the four reversible processes shown in Figure 18.3:

1. (1–2): Adiabatic magnetization of the magnetic refrigerant
2. (2–3): Isothermal heat rejection
3. (3–4): Adiabatic demagnetization
4. (4–1): Isothermal heat addition

In adiabatic magnetization (1–2), the temperature of the material is increased isentropically due to an increase in the magnetic field. The magnetic field is then further increased while the magnetic refrigerant rejects heat to keep the material isothermal (2–3). In actuality, this continuous cooling effect would take a large heat exchanger and a large period of time, both of which are not ideal with respect to cost. With thermal isolation restored, the material is adiabatically cooled (3–4) by partially removing the magnetic field. The final step absorbs heat while simultaneously reducing the magnetic field (4–1).

In Figure 18.3, the net magnetic work, w, required in the cycle is equal to the area of the T-s diagram bounded by (1–2–3–4). Similarly, the cooling capacity, q_c, is the area underneath

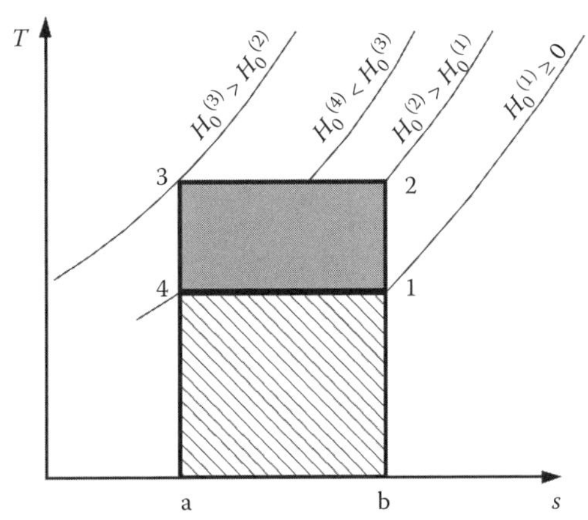

FIGURE 18.3
The Carnot cycle for a magnetic refrigeration application. (From Kitanovski, A. and Egolf, P.W., *Int. J. Refrig.*, 29, 3, 2006. With permission.)

the Carnot cycle, which is (a–b–1–4). The coefficient of performance (COP) that describes the *efficiency* of a refrigerator can then be found:

$$\mathrm{COP}_R = \frac{q_c}{w} = \frac{T_1}{T_2 - T_1}. \tag{18.12}$$

This COP acts as an upper bound for the other refrigeration cycles as each of the four processes is reversible. Since isothermal processes are used, the cooling capacity per unit work is maximized. A disadvantage of the isothermal processes, however, is the reduced temperature spans available for a given magnetic field as compared to other cycles. This can be seen by considering Figure 18.2. The maximum span of a Carnot cycle is constrained by the total entropy curves at high field and low field. In the case of gadolinium using a high field, a Carnot cycle could only operate over spans on the order of 10 K.

18.2.2.2 Brayton Cycle

The Brayton cycle is a more practical cycle for substances not utilizing a first-order phase change as, in practice, it is easier to do. This cycle is composed of two adiabatic processes and two processes where heat is transferred under a constant magnetic field. This is summarized as follows:

(1–2): Adiabatic magnetization
(2–3): Heat rejection with a constant magnetic field, $H_0^{(2)}$
(3–4): Adiabatic demagnetization
(4–1): Heat addition with a constant magnetic field, $H_0^{(1)}$

A *T-s* diagram of the ideal magnetic Brayton cycle is provided in Figure 18.4. This ideal cycle ignores the irreversibilities associated with each process.

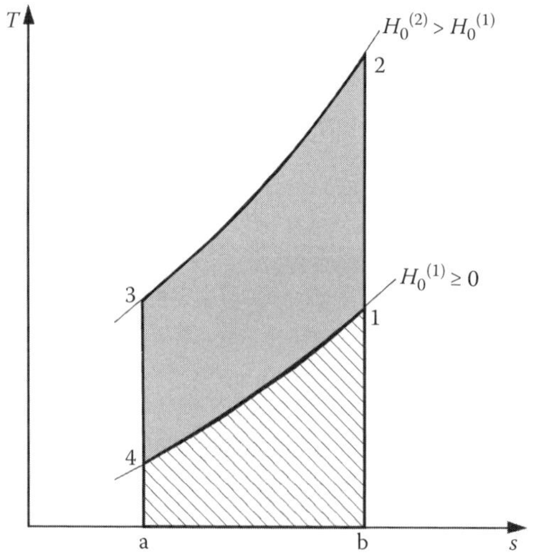

FIGURE 18.4
The Brayton cycle for a magnetic refrigeration process. (From Kitanovski, A. and Egolf, P.W., *Int. J. Refrig.*, 29, 3, 2006. With permission.)

The work done by the magnetic material can be found by calculating the area defined by the process (1–2–3–4) while the heat removed from the working fluid is equal to the area under the cycle (a–b–1–4). In actual cycles, losses of efficiency will occur in the adiabatic magnetization (1–2) and demagnetization (3–4) processes due to an irreversible entropy change.

In theory, with an equivalent effective temperature span, the magnetic Brayton cycle sacrifices efficiency as compared to Carnot. Less heat is absorbed per cycle and more work is inputted; however, this neglects thermal resistances between the heat sinks in a Carnot cycle. For a material like gadolinium with a 10 T field change, the temperature span is about 15–20 K compared to the 5–10 K/10 T temperature change for the Carnot cycle [6]. This is because the heat transfer is done with a constant magnetic field instead of isothermally, resulting in a larger change in the magnetic field in the isentropic processes. A larger temperature range could be obtained by maintaining the isothermal processes of the Carnot cycle but regenerating heat under a constant magnetic field; this is the foundation for the Ericsson and Stirling cycles.

18.2.2.3 Ericsson and Stirling Cycles

The Ericsson and Stirling cycles are very similar to one another from an energy standpoint. These cycles both utilize the isothermal heat transfer of the Carnot cycle and both require heat regeneration to be considered ideal. The adiabatic magnetizing and demagnetizing of the material is however replaced by a constant magnetic field strength, *H*, in the Ericsson cycle and constant magnetization, *M*, in the Stirling cycle. The Ericsson cycle is summarized in the following and a *T-s* diagram of the ideal process with heat regeneration is in Figure 18.5:

(1–2): Heat addition with a constant magnetic field, $H_0^{(1)}$

(2–3): Isothermal heat rejection to working fluid

(3–4): Heat rejection with a constant magnetic field, $H_0^{(3)}$

(4–1): Isothermal heat addition from working fluid

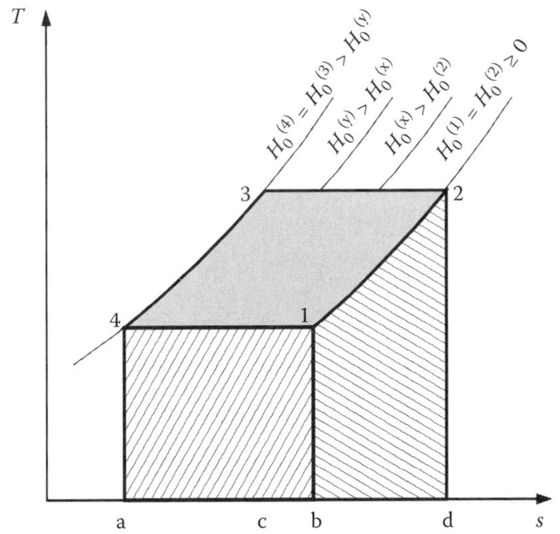

FIGURE 18.5
Magnetic Ericsson cycle for a magnetic refrigeration process. (From Kitanovski, A. and Egolf, P.W., *Int. J. Refrig.*, 29, 3, 2006. With permission.)

The magnetic work in the cycle is again equal to the confined area (1–2–3–4). The area (b–d–2–1) represents the regeneration heat that must be absorbed by the magnetocaloric material. In the ideal case, this area is equal to the heat rejected by the material (a–c–3–4). Since heat transfer is an irreversible process, however, this regeneration of the magnetic material reduces the overall efficiency of the cycles. The area (1–4–a–b) represents the cooling capacity of the Ericsson cycle.

The Ericsson and Stirling cycles are similar; the temperature span is greater than the Carnot cycle and, in the case of ideal regeneration, the Ericsson and Stirling cycles have a COP equivalent to that of the Carnot cycle. An important point to note for all of the cycles discussed so far is that all of the working material follows the same cycle. This is a subtle point but is important in understanding the AMR cycle that follows.

In theory, regeneration can provide high efficiencies and increased temperature spans; however, implementing this with magnetic refrigerants is not easy. Because the working substance is a solid, it is difficult to effectively couple the body to an external thermal storage medium that maintains a temperature distribution between the hot and cold reservoirs. Even if this can be solved, the refrigerants themselves have nonideal properties such that different amounts of heat need to be regenerated between the cooling and warming phases. These considerations were part of the reasoning that lead to the AMR cycle whereby the refrigerant itself is a regenerator linked to hot and cold reservoirs by a heat transfer fluid.

18.2.2.4 Active Magnetic Regenerator Cycle

The process currently employed in magnetic refrigeration cycles above 20 K is the AMR. The Brayton, Ericsson, and Stirling cycles all require high-quality regeneration to be efficient enough for magnetic refrigeration units to be competitive compared to vapor-compression units. The AMR cycle provides this by using the magnetic material itself as a regenerator that is in direct contact with the heat transfer fluid. Passive regenerators are used in various devices to increase the operating span. By using the working material itself as the thermal storage medium, the regenerator is said to be active [22]. A stepwise representation of the AMR process is shown in Figure 18.6. A schematic indicating the cycle-average temperature distribution and heat and work interactions at a location in the regenerator is shown in Figure 18.7.

The cycle is assumed to be in periodic steady state so that the temperature distribution is a stationary periodic function in time. The refrigerant is made to be a porous solid with a large surface area per unit volume. A heat transfer fluid flows in the pores, oscillating between the cold and hot reservoirs in synchronization with the magnetic field waveform. The steps begin with the regenerator being magnetized so that the local temperature of the solid increases due to the local MCE. Fluid is then blown from the cold side to the hot side absorbing energy from the matrix and exiting at a temperature above that of the warm reservoir. As a result, heat is rejected in the warm heat exchanger. The fluid flow stops, the magnetic field is removed, and the temperature decreases at all points due to the MCE. The temperature of solid material at the cold end is now less than it was when the cycle began. Fluid now flows through the bed in the opposite direction, exiting at a lower temperature such that heat will be absorbed from the cold reservoir. The fluid flow stops, the bed is remagnetized, and the cycle begins again.

The solid refrigerant acts as a regenerator to the heat transfer fluid. At one point in the cycle, the regenerator removes heat from the fluid, while at a later point, this heat is returned and an additional amount of energy is added due to magnetic work. This increases the fluid temperature exiting the hot end of the regenerator so that heat

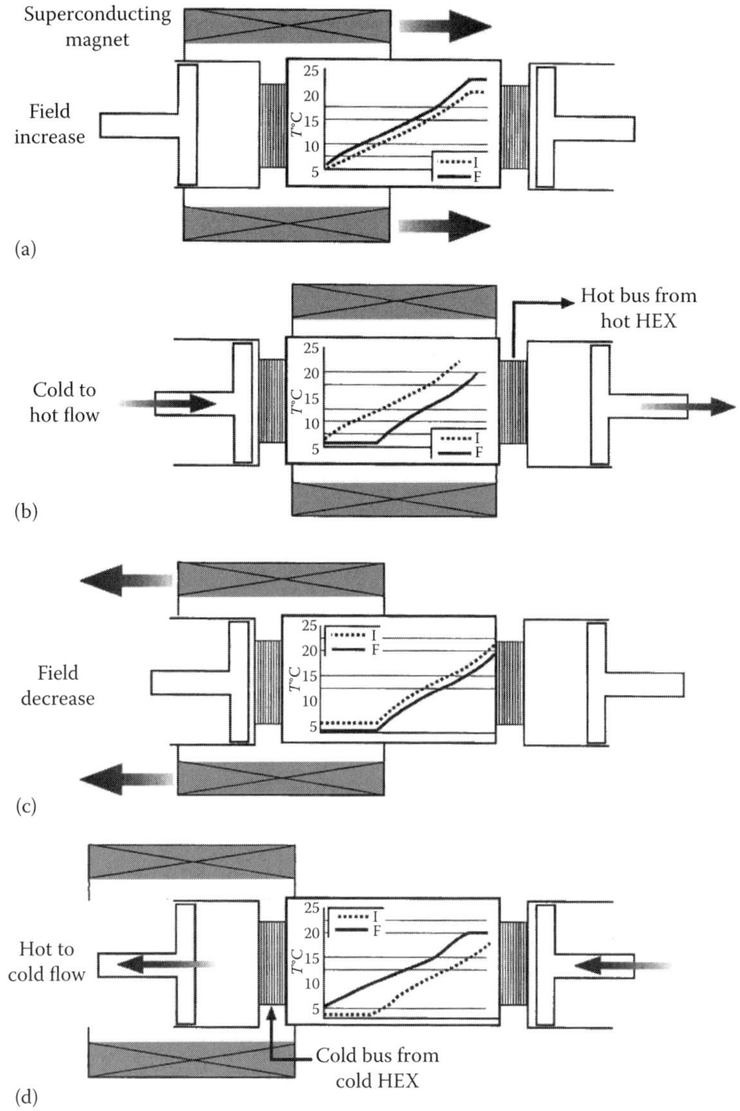

FIGURE 18.6
A stepwise representation of an AMR cycle. (a) Magnetization, (b) magnetized fluid flow from the cold side to the hot side, (c) demagnetization, and (d) fluid flow from the hot to cold side. (From Pecharsky, V.K. and Gschnieder, K.A., Jr., *J. Magn. Magn. Mater.*, 200, 44, 1999. With permission.)

can be rejected to the warm reservoir. The heat flux from the cold end to the warm end, Q, varies with position in the regenerator. Integrated over the entire bed, the total work input will balance the net heat transfer. An AMR can use a single material as the regenerator or a combination of different materials. Using multiple materials complicates the regenerator design process but allows for a larger temperature span to be created between the hot and cold reservoirs.

Determining the optimal composition, structure, matrix geometry, and operating conditions of an AMR is an active area of research. The highly nonlinear properties of

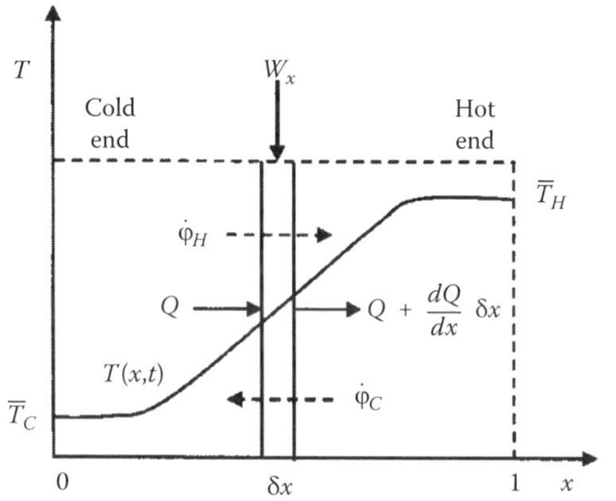

FIGURE 18.7
Temperature distribution in an AMR regenerator. (From Rowe, A.M. and Barclay, J.A., *J. Appl. Phys.*, 93, 1672, 2003. With permission.)

both first- and second-order materials make this a complex problem. Simplified analyses provide some guidance regarding the ideal MCE and flow conditions that satisfy the macroscopic entropy balance on a regenerator [48,49]. When the field-dependent specific heat of a refrigerant is taken into consideration, the ideal MCE as a function of temperature, $\Delta T(T)_{ideal}$, is given by

$$\Delta T(T)_{ideal} = \left(\Delta T_{ref} + T_{ref}\right)\left(\frac{T}{T_{ref}}\right)^{\sigma} - T, \quad (18.13)$$

where σ is the ratio of specific heat at low field to that at high field and the subscript *ref* indicates the MCE at a reference temperature, T_{ref}. For this equation to be valid, the fluid thermal capacity ratio must also equal the specific heat ratio [48].

Although the temperature span is increased significantly with an AMR cycle and multiple refrigerants, a single AMR stage may not be enough to operate between 20 and 300 K. A number of AMR units can be arranged together to span large temperature differences, to increase cooling power, or to better match cooling to the enthalpy change of a fluid like hydrogen. In a series configuration, the cold reservoir of a warmer stage is also the hot reservoir of a lower-temperature stage. In a parallel configuration, one AMR stage precools hydrogen going to another stage with both stages rejecting heat to the same reservoir [50]. Because each stage in an overall liquefaction system may have different cooling loads and operating temperatures, an assortment of magnetocaloric materials are required. This adds a degree of complexity in the optimization of an entire liquefaction unit, in addition to added capital costs for building several AMRs. Irreversibilities resulting from various effects can also compound making minimization of entropy generation and losses in magnetic refrigeration an important exercise [23].

18.2.3 Regenerator Design

Magnetic liquefaction of hydrogen, like any process, contains irreversibilities. These irreversibilities reduce the efficiency of converting magnetic work into refrigeration, making the overall process require more energy. Sources of inefficiency include nonideal materials, imperfect heat transfer, flow losses, eddy currents, and parasitic heat leaks in the device and from the environment.

Assuming suitable magnetic refrigerants are used, magnetic refrigerators using the AMR cycle require effective heat transfer. Regenerative heat transfer between the solid matrix and the fluid is done through direct contact, while transfer to the external reservoirs is through heat exchangers. If a single AMR stage is operating over a large temperature span, the impacts of imperfect convective heat transfer in the regenerator can be significant. In theory, because the cooling power of a device can be increased with the cycle frequency, there is a need for larger convective transfer rates. Optimizing regenerator structure and operating parameters using entropy generation minimization has also been performed on AMRs [51,52].

Regenerative heat transfer also requires careful design of the regenerator matrix so that the characteristic dimension of the structure is sufficiently small to ensure the material is being fully utilized [53]. Transfer of energy from the regenerators depends upon the solid thermal conductivity and specific heat. Higher conductivities allow for faster energy transfer to the working fluid, which reduces the time for parasitic heat leaks. Thermal diffusivity in the solid is linked to the operating frequency of an AMR through the Fourier number [54]. This plays an important role in optimized designs as the cooling capacity and losses of a system depend upon operating frequency. As discussed by Zhang et al., an optimum frequency exists, which minimizes the overall exergy destruction [23].

Other losses that contribute to decreased efficiency are flow losses and eddy currents. For a regenerator to have good thermal effectiveness, a large wetted area per unit volume is desired; however, this can lead to larger pumping powers needed to oscillate fluid through the structure. This problem becomes more significant as operating frequency increases. Eddy currents arise in conductors subjected to time-varying magnetic fields and are dissipated as heat due to resistance in the material. This problem scales with the square of the rate of field change [55]. Careful design choices and the use of materials with low electrical conductivity can minimize this problem. In the case of regenerators, spherical particle beds are common for the matrix, and, because of small size and point contacts between the particles, eddy-current losses can be small for typical operating frequencies [56].

18.3 Magnetocaloric Materials

Materials exhibit the MCE to varying degrees and single materials are only effective within a limited temperature range. This is one of the main challenges to developing commercial magnetic devices. A material with a wide operating range and a large MCE should reduce the required magnetic field strength while maintaining efficiency. This can benefit cost and relax some of the design constraints. For these reasons, developing and characterizing magnetocaloric materials is an integral part of improving magnetic refrigeration and facilitating the use of magnetic cycles for hydrogen liquefaction [57]. This section briefly discusses refrigerants, desirable properties, and materials relevant to hydrogen liquefaction and their operating range. This overview will be brief as the subject of materials is vast; the reader is encouraged to explore the literature for more details.

18.3.1 Requirements for an Effective Refrigerant

The choice of materials to use within a given cycle is still open to debate. In recent years, first-order materials have received a significant amount of attention due to large magnetic entropy changes using materials that are less expensive than rare-earth elements and intermetallic compounds on the basis of equivalent volume. Hysteresis tends to be significant in many of these materials and the width of the ordering transition in terms of temperature can be narrow.

18.3.1.1 Refrigerant Capacity

There are many properties that can be used to characterize refrigerants. Manufacturability, cost of materials and processing, toxicity, ductility, and a range of thermal properties can be considered. The most important are those that determine the ability to produce useful cooling efficiently; if this cannot be substantiated, the remaining properties are irrelevant. In terms of the AMR cycle, the ideal scaling of adiabatic temperature change with temperature has already been discussed. A general parameter used to characterize any refrigerant absorbing and rejecting heat isothermally is the refrigerant capacity (RC) as defined by Wood and Potter [57],

$$\mathrm{RC} \equiv \Delta S \Delta T_{sp},$$

where, in a reversible cycle, the isothermal entropy change, ΔS, is the same for hot and cold heat transfer processes and ΔT_{sp} is the temperature span. For a magnetic refrigerant, the magnetic entropy change as a function of temperature determines the maximum span and feasible operating points. For any magnetic refrigerant and applied field, a maximum value of RC can be determined. This parameter is useful in comparing magnetocaloric materials as it relates useful cooling to field strength on a volumetric basis. All other properties being the same, a material with a broad transition and large entropy change will have better potential as a refrigerant than one with a narrow transition and small entropy change.

A modified version of RC is used by Gschneidner and Pecharsky to compare materials [58]. They define a relative cooling power (RCP) in terms of the magnetic entropy change (RCP(S)) or the adiabatic temperature change (RCP(T)). The method is similar to [57]; however, to simplify things, they use the temperature span defined by the respective curve (entropy change or adiabatic temperature change) at half the maximum value, δT_{FWHM}. The RCP values are normalized by the applied field strength so that RCP(S) has units of energy per unit volume of refrigerant per unit field, that is, J/cm^3/T. A subset of materials for use in the 20–80 K range are listed in Table 18.1 [58].

One interesting fact to note in Table 18.1 is the magnitude of the adiabatic temperature change per unit field change. These values are on the order of 1 K/T and are the primary reason why a pure magnetic stage for hydrogen liquefaction requires superconducting magnets and highly effective regenerator designs.

The shapes of magnetic entropy curves as a function of temperature for various materials active below 100 K are shown in Figure 18.8. As can be seen, there tends to be a trade-off between the maximum entropy change and the width of the transition. While some materials have larger entropy spikes, they are only effective within a short temperature range. For example, $ErCo_2$ produces a large entropy spike around 35 K but becomes ineffective within 5 K in either direction, while $GdNi_2$ has a lower maximum entropy change but is effective

TABLE 18.1

Magnetocaloric Properties and Curie Temperatures for Materials with a Single Magnetic Transition

Material	T_C or T_{max}	$\frac{-\Delta S}{\Delta H}\left(\frac{mJ}{cm^3KT}\right)$	$\frac{-RCP(S)}{\Delta H}\left(\frac{mJ}{cm^3T}\right)$	ΔH(T)	$\frac{\Delta T_{ad}}{\Delta H}\left(\frac{K}{T}\right)$	$\frac{RCP(T)}{\Delta H}\left(\frac{K^2}{T}\right)$	ΔH(T)
$(Dy_{0.1}Er_{0.9})Al_2$	17.7	39.2	785	5	1.55	52.7	10
$DyNi_2$	20.5 ± 1.5	42.4	959	5	1.64	38.5	5
Gd_2PdSi_3	21	13.9	599	8	1.09	40.2	8
$(Gd_{0.14}Er_{0.86})Al_2$	24	—	—	—	1.27	53.4	7
$(Dy_{0.25}Er_{0.75})Al_2$	24.4	28.1	843	5	1.37	58.9	10
25–50 K							
$HoAl_2$	30 ± 3	39.0	1010	4	1.50	—	8
GdNiGa	30	24.0	839	9	1.17	50.4	9
$(Dy_{0.4}Er_{0.6})Al_2$	31.6	25.6	1070	5	1.25	68.8	10
$NdMn_2Si_2$	32	—	—	—	1.40	16.8	6
$(Dy_{0.5}Er_{0.5})Al_2$	38.2	25.2	909	5	1.26	68.0	10
GdPd	39 ± 1	—	—	—	1.36	32.6	5
$(Dy_{0.55}Er_{0.45})Al_2$	40.8	25.1	902	5	1.25	66.2	10
$ErCo_2$	44	45.0	770	7	1.69	21.1	7
$(Dy_{0.5}Ho_{0.5})Al_2$	46	26.6	839	—	—	—	—
$(Dy_{0.7}Er_{0.3})Al_2$	47.5	23.1	900	5	1.2	69.6	10
50–100 K							
$(Dy_{0.85}Er_{0.15})Al_2$	55.7	22.9	963	5	1.15	74.8	10
$DyAl_2$	63.9	22.0	951	5	1.10	73.7	10
GdNi	71	—	—	—	1.06	46.8	7
$GdNi_2$	72	19.6	861	7	0.83	40.1	7
$HoCo_2$	82	32.0	831	7	0.85	14.4	6
GdNiIn	94	13.9	1180	9	0.74	71.4	9

Source: Gschneidner, K.A., Jr. and Pecharsky, V.K., Magnetic refrigeration, Chapter 25 of *Intermetallic Compounds: Vol. 3, Principles and Practice*, John Wiley & Sons, New York, 2002.

over a wide temperature span. Based on the RCP(S) values listed in Table 18.1, $GdNi_2$ has a better potential as a refrigerant. The use of GdNi and $GdNi_2$ in the AMR cycle operating between 20 and 80 K for hydrogen liquefaction was discussed by Zimm et al. [17].

As discussed previously with the AMR concept, one way to utilize materials with larger spikes, but narrow transitions, is to use regenerators that contain multiple materials. This allows for the overall entropy change to remain relatively constant over a broad temperature span. However, regenerators composed of more than one material become difficult to model theoretically and numerically. With this in mind, several test devices and models have been developed with the intent of testing and understanding the properties of AMRs specifically in the 80–20 K range [22,24,26,27,31,33–36,60].

18.3.1.2 Tailoring Properties

Another useful property for a magnetic refrigerant is the ability for its Curie temperature to be shifted by altering its composition. This allows a material to be used in a specific temperature range that may lack quality refrigerants or ones that are easy to work with. Furthermore, a less expensive material might be able to replace the operating region of a more expensive refrigerant.

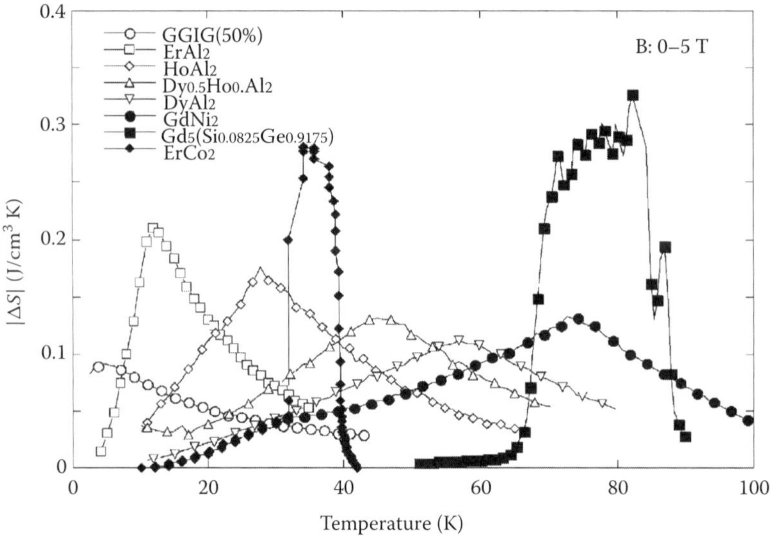

FIGURE 18.8
The entropy change of various materials over a range of temperatures. (From Sakata, K. et al., *J. Power Sources*, 159, 100, 2006. With permission.)

An example of a tunable alloy is $Gd_5(Si_xGe_{1-x})_4$—the ordering temperature increases with Si content. The $x = 0.0825$ composition is shown in Figure 18.8. (Note: the label in the legend is missing the subscript 4.) Matsumoto et al. characterized the magnetization and entropy change of this alloy with a range of compositions specifically for use in hydrogen liquefaction [61]. While large entropy changes were measured, they also noted hysteresis, which is an irreversibility that may have a significant impact on performance. Rare-earth alloys such as Gd_xEr_{1-x} are also tunable while maintaining good magnetocaloric properties.

One other avenue for creating a regenerator with desired properties is to produce a composite material that is a mixture of refrigerants [62–64]. This has been examined experimentally and theoretically for low- and high-temperature AMRs. One study of particular relevance for hydrogen liquefaction involved the creation of a composite material with a relatively uniform and broad magnetic entropy change below 80 K [63]. Creating a composite may create a better magnetic entropy curve; however, it will also create irreversibility due to heat transfer between particles.

18.3.2 AMR Performance

The consequences of material properties, regenerator composition, and operating parameters are discussed in a paper by Matsumoto et al. where an AMR hydrogen liquefaction unit operating from 77 to 20 K is analyzed numerically [26]. In the case of a single material AMR, the impacts of the working fluid's mass flow on the cooling capacity of the refrigerator are determined. It is found that both low and high mass flows reduce the cooling capacity while an intermediate value provides an optimum cooling capacity. The effect of using materials with different Curie temperatures is also assessed while maintaining the hot reservoir at 77 K and varying the overall temperature span of the bed. For materials with Curie temperatures of 60 K (Tc60), 70 K (Tc70), and 80 K (Tc80), each material appears beneficial in a different aspect. The Tc80 material provides the highest COP value, the Tc70 material provides the highest cooling capacity, and the Tc60 material has the smallest COP and cooling capacity.

Evaluating the results for the high cooling capacity of Tc70 reveals that this material exhibits the largest average entropy change within the 60–77 K region, which is similar to the operating temperature region. From this, it is deduced that the highest cooling capacity comes when a material provides the largest average entropy change within the operating region of the AMR. Therefore, although the Tc60 had the largest peak entropy change, its entropy change throughout the entire operating temperature range of the material bed was poor compared to the Tc70 material. This is consistent with the RC method of ranking materials.

These results also support the idea that using multiple materials for the regenerator in the AMR could be more beneficial than one material. Since the temperature of the refrigerant bed is distributed as shown in Figure 18.7, using multiple materials allows for each material in the bed to operate closer to its Curie point and the maximum entropy change. Matsumoto et al. analyzed the effect of two-layered AMR beds with different quantities of the Tc60 and Tc80. It was found that the highest cooling capacity is obtained when the material bed is 30% Tc60. The COP however decreased with an increasing percentage of the lower Curie temperature material giving rise to the notion that the high-temperature side plays an important part in reducing the irreversibilities of the cycle [26].

18.3.3 Refrigerant Selection

A large number of materials have been discovered that are potentially useful as magnetic refrigerants. Table 18.2 shows a list of some magnetic refrigerants and their Curie

TABLE 18.2

Curie Temperatures for a Variety of Magnetocaloric Materials

Material	T_C (K)	Reference	Material	T_C (K)	Reference
$GdMn_2$	300	[65]	$Gd_5(Si_{0.0825}Ge_{0.9175})$	75	[66]
Gd	293	[66]	GdN	65	[65]
$Gd_{0.9}Y_{0.1}$	281	[65]	$DyAl_2$	63	[66]
GdZn	268	[65]	$Gd_{025}Er_{0.75}$	60	[67]
$Gd_{0.73}Dy_{0.27}$	266	[66]	$Gd_{2.8}S_4$	58	[65]
$Gd_{0.8}Y_{0.2}$	254	[65]	$GdAg_{0.7}In_{0.3}$	57	[65]
$Gd_{0.36}Dy_{0.64}$	234	[66]	$Gd_{0.2}Er_{0.8}$	51	[67]
$GdZn_{0.85}In_{0.15}$	224	[65]	$Gd_{0.7}Th_{0.3}Al_2$	50	[65]
$Gd_{0.67}Y_{0.33}$	211	[65]	$Dy_{0.7}Er_{0.3}Al_2$	45	[66]
GdGa	200	[65]	$Gd_{2.76}S_4$	42	[65]
$Gd_{0.18}Dy_{0.82}$	195	[66]	$Gd_{0.139}Er_{0.861}Al_2$	40	[67]
$Gd_{0.7}La_{0.3}$	185	[65]	$GdNi_2$ (amorphous)	38	[65]
Dy	180	[66]	$Dy_{0.5}Er_{0.5}Al_2$	38	[66]
$GdAl_2$	153	[65]	$Gd_{0.1}Er_{0.9}Al_2$	33	[67]
$Gd_{0.8}Au_{0.2}$	150	[65]	$GdNi_5$	32	[65]
Gd_5	135	[66]	$Gd_{0.65}Th_{0.36}Al_2$	30	[65]
$Gd_{0.68}N_{0.32}$	125	[65]	$Gd_{2.73}S_4$	28	[65]
$Gd_5(Si_{0.225}Ge_{0.775})_4$	120	[66]	$GdAg_{0.8}In_{0.2}$	24	[65]
$GdAg_{0.5}In_{0.5}$	111	[65]	$Dy_{0.25}Er_{0.75}Al_2$	24	[66]
$Gd_{0.5}Ag_{0.5}$	100	[65]	$Gd_{2.71}S_4$	21	[65]
$Gd_{0.8}Th_{0.2}Al_2$	90	[65]	$Gd_{0.028}Er_{0.972}Al_2$	20	[67]
$Gd_5(Si_{0.15}Ge_{0.85})_4$	90	[66]	$ErAl_2$	12	[66]
$GdNi_2$ (crystalline)	81	[65]			

temperatures. This list is not exhaustive but provides a good spread of materials from room temperature to the hydrogen boiling point.

Table 18.2 demonstrates the need for a wide range of refrigerants required to liquefy hydrogen from room temperature. Even if one material could be used for every 20 K temperature span, a minimum of 14 different materials would still be required. When considering the different combinations of materials that are possible, the design of a full magnetic liquefaction process from 300 K is complicated. Alternatively, hydrogen can be precooled by liquid natural gas (LNG) (112 K) or LN_2 (77 K) resulting in the use of fewer materials to reach liquid hydrogen. Given the challenge of a magnetic device operating between room temperature and liquid hydrogen, most designs have focused on a hybrid liquefier that used a conventional cycle to LN_2 and a magnetic stage(s) operating below 80 K. These low-temperature magnetic devices will be discussed in the following section.

18.4 20–80 K Magnetic Refrigeration Devices

A number of cryogenic devices have been developed to specifically test materials and operating parameters in the 80–20 K range. While not meant to be liquefiers, these devices were usually part of a liquefier development path, and results were used to assist in determining preferred configurations for magnetic stages operating below a LN_2 upper stage. Other devices have been developed to operate below 20 K using LH_2 as an upper heat sink or even below 4 K with LHe as the hot reservoir. Near-room-temperature MR is another active application area with device development activities. These devices will not be discussed further; instead, we will focus on those devices specifically relevant to LH_2 liquefaction. A discussion of some important design considerations for magnetic devices follows.

18.4.1 Design Considerations

Some unique considerations arise when building devices utilizing magnetic fields. To maximize cooling power, magnetic fields are as high as practically possible. Constraints on this arise due to magnetic forces which, in some cases, also restrict operating frequency. Another consideration is sealing and one must decide if the regenerators are to have static or dynamic seals. Because superconducting magnets can be expensive, simple designs are preferred. The desire to use solenoid coils instead of complex windings leads to device geometries built around the magnet field shape and ability to access the high field region.

18.4.1.1 Forces

Structural forces between current-carrying elements are present in systems using coil magnets (Lorentz forces); however, these are ignored as they typically concern the magnet design. The following discussion focuses on the forces related to magnetic work input.

A magnetic cycle requires cyclic application of a magnetic field waveform. This can be done in three ways: (1) fix a material and then charge and discharge a coil, (2) use a static field and move the AMR in or through the field, and (3) fix the AMR and move the field. The last two nearly are equivalent in theory, but differences arise due to the ability to create a semicontinuous regenerator structure and the appearance of a relatively constant permeability in the field region. Charging a coil is undesirable because of frequency limitations and the additional power flows required due to the changing energy in the field. Some common configurations used are reciprocating AMRs and rotating structures housing regenerator beds. An analysis of force modeling for these two types of geometries was reported by Barclay et al. [68]. It was found that reciprocating geometries using single high field regions resulted in large unbalanced magnetic forces. This leads to an increase in structural and drive system ratings for these geometries. In the case of a rotating wheel regenerator structure, although there is good force balancing due to relatively constant permeability in the field region, there is however a radial force that must be designed for. An analysis of dynamic forces on a reciprocating AMR device is reported in [69]. Part of this work examined the use of a passive energy storage mechanism to reduce the size of a drive motor by damping out the reversing, unbalanced forces on the moving AMR structure. While reductions in torque were possible, the frequency of the torque waveform doubled and dynamic tuning would be needed.

A magnetic material in a field gradient will also have a net force. Magnetocaloric materials are subjected to magnetic forces when they are moved from regions of low field to high field, and the force depends upon the state of the material and the field variation in space. Forces on a refrigerant below the Curie temperature are higher than those above the Curie temperature. Regenerator matrix structures are often particle beds but can take various shapes. The matrix elements will be subjected to magnetic body forces that vary with location in the matrix and time. This can also create shear stresses and cyclic loading. Assuming the refrigerant is not too brittle, one of the main concerns with a particle bed is the movement and migration of particles. It is important that the solid matrix maintain its structure so that preferential flow paths do not arise and that the property distribution through the bed is as intended. Monolithic structures have advantages, but they can be difficult to make with large specific surface areas. One idea for creating a single structural element using particles is to bond them together [70]. This idea has been experimentally tested and does seem to work; however, because of the very small hydraulic diameters of the AMRs, creating the structure without affecting the homogenous nature of the void distribution is difficult. Another approach is to use sintered particle beds; however, maintaining material properties and eddy-current generation are concerns for this technique. Discrete microstructured elements can also be fabricated and used to create a larger regenerator and is a technique now being used.

18.4.1.2 Flow

The AMR concept requires effective regenerators in which the fluid–solid heat transfer occurs with small temperature differences. This must be balanced against pumping power requirements due to pressure drops through the regenerator and remaining fluid system. The regenerator is the element separating warm and cold temperature regions and must act as a thermal resistance between the reservoirs. These requirements come with tradeoffs as good convective heat transfer often comes with small hydraulic diameters, which lead to higher pressure drops. No matter how oscillating flow is created and controlled, there will be a need for good sealing around the regenerator structure.

As with magnetic forces, the choices made in creating a magnetic field waveform for the refrigerant are linked to sealing requirements. Some designs can use static seals while others need dynamic seals. The latter are, of course, more challenging. Ideally, the void space in a regenerator is small and fluid flows uniformly through the matrix. Preferential channeling or leaks around the regenerator between the temperature reservoirs should be prevented. While rotating regenerator structures simplify the magnetic forces and structural and drive system requirements, they usually have dynamic seals between the moving regenerators and the stationary housing. A good dynamic seal may impose additional losses due to friction. Dynamic seals must also be able to operate over long periods at cryogenic temperatures.

When using magnetic refrigeration stages for liquefaction, ideally, sensible heat is removed from hydrogen at a number of intermediate temperatures. However, sensible cooling can also occur using a portion of the cold AMR heat transfer fluid bypassing the regenerators and, instead, absorbing energy from the hydrogen stream through a heat exchanger [71]. Not only does this reduce entropy generation in cooling the hydrogen gas, it can also help to thermally balance the regenerator. This latter effect arises due to the difference in heat capacity of the refrigerant between high and low fields. As will be seen later, bypass flow configurations have been implemented in the design of staged magnetic liquefiers.

18.4.1.3 Parasitic Magnet Load

An unavoidable parasitic load for magnetic refrigeration systems using superconducting magnets arises from the need to maintain the magnet at operating temperature. The needed cooling power can be relatively small as magnets can be operated in persistent mode, which means the power supply for the field is no longer needed. The parasitic load is then dominated by heat leaks to the magnet. The impacts of magnet cooling have been studied and results show that the impact on efficiency depends upon the operating temperature of the magnetic refrigerator and the cooling power of the device [72]. For low temperatures and high cooling powers relative to the magnet cooling requirement—as with hydrogen liquefaction applications—the impact of magnet cooling requirements can be very small. This study is now somewhat dated and had assumed a superconducting magnet operating at 4.2 K. Given recent progress using high-temperature superconductors to build magnets, the impacts of magnet parasitic load may be even less of a concern, particularly for hydrogen liquefaction.

18.4.2 Cryogenic AMR Devices

Although designs for AMR devices for hydrogen liquefaction include reciprocating and rotary configurations, to date, all experimental devices operating between 80 and 20 K have been reciprocating. Experimental and theoretical results for some of these devices will be discussed.

The development of a test device designed to study magnetic refrigeration between 77 and 20 K using a lead recuperator was discussed by Matsumoto et al. [33]. This device did not use the refrigerant itself as a regenerator and, unlike the AMR cycle, the entire mass of refrigerant underwent the same cycle. This piston-type reciprocating device used sintered $DyAl_{2.2}$ as a refrigerant. In the design, the regenerator was moved by a piston throughout the cycle, transferring heat to and from the magnetic material. The refrigeration unit was found to produce a temperature span of between 50.3 and 58.7 K in the regenerator and 48.3 and 59.1 K in the magnetic refrigerant for a 5 T magnetic field and a cycle of 300 s. The primary source of losses noted in the experiment were due to poor heat transfer between the two solid materials of the regenerator and the magnetic refrigerant.

Degregoria et al. [34] describe an AMR refrigeration apparatus, seen in Figure 18.9, using an immersion-cooled superconducting magnet with fields up to 7 T. A novel feature of this design is that the magnet reciprocates while the regenerator beds are stationary. This allows for static sealing and simple flow system design.

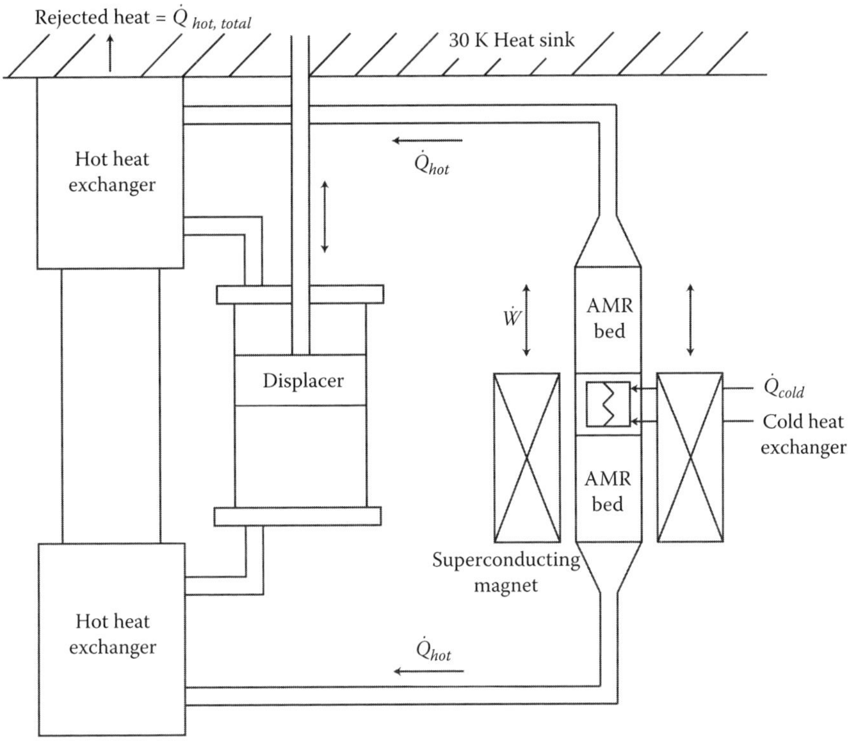

FIGURE 18.9
Component schematic for the Degregoria et al. AMR test apparatus. (From Degregoria, A.J. et al., *Adv. Cry. Eng.*, 37, 875, 1992.)

AMR performance was investigated in the range of 10–20 K using $Er_xGd_{1-x}Al_2$ and in $GdNi_2$ in the 40–80 K range. Tests were performed using fields of 5 and 7 T and results were compared to numerical predictions. The regenerator matrix was crushed irregular particles. Numerical results overestimated the actual test data; however, a zero-load temperature span of ~44 K was achieved at 5 T using $GdNi_2$. Errors in the model predictions were suggested to be a combination of poor correlations and uncertain material properties.

Wang et al. [35] describe experimental results using an AMR test apparatus described in [34] but modified to use a LN_2 heat sink (seen in Figure 18.10) instead of a separate cryocooler. This modification allowed the heat rejection to be quantified by monitoring the boil-off rate. Heat loads could be set using a resistive heater in the cold section.

A large number of tests were performed using $GdNi_2$ with a focus on efficiency measures of the intrinsic AMR cycle, that is, using the magnetic work and pumping work across the regenerators. The sensitivity of efficiency to fluid displaced through the regenerator, temperature span, and cooling load was also measured. The intrinsic efficiency was found to be higher than 50% of Carnot over a range of operating conditions. This work suggests that with appropriate materials, the AMR cycle can produce high efficiency.

Zimm et al. [36] describe a reciprocating test apparatus using a stationary, 7 T superconducting magnet. One of the goals of this device shown in Figure 18.11 was to increase the amount of refrigerant significantly from previous tests and maximize the use of the high field volume. A total of four kilograms of $GdNi_2$ was used in the tests. The use of flex hoses to couple the reciprocating regenerator assembly to the heat exchangers resulted in a

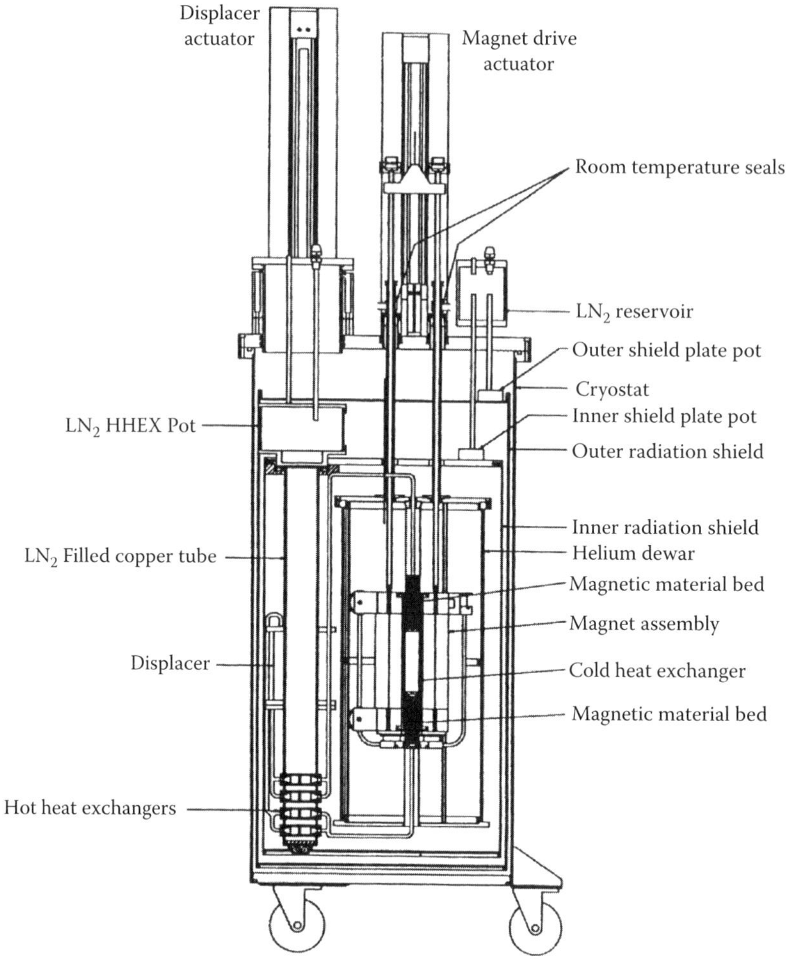

FIGURE 18.10
Cross-sectional view of the AMR test apparatus. (With kind permission from Springer Science+Business Media B.V.: *Cryocoolers*, Experimental results of an efficient active magnetic regenerator refrigerator, 8, 1995, 665–676, Wang, A.A., Johnson, J.W., Niemi, R.W., Sternberg, A.A., and Zimm, C.B.)

significant dead volume in the flow circuit. Numerical predictions suggested that a much larger cooling power could be obtained than was experimentally achieved and the flex hose volume was identified as a serious problem. Inserting Teflon liners to reduce the dead volume nearly doubled the cooling power.

Another reciprocating test apparatus was created by Rowe et al. with the intent of testing different regenerator materials and configurations with little turnover time [27]. This unit has the ability to test refrigerants over a wide range of temperatures to determine their effectiveness and temperature spans in relation to different magnetic fields. Various refrigerant geometries can be used in single or multimaterial regenerator beds. The device uses small quantities of materials, which reduces material costs and design forces and can operate at frequencies as high as 1 Hz. Figure 18.12 shows a schematic of the apparatus configured for operation with LN_2.

Unlike the previous devices, this apparatus was designed to allow testing from room temperature down to 20 K. Most of the reported results using the device are for

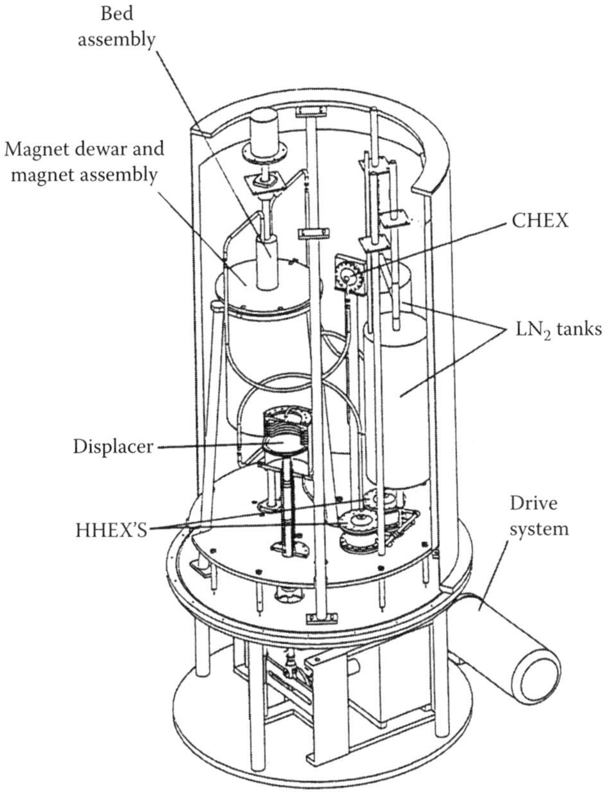

FIGURE 18.11
3D section view of the AMR device. (With kind permission from Springer Science+Business Media B.V.: *Adv. Cry. Eng.*, Test results on a 50 K magnetic refrigerator, 41, 1996, 1675–1681, Zimm, C.B., Johnson, J.W., and Murphy, R.W.)

near-room-temperature operation; however, tests have been carried out using $GdNi_2$, $DyAl_2$, and $Gd_5Si_{0.33}Ge_{3.67}$ alone and in layered structures. Results using $Gd_5Si_{0.33}Ge_{3.67}$ were reported in [23]. Due to an inability to maintain the hot heat sink much below 90 K, the maximum no-load span was 8 K at 5 T. Thus, the operating point was far above the temperature of the peak MCE (~75 K). Near-room-temperature tests using a three-material regenerator and 5 T produced a peak no-load temperature span of ~85 K.

Experience gained with test devices as well as model results has been used to guide the design of magnetic liquefiers of hydrogen. A limited amount of work on magnetic liquefaction is currently being performed in Japan, the United States, and Canada. Some liquefier concepts will be discussed in the following section.

18.5 Magnetic Liquefier Concepts

A wide variety of research has occurred in the field of magnetic refrigeration to determine preferred system configurations for hydrogen liquefaction. Although the AMR cycle is at the heart of system designs, many process configurations are still being assessed.

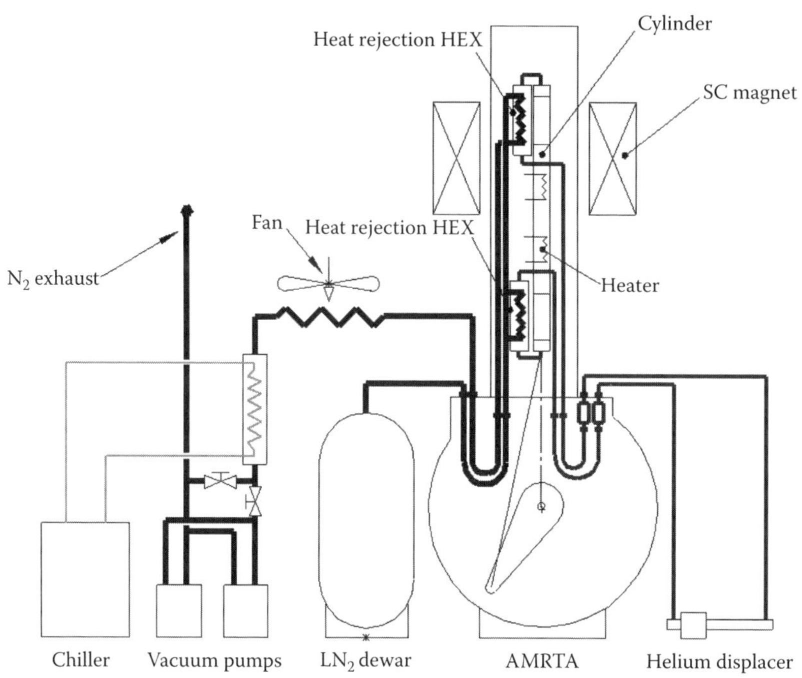

FIGURE 18.12
Picture and schematic for the AMRTA using LN_2 as the hot reservoir. (From Rowe, A. and Tura, A., *AIP Conf. Proc.*, 985, 1292, 2008. With permission.)

These include the number of AMR stages, how they are cascaded together, the type and number of materials to use, and the best operating parameters. Some processes utilize only magnetic cycles to span the 300–20 K temperature gap, while others use LN_2 or LNG to precool hydrogen before being liquefied magnetically. These are areas being investigated to improve the overall efficiency and cost-effectiveness of magnetic hydrogen liquefaction units.

18.5.1 Process Configurations

This section describes process flow configurations proposed for liquefying hydrogen with magnetic cycles. Recently, three different configurations have been analyzed by Utaki et al. [25] using the numerical model by Engelbrecht [73]. All three cases use series cascades of AMR systems with different precooling temperatures. Each configuration uses a magnetic Carnot stage (CMR) for the final removal of latent heat and liquefaction [20]. It is difficult to ascertain, but it appears that bypass flow is assumed in some stages. This study also looks at using hydrogen, propane, and glycol/water mixture as heat transfer fluids.

The starting temperatures for the three cases are 300, 77, and 120 K. Starting points of 77 and 120 K are established by precooling hydrogen with LN_2 and LNG, respectively. The model results for all cases assume a constant magnetic entropy change of 86.2 mJ/cm^3 K at 5 T independent of temperature. This assumes an ideal magnetocaloric material in an ideally layered regenerator. Making this assumption removes the dependency on available materials, allowing for a comparison between the fundamental configurations of each of the three cases. The heat emitted by the conversion from ortho-hydrogen to para-hydrogen

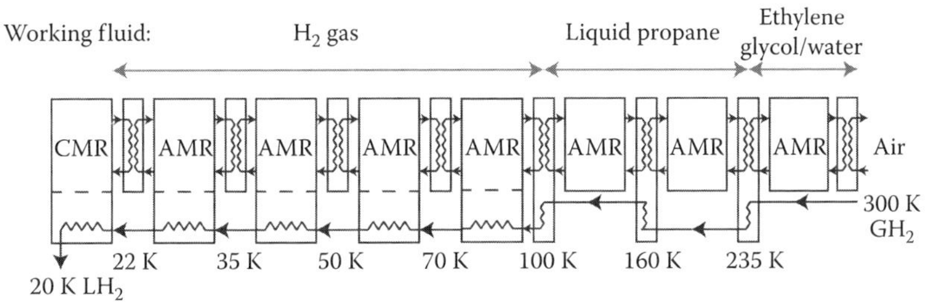

FIGURE 18.13
The aforementioned magnetic refrigerator is case 1, which uses 7 AMR stages and one CMR stage. (From Utaki, T. et al., *Cryocooler*, 14, 645, 2007.)

is also taken into account in the model and is included as a heat load during each of the liquefaction processes. To compare the work, COP, and liquefaction efficiencies, all cases produce the same amount of hydrogen and use optimized regenerator beds and mass flows for each stage.

The first case shown in Figure 18.13 utilizes seven AMR stages from room temperature to 22 K where a CMR stage is used to liquefy hydrogen from 22 to 20 K. The working fluids for the AMR stages are a 56% ethylene glycol/water mix from 300 to 235 K, liquid propane from 235 to 95 K, and gaseous hydrogen from 95 to 22 K. Ethylene glycol/water and liquid propane are used as these fluids were found to have a significantly higher COP per stage. The use of hydrogen as the heat transfer fluid in an AMR offers some potential advantages over helium; however, the ability of known refrigerants to operate in an H_2 environment is unknown.

The second case uses LN_2 to precool hydrogen to 77 K. From this point, hydrogen is cooled to 22 K through three AMR stages and then liquefied using a CMR. In the final case, gaseous hydrogen is first cooled to 120 K by LNG and then run through five AMR stages and a CMR liquefaction stage. Both cases 2 and 3 use hydrogen as the working fluid in the system. Process flow diagrams for cases 2 and 3 are provided in Figure 18.14.

Since the appeal of magnetic refrigeration is its theoretically low energy cost compared to gas systems, the most important results from the three separate cases are the work input and liquefaction efficiency. The work input is described as the work required to liquefy hydrogen from room temperature. In the case where precooling is performed with LN_2, additional energy is added to the AMR and CMR work totals to account for the work input to the nitrogen liquefier. The efficiency of each process is found by dividing the theoretical liquefaction energy by the actual energy required. For liquid hydrogen, which has been converted to ortho–para equilibrium, this theoretical work is 3.92 kWh/kg H_2 [39]. In the paper by Utaki et al., a 0.01 ton H_2/day unit is modeled so the minimum power required is then 1.63 kW. All three units are assumed to produce the same amount of product, making the CMR work identical in all cases. A Carnot efficiency of 50% is assumed based upon previous tests, making the power required for the CMR stage 1.30 kW [21]. Table 18.3 summarizes the work requirements for each part of the three cases in addition to the liquefaction efficiency of hydrogen.

From Table 18.3, case 2 shows the highest liquefaction efficiency at 46.5% followed closely by case 3 at 46.3%. Factors to take into account include the added capital cost for the additional AMR units in the 5-stage process and increased irreversibilities that may occur in using a larger number of stages. Another result to note is the decreased amount of work in case 1 when the number of AMR stages is increased from seven to nine. The added stages allow for each AMR to cover smaller temperature spans resulting in an overall more efficient system.

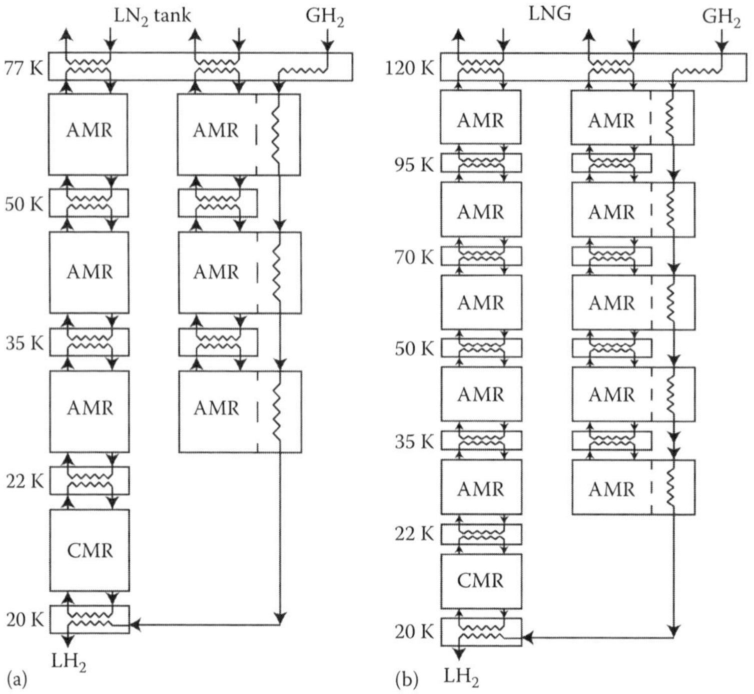

FIGURE 18.14
Case 2 (a) contains 3 AMR stages with a CMR stage and case 3 (b) contains 5 AMR stages with a CMR stage. (From Utaki, T. et al., *Cryocooler*, 14, 645, 2007.)

TABLE 18.3
Summary of the Input Work and Liquefaction Efficiency for Cases 1–3

	Case 1 (7 Stages)	Case 1 (8 Stages)	Case 1 (9 Stages)	Case 2	Case 3
Precooling work (kW)	0	0	0	1.49	2.23
AMR work (kW)	13.09	11.08	9.90	0.72	
CMR work (kW)	1.30	1.30	1.30	1.30	1.30
Total work (kW)	14.39	12.38	11.20	3.51	3.53
Liquefaction efficiency (%)	11.4	13.2	14.6	46.5	46.3

Source: Utaki, T. et al., *Cryocooler*, 14, 645, 2007.

Other process configurations using series and parallel AMR stages have been identified for efficient hydrogen liquefaction [50]. While multistaged processes can reduce operating costs by better matching the cooling requirements of hydrogen, capital costs may ultimately limit the number of stages. Some of the more detailed magnetic liquefier designs reported in the literature are discussed in the next section.

18.5.2 Liquefier Designs

To date, a small number of magnetic liquefier designs have been developed and analyzed in detail. These devices are summarized here and provide a background for the current development activities on magnetic liquefaction discussed in the final section.

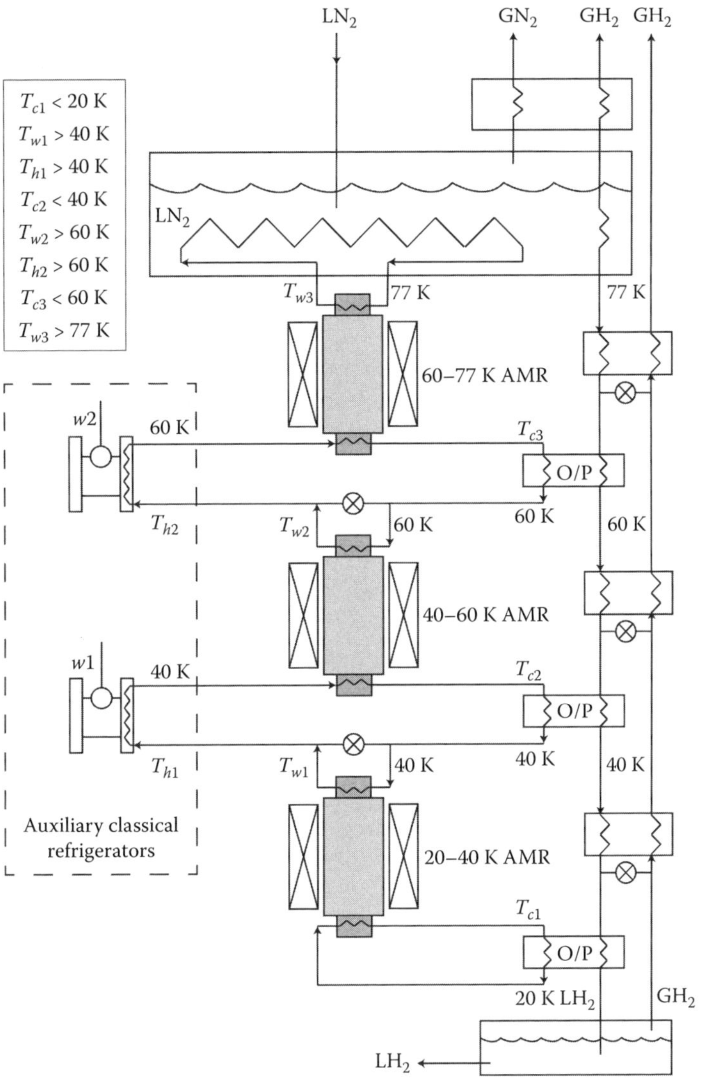

FIGURE 18.15
The aforementioned configuration represents the three-stage AMR liquefier used in the exergy analysis. (From Zhang, L. et al., *Cryogenics*, 33, 667, 1993.)

An early design operating with a LN_2 precooling system was analyzed by Zhang et al. [23,74]. The device depicted in Figure 18.15 uses three magnetic stages in series with ortho–para conversion assumed as a load for each stage. In the analysis, the cooling power and exergetic equivalent losses of the process are examined by altering various system parameters including material type, bed size, and temperature. Each regenerator stage in the design is also assumed to be made of a mixture of alloys from the family $Gd_{1-x}Er_xAl_2$, while the heat transfer fluid in the AMR system is helium at 10 atm. This fluid is then used to cool a 1 atm hydrogen stream through external heat exchangers. Entropy generation and exergy balances are used to determine the impacts of inefficiencies in the regenerators, auxiliary refrigeration units, and external heat exchangers.

Several design features were suggested from the study to improve the efficiency of liquid hydrogen production using this specific configuration and operating space. The first suggestion is to use a large magnetic material bed. A larger material bed reduces the exergy losses relative to the cooling capacity produced. The material bed should also be a shape that maximizes heat transfer to the working fluid. Increasing the mass flow of the helium was identified as a way of reducing losses by decreasing the temperature difference between the working fluid and the stage temperature. This causes the cooling capacity to increase while the exergy destruction remains relatively constant. Alternately, the cycle's frequency can also be increased to achieve the same effect. A novel finding from this research is that using a high magnetic field may not always be advantageous to the overall system.

Following the test results for the AMR reported by Degregoria and described earlier [34], Janda et al. designed a process to liquefy hydrogen at a rate 0.01 ton H_2/day with the intention to eventually scale this up to 1 ton/day. The device uses two stages to span from 80 to 20 K and is shown in a simplified form in Figure 18.16. Helium is used as the working fluid in the AMRR while precooling is again performed using LN_2. The previously tested material $GdNi_2$ is hypothesized to span from 80 to 40 K under a 7 T magnetic field, while GdPd was to be used in the 40 to 20 K stage with a 5 T field. Although this specific unit was never built, it provided a foundation for a number of studies [18,19,23,74]. The most recent activity related to this device was an optimization study by Zhang et al. [19], which studied the impacts of particle size, material bed lengths, and the interstage temperature on the overall process efficiency.

A novel feature of the design is the use of unbalanced flow in the regenerators. This may actually benefit the AMR cycle efficiency due to the variation in specific heat of the magnetic material. If the materials are largely operating below their Curie temperatures,

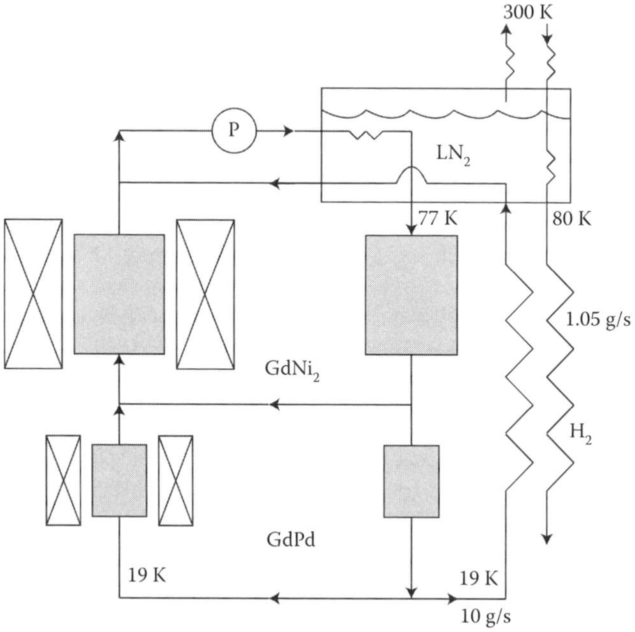

FIGURE 18.16
Simplified AMR schematic by Zhang et al. of Janda's two-stage hydrogen liquefier. (From Zhang, L. et al., *Cryogenics*, 40, 269, 2000.)

the specific heat in the demagnetized state is larger than when magnetized. For a local Brayton-type process to occur with this varying specific heat, the thermal mass of fluid should be different for high and low field blows. The optimized results indicated that the bypass ratio for the low-temperature stage was on the order of 12%, whereas for the high-temperature stage, it was ~2.5%. The optimal interstage temperature of the process was found to be 41 K, and the liquefaction efficiency was determined to be ~20%. This is a good number given the scale of the device.

As part of the Japanese World Energy Network project phase II, the potential of magnetic refrigeration for large-scale liquefaction at high efficiencies was investigated [66]. This study began by analyzing a 10 kg/day laboratory prototype. The schematic of the device is shown in Figure 18.17. One of the main design features is the use of a belt carrying discrete regenerator beds. The reason for this is so that an inexpensive solenoidal superconducting magnet could be used. The belt rotates the regenerators through the high and low field regions where manifolds are used to regenerate pressurized helium flowing through the heat exchange system. This type of design has the benefits of a rotary configuration in terms of force balancing; however, sealing and controlling flow is not a trivial problem. The design calculations assumed operation between room temperature and 20 K with six stages in parallel and the need for 14 different magnetocaloric materials. Overall, liquefaction efficiency was estimated to be ~50%.

Currently, research and development activities are occurring in Japan, Canada, and the United States. The US DOE is funding the development of a magnetic liquefier prototype by Heracles Energy [75]. In Canada, magnetic liquefaction is being studied by our group as part of a strategic research network on hydrogen (H2Can) funded by the Natural Sciences and Engineering Research Council of Canada (NSERC).

In Japan, recent work on hydrogen liquefaction was reported by Matsumoto et al. [21]. Their design is based upon the AMRR system of Utaki (discussed earlier) where hydrogen is precooled by LNG then run through three AMR stages and a final CMR liquefaction stage.

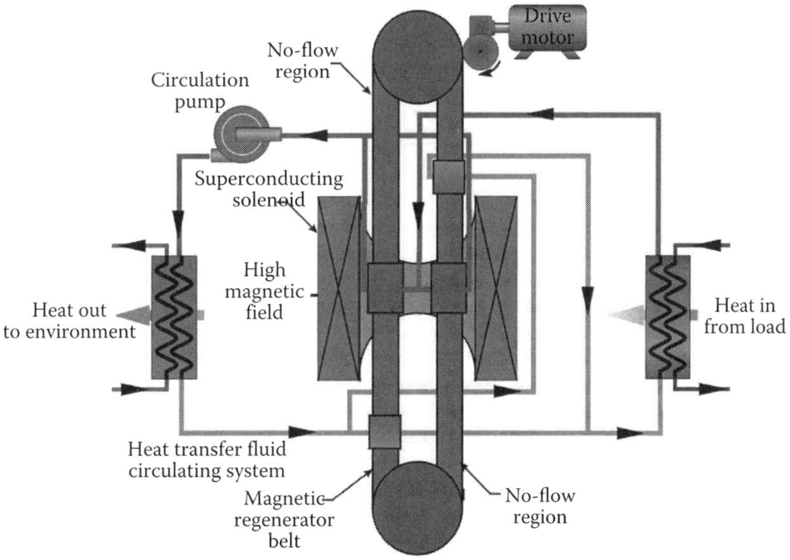

FIGURE 18.17
Schematic of an AMRR device using a belt of regenerators as the magnetocaloric material. (From Iwasaki, W., *J. Hydrogen Energy*, 28, 559, 2003. With permission.)

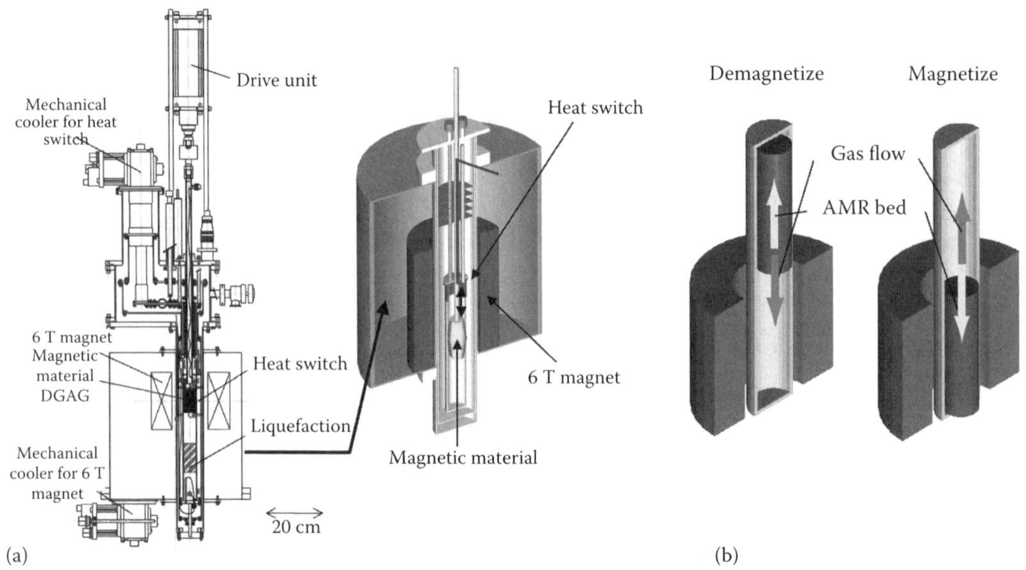

FIGURE 18.18
Cross-sectional view of the magnetic liquefaction test apparatus and CMR (a) and AMR (b) operation diagrams. (From Matsumoto, K., Kondo, T., Yoshioka, S., Kamiya, K., and Numazawa, T., Magnetic refrigerator for hydrogen liquefaction, *J. Phys. Conf. Ser.*, 150, 012028, 2009. With permission from IOP Publishing Ltd.)

The starting point of 112 K was chosen due to the abundance of LNG plants in Japan. Because the AMR stages of this device have already been described, this final section will focus on the CMR stage.

The CMR converts gaseous hydrogen at 22 K to liquid hydrogen at 20 K. This is done through condensing hydrogen directly onto the refrigerant itself. Because most refrigerants are not hydrogen resistant, a ceramic material is used for this stage. This CMR refrigerant is $Dy_{2.4}Gd_{0.6}Al_5O_{12}$, or DGAG, and the thermal efficiency as compared to the ideal Carnot cycle is between 50% and 60% and the liquefaction efficiency is ~90% [21]. The active magnetic refrigerator stages have not yet been fixed in terms of the number of stages and materials used, but the entire process is to be examined in the near future from either a LN_2 or LNG starting point. A cross-sectional view of the test apparatus and CMR and AMR stages are seen in Figure 18.18.

18.6 Summary

When Brown used gadolinium as a magnetic refrigerant in 1976, the seed for magnetic liquefaction of hydrogen was planted. Thirty-five years later, a broad array of research has been performed, improving our understanding of magnetocaloric materials, regenerator design, magnetic devices, and liquefaction processes. Studies addressing design principles, test devices, and magnetic refrigerants have proven that high efficiencies are possible; however, the development of commercial devices is still a challenge. The AMR cycle is the preferred method of generating cooling over significant temperature spans, but the thermodynamic implications of using multiple materials in a single AMR are still being developed. Highly nonlinear material properties and a lack of analytic expressions describing AMR behavior make design a challenge. Devices operating with high magnetic fields must also accommodate for magnetic forces. Depending upon the chosen device geometry, these forces can be high and reversing and may lead to eddy-current heating and other losses. While numerous magnetic refrigerants have been identified, there appears to be a trade-off between the maximum MCE and the width of the magnetic transition. Layered AMR geometries may partially solve this problem, but there are limited experimental results to date.

With all of those mentioned earlier in mind, it should also be pointed out that the cumulative amount of work on magnetic liquefier development is relatively small compared to the effort that has gone into gas cycles. Gas compression and expansion technologies are mature with a diverse set of component manufacturers serving the market. As a result, we are likely to see only modest gains in conventional liquefier efficiency. The capital costs of large-scale conventional systems could easily decrease if repeated builds of standardized designs were to occur. Magnetic refrigeration provides an alternative process that may reduce the costs of liquid hydrogen at device scales that are uneconomical with conventional cycles. Higher efficiencies at smaller scales would allow liquefaction systems to be deployed in a more distributed manner instead of at large central facilities as we currently see. The preferred geometries and process configurations of magnetic liquefaction systems are still unknown, but with continued advances in high-temperature superconductors, materials, and our understanding of magnetic cycles, the prospects for commercialization of magnetic liquefiers are improving.

References

1. A. Smith, Who discovered the magnetocaloric effect? *Eur. Phys. J. H.*, 2013;38(4);507–517.
2. P. Weiss, A. Piccard, Sur un nouveau phénomène magnétocalorique. *Compt. Rend. Ac. Sci.*, 1918;166;352.
3. P. Debye, Einige Bemerkungen zur Magnetisierung bei tiefer Temperatur. *Ann. Phys.*, 1926;81;1154–1160.
4. W.F. Giauque, A thermodynamic treatment of certain magnetic effects. A proposed method of producing temperatures considerably below 18 absolute. *J. Am. Chem. Soc.*, 1927;49;1864–1870.
5. T. Hashimoto, T. Numasawa, M. Shino, T. Okada, Magnetic refrigeration in the temperature range from 10 K to room temperature: The ferromagnetic refrigerants. *Cryogenics*, 1981;21;647–653.
6. G.V. Brown, Magnetic heat pumping near room temperature. *J. Appl. Phys.*, 1976;47;3673.
7. W.A. Steyert, Magnetic refrigerators for use at room temperature and below. *J. de Physique Collogue*, 1978;39;1598.
8. J.A. Barclay, W.A. Steyert Jr., Active magnetic regenerator. United States Patent 4332135, 1982.
9. T. Hashimoto, T. Ikegami, T. Okada, Magnetic refrigeration in the temperature range from 10 K to room temperature: The ferromagnetic refrigerants. *Cryogenics*, 1981;21;647–653.
10. T. Hashimoto, New application of complex magnetic materials to the magnetic refrigerant in an Ericsson magnetic refrigerator. *J. Appl. Phys.*, 1987;62;3873–3878.
11. T. Hashimoto, Recent investigations on refrigerants in magnetic refrigeration. *Adv. Cry. Eng.*, 1986;32;261–270.
12. C.B. Zimm, W.F. Stewart, J.A. Barclay, C.K. Campenni, Measured properties if GdNi for magnetic refrigeration. *Adv. Cry. Eng.*, 1988;33;791–798.

13. J.A. Barclay, W.C. Overton Jr, W.F. Stewart, C.B. Zimm, Magnetic refrigeration for efficient cryogen liquefaction. Department of Energy, Division of Advanced Energy Projects, Final Report (1982–1986).
14. J.A. Waynert, A.J. Degregoria, R.W. Foster, J.A. Barclay, Evaluation of industrial magnetic heat pump/refrigerator concepts that utilize superconducting magnets, Argonne National Laboratory Report, ANL-89/23 June 1989.
15. J.A. Barclay, Prospects for magnetic liquefaction of hydrogen. *Proceedings of the XVIII International Congress of Refrigeration*, Montreal, Quebec, Canada, 1991, pp. 297–305.
16. D. Janda, A.J. Degregoria, J. Johnson, S. Kral, Design of an active magnetic regenerative hydrogen liquefier. *Adv. Cry. Eng.*, 1992;37;891–899.
17. C.B. Zimm, E.M. Ludeman, M.C. Severson, T.A. Henning, Materials for regenerative magnetic cooling. *Adv. Cry. Eng.*, 1992;37;883–891.
18. A.J. Degregoria, Modeling the active magnetic refrigerator. *Adv. Cry. Eng.*, 1992;37;867.
19. L. Zhang, S.A. Sherif, A.J. Degregoria, C.B. Zimm, T.N., Verziroglu, Design optimization of a 0.1 ton/day active magnetic regenerative H2 liquefier. *Cryogenics*, 2000;40;269–278.
20. K. Kamiya, H. Takahashi, T. Numazawa, H. Nozawa, T. Yanagitani, Hydrogen liquefaction by magnetic refrigeration. *Proceedings of the International Cryocooler Conference*, Boulder, CO, 2007.
21. K. Matsumoto, T. Kondo, S. Yoshioka, K. Kamiya, T. Numazawa, Magnetic refrigerator for hydrogen liquefaction. *J. Phys. Conf. Ser.*, 2009;150;012028.
22. K. Matsumoto, T. Hashimoto, Thermodynamic analysis of magnetically active regenerator from 30 to 70 K with a Brayton-like cycle. *Cryogenics*, 1990;30;840–845.
23. L. Zhang, S.A. Sherif, T.N. Verziroglu, J.W. Sheffield, Second law analysis of active magnetic regenerative hydrogen liquefiers. *Cryogenics*, 1993;33;667–674.
24. J. Dikeos, A.M. Rowe, A. Tura, Numerical analysis of an active magnetic regenerator (AMR) refrigeration cycle. *Adv. Cry. Eng.*, 2006;823;993–1000.
25. T. Utaki, K. Kamiya, T. Nakagawa, T.A. Tamamoto, T. Numazawa, Research on a magnetic refrigeration cycle for hydrogen liquefaction. *Cryocooler*, 2007;14;645.
26. K. Matsumoto, T. Kondo, T. Numazawa, Numerical analysis of active magnetic regenerators for hydrogen magnetic refrigeration between 20 and 77 K. *Cryogenics*, 2010;doi: 10.1016/j.cryogenics.2010.06.003.
27. J.A. Barclay, The theory of an active magnetic regenerative refrigerator. NASA Report, NASA-CP-2287, 1983.
28. R.W. Murphy, F.C. Chen, V.C. Mei, G.L. Chen, Analysis of magnetic refrigeration cycles. *Proceedings of the 26th Intersociety Energy Conversion Conference*, Boston, MA, 1991, pp. 482–486.
29. C. Carpetis, Numerical study of magnetic refrigeration including consideration of magnetic nanocomposites. *Proceedings of the XVIII International Congress of Refrigeration*, Montreal, Quebec, Canada, 1991, pp. 331–337.
30. S. Schuricht, A.J. Degregoria, C.B. Zimm, The effects of a layered bed on active magnetic regenerator performance. *Proceedings of the International Cryocooler Conference*, Santa Fe, NM, 1992.
31. L. Zhang, S.A. Sherif, T.N. Veziroglu, J.W. Sheffield, Performance analysis of reciprocating magnetic liquefiers. *Int. J. Hydrogen Energy*, 1994;12;945–956.
32. C. Carpetis, An assessment of the efficiency and refrigeration power of magnetic refrigerators with ferromagnetic refrigerants. *Adv. Cry. Eng.*, 1994;39;1407–1415.
33. K. Matsumoto, T. Ito, T. Hashimoto, An Ericsson magnetic refrigerator for low temperature. *Adv. Cry. Eng.*, 1988;33;743–750.
34. A.J. Degregoria, L.J. Feuling, J.F. Laatsch, J.R. Rowe, J.R. Trueblood, Test results of an active magnetic regenerative refrigerator. *Adv. Cry. Eng.*, 1992;37;875.
35. A.A. Wang, J.W. Johnson, R.W. Niemi, A.A. Sternberg, C.B. Zimm, Experimental results of an efficient active magnetic regenerator refrigerator, *Cryocoolers*, 1995;8;665–676.
36. C.B. Zimm, J.W. Johnson, R.W. Murphy, Test results on a 50 K magnetic refrigerator. *Adv. Cry. Eng.*, 1996;41;1675–1681.
37. A. Rowe, A. Tura, Cryogenic testing of an active magnetic regenerative refrigerator. *AIP Conf. Proc.*, 2008;985;1292–1298.

38. J.A. Barclay, Magnetic refrigeration: A review of a developing technology. *Adv. Cry. Eng.*, 1988;33;719–731.
39. W. Peschka, *Liquid Hydrogen: Fuel of the Future*, Springer-Wein, New York, 1992.
40. V.K. Pecharsky, K.A. Gschnieder Jr., Y. Mudryyk, D. Paudyal, Making the most of the magnetic and lattice entropy changes. *J. Magn. Magn. Mater.*, 2009;321;3541–3547.
41. K.A. Gschneidner Jr., V.K. Pecharsky, A.O. Tsokol, Recent developments in magnetocaloric materials. *Rep. Prog. Phys.*, 2005;68;1479–1539.
42. H.B. Callen, *Thermodynamics and an Introduction to Thermostatistics*, 2nd edn., John Wiley & Sons, New York, 1985.
43. A. Kitanovski, P.W. Egolf, Thermodynamics of magnetic refrigeration. *Int. J. Refrig.* 2006;29;3–21.
44. G. Bisio, G. Rubatto, P. Schiapparelli, Magnetic systems depending on three or two variables; thermodynamic analysis and some existing and possible applications. *Energy Conv. Manage.*, 1999;40;1267–1286.
45. T.H.K. Barron, Gruneisen parameters for the equation of state of solids. *Ann. Phys.*, 1957;1;77–90.
46. V.K. Pecharsky, K.A. Gschnieder Jr., Magnetocaloric effect and magnetic refrigeration. *J. Magn. Magn. Mater.*, 1999;200;44–56.
47. A.M. Rowe, J.A. Barclay, Ideal magnetocaloric effect for active magnetic regenerators. *J. Appl. Phys.*, 2003;93;1672–1676.
48. C.R. Cross, J.A. Barclay, A.J. Degregoria, S.R. Jaeger, J.W. Johnson, Optimal temperature-entropy curves for magnetic refrigeration. *Adv. Cry. Eng.*, 1988;33;767.
49. J.L. Hall, C.E. Reid, I.G. Spearing, J.A. Barclay, Thermodynamic considerations for the design of active magnetic regenerative refrigerators. *Adv. Cry. Eng.*, 1996;41;1653.
50. J.A. Barclay et al. Apparatus and methods for cooling and liquefying a fluid using a magnetic refrigeration. United States Patent 6467274, 2002.
51. J.A. Barclay, S. Sarangi, Selection of regenerator geometry for magnetic refrigerator applications. In *Cryogenic Processes and Equipment*, P.J. Kerney (ed), The American Society of Mechanical Engineers, New York, 1984.
52. P. Li, M. Gong, J. Wu, Geometric optimization of an active magnetic regenerative refrigerator via second-law analysis. *J. Appl. Phys.*, 2008;104;103536.
53. M. Ogawa, R. Li, T. Hashimoto, Thermal conductivities of magnetic intermetallic compounds for cryogenic regenerator. *Cryogenics*, 1991;41;405–410.
54. K. Engelbrecht et al., The effects of internal temperature gradients on regenerator performance. *Trans. ASME*, 2006;128;1060.
55. P. Kittel. Eddy current heating in magnetic refrigerators. *Adv. Cry. Eng.*, 1990;35;1141–1148.
56. A. Tura, J. Roszmann, J. Dikeos, A. Rowe. Cryogenic active magnetic regenerator test apparatus. *Adv. Cry. Eng.*, 2006;51;985–992.
57. M.E. Wood, W.H. Potter, General analysis of magnetic refrigeration and its optimization using a new concept: Maximization of refrigerant capacity. *Cryogenics*, 1985;25;667–683.
58. K.A. Gschneidner Jr., V.K. Pecharsky. Magnetic refrigeration. Chapter 25 of *Intermetallic Compounds: Vol. 3, Principles and Practice*, J.H. Westbrook and R.L. Fleischer (eds), John Wiley & Sons, New York, 2002.
59. K. Sakata, E. Mizutani, K. Fukuda, A review of topics in hydrogen-related innovative materials in Japan. *J. Power Sources*, 2006;159;100–106.
60. J.A. Barclay, W.F. Stewart, W.C. Overton, R. Chesebrough, M. McCray, D. McMillan, An apparatus to determine the heat capacity and thermal conductivity of a material from 1 to 300 K in magnetic fields up to 9T. *Adv. Cry. Eng.*, 1984;30;425–432.
61. K. Matsumoto, T. Okano, T. Kouen, T. Numazawa, K. Kamiya, S. Nimori, Magnetic entropy change of magnetic refrigerants with first order phase transition suitable for hydrogen refrigeration. *IEEE Trans. Appl. Superconduct.*, 2004;14;1738–1741.
62. C.E. Reid, J.A. Barclay, J.L. Hall, S. Sarangi, Selection of magnetic materials for an active magnetic regenerative refrigerator. *J. Alloys Compounds*, 1994;207–208;366–371.
63. K.A. Gschneidner Jr., V.K. Pecharsky, S.K. Malik, The (Dy1-xErx)Al2 alloys as active magnetic regenerators for magnetic refrigeration. *Adv. Cry. Eng.*, 1996;42;475–483.

64. A. Smaili, R. Chahine, Composite materials for Ericsson-like magnetic refrigeration cycles, *J. Appl. Phys.*, 1997;81(2);824–829.
65. J.A. Barclay, Can magnetic refrigerators liquefy hydrogen at high efficiency. *ASME-Paper 81-HT-82, 20th Joint ASME/AICHE National Heat Transfer Conference*, Milwaukee, WI, August 2–5, 1981.
66. W. Iwasaki, Magnetic refrigeration technology for an international clean energy network using hydrogen energy (WE-NET). *J. Hydrogen Energy*, 2003;28;559–567.
67. L. Zhang, S.A. Sherif, T.N. Verziroglu, J.W. Sheffield, Performance analysis of reciprocating magnetic liquefiers. *J. Hydrogen Energy*, 1994;19;945–956.
68. J.A. Barclay et al., Design limitations on magnetic refrigerators imposed by magnetic forces. *Proceedings of the Fourth International Cryocooler Conference*, Easton, MD, September, 1986.
69. A. Rowe, J. Barclay, Static and dynamic force balancing in reciprocating active magnetic refrigerators, *Adv. Cry. Eng.*, 2002;47;1003–1010.
70. R.W. Merida, J.A. Barclay, Monolithic regenerator technology for low temperature (4 K) Gifford-McMahon cryocoolers. *Adv. Cry. Eng.*, 1998;43;1597–1604.
71. A.J. Degregoria et al. Active magnetic regenerator method and apparatus United States Patent 5249424, 1993.
72. J.A. Barclay, W.F. Stewart, The effect of parasitic refrigeration on the efficiency of magnetic liquefiers. *Proceedings of the 17th Intersociety Energy Conversion Engineering Conference*, Los Angeles, CA, 1982, pp. 1166–1170.
73. K.L. Engelbrecht, A numerical model of an active magnetic regenerator refrigeration system. Master thesis, University of Wisconsin, Madison, WI, 2004.
74. L. Zhang, S.A. Sherif, T.N. Verziroglu, J.W. Sheffield, On exergy losses in AMR hydrogen liquefiers. *J. Hydrogen Energy*, 1994;19;447–452.
75. J. Barclay, Active magnetic regenerative liquefier. *2010 Annual Merit Review Proceedings*, 2010, Available at http://www.hydrogen.energy.gov/annual_review10_production.html#delivery.

19

Compact Hydrogen Storage in Cryogenic Pressure Vessels

Salvador M. Aceves
Lawrence Livermore National Laboratory

Francisco Espinosa-Loza
Lawrence Livermore National Laboratory

Elias Ledesma-Orozco
Universidad de Guanajuato

Guillaume Petitpas
Lawrence Livermore National Laboratory

CONTENTS

19.1 Introduction

As a universal transportation fuel that can be generated from water and any energy source, hydrogen is a leading candidate to supplant petroleum with the potential to ultimately eliminate petroleum dependence, associated air pollutants, and greenhouse gases [1]. The predominant technical barrier limiting widespread use of hydrogen automobiles is storing enough hydrogen fuel on board to achieve sufficient (500+ km) driving range in a compact, lightweight, rapidly refuelable, and cost-effective system.

There are three major conceptual approaches to storing hydrogen on board automobiles: (1) gas compressed to high pressures (e.g., 350–700 atm) [2,3], (2) lower pressure absorption of hydrogen within porous and/or reactive solids [4,5], or (3) cryogenic liquid (LH_2) at temperatures near its boiling point (20.3 K) [6]. Each approach faces fundamental limits. Hydrogen stored as a compressed gas occupies a relatively large volume at ambient temperature, while materials to absorb hydrogen add significant weight, cost, and thermal

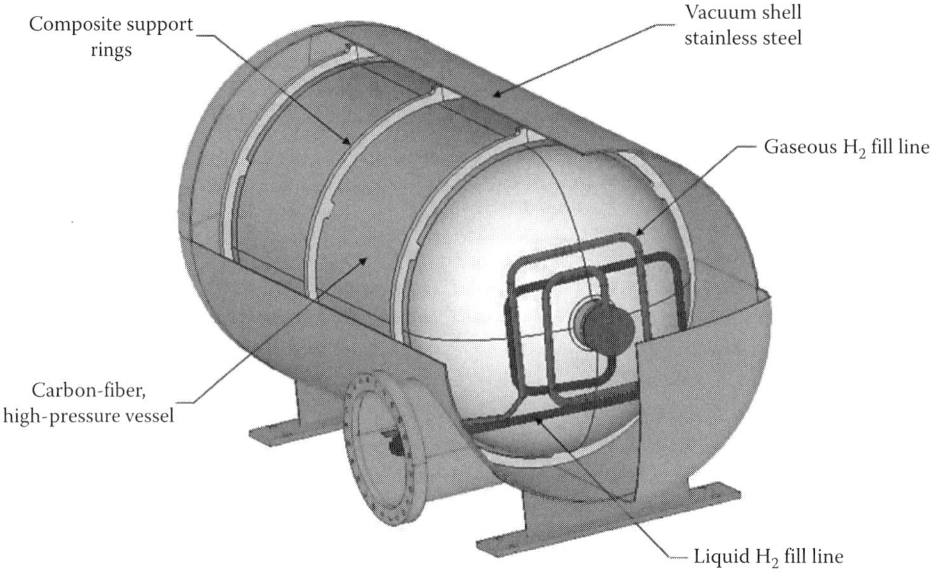

FIGURE 19.1
Cryogenic pressure vessel design. Inner vessel is an aluminum-lined, carbon fiber–wrapped pressure vessel typically used for storage of compressed gases. This vessel is surrounded by a vacuum space filled with numerous sheets of highly reflective metalized plastic (minimizing heat transfer into vessel) and an outer jacket of stainless steel.

complexity to onboard storage systems. Finally, liquid hydrogen (LH_2) evaporates very easily, pressurizing quickly as it absorbs heat from the environment (typically venting after 3 days of inactivity) or from distribution, transfer, and refueling operations.

A new approach has been developed over the last decade that combines existing storage technologies to capture the advantages of both: *cryogenic* pressure vessels [7–9]. Cryogenic vessels comprise a high-pressure vessel enclosed in high-performance vacuum multilayer thermal insulation (Figure 19.1), enabling hydrogen storage at cryogenic temperatures (as cold as 20 K) *and* high pressures (e.g., ~350 atm). For a pressure vessel of given size and cost, a cryogenic pressure vessel stores substantially more hydrogen than a vessel at ambient temperature without the additional weight and cost of hydrogen-absorbent materials but with far greater thermal endurance than conventional (i.e., low pressure) cryogenic LH_2 tanks. Cryogenic pressure vessels can essentially eliminate evaporative losses for practical automotive refueling and driving scenarios. High-density storage enables inexpensive, long-range hydrogen-fueled automobiles, and cryogenic operation reduces isentropic expansion energy, potentially improving safety of operation.

19.2 Historical Perspective

When initially proposed in the early 1970s, hydrogen-fueled vehicles demanded low-pressure LH_2 [10,11]. Pressurized hydrogen storage was not considered viable. Metallic tanks were extremely heavy and bulky, and composite materials were not available for automotive applications due to technical maturity and cost issues.

Progress in cryogenic technology (a factor of 1000 improvement on cryogenic insulation over 50 years [12]) made a compelling case for LH_2 vehicles. BMW, Linde, and Magna Steyr demonstrated LH_2 vessels with world-leading weight, volume, and cost performance [6,13]. However, while viable for large vehicles (>10 kg H_2), LH_2 tanks suffer from considerable evaporative losses when reduced in size due to lower thermal mass that magnifies the effect of environmental heat transfer.

The aerospace industry, more sensitive to weight and less sensitive to cost, did consider composite materials for multiple structural applications. In particular, storage of cryogenic propellants and oxidizers is a key application that was researched from the 1960s [14]. While light and efficient for space travel, cryogenic fluids suffer from evaporative losses during long missions, and minimizing these losses is critical for reducing launch weight. While most efforts have been dedicated to zero boil-off systems where high-performance thermal insulation is complemented by active cryocooling to eliminate evaporative losses [15,16], the concept of storing cryogens in a vessel rated for high pressure (34–120 atm) was also considered for this application [17]. The authors observed the thermodynamic potential for trading-off vessel strength (pressure rating) versus insulation performance: greater pressure rise allows more heat transfer and less insulation. Conversion between the two phases of hydrogen nuclear spin (para to ortho) is endothermic, and the authors also considered the use of catalysts to promote the reaction and further reduce evaporative losses. The results indicate an advantage for cryogenic pressure vessels versus low-pressure cryogenic vessels for Mars exploration due to lower evaporative losses.

Progress in composite pressure vessel technology enabled automotive application starting in the early 1990s and leading to the development of high-performance pressure vessels capable of storing hydrogen at large weight fraction (>10% H_2) [18]. However, volumetric performance—critical for efficient packaging within a vehicle—was still limited by the low density of compressed gas.

The first automotive cryogenic pressure vessels were developed at Concordia University (Montreal, Canada) in the early 1990s [19,20]. The concept for these *thermocontrolled* vessels consisted of filling insulated pressure vessels (rated for 300 atm) with LH_2, which was subsequently evaporated, initially from contact with the warm tank and finally by circulating warm gases (e.g., engine exhaust) through an in-tank heat exchanger. Aside from the capacity advantage of storing high-density LH_2, thermocontrolled vessels provide a source of high-pressure hydrogen that can be used for direct injection into a hydrogen spark-ignited engine, eliminating power loss that would otherwise occur due to hydrogen displacing ambient air [21]. The potential for evaporative losses for infrequent drivers or during long periods of inactivity was not considered and may be an issue when the cryogenic vessel is pressurized through intentional heat transfer.

Detailed thermodynamic modeling of cryogenic pressure vessels was first conducted at Lawrence Livermore National Laboratory [8]. The analysis revealed the potential for high-capacity vessels that eliminate evaporative losses during regular use. The margin afforded by the high-pressure rating combined with the cooling that occurs as gaseous hydrogen is extracted reduces sensitivity to environmental heat transfer by an order of magnitude when compared to LH_2 vessels. It may also enable insulation simplification. While LH_2 tanks demand very-high-performance insulation (<3 W heat transfer), cryogenic pressure vessels can operate at higher heat transfer rates (possibly as much as 10 W), enabling thinner and/or simpler insulations that improve packaging efficiency and reduce system cost.

19.3 Thermodynamics of Cryogenic Pressure Vessels

Cryogenic pressure vessels address three key problems stemming from the high volatility of LH_2: evaporative losses after a short period of inactivity (dormancy), cumulative evaporative losses for short daily driving distances, and risk of being stranded due to fuel evaporation after long-term parking.

The dormancy (period of inactivity before a vessel releases hydrogen to reduce pressure buildup) is an important parameter for LH_2 vehicle acceptability. Dormancy can be calculated from the first law of thermodynamics [22] and the properties of H_2 [23] and can be illustrated with a diagram of hydrogen thermodynamic properties (Figure 19.2) to simplify visualization and graphical calculation of dormancy for hydrogen vessels.

Figure 19.2 uses axes of internal energy and density instead of more traditional temperature and pressure. A dormancy calculation begins by identifying the initial thermodynamic state in Figure 19.2 (density and internal energy) of the hydrogen contained in the vessel. From this initial point (e.g., point A), the thermodynamic state of hydrogen fuel on board a parked vehicle moves horizontally to the right (warming at constant density) as

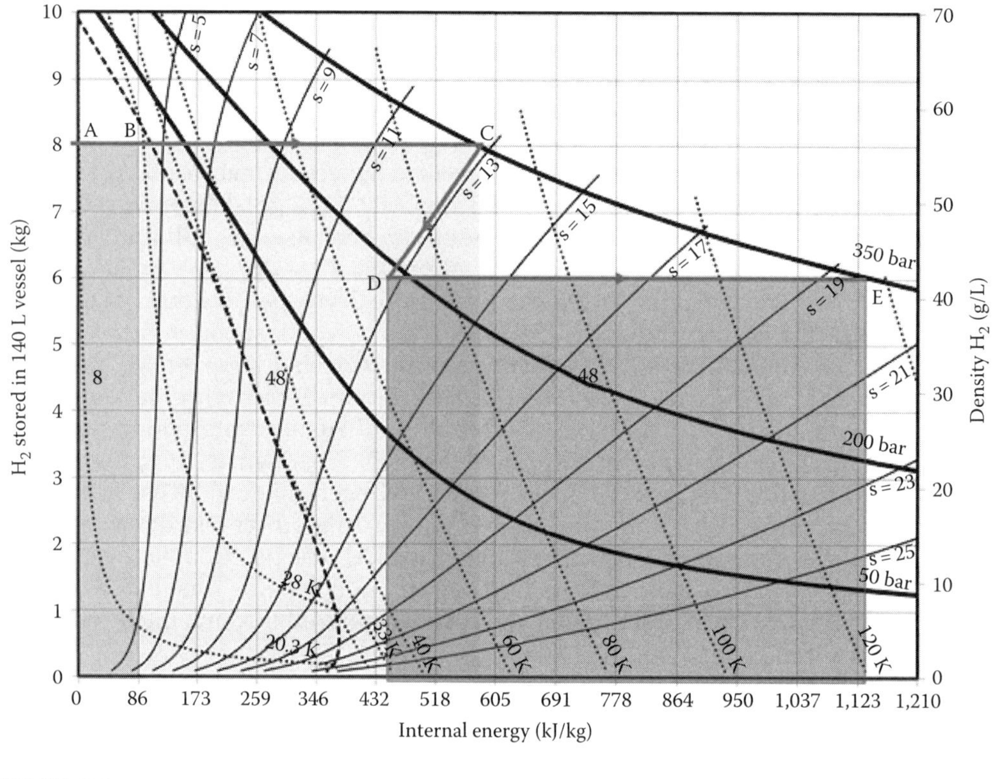

FIGURE 19.2
Phase diagram for hydrogen showing density (right vertical axis) and internal energy (horizontal axis), with lines for constant pressure (thick lines), temperature (dotted lines), and entropy (thin lines). The figure also shows a thick dashed saturation line that separates the supercritical phase (right) from the two-phase (liquid and vapor) region (left). A second vertical axis in the left shows the mass of hydrogen contained in a vessel with 140 L internal volume, which would store 10 kg of LH_2 at 20 K and 1 atm. The figure also shows letters and areas representing dormancy (in Watt days) of conventional LH_2 tanks (AB, medium-gray area) and cryogenic pressure vessels (B to E, medium + light + dark-gray areas).

heat enters from the environment, until the hydrogen pressure reaches the vessel maximum and some hydrogen needs to be used or vented. The cumulative thermal energy absorbed while a car is parked can be calculated by multiplying the amount of hydrogen in the vessel by the total change in its specific internal energy. This total thermal energy is shown as the *area* in Figure 19.2 under the horizontal line joining the initial and final points in the process (neglecting temperature stratification and vessel thermal capacity). Dormancy is then equal to the total heat absorbed (the area under the line) divided by the heat entry rate.

An appropriate choice of scales in Figure 19.2 radically simplifies dormancy calculations. The grid scale in the internal energy (horizontal) axis is set at 86.4 kJ/kg H_2, which converts to 1 W-day/kg H_2 (1 day = 86,400 s). The grid scale in the vertical axis represents 1 kg H_2. Therefore, the area of a grid square represents 1 W-day of heating. The total change in internal energy (in Watt-days) can be easily calculated by counting the squares under the horizontal line representing the parking process. Dormancy is calculated by dividing the internal energy change (in Watt-days) by the rate of heat transfer (in Watts).

As an illustration, consider a parked hydrogen automobile with a conventional LH_2 tank with 140 L internal volume and 6 atm maximum working pressure, which is 80% full with 8 kg LH_2 at 20 K and 1 atm (point A in Figure 19.2). Once the vehicle is parked, heat entry warms the hydrogen, increasing both its temperature and pressure. Dormancy ends in this case when the pressure reaches 6 atm (point B in Figure 19.2), when hydrogen venting becomes necessary to maintain pressure within the vessel's limits. Total heat absorbed during this process from point A to point B can be calculated by counting the number of squares (8 W-days) in the area marked in green. Dormancy can then be calculated by dividing 8 W-days by the heat transfer rate (e.g., 2 days for a vessel absorbing heat at a rate of 4 W).

Figure 19.2 illustrates the dramatic dormancy advantage of automobiles with cryogenic pressure vessels. An auto initially filled with 8 kg LH_2 at 1 atm and 20 K can remain parked until the pressure reaches 340 atm (point C in the figure) without venting any hydrogen. Counting squares under the line joining point A and point C, we obtain 8 + 48 = 56 W-days, *seven times* greater thermal endurance than a conventional LH_2 tank.

Furthermore, unlike conventional LH_2 vessels, cryogenic pressure vessels dramatically extend dormancy as the vehicle is driven. For example, if the parked vehicle is driven when the hydrogen is at state C (Figure 19.2) consuming 2 kg of H_2 fuel, the remaining hydrogen in the vessel expands and cools following a constant entropy line from point C to point D, extending the thermal endurance of the vessel by an additional 48 W-days before any losses occur (at point E in Figure 19.2). Further driving substantially extends dormancy, essentially eliminating fuel evaporation for even very moderate driving patterns.

In principle, Figure 19.2 enables simple analyses of arbitrary cycles of driving and parking periods. Evaporative losses and dormancy are easily calculable, given a driving schedule, vessel volume, and thermal performance (i.e., heat transfer leak rate).

Figure 19.2 is somewhat conservative because it neglects secondary effects such as vessel heat capacity and heat potentially absorbed by conversion between the two states of nuclear spin arrangement (i.e., *para*-hydrogen conversion to *ortho*-hydrogen) of hydrogen molecules. Both of these effects tend to increase the dormancy of the vessels but are most significant only for warmer temperatures ($T > 77$ K) and partially full vessels where thermal endurance is a substantially easier challenge. Both effects are negligible at the very low (i.e., 20–30 K) temperatures where conventional LH_2 tanks operate.

From Figure 19.2, it is clear that insulation performance can be traded off versus vessel pressure rating: dormancy can be increased by either improving insulation or by

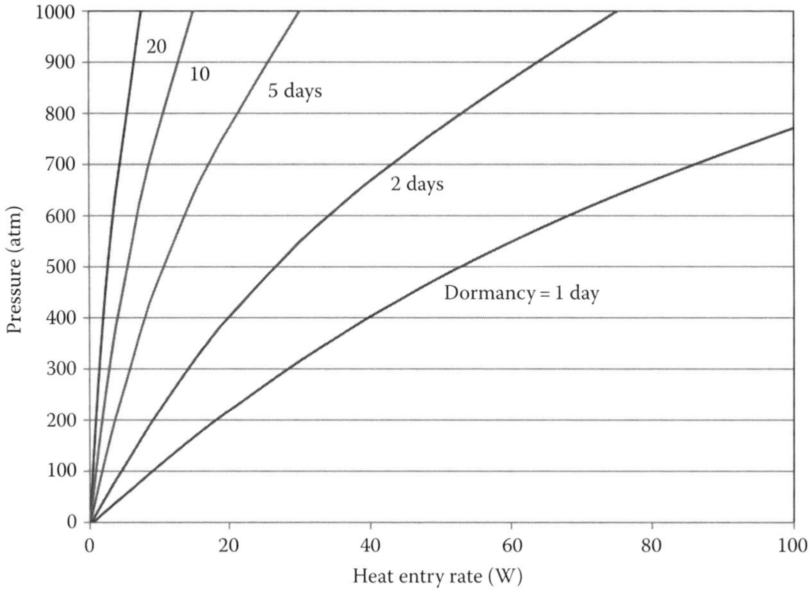

FIGURE 19.3
Contour lines of dormancy versus heat transfer rate (x-axis) and vessel rated pressure (y-axis) for an initially full vessel (140 L internal volume and 10 kg LH_2 at 1 atm).

strengthening the vessel. This trade-off is better illustrated in Figure 19.3, which shows contour lines of dormancy versus heat transfer rate (x-axis) and vessel rated pressure (y-axis) for an initially full vessel (140 L internal volume and 10 kg LH_2 at 1 atm). The figure shows that a pressure vessel may reach acceptable dormancy (5 days) through either (1) high-performance insulation (~3 W) at low rated pressure (~167 atm) or (2) high rated pressure (700 atm) at low-performance insulation (~17 W). Even larger heat transfer rates would be allowable in continuously driven vehicles (e.g., taxis, buses) since strong vessels (>500 atm rated pressure) deliver 1 day of dormancy at heat entry rates over 60 W. The optimum design point depends on the relative cost of fiber and insulation, as well as the particular mission requirements.

It should also be noted that Figure 19.3 assumes a completely full vessel—the best case for low-pressure vessels and the worst case for high-pressure vessels. Low-pressure vessels (<10 atm) have maximum dormancy when full because the thermal inertia of the evaporating LH_2 slows down pressure rise. On the other hand, cryogenic pressure vessels gain most of their dormancy from containing the hydrogen as they heat up and therefore have longest dormancy at low fill levels where the vessel can heat up more before reaching the rated pressure. Dormancy is infinite when ambient temperature hydrogen at rated pressure is denser than the cryogenic hydrogen stored in the vessel.

Considering that (1) vehicles are typically not filled to full capacity before extended parking and (2) even short distance driving plays a considerable role in cryogenic pressure vessel dormancy (Figure 19.2), it is important to consider all factors for enabling quick visualization of thermal performance and dormancy. Figures 19.4 (vessel rated for 135 atm) and 19.5 (vessel rated for 350 atm) assume a 140 L vessel (10 kg LH_2 when full) with 5 W heat transfer rate, installed on board an efficient vehicle (100 km/kg H_2). For any given vessel pressure (x-axis) and hydrogen mass (y-axis), the figures give (1) dormancy, (2) daily driving distance necessary to eliminate evaporative losses, (3) temperature, and (4) entropy.

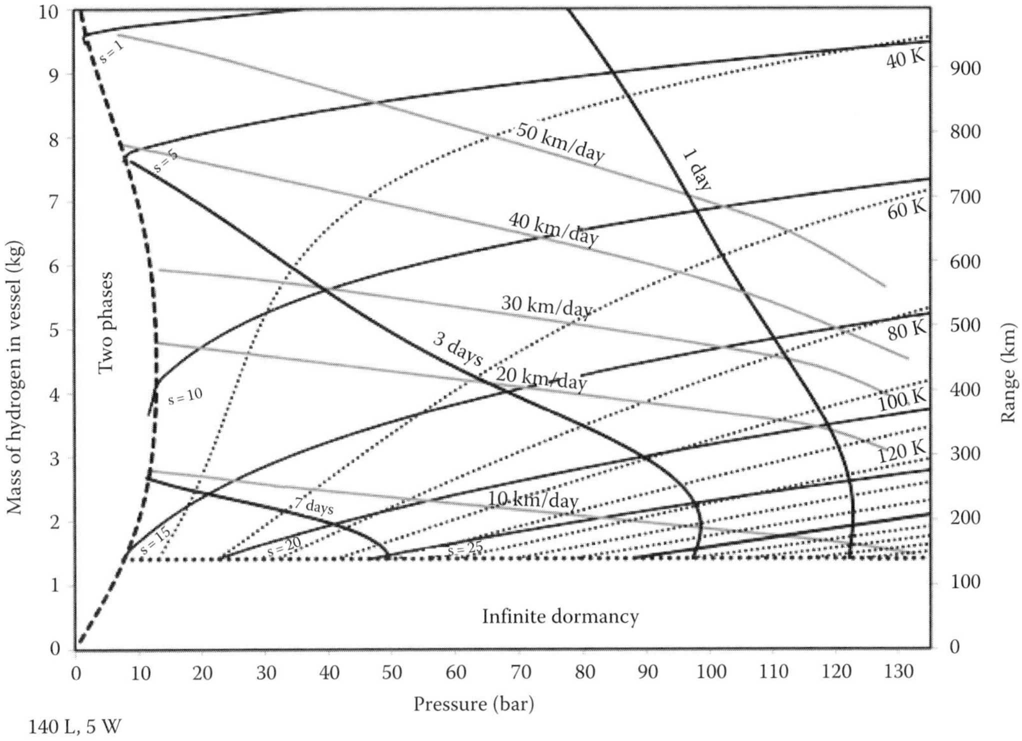

FIGURE 19.4
Contour lines of entropy (thin, continuous), temperature (dotted), dormancy (thick), and minimum daily driving distance necessary for avoiding evaporative losses (gray), as a function of vessel rated pressure (x-axis) and hydrogen mass in vessel (y-axis). The figure assumes a 140 L vessel (10 kg LH_2 when full) with 5 W heat entry, installed on board an efficient vehicle (100 km/kg H_2), and 135 bar vessel rated pressure.

As an example of the application of these figures, assume that a vessel with a 140 L internal volume and 135 atm rated pressure contains 6 kg H_2 at a pressure of 105 atm. Finding the location of this point in Figure 19.4, we determine that the vessel contains hydrogen at ~60 K and will start venting in ~1 day if left unused and driving ~45 km/day will suffice for eliminating all evaporative losses.

A comparison between Figures 19.4 and 19.5 illustrates the thermal performance advantages of stronger vessels. The infinite dormancy region more than doubles in size as the pressure rating is increased to 350 atm (Figure 19.5) due to the higher ambient temperature storage capacity. Even in the regions where evaporative losses are possible, dormancy is considerably longer, and minimum daily driving distance is much shorter as the rated pressure is raised, eliminating evaporative loses under practical driving scenarios. These advantages of cryogenic pressure vessels must be balanced against the added cost and weight of structural materials and the cost of high-performance insulation.

It is worth noting that there is no risk of being stranded due to evaporative losses during cryogenic pressure vessel long-term parking. Even when thermal equilibrium with the environment is reached (after months of parking), a substantial fraction of the original hydrogen remains (15% at 135 atm and 30% at 350 atm, infinite dormancy limit in Figures 19.4 and 19.5). This is sufficient for reaching a refueling station even in remote locations.

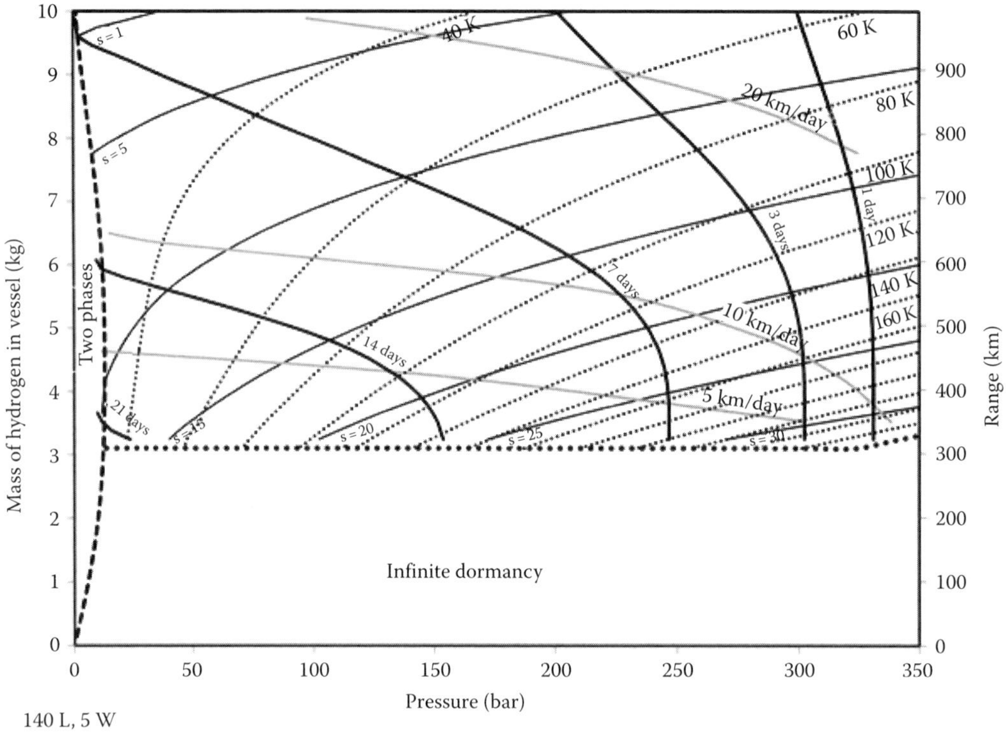

FIGURE 19.5
Contour lines of entropy (thin, continuous), temperature (dotted), dormancy (thick), and minimum daily driving distance necessary for avoiding evaporative losses (gray), as a function of vessel rated pressure (*x*-axis) and hydrogen mass in vessel (*y*-axis). The figure assumes a 140 L vessel (10 kg LH_2 when full) with 5 W heat entry, installed on board an efficient vehicle (100 km/kg H_2), and 350 bar vessel rated pressure.

19.4 Potential Safety Advantages

Cryogenic pressure vessels offer a number of potential safety advantages. The most dramatic and perhaps counterintuitive is the radically lower theoretical burst energy of *low-temperature* H_2. Figure 19.6 shows the theoretical maximum mechanical energy released by a sudden adiabatic expansion to atmospheric pressure (e.g., in a vessel rupture) of high-pressure hydrogen gas from three temperatures (80, 150, and 300 K). H_2 stored at 70 atm and 300 K will release a maximum mechanical energy of 0.55 kWh/kg H_2 if suddenly (i.e., adiabatically) expanded to atmospheric pressure (cooling substantially in the process). Counterintuitively, this maximum energy release increases only slightly with *much* higher H_2 pressures. Raising vessel pressure to 1000 atm (1400% increase from 70 atm) increases maximum mechanical energy release by only 10% while shrinking vessel volume and strengthening (thickening) vessel walls many times over. The low burst energy and high hydrogen storage density of cryogenic temperatures combine synergistically, permitting smaller vessels that can be better packaged on board to withstand automobile collisions. The vacuum jacket surrounding a cryogenic pressure vessel (Figure 19.1) offers a second layer of protection, eliminating environmental impacts over the life of the pressure vessel. Vacuum jacketing also provides expansion

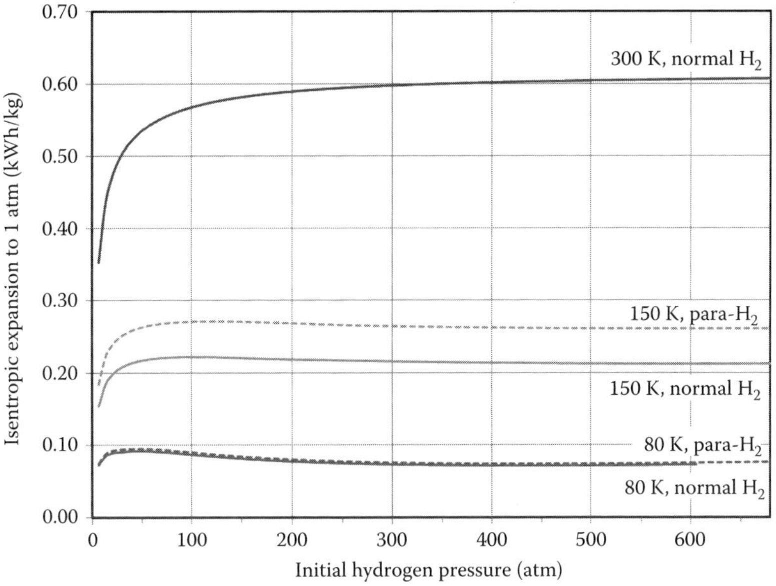

FIGURE 19.6
Maximum mechanical energy (per kilogram of hydrogen) released upon instantaneous expansion of H_2 gas (e.g., from a pressure vessel) as a function of initial storage pressure at 80, 150, and 300 K. For comparison, note that the chemical energy content of hydrogen is 33.3 kWh/kg. This mechanical energy is the theoretical maximum available work based on reversible adiabatic expansion from the pressure shown to 1 atm, calculated from internal energy differences of H_2 gas before and after isentropic expansion. The figure reports normal-hydrogen expansion energy at 300 K and para-hydrogen and normal-hydrogen expansion energy at 80 and 150 K. In the latter two cases, the real adiabatic expansion energies are slightly lower because the para content is less than 100% before expansion (between 99.8% para and equilibrium, depending on the previous use of the storage and the actual ortho-to-para transition rate).

volume to mitigate shocks from hydrogen release. Cryogenic vessels avoid the fast fill heating and overpressures (up to 25%) typical of ambient temperature vessels, consequently operating at higher safety factors, especially as driving the automobile cools the remaining hydrogen fuel and reduces average hydrogen pressures further over typical driving and refueling cycles. Finally, due to the high density of cryogenic hydrogen and the relatively low refueling pressure, the number and amplitude of pressure peaks in a cryogenic pressure vessel can be lower than in ambient high-pressure vessels.

19.5 Technology Validation

Three generations of cryogenic pressure vessels incorporating aluminum-lined, composite-wrapped vessels, an outer vacuum vessel, and multilayer vacuum insulation to minimize heat transfer have been designed, assembled, and demonstrated at LLNL (Figure 19.7). The designs also included instrumentation for pressure and temperature as well as safety devices to prevent vessel failure. The figure shows specific weight and volume performance for the three vessels (weight fraction and grams of H_2 per liter), as well as the Department of Energy (DOE) 2010, 2015, and ultimate weight and volume targets [24].

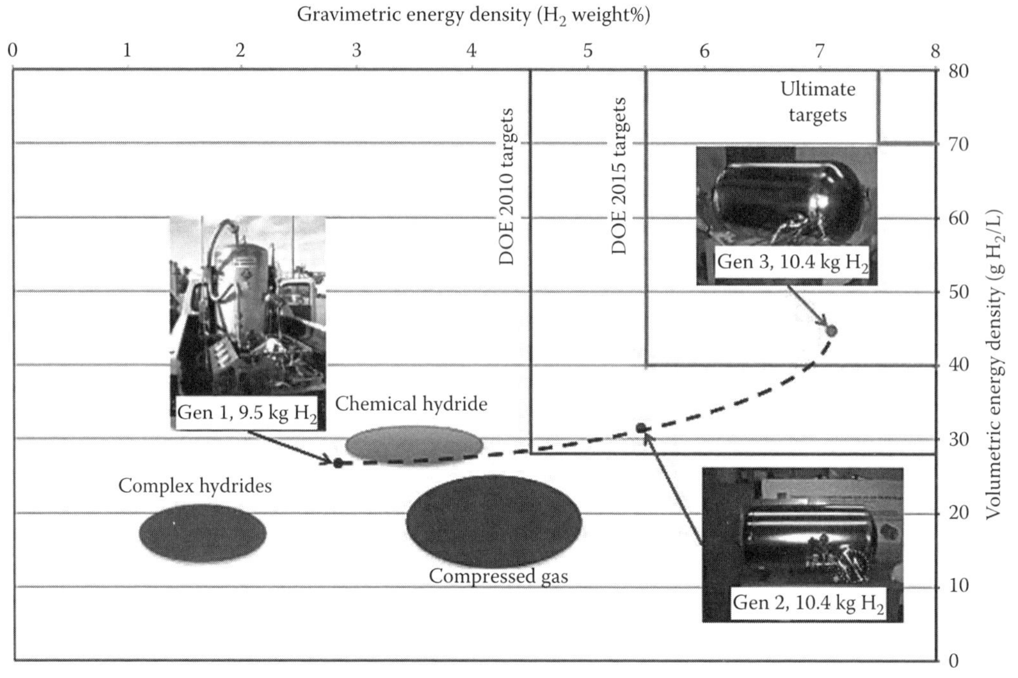

FIGURE 19.7
Weight and volume performance of three generations of cryogenic pressure vessel prototypes. The figure also shows weight and volume performances specified by the US DOE 2010 target (4.5% weight fraction and 30 g H_2/L), 2015 target (6% weight fraction and 40 g H_2/L), and ultimate target (7.5% weight fraction and 70 g H_2/L). (From Department of Energy, Office of Energy Efficiency and Renewable Energy Hydrogen, Fuel Cells & Infrastructure Technologies Program Multi-Year Research, Development and Demonstration Plan, http://www.eere.energy.gov/hydrogenandfuelcells/mypp.)

The latest prototype cryogenic pressure vessel (generation 3) takes advantage of the insensitivity to environmental heat transfer to considerably reduce insulation thickness (to ~1.5 cm vs. ~3 cm typical of LH_2 vessels), significantly reducing system volume. This more compact vessel is expected to still provide acceptable dormancy.

Table 19.1 lists detailed component weight and volume breakdown for the generation 3 cryogenic pressure vessel. Thin insulation produces a very compact and light design that meets the very challenging DOE 2015 weight and volume targets and is within 3 kg of meeting the DOE ultimate weight target. Further gains in packaging density may be obtained by pressurizing LH_2, theoretically increasing hydrogen storage capacity by ~20% for isothermal compression (84 kg/m^3 hydrogen density at 20 K and 250 atm), although 5%–10% density increase is more likely once pumping inefficiencies and compression heating are considered.

The US DOE has also established *cost* targets for hydrogen storage systems [24]. While cryogenic pressure vessels are projected to be cost competitive with respect to alternatives [25], they still cost three to four times more than the DOE 2010 target ($13.6/kWh vs. $4/kWh). Cost may therefore be the most challenging of DOE targets, even though future improvements on vessel materials, manufacture, and insulation may reduce expense below projected values.

TABLE 19.1

Weight and Volume of the Main System Components for the Generation 3 Cryogenic Pressure Vessel and for the Dual-Volume System

	Generation 3		Dual Volume (Projected)	
Components	**Weight (kg)**	**Volume (L)**	**Weight (kg)**	**Volume (L)**
Hydrogen inside cylindrical vessel(s)	10.7	151	6.0	86.6
Hydrogen within low-pressure vessel	0	0	5.4	76.7
Total hydrogen stored	10.7	151	11.4	163.3
High-pressure vessel(s)	61	28	11.9	7.2
Low-pressure (interstitial) vessel	0	0	81.7	10.2
Insulation and vacuum shell	57	46	53.6	24.56
Total, vessel and vacuum shell	118	74	147.2	41.96
Total, vessel(s), vacuum shell, and hydrogen	129	225	158.6	205.26
Total for accessories	16	10	16	10
Total, storage system	145	235	174.6	215.26
H_2 weight fraction	7.38%		6.53%	
Volume performance, g/L		45.49		52.95

The generation 3 vessel was built and installed on board a Toyota Prius experimental vehicle. The dual-volume system has been designed and analyzed, but it has not been built. For comparison, the DOE 2010 targets are 4.5% weight fraction and 30 g H_2/L; DOE 2015 targets are 6% weight fraction and 40 g H_2/L; ultimate DOE targets are 7.5% weight fraction and 70 g H_2/L [24]. Storage calculations assume 70.7 kg/m^3 hydrogen density (LH_2 at 20 K and 1 atm). The table assumes that all stored hydrogen is usable, although driving at high power for a long time may drop the vessel pressure below the necessary level to run the engine or fuel cell, reducing the net hydrogen storage density by a few percent [26].

19.6 Increasing Packaging Efficiency with Dual-Volume Vessels

Cryogenic pressure vessels store high-density LH_2 while avoiding evaporative losses with thin insulation resulting in high *system* storage density. However, even LH_2 is only a fourth as energetically dense as gasoline, and the generation 3 vessel is approximately six times larger than a gasoline tank with the same energy storage. While there is still potential for improving cryogenic pressure vessels through more compact construction, we also see the possibility of using *conformable* designs that may package better inside the vehicle than conventional cylindrical shapes.

Conformable vessels hold the promise of better filling *box*-shaped or even irregularly shaped spaces, improving packaging efficiency by 20%–40% over conventional cylindrical vessels. However, conformable vessels remain a structural challenge [27,28]. Noncylindrical, nonspherical shapes suffer from bending stresses that quickly exceed the strength of structural materials—even ultrastrong carbon fiber. Applicability of conformable box-shaped vessels is therefore limited to low pressures (~6 atm), typically too low for compact hydrogen storage.

We can, however, take advantage of low-pressure conformable vessels through a new concept: *dual-volume containers*. The concept of a dual-volume container (Figure 19.8)

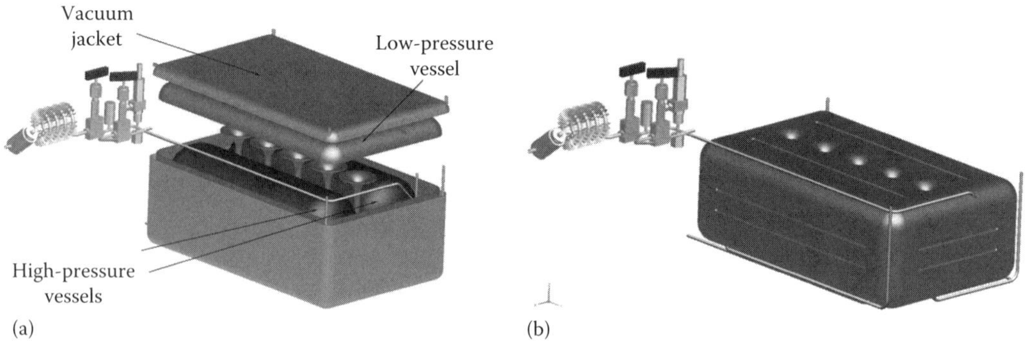

FIGURE 19.8
Geometry of dual-volume conformable pressure vessel comprising two cylindrical high-pressure (350 atm) vessels surrounded by a low-pressure (6 atm) interstitial box-shaped vessel. Finally, a vacuum vessel and multilayer insulation surround the outer vacuum vessel. (a) Vessel with the upper vacuum and interstitial vessels removed for better visualization. (b) Fully assembled dual-volume vessel.

combines two high-pressure (~250 atm) cylindrical vessels enclosed within a low-pressure (~6 atm) box-shaped conformable container that can be filled with low-pressure LH_2, enabling utilization of interstitial spaces and corners that are normally wasted in cylindrical vessels. The low-pressure container is then surrounded with thin (6.5 mm) vacuum multilayer insulation for reducing heat transfer into the vessel.

The geometry of dual-volume vessels is synergistic with hydrogen properties because the interstitial space can store LH_2 at low pressure and high density. While the low-pressure container will have reduced thermal endurance (~2 days), it can be used for extending vehicle range when most needed—during long trips. Continuous driving quickly consumes hydrogen before any evaporative losses may occur. The driver may then fill only the cylindrical pressure vessels with LH_2 and leave the interstitial vessel empty when little driving is anticipated, avoiding evaporative losses while still maintaining a reasonable driving range.

The dual-volume design in Figure 19.8 has improved packaging efficiency due to interstice utilization. A system with 11.4 kg H_2 storage capacity would store 5.4 kg LH_2 in 76.7 L of interstitial space, often wasted in cylindrical vessels. Each of the inner high-pressure linerless vessels stores 3 kg LH_2 in 43.3 L of internal volume.

The dual-volume vessel in Figure 19.8 has been analyzed and designed by finite element. Table 19.1 lists the projected weight and volume of the vessel and accessories. The table shows that the dual-volume vessel has potential for producing very high weight and volume performance, and it is projected to meet or exceed the DOE 2015 targets.

Vessel volume listed in Table 19.1 is the total external volume (i.e., the *shrink-wrap* volume). However, considering that spaces in vehicles are typically box-shaped rather than cylindrical, it may be appropriate to define vessel's *effective* volume performance not in terms of total external volume but rather in terms of the volume of a box that encloses the system. It would therefore make sense to define a conformability credit as the hydrogen volume stored in a conformable vessel that occupies a box-shaped space divided by the hydrogen volume in cylindrical vessels that can be packaged within the same box-shaped space. This criterion would give the dual-volume vessel a 22% conformability credit that would increase the volumetric performance from 53 to 64 g H_2/L—over 90% of the density of pure LH_2 at ambient pressure (70.7 g/L).

It may be possible to reach the DOE ultimate volume target (70 g/L) by filling the cylindrical vessels with pressurized LH_2 and the interstitial vessel with low-pressure LH_2. At 84 g/L (LH_2 at 20 K and 250 atm), the total amount of hydrogen in the cylindrical vessels would grow to 7.3 kg, for a total of 12.7 kg once the interstitial vessel is considered. This would raise weight and volume performance metrics to 7.3% H_2 and 72 g H_2/L (assuming 22% conformability credit). If packaged into a vehicle, this dual-volume system would provide an unprecedented level of storage performance, potentially enabling practical range in most hydrogen vehicles.

19.7 Conclusions

Unlike other fuels, hydrogen can be produced and consumed without generating carbon dioxide. If generated using renewable energy, hydrogen becomes a versatile, storable, and universal carbonless energy carrier that does not pollute. The greatest engineering challenge associated with hydrogen fuel is storing enough hydrogen on board the vehicle for a reasonable range (500 km).

First developed for aerospace applications, cryogenic pressure vessels are now being researched for automotive use. Cryogenic pressure vessels can store high-density hydrogen, similar to LH_2 vessels, without the evaporative losses. They are more compact than compressed hydrogen, therefore reducing the need for expensive carbon fiber. They are light and offer potential safety advantages due to reduced isentropic expansion energy at low temperature.

The versatility and flexibility of cryogenic pressure vessels enable high hydrogen storage density while avoiding evaporative losses, maintaining reasonable cost, and allowing fast refueling. The recent generation 3 prototype is the first automotive hydrogen vessel to meet the DOE 2015 weight and volume targets.

Conformable dual-volume vessels store hydrogen in interstitial and corner spaces that typically go unused in cylindrical vessels, thereby improving packaging efficiency and providing even higher hydrogen storage performance, leading to unprecedented levels of volume storage capacity. Predicted storage packaging density would enable long-range hydrogen transportation in a broad line of today's vehicles. Ongoing and planned work is addressing the technical challenge of simultaneously improving insulation performance and packaging efficiency while maintaining low cost.

Acknowledgments

This project was funded by DOE's Fuel Cell Technologies Program, with Monterey Gardiner as the technology development manager. This work was performed under the auspices of the US DOE by Lawrence Livermore National Laboratory under contract DE-AC52-07NA27344.

References

1. Berry, G.D. and Aceves, S.M. The case for hydrogen in a carbon constrained world, *Journal of Energy Resources Technology*, 127, 89–94, 2005.
2. von Helmolt, R. and Eberle, U. Fuel cell vehicles: Status 2007, *Journal of Power Sources*, 165, 833–843, 2007.
3. Ciancia, A., Pede, G., Brighigna, M., and Perrone, V. Compressed hydrogen fuelled vehicles: Reasons of a choice and developments in ENEA, *International Journal of Hydrogen Energy*, 21, 397–406, 1996.
4. Ahluwalia, R.K. and Peng, J.K. Automotive hydrogen storage system using cryo-adsorption on activated carbon, *International Journal of Hydrogen Energy*, 34, 5476–5487, 2009.
5. Forde, T., Eriksen, J., Pettersen, A.G., Vie, P.J.S., and Ulleberg, O. Thermal integration of a metal hydride storage unit and a PEM fuel cell stack, *International Journal of Hydrogen Energy*, 34, 6730–6739, 2009.
6. Amaseder, F. and Krainz, G. Liquid Hydrogen Storage Systems developed and manufactured for the first time for Customer Cars, SAE Paper 2006-01-0432, SAE International, Warrendale, PA, 2006.
7. Aceves, S.M., Berry, G.D., and Rambach, G.D. Insulated pressure vessels for hydrogen storage on vehicles, *International Journal of Hydrogen Energy*, 23(7), 583–591, 1998.
8. Aceves, S.M. and Berry, G.D. Thermodynamics of insulated pressure vessels for vehicular hydrogen storage, *ASME Journal of Energy Resources Technology*, 120(2), 137–142, 1998.
9. Aceves, S.M., Berry, G.D., Martinez-Frias, J., and Espinosa-Loza, F. Vehicular storage of hydrogen in insulated pressure vessels, *International Journal of Hydrogen Energy*, 31, 2274–2283, 2006.
10. Williams, L.O. Hydrogen powered automobiles must use liquid hydrogen, *Cryogenics*, 13, 693–698, 1973.
11. Gregory, D.P. The hydrogen economy, *Scientific American*, 228(1), 13–21, 1973.
12. Scurlock, R.G. Development of low-loss storage of cryogenic liquids over the past 50 years, in Timmerhaus, K.D. and Reed, R.P. (eds). *Cryogenic Engineering, Fifty Years of Progress*, Springer, New York, 2007.
13. Braess, H.H. and Strobl, W. Hydrogen as a fuel for road transport of the future: Possibilities and prerequisites, *Proceedings of the 11th World Hydrogen Energy Conference*, Stuttgart, Germany, 1996.
14. Hanson, M.P. Glass, boron and graphite filament-wound resin composites and liners for cryogenic pressure vessels, *Journal of Macromolecular Science, Part B*, 1, 651–665, 1967.
15. Plachta, D. and Kittel, P. An updated zero boil-off cryogenic propellant storage analysis applied to upper stages or depots in an LEO environment, Paper AIAA-2002–3589, *Proceedings of the 38th Joint Propulsion Conference and Exhibit*, Indianapolis, IN, 2002.
16. Ho, S.H. and Rahman, M.M. Three-dimensional analysis for liquid hydrogen in a cryogenic storage tank with heat pipe-pump system, *Cryogenics*, 48, 31–41, 2008.
17. Mueller, P.J., Batty, J.C., and Zubrin, R.M. High-pressure cryogenic hydrogen storage system for a Mars sample return mission, *Cryogenics*, 36, 815–822, 1996.
18. Mitlitsky, F., Weisberg, A.H., and Myers, B. Vehicular hydrogen storage using lightweight tanks (regenerative fuel cell systems), *Proceedings of the U.S. DOE Hydrogen Program Annual Review Meeting*, Lakewood, CO, 1999.
19. Krepec, T., Miele, D., and Lisio, C. Improved concept of hydrogen on-board storage and supply for automotive applications, *International Journal of Hydrogen Energy*, 15, 27–32, 1990.
20. Tummala, M., Krepec, T., and Ahmed, A.K.W. Optimization of thermocontrolled tank for hydrogen storage in vehicles, *International Journal of Hydrogen Energy*, 22, 525–530, 1997.
21. White, C.M., Steeper, R.R., and Lutz, A.E. The hydrogen-fueled internal combustion engine: A technical review, *International Journal of Hydrogen Energy*, 31, 1292–1305, 2006.
22. Moran, M.J. *Availability Analysis: A Guide to Efficient Energy Use*, Prentice Hall, Inc., Englewood Cliffs, NJ, 1982.

23. McCarty, R.D. *Hydrogen: Its Technology and Implications, Hydrogen Properties, Volume III*, CRC Press, Cleveland, OH, 1975.
24. Department of Energy, Office of Energy Efficiency and Renewable Energy Hydrogen. Fuel Cells & Infrastructure Technologies Program Multi-Year Research, Development and Demonstration Plan, http://www.eere.energy.gov/hydrogenandfuelcells/mypp
25. Lasher, S., McKenney, K., Sinha, J., and Rosenfeld, J. Analyses of hydrogen storage materials and on-board systems, *Proceedings of the DOE Hydrogen and Fuel Cell Annual Merit Review*, Crystal City, VA, 2008, http://www.hydrogen.energy. gov/pdfs/progress08/iv_e_1_lasher.pdf
26. Ahluwalia, R.K. and Peng, J.K. Dynamics of cryogenic hydrogen storage in insulated pressure vessels for automotive applications, *International Journal of Hydrogen Energy*, 33, 4622–4633, 2008.
27. Haaland, A. High-pressure conformable hydrogen storage for fuel cell vehicles, *Proceedings of the US Department of Energy 2000 Hydrogen Program Review*, National Renewable Energy Laboratory Report NREL/CP-570-28890, Golden, CO, 2000. Available online at: http://www1.eere.energy.gov/hydrogenandfuelcells/pdfs/28890cc.pdf.
28. Aceves, S.M., Berry, G.D., Weisberg, A.H., Espinosa-Loza, F., and Perfect, S.A., Advanced concepts for vehicular containment of compressed and cryogenic hydrogen, *Proceedings of the 16th World Hydrogen Energy Conference*, Lyon, France, July 10–15, 2006.

20

Metal Hydrides

Sesha S. Srinivasan
Tuskegee University

Prakash C. Sharma
Tuskegee University

Elias K. Stefanakos
University of South Florida

D. Yogi Goswami
University of South Florida

CONTENTS

20.1 Metal Hydrides

Hydrogen storage by metal hydrides comprises an intermetallic alloy phase that has the capability to absorb and hold vast amounts of hydrogen by chemical bonding.[1] An appropriate hydrogen storage matrix should have the capacity to absorb and release hydrogen without compromising the matrix structure. Metal hydrides are prepared by reaction between a metallic phase and hydrogen. When exposed to hydrogen at certain pressures and temperatures, these phases absorb large quantities of hydrogen gas and form the corresponding metal hydrides. If this is the scenario, the hydrogen is distributed compactly throughout the intermetallic lattice. Metal hydrides represent an exciting process of hydrogen storage, which is inherently safer than the compressed gas or liquid hydrogen storage. Additionally, some intermetallics (including metals and alloys) store hydrogen at a higher volume density than liquid hydrogen.

The qualities required to make these intermetallics useful include the facility to absorb and release large amounts of hydrogen gas, many times without damaging the storage material and with good selectivity (only hydrogen absorption). Moreover, suitable metal hydrides absorb and release hydrogen at rates that can be controlled by adjusting temperature and/or pressure.

As mentioned in the preceding text, a typical relation of metal hydrides with hydrogen can be expressed as $M + x/2\ H_2 \leftrightarrow MHx + \text{Heat}$, where M represents the intermetallic matrix and H is hydrogen. The effect is reversible and the direction is determined by the pressure (and temperature) of the hydrogen gas. If the pressure is above a certain level (the equilibrium pressure), the effect proceeds to the right to form the hydride, whereas below the equilibrium pressure, hydrogen is liberated and the intermetallic matrix returns to its original state. The equilibrium pressure, itself, depends upon temperature. It increases due to expansion with an increment in temperature (and vice versa). The molecular hydrogen first dissociates into hydrogen atoms due to the available metal catalysts and interacts with the host metal lattice to form metal hydride as shown in the schematics of Figure 20.1a. Similarly, the reverse reaction yields hydrogen in gaseous form as represented in Figure 20.1b.

In general, hydrides are classified according to the nature of the bonding of hydrogen to the host lattice such as covalent, saline or ionic, and metallic, all of these classes bearing different bonding characteristics.[2] Such a classification doesn't always clarify the characteristic features of the compound in question. For example, lithium hydride is classified as a saline hydride, where in reality, it actually exhibits some covalent characteristics. The rare-earth hydrides are normally classified in the metallic hydride group, whereas they exhibit some characteristics similar to those of volatile and saline hydrides (e.g., high heat of formation). The ionic hydrides are usually crystalline and show high heats of formation and high melting points. The ionic alkali and alkaline-earth hydrides have a higher density than the pure alkali (45%–75%) and alkaline-earth metals (20%–25%).[3] Covalent hydrides are generally thermally unstable, and this instability goes up with

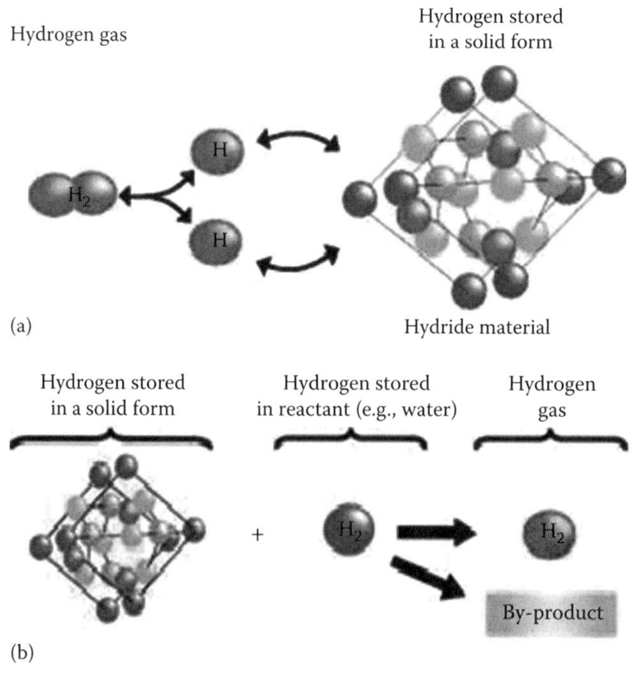

FIGURE 20.1
(a) Metal hydride formation during H_2 absorption in host metal lattice. (b) Hydrogen gas release during the dehydrogenation of metal hydride.

increasing atomic weight of the nonhydrogen element(s). Typical covalent hydrides are carbon hydrides, boron hydrides, germanium hydrides, etc. Covalent hydrides normally exhibit low symmetric structures. Metallic complex hydrides are normally formed by the transition metals, for example, ScH_2.[17] They generally exhibit metallic characteristic properties such as high thermal and electrical conductivity, hardness, luster, etc. Due to the wide homogeneity ranges adopted by metallic hydrides, they have occasionally been considered as solid solutions of hydrogen in the interstitials of metal, alloy, or intermetallic matrices. Yet another classification based on the carbon-based sorbents, for example, single-wall or multiwall carbon nanotubes, polymer nanostructures, zeolites, and metal organic frameworks, possesses weak van der Waals attraction toward hydrogen. The classification of hydrides on the basis of temperature of operation and thermodynamic values such as ΔH of reaction is schematically shown in Figure 20.2. It is clearly understood that the metal or complex hydrides operate in the region of moderate temperatures 0°C–100°C with appropriate ΔH of ~30–55 kJ/mol H_2. Whereas the ionic or covalent hydrides require high temperature, ΔH of ~70–80 kJ/mol H_2 and the carbon-based sorbent systems mandate cryogenic operating temperatures with ΔH <20 kJ/mol H_2.

Hydrogen can be packed and stored in a solid state by forming metal hydride.[4–12] During the formation of the metal hydride, hydrogen molecules are dissociated into hydrogen atoms, which insert themselves into interstitial spaces inside the lattice of intermetallic compounds and/or alloys (Figure 20.1).

At a large distance from the metal surface, the energy difference between a hydrogen molecule and two separate hydrogen atoms is the dissociation energy ($H_2 \rightarrow 2H$, E_D = 435.99 kJ/mol). The first attractive interaction of the hydrogen molecule approaching the

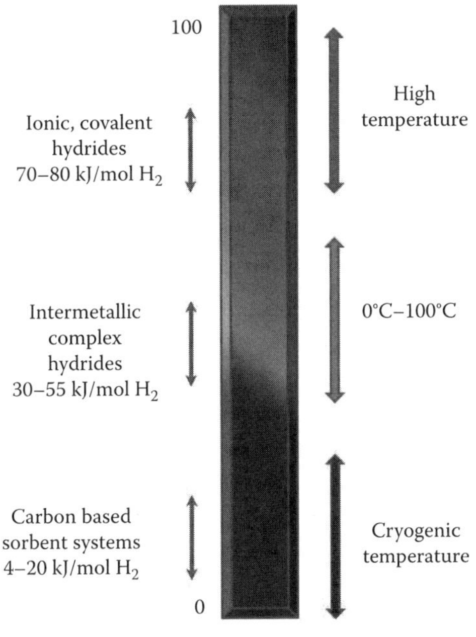

FIGURE 20.2
Classification of hydrogen storage systems on the basis of temperature of operation and the thermodynamic values such as heats of reaction.

metal surface is the van der Waals force leading to the physisorbed state ($E_{Phys} \approx 10$ kJ/mol) more or less one hydrogen molecule radius (≈0.2 nm) from the metal surface. Closer to the surface, the hydrogen has to overcome an activation barrier in order to dissociate and create the hydrogen metal bond. The height of the activation barrier depends on the surface elements involved. Hydrogen atoms sharing their electrons with the metal atoms at the surface are in the chemisorbed state ($E_{Chem} \approx 50$ kJ/mol·H_2). The chemisorbed hydrogen atoms may have a high surface mobility, interact with each other, and form surface phases at considerably high coverage. In the following step, the chemisorbed hydrogen atoms can jump into the subsurface layer and finally diffuse to interstitial sites through the host metal lattice as depicted in Figure 20.3. An interstitial hydrogen atom contributes its electron to the band structure of the metal.

In the α-phase solid solution, the hydrogen to metal ratio is small ($H/M < 0.1$), and hydrogen is exothermically dissolved in the metal. The metal lattice expands relative to the hydrogen concentration by approximately 2–3 Å^3 per hydrogen atom.[13] At greater hydrogen concentrations in the host metal ($H/M > 0.1$), a strong H–H interaction due to the lattice expansion becomes significant, and the hydride phase (β-phase) nucleates and grows. The hydrogen concentration in the hydride phase is commonly found to be $H/M = 1$.

The volume expansion between the coexisting α- and the β-phase[14] corresponds in many cases 10%–20% of the metal lattice. Therefore, at the phase boundary, large stresses are created that frequently lead to a decrepitation of brittle host metals including intermetallic compounds resulting in a final hydride powder with a representative particle size of 10–100 μm.

An idealized representation of pressure–concentration–temperature (PCT) isotherms for the α-phase solid solution and β-phase hydride is shown in Figure 20.4. A plateau is observed on the pressure versus hydrogen/metal ratio for fixed temperature. When the reaction is complete, another sharp pressure rise is seen if more hydrogen is added.

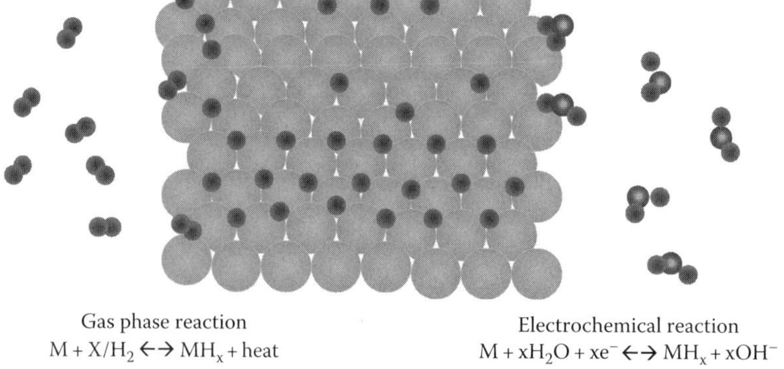

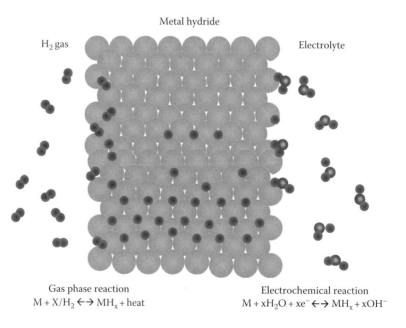

FIGURE 20.3
Schematic representation of the dissociation of hydrogen gas molecule.

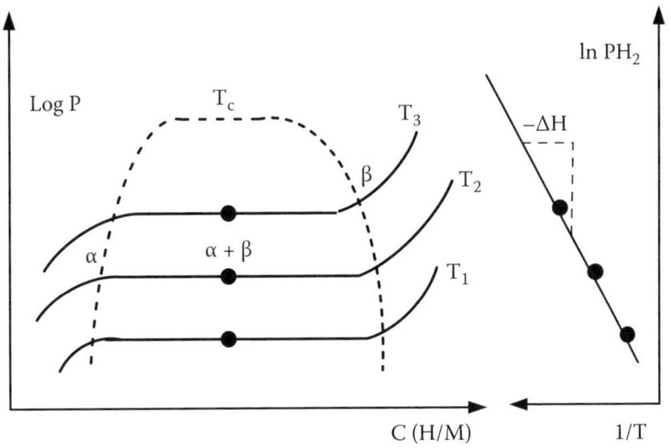

FIGURE 20.4
Effect of temperature on the general features of the PC isotherms.

At a certain temperature, the plateau pressure reflects one point on a pressure–temperature (van't Hoff) plot. The log of pressure versus reciprocal temperature is linear with a negative slope and is related to the heat of reaction (heat of hydriding); the plateau pressures must increase with temperature due to the linear relationship between pressure and temperature.

An increment in isothermal temperature causes the plateau pressure to increase and, at the same time, reduces the width of the plateau that represents the miscibility regime of the α- and β-phase. This narrowing process of the plateau with increasing of temperature continues until, eventually, at a certain *critical temperature*, T_c, the plateau disappears totally, reducing the miscibility of the two phases to zero, and the α-phase converts continuously into the β-phase. The slope and length of the equilibrium plateau are of particular importance for hydrogen storage application; a flat plateau enables the reversible absorption and desorption of hydrogen from a metal simply by raising or lowering the surrounding hydrogen pressure above or below the plateau pressure.

The equilibrium pressure as a function of temperature is related to the changes of ΔH and ΔS, in that order, by the van't Hoff equation:[15–17]

$$\ln\left[\frac{p(H_2)}{p^\circ}\right] = \frac{\Delta H}{RT} - \frac{\Delta S}{R} \tag{20.1}$$

where

$p(H_2)$ is the hydrogen equilibrium pressure at α- to β-phase hydride conversion, determined from PCT isotherms
p° is the standard pressure
R is the gas constant
T is the temperature

The ΔH, of hydride formation is an important parameter characterizing the alloy as a proper hydrogen absorber for various applications. In general, the difference in standard ΔS between a metal hydride alloy and its hydride is small and is on the order of

10 J/mol/K. The modification in ΔS with hydride formation is mainly provided by the loss of the standard ΔS of hydrogen gas (130.858 J/mol/K at 298 K), which means that ΔS can be assumed to be a constant and does not depend on the nature of the metal hydride alloy. The knowledge of ΔH particularly is significant to the heat management needed for practical engineering devices and is a fundamental measure of the M–H bond strength. The van't Hoff plot (lnP vs. 1/T) is a practical way to compare hydrides due to their thermal stabilities.

An overview of hydrogen storage alloys has been discussed from the solid–gas reaction point of view.[18] There are a number of important properties that must be considered in metal hydride storage. Some of the most important ones are (1) ease of activation, (2) heat transfer rate, (3) kinetics of hydriding and dehydriding, (4) resistance to gaseous impurities, (5) cyclic stability, (6) safety, (7) weight, and (8) cost. Although metal hydrides can theoretically store large amounts of hydrogen in a safe and compact way, the practical gravimetric hydrogen density is limited to <3 mass%. It is still a challenge to explore the properties of lightweight metal and complex hydrides.

Hydrogen is a highly reactive element and has been found to form hydrides and solid solutions with thousands of metals and alloys.[13] A hydride *family tree* of the elements, alloys, and complexes is shown in Figure 20.5.

The majority of the 91 natural elements below hydrogen in the periodic table will form hydrides under appropriate conditions such as VH_2, NaH, LaH_2, and ZrH_2. Unfortunately, the PCT properties are not very suitable for the 1–10 atm and 0°C–100°C range used for practical hydrogen storage. Only V is in the range, and there has been extensive study of solid solutions of V and other metals. Pd has been used for over 100 years for H storage,

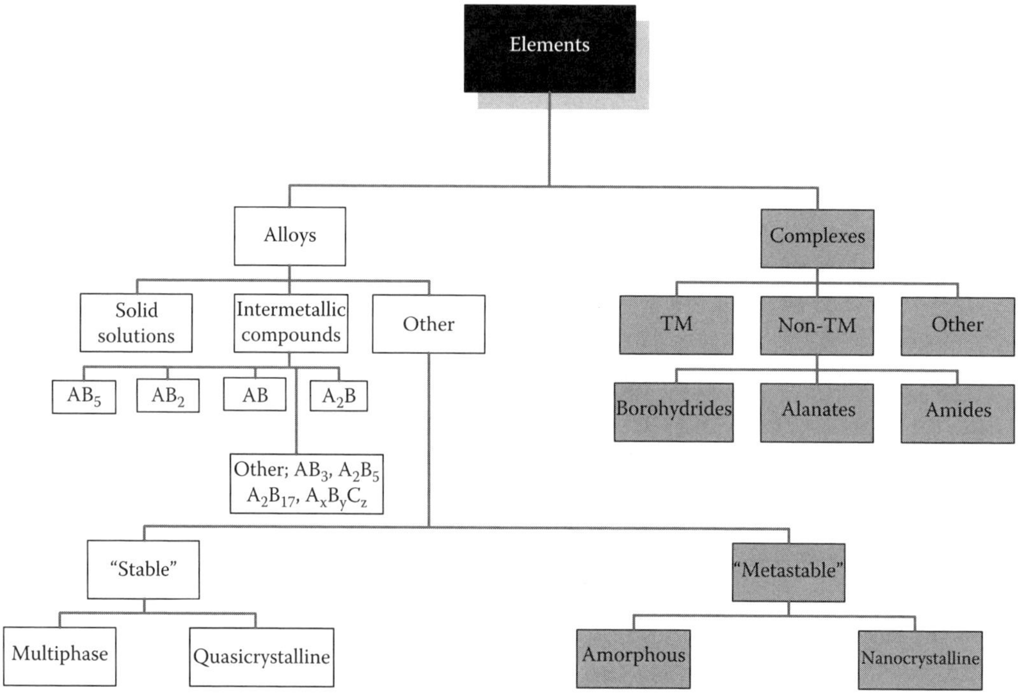

FIGURE 20.5
Family tree of the elements, alloys, and complexes.[13]

but it is very costly, it doesn't hold much hydrogen, and it requires heating temperatures above 100°C to release atomic hydrogen.

20.1.1 Intermetallic Alloys

To prepare useful reversible hydrides, strong hydride-forming (A) elements are combined with weak hydriding (B) elements to form alloys (especially intermetallic compounds) that have the desired intermediate thermodynamic affinities for hydrogen. A good example of this characteristic is the combination of La (forming LaH_2 with 25°C, $P_d = 3 \times 10^{-29}$ atm, and $\Delta H_f = -208$ kJ/mol H_2) and Ni (NiH with 25°C, $P_d = 3400$ atm, and $\Delta H_f = -8.8$ kJ/mol H_2) to form the intermetallic compound $LaNi_5$ ($LaNi_5H_6$ with 25°C, $P_d = 1.6$ atm, and $\Delta H_f = -30.9$ kJ/mol H_2). This ability to *interpolate* between the extremes of elemental hydriding behavior opened the door to the modern world of reversible hydrides.

The AB_5 family has an exceptional versatility because many different elemental species can be substituted (at least partially) into the A and B lattice sites. The A elements tend to be one or more of the lanthanides (elements 57–71), Ca, or other elements such as Y and Zr. The B elements are based on Ni with many other possible substitutional elements such as Co, Al, Mn, Fe, Cu, Sn, Si, and Ti. Modern commercial AB_5 hydriding alloys are for the most part based on the use of the lanthanide mixture or mischmetal (Mm = Ce + La + Nd + Pr) for the A site and Ni + Al + Mn + Co + … for the B site.

Hydrogen storage capacity is on the uncomfortably low side, not passing 1.3 wt.% on the plateau basis. Alloy raw material cost is high; at least by comparison to other systems (AB_2 and AB) that are discussed in the following, the AB_5 alloys are easy to activate, hardly requiring any heating. They decrepitate on the first H/D cycle to fine powder, which is mildly pyrophoric if suddenly exposed to air, a well-known property that must be included in safety considerations. Both easy activation and pyrophoricity indicate the AB_5 alloys do not form defensive oxide layers. Intrinsic kinetics of the AB_5 alloys is generally very good. In an attempt to increase the kinetics and hydrogen storage capacity in AB_5-type alloys such as $MmNi_{4.6}Fe_{0.4}$, mechanochemical pulverization at room temperature was carried out.[19] Figure 20.6 represents the x-ray diffraction (XRD) profiles of $MmNi_{4.6}Fe_{0.4}$ prepared using radio-frequency induction melting followed by mechanochemical pulverization.

The nanocrystallization of peaks was observed with broadening of full width at half maximum (FWHM), thus enhancing the kinetics of hydrogen release as shown in Figure 20.7.

The scanning electron microscope (SEM) micrographs (see Figure 20.8) further justify the smaller and homogeneous crystallites with well-defined grains that enable for the higher hydrogen absorption and desorption cycling capacity with twofold increase of kinetics at room temperature.

Like the AB_5 compounds, the AB_2 intermetallics represent a large and versatile group of hydriding materials with good PCT properties at ambient temperature. The A elements are often from the IVA group (Ti, Zr, Hf) and/or rare-earth series (no. 57–71). The B elements can be a variety of transition or nontransition metals with somewhat of a preference for atomic numbers 23–26 (V, Cr, Mn, Fe). A large variety of substitutions are possible for both A and B elements, thus providing a high degree of fine tuning of PCT properties.

The decisive advantages in regard to hydrogen storage and related thermal applications are high hydrogen capacity, ease of activation, very rapid rates of absorption and desorption, long cycling life, and low cost of materials. The main disadvantages for hydrogen storage are high ΔH values (moderate stabilities of hydrides). These properties make these compounds good for hydrogen compression applications. An excellent example is $ZrMn_2$ that

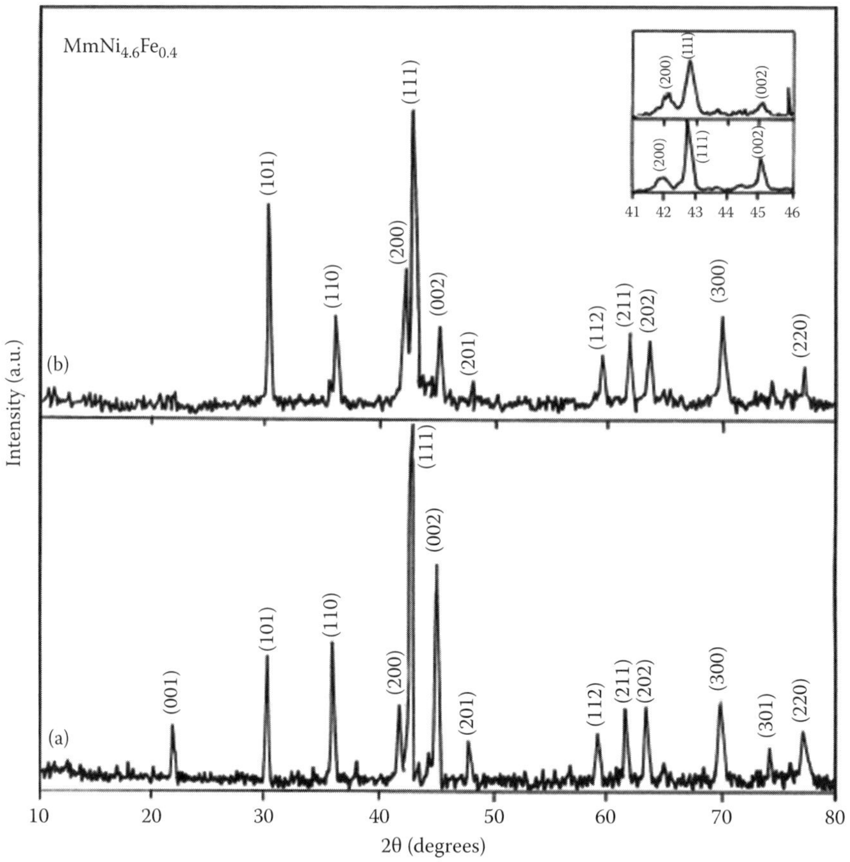

FIGURE 20.6
X-ray diffractograms of (a) as-synthesized $MmNi_{4.6}Fe_{0.4}$ using radio-frequency induction melting and (b) induction melting followed by mechanical pulverization.

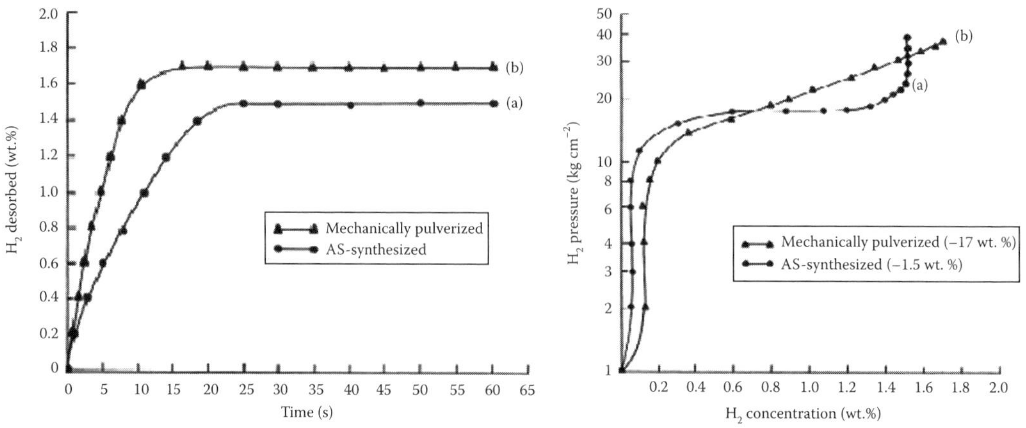

FIGURE 20.7
(a) Hydrogen desorption kinetic curves and (b) PCT of as-synthesized and mechanically pulverized $MmNi_{4.6}Fe_{0.4}$.

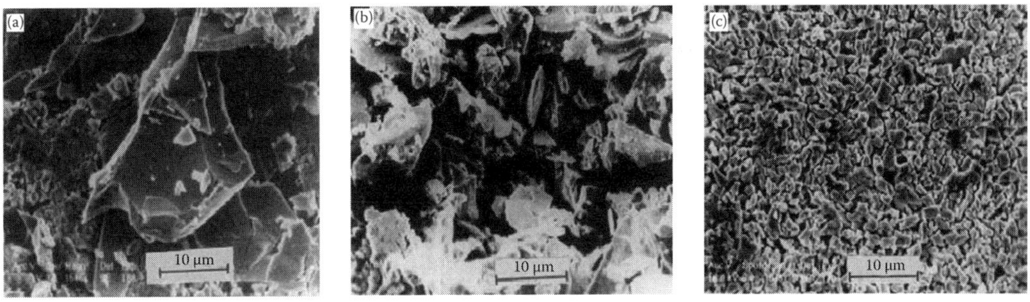

FIGURE 20.8
SEM micrographs of (a) as-synthesized (b) mechanically pulverized and (c) hydrogenated $MmNi_{4.6}Fe_{0.4}$ alloy.

contents 3.6 atoms of hydrogen and ΔH = −53.0 kJ/mol H_2, showing a high hydrogen content and a high T_{dec}, related to the high ΔH.[20] The high T_{dec} and the high hydrogen content make this compound good for hydrogen compression.

Method of preparing an alloy of zirconium and manganese suitable for storing hydrogen consists in two steps[21]: make an intimate mixture of Zr and Mn in correct stoichiometric ratio and heat the sample between 900°C and 1150°C. The intimate mixture is done using a straightforward, repeatable, and inexpensive procedure, ball milling under a H_2 atmosphere. The annealing was applied for 10 h in a helium atmosphere at 1100°C.

A is typically of the group IVA elements Ti, Zr, or Hf, and B is a transition metal, typically Ni. Another family is based on Mg_2Ni.[22] Unfortunately, the A_2B's offer little in the 0°C–100°C and 1–10 atm range, at least with the present state of the art. They are definitely more stable. H storage capacity and cost properties of Mg_2Ni are attractive, but desorption temperatures are too elevated for most applications. Mg_2Ni is not very amenable to modification of PCT properties by ternary and higher-order substitutions. Various attempts to lower desorption temperatures have not been especially successful.

20.1.2 Solid Solutions

The first example of a reversible intermetallic hydride was demonstrated with the AB compound, ZrNi, by Libowitz in 1957.[23] The hydride $ZrNiH_3$ has a 1 atm desorption temperature of about 300°C, too high for hydrogen storage applications but suitable for hydrogen compression.

These intermetallic alloys show good volumetric and gravimetric reversible H storage capacities, competitive with the best of the AB_5's and AB_2's. Activation is pretty slow and hard for the ZrNi-based alloys. In any event, it may take a day or so and high pressures (50+ atm) for total activation. The passive oxide films that can form on ZrNi (and its derivatives) don't result in a high degree of sensitivity to gaseous impurities in the H_2 used; hence, these alloys are resistant to impurities. Cyclic stability of the lower plateau is great, but the upper plateau tends to drift higher with H/D cycling. Although there is vast information available about ZrNi alloys, the intermediate compositional alloys ZrNi 70/30 and ZrNi 30/70 have been sparsely reported for hydrogen storage.[24,25] The XRD analysis of ZrNi70/30 before and after hydrogenation is shown in Figure 20.9. The XRD spectra show a significant change in the intensity of the peaks in ZrNi 70/30 hydride compared with the unhydrided sample (33.2°, 34.2°, 36.5°, and 43°) and the presence of a new peak (36.7°) attributed to the ZrNi 70/30 hydride phase. The crystallite

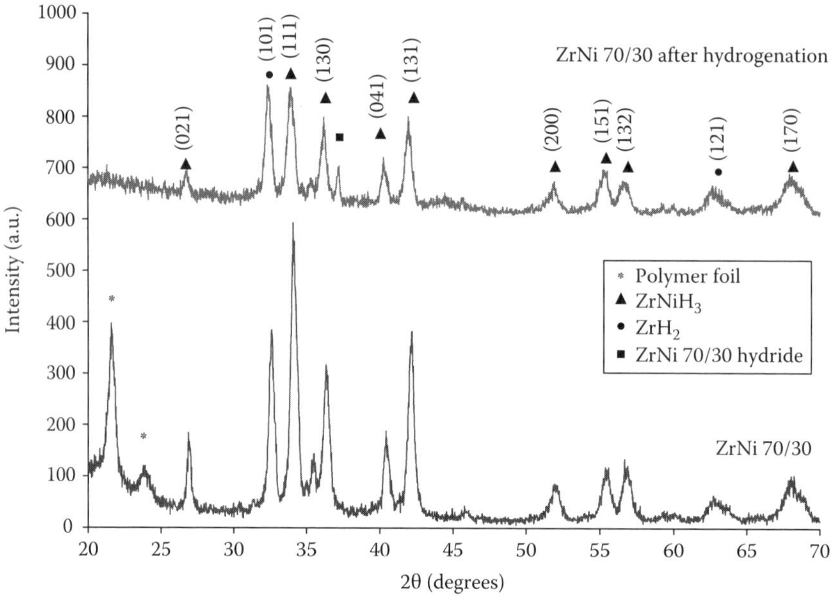

FIGURE 20.9
XRD spectra of ZrNi 70/30 before and after hydrogenation.

sizes of the $ZrNiH_3$ (~48 nm) and ZrH_2 (~39 nm) are calculated and compared with the unhydrided sample. The scanning micrographs of the ZrNi 70/30 after a number of hydrogenation cycles show the development of microcracks due to hydrogen interaction with the host lattice at high temperatures (Figure 20.10).

Figure 20.11 illustrates the absorption–desorption PCT for the ZrNi 70/30 alloy, giving evidence of a typical metal hydride behavior. The plateau region is the most important segment of a PCT plot, which represents the pressure of hydrogen in equilibrium with the metal–metal hydride ($\alpha + \beta$) phases, representing the dissociation pressure of the metal hydride at the desired temperature; the temperature indicates the thermal stability of the metal hydride. Moreover, when the temperature goes from 300°C, 325°C, 350°C, 375°C, and 390°C, it causes the plateau pressure to increase and, at the same time, reduces the width of the plateau that represents the miscibility regime of the α- and β-phase. As an example, at 375°C, hydrogen content of 1 wt.% is reached at 13 bars. According to the investigations, ZrNi 70/30 alloy presents a metal hydride behavior: low kinetics, moderate to high equilibrium pressures, and moderate to high hydrogen content, conditions that make this alloy the most convenient for hydrogen compression applications. From the PCT isotherms, the ΔH of reaction has been calculated to be ~39 kJ/mol H_2 for the ZrNi70/30.

In addition to the AB_5, AB_2, AB, and A_2B intermetallic compounds discussed previously, various other families of intermetallics have been shown capable of reversible hydriding/dehydriding reactions.[26,27] Examples include AB_3, A_2B_7, A_6B_{23}, A_2B_{17}, and A_3B. Most structures involve long-period AB_5 and AB_2 stacking sequences and are thus crystallographically related to these two classic families. Most have narrow plateaux with long sloping upper legs (e.g., $GdFe_3$) or multiple plateaux (e.g., $NdCo_3$ or Pr_2Ni_7).

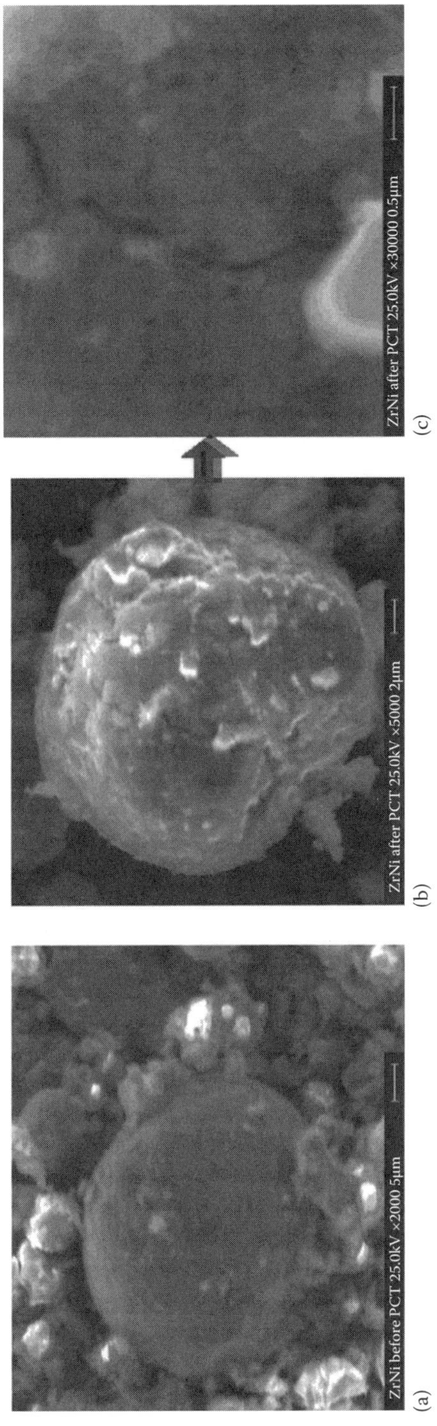

FIGURE 20.10
SEM micrographs of (a) pristine, (b) hydrogenated, and (c) hydrogenated in higher magnification showing microcracks of ZrNi 70/30 material.

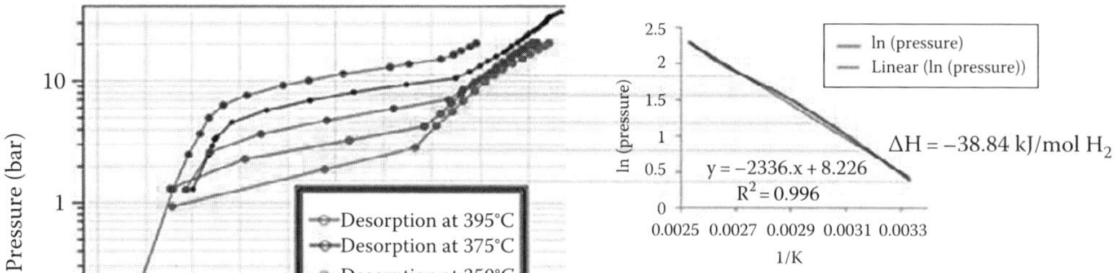

$$\ln \frac{P_2}{P_1} = \ln P_2 - \ln P_1 = \frac{-\Delta H}{R}\left(\frac{1}{T_2} - \frac{1}{T_1}\right) \quad \text{Van't Hoff equation}$$

$$\Delta H = -R\frac{(\ln P_2 - \ln P_1)}{\left(\frac{1}{T_2} - \frac{1}{T_1}\right)}$$

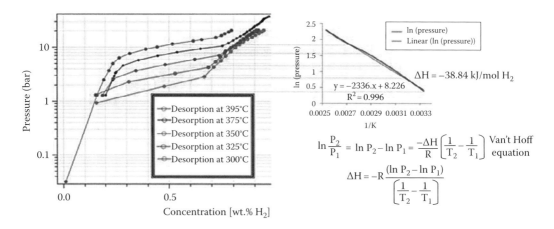

FIGURE 20.11
Absorption and desorption PCT of ZrNi70/30 at 300°C, 325°C, 350°C, 375°C, and 390°C.

TABLE 20.1
Theoretical Capacities of Hydriding Substances as Hydrogen Storage Media

Medium	Hydrogen Content (kg/kg)	Hydrogen Storage Capacity (kg/L of Vol.)	Energy Density (kJ/kg)	Energy Density (kJ/L of Vol.)
MgH_2	0.070	0.101	9,933	14,330
Mg_2NiH_4	0.0316	0.081	4,484	11,494
VH_2	0.0207		3,831	
$FeTiH_{1.95}$	0.0175	0.096	2,483	13,620
$TiFe_{0.7}Mn_{0.2}H_{1.9}$	0.0172	0.090	2,440	12,770
$LaNi_5H_{7.0}$	0.0137	0.089	1,944	12,630
$R.E.Ni_5H_{6.5}$	0.0135	0.090	1,915	12,770
Liquid H_2	1.00	0.071	141,900	10,075
Gaseous H_2 (100 bar)	1.00	0.0083	141,900	1,170
Gaseous H_2 (200 bar)	1.00	0.0166	141,900	2,340
Gasoline	—	—	47,300	35,500

20.1.3 Composites

When the mass of the metal or alloy is taken into account, the metal hydride gravimetric storage density is comparable to storage of pressurized hydrogen. The best achievable gravimetric storage density is about 0.07 kg of H_2/kg of metal, for a high-temperature hydride such as MgH_2 as shown in Table 20.1, which gives a comparison of some hydriding substances with liquid hydrogen, gaseous hydrogen, and gasoline.[28,29] MgH_2 can effectively store hydrogen due to its thermodynamic stability; however, reaction kinetics are too slow and the decomposition temperature is high, at approximately 330°C.[20] A possible way to achieve Mg-like storage capacity but with reversible hydrogenation characteristics is to form composites with Mg as one of the components as discussed earlier. The other component may be one of the known hydrogen storage intermetallic alloys.[30]

Since the $MmNi_{4.6}Fe_{0.4}$ alloy discussed in Section 20.1.1 exhibited hydrogen storage capacity of less than 2.0 wt.%, a novel strategy of forming a composite phase Mg-Xwt.% $MmNi_{4.6}Fe_{0.4}$ was carried out.[31] This system has dual advantages in a sense that it reduces the weight penalty of the system for hydrogen storage applications and also enhances the hydrogen storage capacity >5 wt.% at moderate temperatures. The flowchart for the synthesis and characterization of Mg-Xwt.% $MmNi_{4.6}Fe_{0.4}$ (X = 10, 20, 30, 40, and 50) is shown in Figure 20.12.

The dehydrogenation kinetic and PCT curves as demonstrated in Figure 20.13a and b reveal the high hydrogen storage capacity (~5 wt.%) at temperatures >300°C.

The XRD pattern of the composite material shows the presence of both Mg and $MmNi_{4.6}Fe_{0.4}$ phases; additional peak pertaining to MgH_2 was observed in the hydrogenated materials (see Figure 20.14). The effective hydrogenation and dehydrogenation in these composite materials have been confirmed through the hydrogen-induced microchannels as exhibited in SEM micrographs and are shown in Figure 20.15. The Mg in the host matrix of $MmNi_{4.6}Fe_{0.4}$ thus facilitates lower desorption temperatures due to the available catalytic nanoparticles of Ni and Fe, which surpassed the hydrogen storage characteristics of Mg_2Ni or other Mg-based intermetallic systems.

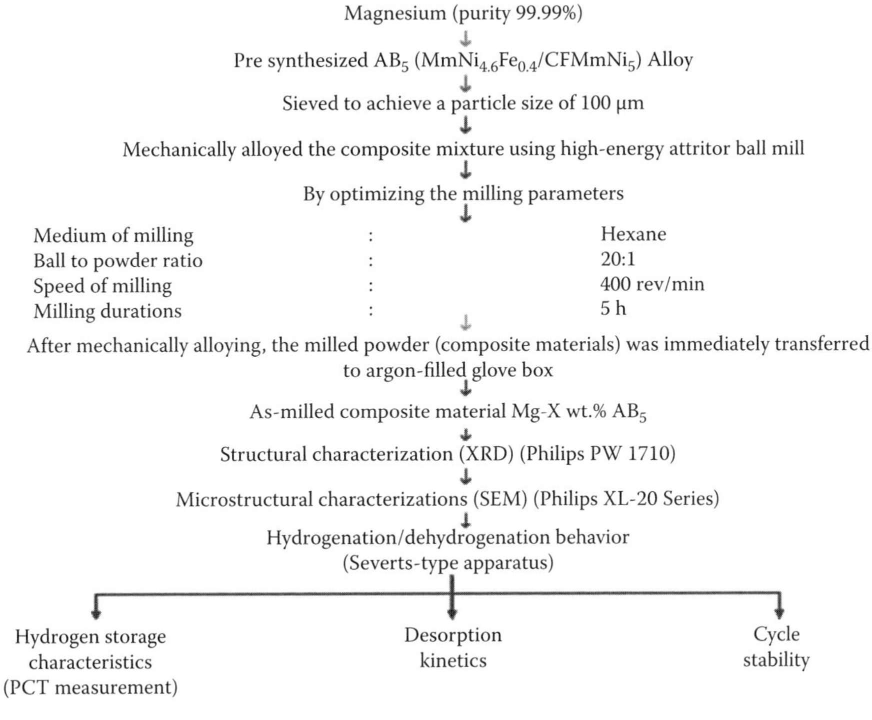

FIGURE 20.12
Flowchart diagram of synthesis and characterization of Mg-Xwt.% $MmNi_{4.6}Fe_{0.4}$ composite material.

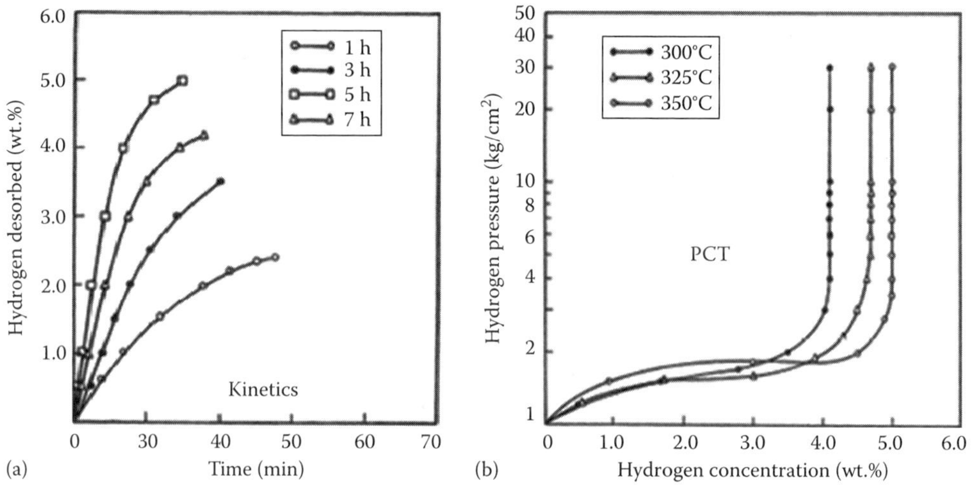

FIGURE 20.13
(a) Hydrogen desorption kinetics and (b) PCT characteristics of Mg-Xwt.% $MmNi_{4.6}Fe_{0.4}$ at moderate temperatures.

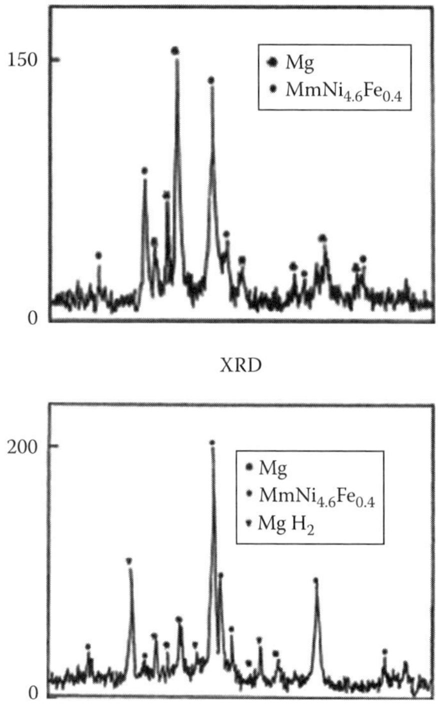

FIGURE 20.14
XRD pictures of Mg-Xwt.% $MmNi_{4.6}Fe_{0.4}$ before and after hydrogenation at moderate temperatures.

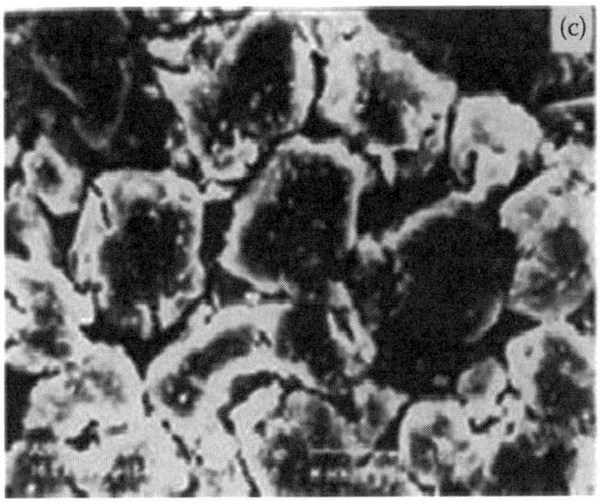

SEM

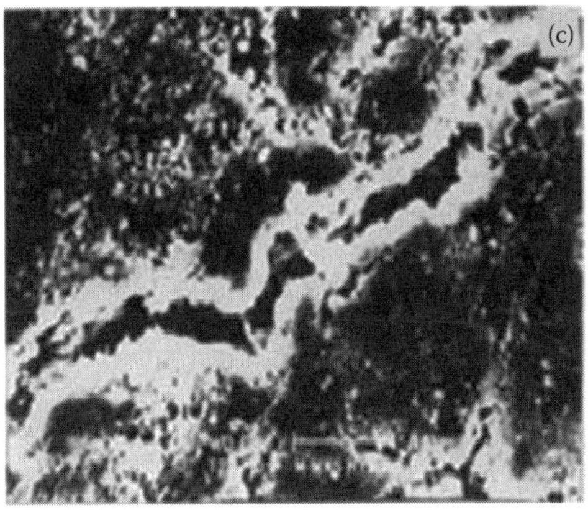

FIGURE 20.15
SEM pictures of Mg-Xwt.% $MmNi_{4.6}Fe_{0.4}$ before and after hydrogenation at moderate temperatures.

References

1. J.H.N. van Vucht, F.A. Kuipers, and H.C.A.M. Bruning (1970). Reversible room-temperature absorption of large quantities of hydrogen by intermetallic compounds, Phillips Research Reports. 25, 133–140.
2. Z.P. Li, B.H. Liu, K. Arai, N. Morigazaki, and S. Suda (2003). Protide compounds in hydrogen storage systems, *Journal of Alloys and Compounds*, 356–357, 469–474.
3. P. Vajeeston (2004). Theoretical modeling of hydrides, Dissertation presented for the degree of Doctor Scientiarum, University of Oslo, Oslo, Norway.
4. G.G. Libowitz and A.J. Maeland (1979). In *Handbook on the Physics and Chemistry of Rare Earths*, K.A.J. Gschneidner and L. Eyring, eds. North Holland Publishing, Amsterdam, the Netherlands.
5. Y. Fukai (1993). *The Metal-Hydrogen System, Basic Bulk Properties*, 1st edn., Springer-Verlag, Berlin, Germany
6. W.M. Muller, J.P. Blackledge, and G.G. Libowitz (1968). *Metal Hydrides*, Academic Press, New York.
7. B.L. Shaw (1967). *Inorganic Hydrides*, Pergamon Press, Oxford, U.K.
8. G.D. Sandrock (1978). In *Hydrides for Energy Storage, The Metallurgy and Production of Rechargeable Hydrides*, A.F. Anderson and A.J. Maeland, eds. Pergamon Press, Oxford, U.K., p. 353.

9. J.J. Reilly (1977). *Hydrogen: Its Technology and Implications*, 2, 13.
10. L. Schlapbach, A. Seiler, F. Stucki, and H.C. Siegman (1980). Surface effects and the formation of metal hydrides, *Journal of Less Common Metals*, T3, 145.
11. K.J. Gross (1998). Intermetallic materials for hydrogen storage, PhD thesis, University of Fribourg, Fribourg, Switzerland.
12. S.S. Sai Raman (2000). Hydrogen storage materials, PhD thesis, Banaras Hindu University, Varanasi, India.
13. G. Sandrock (1999). A panoramic overview of hydrogen storage alloys from a gas reaction point of view, *Journal of Alloys and Compounds*, 293–295, 877–888.
14. S. Bliznakov, E. Lefterova, L. Bozukov, A. Popov, and P. Andreev (2005). Techniques for characterization of hydrogen absorption/desorption in metal hydride alloys, *Proceedings of the International Workshop: Advanced Techniques for Energy Sources Investigation and Testing*, Sofia, Bulgaria.
15. L. Schlapbach, I. Anderson, J. Burger, and K.H.J. Buschov (1994). *Electronic and Magnetic Properties of Metals and Ceramics*, VCH, Weinheim, Germany, Part II, Vol. 331, p. 271.
16. M. Yamaguchi, E. Akiba, and K.H.J. Buschov (1994). *Electronic and Magnetic Properties of Metals and Ceramics*, VCH, Weinheim, Germany, Part II, p. 334.
17. J. Huot and C. Julien (2002). *Proceedings of New Trends in Intercalation Compounds for Energy Storage*, Kluwer Academic Publishers, Dordrecht, the Netherlands, Vol. 61, p. 109.
18. G. Sandrock (1999). A panoramic overview of hydrogen storage alloys from a gas reaction point of view, *Journal of Alloys and Compounds*, 293–295, 877–888.
19. V.V. Sarma, S.S. Sairaman, D.J. Davidson, and O.N. Srivastava (2001). On the mechanically pulverized $MmNi_{4.6}Fe_{0.4}$ as a viable hydrogen storage material, *International Journal of Hydrogen Energy*, 26, 231–236.
20. Hydrogen Storage Materials Database, US Department of Energy, http://www.hydrogenmaterialssearch.govtools.us/.
21. I.I. Bulyk, Yu.B. Basaraba, and A.M. Trostianchyn (2004). Features of the HDDR process in ZrT_2 (T = Cr, Mn, Fe, Co) compounds, *Journal of Alloys and Compounds*, 367, 283–288.
22. A. Maddalena, M. Petris, P. Palade, S. Sartori, G. Principi, E. Settimo, B. Molinas, and S.L. Russo (2006). Study of Mg-based materials to be used in a functional solid state hydrogen reservoir for vehicular applications, *International Journal of Hydrogen Energy*, 31(14), 2097–2103.
23. G.G. Libowitz, H.F. Hayes, and R. Thomas (1957). The system zirconium-nickel and hydrogen, *Journal Chemistry and Physics*, 27, 514.
24. L.E.A. Berlouis, N. Comisso, and G. Mengoli (2006). Changes in hydrogen storage properties of binary mixtures of intermetallic compounds submitted to mechanical milling, *Journal of Electroanalytical Chemistry*, 586, 105–111.
25. D. Escobar, S. Srinivasan, Y. Goswami, and E. Stefanakos (2008). Hydrogen storage behavior of ZrNi 70/30 and ZrNi 30/70 composites, *Journal of Alloys and Compounds*, 458(1–2), 223–230.
26. G. Sandrock (1997). State-of-the-art review of hydrogen storage in reversible metal hydrides for military fuel cell applications, Report Prepared for Department of the Navy, Office of Naval Research, Contract Number N00014-97-M-0001.
27. G. Sandrock (1995). *Hydrogen Energy System—Production and Utilization of Hydrogen and Future Aspects*, Kluwer Academic, Dordrecht, the Netherlands, p. 135.
28. T.N. Veziroglu (1987) Hydrogen technology for energy needs of human settlements, *International Journal of Hydrogen Energy*, 12(2), 99–129.
29. W. Grochala and P. Edwards (2004). Hydrides of the chemical elements for the storage and production of Hydrogen, *Chemical Review*, 104, 1283–1315.
30. S.S.S. Raman and O.N. Srivastava (1996). Hydrogenation behavior of the new composite storage materials, *Journal of Alloys and Compounds*, 241, 167–174.
31. D.J. Davidson, S.S. Sairaman, and O.N. Srivastava (1999). Investigation on the synthesis, characterization and hydrogenation behavior of new Mg-based composite materials Mg-Xwt.% $MmNi_{4.6}Fe_{0.4}$ prepared through mechanical alloying, *Journal of Alloys and Compounds*, 292, 194–201.

21

Complex Hydrides

Sesha S. Srinivasan
Tuskegee University

Prakash C. Sharma
Tuskegee University

Elias K. Stefanakos
University of South Florida

D. Yogi Goswami
University of South Florida

CONTENTS

21.1 Complex Hydrides

In a simple definition, complex metal hydrides are metallic elements (one or more metal elements) bonded with hydrogen. There are various types of complex metal hydrides that can be classified by the amount of different species present in the molecule: binary, ternary, and quaternary hydrides (see Table 21.1).

The general reactions for hydrogen desorption and absorption for binary metal hydrides are as follows:

$$MH_n \rightarrow M + \frac{n}{2} H_{2(g)} \tag{21.1}$$

$$M + \frac{n}{2} H_{2(g)} \rightarrow MH_n \tag{21.2}$$

Ternary and quaternary metal hydrides follow complex reactions path that are specific for each compound.

TABLE 21.1

Classification of Complex Hydrides

Classification	Example
Binary[3]	LiH, MgH_2, etc.
Ternary[3]	$LiBH_4$, $Zn(BH_4)_2$, etc.
Quaternary	$LaMg_2NiH_7$,[29] $LiMg_2RuH_7$,[30] etc.

The group one, two, and three light elements, for example, Li, Mg, B, Al, can form a large variety of metal–hydrogen complexes. They are very interesting due to their light weight and the number (often 2) of hydrogen atoms per metal atom. The principal distinction between the complex hydrides and the previously described metallic hydrides is the transition to an ionic or covalent compound of the metals upon hydrogen absorption. The hydrogen in the multifaceted hydrides is often located in the corners of a tetrahedron with boron or aluminum in the center. The negative charge of the anion, $[BH_4]^-$ and $[AlH_4]^-$, is compensated by a cation, for example, Li^+ or Na^+.[1]

There have been reports that there are about 70 known complex hydrides; some of those, such as $BaReH_9$ with 2.7 wt.% of hydrogen,[2] have been reported to dehydride below 100°C, but the low hydrogen content renders them ineffective as storage materials. The one family of complex hydrides that contains as much hydrogen as the aluminum hydrides is the borohydrides. The borohydrides vary greatly in hydrogen content, up to a maximum of 20.8 wt.% for $Be(BH_4)_2$. This compound is normally not considered because of the toxicity of beryllium; however, there are numerous other known borohydrides. These hydrides release hydrogen slowly. Unlike the metallic hydrides, hydrogen is released via cascade decompositions in complex hydrides, and the step reactions call for different circumstances. Therefore, there is a big difference between the theoretical and the practically attainable hydrogen capacities.

There are about 234 complex chemical hydrides that have been reported with theoretical hydrogen storage capacity.[3] A list of various complex chemical hydrides presently under investigation with their available capacities and operating temperature is given in Table 21.2.

21.1.1 Alanates

Complex hydrides, $MAlH_4$, MBH_4, and $N(AlH_4)_2$ (M = Na, Li, K; N = Mg), are emerging as promising hydrogen storage materials because of their high potential storage capacity.[3] However, they are generally characterized by irreversible dehydriding or extremely slow hydrogen cycling kinetics. The breakthrough discovery of doping with a few mole percent of Ti catalyst has enhanced the dehydrogenation kinetics of $NaAlH_4$ at low operating temperatures (<150°C) and was the starting point in reinvestigating these complex hydride systems for hydrogen storage.[4] However, reduced availability of reversible hydrogen (~4 to 5 wt.%), poor cyclic stability, and loss of the catalytic function of Ti species necessitate the search for new and efficient complex hydride systems. The hydride complexes such as $NaAlH_4$ and $NaBH_4$ are known to be stable and decompose only at elevated temperatures, often above the melting point of the complex. However, the addition of a few mole concentrations of titanium species to $NaAlH_4$ eases the release of hydrogen at moderate temperatures and ambient pressure.[4] The decomposition of Ti-doped $NaAlH_4$ proceeds in two

TABLE 21.2

Theoretical Hydrogen Storage Capacities of Complex Hydrides

No.	Complex Chemical Hydride	Theoretical Capacity (wt.%)	Reversible Capacity (wt.%)	Operating Temperature (°C)	Remarks
1	Ti-doped $NaAlH_4$	7.5	5.5	100–150	High rehydrogenation pressure, poor cycle life, loss of catalytic activity, less available capacity
2	Undoped and Ti-doped $LiAlH_4$	10.5	6.3	120–170	Problems with reversibility and reduced thermodynamic stability
3	Undoped and doped $LiBH_4$	18.2	9.0	200–400	High operating temperature, rehydrogenation problem, possible borane gas evolution
4	$Mg(AlH_4)_2$	9.3	6.6	200–250	High operating temperature, thermodynamic stability
5	$NaBH_4/H_2O$	10.5	9.2	Ambient	Hydrolysis reaction, irreversibility, one-time use
6	Li_3N ($LiNH_2$/ LiH)	11.3	6.5–7.0	255–285	High operating temperature, possible ammonia evolution
7	B-H-Li-N	10.0		80–150	Rehydrogenation problem
8	AlH_3	10.5		150	Ball-milling-induced decomposition, irreversible
9	H_3BNH_3	18.3	12.6		Ammonia evolution possibility, irreversible

steps with the total released hydrogen of ~5.5 wt.% at 100°C–150°C as given in Equations 21.4 and 21.5:

$$3NaAlH_4 \rightleftarrows Na_3AlH_6 + 2Al + 3H_2\left(3.72 \text{ wt.\% } H_2\right) \tag{21.3}$$

$$Na_3AlH_6 \rightleftarrows 3NaH + Al + \frac{3}{2}H_2\left(1.8 \text{ wt.\% } H_2\right) \tag{21.4}$$

Figure 21.1 represents the TGA profile of two-step hydrogen decomposition from $NaAlH_4$ with hydrogen release in weight percentage.[5] The differential seanning calorimetry (DSC) curves at different ramping rates such as 1°C–3°C/min also demonstrated two-step hydrogen decomposition by endothermic transitions as exhibited in Figure 21.1. It is also discernible that the heats of decomposition obtained from these endothermic profiles (ΔH = 30–40 kJ/mol H_2) are well matched with the theoretical limits.

Another advancement pertaining to the successful synthesis of $NaAlH_4$ from ball milling of NaH and Al in presence of Ti catalyst was reported.[6] Long-term hydrogenation and dehydrogenation cycling on this system reveals the complete reversibility of the Na_3AlH_6 phase; however, the partial regeneration of $NaAlH_4$ phase was obtained due to the lack of available aluminum and the catalytic activity of titanium in the host matrix (see Figure 21.2).

Following this breakthrough discovery, an effort was initiated in the US DOE hydrogen program to develop $NaAlH_4$ and related alanates as hydrogen storage materials.[7,8] Another complex

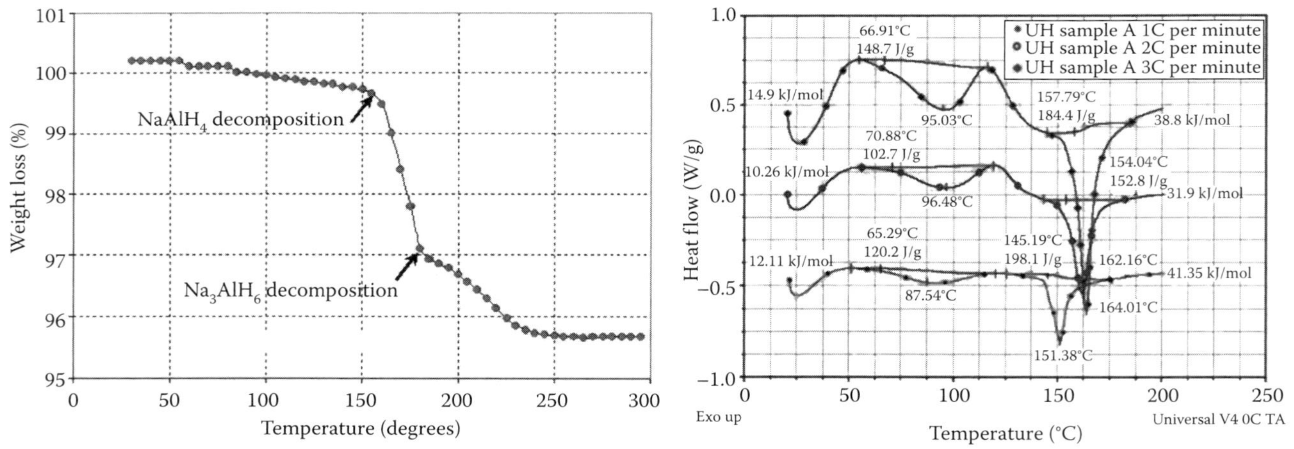

FIGURE 21.1
TGA and DSC profiles of $NaAlH_4$ with two steps of hydrogen decomposition.

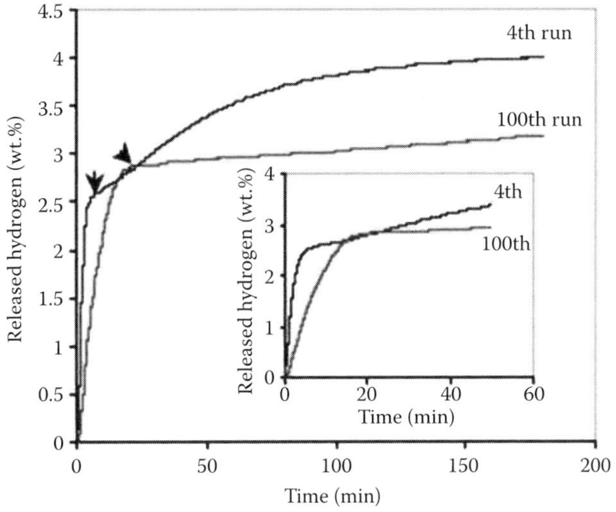

FIGURE 21.2
Hydrogen decomposition in $NaAlH_4$ during the 4th and 100th cycles.

TABLE 21.3
New Complex Hydrides and Their Hydrogen Storage Capacity

S. No.	Complex Hydride	Theoretical Capacity (wt.%)	Decomposition Temperature, T_{dec} (°C)
1	$LiAlH_2(BH_4)_2$	15.2	
2	$Mg(BH_4)_2$	14.8	260–280
3	$NH_4Cl + LiBH_4$	13.6	>ambient
4	$Ti(BH_4)_3$	12.9	ca. 25
5	$Fe(BH_4)_3$	11.9	−30 to −10
6	$Ti(AlH_4)_4$	9.3	−85
7	$Zr(BH_4)_3$	8.8	<250
8	$Zn(BH_4)_2$	8.4	85

hydride, $Mg(AlH_4)_2$, contains 9.6 wt.% of hydrogen that decomposes below 200°C.[9] Some of the new complex hydrides and their theoretical capacities are listed in Table 21.3.

21.1.2 Borohydrides

Borohydride complexes with suitable alkali or alkaline earth metals are a promising class of compounds for hydrogen storage. $LiBH_4$ is one of the lightweight and high hydrogen storage capacity materials. It has a theoretical gravimetric and volumetric hydrogen storage capacities of 18.5 wt.% and 121 kg H_2/m^3,[10] which surpasses the DOE and FreedomCAR targets. Initial investigations have demonstrated that the hydrolysis reaction (21.5)[11–13] of $LiBH_4$ at room temperatures effectively releases hydrogen exothermically; however, the process is irreversible with $LiBO_2$ as a by-product. Hydrolysis reaction proceeds as follows:

$$LiBH_4 + 2H_2O \rightarrow LiBO_2 + 4H_2 \tag{21.5}$$

TABLE 21.4

Thermal Analysis of $LiBH_4$

Temperature Range (°C)	Description
108–112	Endothermic peak. Structural transition.
268–286	Fusion process with a slight weight loss.
380	Main weight loss due to H_2 decomposition.
483–492	Authors are not certain; however, it coincides with a weight loss.

TABLE 21.5

Thermal Studies by Zuttel et al. on $LiBH_4$

Temperature Range (°C)	Description
100	Structural transition with a slight weight loss
270	Fusion phase
320	First significant weight loss
400–500	Second significant weight loss

TABLE 21.6

Thermal Studies by Orimo et al. on $LiBH_4$[14]

Temperature Range (K)	Description
380 (~107°C)	Structural transition
550 (~277°C)	Melting phase
600 to 700 (~327°C to 427°C)	Dehydriding reaction

An alternate route that has been previously investigated is to thermally destabilize $LiBH_4$. $LiBH_4$ has a known melting point around 275°C.[18] For the thermal properties of $LiBH_4$, refer to Table 21.4.

Recent thermal studies by Zuttel et al.[18] and Orimo et al.[14] further expand the understanding regarding $LiBH_4$ behavior (see Tables 21.5 and 21.6).

As an additional note, the structural transition of $LiBH_4$ has been found to be from orthorhombic[15] to hexagonal structure.[16]

According to Stasinevich and Egorenko,[17] the decomposition of alkali metal tetrahydroborides can proceed as reactions:

$$MBH_4 \rightarrow M + B + 2H_2 \tag{21.6}$$

$$MBH_4 \rightarrow MH + B + \frac{3}{2}H_2 \tag{21.7}$$

For $LiBH_4$, the dehydriding reaction (21.8) and rehydriding reaction (21.9) can be generally described as

$$LiBH_4 \rightarrow LiH + B + \frac{3}{2}H_2 \tag{21.8}$$

$$LiH + B + \frac{3}{2}H_2 \rightarrow LiBH_4 \tag{21.9}$$

Theoretically, the reaction (21.8) releases around 13.8 wt.% of hydrogen. According to Orimo et al.,[14] the rehydrogenation of $LiBH_4$ was achieved; however, the energy levels consumed during this process, 35 Mpa of hydrogen and 873 K, might not make it cost-effective. Additional investigations performed with $LiBH_4$ included the doping with SiO_2 as a catalyst, thus enabling the decrease of $LiBH_4$ dehydriding temperature to 300°C.[18] A recent report indicates a dehydrogenation–rehydrogenation cycle improvement and reducing the reaction enthalpy of $LiBH_4$ by the addition of MgH_2 in a ratio of $2LiBH_4 + MgH_2$.[19] The addition of MgH_2 reversibly destabilizes $LiBH_4$, which in consequence increases the hydrogen equilibrium pressure as reaction (21.10) and also the scheme of Figure 21.3:

$$LiBH_4 + \frac{1}{2}MgH_2 \leftrightarrow LiH + \frac{1}{2}MgB_2 + 2H_2 \tag{21.10}$$

$Zn(BH_4)_2$ is a ternary complex metal borohydride with a decomposition temperature of around 85°C.[3] Its theoretical hydrogen capacity is about 8.5wt.%, and it can be synthesized by metathesis reaction of $NaBH_4$ and $ZnCl_2$ in diethyl ether.[20] A recent report from Eun Jeon et al.[21] indicates that zinc borohydride was successfully synthesized by ball milling $ZnCl_2$ and $NaBH_4$ without the use of a solvent; refer to the following reaction:

$$ZnCl_2 + 2NaBH_4 \rightarrow Zn(BH_4)_2 + 2NaCl \tag{21.11}$$

An example of the decomposition of a borohydride ($Zn(BH_4)_2$) is as follows:

$$Zn(BH_4)_2 \leftrightarrow Zn + 2B + 4H_2 \tag{21.12}$$

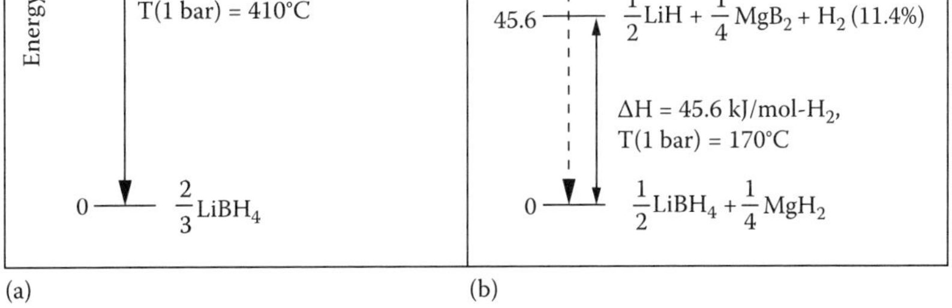

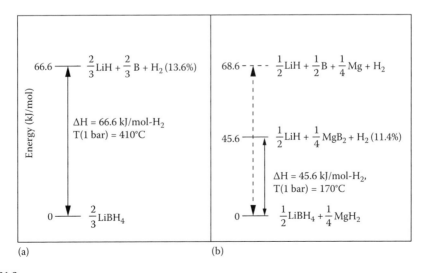

FIGURE 21.3
Dehydrogenation reaction mechanism with thermodynamic values for (a) pure $LiBH_4$ and (b) destabilized $LiBH_4$ with MgH_2.[25]

which shows $Zn(BH_4)_2$ thermally decomposing into the constituent elements with release of hydrogen. The main reason that complex hydride compounds have not been considered for hydrogen storage before is their reported lack of reversibility.

21.1.3 Amides

Recently, a new hydrogen storage system, Li_3N, which absorbs 11.5 wt.% of hydrogen reversibly, has been reported.[22,23] The hydrogenation of Li_3N is a two-step reaction as shown in the schematics of Figure 21.4.

Li_3N absorbs 5.74 wt.% of hydrogen for the first step and 11.5 wt.% for second step. Since the hydrogen pressure for the reaction corresponding to the first step is very low (about 0.01 bar at 255°C), only the second step reaction of Li_2NH with H_2 leads to the reversible storage capacity. The plateau pressure for imides hydrogenation is 1 bar at a relatively high temperature of 285°C.[24] However, the lower temperature of reaction at 220°C and the increase of plateau pressure (27–32 bars) was achieved with MgH_2 addition as demonstrated in Figure 21.5.[25,26] Further research on this system may lead to additional improvements in operating conditions with improved capacity.

21.1.4 Quaternary Hydrides

There have been numerous works in finding complex hydrides for hydrogen storage, the most important and recent of which are summarized in Figure 21.6. While this figure certainly represents only a small portion of research performed on these systems, it nevertheless clearly illustrates that most materials either require temperatures that are too high for practical use or simply have a capacity that is too low. This figure summarizes mainly borohydride- and amide-based materials for hydrogen storage, as mentioned in the previous sections.

While it appears that there are several materials that would meet the DOE guidelines, the amidoborane samples,[27] such as $LiNH_2BH_3$ or $NaNH_2BH_3$, are nonreversible, thereby

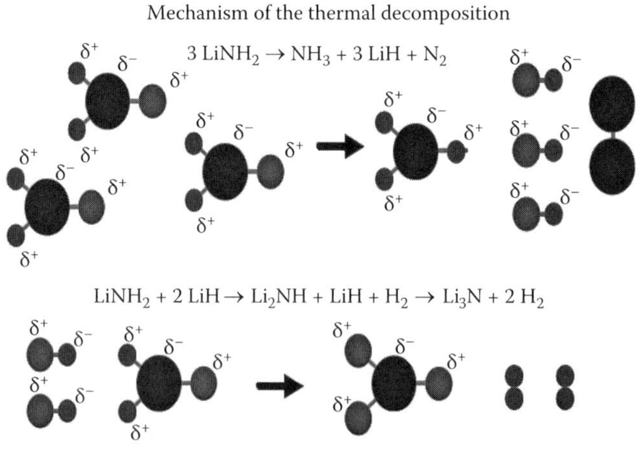

FIGURE 21.4
Schematics and the stepwise decomposition reaction of Li-amide system.[23]

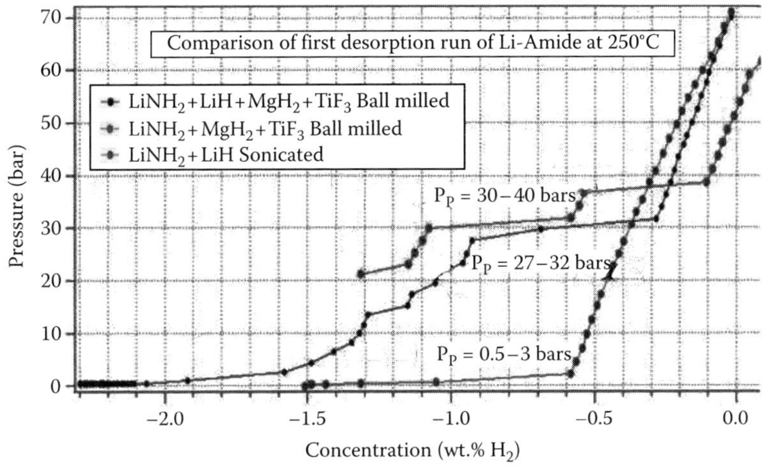

FIGURE 21.5
Pressure-composition isotherms of Li amide and Li amide/Mg hydride.

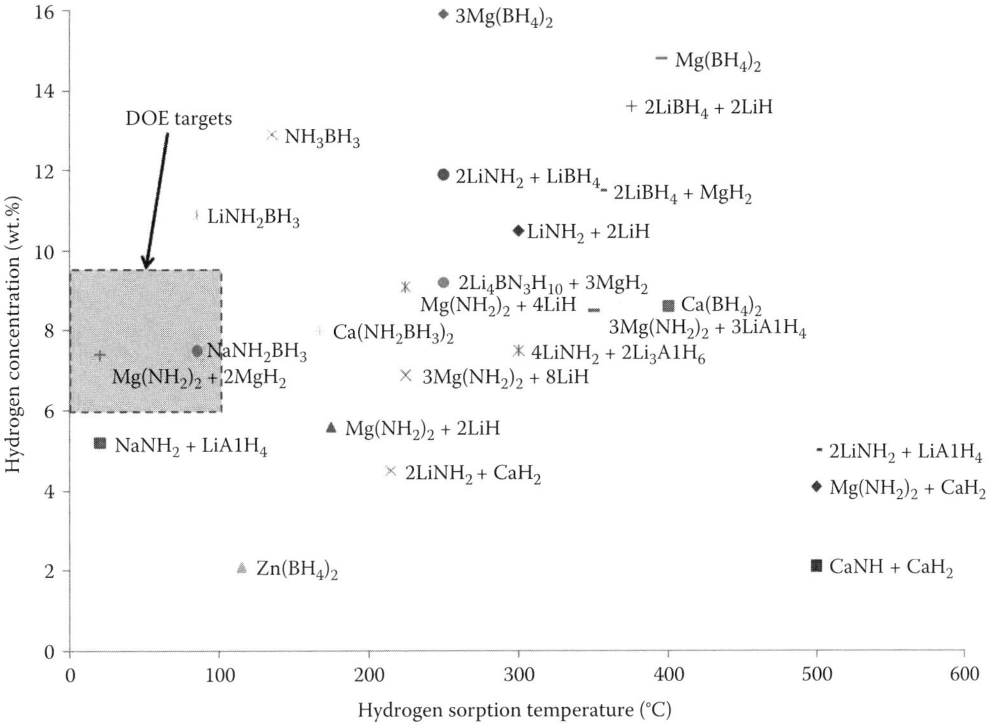

FIGURE 21.6
Hydrogen sorption capacity and temperature of selected complex hydrides and chemical hydrides with DOE target range highlighted.

making the systems impractical for mobile use, as required by the DOE. $Mg(NH_2)_2$ + $2MgH_2$ was found to release 7.6wt.% around room temperature during ball milling[28] but exhibited such a low enthalpy that a high pressure (much higher than that required by the DOE) would be required to rehydrogenate the material, thereby also making the material impractical for use as a reversible hydrogen storage system. A promising system for hydrogen storage has been magnesium amide ($Mg(NH_2)_2$) with a capacity of between 5.6 and 9.2 wt.%,[14,28,29] though all of these systems require temperatures of close to 200°C with a reduction in capacity directly proportional to the reduction in temperature.

By combining the advantages of some of these systems, namely, the borohydride family of materials with the magnesium amide systems, it is thought that a combinatorial effect can be achieved with a reduction in hydrogen sorption temperature, reversibility, as well as a high hydrogen capacity. The overall goal of the investigation of complex hydrides for hydrogen storage is to reduce the hydrogen release temperature, which can be accomplished either by reducing the particle size, as is the case for MgH_2, or by destabilizing the material through the addition of catalysts or other additives. Ball milling is the chosen processing technique, as this combines both chemical and mechanical synthesis of the material. By ball milling, a homogenous mixture with reduced particle size can be achieved, as schematically indicated by Figure 21.7. The parent compounds are combined in the ball mill container, and through milling at high speeds, the materials grind each other down to

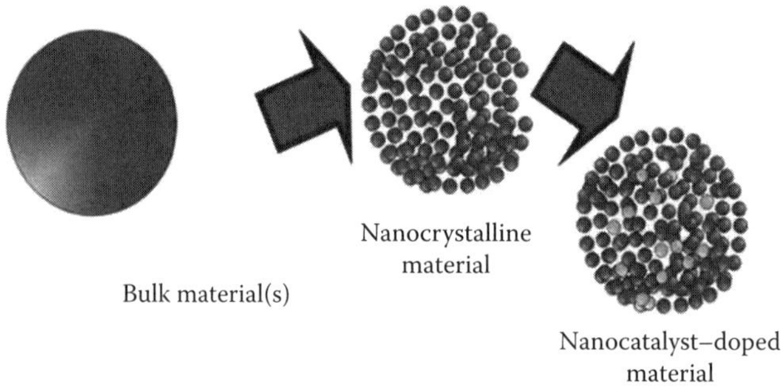

FIGURE 21.7
Schematic mechanochemical synthesis approach for complex hydrides to reduce particle size and achieve a homogenous mixture of parent compounds.

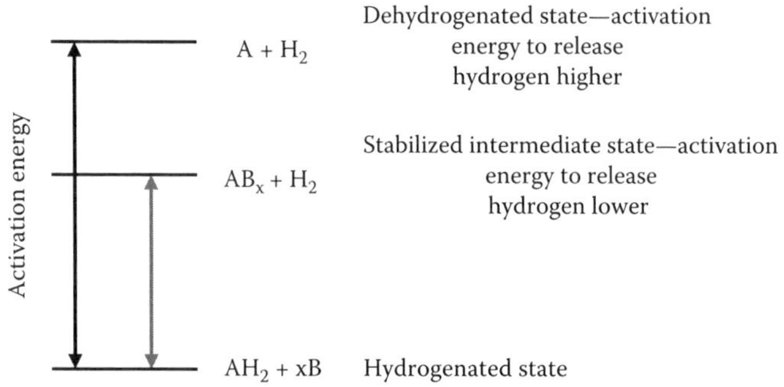

FIGURE 21.8
Destabilization of complex hydride through additives or catalysts.

a smaller particle size and produce a homogenous mixture, possibly with a new chemical composition.

Furthermore, by combining several materials, whether they are in small quantities, so as to count as a catalyst, or in larger quantities, to be considered destabilizers, the activation energy for hydrogen release or absorption can be altered and ideally brought to a point where a low temperature is enough to release the hydrogen. When a material is destabilized, it can react with the additive during dehydrogenation to form a new compound, one that requires a lower energy, as schematically shown in Figure 21.8. An additive, B, can allow the hydrogenated material, A, to form an intermediate compound, AB, while releasing hydrogen, thereby lowering the activation energy required for desorption. Various nano-additives are investigated for the complex hydride LiBNH as well as MgH_2, which is added in larger quantities. The details are described in the specific upcoming sections, as appropriate.

References

1. Z.P. Li, B.H. Liu, K. Arai, and S. Suda (2005). Development of the direct borohydride fuel cell, *Journal of Alloys and Compounds, 404–406*, 648–652.
2. E. Orgaz and M. Gupta (1999). Electronic structure of BaReH9, *Journal of Alloys and Compounds, 293–295*, 217–221.
3. Hydrogen Storage Materials Database, US Department of Energy, http://www.hydrogenmaterialssearch.govtools.us/
4. B. Bogdanovic and M. Schwickardi (1997). Ti-doped alkali metal aluminum hydrides as potential novel reversible hydrogen storage materials, *Journal of Alloys and Compounds, 253–254*, 1.
5. D. Sun, S.S. Srinivasan, T. Kiyobayashi, N. Kuriyama, and C.M. Jensen (2003). Rehydrogenation of dehydrogenation $NaAlH_4$ at low temperature and pressure, *Journal of Physical Chemistry B, 107*(37), 10176–10179.
6. S.S. Srinivasan, H.W. Brinks, B.C. Hauback, C.M. Jensen, and D. Sun (2004). Long term cycling behavior of Titanium doped $NaAlH_4$ prepared through solvent mediated milling of Titanium ad-mixed NaH and Al, *Journal of Alloys and Compounds, 377*, 283–289.
7. C.M. Jensen and R.A. Zidan (2002). Hydrogen storage materials and method of making by dry homogenation, US Patent, 6471935.
8. C.M. Jensen and K.J. Gross (2001). Development of catalytically enhanced sodium aluminum hydride as a hydrogen-storage material, *Applied Physics A, 72*, 213.
9. M. Fichtner (2002). Synthesis and structure of magnesium alanates and two solvent adducts, *Journal Alloys and Compounds, 345*(1–2), 286–296.
10. E.M. Fedneva, V.L. Alpatova, and V.I. Mikheeva (1964). $LiBH_4$ complex hydride materials, *Russian Journal of Inorganic Chemistry, 9*, 826.
11. H.I. Schlesinger, H.C. Brown, A.E. Finholt, J.R. Gilbreath, H.R. Hoekstra, and E.K. Hyde (1953). Sodium borohydride, its hydrolysis and its use as a reducing agent and in the generation of hydrogen, *Journal of American Chemical Society, 75*, 215–219.
12. V.C.Y. Kong, F.R. Foulkes, D.W. Kirk, and J.T. Hinatsu (1999). Development of hydrogen storage for fuel cellgenerators. i: Hydrogen generation using hydrolysishydrides, *International Journal of Hydrogen Energy, 24*, 665–675.
13. Y. Kojima, Y. Kawai, M. Kimbara, H. Nakanishi, and S. Matsumoto (2004). Hydrogen generation using sodium borohydride solution and metal catalyst coated on metal oxide, *International Journal of Hydrogen Energy, 29*, 1213–1217.
14. S. Orimo, Y. Nakamori, G. Kitahara, K. Miwa, N. Ohba, S. Towata, and A. Zuttel (2005). Dehydriding and rehydriding reactions of $LiBH_4$, *Journal of Alloys and Compounds, 404–406*, 427–430.

15. P.M. Harris and E.B. Meibohm (1947). Structure of lithium borohydride $LiBH_4$, *Journal of American Chemical Society*, *69*, 1231.
16. S. Gomes, H. Hagemann, and K. Yvon (2002). Lithium boro-hydride $LiBH_4$: II. Raman spectroscopy, *Journal of Alloys and Compounds*, *346*, 206.
17. D.S. Stasinevich and G.A. Egorenko (1968). Thermographic investigation of alkali metal and magnesium tetrahydroborates at pressure up to 10 atm, *Russian Journal of Inorganic Chemistry*, *13*, 341–343.
18. A. Zuttel, S. Rentsch, P. Fischer, P. Wenger, P. Sudan, Ph. Mauron, and Ch. Emmenegger (2003). Hydrogen storage properties of $LiBH_4$, *Journal of Alloys and Compounds*, *356–357*, 515–520.
19. J.J. Vajo, S.L. Skeith, and F. Mertens (2005). Reversible storage of hydrogen in destabilized $LiBH_4$, *Journal of Physical Chemistry B*, *109*, 3719–3722.
20. T.J. Marks and J.R. Kolb (1977). Covalent transition metal, lanthanide, and actinide tetrahydroborate complexes, *Chemical Review*, *77*, 263.
21. E. Jeon and Y.W. Cho (2006). Mechanochemical synthesis and thermal decomposition of zinc borohydride, *Journal of Alloys and Compounds*, *422*, 273–275.
22. P. Chen, Z. Xiong, J. Luo, J. Lin, and K.L. Tan (2002). Interaction of hydrogen with metal nitrides and imides, *Nature*, *420*(2), 302.
23. P. Chen, Z. Xiong, J. Luo, J. Lin, and K.L. Tan (2003). Interaction between lithium amide and lithium hydride, *Journal of Physical Chemistry B*, *107*, 10967.
24. H.I. Schlesinger and H.C. Brown (1940). Metallo borohydrides. III. Lithium borohydride, *Journal of American Chemical Society*, *62*, 3429–3435.
25. W.F. Luo (2004). $LiNH_2$-MgH_2: A viable hydrogen storage system, *Journal of Alloys Compounds*, *381*, 284–287.
26. S.S. Srinivasan, E. Stefanakos, Y. Goswami, M. Jurcyzk, and M. Smith (2006). *16th World Hydrogen Energy Conference*, Leon, France, June'06, Transition metal assisted new complex hydrides for hydrogen storage, *WHEC Proceedings*.
27. Z. Xiong, C.K. Yong, G. Wu, P. Chen, W. Shaw, A. Karkamkar, T. Autrey, M.O. Jones, S.R. Johnson, P.P. Edwards, and W.I.F. David (2008). High-capacity hydrogen storage in lithium and sodium amidoboranes, *Nature Materials*, *7*, 138–141.
28. J. Hu, G. Wu, Y. Liu, Z. Xiong, P. Chen, K. Murata, K. Sakata, and G. Wolf (2006). Hydrogen release from $Mg(NH_2)_2$-MgH_2 through mechanochemical reaction, *Journal of Physical Chemistry B*, *110*, 14688–14692.
29. G. Renaudin, L. Guénée, and K. Yvon (2003). La_2MgNiH_7, a novel quaternary metal hydride containing tetrahedral $[NiH]^{4-}$ complexes and hydride anions, *Journal of Alloys and Compounds*, *350*, 145–150.
30. B. Huang, K. Yvon, and P. Fischer (1994). $LiMg_2RuH_7$, a new quaternary metal hydride containing octahedral $[Ru(H)H_6]^{4-}$ complex anions, *Journal of Alloys and Compounds*, *210*, 243–246.

22

Nanomaterials for Hydrogen Storage

Sesha S. Srinivasan
Tuskegee University

Prakash C. Sharma
Tuskegee University

Elias K. Stefanakos
University of South Florida

D. Yogi Goswami
University of South Florida

CONTENTS

22.1 Nanomaterials for Hydrogen Storage

22.1.1 Nanoparticles/Nanocrystalline

It is generally known that pristine MgH_2 theoretically can release ~7.6 wt.% H_2 at decomposition temperatures >300°C at a H_2 pressure of ~1 bar.[1–3] However, so far, MgH_2 based materials have limited practical applications because both hydrogenation and dehydrogenation reactions are very slow, and hence, relatively high temperatures are required.[4] The phenomenon of mechanical milling helps to pulverize the particles of MgH_2 into micro- or nanocrystalline phases and thus leads to lowering the activation energy of desorption. Without using catalysts, the activation energy of absorption corresponds to the activation barrier for the dissociation of the H_2 molecule and the formation of hydrogen atoms. The activation energies of the H_2 sorption for the bulk MgH_2, mechanically milled MgH_2, and nanocatalyst-doped MgH_2 are 162, 144, and 71 kJ/mol, respectively. It is undoubtedly seen that the activation barrier has been drastically lowered by nanocatalyst doping. Figure 22.1 represents the thermogravimetric profiles of bulk MgH_2, which exhibits weight loss due to H_2 decomposition at 415°C.

However, mechanochemical milling of MgH_2 introduces defects and reduces the particle size. Thus, obtained micro-/nanocrystalline MgH_2 grains show endothermic H_2 decomposition (see Figure 22.2) at an earlier temperature of 340°C. In addition to nanoscale formation, doping by a nanocatalyst certainly decreases the onset transition temperature by as much as 100°C (Figures 22.1 and 22.2).

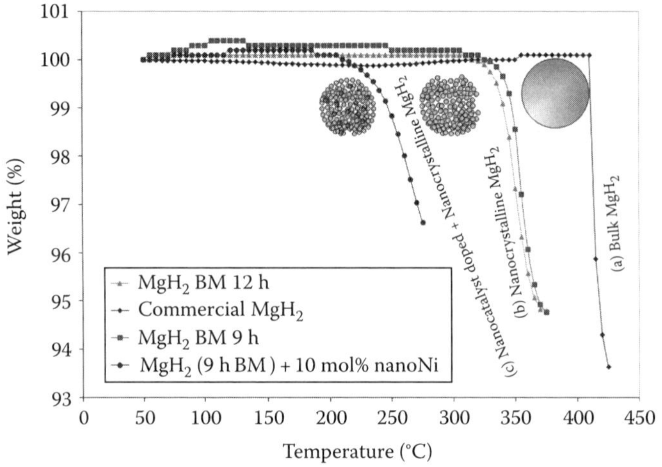

FIGURE 22.1
Thermogravimetric profiles of (a) bulk MgH_2, (b) nanocrystalline MgH_2, and (c) nanocatalyst-doped nanocrystalline MgH_2.

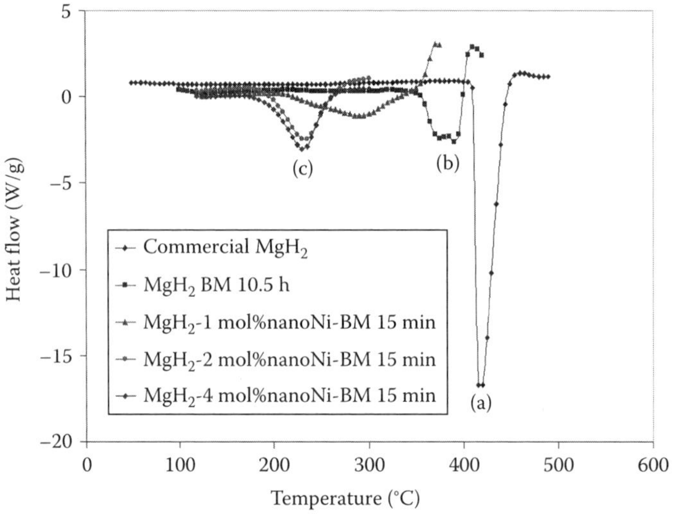

FIGURE 22.2
DSC profiles of (a) bulk MgH_2, (b) nanocrystalline MgH_2, and (c) nanocatalyst-doped nanocrystalline MgH_2.

Moreover, the enthalpy of decomposition, ΔH, obtained from these profiles is given in Table 22.1.

The structural characterization of the bulk, mechanochemically prepared, and nanocatalyst-doped MgH_2 has been carried out using XRD and is shown in Figure 22.3. The Bragg reflection of bulk MgH_2 shows sharp crystalline peaks. However, in the mechanochemically prepared MgH_2, the crystallite size is reduced to nanodimension (approximately five times reduced intensity) with broadening of the FWHM. No additional peaks correspond to MgO, or no other impurities were observed in the ball-milled samples. For the nanocatalyst-doped MgH_2, a small inflection peak at an angle of 45° represents the nano-Ni. A lattice strain was found in the mechanochemically milled samples in comparison to the bulk MgH_2.

TABLE 22.1

Enthalpy of H_2 Decomposition (ΔH_{dec})

Sample	T_{dec} (°C)	ΔH_{dec} (kJ/mol H_2)
Commercial MgH_2	415	55.98 ± 0.50
MgH_2(MC)	375	46.35 ± 0.50
MgH_2(MC)–nano-Ni	225	36.50 ± 0.50

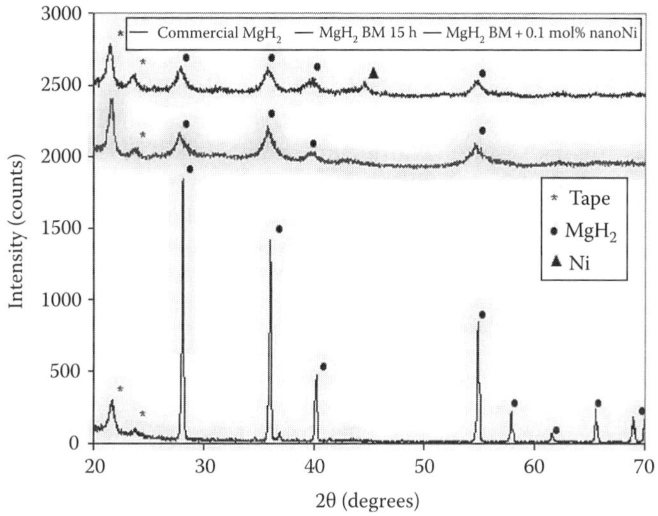

FIGURE 22.3
X-ray diffraction profiles of bulk, mechanochemically processed, and nanocatalyst-doped MgH_2 show the presence of MgH_2 structural phase.

FIGURE 22.4
SEM micrographs of (a) bulk and (b) nanocrystalline MgH_2.

Another significant improvement in the nanocrystalline MgH_2, over bulk samples, is the reduced particle size as confirmed by SEM microstructural observations indicated in Figure 22.4. Bulk MgH_2 exhibits a larger crystallite size of 212 nm, whereas the nanocrystalline counterpart is shown to be about 27 nm, almost an order of magnitude less. The reduction in the particle size is thought to be the cause of the associated reduction of the decomposition temperature by at least 100°C (see Figures 22.1 and 22.2).

22.1.2 Nanocatalytic Dopants

Transition metal complexes, TMH_x (T = Mg; M = Fe, Co, Ni), have been identified as the potential candidates for H_2 storage. These hydrides, especially Mg_2FeH_6, have shown excellent cyclic capacities even without a catalyst.[5] However, at low temperatures, the cyclic nature of the hydride is limited. Also Mg_2FeH_6 shows the highest known volumetric H_2 density of 150 kg/m^3, which is more than double that of liquid H_2 as shown in Figure 22.5.[6] Its gravimetric H_2 density exceeds 5 wt.%, at ~400°C, and is only slightly lower than that of MgH_2 (~7.6 wt.%).

The Mg_2FeH_6 hydride belongs to the family of Mg-transition metal complex hydrides that have TMH_x complexes analogous to the $[AlH_4]^{1-}$ alanate complexes, for example, the $[FeH_6]^{4-}$ complex in Mg_2FeH_6 or $[NiH_4]^{4-}$ in Mg_2NiH_4.[7] MgH_2 forms ternary and quaternary hydride structures by reacting with various transition metals (Fe, Co, Ni, etc.) and thus leads to improved kinetics.[8] Moreover, the nanoscale version of these transition metal particles offers an additional H_2 sorption mechanism via its active surface sites.[9] In a similar way, the synergistic approach of doping nanoparticles of Fe and Ti with a few mol% of carbon nanotubes (CNTs) on the sorption behavior of MgH_2 has recently been investigated.[10] The addition of CNTs significantly promotes H_2 diffusion in the host metal lattice of MgH_2 due to the short pathway length and creation of fast diffusion channels.[11]

Fe does not form intermetallic compounds with Mg but instead readily combines with H_2 and Mg to form the ternary Mg_2FeH_6 hydride[5] according to the following reactions:

$$2Mg + Fe + 3H_2 \leftrightarrow Mg_2FeH_6 \quad (22.1)$$

$$2MgH_2 + Fe + H_2 \leftrightarrow Mg_2FeH_6 \quad (22.2)$$

The synthesis of nanostructured Mg_2FeH_6 has been carried out by an inexpensive mechanochemical process (Fritsch Pulverisette 6) under reactive gas (H_2) milling. The appropriate stoichiometries of elemental Mg (–325 mesh, 99.9% pure) and fine Fe powder

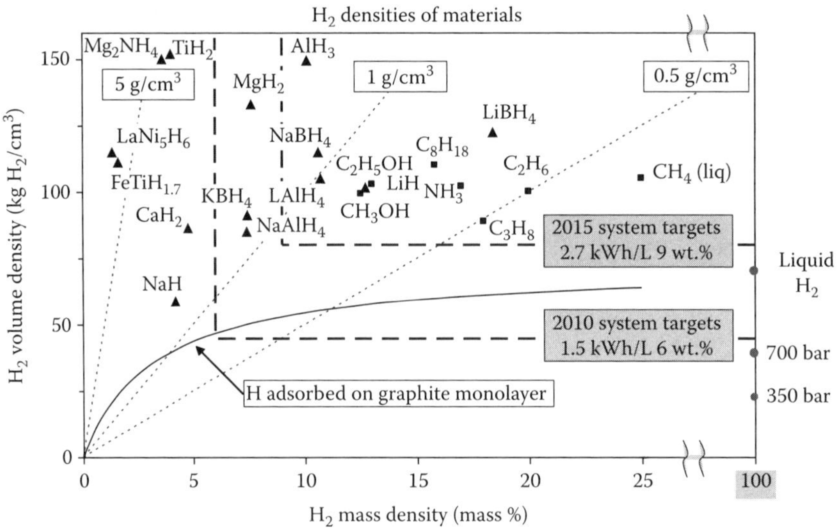

FIGURE 22.5
Volumetric and gravimetric H_2 densities of various hydrogen storage systems.

(–100 mesh, 99.8% pure) were taken in a 2:1 mole ratio and milled under continuous flow of H_2 (~2–3 atm). The milling parameters, such as speed (300 rpm), mode of rotations (forward and reverse motion), and time duration (5 h), were optimized to control the nanostructure grain size of the 2Mg + Fe mixture. After 5 h of milling, the sample was collected in a N_2-filled glove box for further characterization. X-ray diffraction profiles of bulk and nanocrystalline Mg_2FeH_6 are shown in Figure 22.6. It is clearly discernible from this figure that the crystallite size of the nanocrystalline sample is reduced by a factor of 3 when compared to the bulk counterparts. Powder x-ray diffraction of the mixture before and after mechanochemical processing under H_2 ambient shows the disappearance of sharp crystalline peaks corresponding to Mg and formation (FWHM calculations) of nanocrystalline reflections as exhibited in Figure 22.6.

Figure 22.7 represents the comparative differential scanning calorimetric (DSC) patterns of bulk and nanocrystalline Mg_2FeH_6. It is interesting to note that the endothermic peak

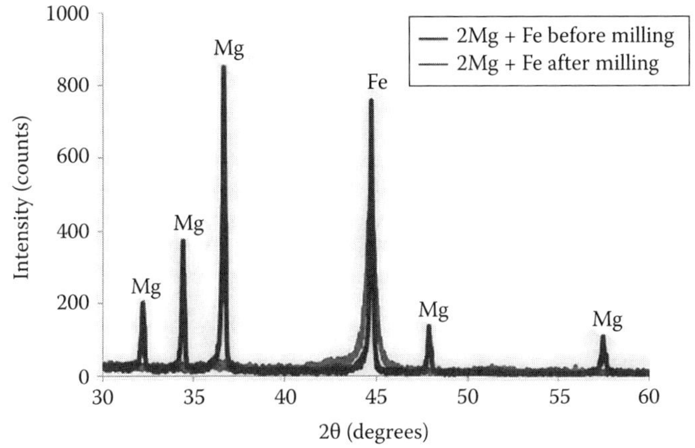

FIGURE 22.6
XRD profiles of bulk and nanocrystalline Mg_2FeH_6.

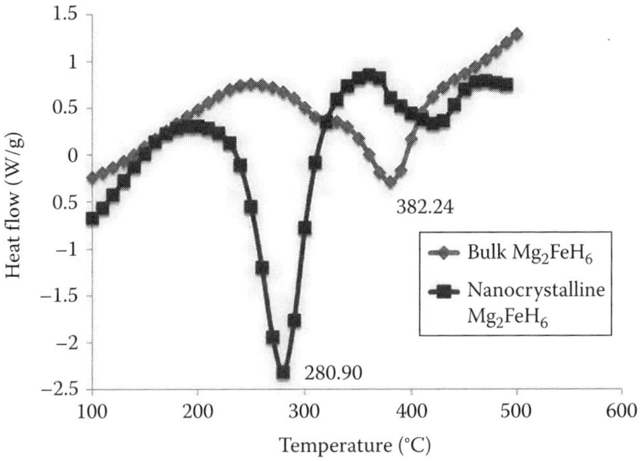

FIGURE 22.7
DSC curves of bulk and nanocrystalline Mg_2FeH_6.

due to H_2 desorption occurs at a lower temperature, 280°C, for the nanocrystalline structure, whereas the bulk sample exhibits an endothermic peak at 383°C. An early onset at around 250°C is observed for the nanocrystalline material (n-Mg_2FeH_6).

We have also presented in the present review that the kinetics of absorption and desorption increase upon nanocrystallization and Ti-catalyst doping into the host lattice structure of Mg_2FeH_6. Figure 22.8 clearly shows at least a threefold–fourfold improvement in the hydrogenation kinetics of Mg_2FeH_6 by the simultaneous nanocrystallization and nanocatalytic doping.

The absorption kinetics of Ti-doped nanocrystalline Mg_2FeH_6 increases with increasing cycle number as depicted in Figure 22.9. This is due to the fact that during each and

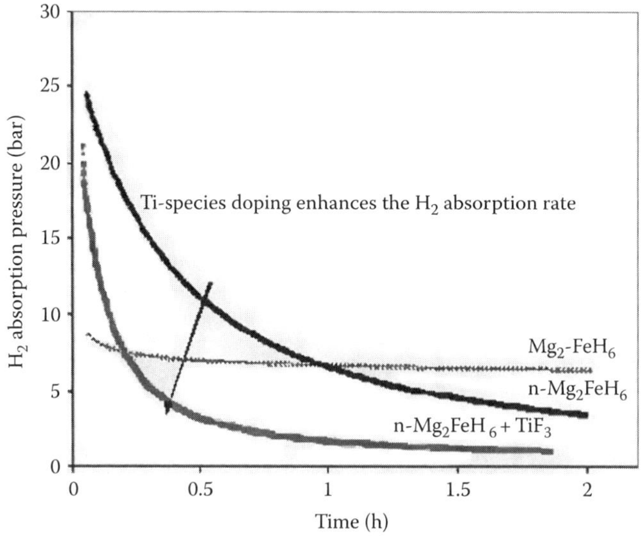

FIGURE 22.8
Absorption kinetics of bulk Mg_2FeH_6, nanocrystalline Mg_2FeH_6, and Ti-doped nanocrystalline Mg_2FeH_6 at 250°C.

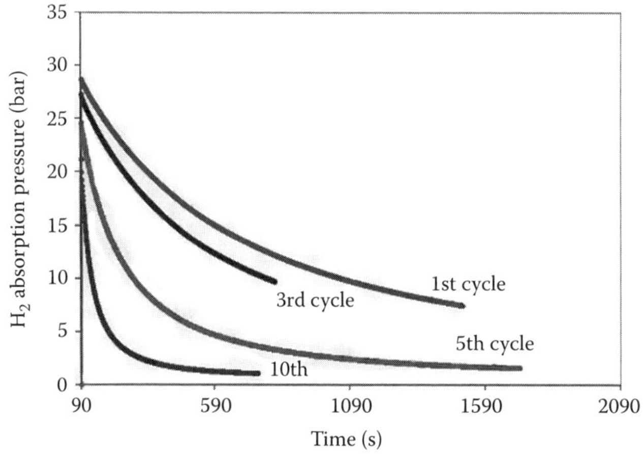

FIGURE 22.9
H_2 absorption kinetics of Ti-doped nanocrystalline Mg_2FeH_6; rate increases with cycling (sample mass of 1.5 g and temperature of desorption of 250°C).

every cycle, greater number of H atoms is being absorbed by the host lattice structure. Additionally, Ti addition only influences the sorption kinetics of nanocrystalline Mg_2FeH_6 and does not improve the overall H capacity.

References

1. Zaluska A, Zaluski L, Strom-Olsen JO. Structure, catalysis and atomic reactions on the nanoscale: A systematic approach to metal hydrides for hydrogen storage, *Appl. Phys. A* 2001;72:157.
2. Price TCE, Grant DM, Walker GS. Synergistic effect of $LiBH_4$ + MgH_2 as a potential reversible high capacity hydrogen storage material, *Ceram. Trans.* 2009;202:97.
3. Niemann MU, Srinivasan SS, McGrath K, Kumar A, Goswami DY, Stefanakos EK, Nanocrystalline effects on the reversible hydrogen storage characteristics of complex hydrides, *Ceram. Trans.* 2009;202:111.
4. Huot J, Liang G, Boily S, Van Neste A, Schulz R. Structural study and hydrogen sorption kinetics of ball-milled magnesium hydride, *J. Alloys Comp.* 1999;293/295:495.
5. Bogdanovic B, Schlichte K, Reiser A. Thermodynamic properties and cyclic-stability of the system Mg_2FeH_6. In: *Hydrogen Power: Theoretical and Engineering Solutions*, T.O. Saetre, ed., Kluwer Academic, Dordrecht, the Netherlands, 1991, p. 291.
6. Zuttel A, Wenger P, Rentsch S, Sudan S, Mauron P, Emmenegger C. $LiBH_4$ a new hydrogen storage material, *J. Power Sources* 2003;118:1.
7. Sandrock G. A panoramic overview of hydrogen storage alloys from a gas reaction point of view, *J. Alloys Comp.* 1999;293–295:877.
8. Yvon K, Bertheville B. Magnesium based ternary metal hydrides containing alkali and alkaline-earth elements, *J. Alloys Comp.* 2006;425:101.
9. Jeon K-J, Theodore A, Wu C-Y, Cai M. Hydrogen absorption/desorption kinetics of magnesium nano-nickel composites synthesized by dry particle coating technique, *Int. J. Hydrogen Energy* 2007;32(12):1860–1868.
10. Yao X, Wu CZ, Wang H, Cheng HM, Lu GQ. Effects of carbon nanotubes and metal catalysts on hydrogen storage in magnesium nanocomposites, *J. Nanosci. Nanotechnol.* 2006;6(2):494.
11. Wu CZ, Wang P, Yao X, Liu C, Chen CM, Lu GQ, Cheng HM. Hydrogen storage properties of MgH_2/SWNT composite prepared by ball milling, *J. Alloys Comp.* 2006;420:278.

23

Chemical Hydrogen Storage

Sesha S. Srinivasan
Tuskegee University

Prakash C. Sharma
Tuskegee University

Elias K. Stefanakos
University of South Florida

D. Yogi Goswami
University of South Florida

CONTENTS

23.1 Chemical Hydrogen Storage

The term *chemical hydrogen storage* is used to describe storage technologies in which hydrogen is generated through a chemical reaction. Common reactions involve chemical hydrides with water or alcohols. Typically, these reactions are not easily reversible on-board a vehicle. Hence, the *spent fuel* or by-products must be removed from the vehicle and regenerated off-board.

Hydrolysis reactions involve the oxidation reaction of chemical hydrides with water to produce hydrogen. The reaction of sodium borohydride has been the most studied to date.[1] This reaction is

$$NaBH_4 + 2H_2O \rightarrow NaBO_2 + 4H_2 \tag{23.1}$$

In the first embodiment, slurry of an inert stabilizing liquid protects the hydride from contact with moisture and makes the hydride pumpable. At the point of use, the slurry is mixed with water and the consequent reaction produces high-purity hydrogen. The reaction can be controlled in an aqueous medium via pH and the use

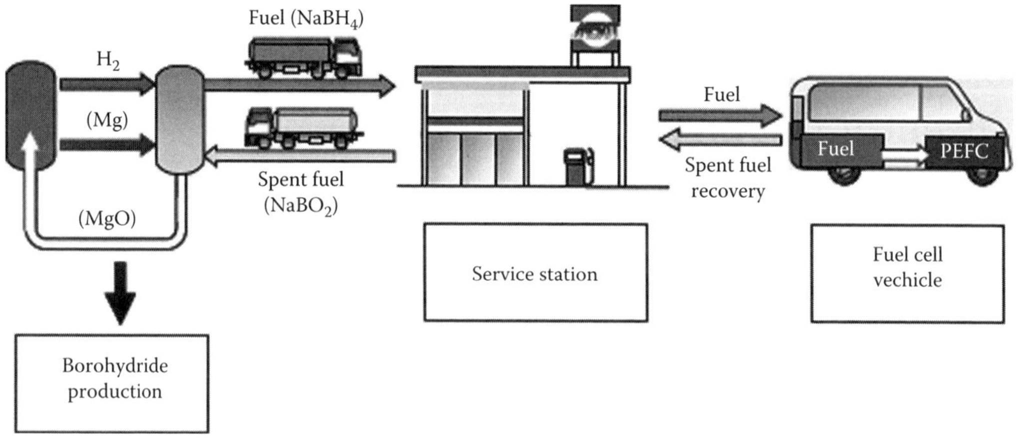

FIGURE 23.1
Hypothetical hydrogen economy using $NaBH_4$ as hydrogen storage.

of a catalyst. While the material hydrogen capacity can be high and the hydrogen release kinetics fast, the borohydride regeneration reaction must take place off-board. Regeneration energy requirements cost and life-cycle impacts are key issues currently being investigated.

Millennium Cell has reported that their $NaBH_4$-based Hydrogen on Demand™ system possesses a system gravimetric capacity of about 4 wt.%.[1] Similar to other material approaches, the issues include system volume, weight and complexity, and water availability.

Another hydrolysis reaction that is presently being investigated by Safe Hydrogen is the reaction of MgH_2 with water to form $Mg(OH)_2$ and H_2.[2] In this case, particles of MgH_2 are contained in nonaqueous slurry to inhibit premature water reactions when hydrogen generation is not required. Material-based capacities for the MgH_2 slurry reaction with water can be as high as 11 wt.%. However, similar to the $NaBH_4$ approach, water must also be carried on-board the vehicle in addition to the slurry, and the $Mg(OH)_2$ must be regenerated off-board.

An idea on how the complex metal hydrides would be processed during the hydrogen economy is depicted in Figure 23.1 using sodium borohydride ($NaBH_4$).

A new chemical approach may be hydrogen generation from ammonia borane materials by the following reactions:

$$NH_3BH_3 \rightleftarrows NH_2BH_2 + H_2 \rightleftarrows NHBH + H_2 \tag{23.2}$$

The first reaction, which occurs at less than 120°C, releases 6.1 wt.% hydrogen, while the second reaction, which occurs at approximately 160°C, releases 6.5 wt.% hydrogen.[3] Recent studies indicate that hydrogen release kinetics and selectivity are improved by incorporating ammonia borane nanosized particles in a mesoporous scaffold.

References

1. Wu C, Bai F, Yi Y, Zhang B. Cobalt boride catalysts for hydrogen generation from alkaline $NaBH_4$ solution. *Materials Letters* 2005;59:1748–1751.
2. Mcclaine AW, Tullmann S, Brown K. Chemical hydride slurry for hydrogen production and storage, *Proceedings of the US DOE Hydrogen Energy Program*, Progress Report, 2004:204–209.
3. Karkamkar A, Aardahl C, Autrey T. Recent developments on hydrogen release from ammonia borane, *Material Matter* 2007;2(2):6–9.

24

Hydrogen Adsorption and Storage on Porous Materials

K. Mark Thomas
Newcastle University

CONTENTS

24.1 Introduction

The current interest in hydrogen as a sustainable fuel is related to strategic and environmental concerns. The former includes security of supply with the increasing worldwide demand for energy, concerns that world oil production is close to reaching a peak,[1] and the long-term availability of oil supplies. The latter is related to the need for more environmentally friendly alternative fuel due to pollution from fossil fuel combustion leading to poor air quality and global warming due to carbon dioxide emissions.

There are major scientific and engineering challenges, which need to be overcome before the development of a hydrogen-based economy and widespread use of hydrogen as an alternative fuel to oil can be achieved. The storage of hydrogen for use in vehicles is perhaps the most difficult problem to overcome. Handling and storage technologies for hydrogen in an industrial situation, where safety issues can be closely controlled, are well established. However, the use of hydrogen for onboard transport applications involving the general public is much more difficult because of safety, scientific, engineering, and economic constraints.

Hydrogen storage requirements for vehicles are extremely challenging because of safety issues and are determined by storage volume requirements that should not compromise interior space and performance requirements. The US DOE hydrogen onboard storage system targets[2] for vehicles include the weight of the storage tank and associated cooling, pressure, and delivery control equipment in the calculations for real vehicle applications. The target for 2015 is a hydrogen-powered average corporate vehicle with storage capacity of 5–13 kg of hydrogen giving a refueling range >300 miles (483 km) with a refueling time < 5 min. Smaller more fuel-efficient cars are likely to be used in the future leading to slightly less demanding hydrogen storage requirements.

When hydrogen is used as a fuel for transport applications, a much larger storage tank is required than used for current hydrocarbon-fuelled vehicles. The reason for this is that while hydrogen has a higher combustion enthalpy than hydrocarbon fuels on a weight basis, it has a much lower value on a volume basis. Safety issues and convenience are paramount for storage applications and there are severe practical limitations for tank size, refueling range, etc. The limitations on storage space for hydrogen in vehicles lead to the requirement for greater efficiency, and this could be provided by fuel-cell-powered vehicles, which have greater efficiency than internal combustion engines. However, larger storage volumes are still required compared with current vehicles. The durability of fuel cells and their current reliance on precious metal catalyst are issues. The development of a hydrogen economy has potential strategic and environmental benefits provided that hydrogen can be produced in an environmentally sustainable manner leading to reduced reliance on imported oil, air pollution, and release of carbon dioxide.

The development of a suitable hydrogen storage method is the major scientific challenge to overcome, before the technology necessary for change from petroleum to sustainable hydrogen, as an energy carrier for transport applications can be implemented. The storage methods currently being considered include high-pressure gas, liquid hydrogen, adsorption on porous materials at lower pressures, complex hydrides, and hydrogen intercalation in metals.[3] The main constraint for vehicles is the volume required to store sufficient hydrogen for acceptable refueling ranges. Although hydrogen storage tanks up to 70 MPa have been used, there is still a limitation on the amount of hydrogen that can be delivered from a given size of tank. The energy required to compress the gas is a consideration in relation to total energy use. Liquid hydrogen provides a higher hydrogen storage density ($\sim \times 2$) that of compressed hydrogen gas tank, but there are limitations due to the extremely low temperatures required (critical temperature 32.98 K),[4] and significant losses of hydrogen may occur due to evaporation during storage. Liquefaction costs are also significant. Compressed gas and liquid hydrogen storage methods currently available do not fully satisfy the safety and refueling range criteria for hydrogen storage for vehicles. Therefore, novel hydrogen storage methods involving high-pressure vessel containing porous materials, metal hydrides, or other materials are being considered. This approach combines the advantages of both compressed gas and solid state material storage to increase the total hydrogen storage capacity at lower pressures than compressed gas alone.

Porous materials have nanometer-sized pores that provide large internal surface areas and pore volumes. These materials are widely used for gas storage,[3,5] purification and separation,[6,7] removal of trace pollutant species,[8,9] and as catalysts and catalyst supports.[10,11] Hydrogen physisorption on range porous materials is one of the storage methods being investigated for use in transport applications.[12–22] A wide range of carbons, aluminosilicates, aluminophosphates, silicas, microporous polymers, covalent organic frameworks (COFs), and porous metal–organic framework (MOF) materials with a wide variety of porous structures, surface chemistry, and structural flexibility have been studied. Porous carbons have relatively rigid structures and the surface sites are limited to hydrophobic graphene layers and oxygen and nitrogen surface functional groups. The structures of MOF coordination polymers were reported[23,24] but the first gas adsorption studies were not published until 1997.[25] These materials may have diverse surface chemistry,[26] pore architectures,[27] framework flexibility,[12] and very large internal surface areas and pore volumes.[28] Porous coordination polymers may be prepared from a wide range of multidentate ligands and metals or metal clusters (secondary building units [SBUs]) to give materials with diverse surface chemistry and porous structure characteristics.[27,29,30] The pore structure and surface chemistry of MOFs may be tailored to specific applications and have the greatest range of surface chemistry for optimizing hydrogen interactions with surface sites varying from unsaturated metal centers to hydrophobic ligand surfaces. MOFs with zeolite-type structures have also been synthesized and these may also have applications for gas storage.[31,32]

Hydrogen adsorption characteristics have good correlations with a variety of pore structural characteristics determined by gas adsorption methods for a wide range of porous materials.[13,17,19] The current status of experimental studies of hydrogen physisorption and storage on porous materials is reviewed and future prospects are discussed.

24.2 Experimental Methods for Hydrogen Adsorption

Hydrogen adsorption measurements for porous materials require stringent experimental protocols and validation using extensive reproducibility studies in order to establish the accuracy.[13,29,33] These measurements can be obtained using either (a) direct gravimetric instruments of adsorption or (b) indirect volumetric (or manometric) measurements using a Sievert apparatus. Gravimetric measurements are usually used up to ~20 bar and volumetric measurements can be used up to 200 bar pressure although results have usually only been reported up to 100 bar. Volumetric instruments have a simpler design, which is cheaper to construct for high-pressure instruments. Adsorption/desorption in volumetric instruments is driven by pressure differences. Gravimetric measurements are carried out under isobaric conditions (constant chemical potential) and, hence, may provide accurate adsorption/desorption kinetics.[6,34] Both measurement techniques require ultraclean high-vacuum systems with all metal seals with diaphragm and turbo pumps, which can be evacuated to 10^{-10} bar to completely degas the material, and a gas purification system is needed to remove trace impurities from hydrogen. The buoyancy corrections for gravimetric measurements and the sample volume measurement for volumetric measurements are equivalent since helium is used to determine both corrections.[33] The equilibrium degassed state for porous material is easily determined for gravimetric instruments, but establishing that the sample is completely degassed is more difficult in the case of volumetric measurements. In the latter, errors associated with introduction of doses of gas are cumulative for each isotherm pressure step and increase with increasing pressure.[33,35] Also, the importance of the amount remaining unadsorbed in the dead space increases with increasing pressure.[36] The relative magnitudes of the corrections and amounts adsorbed lead to a major source of uncertainty influencing the accuracy of high-pressure hydrogen adsorption measurements. Broom et al. have provided a detailed discussion of sources of errors for hydrogen adsorption measurements.[33,37] Reproducibility studies carried out by various laboratories have been reported[38–40] and reasonable agreement has been obtained.

Adsorption isotherms are presented on either surface excess or absolute basis in moles/g of outgassed adsorbent.[36,41] Sometimes, for gas storage applications, total wt% loading is quoted.[42–44] The adsorbed hydrogen phase is similar to an incompressible fluid, and the adsorbate densities for various hydrogen/adsorbent systems estimated from the maximum amount of hydrogen adsorbed with the total pore volumes obtained from gas adsorption studies or crystallographic pore volumes determined using the PLATON software package are typically in the range of 0.045–0.07 g cm^{-3}.[45–47] In comparison, the density values for liquid hydrogen are 0.0708 g cm^{-3} at 20.28 K and 0.077 g cm^{-3} at 13.8 K.[4] The density of liquid H_2 probably represents an upper limit for the density of hydrogen adsorbed in pores at 77 K and higher temperatures.

24.3 Activation of Porous Materials

The stability of porous materials and the conditions to obtain fully degassed porous structures for adsorption studies vary considerably. Porous carbons usually require heating to ~150°C under high vacuum to remove adsorbed material. However, in the case of MOFs, the activation procedure is often more complex. The stability,[43] purity,[48] and activation procedures[49] for MOFs are sometimes issues resulting in difficulties in obtaining repeatable adsorption results. Comparison of x-ray powder diffraction data with single-crystal data allows the purity of MOF samples prior to activation to be confirmed. The guests (templates, solvent, or coordinated solvent trapped inside MOFs) must be removed to form the porous structure and this may result in structural change that needs to be characterized. Activation of MOFs to form porous structures can be achieved by heating in ultrahigh vacuum, exchanging with a more volatile solvent to facilitate removal and extraction with supercritical carbon dioxide. Comparison of the methods for four MOFs shows that supercritical carbon dioxide drying results in large increases in the surface area as determined by gas adsorption measurements.[50] It was proposed[50] that supercritical carbon dioxide extraction inhibits mesopore collapse leaving micropores accessible to gases. It is evident that supercritical carbon dioxide drying has advantages in activating the porous structures of some MOFs.

The removal of coordinated solvent may lead to the formation of unsaturated or *open* metal centers, which are high-energy sites for adsorption. Therefore, thermal and chemical stability and framework flexibility of desolvated MOF porous material structures and activation procedures are important considerations when comparing adsorption characteristics.

24.4 Characterization of Porous Materials

Pores are classified by the International Union of Pure and Applied Chemistry (IUPAC) on the basis of size as micropores (<2 nm), mesopores (2–50 nm), and macropores (>50 nm).[36] Adsorption of hydrogen in the narrowest pores (<0.5 nm), where hydrogen is similar in size to pore dimensions, has the complication of quantum molecular seiving effects, giving rise to slow rates of adsorption and desorption due to activated diffusion effects.[51,52] Rigorous distinction of adsorption characteristics based on the classes of microporosity is not possible because they depend on adsorbate and adsorbent interactions rather than dimensions.[53] The terms *nanopore and nanoporosity* are not defined in the IUPAC scheme but generally refers to nanometer-sized pores. Characterization of the porous structures of materials is difficult, for example, MOF materials are flexible and the structure may partially collapse on desolvation. A variety of isotherm equations and adsorptives have been used to characterize porous structures using gas adsorption techniques. Porous structures are characterized by surface areas (determined using Langmuir, Brunauer–Emmett–Teller [BET], Dubinin–Radushkevich [DR], etc., equations), pore volumes (total, micropore [DR], etc.), and pore size distributions. Pore volumes, surface areas, and pore size distributions are estimates for relative comparisons for characterization purposes.

BET surface areas obtained from nitrogen adsorption at 77 K are frequently used as a standard procedure for characterization of porous materials.[54] Studies suggested that BET surface area correlates with MOF surface areas determined from crystallographic data.[55] However, heterogeneous surfaces are found in MOFs and the BET surface area is best described as an *apparent* surface area. If a Type I isotherm has a plateau at high relative pressure, the pore volume is given by the amount adsorbed (converted to a liquid volume) since the mesopore volume and external surface are both relatively small. Porous structure characterization parameters (Langmuir and BET surface areas and pore volumes) obtained from gas adsorption studies of MOFs have interrelated correlations.[19]

The surface excess obtained from the high-pressure adsorption isotherm is the amount present in the interfacial layer over the gas concentration present at the same equilibrium gas pressure, in which the gas-phase concentration is constant up to the Gibbs surface.[36] High-pressure adsorption studies usually report the surface excess, but in the case of gas storage, the total amount of gas that can be stored by a porous material is also relevant. Accurate measurements of the absolute amounts of adsorbed hydrogen are more difficult.[56] The total amount adsorbed is often estimated from the following equation[57]:

$$n_{tot} = n_e + p_v \times \rho_{gas}$$

where

n_{tot} is the total absolute uptake
n_e is the surface excess uptake
ρ_{gas} is the bulk density of gas (hydrogen)
p_v is the pore volume

The use of crystal density, which is mainly used for reporting hydrogen storage on MOFs may give unrealistically high volumetric storage values.[58] It was suggested that tap or packing densities were the most suitable. A comparison of two carbons and MOF-5 showed that the carbons gave the highest volumetric capacity.[58]

The amount adsorbed on a volumetric basis is the most appropriate for hydrogen storage for transport applications. The gas uptake on a volumetric basis for real systems is also subject to packing effects, which are determined by particle size and shape. Packing effects will reduce the uptake of real systems and further work is required to establish limitations associated with these effects.

24.5 Hydrogen Adsorption Measurements

Hydrogen adsorption has been investigated for a wide range of porous materials ranging from amorphous materials (activated carbons [ACs], porous polymers, etc.) to crystalline materials (MOF materials, zeolites, etc.) and also, metal-doped materials. The hydrogen adsorption isotherms for porous materials at 77 K have been studied extensively and are related to pore structure characterization parameters. Hydrogen adsorption at room temperature is much lower. Similarities are observed for the all types of porous materials.

Hydrogen adsorption on MOFs has received much recent attention because of the wide range of surface chemistry and crystalline structures with some of the largest pore volumes and surface areas available. Figure 24.1 shows the structure of $[Cu_3(dcbpyb) \cdot (H_2O)_3] \cdot 8dmso \cdot 15dmf \cdot 3H_2O$ (NOTT-112) and hydrogen adsorption isotherms on a surface excess and total hydrogen basis over the pressure range of 0–77 bar at 77 K.[59] The maximum surface excess is 7.07 wt% at 35–45 bar while the total hydrogen adsorption was 10 wt% at 77 bar. The polyhedral framework material has one the highest total volumetric capacities (54 g L^{-1}) at 80 bar and 77 K.[57] The surface excess adsorption isotherm has a maximum, and this is the amount of hydrogen present in the interfacial layer over the gas concentration present at the same equilibrium gas pressure, in which the gas-phase concentration is constant up to the Gibbs surface.[36] The total amount present takes into account the high-pressure gas present in the pores at equilibrium. NOTT-112 has a structure containing three types of cages: (1) cage A (12 open copper sites and diameter 13 Å), (2) cage B (diameter 13.9 Å), and (3) cage C (diameter 20 Å). NOTT-112 has a substantial uptake of 2.3 wt% at 1 bar and this is attributed to open Cu(II) centers present in relatively small cage A (see Figure 24.1a).

24.6 MOF Series Based on Dicarboxylate and Tetracarboxylate Ligands

Hydrogen adsorption has been investigated for two series of MOFs where there is a systematic change in the length of the ligand framework linker to change the pore volume and surface area without changing the structural topology: (1) isoreticular metal–organic frameworks (IRMOFs) of Yaghi et al.[5,60–66] based on Zn_4OL_3 stoichiometry (L = various dicarboxylate linkers) (see Figure 24.2) and (2) the NOTT homologous series with stoichiometry Cu_2L' (L′ = various tetracarboxylate ligands) of Schroeder et al. (see Figure 24.3).[44,45] The IRMOF series of MOF materials were used in the identification of a correlation between H_2 saturation amount and Langmuir surface area determined from N_2 adsorption at 77 K.[65]

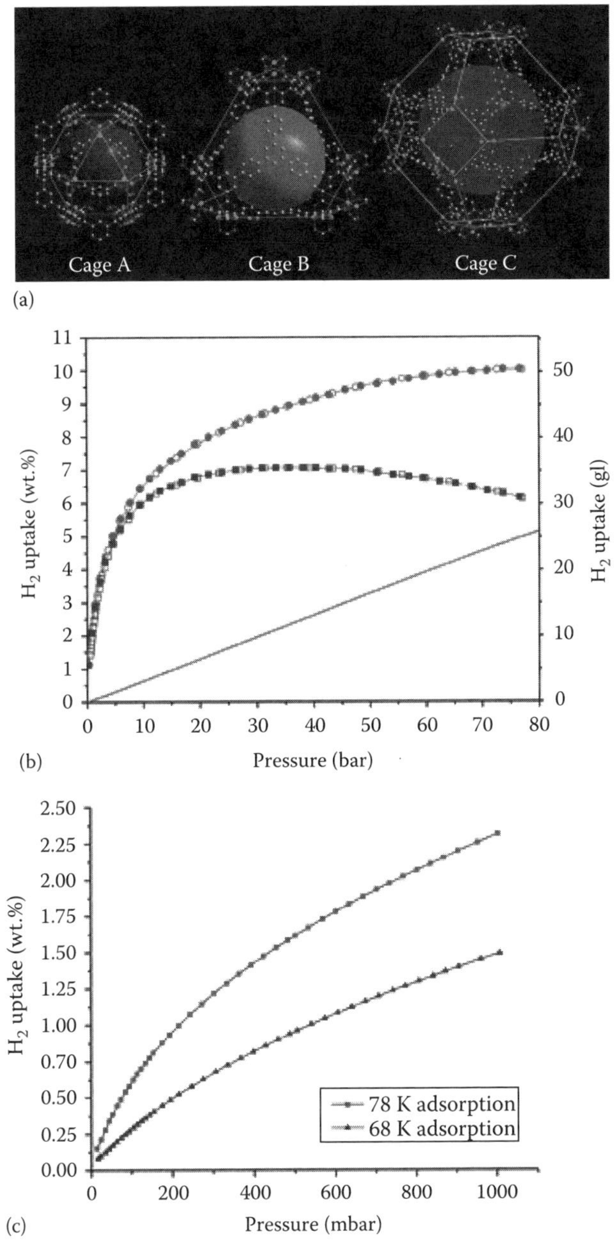

FIGURE 24.1
Structure and high-pressure hydrogen sorption excess adsorption isotherms for NOTT-112[59]. (a) Structure showing cages (b) surface excess (squares), total uptake (circles), and density of gas (line)—filled symbols (adsorption) and open symbols (desorption)—and (c) gravimetric total H_2 uptakes at 78 and 88 K up to 1 bar. (Yan, Y., Lin, X., Yang, S. H., Blake, A. J., Dailly, A., Champness, N. R., Hubberstey, P., Schroder, M., *Chem. Commun.*, 1025–1027, 2009. Reproduced by permission of the Royal Society of Chemistry.)

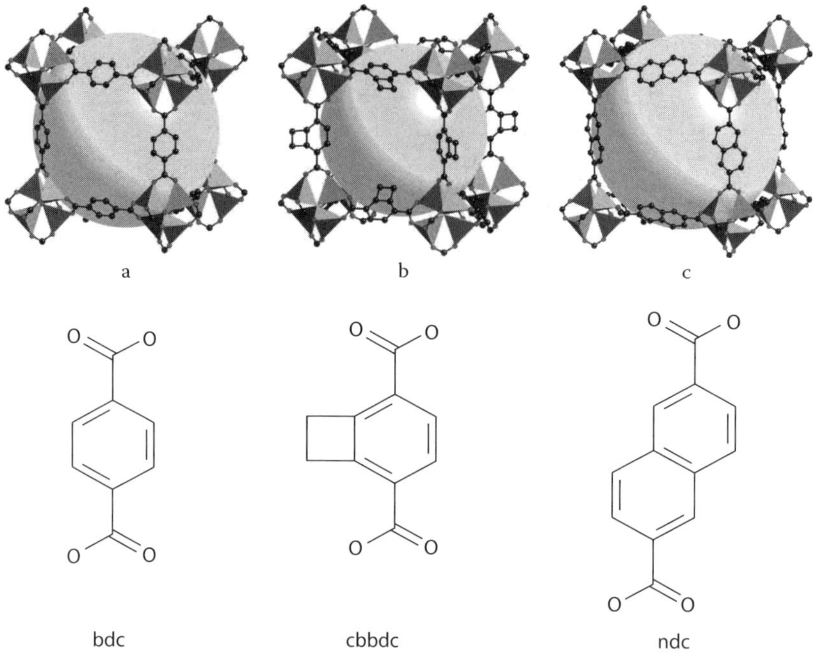

FIGURE 24.2
Structure of some of the IRMOF series of MOFs and the corresponding ligands: (a) IRMOF-1, (b) IRMOF-6, and (c) IRMOF-8. (Rosi, N. L., Eckert, J., Eddaoudi, M., Vodak, D. T., Kim, J., O'Keeffe, M., Yaghi, O. M., *Science*, 300, 1127–1129, 2003. Reproduced with permission from Science.)

Figure 24.4 shows a comparison of the amount of hydrogen adsorbed at 1 bar and at high pressure (at or close to saturation) with BET and Langmuir surface areas and pore volume for the IRMOF series. The results show that good correlations are observed for the hydrogen uptake at high pressure and pore characterization parameters. A correlation between hydrogen uptake at 1 bar and BET surface area and pore volume was not observed. Figure 24.5 shows the variation amount of hydrogen adsorbed at 1 bar and high pressure with BET and total pore volume for the NOTT tetracarboxylate series of MOFs. Good correlations were observed between the hydrogen uptake at high pressure on a weight basis and the pore characterization parameters. This series also does not show any correlations between the hydrogen uptakes on a weight basis at 1 bar pressure and pore structure characterization parameters. Both these series of MOF structures show progressive increases in hydrogen–surface excess adsorption at or close to saturation at high pressure with increasing pore size, pore volume, and surface area characterization parameters. Similar correlations were not observed for hydrogen adsorption at 1 bar pressure.

Hydrogen adsorption isotherms vary subtly with pore size. Adsorption potential is higher in narrow pores because of overlap of potential energy fields from the pore walls and this increases adsorption at low pressure. Hydrogen interactions with surfaces can also be enhanced by changing the surface chemistry. The inclusion of open or unsaturated metal centers is a possible method of enhancing hydrogen–surface interactions.[26,67,68] Henry's law constant (K_H) and isosteric enthalpies of adsorption

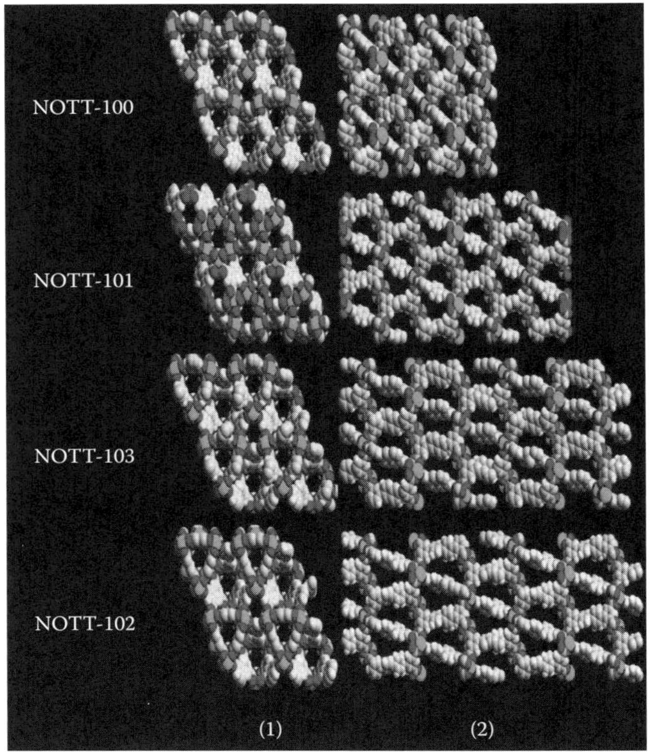

(a)

FIGURE 24.3
(a) Structures of some of the NOTT series of Cu_2L (L, tetracarboxylate ligand) MOFs: (1) view along the *c* axis and (2) view along the *a* axis. (Reproduced with permission from Lin, X., Telepeni, I., Blake, A. J., Dailly, A., Brown, C. M., Simmons, J. M., Zoppi, M., Walker, G. S., Thomas, K. M., Mays, T. J., Hubberstey, P., Champness, N. R., Schroder, M., *J. Am. Chemi. Soc.*, 131, 2159–2171. Copyright 2009, American Chemical Society; Lin, X., Jia, J., Zhao, X., Thomas, K. M., Blake, A. J., Walker, G. S., Champness, N. R., Hubberstey, P., Schroder, M., *Angew. Chem. Int. Ed.*, 45, 7358–7364. Copyright 2006, American Chemical Society.)

(continued)

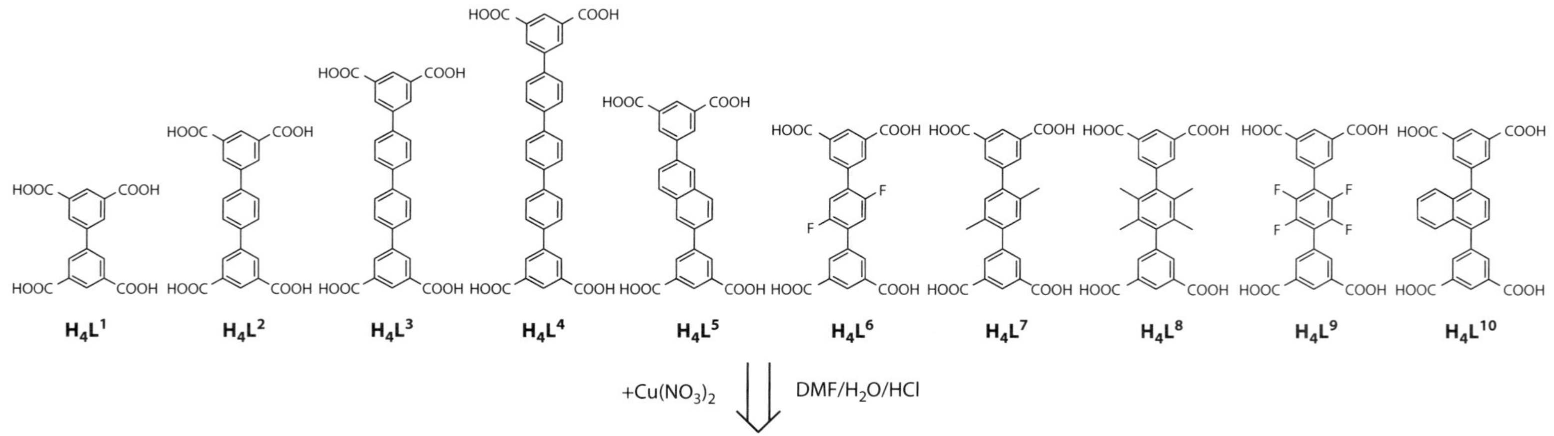

(b) **NOTT-100** **NOTT-101** **NOTT-102** **NOTT-103** **NOTT-104** **NOTT-105** **NOTT-106** **NOTT-107** **NOTT-108** **NOTT-109**

FIGURE 24.3 (continued)
(b) Structures of the series of tetracarboxylic acid ligands in the NOTT series of MOFs. (Reproduced with permission from Lin, X., Telepeni, I., Blake, A. J., Dailly, A., Brown, C. M., Simmons, J. M., Zoppi, M., Walker, G. S.,Thomas, K. M., Mays, T. J., Hubberstey, P., Champness, N. R., Schroder, M., *J. Am. Chemi. Soc.*, 131, 2159–2171. Copyright 2009, American Chemical Society.)

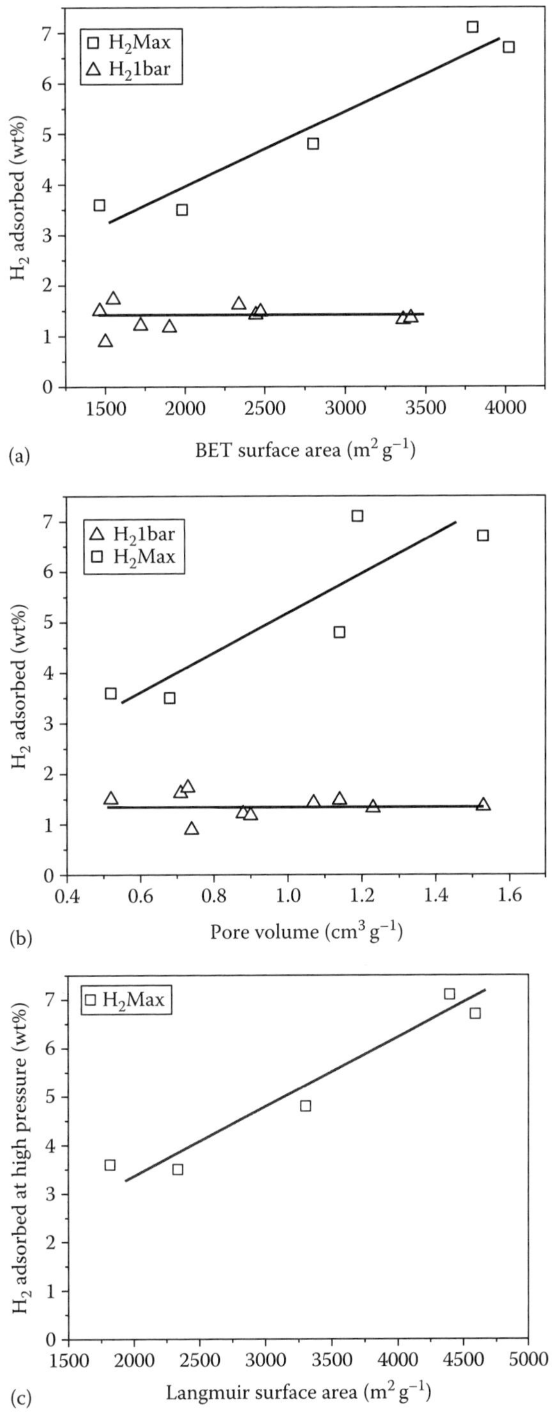

FIGURE 24.4
The variation of H_2 adsorbed at high pressure at 77 K with pore characterization parameters for the IRMOF series of MOFs.[20,43,62,65,69,191,193] (a) BET surface area, (b) pore volume, and (c) Langmuir surface area.

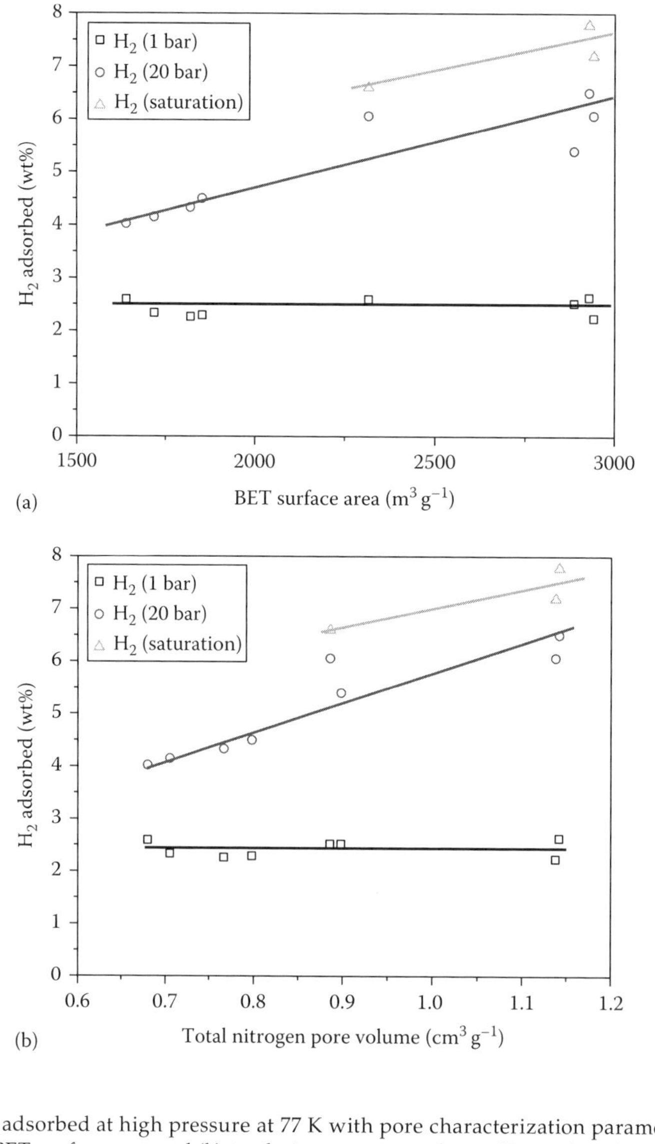

FIGURE 24.5

The variation of H_2 adsorbed at high pressure at 77 K with pore characterization parameters for the NOTT series of MOFs. (a) BET surface area and (b) total nitrogen pore volume. (Reproduced with permission from Lin, X., Telepeni, I., Blake, A. J., Dailly, A., Brown, C. M., Simmons, J. M., Zoppi, M., Walker, G. S.,Thomas, K. M., Mays, T. J., Hubberstey, P., Champness, N. R., Schroder, M., *J. Am. Chemi. Soc.*, 131, 2159–2171. Copyright 2009, American Chemical Society; Lin, X., Jia, J., Zhao, X., Thomas, K. M., Blake, A. J., Walker, G. S., Champness, N. R., Hubberstey, P., Schroder, M., *Angew. Chem. Int. Ed.*, 45, 7358–7364. Copyright 2006, American Chemical Society.)

at zero surface coverage are measures of H_2–surface interactions. Materials with larger pores and total pore volumes provide increased adsorption capacity at high pressures.[69] The influence of these factors can be seen in the crossover of isotherms observed with changing pore size for the NOTT copper tetracarboxylate series of MOFs (see Figure 24.6). Cu_2(bptc) has a higher affinity with H_2 than Cu_2(tptc) at low pressure,

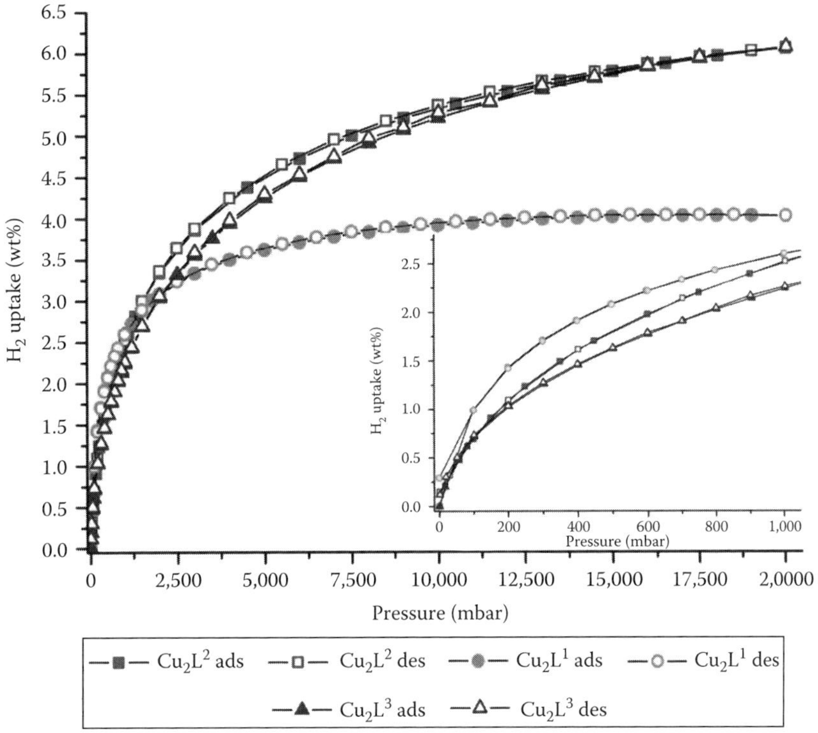

FIGURE 24.6
Adsorption isotherms for Cu_2L (L^1 = bptc, L^2 = tptc, and L^3 = qptc); filled symbols (adsorption) and open symbols (desorption). (Lin, X., Jia, J., Zhao, X., Thomas, K. M., Blake, A. J., Walker, G. S., Champness, N. R., Hubberstey, P., Schroder, M.: *Angew. Chem. Int. Ed.* 2006. 45, 7358–7364. Copyright Wiley-VCH Verlag GmbH & Co. KGaA. Reproduced with permission.)

but the lower pore volume limits the maximum H_2 adsorption to 4.02 wt% at 20 bar, while Cu_2(tptc) has higher hydrogen adsorption at >1.2 bar and 6.06 wt% at 20 bar.[44] At low pressure and surface coverage (or loadings), the enthalpy of hydrogen adsorption is important, whereas at high pressure, the total pore volume provides a limit for hydrogen adsorption. It is evident that there is a compromise for the structure of porous materials for optimum hydrogen adsorption characteristics, between adsorption saturation capacity at high pressure and increased adsorption at low pressure. The density of the porous material is also a factor since this defines the mass that can be put in a storage tank of a specific volume.

24.7 Hydrogen Adsorption Studies up to 1 Bar Pressure

Initially, most hydrogen adsorption studies for all porous materials were carried out at pressures up to 1 bar pressure because of the lack of availability of high-pressure instruments. These measurements provide data for determining the isosteric enthalpies of

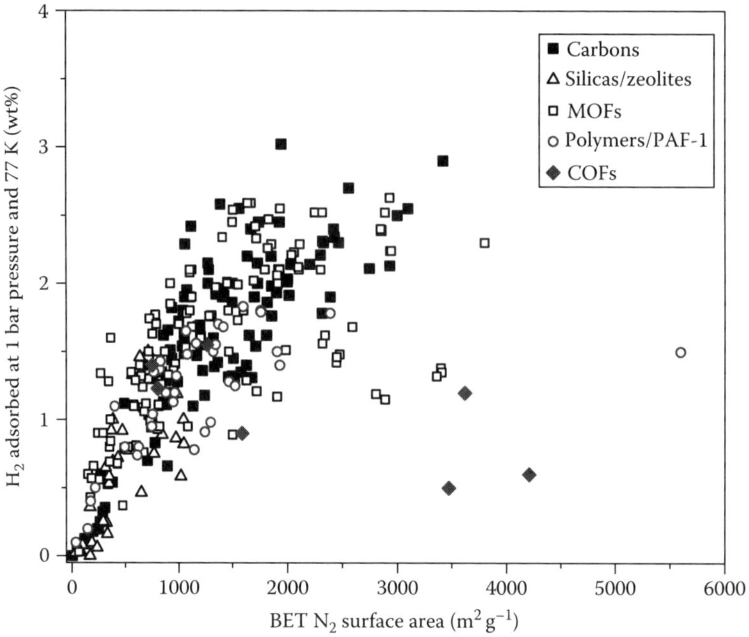

FIGURE 24.7
The variation of selected H_2 amounts adsorbed at 1 bar and 77 K with BET surface area for porous adsorbents: carbon materials,[46,51,70,72–77,80,82–84] silicas/zeolites,[72,86–88] polymers,[91–96] COFs,[98–102] and MOFs.[26,32,42,44,45,59,65–67,104–135,137,143–145]

adsorption at zero surface coverage, which is a fundamental measure of the hydrogen–surface interaction and a significant factor in determining the hydrogen desorption characteristics as a function of temperature. As discussed earlier, the uptakes at 1 bar pressure are not related to the maximum surface capacity at high pressure. The smallest micropores contribute more to the hydrogen adsorption uptake at 1 bar than larger pores.[46,70,71] Figure 24.7 shows the variation of H_2 uptakes at 1 bar and 77 K versus BET surface area for carbons,[46,51,70,72–84] silicas, aluminas, zeolites,[72,79,85–90] porous polymers,[91–97] COFs,[98–103] and MOFs.[26,32,42,44,45,59,62,65–67,96,104–145] It is evident that similarities exist for H_2 adsorption at 1 bar for all porous materials with surface areas below ~2000 $m^2\ g^{-1}$. However, a distribution of results is observed and this is much wider for high (>2000 $m^2\ g^{-1}$) surface area materials. It is apparent that measurement of hydrogen adsorption uptakes at 1 bar is not a guide to ultimate hydrogen capacity.

Pressure is a significant parameter for storage applications and high adsorption at low pressure is of interest in this respect. Recently, there has been a report of 6.1 wt% reversible hydrogen adsorption on lithium-doped conjugated microporous polymer (Li-CMP).[146] This is at least twice that of the highest amounts adsorbed at 1 bar observed for a wide variety of porous materials as shown in Figure 24.7. Lithium doping in the range up to 7 wt% was investigated with the maximum uptake being for 0.5 wt%. The isosteric enthalpy of adsorption at zero surface coverage was 8.1 kJ mol^{-1} and this decreased with increasing surface coverage. This enthalpy of adsorption is within the typical range of values obtained for nondoped materials. The hydrogen sorption of Li-CMP with optimal Li content (0.5 wt%) at 273 K and 0.1 MPa was very low.

24.8 Hydrogen Adsorption Capacity Studies at High Pressure

Hydrogen adsorption has been studied for a wide range of porous materials including carbons,[46,51,70,72–84,147–173] carbon nanotubes,[174,175] silicas,[72,176] aluminas[72], zeolites,[85–88,177–179] porous polymers,[91–95,180–188] COFs,[98–103] PAFs,[97] and MOFs.[26,32,42,45,57,59,62,65–67,104–137,143,144,189–246] Selected hydrogen adsorption data at high pressure and 77/298 K and corresponding pore structural characterization data are given in Table 24.1.

Studies of adsorption of gases on porous materials for storage applications have been directed at the synthesis of materials with ultrahigh porosities. Investigation of hydrogen adsorption on MOFs allows comparison with crystallographic void volumes and pore volumes and surface areas determined from gas adsorption studies. However, the densities of materials with very high crystallographic void volumes are very low. Several measurements of hydrogen storage capacity are used to compare the relative merits of materials. These capacity measurements are the surface excess (mass%) and total adsorption (mass%) and the total volumetric capacity (g L^{-1}). The relative merits of a set of materials differ depending on the hydrogen capacity measurement used. The total uptake is more relevant for practical applications of H_2 as a fuel, but it cannot be measured easily by experimental methods and, therefore, is calculated using the pore volume and the gas density. The hydrogen isotherms for these MOFs reach saturation uptakes, and the saturation pressure increases with an increase in cavity size.

MOFs are the porous materials with the highest void volumes, intrinsic surface areas, and pore volumes that have been prepared to date. The materials with the lowest densities are MOF-200 ($Zn_4O(bbc)_2(H_2O)_3{\cdot}H_2O$) and MOF-210 ($Zn_4O(bte)_{4/3}(bpdc)$) and have void volumes (densities) determined from crystallographic analyses of 90% (0.22 g cm^{-3}) and 89% (0.25 g cm^{-3}), respectively. The ultrahigh BET surface areas (10,400 m^2 g^{-1}) obtained for both MOF-200 and MOF-210 are close to the ultimate limit for solid materials. Comparison of hydrogen adsorption capacities gives the relative merits for the best performing materials. Hydrogen excess capacities on a mass% basis are

$$\text{MOF-210}(8.6) > \text{NOTT-112}(7.6) > \text{MOF-200}(7.4) > \text{MOF-177}(7.3)$$
$$> \text{MOF-205}(7) > \text{UCMC-2}(6.9) > \text{PCN-14}(4.4)$$

A range of values (5.3%–7.6%) have been reported for MOF-5.[43,57]

Hydrogen total capacities on a mass% basis are

$$\text{MOF-210}(17.6) > \text{MOF-200}(16.3) > \text{UCMC-2}(12.4) > \text{MOF-205}(12)$$
$$> \text{MOF-177}(11.7) > \text{NOTT-112}(10.7) > \text{PCN-14}(6.7)$$

A range of values (8.2%–10.6%) have been reported for MOF-5.

On the amount adsorbed per volume basis (g L^{-1}),

$$\text{PCN-14}(55) > \text{NOTT-112}(54) > \text{UCMC-2}(50) > \text{MOF-177}(50)$$
$$> \text{MOF-205}(46) > \text{MOF-210}(44) > \text{MOF-200}(36)$$

TABLE 24.1

Selected High-Pressure Hydrogen Adsorption and Porous Structure Characterization Data for Porous Materials

Porous Materials	Surface Area ($m^2\ g^{-1}$)	Pore Volume ($cm^3\ g^{-1}$)	Amount Adsorbed (77 K/wt%)	Amount Adsorbed Ambient Temperature (wt%)	References
MOFs					
$Zn_4O(bdc)_3$ (MOF-5, IRMOF-1)	4171^{L}		5.2 (48 bar)		[65]
	3534^{B}				[65]
		1.19			[69]
	3080^{L}				[191]
	3362^{L}		4.3 (30 bar)	0.45 (60 bar)	[62]
	2296^{B}		5.1 (65 bar)	0.28 (65 bar)	[192]
	3840^{L}			0.9 (60 bar/200 K)	[192]
	2295^{B}				[193]
	2833^{B}				[194]
	4400^{L}		7.1 (40 bar)		[43]
	3800^{B}				[43]
	2885^{B}	1.18			[137]
			5.75 (35 bar)	0.3 (60 bar)	[210]
	2860^{B}		5.3^{E} (80 bar)		[57]
	3480^{L}		8.2^{Tot} (80 bar)		[57]
		1.22	7.6^{E} (80 bar)		[43,57]
		1.55	10.6^{Tot} (80 bar)		[43,57]
MOF-5	3917^{L}	1.39	6.9^{E} (100 bar)		[138]
	2449^{B}		11.8^{Tot} (120 bar)		[138]
$Zn_4O(cbbdc)_3$ (IRMOF-6)	3305^{L}		4.8 (50 bar)		[65]
	2804^{B}				[65]
	2516^{B}				[194]
	3263^{L}	1.14			[66]
	2476^{B}				[66]
$Zn_4O(ndc)_3$ (IRMOF-8)	1818^{L}		3.6 (10–15 bar)	0.4 (30 bar)	[62,191]
	1466^{L}				[62,191]
	1922^{L}				[193]

	1215^{B}	0.52	3.5 (~20 bar)		[69]
$Zn_4O(hpdc)_3$ (IRMOF-11)	2337^{L}		3.5 (34 bar)		[65]
	1984^{B}	0.68			[69]
	1911^{L}				[62]
	2096^{B}				[194]
$Zn_4O(ttdc)_3$ (IRMOF-20)	4593^{L}		6.7 (70–80 bar)		[65]
	4024^{B}				[65]
	4346^{L}				[66]
	3409^{B}	1.53			[69]
$Zn_3O_3(dhbdc)$(MOF-74)	1072^{L}		2.3 (26 bar)		[65]
	950^{B}				[65]
	816^{B}				[194]
	1132^{L}	0.39			[66]
	783^{B}				[66]
	870^{B}		2.8 (30 bar)		[208]
$Zn_4O(btb)$ (MOF-177)	5640^{L}		7.5 (70 bar)		[65]
	4746^{B}				[39]
	4526^{L}				[62]
	3100^{B}	1.58		0.62 (100 bar, 298 K)	[196]
	4300^{L}				[196]
	4508^{B}				[194]
	4500^{B}	1.89	7.3^{E} (80 bar)		[39,57]
	5340^{L}		11.6^{Tot} (80 bar)		[39,57]
$Zn_4O(fma)_3$	1120^{B}		5.2 (39 bar)	0.32 (40 bar)	[243]
	1618^{L}				[243]
$Zn_4O(bdc)(btb)_{4/3}$	2932^{B}	1.47	5.43^{E} (20 bar)		[231]
	3950^{L}		6.23^{A} (20 bar)		[231]
$ZnO_4(L1)_3$	502^{B}	0.2^{DR}		1.12 (48 bar)	[197]
$ZnO_4(L2)_3$	396^{B}	0.13^{DR}		0.98 (48 bar)	[197]
$Cr_3OF(btc)_2$ (MIL-100(Cr))	2700^{L}	1.0	3.28 (26.5 bar)	0.15 (73.3 bar)	[69]
MIL-101(Cr)TM	3197^{B}	1.73	4.01 (10 bar)		[246]
	4546^{L}				[246]

(continued)

TABLE 24.1 (continued)

Selected High-Pressure Hydrogen Adsorption and Porous Structure Characterization Data for Porous Materials

Porous Materials	Surface Area ($m^2\ g^{-1}$)	Pore Volume ($cm^3\ g^{-1}$)	Amount Adsorbed (77 K/wt%)	Amount Adsorbed Ambient Temperature (wt%)	References
$Cr_3OF(ntc)_{3/2}$ (MIL-102 (Cr))	42.1^L	0.12	1.0 (35 bar)	0.05 (35 bar)	[206]
$Al(OH)(bdc)_3$ (MIL-53(Al))	1100^B		3.8 (16 bar)		[125]
	1590^L	0.59			[69]
	1779^L		2.9 (~20 bar)		[193]
	933^B				[193]
$Cr(OH)(bdc)_3$, MIL-53(Cr)	1100^B		3.1 (16 bar)		[125]
	1500^L	0.56			[69]
$Cr_3OF(bdc)_3$ (MIL-101b(Cr))	5500	1.9	6.1 (60 bar)	0.43 (80 bar)	[69]
$Cu_3(btc)_2$ (HKUST-1)	2257^L		3.2 (~50 bar)		[65]
	1944^B				[65]
	1154^B		3.6 (~30 bar)	0.35 (65 bar)	[192]
	1958^L				[193]
	1781^B				[194]
	2175^L	0.75			[66]
	1507^B				[66]
		0.684^{N2}	3.6 (10 bar)		[47]
		0.703^{CO2}			[47]
		0.72^X			[47]
	1482^B	0.828	4.1 (26 bar)		[49]
	2302^L				[49]
$Cu_2(bptc)$ (NOTT-100, MOF-505)	1830^L	0.63^{DR}			[26]
	1547^B				[194]
	1670^B	0.680^{N2}	4.02 (20 bar)		[44,45]
		0.683^X	4.2^e		[44,45]
$Cu_2(tptc)$ (NOTT-101)	2247^B	0.886^{N2}	6.06 (20 bar)		[44,45]
	2316^B	1.083^X	6.7^e		[44,45]
			6.6^{Tot} (60 bar)		[44,45]

Cu_2(qptc) (NOTT-102)	2942^{B}	1.138^{N2}	6.07 (20 bar)		[44,45]
		1.284^{X}	7.01^{e}		[44,45]
			7.2^{Tot} (60 bar)		[44,45]
$Cu_2(C_{26}O_8H_{12})$ (NOTT-103)	2929^{B}	1.142^{N2}	6.51^{Tot} (20 bar)		[44]
			7.78^{Tot} (60 bar)		[44]
$Cu_2(C_{22}O_8H_8F_2)$ (NOTT-105)	2387^{B}	0.898^{N2}	5.40^{Tot} (20 bar)		[44]
$Cu_2(C_{24}H_{14}O_8)$ (NOTT-106)	1855^{B}	0.798^{N2}	4.50^{Tot} (20 bar)		[44]
$Cu_2(C_{26}H_{20}O_8)$ (NOTT-107)	1822^{B}	0.767^{N2}	4.46^{Tot} (20 bar)		[44]
$Cu_2(C_{26}O_8H_{12})$ (NOTT-109)	1718^{B}	0.705^{N2}	4.15^{Tot} (20 bar)		[44]
Cu_2(pddip)$(H_2O)_2(DMF)_{7.5}$ $(H_2O)_5$ (NOTT-110)	2960^{B}	1.22^{N2}	5.43^{E} (55 bar)		[244]
			7.62^{Tot} (55 bar)		[244]
			8.5^{Sat}		[244]
[Cu_2(dhpddip)$(H_2O)_2$]$(DMF)_{7.5}$ $(H_2O)_5$ (NOTT-111)	2930^{B}	1.19^{N2}	5.47^{E} (48 bar)		[244]
			7.36^{Tot} (48 bar)		[244]
Cu_3(dcbpyb) (NOTT-112)	3800^{B}	1.62^{N2}	7.07^{E} (35–40 bar)		[59]
		1.69^{Ar}	10.0^{Tot} (77 bar)		[59]
			10.7^{Tot} (80 bar)		[57]
Cu_2(bdi) (PCN-46)	2500^{B}	1.012	7.16^{Tot} (60 bar)		[237]
$Cu_3(tatb)_2$ catenated (PCN-6)	3800^{L}		7.2^{E} (50 bar)		[240]
			9.5^{Tot} (50 bar)		[240]
$Cu_3(tatb)_2$ noncatenated (PCN-6′)	2700^{L}		4.2^{E} (50 bar)		[240]
			5.8^{Tot} (50 bar)		[240]
Mn(btt)	2100^{B}	0.795^{X}	5.1 (40 bar)	0.95 (90 bar)	[42]
			6.9 (90 bar)		[42]
$Zn_7O_2(pda)_5$		0.17^{H2O}		1.01 (71.43 bar)	[198]
		0.21^{M}			[198]
$Zn(mim)_2$ (ZIF-8)	1810^{L}	0.663^{X}	3.1 (55 bar)		[32]
	1630^{B}	0.636μp			[32]
	1400^{L}				[31]
	1030^{B}				[31]
			3.3 (30 bar)	0.13 (60 bar)	[210]
Zn(ndc)$(bpe)_{1/2}$	303^{L}	0.2	2.0 (40 bar)	0.3 (65 bar)	[199]
Ni_2(dhtp)	1083^{L}	0.41	1.8 (50 bar)	0.3 (65 bar)	[110]

(continued)

TABLE 24.1 (continued)

Selected High-Pressure Hydrogen Adsorption and Porous Structure Characterization Data for Porous Materials

Porous Materials	Surface Area (m^2 g^{-1})	Pore Volume (cm^3 g^{-1})	Amount Adsorbed (77 K/wt%)	Amount Adsorbed Ambient Temperature (wt%)	References
$Ni_3(btc)_2(3\text{-pic})_6(pd)_3$		0.63	2.6 (15 bar)	0.15 (15 bar)	[200]
$Cu_2(hfipbb)_2(H_2hfipbb)$				1.0 (48 bar)	[201]
$Fe_3(OH)(pbpc)_3$	1200^B		3.05 (20 bar)		[143]
$Ni_3(OH)(pbpc)_3$	1553^B		4.15 (20 bar)		[143]
$Cu_2(abtc)$ (SNU-5)	2850^L	1.0	5.22^E (50 bar)		[245]
			6.76^{Tot} (50 bar)		[245]
$Zn_2(abtc)(dmf)_2$ (SNU-4)	1460^L	0.53	3.70^E (50 bar)		[245]
			4.49^{Tot} (50 bar)		[245]
$Cu_2(abtc)(H_2O)_2(dmf)_2 . H_2O$ (JUC-62)		0.888^X	4.71 (40 bar)		[203]
$Cu_2(abtc)(H_2O)_2 \cdot 3dma$ (PCN-10)	1779^L	0.67	4.33^E (20 bar)	0.25 (45 bar)	[118]
	1407^B				[118]
$Cu_2(sbtc)(H_2O)_2 \cdot 3dma$ (PCN-11)	2442^L	0.91	5.05^E (20 bar)	0.34 (45 bar)	[118]
	1931^B				[118]
$Cu_2(adip)$ (PCN-14)	1753^B	0.87	4.42^E (45 bar)		[241]
	2176^L		6.7 (80 bar)		[57,241]
$[Ni_3O(H_2O)_3(adc)_3] \cdot (dma)_2\}n$, (PCN-19)	723B	0.38	1.67 (48 bar)		[234]
	823L				[234]
$Cu_3(ttca)_2$(PCN-20)	3525^B		6.2^E (50 bar)		[242]
	4237^L				[242]
$Zn(bdc)(ted)_{0.5}$	1794^B	0.73	4.1 (20 bar)	0.2 (50 bar)	[128,204]
$Zn_3(bdc)_3Cu(pyen)$ (M′MOF1)		0.257^M	1.25 (10 bar)		[52]
		0.29^X	1.35^e		[52]
$Co(bdc)(dabco)_{0.5}$	1595^B		4.11 (40 bar)	<0.5 (100 bar)	[205]
	2120^L				[205]
$Cu(bdc)(dabco)_{0.5}$	1300^B		2.7 (40 bar)		[205]
	1703^L				[205]
$Zn(bdc)(dabco)_{0.5}$	1165^B		3.17 (40 bar)		[205]
	1488^L				[205]

$Mn_3[(Mn_4Cl)_3(tpt\text{-}3tz)_8(dmf)_{12}]_2$	1580^{B}		3.7 (25 bar)	0.5 (68 bar)	[217]
	1700^{L}				[217]
$Cu_3[(Cu_4Cl)_3(tpb\text{-}3tz)_8]_2 1 \cdot 1CuCl_2$	1120^{B}		2.8 (30 bar)	0.5 (68 bar)	[217]
	1200^{L}				[217]
$Fe_3[(Fe_4Cl)_3(btt)_8(MeOH)_4]_2$ (Fe-btt)	2010^{B}			1.1 (100 bar, 298 K)	[273]
Co(bdp)	2670^{L}	0.93	3.1 (30 bar)		[207]
Cu(dccptp)(NO_3)	268^{B}	0.113	1.91 (20 bar)		[144]
$Cd_3(bpdc)_3$ (JUC-48)	880^{L}	0.19	2.8 (40 bar)		[211]
$Ni_3(OH)(pbpc)_3$	1553^{B}		4.15 (20 bar)		[143]
$Fe_3(OH)(pbpc)_3$	1200^{B}		3.05 (20 bar)		[143]
$Zn_2(bpytc)$	312.7^{B}	0.187^{X}	1.08 (4 bar)	0.057 (4 bar)	[212]
	423.7^{L}				[212]
$Cu_3(bhtc)_2$ (UMCM-150)	2300^{B}		5.7 (45 bar)		[127]
	3100^{L}				[127]
$Zn_4O(t^2dc)(btb)_{4/3}$(UCMC-2)	5200^{B}	2.32	6.9^{Ex} (46 bar),		[57,232]
	6060^{L}		12.4^{Tot} (80 bar)		[57,232]
	6060^{L}				[57,232]
MOF-200, $Zn_4C_{90}H_{60}O_{16}$ Bbc	4530^{B}	3.59	7.4^{Ex}, 16.3^{Tot} (80 bar)		[57]
	10400^{L}				[57]
MOF-205 $Zn_4C_{48}H_{26}O_{13}$ bttb, ndc	4460^{B}	2.16	7.0^{Ex}, 12.0^{Tot} (80 bar)		[57]
	6170^{L}				[57]
MOF-210 $Zn_4C_{58}H_{28}O_{13}$ bte, bpdc	6240^{B}	3.60	8.6^{Ex}, 17.6^{Tot} (80 bar)		[57]
	10400^{L}				[57]
$Cu_3(bipy)_{1.5}(ndc)_3$	113		1 (15 bar)		[233]
$[Cu(bpe)_{0.5}(ndc)] \cdot 0.5H_2O$	337		1 (15 bar)		[233]
$Zn_9O_3(2,7\text{-}ndc)_{12}(dmf)_3$	901^{B}	0.458	2.32 (20 bar)^{E}		[135]
	1281^{L}				[135]
$Cu_2(bpndc)_2(bpy)$ (SNU-6)	2590^{B}	1.05	4.87^{E} (70 bar)		[134]
	2910^{L}		10.0^{Tot} (70 bar)		[134]
$Zn_2(bpndc)_2(bpy)$ (SNU-9)	1034^{L}	0.366	3.63^{E} (90 bar)		[236]
	824^{B}		6.23^{Tot} (90 bar)		[134]
$Y(btc)(H_2O)$			2.1 (10 bar)		[202]

(continued)

TABLE 24.1 (continued)

Selected High-Pressure Hydrogen Adsorption and Porous Structure Characterization Data for Porous Materials

Porous Materials	Surface Area ($m^2 g^{-1}$)	Pore Volume ($cm^3 g^{-1}$)	Amount Adsorbed (77 K/wt%)	Amount Adsorbed Ambient Temperature (wt%)	References
$Ag_2[Ag_4(trz)_6]$ FMOF-1	810^B	0.324	2.33 (77 bar)		[238]
			4.1^{vol} (64 bar)		[238]
$[Sm_2(oxdaa)_6Zn_3(H_2O)_6]\cdot 1.5H_2O$	719^B		1.19 (34 bar)	0.54 (35 bar, 298 K)	[239]
$(Me_2NH_2)[In(bptc)]$	820^B	0.325	2.36 (20 bar)		[228]
$(Me_2NH_2)[In(bptc)]^-Li^+$	1024^B	0.419	2.88 (20 bar)		[228]
Carbon materials					
Carbon nanotubes					
Multiwalled carbon nanotubes (MWNTs)	$165–406^B$	$0.588–1.063^{TPV}$		<0.78 (100 bar, 298 K)	[174]
		0.097–0.178μp			[174]
Nanotubes	800		1.7 (35 bar)	0	[175]
MWNTs	246–260	0.08–0.12		0.2–0.34 (90 bar, 298 K)	[82]
Activated MWNT	1220^B	0.29μp	2^E (40 bar)	0.16 (200 bar, 298 K)	[162]
Nanofibers					
Nanofibers and activated nanofibers	$150–265^B$	0.03–0.1μp	0.8^E (40 bar)	0.15–0.2 (200 bar, 298 K)	[162]
Activated (steam/KOH) rayon carbon fibers	3144^B	0.744μp	7.01 (40 bar)	1.46 (40 bar, 298 K)	[170]
Carbon nanofibers (phenol formaldehyde resin/ KOH activated)	$520–1765^B$	$0.26–1.0^{TPV}$	1.65–3.45 (40 bar)		[172]
		$0.26–0.48^{\mu pCO2}$			[172]
Graphene					
Graphene-like nanosheets	640		1.2 (10 bar)	0.1 (10 bar)	[250]
Porous Carbons					
CVD carbons from acetonitrile using zeolite β-template	$2191–3150^B$	$1.74–2.41^{TPV}$	3.9–6.9 (20 bar)		[161]
	1150–2397μp	$0.56–1.13^{\mu p}$	$4.67–8.33^e$		[161]
CVD carbons from acetonitrile using zeolite β-template	$1721–2535^B$	$1.09–1.56^{TPV}$	3.3–5.3 (20 bar)		[157]
CVD carbons from acetonitrile using Mg–Al (Mg/Al = 2) layered double hydroxides (LDH) template	$800–1268^B$	$1.2–1.5^{TPV}$	1.8–2.8 (20 bar)		[158]

CVD carbons from acetonitrile or ethylene using zeolite 13X or zeolite Y template	$808–1918^{B}$	$0.49–1.38^{TPV}$	2.0–4.5 (20 bar)		[159]
	$194–1077^{\mu p}$	$0.1–0.5^{\mu p}$			[159]
Porous Carbons and carbon nanotubes	$22–2564^{B}$	$0.0065–0.75^{dB}$	$0–4.4^{I}$	$0–0.54^{I}$	[156]
Carbons from biomass	$2000–3100^{B}$	$1.11–1.68^{TPV}$	3.99–5.05 (10 bar)		[81]
ACs	890–3000	$0.48–1.6^{TPV}$	1.9–4.5 (10 bar)		[75]
Ordered mesoporous carbon materials using silica	$711–2390^{B}$	$0.5–1.22^{TPV}$	1.4–3.5 (10 bar)		[80]
KOH superactivated carbon KAU5	3183^{B}	$0.72^{\mu pCO2}$		1.2 (200 bar)	[162]
ACs and superactivated carbons	$1058–3808^{B}$	$0.43–0.68^{CO2\mu p}$	2.2–5.6 (40 bar)	0.6–1.1 (200 bar)	[162]
AC and carbon cloths	$1049–1730^{B}$	$0.79–0.39^{TPV}$		0.24–0.56 (90 bar, 298 K)	[82,83]
		$0.27–0.5^{CO2\mu p}$			[82,83]
Chemically activated phenolic resin carbons	$1424–2811^{B}$		$2.81–5.75^{E}$ (40 bar)	0.5 (RT, 40 bar)	[166]
Carbon (wood derived + KOH activation)	2450^{B}	1.19^{TPV}		0.8 (298 K, 2 MPa)	[248]
		$0.63^{\mu PCO2}$			[248]
		$0.83^{\mu PN2}$			[248]
Carbide-derived carbons	$1150–3038^{B}$	0.48–1.34	$2.9–4.7^{E}$ (60 bar)		[167]
B-doped carbon	780^{B}		3.2 (40 bar)	0.4 (56 bar)	[247]
BC_{11}	780^{B}		3 (80 bar)	0.37 (80 bar, ambient)	[256]
BC_6	609^{B}		3.8 (80 bar)	0.54 (80 bar, ambient)	[256]
Polycarbonate char	$320–2096^{B}$	$0.13–0.87^{TPV}$	2.27–4.68 (26 bar)	0.17–0.29 (26 bar, 273 K)	[164]
	$370–2456^{N2DR}$				[164]
	$713–1400^{CO2\mu p}$				[164]
Nitrogen-containing carbons	$1912–3698^{B}$	$0.99–1.71^{TPV}$	$3.4–5.3^{E}$ (20 bar)		[163]
	$1510–2838^{\mu P}$	$0.65–1.24^{\mu PV}$	$4.1–6.9^{Tot}$ (20 bar)		[163]
Nanoporous carbon particles	2711^{L}	0.87^{TPV} $0.38^{\mu P}$	4.2 (50 bar)	0.25 (50 bar)	[160]
	1995^{B}				[160]
Carbon monoliths	$2147–2647^{B}$	$0.894–1.512^{TPV}$		0.40–0.46 (50 bar)	[165]
		$0.422–0.515^{DR\mu PCO2}$			[165]
Nitrogen-doped carbons, KOH activated	$334–2745^{B}$	$0.36–2.12^{TPV}$	1.27–6.84 (20 bar), 8.24^{Le}		[168]
Carbon CO_2/KOH activated	$2009–3190^{B}$	$1.0–1.69^{TPV}$	5.63–7.08 (20 bar)		[169]

(continued)

TABLE 24.1 (continued)

Selected High-Pressure Hydrogen Adsorption and Porous Structure Characterization Data for Porous Materials

Porous Materials	Surface Area ($m^2\ g^{-1}$)	Pore Volume ($cm^3\ g^{-1}$)	Amount Adsorbed (77 K/wt%)	Amount Adsorbed Ambient Temperature (wt%)	References
Carbon KOH activated	1579^B	0.9^{TPV}	4.54 (20 bar)		[84]
	2930^B	1.52^{TPV}	6.24 (20 bar)		[84]
Carbon AX21	2780^B	0.86^{TPV}	5.2 (29 bar)		[171]
Carbide-derived carbons	1908–2301	0.6–1.06	4.61^E (28 bar)	$0.67–0.78^E$ (100 bar)	[173]
			4.0^E (25 bar)		[173]
MCM materials					
MCM-41	$916–1060^B$		$1.68–2.01^E$ (30–45 bar)	0.1 (45 bar, 296 K)	[176]
Si, Zn–doped MCM-41	$516–875^B$		$1.07–1.83^E$ (40 bar)		[176]
Porous Polymers					
Polyporphyrins	1522^B	0.85^{TPV}	5.0 (65 bar)		[186]
	2030^L				[186]
Hyper-cross-linked polymer 4,4′-bis(chloromethyl)-1,1′-biphenyl (BCMBP),	2992^L	$0.633^{\mu P}$	3.68 (15 bar)		[184]
	1904^B	0.54^{DFT}			[184]
Hyper-cross-linked polymer dichloroxylene (p-DCX)	2096^L	$0.494^{\mu P}$	3.18 (15 bar)		[184]
	1370^B	0.46^{DFT}			[184]
Hyper-cross-linked polyanilines	630^B	0.94^{TPV}	2.2 (30 bar)		[182]
Triptycene-based polymer (Trip-PIM)	1416^B		2.71 (10 bar)		[94,95]
Hyper-cross-linked BCMBP	1366^B		2.8 (15 bar)		[187]
	2096^L	0.55^{DFT}			[187]
PIMs PIM-1, HATN,CTC	$760–830^B$		1.44–1.70 (10 bar)		[92]
Hyper-cross-linked polymer vinylbenzyl chloride	1466^B	0.48	3.04 (15 bar)		[91]
Triptycene-based polymers (Me and iPr)	1760^B		3.4 (18 bar)		[95]
	1601^B		3.3 (18 bar)		[95]
Porph-PIM	960^B		2.0 (18 bar)		[251]
CTC-PIM	770^B		1.7 (10 bar)		[92]

PIM-1	750^{B}		1.45 (18 bar)		[92]
Polyimide framework OFP3	1159^{B}		3.94 (10 bar)		[252]
Polymer PS4AC2, PS4AC1, Pt4AC, PS4TH	$762–1043^{B}$	0.425–0.757	2.2–3.7 (60 bar)	0.43–0.50 (298 K, 70 bar)	[253]
	$1122–1412^{L}$				[253]
Polymers	1246^{B}	0.729	3.0 (75 bar)		[188]
	1515^{L}				[188]
Porous aromatic framework (PAF-1)	5600^{B}	3.1	7.0^{E} (48 bar)		[97]
	7100^{L}				[97]
COFs					[98]
COF-1, C_3H_2BO	750^{B}		1.48^{Sat}		[103]
	970^{L}				[103]
COF-5, $C_9H_4BO_2$	$1670(2050)^{B}$		3.58^{Sat}		[103]
	$1990(3300)^{L}$				[103]
COF-6, $C_8H_3BO_2$	750^{B}		2.26^{Sat}		[103]
	980^{L}				[103]
COF-8, $C_{14}H_7BO_2$	$1350(1710)^{B}$		3.5^{Sat}		[103]
	$1400(2110)^{L}$				[103]
COF-10, $C_{12}H_6BO_2$	$1760(1980)^{B}$		3.92^{Sat}		[103]
	$2080(4620)^{L}$				[103]
COF-102, $C_{25}H_{24}B_4O_8$	3620^{B}		7.24^{Sat}		[103]
	$4650^{L,}$				[103]
COF-103, $C_{24}H_{24}B_4O_8Si$	3530^{B}		7.05^{Sat}		[103]
	$4630^{L,}$				[103]
Zeolites					
Zeolites, Na-LEV, H-OFF, Na-MAZ, and Li-ABW			1.02–2.07 (16 bar)		[179]
Zeolites Na–A, Na–Li–A, Li–A, K–A				0.42–0.49 (60 bar, 293 K)	[178]
Ion-exchanged zeolites	$384–725^{B}$		1.32–2.19 (15 bar)		[85]
Zeolites A, X, Y, and RHO	$3–725^{B}$		0–1.81 (15 bar)		[86]
Low silica Type X zeolite	717^{B}	0.568		0.6 (100 bar)	[87]

^{B} determined using BET equation; ^{L} determined using Langmuir equation; ^{DR} determined using Dubinin–Radushkevich equation; ^{e} estimate of saturation; ^{E} surface excess; M determined from methanol adsorption; ^{X} determined from x-ray diffraction studies; ^{H_2O} determined from water vapor adsorption; ^{dB} determined an extension of de Boer's *t*-method for the micropore volume; ^{I} interpolated from graph in paper; $^{\mu P}$ micropore volume or surface area; DFT density functional theory calculation.

The different orders for hydrogen saturation amounts on the basis of weight and volume mainly arise because of the low density of some of the materials with the largest pore volumes. Examples of materials with low densities are MOF-177 (0.427 g cm^{-3}), MOF-200 (0.22 g cm^{-3}), MOF-205 (0.38 g cm^{-3}), and MOF-210 (0.25 g cm^{-3}). The best materials on a volumetric basis are PCN-14 (0.83 g cm^{-3}), NOTT-112 (0.48 g cm^{-3}), and UCMC-2 (0.40 g cm^{-3}). The results indicate that a highly porous material with an ultrahigh gravimetric storage capacity may not necessarily be an optimum volumetric H_2 storage material. Therefore, it is evident that comparisons of hydrogen storage capacity on a volumetric basis with porous structure characteristics are more complex. Hydrogen adsorption on amorphous porous materials such as ACs on a surface excess gives similar or slightly lower values. Comparisons of hydrogen volumetric storage capacities are needed on precisely the same basis.

The amounts of hydrogen adsorbed under high pressure at 77 and 298 K and corresponding surface areas and pore volumes have been investigated for MOFs,[32,44,59,62,65,66,69,125,137,143,144,179,191–208,210,211,212,228,238–244,246] carbons,[81,156,160–162,247–250] zeolites,[85–87,178,179] silicas,[176] polymers,[91,92,94,95,182,184,251–253] and COFs.[98,103] Correlations have been observed between the amount of hydrogen at saturation (or close to saturation capacity) on a wt% adsorbent basis at 77 K and Langmuir surface area, BET surface area, and total pore volumes, as shown in Figures 24.8 through 24.10 for all types of porous materials. A similar correlation is observed between the amount of hydrogen at saturation (or close to saturation capacity) and micropore volume obtained from carbon dioxide adsorption data for carbon materials.[13,19] Porous materials for storage applications have wider pores and the BET N_2 surface area, Langmuir surface area, and total pore volume have good correlations for all porous materials, as shown in Table 24.1. The correlation for hydrogen high-pressure uptake with BET surface area is observed for porous materials with Type I and II nitrogen adsorption isotherms, and therefore, the correlation is more general (see Figure 24.9). Figure 24.10 shows the

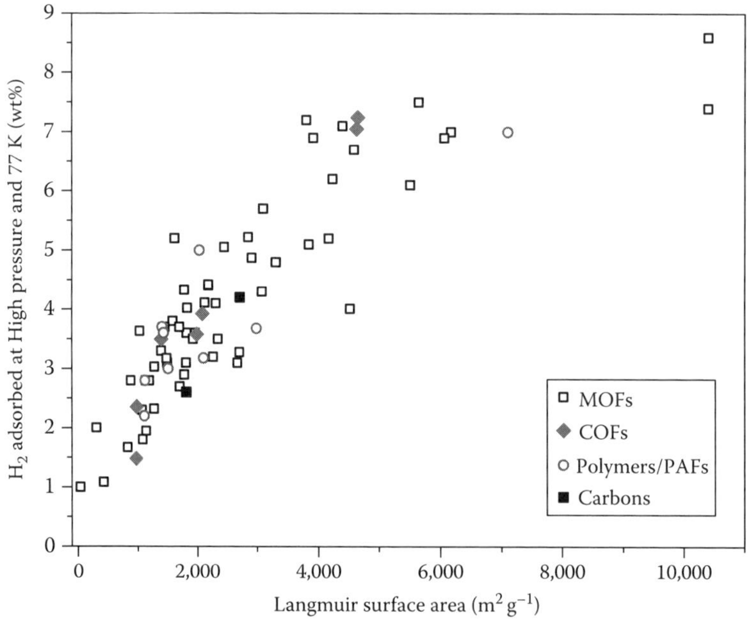

FIGURE 24.8
The variation of H_2 adsorbed (wt%) at saturation at 77 K with Langmuir surface area (m^2 g^{-1}) for porous MOF materials,[26,32,39,43,45,49,57,62,65,66,69,110,118,125,127,134,135,138,191–193,196,199,205–208,210–212,217,231,232,234,236,240–242,245] polymers,[97,184,186,188,253] COFs,[103] and carbon materials.[160] (Data are taken from the references in Table 24.1.)

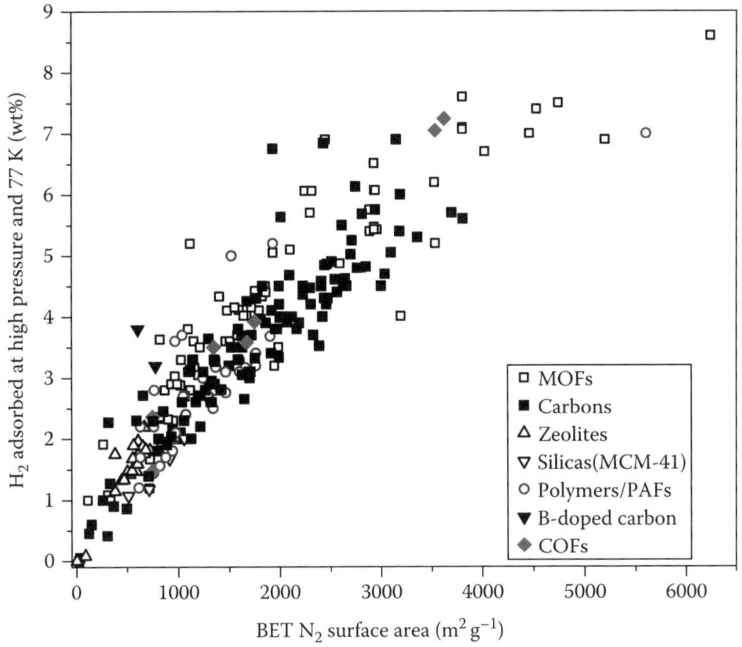

FIGURE 24.9
The variation of H_2 adsorbed (wt%) at saturation at 77 K with BET surface area ($m^2\ g^{-1}$) for porous MOF materials,[26,32,39,42–45,47,49,57,59,62,65,66,69,118,125,127,134,135,137,138,143,144,191–193,196,204,205,208,210,212,217,228,231,232,234,236–239,241–244] carbons,[75,80–84,156–174,248,250] boron-doped carbon,[247,256] MCM-41,[176] zeolites,[85,86] polymers,[91,92,94,95,97,182,184,186–188,251–253] and COFs.[98] (Data are taken from the references in Table 24.1.)

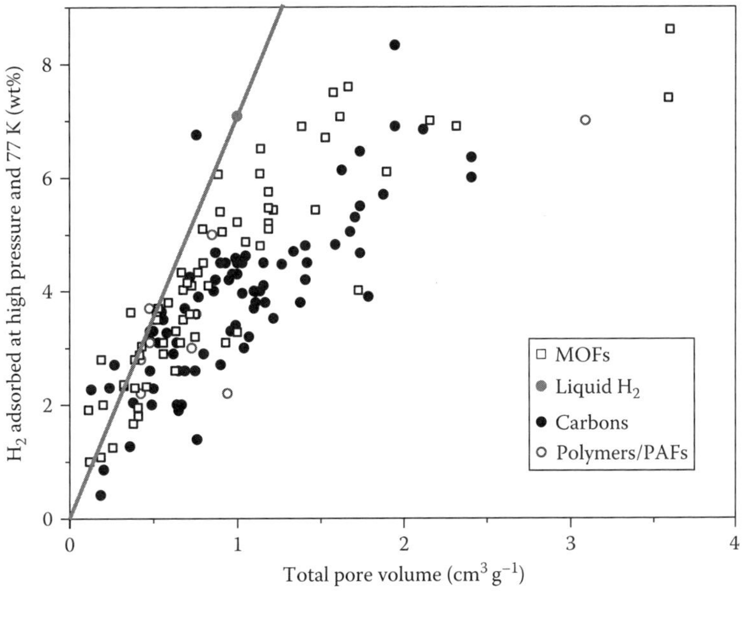

FIGURE 24.10
The variation of H_2 adsorbed (wt%) at saturation at 77 K with total pore volume ($cm^3\ g^{-1}$) for porous MOFs,[32,42,44,45,47,52,57,59,62,65,66,69,110,118,125,128,137,144,191–193,196,199,200,204,206–208,210,211,234,236,238,239,241,244,245] carbons,[75,80–84,157–159,161,163,165,168–174] and polymer[91,97,182,186,188,253] materials. (Data are taken from the references in Table 24.1.)

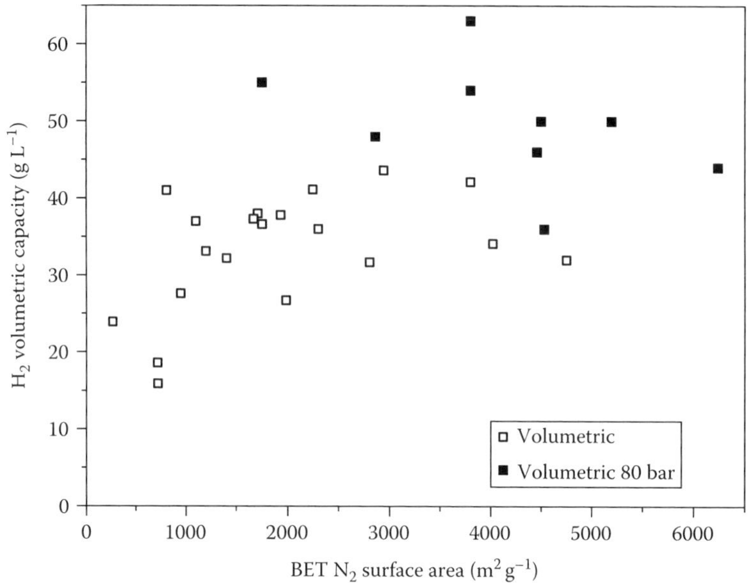

FIGURE 24.11
The variation of H_2 excess volumetric capacity[39,43,45,57,65,118,122,125,127,143,144,205,234,239,241] and volumetric capacity at 80 bar (crystal density was applied to the gravimetric density)[57] versus BET surface area for selected MOFs.

line corresponding to liquid hydrogen filling the pores for comparison with hydrogen uptake data at or close to saturation with total pore volume. The hydrogen saturation uptake values are usually lower than the line for liquid hydrogen filling the pores. The adsorbed hydrogen phase at 77 K is similar to an incompressible fluid with a slightly lower density than liquid hydrogen (critical temperature 32.98 K).[4] MOFs usually have narrower pore size distributions than other porous materials and are usually closer to the line for liquid hydrogen.

Hydrogen storage on a volumetric basis is generally considered to be the most significant characteristic for vehicle applications. These data are mainly available for MOFs. The clear correlations of H_2 excess adsorption on a mass basis with porous structure characterization parameters are no longer observed when the H_2 excess adsorption on a volumetric basis is compared with porous structure characteristics. Figure 24.11 shows the variation of hydrogen adsorbed versus BET surface area. There is an initial improvement up to ~2000 m^2 g^{-1}, but above these values, there is a great deal of scatter and there is no clear correlation with BET surface area.

24.9 Temperature Dependence of Hydrogen Physisorption

Physisorption decreases markedly with increasing temperature and H_2–surface interactions are weak, limiting hydrogen adsorption above cryogenic temperatures. This is the major limitation to the use of hydrogen storage using physisorption on porous materials. The temperature dependence of hydrogen physisorption is illustrated in Figure 24.12, which shows normalized hydrogen adsorption isobars for 1 bar pressure for an AC, MOF $Ni_3(btc)_2(3\text{-pic})_6(pd)_3$ (C), and the ethanol (**E**)- and methanol (**M**)-templated polymorphs of $Ni_2(bpy)_3(NO_3)_4$.[200] The shapes of the isobars in the low-pressure region for **E** and **M** are

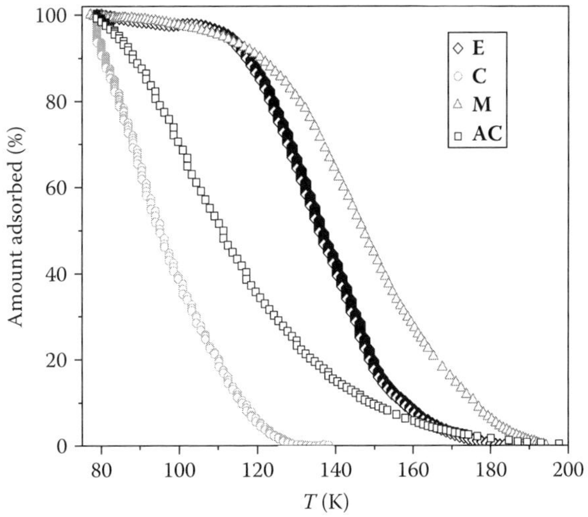

FIGURE 24.12
Isobars for desorption of hydrogen at 1 bar with heating rate 0.3 K min (only every 45th point included for clarity) from AC (∀), $Ni_3(btc)_2(3\text{-pic})_6(pd)_3$ (o), **E** polymorph of $Ni_2(bpy)_3(NO_3)_4$ (M), and **M** polymorph of $Ni_2(bpy)_3(NO_3)_4$ (Δ). (Zhao, X., Xiao, B., Fletcher, A. J., Thomas, K. M., Bradshaw, D., Rosseinsky, M. J., *Science*, 306, 1012, 2004. Copyright Science.)

different from the chiral framework; $Ni_3(btc)_2(3\text{-pic})_6(pd)_3$ and AC are due to hysteretic adsorption in **E** and **M** and this is discussed later. There is little or no conclusive evidence for significant quantities of H_2 being physisorbed above temperature of 195 K for a hydrogen pressure of 1 bar. The amounts of hydrogen adsorbed at ambient temperature conditions and high pressure reported in the literature[42,62,69,110,118,128,191,192,196–201,204–206,210,212,217] are usually < 1% and average ~0.5 wt% as shown in Table 24.1.[19] Hydrogen–surface interactions and surface chemistry are major factors under ambient temperature conditions. However, experimental uncertainties in high-pressure measurements under ambient temperature conditions make it difficult to establish correlations with characterization parameters. The low uptakes under these conditions are consistent with low enthalpies for hydrogen adsorption on porous materials.[13,19,200]

24.10 Hydrogen–Surface Interactions in Pores

Adsorption is enhanced by interactions with specific surface sites and in narrow pores by overlap of potential energies from pore walls. Smaller pore sizes enhance adsorption at low pressures, whereas larger pores are necessary for high hydrogen capacity. Hydrogen–surface interactions may be electrostatic and dispersive (van der Waals) forces and Kubas coordination to a transition metal.[254,255] The possible hydrogen interactions with surface sites are

1. Hydrogen coordination to unsaturated metal centers via Kubas coordination
2. Quadrupole moment interaction with the local electric field gradient

3. Electron cloud polarization by a charge center
4. Dispersive forces (van der Waals)
5. Quadrupole moment between neighboring hydrogen molecules

The interaction between hydrogen and surface sites is a critical aspect for improving hydrogen adsorption at temperatures above cryogenic temperatures. MOFs have a wider range of possible surface chemistry than other porous materials since the surface chemistry may vary from open metal centers to hydrophobic surface associated with linker ligands and also may include functional groups on the ligands. It has been estimated that an interaction energy in the range of 20–40 kJ mol^{-1} is required for hydrogen storage applications.[254] Porous carbons with high oxygen, nitrogen, and boron contents have also been synthesized to investigate the effect on hydrogen adsorption characteristics.[46,157–159,161,163,168,247,256]

The synthesis and structure of the first stable organometallic complex with a H_2-coordinated side-on (η^2) to the metal were reported by Kubas et al. in 1984.[257] Subsequently, many similar complexes have been reported, mainly for second- and third-row transition metals, and some of these complexes are stable under ambient conditions.[258,259] H_2-coordinated side-on (η^2) to metals involves nonclassical two-electron three-center bonds with longer H-H bond length than gas-phase H_2 bond length (0.74 Å). Kubas[258,259] has divided the η^2–hydrogen coordination to metals into the following categories based on the H-H bond length:

1. True H_2 complexes (H-H, 0.8–1.0 Å)
2. Elongated H_2 complexes (H-H, 1.0–1.3 Å)
3. Compressed dihydrides (H-H, 1.3–1.6 Å)
4. True dihydrides (H-H, ~1.6 Å)[258]

Hydrogen binding in Kubas complexes is often reversible. Changes in substitution on diphosphine ligands in [RuCl (η^2–H_2)(pp_2)][BF_4] result in the H-H bond length increasing systematically from 0.97 to 1.03 Å with increasing substituent electron-donor ability indicating that varying the substitution in ligands influences the coordination of H_2 to metals.[260]

These η^2–hydrogen–metal complexes are of interest in relation to a strategy of changing surface chemistry in MOFs by the inclusion of open or unsaturated metal centers for Kubas coordination of hydrogen. In principle, the hydrogen–metal interaction strength in MOFs can be varied from very weak physisorption interactions, through elongated coordinated H_2 to dihydrides by changing the metal, ligand, and substitution on the ligand. The aim is to bind hydrogen weakly to metals via nondissociative chemisorption so that enhanced hydrogen adsorption occurs at higher temperatures well above cryogenic temperatures. If the hydrogen–metal interaction is sufficiently strong so that stable hydrides are formed, it will be detrimental for hydrogen storage applications because of difficulty in desorbing the H_2. In principle, fine-tuning the hydrogen–metal–η^2 coordination interaction may provide materials with facile hydrogen adsorption and desorption characteristics. A correlation between coordinatively unsaturated metal centers and enhanced hydrogen–surface density in many MOFs has been proposed.[208]

Studies of η^2–hydrogen coordination in Kubas organometallic molecules indicate that bulky ligands inhibit H_2 splitting. This suggests confinement in pores may have an effect on coordination.[259] The interaction of η^2–H_2 coordinated to a metal with the surrounding pore structure is likely to have a significant influence on stability and bonding. Hydrogen interaction with open metal sites in MOFs has been observed but, for the systems studied so far, it is much weaker than found in Kubas complexes. Only a relatively small number of Kubas complexes of first-row transition metals are known, whereas the MOFs that have been synthesized mainly contain first-row transition metals.[258,259] Further work is required to develop new MOFs incorporating unsaturated metal centers with stronger interactions with hydrogen for storage applications above cryogenic temperatures.

Theoretical studies of H_2 adsorption on porous materials suggest that inclusion of metal ions, such as Li^+, Na^+, or Mg^{2+}, might increase hydrogen adsorption by nondissociative H_2 binding. Mulfort et al. observed increases in both the overall H_2 adsorbed and the isosteric enthalpy for a Li-doped interpenetrated MOF $Zn_2(ndc)_2$(diPyNi) via chemical reduction with lithium metal.[130] The increases were ascribed to H_2–lithium interactions most likely enhanced by increased ligand polarizability and framework structural changes during chemical reduction.

H_2 adsorption in (Me_2NH_2)[In(bptc)] is enhanced by exchange of $Me_2NH_2^+$ for Li^+ cations.[261] The Li^+-exchanged material had a lower isosteric enthalpy for H_2 adsorption than the parent material. Cation exchange results in increased H_2 capacity due to increased accessible pore volume and lower adsorption enthalpy due to increased pore size.[261]

24.10.1 Isosteric Enthalpies of Hydrogen Adsorption

The isosteric enthalpy of hydrogen adsorption (Q_{st}) at zero surface coverage is a fundamental measurement of the interaction of hydrogen with surfaces.[46,262] Adsorbate–adsorbate and other interactions are factors that influence the adsorption enthalpies at higher surface coverage. The isosteric enthalpy of adsorption is very sensitive to the accuracy of experimental data, the methods used for interpolating between isotherm points to calculate ln(p) values for specific amounts adsorbed for the temperatures used, and the methods for determining Henry's law constant from the low-pressure data. Hydrogen adsorption is normally measured above the critical temperature (33 K) where a saturated vapor pressure cannot be used to compare isotherms at different temperatures. Therefore, isotherm equations that do not use a saturated vapor pressure, for example, virial equation methods[263,264] and Langmuir–Freundlich isotherm methods, have been used. The latter does not reduce to Henry's law and therefore cannot be used at low pressure. Virial parameters can be used for comparisons of H_2–surface and H_2–H_2 interactions in pores.[46,51,52] Comparison of these methods for hydrogen and deuterium adsorption on $Zn_3(bdc)_3$Cu(pyen) has shown that the virial methods give a more accurate description of hydrogen and deuterium isotherms at low surface coverage due to the fact that the Langmuir–Freundlich isotherm does not reduce to Henry's law at zero surface coverage.[52]

The isosteric enthalpies of adsorption of hydrogen on a wide range of porous materials have been reported.[39,42,44,46,52,59,66,67,69,98,110–112,118,120,121,147–149,161,177,182–184,191,192,195,196,202,206,214–226,228,244,247–249,256,265–280] The hydrogen adsorption enthalpies for porous materials at cryogenic temperatures are given in Table 24.2 and have the following ranges:

1. Carbons: 1.4–12.5 kJ mol^{-1} [46,147–149,161,247–249]; Boron-doped carbon, 10–20 kJ mol^{-1} [247,256]
2. Zeolites: 5.9–18.2 kJ mol^{-1} [148,177,267–271,274–280]
3. Silicas, siloxanes: 5.4–10.4 kJ mol^{-1} [89,90,148]
4. MOFs: 5.1–13.5 kJ mol^{-1} [39,42,44,52,59,66,67,69,110–112,118,120–123,127,128,133,190–192,195,196,202,214–226,228,235,237,244,245,247,271,281,282]
5. COFs: 2.7–8.8 kJ mol^{-1} [98,103]
6. Porous polymers: 5.7–10.0 kJ mol^{-1} [182–184]

TABLE 24.2

Selected Literature Values for Experimental Values for Isosteric Enthalpies of Adsorption of H_2 on Porous Materials from Gas Adsorption Measurements over a Range of Temperatures and Microcalorimetry

Porous Materials	Isosteric Enthalpy of Adsorption (Q_{st}/kJ mol^{-1})	Temperature Range (K)	References
Metal organic framework			
$Zn_4O(bdc)_3$ (MOF-5, IRMOF-1)	4.9–4.4	77–87	[66]
	4.1	77–298	[191]
	3.8	77–298	[192]
	5.2–4.7	77–87	[215]
$Zn_4O(ndc)_3$ (IRMOF-8)	6.1	77–298	[191]
$Zn_4O(hpdc)_3$ (IRMOF-11)	9.1–5.1	77–87	[66]
$Zn_3O_3(dhbdc)$ (MOF-74)	8.3–5.6	77–87	[66]
$Zn_4O(btb)$ (MOF-177)	11.3–5.8	273–323	[196]
	4.4	77–87	[39]
$Zn_3(bpdc)_3(bpy)$	7.1–5.1	77–87	[120]
$Zn_3(bdt)_3$	8.7–6.8	77–87	[67]
$Zn_2(abtc)(dmf)_2$ (SNU-4)	7.24–5.91	77–87	[245]
$Cu_2(bptc)$ (NOTT-100)	6.3	77–87	[44,45]
$Cu_2(tptc)$ (NOTT-101)	5.35	77–87	[44,45]
$Cu_2(qptc)$ (NOTT-102)	5.4	77–87	[45,244]
$Cu_2(C_{26}O_8H_{12})$ (NOTT-103)	5.7	77–87	[44]
$Cu_2(C_{22}O_8H_8F_2)$ (NOTT-105)	5.77	77–87	[44]
$Cu_2(C_{24}H_{14}O_8)$ (NOTT-106)	6.34	77–87	[44]
$Cu_2(C_{26}H_{20}O_8)$ (NOTT-107)	6.70	77–87	[44]
$Cu_2(pddip)(H_2O)_2(DMF)_{7.5}(H_2O)_5$ (NOTT-110)	5.68	77–87	[244]
$[Cu_2(dhpddip)(H_2O)_2](DMF)_{7.5}(H_2O)_5$ (NOTT-111)	6.21	77–87	[244]
$Cu_3(dcbplb)$ (NOTT-112)	5.64–4.74	77–87	[59]
$Cu_3(btc)_2$ (HKUST-1)	6.6–6.0	77–87	[66]
	4.5	77–87	[192]
	7–6	77–87	[216]
$Cu_2(abtc)(H_2O)_2{\cdot}3dma$ (PCN-10)	7–4	77–300	[118]
$Cu_2(abtc)(dmf')_2$ (SNU-5′)	6.53–5.91	77–87	[245]
$Cu_2(abtc)$ (SNU-5)	11.6–4.43	77–87	[245]
$Cu_2(sbtc)(H_2O)_2{\cdot}3dma$ (PCN-11)	7–4	77–300	[118]
$Cu_2(adip)$ (PCN-14)	8.6–5.9	77–87	[241]
$Cu_2(bdi)$ (PCN-46)	7.2	77–87	[237]
$[Cu_2(bpndc)(bpy)]n$ (SNU-6)	7.74	77–87	[134]
$Cr_3OF(btc)_2$ (MIL-100(Cr))	6.3–5.6^{MC}	77	[69]
$Cr_3OF(bdc)_3$ (MIL-101(Cr))	10.0–9.3^{MC}	77	[69]
$H_2[Co_4O(tatb)_{8/3}]$ (PCN-9)	10.1	77–87	[214]
$Co_3(bpdc)_3(bpy)$	6.8–5.7	77–87	[120]
$Mg_3(ndc)_3$	9.5–7.0	77–87	[195]
$Mn_3(bdt)_3$	8.4–6.3	77–87	[67]
$Mn_2(bdt)Cl_2$	8.8–6.0	77–87	[67]

TABLE 24.2 (continued)

Selected Literature Values for Experimental Values for Isosteric Enthalpies of Adsorption of H_2 on Porous Materials from Gas Adsorption Measurements over a Range of Temperatures and Microcalorimetry

Porous Materials	Isosteric Enthalpy of Adsorption (Q_{st}/kJ mol^{-1})	Temperature Range (K)	References
$Mn_3[(Mn_4Cl)_3(btt)_8]_2$	10.1–5.5	77–87	[42,121]
$Cu_3[(Cu_4Cl)_3(btt)_8]_2$	9.5–6	77–87	[122]
$Fe_3[(Mn_4Cl)_3(btt)_8]_2$ $FeCl_2$	10.2–5.5	77–87	[121]
$Co_3[(Mn_4Cl)_3(btt)_8]_2{\cdot}1.7\ CoCl_2$	10.5–5.6	77–87	[121]
$Ni_{2.75}Mn_{0.25}[(Mn_4Cl)_3(btt)_8]_2$	9.1–5.2	77–87	[121]
$Cu_3[(Cu_{2\cdot 9}Mn_{1.1}Cl)_3(btt)_8]_2{\cdot}2CuCl_2$	8.5–6.0	77–87	[121]
$Zn_3[(Zn_{0\cdot 7}Mn_{3.3}Cl)_3(btt)_8]_2{\cdot}2ZnCl_2$	9.6–5.5	77–87	[121]
$Li_{3.2}Mn_{1.4}[(Mn_4Cl)_3(btt)_8]_2{\cdot}0.4LiCl$	8.9–5.4	77–87	[121]
$Mn_3[(Mn_4Cl)_3(btt)_8]_2{\cdot}0.75CuPF_6$	9.9–5.6	77–87	[121]
$Fe_3[(Fe_4Cl)_3(btt)_8(MeOH)_4]_2{\cdot}$(Fe-btt)	11.9	77–87	[273]
$Mn_3[(Mn_4Cl)_3$(tpt-3tz)$_8(DMF)_{12}]_2$	7.6	77–87	[217]
$Cu_3[(Cu_4Cl)_3$(tpb-3tz)$_8]_2{\cdot}1.1CuCl_2$	8.2	77–87	[217]
$Mn_3[Co(CN)_6]_2$	5.9–5.3	77–87	[215]
$Fe_3[Co(CN)_6]_2$	6.6–6.3	77–87	[215]
$Co_3[Co(CN)_6]_2$	6.8–6.5	77–87	[215]
$Ni_3[Co(CN)_6]_2$	7.4–6.9	77–87	[215]
$Cu_3[Co(CN)_6]_2$	7.0–6.7	77–87	[215]
	7.2	77–87	[218]
$Zn_3[Co(CN)_6]_2$	6.5–6.3	77–87	[215]
$Ni[Fe(CN)_5NO]$	7.5–5.7	77–87	[218]
$Co[Fe(CN)_5NO]$	6.5–5.0	77–87	[218]
	8.5–5.5		[219]
$Zn_3K_2[Fe(CN)_6]_2$	8.3–7.1	75–85	[220]
	9.0–7.9	77–87	[221]
$Zn_3Rb_2[Fe(CN)_6]_2$	6.8	75–85	[220]
	7.9–7.3	77–87	[221]
$Zn_3Cs_2[Fe(CN)_6]_2$	6.3	75–85	[220]
$Zn_3Na_2[Fe(CN)_6]_2$	7.7–7.0	77–87	[221]
$Zn_3[Co(CN)_6]_2$-R	6.1	75–85	[220]
$Zn_3[Co(CN)_6]_2$	6.5–6.3	77–87	[221]
$Zn_3Li_2[Fe(CN)_6]_2{\cdot}2H_2O$	7.9–6.1	77–87	[221]
$H_2\ Zn_3\ [Fe(CN)_6]_2{\cdot}2H_2O$	8.2–7.8	77–87	[221]
$NaNi_3(OH)(sip)_2$	10.4–9.4	77–87	[112]
Co(pyz)$[Ni(CN)_4]$	7.2	77–87	[222]
Co(pyz)$[Pd(CN)_4]$	7.8	77–87	[222]
Co(pyz)$[Pt(CN)_4]$	7.6	77–87	[222]
Ni(pyz)$[Ni(CN)_4]$	7.2	77–87	[222]
Ni(bpy)$[Ni(CN)_4]$	7.5	77–87	[222]
Ni(bpy)$[Pd(CN)_4]$	7.0	77–87	[222]
Ni(dpac)$[Ni(CN)_4]$	6.0	77–87	[222]
Ni_2(dhtp)$(H_2O)_2{\cdot}8H_2O$	13.5^{VTIR}	92–150	[110,223]

(continued)

TABLE 24.2 (continued)

Selected Literature Values for Experimental Values for Isosteric Enthalpies of Adsorption of H_2 on Porous Materials from Gas Adsorption Measurements over a Range of Temperatures and Microcalorimetry

Porous Materials	Isosteric Enthalpy of Adsorption (Q_{st}/kJ mol^{-1})	Temperature Range (K)	References
Cd(2-pmc)$_2$ *rho*-ZMOF	8.7	77–87	[190]
$Cu_6O(tzi)_3(NO_3)$	9.5–4.7	77–87	[123]
Cu(bdc)(ted)$_{0.5}$	6.1–4.9	77–87	[128]
$Cu_3(bhtc)_2$ (UMCM-150)	7.2–4.9	77–87	[127]
$In_3O(abtc)_{1.5}(NO_3)$	6.5	77–87	[225]
In(4,6-pmdc)$_2$ *sod*-ZMOF	8.4	77–87	[190]
$((CH_3)_2NH_2)$[In(bptc)]	7.6	77–87	[228]
$((CH_3)_2NH_2)$[In(bptc)]Li^+	6.1	77–87	[228]
Y(btc)(H_2O)	7.3	77–87	[202]
$Zn_3(bdc)_3$Cu(pyen)(M′MOF1)	12.29–9.5	77–87	[52]
	12.44–9.5 (D_2)	77–87	[52]
Zn(bdc)(ted)$_{0.5}$	5.3–5	77–87	[128]
$Zn_2(ndc)_2$(diPyNi)	5.7–3.3	77–87	[130]
$Zn_2(ndc)_2$(diPyNi) $^-Li^+$	6.1–5.1	77–87	[130]
Zn(mim)$_2$ (ZIF-8)	4.5	30–300	[210]
Zn(tbip)	6.7–6.4	77–87	[226]
Zn_3(OH) (*p*-cdc)$_{2.5}$	7.2–3	77–87	[224]
$Zn_4O(t^2dc)(btb)_{4/3}$(UCMC-2)	6.4–4.2	77–87	[232]
$Zn_2(cnc)_2$(dpt)	7.85–6.4	77–87	[243]
$Co_4O(tatb)_{8/3}\cdot 2H_3O^+\cdot 5H_2O\cdot 8$dmso (PCN-9(Co))	10.1–6	77–87	[235]
$Fe_4O(tatb)_{8/3}\cdot 5H_2O\cdot 10$dmso (PCN-9(Fe))	6.4–4	77–87	[235]
$Mn_4O(tatb)_{8/3}\cdot 2H_3O^+\cdot 5H_2O\cdot 8$dmso (PCN-9(Mn))	8.7–5.5	77–87	[235]
Mg formate	7–6.5	77–87	[272]
$[Cd_5(Tz)_9](NO_3)_3\cdot 8H_2O$	13.3–6	77–87	[281]
Mg-MOF-74	9.4^{VTIR}	77–97	[271]
Co-MOF-74	11.2^{VTIR}	77–97	[271]
Cr-bdc nanostructures	8	77–87	[282]
Cu-bdc nanostructures	4	77–87	[282]
Carbon materials			
Graphene-like nanosheets	5.9–4	77–87	[250]
Functionalized ACs from coconut and polyacrylonitrile	5.2–3.9	77–114	[46]
Carbon (charcoal)	7.3–1.4	17–90	[147]
Carbons (charcoal)	7.9–6	77–90	[148]
AC AX-21	6.4–5.0	77–273	[149]
Carbons from biomass	4.1–7.5	77–87	[81]
Carbon (wood derived + KOH activation)	10.9	298–333	[248,249]
	17–20	298–333	[248,249]
Carbide-derived carbons	11–6	77–87	[265]
Carbons zeolite β-templated	8.2–4	77–87	[161]

TABLE 24.2 (continued)

Selected Literature Values for Experimental Values for Isosteric Enthalpies of Adsorption of H_2 on Porous Materials from Gas Adsorption Measurements over a Range of Temperatures and Microcalorimetry

Porous Materials	Isosteric Enthalpy of Adsorption (Q_{st}/kJ mol^{-1})	Temperature Range (K)	References
Boron-doped carbon	12.47–10.8	77–298	[247]
BC_6	20–10	77–87	[256]
Zeolites			
Zeolite Na-A	10.7–6.2	40–120	[177]
Zeolites 4A, 5A, 13X	7.9–5.9	75–90	[148]
Aluminophosphates	9.2–3.7	77–87	[266]
Mg exchanged faujasite, Na/K-ZSM-5, (Mg,Na)-Y, (Ca,Na)-Y, Ca-X, Mg-X, Li-FER, Na-FER, K-FER, Li-ZSM-5, Na-ZSM-5	18.2–3.5^{VTIR}	120–150 79–95	[267–269,271,274–280]
Silicas			
Silica gel	7.3–5.4	75–90	[148]
Functionalized cubic	7.2		[89]
Siloxane cages			
Poly(ethynylene aryleneethenylene silsesquioxane	8	77–87	[90]
Polymers			
Hyper-cross-linked porous polymers	7.5–6	77–87	[184]
Hyper-cross-linked polyaniline polymers	9.3–5.7	77–87	[182]
Hyper-cross-linked materials	7.8–5.2	77–87	[93]
Polyporphyrins	7.6–5.8	77–298	[186]
Polymers POPs1–4	9–5	77–87	[188]
PAF-1	4.6	77–87	[97]
COFs 1,5,6,8,10, 102, 103	8–3.9	77–87	[103]

MC, Low surface coverage microcalorimetry method; VTIR, variable temperature IR method. All other measurements are from hydrogen adsorption isotherm studies.

A correlation has been observed between ΔH_0 and ΔS_0 for hydrogen adsorption on Mg-MOF-74, Co-MOF-74, and cation-exchanged zeolites.[271] This suggests that the optimum value of ΔH_0 for hydrogen delivery at room temperature is likely to be in the range of 22–25 kJ mol^{-1}. Other estimates have suggested that an enthalpy of adsorption of 15 kJ mol^{-1} is optimum for hydrogen storage applications,[283] while a range of 20–40 kJ mol^{-1} has also been proposed for hydrogen storage applications.[254] The Q_{st} values obtained for hydrogen adsorption on porous materials are much higher than the enthalpy of vaporization of 0.9 kJ mol^{-1} for H_2 at 20.28 K.[4] The surface chemistry in MOFs can be varied to a greater extent than in case of other porous materials. Some MOFs have open metal centers that have been proposed as sites for η^2–H_2 coordination similar to that found in Kubas compounds. There is some evidence for stronger interactions, but the enthalpies of hydrogen adsorption at zero surface coverage for MOFs with open metal centers are often not appreciably higher than for other porous materials (see Table 24.2). This is shown by enthalpies of hydrogen adsorption in the range of 5.3–6.7 kJ mol^{-1} observed for the copper tetracarboxylate series.[44] The highest enthalpies of hydrogen adsorption so far obtained for MOFs are 12.3 kJ mol^{-1} for $Zn_3(bdc)_3 \cdot Cu(pyen)$ (gas adsorption method, 77–87 K), 13.5 kJ mol^{-1} for desolvated $Ni_2(dhtp)(H_2O)_2 \cdot 8H_2O$ (variable temperature infrared (IR) method), and 13.3 kJ mol^{-1} for $[Cd_5(Tz)_9](NO_3)$. These enthalpies of hydrogen adsorption at cryogenic temperatures compare with the following maximum values: boron-doped carbon (>7%) (~12.5 kJ mol^{-1} and (BC_6) 20 kJ mol^{-1}),[247,256] ~11 kJ mol^{-1} for carbide-derived carbons,[265] <8.2 kJ mol^{-1} for other carbons,[161] 18.2 kJ mol^{-1} for zeolites,[268] 8 kJ mol^{-1} for COFs,[103] and 10.0 kJ mol^{-1} for porous polymers.

Recent results for H_2 adsorption on an ultramicroporous carbon showed that the main process, physisorption with an enthalpy of adsorption of 9–11 kJ mol^{-1}, was accompanied by a slow process. The latter involves slow uptake at high pressures with a high enthalpy of adsorption (17–20 kJ mol^{-1}) and hysteresis on desorption. The combined result is relatively high levels of hydrogen uptake at near-ambient temperatures and pressures (e.g., up to 0.8 wt% at 298 K and 2 MPa). These unusual adsorption characteristics were attributed to contributions from polarization-enhanced physisorption induced by residual traces of alkali metals from chemical activation.

Currently, the Q_{st} values reported for hydrogen adsorption on porous materials are too low for significant adsorption at ambient temperatures. An intermediate weak nondissociative η^2 coordination or other weak chemisorption with Q_{st} values in excess of values observed so far and probably in the range of 20–40 kJ mol^{-1} with facile adsorption and desorption characteristics is required for storage applications. If the hydrogen is strongly chemisorbed H_2 desorption, it would require desorption temperatures that are too high.

24.10.2 Diffraction Studies of Hydrogen in Pores

X-ray diffraction studies have provided direct evidence of hydrogen adsorption.[284,285] Dense aggregates of adsorbed H_2 were observed in MOF $[Rhbza)_4(pyz)]_n$ at 90 K.[285] Hydrogen adsorbate molecules were located near a corner of rectangular nanochannels in $[Cu_2(pzdc)_2(pyz)]n$ coordination polymer. These results are consistent with adsorption due to overlap of potential energy fields from the pore walls.[284]

Neutron diffraction provides direct evidence for the position of hydrogen adsorption sites. Spencer et al. used variable-temperature single-crystal neutron diffraction to investigate hydrogen-loaded MOF-5 $[Zn_4O(bdc)_3]$.[286] The sites for hydrogen adsorption were the nodal regions with one site located over the shared vertex of the ZnO_4 units at the center of the node and the other over the face of ZnO_4. The hydrogen adsorption sites of MOF-5 were also determined using deuterium loading and powder neutron diffraction methods.[287] The initial adsorption sites were identified as the center of three ZnO_3 triangular faces followed by the top of single ZnO_3 triangles. At higher loading, ZnO_2 sites and sites on the hexagonal linkers were occupied. 3D interlinked nanoclusters of hydrogen molecules were observed at high loading.

Six D_2 adsorption sites were observed in HKUST-1 ($Cu_3(btc)_2$) using neutron powder diffraction techniques. Adsorption occurred initially on coordinatively unsaturated axial sites of the dinuclear Cu center [Cu–D_2 = 2.39(1) Å].[288] This distance is much larger than that observed for σ-bonded η^2–H_2 complexes, but it represents a significant interaction. Competitive adsorption on other D_2 sites proceeded with increasing pore size. Neutron powder diffraction studies showed that adsorbed D_2 is associated with the Mn center (Mn–D_2 = 2.27 Å) in $Mn_3[(Mn_4Cl)_3(btt)_8.(CH_3OH)_{10}]_2$.[42] One of the strongest D_2 adsorption sites was 2.47 Å from the Cu^{2+} ions in $HCu_3[(Cu_4Cl)_3(btt)_8]_2 \cdot 3.5HCl$.[122]

Neutron powder diffraction studies on D_2-loaded $Cu_3(dcbplb)$ (NOTT-112) showed that the axial sites of exposed Cu(II) ions in the smallest cuboctahedral cages are the initial, strongest binding sites with a Cu–D_2 distance of 2.23 Å. This allows discrimination between the two types of exposed Cu(II) sites at the paddle wheel nodes.[289]

Studies have shown that in Zn_2(dhbdc) (MOF-74), the first site occupied had a longer Zn–D_2 distance (2.6 Å).[208] Powder neutron diffraction studies[273] of $Fe_3[(Fe_4Cl)_3(btt)_8(MeOH)_4]_2$ (Fe-btt), under various deuterium loadings at 4 K, showed 10 different adsorption sites. The strongest binding site was 2.17(5) Å from the Fe^{2+} in the framework.

Neutron diffraction and inelastic neutron scattering (INS) studies have shown that hydrogen (deuterium) adsorption on Cr MIL-53 at low temperature involves the occupation of four different adsorption sites. Adsorption sites near the Cr–O clusters had the highest occupations, while adsorption occurs in the corners formed by the linker ligands. Significant breathing of the framework occurred and this was pronounced above the boiling point of hydrogen.[290]

High-resolution neutron powder diffraction studies of hydrogen adsorption on $Cu_3[Co(CN)_6]_2$ at 1, 2, and 2.3 H_2/Cu loadings showed that adsorption occurred on two sites.[291] The strongest adsorption site was an interstitial location within the structure. The second adsorption type of site was exposed Cu^{2+} ion coordination sites resulting from the presence of $[Co(CN)_6]^{3-}$ vacancies.

Neutron diffraction studies revealed four distinct D_2 adsorption sites in Y(btc). The strongest adsorption sites identified were the aromatic rings in the organic linkers rather than the open metal sites, as reported for other MOFs.[202] At high H_2 loadings, highly symmetric novel nanoclusters of H_2 molecules with relatively short H_2–H_2 contact distances were formed.

Hydrogen interactions in MOF materials are complex and a variety of factors influence adsorption interactions when H_2 is confined in pores. The metal–H_2 and D_2 distances observed earlier are in the range of 2.17–2.6 Å, which is significantly longer than the range of 1.6–1.92 Å reported[257,260,292] for Kubas compounds, but the shorter distances represent a significant interaction. Comparison of the metal–H_2 (or M–D_2) distances with corresponding isosteric enthalpy data suggests that other factors may also influence H_2–surface interactions.

24.10.3 Spectroscopic Studies of Hydrogen–Surface Interactions

INS measurements on porous nickel phosphateVSB-5[293] and porous hybrid inorganic/organic framework material $NaNi_3(OH)(sip)_2$[112,293] showed that initially, adsorption occurred on unsaturated Ni(II) sites in the latter. Binding sites consistent with physisorption were occupied as hydrogen loading increased and finally, at higher loading interactions between H_2 adsorbate molecules occurred. Studies of unsaturated metal centers in metal cyano[68,107,215,294] and MOF materials suggest that H_2 interactions with surface sites in these materials are weaker.[26,291,294]

INS studies[295] of D_2 adsorption on HKUST-1 were consistent with previous neutron powder diffraction studies. Three binding sites for H_2 loading <2.0 H_2/Cu were identified and these were progressively populated with increasing loading. The variation of INS peaks with temperature showed that the peak corresponding to population at H_2/Cu ratio <0.5 had the highest enthalpy (6–10 kJ mol^{-1}), while the peaks for the other sites had lower enthalpies.

Zhou et al. proposed[296] that hydrogen binding in $(Mn_4Cl)_3(btt)_8$-MOF[42] is not of the expected Kubas type because there is (a) no significant charge transfer from metal to H_2, (b) no evidence of any H_2–σ* Mn-d orbital hybridization, (c) no significant H-H bond elongation, and (d) no significant shift in ν(H-H) stretching vibration.[296] The short metal–H_2

distances and relatively high binding energies were explained by an enhanced classical coulombic interaction. This effect is unlikely to be sufficient to increase the hydrogen storage temperature to ambient conditions.

INS spectra of $Fe_3[(Fe_4Cl)_3(btt)_8(MeOH)_4]_2$ (Fe-btt) were consistent with the strong rotational hindering of adsorbed H_2 molecules at low loadings.[273] Catalytic conversion of *ortho* to *para* hydrogen by the paramagnetic iron centers was also observed.

IR spectroscopic studies of H_2 adsorption on various MOF materials at 300 K and high pressures (27–55 bar) have shown that ν(H-H) stretching vibration at 4155 cm^{-1} for *ortho* H_2 is shifted (30–40 cm^{-1}) to lower energy.[297] H_2 molecules interact with the organic ligands of MOFs $M(bdc)(ted)_{0.5}$ (M = Ni, Cu, Zn.) instead of the saturated metal centers located at the corners of the unit cell. Density functional calculations showed that for the induced dipole associated with the trapped H_2 in $M(bdc)(ted)_{0.5}$ systems, the strongest dipole moment is for the site in the unit cell corner dominated by the interaction with the benzene ligand rather than the metal center. A weak dependence of ν(H-H) on cations was observed for 1D pore structures $M_3[HCOO]_6$ (M = Mn, Co, Ni)-type formate structures, and this was attributed to a small change in pore size rather than a direct interaction with the metal centers. No correlation was observed between H_2 binding energies (determined by isotherm measurements) and the ν(H-H) stretch shift, indicating that these shifts are dominated by the environment (organic ligand, metal center, and structure) rather than the strength of the interaction.[297]

Interaction between adsorbed hydrogen and the coordinatively unsaturated Mg^{2+} and Co^{2+} cationic centers in Mg-MOF-74 and Co-MOF-74, respectively, has been investigated using variable temperature infrared (VTIR) spectroscopy.[271] The H_2 molecule is perturbed by the cationic adsorption center making the ν(H-H) stretching mode IR active at 4088 and 4043 cm^{-1} for Mg-MOF-74 and Co-MOF-74, respectively. The corresponding enthalpies of adsorption were 9.4 and 11.2 kJ mol^{-1} for Mg-MOF-74 and Co-MOF-74, respectively. The ν(H-H) IR bands for zeolites were in the range of 4056–4111 cm^{-1}. These IR band shifts are consistent with only very weak interactions with unsaturated metal centers.

The ν(H-H) stretching vibration in H_2 becomes IR active when adsorbed on surfaces due to lowering of symmetry. However, the IR spectra are complex because of the presence of *ortho* and *para* forms of H_2, and in situ evidence for a single-site catalyzed conversion of adsorbed H_2 was obtained.[298] The IR bands shift to lower energy relative to the gas-phase spectrum and the intensity is proportional to the strength of the H_2–surface interaction. IR spectroscopy studies have shown that unsaturated copper sites in HKUST-1 are preferential adsorption sites for H_2 and the ν(H-H) band is ~70 cm^{-1} lower than the corresponding gas-phase band.[298] Smaller shifts (37–45 cm^{-1}) in ν(H-H) have been observed for H_2 adsorption on a cross-linked polymer. It was proposed that this shift was primarily due to specific interactions of the H_2 molecule with the electron-rich part of the polymer.[299] The H_2 adsorption energy estimated from the temperature dependence of IR spectral features was 10 kJ mol^{-1}. This is similar to the highest value obtained from INS measurements.[295]

The shifts to lower energy in low-temperature-stable η^2–H_2 complexes are much larger with ν(H-H) typically in the range of 2600–3250 cm^{-1} [258] (ν(H-H) for H_2 in gas phase is 4160 cm^{-1} [298]). The IR spectrum of $W(CO)_3(PCy_3)_2(\eta^2–H_2)$ gives ν(H-H) of 2690 cm^{-1}, while the H-H bond length in $W(CO)_3(P\mathit{i}Pr_3)_2(\eta^2–H_2)$ was 0.84 Å as determined by neutron diffraction studies compared with 0.74 Å in H_2 gas.[257] Raman spectroscopy has also been used to study the adsorbed H_2 phase in porous materials at room temperature and under cryogenic conditions.[300–302] The results confirm that the interaction strength for adsorption of molecular hydrogen is small and consistent with physisorption for single-walled nanotubes, MOF-5, and HKUST-1.

Overall, the spectroscopic evidence currently available indicates that interactions of molecular hydrogen with surfaces in pores are weak. This is consistent with crystallographic diffraction studies that show longer distances than those observed for Kubas compounds and low enthalpies for hydrogen adsorption on porous materials. The inclusion of Kubas-type motifs in MOFs has the greatest potential for developing materials for hydrogen storage under ambient conditions.

24.10.3.1 Framework Flexibility and Hysteretic Adsorption in MOFs

Some MOFs have framework flexibility and structural change may occur during the adsorption process. MOF framework flexibility may occur either without any bond breaking or some materials undergo bond breaking and reforming reactions. The latter are much rarer and the former have possible applications in storage applications. The flexibility in the MIL series is an example of large structural changes during adsorption.[12]

Structural change without bond breaking occurs with a scissoring motion leading to an increase in crystallographic cell volume of ethanol and methanol templated phases E and M of on $Ni_2(bpy)_3.(NO_3)_4$ during adsorption of the templates.[303,304] Hysteretic adsorption was observed initially for hydrogen adsorption on $Ni_2(bpy)_3.(NO_3)_4$[200] (see Figure 24.13) and several further systems have been identified.[207,227,305] This is thought to be a kinetic trapping effect. The corresponding isobars are shown in Figure 24.12. It is apparent that hydrogen desorption does not start until higher temperatures (~110 K).[200] These materials have windows and pore cavities and the diffusion through the very narrow windows and framework flexibility are probably factors in hysteretic adsorption. Hysteretic hydrogen adsorption was subsequently observed in a related system for desolvated $[Co_2(bpy)_3(SO_4)_2(H_2O)_2](bpy)(CH_3OH)$.[306]

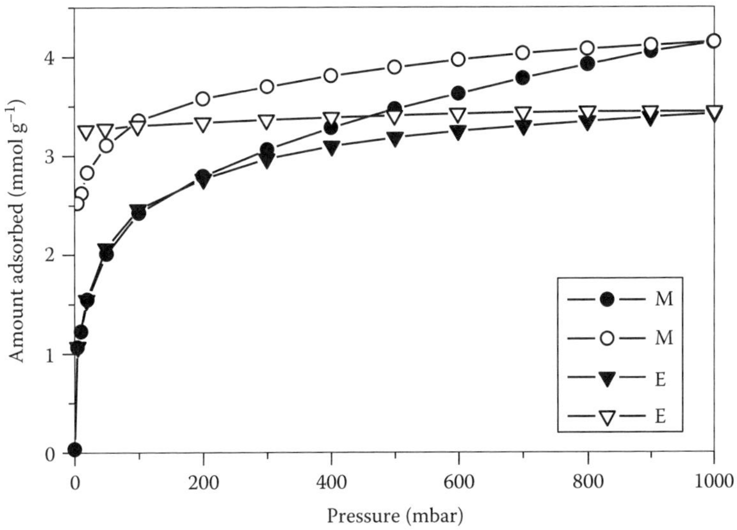

FIGURE 24.13
Adsorption/desorption isotherms for H_2 adsorption (filled symbols) and desorption (open symbols) isotherms on MOF adsorbents at 77 K: **M** polymorph of $Ni_2(bpy)_3(NO_3)_4$ (●, ○); **E** polymorph of $Ni_2(bpy)_3(NO_3)_4$ ((▼, ▽). (Zhao, X., Xiao, B., Fletcher, A. J., Thomas, K. M., Bradshaw, D., Rosseinsky, M. J., *Science*, 306, 1012, 2004. Copyright Science.)

Cation-induced kinetic trapping and enhanced hydrogen adsorption in a modulated anionic MOF (In(III) centers and tetracarboxylic acid) have been observed.[307] The framework exhibits hysteretic hydrogen adsorption with piperazinium dications in pores, but on exchange of these dications with lithium cations, no hysteresis is observed.

Ferey et al. have developed the MIL series of MOFs where the changes in structure during adsorption are much larger.[229,308] Hysteresis has been observed for hydrogen adsorption on the MIL-53, (M = Al^{3+},Cr^{3+}) at high pressure and 77 K.[125]

Choi et al. studied[207] H_2 adsorption on a MOF Co(bdp) that showed temperature-dependent broad hysteresis loops over the temperature range of 50–87 K. The results suggested a pressure-, temperature-, and gas-dependent pore-opening mechanism with the hysteretic behavior governed by phase transitions with energies comparable to the H_2 adsorption enthalpy. Some high-pressure H_2 hysteresis has also been observed for FMOF-1, which was formed by reaction of Ag(I) with 3,5-bis(trifluoromethyl)-1,2,4-triazolate.[305] The fluoro-lined channels and cavities of the framework show hysteretic adsorption of H_2.

There are now a number of examples of hysteretic adsorption of hydrogen[125,200,207,305] and other gases.[309–311] Hysteretic H_2 adsorption has been observed in several MOFs and may have applications for hydrogen storage. A microscopic theory has been proposed showing that enhanced adsorption hysteresis of H_2-activated diffusion along the small-pore channels is a dominant equilibration process. The sensitivity of gas adsorption to temperature changes is explained by thermodynamics. Analysis of transient adsorption dynamics reveals that the hysteretic H_2 adsorption is an intrinsic adsorption characteristic in diffusion-controlled small-pore systems, but this has not yet been realized.[312]

24.10.3.2 Kubas-Type Compounds Doped in Porous Materials

Kubas-type compounds have been doped into mesoporous materials to produce potential hydrogen storage materials. The problem is getting high enough loadings. Low-coordinate Ti(III) fragments with controlled geometries designed specifically for σ–H_2 binding have been dispersed on mesoporous silica using tri- and tetrabenzyl Ti precursors. The optimal loading level of 0.2 mol equivalents of tetrabenzyl Ti gave a material with total hydrogen storage capacity at 77 K of 21.45 wt% (including compressed gas at 100 atm). The binding adsorption enthalpies increased with increasing surface coverage to a maximum of 22.15 kJ mol^{-1}. A quantum increasing chemical computational model of the titanium–H_2 binding sites of a mesoporous silica is in agreement with experimental results that average Ti–H_2 interaction energies increase as the number of bound H_2 molecules. These findings suggest that η^2–H_2 binding to unsaturated metal centers may possibly be exploited for hydrogen storage. The silica-supported systems have 5–10 wt% of organometallic Ti, V, and Cr grafted onto mesoporous silica.[313–315] These materials are of interest because high H_2/metal ratios and enthalpies increase with increasing adsorption. In general, the H_2 storage ability of the materials decreases in the order Ti > V > Cr and $M^{2+} > M^{3+}$.

24.10.3.3 Hydrogen Spillover on Porous Materials

Hydrogen spillover is well established in catalysis where hydrogen atoms are formed on metal particles by dissociative chemisorption and diffuse from the particles onto pore surfaces where further diffusion may occur. The use for hydrogen storage on porous materials was first proposed by Yang and coworkers.[316,317] The hydrogen atoms react with surface sites, and therefore, surface chemistry and surface area are important. Studies have been undertaken for hydrogen spillover on carbon nanotubes,[316,318,319] graphite nanofibers,[318,320]

ACs,[321–323] nitrogen-doped carbons,[324] zeolites,[87] Al-doped MCM-41 mesoporous silica,[325] COFS,[326] and MOFs.[327,328] The adsorption isotherms are reversible with acceptable desorption characteristics at ambient temperature. Isotope tracer studies have been undertaken on IRMOF-8 bridged to Pt/C to understand the mechanism. Studies of hydrogen spillover on MOF-5,[329] HKUST-1, and MIL-101[326] showed that H_2 storage capacity at 298 K and high pressure correlated with the surface area and pore volume. Hydrogen uptake at room temperature in composite MIL-100(Al)/Pd(10%) is ~2× that of the MIL-100. This is accounted for by Pd hydride formation and a spillover mechanism.[330]

Sample-to-sample consistency in storage capacity on the bridged MOF samples is difficult to achieve unlike Pt/C sorbents. However, significant storage enhancements are still observed.[331] The effect of Pd nanoparticles on the rate of hydrogen adsorption provides supporting evidence for the hydrogen spillover mechanism in Pd/ultramicroporous carbon systems.[249] Although this method may enhance storage capacities by ~3 compared with physisorption under ambient temperature conditions, it is still well below the hydrogen storage capacities required for applications.

24.10.3.4 Comparison of H_2 and D_2 Adsorption

Both H_2 and D_2 have bond lengths of 0.7416 Å, but different zero point energies.[332] Small differences in physical properties (density, boiling point, amplitudes of vibration, dissociation energy, etc.) and adsorption characteristics are observed.[46,51,52,333–335] Comparison of H_2 and D_2 adsorption at 77 and 87 K shows that D_2/H_2 molar ratios in the range of 1.06–1.15 are typical for porous MOF,[45,47,52,143] carbons,[46,51] and zeolite materials.[177]

When differences between H_2 or D_2 sizes and pore size are similar to the de Broglie wavelength, quantum effects occur and this leads to molecular sieving of H_2 and D_2.[336–338] Quantum effects lead to higher amounts of deuterium adsorbed than hydrogen under the same experimental conditions.[46,47,51,52,334,335,338] Kinetic quantum molecular sieving was first observed experimentally for adsorption and desorption of H_2 and D_2 on a carbon molecular sieve (Takeda 3a), and a microporous carbon[51] and typical adsorption/desorption kinetic profiles for D_2 and H_2 are shown in Figure 24.14. This figure shows that D_2 adsorption is faster than H_2 for both adsorption and desorption contrary to the trend expected based on mass. Analysis of the H_2 and D_2 adsorption isotherms using a virial equation shows that the D_2–D_2 interactions are smaller than the H_2–H_2 interactions, and this is consistent with the smaller amplitude of vibration of D_2 compared with H_2.[51,52]

Similar observations of faster deuterium adsorption kinetics compared with hydrogen were reported subsequently for $Zn_3(bdc)_3 \cdot Cu(pyen)$[52] and zeolites.[88] The MOF $Zn_3(bdc)_3 \cdot Cu(pyen)$ has two types of pores present in the material and independent quantum effects were observed for these pores.[52] The activation energies for hydrogen were slightly greater than for deuterium for both types of pores. The difference between the adsorbate dimensions and the minimum pore dimension needs to be close to the de Broglie wavelength for quantum effects to be observed. The faster adsorption kinetics for D_2 compared with H_2 is contrary to that expected on a mass basis. The larger amplitude of vibration of hydrogen atoms leads to an effective larger cross section for hydrogen leading to slower hydrogen adsorption and desorption kinetics when the pore size is similar to the de Broglie wavelength. Theoretical studies of quantum kinetic molecular sieving have been reported.[339,340] These kinetic effects have only been observed at low temperature in materials with narrow pores. This effect would not be an issue for storage in materials with larger pores. Kinetic quantum molecular sieving has possible applications for isotope separation.

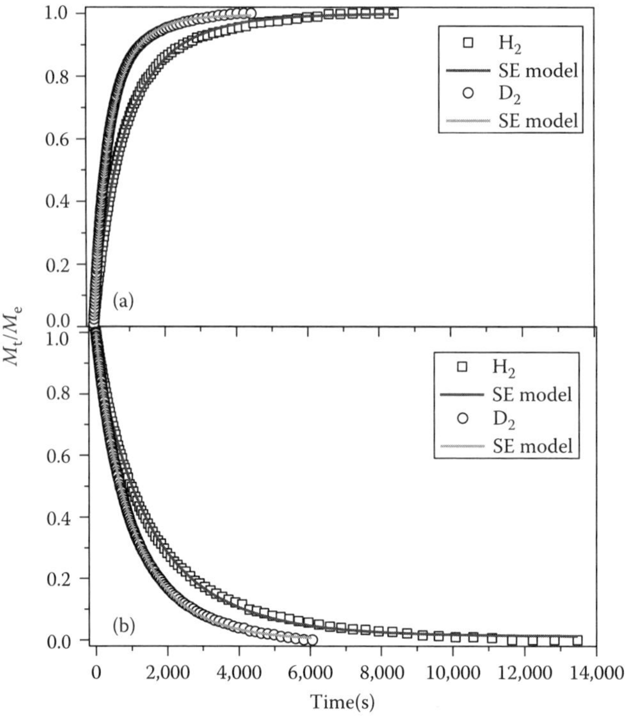

FIGURE 24.14
Typical H_2 and D_2 adsorption and desorption kinetic profiles and fitting of the stretched exponential kinetic model for pressure increments (1–2 kPa) and decrements (2–1 kPa) on CMS T3A at 77 K (a) adsorption and (b) desorption. (Reproduced with permission from the American Chemical Society.)

24.11 Applied Electric Field and Hydrogen Adsorption

It has been proposed that an applied electric field may increase hydrogen adsorption. Density functional theory has been used to show that an applied electric field may improve the hydrogen storage properties of polarizable substrates. The adsorption of a layer of hydrogen molecules on several nanomaterials was investigated.[341]

The effect of an applied electric field on the hydrogen physisorption isotherm of platinum-supported carbon sorbents has shown that sorption enhancement was obtained by applying a positive electrical potential of 2000 V, and this was ascribed to stronger hydrogen–surface interactions. The interaction was van der Waal type between the σ-bonds of hydrogen and the π-bonds of the aromatic rings. However, the interaction between electrically charged carbon and hydrogen might involve stronger orbital interactions between hydrogen and carbon.[342]

24.12 Conclusions

The synthesis of porous materials for hydrogen adsorption and desorption with capacities at cryogenic temperatures that meet the US DOE targets has recently been the focus of a great deal of research. Several groups have reported experimental measurements showing maximum excess adsorption of ~6.5 to 8.6 wt% of hydrogen at 77 K and < 100 bar on MOF (MOF-200, MOF-210, NOTT-112),[57,59] polymers (PAF-1),[97] and various porous carbon materials.[168,169] It is apparent that similar adsorption uptakes are observed for the best performing porous materials. Total hydrogen uptakes for MOF-200 (16.3 wt%), MOF-210(17.6 wt%),[57] NOTT-112(10 wt%),[59] and UMCM-2(12.4 wt%)[232] at 80 bar have been observed at 77 K.[57] However, for storage applications, hydrogen capacity on a volume basis is more important. The materials with highest adsorption capacities on a volumetric basis are PCN-14(55 g L^{-1}) and NOTT-112 (54 g L^{-1}).[57,59,241]

Hydrogen maximum surface excess capacity (wt%) and BET N_2 surface area at 77 K have a strong correlation for a wide range of porous materials. Similar correlations are observed for all porous materials for maximum surface excess hydrogen adsorbed with other pore structure characterization parameters, such as pore volume. However, when the hydrogen maximum excess adsorption is expressed on a volumetric basis for the adsorbent, the correlations are much less clear for surface areas >2000 m^2 g^{-1}. Hydrogen adsorbate can be considered as an incompressible fluid with a density similar to liquid hydrogen, and the maximum surface excess adsorbed is limited by the available total pore volume of the adsorbent. The amounts of hydrogen adsorbed on a wide range of porous materials under high pressure and ambient temperature conditions are small (less than ~1 wt%).

The maximum enthalpies of hydrogen adsorption obtained experimentally over the temperature range of 77–87 K are MOFs (12–13.5 kJ mol^{-1}), boron-doped carbon (20 kJ mol^{-1}), carbons (20 kJ mol^{-1}), polymers (9 kJ mol^{-1}), and zeolites (18 kJ mol^{-1}). However, most reports have isosteric enthalpy adsorption measurements <13 kJ mol^{-1} (see Table 24.2). Currently, the highest isosteric enthalpies of hydrogen adsorption for MOFs have been observed for $Zn_3(bdc)_3{\cdot}Cu(Pyen)$,[52] $Ni_2(dhtp)$,[223] and $[Cd_5(Tz)_9](NO_3)_3$.[281] These materials have open metal centers. The thermodynamic results and structural and spectroscopic observations of H_2 adsorbed in pores with open metal centers in MOFs show that the interactions observed so far are much weaker than those observed for η^2–H_2 bonding in Kubas compounds. Higher enthalpies of adsorption probably in the range of 20–40 kJ mol^{-1} are required for increased adsorption for storage applications at temperatures higher than 77 K.

Adsorption of hydrogen on porous materials is currently limited to cryogenic temperatures (77 K), and this limits the use of the technique in real practical transport application situations. The design of porous materials with enhanced adsorption capacity under ambient conditions involves increasing hydrogen–surface interactions. The stability of η^2-type hydrogen–metal complexes suggests that, in principle, unsaturated metal centers in MOFs may act as sites for η^2-type hydrogen coordination and this is potential development route for materials with enhanced hydrogen–surface interactions. Also, the doping of Kubas compounds into porous materials is also a possible option. The effect of confinement in pores on hydrogen–metal interactions is less clear. Structural studies of Kubas (η^2–H_2) coordination compounds provide evidence for a wide variation in H_2 interaction strength with metals. This suggests that modification of the hydrogen–metal interaction can be achieved by varying the metal, ligand, and ligand substitution in MOFs. The ability to fine-tune the hydrogen–metal specific interactions involving weak nondissociative chemisorption rather than nonspecific physisorption is important. However, open metal centers only represent a small fraction of the internal surface area of MOFs. Coordination of multiple H_2 molecules per unsaturated metal is necessary for storage applications. However, only a few stable solid bis-H_2 organometallic complexes are known.[258,259] Multiple coordination of H_2 to copper centers has been proposed for $Zn_3(bdc)_3{\cdot}Cu(Pyen)$ based on the variation in enthalpy of adsorption with surface coverage. Interaction of H_2 molecules with both sides of copper in the Cu(Pyen) pillar increases the influence of open metal centers.[52] However, confinement in pores may influence the primary H_2 interactions with open metal sites by interactions from other pore surfaces. Secondary interactions influence framework flexibility and window opening in MOFs, which control hysteretic adsorption characteristics, and these may also be useful in the design strategies for storage applications. An improved understanding of surface chemistry and framework flexibility will lead to porous materials with improved hydrogen adsorption characteristics for storage applications under high pressure and ambient temperature conditions.

Acknowledgments

The author would like to thank collaborators for many useful discussions and the Carbon Trust, EPSRC, and Leverhulme Trust for financial support.

Abbreviations

abtc	azobenzene-3,3′,5,5′-tetracarboxylate
adc	9,10-anthracenedicarboxylate
bbc	4,4′,4″-(benzene-1,3,5-triyl-tris(benzene-4,1-diyl))tribenzoate
bdc	benzene-1,4-dicarboxylate
bdp	1,4-benzenedipyrazolate
bdt	1,4-benzeneditetrazolate
bhtc	biphenyl-3,4′,5-tricarboxylate
bpdc	biphenyldicarboxylate
bpe	4,4′-trans-bis(4-pyridyl)-ethylene
bpndc	benzophenone 4,4′-dicarboxylic acid
bptc	3,3′,5,5′-biphenyltetracarboxylate
bpy	4,4′-bipyridine
bpydc	2,2′-bipyridyl-5,5′-dicarboxylate
btb	1,3,5-benzenetribenzoate
btc	1,3,5-benzenetricarboxylate
bte	4,4′,4″-[benzene-1,3,5-triyl-tris (ethyne-2,1-diyl)]tribenzoate
btt	1,3,5-benzenetristetrazolate
bttb	4,4,44″-benzene-1,3,5-triyltribenzoate
bza	benzoate
cbbdc	1,2-dihydrocyclobutylbenzene-3,6-dicarboxylate
cnc	4-carboxycinnamic
Cy	cyclohexane
dabco	1,4-diazabicyclo[2.2.2]octane
dcbpyb	tris(3′,5′-dicarboxy[1,1′-biphenyl]-4-yl)benzene
def	*N,N′* diethylformamide
dhbdc	2,5-dihydroxyl-1,4-denzenedicarboxylate

dhtp	2,5-dihydroxyterephthalate
diPyNi	*N,N′*-di-(4-pyridyl)-1,4,5,8-naphthalenetetracarboxydiimide
dma	dimethylacetamide
dmf	dimethylformamide
dpac	dipyridylacetylene
dpt	3,6-di-4-pyridyl-1,2,4,5-tetrazine
fma	fumarate
Hdccptp	3,5-dicyano-4-(4-carboxyphenyl)-2,2′:6′4″-terpyridine
H_2t_2dc	thieno[3,2-*b*]thiophene-2,5-dicarboxylic acid
H_3tzi	5-tetrazolylisophthalic acid
hfipbb	4,4′-(hexafluoroisopropylidene) bis benzoate
hpdc	4,5,9,10-tetrahydropyrene-2,7-dicarboxylate
H_2Oxdaa	oxydiacetic acid
H_2pbpc	LH2 = pyridine-3,5-bis(phenyl-4-carboxylic acid)
H_2tbip	5-tert-butyl isophthalic acid
H_2t^2dc	thieno[3,2-*b*]thiophene-2,5-dicarboxylic acid
H_3tpb-3tz	1.3,5-tri-*p*-(tetrazol-5-yl)-phenylbenzene
H_3tpt-3tz	2,4,6-tri-*p*-(tetrazol-5-yl)-phenyl-*s*-triazine
H_4adip	5,5′-(9,10-anthracenediyl)diisophthalic acid
H_4bdi	5,5′-(buta-1,3-diyne-1,4-diyl)diisophthalic acid
$H_4dhpddip$	2,7-(9,10-dihydrophenanthrenediyl)diisophthalic acid
H_4pddip	2,7-(phenanthrenediyl)diisophthalic acid
$H_6tdcbpyb$	1,3,5-tris(3′,5′-dicarboxy[1,1′-biphenyl]-4-yl)benzene
H_6tdbb	1,3,5-tris(3′,5′-dicarboxy[1,1′-biphenyl]-4-yl-)benzene
L^1	6,6′-dichloro-2,2′-diethoxy-1,1′-binaphthyl-4–4′-dibenzoate
L^2	6,6′-dichloro-2,2′-dibenzyloxy-1,1′-binaphthyl-4,4′-dibenzoate
mim	2-methylimidazole
ndc	naphthalene-2,6-dicarboxylate
ntc	naphthalene tetracarboxylate
p-$cdcH_2$	1,12-dihydroxycarbonyl 1–1,12-dicarba-*closo*-dodecaborane
pd	1,2-propanediol
$pdaH_2$	*p*-phenylenediacrylic acid
3-pic	3-picoline
pp	bis-1,2(diarylphosphino)ethane
$pyenH_2$	5-methyl-4-oxo-1,4-dihydro-pyridine-3-carbaldehyde
pyz	pyrazine
4,6-pmdc	4,6-pyrimidicarboxylate
2-pmc	2-pyrimidinecarboxylate
pzdc	pyrazine-2,3-dicarboxylate
qtpc	quaterphenyl-3,3‴,5,5‴ tetracarboxylic acid
sbtc	*trans*-stilbene-3,3′,5,5′-tetracarboxylate
sip	5-sulfoisophthalate
t^2dc	thieno[3,2-b]thiophene-2,5-dicarboxylate
tatb	4,4′,4″-*s*-triazine-2,4,6-triyltribenzoate
ted	triethylenediamine
tmbdc	2,3,5,6-tetramethylbenzene-1,4-dicarboxylate
tptc	terphenyl-3,3″″,5,5″ tetracarboxylate
trz	1,3,5-bis(trifluoromethyl)-1,2,4-triazolate ion

References

1. Hirsch, R. L., Bezdek, R., Wendling, R. *Peaking of World Oil Production*. www.netl.doe.gov/publications/others/pdf/Oil_Peaking_NETL.pdf, 2005.
2. phttp://www.eere.energy.gov/hydrogenandfuelcells/hydrogen/storage. html#objectives.
3. Schlapbach, L., Zuttel, A. *Nature* **2001**, *414*, 353–358.
4. David, R. L. (ed). *CRC Handbook of Chemistry and Physics*, 74th edn., The Chemical Rubber Co., Boca Raton, FL, **1993**.
5. Eddaoudi, M., Kim, J., Rosi, N., Vodak, D., Wachter, J., O'Keeffe, M., Yaghi, O. M. *Science* **2002**, *295*, 469–472.
6. Reid, C. R., Thomas, K. M. *J. Phys. Chem. B* **2001**, *105(33)*, 10619–10629.
7. Chagger, H. K., Ndaji, F. E., Sykes, M. L., Thomas, K. M. *Carbon* **1995**, *33*, 1405–1411.
8. Jiang, S. Y., Gubbins, K. E., Balbuena, P. B. *J. Phys. Chem.* **1994**, *98*, 2403–2411.
9. Britt, D., Tranchemontagne, D., Yaghi, O. M. *Proc. Natl. Acad. Sci. (USA)* **2008**, *105(33)*, 11623–11627.
10. Davis, M. E. *Nature* **2002**, *417*, 813–821.
11. Corma, A. *Chem. Rev.* **1997**, *97*, 2373–2419.
12. Fletcher, A. J., Thomas, K. M., Rosseinsky, M. J. *J. Solid State Chem.* **2005**, *178*, 2491–2510.
13. Thomas, K. M. *Catalysis Today* **2007**, *120*, 389–398.
14. Felderhoff, M., Weidenthaler, C., von Helmolt, R., Eberle, U. *Phys. Chem. Chem. Phys.* **2007**, *9*, 2643–2653.
15. Morris, R. E., Wheatley, P. S. *Angew. Chem. Int. Ed.* **2008**, *47*, 4966–4981.
16. van den Berg, A. W. C., Arean, C. O. *Chem. Commun.* **2008**, 668–681.
17. Collins, D. J., Zhou, H. C. *J. Mater. Chem.* **2007**, *17*, 3154–3160.
18. Dinca, M., Long, J. R. *Angew. Chem. Int. Ed.* **2008**, *47*, 6766–6779.
19. Thomas, K. M. *Dalton Trans.* **2009**, 1487–1505.
20. Lin, X., Jia, J., Hubberstey, P., Schroder, M., Champness, N. R. *Cryst. Eng. Commun.* **2007**, *9*, 438–448.
21. Ma, S. Q., Zhou, H. C. *Chem. Commun.* **2010**, *46*, 44–53.
22. Thomas, K. M. In *Energy Materials*, Bruce, D. W., O'Hare, D., Walton, R. I., Eds., Wiley, New York, **2011**, pp. 245–281.
23. Kinoshita, Y., Matsubara, I., Hibuchi, T., Saito, Y. *Bull. Chem. Soc. Jpn.* **1959**, *32*, 1221–1226.
24. Hoskins, B. F., Robson, R. *J. Am. Chem. Soc.* **1989**, *111*, 5962–5964.
25. Kondo, M., Yoshitomi, T., Seki, K., Matsuzaka, H., Kitagawa, S. *Angew. Chem. Int. Ed.* **1997**, *36*, 1725–1727.
26. Chen, B., Ockwig, N. W., Millward, A. R., Contreras, D. S., Yaghi, O. M. *Angew. Chem. Int. Ed.* **2005**, *44*, 4745–4749.
27. Kitagawa, S., Kitaura, R., Noro, S. *Angew. Chem. Int. Ed.* **2004**, *43*, 2334–2375.
28. Ferey, G., Mellot-Draznieks, C., Serre, C., Millange, F., Dutour, J., Surble, S., Margiolaki, I. *Science* **2005**, *309*, 2040; *Science* **2005**, *310*, 1119.
29. Ferey, G. *Chem. Soc. Rev.* **2008**, *37*, 191–214.
30. Kitagawa, S., Matsuda, R. *Coord. Chem. Rev.* **2007**, *251*, 2490–2509.
31. Huang, X.-C., Lin, Y.-Y., Zhang, J.-P., Chen, X.-M. *Angew. Chem. Int. Ed.* **2006**, *45*, 1557–1559.
32. Park, K. S., Ni, Z., Cote, A. P., Choi, J. Y., Huang, R., Uribe-Romo, F. J., Chae, H. K., O'Keeffe, M., Yaghi, O. M. *Proc. Natl. Acad. Sci. (USA)* **2006**, *103*, 10186–10191.
33. Broom, D. P. *Int. J. Hydrogen Energy* **2007**, *32*, 4871–4888.
34. Harding, A. W., Foley, N. J., Norman, P. R., Francis, D. C., Thomas, K. M. *Langmuir* **1998**, *14*, 3858–3864.
35. Fuller, E. L., Poulis, J. A., Czanderna, A. W., Robens, E. *Thermochim. Acta* **1979**, *29*, 315–318.
36. Sing, K. S. W., Everett, D. H., Haul, R. A. W., Moscou, L., Pierotti, R. A., Rouquerol, J., Siemieniewska, T. *Pure Appl. Chem.* **1985**, *57*, 603–619.

37. Broom, D. P., Moretto, P. *J. Alloys Compd.* **2007**, *446–447*, 687–691.
38. Anson, A., Benham, M., Jagiello, J., Callejas, M. A., Benito, A. M., Maser, W. K., Zuettel, A., Sudan, P., Martinez, M. T. *Nanotechnology* **2004**, *15*, 1503–1508.
39. Furukawa, H., Miller, M. A., Yaghi, O. M. *J. Mater. Chem.* **2007**, *17*, 3197–3204.
40. Zlotea, C., Moretto, P., Steriotis, T. *Int. J. Hydrogen Energy* **2009**, *34*, 3044–3057.
41. Sircar, S. *Ind. Eng. Chem. Res.* **1999**, *38*, 3670–3682.
42. Dinca, M., Dailly, A., Liu, Y., Brown, C. M., Neumann, D. A., Long., J. R. *J. Am. Chem. Soc.* **2006**, *128*, 16876–16883.
43. Kaye, S. S., Dailly, A., Yaghi, O. M., Long, J. R. *J. Am. Chem. Soc.* **2007**, *129*, 14176–14177.
44. Lin, X., Telepeni, I., Blake, A. J., Dailly, A., Brown, C. M., Simmons, J. M., Zoppi, M. et al. *J. Am. Chem. Soc.* **2009**, *131*, 2159–2171.
45. Lin, X., Jia, J., Zhao, X., Thomas, K. M., Blake, A. J., Walker, G. S., Champness, N. R., Hubberstey, P., Schroder, M. *Angew. Chem. Int. Ed.* **2006**, *45*, 7358–7364.
46. Zhao, X. B., Xiao, B., Fletcher, A. J., Thomas, K. M. *J. Phys. Chem. B* **2005**, *109*, 8880–8888.
47. Xiao, B., Wheatley, P. S., Zhao, X., Fletcher, A. J., Fox, S., Rossi, A. G., Megson, I. L., Bordiga, S., Regli, L., Thomas, K. M., Morris, R. E. *J. Am. Chem. Soc.* **2007**, *129*, 1203–1209.
48. Hafizovic, J., Bjorgen, M., Olsbye, U., Dietzel, P. D. C., Bordiga, S., Prestipino, C., Lamberti, C., Lillerud, K. P. *J. Am. Chem. Soc.* **2007**, *129*, 3612–3620.
49. Liu, J., Culp, J. T., Natesakhawat, S., Bockrath, B. C., Zande, B., Sankar, S. G., Garberoglio, G., Johnson, J. K. *J. Phys. Chem. C* **2007**, *111*, 9305–9313.
50. Nelson, A. P., Farha, O. K., Mulfort, K. L., Hupp, J. T. *J. Am. Chem. Soc.* **2009**, *131*, 458–460.
51. Zhao, X., Villar-Rodil, S., Fletcher, A. J., Thomas, K. M. *J. Phys. Chem. B* **2006**, *110*, 9947–9955.
52. Chen, B., Zhao, X., Putkham, A., Hong, K., Lobkovsky, E. B., Hurtado, E. J., Fletcher, A. J., Thomas, K. M. *J. Am. Chem. Soc.* **2008**, *130*, 6411–6423.
53. Marsh, H. *Carbon* **1987**, *25*, 49–58.
54. Rouquerol, F., Rouquerol, J., Sing, K. *Adsorption by Powders and Porous Solids*, Academic Press, London, U.K., 1999.
55. Walton, K. S., Snurr, R. Q. *J. Am. Chem. Soc.* **2007**, *129*, 8552–8556.
56. Mertens, F. O. *Surf. Sci.* **2009**, *603*, 1979–1984.
57. Furukawa, H., Ko, N., Go, Y. B., Aratani, N., Choi, S. B., Choi, E., Yazaydin, A. O. et al. *Science* **2010**, *329*, 424–428.
58. Juan-Juan, J., Marco-Lozar, J. P., Suarez-Garcia, F., Cazorla-Amoros, D., Linares-Solano, A. *Carbon* **2010**, *48*, 2906–2909.
59. Yan, Y., Lin, X., Yang, S. H., Blake, A. J., Dailly, A., Champness, N. R., Hubberstey, P., Schroder, M. *Chem. Commun.* **2009**, 1025–1027.
60. Rosi, N. L., Eckert, J., Eddaoudi, M., Vodak, D. T., Kim, J., O'Keeffe, M., Yaghi, O. M. *Science* **2003**, *300*, 1127–1129.
61. Yaghi, O. M., O'Keeffe, M., Ockwig, N. W., Chae, H. K., Eddaoudi, M., Kim, J. *Nature* **2003**, *423*, 705–714.
62. Rowsell, J. L. C., Millward, A. R., Park, K. S., Yaghi, O. M. *J. Am. Chem. Soc.* **2004**, *126*, 5666–5667.
63. Rowsell, J. L. C., Spencer, E. C., Eckert, J., Howard, J. A. K., Yaghi, O. M. *Science* **2005**, *309*, 1350–1354.
64. Rowsell, J. L. C., Yaghi, O. M. *Angew. Chem. Int. Ed.* **2005**, *44*, 4670–4679.
65. Wong-Foy, A. G., Matzger, A. J., Yaghi, O. M. *J. Am. Chem. Soc.* **2006**, *128*, 3494–3495.
66. Rowsell, J. L. C., Yaghi, O. M. *J. Am. Chem. Soc.* **2006**, *28*, 1304–1315.
67. Dinca, M., Yu, A. F., Long, J. R. *J. Am. Chem. Soc.* **2006**, *128*, 8904–8913.
68. Prestipino, C., Regli, L., Vitillo, J. G., Bonino, F., Damin, A., Lamberti, C., Zecchina, A., Solari, P. L., Kongshaug, K. O., Bordiga, S. *Chem. Mater.* **2006**, *18*, 1337–1346.
69. Latroche, M., Suble, S., Serre, C., Mellot-Draznieks, C., Llewellyn, P. L., Lee, J. H., Chang, J. S., Jhung, S. H., Ferey, G. *Angew. Chem. Int. Ed.* **2006**, *45*, 8227–8231.
70. Gogotsi, Y., Dash, R. K., Yushin, G., Yildirim, T., Laudisio, G., Fischer, J. E. *J. Am. Chem. Soc.* **2005**, *127*, 16006–16007.

71. Armandi, M., Bonelli, B., Arean, C. O., Garrone, E. *Microporous Mesoporous Mater.* **2008**, *112*, 411–418.
72. Nijkamp, M. G., Raaymakers, J. E. M. J., van Dillen, A. J., de Jong, K. P. *Appl. Phys. A Mater. Sci. Process.* **2001**, *72*, 619–623.
73. Pang, J., Hampsey, J. E., Wu, Z., Hu, Q., Lu, Y. *Appl. Phys. Lett.* **2004**, *85*, 4887–4889.
74. Parra, J. B., Ania, C. O., Arenillas, A., Rubiera, F., Palacios, J. M., Pis, J. J. *J. Alloys Compd.* **2004**, *379*, 280–289.
75. Texier-Mandoki, N., Dentzer, J., Piquero, T., Saadallah, S., David, P., Vix-Guterl, C. *Carbon* **2004**, *42*, 2744–2747.
76. Schimmel, H. G., Nijkamp, G., Kearley, G. J., Rivera, A., de Jong, K. P., Mulder, F. M. *Mater. Sci. Eng. B Solid State Mater.* **2004**, *B108*, 124–129.
77. Schimmel, H. G., Kearly, G. J., Nijkamp, M. G., Visser, C. T., de Jong, K. P., Mulder, F. M. *Chem. Eur. J.* **2003**, *9*, 4764–4770.
78. Takagi, H., Hatori, H., Yamada, Y., Matsuo, S., Shiraishi, M. *J. Alloys Compd.* **2004**, *385*, 257–263.
79. Takagi, H., Hatori, H., Soneda, Y., Yoshizawa, N., Yamada, Y. *Mater. Sci. Eng. B* **2004**, *108*, 143–147.
80. Gadiou, R., Saadallah, S. E., Piquero, T., David, P., Parmentier, J., Vix-Guterl, C. *Microporous Mesoporous Mater.* **2005**, *79*, 121–128.
81. Cheng, F., Liang, J., Zhao, J., Tao, Z., Chen, J. *Chem. Mater.* **2008**, *20*, 1889–1895.
82. Zubizarreta, L., Gomez, E. I., Arenillas, A., Ania, C. O., Parra, J. B., Pis, J. J. *Adsorption* **2008**, *14*, 557–566.
83. Zubizarreta, L., Arenillas, A., Pis, J. J. *Int. J. Hydrogen Energy* **2009**, *34*, 4575–4581.
84. Guo, H. L., Gao, Q. M. *Int. J. Hydrogen Energy* **2010**, *35*, 7547–7554.
85. Langmi, H. W., Book, D., Walton, A., Johnson, S. R., Al-Mamouri, M. M., Speight, J. D., Edwards, P. P., Harris, I. R., Anderson, P. A. *J. Alloys Compd.* **2005**, *404–406*, 637–642.
86. Langmi, H. W., Walton, A., Al-Mamouri, M. M., Johnson, S. R., Book, D., Speight, J. D., Edwards, P. P., Gameson, I., Anderson, P. A., Harris, I. R. *J. Alloys Compd.* **2003**, *356–357*, 710–715.
87. Li, Y., Yang, R. T. *J. Phys. Chem. B* **2006**, *110*, 17175–17181.
88. Chu, X. Z., Zhou, Y. P., Zhang, Y. Z., Su, W., Sun, Y., Zhou, L. *J. Phys. Chem. B* **2006**, *110*, 22596–22600.
89. Chaikittisilp, W., Sugawara, A., Shimojima, A., Okubo, T. *Chem. Mater.* **2010**, *22*, 4841–4843.
90. Chaikittisilp, W., Sugawara, A., Shimojima, A., Okubo, T. *Chem. Eur. J.* **2010**, *16*, 6006–6014.
91. Lee, J. Y., Wood, C. D., Bradshaw, D., Rosseinsky, M. J., Cooper, A. I. *Chem. Commun.* **2006**, 2670–2672.
92. McKeown, N. B., Gahnem, B., Msayib, K. J., Budd, P. M., Tattershall, C. E., Mahmood, K., Tan, S., Book, D., Langmi, H. W., Walton, A. *Angew. Chem. Int. Ed.* **2006**, *45*, 1804–1807.
93. Germain, J., Hradil, J., Frechet, J. M. J., Svec, F. *Chem. Mater.* **2006**, *18*, 4430–4435.
94. Ghanem, B. S., Msayib, K. J., McKeown, N. B., Harris, K. D. M., Pan, Z., Budd, P. M., Butler, A., Selbie, J., Book, D., Walton, A. *Chem. Commun.* **2007**, *1*, 67–69.
95. Ghanem, B. S., Hashem, M., Harris, K. D. M., Msayib, K. J., Xu, M. C., Budd, P. M., Chaukura, N. et al. *Macromolecules* **2010**, *43*, 5287–5294.
96. Zhang, W., Li, C., Yuan, Y. P., Qiu, L. G., Xie, A. J., Shen, Y. H., Zhu, J. F. *J. Mater. Chem.* **2010**, *20*, 6413–6415.
97. Ben, T., Ren, H., Ma, S. Q., Cao, D. P., Lan, J. H., Jing, X. F., Wang, W. C. et al. *Angew. Chem. Int. Ed.* **2009**, *48*, 9457–9460.
98. Han, S. S., Furukawa, H., Yaghi, O. M., Goddard, W. A. *J. Am. Chem. Soc.* **2008**, *130*, 11580–11581.
99. Tilford, R. W., Mugavero, S. J., Pellechia, P. J., Lavigne, J. J. *Adv. Mater.* **2008**, *20*, 2741–2746.
100. Weber, J., Antonietti, M., Thomas, A. *Macromolecules* **2008**, *41*, 2880–2885.
101. El-Kaderi, H. M., Hunt, J. R., Mendoza-Cortes, J. L., Cote, A. P., Taylor, R. E., O'Keeffe, M., Yaghi, O. M. *Science* **2007**, *316*, 268–272.
102. Cote, A. P., El-Kaderi, H. M., Furukawa, H., Hunt, J. R., Yaghi, O. M. *J. Am. Chem. Soc.* **2007**, *129*, 12914–12915.
103. Furukawa, H., Yaghi, O. M. *J. Am. Chem. Soc.* **2009**, *131*, 8875–8883.

104. Lee, E. Y., Suh, M. P. *Angew. Chem. Int. Ed.* **2004**, *43*, 2798–2801.
105. Dybtsev, D. N., Chun, H., Yoon, S. H., Kim, D., Kim, K. *J. Am. Chem. Soc.* **2004**, *126*, 32–33.
106. Dybtsev, D. N., Chun, H., Kim, K. *Angew. Chem. Int. Ed.* **2004**, *43*, 5033–5036.
107. Chapman, K. W., Southon, P. D., Weeks, C. L., Kepert, C. J. *Chem. Commun.* **2005**, *26*, 3322–3324.
108. Perles, J., Iglesias, M., Martin-Luengo, M. A., Monge, M. A., Ruiz-Valero, C., Snejko, N. *Chem. Mater.* **2005**, *17*, 5837–5842.
109. Chun, H., Dybtsev, D. N., Kim, H., Kim, K. *Chem. Eur. J.* **2005**, *11*, 3521–3529.
110. Dietzel, P. D. C., Panella, B., Hirscher, M., Blom, R., Fjellvag., H. *Chem. Commun.* **2006**, *9*, 959–961.
111. Guo, X., Zhu, G., Li, Z., Sun, F., Yang, Z., Qiu, S. *Chem. Commun.* **2006**, 3172–3174.
112. Forster, P. M., Eckert, J., Heiken, B. D., Parise, J. B., Yoon, J. W., Jhung, S. H., Chang, J. S., Cheetham, A. K. *J. Am. Chem. Soc.* **2006**, *128*, 16846–16850.
113. Jia, J., Lin, X., Blake, A. J., Champness, N. R., Hubberstey, P., Shao, L., Walker, G., Wilson, C., Schroeder, M. *Inorg. Chem.* **2006**, *45*, 8838–8840.
114. Moon, H. R., Kobayashi, N., Suh, M. P. *Inorg. Chem.* **2006**, *45*, 8672–8676.
115. Navarro, J. A. R., Barea, E., Salas, J. M., Masciocchi, N., Galli, S., Sironi, A., Ania, C. O., Parra, J. B. *Inorg. Chem.* **2006**, *45*, 2397–2399.
116. Rood, J. A., Noll, B. C., Henderson, K. W. *Inorg. Chem.* **2006**, *45*, 5521–5528.
117. Ma, S., Wang, X. S., Yuan, D., Zhou, H. C. *Angew. Chem. Int. Ed.* **2008**, *47*, 4130–4133.
118. Wang, X. S., Ma, S. Q., Rauch, K., Simmons, J. M., Yuan, D. Q., Wang, X. P., Yildirim, T., Cole, W. C., Lopez, J. J., Meijere, A. d., Zhou, H. C. *Chem. Mater.* **2008**, *20*, 3145–3152.
119. Chun, H., Jung, H., Koo, G., Jeong, H., Kim, D. K. *Inorg. Chem.* **2008**, *47*, 5355–5359.
120. Lee, J. Y., Pan, L., Kelly, S. P., Jagiello, J., Emge, T. J., Li, J. *Adv. Mater.* **2005**, *17*, 2703–2706.
121. Dinca, M., Long, J. R. *J. Am. Chem. Soc.* **2007**, *129*, 11172–11176.
122. Dinca, M., Han, W. S., Liu, Y., Dailly, A., Brown, C. M., Long, J. R. *Angew. Chem. Int. Ed.* **2007**, *46*, 1419–1422.
123. Nouar, F., Eubank, J. F., Bousquet, T., Wojtas, L., Zaworotko, M. J., Eddaoudi, M. *J. Am. Chem. Soc.* **2008**, *130*, 1833–1835.
124. Choi, S. B., Seo, M. J., Cho, M., Kim, Y., Jin, M. K., Jung, D. Y., Choi, J. S., Ahn, W. S., Rowsell, J. L. C., Kim, J. *Cryst. Growth Des.* **2007**, *7*, 2290–2293.
125. Férey, G., Latroche, M., Sérre, C., Millange, F., Loiseau, T., Percheron-Guegan, A. *Chem. Commun.* **2003**, *24*, 2976–2977.
126. Kramer, M., Schwarz, U., Kaskel, S. *J. Mater. Chem.* **2006**, *16*, 2245–2248.
127. Wong-Foy, A. G., Lebel, O., Matzger, A. J. *J. Am. Chem. Soc.* **2007**, *129*, 15740–15741.
128. Lee, J. Y., Olson, D. H., Pan, L., Emge, T. J., Li, J. *Adv. Funct. Mater.* **2007**, *17*, 1255–1262.
129. Chun, H., Moon, J. *Inorg. Chem.* **2007**, *46*, 4371–4373.
130. Mulfort, K. L., Hupp, J. T. *J. Am. Chem. Soc.* **2007**, *129*, 9604–9605.
131. Park, H., Britten, J. F., Mueller, U., Lee, J. Y., Li, J., Parise, J. B. *Chem. Mater.* **2007**, *19*, 1302–1308.
132. Lee, E. Y., Jang, S. Y., Suh, M. P. *J. Am. Chem. Soc.* **2005**, *127*, 6374–6381.
133. Humphrey, S. M., Chang, J. S., Jhung, S. H., Yoon, J. W., Woo, P. T. *Angew. Chem. Int. Ed.* **2007**, *46*, 272–275.
134. Park, H. J., Suh, M. P. *Chem. Eur. J.* **2008**, *14*, 8812–8821.
135. Park, M., Moon, D., Yoon, J. W., Chang, J. S., Lah, M. S. *Chem. Commun.* **2009**, *15*, 2026–2028.
136. Ma, S. Q., Yuan, D. Q., Wang, X. S., Zhou, H. C. *Inorg. Chem.* **2009**, *48*, 2072–2077.
137. Sabo, M., Henschel, A., Froede, H., Klemm, E., Kaskel, S. *J. Mater. Chem.* **2007**, *17*, 3827–3832.
138. Saha, D. P., Wei, Z. J., Deng, S. G. *Sep. Purif. Technol.* **2009**, *64*, 280–287.
139. Zhang, Y. B., Zhang, W. X., Feng, F. Y., Zhang, J. P., Chen, X. M. *Angew. Chem. Int. Ed.* **2009**, *48*, 5287–5290.
140. Shu-Hao, H., Chia-Her, L., Wei-Chang, W., Sue-Lein, W. *Angew. Chem. Int. Ed.* **2009**, *48*, 6124–6127.
141. Duriska, M. B., Neville, S. M., Lu, J. Z., Iremonger, S. S., Boas, J. F., Kepert, C. J., Batten, S. R. *Angew. Chem. Int. Ed.* **2009**, *48*, 8919–8922.
142. Gedrich, K., Senkovska, I., Baburin, I. A., Mueller, U., Trapp, O., Kaskel, S. *Inorg. Chem.* **2010**, *49*, 4440–4446.

143. Jia, J. H., Lin, X., Wilson, C., Blake, A. J., Champness, N. R., Hubberstey, P., Walker, G., Cussen, E. J., Schroder, M. *Chem. Commun.* **2007**, 840–842.
144. Yang, W., Lin, X., Jia, J., Blake, A. J., Wilson, C., Hubberstey, P., Champness, N. R., Schroeder, M. *Chem. Commun.* **2008**, 359–361.
145. Huang, S.-H., Lin, C.-H., Wu, W.-C., Wang, S.-L. *Angew. Chem. Int. Ed.* **2009**, *48*, 6124.
146. Li, A., Lu, R. F., Wang, Y., Wang, X., Han, K. L., Deng, W. Q. *Angew. Chem. Int. Ed.* **2010**, *49*, 3330–3333.
147. van Dingenen, W., van Itterbeek, A. *Physica (Hague)* **1939**, *6*, 49–58.
148. Basmadjian, I. D. *Can. J. Chem.* **1960**, *38*, 141–148.
149. Benard, P., Chahine, R. *Langmuir* **2001**, *17*, 1950–1955.
150. Gundiah, G., Govindaraj, A., Rajalakshmi, N., Dhathathreyan, K. S., Rao, C. N. R. *J. Mater. Chem.* **2003**, *13*, 209–213.
151. Zhou, Y., Feng, K., Sun, Y., Zhou., L. *Chem. Phys. Lett.* **2003**, *380*, 526–529.
152. Kayiran, S. B., Lamari, F. D., Levesque, D. *J. Phys. Chem. B* **2004**, *108*, 15211–15215.
153. Shiraishi, M., Takenobu, T., Kataura, H., Ata, M. *Appl. Phys. A* **2004**, *78*, 947–953.
154. Zhou, L., Zhou, Y., Sun, Y. *Int. J. Hydrogen Energy* **2004**, *29*, 475–479.
155. Poirier, E., Chahine, R., Tessier, A., Bose, T. K. *Rev. Sci. Inst.* **2005**, *76*, 055101.
156. Panella, B., Hirscher, M., Roth, S. *Carbon* **2005**, *43*, 2209–2214.
157. Pacula, A., Mokaya, R. *J. Phys. Chem. C* **2008**, *112*, 2764–2769.
158. Pacula, A., Mokaya, R. *Microporous Mesoporous Mater.* **2007**, *106*, 147–154.
159. Yang, Z. X., Xia, Y. D., Sun, X. Z., Mokaya, R. *J. Phys. Chem. B* **2006**, *110* 18424–18431.
160. Hu, Q. Y., Lu, Y. F., Meisner, G. P. *J. Phys. Chem. C* **2008**, *112*, 1516–1523.
161. Yang, Z., Xia, Y., Mokaya, R. *J. Am. Chem. Soc.* **2007**, *129*, 1673–1679.
162. Jorda-Beneyto, M., Suarez-Garcia, F., Lozano-Castello, D., Cazorla-Amoros, D., Linares-Solano, A. *Carbon* **2007**, *45*, 293–303.
163. Xia, Y., Walker, G. S., Grant, D. M., Mokaya, R. *J. Am. Chem. Soc.* **2009**, *131*, 16493–16499.
164. Mendez-Linan, L., Lopez-Garzon, F. J., Domingo-Garcia, M., Perez-Mendoza, M. *Energy Fuels* **2010**, *24*, 3394–3400.
165. Balathanigaimani, M. S., Shim, W.-G., Kim, T.-H., Cho, S.-J., Lee, J.-W., Moon, H. *Catalysis Today* **2009**, *146*, 234–240.
166. Meisner, G. P., Hu, Q. *Nanotechnology* **2009**, *20*, 204023/1–204023/10.
167. Gogotsi, Y., Portet, C., Osswald, S., Simmons, J. M., Yildirim, T., Laudisio, G., Fischer, J. E. *Int. J. Hydrogen Energy* **2009**, *34*, 6314–6319.
168. Zheng, Z., Gao, Q., Jiang, J. *Carbon* **2010**, *48*, 2968–2973.
169. Wang, H., Gao, Q., Hu, J. *J. Am. Chem. Soc.* **2009**, *131*, 7016–7022.
170. Gao, F., Zhao, D. L., Li, Y., Li, X. G. *J. Phys. Chem. Solids* **2010**, *71*, 444–447.
171. Weinberger, B., Lamari, F. D. *Int. J. Hydrogen Energy* **2009**, *34*, 3058–3064.
172. Suarez-Garcia, F., Vilaplana-Ortego, E., Kunowsky, M., Kimura, M., Oya, A., Linares-Solano, A. *Int. J. Hydrogen Energy* **2009**, *34*, 9141–9150.
173. Bonilla, M. R., Bae, J. S., Nguyen, T. X., Bhatia, S. K. *J. Phys. Chem. C* **2010**, *114*, 16562–16575.
174. Lee, S.-Y., Park, S.-J. *Int. J. Hydrogen Energy* **2010**, *35*, 6757–6762.
175. Bianco, S., Giorcelli, M., Musso, S., Castellino, M., Agresti, F., Khandelwal, A., Russo, S. L., Kumar, M., Ando, Y., Tagliaferro, A. *J. Nanosci. Nanotechnol.* **2010**, *10*, 3860–3866.
176. Sheppard, D. A., Buckley, C. E. *Int. J. Hydrogen Energy* **2008**, *33*, 1688–1692.
177. Stephanie-Victoire, F., Goulay, A. M., de Lara, E. C. *Langmuir* **1998**, *14*, 7255–7259.
178. Kayiran, S. B., Darkrim, F. L. *Surf. Interface Anal.* **2002**, *34*, 100–104.
179. Dong, J., Wang, X., Xu, H., Zhao, Q., Li, J. *Int. J. Hydrogen Energy* **2007**, *32*, 4998–5004.
180. McKeown, N. B., Budd, P. M. *Chem. Soc. Rev.* **2006**, *35*, 675–683.
181. Budd, P. M., Butler, A., Selbie, J., Mahmood, K., McKeown, N. B., Ghanem, B., Msayib, K., Book, D., Walton, A. *Phys. Chem. Chem. Phys.* **2007**, *9*, 1802–1808.
182. Germain, J., Frechet, J. M. J., Svec, F. *J. Mater. Chem.* **2007**, *17*, 4989–4997.
183. Jiang, J. X., Su, F., Trewin, A., Wood, C. D., Niu, H., Jones, J. T. A., Khimyak, Y. Z., Cooper, A. I. *J. Am. Chem. Soc.* **2008**, *130*, 7710–7720.

184. Wood, C. D., Tan, B., Trewin, A., Niu, H., Bradshaw, D., Rosseinsky, M. J., Khimyak, Y. Z. et al. *Chem. Mater.* **2007**, *19*, 2034–2048.
185. Germain, J., Svec, F., Frechet, J. M. J. *Chem. Mater.* **2008**, *20*, 7069–7076.
186. Xia, J., Yuan, S., Wang, Z., Kirklin, S., Dorney, B., Liu, D.-J., Yu, L. *Macromolecules* **2010**, *43*, 3325–3330.
187. Wood, C. D., Tan, B., Trewin, A., Su, F., Rosseinsky, M. J., Bradshaw, D., Sun, Y., Zhou, L., Cooper, A. I. *Adv. Mater.* **2008**, *20*, 1916-+.
188. Yuan, S. W., Dorney, B., White, D., Kirklin, S., Zapol, P., Yu, L. P., Liu, D. J. *Chem. Commun.* **2010**, *46*, 4547–4549.
189. Yoon, J. H., Choi, S. B., Oh, Y. J., Seo, M. J., Jhon, Y. H., Lee, T. B., Kim, D., Choi, S. H., Kim, J. *Catalysis Today* **2007**, *120*, 324–329.
190. Sava, D. F., Kravtsov, V. C., Nouar, F., Wojtas, L., Eubank, J. F., Eddaoudi, M. *J. Am. Chem. Soc.* **2008**, *130*, 3768–3770.
191. Dailly, A., Vajo, J. J., C. Ahn, C. *J. Phys. Chem. B* **2006**, *110*, 1099–1101.
192. Panella, B., Hirscher, M., Puetter, H., Mueller, U. *Adv. Funct. Mater.* **2006**, *16*, 520–524.
193. Panella, B., Hones, K., Muller, U., Trukhan, N., Schubert, M., Putter, H., Hirscher, M. *Angew. Chem. Int. Ed.* **2008**, *47*, 2138–2142.
194. Millward, A. R., Yaghi, O. M. *J. Am. Chem. Soc.* **2005**, *127*, 17998–17999.
195. Dinca, M., Long, J. R. *J. Am. Chem. Soc.* **2005**, *127*, 9376–9377.
196. Li, Y., Yang, R. T. *Langmuir* **2007**, *23*, 12937–12944.
197. Kesanli, B., Cui, T., Smith, M. R., Bittner, E. W., Brockrath, B. C., Lin, W. *Angew. Chem. Int. Ed.* **2005**, *44*, 72–75.
198. Fang, Q. R., Zhu, G. S., Xue, M., Zhang, Q. L., Sun, J. Y., Guo, X. D., Qiu, S. L., Xu, S. T., Wang, P., Wang, D. J., Wei, Y. *Chem. Eur. J.* **2006**, *12*, 3754–3758.
199. Chen, B., Ma, S., Zapata, F., Lobkovsky, E. B., Yang, J. *Inorg. Chem.* **2006**, *45*, 5718–5720.
200. Zhao, X., Xiao, B., Fletcher, A. J., Thomas, K. M., Bradshaw, D., Rosseinsky, M. J. *Science* **2004**, *306*, 1012–1015.
201. Pan, L., Sander, M. B., Huang, X., Li, J., Smith, M., Bittner, E., Bockrath, B., Johnson, J. K. *J. Am. Chem. Soc.* **2004**, *126*, 1308–1309.
202. Luo, J., Xu, H., Liu, Y., Zhao, Y., Daemen, L. L., Brown, C., Timofeeva, T. V., Ma, S., Zhou, H. C. *J. Am. Chem. Soc.* **2008**, *130*, 9626–9627.
203. Xue, M., Zhu, G., Y. Li, X. Zhao, Jin, Z., Kang, E., Qiu, S. *Cryst. Growth Des.* **2008**, *8*, 2478–2483.
204. Liu, J., Lee, J. Y., Pan, L., Obermyer, R. T., Simizu, S., Zande, B., Li, J., Sankar, S. G., Johnson, J. K. *J. Phys. Chem. C* **2008**, *112*, 2911–2917.
205. Takei, T., Kawashima, J., Li, T., Maeda, A., Hasegawa, M., Kitagawa, T., Ohmura, T., Ichikawa, M., Hosoe, M., Kanoya, I., Mori, W. *Bull. Chem. Soc. Jpn.* **2008**, *81*, 847–856.
206. Surble, S., Millange, F., Serre, C., Duren, T., Latroche, M., Bourrelly, S., Llewellyn, P. L., Ferey, G. *J. Am. Chem. Soc.* **2006**, *128*, 14889–14896.
207. Choi, H. J., Dincă, M., Long, J. R. *J. Am. Chem. Soc.* **2008**, *130*, 7848–7850.
208. Liu, Y., Kabbour, H., Brown, C. M., Neumann, D. A., Ahn, C. C. *Langmuir* **2008**, *24*, 4772–4777.
209. Autie-Castro, G., Autie, M., Reguera, E., Santamaria-Gonzalez, J., Moreno-Tost, R., Rodriguez-Castellon, E., Jimenez-Lopez, A. *Surf. Interface Anal.* **2009**, *41*, 730–734.
210. Zhou, W., Wu, H., Hartman, M. R., Yildirim, T. *J. Phys. Chem. C* **2007**, *111*, 16131–16137.
211. Fang, Q. R., Zhu, G. S., Jin, Z., Ji, Y. Y., Ye, J. W., Xue, M., Yang, H., Wang, Y., Qiu., S. L. *Angew. Chem. Int. Ed.* **2007**, *46*, 6638–6642.
212. Lin, X., Blake, A. J., Wilson, C., Sun, X. Z., Champness, N. R., George, M. W., Hubberstey, P., Mokaya, R., Schroder, M. *J. Am. Chem. Soc.* **2006**, *128*, 10745–10753.
213. Comotti, A., Bracco, S., Sozzani, P., Horike, S., Matsuda, R., Chen, J., Takata, M., Kubota, Y., Kitagawa, S. *J. Am. Chem. Soc.* **2008**, *130*, 13664–13672.
214. Ma, S., Zhou, H. C. *J. Am. Chem. Soc.* **2006**, *128*, 11734–11735.
215. Kaye, S. S., Long, J. R. *J. Am. Chem. Soc.* **2005**, *127*, 6506–6507.
216. Lee, J. Y., Li, J., Jagiello, J. *J. Solid State Chem.* **2005**, *178*, 2527–2532.
217. Dinca, M., Dailly, A., Tsay, C., Long, J. R. *Inorg. Chem.* **2008**, *47*, 11–13.

218. Culp, J. T., Matranga, C., Smith, M., Bittner, E. W., Bockrath, B. *J. Phys. Chem. B* **2006**, *110*, 8325–8328.
219. Reguera, L., Balmaseda, J., Krap, C. P., Reguera, E. *J. Phys. Chem. C* **2008**, *112*, 10490–10501.
220. Reguera, L., Balmaseda, J., Castillo, L. F. d., Reguera, E. *J. Phys. Chem. C* **2008**, *112*, 5589–5597.
221. Kaye, S. S., Long, J. R. *Chem. Commun.* **2007**, 4486–4488.
222. Culp, J. T., Natesakhawat, S., Smith, M. R., Bittner, E., Matranga, C., Bockrath, B. *J. Phys. Chem. C* **2008**, *112*, 7079–7083.
223. Vitillo, J. G., Regli, L., Chavan, S., Ricchiardi, G., Spoto, G., Dietzel, P. D. C., Bordiga, S., Zecchina, A. *J. Am. Chem. Soc.* **2008**, *130*, 8386–8396.
224. Farha, O. K., Spokoyny, A. M., Mulfort, K. L., Hawthorne, M. F., Mirkin, C. A., Hupp, J. T. *J. Am. Chem. Soc.* **2007**, *129*, 12680–12681.
225. Liu, Y., Eubank, J. F., Cairns, A. J., Eckert, J., Kravtsov, V. C., Luebke, R., Eddaoudi, M. *Angew. Chem. Int. Ed.* **2007**, *46*, 3278–3283.
226. Pan, L., Parker, B., Huang, X., Olson, D. H., Lee, J. Y., Li, J. *J. Am. Chem. Soc.* **2006**, *128*, 4180–4181.
227. Burrows, A. D., Cassar, K., Duren, T., Friend, R. M. W., Mahon, M. F., Rigby, S. P., Savarese, T. L. *Dalton Trans.* **2008**, 2465–2474.
228. Yang, S., Lin, X., Blake, A. J., Thomas, K. M., Hubberstey, P., Champness, N. R., Schroder, M. *Chem. Commun.* **2008**, 6108–6110.
229. Liu, Y., Her, J. H., Dailly, A., Ramirez-Cuesta, A. J., Neumann, D. A., Brown, C. M. *J. Am. Chem. Soc.* **2008**, *130*, 11813–11818.
230. Gao, C. Y., Liu, S. X., Xie, L. H., Sun, C. Y., Cao, J. F., Ren, Y. H., Feng, D., Su, Z. M. *CrystEng. Comm.* **2009**, *11*, 177–182.
231. Xiang, Z., Lan, J., Cao, D., Shao, X., Wang, W., Broom, D. P. *J. Phys. Chem. C* **2009**, *113*, 15106–15109.
232. Koh, K., Wong-Foy, A. G., Matzger, A. J. *J. Am. Chem. Soc.* **2009**, *131*, 4184–1485.
233. Kanoo, P., Matsuda, R., Higuchi, M., Kitagawa, S., Maji, T. K. *Chem. Mater.* **2009**, *21*, 5860–5866.
234. Ma, S. Q., Simmons, J. M., Yuan, D. Q., Li, J. R., Weng, W., Liu, D. J., Zhou, H. C. *Chem. Commun.* **2009**, 4049–4051.
235. Ma, S., Yuan, D., Chang, J.-S., Zhou, H.-C. *Inorg. Chem.* **2009**, *48*, 5398–5402.
236. Park, H. J., Suh, M. P. *Chem. Commun.* **2010**, *46*, 610–612.
237. Zhao, D., Yuan, D. Q., Yakovenko, A., Zhou, H. C. *Chem. Commun.* **2010**, *46*, 4196–4198.
238. Yang, C., Wang, X. P., Omary, M. A. *J. Am. Chem. Soc.* **2007**, *129*, 15454-+.
239. Wang, Y., Cheng, P., Chen, J., Z. Liao, D., Yan, S. P. *Inorg. Chem.* **2007**, *46*, 4530–4534.
240. Ma, S. Q., Eckert, J., Forster, P. M., Yoon, J. W., Hwang, Y. K., Chang, J. S., Collier, C. D., Parise, J. B., Zhou, H. C. *J. Am. Chem. Soc.* **2008**, *130*, 15896–15902.
241. Ma, S. Q., Simmons, J. M., Sun, D. F., Yuan, D. Q., Zhou, H. C. *Inorg. Chem.* **2009**, *48*, 5263–5268.
242. Wang, X. S., Ma, S. Q., Yuan, D. Q., Yoon, J. W., Hwang, Y. K., Chang, J. S., Wang, X. P., Jorgensen, M. R., Chen, Y. S., Zhou, H. C. *Inorg. Chem.* **2009**, *48*, 7519–7521.
243. Xue, M., Liu, Y., Schaffino, R. M., Xiang, S. C.,, Zhao, X. J., Zhu, G. S., Qiu, S. L., Chen, B. L. *Inorg. Chem.* **2009**, *48*, 4649–4651.
244. Yang, S. H., Lin, X., Dailly, A., Blake, A. J., Hubberstey, P., Champness, N. R., Schroder, M. *Chem. Eur. J.* **2009**, *15*, 4829–4835.
245. Lee, Y. G., Moon, H. R., Cheon, Y. E., Suh, M. P. *Angew. Chem. Int. Ed.* **2008**, *47*, 7741–7745.
246. Yang, J., Zhao, Q., Li, J., Dong, J. *Microporous Mesoporous Mater.* **2010**, *130*, 174–179.
247. Chung, T. C. M., Jeong, Y., Chen, Q., Kleinhammes, A., Wu, Y. *J. Am. Chem. Soc.* **2008**, *130*, 6668–6669.
248. Bhat, V. V., Contescu, C. I., Gallego, N. C., Baker, F. S. *Carbon* **2010**, *48*, 1331–1340.
249. Bhat, V. V., Contescu, C. I., Gallego, N. C. *Carbon* **2010**, *48*, 2361–2364.
250. Srinivas, G., Zhu, Y. W., Piner, R., Skipper, N., Ellerby, M., Ruoff, R. *Carbon* **2010**, *48*, 630–635.
251. McKeown, N. B., Budd, P. M., Book, D. *Macromol. Rapid Commun.* **2007**, *28*, 995–1002.
252. Makhseed, S., Samuel, J. *Chem. Commun.* **2008**, *36*, 4342–4344.
253. Yuan, S. W., Kirklin, S., Dorney, B., Liu, D. J., Yu, L. P. *Macromolecules* **2009**, *42*, 1554–1559.
254. Lochan, R. C., Head-Gordon, M. *Phys. Chem. Chem. Phys.* **2006**, *8*, 1357–1370.
255. Reguera, E. *Int. J. Hydrogen Energy* **2009**, *34*, 9163–9167.

256. Jeong, Y., Chung, T. C. M. *Carbon* **2010**, *48*, 2526–2537.
257. Kubas, G. J., Ryan, R. R., Swanson, B. I., Vergamini, P. J., Wasserman, H. J. *J. Am. Chem. Soc.* **1984**, *106*, 451–452.
258. Kubas, G. J. *Chem. Rev.* **2007**, *107*, 4152–4205.
259. Kubas, G. J. *Proc. Natl. Acad. Sci. (USA)* **2007**, *104*, 6901–6907.
260. Dutta, S., Jagirdar, B. R., Nethaji, M. *Inorg. Chem.* **2008**, *47*, 548–559.
261. Yang, S., Lin, X., Blake, A. J., Thomas, K. M., Hubberstey, P., Champness, N. R., Schröder, M. *Chem. Commun.* **2008**, 6108–6110.
262. Cerny, S. *Chem. Phys. Solid Surf. Heterog. Catal.* **1983**, *2*, 1–57.
263. Cole, J. H., Everett, D. H., Marshall, C. T., Paniego, A. R., Powl, J. C., Rodriguez-Reinoso, F. *J. Chem. Soc. Faraday Trans.* **1974**, *70*, 2154–2169.
264. Czepirski, L., Jagiello, J. *Chem. Eng. Sci.* **1989**, *44*, 797–801.
265. Yushin, G., Dash, R., Jagiello, J., Fischer, J. E., Gogotsi, Y. *Adv. Funct. Mater.* **2006**, *16*, 2288–2293.
266. Jhung, S. H., Kim, H. K., Yoon, J. W., Chang, J. S. *J. Phys. Chem. B* **2006**, *110*, 9371–9374.
267. Palomino, G. T., Carayol, M. R. L., Arean, C. O. *J Mater. Chem.* **2006**, *16*, 2884–2885.
268. Arean, C. O., Palomino, G. T., Carayol, M. R. L. *Appl. Surf. Sci.* **2007**, *253*, 5701–5704.
269. Arean, C. O., Delgado, M. R., Palomino, G. T., Rubio, M. T., Tsyganenko, N. M., Tsyganenko, A. A., Garrone, E. *Microporous Mesoporous Mater.* **2005**, *80*, 247–252.
270. Palomino, G. T., Carayol, M. R. L., Arean, C. O. *Catalysis Today* **2008**, *138*, 249–252.
271. Areán, C. O., Chavan, S., Cabello, C. P., Garrone, E., Palomino, G. T. *ChemPhysChem* **2010**, *11*, 3237–3242.
272. Schmitz, B., Krkljus, I., Leung, E., Hoeffken, H. W., Mueller, U., Hirscher, M. *ChemSusChem* **2010**, *3*, 768–761.
273. Sumida, K., Horike, S., Kaye, S. S., Herm, Z. R., Queen, W. L., Brown, C. M., Grandjean, F., Long, G. J., Dailly, A., Long, J. R. *Chem. Sci.* **2010**, *1*, 184–191.
274. Arean, C. O., Manoilova, O. V., Bonelli, B., Delgado, M. R., Palomino, G. T., Garrone, E. *Chem. Phys. Lett.* **2003**, *370*, 631–635.
275. Palomino, G. T., Delgado, M. R., Tsyganenko, N. M., Tsyganenko, A. A., Garrone, E., Bonelli, B., Manoilova, O. V., Arean, C. O. In *Molecular Sieves: From Basic Research to Industrial Applications, Pts a and B*, Cejka, J., Zilkova, N., Nachtigall, P., Eds., Elsevier Science Bv, Amsterdam, the Netherlands, **2005**, Vol. 158, pp. 853–860.
276. Arean, C. O., Palomino, G. T., Garrone, E., Nachtigallova, D., Nachtigall, P. *J. Phys. Chem. B* **2006**, *110*, 395–402.
277. Nachtigall, P., Garrone, E., Palomino, G. T., Delgado, M. R., Nachtigallova, D., Arean, C. O. *Phys. Chem. Chem. Phys.* **2006**, *8*, 2286–2292.
278. Arean, C. O., Palomino, G. T., Carayol, M. R. L., Pulido, A., Rubes, M., Bludsky, O., Nachtigall, P. *Chem. Phys. Lett.* **2009**, *477*, 139–143.
279. Palomino, G. T., Bonelli, B., Arean, C. O., Parra, J. B., Carayol, M. R. L., Armandi, M., Ania, C. O., Garrone, E. *Int. J. Hydrogen Energy* **2009**, *34*, 4371–4378.
280. Palomino, G. T., Arean, C. O., Carayol, M. R. L. *Appl. Surf. Sci.* **2010**, *256*, 5281–5284.
281. Zhong, D. C., Lin, J. B., Lu, W. G., Jiang, L., Lu, T. B. *Inorg. Chem.* **2009**, *48*, 8656–8658.
282. Xin, Z. F., Bai, J. F., Shen, Y. M., Pan, Y. *Crystal Growth Des.* **2010**, *10*, 2451–2454.
283. Bhatia, S. K., Myers, A. L. *Langmuir* **2006**, *22*, 1688–1700.
284. Kubota, Y., Takata, M., Matsuda, R., Kitaura, R., Kitagawa, S., Kato, K., Sakata, M., Kobayashi, T. C. *Angew. Chem. Int. Ed.* **2005**, *44*, 920–923.
285. Takamizawa, S., Nakata, E. *Cryst. Eng. Commun.* **2005**, *7*, 476–479.
286. Spencer, E. C., Howard, J. A. K., McIntyre, G. J., Rowsell, J. L. C., Yaghi, O. M. *Chem. Commun.* **2006**, 278–280.
287. Yildirim, T., Hartman, M. R. *Phys. Rev. Lett.* **2005**, *95*, 215504.
288. Peterson, V. K., Liu, Y., Brown, C. M., Kepert, C. J. *J. Am. Chem. Soc.* **2006**, *128*, 15578–15579.
289. Yan, Y., Telepeni, I., Yang, S. H., Lin, X., Kockelmann, W., Dailly, A., Blake, A. J. et al. *J. Am. Chem. Soc.* **2010**, *132*, 4092-+.
290. Mulder, F. M., Assfour, B., Huot, J., Dingemans, T. J., Wagemaker, M., Ramirez-Cuesta, A. J. *J. Phys. Chem. C* **2010**, *114*, 10648–10655.

291. Hartman, M. R., Peterson, V. K., Liu, Y., Kaye, S. S., Long, J. R. *Chem. Mater.* **2006**, *18*, 3221–3224.
292. Kubas, G. J., Burns, C. J., Eckhert, J., Johnson, S. W., Larson, A. C., Vergamini, P. J., Unkefer, C. J., Khalsa, G. R. K., Jackson, S. A., Eisenstein, O. *J. Am. Chem. Soc.* **1993**, *115*, 569–581.
293. Forster, P. M., Eckert, J., Chang, J.-S., Park, S.-E., Férey, G., Cheethem, A. K. *J. Am. Chem. Soc.* **2003**, *125*, 1309–1312.
294. Chapman, K. W., Chupas, P. J., Maxey, E. R., Richardson, J. W. *Chem. Commun.* **2006**, 4013–4015.
295. Liu, Y., Brown, C. M., Neumann, D. A., Peterson, V. K., Kepert, C. J. *J. Alloys Compd.* **2007**, *446–447*, 385–388.
296. Zhou, W., Yildirim, T. *J. Phys. Chem. C* **2008**, *112*, 8132–8135.
297. Nijem, N., Veyan, J.-F. o., Kong, L., Li, K., Pramanik, S., Zhao, Y., Li, J., Langreth, D., Chabal, Y. J. *J. Am. Chem. Soc.* **2010**, *132*, 1654–1664.
298. Bordiga, S., Regli, L., Bonino, F., Groppo, E., Lamberti, C., Xiao, B., Wheatley, P. S., Morris, R. E., Zecchina, A. *Phys. Chem. Chem. Phys.* **2007**, *9*, 2676–2685.
299. Spoto, G., Vitillo, J. G., Cocina, D., Damin, A., Bonino, F., Zecchina, A. *Phys. Chem. Chem. Phys.* **2007**, *9*, 4992–4999.
300. Panella, B., Hirscher, M. *Phys. Chem. Chem. Phys.* **2008**, *10*, 2910–2917.
301. Centrone, A., Siberio-Perez, D. Y., Millward, A. R., Yaghi, O. M., Matzger, A. J., Zerbi, G. *Chem. Phys. Lett.* **2005**, *411*, 516–519.
302. Centrone, A., Brambilla, L., Zerbi, G. *Phys. Rev. B* **2005**, *71*, 245406.
303. Fletcher, A. J., Cussen, E. J., Prior, T. J., Rosseinsky, M. J., Kepert, C. J., Thomas, K. M. *J. Am. Chem. Soc.* **2001**, *123*, 10001–10011.
304. Fletcher, A. J., Cussen, E. J., Bradshaw, D., Rosseinsky, M. J., Thomas, K. M. *J. Am. Chem. Soc.* **2004**, *126*, 9750–9759.
305. Yang, C., Wang, X., Omary, M. A. *J. Am. Chem. Soc.* **2007**, *129*, 15454–15455.
306. Bradshaw, D., Warren, J. E., Rosseinsky, M. J. *Science* **2007**, *315(5814)*, 977–980.
307. Yang, S., Lin, X., Blake, A. J., Walker, G. S., Hubberstey, P., Champness, N. R., Schröder, M. *Nat. Chem.* **2009**, *1*, 487–493.
308. Llewellyn, P. L., Maurin, G., Devic, T., Loera-Serna, S., Rosenbach, N., Serre, C., Bourrelly, S., Horcajada, P., Filinchuk, Y., Férey, G. *J. Am. Chem. Soc.* **2008**, *130* 12808–12814.
309. Humphrey, S. M., Oungoulian, S. E., Yoon, J. W., Hwang, Y. K., Wise, E. R., Chang, J. S. *Chem. Commun.* **2008**, 2891–2893.
310. Ma, S., Wang, X. S., Manis, E. S., Collier, C. C., Zhou, H. C. *Inorg. Chem.* **2007**, *46* 3432–3434.
311. Culp, J. T., Smith, M. R., Bittner, E., Bockrath, B. *J. Am. Chem. Soc.* **2008**, *130*, 12427–12434.
312. Kang, J., Wei, S.-H., Kim, Y.-H. *J. Am. Chem. Soc.* **2010**, *132*, 1510–1511.
313. Hamaed, A., Trudeau, M., Antonelli, D. M. *J. Am. Chem. Soc.* **2008**, *130*, 6992–6999.
314. Hamaed, A., Hoang, T. K. A., Trudeau, M., Antonelli, D. M. *J. Organometal. Chem.* **2009**, *694*, 2793–2800.
315. Hamaed, A., Van Mai, H., Hoang, T. K. A., Trudeau, M., Antonelli, D. *J. Phys. Chem. C* **2010**, *114*, 8651–8660.
316. Lueking, A., Yang, R. T. *J. Catal.* **2002**, *206*, 165–168.
317. Yang, F. H., T. Yang, R. *Carbon* **2002**, *40*, 437–444.
318. Lueking, A. D., Yang, R. T. *Appl. Catal. A General* **2004**, *265*, 259–268.
319. Yang, F. H., Lachawiec, A. J., Yang, R. T. *J. Phys. Chem. B* **2006**, *110*, 6236–6244.
320. Lueking, A. D., Yang, R. T., Rodriguez, N. M., Baker, R. T. K. *Langmuir* **2004**, *20*, 714–721.
321. Lachawiec, A. J., Qi, G. S., Yang, R. T. *Langmuir* **2005**, *21*, 11418–11424.
322. Li, Y. W., Yang, R. T. *J. Phys. Chem. C* **2007**, *111*, 11086–11094.
323. Wang, L. F., Yang, R. T. *J. Phys. Chem. C* **2008**, *112*, 12486–12494.
324. Wang, L. F., Yang, R. T. *J. Phys. Chem. C* **2009**, *113*, 21883–21888.
325. Ramachandran, S., Ha, J. H., Kim, D. K. *Catal. Commun.* **2007**, *8*, 1934–1938.
326. Li, Y., Yang, R. T. *AIChE J.* **2008**, *54*, 269–279.
327. Li, Y., Yang, R. T. *J. Am. Chem. Soc.* **2006**, *128*, 726–727.
328. Li, Y., Yang, R. T. *J. Am. Chem. Soc.* **2006**, *128*, 8136–8137.
329. Lachawiec, A. J., Yang, R. T. *Langmuir* **2008**, *24*, 6159–6165.

330. Zlotea, C., Campesi, R., Cuevas, F., Leroy, E., Dibandjo, P., Volkringer, C., Loiseau, T., Férey, G. r., Latroche, M. *J. Am. Chem. Soc.* **2010**, *132*, 2991–2997.
331. Stuckert, N. R., Wang, L., Yang, R. T. *Langmuir* **2010**, *26*, 11963–11971.
332. Stoicheff, B. P. *Can. J. Phys.* **1957**, *35*, 730–741.
333. Fletcher, A. J., Thomas, K. M. *J. Phys. Chem. C* **2007**, *111*, 2107–2115.
334. Tanaka, H., Kanoh, H., El-Merraoui, M., Steele, W. A., Yudasaka, M., Iijima, S., Kaneko, K. *J. Phys. Chem. B* **2004**, *108*, 17457–17465.
335. Noguchi, D., Tanaka, H., Kondo, A., Kajiro, H., Noguchi, H., Ohba, T., Kanoh, H., Kaneko, K. *J. Am. Chem. Soc.* **2008**, *130*, 6367–6372.
336. Beenakker, J. J. M., Borman, V. D., Krylov, S. Y. *Phys. Rev. Lett.* **1994**, *72*, 514–517.
337. Beenakker, J. J. M., Borman, V. D., Krylov, S. Y. *Chem. Phys. Lett.* **1995**, *232*, 379–382.
338. Yaris, R., Sams, J. R. J. *J. Chem. Phys.* **1962**, *37*, 571–576.
339. Kumar, A. V. A., Bhatia, S. K. *Phys. Rev. Lett.* **2005**, *95*, 245901.
340. Kumar, A. V. A., Jobic, H., Bhatia, S. K. *J. Phys. Chem. B* **2006**, *110*, 16666–16671.
341. Zhou, J., Wang, Q., Sun, Q., Jena, P., Chen, X. S. *Proc. Natl. Acad. Sci. (USA)* **2010**, *107*, 2801–2806.
342. Shi, S. Z., Hwang, J. Y., Li, X., Sun, X., Lee, B. I. *Int. J. Hydrogen Energy* **2010**, *35*, 629–631.

25

Hydrogen Storage in Hollow Microspheres

Laurent Pilon
University of California, Los Angeles

CONTENTS

25.1 Introduction

The element hydrogen is the most commonly found element in the universe. However, hydrogen molecules (H_2) are not readily available. As such, it is an energy carrier as opposed to a fuel. It can be used in various mobile applications such as (1) in proton exchange membrane (PEM) fuel cell for transportation systems or mobile devices (e.g., laptops and cell phones) where it catalytically reacts with oxygen to produce water and electricity, (2) in internal combustion engines for surface transportation where it can be mixed with liquid fuel [1,2], or (3) in rocket propulsion [3]. Akunets et al. [2] also suggested using a mixture of liquid oxygen and hydrogen in polymer microballoons for jet engine fuel [2].

Hydrogen storage for such mobile applications is arguably one of the main technological challenges for a viable hydrogen economy. This chapter focuses on hydrogen storage in hollow glass microspheres or microcapsules in general. First, various power sources and fuels for mobile applications are compared based on their energy densities. Then, competing hydrogen storage technologies are reviewed. Moreover, principles, design parameters, material considerations, and performances associated with hydrogen storage in hollow glass microspheres are discussed in detail. Processes for synthesizing hollow glass microspheres are also reviewed.

25.1.1 Hydrogen Storage Technologies

In order to compare different fuels and energy carriers as well as hydrogen storage solutions, it is useful to remember that (1) energy contained in 1 kg of hydrogen is equivalent to that in 1 gal (3.78 L) of gasoline and that (2) current gasoline or diesel tanks contain between 10 and 30 gal. However, fuel cells are more efficient than internal combustion engines, thus reducing the required amount of hydrogen on board to 5–8 kg. In addition, a midsize fuel cell car cruising at 100 km/h (62 mph) consumes about 400 mg of H_2/s, which needs to be delivered on demand [4].

The US Department of Energy (DOE) has developed a technology roadmap for hydrogen storage. It sets quantitative targets that would ensure vehicle autonomy greater than 300 miles as well as safe and flexible driving, fast refueling time, procedure, and retail sales comparable to existing ones [5]. The technical criteria for selecting a specific hydrogen storage technology for transportation applications and the associated 2015 US DOE targets have been identified as follows [5]:

1. *Safe operation under all circumstances* including road accidents since hydrogen reacts explosively with oxygen above the ignition temperature of 450°C at atmospheric pressure
2. *Large gravimetric energy density* defined as the mass of H_2 (or energy) stored per unit mass of storage systems, which should be larger than 9 wt.% or 10.8 MJ/kg
3. *Large volumetric energy density* defined as the mass of H_2 or energy stored per unit volume of storage systems, which should be larger 81 g of H_2/L or 9.72 MJ/L
4. *Refueling time* of hydrogen tanks as short as refueling with current fuels such as diesel or gasoline (2.5 min for 5 kg of H_2)
5. *Reversibility of uptake and release* so that the storage system can be used numerous times (1500 cycles from 1/4 to full) for reliability and cost-effectiveness

6. On-demand availability with a minimum mass flow rate of 0.02 g/s/kW
7. Operating temperatures between −40°C and 60°C to work in any weather conditions
8. *Low energy requirements* for loading and unloading H_2 in order to achieve maximum energy efficiency
9. *Safe dormancy properties*, that is, at room temperature, very little hydrogen should leak out from the storage solution making the storage safe while not in use for extended periods of time
10. Low fabrication and operation costs for a pretax cost of $2/kW/h

Gaseous hydrogen at room temperature without its storage tank has gravimetric energy density of 143 MJ/kg and therefore exceeds by far DOE's target. Unfortunately, gaseous hydrogen at standard temperature and pressure has volumetric energy density of 0.01079 MJ/L. Such a very low value requires that hydrogen be compressed drastically. Numerous techniques have been proposed and can be grouped as (1) compressed hydrogen gas, (2) liquid hydrogen, (3) cryoadsorption, and (4) [6–8]. High-pressure H_2 storage solutions at room temperature include gas cylinders, underground reservoirs, and hollow glass microspheres [9–11]. Hydrogen can also be stored in liquid phase at cryogenic temperatures (≤20 K). Low-temperature cryoadsorption of hydrogen is achieved in (1) carbon nanotubes [12], (2) activated carbon [13], (3) carbon aerogels [13], (4) metal-organic framework [14], and (5) zeolite. Chemical hydrogen storage includes (1) metal hydrides [8,15], (2) liquid (organic) hydrides, and (3) ammonia and methanol. Unfortunately, none of the current technologies meet the aforementioned 2015 performance and cost targets set by the US DOE by a wide margin [16,17].

As an energy carrier for transportation systems and portable applications, hydrogen storage performances should be compared with other fuels or power systems including (1) liquid fuels (e.g., gasoline, diesel, biodiesel, ethanol, and methanol), (2) batteries (e.g., lithium ion, fluoride ions, and zinc–air) used in hybrid and all-electric cars, (3) solid fuel (e.g., Al, Zn, and Li) used in rocket propulsion, (4) compressed or liquified natural gas already widely used for public transportation, and (5) compressed air. To do so, it is useful to define the gravimetric and volumetric energy densities representing the amount of energy stored in a system per unit mass or unit volume, respectively. Figure 25.1 plots the gravimetric versus volumetric energy densities for each fuel or energy storage technologies and the associated container as reported in the literature [11,18]. It is evident that gasoline or diesel or biodiesel has outstanding performances that explain their widespread use. These performances are only matched by those of solid fuels, which cannot be considered for surface transportation. Performances of battery technologies, however, remain an order of magnitude smaller than liquid fuels.

Figure 25.2 also compares the best reported performances of the various hydrogen storage technologies [8,11]. It is evident that metal hydrides offer large volumetric energy density but relatively low gravimetric density. Liquid H_2 storage at 20 K and atmospheric pressure as well as cryoadsorption and cryocompressed storages show good performance but require large energy for liquefaction (27.9% of the stored energy [8]). They also feature poor dormancy properties.

25.1.2 Hydrogen Storage in Hollow Glass Microspheres

As previously discussed, the volumetric energy density of gaseous hydrogen at room temperature is very small and falls short of the DOE targets [5]. One strategy to increase its

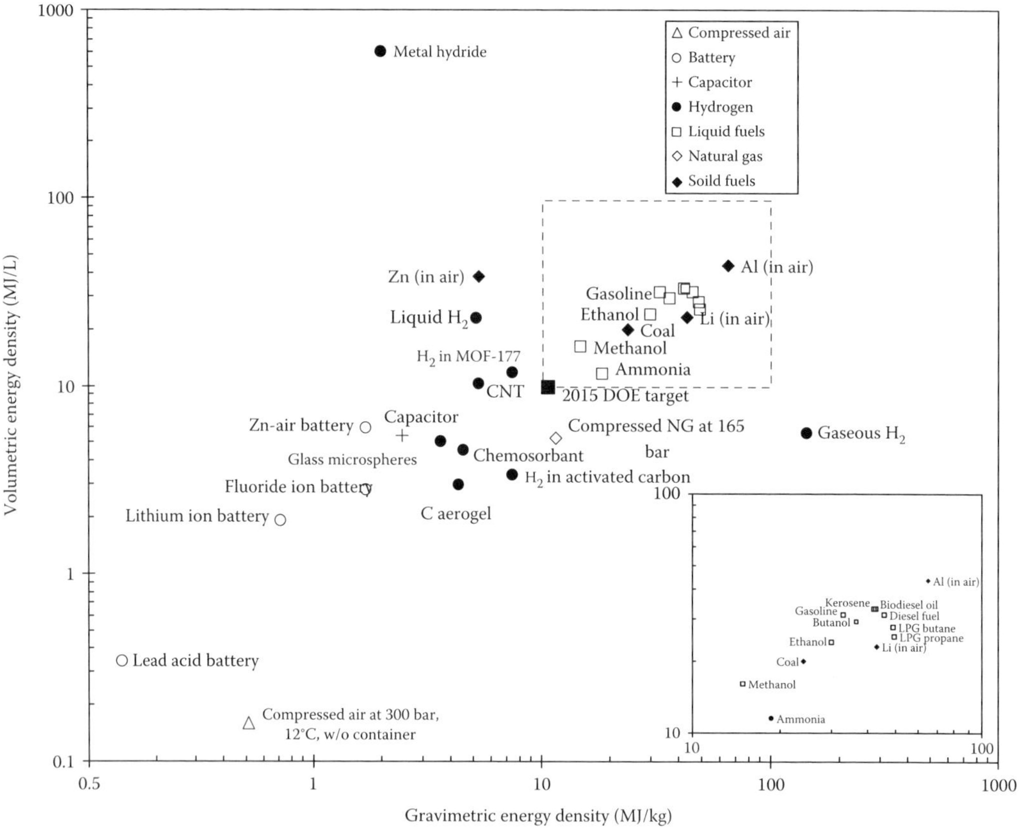

FIGURE 25.1
Comparison of gravimetric and volumetric energy densities for different fuels and hydrogen storage technologies. (From Yartys, V.A. and Lototsky, M.V., An overview of hydrogen storage methods, in *Hydrogen Materials Science and Chemistry of Carbon Nanomaterials*, Veziroglu, T.N., Zaginaichenko, S.Y., Schur, D.V., Baranowski, B., Shpak, A.P., Skorokhod, V.V., and Kale, A., eds., Kluwer Academic Publishers, Dordrecht, the Netherlands; Herr, M. and Lercher, J.A., Hydrogen storage in microspheres—Final report, ET-TN-03-628, September 9, 2003; Teitel, R.J., Microcavity hydrogen storage final progress report, Prepared for US Department of Energy and Environment, Report BNL 51439, 1981.)

volumetric energy density is to compress H_2 inside hollow glass microspheres or microcapsules in general. Loading and unloading of hydrogen gas in and out is based on the fact that gas permeation through the solid shell is a thermally activated process, that is, gas permeation increases exponentially with temperature. Hollow microspheres are also called microcapsules, microcavities, microbubbles, or microballoons. This was first proposed by Teitel in 1980 [9,10]. The initial thought was to use it in combination with a metal hydride storage system designed to store hydrogen released by hollow glass microspheres during cooldown and able to provide hydrogen during cold starts and accelerations. Moreover, "the addition of a metal hydride to the system would increase its hydrogen volumetric energy density and reduce the gravimetric energy density" [9]. Since then, hollow glass microspheres have been considered as stand-alone H_2 storage solutions [19]. Geometries other than microspheres have also been considered including microcylinders [4,20] and foams [21].

Figure 25.3 illustrates the typical life cycle of hollow microspheres for H_2 storage including (1) hydrogen loading, (2) storage and distribution, (3) onboard H_2 unloading,

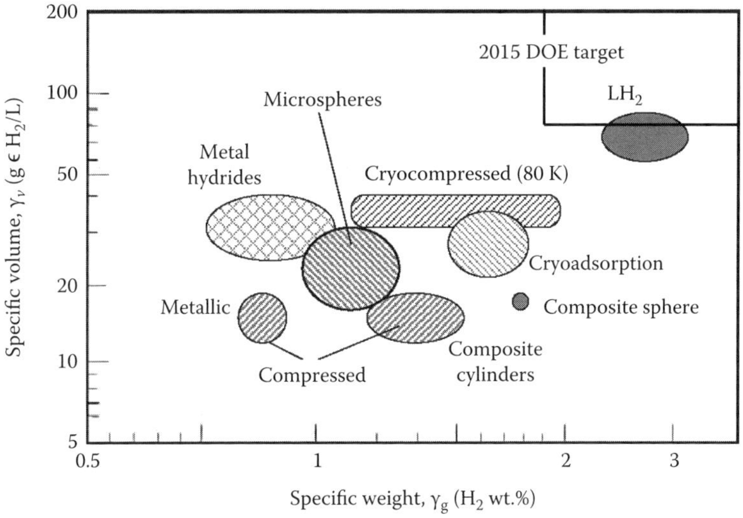

FIGURE 25.2
Comparison of gravimetric and volumetric energy densities for hydrogen storage technologies. (From Herr, M. and Lercher, J.A., Hydrogen storage in microspheres—Final report, ET-TN-03-628, September 9, 2003.)

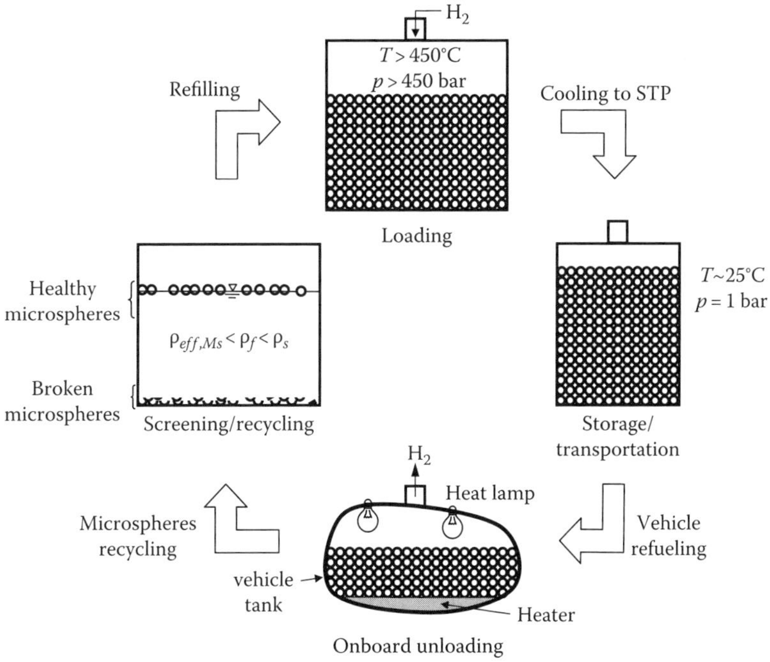

FIGURE 25.3
Life cycle of hollow microspheres from hydrogen loading and microsphere storage to vehicle refueling and microsphere recycling.

and (4) screening and recycling of healthy microcapsules. Practically, hydrogen loading in hollow microspheres can be performed at industrial scale in a batch process off vehicle in an autoclave at high temperatures (≈400°C) and pressures (>450 bar) to accelerate hydrogen permeation through the container's shell. Then, the hollow microspheres are cooled to room temperature so that hydrogen gas remains trapped inside due to significant reduction in the H_2 permeation with decreasing temperature (see Section 25.2.7.2). The hollow microspheres, then, can be safely transported to distribution points at room temperature and atmospheric pressure. Hydrogen refueling of a vehicle would consist of sucking the spent microspheres out of the tank and pouring loaded ones in the tank in a manner very similar to current gasoline or diesel refueling. On-demand hydrogen release from the microspheres would be induced by an onboard electric heater or a heat lamp integrated in the tank. Alternatively, unloading could also be achieved by mechanically or thermally destroying the microspheres, thus releasing their H_2 content. This strategy would still require removing the broken microspheres before tank refueling to be remelted or discarded. However, small pieces of broken microspheres could constitute a health hazard [22]. Empty microspheres removed from the tank would be screened for cracks and sorted by size and recycled to undergo the same sequence of H_2 loading, distribution, and unloading. Separation of broken microspheres from reusable ones can be achieved by the sink–float method [23,24]. This consists of placing all microspheres in a fluid whose density ρ_f is smaller than that of the shell materials ρ_s but larger than the effective density of the microspheres ρ_{MS}, that is, $\rho_{MS} < \rho_f < \rho_s$. Thus, pieces of broken microspheres will sink while *healthy* microspheres will float.

Hydrogen storage in hollow glass microspheres presents the following advantages over the previously reviewed storage technologies. First, hollow microspheres have high gravimetric energy density [11]. Hydrogen can be stored under internal pressure higher than that inside conventional cylinders [11]. Hydrogen-filled hollow glass microspheres are also easy and safe to handle at atmospheric pressure and ambient temperature and can be poured or pumped in tanks of any arbitrary geometries and made of lightweight materials (e.g., plastic) [11,19]. The technology is inexpensive and requires low energy consumption for producing large quantities of microcontainers [7]. In addition, this technology has good dormancy characteristics [7]. It is also resistant to contamination by atmospheric gases, unlike metal hydrides. Similarly, the microspheres are expected to remain stable and intrinsically safe under accident or fire conditions thanks to the small volume of hydrogen stored in each microspheres and their conformation or ability of the bed to change shape caused by potential deformation of the outside container [7,19,25]. Finally, the technical risks are minimal [7] and scaling from benchtop to full-scale utilization appears to be straightforward as hollow microspheres are already produced at industrial scale albeit for other applications and without the desired mechanical properties required for H_2 storage [26,27].

The challenges of hydrogen storage in microcontainers are [11,28] (1) the low volumetric energy density, (2) the relatively low fraction of recoverable hydrogen [29], (3) the fact that one needs to heat the microcapsules at temperatures above the operating temperature of PEM fuel cells, (4) the small H_2 release rate, (5) the large amount of energy required to compress hydrogen to very high pressures used during H_2 loading (25% of storage energy [8]), and (6) the cost of disposing or recycling glass microspheres. Table 25.1 summarizes the different advantages and disadvantages of hydrogen storage in microcapsules.

TABLE 25.1

Established and Potential Advantages and Disadvantages Associated with Current Hydrogen Storage in Hollow Microcapsules

Advantages	Disadvantages
High gravimetric energy density	Low volumetric energy density
Handling at room temperature and atmospheric pressure	Requires hydrogen loading under very high pressures
Inexpensive and high-throughput manufacturing process	Requires heating for loading or unloading
Intrinsically safe	Potentially high loading energy cost
Good dormancy characteristics	Requires reprocessing of spent microcapsules
Resistant to contamination/poisoning by atmospheric gases	Potential health hazard from broken microcapsules
Can be poured and fit in any container solution	Uncertain consumer acceptance of refueling conditions
Easily scalable from lab to industrial scale	Large uncertainty in achieving desired performances
Minor technical risks	

Source: Robinson, S.L. and Handrock, J.L., Hydrogen storage for vehicular applications: Technology status and key development areas, Sandia National Laboratory Report SAND94-8229-UC-406, 1994.

25.2 Design Parameters for Hydrogen Storage in Hollow Microspheres

25.2.1 Principles

The amount of hydrogen gas stored inside a capsule and its permeation rate depends on the inner and outer pressures, the temperature, and the shell material and thickness. To achieve the maximum storage capacity, pressure inside the microcapsules should be as large as possible. Similarly, for fast H_2 release, the outside pressure should be reduced and the temperature increased while the shell should be as thin as possible. However, the operating temperatures and pressures during loading and unloading and the handling of the hollow microcontainers should not threaten their mechanical integrity. The shell material should have the proper permeation properties but also the mechanical properties able to stand large pressures and temperatures as well as the thermal cycling associated with successive loading and unloading. Consequently, microcapsules' geometry and material as well as the operating temperatures and pressures must be optimized to minimize the loading and unloading times and to maximize the energy densities, the permeation rate, the hold time, and the number of life cycles. To do so, concepts of burst and buckling pressures, gas permeation, as well as geometric and material considerations are reviewed.

25.2.2 Hydrogen Properties

25.2.2.1 Density

At temperatures and pressures of interest for hydrogen storage, gaseous hydrogen cannot be considered as an ideal gas. Instead, a more complex equation of state must be used to account for repulsion forces between H_2 molecules. The Beattie–Bridgeman equation gives the hydrogen pressure P as a function of temperature T and density ρ [30]:

$$P = \rho^2 RT\left(1 - \frac{c\rho}{T^3}\right)\left[\frac{1}{\rho} + B_0(1 - b\rho)\right] - A_0\rho^2(1 - a\rho) \qquad (25.1)$$

where
- R is the ideal gas constant (=8.314 J/mol K)
- the pressure P is expressed in Pa
- ρ is expressed in kg/m^3
- T is expressed in K
- the parameters a, b, and c are equal to -5.06×10^{-6} m^3/mol, -43.56×10^{-6} m^3/mol, and 5.04 m^3 K^3/mol, respectively
- the coefficients A_0 and B_0 are equal to 0.02 Pa m^3/mol^2 and 20.96×10^{-6} m^3/mol, respectively [30]

Figure 25.4 compares the actual density of hydrogen ρ with that predicted by the ideal gas law $\rho = PM/RT$ for temperature and pressure ranging 48–398 K and 1–1000 bar (14.5–14,500 psi), respectively. It shows that H_2 deviates from ideal gas law for values of PM/RT larger than 10, that is, for low temperatures and/or high pressures.

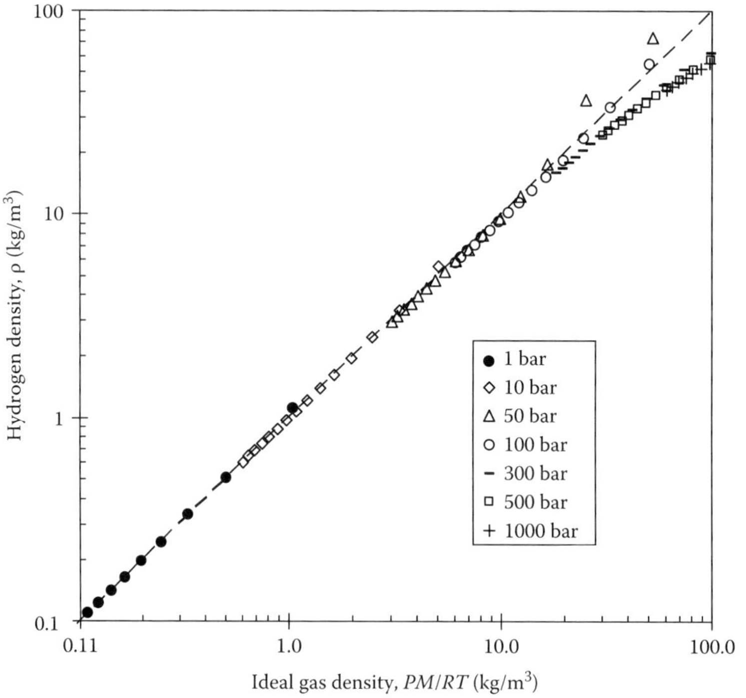

FIGURE 25.4
Actual hydrogen density versus ideal gas density ($\rho = PM/RT$) for pressure ranging from 1 to 1000 bar and temperatures between 48 and 398 K. (From Lemmon, E.W. et al., NIST reference fluid thermodynamic and transport properties database (REFPROP): Version 8.0, National Institute of Standards and Technology, Boulder, CO, http://www.nist.gov/srd/nist23.htm, 2007.)

25.2.2.2 Compressibility Factor

Alternative to the previous equation of state, one can define the compressibility factor as

$$Z(T,P) = \frac{PM}{\rho RT} \tag{25.2}$$

Figure 25.5 shows the compressibility factor of hydrogen as a function of pressure P between 1 and 1000 bar for different values of temperature T between 48 and 398 K. Within this range, the compressibility factor $Z(T,P)$ of H_2 varies between 0.68 and 2.23. It is nearly equal to unity (ideal gas) for pressure less than 10 bar and temperatures larger than 48 K.

25.2.3 Burst Pressure

The volumetric energy density of hollow microspheres filled with hydrogen depends strongly on the differential pressure it can sustain. The maximum pressure of H_2 inside a hollow microsphere depends on the inner and outer radii denoted by r_i and r_o, respectively and on the biaxial tensile strength of the shell materials denoted by $\sigma_{s,max}$. For thin-walled microspheres (i.e., $r_o > 5(r_o - r_i)$), the burst pressure denoted by P_{max} is expressed as [31]

$$P_{max} = \frac{2\sigma_{s,max}(r_o - r_i)}{S_f r_0} \quad \text{for hollow microspheres} \tag{25.3}$$

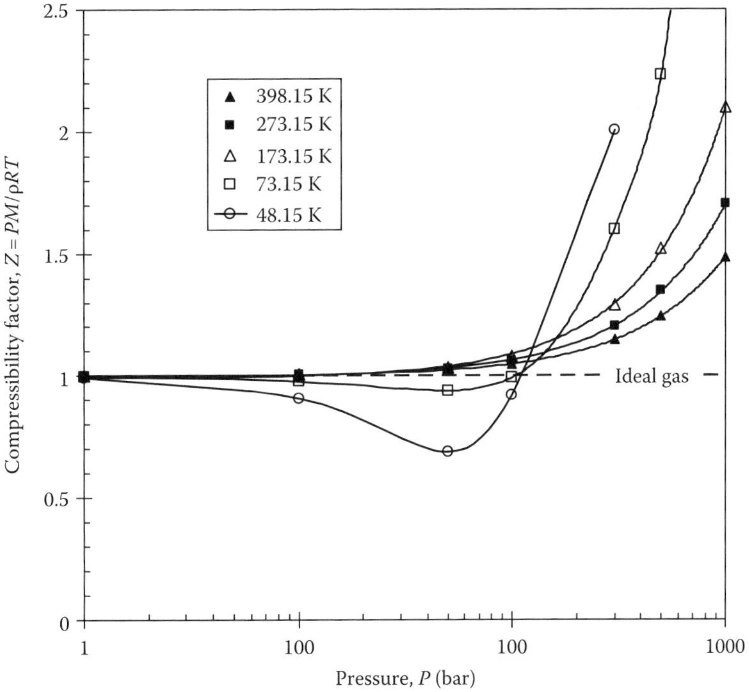

FIGURE 25.5
Compressibility factor of hydrogen as a function of pressure for temperature ranging from 48 to 398 K and pressure between 1 and 1000 bar. (From Lemmon, E.W. et al., NIST reference fluid thermodynamic and transport properties database (REFPROP): Version 8.0, National Institute of Standards and Technology, Boulder, CO, http://www.nist.gov/srd/nist23.htm, 2007.)

A safety factor S_f, typically ranging from 1.5 to 10, must also be considered for the actual design and fabrication of the storage system [32]. It is often taken as 1.5 for H_2 storage applications since the breaking of one microsphere does not endanger the integrity of the entire hydrogen storage solution [32]. Equation 25.3 indicates that the internal pressure increases as the tensile strength $\sigma_{s,max}$ and the shell thickness $(r_o - r_i)$ increase and as the outer radius r_o decreases.

For cylindrical thin-walled containers, the burst pressure is expressed as [31]

$$P_{max} = \frac{\sigma_{s,max}(r_o - r_i)}{S_f r_0} \quad \text{for cylinders} \tag{25.4}$$

Here, $\sigma_{s,max}$ is the maximum strength in the radial direction also called hoop stress σ_h. For cylindrical geometry, a longitudinal stress σ_l exists but is only half the hoop (radial) stress, that is, $\sigma_l = \sigma_h/2$ [31]. It is evident from Equations 25.3 and 25.4 that hollow microspheres are, a priori, preferable to cylinders with the same inner and outer diameters as they can sustain burst pressures twice as large. However, synthesis of hollow microspheres is more challenging than microcylinders [4] (see Section 25.5).

Achieving large burst pressure first guided the search for the best shell materials that could stand large tensile stress. Teitel [9,10] suggested using glass for their outstanding mechanical properties under a wide range of temperatures. Indeed, the tensile strength of glasses is reported to be above 4 GPa and can reach up to 6.9 GPa for quartz (SiO_2) at room temperature compared with 460 MPa for steel [33,34]. Figure 25.6 shows the burst pressure predicted by Equation 25.3 for thin-walled microspheres as a function of radii ratio r_o/r_i

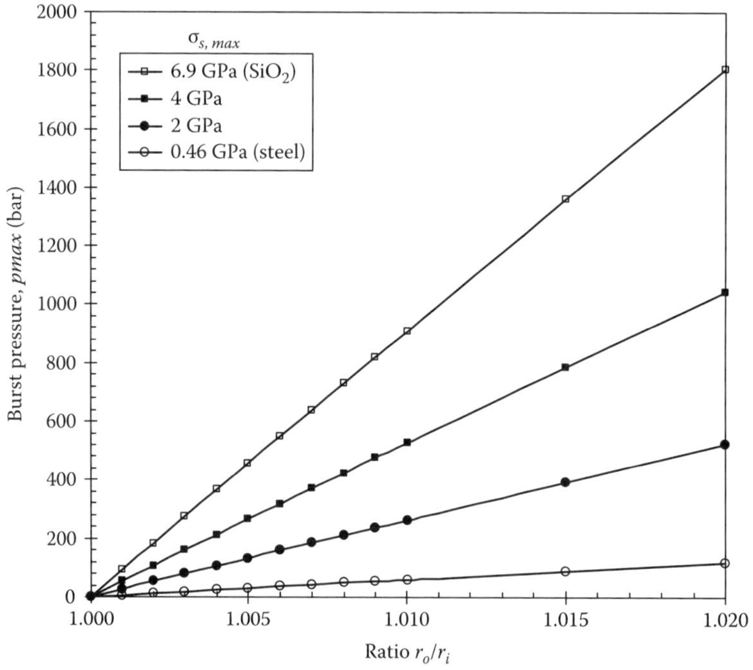

FIGURE 25.6
Burst pressure of thin-walled microspheres predicted by Equation 25.3 for different values of tensile strength $\sigma_{s,max}$ and $S_f = 1.5$. Burst pressure for cylinders with the same r_o/r_i ratio is half of that shown.

for tensile strength σ_{max} varying from 460 MPa (steel) to 6.9 GPa (quartz) and safety factor equal to 1.5. The radii ratio r_o/r_i is kept smaller than 1.02, beyond which Equation 25.3 is no longer valid [31].

For either microspheres or microcylinders, the absence of defects in the sphere membrane (or shell) and the close-to-perfect sphericity or cylindricity are essential elements in order to achieve large burst pressure. Imperfection may exist and scratches can appear at the surface of the microcapsules during manufacturing and handling, which can result in decrease in the tensile strength [35]. To prevent scratches from occurring, the microcapsule surface can be acid etched or coated with a hard coating material as discussed in Sections 25.4 and 25.5.7.

25.2.4 Buckling Pressure

During hydrogen loading, the external pressure should be controlled to prevent the microspheres from collapsing when the difference between inner and outer pressures exceeds the buckling pressure. The classical expression for the static buckling pressure P_{cr} for isotropic thin-walled and shallow spherical shell under uniform pressure is given by [36,37]

$$P_{cr} = \frac{2E_s(r_o - r_i)^2}{r_o^2\sqrt{3(1-\nu_s^2)}} \tag{25.5}$$

where E_s and ν_s are the Young's modulus and Poisson's ratio of the shell material, respectively. The previous expression provides an upper limit for the pressure under which the hollow microspheres can be exposed [27]. In general, thin-walled microspheres can stand very large buckling loads [36]. In addition, numerous studies have also investigated dynamic buckling and how it is affected by shell defects, nonsphericity, nonuniformity in shell thickness, as well as coaxial loads [37–39]. However, experimental study for hollow microspheres were found to be more reliable to estimate the buckling pressure [37,39].

Rambach and Hendricks [27] experimentally investigated the buckling pressure of silicate glass microspheres 45 μm in average diameter and wall thickness of 0.9 μm. The buckling pressure predicted by Equation 25.5 was 1172 bar (17,000 psi) using $E_s = 62.0$ GPa and $\nu_s = 0.22$. Experimentally, however, more than 20% of the microspheres collapsed for loading pressures larger than 500 bar at room temperature. Failure rate exceeded 80% for external pressure larger than 800 bar. No significant failure occurs for external pressure smaller than 414 bar (6000 psi) at the loading temperature up to 350°C thus setting practical limits of operations.

25.2.5 Hydrogen Permeation Processes

Gas permeability refers to the steady flow rate of gas across a specimen per unit of differential pressure and per unit thickness [40]. It is denoted by K and expressed in mol/Pa m s. The mass transfer rate of hydrogen entrapped in a single microsphere is expressed as [29]

$$\frac{dm_i}{dt} = \pm\frac{4\pi r_i^2 MK(T)}{(r_o - r_i)}(P_o - P_i) \tag{25.6}$$

where P_i and P_o are the inner and outer pressures, respectively. The ± sign depends whether hydrogen is loaded (+) ($P_o > P_i$) or released (−) ($P_o < P_i$) from the microspheres. Here, changes in the surface area for diffusion were ignored, and the inner surface area $4\pi r_i^2$ was chosen since it is the smallest and therefore controls the hydrogen permeation [29]. The mass of H_2 contained inside the microspheres is given by [41,42]

$$m_i = \rho \frac{4\pi r_i^3}{3} = \frac{4\pi r_i^3}{3} \frac{P_i M}{Z(T, P_i)RT} \tag{25.7}$$

Similarly, for a single microcylinder, the mass transfer rate and the mass of H_2 entrapped are, respectively, expressed as

$$\frac{dm_i}{dt} = \pm \frac{2\pi r_i LMK(T)}{(r_o - r_i)}(P_o - P_i) \quad \text{and} \quad m_i = \frac{P_i M}{Z(T, P_i)RT} \pi r_i^2 L \tag{25.8}$$

where
L is the length of the cylinder
r_i and r_o are the inner and outer diameters, respectively

25.2.6 Geometric Considerations

Several geometric parameters should be optimized for hydrogen storage in microcapsules, namely, (1) the shape (spherical or cylindrical), (2) the inner and outer diameters r_i and r_o of the microspheres or microcylinders, (3) the shell thickness ($r_o - r_i$), and (4) the volume fraction or packing fraction ϵ of microcapsules in the storage tank, which depends on their size distribution. Each of these design parameters are discussed in the following sections.

25.2.6.1 Shape

Most studies have focused on hydrogen in hollow glass microspheres by virtue of the fact that microspheres have the highest burst pressure and smaller release time constant. In addition, glass was the material of choice for its high tensile strength and permeability. However, their synthesis can be more complex and costly than those of cylinders [4,20]. Yan et al. [20] manufactured hollow silica fibers with outer diameter ranging from 160 to 260 μm and wall thickness between 16 and 35 μm. The authors also discussed sealing and strengthening of the microcylinders (see Section 25.4)

25.2.6.2 Size

Commercially available hollow glass microspheres feature outer diameter between 5 and 500 μm and shell thickness between 0.5 and 20 μm [11]. Increasing the shell thickness increases the burst and buckling pressures and extends the hold time of the microspheres. However, it also reduces the volume available for hydrogen storage and the volumetric energy density while increasing the loading and unloading times. Thus, there exist an optimum values for outer and inner radii associated with operating temperature and pressure.

25.2.6.3 Packing Fraction

The maximum theoretical packing of monodisperse spheres is 74% corresponding to a body-centered (BCC) and face-centered (FCC) cubic face-centered arrangement as illustrated in Figure 25.7. However, this highly ordered arrangement is not achieved in practice

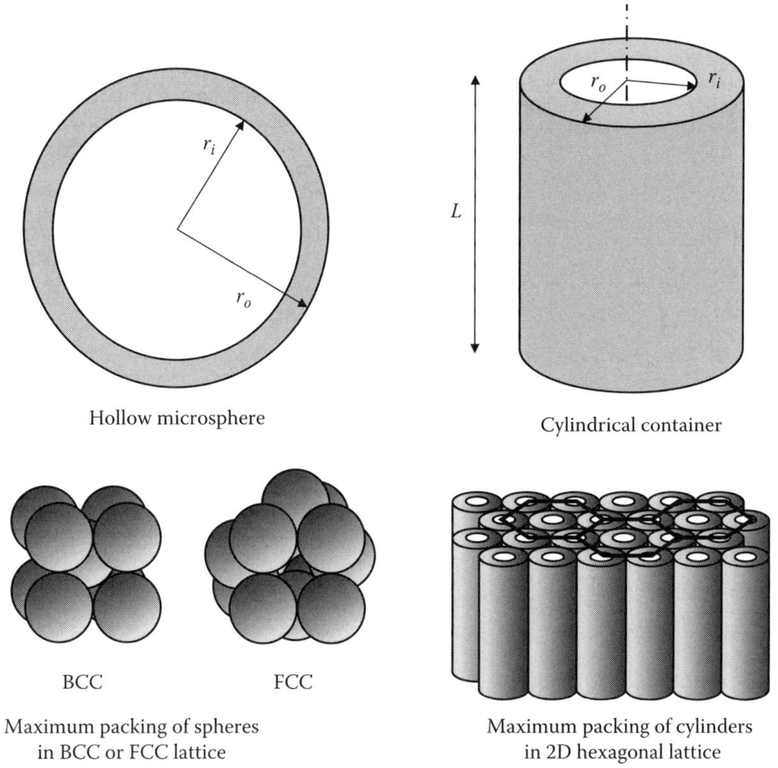

FIGURE 25.7
Schematic and dimensions of microspheres and microcylinders along with their spatial arrangement.

and maximum random packing of 63.7% is more realistic [43]. Larger packing can be achieved if polydisperse spheres are used [43], but this can also complicate the handling, sorting, and recycling of the microspheres [11] since the microsphere density will vary with size, which is incompatible with the sink–float method [23,24].

Further increase in packing fraction can be obtained if the spheres are deformed or if polymeric or glass foams are used [21,44]. Pientka et al. [44] showed that extruded polystyrene foams can be used for hydrogen separation from CO_2 and N_2 as well as for hydrogen storage. Metallic foams can stand larger pressure than polymeric foams but have much larger density and may become brittle when exposed to H_2. Banyay et al. [21] concluded that metallic foams meet the volumetric energy density DOE target but not the gravimetric one and vice versa for polymeric foams. The authors suggested that composite foams might satisfy both targets. In the case of closed-cell foams for H_2 storage, Equation 25.6 is not valid and modeling mass transfer through foams has been discussed in the literature [45].

Finally, the maximum packing fraction of monodisperse and aligned cylinders ordered in 2D hexagonal lattice is 90.7% [4], while it is 0.82–0.83 if they are randomly packed [46].

25.2.7 Material Considerations

The shell material is selected for its tensile strength and permeability as a function of temperature along with the specific heat and optical properties depending on the heating process. These properties are reviewed in the following sections.

25.2.7.1 Mechanical and Thermal Properties

Mechanical and thermal properties of glass depend on compositions, sample history, and temperature. Glass is brittle and much stronger under compression than tensile loads [47,48]. Table t-mechanics provides of range of realistic values for $\sigma_{s,max}$, E_s, ν_s, ρ_s, k_s, and $c_{p,s}$ [47]. Note that tensile strength is significantly affected by surface imperfections or flaws [47–49]. Strength of glass decreases as temperature increases [49]. It is also size dependent and strength of glass fibers increases sharply as the fiber diameter decreases, for example [48]. Attempts to correlate glass strength and composition were typically unsuccessful [48] and experimental data showed very large variations mainly due to the fact that surface flaws determine the glass strength and are difficult to observe directly [49]. The ultimate tensile strength $\sigma_{s,max}$ is expressed as [49,50]

$$\sigma_{s,max} = \sqrt{\frac{E_s \alpha}{a}} \tag{25.9}$$

where
- E_s is the shell material Young's modulus
- a is the interatomic distance
- α is the surface energy

For commercial oxide glass, a is not significantly affected by the compositions [50] and is of the order of 2×10^{-10} m [49,50] while α varies from 3.5 J/m^2 for soda-lime glass to 5.2 J/m^2 for vitreous silica [49]. Thus, the ultimate tensile strength is 16 GPa for soda-lime silicate and 24 GPa for vitreous silica, for example. These values are much higher (up to 60% for silica [49]) than experimental measurements due to the unavoidable presence of surface flaws [49,50]. Thus, experimentally measured values of $\sigma_{s,max}$ should be used in estimating the burst pressure from Equation 25.3 (Table 25.2).

25.2.7.2 Hydrogen Permeability

The first mechanism for hydrogen dissolution in glass consists of physical dissolution when H_2 molecules occupy the interstices of the glass. The second mechanism involves

TABLE 25.2

Summary of Mechanical and Thermal Properties of Glasses Considered for Hydrogen Storage

Property	Vitreous Silica	Soda-Lime Glass	Borosilicate Pyrex
$\sigma_{s,max}$ (GPa)	24	16	14
E_s (GPa)	72.9–77.2	70–72	64
ν_s (-)	0.165–0.177	0.25	0.2
ρ_s (kg/m^3)	2200	2520	2230
$c_{p,s}$ (J/kg K)	712	754	750
$k_{c,s}$ (W/m K)	1.46–1.71	1.2–1.78	1.2–1.78

Source: Bansal, N.P. and Doremus, R.H., *Handbook of Glass Properties*, Academic Press, Inc., Orlando, FL, 1986.

chemical reactions between H_2 and the glass resulting in the formation of OH groups. For example, hydrogen causes the reduction of variable-valence ions such as Fe^{3+}, Ce^{4+}, and Sn^{4+} [51]. In particular, Fe^{3+} can be reduced almost completely to Fe^{2+} according to the following reaction [51]:

$$H_2 + 2(\equiv \mathrm{Si} - \mathrm{O})^- + 2\mathrm{Fe}^{3+} \rightleftharpoons 2(\equiv \mathrm{Si} - \mathrm{OH}) + 2\mathrm{Fe}^{2+} \tag{25.10}$$

The reaction rate is much faster than the diffusion rate. Thus, the process is diffusion limited and can be accounted for through standard mass diffusion model with some effective diffusion coefficient, permeability, and solubility [52]. Several empirical relationships for temperature dependence of gas permeability $K(T)$ have been suggested and are of the following general form:

$$K(T) = K_0 T^n \exp\left(\frac{-Q_K}{T}\right) \tag{25.11}$$

where K_0, n, and Q_K are constants determined empirically. Souers et al. [53] identified K_0 and Q_K while n was set equal to 1 as suggested by Doremus [54] and Shelby [40]. Alternatively, other researchers [2,4,29,42,55–59] used the standard Arrhenius law for which $n = 0$. The values of K_0 and Q_K are sometimes expressed in terms of (1) the content of glass network formers such as SiO_2, B_2O_3, and P_2O_5 denoted by G and expressed in mol.% or (2) the content of modifier oxide such as CaO, Na_2O, MgO, SrO, and NaO denoted by M. Table 25.3 summarizes values of K_0 and Q_K as reported in literature for vitreous silica, silicate, soda-lime, and borosilicate glasses [29,53,55–58,60,61]. In general, hydrogen permeability increases with G and decreases with M as suggested by the expressions for Q_K reported in Table 25.3. In addition, the permeability has been reported in different units such as molecules/cm s atm and mol/cm s when measured with 1 atm pressure difference across the sample. Thus, conversion was sometimes necessary to compare the different studies according to

$$K\,(\text{in mol/Pa m s}) = \frac{100}{P_{atm} N_A} K\,(\text{in molecules/cm s atm}) = \frac{100}{P_{atm}} K\,(\text{in mol/cm s}) \tag{25.12}$$

where
- N_A is the Avogadro number (=6.022 × 10^{23} molecules/mol)
- P_{atm} is the atmospheric pressure (=1.013 × 10^5 Pa) or the pressure difference at which the measurements were performed

Note that H_2 permeability in glass can be considered to be independent of pressure up to 1000 bar [47]. Figure 25.8 shows the evolution of permeability as a function of temperature for the different glasses and references summarized in Table 25.3. It shows that H_2 permeability increases significantly with temperature and can vary widely with glass composition. It is also evident that permeability is systematically larger in vitreous silica or quartz than in glasses with other compositions.

TABLE 25.3

Empirical Constants for the Permeability of Various Glasses Used in Equation 25.11

Composition	n	$K_0 \times 10^{17}$ (mol/Pa m s)	Q_K(K)	References	Reported in
Soda-lime glass ($72\% \leq G \leq 100\%$)	0	8,100	17,330 − 127.8 G	[29]	mol/Pa m s
Silicate glasses	1	$3.4 + 8 \times 10^{-4}$ M^3	3,600 + 165 M	[53]	mol/Pa m s
Soda-lime glasses (M = 19%)	1	20	5,600	[53]	mol/Pa m s
Vitreous silica (see Ref. [40])	0	58.1	4,234	[55]	[a]
Vitreous silica (quartz)	0	8,688	4,469	[56]	molecules/cm s
Vitreous silica (quartz)	0	7,409	4,539	[57]	molecules/cm s
Pyrex (borosilicate) (D_2)	0	10.7	4,529	[58]	mol/cm s
Pyrex (borosilicate)	1	2.8	4,026	[60]	molecules/cm s atm
Pyrex (borosilicate) (D_2)	0	23.1	4,199	[61]	mol/cm s

[a] cc(STP)/s cm² area/mm thickness/cm Hg.

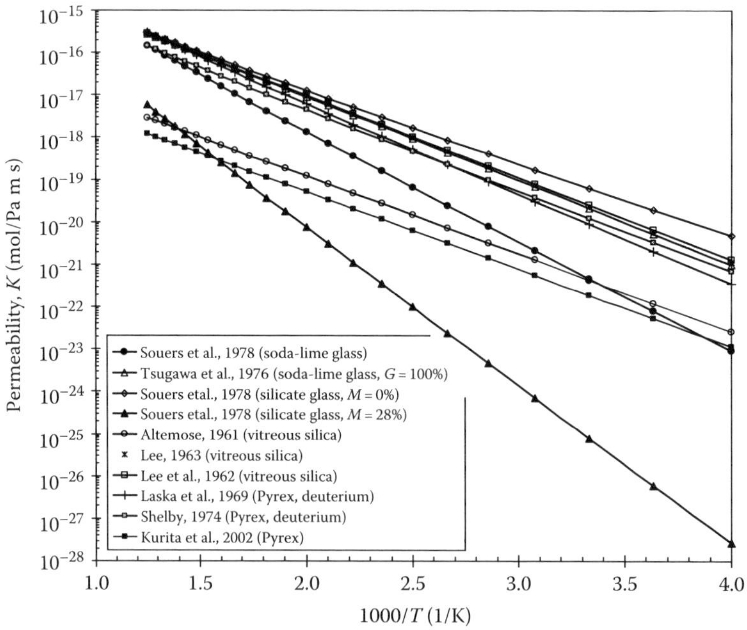

FIGURE 25.8
Hydrogen permeation through soda-lime, silicate, vitreous silica, and Pyrex as a function of 1000/T as reported in the literature and summarized in Table 25.3.

25.2.7.3 Hydrogen Diffusion Coefficient and Solubility in Glass

Diffusion coefficient $D(T)$ (in m^2/s) and solubility $S(T,P)$ (in mol/m^3 Pa) of hydrogen have also been reported in the literature for Pyrex [58,60], quartz [56–58], and silicate glass [59]. They are related to the permeability through [40]

$$K(T) = D(T)S(T) \tag{25.13}$$

They can be assumed to be independent of pressure below 1000 bar [47]. To be consistent with Equations 25.11 and 25.13, diffusion coefficient and solubility are expressed as

$$D(T) = D_0 T^n \exp\left(\frac{-Q_D}{T}\right) \quad \text{in m}^2/\text{s} \tag{25.14}$$

$$S(T,P) = \frac{K(T,P)}{D(T)} = S_0 \exp\left(\frac{Q_S}{T}\right) \quad \text{in mol/Pa m}^3 \tag{25.15}$$

where D_0, Q_D, S_0, and Q_S are empirical constants such that $K_0 = D_0 S_0$ and $Q_K = Q_D - Q_S$. Figure 25.9 shows the diffusion coefficient of vitreous silica and borosilicate Pyrex as a function of temperature as reported in the literature [40,57,61] and summarized in Table 25.4.

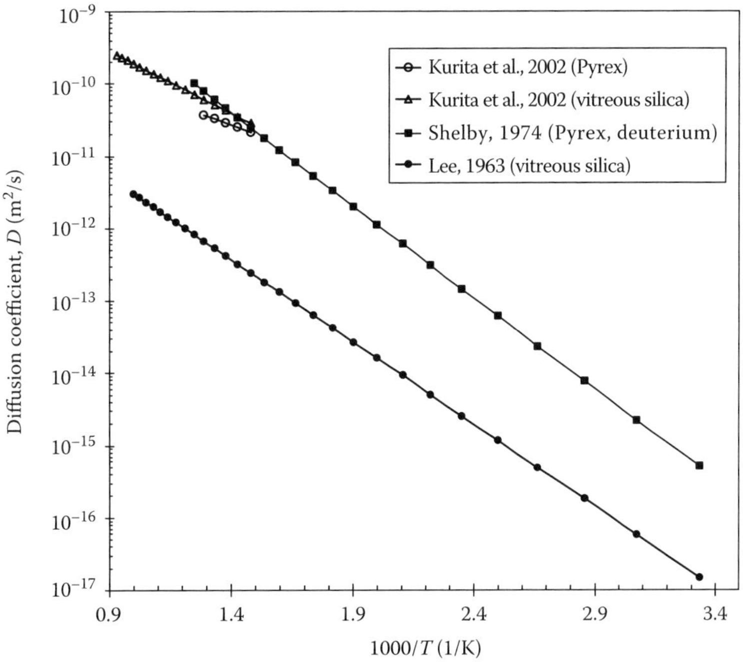

FIGURE 25.9
Hydrogen diffusion coefficient through vitreous silica and borosilicate Pyrex as a function of $1000/T$ as reported in the literature [40,57,61] and summarized in Table 25.3.

TABLE 25.4

Empirical Constants for the Permeability of Various Glasses Used in Equation 25.14

Composition	n	$D_0 \times 10^{10}$ (m²/s)	Q_D (K)	References	Gas
Vitreous silica	0	5.65	5237	[57]	H_2
Pyrex 7740	1	1.06	5385	[60]	D_2
Borosilicate Pyrex	0	14.0	2820	[61]	H_2
Vitreous silica	0	96.0	3928	[61]	H_2

Recently, Shelby et al. [62–65] presented experimental study of what they called *photo-induced hydrogen outgassing* from borosilicate glass doped with various metal oxides especially Fe_3O_4. The authors observed that [62–65] (1) hydrogen release from a slab of doped borosilicate glass placed in a vitreous silica tube was accelerated when exposed to an incandescent heating lamp compared with heating in a furnace at 400°C, (2) the onset of outgassing was observed immediately with lamp heating but was slower for furnace heating, (3) increasing the lamp intensity accelerated the H_2 release rate and the overall H_2 released from the sample, (4) borosilicate glass CGW 7070 demonstrated the best H_2 release response, and (5) increasing the Fe_3O_4 doping level increases the H_2 release rate. The authors suggested that "infrared radiation is contributing the activation energy necessary for hydrogen diffusion." To the best of our knowledge, this would constitute a new physical phenomenon [66]. It remains unclear, however, how the total irradiance and the spectral nature of radiation would be accounted for in an Arrhenius type of relation for the diffusion coefficient or permeability. Moreover, the reported experimental data do not isolate the proposed mechanism from the well-known thermally activated gas diffusion. In fact, Kitamura and Pilon [67] numerically showed that the experimental observations can be qualitatively explained based on conventional thermally activated gas diffusion and by carefully accounting for the participation of the silica tube to radiation transfer along with the spectral properties of the silica tube and the glass samples. In brief, the radiation emitted by the incandescent lamp has a peak emission between 1 and 2 μm and reaches directly the sample since the silica tube is nearly transparent up to 3.5 μm. On the contrary, for furnace heating at 400°C, the silica tube absorbs a large fraction of the incident radiation, which reduces the heating rate and the H_2 release rate. However, between 0.8 and 3.2 μm, undoped borosilicate does not absorb significantly. Coincidentally, Fe_3O_4 doping increases the absorption coefficient and also reacts with H_2 to form ferrous ions, which increase the absorption coefficient of the sample by two orders of magnitude. Thus, doped and reacted samples heat up much faster when exposed to the heating lamp resulting in the observed faster response time and larger H_2 release rate.

25.2.7.4 Optical Properties

In the case when heating of microcapsules is achieved by infrared lamp heating, the optical properties of the glass shell are essential to predict the heating rate and the temperature on which hydrogen permeability depends. Optical properties of fused quartz (vitreous silica) from ultraviolet to infrared at room temperature have been reviewed by

Kitamura et al. [68]. Those of soda-lime silicate glass can be found in Ref. [69] while De Sousa Meneses et al. [70] reported the optical properties of borosilicate glass at room temperature over the spectral range beyond 4.0 μm.

As a first-order approximation, one can assume that glass optical properties do not change significantly with temperature (see Ref. [69], Fig. 2). However, they may change with ions doping and due to reaction with hydrogen. For example, in soda-lime silicate glass, Johnston and Chelko [51] established that reduction of Fe^{3+} by H_2 into Fe^{2+} results in significant increase in the absorptance of the glass sample in the spectral range from 0.4 to 2.5 μm. The changes were apparent with the unaided eye. Similarly, Shelby and Vitko [71] observed (1) an increase in absorptance beyond 0.8 μm and (2) a reduction in absorptance between 0.4 and 0.8 μm for soda-lime silicate. Rapp [65] confirmed Shelby and Vitko's results for Fe_3O_4-doped borosilicate glass showing an increase in the Fe^{2+}/Fe^{3+} ratio as the duration of exposure to hydrogen gas increases [64]. This was attributed to the fact that the absorption band around 380 nm corresponds to the ferric state (Fe^{3+}) while a peak around 1.1 μm corresponds to the ferrous state Fe^{2+}. In addition, the formation of OH groups results in a strong absorption band at wavelengths around 2.73–2.85, 3.5, and 4.5 μm [64,72,73]. The refractive and absorption indices of vitreous silica and borosilicate glass used in this study are presented in Figure 25.10.

25.3 Performance Assessment

Several parameters are useful in assessing and comparing the performances of hydrogen storage solutions including hollow glass microspheres. Parameters of particular interest are (1) gravimetric and volumetric energy densities, (2) loading and unloading times, as well as (3) the energy required to store the hydrogen gas.

25.3.1 Gravimetric and Volumetric Energy Densities

The gravimetric energy density η_g of a bed of monodisperse hollow microspheres is expressed in MJ/kg and given by

$$\eta_g = \frac{N_T m_{H_2} \Delta H_L}{N_T(m_s + m_{H_2})} = \frac{\rho r_i^3 \Delta H_L}{\rho r_i^3 + \rho_s\left(r_o^3 - r_i^3\right)} \tag{25.16}$$

where

N_T is the total number of microspheres in the container

m_s and m_{H_2} are the mass of the solid shell and of the entrapped H_2 in a single microsphere

The densities of H_2 and of the glass shell are denoted by ρ and ρ_s. The lower heating value (LHV) denoted by ΔH_L represents the amount of heat released from combusting a unit mass of H_2 at 25°C and returning the combusting products (H_2O) to 150°C. It is equal to 120 MJ/kg, which exceeds that of all conventional fuels including gasoline (42 MJ/kg), ethanol (27 MJ/kg), natural gas (47 MJ/kg), and coal (23 MJ/kg) [8]. Note that the mass of the container is not considered here since it can be a lightweight plastic whose mass is negligible compared with that of the microspheres.

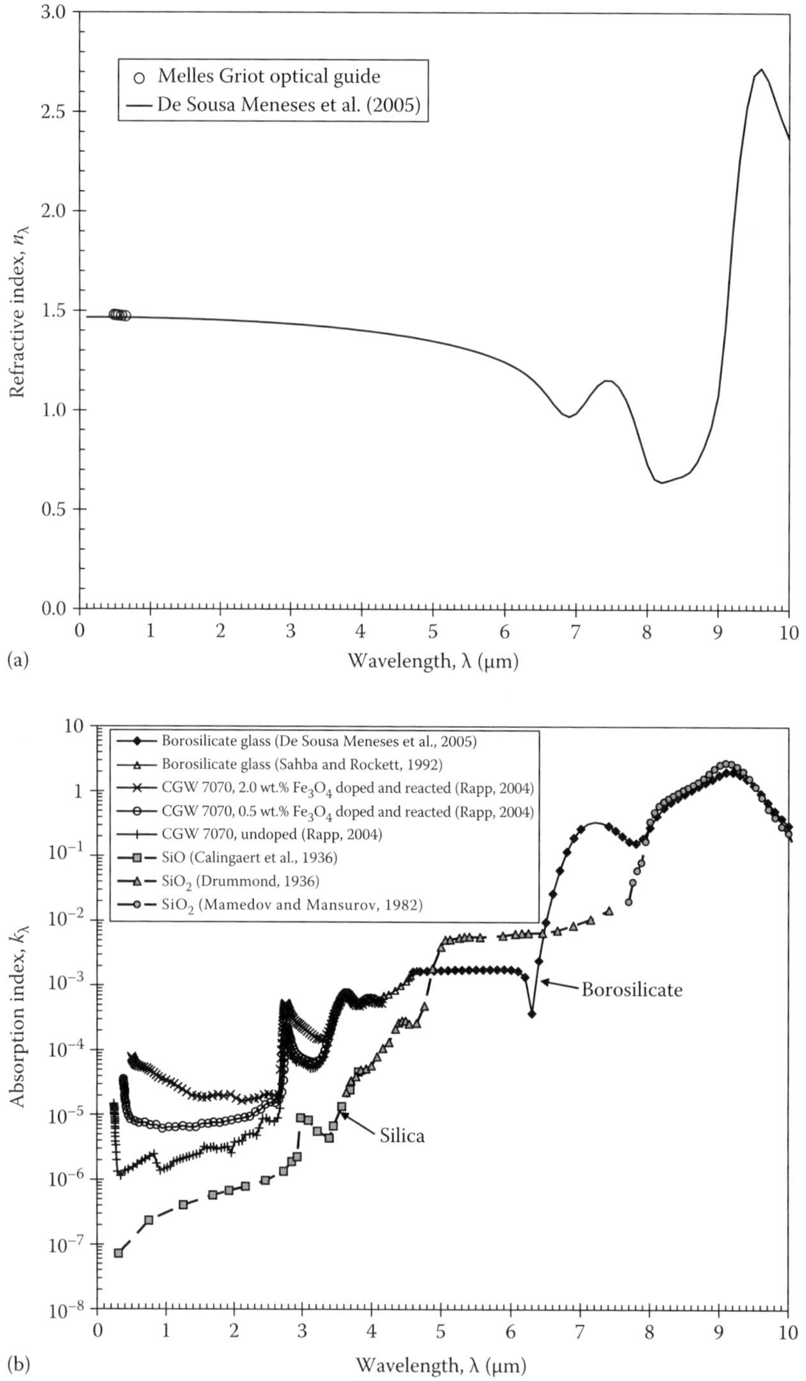

FIGURE 25.10
The (a) refractive and (b) absorption indices of undoped and Fe_3O_4 borosilicate glass obtained or retrieved from various sources. (From Rapp, D.B., Photo-induced hydrogen outgassing of glass, PhD thesis, Alfred University, Alfred, NY, 2004; De Sousa Meneses, D. et al., *J. Non-Crystal. Solids*, 351, 124, 2005; Sahba, N. and Rockett, T.J., *J. Am. Ceramic Soc.*, 75(1), 209, 1992.)

Similarly, the volumetric energy density η_v for monodisperse spheres is expressed in MJ/m^3 and given by

$$\eta_v = \frac{N_T m_{H_2} \Delta H_L}{N_T V_{total}} = \rho\epsilon\Delta H_L \left(\frac{r_i}{r_o}\right)^3 \tag{25.17}$$

where
V_{total} is the total volume of the storage solution
ϵ is the packing fraction of the hollow glass microspheres

Alternatively, the storage capacity and the hydrogen relative weight content have been used extensively to assess the performance of storage solutions [2,11]. The relative weight percent of hydrogen in the microsphere, expressed in kg of H_2 per kg of microspheres or wt.%, is defined as

$$\gamma_g = \frac{m_{H_2}}{m_s} = \frac{\rho r_i^3}{\rho_s(r_o^3 - r_i^3)} \tag{25.18}$$

Making use of the equality $(r_o^3 - r_i^3) = (r_o - r_i)(r_o^2 + r_i r_o + r_i^2) \approx 3r_i^2(r_o - r_i)$, γ_g simplifies to [2]

$$\gamma_g \approx \frac{\rho r_i}{3\rho_s(r_o - r_i)} \tag{25.19}$$

Based on Equation 25.19, γ_g can be expressed in terms of burst pressure and tensile strength of the shell materials:

$$\gamma_g = \frac{2\rho\sigma_{s,max}}{3\rho_s S_f P_{max}} \tag{25.20}$$

On the other hand, the effective density of hydrogen stored in a bed of monodisperse hollow microspheres, expressed in kg of H_2/m^3 of bed, is given by [35]

$$\gamma_v = \frac{\eta_v}{\Delta H_L} = \rho\epsilon\left(\frac{r_i^3}{r_o^3}\right) \tag{25.21}$$

Note that because the shell thickness is relatively thin, $r_i^3 \approx r_o^3$ and γ_v depend essentially on the internal hydrogen pressure.

Figure 25.11 shows the gravimetric and volumetric energy densities of hydrogen stored at 300 K in randomly packed $(\epsilon = 0.63)$ monodisperse hollow glass microspheres 50 μm in diameter with shell thickness of 1 μm. The glass tensile strength $\sigma_{s,max}$ was taken as 1 GPa. It indicates that even for a conservative value of $\sigma_{s,max}$, hollow glass microspheres can achieve the 2015 DOE target for gravimetric energy density. Unfortunately, the volumetric energy density falls short of the 2015 DOE target by a wide margin unless the internal pressure greatly exceeds 1000 bar, which would require a large amount of energy for compressing H_2.

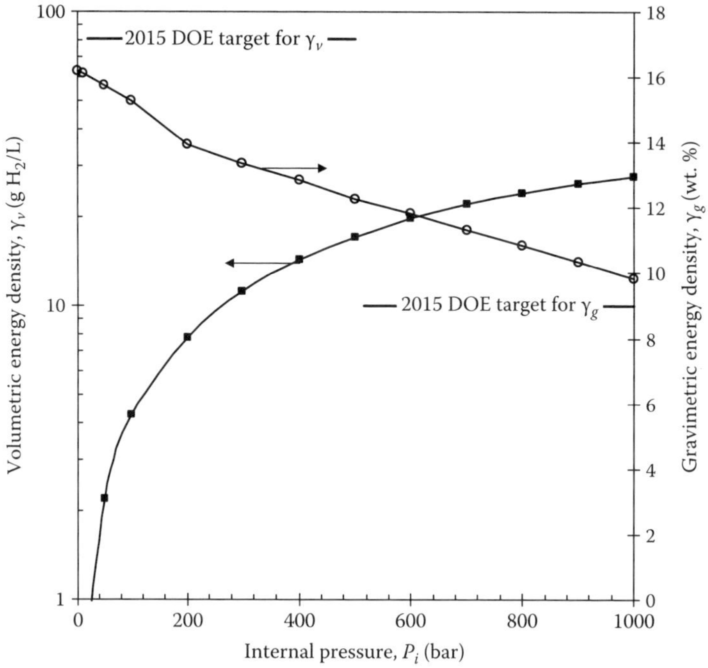

FIGURE 25.11
Gravimetric and volumetric energy densities of hydrogen stored in monodisperse hollow glass microspheres with $r_o = 25$ μm, $r_o - r_i = 1$ μm, $S_f = 1.5$, $\sigma_{s,max} = 1$ GPa, $\varepsilon = 0.63$, and $\rho_s = 2230$ kg/m^3.

Finally, if the storage bed consists of monodisperse microcylinders with inner and outer diameters r_o and r_i, the gravimetric and volumetric energy densities are expressed as

$$\gamma_g = \frac{\rho r_i^2}{\rho_s(r_o^2 - r_i^2)} \quad \text{and} \quad \gamma_v = \rho\epsilon\left(\frac{r_i^2}{r_o^2}\right) \tag{25.22}$$

25.3.2 Loading and Unloading Times

Combining Equation 25.6 with equation of state (25.2) yields the following governing equation for the inner H_2 pressure:

$$\frac{d}{dt}\left(\frac{P_i}{Z(T,P_i)T}\right) = -\frac{3RK(T)}{r_o(r_o - r_i)}(P_o - P_i) \tag{25.23}$$

Then, assuming that the temperature T and the outer pressure P_o remain constant during the loading or unloading processes while the compressibility factor $Z(T,P_i)$ is approximately constant for pressure varying between $P_i(0)$ and $P_i(t)$ yields expression for the pressure inside the microcontainers:

$$P_i(t) = P_i(0) \pm [P_i(0) - P_o]exp\left(-\frac{t}{\tau}\right) \tag{25.24}$$

where the time constant τ is expressed as

$$\tau = \frac{r_i(r_0 - r_i)}{3Z(T,P_i)RTK(T)} \quad \text{for hollow spheres} \tag{25.25}$$

The time needed for the pressure difference ($P_i - P_o$) to decrease by 63% and 95% from its initial value is equal to τ and 3τ, respectively. The presence of the compressibility factor Z in the denominator suggests that H_2 release is faster for large temperatures and pressures when Z is greater than 1 (see Figure 25.5). Treating hydrogen as an ideal gas at temperature T and pressure P_i gives $Z(T,P_i) = 1.0$ and $\tau = r_o(r_o - r_i)/3RTK(T)$ as often encountered in the literature [11,29,53]. Similarly, Equation 25.24 is valid for cylindrical containers with a time constant given by

$$\tau = \frac{r_i(r_o - r_i)}{2Z(T,P_i)RTK(T)} \quad \text{for cylinders} \tag{25.26}$$

Note that τ is independent of the cylinder length L [4]. In addition, assuming that the container wall is thin, the time constant for a sphere is 33% smaller than that for a cylinder with the same inner and outer diameters. In other words, loading and unloading times are shorter for spheres than for cylinders. The maximum shelf time of loaded microcapsules can be estimated from Equations 25.25 and 25.26 using $T = 300$ K. If the temperature of the microcontainer is not constant during the loading or unloading processes, the energy conservation equation must also be solved giving the temperature distribution within the container shell or within the packed bed. Heat transfer in packed beds has been studied extensively [74–85]. Models accounted for combined heat conduction through the wall of the hollow microspheres as well as for absorption, scattering, and emission of radiation models found in the literature can be adapted to predict the temperature inside a bed of hollow glass microspheres heated by a resistive heater or an incandescent lamp.

Figure 25.12 shows the evolution of the time constant τ as a function of temperature predicted by Equation 25.25 using permeation $K(T)$ reported by Lee et al. [56] for hollow vitreous silica microspheres 50 μm in diameter with shell thickness equal to 1 μm, that is, $r_o = 25$ μm and $r_o - r_i = 1$ μm. It also shows the time constant τ if the gas is assumed to be ideal, that is, $Z(T,P) = 1$. Note that for pressure less than 100 bar (1450 psi), the loading and unloading time is independent of pressure as observed experimentally with soda-lime glass [29]. However, when the initial hydrogen pressure is large ($Z(T,P) > 1$), the time constant depends on pressure and is shorter than that predict assuming ideal gas behavior.

Figure 25.13 shows the loading and unloading characteristic time τ as a function of the microsphere radius r_o at temperatures corresponding to (1) storage and handling temperature (25°C) and (2) the loading and unloading temperatures of 200°C and 400°C assuming that the shell is 1 μm thick and made of vitreous silica. It establishes that the loading and unloading time decreases significantly with decreasing microsphere radius and increasing temperature.

As previously mentioned, permeation is associated with the steady-state transport of hydrogen through the glass shell. Thus, it does not account for the time necessary for H_2 to diffuse through the shell of the microspheres before reaching a steady-state mass flux. This lag time denoted by τ_l corresponds to the time needed to establish the glass solubility in the microspheres wall. As a first-order approximation, τ_l can be expressed as [29]

$$\tau_l = \frac{(r_o - r_i)^2}{6D(T)} \tag{25.27}$$

where the lag time depends only on the shell thickness. It is typically much smaller than the unloading or loading time at low temperatures (i.e., $\tau_l \ll \tau$). For example, for hollow

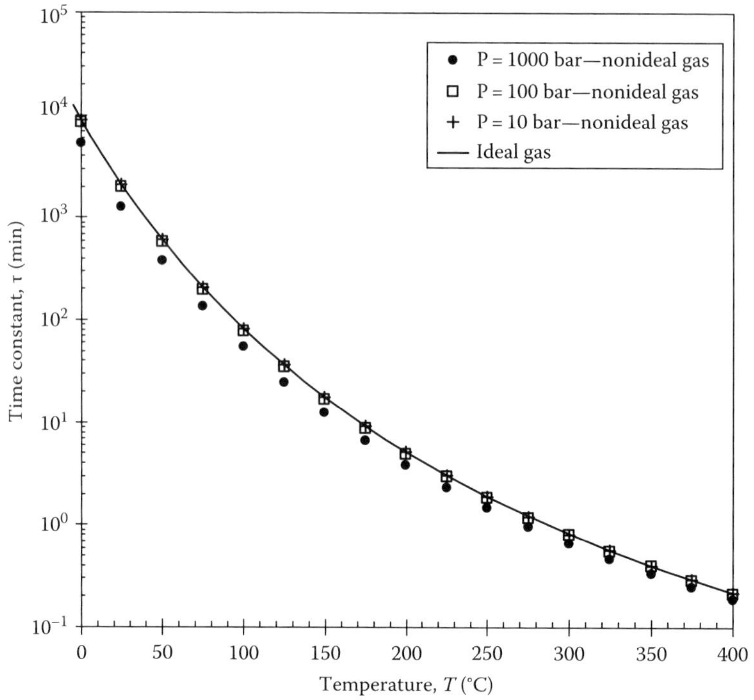

FIGURE 25.12
Loading and unloading time constant τ predicted by Equation 25.25 for vitreous silica glass microspheres with r_o = 25 μm, $(r_o - r_i)$ = 1 μm, $Z(T,P)$ = 1.0, $K_0 = 8.688 \times 10^{-14}$ mol/Pa m s, and Q_K = 4469 K. Time constant for microcylinders can be found by multiplying the results shown by 3/2. (From Lee, R.W. et al., *J. Chem. Phys.*, 36(4), 1062, 1962.)

vitreous silica microspheres with r_o = 25 μm and $(r_o - r_i)$ = 1 μm, the time constant τ is 0.9 days, 3.8 min, and 11 s compared with the corresponding time lag τ_l of 6.2 min, 0.3 s, and 7 ms at temperature of 25°C, 200°C, and 400°C, respectively. Nonetheless, this suggests that 95% of the hollow glass microspheres content is released in 33 s at 400°C and in 2.7 days at 25°C.

25.3.3 Filling and Discharging Energy Requirements

Another key element is assessing that the performance of hydrogen storage solution is (1) the energy required to compress the hydrogen gas to pressures in excess of 450 MPa during loading and (2) the thermal energy provided to heat up the microspheres and their content around 400°C to achieve high permeation rates.

Because of repulsion forces between H_2 molecules, hydrogen gas deviates significantly from an ideal gas under loading conditions and requires large compression work. Energy-efficient and cost-effective hydrogen compressors operating at high pressure are critical components of the envisioned hydrogen economy [86]. Hydrogen gas has also small molecules and low viscosity, which renders sealing of the compressors challenging. In addition, at high temperatures and pressures, hydrogen permeates through steel and causes embrittlement resulting in material failure, high maintenance costs, and safety concerns.

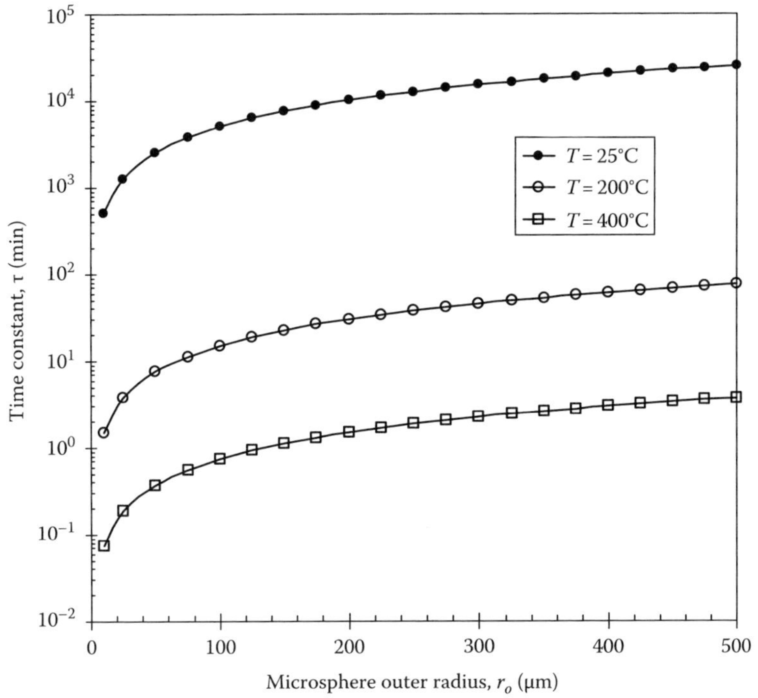

FIGURE 25.13
Loading and unloading time constant τ predicted by Equation 25.25 for vitreous silica glass microspheres as a function of r_o assuming $(r_o - r_i) = 1$ μm, $Z(T,P) = 1.0$, $K_0 = 8.688 \times 10^{-14}$ mol/Pa m s, and $Q_K = 4469$ K. (From Lee, R.W. et al., *J. Chem. Phys.*, 36(4), 1062, 1962.)

Material embrittlement can be addressed by using more expensive material alloys such as chromium/molybdenum/vanadium alloys.

The total specific energy E_t required per loading and unloading cycle of 1 kg of microspheres with H_2 is the sum of (1) the specific work done for pressurizing the hydrogen during loading denoted by W_p and (2) the specific thermal energy for heating both the microspheres and the hydrogen to desired temperatures denoted by Q_u, that is,

$$E_t = W_p + 2Q_u \tag{25.28}$$

where the factor 2 accounts for thermal energy required for both loading and unloading.

The compression work required for pressurizing H_2 depends on the thermodynamic process, the number of stages, the compressor efficiency β, and the initial and final hydrogen pressures denoted by P_1 and P_2, respectively. For isothermal compression and assuming ideal gas behavior, the specific compression work W_p (in J/kg) is given by [86]

$$\beta W_p = -\int_{\rho_1}^{\rho_2} \frac{P(T,\rho)}{\rho^2} d\rho = ZRTln\left(\frac{\rho_1}{\rho_2}\right) \tag{25.29}$$

For isentropic compression, the specific compression work is expressed as [86]

$$\beta W_p = \frac{k}{(k-1)}\frac{P_1}{\rho_1}\left[\left(\frac{P_2}{P_1}\right)^{(k-1)/k} - 1\right] \tag{25.30}$$

where
k is the specific heat ratio ($k = c_p/c_v$)
ρ_1 is the density at T_1 and P_2

Assuming hydrogen is compressed from standard temperature and pressure $\rho_1 = 0.09$ kg/m^3 and $k = 1.41$. Multistage compressors with cooling between stages operate between these two limiting cases albeit closer to isothermal [86]. Figure 25.14 shows the work W_p required for adiabatic and isothermal hydrogen compressions as a fraction of the hydrogen LHV as a function of final pressure P_2. For example, multistage compression from atmospheric pressure to 200 and 800 bar requires 11.9% and 19.1% of the LHV of hydrogen, respectively. It can reach 20% if one accounts for mechanical and electrical losses. These results are in good agreement with analysis by Ramback et al. [26,27,32]. The authors estimated the specific compression work for temperatures up to 400°C and pressure less than 620 bar to range between 10% and 20% of the LHV of H_2 assuming a three-stage compressor with 75% efficiency (Fig. 7 in Ref. [32]). However, Ramback [32] noted that it is less than the energy needed to liquify hydrogen, and therefore, hydrogen storage in hollow glass

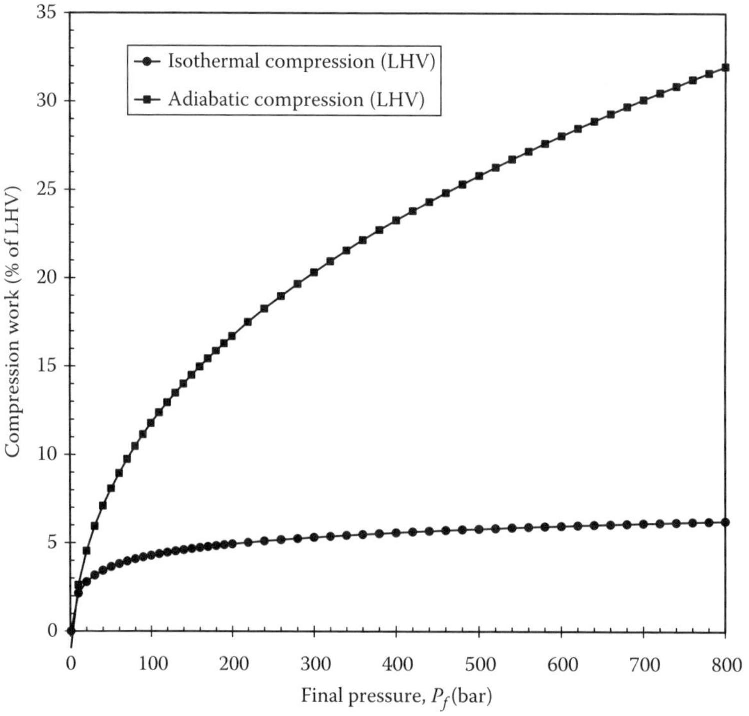

FIGURE 25.14
Energy required for adiabatic and isothermal compressions as a fraction of the hydrogen LHV.

TABLE 25.5

Expression of Design Parameters for Thin-Walled Monodisperse Hollow Microspheres and Microcylinders

Parameters	Symbol	Microspheres	Microcylinders
Burst pressure	P_{max}	$P_{max} = \dfrac{2\sigma_{s,max}(r_o - r_i)}{S_f r_0}$	$P_{max} = \dfrac{\sigma_{s,max}(r_o - r_i)}{S_f r_0}$
Buckling pressure	P_{cr}	$P_{cr} = \dfrac{2E_s(r_o - r_i)}{r_o\sqrt{3(1-v_s^2)}}$	$P_{cr} = \dfrac{E_s(r_o - r_i)}{r_o\sqrt{3(1-v_s^2)}}$
Maximum packing	ϵ	74%	90.7%
Random packing	ϵ	63%	82%–83%
Loading time constant	T	$\tau = \dfrac{r_i(r_o - r_i)}{3Z(T,P_i)RTK(T)}$	$\tau = \dfrac{r_i(r_o - r_i)}{2Z(T,P_i)RTK(T)}$
Gravimetric density	γ_g	$\gamma_g \approx \dfrac{\rho r_o^3}{\rho_s(r_o^3 - r_i^3)}$	$\gamma_g = \dfrac{\rho r_i^2}{\rho_s(r_o^2 - r_i^2)}$
Volumetric density	γ_v	$\gamma_v = \rho\epsilon r_i^3/r_o^3$	$\gamma_v = \rho\epsilon r_i^2/r_o^2$

microsphere can be economically competitive. On the contrary, Robinson and Iannicci [87] concluded that compressing H_2 at high pressures in excess of 1000 bar appears unpractical from an energy consideration point of view despite the fact that hollow glass microspheres could sustain such pressures [11]. Similar conclusions were reached by Bossel [86] and was identified as one of the reasons why the hydrogen economy *will never make sense.*

The specific thermal energy required to heat up the microspheres and the hydrogen from T_1 to T_2 can be expressed as

$$Q_u\,(\mathrm{J/kg}) = \int_{T_1}^{T_2} c_{eff}(T)\,dT \tag{25.31}$$

where c_{eff} is the effective specific heat of the microcontainer bed defined as $c_{eff} = \epsilon[c_p r_i^3/r_o^3 - c_{s,p}(1 - r_i^3/r_o^3)]$. As a first-order approximation, one can assume c_{eff} to be constant so Q_u is proportional to the temperature difference $(T_1 - T_2)$. Ramback [27] showed that it represents between 2% and 6% for monodisperse soda-lime silicate hollow microspheres 50 μm in diameter with a 0.9 μm thick shell for $\epsilon = 0.63$ and $(T_1 - T_2)$ between 100°C and 400°C and pressures between 248 and 620 bar.

Table 25.5 summarizes expressions for design parameters and assessing performances of thin-walled monodisperse hollow microspheres and microcylinders.

25.4 Experiments

Only a few experimental studies focusing on H_2 storage in glass microcontainers have been performed over the last three decades and are summarized in Table 25.6. Teitel [9,10] performed crush tests and loading tests with different commercial hollow microspheres made of soda-lime silica glass with outer radius ranging from 14 to 55 μm and wall thickness from 0.4 to 2.2 μm. They showed that the survival rate varied from 30% to 90% depending

TABLE 25.6

Summary of Experimental Studies on Hydrogen Storage in Microcapsules

References	Composition	Shape	r_o (μm)	$(r_o - r_i)$ (μm)	Loading P (bar)/T (°C)	Outgassing T(°C)	η_g (wt.%)	η_v (g/L)
[29]	Soda-lime	Microspheres	40–45	1	104	100–300	2	
[18]	Soda-lime	Microspheres	14–55	0.4–2.2	551/350	N/A	5.3	12
[20]	Fused quartz	Microcylinders	160–260	16–35	N/A	N/A	2	N/A
[35]	S60/1000 (3M)	Microspheres	5–15	200	350	4	14	
[27]		Microspheres	25.5	0.9	581/350			
[34]	Soda-lime	Microcylinders	100–500	2–10	100–250	200–250	15–17	10–15

2015 DOE targets are η_g = 9 wt.% and η_v = 81 g H_2/L.

on the glass composition, the microsphere radii, and the loading gas. The authors suggested an aspect ratio $r_0/(r_0 - r_i)$ of at least 30. Loading tests were performed at up to 551 bar (8000 psi) at 350°C. The author reported a gravimetric energy density of 4%–5% and a volumetric energy density of 20 g of H_2/L. Pressure would need to be significantly increased to reach the 2015 DOE target of 9 wt.% and 81 g of H_2/L. Microspheres breakage rate during the loading and unloading cycle was about 15%.

Tsugawa et al. [29] loaded soda-lime silicate microspheres 40–45 μm in outer diameter with a 1 ± 0.1 μm thick shell. The microspheres were placed in a highly impermeable beryllium container and heated at 763 K. Then, hydrogen was introduced and pressurized for 24 h until pressure reached 103 atm. The microspheres were quenched for 1 min in liquid nitrogen. The gravimetric energy density was reported as 2 wt.%. Finally, outgassing was performed at constant temperature ranging from 373 to 573 K. The authors also established experimentally that soda-lime glass microspheres pressurized with H_2 fill to only 82 vol.% instead of 100% with helium and 2% of H_2 remained inside the microspheres after outgassing.

Duret and Saudin [35] performed crush and burst tests on commercial hollow glass microspheres, provided from 3M and Saint Gobain, with outer diameter between 2 and 80 μm and a mean diameter between 12 and 30 μm. Hollow glass microsphere effective density was between 0.57 and 1.5 g/cm^3. The crush test was performed in water at 1000 bar for a few hours followed by a float test. Survival rate as low as 10% was recorded. Unbroken microspheres underwent a burst test at 200 bar and 400°C. 3M microspheres could sustain the crush test and achieved gravimetric and volumetric energy density of 4 wt.% and 14 g/L, respectively.

Recent studies have explored the use of hydrogen storage in microcylinders. First, Yan et al. [20] drew fused quartz hollow fibers (or microcylinders) 15 cm in length and with outer diameter ranging from 160 to 260 μm and wall thickness between 16 and 35 μm. They also subjected the fibers to chemical strengthening through hydrofluoric acid (HF) etching [88]. In addition, fiber coating with Shellac was performed. This reduced significantly the frequency of breakage and protected the fibers' surfaces from the effects of aging and handling. Different end sealing techniques were also tested including (1) heat treating the tip of fiber bundles in a furnace, (2) individual fiber sealing with an open flame, and (3) end plugging the fibers with colloidal silica. Flame sealing method provided the best sealing. The authors also performed hot static and dynamic fatigue tests and burst tests as well as hydrogen loading and unloading tests. They established that liquid-phase HF etching significantly increases the tensile strength of the microcylinders and enables further control of the wall thickness. Finally, a gravimetric energy density of 2 wt.% was reported.

Kohli et al. [34] loaded soda-lime silicate glass hollow microspheres with aspect ratio $r_o/(r_o - r_i)$ ranging from 80 to 180 at outside pressure up to 250 bar and temperature of 200°C–250°C. They estimated the gravimetric energy density at 15–17 wt.% and volumetric energy density between 10 and 15 g/L (Fig. 4 in Ref. [34]).

Based on the previous discussion, it is evident that hollow glass microspheres are the microcapsules offering the best mechanical and gas permeation properties, which maximize gravimetric and volumetric energy densities and minimize loading and unloading times. The next section presents manufacturing processes for producing hollow glass microspheres.

25.5 Synthesis of Hollow Glass Microspheres

Extensive efforts have been devoted to the fabrication of hollow glass microspheres for applications ranging from thermal insulation and fire retardant to lightweight composite materials for flotation and reinforced materials. They have also been used for viscosity modification, shrinkage reduction, and chemical resistance enhancement of plastics or paints, as well as for radiation shield [89]. Hollow microspheres made of materials other than glass have also been used to encapsulate drugs for controlled drug delivery or colorant or artificial flavor in food science. Thus, there exist numerous synthesis methods with different levels of control in the size distribution, shell thickness, material composition, mechanical properties, and throughput. The different synthesis processes have been reviewed extensively elsewhere [90–93] and this section is meant to give a relatively brief overview. In addition, numerous patents have been claimed for manufacturing, functionalizing, or strengthening hollow glass microspheres [94–110].

25.5.1 Hollow Microspheres in Fly Ash

Hollow glass microspheres have been observed in fly ash emanating from coal-fired power plants [90,111]. Indeed, coal particles injected in the furnace contain silica, sulfates, and other inorganic matter. As coal particles burn from the outside inward, carbon gets oxidized, and the remaining matter forms a molten glassy outer shell. The inner coal continues burning and releases gases such as carbon dioxide and sulfur oxides resulting from the oxidation of carbon and sulfates at high enough temperatures. These gases are entrapped in the molten glassy shell and blow it to produce a hollow microsphere, which solidifies as it exits the furnace. However, the hollow microspheres formed in this process (1) may not be perfectly spherical, (2) feature a wide size distribution (from a few to hundreds of microns), and (3) have nonuniform wall thickness [90]. Moreover, impurities and particles are often present in the shell along with tiny bubbles [90]. The nature of the process offers very little control except for the size of the initial coal particles [111].

25.5.2 Spray Pyrolysis Process

Several processes aimed to emulate hollow microspheres formation in coal burning but in a controlled manner have been developed over the last 50 years. The most widely used processes can be reduced to two consecutive stages: (1) fabrication as semiproducts in the form of coarse or fine irregular particles or spheres and (2) blowing (or molding) of the semiproducts into hollow glass microspheres [92]. The main differences in the fabrication of hollow microspheres lie in the fabrication of the micropowder, which determines the dimensions and properties of the hollow glass microspheres. Formation of the semiproduct can include (1) glass frit, (2) liquid-droplet process, (3) solgel process, and (4) rotating and electric arc methods. The second stage consists of spray pyrolysis in flame or hot gas or a rotating electric arc plasma. Microspheres with porous walls have also been synthesized using spray pyrolysis [112,113].

25.5.2.1 Blowing Process

Glass frits, microspheres, or micropowders can be transformed in hollow microspheres using flame spray pyrolysis process [114]. It consists of spraying the semiproducts containing a blowing agent (e.g., sulfates) into an oxy-fuel flame at temperature between 1000°C and 1200°C. Upon rapid heating, the frit assumes spherical shape thanks to reduced viscosity and large surface tension. Beyond a certain temperature, the blowing agent undergoes thermal decomposition releasing gases (e.g., SO_2 and O_2), which results in the formation of a cavity that expands over time. Then, the resulting spherical hollow glass microspheres are rapidly cooled from the outside, which provides enough mechanical strength to retain their sphericity. The glass composition as well as the microsphere size and wall thickness can be controlled by changing the size, shape, and composition of the initial glass frit including the blowing agent [101,115,116]. Further control can be achieved through process parameters such as the residence time and the temperature history. Figure 25.15 shows the history of the semiproduct (e.g., glass frit or spherical particles) undergoing flame spray pyrolysis.

Other blowing processes have been proposed such as replacing flames by high-temperature gases. Similarly, Bica [117] formed hollow glass microspheres from glass frit introduced in a rotating electric arc plasma. The author reported that 85% of the glass frit was transformed into hollow microspheres with diameters between 2 and 24 μm and wall thickness between 0.4 and 1.6 μm.

25.5.2.2 Semiproduct Production

Glass frits are the simplest semiproducts used to produce hollow glass microspheres. They are obtained by mechanically crushing glass or by rapidly cooling hot glass, which shatters under thermal stress. Glass beads can also be produced from a stream of molten glass using (1) a horizontal rotating paddle wheel [118] to mechanically break

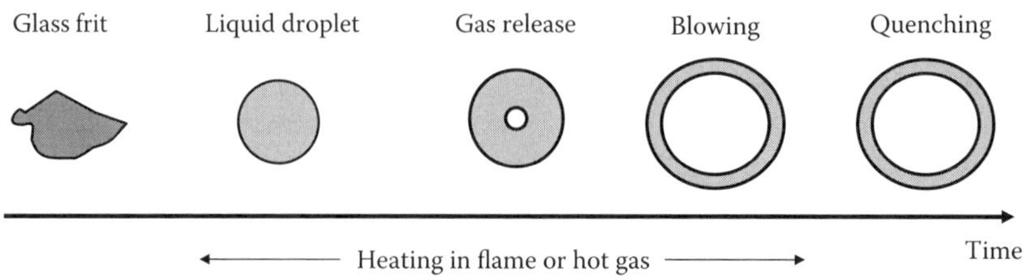

FIGURE 25.15
Schematic of a spray pyrolysis process to produce hollow glass microspheres.

the vertically flowing stream or (2) under electric field in the longitudinal direction of the stream and a temporally variable magnetic field perpendicular to the stream [119]. However, the resulting glass frit has irregular shape and wide mass and size distributions. Consequently, the resulting hollow microspheres produced after the blowing process have also a widespread size distribution and shell thickness. Thus, a more advanced process has been conceived to produce semiproducts with controlled size and composition.

In dried-gel process, a gel having the desired glass composition is formed, dried, and ground into fine particles. Then, spray pyrolysis of the sieved gel powder (as opposed to glass frit) is performed. For example, Downs and Miller [120] claimed a solgel process resulting in a glass shell composed of at least 99% silica that does not suffer from inhomogeneities and phase separation associated with multiple component glass shells. A blowing agent such as urea may also be added to produce large hollow microspheres with thin wall [121]. Figure 25.16 shows micrographs of dried-gel particles and the resulting hollow glass microspheres after

However, dried-gel process tends to produce hollow microspheres with a wide range of diameter and wall thickness as illustrated in Figure 25.16. This drawback has been addressed by pressing the gel powder into mold to form cylindrical pellets of uniform and arbitrary size [122]. Moreover, Schmitt et al. [123] produced iron-doped sodium borosilicate

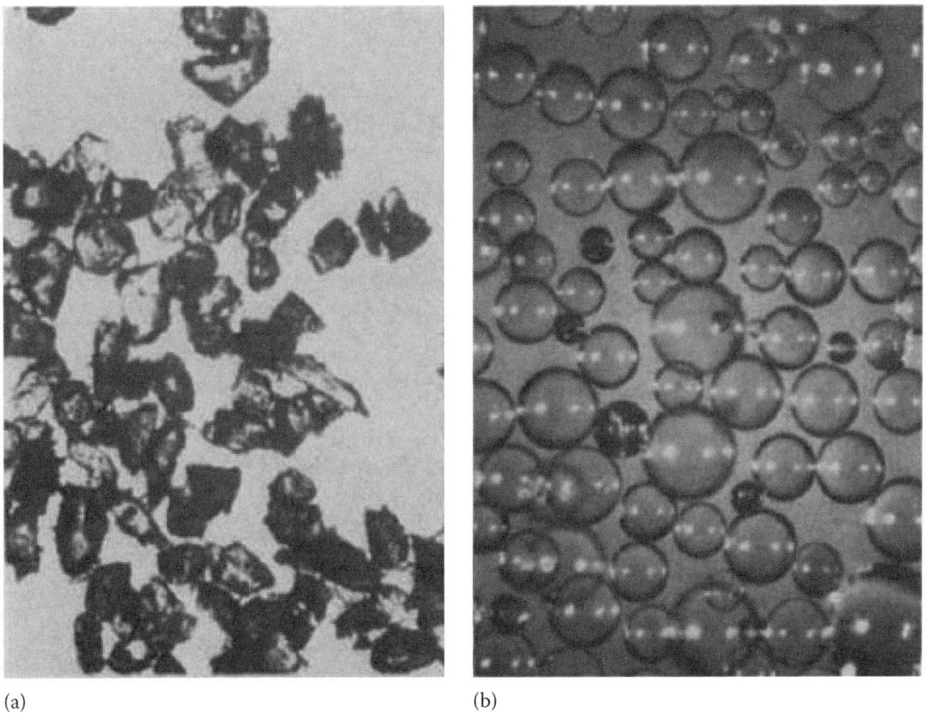

(a) (b)

FIGURE 25.16
Micrographs of (a) dried-gel particles and (b) resulting hollow glass microspheres after spray pyrolysis. (Reproduced from Hendricks, C.D., *J. Nuclear Mater.*, 85–86(1), 107–111, 1979; Rosencwaig, A. et al., Laser fusion hollow glass microspheres by the liquid-droplet method, Lawrence Livermore Laboratory Report UCRL-81421, June 5, 1978.)

granules doped by spray drying an aqueous suspension of crushed xerogel powder with iron chloride or iron sulfate and surfactants. They investigated the effects of heat treatment of the gel, spray drying parameters, and blowing agents.

25.5.3 Liquid-Droplet Method

Liquid-droplet process to produce glass beads to be blown into hollow microspheres was developed for producing laser fusion targets [121,124–126]. Figure 25.17 shows a schematic of the vertical-drop tower along with the different zones and thermal stages undergone by the liquid droplets. First, glass-forming components are dissolved and mixed in an aqueous solution. An in-line stream made of a large quantity of uniform-size droplets is generated by a Rayleigh–Taylor droplet generator [121,124,125] or a vibrating nozzle [121] at the top of a vertical-drop furnace. Air is drawn in the column in a controlled manner from the top to the bottom of the furnace. The generated droplets travel downward through a region of moderate temperature (region 1) to remove the water from the outer surface and to create an elastic gel membrane that encapsulates the rest of the droplet. Then, in region 2, water evaporates from the droplet and diffuses through the membrane resulting in dry solid gel particles. Upon further heating at higher temperatures (region 3), the gel

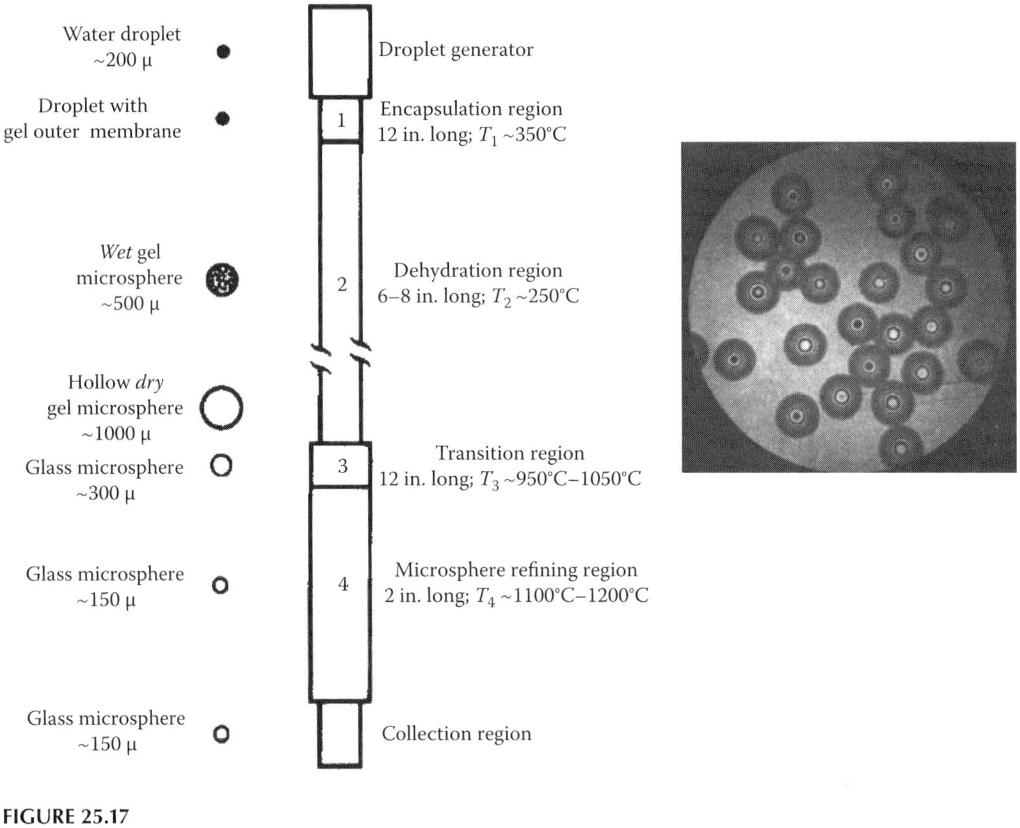

FIGURE 25.17
Schematic of a vertical-drop furnace used in liquid-droplet process and interference micrograph of the produced hollow glass microspheres. The column was 5 m high and made of quartz tube 7.5 cm in diameter. (Reproduced from Hendricks, C.D., *J. Nuclear Mater.*, 85–86(1), 107–111, 1979; Rosencwaig, A. et al., Laser fusion hollow glass microspheres by the liquid-droplet method, Lawrence Livermore Laboratory Report UCRL-81421, June 5, 1978.)

microspheres fuse to glass and collapse to form denser glass microspheres. In region 4, gases are released and diffuse through the shell whose wall becomes uniform thanks to lower viscosity caused by higher temperatures and to the action of surface tension forces. The hollow glass microspheres are then rapidly quenched and collected. This process produces nearly monodisperse microspheres with excellent sphericity, wall uniformity, and concentricity. It also offers control of the microspheres' diameter and shell thickness by controlling the liquid solution composition, the droplet formation process, the temperature profile in the column, and the liquid flow rate [91]. The microsphere size can be increased by increasing the drop tower height [121]. However, it requires that the glass-forming compounds be in solution. Unfortunately, some compositions form a gel so rapidly that droplets cannot be produced. In addition, the produced microspheres are small, spherical, monodisperse, and concentric as illustrated in Figure 25.17 [121]. However, the throughput is relatively small. Dried-gel process addresses the limitations in terms of composition, microsphere size, and throughput but has its own drawbacks as previously discussed [121,125].

25.5.4 Hollow Glass Microspheres by Solgel Process

The synthesis of hollow glass microspheres by solgel process has been described by Ding and Day [127]. The starting aqueous solution is prepared with a precursor solution of metal alkoxide (e.g., tetraethyl orthosilicate [TEOS]) with ethanol and hydrochloric acid (HCl). Partial hydrolyzation (or polymerization) of the metal alkoxide, catalyzed by the acid, takes place over time. The solution is used to produce droplets with a droplet generator such as a vibrating nozzle [127]. The falling droplets are rapidly transformed into solid gel microspheres upon drying at temperatures less than 250°C. Finally, water, ethanol, and HCl are removed from the gel microspheres by thermal treatments at 500°C to 700°C to form hollow glass microspheres. Ding and Day [127] produced microspheres 60–81 μm in diameter. The advantages of this process include uniform sphere size, high chemical purity, low processing temperatures, and controllable microstructures along with the possibility of achieving compositions that cannot be made by conventional melting. Finally, hollow glass microspheres can also be synthesized using colloidal templating of sacrificial polystyrene latex microspheres removed by calcination [128] as well as miniemulsions [129].

25.5.5 Hollow Silica Aerogel Spheres

Hollow silica aerogel spheres have been synthesized by combining a droplet-generation method with solgel processing [130,131]. The process starts with preparation of TEOS, ethanol, and HNO_3 aqueous solution at pH 2. The solution is flown through the outer nozzle of a dual-nozzle hollow droplet generator. The hollow droplets are briefly injected in a gelation chamber where, upon contact with the gelation agent, they transform into rigid alcogel spheres. Two methods were proposed, namely, [130] (1) the levitation method where the droplets fall in an upward stream of NH_3–N_2 gas mixture acting as a gelation agent and (2) the NH_4OH vapor column techniques where the droplets fall through the saturated vapor above liquid NH_4OH. Both approaches are illustrated in Figure 25.18.

They are then collected and aged for 24 h in ethyl alcohol before being dried with supercritical CO_2. The process results in hollow silica aerogel spheres with a hollow core surrounded by a shell of silica aerogel and a thin outer membrane as shown in Figure 25.18. Spheres thus formed were several hundreds of microns in diameters and as large as 2 mm.

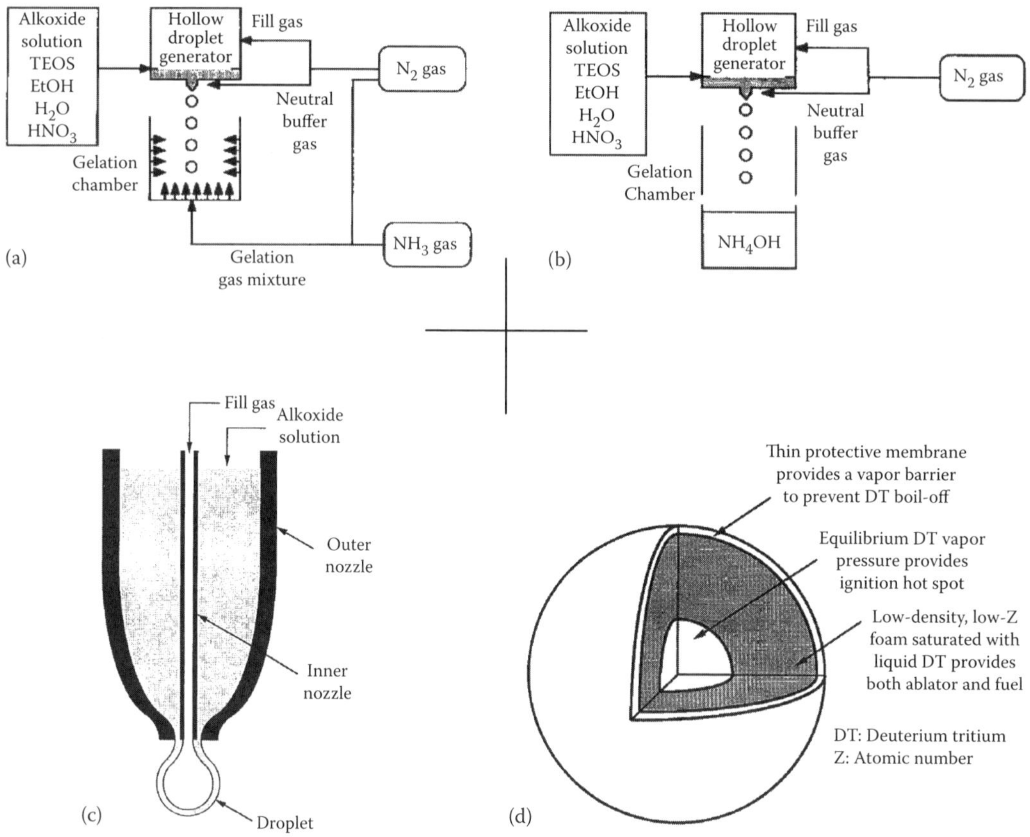

FIGURE 25.18
Illustration of hollow silica aerogel spheres synthesis method by (a) levitation in NH_3–N_2 gas mixture or (b) free fall in NH_4OH vapor column along with (c) dual-nozzle hollow droplet generator and resulting (d) hollow silica aerogel spheres. (Reproduced from Kim, K.K., *J. Am. Ceramic Soc.*, 74(8), 1987, 1991.)

Porosity of the aerogel ranged from 90% to 97% with pore diameters between 9 and 60 nm. The porosity and pore size could be controlled by varying (1) the solution composition (and its viscosity), (2) the gelation time, and (3) the gelation method. Optimization of the process parameters were later investigated [131]. The overall diameter can be controlled through the design of the dual nozzle and the solution viscosity.

25.5.6 Porous Wall Glass Hollow Microspheres

More recently, porous wall glass microspheres made of silica glass containing boron oxide, alkaline earths, and alkali have been synthesized with diameters from 2 to 100 μm and shell thickness from 10 to 300 nm [112]. The first step of the process consists of flame spray pyrolysis to form conventional hollow glass microspheres. The shell is made porous through heat treatment and acid leached [112,113]. The heat treatment induces phase separation between a continuous silica-rich phase interconnected with an alkali- and borate-rich phase. Acid leaching with 3 M HCl at 80°C–85°C dissolves away the alkali–borate phase, thus making the wall porous with pore size between 1 and 100 nm. Figure 25.19 shows a schematic of the porous wall glass hollow microspheres along with an SEM

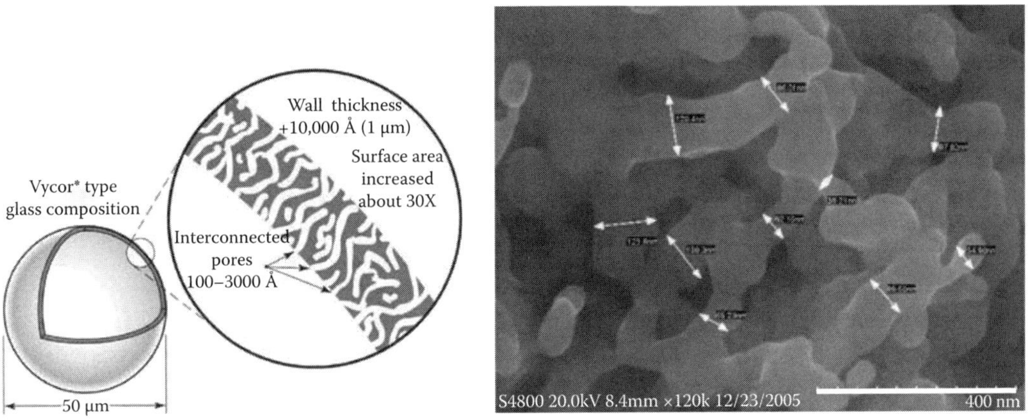

FIGURE 25.19
Schematic and SEM image of porous wall glass hollow microspheres. (Reproduced from Heung, L.K. et al., Hollow porous-wall glass microspheres for hydrogen storage, Patent Application No. US 2006/0059953 (October 21, 2005) and International Application No. PCT/US2006/040525 (October 17, 2006); Wicks, G.G. et al., *ACerS Bull.*, 23–28, June 2008.)

picture of the porous wall. The porosity of the wall can be controlled by selecting different glass compositions and silica/alkali/borate ratios. The hollow microspheres can then be sorted by flotation techniques. Their density ranged from 0.1 to 0.7 g/cm^3. The authors were also able to fill the microspheres with palladium through the porous wall [113]. Thus, the microspheres could potentially be filled with various hydrogen adsorbent materials [113]. The porous wall would enable faster loading and release of the hydrogen.

Finally, alternative processes have been developed to synthesize hollow glass microspheres with porous wall combining (1) colloidal templating of polystyrene latex microspheres, to form a large microcavity with (2) solgel method to form the mesoporous wall after calcination of both the micelles and the microsphere [132–135]. Similarly, a combination of emulsions (e.g., oil/water) and the solgel method has been demonstrated by the self-assembly of silica and surfactants at the oil/water droplet interface [136,137]. In addition to pursuing these synthesis efforts, mechanical and thermal testing of these novel microspheres are needed including cycling under realistic pressures and temperatures expected during hydrogen loading and unloading.

25.5.7 Surface Treatment and Additional Functionality

As discussed in Section 25.2.7.1, tensile strength of the glass shell strongly depends on imperfection or flaws at the shell surface. Such flaws must be removed to strengthen and to improve the durability of the glass shell. Surface treatment by heat polishing [138] and chemical hardening [88] can be used to smoothen the microspheres' surface and strengthen their shell. Hendrick et al. [121] used a mixture of HNO_3 and NH_4F at 90°C followed by washes with water, acetone, and alcohol. Figure 25.20 compares SEM images of the surface of hollow glass microsphere commercially available with those synthesized and acid etched by Hendricks et al. [121,125].

The authors established that this etching process results in a 100–200 Å surface finish and no sign of surface deterioration after exposure to humid air compared with a few days when using conventional HCl or HF wash. Acid etching also provides further control of the shell thickness. In addition, processes have been developed for coating the hollow glass

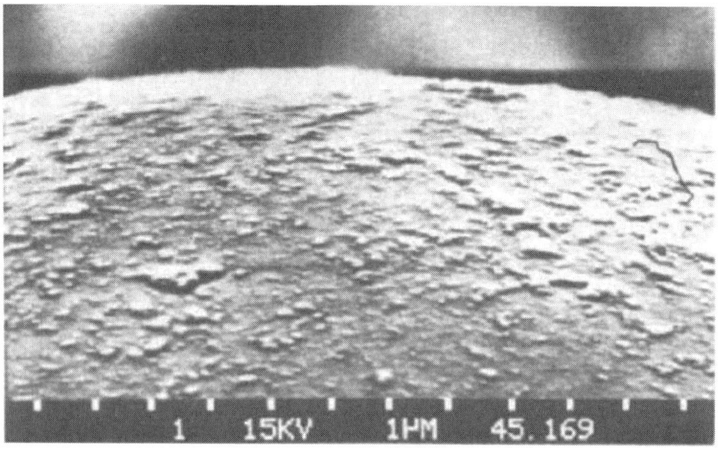

(a)

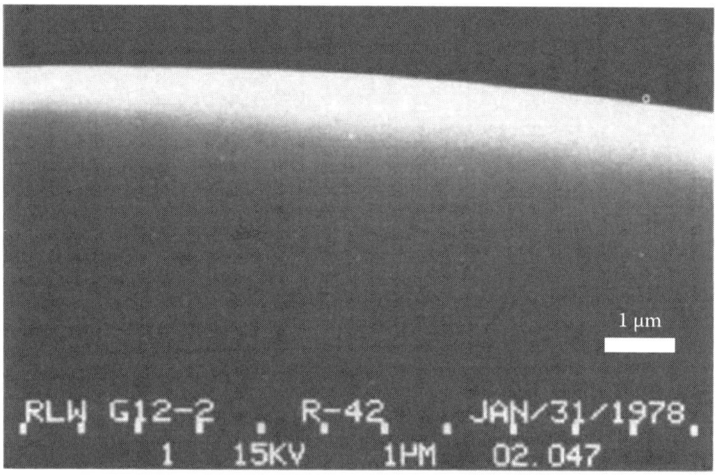

(b)

FIGURE 25.20
SEM images of the surface of hollow glass microsphere (a) commercially available and (b) synthesized by liquid droplet method and chemically etched. (From Hendricks, C.D., *J. Nuclear Mater.*, 85–86(1), 107–111, 1979.)

microspheres with a thin layer of materials to strengthen the thin glass shell [20,34,125,139]. For example, hollow glass microspheres have been coated with nickel [140], TiO_2, AlN [141], or with a gel [142] to name a few. Processes have also been developed to form a multilayer shell [143,144]. However, this protective coating should be thin and nonreactive with H_2 to limit potential barrier to hydrogen permeation. Acid etching and coatings of the shell of the hollow glass microspheres could be repeated between consecutive uses albeit at an additional cost.

25.5.8 Other Materials and Shapes

Hollow microspheres have also been made of polymers. They can be produced by emulsion processes as well as by coating solid template particles (e.g., polystyrene or silica) in liquid suspensions or in fluidized beds [91]. The solid particles act as sacrificial cores and are removed by thermolysis [91] or chemical etching. Hollow microspheres have also been synthesized with various polymeric materials and various size distributions [145–148].

As previously discussed, synthesis of cylindrical glass capillaries or hollow glass fibers is much more straightforward as they can be produced by simply drawing hollow fibers [149–151]. These products are inexpensive and commercially available as capillary tubes. For hydrogen storage, both cylinder ends must be sealed as described in details in Ref. [20].

25.6 Conclusions and Perspectives

This chapter has reviewed the principles, advantages, and challenges of hydrogen storage in hollow glass microspheres. Glass, and in particular vitreous silica, appears to be the material of choice for its superior tensile strength and good hydrogen permeability. In addition, hollow microspheres can sustain larger pressures and feature larger hydrogen permeation rate than cylinders with the same inner and outer radii and internal pressure. Significant progress has been made in synthesizing monodisperse hollow glass microspheres with controlled diameter and shell thickness. Strategies have been developed to further strengthen the shell through acid etching, heat polishing, or coating with hard material. Hollow glass microspheres feature excellent gravimetric energy density and can be safely handled at room temperature and atmospheric pressure and poured in any container solution. They are relatively inexpensive to make and the associated technology can be scaled up and present little risks. However, their volumetric energy density falls short of the 2015 DOE target. Moreover, the energy cost for compressing hydrogen at high pressure during loading is very high. Detailed cost analysis for this technology could not be found in the open literature. However, it has been stated that "the use of commercial grade hollow-glass microspheres for high pressure hydrogen storage has been shown to be cost ineffective" [7]. This conclusion was reached based on experimental tests performed by Teitel [18] and due to (1) material and synthesis cost, (2) the low volumetric energy density achieved, (3) high microsphere breakage rate, (4) the low fraction of recoverable hydrogen, and (5) the fill and release energy requirements. On the other hand, Akunets et al. [2] suggested that hydrogen storage in hollow glass microspheres could be used in applications when storage lifetime, safety, and weight are major concerns and where cost is not the determining factor. Similarly, Robinson and Handrock [7] suggested the use of hollow glass microspheres for fleet applications where volume and refueling concerns are not critical.

Based on the previous discussion and in order to make this technology competitive with other hydrogen storage technologies, the following technical recommendations are made [11,28]:

1. Develop energy-efficient, reliable, cost-effective, and clean 1000 bar hydrogen compressors [28]. Technologies to reduce energy requirement for H_2 compression would have no moving parts and be oil- and lubricant-free thus reducing the risk for contamination of hydrogen. They include metal hydride hydrogen compressors [152,153], for example. This effort would also benefit other high-pressure hydrogen storage technologies.
2. Develop inexpensive and high-throughput processes to produce high-quality monodisperse microspheres with large tensile strength and uniform shell thickness

able to withstand high internal pressures. The shell material should also feature adequate permeability to achieve long dormancy time as well as short unloading times for on-demand delivery.

3. Optimize existing or develop new coating and etching techniques to strengthen the shell and reduce breakage rate during loading and handling. This would increase reliability and the number of loading and unloading cycles hollow glass microspheres can withstand.
4. Develop materials and strategies to unload H_2 at temperature below 100°C. Then, heating could be achieved by using waste heat generated by the fuel cell or a simple heater powered by a battery during fuel cell start-up.
5. Develop novel approaches to control H_2 permeability through nonthermal methods such as magnetic, electric, or electromagnetic [28].
6. Assess the financial and energy costs associated with reprocessing the spent microspheres.

Nomenclature

a, b, c	Parameters in Bettie–Bridgeman equation
A_0, B_0	Parameters in Bettie–Bridgeman equation (=0.02 Pa m^3/mol^2)
c_p	Specific heat at constant pressure (J/kg K)
c_v	Specific heat at constant volume (J/kg K)
D	Diffusion coefficient (m^2/s)
D_0	Maximum diffusion coefficient (m^2/s)
E	Young's modulus of the shell material (Pa)
E_t	Specific total energy required for loading and unloading (J/kg)
ΔH_L	Hydrogen LHV (J/kg)
k	Ratio of specific heat, $\gamma = c_p/c_v$
k_c	Thermal conductivity (W/m K)
K	Permeability (mol/Pa m s)
K_0	Maximum permeability (mol/Pa m s)
L	Cylinder length (m)
M	Molar mass (g/mol) (=2.0158 g/mol)
P_{cr}	Buckling pressure (Pa or bar)
P_i	Inner shell pressure (Pa or bar)
P_{max}	Burst pressure (Pa or bar)
P_o	Outer shell pressure (Pa or bar)
Q_D	Specific thermal energy required for diffusivity (K)
Q_K	Specific thermal energy required for permeation (K)
Q_S	Specific thermal energy required for solubility (K)
Q_u	Thermal energy required for H_2 unloading (J/kg)
R	Universal gas constant (R = 8.314 J/mol K)
r_i	Inner shell diameter (m)
r_o	Outer shell diameter (m)
S	Solubility of hydrogen in shell material (mol/m^3 Pa)
t	Time (s)

T	Temperature (K)
W	Specific compression work (J/kg)
Z	Compressibility factor

Greek Symbols

α	Surface energy (J/m^2)
β	Compressor efficiency
γ_g	Gravimetric energy density (in wt.%)
γ_g	Volumetric energy density (in kg/L)
η_g	Gravimetric energy density (in MJ/kg)
η_v	Volumetric energy density (in MJ/L)
ν	Poisson's ratio of the shell material
ρ	Density of hydrogen gas (kg/m^3)
ρ_{eff}	Effective density of microcapsules (kg/m^3)
ρ_f	Density of the fluid in sink–float method (kg/m^3)
ρ_s	Density of the shell materials (kg/m^3)
σ_{max}	Tensile strength (Pa)
σ_h	Hoop stress (Pa)
σ_l	Longitudinal stress (Pa)
τ	Permeation time constant (s)
T_ℓ	Lag time constant (s)

Subscripts

i	Refers to initial state
f	Refers to final state
s	Refers to the shell

References

1. J.M. Norbeck, J.W. Heffel, T.D. Durbin, B. Tabbara, J.M. Bowden, and M.C. Montano, *Hydrogen Fuel for Surface Transportation*, SAE International, Warrendale, PA, 1996.
2. A.A. Akunets, N.G. Basov, V.S. Bushuev, V.M. Dorogotovtsev, A.I. Gromov, A.I. Isakov, V.N. Kovylnikov, Y.A. Merkul'ev, A.I. Nikitenko, and S.M. Tolokonnikov, Super-high-strength microballoons for hydrogen storage, *International Hydrogen Energy*, 19(8), 697–700, 1994.
3. G.P. Sutton and O. Biblarz, *Rocket Propulsion Elements*, 7th edn. Wiley & Sons, New York, 2000.
4. N.K. Zhevago and V.I. Glebova, Hydrogen storage in capillary arrays, *Energy Conversion and Management*, 48(5), 1554–1559, 2007.
5. US Department of Energy Hydrogen, Targets for on-board hydrogen storage systems, last accessed on December 20, 2008, http://www.eere.energy.gov/hydrogenandfuelcells/pdfs/freedomcar_tar gets_explanations.pdf.
6. L.M. Das, On-board hydrogen storage systems for automotive application, *International Journal of Hydrogen Energy*, 21(9), 789–800, 1996.
7. S.L. Robinson and J.L. Handrock, Hydrogen storage for vehicular applications: Technology status and key development areas, Sandia National Laboratory Report SAND94-8229-UC-406, 1994.
8. V.A. Yartys and M.V. Lototsky, An overview of hydrogen storage methods, in *Hydrogen Materials Science and Chemistry of Carbon Nanomaterials*, T.N. Veziroglu, S.Y. Zaginaichenko, D.V. Schur, B. Baranowski, A.P. Shpak, V.V. Skorokhod, and A. Kale, eds. Kluwer Academic Publishers, Dordrecht, the Netherlands.

9. R.J. Teitel, Hydrogen supply method, US Patent No. 4211537, July 8, 1980.
10. R.J. Teitel, Hydrogen supply system, US Patent No. 4302217, November 24, 1981.
11. M. Herr and J.A. Lercher, Hydrogen storage in microspheres—Final report, ET-TN-03-628, September 9, 2003.
12. A.C. Dillon, K.M. Jones, T.A. Bekkedahl, C.H. Kiang, D.S. Bethune, and M.J. Heben, Storage of hydrogen in single-walled carbon nanotubes, *Nature*, 386(6623), 377–379, 1997.
13. M. Jordá-Beneytoa, F. Suárez-García, D. Lozano-Castellóa, D. Cazorla-Amorós, and A. Linares-Solano, Hydrogen storage on chemically activated carbons and carbon nanomaterials at high pressures, *Carbon*, 45(2), 293–303, 2007.
14. J. Rowsell and O.M. Yaghi, Strategies for hydrogen storage in metal-organic frameworks, *Angewandte Chemie International Edition*, 44, 4670–4679, 2005.
15. B. Sakintuna, F. Lamari-Darkrim, and M. Hirscher, Metal hydride materials for solid hydrogen storage: A review, *International Journal of Hydrogen Energy*, 32(9), 1121–1140, 2007.
16. A. Bouza, C.J. Read, S. Satyapal, and J. Milliken, Annual DOE hydrogen program review hydrogen storage, US Department of Energy, Energy Efficiency and Renewable Energy, Office of Hydrogen, Fuel Cells and Infrastructure Technologies, 2004.
17. S.G. Chalk and J.F. Miller, Key challenges and recent progress in batteries, fuel cells, and hydrogen storage for clean energy systems, *Journal of Power Sources*, 159, 73–80, 2006.
18. R.J. Teitel, Microcavity hydrogen storage final progress report, US Department of Energy and Environment, Report BNL 51439, 1981.
19. L. Yang, B.Z. Jang, and J. Guo, Micro-spheres for storing and delivering hydrogen to fuel cells, in *Proceedings of the International Hydrogen Energy Congress and Exhibition IHEC 2005*, Istanbul, Turkey, July 13–15, 2005, pp. 1–6.
20. K.L. Yan, B.G. Sellars, J. Lo, S. Johar, and M.K. Murthy, Storage of hydrogen by high-pressure microencapsulation in glass, *International of Journal of Hydrogen Energy*, 10, 517–522, 1985.
21. G.A. Banyay, M.M. Shaltout, H. Tiwari, and B.V. Mehta, Polymer and composite foam for hydrogen storage application, *Journal of Materials Processing Technology*, 191(1–3), 102–105, 2007.
22. R.G. Zalosh and S.N. Bajpai, Hydrogen microsphere hazard evaluation, in *Proceedings of the DOE Thermal and Chemical Storage Annual Contractor's Review Meeting*, Tysons Corner, VA, 1981, pp. 211–213.
23. W. Paul and R.V. Jones, Production and separation of small glass spheres, *British Journal of Applied Physics*, 3, 311–314, 1952.
24. E.H. Farnum, J.R. Fries, J.W. Havenhill, M.L. Smith, and D.L. Stoltz, Method for selecting hollow microspheres for use in laser fusion targets, US Patent No. 3997435, December 14, 1976.
25. R.J. Teitel, Development status of microcavity hydrogen storage for automotive applications, in *Proceedings of the DOE Chemical/Hydrogen Energy Systems Contractor Review*, DOE CONF-791127, Reston, VA, 1979, US Department of Energy, Washington, DC.
26. G.D. Rambach, Hydrogen transport and storage in engineered glass microspheres, in *Proceedings of the Sixth Annual US Hydrogen Meeting—Hydrogen Technologies: Moving toward Commercialization*, Alexandria, VA, March 7–9, 1995, National Hydrogen Association, pp. 163–172.
27. G.D. Rambach and C. Hendricks, Hydrogen transport and storage in engineered microspheres, in *Proceedings of the 1996 US DOE Hydrogen Program Review*, NREL/CP-430-21938, Miami, FL, May 1–2, 1996, Vol. 2, pp. 765–772.
28. T. Riis, E.F. Hagen, P.J.S. Vie, and Ø. Ulleberg, Hydrogen production and storage—R&D priorities and gaps, International Energy Agency, 2006, www.iea.org/Textbase/papers/2006/hydrogen.pdf.
29. R.T. Tsugawa, I. Moen, P.E. Roberts, and P.C. Souers, Permeation of helium and hydrogen from glass-microsphere laser targets, *Journal of Applied Physics*, 47(5), 1987–1993, 1976.
30. Y.V.C. Rao, *An Introduction to Thermodynamics*, Universities Press, Hyderabad, India, 2004.
31. J.P. Den Hartog, *Strength of Materials*, Dover Publications, New York, 1961.
32. G.D. Rambach, Hydrogen transport and storage in engineered glass microspheres, in *Proceedings of the US DOE Hydrogen Program Review Meeting, Sandia National Laboratory*, UCR-JC-11708, April 18–21, 1994.

33. G.M. Bartenev and D.S. Sanditov, The strength and some mechanical and thermal characteristics of high-strength glasses, *Journal of Non-Crystalline Solids*, 48(2–3), 405–421, 1982.
34. D.K. Kohli, R.K. Khardekar, R. Singh, and P.K. Gupta, Glass micro-container based hydrogen storage scheme, *International of Hydrogen Energy*, 33(1), 417–422, 2008.
35. D. Duret and A. Saudin, Microspheres for on-board hydrogen storage, *International of Journal of Hydrogen Energy*, 19, 757–764, 1994.
36. H. Kunieda, Classical buckling load of spherical domes under uniform pressure, *Journal of Engineering Mechanics*, 118(8), 1513–1525, 1992.
37. N.F. Morris, Shell stability: The long road from theory to practice, *Engineering Structures*, 18(10), 801–806, 1996.
38. W. Wunderlich and U. Albertin, Buckling behaviour of imperfect spherical shells, *International Journal of Non-Linear Mechanics*, 37(4–5), 589–604, 2002.
39. M. Deml and W. Wunderlich, Direct evaluation of the 'worst' imperfection shape in shell buckling, *Computer Methods in Applied Mechanics and Engineering*, 149(1–4), 201–222, 1997.
40. J.E. Shelby, *Handbook of Gas Diffusion in Solids and Melts*, ASM International, Materials Park, OH, 1996.
41. F.J. Norton, Helium diffusion through glass, *Journal of the American Ceramic Society*, 36(3), 90–96, 1953.
42. F.J. Norton, Permeation of gases through solids, *Journal of Applied Physics*, 28(1), 34–39, 1957.
43. H.J. Frost and R. Raj, Limiting densities for dense random packing of spheres, *Journal of the American Ceramic Society*, 65(2), C-19–C-21, 1982.
44. Z. Pientka, P. Pokorný, and K. Belafi-Bako, Closed-cell polymeric foam for hydrogen separation and storage, *Journal of Membrane Science*, 304(1–2), 82–87, 2007.
45. L. Pilon, A.G. Fedorov, and R. Viskanta, Gas diffusion in closed-cell foams, *Journal of Cellular Plastics*, 36, 451–474, 2000.
46. D.N. Sutherland, Random packing of circles in a plane, *Journal of Colloid and Interface Science*, 60(1), 96–102, 1977.
47. N.P. Bansal and R.H. Doremus, *Handbook of Glass Properties*, Academic Press, Inc., Orlando, FL, 1986.
48. G.W. McLellan and E.B. Shand, *Glass Engineering Handbook*, 3rd edn. Mc Graw Hill, New York, 1984.
49. R.H. Doremus, *Glass Science*, 2nd edn. John Wiley & Sons, New York, 1994.
50. H. Rawson, *Properties and Applications of Glass*, Elsevier, New York 1980.
51. W.D. Johnston and A.J. Chelko, Reduction of ions in glass by hydrogen, *Journal of American Ceramics Society*, 53(6), 295–301, 1970.
52. J.E. Shelby and J. Vitko Jr., Hydrogen transport in a machinable glass-ceramic, *Journal of Non-Crystalline Solids*, 45, 83–92, 1981.
53. P.C. Souers, I. Moen, R.O. Lindhal, and R.T. Tsugawa, Permeation eccentricities of He, Ne, and D-T from soda-lime glass microbubbles, *Journal of the American Ceramic Society*, 61(1–2), 42–46, 1978.
54. R.H. Doremus, Pre-exponential factor of temperature in the diffusion equation, *Journal of Chemical Physics*, 34(6), 2186–2187, 1961.
55. V.O. Altemose, Helium diffusion through glass, *Journal of Applied Physics*, 32(7), 1309–1316, 1961.
56. R.W. Lee, R.C. Frank, and D.E. Swets, Diffusion of hydrogen and deuterium in fused quartz, *The Journal of Chemical Physics*, 36(4), 1062–1071, 1962.
57. R.W. Lee, Diffusion of hydrogen in natural and synthetic fused quartz, *The Journal of Chemical Physics*, 38(2), 448–455, 1963.
58. H.M. Laska, R.H. Doremus, and P.J. Jorgensen, Permeation, diffusion, and solubility of deuterium in pyrex glass, *Journal of Chemical Physics*, 50(1), 135–137, 1969.
59. J.L. Barton and M. Morain, Hydrogen diffusion in silicate glass, *Journal of Non-Crystalline Solids*, 3, 115–126, 1970.
60. J.E. Shelby, Helium, deuterium, and neon migration in a common borosilicate glass, *Journal of Applied Physics*, 45(5), 2146–2149, 1974.

61. N. Kurita, N. Fukatsu, H. Otsuka, and T. Ohashi, Measurements of hydrogen permeation through fused silica and borosilicated glass by electrochemical pumping using oxide protonic conductor, *Solid State Ionics*, 146, 101–111, 2002.
62. J.E. Shelby and B.E. Kenyon, Glass membrane for controlled diffusion of gases, US Patent No. 6231642 B1, May 15, 2001.
63. B.E. Kenyon, Gas solubility and accelerated diffusion in glasses and melts, Master's thesis, Alfred University, Alfred, NY, 1998.
64. D.B. Rapp and J.E. Shelby, Photo-induced hydrogen outgassing of glass, *Journal of Non-Crystalline Solids*, 349, 254–259, 2004.
65. D.B. Rapp, Photo-induced hydrogen outgassing of glass, PhD thesis, Alfred University, Alfred, NY, 2004.
66. R.K. Brow and M.L. Schmitt, A survey of energy and environmental applications of glass, *Journal of the European Ceramic Society*, 29(7), 1193–1201, 2009.
67. R. Kitamura and L. Pilon, Radiative heat transfer in enhanced hydrogen outgassing of glass, *International Journal of Hydrogen Energy*, 34(16), 6690–6704, 2009.
68. R. Kitamura, L. Pilon, and M. Jonasz, Optical constants of silica glass from extreme ultraviolet to far infrared at near room temperature, *Applied Optics*, 46(33), 8118–8133, 2007.
69. R. Viskanta and J. Lim, Transient cooling of a cylindrical glass gob, *Journal of Quantitative Spectroscopy and Radiation Transfer*, 73, 481–490, 2002.
70. D. De Sousa Meneses, G. Gruener, M. Malki, and P. Echegut, Causal Voigt profile for modeling reflectivity spectra of glasses, *Journal of Non-Crystalline Solids*, 351, 124–129, 2005.
71. J.E. Shelby and J. Vitko Jr., The reduction of iron in soda-lime-silicate glasses by reaction with hydrogen, *Journal of Non-Crystalline Solids*, 53, 155–163, 1982.
72. S.P. Faile and D.M. Roy, Dissolution of hydrogen in fused quartz, *Journal of American Ceramics Society*, 54(10), 533–534, 1971.
73. I. Fanderlik, *Glass Science and Technology, Vol. 5: Optical Properties of Glass*, Elsevier Science, New York, 1983.
74. W. Schotte, Thermal conductivity of packed beds, *AIChE Journal*, 6(1), 63–67, 1960.
75. J.C. Chen and S.W. Churchill, Radiant heat transfer in packed beds, *AIChE Journal*, 8, 35–41, 1963.
76. M.Q. Brewster and C.L. Tien, Radiative transfer in packed fluidized beds: Dependent versus independent scattering, *ASME Journal of Heat Transfer*, 104, 573–579, 1982.
77. K. Kamiuto, Correlated radiative transfer in packed-sphere systems, *Journal of Quantitative Spectroscopy and Radiative Transfer*, 43, 39–43, 1990.
78. K. Kamiuto, M. Iwamoto, M. Sato, and T. Nishimura, Radiation-extinction coefficients of packed-sphere systems, *Journal of Quantitative Spectroscopy and Radiative Transfer*, 45, 93–96, 1991.
79. K. Kamiuto, Analytical expression for total effective thermal conductivities of packed beds, *Journal of Nuclear Science and Technology*, 28(12), 1153–1156, 1991.
80. B.P. Singh and M. Kaviany, Independent theory versus direct simulation of radiative heat transfer in packed beds, *International Journal of Heat and Mass Transfer*, 34, 2869–2882, 1991.
81. B.P. Singh and M. Kaviany, Modelling radiative heat transfer in packed beds, *International Journal of Heat and Mass Transfer*, 35, 1397–1405, 1992.
82. K. Kamiuto, Radiative properties of packed-sphere systems estimated by the extended emerging-intensity fitting method, *Journal of Quantitative Spectroscopy and Radiative Transfer*, 47, 257–261, 1992.
83. K. Kamiuto, Combined conduction and correlated-radiation heat transfer in packed beds, *Journal of Thermophysics and Heat Transfer*, 7(3), 496–501, 1993.
84. M. Kaviany and B.P. Singh, Radiative heat transfer in porous media, *Advances in Heat Transfer*, 23(23), 133–186, 1993.
85. K. Nasr, R. Viskanta, and S. Ramadhyani, An experimental evaluation of the effective thermal conductivities of packed beds at high temperatures, *ASME Journal of Heat Transfer*, 116(4), 829–837, 1994.

86. U. Bossel, Does a hydrogen economy make sense? *Proceedings of the IEEE*, 94(10), 1826–1837, 2006.
87. S.L. Robinson and J.J. Iannucci, Technologies and economics of small-scale hydrogen storage, Sandia National Laboratory Report SAND-79–8646, 1979.
88. B.A. Proctor, Strengths of acid-etched glass rods, *Nature*, 187, 492–493, 1960.
89. A.S. Geleil, M.M. Hall, and J.E. Shelby, Hollow glass microspheres for use in radiation shielding, *Journal of Non-Crystalline Solids*, 352(6–7), 620–625, 2006.
90. C.D. Hendricks, Glass spheres, in *Materials Handbook, Ceramics and Glasses*, Vol. 4, S.J. Schneider Jr., ed. ASM International, Materials Park, OH.
91. J. Bertling, J. Blömer, and R. Kümmel, Hollow microspheres, *Chemical Engineering Technology*, 27(8), 829–837, 2004.
92. V.V. Budov, Hollow glass microspheres. use, properties, and technology (review), *Glass and Ceramics*, 51(7–8), 230–235, 1994.
93. E.F. Medvedev, Use of structural criteria for calculating oxide glass compositions for hydrogen microcontainers (a review), *Glass and Ceramics*, 63(7–8), 222–226, 2006.
94. F. Veatch and R.W. Burhans, Process of producing hollow particles and resulting product, US Patent 2797201, June 25, 1957.
95. F. Veatch, H.E. Alford, and R.C. Croft, Hollow glass particles and method of producing the same, US Patent No. 2978340, April 4, 1961.
96. F. Veatch, H.E. Alford, and R.C. croft, Apparatus for producing hollow glass particles, US Patent No. 3129086, April 14, 1964.
97. W.R. Beck and D.L. O'Brien, Glass bubbles prepared by reheating solid glass particles, US Patent No. 3365315, January 23, 1968.
98. H.E. Alford and F. Veatch, Microsphere glass agglomerates and method for making them, US Patent No. 3458332, July 29, 1969.
99. C.D. Hendricks, Method for producing small hollow spheres, US Patent No. 4133854, January 9, 1979.
100. A. Rosencwaig, J.C. Koo, and J.L. Dressler, Method for producing small hollow spheres, US Patent No. 4257799, March 24, 1981.
101. P.A. Howell, Glass bubbles of increased collapse strength, US Patent No. 4391646, July 5, 1983.
102. P. Garnier, D. Abriou, and M. Coquillon, Process for producing glass microspheres, US Patent No. 4661137, April 28, 1987.
103. H.J. Marshall, Glass microbubbles, US Patent No. 4767726, August 30, 1988.
104. P. Garnier, D. Abriou, and J.-J. Gaudiot, Production of glass microspheres, US Patent No. 4778502, October 18, 1988.
105. R.W.J. Lencki, R.J. Neufeld, and T. Spinney, Method of producing microspheres, US Patent No. 4822534, April 18, 1989.
106. K.E. Goetz, J.A. Hagarman, and J.P. Giovene Jr., Hollow glass spheres, US Patent No. 4983550, January 8, 1991.
107. J. Block, N.J. Tessier, and A.J. Colageo, Method of making small hollow glass spheres, US Patent No. 5069702, December 3, 1991.
108. R.W. Rice A.J. Colageo J. Block, J.W. Lau, Method for making low sodium hollow glass microspheres, US Patent No. 5176732, January 5, 1993.
109. P. Garnier, D. Abriou, and J.-J. Gaudiot, Method for selecting hollow microspheres for use in laser fusion targets, US Patent No. 5256180, October 26, 1993.
110. S. Kawachi and Y. Sato, Glass composition for glass bubbles with increased compressive strength, US Patent No. 5292690, March 8, 1994.
111. E.V. Sokol, N.V. Maksimova, N.I. Volkova, E.N. Nigmatulina, and A.E. Frenkel, Hollow silicate microspheres from fly ashes of the Chelyabinsk brown coals (South Urals, Russia), *Fuel Processing Technology*, 67(1), 35–52, 2000.
112. L.K. Heung, R.F. Schumacher, and G.G. Wicks, Hollow porous-wall glass microspheres for hydrogen storage, Patent Application No. US 2006/0059953 (October 21, 2005) and International Application No. PCT/US2006/040525 (October 17, 2006).

113. G.G. Wicks, L.K. Heung, and R.F. Schumacher, Microspheres and microworlds, *ACerS Bulletin*, 23–28, June 2008.
114. J.R. Fisher and M.D. Rigterink, Formation of expanded silica spheres, US Patent No. 2883347, April 21, 1959.
115. V.V. Budov and V.Y. Stetsenko, Choice of glass composition for producing hollow microspheres, *Glass and Ceramics*, 45(8), 289–291, 1988.
116. V.V. Budov, Physicochemical processes in producing hollow glass microspheres, *Glass and Ceramics*, 47(3), 77–79, 1990.
117. I. Bica, Formation of glass microspheres with rotating electrical arc, *Materials Science and Engineering B*, 77, 210–212, 2000.
118. P.D. Law, Method and apparatus for production of glass beads by use of a rotating wheel, US Patent No. 3310391, March 21, 1967.
119. E.M. Guyer and J.E. Nitsche, Method and apparatus for manufacturing glass beads, US Patent No. 3313608, April 11, 1967.
120. R.L. Downs and W.J. Miller, Hollow microspheres of silica glass and method of manufacture, US Patent No. 4336338, June 22, 1982.
121. C.D. Hendricks, A. Rosencwaig, R.L. Woerner, J.C. Koo, J.L. Dressler, J.W. Sherohman, S.L. Weinland, and M. Jeffries, Fabrication of glass sphere laser fusion targets, *Journal of Nuclear Materials*, 85–86(1), 107–111, 1979.
122. T.P. O'Holleran, R.L. Nolen, R.L. Downs, D.A. Steinman, and R.L. Crawley, Summary abstract: Shells from compacted powders, *Journal of Vacuum Science and Technology*, 18(3), 1242–1243, 1981.
123. M.L. Schmitt, J.E. Shelby, and M.M. Hall, Preparation of hollow glass microspheres from sol-gel derived glass for application in hydrogen gas storage, *Journal of Non-Crystalline Solids*, 352(6–7), 626–631, 2006.
124. A. Rosencwaig, J.L. Dressler, J.C. Koo, and C.D. Hendricks, Laser fusion hollow glass microspheres by the liquid-droplet method, Lawrence Livermore Laboratory Report UCRL-81421, June 5, 1978.
125. C.D. Hendricks, Fuel pellets and optical systems for inertially confined fusion, *Journal of Nuclear Materials*, 85–86(1), 79–86, 1979.
126. J. Koo, J. Dressler, and C. Hendricks, Low pressure gas filling of laser fusion microspheres, *Journal of Nuclear Materials*, 85–86(1), 113–115, 1979.
127. J.Y. Dinga and D.E. Day, Preparation of silica glass microspheres by sol-gel processing, *Journal of Materials Research*, 6(1), 168–174, 1991.
128. F. Caruso, R.A. Caruso, and H. Möhwald, Nanoengineering of inorganic and hybrid hollow spheres by colloidal templating, *Science*, 282(5391), 1111–1114, 1998.
129. B. Peng, M. Chen, S. Zhou, L. Wu, and X. Ma, Fabrication of hollow silica spheres using droplet templates derived from a miniemulsion technique, *Journal of Colloid and Interface Science*, 321(1), 67–73, 2008.
130. K.K. Kim, K.Y. Jang, and R.S. Upadhye, Hollow silica spheres of controlled size and porosity by sol-gel processing, *Journal of the American Ceramic Society*, 74(8), 1987–1992, 1991.
131. K.Y. Jang and K. Kim, Evaluation of sol-gel processing as a method for fabricating spherical-shell silica aerogel inertial confinement fusion targets, *Journal of Vacuum Science & Technology A*, 10, 1152–1157, 1992.
132. B. Tan and S.E. Rankin, Dual latex/surfactant templating of hollow spherical silica particles with ordered mesoporous shells, *Langmuir*, 21(18), 8180–8187, 2005.
133. S. Zhang, L. Xu, H. Liu, Y. Zhao, Y. Zhang, Q. Wang, Z. Yu, and Z. Liu, A dual template method for synthesizing hollow silica spheres with mesoporous shells, *Materials Letters*, 63(2), 258–259, 2009.
134. Y.-Q. Yeh, B.-C. Chen, H.-P. Lin, and C.-Y. Tang, Synthesis of hollow silica spheres with mesostructured shell using cationic-anionic-neutral block copolymer ternary surfactants, *Langmuir*, 22(1), 6–9, 2006.

135. N. Kato, T. Ishii, and S. Koumoto, Synthesis of monodisperse mesoporous silica hollow microcapsules and their release of loaded materials, *Langmuir*, 26(17), 14334–14344, 2010.
136. S. Schacht, Q. Huo, I.G. Voigt-Martin, G.D. Stucky, and F. Schth, Oil-water interface templating of mesoporous macroscale structures, *Science*, 273(5276), 768–771, 1996.
137. Z. Teng, Y. Han, J. Li, F. Yan, and W. Yang, Preparation of hollow mesoporous silica spheres by a sol-gel/emulsion approach, *Microporous and Mesoporous Materials*, 127(1–2), 67–72, 2010.
138. B.A. Proctor, I. Whitney, and J.W. Johnson, The strength of fused silica, *Proceedings of the Royal Society of London. Series A, Mathematical and Physical Sciences*, 297(1451), 534–557, 1967.
139. C.S. Chamberlain, G.F. Vesley, P.G. Zimmerman, and J.W. McAllister, Coated glass microbubbles and article incorporating them, US Patent No. 4618525, October 21, 1986.
140. Zhang, Q., Wu, M., and Zhao, W., Electroless nickel plating on hollow glass microspheres, *Surface and Coatings Technology*, 192(2–3), 213–219, 2005.
141. K.H. Moh, Glass and glass-ceramic bubbles having an aluminum nitride coating, US Patent No. 5691059, November 25, 1997.
142. M.E. Holman, Gel-coated microcapsules, US Patent No. 6099894, August 8, 2000.
143. A.G. Baker and A.J. Baker, Process for producing hollow, bilayered silicate microspheres, US Patent No. 4549892, October 29, 1985.
144. Q. Sun, P.J. Kooyman, J.G. Grossmann, P.H.H. Bomans, P.M. Frederik, P.C.M.M. Magusin, T.P.M. Beelen, R.A. van Santen, and N.A.J.M. Sommerdijk, The formation of well-defined hollow silica spheres with multilamellar shell structure, *Advances in Materials*, 15, 1097–1100, 2003.
145. M. Okubo and H. Minami, Control of hollow size of micron-sized monodispersed polymer particles having a hollow structure, *Colloid & Polymer Science*, 274(5), 433–438, 1996.
146. W. Schmidt and G. Roessling, Novel manufacturing process of hollow polymer microspheres, *Chemical Engineering Science*, 61(15), 4973–4981, 2006.
147. G. Crotts and T.G. Park, Preparation of porous and nonporous biodegradable polymeric hollow microspheres, *Journal of Controlled Release*, 35, 91–105, 1995.
148. T. Brandau, Preparation of monodisperse controlled release microcapsules, *International Journal of Pharmaceutics*, 242, 179–184, 2002.
149. J.A. Burgman and L.L. Margason, Method and apparatus for forming hollow glass fibers, US Patent No. 3268313, August 23, 1966.
150. T.H. Jensen, Hollow glass fiber bushing, method of making hollow fibers and the hollow glass fibers made by that method, US Patent No. 4758259, July 19, 1988.
151. T. Maruyama, H. Matsumoto, and Y. Miyake, Glass capillary tube and method for its production, US Patent No. 4882209, November 21, 1989.
152. P. Muthukumar, M.P. Maiya, and S.S. Murthy, Performance tests on a thermally operated hydrogen compressor, *International Journal of Hydrogen Energy*, 33(1), 463–469, 2008.
153. P. Muthukumar, M.P. Maiya, and S.S. Murthy, Parametric studies on a metal hydride based single stage hydrogen compressor, *International Journal of Hydrogen Energy*, 27(10), 1083–1092, 2002.
154. E.W. Lemmon, M.L. Huber, and M.O. McLinden, NIST reference fluid thermodynamic and transport properties database (REFPROP): Version 8.0, National Institute of Standards and Technology, Boulder, CO, 2007, http://www.nist.gov/srd/nist23.htm.
155. N. Sahba and T.J. Rockett, Infrared absorption coefficients of silica glasses, *Journal of the American Ceramic Society*, 75(1), 209–212, 1992.

26

Slush Hydrogen Storage

S.A. Sherif
University of Florida

S. Gursu
OMV Petrol Ofisi AS

T. Nejat Veziroglu
International Association for Hydrogen Energy

CONTENTS

This chapter provides an overview of slush hydrogen production methods and reviews the current state of knowledge regarding slush hydrogen utilization.

26.1 Introduction

Some of the problems associated with the utilization of hydrogen include its low density, temperature stratification, short holding time due to its low latent heat, hazards associated with high vent rates, and unstable transport conditions caused by sloshing of the liquid in the fuel tank. Rapial and Daney [1] report that there are two techniques that can completely eliminate some of these undesirable features and partially eliminate the others. One is the production of liquid–solid hydrogen mixtures, *slush hydrogen*, by further refrigeration of the liquid. The other is the gelation of liquid hydrogen by the addition of a fine particle gelling substance.

Slush hydrogen is a mixture of liquid and frozen hydrogen in equilibrium with the gas at the triple point, 13.8 K. The density of the icelike form is about 20% higher than that of the boiling liquid. To obtain the icelike form, one has to remove the heat content of the liquid at 20.3 K until the triple point is reached and then remove the latent heat of fusion. The *cold content* of the icelike form of hydrogen is some 25% higher than that of the saturated vapor at 20.3 K. Justi [2] reports that in producing the slush hydrogen, in comparison with the icelike form, one loses part of the increase in density and cold storage, but in mixing the icelike form with liquid hydrogen, one obtains a medium that is transportable in pipelines. Slush hydrogen can be produced from liquid hydrogen with the help of a vacuum pump and expansion through a jet [2].

Baker and Matsch [3] gave a tabulated comparison among the properties of liquid and slush hydrogen. According to them, the smaller enthalpy of slush hydrogen reduces the evaporation losses during storage and transport and as a consequence permits longer storage without venting. The increased density permits ground transport equipment to carry 15% greater loads; for space vehicles, a given fuel load can be carried in tankage, which are 13% smaller in volume.

Kandebo [4] reported that slush hydrogen is expected to be the primary propellant for the National Aero-Space Plane (NASP). He mentioned that using slush instead of liquid hydrogen reduces the physical size of the NASP and cuts the projected gross liftoff weight by up to 30%. The freeze–thaw process, which is the most used slush production process, and the auger process, which is relatively new, were discussed and compared. One expensive alternative method to pressurize tanks was also mentioned. This alternative is to pressurize tanks with a combination of helium and hydrogen, because helium does not condense at slush hydrogen storage temperatures and is not very dense.

Justi [2] reported that the slush-like form of hydrogen has the advantage of a greater *cold content* compared to liquid hydrogen. The disadvantage of slush hydrogen lies in the present state of the technology for its production, which is not yet well advanced, so that it is impractical to calculate a realistic price. Justi argues, however, that the production cost of slush hydrogen now and in the near future is greater than the liquefaction costs of hydrogen because of the larger energy use involved.

Sindt [5] studied the characteristics of slush hydrogen preparation, storage, transfer, and instrumentation. In that study, slush preparation by intermittent vacuum pumping was discussed. Slush was aged 100 h, during which time the solid particle size and structure were observed. The solid particle structure was found to change dramatically during aging, even though the changes in the particle size were insignificant. Sindt also mentioned that slush with over 50% solid content could be transferred and pumped with losses similar to those in triple-point liquid hydrogen if the Reynolds numbers were high enough.

Voth [6] reported on a slush hydrogen production method with an auger rotating inside a brass tube refrigerated with liquid helium. The auger was reported to produce small particles from the cryogens on a continuous basis so that the resulting slush mixture could be transferred and stored. He also reported that the auger is capable of producing slush at pressures higher than the triple-point pressure of the cryogen.

Methods for determining the quality of liquid–solid hydrogen mixtures, along with relevant thermodynamic analyses were discussed by Daney and Mann [7] and Mann et al. [8]. Properties of liquid and slush hydrogen and their applications were discussed by Smith [9] and Angus [10]. The flow characteristics of slush hydrogen with different solid contents were examined by Sindt and Ludtke [11] and by Sindt et al. [12]. The latter study reported on experimental results in which a second component was added to liquid hydrogen to determine the effect produced by the freeze–thaw process on the size of the solid particles. Perrel and Haase [13] developed a computer simulation method to describe the self-catalysis of ortho-hydrogen molecules. The simulations were performed on a small computer using Monte Carlo techniques. Roder [14] studied the phase transitions in solid hydrogen and suggested that, except for the lambda transition, which occurs at low temperatures in solid hydrogen and involves a change from hexagonal close-packed (hcp) to face-centered cubic (fcc), there may be an additional phase transition in solid hydrogen.

Although current interest in liquid–solid mixtures of hydrogen as a potential rocket propellant has led to theoretical and experimental investigations of its characteristics, there is still a lot of potential for additional investigations. For example, the problem of transferring heat to slush hydrogen along with solid density stratification in large containers needs to be investigated more thoroughly.

26.2 Production of Slush Hydrogen

Several successful methods for producing solids from liquid–solid mixtures were reported in the literature. However, if the preparation method is to be useable, it must work in systems of all sizes and still meet certain basic criteria that create a process, which is thermodynamically efficient and reproducible. The process should also be able to produce a mixture with enough fluidity to transfer through existing systems and with sufficient solid content so that the density and heat capacity are significantly greater than those for the normal boiling-point liquid.

26.2.1 Freeze–Thaw Method

Sindt [5] summarized the basic principles of the freeze–thaw method and its feasibility. Two of the methods reported by Sindt involve the formation of solid particles by spraying precooled liquid into an environment well below the triple-point pressure and inducing solid formation by allowing precooled helium gas to flow through a triple-point temperature liquid. The spray technique produced a very-low-density solid particle, which had to be partially melted to form a liquid–solid mixture. The technique of allowing helium to flow through the liquid produced solid by reducing the partial pressure of hydrogen below its triple-point pressure, which resulted in a clear, rigid tube of solid forming around the helium bubble train. The solid was found not to be convertible to a mixture without further treatment such as crushing. Sindt found that a more desirable preparation method was to evacuate the ullage over the liquid, which created a solid layer at the liquid surface. The solid layer produced had a texture that was very dependent upon the rate of vacuum pumping. Low pumping rates produced a dense, nearly transparent layer of solid. Subsequent breaking of the solid left large rigid pieces that formed inhomogeneous mixtures even with vigorous mixing. Higher pumping rates produced a solid layer that was porous and consisted of agglomerates of loosely attached fine solid particles. The solid layer was found to be easily broken and mixed with the liquid to form a homogeneous liquid–solid mixture that could be defined as slush. Sindt discovered, however, that solids could not be continuously generated at the liquid surface without vigorous mixing.

Sindt [5] also reported on a continuous preparation method, which he developed, capable of forming solid layers in cycles by periodic vacuum pumping. During the pumping (pressure reduction) portion of the cycle, the solid layer formed as previously described. When the pumping was stopped, the solid layer melted and settled into the liquid, forming slush. Although mixing was not necessarily required, the preparation method was accelerated by some mechanical mixing, which helped break the layer of solids. This technique has been called the *freeze–thaw* process.

In the same study, Sindt was able to determine the acceptable limits of vacuum pumping rates for freeze–thaw slush production. He found that low pumping rates were unacceptable because of the texture of the solid produced. Very high pumping rates were also found to be undesirable because liquid droplets were carried into the warm regions of the system by the velocity of the evolving gas. The evaporation of these droplets was, therefore, considered a loss to the preparation of solid. Sindt also determined the rate that produces solids that are easily transformable into slush by measuring the evolved gas during many cycles of freeze–thaw preparation. This rate was determined to be 0.9 $m^3\ s^{-1}\ m^{-2}$ of liquid surface area at vacuum pump inlet conditions of 300 K and 6.9 kN m^{-2}. This was also found to be the pumping rate during the evacuation phase of the freeze–thaw cycle. He concluded that the range of the vacuum pumping rate for satisfactory slush production was 0.8–1.2 $m^3\ s^{-1}\ m^{-2}$ of liquid surface area. Pumping rates in that range were found to result in an efficient exchange of energy between the evaporating and freezing liquids. The efficiency of this exchange of energy is best illustrated by the experimentally determined irreversibility reported by Daney and Mann [7], which was found to be less than 3.4%. Pumping rates within this range have been used to produce slush of solid fractions to 0.45 in vessels 10–76 cm in diameter.

According to Voth [6], although the freeze–thaw production method was proved to be technically feasible and fully developed, it was shown to have disadvantages. These include the fact that the freeze–thaw process is a batch process and that it operates at the triple-point pressure of the cryogen. For hydrogen, this pressure was estimated to be 0.07 bar absolute, a pressure capable of drawing air through inadvertent leaks in the system. The air could be thought of as presenting a potential safety problem in a hydrogen system. Voth also showed that the freeze–thaw process required either costly equipment to recover the generated triple-point vapor or the loss of approximately 16% of the normal boiling-point liquid hydrogen and approximately 24% of the normal boiling-point liquid oxygen if the vapor was discarded.

26.2.2 Auger Method

Voth [6] reported on a study for producing liquid–solid mixtures of oxygen or hydrogen using an auger. He described an auger used to scrape frozen solid from the inside of a refrigerated brass tube in order to produce slush hydrogen. He showed that slush hydrogen could be continuously produced by this method, and since it could be immersed in liquid, slush was produced at pressures above the triple-point pressure. Voth was also able to produce the increased pressure pneumatically or by generating temperature stratification near the surface of the liquid. He observed that the auger system produced particles in the size realm of the particle produced by the freeze–thaw method so that, like the freeze–thaw-produced slush, the auger-produced slush could be readily transferred and stored.

According to Voth [6], the surface freezing occurring in the freeze–thaw production process is thermodynamically reversible. In contrast, the freezing process in the auger is irreversible since a temperature difference must exist between the refrigerant and the freezing cryogen, and the energy added to scrape the solid cryogen out of the brass tube must be removed by the refrigerant. Voth argued that in spite of these irreversibilities, the auger system was found to require less energy to produce slush hydrogen than a particle freeze–thaw system. He reported that the temperature difference required refrigeration temperatures below the triple-point temperature of hydrogen so a gaseous or liquid helium refrigerator was required for the auger.

Although this method was determined to offer the advantage of operating above atmospheric pressures (an important consideration in lowering the possibility of oxygen intrusion into the slush), it was shown to have the disadvantage of using expensive helium coolant.

26.3 Quality Determination

A thermodynamic analysis of the freeze–thaw production process for forming mixtures of slush developed by Mann et al. [8] showed that the measurement of two quantities, (1) the mass fraction pumped during the production process and (2) the heat leak per unit mass into the production Dewar, was sufficient to determine the quality of the resulting mixture. Since irreversibilities were observed to exist in the freeze–thaw process, a comparison of the predicted quality with the qualities determined by the measurement of thermodynamic properties was found necessary.

According to Daney and Mann [7], in the freeze–thaw method of forming a triple-point mixture of hydrogen, the latter was partially evaporated under the reduced pressure obtained by a vacuum pump, while a refrigeration effect (approximately equal to the latent heat of vaporization) was experienced by the remaining liquid or liquid–solid mixture. They suggested, by specifying the initial state and the process or path to be followed, that the quality of the liquid–solid mixture could be predicted as a function of the mass of the vapor removed. According to them, therefore, the problem of quality prediction consisted of two parts. The first was the determination of when a specific initial state, triple-point liquid, was reached, while the second was the prediction of the quality once this standard initial state was achieved. They also reported that the arrival at the triple-point liquid condition may be determined in several ways: (1) vapor pressure measurement, (2) visual observation of the formation of a few solid particles, and (3) prediction from the mass fraction of the vapor pumped off in the cooling-down process. Daney and Mann [7] argued that the first two methods provided a more accurate means of determining the arrival at triple-point liquid conditions, and hence, a more accurate quality determination could be realized. However, they also argued that it may be of interest to see what accuracy in quality determination could be expected by starting the mass accounting with a normal boiling-point liquid instead of a triple-point liquid.

26.4 Energy Requirements

A study to determine the energy requirements of the freeze–thaw and auger methods was reported by Voth [6] who mentioned that producing slush appears to be a reasonable way of increasing the density and heat capacity of cryogens. He observed that when large quantities of slush were produced, the energy required to produce them became significant. He therefore concluded that the production energy depended on both the thermodynamic reversible energy requirements to produce slush and the energy required to overcome irreversibilities in a practical system. He determined the energy required to produce slush with a 0.5 solid mass fraction for four practical slush hydrogen producing systems. Three of the systems employed the freeze–thaw production method, while the fourth used an auger and a helium refrigerator to produce slush from normal boiling-point liquid hydrogen.

According to Voth [6], the thermodynamic reversible energy required to produce slush hydrogen was equal to the thermodynamic availability of the slush. Also using normal hydrogen at a temperature of 300 K and a pressure of 1.013 bar as the base fluid, the reversible energy required to produce normal boiling-point para-hydrogen was 3971.4 W h kg^{-1} while the reversible energy required to produce slush with a solid mass fraction of 0.5 was 4372.8 W h kg^{-1}. Voth explained that the energy required by practical systems was higher because of component inefficiencies. The calculated energies for the four cases were based on liquefier and refrigerator efficiencies of 40% of that of Carnot. He assumed that the vacuum pumps required by the freeze–thaw production had an efficiency of 50% of the isothermal efficiency. However, he did not include the increased production energy due to heat leak into the containers and transfer lines because it was difficult to estimate without a firm system definition and because the heat leak would have been nearly equal for all of the systems studied. Because heat leak was not included, the calculated energies for the four cases were lower than those of an actual system, but the results did allow a comparison among the various slush production systems.

Voth also found that the auger method introduced additional irreversibilities because of the rotational power added to the slush generator and the temperature difference between the refrigerant and the freezing hydrogen. He concluded that a value of 1.05 for the ratio of total supplied refrigeration to the refrigeration available for the freezing hydrogen in the auger system was usually needed to determine the refrigeration required to produce slush with a solid fraction of 0.5. The input energy to the refrigerator connected to the auger was observed to depend on the refrigeration temperature; the lower the temperature, the higher the input energy. For the auger system, he assumed that the lowest refrigeration temperature was 10 K, and since a refrigerator was required to cool the normal boiling-point liquid to a triple-point liquid, the highest refrigeration temperature was assumed to be 19.76 K. He noted that these refrigeration temperatures were within the capabilities of a closed-cycle helium refrigerator. Voth also reported that while the power required to produce slush hydrogen using the auger was relatively easy to calculate, the power required to produce slush hydrogen using the freeze–thaw production method was very system dependent.

26.5 Solid Particle Characteristics

Since applications of slush hydrogen always include some storage, Mann et al. [8] observed the aging effects on solid particles in slush as solids aged as much as 100 h. Particles of slush in freshly prepared solids were observed to be in the form of agglomerates of very small subparticles loosely attached. These agglomerates were remnants of the porous layer that forms during the pumping phase of preparation. They observed that as slush aged, the solid particle configuration changed dramatically, and the loosely attached subparticles filled in with solids to form a more smoothly rounded particle of higher density. They asserted that the exact mechanism of the change was not completely understood but suggested that it involved a mass energy exchange, which occurred in such a way as to assume a more stable spherical shape with less surface area. According to them, the major portion of the change occurred within the first 5 h. They remarked that accompanying the change in particle configuration was an increase in the density of slush that had settled to the bottom of the vessel.

This tended to increase density partly from the increase in the bulk density of each particle and partly from a closer packing of particles due to their more rounded shape after aging. They thus concluded that the aging phenomenon of the particles and the accompanying increase in settled slush density were apparently a result of heat flow to the slush. Freshly prepared slush was concluded to have a maximum settled solid fraction of 0.35–0.45.

During the experimental study of Mann et al. [8], the slush particle size and particle settling velocity were measured using photo instrumentation. They noted that these characteristics of particles should be determined because they generally affect the fluid mechanics of liquid–solid mixtures. They also reported that particle size determines the minimum restrictions permissible in systems using the mixtures. They concluded that although particle configuration changed with age, particle size did not change significantly and that the distribution of particle size could be satisfactorily described by a modified logarithmic function. They reported that size ranged from 0.5 to 10 mm, with 2 mm being the most frequently occurring size. They also reported over 170 settling velocity measurements on both fresh and aged slush particles and concluded that the settling velocity of fresh particles was lower than that of aged particles and was highly dependent on the shape of individual particles.

Sindt [5] reported that mixing of slush hydrogen to maintain homogeneity was required in most of the potential applications, and if mixers were to impart velocities to the fluid greater than the settling velocities, particles would have been carried along and mixing would have resulted. Sindt used several methods to mix the slush in a vessel. One method used a simple propeller with blades designed to impact velocity to the fluid with lift. Another method used high-velocity streams of fluid expelling from openings in a duct. According to him, both methods of mixing had been used to mix slush of a solid fraction as high as 0.6. Also, the solid content in slush could be maintained and even increased by transferring slush into the vessel and removing liquid through a screen with 0.6 mm openings. Sindt explained that using this method, the solid fraction could be increased until the solids protrude above the liquid–gas interface.

26.6 Transfer Characteristics

Most of the studies on transfer characteristics of slush hydrogen have been carried out at the Cryogenics Division of the National Institute of Standards and Technology (NIST). Experimental results pertaining to slush hydrogen fluidity will be reported in this section.

26.6.1 Slush Hydrogen Flow in Pipes

Sindt [5] reported that transfer and pumping characteristics of slush hydrogen should be similar to those of liquid hydrogen if slush is to be used in the existing liquid hydrogen system and if slush is to be handled as a fluid. During the experimental study of Sindt et al. [12], slush of solid fractions as high as 0.6 was transferred through vacuum-insulated pumps of 16.6–25 mm in diameter. Transfer pressure losses were determined in triple-point liquid and in slush of solid fractions up to 0.55 in a 16.6 mm diameter pipe. They estimated slush data to have an uncertainty occurring at the high solid fractions and the lowest flow rates. According to them, pressure loss data for flowing slush revealed that at low flow rates, pressure losses in flowing slush were double those of triple-point liquid if solid fractions were high.

The data also revealed that at high flow rates, pressure losses in flowing slush were less than the losses in triple-point liquid if the solid fraction was near 0.35. They also carried out a more complete analysis of the losses in flowing slush hydrogen by calculating a friction factor and plotting it versus the Reynolds number. In the calculations, they used triple-point liquid viscosity because the viscosity of slush hydrogen was generally unknown.

Sindt and Ludtke [11] observed that the characteristics of slush hydrogen flow losses at the higher Reynolds numbers were more evident when displayed in terms of the friction factors than if displayed as pressure losses. They reported that friction losses in flowing slush of solid fractions to 0.4 were as much as 10% less than in liquid and, that, the friction-loss curve for 0.5 solid fraction slush appeared to be following a slope such that it would be below triple-point liquid losses at high Reynolds numbers. They reported these data for a 16.6 mm diameter pipe and concluded that low losses in slush aged 10 h were 4%–10% higher than losses in 1 h old slush.

The experiments of Sindt et al. [12] revealed that a gradient in the solid concentration may develop when the slush velocity is 0.5 m s^{-1}, while at a velocity of 0.15 m s^{-1}, the solids were likely to settle forming a sliding bed. Considering these observations, it was concluded that the critical velocity of slush in the pipe was about 0.5 m s^{-1}.

26.6.2 Slush Hydrogen Flow in Restrictions

Sindt et al. [12] experimentally determined flow losses in restrictions such as valves, orifices, and venturis using slush hydrogen of solid fractions to 0.5. They determined the losses in a 3/4 in. nominal copper globe valve with the valve fully opened and in cases when it was partially opened and expressed the losses as a resistance coefficient independent of the slush solid fraction. According to their study, with the valve partially opened, the losses seemed to be less with increasing the solid fraction, while with the valve fully opened, the trend was reversed. Also, the loss coefficients compared favorably with the typical valve coefficient for water.

Sindt and Ludtke [11] remarked that liquid–solid mixtures of hydrogen flow through orifices of 6.4 and 9.5 mm in diameter with no plugging and with a pressure drop of the same magnitude as that of the triple-point liquid. They observed that the slush mass flow rate was dependent on the pressure differential across the orifice as would have been expected for a nearly incompressible, Newtonian fluid, and that when the minimum pressure in the orifice reached the triple-point pressure, cavitation with accompanying choking occurred. At that point the mass flow became dependent on the upstream pressure and independent of the downstream pressure. Flow characteristics of slush flowing through a venturi have also been determined by Sindt et al. [12]. According to them, flow losses through the venturi were not affected by the solid fraction of the slush.

26.7 Conclusions

This chapter presented an overview of the state of the art of slush hydrogen production and utilization technologies. It is apparent that the use of slush hydrogen should be considered only for cases in which higher density and greater solid content are really needed. This is mainly because the production costs of slush hydrogen now and in the near future are greater than the liquefaction costs of hydrogen due to the larger energy use involved.

References

1. A.S. Rapial and D.E. Daney, Preparation and characterization of slush and nitrogen gels, NBS Technical Note 378, National Bureau of Standards, Boulder, CO (1969).
2. E.W. Justi, *A Solar Hydrogen Energy System*, Plenum Press, New York (1987).
3. C.R. Baker and L.C. Matsch, Production and distribution of liquid hydrogen, *Advances in Petroleum Chemistry and Refining* **10** 37–81 (1965).
4. S.W. Kandebo, Researchers explore slush hydrogen as fuel for national aero-space plane, *Aviation Week & Space Technology* 37–38 (1989).
5. C.F. Sindt, A summary of the characterization study of slush hydrogen, *Cryogenics* **10** 372–380 (1970).
6. R.O. Voth, Producing liquid–solid mixtures (slushes) of oxygen or hydrogen using an auger, *Cryogenics* **25** 511–517 (1985).
7. D.E. Daney and D.B. Mann, Quality determination of liquid–solid hydrogen mixtures, *Cryogenics* **7** 280–285 (1967).
8. D.B. Mann, P.R. Ludtke, C.F. Sindt, and D.B. Chelton, Liquid–solid mixtures of hydrogen near the triple point, *Advances in Cryogenic Engineering*, Vol. 11, Plenum Press, New York, pp. 207–217 (1966).
9. E.M. Smith, Slush hydrogen for aerospace applications, *International Journal of Hydrogen Energy* **14**(3) 201–213 (1989).
10. H.C. Angus, Storage distribution and compression of hydrogen, *Chemistry and Industry*, **16** 68–72 (1984).
11. C.F. Sindt and P.R. Ludtke, Slush hydrogen flow characteristics and solid fraction upgrading, *Advances in Cryogenic Engineering*, Vol. 15, Plenum Press, New York, pp. 382–390 (1970).
12. C.F. Sindt, P.R. Ludtke, and D.E. Daney, Slush hydrogen fluid characterization and instrumentation, NBS Technical Note 377, National Bureau of Standards, Boulder, CO (1969).
13. L.R. Perrel and D.G. Haase, Self-catalysis in solid hydrogen, a computer simulation, *American Journal of Physics*, **52**(9) 831–833 (1984).
14. H.M. Roder, A new phase transition in solid hydrogen, *Cryogenics* **13**(7) 439–440 (1973).

Section VIII

Hydrogen Conversion and End Use

27

*Hydrogen-Fueled Internal Combustion Engines**

Sebastian Verhelst
Ghent University

Thomas Wallner
Argonne National Laboratory

Roger Sierens
Ghent University

CONTENTS

* Reprinted from *Progress in Energy and Combustion Science 35*, Verhelst, S. and Wallner, T., Hydrogen-fueled internal combustion engines, 490–527, Copyright 2009, with permission from Elsevier.

27.1 Introduction

27.1.1 Incentives and Drawbacks for Hydrogen as an Energy Carrier

The current way of providing the world's energy demand, based primarily on fossil fuel, is becoming increasingly untenable. Fossil fuel reserves, once hardly ever given a second thought, now are clearly exhaustible. Fossil fuel prices have never been more volatile, influenced first by economic acceleration mostly in China and India and subsequently by economic recession. The difficulty of controlling prices and the uncertain reserves are strong incentives for pursuing energy security. Global warming and local pollution hot spots associated with fossil fuel usage are further significant environmental and societal problems.

These are strong drivers for research, development and demonstrations of alternative energy sources, energy carriers, and in the case of transportation, power trains. The use of hydrogen as an energy carrier is one of the options put forward in most governmental strategic plans for a sustainable energy system. The US Department of Energy (DOE); the European Commission's Directorate-General for Research; the Japanese Ministry of Economy, Trade and Industry; the Indian Ministry of New and Renewable Energy; and many others have formulated vision reports and published funding calls for hydrogen programs [1–4].

The attractiveness of hydrogen lies in the variety of methods to produce hydrogen as well as the long-term viability of some of them (from fossil fuels, from renewable energy [biomass, wind, solar [5], from nuclear power, etc.]), the variety of methods to produce energy from hydrogen (internal combustion engines, gas turbines, fuel cells), virtually zero harmful emissions, and potentially high efficiency at the point of its use. Compared to biofuels, a recent study reported the yield of final fuel per hectare of land for different biomass-derived fuels and of hydrogen from photovoltaics or wind power [6]. The results show that the energy yield of land area is much higher when it is used to capture wind or solar energy. Compared to electricity, using hydrogen as an energy carrier is advantageous in terms of volumetric and gravimetric energy storage density. However, there are also serious challenges to overcome when hydrogen is to be used as an energy carrier. Although better than batteries in storage terms, its very low density implies low energy densities compared to the fuels in use today, even when compressed to 700 bar or liquefied, both of which incur substantial energy losses. Thus, distribution, bulk storage, and onboard vehicle storage are heavily compromised. Also, in case of hydrogen-fueled vehicles, care must be taken to ensure that the well-to-wheel greenhouse gas emission reduction compared to hydrocarbon fuel turns out to be positive [7]. Nevertheless, the advantages offered by hydrogen are significant enough to warrant the exploration of its possibilities.

27.1.2 Using Hydrogen in Internal Combustion Engines: Sense or Nonsense?

There are numerous works and opinions as to what constitutes the *best* fuel/energy carrier as well as power train: heavily optimized hydrocarbon fueled engines, biofuels, electricity, hydrogen, etc. However, there are always a multitude of aspects to be taken into account, ranging from well-to-wheel (or cradle-to-grave) primary energy use, greenhouse gas emissions, tailpipe emissions relating to local pollution, cost, practicality, to customer acceptance, etc., which clearly are not all easily scored and ranked, making it very hard to predict the winner(s). Moreover, there appears to be no silver bullet, and numerous choices merit a detailed study in exploring the possibilities and drawbacks.

Concerning hydrogen as an energy carrier, Shelef and Kukkonen [7] have compared hydrogen fuel cell (H_2FC) vehicles and H_2ICE vehicles to gasoline vehicles and electric vehicles, on the basis of well-to-wheel carbon dioxide (CO_2) emissions and primary energy use. Fifteen years have passed since that study, but overall, the arguments and challenges for the implementation of hydrogen as a vehicle fuel largely hold. However, the data that were used clearly need updating to see how this affects the conclusions. Shelef and Kukkonen concluded that the H_2FC vehicle could decrease primary energy use and greenhouse gas emissions compared to gasoline and natural gas vehicles, but that H_2ICE vehicles increased both. However, a recent study at Argonne National Laboratory that compares the fuel economy potential of hydrogen power trains to conventional gasoline vehicles concludes that by 2045, a H_2ICE hybrid-electric vehicle (HEV) would only consume 9% more than a H_2FC HEV, as a result of the recent and expected future significant improvements in hydrogen engine technology [8].

Thus, the perceived large difference in fuel economy between a H_2FC HEV and H_2ICE HEV is lower than frequently reported and will decrease over time. Currently, H_2ICEs are much cheaper than H_2FCs, both directly and in terms of fuel cost (with high fuel purity requirements for the H_2FCs). Furthermore, using ICEs allows bifuel operation (e.g., the engine can run on gasoline as well as on hydrogen), alleviating fuel station density and autonomy requirements. This could facilitate the start-up of a hydrogen infrastructure, where the experience gained with transport, fueling, and storage directly translates to fuel cell vehicles.

This explains the US DOE's position that while FC vehicles consistently achieve the highest fuel efficiency, the H_2ICE can serve as a bridging technology and might help in the development of the infrastructure needed for hydrogen fuel [1]. Similarly, the well-to-wheels study of the European Commission's Directorate General Joint Research Center, in cooperation with CONCAWE and EUCAR, concludes that H_2ICE vehicles are available in the near term at a lower cost than fuel cell vehicles [9].

27.1.3 Motivation and Outline

This chapter sets out to review hydrogen engine technology. It does not discuss where, when, and if hydrogen should be used as an energy carrier. The previous sections provide a brief summary of the attractiveness of using hydrogen, the interest in H_2ICEs, and a warning about the challenges.

Previous articles have already discussed hydrogen engine technology, with excellent reviews by Das [10,11], Eichlseder et al. [12], and White et al. [13]. Most of these papers focus on the application without attention to the fundamentals of hydrogen combustion and mostly cover only certain aspects of H_2ICEs. Given the multitude of recent studies as well as ongoing efforts in developing H_2ICEs and H_2ICE vehicles, it is considered timely to offer a comprehensive review.

The following starts out by considering the fundamental properties of hydrogen and hydrogen combustion. Some important H_2ICE features can already be expected from the physical and chemical properties of hydrogen compared to hydrocarbons; these are discussed in Section 27.2.1. Sections 27.2.2 and 27.2.3 cover the laminar and turbulent burning velocities of hydrogen mixtures at engine conditions, to set the stage for Section 27.3 on modeling of H_2ICEs and subsequent chapters. In most of the initial work on H_2ICEs, abnormal combustion encountered with operation on hydrogen was the main subject, so this is covered in Section 27.4, after which engine features are discussed (Section 27.5), which allow the operation of a hydrogen-fueled engine. For these designated or converted hydrogen engines, Section 27.6 discusses the possibilities of controlling the power output while striving for maximum efficiency and minimum emissions. Section 27.7 reviews some safety aspects (those relevant to experimental H_2ICE work), after which the performance of H_2ICEs in vehicles is evaluated using selected examples in Section 27.8. Section 27.9 presents the use of mixtures of other fuels with hydrogen. Concluding Sections 27.10 and 27.11 summarize the gaps remaining in current knowledge and the most important points from the current work.

Experimental data presented to clarify the effects of heat flux (Section 27.3.4), combustion anomalies (Section 27.4), and mixture formation (Sections 27.3.1 and 27.6) on combustion were collected on automotive-size single-cylinder research engines operated at the Center for Transportation Research at Argonne National Laboratory, the Institute for Internal Combustion Engines at Graz University of Technology, and the Department of Flow, Heat and Combustion Mechanics at Ghent University.

27.2 Fundamentals

27.2.1 Physical and Chemical Properties of Hydrogen Relevant to Engines

Starting from some physical and chemical properties of hydrogen and hydrogen–air mixtures, a number of H_2ICE features can already be defined or expected.

Table 27.1 lists some properties of hydrogen compared with methane and iso-octane [14–17], which are taken here as representing natural gas and gasoline, respectively, as it is easier to define properties for single-component fuels. The small and light hydrogen molecule is very mobile (high mass diffusivity) and leads to a very low density at atmospheric conditions.

The wide range of flammability limits, with flammable mixtures from as lean as $\lambda = 10$ to as rich as $\lambda = 0.14$ ($0.1 < \varphi < 7.1$), allows a wide range of engine power output through changes in the mixture equivalence ratio. The flammability limits widen with increasing temperature, with the lower flammability limit dropping to 2 vol.% at 300°C (equivalent to $\lambda = 20/\varphi = 0.05$) [18]. The lower flammability limit increases with pressure [18], with the upper flammability limit having a fairly complex behavior in terms of pressure dependence [19] but of lesser importance to engines. As will be discussed later, in practice, the lean limit of H_2ICEs

TABLE 27.1

Hydrogen Properties Compared with Methane and Iso-Octane Properties

Property	Hydrogen	Methane	Iso-Octane
Molecular weight (g/mol)	2.016	16.043	114.236
Density (kg/m^3)	0.08	0.65	692
Mass diffusivity in air (cm^2/s)	0.61	0.16	~0.07
Minimum[a] ignition energy (mJ)	0.02	0.28	0.28
Minimum[a] quenching distance (mm)	0.64	2.03	3.5
Flammability limits in air (vol%)	4–75	5–15	1.1–6
Flammability limits (λ)	10–0.14	2–0.6	1.51–0.26
Flammability limits (φ)	0.1–7.1	0.5–1.67	0.66–3.85
Lower heating value (MJ/kg)	120	50	44.3
Higher heating value (MJ/kg)	142	55.5	47.8
Stoichiometric air-to-fuel ratio (kg/kg)	34.2	17.1	15.0
Stoichiometric air-to-fuel ratio (kmol/kmol)	2.387	9.547	59.666

Note: Data given at 300 K and 1 atm.

[a] Corresponding equivalence ratios given in text.

is reached for lower air-to-fuel equivalence ratios than mentioned earlier, in the vicinity of $\lambda = 4/\varphi = 0.25$. The lower flammability limit is mostly determined by the classical method of flame propagation in a tube. As mentioned earlier, the mass diffusivity of hydrogen is high, and this causes a difference in the limit for upward or downward propagating flames, due to preferential diffusion in the presence of buoyancy [20,21] (see Section 27.2.2). For upward propagating flames, mixtures as lean as 4% hydrogen in air are still flammable but are noncoherent and burn incompletely. The value of 4% pertains to one particular experimental configuration, so in real-world situations, the limit may well be below 4% (or above, depending on conditions). The absolute limit is thus not well known even today. However, this limit is important for safety considerations but less so for engine combustion.

The minimum ignition energy of a hydrogen–air mixture at atmospheric conditions is an order of magnitude lower than for methane–air and iso-octane–air mixtures. It is only 0.017 mJ, which is obtained for hydrogen concentrations of 22%–26% ($\lambda = 1.2$–$1.5/\varphi = 0.67$–0.83) [22]. The minimum ignition energy is normally measured using a capacitive spark discharge and thus is dependent on the spark gap. The figure quoted earlier is for a gap of 0.5 mm. Using a 2 mm gap, the minimum ignition energy is about 0.05 mJ and more or less constant for hydrogen concentrations between 10% and 50% ($\lambda = 0.42$–3.77, $\varphi = 0.27$–2.38), with a sudden increase when the concentration of H_2 is below 10% [22].

The quenching distance can be experimentally derived from the relation between the minimum ignition energy and the spark gap size [23] or directly measured [24]. It is minimal for mixtures around stoichiometry and decreases with increasing pressure and temperature. As can be seen in Table 27.1, it is about one third that for methane and iso-octane. This affects crevice combustion and wall heat transfer, as will be discussed later.

Finally, note the large difference between the lower and higher heating values of hydrogen compared with methane and iso-octane, which is easily explained as H_2O is the sole combustion product of hydrogen. Also note the large difference in stoichiometric air-to-fuel ratio of hydrogen compared with methane and iso-octane, as well as the large difference in stoichiometric air-to-fuel ratio in mass terms versus in mole terms.

The properties of the different fuels also determine their ability to efficiently store enough energy onboard a vehicle. The US DOE has set goals for volumetric as well as

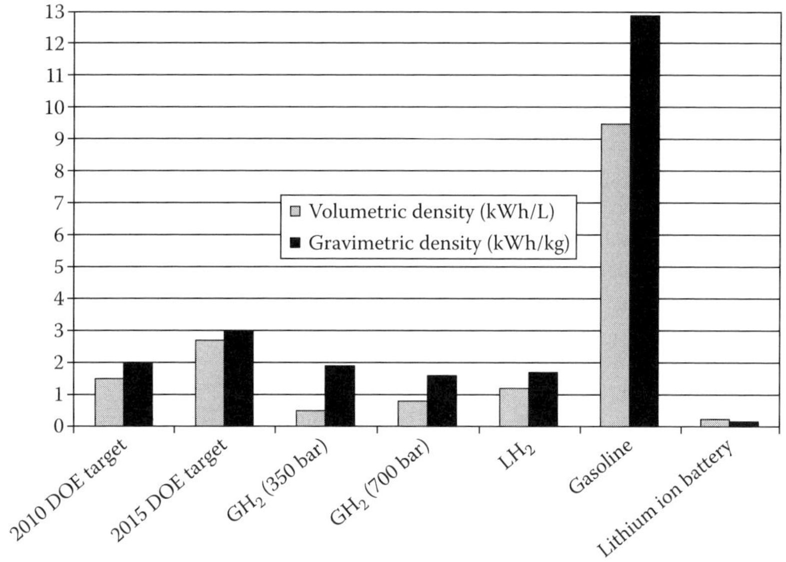

FIGURE 27.1
Volumetric and gravimetric energy storage densities of compressed gaseous hydrogen (GH_2) and liquefied hydrogen (LH_2): DOE targets and current levels compared with gasoline and lithium ion batteries.

gravimetric storage densities of hydrogen storage systems. Figure 27.1 shows a comparison of DOE targets, current energy density levels of gaseous as well as liquid hydrogen storage systems as well as the energy density of gasoline and a lithium ion battery system for comparison. Neither compressed nor liquid hydrogen storage can currently meet the 2010 DOE goals of 1.5 kWh/L and 2 kWh/kg, respectively [25].

Table 27.2 lists the properties of hydrogen–air mixtures, at stoichiometric and at the lean limit mentioned earlier, compared with stoichiometric methane–air and iso-octane–air mixtures [14–17]. The volume fraction of fuel in the fuel–air mixture can be directly calculated from the molar stoichiometric air-to-fuel ratio listed in Table 27.1. The large volume fraction occupied by hydrogen has consequences for the attainable engine power density (see further below). Combined with the wide flammability limits, it also has an important effect on mixture properties such as the kinematic viscosity and thermal conductivity. These properties vary much more than in conventionally fueled engines. This affects, for example, nondimensional numbers used in modeling (see Section 27.3), such as Reynolds numbers, which can substantially differ from the numbers for hydrocarbon combustion. The comparatively large variation in mixture density and thus the speed of sound affects the gas dynamics in engines with external mixture formation.

An increased ratio of specific heats results in an increased amount of compression work. However, the actual compression work, particularly for direct-injection (DI) operation, strongly depends on the injection strategy (see Section 27.6). Calculations have shown that injection timing and duration are the dominating factors compared with the fuel properties, and efficiency benefits of up to 4% can be gained when employing an optimized injection strategy [26].

There is some ambiguity concerning the autoignition temperature of fuels in general and hydrogen in particular. For instance, for methane, values have been found ranging from 810 K [27] to 868 K [28]. For hydrogen, values were found from 773 K [29] to 858 K [30]. Some sources list the autoignition temperature for hydrogen as lower than that for

TABLE 27.2

Mixture Properties of Hydrogen–Air, Methane–Air, and Iso-Octane–Air

Property	H_2–Air $\lambda = 1$ $\varphi = 1$	H_2–Air $\lambda = 4$ $\varphi = 0.25$	CH_4–Air $\lambda = 1$ $\varphi = 1$	C_8H_{18}–Air $\lambda = 1$ $\varphi = 1$
Volume fraction fuel (%)	29.5	9.5	9.5	1.65
Mixture density (kg/m^3)	0.850	1.068	1.123	1.229
Kinematic viscosity (mm^2/s)	21.6	17.4	16	15.2
Autoignition temperature (K)	858[a]	>858[a]	813[a]	690[a]
Adiabatic flame temperature (K)	2390	1061	2226	2276
Thermal conductivity (10^{-2} W/m K)	4.97	3.17	2.42	2.36
Thermal diffusivity (mm^2/s)	42.1	26.8	20.1	18.3
Ratio of specific heats	1.401	1.400	1.354	1.389
Speed of sound (m/s)	408.6	364.3	353.9	334.0
Air-to-fuel ratio (kg/kg)	34.2	136.6	17.1	15.1
Mole ratio before/after combustion	0.86	0.95	1.01	1.07
Laminar burning velocity, ~360 K (cm/s)	290	12	48	45
Gravimetric energy content (kJ/kg)	3758	959	3028	3013
Volumetric energy content (kJ/m^3)	3189	1024	3041	3704

Note: Data given at 300 K and 1 atm (with the exception of the laminar burning velocity, given at 360 K and 1 atm).

[a] See text.

methane; other sources list the opposite. This ambiguity can be at least partly explained by the sensitivity of autoignition temperatures to the experimental apparatus, the experimental procedure, and the criterion used for defining the value [31].

For spark-ignition (SI) engines, with a propagating flame front, autoignition of the unburned mixture ahead of the flame front is unwanted, as it can result in knocking* combustion. The efficiency of an SI engine is influenced by the compression ratio and the ignition timing (among others), the choices of which are dependent on the autoignition temperature of the fuel–air mixture, so this is an important parameter. For liquid hydrocarbons, the octane rating is more commonly used as a measure of the propensity of a fuel–air mixture to undergo preflame reactions. For hydrogen, a research octane number (RON) in excess of 130 and a motor octane number (MON) of 60 have been reported [33,34]. It is also noteworthy that for the determination of the *methane number* (MN) of a gaseous fuel, hydrogen is taken as a reference fuel, having an MN of zero [35], giving the impression that it is very prone to knock (for details, see Section 27.4).

The laminar burning velocity of stoichiometric hydrogen–air mixtures is much higher than that of methane and iso-octane. However, if lean-burn strategies are used, the burning velocity can be much lower (see value for $\lambda = 1/\varphi = 0.25$). For mixtures around stoichiometry, the high burning velocity and high adiabatic flame temperature point to high nitrogen oxide (NO_X) emissions (see later). The laminar burning velocity of hydrogen mixtures is extensively discussed in Section 27.2.2.

* The term *knock* is commonly used to denote end-gas autoignition but in itself is a poor term, as it denotes the physical manifestation of abnormal oscillations in the cylinder pressure [32] and can also result from, for example, surface ignition.

TABLE 27.3

Theoretical Power Densities of Hydrogen-, Methane- and Iso-Octane-Fueled Engines

	Hydrogen (%)	Methane (%)	Iso-Octane (%)
PFI	86	92	100
DI	119	100	100

Combining the lower heating value (LHV) of hydrogen, its density, and the stoichiometric air requirement, one can calculate the maximum theoretical power density of different engine concepts. Table 27.3 compares the theoretical power density of hydrogen-and methane-fueled engines with an iso-octane-fueled engine as the reference. Values for both port-fuel-injection (PFI) engines and DI engines are quoted. Note the large difference for the gaseous fuels, with a (theoretical) power density increase of 38% for hydrogen when switching from PFI to DI.

Finally, it is noteworthy that the possibility of qualitative load control (changing the mixture richness at wide open throttle [WOT]), the tolerance for substantial mixture dilution (either through excess air or exhaust gas recirculation [EGR]), the high autoignition temperature (allowing high compression ratios), and the generally fast burn rate are all factors contributing to potentially high engine efficiencies. As will be discussed in Section 27.6, this has been experimentally confirmed. As will be briefly discussed in Section 27.3.4, however, heat losses from cylinder gases to the combustion chamber walls can be higher with hydrogen compared with conventional fuels, negatively affecting efficiencies.

27.2.2 Laminar Burning Velocity: Influence of Preferential Diffusion

The laminar burning velocity, u_l, of a fuel–air mixture is an important physicochemical property due to its dependence on pressure, temperature, mixture equivalence ratio, and diluent concentration. It affects the combustion rate in an engine, the equivalence ratio limits for stable combustion, the tolerance for EGR, etc. Most engine combustion models assume the flame structure to be that of a (stretched) laminar flame, with the effect of the in-cylinder turbulence to be one of stretching and wrinkling the flame, thereby increasing the flame area. Consequently, data on the laminar burning velocity and its dependence on pressure, temperature, mixture composition, and stretch rate are a prerequisite. In the following, the available data are discussed, along with the effects of preferential diffusion. These effects have long been known [36] but are not always familiar to engine researchers. As the effects are very pronounced for the case of hydrogen, the issue is discussed here at some length.

27.2.2.1 Flame Front Instabilities

Several mechanisms exist that can trigger instability of a laminar flame. As these instabilities have important implications on hydrogen combustion, this section gives a brief overview of the effects a disturbance (perturbation) can have on a flame front, mainly from a phenomenological point of view [37–39].

When the laminar flame is regarded as a passive surface (an infinitely thin interface separating low-density burned gases from higher density unburned gases), a wrinkling of the flame front will not affect the flame intensity but will increase the volumetric burning rate through increased flame area. The discontinuity of density ($\rho_u \rightarrow \rho_b$) causes a hydrodynamic instability known as the Darrieus–Landau instability [37,38]. Simply speaking, a wrinkle of the flame front will cause a widening of the streamtube to the protrusion of the flame front into the unburned gases, resulting in a locally decreased gas velocity. This will cause a further protrusion of this flame segment as the flame speed remains unchanged (because the flame structure is not affected). Thus, a flame is unconditionally unstable when only considering hydrodynamic stretch and neglecting the effect of flame stretch (see later) on the structure of the flame.

The lower density of the burned gases compared to the unburned gases is also the cause for a second instability arising from gravitational effects. This body-force or buoyant instability, also known as the Rayleigh–Taylor instability, arises when a less-dense fluid is present beneath a more-dense fluid, such is the case in, for example, an upwardly propagating flame.

Finally, flame instability can be caused through unequal diffusivities [37,38]. As the flame propagation rate is largely influenced by the flame temperature, and this is in turn influenced by the conduction of heat from the flame front to the unburned gases and the diffusion of reactants from the unburned gases to the flame front, a perturbation of the balance between diffusivities can have important effects. Three diffusivities are of importance: the thermal diffusivity of the unburned mixture (D_T), the mass diffusivity of the so-called deficient* reactant ($D_{M,lim}$), and the mass diffusivity of the so-called excess† reactant ($D_{M,exc}$). The ratio of two diffusivities can be used to judge the stability of a flame when subjected to a perturbation or flame stretch.

The Lewis number Le of the deficient reactant is defined as the ratio of the thermal diffusivity of the unburned mixture to the mass diffusivity of the deficient reactant:

$$Le = \frac{D_T}{D_{M,lim}} \tag{27.1}$$

If this Lewis number is greater than unity, the thermal diffusivity exceeds the mass diffusivity of the limiting reactant. When this is the case, a wrinkled flame front will have parts that are *bulging* toward the unburned gases that lose heat more rapidly than diffusing reactants can compensate for. The parts that recede in the burned gases, on the contrary, will increase in temperature more rapidly than being depleted of reactants. As a result, the flame speed of the *crests* will decrease and the flame speed of the *troughs* will increase, which counteracts the wrinkling and promotes a smooth flame front. The mixture is then called thermodiffusively stable. When the Lewis number is smaller than unity, similar reasoning shows that a perturbation is amplified, which indicates unstable behavior.

Another mechanism involving unequal diffusivities is the following: When the limiting reactant diffuses more rapidly than the excess reactant ($D_{M,lim} > D_{M,exc}$), it will reach a bulge of the flame front into the unburned gases more quickly and cause a local shift in mixture ratio. As in this case, the more diffusive reactant is the limiting reactant, the local mixture ratio will shift so that it is nearer to stoichiometry, and the local flame speed will increase. Thus, a perturbation is amplified and the resulting instability is termed a preferential diffusion instability. This mechanism is easily illustrated by the propensity of rich heavier-than-air fuels (e.g., propane/air [40], iso-octane/air [41]) and lean lighter-than-air

* This refers to the reactant limiting the rate of reaction. Thus, in a lean flame, the deficient reactant is the fuel; in a rich flame, it is oxygen.

† For a lean flame, this is oxygen; for a rich flame, this is the fuel component.

fuels (e.g., methane/air [42,43], hydrogen/air [40]) to develop cellular flame fronts (see also the review paper by Hertzberg [21]). The selective diffusion of reactants can be viewed as a stratification of the mixture [44].

Both mechanisms involving unequal diffusivities are sometimes called differential diffusion instabilities, or instabilities due to non-equidiffusion.

In reality, all mechanisms described earlier are simultaneously present. Disturbances of a flame front causing it to deviate from a steady planar flame can be summarized in one scalar parameter, the rate of flame stretch, a, which is defined as the normalized rate of change of an infinitesimal area element of the flame:

$$\alpha = \frac{1}{A}\frac{dA}{dt} \tag{27.2}$$

The combined effect of the instability mechanisms is dependent on the magnitude of the stretch rate. For instance, thermodiffusively stable spherically expanding flames start out smooth, as the stretch rate is initially high enough for thermodiffusion to stabilize the flame against hydrodynamic instability. For small to moderate rates of stretch, the effect of stretch on the burning velocity can be expressed to first order [37] by

$$u_l - u_n = L_\alpha \tag{27.3}$$

where

The subscript "n" denotes the stretched value of the normal burning velocity

L is a Markstein length

Depending on the sign of L and whether the flame is positively or negatively stretched, the actual burning velocity can be increased or decreased compared to the stretch-free burning velocity, u_l. A positive Markstein length indicates a diffusionally stable flame, as flame stretch decreases the burning velocity. Any disturbances (wrinkles) of the flame front will thus tend to be smoothed out. A negative Markstein length indicates an unstable flame. A perturbation of the flame front will then be enhanced, and such flames quickly develop into cellular structures. The Markstein length is also a physicochemical parameter that embodies the effect of a change in flame structure when the flame is stretched. Thus, when measuring burning velocities, it is important that this is done at a well-defined stretch rate and the Markstein length is simultaneously measured so that the stretch-free burning velocity can be calculated. It has taken a while for the effects of stretch to be understood and for measuring methodologies to be developed that could take the effects into account. As illustrated in the following section, this is the main reason for the large spread in the reported data on hydrogen mixture burning velocities throughout the years.

27.2.2.2 Laminar Burning Velocity at Atmospheric Conditions

Contemporary reviews of data and correlations for the laminar burning velocity of hydrogen–air mixtures show a wide spread of experimental and numerical results [45,46]. Figure 27.2 plots laminar burning velocities against the equivalence ratio for hydrogen–air mixtures at normal temperature and pressure (NTP). Note the large difference in burning velocities, with stoichiometric burning velocities varying from 2.1 up to 2.5 m/s, with even larger differences for the lean mixtures (e.g., for $\lambda = 2$ from 56 to 115 cm/s).

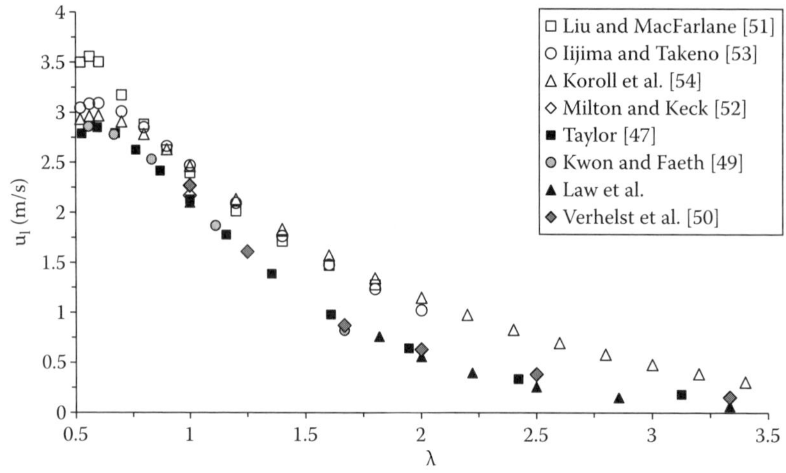

FIGURE 27.2
Laminar burning velocities plotted against air-to-fuel equivalence ratio, for NTP hydrogen–air flames. Experimentally derived correlations from Liu and MacFarlane [51], Milton and Keck [52], Iijima and Takeno [53], and Koroll et al. [54]. Other experimental data from Taylor [47], Vagelopoulos et al. [48], Kwon and Faeth [49], and Verhelst et al. [50].

The cause of this large spread can be found in the influence of the flame stretch rate on experimentally observed burning velocities.

The solid symbols in Figure 27.2 denote stretch-free burning velocities (or rather, burning velocities that were corrected to account for the effects of the flame stretch rate), as measured by Taylor [47], Vagelopoulos et al. [48], Kwon and Faeth [49], and Verhelst et al. [50]. The open symbols denote other measurements that did not take stretch rate effects into account, as reported by Liu and Mac-Farlane [51], Milton and Keck [52], Iijima and Takeno [53], and Koroll et al. [54]. These experiments result in consistently higher burning velocities, with the difference increasing for leaner mixtures.

Because of the very high mass diffusivity of hydrogen (the highest of all fuels), a lean to stoichiometric hydrogen/air flame (i.e., for equivalence ratios such as used in homogeneous charge hydrogen engines) will be diffusionally unstable, both from the Lewis number ($D_T \ll D_{M;H_2}$) and from the preferential diffusion ($D_{M;H_2} > D_{M;O_2}$) point of view. Thus, these flames are very sensitive to the flame stretch rate. Figure 27.3 shows schlieren photographs of a centrally ignited $\lambda = 1.43/\varphi = 0.7$, 365 K, 1 bar H_2–air flame propagating in a constant-volume *bomb* [46].

The flame starts out smooth but quickly evolves into a fully cellular flame. From image processing of the schlieren photographs, the flame radius, r, can be obtained as a function of time, t. Figure 27.4 plots the flame speed, $S_n = dr/dt$, versus the flame stretch rate, α, which can be calculated as $2S_n/r$ for spherically expanding flames. In spherically expanding flames, the flame starts out highly stretched and as the flame grows,

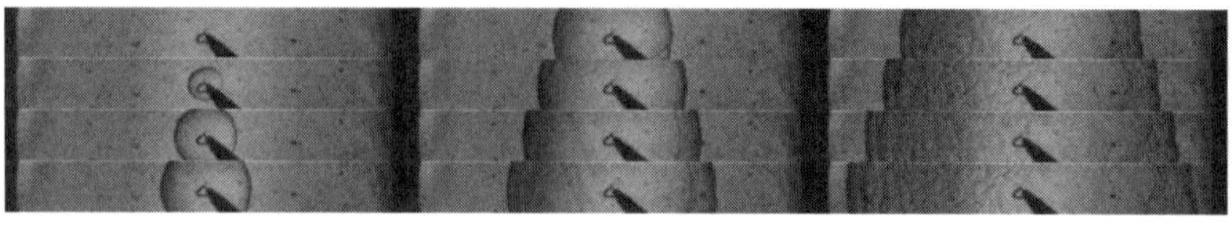

FIGURE 27.3
Schlieren photographs of a $\lambda = 1.43/\varphi = 0.7$, 365 K, 1 bar H_2–air flame. Time interval between frames: 0.641 ms.

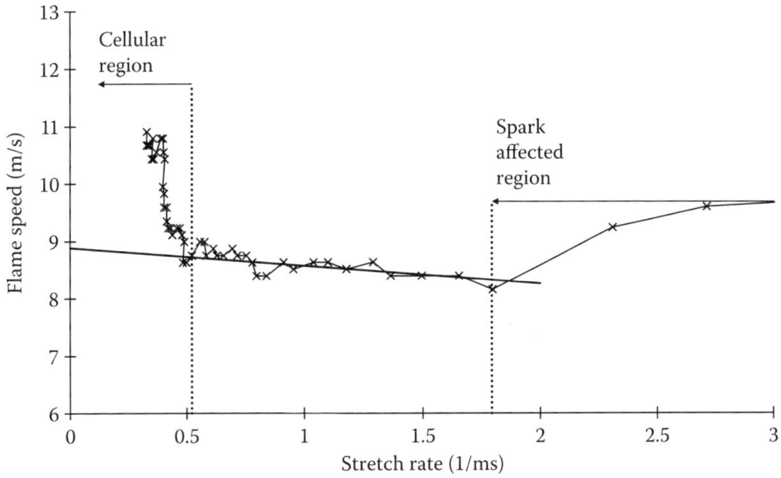

FIGURE 27.4
Development of a $\lambda = 1.43/\varphi = 0.7$, 365 K, 1 bar H_2–air flame. (From Verhelst, S., A study of the combustion in hydrogen-fuelled internal combustion engines, PhD thesis, Ghent University, Ghent, Belgium, 2005, http://hdl.handle.net/1854/3378.)

the stretch rate decreases; thus, Figure 27.4 should be read from right to left. Such a plot demonstrates a number of things. First, the methodology of obtaining stretch-free flame speeds. This can be seen from the data points between the *spark affected* and *cellular* regions. After the effects of the spark *boost* have decayed, a regime is found where the flame speed varies linearly with the flame stretch rate. This can be used to extrapolate toward zero stretch and obtain a stretch-free flame speed, S_s (as illustrated in Figure 27.4, in this case $S_s = 8.9$ m/s). After dividing S_s by the density ratio of unburned to burned gases, ρ_u/ρ_b, one obtains the laminar burning velocity, u_l (for this particular case, $u_l = 1.77$ m/s).

In principle, this procedure does not produce the laminar burning velocity that would be found by a steady, planar ideal computation with perfect thermodynamics, transport, and chemical kinetics. Equation 27.3 applies only in the linear range that actually occurs at low stretch (the cellular range in the figure). However, the procedure illustrated in Figure 27.4 is a convenient one leading to a very good approximation of u_l.

A second property that can be derived from Figure 27.4 is the slope of the linear relation between S_n and α, which is called the burned gas Markstein length, L_b, and is a measure of the sensitivity of the flame to flame stretch. The sign of L_b is indicative of the thermodiffusive stability of the flame: If L_b is positive, S_n decreases for increasing α, so that flame perturbations (causing an increase in local stretch rate) are smoothed out, leading to a stable flame. In the case of Figure 27.4, L_b is positive, so the flame is diffusively stable. In this case, thermodiffusion initially stabilizes the flame against the inherent hydrodynamic stability. When the stretch rate falls below a critical stabilizing value, the flame becomes cellular [55,56]. This can be seen clearly in Figure 27.4 from the sudden acceleration of the flame once the flame stretch rate drops below roughly 500 s^{-1}.

This behavior is probably the reason for the large spread in burning velocities shown in Figure 27.2: The correlations by Liu and MacFarlane [51], Milton and Keck [52], Iijima and Takeno [53], and Koroll et al. [54] were derived from data that did not take the effects of the stretch rate into account. As just explained, the flame stretch rate can cause an increase in burning velocity or flame acceleration due to cellularity.

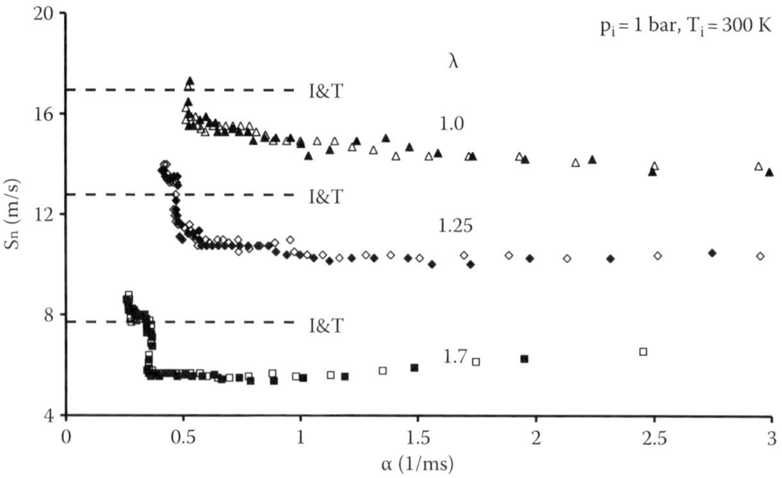

FIGURE 27.5
Flame speed predictions obtained from the correlation by Iijima and Takeno (I&T) [53] plotted on S_n versus α graph [46]. Two different experiments (open and solid symbols) plotted for each condition.

When comparing the burning velocities predicted by, for example, the experimental correlation of Iijima and Takeno [53] with the measurements performed by Verhelst et al. [50], it can clearly be seen that the burning velocities of Iijima and Takeno are all in the cellular region. As Iijima and Takeno calculated burning velocities from pressure records obtained from bomb explosions, flame instabilities could not be seen. If the larger radii were used in the derivation of burning velocities, the flames would have developed cellularity. Figure 27.5 illustrates this: Burning velocity predictions obtained with the correlation of Iijima and Takeno were multiplied by the density ratio to give flame speeds and were added to the S_n versus a plot for some NTP hydrogen–air flames [46]. The predictions all fall in the cellular region, which explains the consistently higher values. The burner measurements of Liu and MacFarlane [51] and the double kernel measurements of Koroll et al. [54] also report higher burning velocities. The deviations with the stretch-corrected measurements increase when going leaner, which could be explained by the decreasing Markstein length (with L_b negative and thus becoming larger in absolute value), resulting in a larger increase in the burning velocity when the flame is positively stretched. The measurements of Liu and MacFarlane are highly stretched due to the very small nozzle used in their measurements [57].

The burning velocity for a stoichiometric hydrogen–air mixture predicted by the correlation by Milton and Keck [52] is lower than the values obtained with stretch rate correction (see Figure 27.2), which could also be due to stretch (a stoichiometric hydrogen–air flame is stable and will thus propagate slower when subjected to positive stretch), if the burning velocity was taken at a small flame radius (i.e., before the onset of cellularity).

The stretch-free measurements show reasonably good correspondence, although the values reported by Vagelopoulos et al. [48] are lower than the others. All bomb-derived data (Taylor [47], Kwon and Faeth [49], and Verhelst et al. [50]) correspond closely.

Note that the very rich equivalence ratio at which the laminar burning velocity peaks (see Figure 27.2: u_l peaks at $\lambda = 0.6/\varphi = 1.7$) can also be explained by the high mass diffusivity of hydrogen [21]. It is noteworthy that the equivalence ratio at which u_l peaks is much richer than the equivalence ratio at which the flame temperature peaks (around stoichiometry).

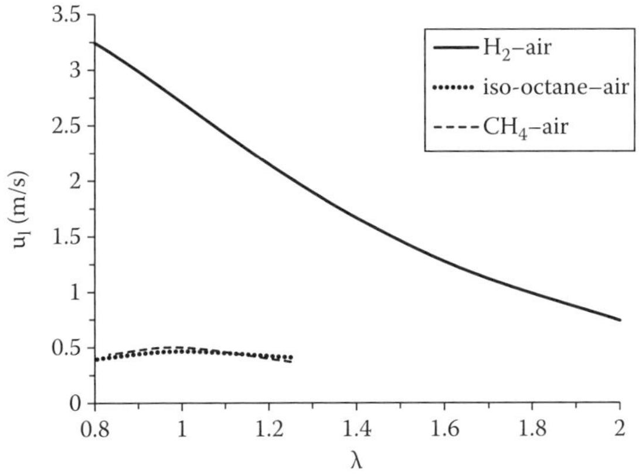

FIGURE 27.6
Laminar burning velocities for hydrogen, iso-octane, and methane–air mixtures, at 1 bar and 360 K, as a function of the air-to-fuel equivalence ratio.

To conclude this section, Figure 27.6 compares stretch-free burning velocities at atmospheric pressure and ~360 K [41,43,46,58] for hydrogen, methane, and iso-octane as a function of equivalence ratio. The figure demonstrates the much higher laminar burning velocity of hydrogen–air mixtures and its strong dependence on the equivalence ratio.

27.2.2.3 Laminar Burning Velocity at Engine Conditions

The previous section demonstrated that care must be taken in using published data for the laminar burning velocity of hydrogen mixtures. The section highlighted this for atmospheric conditions. The current section discusses the laminar burning velocity at engine conditions.

There are very few data available at engine conditions. The range of conditions covered by the correlations of Liu and Mac-Farlane [51], Milton and Keck [52], Iijima and Takeno [53], and Koroll et al. [54], mentioned earlier, include lean to rich mixtures and elevated temperatures (up to 550 K) and pressures (up to 25 atm). However, as discussed previously, they did not account for the effects of stretch and instabilities, which grow stronger with pressure as the flame thickness decreases [50].

Experimental data [59] show hydrogen/air flames at atmospheric conditions to have positive Markstein lengths close to stoichiometric, but all mixture ratios with $\lambda \geq 1/\varphi \leq 1$ have negative Markstein lengths as soon as the pressure exceeds about 4 bar. The consequence of this increasing instability with pressure is illustrated in Figure 27.7, which shows schlieren photographs of a $\lambda = 1.25/\varphi = 0.8$, 300 K, 5 bar H_2–air flame.

In this case, the flame is cellular from inception onward, accelerating throughout its growth. The flame speed increases faster than linearly with decreasing flame stretch rate;

FIGURE 27.7
Schlieren photographs of a $\lambda = 1.25/\varphi = 0.8$, 300 K, 5 bar H_2–air flame, time interval: 0.385 ms.

consequently, the methodology of obtaining stretch-free burning velocities u_l (and its dependence on stretch rate), described in the previous section, is no longer applicable [50,60]. To study the influence of temperature, pressure, and residual gas content, Verhelst et al. [46,50] determined the burning velocity of a spherically expanding flame at a flame radius of 10 mm, for $1 \leq \lambda \leq 3.3/0.3 \leq \varphi \leq 1$, 300 K $\leq T \leq$ 430 K, 1 bar $\leq p \leq$ 10 bar, and 0% $\leq f \leq$ 30% (with f, the residual gas content, in vol.%). This burning velocity is not a fundamental parameter but, as the authors claim, is indicative of the burning rate at a fixed, repeatable condition, representing a compromise that involves a sufficiently large radius to minimize the effects of the SI, while being small enough to limit the acceleration due to the instabilities. It is noteworthy that these are the only data that include the effects of residual gas content, an important parameter, given the operating strategies that are proposed for H_2ICEs (see Section 27.6). A correlation for the burning velocity was derived from this experimental data and partly validated using an engine code [61].

An alternative methodology has been proposed to obtain u_l and Markstein lengths at higher pressures, also from freely expanding spherical flames. The laminar burning velocity u_l and Markstein numbers have been reported for equivalence ratios from $\lambda = 3.3/\varphi = 0.3$ up to stoichiometric, for pressures of 1, 5, and 10 bar [60]. However, this involved numerous experiments and very high camera frame rates, but experimental uncertainty is rather high, especially on the Markstein lengths.

An alternative to experimental determination is the use of a 1D chemical kinetics code to calculate u_l. The H_2/O_2 system is one of the simplest reaction mechanisms, it is fairly well known (with more than 100 mechanisms reported in the literature, e.g., [62]), and computations of u_l are reasonably fast. However, it is perhaps surprising to learn that even for this simple system, there still exist a number of uncertainties, as recently reviewed by Konnov [63]. As the previous discussion has shown, stretch-free data are scarce, especially at engine-like conditions. Thus, validation of reaction mechanisms is very limited at best.

Results were reported using a chemical kinetics code to calculate u_l, using several published reaction mechanisms [46,64]. First, from initial results, the reaction mechanism of Ó Conaire et al. [65] was chosen as it gave the best correspondence to the selected experimental data at atmospheric conditions. Secondly, calculation results were compared with the experimental results from Verhelst et al. [46,50] for a range of pressures, temperatures, equivalence ratios, and residual gas fractions. Note that these experimental results are not stretch-free burning velocities (see above). The authors report that the calculations break down for (very) lean mixtures and higher pressures. The inability of steady, planar calculations to predict burning velocities at very lean mixtures in correspondence to experimentally observed values has recently been elucidated by Williams and Grcar [66]. For moderately lean to stoichiometric mixtures, the effect of temperature and dilution with residuals is reported to be predicted reasonably well. The authors conclude that simulations of the effect of residuals could thus be considered to replace experiments with residuals, which are rather cumbersome.

Bradley et al. [60] compare their stretch-free data at 5 and 10 bar to calculations using the reaction mechanisms of Ó Conaire et al. [65] and Konnov [67]. The results using Konnov's scheme are reported to correspond to the experimental results within the rather large uncertainty bands. Knop et al. [58] propose a correlation for u_l for use in an engine code, based on published experimental results and chemical kinetic calculations. The comparison between simulated and measured engine cycles reported in the paper represents a limited validation of the correlation. Other studies reporting correlations based on chemical kinetic calculations are cited in Section 27.3 on modeling.

To conclude this section, the current state is that experimental data of laminar burning velocities of hydrogen–air mixtures at engine conditions are mostly nonexistent and (consequently) numerical data are nonvalidated. Clearly, this is an area requiring further study. Konnov [63] wrote, "new accurate measurements of hydrogen burning velocities are therefore extremely important for (reaction mechanism) validation."

However, as discussed earlier, accurate burning velocity measurements at lean conditions are next to impossible because of instability. An alternative approach is to test reaction mechanisms on the basis of measured autoignition times [68].

27.2.3 Turbulent Burning Velocity

The turbulent burning velocity, u_t, of hydrogen mixtures is a convenient parameter to calculate the fuel mass burning rate in a hydrogen engine. Contrary to the laminar burning velocity, it depends not only on the mixture properties but also on the flow, the geometry, and the history of the flame [69]. In light of the previous section, an important question is to what extent flame stretch and instabilities influence the turbulent burning velocity of hydrogen flames.

27.2.3.1 Experimental and Numerical Work on the Role of Instabilities and the Effects of Stretch on u_t

First, an overview is given on work evaluating the effect of stretch and laminar flame instabilities on turbulent combustion. A lot of work is devoted to this subject (see the excellent review paper by Lipatnikov and Chomiak [70]). The following overview is limited to work including hydrogen mixtures:

- Bradley and coworkers: Abdel-Gayed et al. [71] investigated the effect of the Lewis number Le (and thus, the laminar flame stability) by measuring the turbulent burning velocity in a fan-stirred bomb using a double kernel method. The measurements comprised hydrogen, propane, and iso-octane, mixed with air at various equivalence ratios and indicated an increase in u_t for unstable mixtures. Later, Abdel-Gayed et al. [72] compiled all contemporary data on the turbulent burning velocity and found a confirmation of this trend; additional measurements by the authors using the fan-stirred bomb also revealed the existence of cellular structures in the turbulent flame for thermodiffusively unstable flames. Recent measurements by Bradley et al. [73] of the turbulent burning velocity in a (different) fan-stirred bomb of statistically spherical explosion flames showed an increase in the ratio u_t/u_l for decreasing Markstein number for mixtures with the same ratio of root mean square (rms) turbulent velocity, u′ to u_l.
- Faeth and coworkers: Wu et al. [74,75] used a turbulent jet burner to measure turbulent burning velocities for hydrogen–air mixtures with various equivalence ratios. The measurements not only showed an increase in u_t for unstable mixtures but also a *dampening* of turbulent fluctuations and thus a decrease of u_t for stable mixtures. Later, Aung et al. [76] reported measurements of statistically spherical explosion flames in a fan-stirred bomb. Mixtures of hydrogen, nitrogen, and oxygen were prepared with almost identical laminar burning velocities but different thermodiffusive stability (stable/neutral/unstable). Again, u_t clearly increased for unstable mixtures. Both measurements on the burner and the bomb showed a strong dependence of u_t on mixture stability even for strong turbulence.

- Koroll et al. [54] recorded schlieren photographs of double kernel explosions in a fan-stirred bomb using hydrogen–air mixtures of varying equivalence ratios. They found a dependence of u_t/u_l on the equivalence ratio, with the ratio being much higher for lean mixtures.
- Goix and Shepherd [77] used a stagnation point flame burner to compare u_t for lean hydrogen–air and propane–air flames with similar laminar burning velocities. At similar rms turbulent velocities, the turbulent burning velocity for the hydrogen mixture was higher than that for the propane mixture. The fractal dimension of the flame surface was compared and was found to be larger for the hydrogen mixture. The ratio u_t/u_l was much higher than the surface area ratio wrinkled/smooth, indicating a substantial difference in local flame speeds.
- Renou et al. [78] measured local flame speeds of spark-ignited flames in a vertical wind tunnel, for stoichiometric methane–air and propane–air mixtures and lean hydrogen–air mixtures. The probability density function (pdf) of local flame speeds was strongly dependent on the Lewis number. For the lean hydrogen/air mixtures, the pdf was much broader indicating the strong effect of stretch on the local flame speed.
- Kido et al. [79] measured local flame speeds of methane, propane, and hydrogen turbulent flames in the weak turbulence region (low u′) with identical laminar burning velocities using a fan-stirred bomb. For a fixed u′, the surface area ratio turbulent/laminar was almost constant but the turbulent burning velocity was very different, caused by a strong difference in local flame speeds. The variation in local flame speeds could be qualitatively explained from the preferential diffusion concept. Later, Kido et al. [80] measured turbulent lean hydrogen flames of different equivalence ratio but similar laminar burning velocity, again for low u′, in the fan-stirred bomb. The turbulent burning velocity increased strongly for leaner mixtures, although the relative increase was found to be much smaller beyond an equivalence ratio of $\lambda = 2.0$. Again, this was found to be qualitatively consistent with changes in local flame speeds due to changes in the local equivalence ratio caused by the preferential diffusion effect.
- Lipatnikov et al. [81] reviewed measurements using a fan-stirred bomb on turbulent burning velocities for lean hydrogen–air mixtures with similar u_l but different Lewis numbers and found a difference in du_t/du_0 amounting to an order of magnitude. The turbulent burning velocity was found to be strongly dependent on the Lewis number even for strong turbulence (u'/u_l 1). The authors also reported a decrease in the smallest wrinkling scale for decreasing Le.
- Chen and Im [82] looked at the correlation of flame speed with stretch in turbulent methane/air flames using 2D direct numerical simulation (2D DNS) with detailed chemistry. Lean and stoichiometric flames were simulated and it was shown that for moderate stretch rates, the local correlation between flame speed and stretch was approximately linear. However, large negative stretch rates (compression) were also found, obtained solely through curvature effects, causing an overall nonlinear correlation of flame speed with stretch. Changes in flame speed were consistent with preferential diffusion theory. Chen and Im [83] also looked at hydrogen–air flames, for equivalence ratios ranging from lean to rich, again using 2D DNS with detailed chemistry. Strong interactions between stretch and preferential diffusion were found to exist in the turbulent flames; the local

correlations between burning velocities and strain and curvature were according to expected diffusive-thermal effects. Im and Chen [84] expanded the work on hydrogen–air flames studying the interaction of twin premixed hydrogen–air flames with 2D DNS and detailed chemistry; the interaction of both rich–rich and lean–lean flames were studied. The local flame front response to turbulence was according to the preferential diffusion mechanism. This resulted in a significant burning rate enhancement for the lean–lean case. This was caused by the global positive stretch on the flame surface increasing the local flame speed, as well as by a *self-turbulization* and increased flame wrinkling.

27.2.3.2 Implications for the Combustion of Hydrogen–Air Mixtures in Engines

The measurements and simulations reviewed in the previous section clearly indicate the existence of an effect of flame instabilities on the turbulent burning velocity that can be very strong in some cases. The influence of pressure was not discussed earlier. One could assume that the effect of local stretch on u_t will decrease with pressure [85,86], as Markstein numbers have been shown to decrease with pressure (see, e.g., [41,43,87]) and the flame thus gets less sensitive to stretch. However, flames at higher pressure have also been shown to get increasingly unstable, as demonstrated in Section 27.2.2, with the dependence of the onset of cellularity on pressure. Turbulent flames have been shown to be more finely wrinkled at higher pressures [86,88,89], which suggest an increasing instability and a larger effect on u_t as pressures increase [90]. Flame instability effects can thus be expected to be relevant to turbulent combustion in SI engines.

Due to the very high mass diffusivity of hydrogen, hydrogen mixtures show a very pronounced preferential diffusion effect. A majority of the work reported in the previous section used hydrogen mixtures exactly for that reason, and several authors advanced turbulent lean hydrogen combustion as the most challenging test for turbulent combustion models [74,81]. Practical mixtures in hydrogen engines will most probably show increased turbulent burning velocities because of instability effects. Heywood and Vilchis [91] compared SI engine operation on hydrogen and propane, with stoichiometric mixtures, by recording schlieren photographs of the flame development in an optically accessible square piston engine. The turbulent flame speed for the propane mixture was an order of magnitude larger than the laminar flame speed, whereas for hydrogen, it was of the same order* (though larger). The characteristic wrinkling scale was found to be smaller for the hydrogen flames.

27.2.3.3 Implications for Modeling of Turbulent Combustion of Hydrogen–Air Mixtures

This section presents some of the current knowledge on modeling of turbulent combustion. Most of the issues discussed in the following are not hydrogen specific. However, these were included to highlight the remaining uncertainties in modeling turbulent combustion so that the reader has a feeling for the extent to which the assumptions of the various engine modeling work presented in Section 27.3 are justified. As discussed in the previous section, the combustion of hydrogen is quite particular and consequently stresses many of the usual assumptions made for turbulent combustion modeling.

* Note that the laminar flame speeds used in their work were taken from [52] and included stretch effects.

A number of turbulent combustion models, dating back to the seminal work by Damköhler (1940), assume the sole effect of turbulence to be an increase in flame surface area through turbulent wrinkling, based on the observations showing an increase in u_t with u′. These models only implement a dependence of u_t on u′ and u_l; these models are numerous and are still popular today.

As more measurements were published, more phenomena became apparent that could not be explained when only considering flame surface wrinkling (e.g., the bending of the u_t vs. u′ curve [72,86,92], qualitatively illustrated in Figure 27.8, with quantitative figures shown in, e.g., [70]). Models were proposed that included quenching effects at excessive flame stretch [92,93]. Later, various u_t correlations based on experimental data or obtained through theoretical work were proposed with additional dependencies: mostly a length scale of turbulence and a transport property of the unburned mixture, for a better correspondence to measurements or resulting from explicit inclusion of stretch effects (e.g., [92,94,95] and others). Also, the effects of turbulence on surface wrinkling as well as on local flame speeds, assuming a linear relation between flame speed and flame stretch (see Section 27.2), were modeled [96].

However, for increasing stretch, a linear relation between flame speed and flame stretch no longer applies, as demonstrated through simulations of highly perturbed laminar flames [97], DNS of turbulent flames [82], as well as through experimental data [81], despite the larger range of applicability of the linear relation than could be assumed from theoretical considerations [82]. Models accounting for this observation are few and still in their infancy. Lipatnikov et al. [81] and Lipatnikov and Chomiak [97] propose a model based on the leading point concept, which (in simplified terms) assumes the global reaction rate to be primarily dominated by the faster burning parts of the flame. Measurements by Kido et al. [79,80] seem to confirm this mechanism with observations of *active* (in the case of lean hydrogen, the convex parts) and *inactive* parts of a flame front. The model proposed by Lipatnikov et al. starts from an experimental u_t correlation and substitutes the chemical (laminar) time scale with a time scale calculated from the *laminar consumption velocity* of *critically perturbed flamelets*, obtained through simulations of stretched laminar flames [97]. Other approaches are suggested by Kobayashi and Kawazoe [88] and Bradley et al. [98].

Ultimately, models that include flame front wrinkling as well as stretch-dependent local flame speeds through a stretched laminar flamelets library are envisaged, but such a library asks for much more data than is currently available (namely, laminar burning velocities, strain, and curvature Markstein lengths for a large range of pressures, temperatures, and mixture compositions). Areas requiring more research in order to work toward such models are suggested by Bradley [90] (Peters' book [99] also is a valuable reference

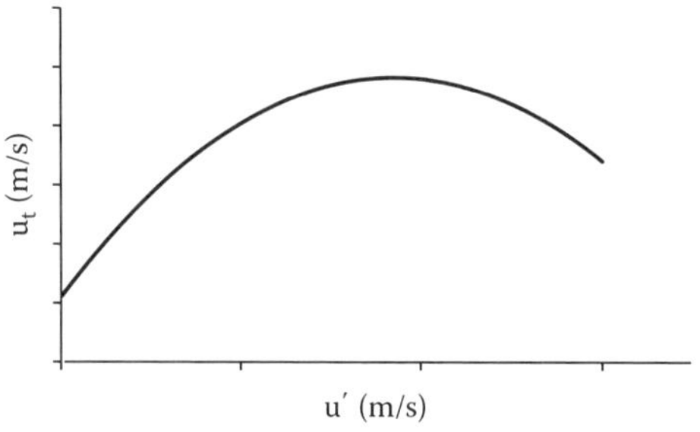

FIGURE 27.8
Illustration of the *bending* of u_t versus u′: u_t initially increases with increasing u′, reaches a maximum, and then decreases again until flame quenching occurs.

on modeling turbulent combustion). There is also uncertainty on this approach, however [100], as these models use a library of steady stretched flamelets, whereas flamelets have been shown to behave differently under transient stretch [44] (e.g., being more resistant to quenching under transient than under steady stretch).

Also noteworthy is that for typical hydrocarbon combustion, u'/u_l is quite high, and therefore many turbulent combustion models start from this assumption. For hydrogen, however, this ratio is much lower at near-stoichiometric conditions due to the high laminar burning velocities, and this assumption is thus invalid.

27.2.4 Hydrogen Jets

As described later, most modern mixture formation systems for hydrogen engines rely on injection of gaseous hydrogen into air using pulsed fuel-injection devices. This section discusses some relevant features of hydrogen jets.

27.2.4.1 Unignited Jets

Regardless of the mixture formation system, the injection process of gaseous hydrogen into air consists of several steps. The actual injection process can be divided into the flow of hydrogen in the injector and nozzle, the transitional flow of the fuel from the injector tip into the downstream volume (intake manifold or combustion chamber), the interaction of the gas jet with the surrounding media, and the consequent generation of turbulence and heat exchange, mixing, and eventual combustion of the fuel. Although the injection event of a gaseous fuel into another gaseous medium may appear simple compared to liquid fuel injection, the disparate length scales in the gas flow present a considerable challenge to the computational modeling of the injection event. Under the assumption of a choked flow and hydrogen being a nearly perfect gas that flows isentropically through the nozzle, the pressure ratio between upstream fuel-supply pressure, p_k, and back pressure, p_0, is defined as

$$\frac{p_k}{p_0} = \left(\frac{2}{\gamma+1}\right)^{\frac{\gamma}{\gamma-1}} \tag{27.4}$$

For hydrogen with a polytropic coefficient, γ, of approximately 1.4, the critical pressure ratio results to be around 0.53, requiring the fuel-supply pressure to be roughly twice the back pressure to obtain critical conditions. Injection is usually designed to be sonic to allow for high mass flow rates and short injection durations even at high engines speeds [13]. Critical injection conditions are also beneficial to the engine control strategy because the amount of injected fuel is independent of back pressure at critical conditions.

The generalized configuration of the near field of an underexpanded jet is well documented, a simplified model of which is presented in Figure 27.9 [101]. The flow at the exit plane of the nozzle is assumed to be choked, that is, Ma = 1. Immediately upon exit from the nozzle, the high ratio of exit pressure to surrounding pressure causes expansion of the gas. Flow subsequently accelerates, thereby generating expansion waves. When these expansion waves meet at the outer boundary, they are reflected as compression waves. Finally, these compression waves coalesce to form an oblique shock structure of barrel shape. This structure encloses a supersonic flow region termed the *zone of silence*, wherein

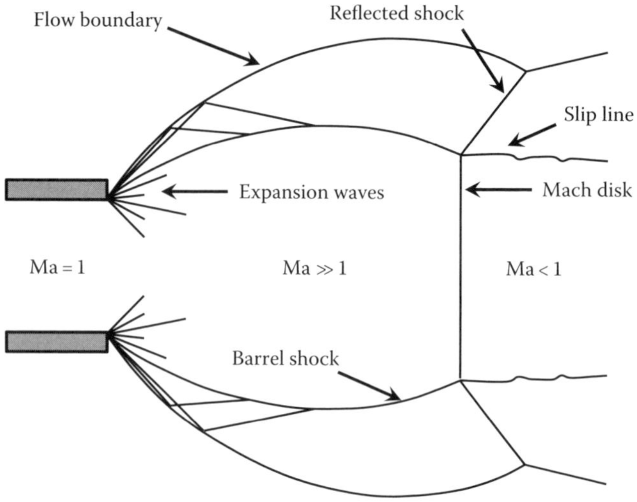

FIGURE 27.9
Principal regions around nozzle exit for an underexpanded jet. (From Owston, R. et al., Fuel–air mixing characteristics of DI hydrogen jets, SAE Paper No. 2008-01-1041, 2008.)

it is assumed that no entrainment takes place, that is, the mass flux at the nozzle exit is conserved throughout this region. For an exit pressure to ambient pressure ratio above approximately 2, the barrel shock culminates in a disk-shaped normal shock called a *Mach disk* and a reflected shock. At sufficiently high-pressure ratios, this process may repeat several times, resulting in a succession of barrel and normal shock structures. Downstream of the final Mach disk, the flow is subsonic, with the adjacent flow from the reflected shock remaining supersonic. Following the abrupt change of density after the Mach disk, pressure in the subsonic flow is assumed to be consistent with that of the surroundings.

The ratio of the largest length scale, the penetration of the gas jet, to the smallest length scale, the flow variations within the orifice, is estimated to be around 4000 [102], precluding the straightforward approach of treating the entire process in a single computational mesh that resolves all length scales, because it would result in impractical computational times.

Various approaches to solve the aforementioned problem have been developed and tested, including dividing the flow problem into tractable portions [102], scaling theories based on conservation of energy equations [103], and combinations of scaling theories with free jet theories [101]. The overarching goal of the aforementioned approaches is to accurately predict the injection characteristics like penetration depth, mixing, and spreading of the jet. Past and ongoing research focuses on the general prediction of these jet characteristics and was widely based on a comparison to results from injection test rigs or pressure chambers [104].

Additional factors that influence the injection and mixture formation event in an actual engine include, but are not limited to, in-cylinder charge motion, changes in supply as well as back pressure during the injection event, surface interaction of the injection jet, and effects between adjacent injection jets. The effect of charge motion and in particular swirl on the development of transient hydrogen jets has been studied experimentally [105]. Experimental investigations on an optical engine also documented effects between adjacent injection jets as well as injection jets and in-cylinder surfaces [106]. It was found that neighboring injection jets from multihole nozzles tend to collapse to one jet. Also, jets close to an in-cylinder surface tend to get drawn toward that surface. Both phenomena are attributed to the so-called Coanda effect [106]. Development and application of simulation tools as well as application and improvements of optical techniques for hydrogen engines have allowed considerable progress and yielded further understanding of the dominant processes for mixing of unignited hydrogen jets, for example [107].

27.2.4.2 Ignited Jets

The development of advanced mixture formation concepts like late injection just before spark timing, multiple injection strategies with injection pulses during compression, as well as combustion or compression ignition (CI) require consideration of flame propagation processes in ignited hydrogen jets. Similar considerations related to hydrogen safety have also been employed to analyze the effects of jet flames resulting from unintended hydrogen releases. Houf et al. [108] emphasized that the knowledge of the flame length and thermal radiation heat flux distribution is important for the safety aspect of hydrogen. Brennan et al. [109] compared modeling results of high-pressure hydrogen jet fires with experimental results. This work was driven by the need to develop contemporary tools for safety assessment of real-scale underexpanded hydrogen jet fires and to study large eddy simulation (LES) model performance to reproduce such large-scale jet fires in an industrial safety context. Mohammadi et al. [110] experimentally investigated the ignition, combustion, and flame behavior of high-pressure and intermittent hydrogen jets in a constant-volume combustion chamber. With parametric studies of the effects of injection pressure, nozzle size, ambient pressure and spark location for various spark timings, and equivalence ratios, they revealed that stable ignition can be achieved even during injection. They also concluded that ignition of hydrogen jets at the end of injection offers the shortest combustion duration, while combustion control is much easier when the jet is ignited just at the time when the jet tip reaches the spark gap. In addition, igniting hydrogen jets at the boundary of the jet is very effective for stabilizing the ignition, and that in combination with advanced spark timing, the intense combustion of hydrogen can be controlled.

Kawanabe et al. [111] confirmed that the flame propagation process qualitatively agreed with experimental data for a wide variety of injection conditions and ignition timings using incompressible-flow-type computational fluid dynamics (CFD) with a k-e turbulence model and the flamelet concept. Differences in flame propagation were reported for jet-tip ignition cases in which the flame mainly propagates along the jet-tip edge and then the burned area grows compared to inner jet ignition cases in which the start of flame propagation is slightly delayed after ignition and the burned area spreads from the inner region of the jet. They also concluded that for the flame propagation process in a hydrogen jet, a local value of the laminar burning velocity significantly affects the turbulent burning velocity. Development of improved jet combustion models also benefits the accuracy of combustion and ultimately emissions prediction of in-cylinder combustion events.

27.3 Modeling Regular Combustion

The following sections discuss the efforts undertaken to analyze and model the regular combustion of hydrogen in engines, that is, where combustion is initiated by the spark after which a turbulent flame develops and propagates throughout the combustion chamber, consuming all of the fuel–air mixture [15]. The discussion is limited to the combustion of pure hydrogen. The interested reader is referred elsewhere for works modeling the effects of hydrogen in addition to other fuels [112] (see also the references cited in Section 27.9). Also, the focus is on work devoted to the development of hydrogen engines and not on work where hydrogen was mainly chosen as a convenient fuel in terms of CPU time, for example, to come to a better understanding of homogeneous charge compression ignition (HCCI) combustion [113,114] by including the detailed chemical kinetics without this leading to prohibitive computing times.

From Sections 27.2.2 and 27.2.3, it is clear that many uncertainties remain concerning both fundamental data on hydrogen–air combustion properties and modeling approaches. This has to be borne in mind in the following discussion (Sections 27.3.2 and 27.3.3, in particular) of the works reporting modeling of the engine cycle of hydrogen engines.

27.3.1 Thermodynamic Analysis of the Working Cycle

The engine efficiencies and losses of the working cycle can be calculated using test data from different operating modes. Starting from a theoretical efficiency of an engine without losses, the individual losses due to incomplete combustion, actual rate of heat release, wall heat losses, and gas exchange can be calculated [15]. Figure 27.10 shows the efficiencies and losses for gasoline and hydrogen operation with both port injection and DI. The data for all fuels were collected on a single-cylinder research engine at an engine speed of 2000 RPM and indicated mean effective pressures (IMEPs) of 2 and 6 bar. In hydrogen DI, the fuel was injected early during the compression stroke (with the start of injection [SOI] at 120°CA before top dead center [BTDC] and an injection pressure of 150 bar), resulting in a fairly homogeneous mixture at spark timing.

At the low load point of 2 bar IMEP, the efficiency of the ideal engine calculated using actual gas properties in gasoline operation is significantly lower than in hydrogen operation. Since this value is mainly influenced by the compression ratio and the air/fuel ratio,

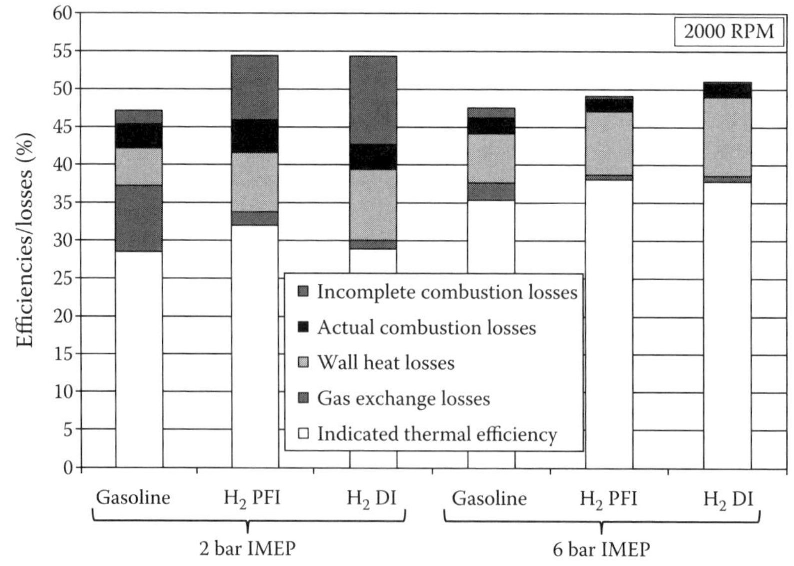

FIGURE 27.10
Analyses of losses compared to the theoretical engine cycle: gasoline versus hydrogen (PFI and DI), at two loads.

hydrogen operation results in higher values, due to the lean operation (see Section 27.6). Losses due to incomplete combustion can be determined by measuring unburned fuel components in the exhaust. These consist of hydrocarbons and carbon monoxide in gasoline operation and unburned hydrogen when fueled with hydrogen. Due to the extremely lean conditions in hydrogen operation ($\lambda = 5.3/\varphi = 0.19$), the losses due to incomplete combustion are significantly higher than in stoichiometric gasoline operation. The loss due to actual combustion rather than an ideal constant-volume combustion at top dead center (TDC) in gasoline operation is around 3%. Due to the lean combustion in hydrogen operation, the losses due to actual combustion are slightly higher than in gasoline operation. Due to the higher pressure levels in hydrogen operation resulting from unthrottled operation, the wall heat losses are considerably higher than in gasoline operation. The higher in-cylinder charge motion in hydrogen DI operation results in higher wall heat losses compared to port injection. Finally, the gas exchange losses in hydrogen operation are only a fraction compared to gasoline, since the engine is operated unthrottled. Overall, this results in an advantage in indicated thermal efficiency for hydrogen port injection over hydrogen DI and gasoline operation. The indicator diagrams for the 2 bar IMEP case are shown in Figure 27.11.

At the medium load point of 6 bar IMEP, the efficiency of the ideal engine in gasoline operation is lower than in hydrogen operation. This is again due to the lean operation. The air–fuel ratio, and hence the theoretical efficiency in hydrogen port injection, is lower than in hydrogen DI due to the air displacement effect with port-injection operation (DI: $\lambda = 2.3/\varphi = 0.43$; PFI: $\lambda = 1.8/\varphi = 0.56$). For this operating point, the loss due to incomplete combustion in gasoline operation is slightly more than 1%. In hydrogen operation both with port and DI, the loss accounts for less than 0.5%. The very complete combustion in hydrogen operation is mainly due to the fast flame speed and small quenching distance at

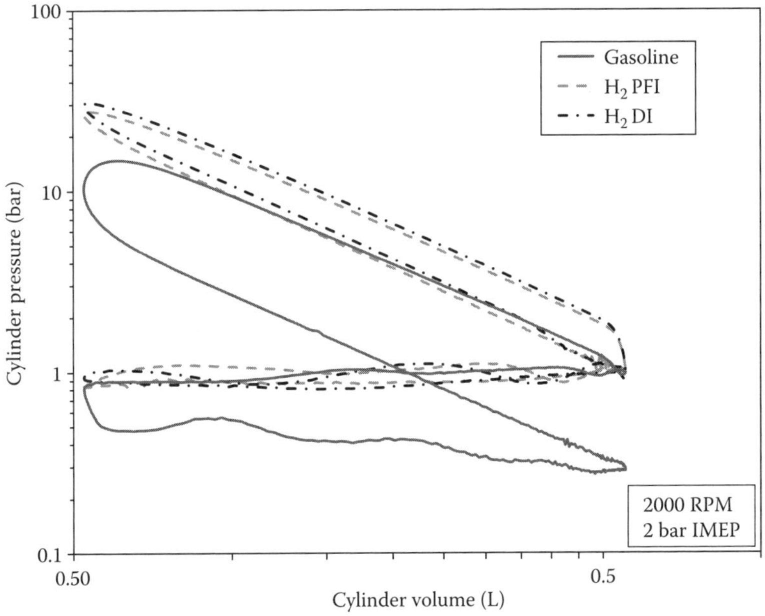

FIGURE 27.11
Pressure versus volume diagrams at 2 bar IMEP for gasoline (stoichiometric throttled), hydrogen PFI, and hydrogen DI (lean WOT).

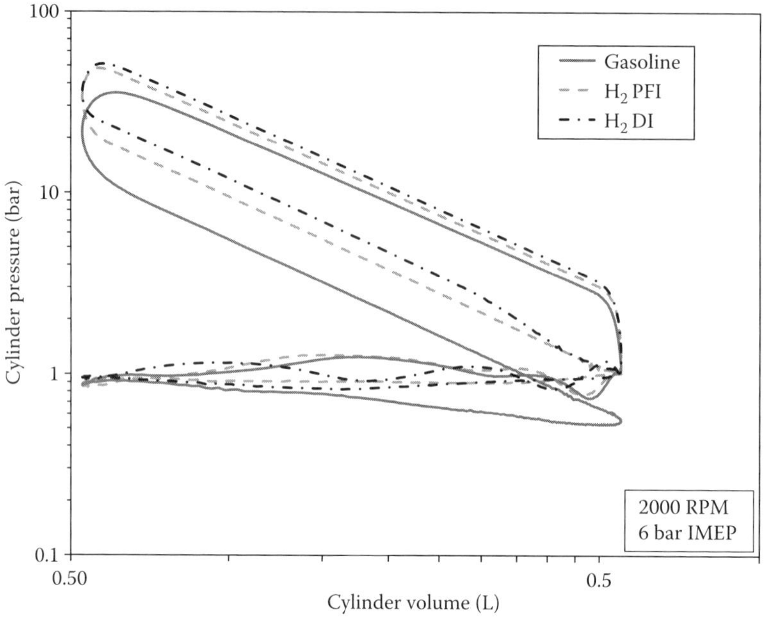

FIGURE 27.12
Pressure versus volume diagrams at 6 bar IMEP for gasoline, hydrogen PFI, and hydrogen DI.

this air–fuel ratio. The loss due to actual combustion in gasoline operation is around 2%. Although the engine is operated unthrottled and therefore lean in hydrogen operation, the combustion is still faster than in gasoline operation. This results in lower losses due to actual combustion in hydrogen operation. However, the higher flame speeds and the smaller quenching distance, which positively affected previous losses, are the main reasons for the increased wall heat losses in hydrogen operation compared to gasoline operation. DI shows even higher wall heat losses compared to port injection, which is likely due to the higher level of in-cylinder charge motion and turbulence caused by the DI event. Finally, the gas exchange losses in hydrogen operation are significantly lower than in gasoline operation since the engine is operated unthrottled. At this operating point, this results in an overall advantage in indicated thermal efficiency of approximately 2.5% in hydrogen operation, both with port and DI, compared to gasoline operation. The indicator diagrams for the 6 bar IMEP case are shown in Figure 27.12.

In both cases, further improvements with hydrogen DI could be gained with stratification resulting from an efficiency-optimized injection strategy (see Section 27.6).

27.3.2 Thermodynamic Models

The earliest attempts at modeling the combustion in hydrogen-fueled engines date from the 1970s. Fagelson et al. [115] used a two-zone quasidimensional* model to calculate power output and NO_X emissions from a hydrogen SI engine. They used a semiempirical turbulent combustion model of the form $u_t = ARe^B u_l$, where A and B are constants; Re is the Reynolds number based on piston diameter, mean piston speed, and burned gas

* A term used to denote multizone thermodynamic engine models, in which certain geometrical parameters are included in the basic thermodynamic approach, mostly the radius of a thin interface (the flame) separating burned from unburned gases, resulting in a *two-zone* formulation [116].

properties; and u_t and u_l are the turbulent and laminar burning velocities, respectively. Spherical flame propagation was assumed, heat transfer was neglected, and NO_X formation was calculated using 10 constant mass zones in the burned gases and the extended Zeldovich mechanism. The laminar burning velocity was calculated from an overall second order reaction with an estimated activation energy. The model was validated against measurements with varying equivalence ratio and ignition timing only. Prabhu-Kumar et al. [117] used this model to predict the performance of a supercharged hydrogen engine, with no changes made to the original model. They reported an overestimation of the rate of pressure rise (and thus of the burning velocity). Sadiq Al-Baghdadi and Al-Janabi [118] and Baghdadi [119] also used the model by Fagelson et al. [115]. They compared simulations to measured power output, NO_X emissions, and brake thermal efficiency for varying compression ratio and supercharging pressure [118]; no pressure diagrams were shown in the paper. In another paper [119], the same model was used and experimental values for varying compression ratio, equivalence ratio, and engine speed were compared to simulations. Again, only brake quantities were compared without detailing the correspondence in terms of burning rate and cylinder pressure development.

Keck [120] reported measurements in an optically accessible engine operated on propane as well as hydrogen and used a turbulent entrainment model to compare predicted trends with experimentally observed trends. Excellent agreement was found between predictions and measurement, indicating a much smaller flame brush thickness for hydrogen operation compared to propane and a much higher initial flame expansion speed (with the difference in flame speeds decreasing throughout the combustion duration but with the flame speed always remaining higher for hydrogen).

Zero- and multidimensional models have been used for hydrogen engine simulation at the Czech Technical University [121,122]. A zero-dimensional model (i.e., models where the thermodynamic state of a single zone, encompassing the complete cylinder contents, is calculated) based on the GT-Power code was used with Wiebe's law fitted to measured rates of heat release. The extended Zeldovich mechanism [123] was used to calculate NO_X emissions. The so-called Advanced Multizone Eulerian Model was developed for multidimensional simulation. This model was a combination of zero-dimensional and multidimensional elements: The combustion chamber seemed to be limited to simple geometries because of limitations to grid generation, and the heat transfer was modeled for the cylinder contents as a bulk volume. The combustion model was a *semiempirical pdf-like* model that relied on a measured rate of heat release and the assumption of a hemispherical flame front to track flame propagation.

Ma et al. [124] used a zero-dimensional model using Wiebe's law. It is not clear to what data this law was fitted. The model was used to calculate the effects of varying compression ratio and ignition timing and to determine an *optimum cylinder diameter* for a fixed equivalence ratio. No validation against experimental data or any justification for extrapolating outside the conditions for which the fit is valid was given, so the quality of the reported results is doubtful.

D'Errico et al. [125,126] reported full-cycle simulations using 1D gas dynamic calculations combined with a quasidimensional combustion model, for a hydrogen engine with cryogenic port injection. The gas dynamic algorithm was adapted so that the injection and transport of cryogenic hydrogen along the intake ducts were incorporated. The methodology proposed by Verhelst and Sierens [45] was used to construct a correlation for the laminar burning velocity from chemical kinetic calculations using an in-house reaction scheme. The burning rate was modeled using a fractal approach in Ref. [125] and Zimont's model in Ref. [126]. Nitric oxide (NO) emissions were calculated using the *super extended*

Zeldovich mechanism. Simulations were run for varying engine speed and equivalence ratio and compared to experiments. The simulated and measured pressure traces in the intake and exhaust duct systems showed very good agreement. The combustion pressure was well predicted for stoichiometric and moderately lean mixtures but was less satisfactory for (very) lean conditions at medium to high engine speeds. The authors pointed to the effects of differential diffusion and instabilities for these (very lean) conditions and the high ratios of turbulent to laminar burning velocities reported for these mixtures [70], which were unaccounted for in the combustion model.

Verhelst and Sierens [61] reported calculations of the power cycle of a hydrogen-fueled engine using a quasidimensional two-zone combustion model framework. They reported the difficulties in obtaining stretch-free laminar burning velocities and proposed a correlation based on measurements of cellular flames [46,50]. This correlation was then used with a number of turbulent burning velocity models, comparing simulations to measurements on a hydrogen-fueled Cooperative Fuel Research (CFR) engine for varying compression ratio, ignition timing, and equivalence ratio. All models predicted the effects of compression ratio and ignition timing well, but some did not predict the effects on equivalence ratio well.

Safari et al. [127] used a similar approach, taking the laminar burning velocity correlation from Verhelst and Sierens [61], with the fractals turbulent combustion model that was one of the models performing reasonably well in Ref. [61]. They included the reaction kinetics of the H_2–O_2–N_2 system to allow calculations of NO emissions. The wall heat transfer model by Woschni [128] was empirically adapted for hydrogen. After validation, lean-burn, *cooled dry* EGR, and *hot wet* EGR strategies were examined numerically.

27.3.3 CFD Models

Johnson [129] used the Kiva-3V engine simulation code developed at Los Alamos National Laboratory with the standard eddy-turnover model to simulate a hydrogen engine at a fixed equivalence ratio and volumetric efficiency. The standard model contains one free parameter that was adapted for hydrogen and held constant for varying ignition timing and engine speed. The model was validated for bulk quantities (NO emissions and brake thermal efficiency) against the experiments reported in Ref. [130]. Fontana et al. [131] modified the Kiva-3V code to simulate an SI engine fueled with a hydrogen/gasoline mixture. They used a hybrid model in which the global reaction rate is either given by the standard eddy-turnover model or a weighed reaction rate based on two global reaction rate expressions, one for hydrogen combustion and one for gasoline. They validated the model for gasoline operation and then calculated the effects of adding various hydrogen concentrations to gasoline.

Shioji et al. [132] calculated flame propagation and NO_X formation for two hydrogen-fueled engines using an in-house CFD code, with a flame area evolution model and the laminar burning velocity correlation by Liu and MacFarlane [51]. They also evaluated the use of laminar burning velocities obtained from chemical kinetic calculations. Agreement with experimental results was fair, apart from high engine speed, lean mixture conditions. The authors quote these findings to "suggest the requirement for the consideration of an increase in local laminar burning velocity due to the selective diffusion of hydrogen."

Adgulkar et al. [133] used the AVL Fire software for a CFD simulation of the power cycle in a simplified engine geometry. Several combustion models were evaluated by comparing calculations to the results from Ma et al. [124]. As stated in the previous section, the results reported by Ma et al. [124] are doubtful, so this is a poor validation of these calculations.

CFD simulations have been used by a team from TU Graz and BMW to investigate the mixture formation and combustion in DI engines [134–136]. The Fluent code was used with the turbulent burning velocity model by Zimont [95]. The laminar burning velocity was obtained from chemical kinetic calculations using the reaction scheme of Ó Conaire et al. [65], neglecting the influence of residual gas. The prediction of the flame propagation and rate of heat release corresponds well to measurements obtained on an optical engine.

At the Institut Français du Pétrole (IFP), the Extended Coherent Flame Model has been adapted for a better representation of hydrogen combustion in engines by adding a source term to the flame surface density transport equation [137]. The terms representing laminar flame propagation in the flame surface density transport equation are usually neglected. However, in the case of hydrogen, due to the high laminar burning velocity of mixtures around stoichiometry, this cannot be justified, so the laminar term was added to the model. Knop et al. [58] discussed the problems in finding a suitable correlation for the laminar burning velocity and proposed a correlation based largely on the correlation of Verhelst [46,50,61] but extended to rich mixtures (presumably through chemical kinetic calculations but not detailed in the paper) to allow computations of stratified combustion in DI engines. Another important contribution in the paper by Knop et al. [58] is an extended Zeldovich model of which the reaction rate constants were adapted for hydrogen, based on the work of Miller and Bowman [123]. The resulting CFD model was validated both for an engine with cryogenic port injection and a DI engine. The detailed mixture distribution obtained from the CFD simulations was used to explain the sensitivity of flame propagation and NO_X formation to mixture heterogeneity.

27.3.4 Heat Transfer Submodel

Most of the works cited in the engine combustion modeling sections mentioned earlier use a heat transfer submodel to calculate the heat transfer between the cylinder gases and the combustion chamber walls. For SI engines, heat transfer due to radiation is small (<10%) compared to convection and can be neglected given the current uncertainties in modeling convection [138]. The instantaneous heat transfer can be modeled as [134,138]

$$\frac{dQ}{dt} = hA(T - T_{wall}) \tag{27.5}$$

where

h is the convection coefficient averaged over the heat transfer surface
A is the total wall surface area
T is the bulk gas temperature
T_{wall} is the wall temperature averaged over the heat transfer surface

Several models exist for evaluating the heat transfer coefficient h, of which the correlations of Woschni [128] and Annand [139] are the most widely used.

Wei et al. [140], Shudo and Nabetani [141], and Shudo and Suzuki [142] have measured instantaneous heat transfer coefficients in hydrogen-fueled engines. Wei et al. found transient heat transfer coefficients during hydrogen combustion to be twice as high as during gasoline combustion. They evaluated heat transfer correlations and found Woschni's equation to underpredict the heat transfer coefficient by a factor of two, whereas Annand's equation gave reasonable results. Shudo et al. compared the heat transfer coefficients during stoichiometric hydrogen and methane combustion, finding them to be larger in the case of hydrogen. The correlation by Woschni was tested and found to be inadequate.

The shorter quenching distance of a hydrogen flame (see Section 27.2) is put forward as the cause of this increased heat transfer, leading to a thinner thermal boundary layer. Furthermore, for near-stoichiometric combustion, flame speeds are high and cause intensified convection. Hydrogen also has a higher thermal conductivity compared to hydrocarbons, greatly affecting mixture thermal conductivity (see Table 27.2). Shudo and Suzuki [142] construct an alternative heat transfer correlation with an improved correspondence to their measurements. However, the correlation contains two calibration parameters dependent on ignition timing and equivalence ratio.

Recent work reported by Nefischer et al. [143] proposes an adaptation to Schubert's formula [144], using a more detailed description of the turbulent kinetic energy from which the characteristic velocity used in the formula is calculated. For hydrogen engines, the turbulent kinetic energy is reported to be affected through the change (decrease) during the combustion of the number of moles (see Table 27.2). Also, for DI engines, the direct gaseous injection affects the turbulent kinetic energy.

Figure 27.13 shows recent results of heat flux measurements on a CFR engine operated on hydrogen and on methane [145]. The figure plots local heat flux (measured at the cylinder liner) for three conditions that produce the same indicated power: methane–air at a lean equivalence ratio of $\lambda = 1.25/\varphi = 0.8$ and WOT, stoichiometric methane–air using throttling, and stoichiometric hydrogen–air WOT. All measurements were performed at MBT spark timing. The much higher peak heat flux on hydrogen confirms the works cited earlier [140,141].

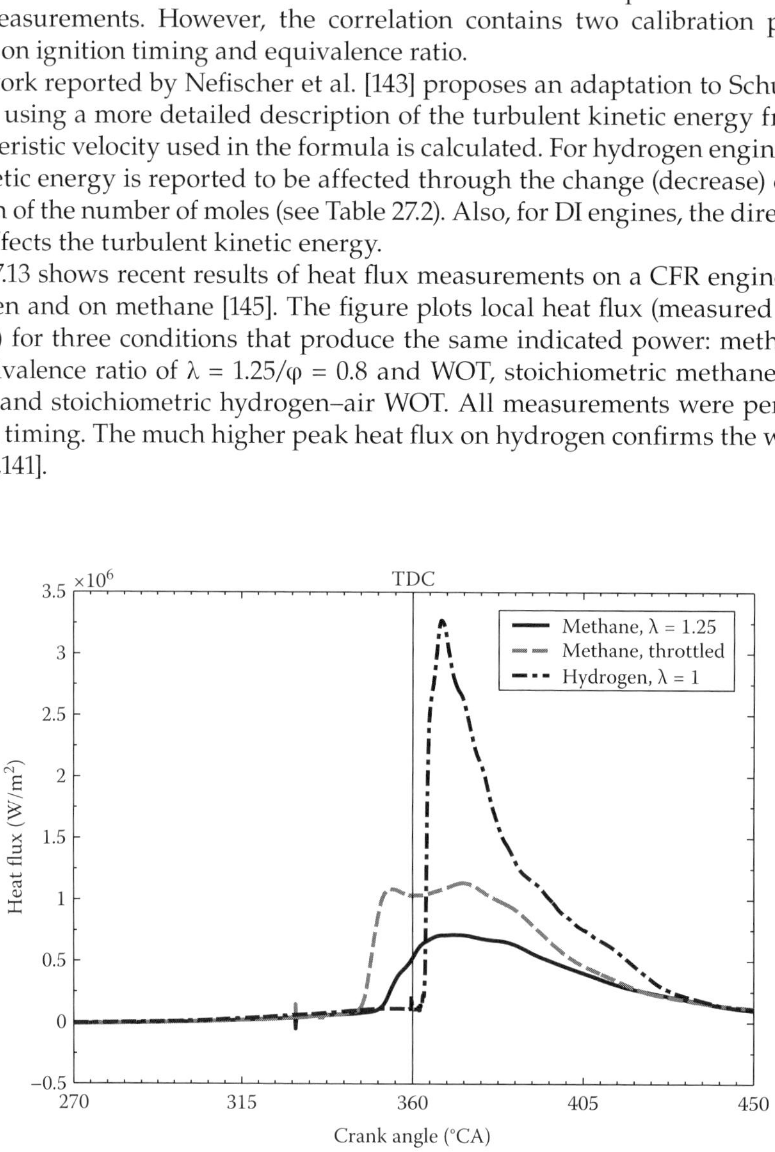

FIGURE 27.13
Measured local heat flux (ensemble average over 35 cycles) for equal indicated power: hydrogen stoichiometric WOT, methane lean WOT, and methane stoichiometric throttled. 600 RPM, compression ratio 8:1.

27.4 Abnormal Combustion

The same properties that make hydrogen such a desirable fuel for internal combustion engines also bear responsibility for abnormal combustion events associated with hydrogen. In particular, the wide flammability limits, low required ignition energy, and high flame speeds can result in undesired combustion phenomena generally summarized as combustion anomalies. These anomalies include surface ignition and backfiring as well as autoignition [146].

Surface ignition is used here to denote uncontrolled ignition induced by a hot spot in the combustion chamber. Preignition generally describes combustion events occurring inside the combustion chamber during the engine compression stroke with the actual start of combustion prior to spark timing. Backfiring, also called back-flash, refers to events in which the hydrogen–air charge combusts during the intake stroke, usually in an intake runner or intake manifold. Engine *knock* is the term used for typical SI engines to describe autoignition of the remaining end gas during the late part of the combustion event with high-pressure oscillations and the typical pinging noise. As already briefly discussed in Section 27.2, care must be taken in using the term knock with hydrogen engines, as further explained below.

27.4.1 Preignition

The typical premature combustion during the engine compression stroke with closed intake valves resulting from preignition, a surface ignition anomaly, can have numerous causes. Because preignition is a stochastic event, detailed investigations of preignition are complicated, and the actual cause of preignition is often nothing more than speculation. Sources for the fresh charge to combust during the compression stroke include hot spark plugs or spark plug electrodes, hot exhaust valves or other hot spots in the combustion chamber, residual gas or remaining hot oil particles from previous combustion events [147], as well as residual charge of the ignition system. In general, both high temperatures and residual charge can cause preignition. Due to the dependence of minimum ignition energy on the equivalence ratio, preignition is more pronounced when the hydrogen–air mixtures approach stoichiometric levels. Also, operating conditions at increased engine speed and engine load are more prone to the occurrence of preignition due to higher gas and component temperatures.

Figure 27.14 shows the in-cylinder pressure trace as well as the crank-angle resolved intake manifold pressure for a combustion cycle in which preignition occurred. A regular combustion event is shown for comparison. The data were taken on an automotive-size single-cylinder hydrogen research engine at an engine speed of 3200 RPM and an IMEP of 7 bar for the regular combustion case (dotted line). The almost symmetrical phasing of the cylinder pressure caused by the preignition results in the IMEP to drop almost to 0. It is interesting to note that the peak pressure for the preignition case is higher than the regular combustion cycle. However, due to the early pressure rise that starts around 80°CA BTDC, the indicated intake pressure trace for the preignition case does not show any significant difference from the regular trace, because the preignition occurred after the intake valves closed.

Measures to avoid preignition include proper spark plug design, design of the ignition system with low residual charge, specifically designed crankcase ventilation, sodium-filled exhaust valves, as well as optimized design of the engine cooling passages to avoid

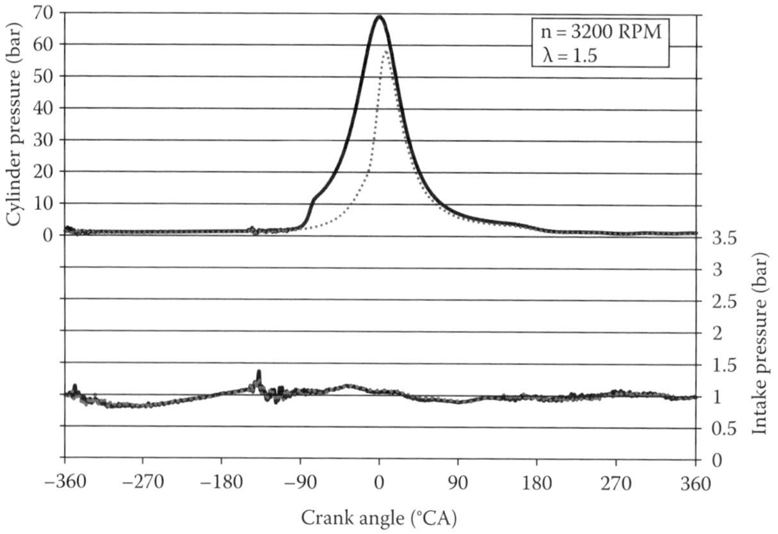

FIGURE 27.14
Typical cylinder and intake manifold pressure traces with preignition (solid lines), compared to regular pressure traces (dotted lines).

hot spots (see Section 27.5). Hydrogen DI into the combustion chamber is another measure to effectively reduce or eliminate the occurrence of preignition depending on the injection strategy (see Section 27.6).

27.4.2 Backfire

Backfiring, or flash-back, describes combustion of fresh hydrogen–air charge during the intake stroke in the engine combustion chamber and/or the intake manifold. With the opening of the intake valves, the fresh hydrogen–air mixture is aspirated into the combustion chamber. When the fresh charge is ignited at combustion chamber hot spots, hot residual gas or particles, or remaining charge in the ignition system, backfiring occurs, similar to preignition. The main difference between backfiring and preignition is the timing at which the anomaly occurs. Preignition takes place during the compression stroke with the intake valves already closed, whereas backfiring occurs with the intake valves open. This results in combustion and pressure rise in the intake manifold, which is not only clearly audible but can also damage or destroy the intake system. Due to the lower ignition energy, the occurrence of backfiring is more likely when mixtures approach stoichiometry. Because most operation strategies with hydrogen DI (see Section 27.6) start injection after the intake valves close, the occurrence of backfiring is generally limited to external mixture formation concepts.

Figure 27.15 shows the cylinder and intake pressure traces for a backfiring cycle measured on an automotive-size hydrogen single-cylinder engine at an engine speed of 3200 RPM and an IMEP of 7 bar. A regular intake and combustion pressure trace is shown as a reference. As soon as the intake valves open, the fresh charge is ignited and combusts in the intake manifold, resulting in an increase of intake pressure of up to 3 bar. The pressure rise is also reflected in the cylinder pressure trace. Once the entire fresh charge is burned, the pressure in the intake manifold decreases; the cylinder pressure at intake valve closing is increased compared to the regular trace. The peak cylinder pressure for this backfiring

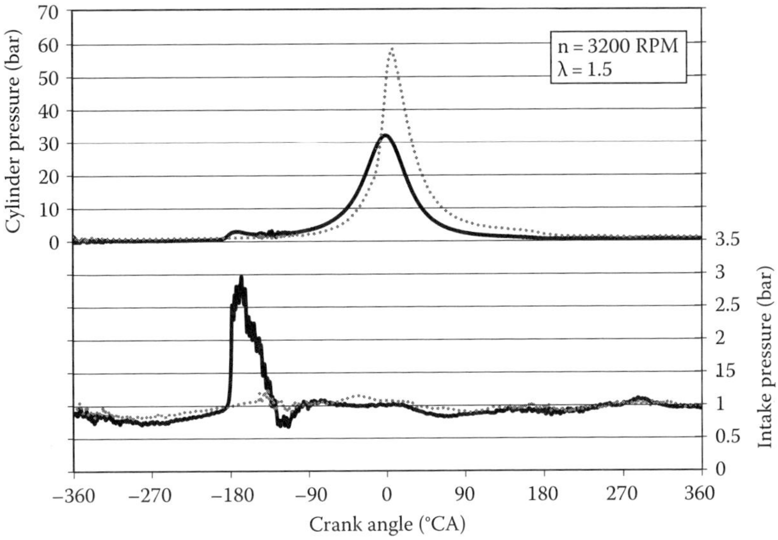

FIGURE 27.15
Typical cylinder and intake pressure traces for backfiring cycle (solid lines), compared to regular pressure traces (dotted lines).

cycle is only about 30 bar and the IMEP is actually negative. For comparison, the peak pressure in motored operation is approximately 21 bar.

Limited information available on combustion anomalies also indicates that preignition and backfiring are closely related with preignition as the predecessor for the occurrence of backfiring. Preignition thereby heats up the combustion chamber, which ultimately leads to backfiring in a consecutive cycle [148–150].

Consequently, any measures that help avoid preignition also reduce the risk of backfiring. In addition, work has been done on optimizing the intake design and injection strategy to avoid backfiring [151].

Injection strategies that allow pure air to flow into the combustion chamber to cool potential hot spots before aspirating the fuel–air mixture were proposed as one potential approach. As a result of experimental and simulation work on a hydrogen PFI engine, a predictive model and guidelines for backfire-free operation were derived. These guidelines were based on the finding that the possibility of backfire mainly depends on the concentration of H_2 residual in the intake ports in PFI hydrogen engines; thus, the leaner the concentration of the residual, the lower the possibility of backfire. Based on this conclusion, it is suggested to limit the end of injection in a fixed range based on engine operation conditions with an earlier end of injection at lower engine speeds and richer hydrogen mixtures [152]. Although trends identified on hydrogen research engines indicated that combustion anomalies significantly limit the operation regime [153], optimization of the fuel-injection strategy in combination with variable valve timing for both intake and exhaust valves allowed operating of a port injected hydrogen engine at stoichiometric mixtures over the entire speed range [154].

27.4.3 Autoignition

When the end-gas conditions (pressure, temperature, time) are such that the end gas spontaneously autoignites, there follows a rapid release of the remaining energy generating high-amplitude pressure waves, mostly referred to as engine knock. The amplitude of the pressure waves of heavy engine knock can cause engine damage due to increased mechanical and thermal stress. The tendency of an engine to knock depends on the engine design as well as the fuel–air mixture properties.

A standard measure to define the knock characteristics of liquid fuels is the octane rating. A CFR engine is used to determine the knock behavior of a specific fuel by comparing its knock resistance to that of a mixture of normal heptane and iso-octane. The most common standardized tests to determine knock resistance on a CFR engine are the research method resulting in a RON [155] and the motor method resulting in a MON [156]. Although these methods were developed and are applicable only to liquid SI engine fuels, octane ratings for hydrogen fuel have been reported in the literature. Reported values range from RON < 88 [157] to RON ¼130 [158] and RON of 130þ for lean mixtures [159]. It is unclear how these values were determined; they must either be estimated values or measured with methods resembling but not according to the ASTM methods. Work has also been performed on emulating the knock measurement on the CFR engine by using low-pass-filtered rate of change of the pressure signal; so far, this work is limited to primary reference fuels [160].

The determination of octane ratings is performed at constant spark advance (13°CA BTDC for RON and 19°CA–26°CA BTDC, depending on the compression ratio for MON). The discrepancies in nominal knock resistance of hydrogen are mainly due to the extremely high flame speeds around stoichiometry, with the strong dependence on the air–fuel ratio, which makes application of standard methods for the determination of knock resistance questionable.

Because of the high knock resistance of methane (115 < MON < 130), the MN was defined to determine the knock characteristics of gaseous fuels. The MN uses a reference fuel blend of methane, with an MN of 100, and hydrogen, with an MN of 0 [161]. Per definition, the MN of hydrogen is 0, which would suggest that hydrogen has a very low knock resistance. This clearly contradicts some of the octane numbers reported in the works cited earlier [158,159].

Work has been reported on attempts to predict the knock behavior of hydrogen-fueled engines. A comparison to experimental results showed good agreement for variation of compression ratio, air–fuel equivalence ratio, and intake air temperature [162]. These results suggest that the operating regime of a hydrogen engine is strongly limited by the occurrence of knocking combustion. However, based on work performed on a multicylinder hydrogen engine at compression ratios of up to 15.3:1, it was stated that knock, as has been observed on gasoline engines, was not observed in any of this hydrogen testing regardless of compression ratio [163].

Figure 27.16 shows the cylinder pressure trace as well as a filtered signal for hydrogen DI operation at 2000 RPM and an engine load of 10 bar IMEP recorded on a single-cylinder research engine with a compression ratio of 12:1. The cylinder pressure signal shows pressure oscillations that are typical for knocking combustion; the high-pass filtered signal shows a maximum pressure amplitude of approximately 3.6 bar. For the same engine speed and load, an operating point with heavy knock was recorded, resulting from further advancing the spark timing (Figure 27.17). Although the regular peak pressure for this operating point is about 90 bar, the maximum pressure with knocking operation reaches 150 bar with oscillations in the high-pass-filtered signal of almost 65 bar.

Similar tests performed on a CFR engine at a compression ratio of 12:1 were targeted at determining the knock characteristics of hydrogen and the applicability of standard automotive knock-detection systems. Comparative analysis of knock intensities of gasoline and hydrogen revealed that knocking pressure traces exhibit similar peak amplitudes as well as similar durations and decays of pressure oscillations [164].

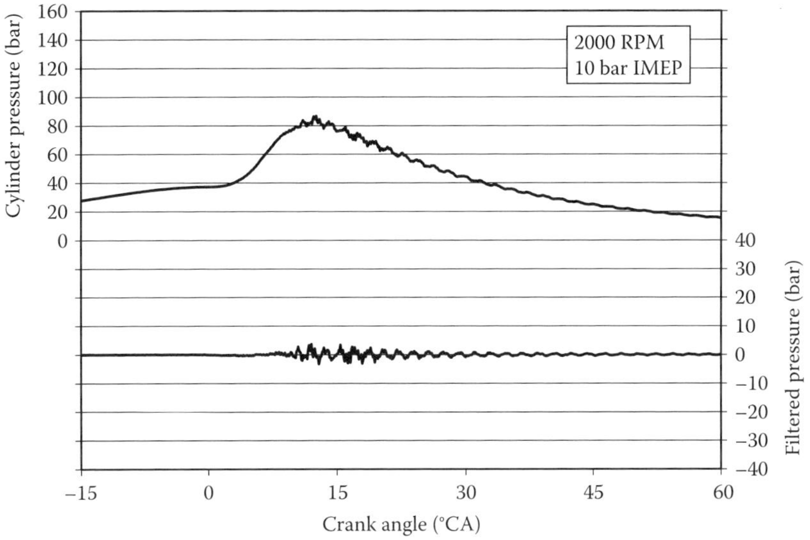

FIGURE 27.16
Typical cylinder pressure trace for light knocking cycle.

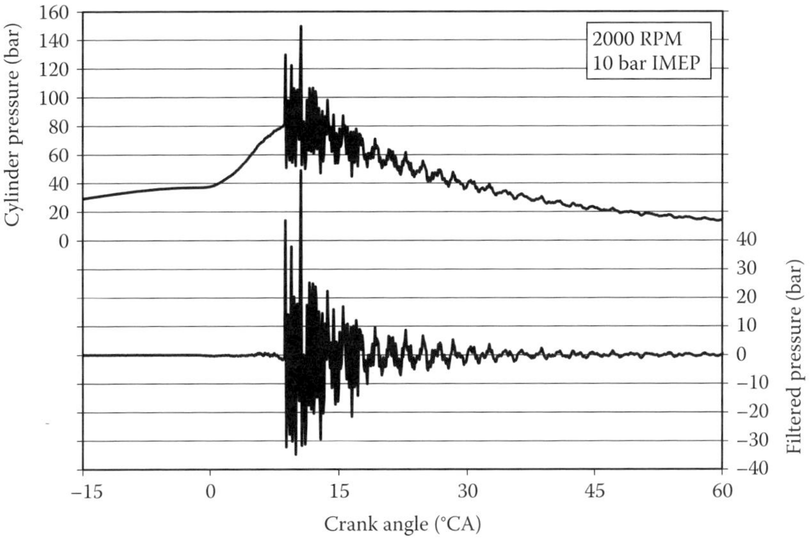

FIGURE 27.17
Typical cylinder pressure trace for heavy knocking cycle.

27.4.4 Modeling Abnormal Combustion Phenomena

The literature on the simulation of abnormal combustion phenomena in a hydrogen engine is quite limited. The model of Fagelson et al. [115], described earlier, was extended by Sadiq Al-Baghdadi [165] with a simple model predicting the occurrence of abnormal combustion, using a *knock integral* type of approach [166]. A graph is shown comparing the simulated and measured cylinder pressure traces for normal combustion. Significant differences can be seen, which casts doubts on the subsequent validation of the *preignition* submodel.

Li and Karim use a two-zone quasidimensional model with a triangular combustion rate law fitted to experimental data coupled with a chemical kinetic scheme [162]. They propose a knock criterion, comparing the energy released by end-gas reactions to the energy released by the normal flame propagation. When this exceeds a critical value, end-gas autoignition occurs. The model is used to predict the knock-limited equivalence ratio as a function of the compression ratio. Similar to the experimental results reported by these authors [167], the knocking regions are very extensive. According to the figures in the paper, which report results for compression ratios between 6:1 and 14:1, stoichiometric operation is impossible without the occurrence of knock. This clearly is contradicted by numerous works cited in the present paper.

Liu et al. [152] use a CFD calculation of the gas dynamics in a port-fueled engine to explain the dependence of backfire occurrence on the injection timing. Although the model is only partly validated, it demonstrates the existence of an optimal timing. Injecting too early leads to a backflow of hydrogen from the cylinder into the intake port, at the end of the intake stroke. Injecting too late results in hydrogen remaining in the intake port because of insufficient time to reach the cylinder. Thus, in both cases, hydrogen is present in the intake port at the time the intake valve opens for the next cycle. This can result in backfire through contact with hot spots (residual gases, exhaust valves, etc.). The optimal injection timing is the one that allows an initial cooling period by freshly aspirated air but also enables all of the injected hydrogen to travel to the cylinder before the intake valve closes (which has been experimentally confirmed [33]).

27.4.5 Avoiding Abnormal Combustion

Limiting the maximum fuel-to-air equivalence ratio is an effective measure for avoiding abnormal combustion in hydrogen operation. Due to the wide flammability limits and fast flame speeds, hydrogen internal combustion engines are usually operated employing a lean-burn strategy and thereby avoid throttle losses (see Section 27.6). The excess air in lean operation acts as an inert gas and effectively reduces combustion temperatures and consequently component temperatures. This significantly reduces the occurrence of abnormal combustion in lean combustion regimes. Although lean operation is also very efficient, it does limit the power output of hydrogen engines. Results from a supercharged and intercooled 1.8 inline four-cylinder engine operated on gasoline as well as hydrogen showed that abnormal combustion can be effectively avoided by limiting the fuel-to-air equivalence ratio. In this particular case, the maximum fuel-to-air equivalence ratio was limited to 0.63 ($\lambda = 1.6$) at 1500 RPM and further reduced as a function of engine speed with a minimum of 0.48 ($\lambda = 2.1$) at 6000 RPM. While effectively avoiding abnormal combustion, this measure also significantly reduces the power output from approximately 120 kW in gasoline operation to approximately 70 kW in hydrogen operation [168].

Further measures to avoid abnormal combustion are given in the next section on hydrogen engine hardware.

27.5 Measures for Engine Design or Conversion

This section discusses some features of engines designed for, or converted to, hydrogen operation. The occurrence of combustion anomalies discussed in the previous section, or more particularly the desire to prevent it, has led to most of the countermeasures put forward in the early work on H_2ICEs.

27.5.1 Spark Plugs

Cold-rated spark plugs are recommended to avoid spark plug electrode temperatures exceeding the autoignition limit and causing backfire [169,170]. Cold-rated spark plugs can be used, since there are hardly any spark plug deposits to burn off [169]. Spark plugs with platinum electrodes are to be avoided, as this can be a catalyst to hydrogen oxidation [11,171] (platinum has been used in the exhaust to oxidize unburned hydrogen [172]).

27.5.2 Ignition System

To avoid uncontrolled ignition due to residual ignition energy, the ignition system should be properly grounded or the ignition cable's electrical resistance should be changed [170,173]. Also, induction ignition in an adjacent ignition cable should be avoided [174], for instance, by using a coil-on-plug system. Somewhat counterintuitively, a high-voltage output ignition system should be provided, as the ignition of hydrogen mixtures asks for an increased secondary ignition voltage (probably because of the lower ion concentration of a hydrogen flame compared to a hydrocarbon flame) [170,173,175,176]; coil-on-plug systems also satisfy this condition. Alternatively, the spark plug gap can be decreased to lower the ignition voltage; this is no problem for hydrogen engines, as there will be almost no deposit formation. Spark plug gaps as small as 0.25 mm have been used [177] (although the gap was subsequently increased to 0.5 mm because of cold-start difficulties due to water condensation at the spark plug tip).

27.5.3 Injection System

It is clear from Section 27.4 that timed injection is a prerequisite. One option is to use port injection and to program the injection timing such that an air cooling period is created in the initial phase of the intake stroke and the end of injection is such that all hydrogen is inducted, leaving no hydrogen in the manifold when the intake valve closes. The timing described here might not be necessary, as work has been reported in which no relation between injection timing and backfire or surface ignition limited equivalence ratio was found [163]. The second option is to use DI during the compression stroke. High flow rate injectors with instantaneous flows around 4–6 g/s at 100 bar supply pressure are needed for DI [178]. With PFI engines, the high flow requirements can be alleviated by using multiple injectors. Timed injection also decreases the amount of unburned fuel in the intake manifold at any given time, limiting the severity of a backfire, should it occur.

27.5.4 Hot Spots

Clearly, hot spots in the combustion chamber that could initiate surface ignition or backfire are to be avoided or minimized. Measures include the use of cooled exhaust valves, multivalve engine heads to further lower the exhaust valve temperature [171,173,179], a proper oil control [147], additional engine coolant passages around valves and other areas with high thermal loads [180] (if possible), the delay of fuel introduction to create a period of air cooling (using timed manifold or DI), and adequate scavenging (e.g., using variable valve timing [163,181]) to decrease residual gas temperatures.

27.5.5 Piston Rings and Crevice Volumes

Experiments have been conducted in which all hot spots were eliminated (careful cleaning of the engine, enhanced oil control or even nonlubricated operation, scavenging of the residual gases, cold spark plugs, cooled exhaust valves, etc.), as well as any uncontrolled spark-induced ignition, and backfire still occurred [177,182]. This suggests that the small quenching distance of hydrogen (together with the wide flammability limits), allowing combustion in the piston top land (the crevice volume above the top piston ring), is a parameter that has been overlooked by many workers. Hydrogen engines have been demonstrated, running on stoichiometric mixtures without any occurrence of backfire, by careful selection of piston rings and crevice volumes, without any need for timed injection or cooled exhaust valves [180]. Workers that have paid attention to increased cooling, enhanced *oil control* by mounting different piston rings, increased scavenging, etc., attribute the resulting wider backfire-free operation region to a reduction of hot spots but have simultaneously (sometimes possibly without realizing it) taken measures to suppress crevice combustion.

Thus, the piston top land clearance can be decreased to prevent hydrogen flames from propagating into the top land; Swain et al. [180] use a clearance of 0.152 mm to quench the hydrogen flame. Some researchers have changed the crevice volumes and/or piston rings with the aim of reducing the reflow of unburned mixture from the second land (the crevice volume between the top two piston rings) to the top land [180,182,183] (preventing *fueling* of a top land flame during exhaust and intake). The smaller quenching distance of a hydrogen flame also implies an increased thermal load for the piston top land; Berger et al. [184] report changes (a special coating) to the top piston ring groove area to account for this.

27.5.6 Valve Seats and Injectors

The very low lubricity of hydrogen has to be accounted for; suitable valve seat materials have to be chosen [171,173] and the design of the injectors should take this into account. This is the case with any dry gaseous fuel (such as natural gas) but can be more critical for hydrogen (compressed natural gas contains small amounts of oil originating from the oil mist in the compressor, whereas hydrogen compressors normally have tighter clearances to limit the leak rate).

27.5.7 Lubrication

An engine lubrication oil compatible with increased water concentration in the crankcase has to be chosen [185]. The report on the hydrogen drive test in Germany by TÜV [173] cites two options: a demulsifying oil and a synthetic oil that forms a solution with water. DeLuchi [186] claims a longer oil lifetime, as the oil is not diluted by hydrogen and there is less formation of acids (perhaps doubtful, given the large quantities of water and NO_X that can be formed during stoichiometric combustion). An ashless oil is recommended to avoid deposit formation (hot spots) [154,169]. Measurements of the composition of the gases in the crankcase at Ghent University [176] showed a very high percentage of hydrogen (+5 vol.%, out of range of the testing equipment) arising from the blowby. Blowby can be expected to be quite high because of the rapid pressure rise (caused by the high flame speed and the resulting fast burn rate) and the low density of hydrogen gas (significantly affecting the mixture density, see Table 27.2). The composition of the lubricating oil was investigated and compared to that of the unused oil. The properties of the oil had severely changed with a strong decrease of the lubricating qualities.

An engine oil specifically developed for hydrogen engines is probably the best solution but currently unavailable. For safety reasons, a forced crankcase ventilation system (see also below) was mounted on the engine to keep the hydrogen concentration well below the lower flammability limit. Air is fed to the crankcase from the lab compressed air net and set to a small overpressure using a pressure regulating valve. A vacuum pump is used to evacuate the crankcase gases, which pass an oil separator first. The crankcase pressure is controlled to a slight underpressure by a balance between the compressed air pressure and a bypass valve on the vacuum pump inlet. The resulting hydrogen concentration in the crankcase with the ventilation system was measured to be below 1 vol.%.

27.5.8 Crankcase Ventilation

Positive crankcase ventilation is generally recommended due to unthrottled operation (high manifold air pressures) and to decrease hydrogen concentrations (from blowby) in the crankcase [171,187]. As will be discussed in Section 27.6, WOT operation is used wherever possible to increase engine efficiency, resulting in high manifold air pressures. Thus, the pressure difference between the crankcase and the intake manifold, such as in throttled gasoline engines, is absent for some operating strategies of hydrogen engines and thus cannot be used as a driving force for crankcase ventilation. This can be solved by, for example, a venturi placed in the intake [171] or other methods used in high manifold pressure engines (such as diesel engines).

27.5.9 Compression Ratio

The choice of the optimal compression ratio is similar to that of any fuel; it should be chosen as high as possible to increase engine efficiency, with the limit given by increased heat losses or the occurrence of abnormal combustion (in the case of hydrogen, primarily surface ignition). The choice may depend on the application, as the optimum compression ratio for highest engine efficiency might be different from the optimum for highest power output [188]. Compression ratios used in H_2ICEs range from 7.5:1 [185] to 14.5:1 [163].

27.5.10 In-Cylinder Turbulence

Because of the high flame speeds of hydrogen, low turbulence combustion chambers (pancake or disk chamber and axially aligned symmetric intake port) can be used, which can be beneficial for the engine efficiency [130,179,180] (increasing the volumetric efficiency and decreasing heat losses). They might even be necessary to avoid excessive rates of pressure rise (possibly even leading to knocking combustion) at stoichiometric operation [180] (where high in-cylinder turbulence could cause very fast flame speeds).

27.5.11 Electronic Throttle

For reasons discussed in Section 27.6, hydrogen engines should be operated at WOT wherever possible, but throttling might be needed at very low loads to maintain combustion stability and limit unburned hydrogen emissions. At medium to high loads, throttling might be necessary to limit NO_X emissions. This can only be realized with a drive-by-wire system, that is, a system in which the throttle position is electronically controlled instead of mechanically linked to the accelerator pedal.

27.5.12 Materials

The effects of hydrogen on the mechanical properties of iron and steel have been widely investigated. Regarding the embrittling effect of hydrogen, it is well known that the dominant effects are a decrease in ductility and true stress at fracture. Hydrogen embrittlement of steels can be classified into three main types [189]:

1. Hydrogen reaction embrittlement arises because of the generation of hydrogen on the surface as a result of a chemical reaction.
2. Environmental embrittlement takes place in the hydrogen-containing atmospheres through adsorption of molecular hydrogen on the surface and its absorption within the lattice after dissociation into atomic form.
3. Internal hydrogen embrittlement, in contrast, takes place in the absence of a hydrogenated atmosphere and is brought about by hydrogen that has entered the lattice during processing or fabrication of steel.

The environmental embrittlement in hydrogen-containing atmospheres results in limitations for material selection of hydrogen storage and fuel systems. Studies have been performed to assess the sensitivity of commonly used stainless steels for hydrogen embrittlement [190] as well as special alloys [191]. In both studies, it was concluded that the tested materials show significant degradation due to the presence of hydrogen. All metallic materials present a certain sensitivity to hydrogen embrittlement, with the sensitivity strongly dependent on the stress level. Materials that can be used for hydrogen applications are brass and copper alloys, aluminum and aluminum alloys, and copper–beryllium. Nickel and high-nickel alloys as well as titanium and titanium alloys are known to be very sensitive to hydrogen embrittlement. For steels, the hydrogen embrittlement sensitivity depends on the exact chemical composition, heat or mechanical treatment, microstructure, impurities, and strength [192]. Negative effects of hydrogen embrittlement have also been documented for certain types of piezomaterials used for hydrogen fuel injectors [178]. Apart from embrittlement effects of onboard hydrogen system components, hydrogen embrittlement testing performed on several grades of high-strength pipeline steels showed a loss in ductility that was, however, recoverable when a charged steel was left for 7 days at ambient temperature after charging. It was concluded that control of cathodic protection systems may be more critical on high-strength steel pipelines [193]. An overview of the hydrogen compatibility of materials can be found in [194].

Concerning specific engine components, intake manifolds of hydrogen internal combustion engines, in particular with PFI, are mostly made of metal to withstand backfire. This measure is mainly taken for development and dynamometer calibration work, since the limits for abnormal combustion have to be established. For vehicle application and demonstration vehicles, calibrations that effectively avoid abnormal combustion have to be employed.

27.6 Mixture Formation and Load Control Strategies

27.6.1 Introduction: Classification of Strategies

The wide ignition limits of hydrogen allow an engine to be operated at extremely lean air–fuel ratios compared to conventional fuels. Concepts for part-load operation of hydrogen engines have been proposed using quantitative control, qualitative control, as well as combined strategies. Quantitative control refers to any approach that limits the amount of fresh charge introduced into the engine in order to limit the power output while keeping the ratio of air to fuel constant. On the other hand, qualitative control is used to describe systems that adjust the air-to-fuel ratio usually by adjusting the amount of fuel introduced to the engine while allowing maximum flow of air. The classification based on part-load operation is closely linked to the air–fuel ratio at which the engine is operated. Reports of hydrogen engines being operated at stoichiometric as well as lean air–fuel ratios have been published. The major advantages of operation at a stoichiometric air–fuel ratio are the increased power output compared to lean-burn strategies as well as the fact that a conventional aftertreatment system can be employed to reduce NO_X emissions. Lean-burn concepts, on the other hand, generally result in a significant increase in achievable engine efficiencies. More recently, strategies that employ both lean-burn and stoichiometric operations have been proposed [195,196].

Besides the aforementioned classification that results from the specific properties of hydrogen as a fuel for internal combustion engines, the charging strategy is another important factor to influence the performance and efficiency of hydrogen internal combustion engines. Supercharged operation has been evaluated as a promising option to mitigate the significant reduction in power output related to hydrogen operation with external mixture formation. From an efficiency standpoint, turbocharging is considered the preferred option; however, reduced throttle response and reduced exhaust energy compared to conventional-fuel operation make the implementation of a turbocharged hydrogen engine more challenging.

Although the self-ignition temperature of hydrogen is considerably higher than that of conventional fuels (H_2 = 585°C, diesel is approx. 250°C), which suggests using a spark plug as the ignition source, attempts have been made to operate hydrogen engines with CI. Successful operation of CI hydrogen engines has been reported for both large displacement stationary engines [197,198] and automotive-size engines [199]. More recently, effort has been reported on hydrogen engines being operated in HCCI mode [200] (see Section 27.6.2.4).

27.6.2 Mixture Formation Strategies

The proper design of the mixture formation process is crucial for achieving high engine efficiencies while meeting more and more stringent emissions targets. Similar to conventionally fueled engines, hydrogen engines have gone through continuing improvement and refinement in terms of mixture formation strategies.

A primary classification of mixture formation strategies can be done based on the location of mixture formation or the location of the hydrogen dosing devices. External mixture formation refers to concepts in which hydrogen and air are mixed outside the combustion chamber, whereas internal mixture formation refers to concepts with hydrogen being introduced directly into the combustion chamber. Some researchers have also proposed combined concepts with a combination of external and internal mixture formation [201–203]. As indicated in Table 27.3, the mixture formation strategy, especially in hydrogen operation, has significant impact on the theoretical power output of the engine. The dramatic difference in theoretical power output is mainly caused by the low density of hydrogen, resulting in a significant decrease in mixture density when external mixture formation is being employed.

Using a classification based on the control of timing/quantity of the induced fuel, one can differentiate systems that use carburetors, mechanically controlled injection devices, and electronically controlled fuel injectors. Modern hydrogen combustion engines almost exclusively use electronically controlled fuel-injection systems; however, the requirements and specifications for these systems change widely based on the injection location and the temperature of the injected fuel. Generally, hydrogen injection systems for external mixture formation are operated at lower injection pressures (2–8 bar) compared to systems for hydrogen DI (5–250 bar). Also, the exposure of injectors to in-cylinder temperatures and pressure in combination with increased injection pressures for internal mixture formation systems still requires further injector development to reach production standards in terms of durability [178]. Research and development have also been performed on external mixture formation concepts with cryogenic hydrogen [126,204–208] posing challenges to the injection system due to extremely low temperatures (boiling temperature of hydrogen is approximately −253°C) and related issues, for example, injector icing.

Finally, a classification based on the resulting or intended mixture homogeneity allows grouping hydrogen combustion engines into homogeneous and stratified concepts. Due to the relatively long time available for mixing of fuel and air, all external mixture formation concepts can be considered homogeneous. However, internal mixture formation concepts with injection of fuel directly into the combustion chamber allow influencing the mixture distribution and homogeneity. This can be of particular interest to reduce the combustion duration and improve the combustion stability at extremely lean conditions or to avoid NO_X emissions critical air–fuel ratios by purposely creating lean and rich zones.

The two most prominent mixture formation strategies for hydrogen engines are hydrogen port injection, which is being used in engine research as well as vehicle demonstrations, and hydrogen DI, which is still at the research stage. The following sections are intended to summarize the most typical characteristics and variations of these mixture formation concepts. First, it is important to reflect on the energy cost of supplying low-pressure (in case of port injection) or high-pressure (DI) hydrogen.

27.6.2.1 Supplying Pressurized Hydrogen

Introduction of hydrogen into the engine, either in the intake manifold or directly into the combustion chambers, requires hydrogen to be supplied at a certain delivery pressure. The most efficient way of providing the required pressure depends on the type of onboard storage as well as the pressure levels required for injection. With compressed hydrogen storage at pressure levels up to 700 bar, sufficient pressure even for high-pressure injection systems operating at pressures above 100 bar is available. However, if no additional compressor is available, the full amount of hydrogen stored in a compressed hydrogen tank cannot be utilized. Assuming a storage pressure of 700 bar and an injection pressure of 100 bar, only 6/7 of the mass of hydrogen stored onboard can be used before the pressure in the tank drops below the required injection pressure. Whether hydrogen is compressed on- or off-board the vehicle, the energy required to compress hydrogen is significant. The minimal work required for compression of hydrogen results from isothermal compression, which is approximated using cooled piston compressors. Nonetheless, even under ideal conditions, the energy required to compress hydrogen from 1 bar to 1000 bar requires more than 7% of the heating value of hydrogen [209]. If hydrogen is stored onboard the vehicle in liquid cryogenic form, compression can be accomplished more efficiently in the liquid state. This allows cutting the compression work by a factor of 5–6 compared to gaseous compression. However, there are several remaining questions in construction and material selection of cryogenic pumps [209]. In addition, liquefaction of hydrogen as employed in current large-scale processes requires about 30% of the energy content of hydrogen [209]. In order to avoid the compression step of liquid hydrogen, ongoing research is performed on cryo-compressed hydrogen storage systems [210,211].

27.6.2.2 Spark-Ignition Port Injection

Hydrogen port injection is probably the most common hydrogen mixture formation strategy and is employed in a variety of variants that differ in multiple aspects including part-load control, air–fuel ratio, and charging strategy. Before evaluating the pros and cons of different mixture formation strategies, it is important to understand the principal correlation between the air–fuel ratio and oxide of nitrogen emissions that is applicable for all homogeneous mixture formation concepts. Figure 27.18 shows a typical trace of oxide of nitrogen emissions as a function of the equivalence ratio for homogeneous port-injection operation. Combustion of lean hydrogen–air mixtures with fuel-to-air equivalence ratios of less than 0.5 ($\lambda > 2$) results in extremely low NO_X emissions. Due to the excess air available in the combustion chamber, the combustion temperatures do not exceed the NO_X critical value of approximately 1800 K [12]. Exceeding the NO_X critical equivalence ratio results in an exponential increase in oxides of nitrogen emissions, which peaks around a fuel-to-air equivalence ratio of 0.75 ($\lambda \sim 1.3$). At stoichiometric conditions, the NO_X emissions are at around 1/3 of the peak value. The highest burned gas temperatures in hydrogen operation occur around a fuel-to-air equivalence ratio near 1.1, but at this equivalence ratio, oxygen concentration is low, so the NO_X concentration does not peak there [212]. As the mixture gets leaner, increasing oxygen concentrations initially offset the falling gas temperatures, and NO_X emissions peak around a fuel-to-air equivalence ratio of 0.75 ($\lambda = 1.3$).

In light of these dependencies of oxide of nitrogen emissions as a function of air–fuel ratio, several operating strategies have been developed that mainly aim at achieving acceptable power densities while simultaneously avoiding excessive NO_X emissions. Conversion engines based on conventional gasoline engines have been operated on hydrogen employing a lean constant air–fuel ratio strategy. Using a conventional throttle and replacing the gasoline fuel system with hydrogen injectors, one can easily implement this strategy [172,213,214]. Selecting an equivalence ratio below the NO_X emissions critical limit of $\varphi \sim 0.5/\lambda \sim 2$ results in extremely low emissions signatures even without the use of any

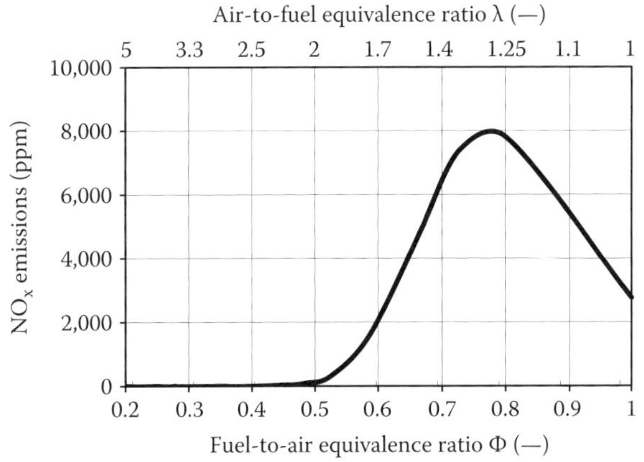

FIGURE 27.18
Correlation of air-to-fuel ratio and NO_X emissions for homogeneous operation. (From Wallner, T. et al., Evaluation of injector location and nozzle design in a direct-injection hydrogen research engine, SAE Paper No. 2008-01-1785, 2009.)

aftertreatment system. Due to the relatively low combustion temperatures, the reduced thermal load of the engine and the increase in required ignition energy for lean hydrogen–air mixtures, this lean operating strategy also effectively avoids combustion anomalies. Implementation of an engine with constant air–fuel ratio of $\varphi \sim 0.55/\lambda = 1.8$ at a compression ratio of 14.5 resulted in an 18% fuel economy improvement compared to the gasoline counterpart while meeting Transitional Low Emission Vehicles (TLEVs) emissions standards without any aftertreatment [215]. However, employing a constant lean air–fuel ratio strategy results in an even more significant loss in power density than shown in Table 27.3 for stoichiometric operation. Assuming a constant fuel-to-air equivalence ratio of 0.5 ($\lambda = 2$) results in a theoretical maximum power output of the hydrogen engine that is only about 50% of a regular gasoline engine in stoichiometric operation (see Figure 27.19). Thus, researchers have investigated the potential of using supercharging in combination with constant lean air–fuel ratio operation to mitigate the significant power loss [216,217]. This concept has also been implemented in several versions of hydrogen-powered pickup trucks running at constant equivalence ratios around $\varphi \sim 0.4/\lambda \sim 2.5$ with compression ratios of up to 12:1 and boost pressures of approximately 0.8 bar [218]. The theoretical maximum power output based on a comparison of calculated mixture calorific values as a function of air–fuel ratio assuming constant efficiencies is in the range of about 80% compared to the naturally aspirated gasoline counterpart (see Figure 27.19).

As has been discussed earlier, a significant increase in engine efficiency (compared to gasoline operation) can be accomplished employing a constant lean air–fuel ratio operating strategy. The air–fuel ratio is generally set as high as possible to achieve acceptable power output while still meeting the emissions targets. However, reducing the equivalence ratio was shown to even further improve engine efficiencies with a peak at around $\lambda = 3.3/\varphi = 0.30$. This peak results from a local minimum of the losses due to increased burn duration and heat transfer as well as the more favorable properties of the working fluid. Also, a sharp decrease in indicated efficiency was observed at $\lambda = 4.5/\varphi = 0.22$ due mostly to greater amounts of unburned hydrogen and a slower burn rate [163].

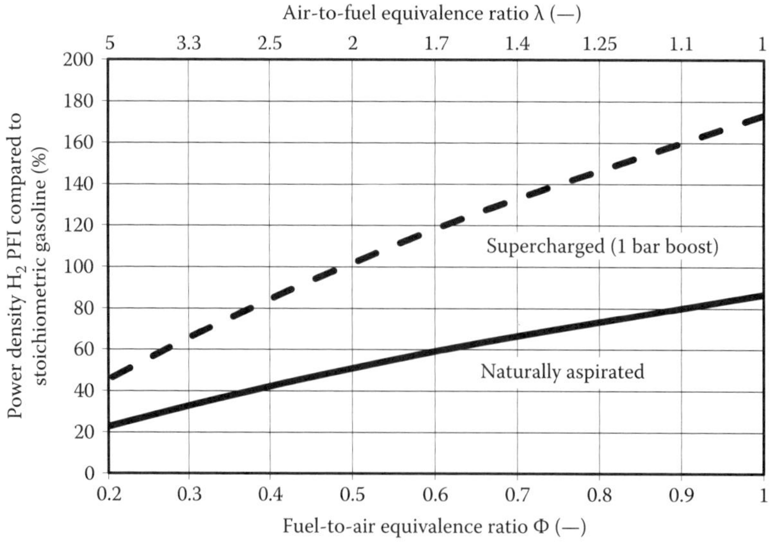

FIGURE 27.19
Theoretical power density of a PFI H_2 engine compared to stoichiometric gasoline operation as a function of equivalence ratio and charging strategy.

As a result, an operating strategy using a variable equivalence ratio as a function of engine load was evaluated for both naturally aspirated engines and supercharged engines and vehicles. Implementation of this variable equivalence ratio strategy in a range from $2 < \lambda < 5/0.2 < \varphi < 0.5$ on a GM 454 spark-ignited PFI engine (commonly known as the Chevrolet big block) showed a more than 20% increase in engine power compared to a carbureted version without increasing the danger of backfiring [219]. The conclusions of this study also include that as a consequence of the wide range of applied mixture composition, the range of ignition timings is also wide. The injection timing shows significant influence in the low load and speed region, and it is not critical in the high load and speed region.

Another significant increase in power density can be achieved with supercharging in combination with a variable equivalence ratio strategy. Drive-cycle simulations based on steady-state engine efficiency and emissions maps of a supercharged 2.3 L hydrogen engine suggested a 3% increase in fuel economy when employing a variable equivalence ratio strategy compared to a constant $\lambda = 2/\varphi = 0.50$ strategy while simultaneously reducing NO_X emissions by more than 80%. Additional simulations simulating the hydrogen engine being operated in a full HEV mode with a pretransmission parallel hybrid architecture suggest another almost 50% increase in fuel economy while reducing NO_X emissions by more than 99% compared to a conventional power train [220]. A similar engine control strategy was used for propelling a hydrogen shuttle bus using a 6.8 L V-10 engine. Due to the lean-burn concept, the engine achieved a more than 12% better brake thermal efficiency compared to the gasoline counterpart at 1500 RPM and 2.62 bar brake mean effective pressure (BMEP) (Ford World-Wide Mapping Point [WWMP]). In-vehicle tests confirmed an increase in power output enabling a more than 30% reduction in 0–35 mph acceleration time compared to the natural gas counterpart while achieving near-zero emissions [221].

An effective way to limit the power loss compared to gasoline or diesel engines is by running hydrogen port-injection engines at stoichiometric air–fuel ratios. However, stoichiometric PFI operation is prone to combustion anomalies and also requires an aftertreatment system to reduce the level of oxide of nitrogen emissions. As indicated in Table 27.3 and Figure 27.19, stoichiometric H_2 PFI operation results in a theoretical power density of approximately 86% compared to gasoline. A six-cylinder 12 L displacement bus engine (MAN H 2866 UH01) converted to bifuel operation was shown to achieve 170 kW in gasoline operation and 140 kW in stoichiometric port-fuel injected hydrogen operation, which is approximately 82% of the gasoline power output, confirming the theoretical considerations. However, in order to avoid combustion anomalies, the compression ratio of the engine had to be reduced to as low as 7.5:1 [185]. With later engine conversions by the same manufacturer, significant improvements could be achieved by using solenoid-driven hydrogen injection valves instead of rotary hydrogen valves. The MAN H2876 UH01, a 12.8 L in-line six-cylinder engine using these improved injectors with sequential injection, achieved a peak brake thermal efficiency of 31% in naturally aspirated stoichiometric hydrogen operation. As can be seen from Figure 27.18, stoichiometric hydrogen operation results in significant amounts of NO_X emissions. The MAN H2876 UH01 engine uses a reducing catalyst with lambda control for emissions aftertreatment. Operated with slight hydrogen surplus, the engine can be operated well below Euro 5 emissions levels, which are mandatory since 2008 [185].

The previous mixture formation concepts employing hydrogen port injection result in compromises either in terms of power density with lean air–fuel ratio approaches or engine efficiency with stoichiometric concepts. A potential solution to this trade-off is combining lean-burn and stoichiometric operating strategies. This concept has been proposed based on engine research results [12,202,222] and has also been implemented in hydrogen demonstration vehicles [154,195]. At low engine loads, the engine is operated at variable lean air–fuel ratios, resulting in good engine efficiencies and extremely low engine-out emissions. Once a certain engine power demand is exceeded, the operating strategy is switched to throttled stoichiometric operation. The NO_X emissions critical operating regime at equivalence ratios 1 < l < 2/0.5 < 4 < 1 is avoided, and a conventional aftertreatment system can be used to reduce oxide of nitrogen emissions in stoichiometric operation. Tests performed on a prototype demonstration vehicle with a 6.0 L V 12 engine employing this operating strategy showed that the emissions add up to only a fraction of the most stringent standards. The test results on a FTP-75 cycle revealed NO_X emissions as low as 0.0008 g/mile (3.9% of the super ultralow emission vehicle [SULEV] standard) and hydrocarbon emissions that were lower than the ambient concentration, indicating that this vehicle actively reduces the concentration of certain emissions components [196].

The low density of hydrogen was shown to significantly reduce the power density with PFI compared to conventional fuels. Calculations based on a verified model have shown that the trapped air mass per cycle could be increased by up to 16% with cryogenic hydrogen port injection (injection temperature around 90 K) compared to ambient hydrogen injection [126]. This leads to a significant increase in power output, making cryogenic injection an effective measure to increase the specific power of hydrogen engines [206]. However, this injection strategy is only feasible in combination with cryogenic hydrogen onboard storage, a technique with promising storage densities but significant challenges due to the complexity of the tank and infrastructure.

27.6.2.3 Spark-Ignition Direct Injection

Efforts to avoid combustion anomalies and increase the power density of hydrogen internal combustion engines while achieving near-zero emissions have led to the development of injection systems for hydrogen DI operation. Similar to common classifications for gasoline engines, hydrogen DI mixture formation strategies have also been grouped in jet-guided, wall-guided, and air-guided concepts [107]. Based on the SOI, one can differentiate early DI and late DI operation; however, no clear threshold between these two categories has been defined. Figure 27.20 shows a schematic of different hydrogen DI strategies and

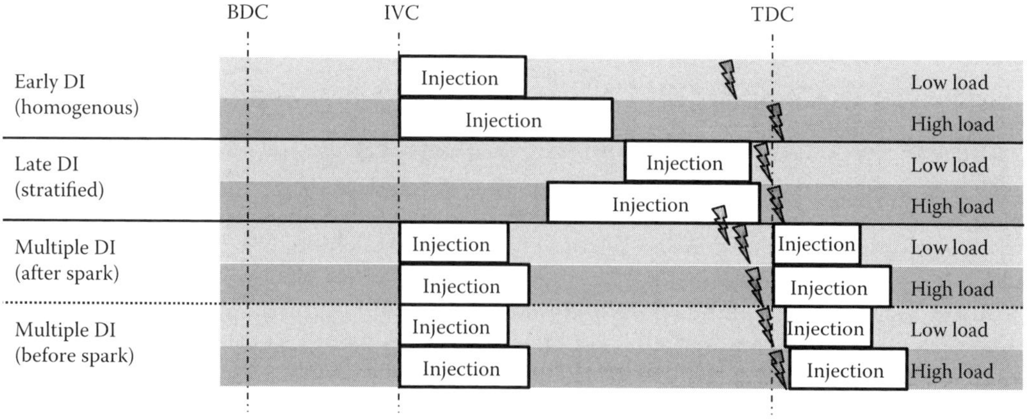

FIGURE 27.20
Schematic of injection strategies for DI. (From Wallner, T. et al., Assessment of multiple injection strategies in a direct injection hydrogen research engine, SAE Paper No. 2009-01-1920, 2009.)

their respective injection timings. Early injection generally refers to any hydrogen DI during the early compression stroke shortly after intake valve closing, whereas late DI refers to strategies with the injection late in the compression stroke generally ending just before spark timing. In order to avoid displacement of fresh charge by hydrogen of low density, the SOI even for early injection is usually set after intake valve closing. Aside from engine operation with one injection pulse per cycle, results of multiple injection strategies with two or more injection pulses per cycle have been reported [134,175,223].

Based on the required injection pressure, the terms low-pressure and high-pressure DI have also been used in the literature. In order to flow hydrogen directly into the combustion chamber, the pressure inside the injection system has to exceed the pressure inside the cylinder. However, only if critical injection conditions are being reached, the amount of hydrogen injected into the engine becomes independent from the cylinder pressure. This is critical for engine calibrations and accurate fuel metering, with the amount of fuel determined only as a function of injection pressure and injection duration. Critical conditions are obtained at a pressure ratio of about 0.53 (see Section 27.2.4 for details) indicating that the injection pressure has to be approximately twice the cylinder pressure to guarantee critical conditions (choked flow). Therefore, operating strategies with early DI require injection pressures in the range of approximately 5–20 bar, late injection strategies up to 100 bar and multiple injection strategies with injection pulses during the actual combustion event of 100–300 bar [224]. The exposure of the injector tip to in-cylinder pressures and temperatures with DI operation poses significant challenges for developing durable injectors with accurate metering capabilities and high flow rates [178].

Injection timing during hydrogen DI operation has crucial influence on the mixture distribution and, therefore, on the combustion characteristics. With early injection, the injected fuel has sufficient time to mix with the air inside the combustion chamber and form an almost homogeneous mixture. With late injection, only limited time for mixing is available, resulting in a stratified charge at spark timing. Those basic trends were also confirmed by using 3D CFD simulation tools (e.g., [225]) and optically accessible engines (e.g., [226]).

The impact of the aforementioned characteristics on NO_X emissions behavior is highly dependent on the engine load or the overall equivalence ratio, respectively. Figure 27.21 shows the NO_X emissions results as a function of SOI for various engine equivalence ratios at an engine speed of 2000 RPM collected on a single-cylinder hydrogen research engine [227,228]. Because the engine is operated without throttling, the equivalence ratio corresponds to engine load. The injector configuration for these investigations is a side-mounted DI injector with a symmetrical 13-hole nozzle configuration (60° included injection angle). At low engine loads, early injection results in extremely low NO_X emissions because the mixture at ignition timing is very likely to be homogeneous [229]. Thus, the lean homogeneous mixture burns without forming NO_X emissions. Late injection at low loads, on the other hand, results in a stratified mixture with hydrogen-rich zones, as well as zones with very lean mixtures or even pure air. Although the overall mixture is still lean, the combustion of rich zones causes a significant increase in NO_X emissions. At high engine loads, this trend appears to be inverted. Early injection results in homogeneous mixtures that approach stoichiometry and produce high NO_X emissions. Late injection is expected to result in stratification, with zones that are even richer than stoichiometric, along with lean zones. This kind of stratification avoids the NO_X critical equivalence ratio regime of $\lambda \sim 1.3/\varphi \sim 0.75$ (see Figure 27.18) and thereby reduces overall NO_X emissions.

Theoretical considerations on mixture calorific values and power densities (Section 27.2) already led to the assumption that hydrogen DI operation results in superior power densities

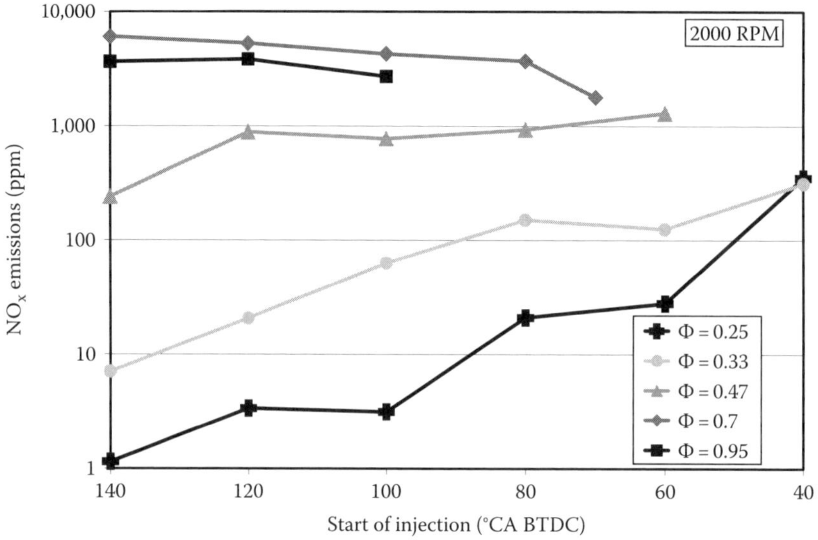

FIGURE 27.21
Typical NO_X emissions pattern for H2 DI operation. (From Wallner, T. et al., Endoscopic investigations in a hydrogen internal combustion engine, *Proceedings of 1st International Symposium on Hydrogen Internal Combustion Engines*, Graz, Austria, 2006, pp. 107–117; Wallner, T. et al., Investigation of injection parameters in a hydrogen DI engine using an endoscopic access to the combustion chamber, SAE Paper No. 2007-01-1464, 2007.)

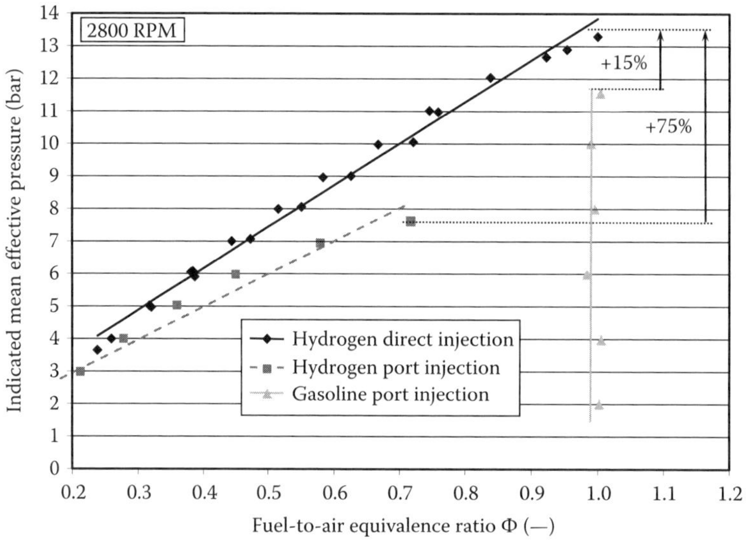

FIGURE 27.22
IMEP comparison for gasoline and hydrogen port injection as well as hydrogen DI operation. (From Eichlseder, H. et al., The potential of hydrogen internal combustion engines in a future mobility scenario, SAE Paper No. 2003-01-2267, 2003.)

compared to port-injection operation with both hydrogen and gasoline. Figure 27.22 shows a comparison of load sweeps as a function of the equivalence ratio measured on a single-cylinder 0.5 L research engine [12]. In the H_2 DI case, an IMEP in excess of 13 bar was achieved, which is approximately 15% higher than the peak IMEP in gasoline operation and more than 75% higher than the peak IMEP in hydrogen port-injection operation.

Because of the displacement of air during the injection of hydrogen into the intake pipe and the combustion anomalies that occur at high engine loads, the maximum achievable mean effective pressure with external mixture formation is distinctly below the values for gasoline. The pattern of the IMEP as a function of equivalence ratio also shows differences between the two H_2 mixture formation methods. Since during the DI of H_2, the air amount remains almost constant with a changing equivalence ratio (displacement effect is precluded because H_2 is not injected before the intake valves close), a leaner equivalence ratio under the same load (or the same IMEP) is established compared with the external mixture formation with H_2. Since both oxides of nitrogen emissions and engine efficiency are strongly dependent on the equivalence ratio, DI operation is capable of achieving higher engine efficiencies and lower NO_X emissions at the same engine load compared to port injection.

The efficiency potential of hydrogen DI operation was demonstrated on a single-cylinder Ford research engine, achieving an estimated peak brake thermal efficiency of more than 45% at an engine speed of 3000 RPM [178]. In order to reflect realistic brake thermal efficiencies, the friction values of the single-cylinder engine were not taken into account, but a friction mean effective pressure of 0.7 bar at 3000 RPM was estimated as typical for a low-friction multicylinder engine. Even at part-load operation, hydrogen DI can be used for optimizing engine efficiency. Although hydrogen's wide flammability limits theoretically allow unthrottled lean operation even at idling conditions, the relatively lean mixtures at those operating conditions result in longer, less efficient combustion durations. A study of basic injection strategies using a single-hole injector on a single-cylinder research engine demonstrated that the combustion duration at low engine loads (IMEP w 2.5 bar) at an engine speed of 2000 RPM could be reduced from more than 50°CA with early injection (SOI = 100°CA BTDC) to around 15°CA with late injection around 60°CA BTDC resulting in an increase in indicated thermal efficiency from 29% to more than 34% [230].

Although the efficiency improvement with hydrogen DI at low and part load as well as high engine loads is significant, a trade-off between optimizing engine efficiency and oxides of nitrogen emissions has been encountered. The aforementioned example with an increase in indicated efficiency from 29% to 34% also led to an increase in NO_X emissions from around 5 ppm with early injection to more than 100 ppm for the efficiency-optimized case [230].

In this respect, multiple injection has been demonstrated as an effective tool to simultaneously achieve high engine efficiencies and low NO_X emissions. Results from a single-cylinder research engine suggest a NO_X emissions reduction potential compared to single-injection strategies in excess of 95% while still achieving acceptable engine efficiencies [134,175,223]. However, due to the short time available for the injection pulse during the combustion phase, multi-injection strategies pose demanding challenges for the hydrogen injection systems in terms of both pressure levels and required injector flow rates. Therefore, the demonstration of the potential of multiple injection strategies has so far been limited to fairly low engine speeds [223].

Hydrogen injectors for DI operation have to reliably operate at hydrogen supply pressures of up to 300 bar while the injector tip heats up to temperatures of 300°C–400°C. It is estimated that a metering accuracy of approximately 2% of the actual flow rate with minimum injection durations as short as 0.1 ms and leakage rates of less than 0.1% of full flow are required for hydrogen DI injectors [224]. Although various injector designs and actuation systems including solenoid, magnetostrictive, and piezoelectric have been considered, currently available injector prototypes still do not meet the durability requirements needed for automotive applications. A major development goal should be to substantially increase injector life from about 200 to 1000 h. This would allow the development and demonstration of an early stage multicylinder engine with advanced features and material technology integrated in the DI fuel system. A longer-term goal is to develop and prove 20,000 h level durability, with a production-oriented design, through further fundamental materials research, accelerated testing on injector rigs, and full-scale engine validation [178].

27.6.2.4 Compression Ignition Direct Injection

For further improvement in engine efficiencies, research has been performed on hydrogen DI operation on CI engines for both stationary applications [197,231,232] and automotive engines [199]. Recently, work on hydrogen HCCI engine research on an optical engine [200] as well as dimethyl ether (DME) assisted hydrogen HCCI operation [233] has been published. In all cases, the high autoignition temperature of hydrogen compared to conventional fuels has resulted in a limited operating range as well as high rates of EGR or intake air preheating as prerequisites to achieve autoignition temperatures. Investigations on an optical engine operated on hydrogen and heptane revealed that for pure hydrogen HCCI, the effect of intake air temperature was relatively small. A 200°C raise in intake temperature for $\lambda = 1.6$–$2.0/\varphi = 0.5$–0.63 resulted in slightly advanced autoignition phasing and also led to a 2.5 bar increase in peak in-cylinder pressure, without causing much difference in the phasing of peak pressure. For comparison, a 20°C increase in intake air temperature for sole heptane HCCI with $\lambda = 1.4/\varphi = 0.7$ advanced the autoignition angle by 10°CA and increased the peak pressure by about 3 bar [200]. Using DME to raise the cetane number and stabilize HCCI combustion in hydrogen engines showed a 13% improvement in indicated thermal efficiency compared to pure hydrogen operation. A peak indicated thermal efficiency of 42%, while achieving exceptionally low NO_X emissions close to zero could be demonstrated [234]. Experimental investigations on a single-cylinder research engine specifically designed for CI showed that stable CI of hydrogen could be achieved employing induction air heating and supercharging without intercooling. However, load ranges were limited to low- and mid-part load to avoid knocking phenomena. These findings led to the development of a dual injection strategy in which a first injection pulse delivered a small amount of fuel into the cylinder. Forming, compressing, and finally igniting an initial mixture around TDC was used to generate temperature in the combustion chamber. A second pulse was subsequently injected and almost immediately converted in a diffusion type of combustion, completely inhibiting knocking phenomena. Making the combustion system more stable and flexible by igniting the first pulse using a spark plug led to an indicated efficiency of the high-pressure cycle of 44% [199].

Whether hydrogen CI operation and hydrogen HCCI operation will be relevant operating modes for hydrogen engines depends on their effective operation ranges as well as efficiency and emissions characteristics. HCCI operation has generally been limited to low engine speeds and loads. In light of the additional complications with stable engine control in HCCI operation, it remains to be seen whether significant improvements compared to conventional hydrogen operation are achievable, since these conventional operating modes already offer excellent engine efficiencies at low engine-out emissions levels at these operating conditions.

27.6.3 Measures for NO_X Control

Due to the absence of carbon in the fuel, the regulated emissions of hydrogen-powered internal combustion engines are theoretically limited to oxides of nitrogen. Carbon monoxide (CO), hydrocarbon (HC), as well as CO_2 emissions are expected to be virtually zero [12].

Measurements on a medium speed, externally supercharged, single-cylinder research diesel-engine-type MAN 1L 24/30 operated on hydrogen showed CO_2 emissions of less than 2% compared to operation on conventional fuel [197]. It is generally assumed that traces of hydrocarbon emissions in the exhaust of hydrogen engines result from the combustion of the lubricating oil.

As shown in Figures 27.18 and 27.21, the oxides of nitrogen emissions of hydrogen engines strongly depend on the engine load and air-to-fuel ratio. Methods for reducing NO_X emissions include engine internal measures, like multiple injections, water injection or EGR, as well as exhaust aftertreatment. Depending on the engine operating strategy, either a conventional three-way catalyst or lean NO_X aftertreatment has to be employed. Regular production-type catalysts were shown to achieve a NO_X conversion efficiency in excess of 99.5% [202]. A BMW Hydrogen 7 vehicle equipped with a catalyst setup consisting of two monoliths—the first one for the stoichiometric operating regime and the second one for reducing NO_X peaks that occur when switching from lean to stoichiometric operation—achieved drive-cycle NO_X emissions that were approximately 0.0008 g/mile, which is equal to 3.9% of the SULEV limit [196].

Although promising results in terms of NO_X emission levels have been achieved with three-way catalysts, research has been performed on more elaborate lean NO_X aftertreatment systems, since the required stoichiometric operation for three-way catalysts to work properly results in a significant loss in engine efficiency compared to lean operation. Measurements on a single-cylinder DI research engine revealed an efficiency loss of 4% with throttled stoichiometric operation compared to unthrottled lean-burn operation at 2000 RPM and an IMEP of 8 bar [26]. Methods to accomplish the reduction in oxides of nitrogen emissions generally consist of a lean NO_X trap as well as other optional components including three-way catalysts and selective catalytic reduction (SCR) converters. Initial results on a straight six-cylinder diesel engine converted to hydrogen external mixture formation operation and equipped with a NO_X absorption three-way catalyst showed NO_X emissions reductions of more than 90% with 3% of the hydrogen fuel injected into the exhaust [235]. An improved system employing hydrogen DI in combination with a NO_X-storage-reduction (NSR) catalyst and an oxidation catalyst showed a NO_X conversion rate of 98% with fuel penalties between 0.2% and 0.5% [236]. Concepts have also been patented in which the additional reducing agent injection in the exhaust is avoided by switching operating modes from lean-burn operation to fuel-rich operation using EGR in order to purge the lean NO_X trap [237,238].

In the mixture formation sections, a number of strategies for NO_X control have already been discussed. Further provisions for engine internal NO_X emission reduction include measures to reduce the in-cylinder temperatures by employing EGR and water injection [239]. Water injection was found to significantly reduce NO_X emissions with only slightly negative impact on engine efficiency [240]. However, although water injection is a very effective measure for NO_X emissions reduction, its practical application will depend on an efficient way of supplying the liquid, for example, by recovering and condensing it from the engine exhaust [241].

EGR is another solution for in-cylinder NO_X emissions reduction. Research on a single-cylinder engine showed that EGR application is an effective technique for reduction of engine knock and NO_X emissions at the expense of engine efficiency above 20% EGR. The combustion knock values decreased substantially by about 85% and the NO_X emissions reduced by about an order of magnitude when EGR levels were increased from 0% to 35% [242]. Although EGR results in an engine efficiency loss compared to unthrottled lean operation, it can still be a viable solution when stringent NO_X emissions limits have to be reached since efficiency improvements compared to throttled stoichiometric operation can be accomplished [216]. An experimental study on a Ford 2.0 L Zetec engine demonstrated that emissions levels of oxides of nitrogen below 1 ppm can be achieved by combining EGR and a three-way catalyst [243,244]. Simulation studies also concluded that the indicated thermal efficiency of cooled EGR is slightly higher than that of a hot EGR strategy. By increasing the EGR percentage (whether cooled or hot), the indicated thermal efficiency increases firstly and then decreases due to the unstable combustion at high EGR rates. The indicated thermal efficiency is mainly influenced by the properties of the cylinder charge, combustion duration, and phasing as well as the wall heat losses. The combination of these factors results in an increased indicated thermal efficiency with moderate EGR levels; however, higher EGR levels result in decreased indicated thermal efficiencies due to inefficient and unstable combustion [127].

27.6.4 Summary of Possible Control Strategies

The preceding sections clearly demonstrate the flexibility of hydrogen as an engine fuel. Consequently, the choices of operating strategies that enable a certain power demand to be met while controlling NO_X emissions are numerous. This is illustrated in summarized form in the following.

At the lowest loads (including idling), the possible strategies, in ascending order of brake thermal efficiency, are

- Fixed stoichiometric operation, with throttling (and/or EGR) and aftertreatment
- Fixed lean equivalence ratio (lean of the NO_X threshold) with throttling, without any need for aftertreatment
- Fixed ultra lean equivalence ratio, with throttling to ensure combustion stability, without any need for aftertreatment

At low loads, again in ascending order of brake thermal efficiency,

- Fixed stoichiometric operation, with throttling (and/or EGR) and aftertreatment
- Fixed lean equivalence ratio (lean of the NO_X threshold) with throttling, without any need for aftertreatment
- Variable equivalence ratio (lean of the NO_X threshold) with WOT, without any need for aftertreatment

At medium loads,

- Fixed stoichiometric operation with throttling (and/or EGR) and aftertreatment
- Fixed lean equivalence ratio (lean of the NO_X threshold) with supercharging, without any need for aftertreatment
- Variable equivalence ratio (between stoichiometric and the NO_X threshold) with WOT and lean NO_X aftertreatment

The latter two generally result in higher efficiencies than the first strategy. Which strategy enables the highest efficiency depends on the supercharging setup and resulting losses of the second strategy compared to the fuel economy penalty incurred by the lean NO_X after treatment for the third strategy.

At the highest loads,

- Fixed stoichiometric DI operation with aftertreatment
- Fixed stoichiometric PFI operation with supercharging and aftertreatment
- Fixed stoichiometric PFI operation with cryogenic fuel injection and aftertreatment
- Fixed lean equivalence ratio (lean of the NO_X threshold) with (high) supercharging, without any need for aftertreatment

Here, the strategy giving the highest efficiency again depends on the systems chosen for injection, supercharging, etc.

This list is not comprehensive; one could also discern, for example, homogeneous from stratified operations.

27.7 Hydrogen Safety

The unique properties of hydrogen require an adapted approach when laying out a safety concept for both hydrogen engine test cells and hydrogen-powered vehicles. The gaseous state of the fuel at ambient conditions in combination with the low density, wide flammability, and invisibility of the gas as well as its flames requires amended measures to guarantee a safety level equivalent to conventional fuels. When properly taking the unique properties into account by facility designers, engineers, and operators, hydrogen can be as safe as, or safer than, gasoline or diesel fuel [188].

Differences in local codes and standards for hydrogen application and use make it impossible to provide an all-inclusive summary of all hydrogen safety aspects. Therefore, this chapter is rather meant as a summary of best-practice recommendations based on the authors' expertise and the limited number of publications in this area.

27.7.1 Test Cell Design

For a hydrogen flame to occur, both an ignitable hydrogen–air mixture and an ignition source need to be present. When operating an internal combustion engine in an enclosed space, it is practically impossible to avoid all ignition sources; therefore, sources of ignition must be expected in the test cell. However, the number of sources and particularly sources where hydrogen likely accumulates (close to the test cell ceiling) should be minimized. It is also common practice to deenergize all electric equipment when increased hydrogen concentrations are detected [245]. Because ignition sources cannot be excluded from the test cell, ignitable hydrogen–air mixtures have to be avoided through proper test cell ventilation.

27.7.1.1 Test Cell Ventilation

The ventilation capabilities of hydrogen test cells are generally designed much larger than those for conventional fuels. For the layout of a hydrogen test cell, two cases with different requirements for ventilation are considered: an enclosed test cell and an open test cell located inside a hi-bay. Due to the natural convection and large amount of air available in hi-bay settings, the hydrogen-specific ventilation requirements are not as critical as for enclosed test cells. As a rule of thumb for enclosed test cells, a minimum ventilation resulting in 1–2 full air exchanges per minute has been established. In order to effectively remove heat created from the test engine, as well as any hydrogen leakage, cross ventilation is widely used for enclosed hydrogen test cells. Fresh air is brought in close to the floor of the test cell and removed from the test cell close to the ceiling after flowing past the experimental setup. This setup results in a controlled flow of fresh air through the test cell and also allows positioning of hydrogen sensors in strategic locations throughout the test cell.

In addition to general test cell ventilation, fume hoods connected to an air blower resulting in a constant stream of flow directly on top of the experimental equipment have been used for hydrogen setups. Due to the number of hydrogen connections in close proximity to the experimental setup and the use of prototype injection equipment, one would expect that a hydrogen leak is most likely to occur in that area. Using a fume hood with constant flow and a hydrogen sensor inside the hood assures fast detection of hydrogen leaks and allows taking countermeasures immediately when a leak occurs [246].

Fume hoods are not only used in enclosed engine test cells but also in hi-bay, open test cell settings. Due to the large volume available for dilution of hydrogen leaks in hi-bay settings, detection of a potential leak could take considerable amount of time. Using hoods on top of the experimental equipment in combination with a hydrogen sensor placed inside the hood is an efficient way of detecting hydrogen leaks, especially in hi-bay settings with no defined air flow pattern [247].

The routing of the crankcase ventilation in hydrogen engine applications also requires special attention since measurements have shown hydrogen crankcase concentrations in excess of 5 vol.% [176]. Possible solutions for test cell setups as well as vehicle applications include venting the crankcase to the atmosphere, routing the crankcase ventilation into the test cell ventilation system, as well as variants of combining forced crankcase ventilation with an oil separator with or without a catalyst to convert the hydrogen to water, after which the gases are routed back into the intake manifold [147,176].

27.7.1.2 Hydrogen Sensors

For detection of hydrogen, several technologies are commercially available including electrochemical, catalytic, thermal conductivity, semiconductor-based, and microelectromechanic sensors. An overview of sensor detection principles and a comparison of sensor performance, including range, cross sensitivity, accuracy, stability, and cost, can be found in [248]. Recommended locations for hydrogen sensors include locations where hydrogen leaks or spills are possible, at hydrogen connections that are routinely separated, where hydrogen could accumulate, as well as in building air intake and exhaust ducts [249]. When designing a hydrogen detection system, one should consider factors including detector response time, detection range, durability/lifetime of the detector, required detector maintenance and calibration, potential cross sensitivity, and area coverage.

A generally accepted and commonly used concentration for alarm activation is around 1 vol.% of hydrogen in air (equivalent to 25% of the lower flammability limit) [221,245]. Other hydrogen detection systems use a progressive approach with several warning and alarm limits that warn the operators at low detection limits (e.g., 10% of the lower flammability limit) and perform automated shutdown of the hydrogen supply system and test equipment if a higher alarm limit is reached. In addition to permanently mounted hydrogen sensors, most experimental facilities also use portable hydrogen detectors for both personal protection and leak checking of hydrogen equipment.

27.7.1.3 **Hydrogen Flame Detectors**

A hydrogen–air flame is colorless, and any visibility is caused solely by impurities. At reduced pressures, a pale blue or purple flame may be present. Severe burns have been inflicted on persons exposed to hydrogen flames resulting from the ignition of hydrogen gas escaping from leaks. Therefore, hydrogen detection systems of various levels of sophistication have been developed and implemented to protect operating personnel. Hydrogen flame detectors can be classified in the following groups [250]:

- Thermal fire detectors classified as rate-of-temperature-rise detectors and overheat detectors have been manufactured for many years and are reliable. Thermal detectors need to be located at or very near the site of a fire.
- Optical sensors for detecting hydrogen flames fall into two spectral regions: ultraviolet (UV) and infrared (IR). UV systems are extremely sensitive; however, they are susceptible to false alarms and can be blinded in foggy conditions. IR systems typically are designed for hydrocarbon fires and are not very sensitive to hydrogen fires.
- Imaging systems mainly are available in the thermal IR region and do not provide continuous monitoring with alarm capability. The user is required to determine if the image being viewed is a flame. UV imaging systems require special optics and are very expensive. Low-cost systems, using low-light silicon charge-coupled device (CCD) video technology with filters centered on the 940 and 1100 nm emission peaks, have been used at some facilities.
- A broom has been used for locating small hydrogen fires, as a dry corn straw or sage grass broom easily ignites as it passes through a flame. A dry fire extinguisher or dust thrown into the air also causes the flame to emit visible radiation. This technique should be used with care in windy, outdoor environments in which the light hydrogen flame can easily be blown around.

The selection of a certain detection system should be based on the ability to detect a flame at sufficient distance as well as the size of flames that can still be detected. Other selection factors include response time, insensitivity to false alarms, as well as the possibility of automatic periodic checkups.

27.7.1.4 **Hydrogen Supply System**

For most research and test applications, hydrogen is stored in cryogenic liquid form or in compressed form. Typical pressure levels for compressed storage range from 138 (2000 PSI) to 414 bar (6000 PSI). Liquid hydrogen is stored at temperatures of –253°C generally in insulated, passive storage systems, meaning that no active cooling is provided. Despite the excessive insulation, the remaining heat input causes liquid hydrogen to evaporate, which increases the pressure. Liquid storage systems require continuous consumption to avoid pressure buildup and ultimately blow off of hydrogen.

The large volume requirements for hydrogen storage systems for both compressed and cryogenic setups usually result in hydrogen being stored outdoors or in dedicated fuel canopies. The hydrogen is then delivered to the test facility via hydrogen supply lines, thereby reducing the amount of energy stored inside the test facility. Also, any leaks resulting from connections of hydrogen cylinders to supply manifolds that have to be opened and retightened when changing out high-pressure cylinders are thus located outside where natural ventilation generally decreases the probability of a buildup of ignitable hydrogen–air mixtures.

Depending on the storage system as well as the application, the hydrogen fuel is delivered to the test facility at widely varying pressures. For compressed storage systems, several high-pressure cylinders are usually hooked up to a manifold to allow extended uninterrupted test runs. For purging and safe leak checking of the system, high-pressure helium is generally used, which is also connected to the delivery system. For most setups, hydrogen or helium passes through manual shutoff valves and check valves and is supplied to a pressure regulator. Once regulated to the appropriate delivery pressure, the hydrogen is fed to the experimental equipment after passing through several manual as well as solenoid-operated safety valves. Using a cascade of solenoid-operated valves allows minimizing the amount of hydrogen that leaks into the test cell in case of a line rupture. In close proximity to the engine, a fast-acting, normally closed three-way valve is used. This valve can be remotely activated by the operator to supply hydrogen to the engine whenever needed. If the engine does not rotate or a hydrogen leak is detected, the three-way valve automatically closes. The normally open path of the three-way valve is connected to a purge line, which allows the engine hydrogen supply line to be depressurized whenever the engine is not operating. The setup with a close-to-the-consumer three-way valve has proven to greatly reduce the risk of unintended hydrogen release into the engine through leaking injectors while minimizing the amount of hydrogen that has to be released to the atmosphere. For long-term shutdowns of the hydrogen supply system or if maintenance is required, hydrogen supply lines are usually purged to the atmosphere and purged with compressed helium.

Accurate metering of the amount of consumed fuel is crucial for the quality of any engine test experiment and provides a necessary baseline for calculation of development-relevant characteristic numbers like engine efficiency or brake-specific fuel consumption. Automated or manual monitoring of the fuel consumption can also be used for leak detection and therefore significantly adds to the safety of hydrogen setups and testing activities. Over the last couple of years, a continuing transition from conventional methods including positive displacement pumps and gravimetric systems toward direct and continuous mass measurement using coriolis meters has been observed [251]. For hydrogen applications, the US Environmental Protection Agency (EPA) has only accepted three methods of hydrogen fuel consumption testing. These three methods are gravimetric measurement; measurement of stabilized pressure, volume, and temperature (PVT); and coriolis mass flow measurement [252]. Recently, another method similar to the carbon balance that is used for fuel consumption calculations for conventional vehicles based on exhaust emissions measurement has been proposed for hydrogen engines, and a detailed comparison of the methods has been provided [253]. Due to the extensive measurement equipment needed for this method, it has so far only found application where a direct fuel consumption measurement is not feasible [196].

Due to the good accuracy over a wide flow range and the direct measurement of fuel mass, a coriolis meter seems to be the preferred method for hydrogen engine testing [246]. For applications that require a range that cannot be covered by a single coriolis meter, systems that switch from a low-range to a high-range coriolis flow meter have been developed and implemented [188].

27.7.2 In-Vehicle Applications

Even more so than for test cell design, it is practically impossible to avoid all ignition sources on hydrogen-powered vehicles. In order to provide a level of safety for hydrogen vehicles comparable to conventionally fueled cars, strategies combining detection of hydrogen leaks and diluting concentrations of hydrogen below the lower ignition limit have been developed. During regular vehicle operation, hydrogen sensors positioned in critical areas throughout the vehicle monitor the hydrogen concentration. In case a hydrogen leak is detected, the measures taken depend on the level of hydrogen concentration. Mitigation measures include, but are not limited to, advisory measures such as various levels of driver warnings, active ventilation using additional fans, as well as executive measures such as disabling the starter relay or hydrogen supply [254]. Hydrogen sensors are used in critical areas of hydrogen vehicles (engine compartment, hydrogen storage area, passenger compartment) to allow the highest level of safety and an early detection of potential hydrogen leaks. Due to the complex design of a vehicle interior and the ventilation systems, 3D CFD simulation in combination with elaborate hydrogen release and detection tests has been performed to determine the most efficient location for hydrogen sensors [255].

Both compressed hydrogen storage systems and cryogenic hydrogen storage systems could potentially allow pressure buildup in case of malfunction or accident. To properly address those risks, extensive tests of hydrogen storage systems including crash tests as well as exposure to fire have been performed. If properly designed, a hydrogen storage system will release hydrogen through overpressure vents strategically positioned in the vehicle.

Measures to increase the safety of hydrogen vehicles also include training of operators as well as potential rescue personnel. Therefore, rescue guidelines have been developed to inform first responders to the different hazards associated with hydrogen-powered vehicles [256].

27.8 Hydrogen Internal Combustion Engine Vehicles

Hydrogen internal combustion engines for automotive application are intended to power vehicles and provide an equivalent level of drivability, range, and safety as conventional-fuel vehicles. However, mainly due to the challenges of onboard hydrogen storage, current hydrogen-powered internal combustion engine vehicles have a limited range and in some cases reduced trunk space available compared to their conventional-fuel counterparts. Nonetheless, due to the immediate availability of hydrogen combustion engines, the extensive knowledge in engine production, durability, and maintenance as well as the capability of combustion engines to run on both hydrogen and conventional fuels (in most cases, gasoline), they are considered a bridging technology toward a widespread hydrogen infrastructure [1]. In this role, hydrogen internal combustion engine vehicles can be considered early adopters to help establishing and expanding a hydrogen infrastructure and building public awareness.

Numerous hydrogen-engine-powered vehicles ranging from two-wheelers to passenger cars, pickup trucks to buses, and off-road equipment have been designed, built, and tested over the last decades. The following section is limited to selected hydrogen internal combustion engine vehicles; design studies and show cars are excluded from this overview.

27.8.1 History

The concept of operating an internal combustion engine on hydrogen is almost as old as the internal combustion engine itself. In 1807, François Isaac de Rivaz of Switzerland invented an internal combustion engine that used a mixture of hydrogen and oxygen for fuel. Rivaz designed a car for this engine—the first internal combustion-powered automobile [257]. Patented by Jean Joseph Etienne Lenoir in 1860, a gas-driven two-stroke engine with horizontal arrangement is considered the first successful internal combustion engine. The engine was powered by hydrogen generated via the electrolysis of water [258]. As early as 1933, Norsk Hydro operated an internal combustion engine vehicle on hydrogen produced from onboard reforming of ammonia [259]. The first hydrogen DI engine dates back to 1933 when Erren Engineering Company proposed injecting slightly pressurized hydrogen into air or oxygen inside the combustion chamber rather than feeding the air–fuel mixture via a carburetor into the engine, a method that commonly resulted in violent backfiring. The patented system required special fuel-injection and control mechanisms but left the other engine components intact. With hydrogen used as a booster, the system eliminated backfiring and achieved much better combustion of hydrocarbons with higher output and lower specific fuel consumption [260]. In 1974, Musashi Institute of Technology introduced the first Japanese hydrogen-fueled vehicle, called Musashi 1, using a four-stroke hydrogen engine and high-pressure storage [261]. The Musashi 2, introduced in 1975, was equipped with hydrogen manifold injection on a four-stroke engine in combination with liquid hydrogen storage [261]. In 1977, Musashi 3 was presented using a spark-ignited two-stroke engine with hydrogen DI [262]. BMW in collaboration with DLR introduced their first hydrogen vehicle in 1979.

27.8.2 Hydrogen Vehicle Characterization

Hydrogen internal combustion engine vehicles can be characterized as either conversion vehicles or dedicated vehicles, with conversion vehicles adapted for hydrogen operation by either a vehicle manufacturer or an aftermarket supplier, whereas dedicated hydrogen cars are specifically designed and built for hydrogen operation by an original equipment manufacturer (OEM). Hydrogen cars have also been built for monofuel operation with hydrogen as the only fuel as well as bifuel solutions with hydrogen as well as gasoline as fuel options. Based on the hydrogen onboard storage system, hydrogen cars can be grouped as compressed hydrogen and cryogenic liquid hydrogen vehicles. Hydrogen as an engine fuel has been applied to reciprocating internal combustion engines as well as rotary engines. The following chapters give a brief overview of selected hydrogen vehicles. Automobiles that use hydrogen as a combustion enhancer in combination with another fuel are not considered in this overview.

27.8.3 Conversion Vehicles

An example for a conversion truck with compressed hydrogen storage is the ETEC H_2ICE Truck Conversion based on a Chevrolet/GMC Truck Silverado/Sierra 1500HD Crew Cab 2WD LS converted to hydrogen operation by Electric Transportation Engineering Corporation. The six-seated light-duty pickup truck is powered by a 6.0 L V-8 engine with hydrogen PFI. A belt-driven supercharger in combination with an intercooler is used to increase the power output of the engine. Hydrogen is stored in three 150 L, Type 3 (aluminum-lined, carbon-fiber-reinforced) tanks at a storage pressure of up to 350 bar, which results in approximately 10.5 kg of usable fuel. The vehicle has an estimated curb weight of 3000 kg [217]. A performance, emissions, and fuel economy study of this vehicle at different air–fuel ratios ($2 < \lambda < 2.85/0.35 < \varphi < 0.50$) showed fuel consumption numbers between 4.1 and 4.5 kg of hydrogen per 100 km, which is energy equivalent to 15.5 and 17 L of gasoline per 100 km (13.8–15.2 mpg) at NO_X emissions levels in the ultralow emissions vehicle (ULEV) and SULEV ranges [263]. So far, about 20 ETEC H_2ICE Truck Conversion vehicles have been built.

Quantum Tecstar has converted over 30 vehicles to hydrogen operation using the Toyota Prius hybrid as a platform. Two compressed hydrogen tanks replace the conventional gasoline tank, leaving the interior of the vehicle unchanged. The converted Prius engine is turbocharged in order to increase the power output in hydrogen operation. With a drivability similar to the gasoline counterpart, the Quantum Hydrogen Prius has an estimated range of 100–130 km per fill while meeting SULEV emissions standards [264].

27.8.4 Bifuel Vehicles

Since 1979, BMW has introduced six generations of hydrogen-powered internal combustion engine vehicles. The latest generation is the BMW Hydrogen 7 bifuel, a luxury sedan powered by a 6.0 LV12 engine. According to the manufacturer's claims, the BMW Hydrogen 7 vehicle has successfully completed the process of series development, meaning that the vehicle and all components have gone through the same design, manufacturing, and quality control processes as any other BMW vehicle. The new hydrogen model is built at BMW's Dingolfing Plant (Germany) parallel to the other models in the BMW 7, 6, and 5 Series, with the drive unit in the BMW Hydrogen 7 coming like all BMW twelve-cylinder engines from the BMW engine production plant in Munich (Germany). The engine is equipped with two separate fuel systems allowing the vehicle to operate on gasoline as well as hydrogen. Gasoline is injected directly into the combustion chambers; hydrogen is injected into the intake manifolds of the naturally aspirated engine [154]. The vehicle is equipped with a cryogenic hydrogen tank located in the trunk of the vehicle in addition to the conventional gasoline tank. The cryogenic tank holds about 8 kg of liquid hydrogen, which allows an estimated range of 200 km in hydrogen operation and another 480 km on gasoline [265]. Approximately 100 BMW Hydrogen 7 bifuel vehicles were built.

Since 1991, Mazda has developed several generations of hydrogen-powered rotary engine vehicles with the Mazda RX-8 Hydrogen RE being the most recent one unveiled in 2003. The hydrogen version of the Renesis engine is equipped with an electric-motor-assist turbocharger that is used to maximize the effectiveness of forced induction throughout the engine speed range [266]. The most recent generation is equipped with two compressed hydrogen tanks with an operating pressure of up to 350 bar, giving the vehicle a range of approximately 100 km in hydrogen operation plus and additional 550 km on gasoline. A combination of lean and stoichiometric hydrogen combustion operation results in a 23% improvement in fuel economy compared to gasoline operation. The performance of the vehicle meeting Japanese SULEV standards is reduced from 154 kW in gasoline to 80 kW in hydrogen operation [201].

27.8.5 Dedicated Hydrogen Vehicles

The BMW Hydrogen 7 monofuel demonstration vehicle was built based on the BMW Hydrogen 7 bifuel car to showcase the emissions reduction potential of a dedicated hydrogen vehicle. On the hardware side, the most significant changes are the removal of the gasoline fuel system including fuel injectors, fuel lines, charcoal filters for tank ventilation, and fuel rail. The two high-pressure fuel pumps were also removed, which reduce the parasitic losses on the engine. For stability reasons, the gasoline fuel tank remains in the vehicle because it is a structural element. The vehicles are equipped with improved catalysts. Independent test results showed that these vehicles achieved emissions levels that were only a fraction of the SULEV standard for NO_X and CO emissions. For nonmethane hydrocarbon (NMHC) emissions, the cycle-averaged emissions were actually 0 g/mile, which required the car to actively reduce emissions compared to the ambient concentration. The fuel economy numbers on the FTP-75 test cycle were 3.7 kg of hydrogen per 100 km, which, on an energy basis, is equivalent to a gasoline fuel consumption of 13.8 L/100 km (17 mpg). Fuel economy numbers for the highway cycle were determined to be 2.1 kg of hydrogen per 100 km, equivalent to 7.8 L of gasoline per 100 km (30 mpg) [196].

Ford Motor Company has been evaluating hydrogen since 1997 as an alternative fuel option for vehicles with internal combustion engines. In 2001, Ford presented the hydrogen engine-powered P2000 vehicle, the first production viable, North American OEM hydrogen internal combustion engine vehicle. The aluminum-intensive five-passenger family sedan was equipped with a highly optimized hydrogen port injection, 14.5:1 compression ratio, 2.0 L engine, gaseous H_2 fuel supply with an operating pressure of up to 250 bar, and a triple-redundant hydrogen safety system consisting of gas sensing as well as active and passive elements. The hydrogen P2000 vehicle met SULEV standards for HC and CO and emitted 0.37–0.74 g/mile of NO_X while showing a metro cycle fuel economy improvement of up to 17.9% relative to gasoline [215]. To demonstrate a commercially viable hydrogen ICE-powered vehicle application, Ford fully engineered a demonstration fleet of 30 E-450 shuttle buses with a 6.8 L Triton engine that runs on hydrogen. The 8–12 passenger shuttle bus with a 4.5 m wheelbase and an estimated gross vehicle weight of 6373 kg is equipped with a compressed hydrogen onboard storage system that holds up to 29.6 kg of hydrogen at a pressure of 350 bar with a resulting vehicle range of 240–320 km. The target specified for the hydrogen-powered shuttle bus is to meet 2010 Phase II heavy-duty emission standards [221,254,255,267].

27.8.6 Overview of Hydrogen Vehicles

Table 27.4 summarizes the most relevant information for the hydrogen-powered vehicles that were described in detail in the previous chapters. The summary includes technological aspects such as the type of engine used or the hydrogen storage system as well as vehicle range for hydrogen and, if bifuel, gasoline and the number of vehicles produced or converted.

27.9 Hydrogen in Combination with Other Fuels

27.9.1 Overview: Motivation

The properties of hydrogen, in particular its wide flammability limits, make it an ideal fuel to combine with other fuels and thereby improve their combustion properties. Based on the mixture formation strategy, one can differentiate between blended operation and dual-fuel operation. Blended operation refers to combinations of hydrogen with one or several other gaseous fuels. Typically, the fuel is already stored and delivered to the engine in

TABLE 27.4
Hydrogen Vehicles Overview

Name	Year	Engine	Tank	Capacity	Range	Units Made
Rivaz	1807	1 cyl	Compressed			Prototype
Lenoir	1860	1 cyl	Water electrolysis			Prototype
Norsk Hydro	1933		Ammonia reforming			Prototype
Musashi 1	1974		Compressed	7 Nm^3		Prototype
Musashi 2	1975		Cryo	230 L		Prototype
Musashi 3	1977	2 stroke	Cryo	65 L		Prototype
BMW	1979	3.5 L	Cryo		300 km	
Ford P2000	2001	2.0 L 14	Compressed	1.5 kg	100 km	
BMW Hydrogen 7	2003	6.0 L V12	Cryo	8 kg	200 + 480 km	~100
Mazda RX-8 Hydrogen RE	2003	2 × 654 cc	Compressed	2.4 kg	100 + 550 km	>30
Ford shuttle bus	2004	6.8 LV10	Compressed	29.6 kg	240–320 km	
ETEC Silverado	2004	6.0 L V8	Compressed	10.5 kg	Up to 335 km	~20
Quantum Prius	2005	1.5 L I4	Compressed	1.6 kg (reg.)/2.4 kg (ext.)	100–130 km	>30

blended form using a single carburetion or fuel-injection system. In this respect, hydrogen is frequently used to improve the lean-burn behavior of natural gas. On the other hand, dual-fuel* operation describes any combination of hydrogen and liquid fuels in which several mixture preparation devices are used. These systems either use separate storage and fuel systems for the different fuels or in some cases hydrogen may be produced onboard.

27.9.2 Blends with Hydrogen as a Constituent

The main motivation for adding hydrogen to natural gas is to extend the lean limit of natural gas. On the other hand, the low gravimetric storage density of compressed hydrogen tanks can be significantly improved by blending hydrogen with methane.

27.9.2.1 Natural Gas Dominated Blends

Hydrogen has a burning velocity that is several times higher than that of methane (see Section 27.2.2). An overall better combustion with the addition of hydrogen to natural gas has been verified, even in a wide range of operating conditions (lambda, compression ratio, etc.), generally showing benefits including a higher efficiency and lower CO_2 production and emissions [268–270]. The addition of hydrogen to natural gas allows extending the lean limit of natural gas without going into the lean misfire region, thereby achieving extremely low emission levels that meet the equivalent zero-emission vehicle (EZEV) requirements [268]. A study on a turbocharged lean-burn SI engine operated on natural gas as well as mixtures of hydrogen and natural gas (20/80 and 30/70 H_2/natural gas by vol.%) demonstrated that it was possible to achieve lower NO_X and total hydrocarbons (THCs) emissions without sacrificing engine torque or fuel economy [271].

* Note that this definition is different from the commonly used one to denote the combustion of a homogeneous gas–air mixture by diesel injection.

Hythane® is a registered fuel referring to mixtures of 20 vol.% H_2 and methane, with the trademark being the property of Eden Innovations Ltd. The Denver Hythane Project in 1991 showed a more than 75% reduction in CO and NO_X emissions when using hythane instead of natural gas [272].

27.9.2.2 Hydrogen-Dominated Blends

Adding methane to hydrogen significantly improves the storage density of compressed storage systems and therefore increases the vehicle range of gaseous fueled vehicles. Blending hydrogen with 5 vol.% of methane increases the stored energy content by 11%, while a 20 vol.% blend of methane with hydrogen increases the stored energy content by 46% compared to neat hydrogen [273]. Tests performed on a single-cylinder research engine operated on hydrogen as well as 5 and 20 vol.% blends of methane showed a slight reduction in NO_X emissions with increased methane content while engine efficiencies decreased with increased methane content especially at low engine loads [274]. Vehicle-level tests on a Mercedes Benz E 200 NGT, a bifuel gasoline—natural gas vehicle that was adapted to operate on gasoline, natural gas, hydrogen, and any H_2/natural gas mixture—showed up to 3% improvement in brake thermal efficiency when operated with hydrogen compared to gasoline [168].

27.9.2.3 Multiple Gas Blends

Multi-gas blends can result from pyrolysis, the carbonization of biomass, thermally utilizable waste substances, or excess gases containing H_2 that arise from chemical processes. Gases containing H_2 help shift the lean-burn limit toward greater amounts of excess air than with natural gas. This effect causes mean combustion chamber temperatures to sink while NO_X emissions are reduced to a very low level. Depending on the amount of hydrogen and other gas components, it is possible to attain NO_X values of under 5 ppm. These H_2-rich gas mixtures also have a neutral influence on the degree of efficiency even with extremely high amounts of excess air. The background of this property lies in the considerably higher laminar burning velocity of hydrogen. In the case of coke gas (60% H_2), the laminar burning velocity at $\lambda = 2/\varphi = 0.5$ is the same as that for natural gas at $\lambda = 1.1/\varphi = 0.9$. Especially in the lower and medium load range, this effect can be utilized directly resulting in an efficiency increase of up to 2% with operation using pure hydrogen compared with natural gas [275]. The power output is limited with turbocharged lean-burn gas engines operating with H_2-rich gas mixtures, especially due to the turbocharging unit.

27.9.3 Dual-Fuel Applications

Dual-fuel application of hydrogen with diesel and biodiesel as well as gasoline and alcohol fuels aims at improving combustion properties, hence reducing emissions and increasing fuel conversion efficiencies.

27.9.3.1 Diesel and Biodiesel

Engine testing as well as chassis dynamometer testing of a GM 1.3 L 53 kW diesel engine operated on 20% bio-derived/80% petroleum-derived diesel fuel (B20) with up to 10% hydrogen addition of the total fuel energy showed a slight decrease in NO_X emissions and slightly increased exhaust temperatures at constant engine efficiencies with no negative impact on engine performance or drivability [276].

The use of biodiesel and vegetables oils has been reported to result in reduced thermal efficiencies and increased smoke numbers. Injection of hydrogen in the intake showed a consistent reduction in smoke, CO, and HC emissions for operation on diesel fuel as well as Jatropha oil. Up to a hydrogen mass share of 5% a simultaneous increase in brake thermal efficiency of up to 2% is observed at full load. The faster heat release rate with addition of 10 mass% of hydrogen results in a 10% increase in NO emissions at full load for diesel as well as Jatropha [277].

Experiments were also conducted on onboard production of hydrogen-rich gas to assist partially premixed charge CI engine operation on ultralow sulfur diesel (ULSD) as well as rapeseed methyl ester (RME). It was concluded that hydrogen-rich gas produced by exhaust gas fuel reforming can promote partially premixed CI and result in improved performance and reduced emissions [278].

27.9.3.2 Gasoline and Alcohol Fuels

Test results on a GMC 2500 Sierra 4WD pickup truck showed a fuel consumption reduction relative to pure gasoline operation of approximately 3% in city driving and 4% in highway driving. The operating strategy for this dual-fuel pickup truck consisted of a variable substitution rate of gasoline with hydrogen ranging from 100% at idle and very light load to 0% substitution at full load to avoid any power loss. The average substitution rate of 40%–50% resulted in a vehicle range in dual-fuel application of 110–180 km using a single 350 bar 150 L compressed hydrogen tank. For a similar vehicle using an identical operating strategy emissions' reduction for CO, NO_X and THC emissions of approximately 20%–28% were observed [279].

Tests carried out on a single-cylinder research engine at compression ratios of 7:1, 9:1, and 11:1 and blend ratios of 0, 20, 40, 60, and 80 vol.% of hydrogen in ethanol showed increasing brake thermal efficiencies of up to 4% with increased compression ratio as well as a hydrogen fraction up to a level of 60 vol.%. The hydrogen addition also resulted in higher peak pressures and faster rates of heat release [280]. However, due to the addition of low-density hydrogen in the intake manifold, the brake power of the engine is reduced by up to 10% compared to operation on neat ethanol.

27.10 Filling in the Blanks

As is apparent from the preceding sections, significant progress in hydrogen-fueled engines has been made lately in terms of achievable power density, efficiencies, and emissions. However, it is also clear that some issues still need to be addressed. Here, an attempt at listing some of these is presented.

Concerning the *fundamentals*, the following questions still remain open:

- *Quenching distance*: In order to substantiate or disprove the crevice combustion hypothesis as a possible cause for backfire occurrence (see Section 27.5), more quantitative data are needed on the dependence of the quenching distance on mixture composition, pressure, and temperature. This also is an important parameter for engine heat transfer, as described in Section 27.3.4.
- *Surface ignition*: It would be good to obtain insights into the specific mechanisms of surface ignition, as this can cause backfire and preignition, which both limit the operational range of hydrogen engines.

- *Autoignition temperature*: As described in Sections 27.2 and 27.4, there is no octane-type measure for hydrogen's knock resistance, and widely varying and sometimes contradictory claims can be found in the literature. A detailed study reporting autoignition behavior of different hydrogen mixtures under engine conditions would probably remove much of the current confusion.
- *Laminar combustion data*: As Section 27.2.2 reviews, there are little to no data available on the laminar burning velocity, the stretch rate dependence and laminar flame stability at engine conditions, of hydrogen mixtures. As this is an important parameter for engine modeling work (Section 27.3), this is an important area for further study. Both experimental data on burning velocities and reaction mechanism validation at engine conditions are needed.
- *Engine cycle computation submodels*: Next to laminar burning velocity data, there is a strong need for a turbulent combustion model that incorporates differential diffusion as a defining characteristic of laminar and turbulent hydrogen combustion (Section 27.2.3). Other submodels needed for a complete engine cycle calculation are jet formation and ignition models as well as an in-cylinder heat transfer model (Sections 27.2.4 and 27.3.4).

Relating the engine hardware, the single most important component needing further development is clearly the hydrogen DI injector, as maximum flow rate and durability are currently the main culprits.

In terms of optimizations, or engine software, lots of conditions are still to be explored for DI operation, given the large flexibility of this mixture formation concept: works reporting the impact of injector location, nozzle design, injection pressure, multiple injections, injection strategy, etc., have just recently started. Accurate but computationally efficient engine models would be of great benefit to this optimization process if they could be developed.

Finally, a number of laboratory experiments have been reported here that look very promising but remain to be proven in the field. Thus, demonstration of these concepts would be very interesting. Current vehicles are limited to some operating strategies, so further testing is required to assess the practicality, durability, and actual performance of the more complex but potentially significantly improved operating strategies demonstrated on engine test benches.

27.11 Conclusion

Hydrogen seems to be a viable solution for future transportation, and the hydrogen internal combustion engine could act as a bridging technology toward a widespread hydrogen infrastructure, since hydrogen combustion engine vehicles can initially be designed for bifuel applications. Although hydrogen is the most abundant element in the universe, it is not readily available in its molecular form and has to be produced using other energy sources. Hydrogen is therefore considered an energy carrier rather than an energy source. In order for hydrogen vehicles to become commercially feasible, challenging tasks in hydrogen production, distribution, and storage have to be addressed.

The unique properties of hydrogen compared to both conventional liquid fuels like gasoline and gaseous fuels like methane make it a challenging yet promising fuel for internal combustion engine applications. In particular, the low density of 0.08 kg/m^3 at 300 K and

1 atm, the wide flammability limits ranging from 4 to 75 vol.% of hydrogen in air, and the low minimum ignition energy of 0.02 mJ require special attention when employing hydrogen as an engine fuel. Due to the low density, the mixture calorific value as a measure for theoretical power output is 14% lower than gasoline for mixture-aspirating operation; however, air-aspirating operation results in a theoretical increase in power output of 19% compared to the gasoline baseline. Also, the laminar burning velocity at stoichiometric conditions at 360 K of approximately 290 cm/s is a factor of 6 higher than that of gasoline or methane.

As the wide flammability limits allow hydrogen engines to be operated with substantial dilution (excess air or EGR), the laminar burning velocity and laminar flame stability can vary widely and consequently are important parameters. The influence of the flame stretch rate on the burning velocity and flame stability is likely significant. Currently, there is a lack of data at engine conditions. Both experimental data and chemical reaction mechanism validation at these conditions are needed.

Most modern hydrogen combustion engines employ fuel injection, either in the intake manifold or directly into the combustion chamber. In order to allow accurate metering of the amount of fuel injected into the engine, critical pressure conditions are generally used for the injection process requiring the hydrogen supply pressure to be approximately twice the back pressure. The ratio of the largest scales versus the smallest scales to be considered during an injection event, of approximately 4000, precludes direct numerical simulation of the injection event, and various approaches to accurately simulate the hydrogen injection process have been developed. More advanced combustion concepts require consideration of ignited hydrogen jets, which have been studied experimentally as well as by using 3D CFD simulations but which could benefit from further study.

The turbulent combustion of hydrogen mixtures has been investigated experimentally and numerically, highlighting some peculiarities of hydrogen combustion, with much larger burning velocity enhancements through turbulence for lean (unstable) mixtures than for stoichiometric or rich (stable) mixtures. Very few models have been proposed to take this into account and remain to be validated. Thus, many uncertainties remain in the modeling of the combustion in hydrogen engines, with additional complexity due to currently inaccurate submodels such as in-cylinder heat transfer.

The same properties that make hydrogen such a desirable fuel for internal combustion engines are also responsible for abnormal combustion events associated with hydrogen. In particular, the wide flammability limits, low required ignition energy, and high flame speeds can result in undesired combustion phenomena generally summarized as combustion anomalies, including surface ignition and backfiring as well as autoignition. Backfiring is limited to external mixture formation operation and can be successfully avoided with DI operation. Proper engine design can largely reduce the occurrence of surface ignition. Autoignition is a topic of controversy, and a wide range of octane ratings for hydrogen as fuel has been reported. Evaluating the guidelines for determining the most common octane ratings, MON, and RON leads to the conclusion that these methods must not be applied to hydrogen as a gaseous fuel. Comparative analysis of knock intensities of gasoline and hydrogen revealed that knocking pressure traces exhibit similar peak amplitudes as well as similar durations and decays of pressure oscillations.

A dedicated design for a hydrogen internal combustion engine should include the ignition system and spark plugs, a hydrogen fuel-injection system, a properly sized engine cooling system, as well as proper design and selection of lubrication and materials. A wide variety of mixture formation strategies have been developed for hydrogen engine applications, with hydrogen PFI and hydrogen DI as the two most common concepts. Inherent disadvantages of PFI include a lower power density compared to gasoline engines as well as operational limitations due to the occurrence of combustion anomalies. However, due to the simplicity of the concept, the availability of dedicated fuel-injection systems, and the straightforward conversion process of conventional engines, hydrogen PFI is widely used mainly for demonstration vehicle applications. The wide ignition limits allow hydrogen port-fuel-injected engines to operate unthrottled, and therefore efficiently, over the entire operating regime. The dependence of the only relevant emissions component in hydrogen operation, oxides of nitrogen, is well documented. At fuel-to-air equivalence ratios, φ, of less than 0.5 ($\lambda > 2$), the engine operates without creation of NO_X emissions; increasing the fuel-to-air equivalence ratio beyond this critical threshold results in a sharp increase of NO_X emissions with a peak around $\varphi \sim 0.75$ and a slight decrease when approaching stoichiometric mixtures. Measures to increase the power density of port-fuel-injected hydrogen operation mainly focus on charging strategies. Hydrogen DI opens up another array of variables for influencing the mixture formation and combustion process. Optimizing operational parameters like fuel injector location and nozzle design, injection pressure, as well as injection strategy can be used to influence the mixture distribution and thereby engine efficiency and NO_X emissions characteristics. A 15% increase of power density in hydrogen DI operation compared to gasoline operation has also been demonstrated, and extrapolations from single-cylinder engine efficiency data suggest that a brake thermal efficiency of 45% is achievable. However, due to the limited availability of high-pressure hydrogen injection equipment, hydrogen DI strategies are still in a research stage. Hydrogen DI using a multiple injection strategy has also been demonstrated as an effective measure for significant NO_X emission reductions of up to 95%. Other emissions' reduction provisions include engine internal measures like EGR and water injection as well as after-treatment concepts including three-way catalysts as well as lean NO_X traps.

The unique properties of hydrogen also require special attention when designing a safety concept for both test cell applications and in-vehicle applications. In practical applications, the presence of ignition sources cannot be excluded; hence, proper ventilation to avoid buildup of ignitable mixtures becomes a necessity. Since hydrogen gas as well as hydrogen flames are invisible to the human eye, hydrogen flame cameras as well as hydrogen detectors have been developed for increased safety. Especially for engine test cell applications, a properly designed hydrogen fuel-supply system can significantly add to the overall safety.

Hydrogen internal combustion engine vehicles have a long history, with the earliest attempts dating back to 1807. Major contributions to the development and demonstration of hydrogen internal combustion engines have been made by Musashi Institute of Technology, BMW, as well as Ford Motor Company. Modern H_2ICE vehicles have shown emissions levels that are only a fraction of the most stringent standards while exceeding the fuel economy numbers of their conventional-fuel counterparts. Apart from its use as a neat fuel, hydrogen is also considered as a combustion enhancer, as a blending agent with gaseous fuels, and bifuel applications with both gasoline- and diesel-type fuels.

Finally, although the H_2ICE has made significant progress recently, there remain many topics requiring further investigation, ranging from fundamentals to demonstrations.

Nomenclature

Abbreviations

BDC	Bottom dead center
BMEP	Brake mean effective pressure
BTDC	Before top dead center

BTE	Brake thermal efficiency
CA	Crank angle
CFD	Computational fluid dynamics
CFR	Cooperative fuel research
CI	Compression ignition
CO	Carbon monoxide
CO_2	Carbon dioxide
CPU	Central processing unit
DI	Direct injection
DME	Dimethyl ether
DNS	Direct numerical simulation
DOE	(US) Department of Energy
EGR	Exhaust gas recirculation
EZEV	Equivalent zero-emission vehicle
FC	Fuel cell
FTP	Federal test procedure
HC	Hydrocarbon
HCCI	Homogeneous charge compression ignition
HEV	Hybrid-electric vehicle
H_2	Hydrogen
H_2FC	Hydrogen fuel cell
H_2ICE	Hydrogen-fueled internal combustion engine
IMEP	Indicated mean effective pressure
IR	Infrared
IVC	Inlet valve closing
LES	Large eddy simulation
LHV	Lower heating value
MBT	Minimum spark advance for best torque
MN	Methane number
MON	Motor octane number
mpg	Miles per (US)gallon
NO	Nitric oxide
NO_X	Nitrogen oxides
NMHC	Nonmethane hydrocarbons
NSR	NO_X storage reduction
NTP	Normal temperature and pressure (300 K, 1 atm)
OEM	Original equipment manufacturer
pdf	Probability density function
PFI	Port-fuel injection
RME	Rapeseed methyl ester
rms	Root mean square
RON	Research octane number
RPM	Revolutions per minute
SCR	Selective catalytic reduction
SI	Spark ignition
SOI	Start of injection
SULEV	Super ultralow emission vehicle
TDC	Top dead center
THC	Total hydrocarbons

TLEV	Transitional low-emission vehicle
ULEV	Ultralow-emission vehicle
ULSD	Ultralow sulfur diesel
UV	Ultraviolet
WOT	Wide open throttle
WWMP	World-Wide Mapping Point
ZEV	Zero-emission vehicle

Symbols

A	(Flame) area
D_M	Mass diffusivity
D_T	Thermal diffusivity
L	Markstein length
Le	Lewis number
p	Pressure
r	(Flame) radius
Re	Reynolds number
S	Flame speed
t	Time
T	Temperature
U	Burning velocity
u'	rms turbulent velocity

Greek Symbols

α	(Flame) stretch rate
γ	Residual gas fraction
λ	Air-to-fuel equivalence ratio
ρ	Density
φ	Fuel-to-air equivalence ratio

Subscripts

b	Burned
exc	Excess reactant
l	Laminar
lim	Limiting reactant
n	Normal
s	Stretch free
t	Turbulent
u	Unburned

Acknowledgment

This work would not have been possible without the help of a number of colleagues, friends, mentors, and financing bodies.

S. Verhelst would like to thank Prof. Roger Sierens, who initiated the H_2ICE work at Ghent University in 1992 and under whose supervision S. Verhelst did his PhD on this fascinating topic. Rene Janssens is also to be mentioned, as the technician whose golden hands were of tremendous help to the experimental work. Part of the research at Ghent was sponsored by the Belgian Science Policy TAP programme in the framework of the CHASM project (contract CP/02/222).

Significant progress in S. Verhelst's understanding of hydrogen combustion was made during his stay with the Combustion Group at Leeds University. Many thanks are due to Dr. Rob Woolley, Dr. Malcolm Lawes, Prof. Derek Bradley, Prof. Chris Sheppard, and the European Commission who funded the stay with a Marie Curie Fellowship (ENK6-CT-2000-57).

Parts of the submitted manuscript have been created by UChicago Argonne, LLC, Operator of Argonne National Laboratory (*Argonne*). Argonne, a US DOE Office of Science laboratory, is operated under Contract No. DE-AC02-06CH11357. The US government retains for itself, and others acting on its behalf, a paid-up nonexclusive, irrevocable worldwide license in the said article to reproduce, prepare derivative works, distribute copies to the public, and perform publicly and display publicly, by or on behalf of the government.

Research referenced in this manuscript was partially funded by DOE's FreedomCAR and Vehicle Technologies Program, Office of Energy Efficiency and Renewable Energy. T. Wallner wishes to thank Gurpreet Singh and Lee Slezak, program managers at DOE, for their support.

A hydrogen engine used to run certain experiments presented in this manuscript was provided by Ford Motor Company. Special thanks to the team from Ford Motor Company for their support.

T. Wallner would also like to express his gratitude to all of the individuals from BMW Research and Development and Graz University of Technology involved in preparing, performing, as well as analyzing hydrogen engine research referenced in this manuscript as well as collaborators from BMW North America LLC and BMW Germany involved in the testing of the BMW Hydrogen 7 vehicle at Argonne National Laboratory.

References

1. U.S. Department of Energy. FreedomCAR and vehicle technologies multi-year program plan 2006–2011. U.S. Department of Energy, Washington, DC, 2006.
2. European Commission. The fuel cells and hydrogen joint technology initiative (FCH JTI).
3. Japanese Ministry of Economy, Trade and Industry. Japan hydrogen & fuel cell demonstration project.
4. National Hydrogen Energy Board, Ministry of New and Renewable Energy, Government of India. National hydrogen energy road map.
5. Yilancia A, Dincer I, Ozturk HK. A review on solar-hydrogen/fuel cell hybrid energy systems for stationary applications. *Prog Energy Combust Sci* 2009;35:231–244.
6. Weindorf W, Altmann M. *Yield of Biofuels versus Hydrogen from Photovoltaics and Wind Power.* Ludwig-Bölkow-Systemtechnik GmbH, Ottobrunn, Germany, 2007.
7. Shelef M, Kukkonen CA. Prospects of hydrogen-fueled vehicles. *Prog Energy Combust Sci* 1994;20:139–148.
8. Delorme A, Rousseau A, Sharer P, Pagerit S, Wallner T. Evolution of hydrogen fueled vehicles compared to conventional vehicles from 2010 to 2045. SAE Paper No. 2009-01-1008. SAE World Congress, Detroit, MI, 2009.
9. European Commission's Directorate General Joint Research Center, CONCAWE, EUCAR. Well-to-wheel analysis of future automotive fuels and powertrains in the European context.

10. Das LM. Hydrogen engines: A view of the past and a look into the future. *Int J Hydrogen Energy* 1990;15:425–443.
11. Das LM. Hydrogen–oxygen reaction mechanism and its implication to hydrogen engine combustion. *Int J Hydrogen Energy* 1996;21:703–715.
12. Eichlseder H, Wallner T, Freymann R, Ringler J. The potential of hydrogen internal combustion engines in a future mobility scenario. SAE Paper No. 2003-01-2267, 2003.
13. White CM, Steeper R, Lutz AE. The hydrogen-fueled internal combustion engine: A technical review. *Int J Hydrogen Energy* 2006;31:1292–1305.
14. Glassman I. *Combustion*. Academic Press, Inc., Orlando, FL, 1987.
15. Heywood JB. *Internal Combustion Engine Fundamentals*. McGraw-Hill, New York, 1988.
16. Morley C. GASEQ, a chemical equilibrium program for windows.
17. Perry RH, Green DW (editors). *Perry's Chemical Engineers' Handbook*. McGraw-Hill, New York, 1997.
18. Molnarne M, Schendler T, Schroeder V. Explosionsbereiche von Gasgemischen. In: *Sicherheitstechnische Kenngroessen*, Band 2. Wirtschaftsverlag NW–Verlag fuer neue Wissenschaft, Bremen, Germany, 2003.
19. Schroeder V, Holtappels K. Explosion characteristics of hydrogen–air and hydrogen–oxygen mixtures at elevated pressures. *International Conference on Hydrogen Safety*, Paper No. 120001, Pisa, Italy, 2005.
20. Drell L, Belles FE. Survey of hydrogen combustion properties. Technical Report 1383. National Advisory Committee for Aeronautics, Cleveland, OH, 1958.
21. Hertzberg M. Selective diffusional demixing: Occurrence and size of cellular flames. *Prog Energy Combust Sci* 1989;15:203–239.
22. Ono R, Nifuku M, Fujiwara S, Horiguchi S, Oda T. Minimum ignition energy of hydrogen–air mixture: Effects of humidity and spark duration. *J Electrostat* 2007;65:87–93.
23. Hong S-W, Shin Y-S, Song J-H, Chang S-H. Performance test of the quenching meshes for hydrogen control. *J Nucl Sci Technol* 2003;40(10):814–819.
24. Potter Jr AE. In: Durcarme J, Gerstain M, Lefebvre AH (editors), *Progress in Combustion and Fuel Technology. Flame Quenching*, Vol. 1. Pergamon Press, New York, 1960, pp. 145–182.
25. U.S. Department of Energy. FreedomCAR and fuel partnership 'hydrogen storage technologies roadmap'. U.S. Department of Energy, Washington, DC.
26. Wallner T. Development of combustion concepts for a hydrogen powered internal combustion engine. PhD thesis, Graz University of Technology, Graz, Austria, 2004.
27. International Programme on Chemical Safety. INCHEM—Chemical safety information from intergovernmental organizations.
28. BOC Gases. Material safety data sheets.
29. The Engineering Toolbox. Fuels and chemicals—Auto ignition temperatures. http://www.engineeringtoolbox.com (accessed February 3, 2009).
30. U.S. Department of Energy. Hydrogen analysis resource center. http://hydrogen.pnl.gov/cocoon/morf/hydrogen (accessed February 3, 2009).
31. Martel B. *Chemical Risk Analysis: A Practical Handbook*. Taylor & Francis, New York, 2000.
32. Konig G, Sheppard CGW. End gas autoignition and knock in a spark ignition engine. SAE Paper No. 902135, 1990.
33. Verhelst S, Sierens R, Verstraeten S. A critical review of experimental research on hydrogen fueled SI engines. SAE Paper No. 2006-01-0430, 2006.
34. Peschka W. *Liquid Hydrogen—Fuel of the Future*. Springer-Verlag, New York, 1992.
35. Herdin G, Gruber F, Klausner J, Robitschko R, Chvatal D. Hydrogen and hydrogen mixtures as fuel in stationary gas engines. SAE Paper No. 2007-010012, 2007.
36. Manton J, von Elbe G, Lewis B. Nonisotropic propagation of combustion waves in explosive gas mixtures and the development of cellular flames. *J Chem Phys* 1952;20:153–157.
37. Clavin P. Dynamic behaviour of premixed flame fronts in laminar and turbulent flows. *Prog Energy Combust Sci* 1985;11:1–59.
38. Williams FA. *Combustion Theory*, 2nd edn. Addison-Wesley, Reading, MA, 1985.

39. Law C. Dynamics of stretched flames. *22nd Symposium (International) on Combustion*, Pittsburgh, PA, 1988, pp. 1381–1402.
40. Kwon S, Tseng LK, Faeth GM. Laminar burning velocities and transition to unstable flames in $H_2/O_2/N_2$ and $C_3H_8/O_2/N_2$ mixtures. *Combust Flame* 1992;90:230–246.
41. Bradley D, Hicks RA, Lawes M, Sheppard CGW, Woolley R. The measurement of laminar burning velocities and Markstein numbers for iso-octane–air and iso-octane–*n*-heptane–air mixtures at elevated temperatures and pressures in an explosion bomb. *Combust Flame* 1998;115:126–144.
42. Tseng L-K, Ismail MA, Faeth GM. Laminar burning velocities and Markstein numbers of hydrocarbon/air flames. *Combust Flame* 1993;95:410–426.
43. Gu XJ, Haq MZ, Lawes M, Woolley R. Laminar burning velocity and Markstein lengths of methane–air mixtures. *Combust Flame* 2000;121:41–58.
44. Law C, Sung C. Structure, aerodynamics and geometry of premixed flamelets. *Prog Energy Combust Sci* 2000;26:459–505.
45. Verhelst S, Sierens R. A laminar burning velocity correlation for hydrogen/air mixtures valid at spark-ignition engine conditions. *ASME Spring Engine Technology Conference*, Paper No. ICES2003-555, Salzburg, Austria, 2003.
46. Verhelst S. A study of the combustion in hydrogen-fuelled internal combustion engines. PhD thesis, Ghent University, Ghent, Belgium, 2005. http://hdl.handle. net/1854/3378.
47. Taylor SC. Burning velocity and the influence of flame stretch. PhD thesis, Leeds University, West Yorkshire, U.K., 1991.
48. Vagelopoulos CM, Egolfopoulos FN, Law CK. Further considerations on the determination of laminar flame speeds with the counterflow twin-flame technique. *25th Symposium (International) on Combustion 1341–1CK347*, Pittsburgh, PA, 1994.
49. Kwon OC, Faeth GM. Flame/stretch interactions of premixed hydrogen-fueled flames: Measurements and predictions. *Combust Flame* 2001;124:590–610.
50. Verhelst S, Woolley R, Lawes M, Sierens R. Laminar and unstable burning velocities and Markstein lengths of hydrogen–air mixtures at engine-like conditions. *Proc Combust Inst* 2005;30:209–216.
51. Liu DDS, MacFarlane R. Laminar burning velocities of hydrogen–air and hydrogen–air–steam flames. *Combust Flame* 1983;49:59–71.
52. Milton B, Keck J. Laminar burning velocities in stoichiometric hydrogen and hydrogen–hydrocarbon gas mixtures. *Combust Flame* 1984;58:13–22.
53. Iijima T, Takeno T. Effects of temperature and pressure on burning velocity. *Combust Flame* 1986;65:35–43.
54. Koroll GW, Kumar RK, Bowles EM. Burning velocities of hydrogen–air mixtures. *Combust Flame* 1993;94:330–340.
55. Bechtold JK, Matalon M. Hydrodynamic and diffusion effects on the stability of spherically expanding flames. *Combust Flame* 1987;67:77–90.
56. Bradley D, Sheppard CGW, Woolley R, Greenhalgh DA, Lockett RD. The development and structure of flame instabilities and cellularity at low Markstein numbers in explosions. *Combust Flame* 2000;122:195–209.
57. Wu CK, Law CK. On the determination of laminar flame speeds from stretched flames. *20th Symposium (International) on Combustion*, Pittsburgh, PA, 1984, pp. 1941–1949.
58. Knop V, Benkenida A, Jay S, Colin O. Modelling of combustion and nitrogen oxide formation in hydrogen-fuelled internal combustion engines within a 3D CFD code. *Int J Hydrogen Energy* 2008;33:5083–5097.
59. Aung KT, Hassan MI, Faeth GM. Effects of pressure and nitrogen dilution on flame/stretch interactions of laminar premixed $H_2/O_2/N_2$ flames. *Combust Flame* 1998;112:1–15.
60. Bradley D, Lawes M, Liu K, Verhelst S, Woolley R. Laminar burning velocities of lean hydrogen–air mixtures at pressures up to 1.0 MPa. *Combust Flame* 2007;149:162–172.
61. Verhelst S, Sierens R. A quasi-dimensional model for the power cycle of a hydrogen fuelled ICE. *Int J Hydrogen Energy* 2007;32:3545–3554.

62. Saxena P, Williams FA. Testing a small detailed chemical–kinetic mechanism for the combustion of hydrogen and carbon monoxide. *Combust Flame* 2006;145:316–323.
63. Konnov AA. Remaining uncertainties in the kinetic mechanism of hydrogen combustion. *Combust Flame* 2008;152:507–528.
64. Verhelst S, Sierens R. A two-zone thermodynamic model for hydrogen-fueled S.I. engines. *7th COMODIA—International Conference on Modeling and Diagnostics for Advanced Engine Systems*, Paper No. FL1-3, Sapporo, Japan, 2008.
65. Ó Conaire M, Curran H, Simmie J, Pitz W, Westbrook C. A comprehensive modeling study of hydrogen oxidation. *Int J Chem Kinet* 2004;36:603–622.
66. Williams FA, Grcar JF. A hypothetical burning-velocity formula for very lean hydrogen–air flames. *Proc Combust Inst* 2009;32:1351–1357.
67. Konnov AA. Refinement of the kinetic mechanism of hydrogen combustion. *J Adv Chem Phys* 2004;23:5–18.
68. Williams FA. Detailed and reduced chemistry for hydrogen autoignition. *J Loss Prev Process Indust* 2008;21:131–135.
69. Driscoll JF. Turbulent premixed combustion: Flamelet structure and its effect on turbulent burning velocities. *Prog Energy Combust Sci* 2008;34:91–134.
70. Lipatnikov A, Chomiak J. Molecular transport effects on turbulent flame propagation and structure. *Prog Energy Combust Sci* 2005;31:1–73.
71. Abdel-Gayed RG, Bradley D, Hamid MN, Lawes M. Lewis number effects on turbulent burning velocity. *20th Symposium (International) on Combustion*, Pittsburgh, PA, 1984, pp. 505–512.
72. Abdel-Gayed RG, Bradley D, Lawes M. Turbulent burning velocities: A general correlation in terms of straining rates. *Proc R Soc Lond* 1987; A-414:389–413.
73. Bradley D, Haq MZ, Hicks RA, Kitagawa T, Lawes M, Sheppard CGW et al. Turbulent burning velocity, burned gas distribution and associated flame surface definition. *Combust Flame* 2003;133:415–430.
74. Wu M-S, Kwon S, Driscoll JF, Faeth GM. Turbulent premixed hydrogen/air flames at high Reynolds numbers. *Combust Sci Technol* 1990;73:327–350.
75. Wu M-S, Kwon S, Driscoll JF, Faeth GM. Preferential diffusion effects on the surface structure of turbulent premixed hydrogen/air flames. *Combust Sci Technol* 1991;78:69–96.
76. Aung KT, Hassan MI, Kwon S, Tseng L-K, Kwon O-C, Faeth GM. Flame/stretch interaction in laminar and turbulent premixed flames. *Combust Sci Technol* 2002;174:61–99.
77. Goix PJ, Shepherd IG. Lewis number effects on turbulent premixed flame structure. *Combust Sci Technol* 1993;91:191–206.
78. Renou B, Boukhalfa A, Puechberty D, Trinité D. Effects of stretch on the local structure of freely propagating premixed low-turbulent flames with various Lewis numbers. *27th Symposium (International) on Combustion*, Pittsburgh, PA, 1998, pp. 841–847.
79. Kido H, Nakahara M, Nakashima K, Hashimoto J. Influence of local flame displacement velocity on turbulent burning velocity. *29th Symposium (International) on Combustion*, Pittsburgh, PA, 2002, pp. 1855–1861.
80. Kido H, Nakahara M, Nakashima K, Kim J-H. Turbulent burning velocity of lean hydrogen mixtures. SAE Paper No. 2003-01-1773, 2003.
81. Lipatnikov AN, Chomiak J, Betev AS, Karpov VP. Effect of Lewis number on mass burning rate in lean hydrogen turbulent flames. *European Combustion Meeting*, Paper No. 169, Orleans, France, 2003.
82. Chen JH, Im HG. Correlation of flame speed with stretch in turbulent premixed methane/air flames. *27th Symposium (International) on Combustion*, Pittsburgh, PA, 1998, pp. 819–826.
83. Chen JH, Im HG. Stretch effects on the burning velocity of turbulent premixed hydrogen/air flames. *28th Symposium (International) on Combustion*, Pittsburgh, PA, 2000, pp. 211–218.
84. Im HG, Chen JH. Preferential diffusion effects on the burning rate of interacting turbulent premixed hydrogen–air flames. *Combust Flame* 2002;131:246–258.
85. Gillespie L, Lawes M, Sheppard CGW, Woolley R. Aspects of laminar and turbulent burning velocity relevant to SI engines. SAE Paper No. 2000-01-0192, 2000.

86. Kobayashi H, Kawabata Y, Maruta K. Experimental study on general correlation of turbulent burning velocity at high pressure. *27th Symposium (International) on Combustion*, Pittsburgh, PA, 1998, pp. 941–948.
87. Bradley D, Gaskell PH, Gu XJ. Burning velocities, Markstein lengths, and flame quenching for spherical methane–air flames: A computational study. *Combust Flame* 1996;104:176–198.
88. Kobayashi H, Kawazoe H. Flame instability effects on the smallest wrinkling scale and burning velocity of high pressure turbulent premixed flames. *28th Symposium (International) on Combustion*, Pittsburgh, PA, 2000, pp. 375–382.
89. Haq MZ, Sheppard CGW, Woolley R, Greenhalgh DA, Lockett RD. Wrinkling and curvature of laminar and turbulent premixed flames. *Combust Flame* 2002;131:1–15.
90. Bradley D. Problems of predicting turbulent burning rates. *Combust Theory Model* 2002;6:361–382.
91. Heywood JB, Vilchis FR. Comparison of flame development in a spark-ignition engine fueled with propane and hydrogen. *Combust Sci Technol* 1984;38:313–324.
92. Bradley D, Lau AKC, Lawes M. Flame stretch rate as a determinant of turbulent burning velocity. *Philos Trans R Soc Lond* 1992; A-338:359–387.
93. Bray KNC. Studies of the turbulent burning velocity. *Proc R Soc Lond* 1990; A-431:315–335.
94. Gülder ÖL. Turbulent premixed flame propagation models for different combustion regimes. *23rd Symposium (International) on Combustion*, Pittsburgh, PA, 1990, pp. 743–750.
95. Zimont V. Gas premixed combustion at high turbulence. Turbulent flame closure combustion model. *Exp Ther Fluid Sci* 2000;21:179–186.
96. Weller HG, Tabor G, Gosman AD, Fureby C. Application of a flame-wrinkling LES combustion model to a turbulent mixing layer. *27th Symposium (International) on Combustion*, Pittsburgh, PA, 1998, pp. 899–907.
97. Lipatnikov A, Chomiak J. Lewis number effects in premixed turbulent combustion and highly perturbed laminar flames. *Combust Sci Technol* 1998;137:277–298.
98. Bradley D, Gaskell PH, Gu XJ, Sedaghat A. Flame instabilities in large scale atmospheric gaseous explosions. *4th International Seminar on Fire and Explosion Hazards*, Londonderry, Northern Ireland, 2003.
99. Peters N. *Turbulent Combustion*. Cambridge University Press, Cambridge, U.K., 2000.
100. Lipatnikov AN, Chomiak J. Turbulent flame speed and thickness: Phenomenology, evaluation, and application in multi-dimensional simulations. *Prog Energy Combust Sci* 2002;28:1–74.
101. Owston R, Magi V, Abraham J. Fuel–air mixing characteristics of DI hydrogen jets. SAE Paper No. 2008-01-1041, 2008.
102. Johnson NL, Amsden AA, Naber JD, Siebers DL. Three-dimensional computer modeling of hydrogen injection and combustion.*'95 SMC Simulation Multiconference*, Phoenix, AZ, 1995.
103. Yüceil BK, Ötügen VM. Scaling parameters for underexpanded supersonic jets. *Phys Fluids* 2002;14:4206–4215.
104. Schüers A, Gerbig F, Wimmer A, Kovac K. Thermodynamic analysis of the working process of hydrogen internal combustion engines with direct injection. *9th Symposium the Working Process of the Internal Combustion Engine*, Institute for Internal Combustion Engines and Thermodynamics, Graz University of Technology, Graz, Austria, 2003.
105. Takeyuki K, Takayoshi K, Hidemi S, Yoshitaka Y, Yasuo M. A study on behavior of a transient hydrogen jet in a high swirl flow. *THIESEL 2002 Conference on Thermo-and Fluid Dynamic Processes in Diesel Engines*, Valencia, Spain, 2002.
106. Kirchweger W, Haslacher R, Hallmannsegger M, Gerke U. Application of the LIF-method for the diagnostics of the combustion process of gas-IC-engines. *13th International Symposium on Application of Laser Techniques to Fluid Mechanics*, Lisbon, Portugal, 2006.
107. Kirchweger W, Eichlseder H, Gerbig F, Gerke U. Optical measurement methods for the optimization of the hydrogen DI combustion. *7th International Symposium on Internal Combustion Engines Diagnostics*, Baden-Baden, Germany, 2006.
108. Houf WG, Evans GH, Schefer RW. Analysis of jet flames and unignited jets from unintended hydrogen release. *International Conference on Hydrogen Safety*, San Sebastian, Spain, 2007.

109. Brennan SL, Makarov DV, Molkov V. LES of high pressure hydrogen jet fire. *J Loss Prev Process Indust* 2009;22:353–359.
110. Mohammadi A, Shioji M, Matsui Y, Kajiwara R. Spark-ignition and combustion characteristics of high-pressure hydrogen and natural-gas intermittent jets. *Trans ASME J Eng Gas Turb Power* 2008;130. 062801–1–7.
111. Kawanabe H, Matsui Y, Kato A, Shioji M. Study on the flame propagation process in an ignited hydrogen jet. SAE Paper No. 2008-01-1035, 2008.
112. Conte E, Boulouchos K. A quasi-dimensional model for estimating the influence of hydrogen-rich gas addition on turbulent flame speed and flame front propagation in IC-SI engines. SAE Paper No. 2005-01-0232, 2005.
113. Noda T, Foster DE. A numerical study to control combustion duration of hydrogen-fueled HCCI by using multi-zone chemical kinetics simulation. SAE Paper No. 2001-01-0250, 2001.
114. Liu C, Karim GA. A simulation of the combustion of hydrogen in HCCI engines using a 3D model with detailed chemical kinetics. *Int J Hydrogen Energy* 2008;33:3863–3875.
115. Fagelson JJ, McLean WJ, de Boer PCT. Performance and NO_X emissions of spark-ignited combustion engines using alternative fuels—Quasi one-dimensional modeling. I. Hydrogen fueled engines. *Combust Sci Technol* 1978;18:47–57.
116. Verhelst S, Sheppard CGW. Multi-zone thermodynamic modelling of spark-ignition engine combustion—An overview. *Energy Convers Manag* 2009;50:1326–1335.
117. Prabhu-Kumar GP, Nagalingam B, Gopalakrishnan KV. Theoretical studies of a spark-ignited supercharged hydrogen engine. *Int J Hydrogen Energy* 1985;10:389–397.
118. Al-Baghdadi MARS, Al-Janabi HAKS. A prediction study of a spark ignition supercharged hydrogen engine. *Energy Convers Manag* 2003;44:3143–3150.
119. Al-Baghdadi MARS. Effect of compression ratio, equivalence ratio and engine speed on the performance and emission characteristics of a spark ignition engine using hydrogen as a fuel. *Renew Energy* 2004;29:2245–2260.
120. Keck J. Turbulent flame structure and speed in spark-ignition engines. *19th Symposium (International) on Combustion*, Pittsburgh, PA, 1982, pp. 1451–1466.
121. Takats M, Macek J, Polasek M, Kovar Z, Beroun S, Schloz C. Hydrogen fueled reciprocating engine as an automotive prime mover? *Fisita World Automotive Congress*, Paper No. F98T/P693, Paris, France, 1998.
122. Polasek M, Macek J, Takats M, Vitek O. Application of advanced simulation methods and their combination with experiments to modeling of hydrogen fueled engine emission potentials. SAE Paper No. 2002-01-0373, 2002.
123. Miller JA, Bowman CT. Mechanism and modeling of nitrogen chemistry in combustion. *Prog Energy Combust Sci* 1989;15:287–338.
124. Ma J, Su Y, Zhou Y, Zhang Z. Simulation and prediction on the performance of a vehicle's hydrogen engine. *Int J Hydrogen Energy* 2003;28:77–83.
125. D'Errico G, Onorati A, Ellgas S, Obieglo A. Thermo-fluid dynamic simulation of a S.I. single-cylinder H_2 engine and comparison with experimental data. *ASME Spring Engine Technology Conference*, Paper No. ICES2003-1311, Aachen, Germany, 2006.
126. D'Errico G, Onorati A, Ellgas S. 1D thermo-fluid dynamic modelling of an SI single-cylinder H_2 engine with cryogenic port injection. *Int J Hydrogen Energy* 2008;33:5829–5841.
127. Safari H, Jazayeri S, Ebrahimi R. Potentials of NO_X emission reduction methods in SI hydrogen engines: Simulation study. *Int J Hydrogen Energy* 2009;34:1015–1025.
128. Woschni G. A universally applicable equation for the instantaneous heat transfer coefficient in the internal combustion engine. SAE Paper No. 670931, 1967.
129. Johnson N. Hydrogen as a zero-emission, high-efficiency fuel: Uniqueness, experiments and simulation. *3rd International Conference ICE97, Internal Combustion Engines: Experiments and Modeling*, Naples, Italy, 1997.
130. Blarigan PV. Development of a hydrogen fueled internal combustion engine designed for single speed/power operation. SAE Paper No. 961690, 1996.

131. Fontana G, Galloni E, Jannelli E, Minutillo M. Numerical modeling of a spark-ignition engine using premixed lean gasoline–hydrogen–air mixtures. *14th World Hydrogen Energy Conference*, Montreal, Quebec, Canada, 2002.
132. Shioji M, Kawanabe H, Taguchi Y, Tsunooka T. CFD simulation for the combustion process in hydrogen engines. *15th World Hydrogen Energy Conference*, Yokohama, Japan, 2004.
133. Adgulkar DD, Deshpande NV, Thombre SB, Chopde IK. 3D CFD simulations of hydrogen fuelled spark ignition engine. *ASME Spring Engine Technology Conference*, Paper No. ICES2008-1649, Chicago, IL, 2008.
134. Wimmer A, Wallner T, Ringler J, Gerbig F. H_2-direct injection—A highly promising combustion concept. SAE Paper No. 2005-01-0108, 2005.
135. Messner D, Wimmer A, Gerke U, Gerbig F. Application and validation of the 3D CFD method for a hydrogen fueled IC engine with internal mixture formation. SAE Paper No. 2006-01-0448, 2006.
136. Gerke U, Boulouchos K, Wimmer A. Numerical analysis of the mixture formation and combustion process in a direct injected hydrogen internal combustion engine. *Proceedings 1st International Symposium on Hydrogen Internal Combustion Engines*, Graz, Austria, 2006, pp. 94–106.
137. Benkenida A, Colin O, Jay S, Knop V. Adaptation of the ECFM combustion model to hydrogen internal combustion engines. *Proceedings 1st International Symposium on Hydrogen Internal Combustion Engines*, Graz, Austria, 2006, pp. 195–206.
138. Borman G, Nishiwaki K. Internal-combustion engine heat transfer. *Prog Energy Combust Sci* 1987;13:1–46.
139. Annand WJD. Heat transfer in the cylinders of reciprocating internal combustion engines. *Proc Inst Mech Eng* 1963;177(36):973–996.
140. Wei SW, Kim YY, Kim HJ, Lee JT. A study on transient heat transfer coefficient of in-cylinder gas in the hydrogen fueled engine. *6th Korea-Japan Joint Symposium on Hydrogen Energy*, Hiratsuka, Japan, 2001.
141. Shudo T, Nabetani S. Analysis of degree of constant volume and cooling loss in a hydrogen fuelled SI engine. SAE Paper No. 2001-01-3561, 2001.
142. Shudo T, Suzuki H. New heat transfer equation applicable to hydrogen-fuelled engines. *ASME Fall Technical Conference*, Paper No. ICEF2002-515, New Orleans, LA, 2002.
143. Nefischer A, Hallmannsegger M, Wimmer A, Pirker G. Application of a flow field based heat transfer model to hydrogen internal combustion engines. SAE Paper No. 2009-01-1423, 2009.
144. Schubert A, Wimmer A, Chmela F. Advanced heat transfer model for CI engines. SAE Paper No. 2005-01-0695, 2005.
145. Demuynck J, Zuliani M, Raes N, Verhelst S, Paepe MD, Sierens R. Local heat flux measurements in a hydrogen and methane spark ignition engine with a thermopile sensor. *Int J Hydrogen Energy* 2009;34:9857–9868.
146. Ringler J, Gerbig F, Eichlseder H, Wallner T. Insights into the development of a hydrogen combustion process with internal mixture formation. *Proceedings 6th International Symposium on Internal Combustion Diagnostics*, Baden Baden, Germany, 2004.
147. Stockhausen WF, Natkin RJ, Reams L. Crankcase ventilation system for a hydrogen fueled engine. US Patent No. 6,606,982 B1, August 2003.
148. Lee SJ, Yi HS, Kim ES. Combustion characteristics of intake port injection type hydrogen fueled engine. *Int J Hydrogen Energy* 1995;20:317–322.
149. Kirchweger W. Investigations on the use of an alternative fuel in an internal combustion engine. Master's thesis, Graz University of Technology, Graz, Austria, 2002.
150. Sierens R, Rosseel E. Backfire mechanism in a carbureted hydrogen fuelled engine. *12th World Hydrogen Energy Conference*, Buenos Aires, Argentina, 1998, pp. 1537–1546.
151. Swain M, Swain M. Elimination of abnormal combustion in a hydrogen-fueled engine. Technical Report NREL/TP-425-8196. Department of Energy, Washington, DC, 1995.
152. Liu X-H, Liu F-S, Zhou L, Sun B-G, Schock HJ. Backfire prediction in a manifold injection hydrogen internal combustion engine. *Int J Hydrogen Energy* 2008;33:3847–3855.

153. Ciatti SA, Wallner T, Ng H, Stockhausen WF, Boyer B. Study of combustion anomalies of H_2-ICE with external mixture formation. *ASME Spring Technical Conference*, Paper No. ICES 2006-1398, 2006.
154. Kiesgen G, Klüting M, Bock C, Fischer H. The new 12-cylinder hydrogen engine in the 7 series. The H_2 ICE age has begun. SAE Paper No. 2006-01-0431, 2006.
155. ASTM. Standard test method for research octane number of a spark-ignition engine fuel. ASTM Designation: D 2699-04a. ASTM International, West Conshohocken, PA, 2004.
156. ASTM. Standard test method for motor octane number of a spark-ignition engine fuel. ASTM Designation: D 2700-04a. ASTM International, West Conshohocken, PA, 2004.
157. Specht M. Infomaterial: Regenerative kraftstoffe. Zentrum für Sonnenenergie-und Wasserstoff-Forschung Baden-Württemberg (ZSW) Fachgebiet Regenerative Energieträger und Verfahren, Stuttgart, Germany.
158. Saravanan N, Nagarajan G, Dhanasekaran C, Kalaiselvan K. Experimental investigation of hydrogen port fuel injection in DI diesel engine. *Int J Hydrogen Energy* 2007;32:4071–4080.
159. College of the Desert. *Hydrogen Fuel Cell Engines and Related Technologies Course Manual*, 2001.
160. Swarts A, Yates A. Insights into the role of autoignition during octane rating. SAE Paper No. 2007-01-0008, 2007.
161. Leiker M, Cartelliere W, Christoph H, Pfeifer U, Rankl M. Evaluation of anti-knocking property of gaseous fuels by means of methane number and its practical application to gas engines. *Am Soc Mech Eng* 1972;94(7):55.
162. Li H, Karim GA. Hydrogen fueled spark-ignition engines predictive and experimental performance. *Trans ASME J Eng Gas Turb Power* 2006;128:230–236.
163. Tang X, Stockhausen WF, Kabat DM, Natkin RJ, Heffel JW. Ford P2000 hydrogen engine dynamometer development. SAE Paper No. 2002-01-0242, 2002.
164. Szwaja S, Bhandary K, Naber J. Comparisons of hydrogen and gasoline combustion knock in a spark ignition engine. *Int J Hydrogen Energy* 2007;32:5076–5087.
165. Al-Baghdadi MARS. Development of a pre-ignition submodel for hydrogen engines. *Proc IMechE Part D J Automobile Eng* 2005;219:1203–1212.
166. Douaud AM, Eyzat P. Four-octane-number method for predicting the antiknock behavior of fuels and engines. SAE Paper No. 780080, 1978.
167. Li H, Karim G. Knock in spark ignition hydrogen engines. *Int J Hydrogen Energy* 2004;29:859–865.
168. Eichlseder H, Klell M, Sartory M, Schaffer K, Leitner D. Potential of synergies in a vehicle for variable mixtures of CNG and hydrogen. SAE Paper No. 2009-01-1420, 2009.
169. Das LM. Near-term introduction of hydrogen engines for automotive and agricultural application. *Int J Hydrogen Energy* 2002;27:479–487.
170. Kondo T, Iio S, Hiruma M. A study on the mechanism of backfire in external mixture formation hydrogen engines—About backfire occurred by the cause of the spark plug. SAE Paper No. 971704, 1997.
171. Stockhausen WF, Natkin RJ, Kabat DM, Reams L, Tang X, Hashemi S et al. Ford P2000 hydrogen engine design and vehicle development program. SAE Paper No. 2002-01-0240, 2002.
172. Olavson LG, Baker NR, Lynch FE, Meija LC. Hydrogen fuel for underground mining machinery. SAE Paper No. 840233, 1984.
173. Project Coordinator Motor Vehicles and Road Transport. TÜV Rheinland e.V. for the Federal Ministry for Research and Technology. Alternative energy sources for road transport—Hydrogen drive test. Technical Report, TÜV Rheinland, Cologne, Germany, 1990.
174. MacCarley CA. A study of factors influencing thermally induced backfiring in hydrogen fuelled engines, and methods for backfire control. *16th International Energy Conversion Engineering Conference*, Atlanta, GA, 1981.
175. Gerbig F, Strobl W, Eichlseder H, Wimmer A. Potentials of the hydrogen combustion engine with innovative hydrogen-specific combustion process. *Fisita World Automotive Congress*, Paper No. F2004V113, Barcelona, Spain, 2004.
176. Verhelst S, Sierens R. Hydrogen engine—Specific properties. *Int J Hydrogen Energy* 2001;26:987–990.

177. Lucas GG, Morris LE. The backfire problem of the hydrogen engine. Symposium Organized by the University's Internal Combustion Engine Group, London, U.K., 1980.
178. Welch A, Mumford D, Munshi S, Holbery J, Boyer B, Younkins M et al. Challenges in developing hydrogen direct injection technology for internal combustion engines. SAE Paper No. 2008-01-2379, 2008.
179. Swain MR, Swain MN, Adt RR. Consideration in the design of an inexpensive hydrogen-fueled engine. SAE Paper No. 881630, 1988.
180. Swain MR, Schade GJ, Swain MN. Design and testing of a dedicated hydrogen-fueled engine. SAE Paper No. 961077, 1996.
181. Berckmüller M, Rottengruber H, Eder A, Brehm N, Elsässer G, Müller-Alander G et al. Potentials of a charged SI-hydrogen engine. SAE Paper No. 2003-01-3210, 2003.
182. Koyanagi K, Hiruma M, Furuhama S. Study on mechanism of backfire in hydrogen engines. SAE Paper No. 942035, 1994.
183. Lee JT, Kim YY, Lee CW, Caton JA. An investigation of a cause of backfire and its control due to crevice volumes in a hydrogen fueled engine. *ASME Spring Technical Conference*, Paper No. 2000-ICE-284, San Antonio, TX, 2000.
184. Berger E, Bock C, Fischer H, Gruber M, Kiesgen G, Rottengruber H. The new BMW 12-cylinder hydrogen engine as clean efficient and powerful vehicle powertrain. *Fisita World Automotive Congress*, Paper No. F2006P114, Yokohama, Japan, 2006.
185. Prümm W. Hydrogen engines for city buses. *Proceedings 1st International Symposium on Hydrogen Internal Combustion Engines*, Graz, Austria, 2006, pp. 1–11.
186. DeLuchi MA. Hydrogen vehicles: An evaluation of fuel storage, performance, safety, environmental impacts, and cost. *Int J Hydrogen Energy* 1989;14:81–130.
187. Strebig KC, Waytulonis RW. The Bureau of Mines' hydrogen powered mine vehicle. SAE Paper No. 871678, 1987.
188. Natkin RJ, Tang X, Whipple KM, Kabat DM, Stockhausen WF. Ford hydrogen engine laboratory testing facility. SAE Paper No. 2002-01-0241, 2002.
189. Tiwari GP, Bose A, Chakravartty JK, Wadekar SL, Totlani MK, Arya RN et al. A study of internal hydrogen embrittlement of steels. *Mater Sci Eng* 2000;A286:269–281.
190. Herms E, Olive JM, Puiggali M. Hydrogen embrittlement of 316l type stainless steel. *Mater Sci Eng* 1999;A272:279–283.
191. Madina V, Azkarate I. Compatibility of materials with hydrogen. Particular case: Hydrogen embrittlement of titanium alloys. *Int J Hydrogen Energy* 2009; doi:10.1016/j.ijhydene.2009.01.058.
192. Barthélémy H. Compatibility of metallic materials with hydrogen. *International Conference on Hydrogen Safety*. Review of the present knowledge Paper No. 1.4.66, San Sebastian, Spain, 2007.
193. Hardie D, Charles EA, Lopez AH. Hydrogen embrittlement of high strength pipeline steels. *Corrosion Sci* 2006;48:4378–4385.
194. Marchi CS, Somerday BP. Technical reference on hydrogen compatibility of materials. Technical Report SAND2008-1163. Sandia National Laboratories, Albuquerque, NM, 2008.
195. Göschel B. Der Wasserstoff-Verbrennungsmotor als Antrieb für den BMW der Zukunft. *Proceedings 24th Internationales Wiener Motorensymposium*, Vienna, Austria, 2003.
196. Wallner T, Lohse-Busch H, Gurski S, Duoba M, Thiel W, Martin D et al. Fuel economy and emissions evaluation of a BMW hydrogen 7 mono-fuel demonstration vehicle. *Int J Hydrogen Energy* 2008;33:7607–7618.
197. Rottengruber H, Wiebicke U, Woschni G, Zeilinger K. Wasserstoff-Dieselmotor mit Direkteinspritzung, hoher Leistungsdichte und geringer Abgasemission. Part 3: Versuche und Berechnungen am Motor. *Motortechnische Zeitschrift* 2000;61:122–128.
198. Osafune S, Akagawa H, Ishida H, Egashira H, Kuma Y, Iwasaki W. Development of hydrogen injection clean engine. *CIMAC Congress*, Paper No. 207, Kyoto, Japan, 2004.
199. Heindl R, Eichlseder H, Spuller C, Gerbig F, Heller K. New and innovative combustion systems for the H_2-ICE: Compression ignition and combined processes. SAE Paper No. 2009-01-1421, 2009.

200. Aleiferis P, Rosati M. Hydrogen SI and HCCI combustion in a direct-injection optical engine. SAE Paper No. 2009-01-1921, 2009.
201. Wakayama N, Morimoto K, Kashiwagi A, Saito T. Development of hydrogen rotary engine vehicle. *16th World Hydrogen Energy Conference*, Lyon, France, 2006.
202. Rottengruber H, Berckmüller M, Elsässer G, Brehm N, Schwarz C. Operation strategies for hydrogen engines with high power density and high efficiency. *15th Annual U.S. Hydrogen Conference*, Los Angeles, CA, 2004.
203. Yi HS, Min K, Kim ES. The optimised mixture formation for hydrogen fuelled engines. *Int J Hydrogen Energy* 2000;25:685–690.
204. Hallmannsegger M, Fickel H-C. The mixture formation process of an internal combustion engine for zero CO_2-emission vehicles fueled with cryogenic hydrogen. *IFP International Conference*, Rueil-Malmaison, France, 2004.
205. Hallmannsegger M. Potentials of the four-stroke Otto engine with PFI of cryogenic hydrogen. PhD thesis, Graz University of Technology, Graz, Austria, 2005.
206. Heller K, Ellgas S. Optimisation of a hydrogen internal combustion engine with cryogenic mixture formation. *Proceedings 1st International Symposium on Hydrogen Internal Combustion Engines*, Graz, Austria, 2006, pp. 49–58.
207. Ellgas S. Simulation of a hydrogen internal combustion engine with cryogenic mixture formation. PhD thesis, University of Armed Forces Munich, Munich, Germany, 2008.
208. Boretti A, Watson H. Numerical study of a turbocharged, jet ignited, cryogenic, port injected, hydrogen engine. SAE Paper No. 2009-01-1425, 2009.
209. Eichlseder H, Klell M. *Wasserstoff in der Fahrzeugtechnik*. Vieweg/Teubner, Wiesbaden, Germany, 2008.
210. Aceves SM, Weisberg A, Espinosa-Loza F, Berry G, Ross T. Advanced concepts for containment of hydrogen and hydrogen storage materials. Technical Report. DOE Hydrogen Program Annual Progress Report. U.S. Department of Energy, Washington, DC, 2007.
211. Ahluwalia RK, Peng JK. Dynamics of cryogenic hydrogen storage in insulated pressure vessels for automotive applications. *Int J Hydrogen Energy* 2008;33:4622–4633.
212. Salimi F, Shamekhi AH, Pourkhesalian AM. Role of mixture richness, spark and valve timing in hydrogen-fuelled engine performance and emission. *Int J Hydrogen Energy* 2009;34:3922–3929.
213. Davidson D, Fairlie M, Stuart A. Development of a hydrogen-fuelled farm tractor. *Int J Hydrogen Energy* 1986;11:39–42.
214. Knorr H, Held W, Prümm W, Rüdiger H. The MAN hydrogen propulsion system for city buses. *11th World Hydrogen Energy Conference*, Stuttgart, Germany, 1996, pp. 1611–1620.
215. Szwabowski S, Hashemi S, Stockhausen W, Natkin R, Reams L, Kabat D, Potts C. Ford hydrogen engine powered P2000 vehicle. SAE Paper No. 2002-01-0243, 2002.
216. Verhelst S, Maesschalck P, Rombaut N, Sierens R. Increasing the power output of hydrogen internal combustion engines by means of supercharging and exhaust gas recirculation. *Int J Hydrogen Energy* 2009;34:4406–4412.
217. ETEC hydrogen internal combustion engine full-size pickup truck conversion. Hydrogen ICE truck brochure.
218. Francfort JE, Karner D. Hydrogen ICE vehicle testing activities. SAE Paper No. 2006-01-0433, 2006.
219. Verhelst S, Sierens R. Aspects concerning the optimisation of a hydrogen fueled engine. *Int J Hydrogen Energy* 2001;26:981–985.
220. Wallner T, Lohse-Busch H, Shidore N. Operating strategy for a hydrogen engine for improved drive-cycle efficiency and emissions behavior. *Int J Hydrogen Energy* 2008;34:4617–4625.
221. Natkin R, Denlinger A, Younkins M, Weimer A, Hashemi S, Vaught A. Ford 6.8l hydrogen IC engine for the E-450 shuttle van. SAE Paper No. 2007-01-4096, 2007.
222. Verhelst S, Maesschalck P, Rombaut N, Sierens R. Efficiency comparison between hydrogen and gasoline, on a bi-fuel hydrogen/gasoline engine. *Int J Hydrogen Energy* 2009;34:2504–2510.
223. Wallner T, Scarcelli R, Nande A, Naber J. Assessment of multiple injection strategies in a direct injection hydrogen research engine. SAE Paper No. 2009-01-1920, 2009.

224. Steinrück P, Ranegger G. Timed injection of hydrogen for fuel cells and internal combustion engines. *Proceedings 1st International Symposium on Hydrogen Internal Combustion Engines*, Graz, Austria, 2006, pp. 164–177.
225. Kovac G, Wimmer A, Hallmannsegger M, Obieglo A. Mixture formation and combustion in a hydrogen engine—A challenge for the numerical simulation. *International Congress on Engine Combustion Processes—Current Problems and Modern Techniques*, Munich, Germany, 2005.
226. Kaiser SA, White CM. PIV and PLIF to evaluate mixture formation in a direct-injection hydrogen-fueled engine. SAE Paper No. 2008-01-1034, 2008.
227. Wallner T, Ciatti S, Bihari B, Stockhausen W, Boyer B. Endoscopic investigations in a hydrogen internal combustion engine. *Proceedings 1st International Symposium on Hydrogen Internal Combustion Engines*, Graz, Austria, 2006, pp. 107–117.
228. Wallner T, Ciatti S, Bihari B. Investigation of injection parameters in a hydrogen DI engine using an endoscopic access to the combustion chamber. SAE Paper No. 2007-01-1464, 2007.
229. Salazar VM, Kaiser SA, Nande AM, Wallner T. The influence of injector position and geometry on mixture preparation in a DI hydrogen engine. *9th International Congress 'Engine Combustion Process—Current Problems and Modern Technologies*, Munich, Germany, 2009.
230. Wallner T, Nande A, Naber J. Study of basic injection configurations using a direct-injection hydrogen research engine. SAE Paper No. 2009-01-1418, 2009.
231. Vogel C. Wasserstoff-Dieselmotor mit direkteinspritzung, hoher Leistungsdichte und geringer Abgasemission. Part 1: Konzept. *Motortechnische Zeitschrift* 1999;60:704–708.
232. Prechtl P, Dorer F. Wasserstoff-Dieselmotor mit Direkteinspritzung, hoher Leistungsdichte und geringer Abgasemission. Part 2: Untersuchung der Gemischbildung, des Zünd-und des Verbrennungsverhaltens. *Motortechnische Zeitschrift* 1999;60:830–837.
233. Sakashita Y, Suzuki H, Takagi Y. Controlling onset of heat release by assisted spark ignition in hydrogen HCCI engine supported by DME supplement. SAE Paper No. 2009-01-1419, 2009.
234. Narioka Y, Takagi Y, Yokoyama T, Iio S. HCCI combustion characteristics of hydrogen and hydrogen-rich natural gas reformate supported by DME supplement. SAE Paper No. 2006-01-0628, 2006.
235. Fujita T, Ozawa S, Yamane K, Takagi Y, Goto Y, Odaka M. Performance of NO_X absorption 3-way catalysis applied to a hydrogen fueled engine. *15th World Hydrogen Energy Conference*, Yokohama, Japan, 2004.
236. Kawamura A, Yanai T, Sato Y, Naganuma K, Yamane K, Takagi Y. Summary and progress of the hydrogen ICE truck development project. SAE Paper No. 2009-01-1922, 2009.
237. Tang X, Theis JR, Natkin RJ, Hashemi S, Stockhausen WF. Hydrogen fueled spark ignition engine. US Patent No. US 7,059,114B2, August 2004.
238. Tang X, Theis JR, Natkin RJ, Hashemi S, Stockhausen WF. Hydrogen fueled spark ignition engine. US Patent No. US 6,779,337B2, June 2006.
239. Bleechmore C, Brewster S. Dilution strategies for load and NO_X management in a hydrogen fuelled direct injection engine. SAE Paper No. 2007-01-4097, 2007.
240. Subramaniam V, Mallikarjuna JM, Ramesh A. Effect of water injection and spark timing on the nitric oxide emission and combustion parameters of a hydrogen fueled spark ignition engine. *Int J Hydrogen Energy* 2007;32:1159–1173.
241. Nande A, Wallner T, Naber J. Influence of water injection on performance and emissions of a direct-injection hydrogen research engine. SAE Paper No. 2008-01-2377, 2008.
242. Nande A, Swaja S, Naber J. Impact of EGR on combustion processes in a hydrogen fuelled SI engine. SAE Paper No. 2008-01-1039, 2008.
243. Heffel JW. NO_X emission and performance data in a hydrogen fueled internal combustion engine at 1500 rpm using exhaust gas recirculation. *Int J Hydrogen Energy* 2003;28:901–908.
244. Heffel JW. NO_X emission reduction in a hydrogen fueled internal combustion engine at 3000 rpm using exhaust gas recirculation. *Int J Hydrogen Energy* 2003;28:1285–1292.
245. Pehr K. Safety concept of an engine test rig with liquid hydrogen supply. *Int J Hydrogen Energy* 1993;18:773–781.
246. Rossegger W, Posch U. Design criteria and instrumentation of hydrogen test benches. *Proceedings 1st International Symposium on Hydrogen Internal Combustion Engines*, Graz, Austria, 2006, pp. 132–148.

247. Wallner T, Lohse-Busch H. Light duty hydrogen engine application research at ANL. *Seminar Bridging the Technology. Hydrogen Internal Combustion Engines*, Argonne National Laboratory, Lemont, IL, 2006.
248. Kessler A, Bouchet S, Castello P, Perrette L, Boon-Brett L. Chapter V: Hydrogen safety barriers and measures, HySafe: Biennial report on Hydrogen Safety, 2006.
249. Basic considerations for the safety of hydrogen systems. ISO ISO/TR 15916, 2004.
250. National Aeronautics and Space Administration. *Safety Standard for Hydrogen and Hydrogen Systems. Guidelines for Hydrogen System Design, Materials Selection, Operations, Storage, and Transportation*. National Aeronautics and Space Administration NSS 1740.16, Washington, DC, 1997.
251. Ebner H, Koeck K. Coriolis fuel meter—A modern and reliable approach to continuous and accurate fuel consumption measurement. SAE Paper No. 2000-01-1330, 2000.
252. Paulina C. Hydrogen fuel cell vehicle fuel economy testing at the U.S. EPA national vehicle and fuel emissions laboratory. SAE Paper No. 2004-01-2900, 2004.
253. Thiel W, Krough B. Hydrogen fuel consumption correlation between established EPA measurement methods and exhaust emissions measurements. SAE Paper No. 2008-01-1038, 2008.
254. Richardson A, Gopalakrishnan R, Chhaya T, Deasy S, Kohn J. Design considerations for hydrogen management system on Ford hydrogen fueled E-450 shuttle bus. SAE Paper No. 2009-01-1422, 2009.
255. Gopalakrishnan R, Throop MJ, Richardson A, Lapetz JM. Engineering the Ford H_2 IC engine powered E-450 shuttle bus. SAE Paper No. 2007-01-4095, 2007.
256. BMW. BMW Hydrogen 7-rescue guidelines, 2006.
257. Dutton K. A brief history of the car. *New Ideas* 2006;1.
258. Fairbanks JW. Engine maturity, efficiency, and potential improvements. *Diesel Engine Emission Reduction Conference*, Coronado, CA, 2004.
259. Schöffel K. Hydrogen—The energy carrier for the future. *Workshop—Environmental Issues in Theory and Practice*, Porsgrunn, Norway, 2005.
260. Erren RA. Der Erren-Wasserstoffmotor. *Automobiltechnische Zeitschrift* 1933;19.
261. Furuhama S. Problems of forecasting the future of advanced engines and engine characteristics of the hydrogen injection with LH_2 tank and pump. *Trans ASME J Eng Gas Turb Power* 1982;119:227–242.
262. Furuhama S, Kobayashi Y. A liquid hydrogen car with a two-stroke direct injection engine and LH_2-pump. *Int J Hydrogen Energy* 1982;7:809–820.
263. Lohse-Busch HA, Wallner T, Fleming J. Transient efficiency, performance, and emissions analysis of a hydrogen internal combustion engine pick-up truck. SAE Paper No. 2006-01-3430, 2006.
264. Abele AR. Quantum hydrogen prius. *ARB ZEV Technology Symposium*, Sacramento, CA, 2006.
265. Klugescheid A. BMW introduces world's first hydrogen-drive luxury performance car—the BMW HYDROGEN 7. Press release BMW of North America, LLC, 2006.
266. Mazda. Mazda Renesis hydrogen rotary engine, 2003.
267. Lapetz J, Natkin R, Zanardelli V. The design, development, validation and delivery of the Ford H_2ICE E-450 shuttle bus. *1st (International) Symposium on Hydrogen Internal Combustion Engines*, Graz, Austria, 2006, pp. 20–33.
268. Sierens R, Rosseel E. Variable composition hydrogen/natural gas mixtures for increased engine efficiency and decreased emissions. *J Eng Gas Turb Power* 2000;122:135–140.
269. Bauer CG, Forest TW. Effect of hydrogen addition on the performance of methane-fueled vehicles. Part I: Effect on S.I. engine performance. *Int J Hydrogen Energy* 2001;26:55–70.
270. Bauer CG, Forest TW. Effect of hydrogen addition on the performance of methane-fueled vehicles. Part II: Driving cycle simulations. *Int J Hydrogen Energy* 2001;26:71–90.
271. Munshi S. Medium/heavy duty hydrogen enriched natural gas spark ignition IC engine operation. *Proceedings 1st International Symposium on Hydrogen Internal Combustion Engines*, Graz, Austria, 2006, pp. 71–82.
272. Ortenzi F, Chiesa M, Scarcelli R, Pede G. Experimental tests of blends of hydrogen and natural gas in light-duty vehicles. *Int J Hydrogen Energy* 2008;33:3225–3229.

273. Wallner T, Lohse-Busch H, Ng H, Peters RW. Results of research engine and vehicle drive cycle testing during blended hydrogen/methane operation. *Proceedings National Hydrogen Association Annual Conference*, San Antonio, TX, 2007, pp. 71–82.
274. Wallner T, Ng H, Peters RW. The effects of blending hydrogen with methane on engine operation, efficiency, and emissions. SAE Paper No. 2007-01-0474, 2007.
275. Gruber F, Herdin G, Klausner J, Robitschko R. Use of hydrogen and hydrogen mixtures in a gas engine. *Proceedings 1st International Symposium on Hydrogen Internal Combustion Engines*, Graz, Austria, 2006, pp. 34–48.
276. Shirk MG, McGuire TP, Neal GL, Haworth DC. Investigation of a hydrogen-assisted combustion system for a light-duty diesel vehicle. *Int J Hydrogen Energy* 2008;33:7237–7244.
277. Kumar MS, Ramesh A, Nagalingam B. Use of hydrogen to enhance the performance of a vegetable oil fuelled compression ignition engine. *Int J Hydrogen Energy* 2003;28:1143–1154.
278. Tsolakis A, Megaritis A. Partially premixed charge compression ignition engine with on-board H_2 production by exhaust gas fuel reforming of diesel and biodiesel. *Int J Hydrogen Energy* 2005;30:731–745.
279. Sulatisky M, Hill S, Lung B. Dual-fuel hydrogen pickup trucks. *16th World Hydrogen Energy Conference*, Lyon, France, 2006.
280. Yousufuddin S, Mehdi SN, Masood M. Performance and combustion characteristics of a hydrogen-ethanol-fuelled engine. *Energy Fuels* 2008;22:3355–3362.
281. Wallner T, Nande A, Naber J. Evaluation of injector location and nozzle design in a direct-injection hydrogen research engine. SAE Paper No. 2008-01-1785, 2009.

28

Hydrogen Enrichment

David R. Vernon
Humboldt State University

Paul A. Erickson
University of California, Davis

CONTENTS

28.1 Introduction

Hydrogen enrichment is the addition of hydrogen to mixtures of other fuels and air in the combustion processes. Hydrogen enrichment potentially changes the combustion properties of the entire mixture. When performed properly, hydrogen enrichment typically promotes complete combustion and extends the lean or dilute operating limits.

Hydrogen enrichment can be applied to many different types of combustion processes including internal combustion engines (ICEs), turbine systems, and burners. The effects of hydrogen enrichment depend upon the type of combustion process, the mixing characteristics, the design of the specific combustion device, operating parameters, and type of primary fuel. Hydrogen enrichment has been applied in many different types of ICEs including spark ignition (SI), compression ignition (CI), spark ignition–direct injection (SI–DI), as well as homogeneous charge compression ignition (HCCI) or premixed charge compression ignition (PCCI). This chapter focuses on hydrogen enrichment in SI ICEs with brief mention of hydrogen enrichment in other applications.

By enabling the use of lean or dilute mixtures, hydrogen enrichment can increase thermal efficiency and reduce emissions in SI ICEs over a range of operating conditions with a number of different primary fuels. Efficiency increases from hydrogen enrichment are related to increased ratio of specific heats, increased burn rates, reduced cycle to cycle variability, and reduced throttling losses. Emissions reductions from hydrogen enrichment are related to more complete combustion reducing hydrocarbon (HC) emissions, increased lean operation leading to reduced carbon monoxide (CO) and nitrogen oxide (NOx) emissions, and potential for increased exhaust gas recirculation leading to reduced NOx emissions.

Some potential drawbacks of hydrogen enrichment can include significantly reduced power output, flashback in the intake or premix chamber, and preignition leading to knock. The use of any gaseous fuel in premixed ICE applications reduces power output compared to liquid fuels due to the larger volume of air displaced by the gaseous fuel [1].

Hydrogen enrichment of SI ICEs has been explored with different fuels including natural gas (NG), methanol, ethanol, bioethanol, gasoline, landfill gas (LFG), and biogas.

This chapter intends to introduce the concept of hydrogen enrichment, outline the fundamental theory behind the effects of hydrogen enrichment, give a brief overview of the experimental literature, and introduces the discussion of hydrogen production for hydrogen enrichment.

28.1.1 Hydrogen as an Energy Carrier

Hydrogen is an energy carrier unlike traditional fossil fuels but similar to electricity. Hydrogen is not readily available from any natural source, and significant amounts of energy and capital investment are required to generate hydrogen. Like electricity, storage of hydrogen can be difficult but is not impossible especially if quantities are relatively small.

Hydrogen can be renewably produced using electricity from renewable sources, by processing biomass or other means. Traditionally, the largest pathway for production of hydrogen is that of catalytically reforming steam and NG [2].

Hydrogen has many unique physical properties including low molecular weight, large diffusion coefficient, high thermal conductivity, and low viscosity.

28.1.2 Hydrogen Combustion

Hydrogen has many advantageous combustion properties including high flame speed, small quenching distance, low ignition energy, high autoignition temperature, and high octane. Table 28.1 shows a comparison between some basic properties for hydrogen and for other traditional SI ICE fuels.

Combustion of hydrogen–air mixtures generates low criteria pollutants with only NOx emissions forming with high combustion temperatures and HC due to residual oil combustion [1,3]. Due to the high octane number allowing high compression ratios, hydrogen-fueled SI ICE engines can achieve high indicated efficiency of up to 52% [3]. Hydrogen-fueled vehicles are on average 22% more efficient than gasoline vehicles [1]. A naturally aspirated engine with a given displacement and compression ratio using a stoichiometric hydrogen–air mixture as fuel will have a comparable thermal efficiency but will have 20% reduced power compared to the same engine using gasoline as the fuel [1].

TABLE 28.1

Properties of Common Fuels

Property	Hydrogen (H_2)	Methane (CH_4)	Methanol (CH_3OH)	Gasoline (C_6–C_{12})
LHV specific energy (MJ/kg)	120	48	20	42–44
Energy density at STP (MJ/N m^3)	11	35	15,700	≈32,000
Maximum flame speed (m/s) [4]	2.8	0.37–0.45	0.45 [5]	0.37–0.43
Flammability limits, %vol in air	4.1–74	5.3–15	6–36.5	1.4–7.6
Explosion limits, %vol in air	18.2–58.9	5.7–14	6.7–36	1.4–3
Lean limit equivalence ratio in air	0.1 [4]	0.5	0.57[b]	0.58 [4]
Minimum ignition energy (mJ)	0.02 [4]	0.29 [4]	0.21 [8]	0.24 [4]
Autoignition temperature (°C)	571	632	470	220
Octane (RON) [5]	>130 [4,6]	120	106	92–98
Molecular diffusion coefficient (cm^2/s), in air	0.61	0.16	0.13	0.05
Energy density of stoichiometric mixture with air (kJ/L)	3.23	3.31	3.56	≈3.84[a]

Source: Adapted from Verhelst, S. et al., *Proc. Inst. Mech. Eng. Part D-J. Automob. Eng.*, 221(D8), 911, 2007.

[a] For gasoline average composition $C_8H_{14.96}$.

[b] Lean misfire limit [7].

Experimental work has shown that a hydrogen-fueled SI ICE can operate down to 30% power without throttling due to the wide flammability limits of hydrogen [1]. Hydrogen has potential for increases of 15%–50% in thermal efficiency than gasoline in a fully optimized engine, depending on various factors [1].

The low emissions and high efficiency of hydrogen-fueled SI ICEs make replacement of fossil fuels with hydrogen an interesting long-term option.

28.1.3 Hydrogen Displacing Traditional Fuels

In the near term, hydrogen combustion is unlikely to completely replace fossil fuel combustion for most applications. The factors limiting the near-term replacement of fossil fuels by hydrogen include the greater cost of hydrogen compared to most fossil fuels, the cost and difficulty of storing hydrogen with high energy density, the lack of an extensive distribution infrastructure, as well as real and perceived safety issues [4]. The potential for hydrogen enrichment of standard fuel–air mixtures is an interesting alternative to a complete replacement of standard fuels.

28.1.4 Mixing Fuels to Change Combustion Characteristics

Mixing fuels to achieve desired combustion characteristics is a common practice. The most common vehicle fuels, gasoline and diesel, are selected fractions of the distillation and oil refining process. These blends of HCs are tailored for reduced emissions, improved performance, reliability, and reduced cost.

In ICEs, the addition of lead or isooctane has been used to control autoignition/knock properties, and the addition of oxygenates such as MTBE or ethanol has been used to reduce carbon monoxide and HC emissions.

28.1.5 Hydrogen Enrichment

The addition of hydrogen to a standard air–fuel mixture has been referred to as hydrogen enrichment, hydrogen bifueling, hydrogen boosting, hydrogen supplementation, and dual charge engine, when used with NG as Hythane, among other names. In this chapter, we will refer to the addition of hydrogen to an air–fuel mixture as hydrogen enrichment when the hydrogen adds less than 30% of the chemical energy contained in the primary fuel. In cases where hydrogen provides more than this amount of the total fuel energy, the term hydrogen enrichment is not the best classification, and other terms such as reformed fuel engine, bifueling, or hydrogen fuel with supplemental secondary fuel may be more appropriate.

As a gas, hydrogen is virtually insoluble in common liquid fuels so hydrogen enrichment is typically achieved by mixing hydrogen with air during the intake stroke [4]. When used with gaseous fuels, hydrogen enrichment can be achieved by premixing hydrogen and the primary fuel or by simultaneously adding separate streams of hydrogen and the primary fuel to the intake air [4].

28.1.6 Hydrogen Production

Hydrogen can be produced from many different primary energy sources via a number of different processes. Common hydrogen production processes include thermochemical reformation of liquid or gaseous HCs, electrolysis of water, and gasification or pyrolysis of biomass or coal [2,5]. The hydrogen production techniques will be selected to match the application, primary energy sources, and other constraints. A brief description of the types of relevant hydrogen production techniques for hydrogen enrichment follows with more detailed descriptions in Section 28.6.

Reformation processes for hydrogen production react on a primary fuel with water, air, or both to generate hydrogen-rich gas mixtures that can include carbon monoxide, carbon dioxide, nitrogen, methane, and traces of other gases. The hydrogen can be separated from the hydrogen-rich gas mixture, or the entire mixture can be added to the air–fuel mixture to be enriched.

Electrolysis processes use electrical energy to split water into hydrogen and oxygen, both of which can be added to an air–fuel mixture.

Reformation processes integrated with engines can utilize waste heat from the engine to make the reformation process more efficient. This type of thermal integration is possible for both onboard mobile and stationary engine applications. Applications should consider the additional weight and volume required for these processes, and efforts are underway to minimize the weight, volume, and energy requirements of these hydrogen generation systems.

28.1.7 Measuring Hydrogen Amounts

When enriching a fuel–air mixture with hydrogen, the amount of hydrogen added can be described in several different ways. The amount of hydrogen can be described in volume percent ($\%_{vol}$), mass percent ($\%_{mass}$), and energy percent ($\%_{ener}$). The percent descriptive parameters can be referenced to the relevant parameter for the total mixture, that is, air plus primary fuel plus hydrogen, or they can be referenced to the unenriched mixture, that is, air plus primary fuel, or they can be referenced to either the air only or fuel only. This variety of descriptive parameters and variety of references lead to ambiguity in many comparisons, and care must be taken to ensure consistent treatment of data from different sources.

In the combustion chamber or combustion zone, the actual concentration of hydrogen will be different from the concentration in the intake due to mixing with burned gas zones, residual exhaust gases, or gases from other inlets. Since the quantity of these other gases can depend upon the design of the combustion device as well as on operating parameters, the actual in-cylinder hydrogen concentration can be more complicated to quantify. For instance, in ICEs, the amount of residual exhaust gases in the cylinder during the intake stroke depends upon engine design and varies over a range of operating parameters, so the actual concentration of hydrogen in the cylinder will depend upon a number of design and operating condition variables.

28.1.8 Literature Reviews

There are three substantial reviews in the hydrogen enrichment of SI ICEs area. Jamal and Wyszynski review the application of hydrogen enrichment to gasoline-fueled SI ICEs including a discussion of onboard reformation techniques [1]. Pettersson and Sjöström compiled a substantial review of reformed methanol-fueled SI ICE operation including the addition of hydrogen-rich reformate gases to methanol–air mixtures and discussion of onboard reformation with waste heat recovery [6]. Akansu et al. review hydrogen enrichment of NG in SI ICEs [7]. There are large bodies of experimental work covering the large number of combustion applications and associated fuels across a range of operating conditions. This leads to a large number of variables in comparing the results of hydrogen enrichment experiments.

This chapter does not intend to offer an exhaustive review of the full breadth of this body of work but merely a starting point and summary of the major conclusions found.

28.2 Effects of Hydrogen Enrichment

28.2.1 Theory/Fundamental

A number of fundamental experiments and modeling efforts have addressed the effects of hydrogen enrichment on combustion processes [8–35]. The changes in combustion from hydrogen enrichment are due to the fundamental combustion properties of hydrogen including rapid free radical generation, low ignition energy, high diffusivity, and high autoignition temperature. Generally hydrogen enrichment increases flame speed, increases heat release/burn rate, increases flame stability, extends the lean limit, increases exhaust gas circulation (EGR)/dilution tolerance, increases flame penetration toward cylinder walls and into crevices, and reduces cycle to cycle variation.

The effects of hydrogen enrichment depend upon the type of combustion process, the design of the specific combustion device, operating parameters, and type of primary fuel.

The potential beneficial results of hydrogen enrichment of SI ICEs include reduced engine-out emissions. Hydrogen enrichment enables lean or high EGR operating modes with large decreases in NOx emissions, reductions in CO emissions, maintained low HC emissions, and increased efficiency. Some potential drawbacks to hydrogen enrichment can include decreased power, increased NOx (when lean operation is not used), and preignition issues.

The effects of hydrogen enrichment depend upon a combination of operating parameters and design parameters. This dependence upon a combination of a large number of variables leads to a great number of different possible conditions. Within this parameter space, there are regions where hydrogen enrichment is beneficial, regions with little difference, and regions where hydrogen enrichment can be detrimental. For this reason, many research groups have found benefits from hydrogen enrichment over some set of parameters with detriments over others. It is possible to carry out careful experiments that show hydrogen enrichment to not have beneficial effects if the parameter space explored remains in the region of detrimental effects. For instance, if equivalence ratio, spark timing, and compression ratio are not changed to achieve appropriate combustion conditions for the new mixture, the effect of hydrogen enrichment can be detrimental. The full parameter space for practical operation must be fully explored especially with regard to equivalence ratio. There are sufficiently large regions of practical interest where hydrogen enrichment produces beneficial effects.

28.2.2 Effects of Hydrogen Enrichment on SI ICEs

28.2.2.1 Effects on Efficiency

Hydrogen enrichment changes the peak thermal efficiency of SI ICEs with reductions in some cases and increases of up to 50% depending on the operating conditions and engine design.

28.2.2.1.1 Increased Free Radical Generation

Hydrogen combustion has well-known free radical production reactions [36–38]. The composition of the radical pool in HC reaction systems is determined by the kinetics of reactions involving H, O, OH, HO_2, and H_2O_2 [37]. Table 28.2 lists some example reactions involving hydrogen in the initiation, chain branching, chain propagation, and chain termination steps.

Increasing the concentration of hydrogen increases the rate of production of free radicals via reactions 1, 2, 3, and 4, as well as increasing the rate of the chain reaction or chain propagation step, reaction 6.

Hydrogen enrichment provides a larger pool of H, O, and OH radicals [39,40] at an earlier stage in the combustion process. This higher concentration of radicals has significant effects upon the combustion process and overall reaction rate. These effects include extension of the lean limit [20,21,28], increasing EGR/dilution tolerance [41], and shortening of the flame nucleation period [28], thereby increasing the heat release/burn rate.

In ICEs, combustion occurs at high pressure with higher concentrations of all gaseous species and therefore with increased rates of third-body termination reactions that reduce the concentration of free radicals [36,39]. This reduction in free radical concentration leads to lean limits at higher equivalence ratios and lower EGR/dilution tolerance [36,39]. Enrichment with hydrogen shifts this trend with increasing pressure with lean limits at lower equivalence ratios and higher EGR/dilution tolerance compared to an unenriched system [39].

28.2.2.1.2 Shortened Flame Nucleation Period

Hydrogen enrichment has been found to decrease the flame nucleation period in many combustion processes [28,39]. Shortening the flame nucleation period increases the overall burn rate. In ICEs, this increase in burn rate increases efficiency, reduces cycle to cycle variation, extends the lean limit, and increases EGR tolerance [36].

For ICEs, a hypothetical increase in burn rate that leads to a decrease in burn duration from 100° to 60° is predicted to result in a 4% decrease in brake-specific fuel consumption (bsfc) [36].

TABLE 28.2

Example Reactions Involving Free Radicals in the Hydrogen–Air System

Initiation	
$H_2 + M \rightarrow H + H + M$ (very high temperatures)	(1)
$H_2 + O_2 \rightarrow HO_2 + H$ (other temperatures)	(2)
Chain branching	
$H + O_2 \rightarrow O + OH$	(3)
$O + H_2 \rightarrow H + OH$	(4)
$O + H_2O \rightarrow OH + OH$	(5)
Chain reaction	
$H_2 + OH \rightarrow H_2O + H$	(6)
Chain termination reaction	
$H + O_2 + M \rightarrow HO_2 + M$	(7)

Source: Turns, S.R., *An Introduction to Combustion*, 2nd edn., McGraw-Hill, Singapore, 2000.

28.2.2.1.3 *Increased Flame Speed and Complete Combustion*

Hydrogen is well known to have a flame speed several times higher than most common fuels; see Table 28.1.

Hydrogen-enriched mixtures show increased flame speeds in proportion with the amount of hydrogen added. Hydrogen's high diffusivity, high thermal conductivity, and production of higher concentration of free radicals all contribute to the measured increases in flame speed.

As fuel–air hydrogen mixtures are made leaner or more dilute, the overall flame speed decreases but remains higher than the unenriched mixture.

In ICEs, faster flame speed ensures that the flame traverses the entire combustion chamber before the exhaust valve is opened, reduces partial burn or misfire events, and ensures more complete combustion [36].

28.2.2.1.4 *Extension of Lean or Dilute Limit*

Hydrogen enrichment has been shown to extend the lean limit and increase EGR/dilution tolerance in many different engines with many different fuels. The lean limit or EGR tolerance will lie between that of pure hydrogen combustion and that of the primary fuel [1]. This effect is not linear and addition of small amounts of hydrogen can lead to significant extension of the lean limit [1].

The effect of lean or dilute operation depends upon combustion chamber design [36]. Once again, hydrogen enrichment will not change the general trends with different designs but will typically shift the trends toward higher efficiency.

Without hydrogen enrichment, operation at moderately lean equivalence ratios increases efficiency due to more available oxygen supporting more complete combustion [36]. Operation near the lean limit leads to reduced efficiency due to increased cycle to cycle variations and reduced heat release rate [36].

Hydrogen enrichment extends the lean limit enabling stable operation at leaner equivalence ratios than possible with an unenriched mixture. Operation with ultralean equivalence ratios can further increase efficiency if stable, and complete combustion is maintained as is enabled by hydrogen enrichment.

Lean and ultralean operation increases efficiency by reducing heat losses to the cylinder walls, reducing dissociation, and increasing the ratio of specific heats of the gas mixture during the expansion process [1].

Increasing EGR has been shown to reduce the flame speed and reduce the heat release rate [36,39]. Increasing EGR leads to increased efficiency until the EGR tolerance limit is approached and efficiency is reduced due to increased cycle to cycle variations, reduced heat release rate, and misfires or incomplete combustion [36]. Hydrogen enrichment does not change this general trend but increases the initial flame speed, thereby extending the EGR tolerance. Similar to the case with lean operation, hydrogen enrichment enables operation with high EGR rates achieving a higher efficiency than possible without hydrogen enrichment.

As with ultralean operation, highly dilute operating regimes significantly reduce peak combustion temperatures. Reduced peak combustion temperatures can lead to reduced heat losses to chamber walls, decreased exhaust temperatures, and decreased dissociation throughout the process, all of which lead to increased efficiency.

28.2.2.1.5 *Effect on Cycle to Cycle Variation*

Hydrogen enrichment has been shown to reduce cycle to cycle variation in many different types of ICEs with many different fuels in lean or dilute operating conditions.

Spark timing is set for the average cycle, so reducing cycle to cycle variation reduces losses for *nonaverage* cycles and increases engine efficiency [36].

28.2.2.1.6 Load Control/Throttling

In SI ICEs, throttling is used to control load by reducing the total mass charge inducted into the engine. Hydrogen enrichment extends the lean limit or dilution tolerance reducing the need for throttling at part load, thereby reducing pumping losses and increasing efficiency. In real vehicle drive cycles, a reduced need for throttling can significantly increase efficiency and therefore increase vehicle mileage.

Experimental work has shown that hydrogen-fueled ICEs can operate down to 30% power without throttling [1]. Hydrogen-enriched fuel–air mixtures are expected to achieve throttle-free operation down to loads between those posed by the primary fuel and those achieved by a hydrogen-fueled engine.

28.2.2.1.7 Knock Resistance

There are some discrepancies in the findings of different groups as to whether hydrogen enrichment increases octane and improves knock resistance. The unique properties of hydrogen with high flame speeds, low ignition energy, and high autoignition temperatures can sometimes lead to combustion with acoustic presentation similar to knock and rapid pressure rise without the autoignition of unburned *end gas* that signifies real knock phenomenon. These characteristics have led some to claim that octane number is not an appropriate measure of knock resistance for hydrogen combustion. A different measure of the knock resistance of gaseous fuels called the methane number has been developed [42] and is used to predict knock but has been criticized due to the relatively unusual combustion properties of hydrogen–methane mixtures [43]. Karim et al. have explored knock extensively in dual-fueled engines, where the majority of fuel energy is in the form of hydrogen, and direct injection (DI) of liquid fuels is used as an ignition source [8,43–45].

With hydrogen enrichment via hydrogen-rich gas addition with significant concentrations of CO, knock resistance can show different trends compared to enrichment with pure hydrogen [8].

Overall hydrogen enrichment with hydrogen concentrations below 20% by volume has shown similar knock resistance as the primary fuel at the same equivalence ratio and EGR percentage.

Ultralean or ultradilute engine operation enabled by hydrogen enrichment can significantly increase knock resistance allowing higher compression ratios or higher intake pressures leading to significantly increased efficiency [1].

28.2.2.2 Effects on Emissions

There are multiple strategies for reducing emissions from ICEs. Two of the main strategies are exhaust aftertreatment and lean or dilute operation. The most common exhaust aftertreatment technology for SI engines is the three-way catalytic converter that requires near stoichiometric operation to effectively reduce NOx, HC, and CO emissions. An alternative to aftertreatment is to avoid production of emissions through sufficiently lean or dilute combustion while retaining near-complete combustion. Methods to achieve complete combustion with mixtures sufficiently lean or dilute to greatly reduce NOx traditionally require optimized spark timing and combustion chamber design as well as high turbulence [46]. High turbulence leads to increased heat transfer to the cylinder walls and higher intake pressure drop, decreasing volumetric and thermal efficiency [46]. Hydrogen enrichment is an alternative to high turbulence for enabling complete combustion with lean or dilute mixtures.

28.2.2.2.1 Effect on NOx Formation

Hydrogen enrichment of stoichiometric nondilute mixtures can increase peak temperatures causing moderate increases in NOx emissions [1]. For this reason at a fixed equivalence ratio, increasing hydrogen addition leads to increased NOx formation [1].

Hydrogen enrichment enables the use of ultralean or dilute mixtures reducing the peak combustion temperature to the point where NOx formation is greatly reduced [1,6,7]. NOx formation can be reduced by as much as 98% in some ICE applications when the engine is operated close to the dilution limit [47].

28.2.2.2.2 Effect on CO Emissions

There are some discrepancies in the findings of different groups as to whether hydrogen enrichment increases or decreases CO emissions of stoichiometric nondilute mixtures. Results appear to depend upon the type of primary fuel, operating conditions, and engine design parameters.

With stable lean operation enabled by hydrogen enrichment, CO emissions can be reduced compared to the unenriched case for many fuel types over a range of operating conditions [1,6].

With stable dilute operation with high EGR enabled by hydrogen enrichment, CO emissions can be reduced in some cases but once again show different trends depending upon the type of primary fuel, operating conditions, and engine design parameters.

28.2.2.2.3 Effect on HC Emissions

Hydrogen enrichment of stoichiometric nondilute mixtures tends to moderately reduce HC emissions [1]. The reduction in HC emissions has been linked to promotion of more complete combustion and a reduction in quenching distance that allows flame propagation nearer to the cylinder walls and further into crevices.

Hydrogen enrichment reduces the quenching distance for the fuel–air hydrogen mixture. The smaller quenching distance of hydrogen-enriched flames leads to a smaller volume of unburned fuel–air mixture caused by flame quenching at the chamber walls, thus reducing emissions of unburned HC. This reduction in quenching distance can lead to increased heat transfer to the chamber wall and burning of lubricating film in that location that potentially increases HC emissions from the lubricant.

ICEs operating on lean mixtures show reduced HC emissions until equivalence ratios reach low enough levels that incomplete combustion, misfire, and partial burn sharply increase HC emissions [36].

In ICEs operating in lean regimes, hydrogen enrichment reduces the incidence of misfire or partial burn events, thereby reducing unburned HC emissions compared to an unenriched mixture at the same equivalence ratio [1]. Hydrogen enrichment also enables stable ultralean operation with complete combustion achieving lower HC emissions than possible with the unenriched mixture [1,6,7].

For ICEs operating in dilute regimes, hydrogen enrichment reduces the incidence of misfire or partial burn events, thereby significantly reducing unburned HC emissions compared to an unenriched mixture at the same dilution level [1]. Hydrogen enrichment also enables stable ultradilute operation with complete combustion achieving lower HC emissions than possible with the unenriched mixture [1,6,7].

Overall hydrogen enrichment tends to decrease HC emissions.

28.2.2.2.4 Use with Aftertreatment

The most common aftertreatment technology used with SI ICEs is the three-way catalytic converter. Three-way catalytic converters reduce NOx to N_2 using CO or HC, as well as oxidize CO and HC to CO_2 and H_2O using small amounts of residual oxygen in the exhaust [36]. For the simultaneous efficient conversion of emissions, the three-way catalyst requires high temperatures, on the order of 500°C, and exhaust gases with little excess oxygen, that is, near stoichiometric operation [36]. At operating temperatures with small amounts of excess oxygen in the exhaust gases, three-way catalytic converters can achieve conversion efficiencies of more than 85%, significantly reducing tailpipe emissions [36].

Clearly, the use of lean and ultralean mixtures contradicts the use of a three-way catalyst. The large amount of excess oxygen in the exhaust slows the NOx reduction reactions. Oxidation catalysts can still be used to reduce CO and HC emissions. Oxidation catalysts can achieve conversion efficiencies of 99% for CO and 95% for HC at high temperatures[36]. In this way, ultralean operation enabled by hydrogen enrichment could be used with oxidation catalytic aftertreatment to achieve low emissions.

The use of dilution with EGR with near stoichiometric amounts of air allows the continued use of three-way catalyst aftertreatment to achieve low emissions.

In ultralean and ultradilute operating regimes, the exhaust temperatures can be significantly lower than in the stoichiometric case. The lower exhaust temperatures may reduce the efficiency of catalytic aftertreatment depending upon the specific set of conditions and catalyst types.

28.2.2.3 Potential Disadvantages of Hydrogen Enrichment

28.2.2.3.1 Preignition and Flashback

In SI ICEs operating with hydrogen as the primary fuel backfire, preignition and knock can be issues [3]. Preignition of hydrogen can be caused by hot spots or residual hot gases in the cylinder. Cylinder cooling by liquid evaporation can be lower when hydrogen enrichment is used. There is also the potential for flame flashback into the intake manifold, if hot gases enter the manifold from the cylinder or if preignition flames propagate through the intake valve opening during intake. These issues can be greatly reduced by eliminating valve overlap and using lean or dilute mixtures [1].

Hydrogen-enriched mixtures of other primary fuels and air have less risk of these abnormal combustion events due to the fact that they have slower flame speeds, higher ignition energies, and larger quenching distances compared to hydrogen–air mixtures. The abnormal combustion problems can happen in hydrogen enrichment for cases where large amounts of hydrogen are added [48]. In hydrogen enrichment cases using lower concentrations of hydrogen, these abnormal combustion events are unlikely. Furthermore, in hydrogen-enriched ICEs operating in lean or dilute regimes, preignition and flashback are even more unlikely.

28.2.2.3.2 Reduced Power

The use of a gaseous fuel in premixed ICE applications reduces power output compared to liquid fuels due to the larger volume of air displaced by the gaseous fuel [1]. For this reason, even ICEs operating near stoichiometric air–fuel ratios typically produce slightly less power when using hydrogen enrichment. The degree of derating depends upon the amount of hydrogen added and the difference in volumetric energy density between the stoichiometric hydrogen–air mixture and the stoichiometric primary fuel–air mixture. For instance, since a stoichiometric mixture of hydrogen and air has a volumetric energy density approximately 16% lower than a gasoline air mixture, displacing air with hydrogen leads to an overall decrease in the mixture volumetric energy density proportional to the amount of hydrogen added. Enrichment with hydrogen containing 25% of the total fuel energy would reduce the total energy inducted in the charge by approximately 4%.

The increases in efficiency associated with hydrogen enrichment at many operating conditions partially compensate for the reduced overall volume energy density leading to either a net decrease or a net increase in power output in different applications. Clearly, ICEs using ultralean or ultradilute mixtures even with hydrogen enrichment produce significantly less power due to the significant reduction in fuel inducted per cycle. When operating with lean or dilute mixtures to achieve low NOx emissions, the engine power output is reduced proportionally to the amount of excess air or diluent added. Once again, increases in efficiency associated with hydrogen enrichment at many operating conditions partially compensate for the reduced overall volume energy density caused by lean or dilute operation.

28.3 Experimental Results in Different Types of Combustion Applications

There are a number of distinctly different processes by which fuels can undergo combustion. These processes include hot flames, cool flames, premixed flames, diffusion flames, and homogeneous autoignition. Depending upon the application, combustion may occur through different processes and the effects of hydrogen enrichment would therefore be expected to be different as well.

28.3.1 Internal Combustion Engines

Hydrogen enrichment has been explored with many different kinds of ICEs including SI, CI, SI–DI, as well as HCCI or PCCI systems. Combustion processes are fundamentally different in SI, SI–DI, CI, and HCCI/PCCI ICEs. For this reason, the effects of hydrogen enrichment can be distinctly different in these different types of ICEs.

Hydrogen enrichment has been explored with each of these types of engines operating with a number of different primary fuels across a range of operating parameters and using various engine design parameters. This chapter focuses on hydrogen enrichment in SI ICEs with brief mention of hydrogen enrichment in other applications.

In SI ICEs, there are a number of different design parameters that affect engine performance. This chapter will describe influences of each of the major engine design parameters on hydrogen enrichment effects.

The effects of hydrogen enrichment of SI ICEs are sorted by primary fuel type. For more detailed treatment of specific cases with the effects of engine design parameters and operating parameters, please see the cited references.

28.3.2 SI ICE

28.3.2.1 Process Description

In SI ICEs, the air and fuel are premixed, inducted into the cylinder, compressed, and then ignited via a spark. This process leads to a turbulent flame front traversing the cylinder volume extinguishing near the chamber walls and piston surface [36]. In SI ICEs, combustion typically takes place through the initiation and propagation of hot turbulent flames.

In this application, hydrogen is often fumigated or injected into the intake air. The primary fuels, hydrogen and air, are mixed before they enter the combustion chamber. In this way, the unique properties of hydrogen combustion influence the flame nucleation and propagation throughout the entire combustion process.

28.3.2.2 SI ICE Experimental Results

There are complex trade-offs between thermal efficiency, emissions, and power for ICEs. Without hydrogen enrichment, these trade-offs are met by adjusting the operational parameters of spark timing, throttle position, equivalence ratio or EGR, as well as design parameters such as compression ratio, intake pressure, combustion chamber shape, valve timing, and aftertreatment techniques. Hydrogen enrichment opens up new operating regimes not accessible by unenriched mixtures, thereby significantly changing the optimization strategy. For a given emissions level, an engine operating with hydrogen enrichment can often achieve higher efficiency without changes in the engine design. Further increases in efficiency could be achieved by optimizing engine design for hydrogen enrichment with lean or dilute mixtures through increased compression ratio or boost pressure and thermal integration of hydrogen production, among others.

Figure 28.1 shows the new operating regimes accessible by hydrogen enrichment of gasoline with a hydrogen-rich gas, adapted from [49].

Overall hydrogen enrichment leads to increases in flame speed, flame stability, shortening of the flame nucleation period, and extension of the lean or dilute limits in SI ICEs operating over a range of conditions with many different primary fuels. The shortening of the flame nucleation period and increased flame speed from hydrogen enrichment lead to a faster burn rate and faster pressure rise approaching the ideal Otto-cycle constant-volume heat addition step and increasing thermal efficiency.

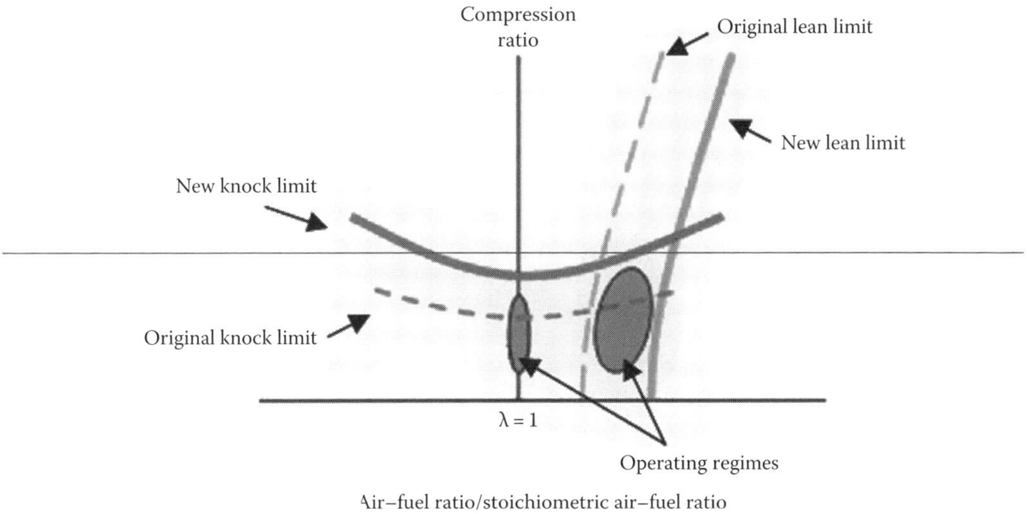

FIGURE 28.1
Hydrogen enrichment enables new operating regimes. (Adapted from Beister, U.-J. and Smaling, R., Hydrogen-enhanced combustion engine could improve gasoline fuel economy by 20% to 30%, MTZ worldwide edition No. 2005-10 2005, October 4, 2005 [cited October 13, 2008], Available from http://www.all4engineers.com/index.php;do=show/alloc=3/lng=en/id=2866/sid=90458fbceb42147dd47b232709a5101c.)

Even within the subset of SI ICEs, the effects of hydrogen enrichment depend upon operating parameters and design parameters. The combination of all of these variables leads to a great number of different possible conditions. Within this parameter space, there are regions where hydrogen enrichment is beneficial, regions with little difference, and regions where hydrogen enrichment can be detrimental.

The complex trade-offs and interactions between operating parameters and design parameters make it impractical to make broad summaries of the effects of hydrogen enrichment. Each experimental case offers a piece of the puzzle that can be fit together to understand the effects of hydrogen enrichment. The following summaries of experimental results are intended to be illustrative of the typical findings for each type of fuel.

28.3.2.2.1 Natural Gas/Methane

NG is a mixture of mostly methane with other short-chain gaseous HCs and a small fraction of inert gases. As a gaseous fuel, NG or pure methane is a promising primary fuel for hydrogen enrichment. A large number of groups have studied the hydrogen enrichment of NG or pure methane [4,7,8,43,46,50–59].

NG is an attractive motor fuel because of the potential for lower emissions, availability, existing distribution infrastructure in some countries, and relatively low cost.

Hydrogen enrichment can enhance the performance characteristics of NG-fueled SI ICEs by increasing power output and efficiency and reducing emissions [46,52].

Gaseous primary fuels have lower total power output in ICEs compared to liquid fuels because the larger volume of the gaseous fuel displaces more intake air. The addition of hydrogen to a mixture of air and gaseous fuel only increases the amount of air displaced by a small amount and therefore does not significantly reduce engine power in this way. The increases in efficiency from hydrogen addition to NG by way of the mechanisms listed previously can more than ameliorate the small reduction in intake mixture energy density, thereby increasing the overall ICE power output [46].

Several studies have confirmed that hydrogen addition of 15–30$\%_{vol}$ of the NG volume increases brake thermal efficiency and reduces bsfc for lean fuel mixtures [7,46,50,52,56]. Under lean conditions, hydrogen enrichment is also shown to increase BMEP [7].

Increasing the concentration of hydrogen enrichment increases the speed of both the flame initiation and flame propagation [46]. The trade-offs between spark timing, thermal efficiency, torque, and emissions are significantly changed by hydrogen enrichment [46].

CO, HC, and CO_2 emissions are reduced by hydrogen enrichment of NG [7].

Hydrogen addition increases NOx concentration at a fixed equivalence ratio but enables significantly leaner operation without penalties to efficiency and therefore can be used to reduce NOx emissions [46,52,56]. In an example case, hydrogen enrichment of 20$\%_{vol}$ extends the lean limit from an equivalence ratio of 0.63–0.54 [7]. Extremely low NOx emissions can be achieved at ultralean conditions enabled by hydrogen enrichment [55,56].

Hydrogen enrichment also increases dilution tolerance enabling high EGR [59]. Allenby et al. show that hydrogen-rich gas addition can enable EGR rates of 25% while maintaining a COV of indicated mean effective pressure (IMEP) below 5% and achieving a reduction in NOx emissions greater than 80% for all cases tested [59].

Overall hydrogen enrichment enables lean or dilute operating conditions that can achieve low emissions across a range of equivalence ratios and operating conditions [52,55].

Large amounts of hydrogen addition to NG can lead to a broader operational region with knock issues [43]. Hydrogen enrichment with hydrogen addition less than 25$\%_{vol}$ maintains the excellent knock-resistant qualities of methane [46,60] (Figure 28.2).

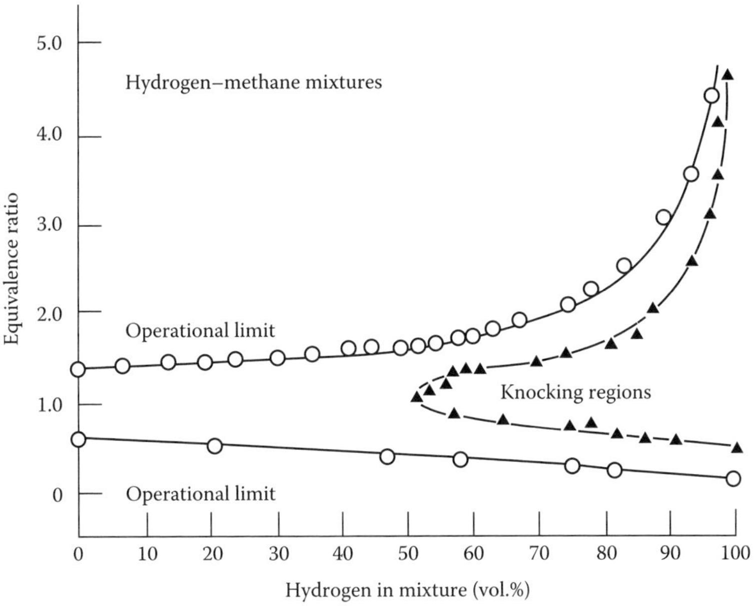

FIGURE 28.2
Knock in hydrogen-enriched NG SI ICE, 6:1 compression ratio. (From Karim, G.A. et al., *Int. J. Hydrogen Energy*, 21(7), 625, 1996.)

28.3.2.2.2 Gasoline

A large number of investigations have been carried out with hydrogen enrichment of gasoline [6,47,61–71]. This is likely due to the popularity of gasoline as a vehicle fuel and the generally good results of hydrogen enrichment.

A substantial review of gasoline hydrogen enrichment including onboard reformation for hydrogen-rich gas production has been compiled by Jamal and Wycinski [1].

Overall hydrogen enrichment of gasoline

- Extends the lean limit enabling reduced NOx and CO emissions and increased thermal efficiency
- Reduces cycle to cycle variation (COV of IMEP)
- Increases the apparent flame speed of the mixture
- Reduces maximum power
- Is still subject to the trend of increased HC emissions with decreasing equivalence ratio for a fixed amount of hydrogen addition
- Decreases HC and increases NOx emissions with increasing hydrogen addition at a fixed equivalence ratio

Without hydrogen, NOx emissions are still unacceptably high when operating near the lean limit for gasoline [1]. Hydrogen enrichment extends the lean limit, thereby enabling operation at ultralean conditions with low NOx emissions and reduced CO emissions [1].

Hydrogen enrichment of gasoline increases the flame speed significantly at all equivalence ratios with large increases near the lean operation limits. In an example case, near the lean limit for gasoline at an equivalence ratio of 0.66, adding 16.8%$_{ener}$ hydrogen increased the apparent flame speed by 61% [62].

Hydrogen enrichment significantly reduced the ignition delay period as measured by the 0%–2% mass fraction burn duration and significantly reduced cycle to cycle variation, particularly with lean equivalence ratios [72].

Lucas and Richards found that hydrogen enrichment increased part load thermal efficiency at wide open throttle resulting in a 30% reduction in bsfc due to reduced pumping losses and reduced heat loss to the cylinder [65].

Sher and Hacohen found that with hydrogen enrichment of 5–15%$_{ener}$, bsfc could be reduced by 10%–20% [69].

Hydrogen enrichment of gasoline using hydrogen-rich gas addition has been explored by a number of groups [64,68,70]. Overall hydrogen-rich gas addition has been found to have similar results to pure hydrogen addition. Beneficial effects on combustion have been attributed to both the hydrogen and CO in the hydrogen-rich gas with inert components such as N_2 or CO_2 acting as small additions of diluent.

Quader et al. from Delphi showed that enrichment of premixed gasoline and air with reformate (simulated partial oxidation reformate 21% H_2, 24% CO) improves combustion initiation and burn rate and extends EGR tolerance and lean limit [68]. When operating at the dilute limit with hydrogen enrichment, NOx emissions were reduced greatly with the same or slight reductions in HC emissions [68]. When operating near the lean limit, emissions of NOx were reduced and efficiency increased by up to 7% with addition of 15%$_{ener}$ reformate [68].

Yüksel et al. show that hydrogen enrichment reduces heat losses to cooling water by up to 36% and unaccounted heat loses by up to 30%, while heat losses to the exhaust remain the same [71].

28.3.2.2.3 Methanol

Pettersson and Sjöström compiled a substantial review of reformed methanol-fueled SI ICE operation including the addition of hydrogen-rich reformate gases to methanol–air mixtures and discussions of onboard reformation with waste heat recovery [6].

The experiments reviewed in this work used large amounts of hydrogen or decomposed methanol addition.

Overall hydrogen enrichment of methanol-fueled SI ICE engines showed an increase in efficiency over operation on pure methanol of up to 18% [6].

Experiments using bottled hydrogen and CO for hydrogen enrichment of methanol found increases in efficiency of up to 25% [6].

Experiments using decomposed methanol (primarily H_2 and CO) for hydrogen enrichment of methanol found increases in system efficiency of up to 42% [6].

The largest efficiency gains were found at part load using lean mixtures [6].

Hydrogen enrichment was shown to enable significant reductions in aldehyde, NOx, CO, and HC emissions, depending upon the specific operating parameters [6].

Tiashen et al. found that hydrogen enrichment of methanol improved combustion characteristics, increased flame propagation rate, shortened the ignition delay period, and increased thermal efficiency [73].

In methanol-fueled engines, cold start at low ambient temperatures can be difficult due to the high heat of vaporization of liquid methanol [6,74]. Incomplete vaporization in the intake leads to lean and less combustible mixtures in the cylinder [6,74]. The addition of hydrogen or a hydrogen-rich gas significantly improves the cold start performance by extending the lean limit in the mixture and greatly enhancing the flammability [6,74].

28.3.2.2.4 Ethanol

Al-baghdadi and Yousufuddin et al. have investigated hydrogen enrichment of ethanol primary fuel [75–77]. Al-baghdadi found hydrogen enrichment of ethanol to improve combustion efficiency, increase thermal efficiency, increase power, reduce bsfc, and reduce toxic emissions [75].

Al-baghdadi found hydrogen enrichment of ethanol to increase NOx emissions and reduce CO emissions, at a fixed compression ratio (CR) and equivalence ratio [75]. The NOx emissions produced by hydrogen-enriched ethanol fuel with up to $3.5\%_{mass}$ H_2 at 12 CR were less than the NOx emissions produced by gasoline fuel at 7 CR [75].

Al-baghdadi found hydrogen enrichment of ethanol to increase brake power up to additions of $2\%_{mass}$ H_2 [75].

Yousufuddin et al. found hydrogen enrichment of ethanol to slightly decrease brake power, increase thermal efficiency, increase pressure rate, and increase heat release rate, across a range of hydrogen addition amounts with a range of compression ratios [76]. Hydrogen enrichment was found to slightly decrease power up to $60\%_{vol}$ H_2 with a decrease in brake power from 3.24 to 3.18 kW with the addition of $20\%_{vol}$ H_2 at a CR of 11, maximum brake torque spark timing, and wide open throttle [76].

Hydrogen enrichment was found to increase brake thermal efficiency up to a maximum with $60\%_{vol}$ H_2. Addition of $20\%_{vol}$ H_2 increased brake thermal efficiency from 23.8% to 25.4%. Hydrogen enrichment was found to slightly decrease power up to $60\%_{vol}$ H_2 with a decrease in brake power from 3.24 to 3.18 kW with the addition of $20\%_{vol}$ H_2 at a CR of 11, maximum brake torque spark timing, and wide open throttle [76].

28.3.2.2.5 Landfill Gas

LFG is comprised of 45%–65% methane with 40%–60% CO_2, as well as nitrogen and other nonmethane organic gases [78].

Hydrogen enrichment of LFG with even small H_2 additions of $3–5\%_{vol}$ extended the operational limits and showed improved performance particularly at the lean and rich limit [78]. Additions of hydrogen also improved the combustion characteristics and reduced cyclic variations of LFG operations especially with the lean and rich mixtures [78].

28.3.2.2.6 SI ICE Summary

There are significant regions of practical SI ICE operating and engine design parameter space where hydrogen enrichment can reduce emissions and increase thermal efficiency with a variety of different fuels.

28.4 Hydrogen Enrichment of Other Types of ICEs

Hydrogen enrichment has been explored with CI ICEs [40,79], HCCI, and PCCI ICEs [80–85], as well as DI–SI ICEs [86,87].

28.4.1 CI ICE Process Description

In CI ICEs, air is inducted into the cylinder, where it is compressed, and then liquid fuel is injected directly into the cylinder. This process leads to a short ignition delay followed by a brief autoignition and premixed flame followed by a longer period diffusion flame surrounding liquid fuel droplets [36]. In standard CI ICEs outside the region containing the liquid fuel droplets, there is only air and no flame can propagate.

Hydrogen enrichment is typically achieved by fumigation of hydrogen into the intake air [40,79].

28.4.1.1 CI ICE Experimental Results

The concentration of hydrogen in the intake air can make a significant difference in the effects of enrichment [40,79].

When the primary fuel is injected into the cylinder containing a hydrogen–air mixture, the effect is dependent on the concentration of hydrogen in the air. If the hydrogen concentration is below the hydrogen–air flammability limit at the cylinder temperature and pressure, then no flame will propagate through the entire cylinder gas volume. Only the hydrogen that interacts directly with the diffusion flame will be combusted during this early period and will affect the combustion properties of the liquid fuel. Hydrogen in the air mixture that does not directly interact with the liquid droplet diffusion flames will be combusted when the cylinder temperature and pressure reach the autoignition temperature of the hydrogen–air-burned gas mixture. This combustion of the hydrogen in the cylinder will increase temperature, decrease oxygen concentration, and increase water concentration, thus affecting the long-duration diffusion flame surrounding the liquid fuel droplets.

For concentrations of hydrogen above the hydrogen–air flammability limit at the cylinder temperature and pressure, it is possible for a premixed flame to propagate through the entire cylinder gas volume after being ignited by the liquid fuel injection. This premixed flame results in higher air temperatures, water concentration, and cylinder pressure, which will influence the liquid droplet diffusion flames.

With large amounts of hydrogen addition, the hydrogen becomes the primary energy source with the directly injected fuel acting as an ignition source or pilot as researched by several groups [43,48,88–95].

The effect of hydrogen enrichment will differ distinctly between these three cases.

Hydrogen enrichment of diesel fuel is selected as an illustrative example in CI ICEs.

28.4.1.1.1 Experimental Results

28.4.1.1.1.1 Diesel In CI engines using diesel fuel and fumigating hydrogen into intake charge air, thermal efficiency is reduced at $10\%_{vol}$ H_2 in intake air and then increases as hydrogen concentration is increased up to $30\%_{vol}$ and to $50\%_{vol}$ [79]. The maximum increase in efficiency is seen with $90\%_{vol}$ H_2 and 65% load giving a specific energy consumption (SEC) of 12.7 MJ/kWh compared to unenriched diesel SEC of 16.7 MJ/kWh [79]. The brake thermal efficiency increases from 22.78% to 27.9% with $30\%_{vol}$ H_2 enrichment [79].

In CI engines, the effect of hydrogen addition is not significant until the hydrogen comprises a large percentage of the total chemical energy input at which point this would be more appropriately called a bifueled engine.

28.4.2 DI–SI ICE

Direct injection–spark ignition (DI–SI) ICEs combine characteristics from traditional SI and CI engines.

28.4.2.1 DI–SI ICE Process Description

In DI–SI engines, fuel is injected directly into the cylinder containing compressed air, and after a short delay, a spark ignites the mixture. In these applications, hydrogen can be fumigated into the air in the intake [86].

Hydrogen enrichment of NG and gasoline fuels are selected as illustrative examples in DI–SI ICEs.

28.4.2.1.1 Experimental Results

28.4.2.1.1.1 Natural Gas/Methane Premixed hydrogen enrichment in a DI stratified methane engine increases efficiency and reduces CO and HC emissions but increases NOx emissions in the part load lean operating region [87]. The increases in NOx can be reduced by retarding ignition timing without reducing the efficiency benefits [87]. Hydrogen enrichment enables lean operation with the same efficiency as nonenriched methane and large reductions in HC, CO, and NOx emissions [87].

28.4.2.1.1.2 Gasoline Hydrogen enrichment (up to $27\%_{ener}$ fuel) to intake air increases combustion efficiency and extends EGR tolerance of gasoline DI with SI [86]. Under some conditions, hydrogen enrichment allows the spark timing to be delayed reducing NOx and HC [86].

28.4.2.2 HCCI/PCCI ICE Process Description

HCCI and PCCI engines aim to achieve autoignition with lower-temperature combustion than SI or CI ICEs. These processes are different than either SI or CI combustion phenomenon.

HCCI and PCCI engines achieve low emissions and high efficiency over a narrow range of operating conditions. HCCI autoignition timing is critical to engine efficiency [83]. Research aims to control the timing of autoignition events over a wider range of operating conditions. One way of controlling autoignition timing is by mixing two fuels with different ignition properties [83]. Hydrogen enrichment has been used to change the autoignition and subsequent cool flameless or cool flame combustion in HCCI and PCCI engines [80–85].

Hydrogen enrichment of dimethyl ether fuel is selected as an illustrative example in HCCI and PCCI ICEs.

28.4.2.2.1 Experimental Results

28.4.2.2.1.1 Dimethyl Ether Shudo et al. have shown that hydrogen enrichment of dimethyl ether leads to improved HCCI engine performance by delaying autoignition [83]. The mechanism they suggest is H_2 consumption of OH radicals during low-temperature combustion delaying the onset of high-temperature combustion [83]. This suggests that the temperature of the combustion process may significantly alter the effects of hydrogen enrichment.

28.4.2.3 Continuous Combustion

Hydrogen enrichment can also be applied to burners [12–35], turbine combustors [2,5,96–99], and other continuous combustion devices.

Hydrogen enrichment extends lean limits and increases dilution tolerance with significant potential benefits in these applications.

28.4.2.3.1 Burners

Burners include small-scale laboratory burners designed to investigate fundamental flame properties to large-scale industrial burners used for process heat in a wide variety of applications. Burners can use premixed, partially premixed, or nonpremixed fuel–air mixtures leading to premixed flames, diffusion flames, or combinations. Burners can use swirl to enhance flame stability with lean or dilute fuel–air mixtures with the potential for reduced NOx, HC, and CO emissions. Each of these types of burner applications is expected to have different effects from hydrogen enrichment.

28.4.2.3.2 Experimental Results

28.4.2.3.2.1 Natural Gas/Methane Higher concentration of OH radicals and concomitant extension of lean stability limits in methane–air burner experiments were found by Schefer using PLIF imaging techniques [28].

28.5 Practical Considerations

28.5.1 Cost of System

Hydrogen enrichment system cost depends significantly on the application and hydrogen source. The costs may include hydrogen storage and delivery or may include a hydrogen production device. Hydrogen could be produced at a central location, distributed to retail outlets, and then delivered to vehicles or stationary applications. Hydrogen could also be produced at the use location by reformation of a fraction of the primary fuel stream or through electrolysis of water using electricity. Centralized versus on-site or integrated hydrogen production strategies involve significantly different costs. Centralized production requires the up-front capital cost for the hydrogen storage equipment and incurs an ongoing variable cost for the delivery of hydrogen. On-site hydrogen production requires an up-front capital cost for the hydrogen production equipment and then most likely a lower ongoing variable cost for the portion of the primary fuel used to generate hydrogen. An integrated hydrogen production strategy such as thermochemical recuperation of waste heat may reduce the variable cost of on-site hydrogen production by increasing the thermal efficiency through the use of waste heat [100]. The potential for lower variable costs with integrated hydrogen production strategy may be offset by potentially higher up-front capital costs due to the need for heat exchangers in addition to the hydrogen production device.

The cost of the hydrogen enrichment system must be weighed against the benefits of primarily reduced emissions and increased combustion energy conversion efficiency.

Currently, for a scenario with hydrogen produced off-site, hydrogen costs more per unit of energy than common bulk fossil fuels and is available in fewer locations.

28.5.2 Impact on Engine Wear

Wear in ICEs is a topic of great debate and little fundamental knowledge [101]. The effect of hydrogen enrichment on engine wear, lubricant lifetime, and engine deposits has yet to be studied in depth.

More rapid pressure rise due to higher burn rates may contribute to accelerated engine wear.

Reduced cylinder temperatures in ICEs are expected to extend lubricant life and reduce engine wear particularly at valves and valve seats. Reduced cylinder temperatures with ultralean or ultradilute operation enabled by hydrogen enrichment may lead to extended lubricant life and reduce engine wear.

Increased water content in the exhaust gases may adversely affect lubricant life, bore wear, and exhaust system life, particularly if the temperature of the combustion products falls below the dew point [101].

Issues such as preignition or knock have been shown to increase engine wear in fossil-fueled systems. It is expected that preignition or knock in hydrogen-enriched systems would have similar effects.

28.5.3 Safety

Hydrogen suffers from both real and perceived safety issues. Due to hydrogen's unique properties, it is prone to leakage, combusts in virtually invisible flames, and is easily ignited over a wide range of concentrations. Also due to hydrogen's unique properties, it dissipates extremely quickly if not confined and creates a flame with low radiant energy that does not heat the surroundings significantly if the flame does not come in direct contact. This combination of properties makes safe use of hydrogen possible but requires different strategies than used with other fuels.

Hydrogen and hydrogen-rich gases are not flammable when not mixed with air or another oxidant. In engine operation, it is important to minimize the volume of fuel that is mixed with air in order to minimize the consequences of unintentional ignition. In hydrogen-enriched fuel mixtures, the reduced ignition energy requires more care to avoid unintentional ignition.

There are safety issues inherent in high-pressure storage of any gas. For this reason, onboard vehicle hydrogen storage systems are likely to be more expensive than low-pressure liquid storage systems.

28.6 Hydrogen Production

Hydrogen can be produced from many different primary energy sources via many different processes. Common hydrogen production processes include thermochemical reformation of liquid or gaseous HCs, electrolysis of water, and gasification or pyrolysis of solid fuels such as biomass or coal [2,5].

Depending upon the application where hydrogen will be used, the available primary energy sources, and other constraints, different hydrogen production techniques will be used.

Reformation processes for hydrogen production generate mixtures of hydrogen and other gases that can include carbon monoxide, carbon dioxide, nitrogen, and methane.

Electrolysis processes generate hydrogen and oxygen, both of which can be added to an air–fuel mixture, but require relatively high-power electrical energy to produce significant quantities of hydrogen.

28.6.1 Thermochemical Reformation

Thermochemical reformation is a process whereby liquid or gaseous HCs are broken up to generate a mixture of hydrogen, carbon monoxide, and carbon dioxide called reformate. Thermochemical reformation processes heat the primary fuel and mix it with steam and/or small quantities of air and then expose the mixture to a catalyst to generate the hydrogen-rich gas or reformate. Depending on the type of fuel being processed and the type of reformation process used, the reformate can include nitrogen, excess water vapor, methane, and traces of other HCs. Hydrogen can be separated and purified for pure hydrogen addition, or the entire reformate gas mixture can be added. When no separation step is taken, the addition of reformate can be referred to as hydrogen-rich gas addition.

With hydrogen enrichment by hydrogen-rich gas addition, relatively inert components such as carbon dioxide and nitrogen are added as well. With small levels of hydrogen-rich gas addition, these inert components may have little effect. With larger levels of hydrogen-rich gas addition, the inert components will act as diluents in many ways similar to exhaust gas recirculation.

28.6.2 Electrolysis

Production of hydrogen via electrolysis can use renewably produced electricity to separate water into hydrogen and oxygen. It would be possible to use renewably produced electricity to produce hydrogen during low-electrical-demand periods, store the hydrogen, and use it to supply electricity when needed and to supply vehicle energy demands. This bold vision for a renewable energy and water cycle is inspiring and one of the drivers for interest in the idea of a hydrogen economy.

Unfortunately, electricity is also an energy carrier requiring significant investment, primary energy use, and emissions to generate. Where an abundance of cheap renewable electricity is available, electrolysis may be able to compete with other sources of hydrogen. If cheap renewable electricity is not available, electrolysis is often more expensive and can be more polluting than other hydrogen production methods on a life cycle basis.

It is unlikely that utilization of electricity produced by ICE generator combinations would be economically or energetically practical to use in electrolysis for hydrogen production [60]. The only cases where using electrolysis to generate hydrogen for enrichment using ICE-generated electricity could be practical are cases where small amounts of hydrogen lead to large increases in efficiency or reductions in emissions.

28.6.3 Alternative Hydrogen Production Methods for Hydrogen Enrichment

There are numerous alternative ways of producing hydrogen either on-board a vehicle, on-site for stationary applications, or in a central location for distribution.

Two alternative options for hydrogen production on-board a vehicle include plasma reformation [47,102] of a portion of the primary fuel stream and the decomposition of a hydrogen containing substance such as sodium borohydride [103].

28.6.4 Hydrogen Production for Vehicle Applications

In vehicle applications, hydrogen production can be carried out on-board the vehicle producing hydrogen as it is needed, or hydrogen can be produced off-board the vehicle and then stored on-board. Onboard hydrogen production requires compact, robust, and reliable hydrogen generation equipment using available energy sources. Off-board hydrogen production requires either centralized hydrogen production with hydrogen distribution infrastructure or distributed hydrogen production infrastructure, in combination with fueling infrastructure and onboard storage.

Although hydrogen has a high specific energy (or mass energy density), due to its low density, it has a low volumetric energy density compared to other common fuels (see Figures 28.3 and 28.4). Note that these figures include the fuel only and not the tank required to carry the fuel. Due to the low volumetric energy density of hydrogen, onboard

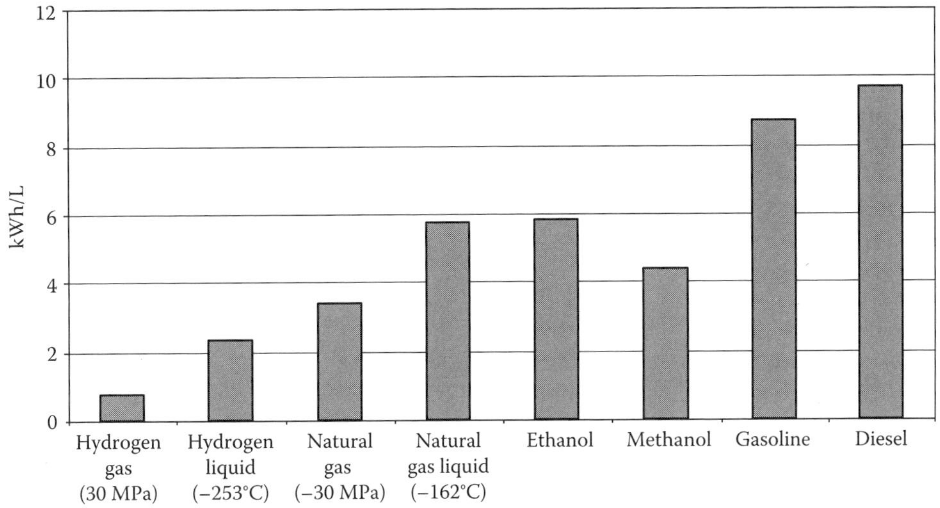

FIGURE 28.3
Volumetric energy density of fuels.

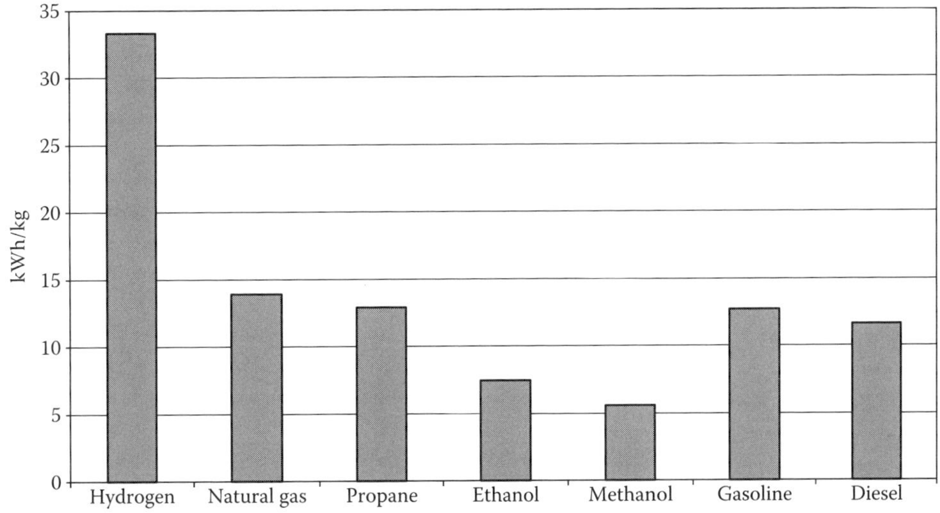

FIGURE 28.4
Specific energy of fuels.

storage of sufficient amounts of hydrogen to allow necessary vehicle range requires significant volumes, high pressures, and/or cryogenic liquids. The difficulty and expense of safe onboard hydrogen storage is a significant hurdle and motivates the development of onboard hydrogen production technologies.

28.6.4.1 Onboard Reformation

Small-scale reformers can achieve maximum thermal efficiencies between 80% and 93% thermal efficiency depending upon the reformation process and feedstock fuel [104]. Generating hydrogen for hydrogen enrichment by reformation of a fraction of the primary fuel stream may be energetically practical since the increase in ICE thermal efficiency may more than offset the losses in the reformation process.

For instance, in order to generate $5\%_{ener}$ hydrogen using a reformation process with a thermal efficiency of 80%, it would require $6.25\%_{ener}$ of the incoming primary fuel stream to be converted. From the experimental results of hydrogen enrichment of gasoline in SI ICEs, hydrogen addition of $5\text{–}15\%_{ener}$ resulted in a reduction in bsfc of 10%–20%. If by using 6.25% of the incoming primary fuel stream it is possible to generate $5\%_{ener}$ hydrogen and achieve a reduction in bsfc of 10%, then this process would result in a net increase in efficiency for the system.

Onboard reformation has challenges associated with the development of small-scale, low-cost, durable systems that can load follow sufficiently well to meet the highly transient demands of vehicle applications.

28.6.4.2 Onboard Electrolysis

Onboard vehicle electrolysis using electricity generated or stored on-board the vehicle is possible. The higher cost of vehicle fuels compared to other sources of primary energy makes this scenario economically impractical. The losses associated with each step in the process from vehicle fuel to electrical energy to hydrogen make this scenario energetically impractical.

28.6.4.2.1 Devices in Popular Media

Recently, many examples of small onboard electrolysis devices for hydrogen enrichment have appeared on the Internet and in the popular media. Even well-designed systems using electrical energy to generate hydrogen have significant efficiency limitations.

For example, a state-of-the-art production automobile engine found in the Toyota Prius converts the primary fuel to mechanical power with a maximum thermal efficiency of approximately 38% [105]. Part of this mechanical power is then converted to electrical power through an alternator or generator [106]. Alternators typically achieve maximum efficiencies in the 60% range with more expensive generators achieving higher efficiency [107]. Commercial large-scale electrolysis systems generate hydrogen with an efficiency between 60% and 73% [108]. Small-scale electrolyzers suitable for onboard vehicle applications are expected to be significantly less efficient than large-scale commercial electrolyzers and may produce hydrogen with a thermal efficiency in the 50% range. The maximum overall efficiency of generating hydrogen with these small onboard electrolysis systems is expected to be less than 11% from fuel energy to hydrogen energy and, under practical operating conditions, may be much lower. If a total hydrogen enrichment of $5\%_{ener}$ was desired, then the production of hydrogen via electrolysis in the manner described would consume approximately 44% of the primary fuel energy. Even with small additions of hydrogen of $1\%_{ener}$, almost 9% of the primary fuel energy would be consumed for hydrogen production. From the experimental results of hydrogen enrichment of gasoline in SI ICEs, hydrogen addition of $5–15\%_{ener}$ resulted in a reduction in bsfc of 10%–20%. This is not a large enough benefit in efficiency to offset the losses in onboard production of hydrogen via electrolysis.

In addition to the fundamental challenges of complex multistep energy conversion systems, the electrolysis devices in the popular media are often designed poorly with inefficient hydrogen production. Overall, most of these systems will not produce enough hydrogen to significantly change the combustion characteristics of the air–fuel mixture being enriched. Often these systems will generate some hydrogen and oxygen along with significant quantities of water vapor due to high overpotential and the associated heat production in the electrolysis unit. Efficiency benefits claimed may be in part due to changing driving conditions or to reduction in power as caused by the water vapor in the intake air stream. Actual electrolysis systems tested informally by the authors in a standard automotive system have shown no reductions in emissions.

28.6.5 Thermochemical Recuperation/Waste Heat Recovery

In thermochemical reformation processes, significant amounts of heat are required to drive the endothermic reformation reactions. This process heat can be supplied by combustion of a portion of the fuel, electric heaters, or by utilization of waste heat sources. In hydrogen enrichment systems with combustion applications, there are often waste heat sources in exhaust streams or coolant loops that can be recovered to supply a portion or all of the heat required by the thermochemical reformation process [100]. By using waste heat instead of burning fuel thermochemical recuperation, the system's overall thermal efficiency can be increased [100]. Under some conditions, the reformate gas can have a higher heating value than the fuel processed due to conversion of a fraction of the heat input into chemical energy [1,100]. The combination of waste heat recovery for supplying heat input and the increase in heating value through the reformation process can potentially lead to an energy benefit from reformation as opposed to a small energy loss if fuel is burned to supply the process heat [1,100].

With hydrogen enrichment of lean or dilute mixtures, the combination of increased engine thermal efficiency, potential for increased heating value through thermochemical recuperation, and the potential for engine efficiency improvements leads to the potential for significant system efficiency improvements.

28.7 Conclusion

Hydrogen enrichment can have significant beneficial effects on many different combustion processes using many different fuels over a range of operating and design parameters. Over some ranges of operating and design parameters, hydrogen enrichment does not show significant benefits and can even be detrimental.

Hydrogen enrichment of fuel–air mixtures changes the combustion properties of the entire mixture. Generally, hydrogen enrichment promotes complete combustion and extends the lean and dilute operating limits. By enabling the use of lean or dilute mixtures, hydrogen enrichment has the potential to increase thermal efficiency and reduce emissions in SI ICEs over ranges of operating parameters found in practical applications.

In SI ICEs, efficiency increases from hydrogen enrichment are related to increased burn rates, reduced cycle to cycle variability, and reduced throttling losses. Hydrogen enrichment promotes more complete combustion leading to reductions in HC emissions. The changed combustion characteristics from hydrogen enrichment enable ultra-lean operation leading to reduced carbon monoxide (CO) and nitrogen oxide (NOx) emissions or allow the use of increased exhaust gas recirculation leading to reduced NOx emissions.

A potential drawback of hydrogen enrichment is slightly reduced power output, due to the fact that the hydrogen gas displaces air reducing the volumetric efficiency of the engine.

In SI ICEs, hydrogen enrichment appears particularly beneficial at part load conditions where it increases efficiency through combustion effects as well as reduced throttling losses.

Overall efficiency gains of 5%–30% are typical, and gains up to 50% have been found using hydrogen enrichment of SI ICEs with a variety of fuels, while maintaining low emissions of CO and reducing NOx and HC emissions.

The use of hydrogen enrichment in commercial applications depends upon the availability of hydrogen either through local hydrogen production or through the development of centralized hydrogen production and distribution infrastructure.

Research continues in the areas of combustion phenomenon, hydrogen production technologies, and engine design.

References

1. Jamal, Y. and M.L. Wyszynski. On-board generation of hydrogen-rich gaseous fuels—A review. *International Journal of Hydrogen Energy* 1994; 19(7): 557–572.
2. Baade, W.F., U.N. Parekh, and V.S. Raman. Hydrogen, in *Kirk-Othmer Encyclopedia of Chemical Technology*. New York: John Wiley & Sons, Inc., 2008, pp. 759–808.
3. Verhelst, S., S. Verstraeten, and R. Sierens. A comprehensive overview of hydrogen engine design features. *Proceedings of the Institution of Mechanical Engineers Part D-Journal of Automobile Engineering* 2007; 221(D8): 911–920.
4. Lynch, F.E. and R. Marmaro, Special purpose blends of hydrogen and natural gas, United States Patent 5139002, 1992.
5. Padro, C.E.G. and J.O. Keller, Hydrogen energy, in *Kirk-Othmer Encyclopedia of Chemical Technology*. New York: John Wiley & Sons, Inc., 2005, pp. 837–866.
6. Pettersson, L. and K. Sjostrom. Decomposed methanol as a fuel—A review. *Combustion Science and Technology* 1991; 80(4–6): 265–303.
7. Akansu, S.O., Z. Dulger, N. Kahraman, and T.N. Veziroglu. Internal combustion engines fueled by natural gas—Hydrogen mixtures. *International Journal of Hydrogen Energy* 2004; 29(14): 1527–1539.
8. Li, H., G.A. Karim, and A. Sohrabi. Knock and combustion characteristics of CH4, CO, H2 and their binary mixtures. SAE Technical Paper 2003-01-3088 2003, doi:10.4271/2003-01-3088.
9. Milton, B.E. and J.C. Keck. Laminar burning velocities in stoichiometric hydrogen and hydrogen—Hydrocarbon gas mixtures. *Combustion and Flame* 1984; 58(1): 13–22.
10. Sher, E. and N. Ozdor. Laminar burning velocities of n-butane/air mixtures enriched with hydrogen. *Combustion and Flame* 1992; 89(2): 214–220.
11. Yu, G., C.K. Law, and C.K. Wu. Laminar flame speeds of hydrocarbon + air mixtures with hydrogen addition. *Combustion and Flame* 1986; 63(3): 339–347.

12. Alavandi, S.K. and A.K. Agrawal. Experimental study of combustion of hydrogen-syngas/methane fuel mixtures in a porous burner. *International Journal of Hydrogen Energy* 2008; 33(4): 1407–1415.
13. Briones, A.M., S.K. Aggarwal, and V.R. Katta. Effects of H2 enrichment on the propagation characteristics of CH4-air triple flames. *Combustion and Flame* 2008; 153(3): 367–383.
14. Burbano, H.J., A.A. Amell, and J.M. García. Effects of hydrogen addition to methane on the flame structure and CO emissions in atmospheric burners. *International Journal of Hydrogen Energy* 2008; 33(13): 3410–3415.
15. Coppens, F.H.V., J. De Ruyck, and A.A. Konnov. Effects of hydrogen enrichment on adiabatic burning velocity and NO formation in methane + air flames. *Experimental Thermal and Fluid Science* 2007; 31(5): 437–444.
16. Coppens, F.H.V. and A.A. Konnov. The effects of enrichment by H2 on propagation speeds in adiabatic flat and cellular premixed flames of CH4 + O2 + CO2. *Fuel* 2008; 87(13–14): 2866–2870.
17. Cozzi, F. and A. Coghe. Behavior of hydrogen-enriched non-premixed swirled natural gas flames. *International Journal of Hydrogen Energy* 2006; 31(6): 669–677.
18. de Ferrières, S., A. El Bakali, B. Lefort, M. Montero, and J.F. Pauwels, Experimental and numerical investigation of low-pressure laminar premixed synthetic natural gas/O2/N2 and natural gas/H2/O2/N2 flames. *Combustion and Flame* 2008; 154(3): 601–623.
19. Di Sarli, V. and A.D. Benedetto. Laminar burning velocity of hydrogen-methane/air premixed flames. *International Journal of Hydrogen Energy* 2007; 32(5): 637–646.
20. Gauthier, S., A. Nicolle, and D. Baillis. Investigation of the flame structure and nitrogen oxides formation in lean porous premixed combustion of natural gas/hydrogen blends. *International Journal of Hydrogen Energy* 2008; 33(18): 4893–4905.
21. Ghoniem, A.F., A. Annaswamy, S. Park, and Z.C. Sobhani. Stability and emissions control using air injection and H2 addition in premixed combustion. *Proceedings of the Combustion Institute* 2005; 30(2): 1765–1773.
22. Halter, F., C. Chauveau, and I. Gökalp. Characterization of the effects of hydrogen addition in premixed methane/air flames. *International Journal of Hydrogen Energy* 2007; 32(13): 2585–2592.
23. Jou, C.-J.G., C.-l. Lee, C.-H. Tsai, H.P. Wang, and M.-L. Lin. Enhancing the performance of a high-pressure cogeneration boiler with waste hydrogen-rich fuel. *International Journal of Hydrogen Energy* 2008; 33(20): 5806–5810.
24. Lafay, Y., B. Renou, G. Cabot, and M. Boukhalfa. Experimental and numerical investigation of the effect of H2 enrichment on laminar methane-air flame thickness. *Combustion and Flame* 2008; 153(4): 540–561.
25. Leung, T. and I. Wierzba. The effect of hydrogen addition on biogas non-premixed jet flame stability in a co-flowing air stream. *International Journal of Hydrogen Energy* 2008; 33(14): 3856–3862.
26. Park, J., S. In Keel, J. Han Yun, and T. Kwon Kim. Effects of addition of electrolysis products in methane-air diffusion flames. *International Journal of Hydrogen Energy* 2007; 32(16): 4059–4070.
27. Ratna Kishore, V., N. Duhan, M.R. Ravi, and A. Ray. Measurement of adiabatic burning velocity in natural gas-like mixtures. *Experimental Thermal and Fluid Science* 2008; 33(1): 10–16. doi:10.1016/j.expthermflusci.2008.06.00
28. Schefer, R.W. Hydrogen enrichment for improved lean flame stability. *International Journal of Hydrogen Energy* 2003; 28(10): 1131–1141.
29. Schefer, R.W., C. White, J. Keller, and D.-R. Derek. *Lean Hydrogen Combustion. Lean Combustion.* Burlington, MA: Academic Press, 2008, pp. 213–254.
30. Shy, S.S., Y.C. Chen, C.H. Yang, C.C. Liu, and C.M. Huang. Effects of H2 or CO2 addition, equivalence ratio, and turbulent straining on turbulent burning velocities for lean premixed methane combustion. *Combustion and Flame* 2008; 153(4): 510–524.
31. Strakey, P., T. Sidwell, and J. Ontko. Investigation of the effects of hydrogen addition on lean extinction in a swirl stabilized combustor. *Proceedings of the Combustion Institute* 2007; 31(2): 3173–3180.
32. Uykur, C., P.F. Henshaw, D.S.K. Ting, and R.M. Barron. Effects of addition of electrolysis products on methane/air premixed laminar combustion. *International Journal of Hydrogen Energy* 2001; 26(3): 265–273.

33. Wang, B., R. Qiu, and Y. Jiang. Effects of hydrogen enhancement in LPG/air premixed flame. *Acta Physico-Chimica Sinica* 2008; 24(7): 1137–1142.
34. Wicksall, D.M. and A.K. Agrawal. Acoustics measurements in a lean premixed combustor operated on hydrogen/hydrocarbon fuel mixtures. *International Journal of Hydrogen Energy* 2007; 32(8): 1103–1112.
35. Wu, Y., I.S. Al-Rahbi, Y. Lu, and G.T. Kalghatgi. The stability of turbulent hydrogen jet flames with carbon dioxide and propane addition. *Fuel* 2007; 86(12–13): 1840–1848.
36. Heywood, J.B. *Internal Combustion Engine Fundamentals*. New York: McGraw-Hill, 1988.
37. Li, J., Z. Zhao, A. Kazakov, and F.L. Dryer, An updated comprehensive kinetic model of hydrogen combustion. *International Journal of Chemical Kinetics* 2004; 36(10): 566–575.
38. Turns, S.R. *An Introduction to Combustion*, 2nd edn. Singapore: McGraw-Hill, 2000.
39. Conte, E. and K. Boulouchos. Influence of hydrogen-rich-gas addition on combustion, pollutant formation and efficiency of an IC-SI engine. SAE Technical Paper 2004; 2004-01-0972, doi: 10.4271/2004-01-0972.
40. McTaggart-Cowan, G.P., H.L. Jones, S.N. Rogak, W.K. Bushe, P.G. Hill, and S.R. Munshi. Direct-injected hydrogen-methane mixtures in a heavy-duty compression ignition engine, in *SAE 2006 World Congress*, Detroit, MI, 2006.
41. Smith, J.A. and G.J.J. Bartley. Stoichiometric operation of a gas engine utilizing synthesis gas and EGR for NOx control. *Journal of Engineering for Gas Turbines and Power* 2000; 122(4): 617–623.
42. Saikaly, K., S. Rousseau, C. Rahmouni, O. Le Corre, and L. Truffet. Safe operating conditions determination for stationary SI gas engines. *Fuel Processing Technology* 2008, doi:10.1016/j.fuproc.2008.05.01.
43. Liu, Z. and G.A. Karim. Knock characteristics of dual-fuel engines fueled with hydrogen fuel. *International Journal of Hydrogen Energy* 1995; 20(11): 919–924.
44. Li, H. and G.A. Karim. Exhaust emissions from an SI engine operating on gaseous fuel mixtures containing hydrogen. *International Journal of Hydrogen Energy* 2005; 30(13–14): 1491–1499.
45. Li, H. and G.A. Karim. An experimental investigation of S.I. engine operation on gaseous fuels lean mixtures. SAE Transactions 2005-01-3765 2005; 114(3): 1600–1608.
46. Karim, G.A., I. Wierzba, and Y. Al-Alousi. Methane-hydrogen mixtures as fuels. *International Journal of Hydrogen Energy* 1996; 21(7): 625–631.
47. Ivanic, Ž., F. Ayala, J. Goldwitz, and J.B. Heywood, Effects of hydrogen enhancement on efficiency and NOx emissions of lean and EGR-diluted mixtures in a SI engine, in *2005 SAE World Congress*. Detroit, MI: SAE, 2005.
48. Saravanan, N., G. Nagarajan, and S. Narayanasamy. An experimental investigation on DI diesel engine with hydrogen fuel. *Renewable Energy* 2008; 33(3): 415–421.
49. Beister, U.-J. and R. Smaling. Hydrogen-enhanced combustion engine could improve gasoline fuel economy by 20% to 30%. MTZ worldwide edition No. 2005-10 2005, October 4, 2005 [cited October 13, 2008]; Available from http://www.all4engineers.com/index.php;do=show/alloc=3/lng=en/id=2866/sid=90458fbceb42147dd47b232709a5101c.
50. Akansu, S.O., N. Kahraman, and B. Çeper. Experimental study on a spark ignition engine fuelled by methane-hydrogen mixtures. *International Journal of Hydrogen Energy* 2007; 32(17): 4279–4284.
51. Munshi, S.R., C. Nedelcu, J. Harris, T. Edwards, J. Williams, F. Lynch et al. Hydrogen blended natural gas operation of a heavy duty turbocharged lean burn spark ignition engine, in *Powertrain & Fluid Systems Conference & Exhibition*. Tampa, FL: SAE, 2004.
52. Bell, S.R. and M. Gupta. Extension of the lean operating limit for natural gas fueling of a spark ignited engine using hydrogen blending. *Combustion Science and Technology* 1997; 123(1–6): 23–48.
53. Han, P., M.D. Checkel, and B.A. Fleck. Hydrogen from reformer gas a novel fuel and bridging technology: A combustion perspective. *International Journal of Hydrogen Energy* 2007; 32(10–11): 1416–1420.
54. Han, P., M. David Checkel, B.A. Fleck, and N.L. Nowicki. Burning velocity of methane/diluent mixture with reformer gas addition. *Fuel* 2007; 86(4): 585–596.
55. Hoekstra, R.L., K. Collier, N. Mulligan, and L. Chew. Experimental study of a clean burning vehicle fuel. *International Journal of Hydrogen Energy* 1995; 20(9): 737–745.

56. Erickson, P.A., D.R. Vernon, E. Jordan, K. Collier, and N. Mulligan. Low NOx operation and recuperation of thermal and chemical energy through hydrogen in internal combustion engines. in *16th Annual Hydrogen Conference of the National Hydrogen Association*, Washington, DC, 2005.
57. Collier, R.K. Reformed natural gas for powering an internal combustion engine, United States Patent 6508209, 2003.
58. Collier, R.K. Low-emission internal combustion engine, United States Patent 6823852, 2004.
59. Allenby, S., W.C. Chang, A. Megaritis, and M.L. Wyszynski. Hydrogen enrichment: A way to maintain combustion stability in a natural gas fuelled engine with exhaust gas recirculation, the potential of fuel reforming. *Proceedings of the Institution of Mechanical Engineers Part D-Journal of Automobile Engineering* 2001; 215(D3): 405–418.
60. Shrestha, S.O.B. and G.A. Karim. Hydrogen as an additive to methane for spark ignition engine applications. *International Journal of Hydrogen Energy* 1999; 24(6): 577–586.
61. Al-Baghdadi, S., M. Abdul-Resul, S. Al-Janabi, and H. Abdul-Kadim. Improvement of performance and reduction of pollutant emission of a four stroke spark ignition engine fueled with hydrogen-gasoline fuel mixture. *Energy Conversion and Management* 2000; 41(1): 77–91.
62. Cassidy, J.F., Emissions and total energy consumption of a multicylinder piston engine running on gasoline and a hydrogen-gasoline mixture, in NASA Technical Note. 1977. Cleveland, OH/Washington DC: NASA Lewis Research Centre.
63. Ecklund, E., Hydrogen enrichment concept preliminary evaluation. Final Report. US Energy Research and Development Administration, Technical Information Centre TEC-75/007, JPL Document 1200-237, Prepared for EPA under interagency Agreement EPA-IAG-D4–0548, December 15, 1975.
64. Kirwan, J.E., A.A. Quader, and M.J. Grieve, Fast start-up on-board gasoline reformer for near zero emissions in spark-ignition engines, in *SAE 2002 World Congress*. Detroit, MI: SAE, 2002.
65. Lucas, G.G. and W.L. Richards. The hydrogen/petrol engine the means to give good part load thermal efficiency. SAE Publication No. 820315 1982.
66. May, H. and D. Gwinner. Possibilities of improving exhaust emissions and energy consumption in mixed hydrogen-gasoline operation. *International Journal of Hydrogen Energy* 1983; 8(2): 121–129.
67. Petkov, T.I. and K.N. Barzev. Some aspects of hydrogen application as a supplementary fuel to the fuel-air mixture for internal combustion engines. *International Journal of Hydrogen Energy* 1987; 12(9): 633–638.
68. Quader, A.A., J.E. Kirwan, and M.J. Grieve, Engine performance and emissions near the dilute limit with hydrogen enrichment using an on-board reforming strategy, in *SAE 2003 World Congress*. Detroit, MI: Society of Automotive Engineers, 2003.
69. Sher, E. and Y. Hacohen. On the modeling of a SI 4-stroke cycle engine fueled with hydrogen-enriched gasoline. *International Journal of Hydrogen Energy* 1987; 12(11): 773–781.
70. Suzuki, T. and Y. Sakurai, Effect of hydrogen rich gas and gasoline mixed combustion on spark ignition engine, in *Powertrain & Fluid Systems Conference & Exhibition*. Toronto, Ontario, Canada: SAE, 2006.
71. Yüksel, F. and M.A. Ceviz. Thermal balance of a four stroke SI engine operating on hydrogen as a supplementary fuel. *Energy* 2003; 28(11): 1069–1080.
72. Rauckis, M.J. and W.J. McLean. The effect of hydrogen addition on ignition delays and flame propagation in spark ignition engines. *Combustion Science and Technology* 1979; 19(5): 207–216.
73. Tiashen, D., L. Jingding, and L. Yingqing, Combustion-supporting fuel for methanol engines: Hydrogen, in *The International Symposium on Hydrogen Systems*, Beijing, China, 1985, pp. 105–113.
74. Pettersson, L. and K. Sjöström. Onboard hydrogen generation by methanol decomposition for the cold start of neat methanol engines. *International Journal of Hydrogen Energy* 1991; 16(10): 671–676.
75. Al-Baghdadi, M. A study on the hydrogen-ethyl alcohol dual fuel spark ignition engine. *Energy Conversion and Management* 2002; 43(2): 199–204.
76. Yousufuddin, S., S.N. Mehdi, and M. Masood. Performance and combustion characteristics of a hydrogen-ethanol-fuelled engine. *Energy & Fuels* 2008; 22(5): 3355–3362.

77. Al-Baghdadi, M.A.S. Hydrogen-ethanol blending as an alternative fuel of spark ignition engines. *Renewable Energy* 2003; 28(9): 1471–1478.
78. Bade Shrestha, S.O. and G. Narayanan. Landfill gas with hydrogen addition—A fuel for SI engines. *Fuel* 2008; 87(17–18): 3616–3626.
79. Saravanan, N. and G. Nagarajan. An experimental investigation of hydrogen-enriched air induction in a diesel engine system. *International Journal of Hydrogen Energy* 2008; 33(6): 1769–1775.
80. Yap, D., S.M. Peucheret, A. Megaritis, M.L. Wyszynski, and H. Xu. Natural gas HCCI engine operation with exhaust gas fuel reforming. *International Journal of Hydrogen Energy* 2006; 31(5): 587–595.
81. Wong, Y.K. and G.A. Karim. An analytical examination of the effects of hydrogen addition on cyclic variations in homogeneously charged compression-ignition engines. *International Journal of Hydrogen Energy* 2000; 25(12): 1217–1224.
82. Tsolakis, A. and A. Megaritis. Partially premixed charge compression ignition engine with on-board H-2 production by exhaust gas fuel reforming of diesel and biodiesel. *International Journal of Hydrogen Energy* 2005; 30(7): 731–745.
83. Shudo, T. and H. Yamada. Hydrogen as an ignition-controlling agent for HCCI combustion engine by suppressing the low-temperature oxidation. *International Journal of Hydrogen Energy* 2007; 32(14): 3066–3072.
84. Shudo, T. An HCCI combustion engine system using on-board reformed gases of methanol with waste heat recovery: Ignition control by hydrogen. *International Journal of Vehicle Design* 2006; 41(1–4): 206–226.
85. Shudo, T. Ignition control in the HCCI combustion engine system fuelled with methanol-reformed gases. *KONES Internal Combustion Engines* 2005; 12: 233–244.
86. Conte, E. and K. Boulouchos. Hydrogen-enhanced gasoline stratified combustion in SI-DI engines. *Journal of Engineering for Gas Turbines and Power* January 22, 2008; 130(2), 022801, 9pp. doi:10.1115/1.2795764.
87. Shudo, T., K. Shimamura, and Y. Nakajima. Combustion and emissions in a methane DI stratified charge engine with hydrogen pre-mixing. *JSAE Review* 2000; 21(1): 3–7.
88. Saravanan, N. and G. Nagarajan. An experimental investigation on optimized manifold injection in a direct-injection diesel engine with various hydrogen flowrates. *Proceedings of the Institution of Mechanical Engineers Part D-Journal of Automobile Engineering* 2007; 221(D12): 1575–1584.
89. Patro, T.N. Burning rate assessment of hydrogen-enriched fuel combustion in diesel-engines. *International Journal of Hydrogen Energy* 1994; 19(3): 275–284.
90. Masood, M., S.N. Mehdi, and P.R. Reddy. Experimental investigations on a hydrogen-diesel dual fuel engine at different compression ratios. *Journal of Engineering for Gas Turbines and Power-Transactions of the ASME* 2007; 129(2): 572–578.
91. Masood, M., M.M. Ishrat, and A.S. Reddy. Computational combustion and emission analysis of hydrogen-diesel blends with experimental verification. *International Journal of Hydrogen Energy* 2007; 32(13): 2539–2547.
92. Lambe, S.M. and H.C. Watson. Optimizing the design of a hydrogen engine with pilot diesel fuel ignition. *International Journal of Vehicle Design* 1993; 14(4): 370–389.
93. Lambe, S.M. and H.C. Watson. Low polluting, energy-efficient CI hydrogen engine. *International Journal of Hydrogen Energy* 1992; 17(7): 513–525.
94. Kumar, M.S., A. Ramesh, and B. Nagalingam. Use of hydrogen to enhance the performance of a vegetable oil fuelled compression ignition engine. *International Journal of Hydrogen Energy* 2003; 28(10): 1143–1154.
95. Banapurmath, N.R., P.G. Tewari, and R.S. Hosmath. Experimental investigations of a four-stroke single cylinder direct injection diesel engine operated on dual fuel mode with producer gas as inducted fuel and Honge oil and its methyl ester (HOME) as injected fuels. *Renewable Energy* 2008; 33(9): 2007–2018.
96. TerMaath, C.Y., E.G. Skolnik, R.W. Schefer, and J.O. Keller. Emissions reduction benefits from hydrogen addition to midsize gas turbine feedstocks. *International Journal of Hydrogen Energy* 2006; 31(9): 1147–1158.

97. Maughan, J.R., J.H. Bowen, D.H. Cooke, and J.J. Tuzson. Reducing gas turbine emissions through hydrogen-enhanced, steam-injected combustion. *Journal of Engineering for Gas Turbines and Power-Transactions of the ASME* 1996; 118(1): 78–85.
98. Juste, G.L. Hydrogen injection as additional fuel in gas turbine combustor. Evaluation of effects. *International Journal of Hydrogen Energy* 2006; 31(14): 2112–2121.
99. Hiroyasu, H., M. Arai, T. Kadota, and J. Yoso. An experimental-study on kerosene-hydrogen hybrid combustion in a gas-turbine combustor. *Bulletin of the JSME-Japan Society of Mechanical Engineers* 1980; 23(184): 1655–1662.
100. Vernon, D.R., E. Jordan, J. Woolley, and P. Erickson. The potential for waste heat recovery by thermochemical recuperation for hydrogen enriched internal combustion. in *ASME Internal Combustion Engine Division Fall Technical Conference*, Charleston, SC, 2007.
101. Becker, E.P. and K.C. Ludema. A qualitative empirical model of cylinder bore wear. *Wear* 1999; 225–229(Part 1): 387–404.
102. Petitpas, G., J.D. Rollier, A. Darmon, J. Gonzalez-Aguilar, R. Metkemeijer, and L. Fulcheri. A comparative study of non-thermal plasma assisted reforming technologies. *International Journal of Hydrogen Energy* 2007; 32(14): 2848–2867.
103. Çakanyildirim, Ç. and M. Gürü. Hydrogen cycle with sodium borohydride. *International Journal of Hydrogen Energy* 2008; 33(17): 4634–4639.
104. Allgeier, T., M. Klenk, T. Landenfeld, E. Conte, K. Boulouchos, and J. Czerwinski. Advanced emission and fuel economy concept using combined injection of gasoline and hydrogen in SI-engines. SAE Technical Paper 2004-01-1270 2004, doi:10.4271/2004-01-1270.
105. Duoba, M., H. Ng, and R. Larsen. In-situ mapping and analysis of the Toyota Prius HEV engine. SAE Technical Paper 2000-01-3096, 2000, doi:10.4271/2000-01-3096.
106. Staunton, R.H., C.W. Ayers, L.D. Marlino, J.N. Chiasson, and T.A. Burress, *Evaluation of 2004 Toyota Prius Hybrid Electric Drive System*. U.S. Department of Energy, 2006, ORNL/TM-2006/423.
107. Bauer, H., ed. *Automotive Handbook*. Stuttgart, Germany: Bosch, 2000.
108. Turner, J.A. Sustainable hydrogen production. *Science* 2004; 305(5686): 972–974.

29

Distribution Networking

Amgad Elgowainy
Argonne National Laboratory

Marianne Mintz
Argonne National Laboratory

Monterey Gardiner
U.S. Department of Energy

CONTENTS

The transition to an alternative transportation fuel requires detailed technical and economic analyses of all aspects of the supporting infrastructure. The previous chapter discussed hydrogen storage. This chapter focuses on the infrastructure required to bring hydrogen fuel from centralized production facilities to the hydrogen vehicles (HVs) that will utilize it. Many believe that, at least in the initial years of a hydrogen transition, the fuel will be supplied primarily via on-site production facilities integrated into stations that in all other respects resemble today's gasoline stations. Although on-site production can be expensive, it eliminates the need for delivery infrastructure. Others believe that hydrogen will be supplied primarily from centrally located hydrogen production facilities. Until demand is sufficient to support dedicated capacity and central production facilities, hydrogen fuel is likely to come from incremental production of existing industrial gas suppliers. Whether from dedicated facilities or as incremental production, cost models suggest that delivery will account for a large portion of hydrogen cost. Understanding the contributors to delivery cost is essential for identifying opportunities to reduce it, as well as for setting appropriate goals and targets. As renewable energy becomes a higher priority, longer delivery distances are expected. There have already been inquiries to compare high-voltage electricity transmission lines to different hydrogen delivery options. At very long distances (greater than several hundred miles), liquid hydrogen appears to be the most attractive option. Liquid hydrogen allows for the lowest cost and highest energy density onboard a vehicle.

29.1 Background

In response to a congressional request in the Energy Policy Act of 2005, the National Research Council (NRC) investigated the maximum practicable number of HVs that could be deployed in the United States by 2020 and beyond, together with the investments, time, and government actions needed to support them. NRC concluded that approximately 2 million HVs could be operating (in a fleet of 280 million light-duty vehicles) in 2020, and about 25 million by 2030, assuming that technical goals are met, consumers readily accept HVs, and policies are established to drive the deployment of hydrogen fuel and fuel cell vehicles. The study found that while other alternatives (e.g., improved fuel economy of conventional vehicles, increased penetration of hybrid vehicles, and biomass-derived fuels) could deliver significantly greater reductions in US oil use and greenhouse gas (GHG) emissions over the next two decades, HVs offer greater long-term potential. Thus, NRC concluded that the development of a hydrogen refueling infrastructure is critical to achieving that potential. Without sales of fuel cell vehicles, fuel providers will be reluctant to invest in fueling capability; conversely, without both actual and perceived fueling capability (convenient station locations, fueling speed, and safety), consumers will be reluctant to purchase fuel cell vehicles [1]. This *chicken and egg* problem increases the risk faced by early purchasers of HVs and investors in hydrogen fuel stations. Thus, the transition to HVs will be challenging.

The cost of hydrogen fuel stations is particularly challenging. The NRC study estimated that a fuel station with a natural gas reformer to produce hydrogen would cost approximately $2.2 million when produced in quantities of 500 or more but $4.4 million if produced in the much smaller quantities likely during the early years of market introduction. Another study showed that the refueling station often accounts for half or more of delivery cost [2].

29.2 Methodology

The delivery costs presented here were computed from the hydrogen delivery scenario analysis model (HDSAM), an Excel-based tool developed by the Argonne National Laboratory with support from the US Department of Energy's (DOE's) Office of Fuel Cell Technologies Office. HDSAM uses a generalized financial and cost engineering framework to estimate the levelized cost (i.e., cost plus a predefined return on investment)* to deliver hydrogen in

* As well as energy use and GHG emissions.

quantities sufficient to meet a given level of market demand for a selected delivery pathway [2,3]. The model links pathway stages or *components* in a systematic market setting to develop capacity/flow parameters for a complete hydrogen delivery infrastructure. Using that systems-level perspective, HDSAM calculates the full, levelized cost (i.e., summed across all components) of hydrogen delivery, accounting for losses and trade-offs among the various component costs. The cost estimates are generated in US dollars using 2005 as the base year.

As with any commodity, distributing hydrogen depends on how it is packaged for dispensing into HVs. Packaging affects not only density (weight/volume) but also the economics of delivery modes and the vehicle's (onboard) storage, a question which has been called the *grand challenge* of the hydrogen economy [3]. The packaging of hydrogen, the distance it must be transported to potential users, and the profile of user demand for hydrogen affect not only the structure of potential delivery systems but also the contribution of delivery to the overall cost of hydrogen motor fuel.

In HDSAM, these parameters are used to define delivery scenarios. Default values contained in the model are discussed in Section 29.3. Briefly, a delivery pathway starts at the hydrogen production plant gate and ends at the vehicle's storage tank. Delivery can be separated into two distinct stages: (1) *transportation* or transmission of hydrogen from the production plant gate to the city gate or other transfer point and (2) *distribution* of hydrogen from the transfer point to the refueling station for dispensing into the vehicle's tank. Thus, the refueling station is considered as a *component* of the distribution system. There are three main delivery *modes* for the transportation or distribution of hydrogen using conventional technologies. These modes are identified here as (1) liquid delivery, (2) tube-trailer delivery, and (3) pipeline delivery. We note that while one mode could be used for transporting the hydrogen to city gate, another mode could be used for the distribution of hydrogen within the city to the refueling stations, thus forming a *mixed-mode* pathway for hydrogen delivery as described later in this chapter. For mixed-mode delivery pathways, the resulting pathway is named for the distribution mode rather than the transportation mode.

Each delivery pathway consists of several major steps to condition and move the hydrogen through the pathway to the HV's onboard storage system. Each major step in the pathway is modeled as a separate *component* or element in the pathway (e.g., liquefiers, trucks, pipelines, terminals, and refueling stations). For this analysis, we consider nine possible pathways, three liquid distribution pathways (Figures 29.1 through 29.3), four tube-trailer distribution pathways (Figures 29.4 through 29.7), and two pipeline distribution pathways (Figures 29.8 and 29.9). Figures 29.1 through 29.9 highlight the components in each of these pathways.

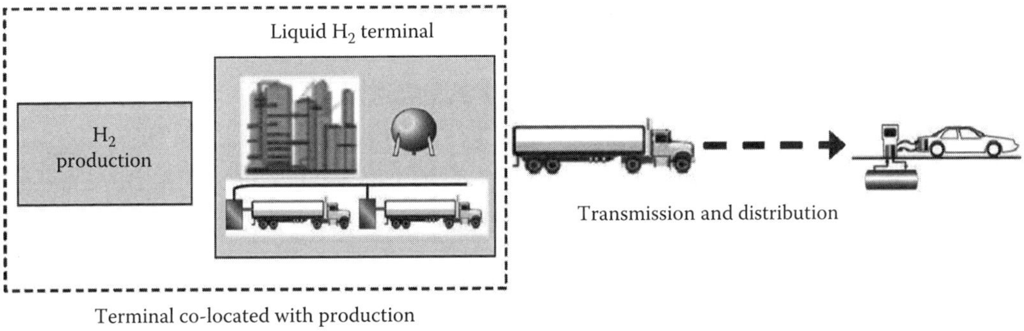

FIGURE 29.1
Liquid delivery pathway with liquid long-term storage.

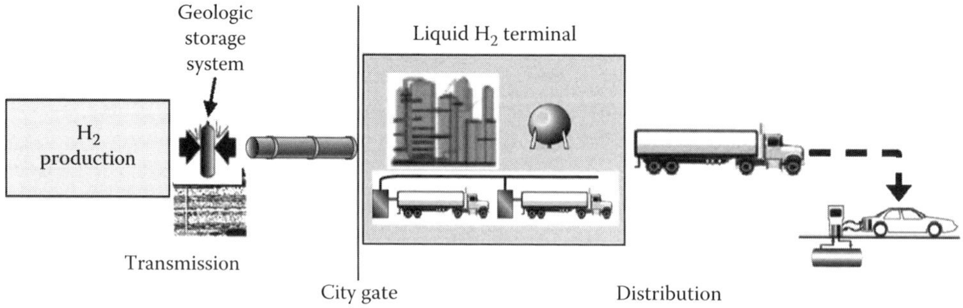

FIGURE 29.2
Mixed-mode pipeline and liquid delivery pathway with long-term geologic storage.

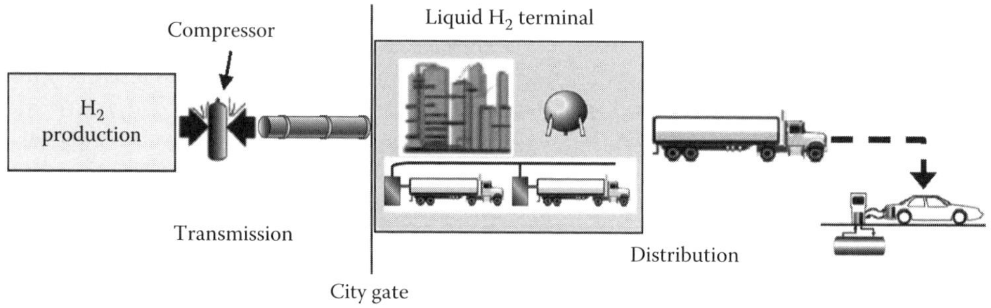

FIGURE 29.3
Mixed-mode pipeline and liquid delivery pathway with liquid long-term storage.

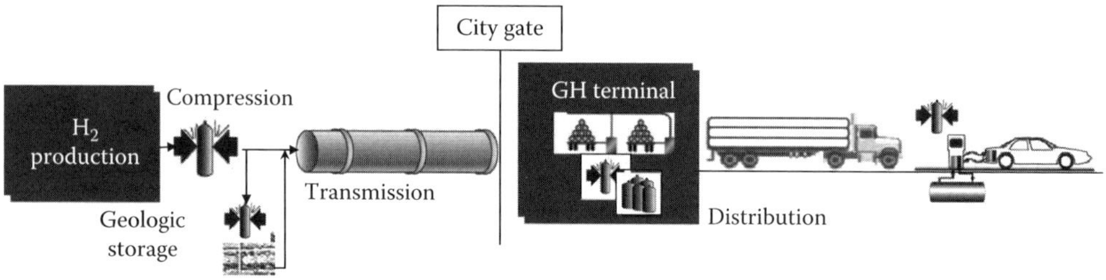

FIGURE 29.4
Mixed-mode pipeline and tube-trailer delivery pathway with long-term geologic storage.

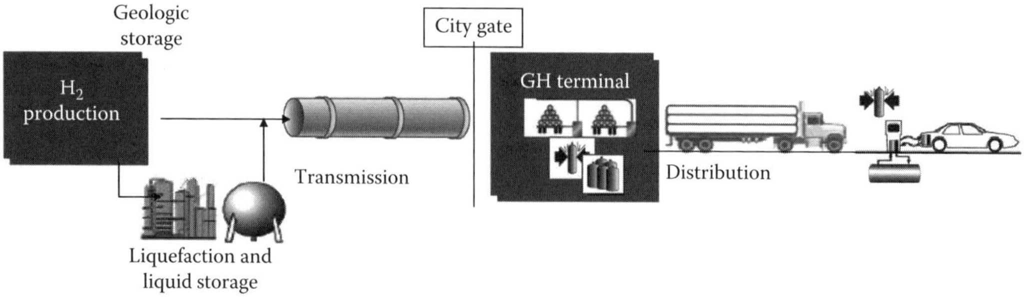

FIGURE 29.5
Mixed-mode pipeline and tube-trailer delivery pathway with liquid long-term storage.

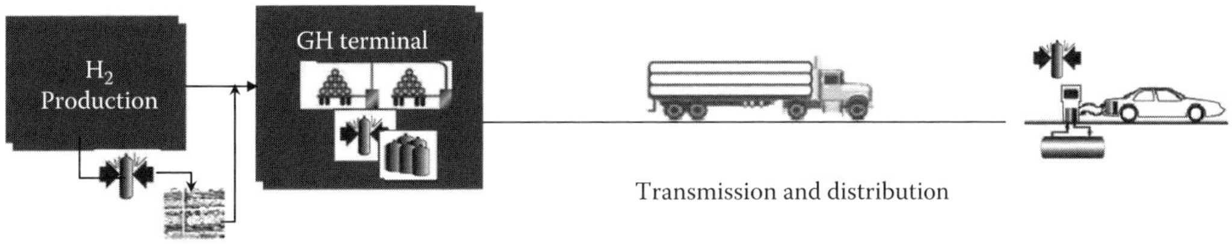

FIGURE 29.6
Tube-trailer delivery pathway with long-term geologic storage.

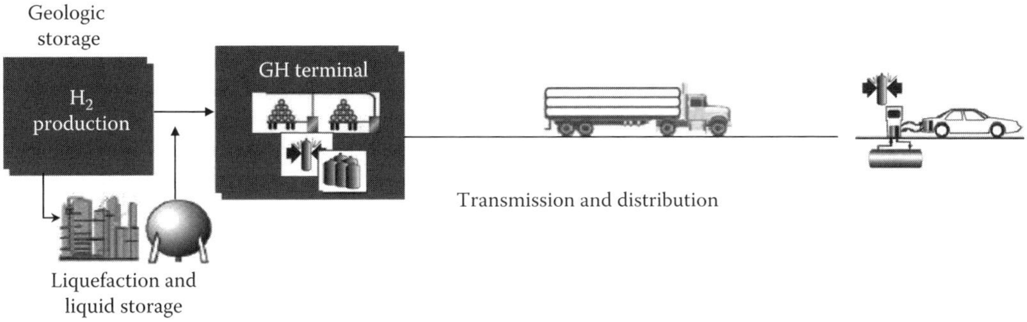

FIGURE 29.7
Tube-trailer delivery pathway with long-term liquid storage.

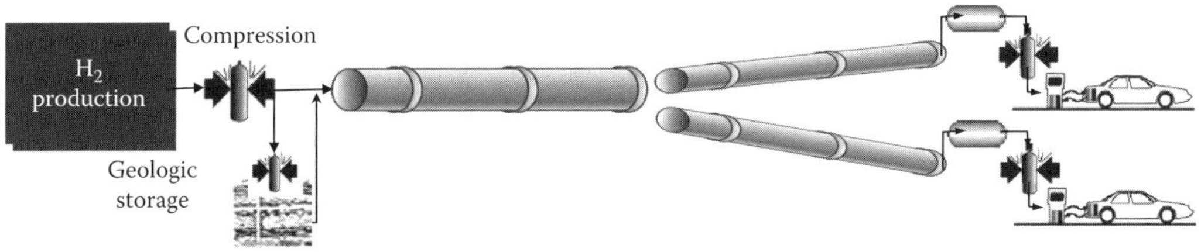

FIGURE 29.8
Pipeline delivery pathway with long-term geologic storage.

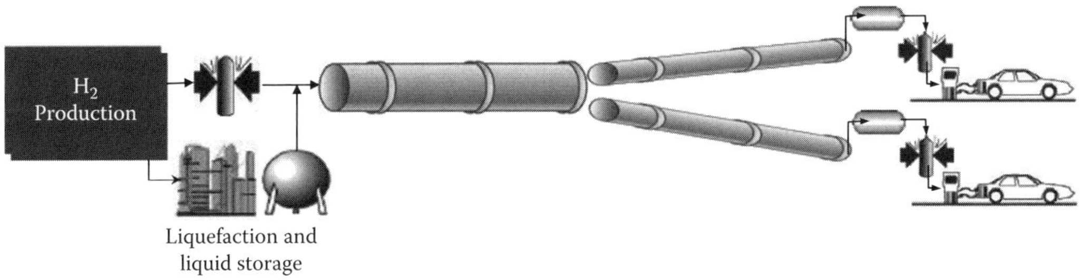

FIGURE 29.9
Pipeline delivery pathway with long-term liquid storage.

29.3 Technical Aspects of Hydrogen Delivery Infrastructure

29.3.1 Infrastructure Storage

Hydrogen storage in the delivery system is necessary to accommodate the temporal imbalances between supply and demand. There are two storage requirements: a short-term capacity to accommodate hourly variations in refueling station demand and a long-term capacity to meet seasonal variations in refueling station demand and scheduled production plant outages.

A representative hourly variation in refueling station demand is illustrated in Figure 29.10, and an illustration of the seasonal variation in demand, together with an annual production plant outage for scheduled maintenance, is shown in Figure 29.11. The seasonal demand variation is a product of annual driving profiles; that is, miles driven in the summer are normally higher than miles driven in the winter.

A series of optimization studies concluded the following for gaseous hydrogen pipeline delivery pathways: (1) long-term storage is most economically provided by compressed gas storage in geologic formations, if geologic storage is available, and in liquid storage, if geologic storage is not available, and (2) for hydrogen delivery by pipeline, short-term storage is most economically provided by low-pressure (~170 bar) compressed gas storage at the refueling station. For the demand profile shown in Figure 29.10, the nominal storage capacity is about 30% of the daily hydrogen dispensed.

For gaseous hydrogen tube-trailer delivery, the tube trailer is dropped off at the refueling site and used to meet the hourly storage needs. Long-term storage is provided by compressed gas geologic storage or liquid hydrogen storage. Several hours of low-pressure gas storage at the terminals are used to fill the tube trailers to ensure smooth loading operations.

For liquefaction and cryogenic liquid delivery of hydrogen, hydrogen storage is not as critical due to the much higher density of liquid hydrogen compared to gaseous hydrogen. However, the high cost of liquefaction dominates the hydrogen delivery cost. The refueling

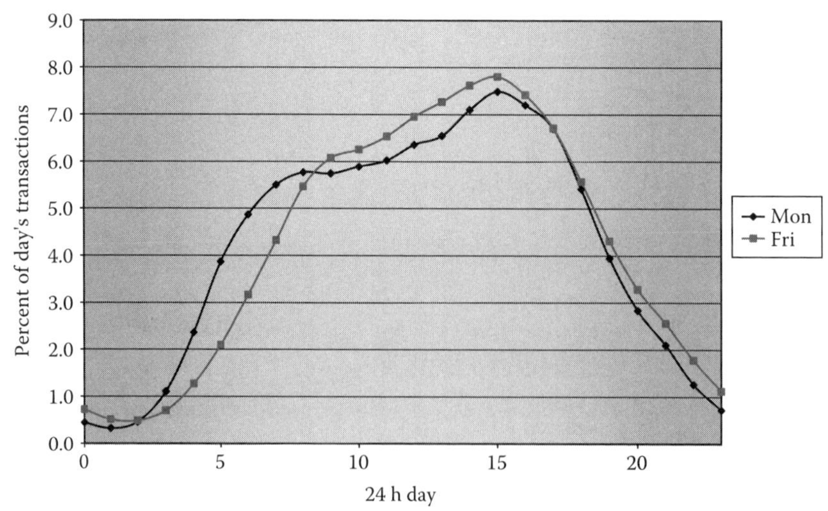

FIGURE 29.10
Hourly variation in refueling station demand (supplied by Chevron based on data from over 400 of their stations).

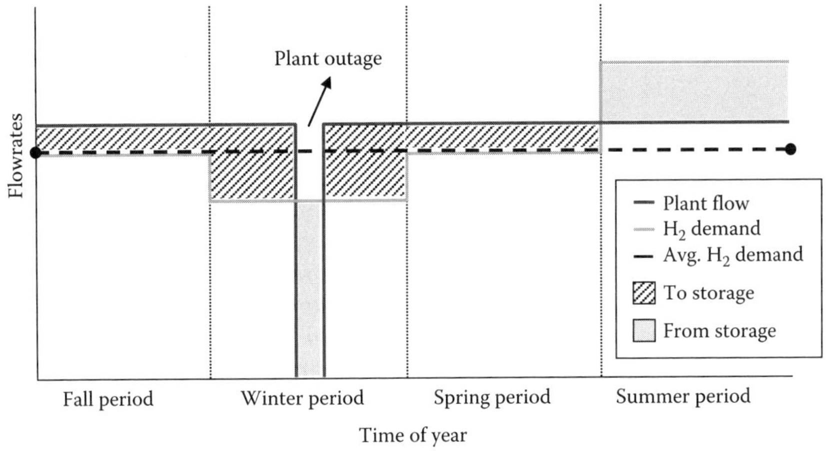

FIGURE 29.11
Seasonal variation in production plant and storage operation.

site hourly storage needs are met by a liquid storage tank. Liquid storage at terminals is needed to handle liquefaction plant outages and the summer peak demand, as well to ensure smooth truck loading operations.

29.3.2 Refueling Stations

Hydrogen fuel stations will serve much the same function as today's gasoline stations. They will dispense hydrogen, gasoline, and perhaps other fuels and will sell various convenience items. Aside from restrictions governing setback and separation distances, their footprint will be comparable to that of conventional gasoline stations, and they will serve similar numbers of vehicles with similar fuel demand profiles.

29.3.2.1 Fueling Profiles

Gasoline-fueled vehicles typically purchase 10–12 gal per *fill* and have an average fuel tank capacity of 16 gal. Thus, a typical *fill* is 62%–75% of tank volume. Assuming a typical gasoline light-duty vehicle (LDV) fuel economy of 22 mpg, a gasoline LDV can travel approximately 265 miles on a 75% fill. Similarly, a hydrogen fuel cell vehicle will typically purchase 4.6 kg of hydrogen, which is about 75% of a 6 kg capacity storage tank onboard the vehicle. If an HV's fuel economy is 58 miles/kg of hydrogen, it has the same range between refueling as a gasoline vehicle (265 miles).

If a typical LDV travels 12,000 miles/year, one fill of 11 gal is required every 7.4 days (22 mpg × 11 gal × 365 days/year/12,000 miles/year), giving an average daily consumption of 1.5 gal (11 gal/7.4 days). This is equivalent to 0.6 kg/day for an HV achieving 2.6 times the fuel economy of a comparable gasoline vehicle.

Peak refueling demand typically occurs on Friday evening as shown in Figure 29.10. In addition to the hourly and daily variations discussed previously, refueling demand is also subject to seasonal fluctuations. The summer driving season typically sees an increase in travel and fuel use in the United States by about 10% above annual average demand. It is assumed that the increase in demand during the 120-day summer driving season corresponds with a decline in demand during the remaining months. This seasonal demand fluctuation is handled upstream of the refueling station by employing either geologic storage or liquid hydrogen storage to supplement production as discussed in Section 29.3.1.

29.3.2.2 Refueling Station Design Parameters

The refueling station includes one or more dispensers, cascade charging units, compressors (for gas delivery), pump/evaporator units (for liquid delivery), precooling unit (for high-pressure gaseous dispensing), and fuel storage units. Station average daily dispensing rates can range from 50 to 6000 kg/day. The lower limit represents a demonstration-scale station visited infrequently by experimental fuel cell vehicles, while the upper limit represents a larger commercial station with capacity similar to large gasoline stations. The station size is specified by its average daily dispensing rate. The vehicle's refueling fill pressure can range from 350 to 700 bar. As such, the maximum cascade charging system pressure is assumed to be 430 bar for the 350 bar fill and 875 bar for the 700 bar fill. The 700 bar fill requires hydrogen precooling to allow refueling at the typical desired fill rate of 1.5 kg/min. The precooling temperature ranges between −20°C and −40°C. The typical configurations of refueling stations receiving gaseous deliveries are shown in Figures 29.12 and 29.13 for 350 bar and 700 bar dispensing, respectively.

29.3.2.2.1 *Refueling Station Cascade Charging System*

The cascade charging system is comprised of several pressure vessels. For example, a typical cascade system for 350 bar filling includes three vessels, each with a 21.3 kg holding

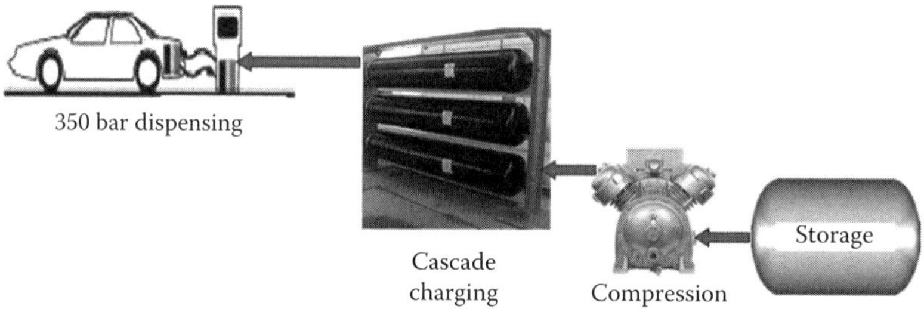

FIGURE 29.12
Refueling station configuration for 350 bar (5000 psi) dispensing from gaseous deliveries.

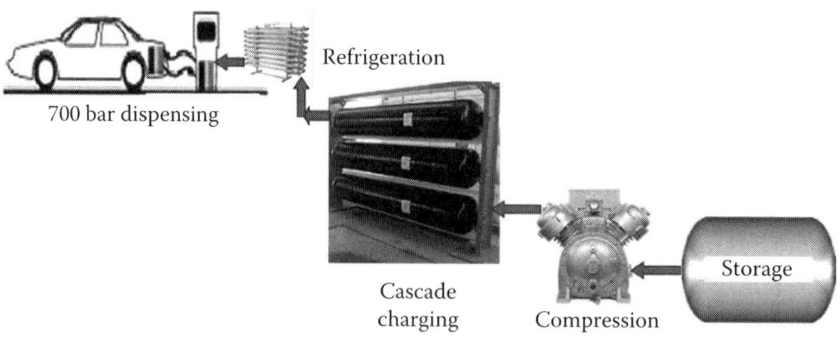

FIGURE 29.13
Refueling station configuration for 700 bar (10,000 psi) dispensing from gaseous deliveries.

capacity and a maximum pressure of 430 bar. There may be more than one bank of 3 cascade charging vessels depending on the size of the refueling station. To satisfy the vehicle filling dynamics, each of the vessels operates under a different minimum pressure, for example, 400, 250, and 100 bar.

29.3.2.2.2 Refueling Station Compressor

For pipeline distribution, the compressor operates in either of the two following modes:

1. During periods of low station demand, the compressor takes suction from the distribution pipeline at approximately 20 bar and delivers intermediate pressure gas to a buffer storage unit at 170 bar.
2. During periods of high station demand, the compressor takes suction from both the distribution pipeline and the buffer storage unit and delivers high-pressure gas to the cascade charging system.

For compressed gas tube-trailer truck distribution, the compressor takes suction from the tube trailer and delivers high-pressure gas to the cascade charging system.

29.3.2.2.3 Liquid Refueling Station Pump, Evaporator, and Storage

While the gaseous refueling stations employ a compressor to charge the cascade system, the liquid refueling stations employ a pump and an evaporator to achieve the same goal. The pump takes suction from the liquid storage tank pressure and raises the pressure to the cascade charging system pressure. The high-pressure liquid is then gasified in the evaporator and charged into the cascade vessels. For vehicles with onboard cryocompressed storage, the pump dispenses hydrogen directly to the vehicle's onboard storage tank at a supercritical state (e.g., 350 bar and 20 K). This eliminates the need for an evaporator and cascade buffer storage, but pumping capacity must be increased to compensate for the lost capacity of the buffer storage. The cryogenic liquid storage tanks at the refueling station are sized to satisfy the station average daily demand. The typical configurations of refueling stations receiving liquid deliveries are shown in Figures 29.14 and 29.15 for gaseous and cryocompressed dispensing, respectively.

29.3.3 Transmission and Distribution Pipelines

There are three stages of pipeline for urban deliveries: transmission pipeline, distribution (trunk) pipeline, and distribution service pipeline. The arrangement is similar to that for

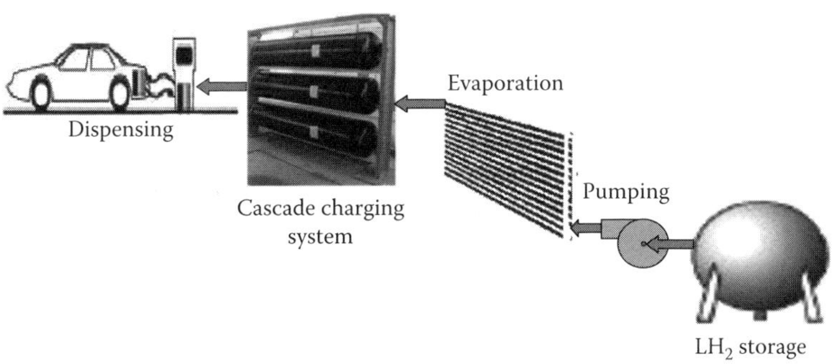

FIGURE 29.14
Refueling station configuration for gaseous dispensing from liquid deliveries.

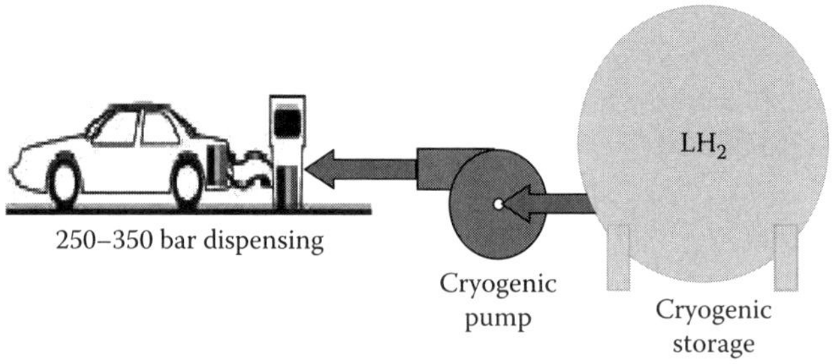

FIGURE 29.15
Refueling station configuration for cryocompressed dispensing from liquid deliveries.

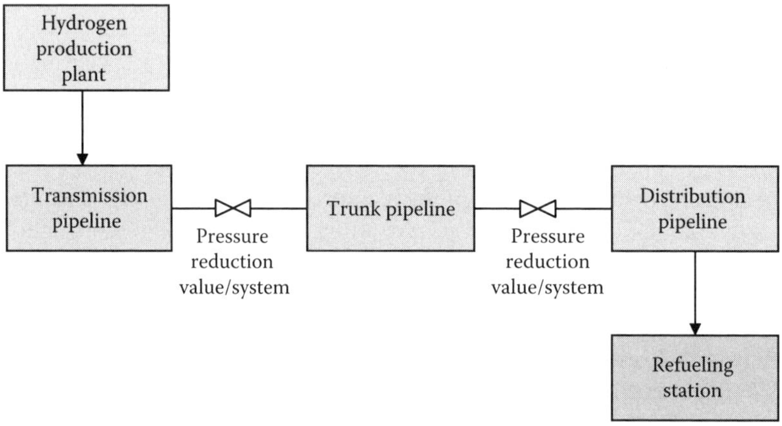

FIGURE 29.16
Transmission and distribution pipeline arrangement.

natural gas transmission and distribution. Hydrogen gas is moved from the production plant to the city gate through large, high-pressure transmission lines. At the city gate, the pressure is reduced, and the gas is moved through trunk pipelines for the distribution system. In the distribution service pipelines, the pressure is once more reduced, and the gas is distributed to the refueling stations. In all cases, the hydrogen pressure is reduced through a combination of pressure drop through the pipeline and a pressure reduction valve and/or system. A flow diagram of a representative system is shown in Figure 29.16.

The current natural gas pipeline system operates with transmission line pressures in the range of 35–80 bar. The current very limited hydrogen transmission pipelines supplying hydrogen to refineries operate in this range as well. Higher transmission pipeline pressures may be feasible and desirable in the future. It is advantageous to keep the hydrogen pipeline pressure as high as deemed practical and safe since vehicle refueling is expected to be at high pressure.

29.3.4 Liquefaction Plants

For small cities or rural communities and during the early market of hydrogen fuel cell vehicles, the construction of transmission and distribution pipelines from the production plant to refueling stations may not be economically feasible. For some combinations of city size and delivery distance, compressed gas tube trailers and liquid hydrogen trucks may be preferred. Liquid hydrogen delivery has added benefits of ensuring the high purity required by fuel cell vehicles and permitting cryocompressed dispensing that can extend HV range.

Hydrogen liquefaction involves gas compression, cooling with water, and then precooling with liquid nitrogen to drop the hydrogen below its inversion temperature.

Hydrogen molecules may exist in para- and orthoforms, depending on the electron configurations in the two atoms in the molecule. At hydrogen's boiling point of −253°C, the equilibrium concentration is primarily parahydrogen; however, at room temperature and above, the equilibrium concentration is about 25% para- and 75% orthohydrogen. If the hydrogen is liquefied without first catalytically converting the ortho- to the para form, the orthohydrogen will slowly convert to parahydrogen in an exothermic reaction releasing about 0.15 kWh/kg of energy. The heat of transformation can cause the evaporation of as much as 50% of the liquid hydrogen over a 10-day period. The ortho- to para-hydrogen conversion is performed during liquefaction by means of a catalyst, with the heat released during conversion removed by cooling with liquid nitrogen and then further cooling with liquid hydrogen. Depending on the plant capacity and the technology employed, the liquefaction electric energy requirement is in the range of 8–18 kWh/kg [4,5].

29.3.5 Gaseous Tube Trailers

A tube trailer incorporates several tubes designed to withstand a specific pressure. For nine tubes, each with a volume of 2.6 m^3, the holding capacity of the trailer is 344 kg with a tube pressure of 180 bar. Since the tube trailer cannot be completely discharged, the delivered capacity is usually reduced (i.e., 280 kg for a heel pressure of 15 bar). The same tube trailer is capable of delivering 650 kg of hydrogen at 350 bar. Composite tubes made of carbon fibers have the potential of delivering hydrogen at higher capacity due to the higher pressure rating and low weight. Composite tube capacity can be increased further (to ~1000 kg) by cooling the compressed gas. The time for loading hydrogen into the tubes depends on the capacity of the loading compressors as well as the temperature limitation imposed by the tube structure. The loading time is typically in the range of 6–10 h.

29.3.6 Liquid Trucks

The typical liquid truck tank capacity is approximately 17,000 gal, with a nominal holding capacity of 4,600 kg of hydrogen. The truck fill time is approximately 2–3 h. The delivered amount is reduced by the boil-off amount during transportation and unloading at refueling stations, which could be as high as 6% of the truck total load. Thus, it is advantageous to minimize the number of stations served per trip to avoid excessive losses.

29.4 Economic Analysis of Hydrogen Delivery Infrastructure

29.4.1 Impact of Refueling Station Cost

Figure 29.17 shows delivery cost estimates for a market with population of approximately one million (e.g., Indianapolis, Indiana) and HV market penetration of 20%, assuming a hydrogen production plant located 100 km (62 miles) from the city boundary and 350 bar

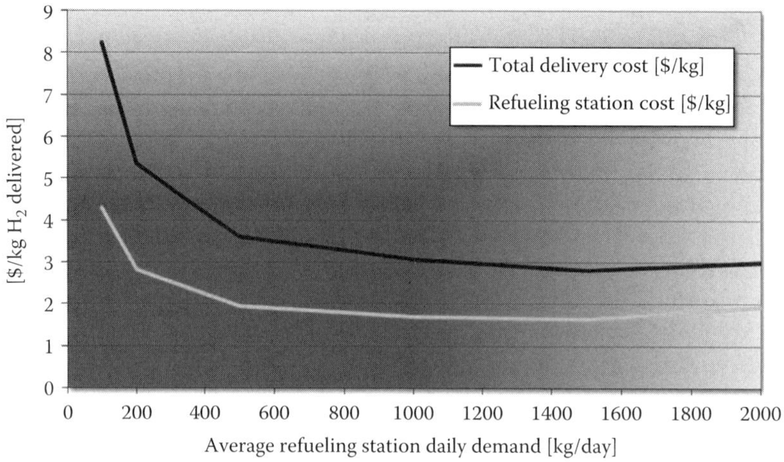

FIGURE 29.17
Total levelized delivery cost and fuel station cost for pipeline delivery.

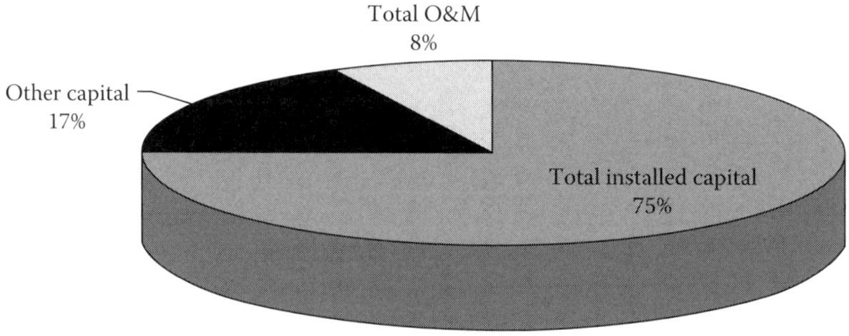

FIGURE 29.18
Breakdown of fuel station cost.

gaseous hydrogen dispensing. Such a scenario generally produces the lowest cost delivery. As shown in the figure, the fuel station is a major contributor to the total cost of hydrogen delivery (which also includes a central compressor, transmission pipeline, distribution pipelines, and, in this scenario, geologic storage), accounting for half or more of that cost. As Figure 29.18 shows, most of the station cost is installed capital. For small stations, installed capital is fairly evenly distributed among compressors, cascade systems, and storage; for larger stations, electrical upgrades become significant (Figure 29.19). These results are robust across a range of station sizes, markets, and daily demand.

29.4.2 Impact of Delivery Mode

Figure 29.20 compares the cost (in $/kg delivered) of delivering hydrogen to Los Angeles via the three main delivery modes assuming 10% market penetration and a station with 300 kg/day average dispensing capacity. For such a market, liquid truck delivery is the least cost option at market penetrations below 10%, while pipeline delivery is preferable for market penetrations above 10%. For all penetrations, tube-trailer delivery is more costly than either liquid truck or pipeline delivery and also poses potential logistics problems by necessitating multiple daily deliveries for larger station sizes. However, since

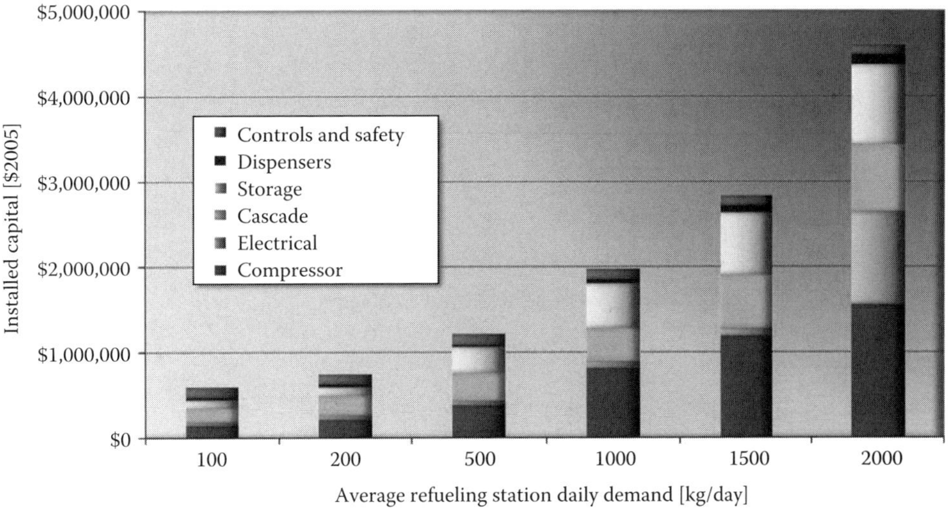

FIGURE 29.19
Gaseous refueling station capital cost.

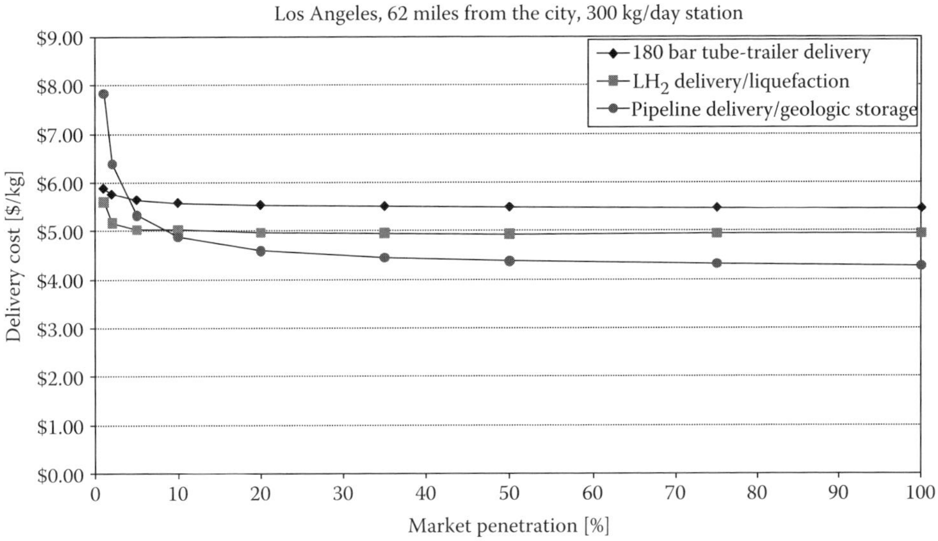

FIGURE 29.20
Comparison of delivery modes for 300 kg/day stations.

higher-pressure tube deliveries (e.g., 350 bar) could reduce daily deliveries and delivery cost, that option could be attractive, especially at lower market penetrations. Table 29.1 contrasts the advantages and disadvantages of each delivery mode for transportation and distribution against several important delivery metrics.

29.4.3 Impact of Distance between Production Plant and City Gate

The location of the production plant with respect to a given market can significantly impact delivery cost. Figure 29.21 shows delivery cost as a function of the distance from production

TABLE 29.1

Relative Advantages and Disadvantages of Delivery Modes for Transportation and Distribution[a]

	Compressed-H_2 Gas in Tube Trailers		Compressed-H_2 Gas via Pipelines		Liquid-H_2 Trucks	
Delivery Mode →	**Transportation (over Long Distance, e.g., >>100 mi)**	**Distribution (and Short Transportation Distance)**	**Transportation (over Long Distance, e.g., >>100 mi)**	**Distribution (and Short Transportation Distance)**	**Transportation (over Long Distance, e.g., >>100 mi)**	**Distribution (and Short Transportation Distance)**
1. Low market demand (<<100 tonne/day)						
Economics (Capital + O&M)	− − −	+ + +	+ +	− − −	+	+ +
Technology readiness	+ +	+ +	+	+	+ + +	+ + +
Logistics of delivery	+	−	+	− − −	+ +	+
Infrastructure storage cost	−	−	−	−	+	+
GHG emissions	+	+	+ +	+ +	− − −	− − −
2. High market demand (>100 tonne/day)						
Economics (Capital + O&M)	− −	+	+ + +	+ +	+	+ +
Technology readiness	+ +	+ +	+	+	+ + +	+ + +
Logistics of delivery	− −	− −	+	− −	+	+
Infrastructure storage cost	− −	− −	− −	− −	+	+
GHG emissions	+	+	+ +	+ +	− − −	− − −

[a] A "+" symbol indicates an advantage, while a "−" symbol indicates disadvantage.

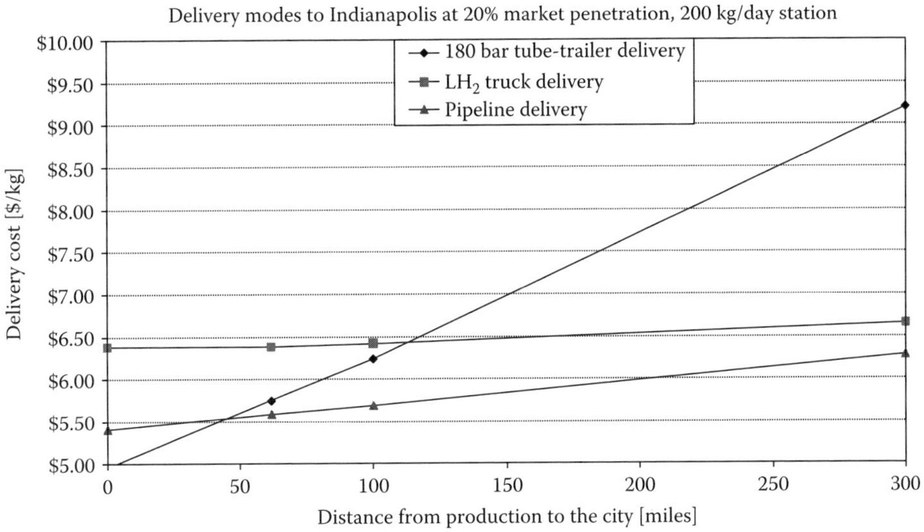

FIGURE 29.21
Comparison of delivery modes at low market penetration and small stations for different distances to city gate.

plant to the city boundary (i.e., transmission distance) for Indianapolis at 10% market penetration and for a station size of 200 kg/day. At low market penetration and short distances between plant and city gate, the figure indicates that tube-trailer delivery is more economical than the other two delivery modes simply because the 200 kg/day station size assumed for that market is matched well with tube-trailer delivered capacity. For greater distances from the city gate, pipeline delivery becomes more economical than tube-trailer delivery due to the rapid increase in the number of tube trailers required as production shifts farther away from the city. The cost of liquid truck delivery exhibits the lowest rate of increase with distance among delivery modes. This is attributed to the high capacity of the liquid trucks and its low cost relative to the other components (e.g., liquefier) in this delivery pathway. This gives liquid truck delivery an advantage over the other two delivery modes at much longer distances (greater than 300 miles) for the assumed market penetration and small station size (which may be typical for early markets). As market penetration and station size increase, pipeline delivery becomes more economical than the other modes regardless of distance from production site to city gate (Figure 29.22). This is because the cost contribution of local service lines (i.e., small-diameter pipelines connecting individual refueling stations with trunk or main distribution lines) to total delivery cost is greatly reduced with increased station capacity and the transmission line cost contribution is reduced by economies of scale at higher HV market penetration. It should be noted that for a given HV market penetration, as station capacity increases, the number of stations decreases proportionately.

29.4.4 Impact of Tube-Trailer Capacity

Figure 29.23 shows the impact of tube-trailer capacity on hydrogen delivery cost. The figure includes the current 180 bar, 280 kg capacity tube trailer; the 350 bar, 500 kg capacity tube trailer currently in demonstration; and a conceptual 1000 kg capacity composite tube trailer. Delivery cost drops significantly by increasing the loading pressure from 180 to 350 bar and could potentially drop further if expectations for the 1000 kg conceptual tube trailer materialize.

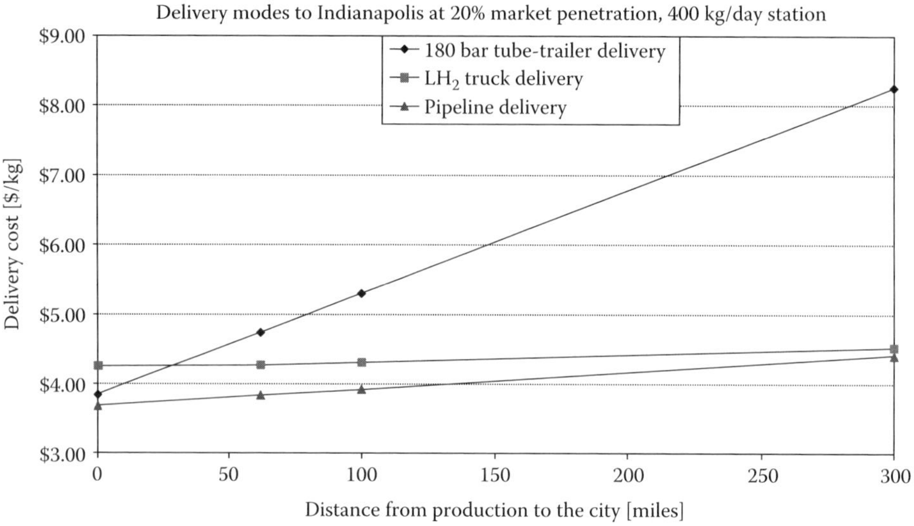

FIGURE 29.22
Comparison of delivery modes at high market penetration and large stations for different distances to city gate.

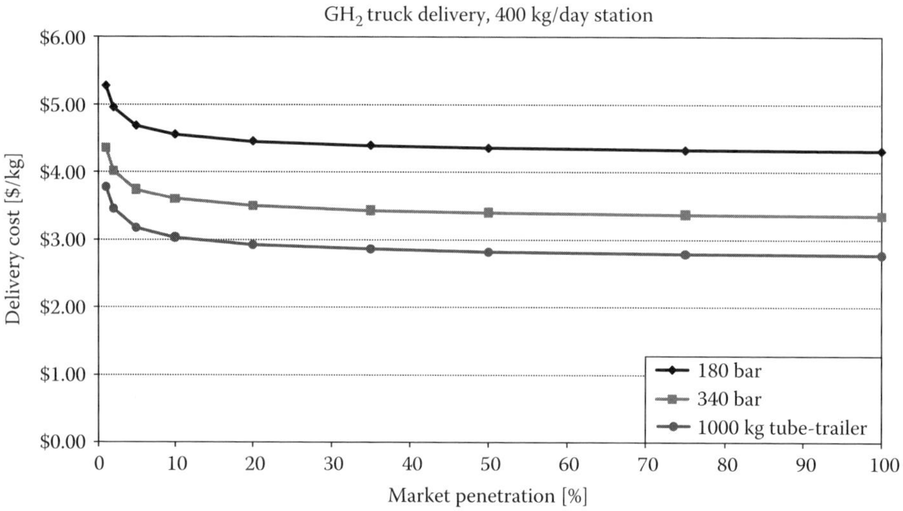

FIGURE 29.23
Comparison of different capacity tube-trailer delivery costs (Indianapolis Market at 62 miles from production plant).

29.4.5 Impact of Dispensing Options

In addition to the transportation and distribution options discussed previously, there are several options for dispensing hydrogen to onboard HV storage tanks. Hydrogen must be compressed or cooled (or a combination of both) to increase its energy density onboard the vehicle and thus extend the vehicle's driving range. At present, consenus is building around 700 bar storage. Honda had favored 350 bar gaseous storage, but recently revised its position to 700 bar, the pressure favored by other automakers (e.g., GM, Daimler, Toyota, and Ford). Some automakers (e.g., BMW) advocate cryocompressed (CcH_2) hydrogen tanks capable of

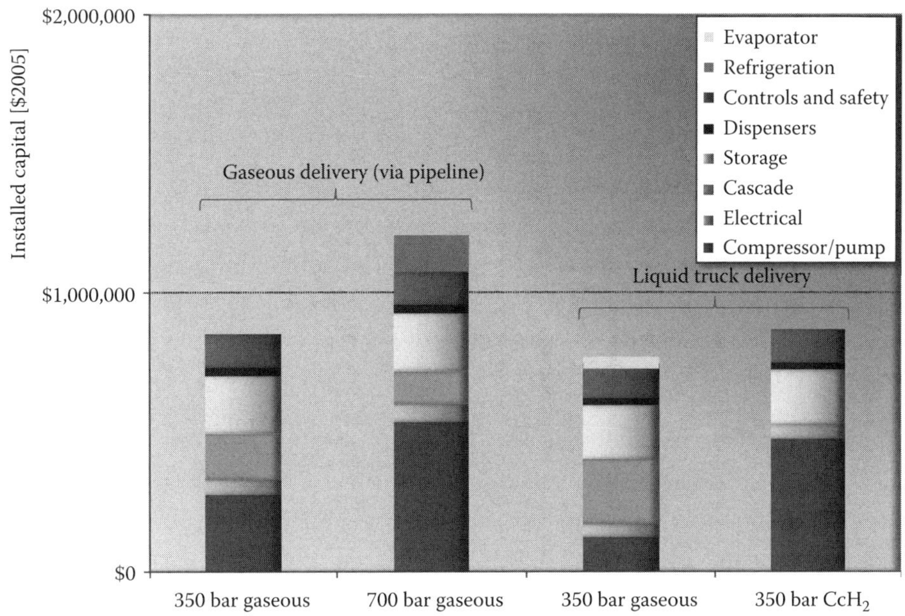

FIGURE 29.24
Comparison of refueling station capital cost for different dispensing options of a 400 kg/day station.

operating at cryogenic temperatures and high pressure. The dispensing option significantly impacts the type and cost of equipment required at the station as well as the station's land requirement or footprint. Since refueling stations are individually owned and operated by small investors in the United States, the refueling station land area, initial capital investment, and the associated risk are a major barrier to automakers' plans to commercialize HVs.

As shown in Figure 29.24, cryocompressed dispensing is much less costly than 700 bar dispensing for a 400 kg/day station. While the cost of cryocompressed dispensing is equivalent to 350 bar gaseous dispensing, the energy density onboard the vehicle is much higher, and the station footprint is much smaller. Table 29.2 contrasts the advantages and disadvantages of each dispensing option in terms of several important refueling station metrics.

29.5 Fuel-Cycle Energy Use and Greenhouse Gas Emissions

Other important aspects of hydrogen delivery are the energy use and GHG emissions associated with hydrogen transmission and distribution from production plants to refueling stations. Figure 29.25 shows on-site energy use and upstream energy consumption (associated with producing and supplying the on-site energy source) for the three main delivery modes. Compression energy is significant for compressed gas delivery via tube trailer or pipeline, consuming the equivalent of 40% of the energy content (lower heating value) of the delivered hydrogen. However, the two options differ with respect to the pathway step or component where compression energy is consumed. While the storage compression takes place at the gaseous (GH_2) terminal for tube-trailer delivery, such compression takes place at the refueling station for pipeline delivery. This difference in the location for storage

TABLE 29.2

Relative Advantages and Disadvantages of Refueling Dispensing Options for H_2 Storage Onboard FCVs[a]

	350 Bar Compressed-H_2 Gas		700 Bar Compressed-H_2 Gas (Cooled to −40°C)		Cryocompressed H_2 (20 K, 350 bar)	
Dispensing Option →	**Small Station (<100 kg/Day)**	**Large Station (>>100 kg/Day)**	**Small Station (<100 kg/Day)**	**Large Station (>>100 kg/Day)**	**Small Station (<100 kg/Day)**	**Large Station (>>100 kg/Day)**
Station economics (Capital + O&M)	−	+	− −	−	+	+ +
Station footprint (land area)		−	−	− −	− −	−
Technology readiness	+	+		−	− −	− −
Vehicle range	−	−	+	+	+ +	+ +
Hydrogen quality	+	+	+	+	+ + +	+ + +
Cost of vehicle's onboard storage	+	+	−	−	+ +	+ +

[a] A "+" symbol indicates an advantage, while a "−" symbol indicates disadvantage.

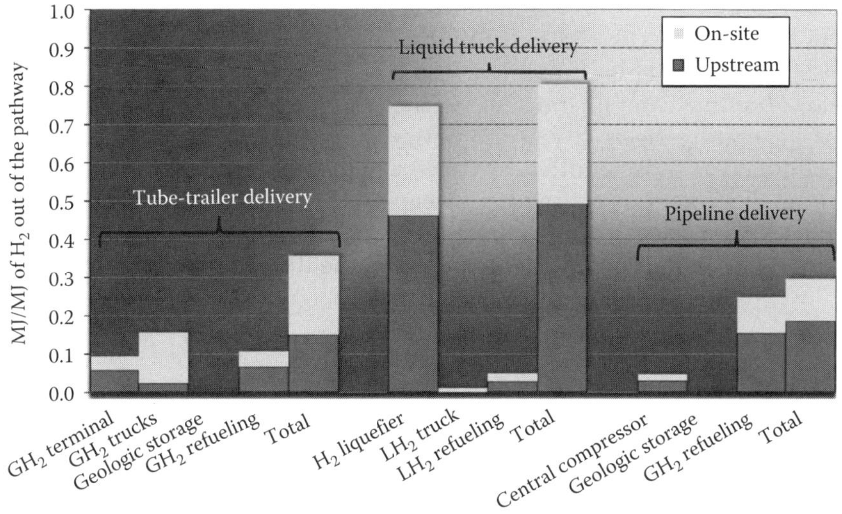

FIGURE 29.25
Comparison of fuel-cycle energy use by delivery mode.

compression results in lower energy consumption at the refueling station for tube-trailer delivery compared to pipeline delivery. Liquid truck delivery consumes significantly more energy than compressed gas delivery, primarily due to the high energy consumption in the liquefaction process. Liquid truck delivery consumes an energy amount, on a life cycle basis, equivalent to 80% of the energy content of the hydrogen. For comparison power plants consume approximately 200% of the energy they produce in generated electricity.

Figure 29.26 shows GHG emissions associated with each of the three delivery modes. Emissions by each component of the delivery pathways are proportional to the energy

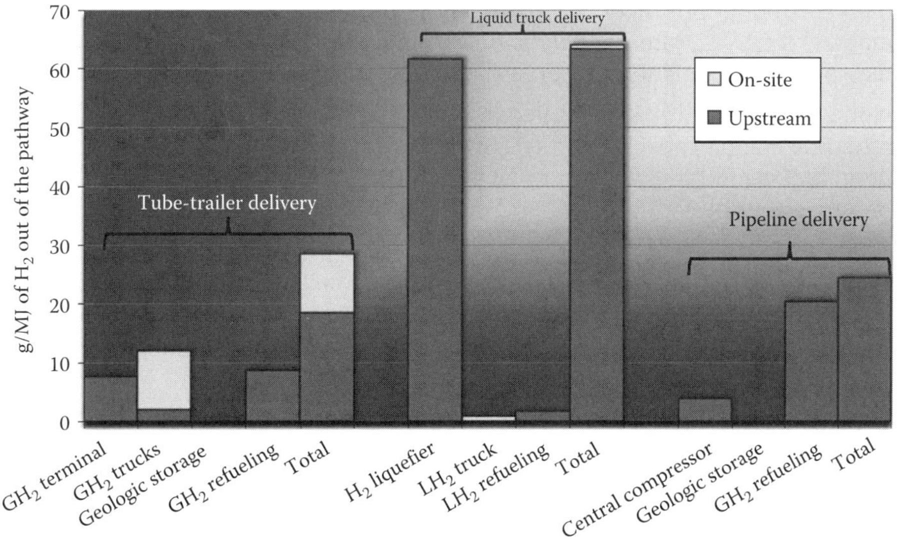

FIGURE 29.26
Comparison of fuel-cycle GHG emissions by delivery mode.

use by that component. The only difference is that the ratios of on-site to upstream emissions do not necessarily correspond with the ratios of on-site to upstream energy use as shown in Figure 29.25. This is mainly because of on-site electricity use, which involves no emissions (for that component) since all the emissions have occurred upstream in the process of generating electricity at the power plants. Other GHG emissions shown in Figure 29.26, such as those emitted by tube trailers and liquid trucks, occur mainly on-site with a smaller fraction occurring upstream in the process of recovering and processing diesel fuel from its petroleum source. As can be seen in the figure, the liquefier is by far the largest GHG emitter of all components in the three delivery pathways.

29.6 Conclusions

The delivery cost of hydrogen is influenced by the key components associated with the transmission and distribution of hydrogen from its point of supply in central plants to the points of demand at refueling stations. The cost of most components in this delivery pathway is primarily impacted by economies of scale, which is a function of the penetration of HVs in a given market. The refueling station is a major cost contributor to the total levelized cost of hydrogen delivery. The transmission distance of hydrogen from its production site to a given market and the delivery mode from the production site to the refueling stations also significantly impact delivery cost.

The following are general conclusions for currently available hydrogen delivery technologies:

- At low market demands (<10% market penetration) with a central plant 100 km (62 miles) or more from the city, the delivery cost of hydrogen to refueling stations is high for all delivery modes ($5–$10/kg of hydrogen or even higher), suggesting that distributed production of hydrogen at refueling stations may serve the early markets for HVs. Alternatively a small semicentral plant located at the city gate may provide sufficiently low delivery cost by tube trailers.
- If the city is small (<400,000 people), the market penetration is low (<10%), the refueling station capacity is small (<400 kg/day), and the distance to the production plant is modest (<100 km or 62 miles), then hydrogen delivery by tube trailer provides the lowest cost option. For early market conditions, delivery costs of $5–$12/kg are anticipated.
- If one or two market conditions move from the *small* to the *large* category, hydrogen delivery by liquid truck may be the lowest-cost approach. However, the energy consumed is up to 40%–80% the energy in the hydrogen delivered due to the energy intensity of hydrogen liquefaction.
- For a maturing hydrogen fuel cell vehicle market (>20% market penetration), hydrogen delivery by pipeline is almost universally preferred, with expected delivery costs in the range of $2–$4/kg of hydrogen depending on the size of the city and market penetration level.
- If the hydrogen production plants are located less than 100 km (62 miles) from the *city gate* and if tube trailers are developed to deliver 1000 kg or more of hydrogen, the cost of tube-trailer delivery drops significantly and approaches the cost of pipeline delivery.

This approach could avoid the required cost, time, disruption, and potential safety concerns of building hydrogen pipeline distribution systems in urban areas.

- The energy use in the delivery of hydrogen can be significant. For pipeline delivery, tube-trailer delivery, and liquid hydrogen delivery, the well-to-vehicle tank energy use is about 30%, 35%, and 80% of the energy in the delivered hydrogen, respectively.
- GHG emissions are lowest with pipeline delivery and moderately higher with tube-trailer delivery but increase dramatically with liquid delivery.

Each of the options examined here—tube-trailer delivery, liquid truck delivery, and pipelines—represents the optimum delivery method at specific maturation of the hydrogen infrastructure and HV market. As such, efforts to reduce the energy requirements and the capital cost of each method can reduce the overall costs of hydrogen delivery in the transition to and widespread use of hydrogen fuel cell vehicles. Possible research efforts include the following:

- Lower-cost composite-based high-pressure storage vessels for hydrogen storage and cascade charging systems at the refueling station. These storage vessels are a major cost for all delivery pathways.
- Composite-based high-pressure (350 bar or higher) tube trailers or other advanced tube trailers (e.g., cold gas tube delivery) with a capacity of 1000 kg of hydrogen.
- Fiber-reinforced polymer (FRP) transmission and/or distribution pipelines to reduce pipeline capital and thus pipeline delivery costs. The distribution lines are the larger portion of the pipeline costs.
- High-efficiency compression, magnetic, or other novel methods for hydrogen liquefaction and the use of renewable energy for liquefaction.

Acknowledgments

This work was supported by the US DOE's Office of Fuel Cell Technologies Office. We gratefully acknowledge the assistance of the team whose efforts led to the initial HDSAM model, most notably Daryl Brown of the Pacific Northwest National Laboratory, Mark Paster (formerly) of the US DOE, Matt Ringer of the National Renewable Energy Laboratory, and Bruce Kelly of Solar Abengoa, as well as a number of industry and laboratory experts who provided valuable inputs and model review.

References

1. National Research Council of The National Academies, 2008, *Transitions to Alternative Transportation Technologies—A Focus on Hydrogen*, The National Academies Press, Washington, DC.
2. A. Elgowainy, M. Mintz, B. Kelly, M. Hooks, and M. Paster, 2008, Optimization of compression and storage requirements at hydrogen refueling stations, *Proceedings of the ASME Pressure Vessels & Piping Conference*, PVP2008-61638, Chicago, IL, July 27–31.

3. M. Mintz, J. Gillette, A. Elgowainy, M. Paster, M. Ringer, D. Brown, and J. Li, 2007, A hydrogen delivery scenario analysis model to analyze hydrogen distribution options: HDSAM, *Transportation Research Record: Journal of the Transportation Research Board*, 1983, 114–120.
4. Hydrogen Delivery Infrastructure Options Analysis, 2008, Mexant, Inc. and partners Air Liquide, Chevron Technology Ventures, Gas Technology Institute, TIAX, LLC, and Argonne National Laboratory, Report Number DE-FG36-05GO1503, http://www/.eere.energy.gov/hydrogenandfuelcells/pdfs/delivery-infrastructure-analysis.pdf (Accessed January 2014).
5. M. Gardiner, S. Satyapal, 2009, Energy requirements for hydrogen gas compression and liquefaction as related to vehicle storage needs, Department of Energy Program Record Number 9013, accessed July 2010 at http://www.hydrogen.energy.gov/pdfs/9013_energy_requirements_for_hydrogen_gas_compression.pdf.

Section IX

Cross-Cutting Topics

30

Development of Hydrogen Safety Codes and Standards in the United States

Chad Blake
National Renewable Energy Laboratory

Carl Rivkin
National Renewable Energy Laboratory

CONTENTS

This chapter provides an overview of hydrogen codes and standards with an emphasis on the national effort supported and managed by the US Department of Energy (DOE). With the help and cooperation of standards and model code development organizations, industry, and other interested parties, DOE has established a coordinated national agenda for hydrogen and fuel cell codes and standards. With the adoption of the research, development, and demonstration (RD&D) roadmap and with its implementation through the Codes and Standards Technical Team (CSTT), DOE helps strengthen the scientific basis for requirements incorporated in codes and standards that, in turn, will facilitate international market receptivity for hydrogen and fuel cell technologies. This work was supported by the US DOE under contract no. DE-AC36-08-GO28308 with the National Renewable Energy Laboratory.

30.1 Introduction

Large quantities of hydrogen have been used safely as a chemical feedstock and industrial gas for many years. Codes, standards, and regulations governing hydrogen storage, distribution, and use at industrial sites are well established. The use of hydrogen as an energy carrier for consumer markets is expected to grow over the next decade, and the development and promulgation of codes and standards for this use are essential to establish a market-receptive environment for commercial hydrogen products and systems.

For hydrogen *energy* use in the United States, the International Code Council (ICC) and the National Fire Protection Association (NFPA) are the two principal model code development organizations. The ICC develops and publishes a family of model codes; the most relevant for hydrogen energy are the International Fire Code (IFC), International Fuel Gas Code (IFGC), International Building Code (IBC), and International Mechanical Code (IMC) [1]. The NFPA develops and publishes both standards and codes [2]. For hydrogen energy, the most widely used of these are NFPA 55 (Standard for the Storage, Use, and Handling of Compressed Gases and Cryogenic Fluids in Portable and Stationary Containers, Cylinders, and Tanks) and NFPA 52 (Vehicular Gaseous Fuel Systems Code). The NFPA has incorporated all of its provisions for hydrogen into a single document, *NFPA 2, Hydrogen Technologies Code*.

Hydrogen standards are typically written under a consensus process by technical committees representing a cross section of interested parties and issued in the United States by organizations such as the American Society of Mechanical Engineers (ASME) for pressure vessels, pipelines, and piping; the Compressed Gas Association (CGA) [6] for pressure vessel operation and maintenance; and the Underwriters Laboratories (UL) for product certification. In the United States, the American National Standards Institute (ANSI) facilitates the development of national standards by accrediting the procedures of standards developing organizations (SDOs) such as those mentioned previously [3] (Figure 30.1).

The adoption and enforcement of codes and standards in the United States take place under the jurisdiction of some 44,000 entities that include city, county, and state governments, as well as special districts such as port and tunnel authorities. Regulations make use of existing standards, either by incorporating appropriate sections of the standards (incorporation by transcription) or by referring to those sections (incorporation by reference). The extremely decentralized enforcement of codes and standards means that the permitting process for hydrogen fuel facilities can be very cumbersome.

The federal government plays a limited role in the development, adoption, and enforcement of codes and standards, but federal safety regulations are incorporated in the Code of Federal Regulations (CFRs). Those that apply to hydrogen are embodied primarily in 49 CFR 171 and 29 CFR 1910, under the jurisdictions of the Department of Transportation (DOT) and Occupational Safety and Health Administration (OSHA), respectively. The DOT regulates the transportation of hydrogen [4]. The OSHA regulates the safe handling of hydrogen in the workplace. OSHA regulations are intended to provide worker safety for hydrogen use [5]. However, neither of these would apply to the public using hydrogen at a retail refueling facility.

While most industrialized countries have adopted regulations, codes, and standards that govern the use of hydrogen, many of these countries also support the development of international standards to facilitate international trade and commerce. For hydrogen energy, the key international SDOs are the International Organization for Standardization (ISO) and the International Electrotechnical Commission (IEC). Information about domestic and international hydrogen codes and standards and current activities of ISO technical committees, including draft standards under preparation or review, can be found at www.fuelcellstandards.com, a website supported by the US DOE. Another useful source of information on hydrogen safety, codes, and standards is the Hydrogen Safety Report, a monthly newsletter published by the National Hydrogen Association (NHA) at www.hydrogensafety.info, also supported by DOE.

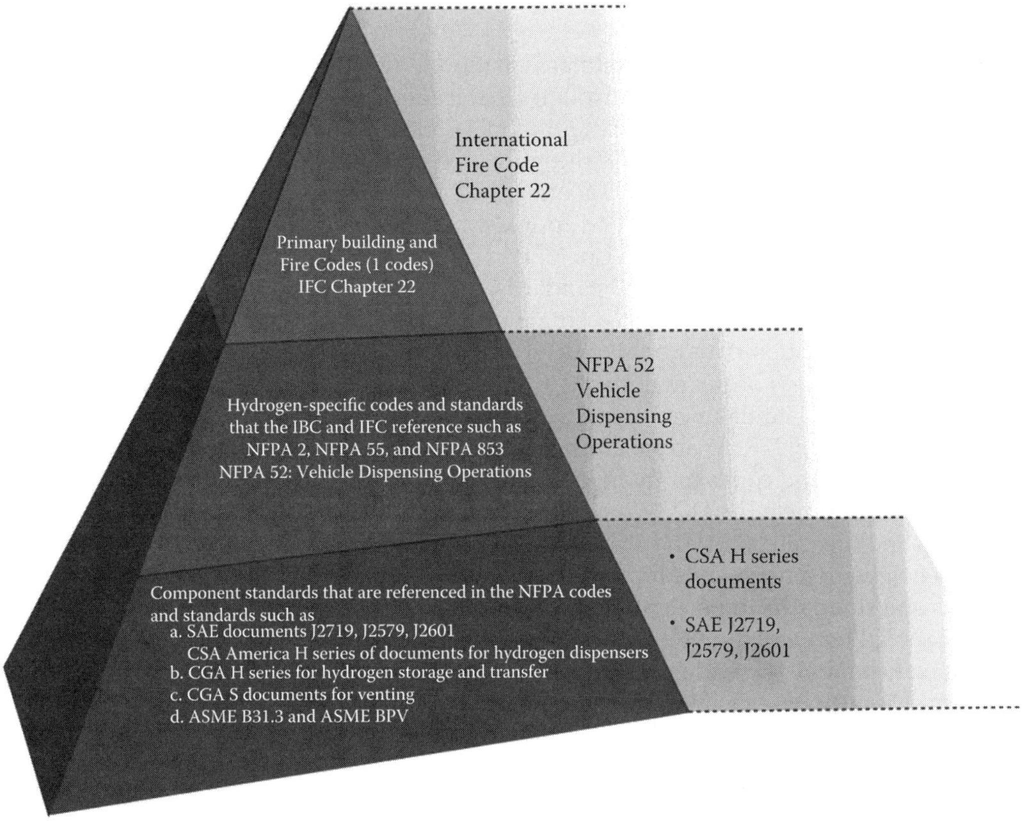

FIGURE 30.1
Codes and standards hierarchy.

30.2 DOE Program for Hydrogen Codes and Standards

For the past decade, the Office of Hydrogen, Fuel Cells and Infrastructure Technologies in DOE has sponsored a collaborative national effort by government and industry to prepare, review, and promulgate codes and standards needed to expedite hydrogen infrastructure development and to help enable the emergence of hydrogen as a significant energy carrier. In addition, DOE has worked to harmonize national and international standards, codes, and regulations that are essential for the safe use of hydrogen by consumers in the United States and throughout the world. The National Renewable Energy Laboratory (NREL) provides technical and programmatic support to DOE for this effort.

DOE has also launched a comprehensive RD&D effort to obtain the data needed to establish a scientific basis for requirements incorporated in hydrogen codes and standards. This RD&D is planned, conducted, and evaluated in collaboration with industry through the US FreedomCAR and Fuel Partnership formed to examine and advance precompetitive research and development of technologies to enable high volume production of affordable hydrogen fuel cell vehicles and the national hydrogen infrastructure necessary to support them. The codes and standards activities of the partnership are conducted through the CSTT that adopted a roadmap to guide the RD&D.

30.2.1 National Templates

Over the past several years, a coordinated national agenda for hydrogen and fuel cell codes and standards has emerged through DOE leadership and the support and collaboration of industry and key standards and model code development organizations (SDOs and CDOs). For example, hydrogen is recognized as a fuel gas, and hydrogen applications have been incorporated in the editions of the ICC model codes since 2003. Provisions for the safe use of hydrogen are included in ICC's International Building, Residential, Fire, Mechanical, and Fuel Gas Codes. Also, NFPA has incorporated hydrogen safety requirements into its family of codes and standards, as noted earlier. The consolidation of all hydrogen safety requirements into a single document, NFPA 2, is a major step toward development of a national hydrogen code.

A key to the success of the national hydrogen and fuel cell codes and standards development efforts to date has been the creation and implementation of national templates through which DOE, NREL, and the major SDOs and CDOs coordinate the preparation of critical standards and codes for hydrogen and fuel cell technologies and applications. The national templates help the DOE to create and maintain a coordinated national agenda for hydrogen and fuel cell codes and standards. DOE leadership has coincided with the emergence of heightened national and international interest in hydrogen energy in general and in codes and standards in particular.

The national templates have been accepted by the major SDOs and CDOs in the United States, the FreedomCAR and Fuel Partnership, key industry associations, and many state and local governments as the guideposts for the coordinated development of standards and model codes. All of the relevant major SDOs and CDOs in the United States are part of this national effort: the ANSI, ASME, American Society of Testing and Materials (ASTM), CGA [6], CSA America [7], ICC, NFPA, Society of Automotive Engineers (SAE), and UL. Industry participants include the FreedomCAR and Fuel Partnership (Chrysler, Ford Motor Company, General Motors, BP, Chevron, ConocoPhillips, ExxonMobil, Shell Hydrogen); other industry members, such as Ballard Power Systems, General Electric, Plug Power, Hydrogenics, UTC Power; and industry associations, such as the American Petroleum Institute (API), NHA, and the US Fuel Cell Council (USFCC). Other federal agencies involved include the DOT and the National Institute of Standards and Technology (NIST). Other organizations participate on an as-needed basis (Figure 30.2).

The objectives of the national templates are to

- Establish by a consensus of the national codes and standards development organizations the CDOs or SDOs that will have the lead in the development of codes and standards for establishing safety requirements for specific components, subsystems, and systems (as shown in the templates) and the organizations that will work collaboratively with (or in support of) the lead organization
- Minimize duplication of efforts in the codes and standards development
- Establish "boundaries" and interfaces among standards for components, subsystems, and systems and identify harmonization requirements *across* such standards
- Identify codes and standards development needs and gaps and identify the organizations that should have responsibility for addressing the gaps

Implementation of the national templates is coordinated through the National Hydrogen and Fuel Cells Codes and Standards Coordinating Committee, created by DOE, NREL, NHA,

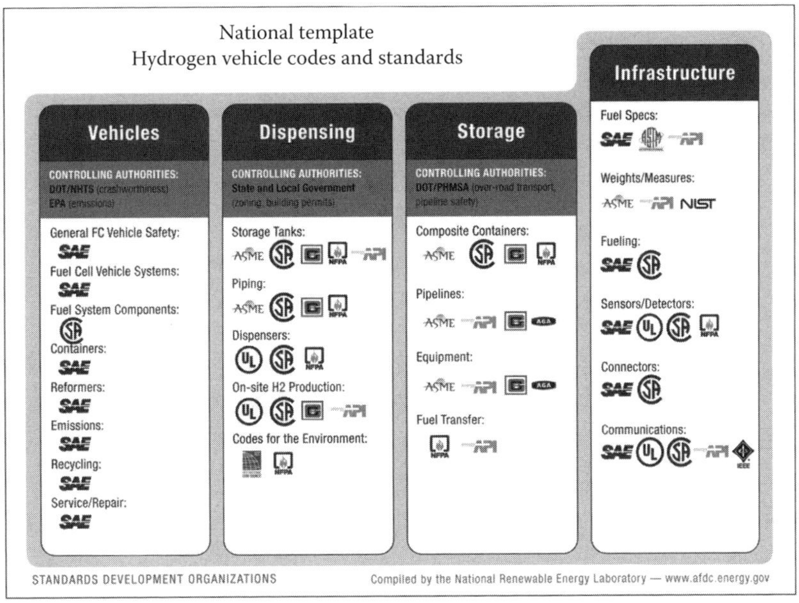

FIGURE 30.2
National template graphic.

and USFCC. The committee conducts monthly conference calls to update participants on current activities and to discuss key issues. In addition, the committee meets quarterly to coordinate codes and standards development and prevent duplication of effort, identify critical deficiencies and gaps in hydrogen codes and standards development that could have an adverse impact on market acceptance, determine a collaborative strategy and action plan to address critical gaps and deficiencies, and identify specific opportunities for organizations to work together in developing codes and standards. The minutes of conference calls and proceedings of meetings are posted at www.hydrogenandfuelcellsafety.info.

DOE supports implementation of the templates through subcontracts with a number of SDOs and CDOs designated for lead roles on the templates. It should be noted that significant work to implement the templates is being done by organizations not funded by DOE. While the templates were not intended to specify which organizations should receive DOE funding, they have helped to solidify the roles of the organizations identified as having a lead role in developing a particular standard.

In summary, the templates continue to function as the seminal documents that help to create a more unified national approach to the development of hydrogen and fuel cell codes and standards. The templates, and the National Hydrogen and Fuel Cells Codes and Standards Coordinating Committee that was formed to manage the templates, have created a *virtual national forum* for SDOs, CDOs, industry, government, and interested parties to address codes and standards issues, both immediate and long term.

30.2.2 Research, Development, and Demonstration for Codes and Standards

The RD&D roadmap helps guide DOE activities that will provide data required for SDOs to develop performance-based codes and standards for a commercial hydrogen-fueled transportation sector in the United States. The roadmap reflects the experience and priorities of the members of the FreedomCAR and Fuel Partnership, which include the DOE, energy companies (BP, Chevron, ConocoPhillips, ExxonMobil, Shell Hydrogen), and the automotive companies (Chrysler, Ford, General Motors) belonging to the US Consortium for Automotive Research (USCAR). The contents of the roadmap are reviewed and revised by the partnership as needed to reflect changing needs and opportunities.

By evaluating specific needs for RD&D, assessing the status of ongoing RD&D, and revising the roadmap as needed, the partnership will ensure new US projects are efficiently leveraged and coordinated with those undertaken internationally. Through the International Partnership for the Hydrogen Economy (IPHE), DOE works with individual countries as well as contributing to global RD&D efforts. Information requirements of international SDOs are considered to help align RD&D projects with needs for code and standard development.

The roadmap includes an assessment of existing hydrogen and fuel cell codes and standards and those that are in the process of being established domestically and internationally and identifies information needs and gaps related to those codes and standards for a hydrogen-based transportation system. The CSTT of the partnership reviews RD&D projects to address gaps and to provide documented research to SDOs on a continuing basis.

The roadmap is organized into four focus areas:

- Hydrogen behavior
- Hydrogen-fueled vehicles
- Hydrogen fuel infrastructure
- Fuel–vehicle interface

The technical goal for each of these focus areas is to gather sufficient information and validating experience on technology applications so that the responsible SDO or CDP can proceed with better data upon which to base requirements incorporated in its codes and standards. Each focus area is subdivided into key target areas, which identify important information needs for which information is required by SDOs and

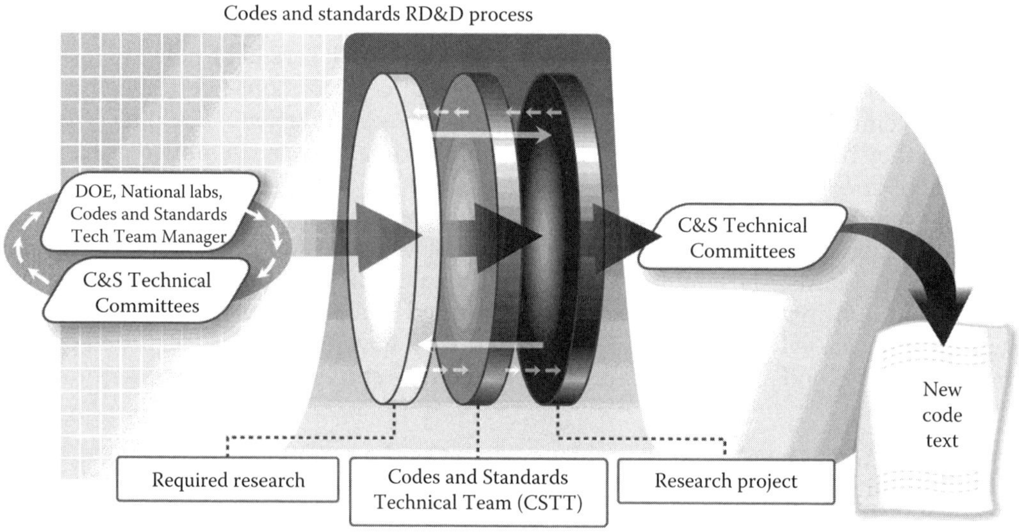

FIGURE 30.3
RD&D process.

CDOs to fully develop codes and standards. The completion of RD&D for the individual technical target area, in conjunction with information distribution, is expected to result in the subsequent development of safe, performance-based codes and standards (Figure 30.3).

30.3 Conclusion

Two key needs for hydrogen safety in consumer applications are the incorporation of data and analysis from RD&D into the codes and standards development process and the application of these codes and standards by state and local code officials. DOE supports a comprehensive program to address both these aspects of hydrogen safety. First, DOE is working with the automobile and energy industries to identify and address high-priority RD&D to establish a sound scientific basis for requirements that are incorporated in hydrogen codes and standards. The high-priority RD&D needs are incorporated and tracked in a roadmap adopted by the CSTT of the FreedomCAR and Fuel Partnership. DOE and its national laboratories conduct critical RD&D and work with key standards and model CDOs to help incorporate RD&D results into the codes and standards process. To address the second aspect, DOE has launched an initiative to facilitate the permitting process for hydrogen fueling stations (HFSs). A key element of this initiative is a web-based information repository that includes toolkits with informational fact sheets, networking charts to encourage information exchange among code officials who have permitted or are in the process of permitting HFSs, case studies of likely near-term HFS configurations, and a database of key codes and standards. The information repository is augmented by workshops for code officials and station developers in jurisdictions that are likely to have HFSs in the near future. This information can be accessed at www.hydrogen.energy.gov/permitting.

The national templates have guided DOE's effort to accelerate the development of key standards and model codes for hydrogen and fuel cell applications. With the help and cooperation of SDOs, CDOs, industry, and other interested parties, DOE has established a coordinated national agenda for hydrogen and fuel cell codes and standards. With the adoption of an RD&D roadmap by the partnership and through its implementation by the CSTT, the DOE will help strengthen the scientific basis for requirements incorporated in these codes and standards that, in turn, will facilitate international market receptivity for hydrogen and fuel cell technologies.

In fiscal year 2010, the DOE's Office of Hydrogen, Fuel Cells and Infrastructure Technologies Program became the Fuel Cell Technologies Program. This programmatic change shifted the focus to fuel cell technologies although it did retain most of the existing project work. The Fuel Cells Technologies Program has responsibility for all potentially commercial fuel cell technologies, which is an expansion beyond the prior program responsibility for PEM hydrogen fuel cells. This new program structure is also more focused on stationary fuel cell applications as opposed to the prior focus on fuel cells for vehicles.

Although this program change will likely not impact the existing national hydrogen codes and standards templates, new templates will likely be developed to address the codes and standards required to support the full range of fuel cell technologies. The CSTT as well as the Multi-Year Plan may be restructured to address the new program responsibilities.

References

1. International Code Council, http://www.iccsafe.org (accessed December 18, 2009).
2. National Fire Protection Association, http://www.nfpa.org (accessed December 18, 2009).
3. American National Standards Institute ANSI, *Overview of the U.S. Standardization System*, 2nd edn., 2007.
4. U.S. Department of Transportation, http://www.hydrogen.dot.gov (accessed December 18, 2009).
5. 29CFR Part 1910.103 and http://www.osha.gov (accessed December 18, 2009).
6. Compressed Gas Association, http://www.cganet.com/isotc197 (accessed December 18, 2009).
7. CSA America, http://www.csa-america.org (accessed December 18, 2009).

31

International Codes, Standards, and Regulations for Hydrogen Energy Technologies

Karen Hall
Technology Transition Corporation, Ltd.

Geoffrey Bromaghim
Technology Transition Corporation, Ltd.

CONTENTS

31.1 Overview

The subject of codes, standards, and regulations for hydrogen energy technologies is vast, and it is a rapidly moving, evolving target. Many dedicated people from industry, government, and trade associations, as well as fire safety specialists, permitting officials, and code and standards development organizations (CDOs and SDOs), are all working together to establish the codes, standards, and regulations that will enable wide-scale deployment of hydrogen energy technologies.

Hydrogen energy safety is based on three primary elements: regulatory requirements, capability of safety technology, and the systematic application of equipment and procedures to minimize risks. Industry currently implements many successful proprietary methodologies for safely handling large amounts of hydrogen. These experiences are making their way into developing standards through direct participation by industry experts. There are several codes, standards, and regulations specifically for hydrogen that are under construction at all levels of government in many countries. There are many efforts underway to standardize hydrogen system components for safety in a variety of potential commercial hydrogen market applications.

Through workshops, working groups, coordination committees, and similar activities, there is also increasingly effective collaboration between industry, CDOs and SDOs, and many other experts. This collaboration makes it possible to identify technical areas of expertise required to produce the codes and standards that stakeholders feel are required to facilitate commercialization of hydrogen and fuel cell technologies and infrastructure. Hydrogen experts have the opportunity to participate directly in technical committees and working groups where issues can be discussed with the appropriate industry groups.

31.2 Background

Before we get into the technical details, it is essential to clarify some of the key terminology.

31.2.1 What Are Codes, Standards, and Regulations?

There are many types of documents of interest to those of us involved in developing codes and standards. What are they and how do they differ? This section provides basic information on the definitions of codes, standards, regulations, technical specifications (TSs), technical reports (TRs), information reports, and recommended practices. It is intended to help the reader understand what is implied by each of these document types and to help assess which type may be most appropriate for a new activity.

31.2.1.1 Codes

The Merriam-Webster's dictionary defines a code as a *systematic statement of a body of law* or *a system of principles or rules*.[1] They generally apply to construction or the built environment. Codes establish minimum requirements for things like offset distances between permanent fixtures, ventilation requirements, plumbing and electrical requirements, and other items relating to a built environment. A code may reference a standard. If you are adding a deck onto your home, or expanding your porch, or installing a hot tub on your deck, you will need to get a permit. To verify the safety and grant approval for such a project, the jurisdiction having authority over your project will want to see proof that the job will be done in a way that conforms to existing codes. The code may reference applicable standards, such as Underwriters Laboratories (UL) standards.

Codes are meaningless unless a state or local jurisdiction adopts them. There are over 44,000 local code enforcement agencies in the United States alone. They have the option of adopting any model code from any year and may make local code amendments. This leads to the potential for a lot of variation in codes. Codes most often encountered in the hydrogen and fuel cell arena include International Code Council (ICC) and National Fire Protection Association (NFPA) codes.

31.2.1.2 Standards

Merriam-Webster defines a standard as "something set up as a rule for measuring or as a model to be followed."[2] This does not appear to be very different from a code. In some cases, they really are not very different. Often, when we talk about a standard for a hydrogen system or component, we are referring to standards for the component or system rather than the standard for installation, although those exist as well. A standard might be performance based, that is, *each unit must meet the following tests*, or it may be a design standard, that is, *the nozzle must be made of the following materials and have the following dimensions*. Standards for manufacturing or testing a unit are independent of where the unit will be used. But standards are not mandatory until they are called out someplace, such as in a code, regulation, procurement contract, or other requirements document. What standards do allow, however, is a consensus process for developing minimum technical requirements to assure uniformity of the product, including safety and performance. Often, when a regulator or code official is unfamiliar with an emerging technology or new equipment design, having a standard gives that official a starting place for evaluating the technology. In addition, it gives the official some confidence that the information is based on best practices and industry consensus. It takes much of the guesswork out of the equation.

The term *International Standards* typically refers to standards published by the International Organization for Standardization (ISO) or the International Electrotechnical Commission (IEC) or both. The capitalization is meant to avoid confusion with other standards developed internationally. An *international standard* is a standard that is adopted by an international standardizing/standards organization and made available to the public.[3] An *International Standard* is one where the international standards organization is ISO or IEC.

Both ISO and IEC have published guides for the preparation of documents prepared in accordance with the ISO/IEC directives. These guides are available on the ISO website (http://www.iso.org) and IEC website (http://www.iec.ch).[4]

31.2.1.3 Regulations

Merriam-Webster defines a regulation as *a rule dealing with details of procedure* or *an order issued by an executive authority of a government and having the force of law*.[5] For example, in the United States, we have regulations from the Department of Transportation that relate to transporting dangerous and hazardous goods. The Federal Aviation Administration (FAA) has regulations regarding what can and cannot be brought onboard an aircraft, how often an aircraft is inspected, how many consecutive hours a pilot can be on duty, etc. These regulations are generally safety-oriented.

31.2.1.4 Technical Specifications

TSs are used by ISO and IEC, among others, when the subject in question is still under development or where for any other reason, the possibility of an agreement to publish a standard is not imminent. In this case, it is used for *prestandardization purposes*.

A TS may also be used when the required support cannot be obtained for a final draft of international standard to pass the approval stage or in case of doubt concerning consensus. In ISO and IEC, TSs are subject to review by the technical committee or subcommittee not later than 3 years after their publication. The purpose of this review is to reexamine the situation that resulted in the publication of a TS and, if it is possible, to achieve the agreement necessary for the publication of an international standard to replace the TS.

31.2.1.5 Publicly Available Specifications

In ISO or IEC, a publicly available specification (PAS) may be developed when a specification is needed prior to all the requirements of a standard being met. A PAS may be an intermediate specification, published prior to the development of a full international standard, or, in IEC, may be a *dual-logo* publication published in collaboration with an external organization. It is a document not fulfilling the requirements for a standard. A PAS remains valid for an initial maximum period of 3 years. The validity may be extended for a single 3-year period, following which it shall be revised to become another type of normative document (such as an international standard) or shall be withdrawn.

31.2.1.6 Technical Reports

When a technical committee or subcommittee has collected data of a different kind from what is normally published as an international standard (this may include data obtained from a survey carried out among the members or data on the *state of the art* in relation to standards of national bodies on a particular subject), the work may be published in the form of a TR. TRs are entirely informative in nature. The technical committee or subcommittee responsible decides on withdrawal of a TR.

31.2.1.7 SAE Documents

The Society of Automotive Engineers (SAE) is an SDO, which is comprised of automotive and other mobility engineering professionals and focuses on developing documents that govern the engineering of powered vehicles. The SAE issues the following types of documents[6]:

- SAE Standards: These TRs are a documentation of broadly accepted engineering practices or specifications for a material, product, process, procedure, or test method.
- SAE Recommended Practices: These TRs are documentations of practice, procedures, and technology that are intended as guides to standard engineering practice. Their content may be of a more general nature, or they may propound data that have not yet gained broad acceptance.
- SAE Information Reports: These TRs are compilations of engineering reference data or educational material useful to the technical community.

SAE has a committee dedicated to developing documents relating to fuel cell vehicles (FCVs).

So while codes, standards, and regulations are each important to protect the public, in most cases, each has a unique niche. Codes generally apply to the built environment. Standards generally apply to components, systems, and testing. And regulations generally apply to transportation or rules of procedure. Each SDO and CDO has specific rules for their process for consensus, and the requirements for different document types vary. For example, the ISO process for a TR requires a less stringent consensus process, and therefore less time, than the process for an International Standard, which has the most stringent requirements to assure consensus.

Now that we have a general feel for the differences, it is important to note one very real similarity. The development of a new code, standard, regulation, or any other document type listed previously takes 2–5 years on average (sometimes *much* longer). That is due to the fact that industry consensus is required, sometimes nationally or internationally. Often, there is too little data available upon which to proceed, and the process may be slow while appropriate testing is conducted or data are gathered.

As regulations are effectively laws, they can typically be obtained through the publishing government agency at no cost. However, as codes and standards are developed by organizations whose business is their development and promulgation, these documents are copyrighted. In addition, many of the pertinent codes and standards are in development. Therefore, the content is subject to change until the final approval stage and publication. Because of this, it is not practical to provide detailed technical information about each code and standard that applies to hydrogen energy technologies.

Instead, this chapter describes the basics of codes, standards, and regulations, highlights the main organizations involved in their development, provides a short overview of the scope of key documents, and provides resources for additional information.

31.3 Why Are Codes, Standards, and Regulations Needed?

For over 50 years, gaseous hydrogen has been used in large quantities as a feedstock in the petroleum refining, chemical, and synthetic fuel industries. Examples include making ammonia for fertilizer and removing sulfur in petroleum refining for such products as *reformulated* gasoline. Hydrogen is also used in the food processing, semiconductor, glass, and steel industries, as well as by electric utilities as a coolant for large turbine generators.

Existing industrial safety rules, regulations, consensus standards, and codes relating to the transporting and utilization of hydrogen are adequate for today's markets. The use of hydrogen has resulted in an admirable safety record. However, in the case of widespread usage of hydrogen for future emerging applications, today's safety rules, consensus standards, codes, etc. may not be adequate. Systematic efforts by local/state/federal government entities, producers of hydrogen products (e.g., automotive industry), codes and standards organizations, users, and others must be devoted to

1. Identifying safety-related issues associated with the production and use of hydrogen-fueled systems
2. Developing or updating and then validating regulations, codes, and standards relating to the safe transportation, use, and servicing of hydrogen-fueled systems

While hydrogen has been used extensively in a number of industrial applications, in energy applications, the only significant use of hydrogen has appeared in space programs. This is beginning to change, given the promise that hydrogen is an efficient energy carrier and an energetic fuel with minimal environmental impact. Systems are being implemented that produce hydrogen from primary energy sources (such as sunlight, wind power, biomass, hydroelectric, and fossil fuels) and use hydrogen in energy applications for home and office heating, for generating electricity, and as a transportation fuel.

Because of the growing applications for hydrogen energy, efforts are underway to create consensus standards for domestic and international use; develop enforceable building, fire, mechanical, plumbing, and other building code provisions; and to harmonize, to a practical extent, requirements from different countries to facilitate international trade.

Widespread hydrogen use requires that safety be intrinsic to all processes and systems. To develop a hydrogen infrastructure that has the public's confidence in its safety and convenience, an industry consensus on safety issues is required. This includes the development of compatible standards and formats (e.g., the same couplings for dispensing the same form of fuel). Product certification protocols are also required.

Codes and standards development is occurring in advance of or in parallel with applications for hydrogen energy and hydrogen-fueled systems. Codes and standards development must be coordinated with technology development so that the technologies can be sited as they enter commercial or precommercial deployment phases. Efforts are also devoted to research and development to validate proposed requirements.

The US Department of Energy (DOE) held a National Hydrogen Energy Roadmap Workshop on April 2–3, 2002, to identify issues surrounding safety, codes, and standards for hydrogen energy systems. The roadmap developed at this seminal event, as well as regular updates in consultation with industry, continues to guide DOE funding priorities for codes and standards.

Following the development of the roadmap, clear roles for the various US SDOs were identified. This allowed work to begin in a coordinated fashion. The creation of National Hydrogen and Fuel Cells Codes and Standards Coordinating Committee (NHFCCSCC) has further increased coordination among US stakeholders by providing a monthly forum to discuss issues and timelines for the development of key codes and standards. This increased coordination helps reduce duplication of effort and provides a mechanism for more timely information exchange, benefiting the national effort.

31.4 Who Is Involved?

A large number of organizations are involved in creating consensus documents in a variety of technical disciplines. CDOs create requirements for the built environment, including building codes, fire codes, mechanical codes, and plumbing codes. SDOs create and maintain standards, TRs, best practices, etc. in the technical discipline for which they have the national or international remit. The best way to understand which organization is responsible for which disciplines is to look at the US national and international templates posted at http://www.hydrogenandfuelcellsafety.info/resources.asp and the matrix of hydrogen and fuel cell codes, standards, and regulations posted at www.fuelcellstandards.com.[7]

31.4.1 United States

In the United States, the development of the codes and standards necessary for commercialization of hydrogen and fuel cell systems is a priority for the US DOE and the hydrogen industry. The US model CDOs—the ICC and the NFPA—both provide processes for public code change proposals. Over the past 5 years, industry has been working closely with these organizations to include changes to facilitate the approval of hydrogen and fuel cell installations. This takes place through the organizations' public code change processes. Members of the public may make recommendations on changes to the US model codes, with supporting justification for the changes. The technical committee responsible for maintaining the specific codes then meets to review all the proposed changes and make a recommendation to the voting members of the CDO (NFPA or ICC). The public then has a chance to review all the proposals and the committee recommendations and provide comments in the case of ICC or exceptions in the form of a notice to make a motion at the annual meeting in the case of the NFPA. Procedures and timelines differ between the ICC and NFPA. Each organization publishes their procedures as well as deadlines for the various code development cycles on their websites:

ICC: www.iccsafe.org

NFPA: www.nfpa.org

Local jurisdictions adopt building and fire codes according to their needs, and they may adopt a model code in its entirety or develop modifications. There are extensive efforts in the United States to identify areas where requirements for hydrogen energy systems may be technically different and work through the open code development processes to harmonize requirements.

Let us now take a look at some of the specific relevant US model code documents promulgated by NFPA and ICC that we need to consider in order to deploy hydrogen energy equipment.

31.4.1.1 National Fire Protection Association

National Fire Protection Association (NFPA) has several existing codes and standards that address the use of hydrogen and hydrogen technologies, including the following:

1. *NFPA 2 Hydrogen Technologies Code, 2011 edition.* This document is NFPA's hydrogen technologies project, which consolidates the requirements of all of the following documents and addresses any areas that the current NFPA hydrogen documents do not address.
2. *NFPA 55 Standard for the Storage, Use, and Handling of Compressed Gases and Cryogenic Fluids in Portable and Stationary Containers, Cylinders, and Tanks, 2013 edition.* This document gives storage, handling, and use requirements for compressed gases including hydrogen. There are two chapters in this document devoted to gaseous and liquefied hydrogen storage systems.
3. *NFPA 52 Vehicular Gaseous Fuel Systems Code, 2013 edition.* This code covers natural gas vehicular fuel systems. These systems include storage of fuels at dispensing facilities and dispensing facility operations, design, maintenance. There are also some onboard vehicle requirements. This code no longer covers hydrogen fueling stations.

4. *NFPA 853 Standard for the Installation of Stationary Fuel Cell Power Plants, 2010 edition.* This document addresses installation of stationary fuel cells and refers to NFPA 55 for the hydrogen storage requirements.
5. *NFPA 30A Code for Motor Fuel Dispensing Facilities and Repair Garages, 2012 edition.* This document covers refueling facilities that use liquid fuels and combinations of liquid and gaseous fuels. It also covers the repair of most gaseous fuel requirements.
6. *NFPA 70 National Electrical Code, 2014 edition.* Article 692 of this document contains requirements for electrical safety for fuel cells.
7. *NFPA 497 Recommended Practice for the Classification of Flammable Liquids, Gases, or Vapors and of Hazardous (Classified) Locations for Electrical Installations in Chemical Process Areas, 2012 edition.* This document gives requirements for the electrical classification of areas where hydrogen would be used or stored. Electrical classification is a critical issue and can have a significant impact on project costs.
8. *NFPA 88A Standard for Parking Structures, 2011 edition.* This document covers the construction and protection of, as well as the control of hazards in, open and enclosed parking structures.
9. *NFPA 86 Standard for Ovens and Furnaces, 2011 edition.* This document contains requirements for hydrogen storage systems that are used for gas quenching operations.
10. *NFPA 5000 Building Construction and Safety Code, 2012 edition.* This document contains construction requirements for buildings storing hydrogen. Hydrogen is not addressed directly, but relevant consideration for hydrogen is addressed generally as a flammable gas in the hazardous material chapter.
11. *NFPA 1 Uniform Fire Code, 2012 edition.* An extract document that contains large pieces of NFPA 52 and 55 (among many other documents). The fire code is meant to address operational safety as opposed to the building code that addresses construction safety.

31.4.1.2 International Code Council

The ICC also promulgates model codes for adoption in the United States and elsewhere. Although the ICC does not have separate standards or codes dedicated to alternative fuels or hydrogen, provisions for hydrogen technologies are predominantly found in the International Fire Code and the International Fuel Gas Code. For example, the International Fire Code includes harmonized separation distance requirements for bulk hydrogen storage, using methodologies developed by Sandia National Laboratories and industry over the past several years.

31.4.1.3 National Standards and Recommended Practices

US SDOs are organizations with a remit to develop standards for specific technology areas. For example, the American Society of Mechanical Engineers (ASME) develops and maintains Boiler and Pressure Vessel Standards. The Compressed Gas Association (CGA) develops and publishes technical information, standards, and recommendations for safe and environmentally responsible practices in the manufacture, storage, transportation, distribution, and use of industrial gases. There are many US SDOs involved in developing requirements for hydrogen energy systems. The US national template for hydrogen vehicle systems and refueling facilities is updated regularly and posted on several websites. An Internet search for the template will also yield a number of presentations made at conferences on the development and maintenance of the template.

The National Renewable Energy Laboratory (NREL) has recently published *Vehicle Codes and Standards: Overview and Gap Analysis*. This report covers six of the most commonly available fuels designated by the DOE as vehicle alternative fuels, including hydrogen. Table 13 in this report provides specific section references in the various codes and standards for hydrogen fuel, equipment, and processes. This 17-page table also provides the edition citation for the model code references for hydrogen. This is a particularly useful reference to help determine which codes and standards may apply for a particular type of equipment or application—particularly in the United States—and when new provisions became available.[8]

As stated previously, there are numerous CDOs and SDOs involved in developing and maintaining the codes and standards that relate to hydrogen energy systems. For example, hydrogen-related standards for onboard vehicles fall under the purview of the SAE. Component documents are developed by the ASME, the National Institute of Standards and Technology (NIST), the American Society for Testing and Materials (ASTM), the CGA, UL, CSA America, and others as appropriate.

An excellent resource for understanding which organizations are involved in standards development and the progress of these efforts is through the NHFCCSCC. This group meets monthly, usually by teleconference, so that all the stakeholders have the regular opportunity to discuss the issues, challenges, and timescales. Minutes of these meetings are posted in the *Hydrogen and Fuel Cell Safety Report*, a monthly online publication that provides news about developing hydrogen and fuel cell codes and standards and related safety information (see Section 31.5 for more information on this resource).* These can be reviewed each month as they are posted or by viewing the related webpages dedicated exclusively to this group where this information is archived.

Here is a brief description of the scope of the national CDOs and SDOs in the development of standards for hydrogen energy systems:

- SAE: FCV standards, including terminology, safety, recyclability, and interface issues between the vehicles and the refueling station.
- ASME: Pipelines, piping, and hydrogen storage containers.
- NIST: Weights and measures for commercial sale of hydrogen.
- ASTM: Sampling and standard test methods—in particular, to measure hydrogen fuel quality.
- CGA: Pipelines, pressure equipment, and serve as the administrator of the US Technical Advisory Group (TAG) for ISO/TC 197 (the ISO Technical Committee on Hydrogen Technologies—described in the next section).
- UL: Hydrogen sensors, flammable gas detectors, and certification.

* The *Hydrogen and Fuel Cell Safety Report* is publicly available at www.hydrogenandfuelcellsafety.info.

- CSA America: Hydrogen dispensing, on-site hydrogen production, and certification. CSA currently has three activities dedicated to hydrogen energy technologies:
  - HPIT 1—Compressed Hydrogen Powered Industrial Truck On-board Fuel Storage and Handling Components
  - HPIT 2—Compressed Hydrogen Station and Components for Fueling Powered Industrial Trucks
  - CHMC 1—Test Method for Evaluating Material Compatibility for Compressed Hydrogen Applications

31.4.2 International Standards

Internationally, there are two main technical committees involved in developing and maintaining international standards for hydrogen and fuel cell technologies:

- IEC/TC 105 is the IEC Committee on Fuel Cell Technologies.
- ISO/TC 197 is the ISO Technical Committee on Hydrogen Technologies.

A description of the scope of the documents developed by these organizations, as well as the status of the documents and points of contact for each work item, is published online at www.fuelcellstandards.com.

The approval process for installation of hydrogen and fuel cell equipment varies between countries. The process may depend on whether the installation is in an industrial or a residential environment, as different authorities have responsibility for the industrial and residential permitting procedures in many countries.

Most countries have regulatory requirements in place for fuel gases. In most cases, hydrogen is not currently covered by existing regulations as a fuel gas; therefore, more time may be required for preparing technical information for the permitting authority and for the review of that information.

Building regulations, codes, and standards describe a set of rules that specify an acceptable level of safety for constructed objects, both buildings and nonbuilding structures. Requirements in these documents may be country-specific and typically deal with issues including

- Design and construction to ensure structural stability of the building and adjoining buildings
- Fire safety, means of escape, prevention of internal and external fire spread, and access and facilities for the fire services
- Ventilation
- Drainage and waste disposal
- Use of combustion appliances and fuel storage
- Protection from falling, collision, and impact
- Energy efficiency and conservation
- Access to and use of the building
- Electrical safety

Some buildings may be exempt from these controls, such as temporary buildings, buildings not frequented by people (unless it is located close to a building that is), small detached buildings (such as garages, garden storage, sheds, and huts), and simple extensions (such as porches, covered ways, and conservatories). However, it is good practice to have an exemption confirmed by the appropriate authority prior to construction.

In Europe, the principal regulations covering hydrogen facilities arise from the national legislation passed to implement the explosive atmosphere (ATEX) directives and the pressure equipment directive. Their requirements are not specific to hydrogen and would equally apply to any fuel that is capable of generating a flammable atmosphere, for example, natural gas or liquefied petroleum gas (LPG), or equipment that contains a fuel under pressure. For some components of the installation, for example, if the hydrogen is produced by internal reformation of natural gas, the requirements of European regulation, the gas appliances directive, may also be applicable.

More detailed information on the applicable European directives can be found in the Installation Permitting Guide. This document was created in response to the growing need for guidance to facilitate small hydrogen and fuel cell stationary installations in Europe. This document is not a standard but is a compendium of useful information for a variety of users with a role in installing these systems. Please see http://www.hyperproject.eu/for the interactive version of this guide.

Please see www.fuelcellstandards.com for information on international standards and regulations applicable to hydrogen and fuel cells. This website contains details on North American, Pacific Rim, European, and international codes, standards, and regulations, including both those already published and those in development.

31.4.2.1 International Standards and ISO/TC 197

The scope of ISO/TC 197 is standardization in the field of systems and devices for the production, storage, transport, measurement, and use of hydrogen. The first meeting of ISO/TC 197 was held in June 1990 in Zurich, Switzerland. The most recent meeting was held on December 5–6, in Paris, France.

ISO/TC 197 has established liaison relationships with other applicable ISO and IEC technical committees, and where appropriate, with subcommittees.

New working groups are formed when a member country submits a new work item proposal (NWIP) to the technical committee, and it is approved through ballot with affirmative votes and agreement to participate in the work. Working groups whose documents are finished are no longer active. As a result, the list of active working groups is ever-changing. ISO/TC 197 working groups are assigned work items. In most cases, a working group is responsible for preparing a single document. In some cases, a working group may be assigned more than one document.

The following is a list of all ISO/TC 197 working groups that have been created, and the documents they were responsible for:

- ISO/TC 197/WG 1 Liquid hydrogen—Land vehicle fuel tanks
- ISO/TC 197/WG 2 Tank containers for multimodal transportation of liquid hydrogen
- ISO/TC 197/WG 3 Hydrogen fuel—Product specification
- ISO/TC 197/WG 4 Airport refuelling facility
- ISO/TC 197/WG 5 Gaseous hydrogen—Land vehicle filling connectors
- ISO/TC 197/WG 6 Gaseous hydrogen and hydrogen blends—Land vehicle fuel tanks
- ISO/TC 197/WG 7 Basic considerations for the safety of hydrogen systems
- ISO/TC 197/WG 8 Hydrogen generators using water electrolysis process

- ISO/TC 197/WG 9 Hydrogen generators using fuel processing technologies
- ISO/TC 197/WG 10 Transportable gas storage devices—Hydrogen absorbed in reversible metal hydride
- ISO/TC 197/WG 11 Gaseous hydrogen—Service stations
- ISO/TC 197/WG 12 Hydrogen fuel: Product specification—Proton exchange membrane (PEM) FCVs
- ISO/TC 197/WG 13 Hydrogen detectors
- ISO/TC 197/WG 14 Hydrogen fuel: Product specification—PEM fuel cell applications for stationary appliances
- ISO/TC 197/WG 15 Gaseous hydrogen—Cylinders and tubes for stationary storage
- ISO/TC 197/WG 16 Basic considerations for the safety of hydrogen systems
- ISO/TC 197/WG 17 Safety of pressure swing adsorption systems for hydrogen separation and purification
- ISO/TC 197/WG 18 Land vehicle fuel tanks and TPRDs
- ISO/TC 197/WG 19 Gaseous hydrogen: Fueling stations—Dispensers
- ISO/TC 197/WG 20 Gaseous hydrogen: Fueling stations—Valves
- ISO/TC 197/WG 21 Gaseous hydrogen: Fueling stations—Compressors
- ISO/TC 197/WG 22 Gaseous hydrogen: Fueling stations—Hoses
- ISO/TC 197/WG 23 Gaseous hydrogen: Fueling stations—Fittings
- ISO/TC 197/WG 24 Gaseous hydrogen: Fueling stations—Part 1: General requirements

An up-to-date listing of which documents have been published, which have been withdrawn, and which projects have been cancelled is maintained on the ISO/TC 197 website. Further information such as the status and target publication date of ISO documents under development can also be found on the ISO website.

In addition, www.fuelcellstandards.com tracks the development of over 200 worldwide standards, allowing the user to search by geographic regions or application. The user can drill down to learn about the scope, stage of development, and contact points for developing documents, as well as ordering information for published documents.

For details on progress and technical issues being addressed by those developing codes and standards, the Fuel Cell and Hydrogen Energy Association publishes regular, timely updates of developing documents in their Hydrogen and Fuel Cell Safety Report (see Section 31.5 for more information).

31.4.2.2 International Standards and IEC/TC 105

The International Electrotechnical Commission (IEC) has a technical committee for fuel cells, which is labeled as IEC/TC 105. The work of this technical committee applies to fuel cells of all types, and therefore, it is not limited to only fuel cells that use hydrogen as its fuel. The scope of IEC/TC 105 is to prepare international standards regarding fuel cell technologies for all fuel cell applications, such as stationary power systems, for transportation both as propulsion systems and auxiliary power units; portable power systems; and micropower systems.

The following documents have been published:

- IEC/TC 105 WG 1 Fuel cell technologies—Part 1: Terminology
- IEC/TC 105 WG 2 Fuel cell technologies—Part 2: Fuel cell modules
- IEC/TC 105 WG 3 Fuel cell technologies—Part 3-1: Stationary fuel cell power systems—Safety
- IEC/TC 195 WG 4 Fuel cell technologies—Part 3-2: Stationary fuel cell power systems—Performance test methods
- IEC/TC 105 WG 5 Fuel cell technologies—Part 3-3: Stationary fuel cell power systems—Installation
- IEC/TC 105 WG 6 Fuel cell system for propulsion and auxiliary power units
- IEC/TC 105 WG 7 Fuel cell technologies—Part 5-1: Portable fuel cell power systems—Safety
- IEC/TC 105 WG 8 Fuel cell technologies—Part 6-100: Micro fuel cell power systems—Safety
- IEC/TC 105 WG 9 Fuel cell technologies—Part 6-200: Micro fuel cell power systems—Performance test methods
- IEC/TC 105 WG 10 Fuel cell technologies—Part 6-300: Micro fuel cell power systems—Fuel cartridge interchangeability
- IEC/TC 105 WG 11 Fuel cell technologies—Part 7-1: Single cell test methods for polymer electrolyte fuel cell (PEFC)
- IEC/TC 105 WG 12 Stationary fuel cell power systems—small stationary fuel cell power systems with combined heat and power output.

The program of work for IEC/TC 105 is available on the IEC website http://www.iec.ch/dyn/www/f?p=103:23:0::::FSP_ORG_ID,FSP_LANG_ID:1309,25

31.4.2.3 National Input to Developing International Standards

Individual countries that participate in ISO and IEC provide input to the developing international standards through their national standard body. In the United States, this is the American National Standards Institute (ANSI). The management of the national TAG is assigned to a relevant SDO. In the United States, the secretariat for the ISO/TC 197 US TAG is the Compressed Gas Association (CGA), while the secretariat for the IEC/TC 105 US TAG is CSA International.

31.4.2.4 Global Technical Regulations

As described earlier, regulations carry the force of law. Many countries have regulations to cover topics covered by the codes in the United States, such as fire safety, gas use, buildings, and electrical safety.

In the case of hydrogen, there has been recent success in the development of a global technical regulation (GTR).

A GTR is (1) a process for developing and promulgating motor vehicle safety standards and/or regulations for motor vehicles by the participating counties and (2) the standards and/or regulations emanating from that process.

The World Forum for Harmonization of Vehicle Regulations of the United Nations Economic Commission for Europe (UN/ECE/WP29) has the lead role in the global harmonization of automotive regulations—focusing on vehicles at the time of manufacturing. The GTR concept was created by the 1998 UN global agreement to internationally harmonize vehicle regulations and make vehicles and vehicle parts produced under GTRs available for sale in any country. The signatories to the global agreement include

- United States
- European Community
- Canada
- Japan
- France
- Germany
- Russian Federation
- Republic of Korea
- People's Republic of China

With respect to hydrogen and fuel cells, UNECE recently adopted a UN GTR governing the safety of hydrogen and fuel cell vehicles.

Additional information regarding GTRs and the process can be found at the following website: www.unece.org.

31.4.3 Other Regulations

We have previously discussed the US model codes promulgated by the ICC and NFPA in the United States. In addition, one must comply with environmental regulations and Occupational Safety and Health Administration (OSHA) regulations. These are not, however, particular to hydrogen and will therefore not be covered in this chapter.

Many other countries cover similar safety principles in national regulations. There are regulations for gas safety, construction, hazardous substances, health and safety at work, pressure systems, electricity, and many others. In most cases, these regulations are also not specific to hydrogen.

In Japan, there is a national regulation for hydrogen refueling stations. As the international standards for hydrogen energy systems are published, there is an increased opportunity to reference these consensus documents in national codes and regulations. This approach allows harmonized requirements for global commercialization.

31.5 Resources Available in the Hydrogen and Fuel Cell Safety Report

The US hydrogen and fuel cell industry works closely with the DOE, the US UL, national CDOs and SDOs, the international SDOs, and other key stakeholders to keep interested parties informed about developments in hydrogen and fuel cell requirements through a bimonthly electronic newsletter, the Hydrogen and Fuel Cell Safety Report, which is available online at www.hydrogenandfuelcellsafety.info. Interested parties may sign up to receive an e-mail announcing when the new issue is posted.

In addition, this website posts minutes from the meetings of the National Hydrogen and Fuel Cells Codes & Standards Coordinating Committee.

The website also collects technical resources, including information on hydrogen properties; proceedings from workshops, including permitting workshops; case studies; the US national and international codes and standards templates; educational websites; R&D data; links to CDO and SDO websites and activities; and anything else that can be shared to facilitate the development of safety, codes, and standards and installation permitting.

References

1. Merriam-Webster Incorporated. (2010). Code. Merriam-Webster online dictionary: http://www.merriam-webster.com/dictionary/code?show=0&t=1285586006 (accessed September 27, 2010).
2. Merriam-Webster Incorporated. (2010). Standard. Merriam-Webster online dictionary: http://www.merriam-webster.com/dictionary/standard?show=0&t=1287425346 (accessed September 27, 2010).
3. ISO/IEC. (2004). ISO/IEC Directives: Part 2—Rules for the structure and drafting of International Standards. Definition 3.2.1.1. Fifth Edition.
4. Available at http://www.iso.org and http://www.iec.ch.
5. Merriam-Webster Incorporated. (2010). Regulation. Merriam-Webster online dictionary: http://www.merriam-webster.com/dictionary/regulation (accessed September 27, 2010).
6. The descriptions of each type are available from the SAE website at www.sae.org.
7. Available at http://www.hydrogenandfuelcellsafety.info/resources.asp and http://www.fuelcellstandards.com/.
8. National Renewable Energy Laboratory. (2010). Vehicle codes and standards: Overview and gap analysis, Technical Report NREL/TP-560-47336. Available at http://www.nrel.gov/docs/fy10osti/47336.pdf and http://www.osti.gov/bridge.

32

Sensors for the Hydrogen Economy

Gary G. Ihas
University of Florida

Neil S. Sullivan
University of Florida

CONTENTS

Four types of hydrogen detectors are used by researchers, engineers, and manufacturers today, and if hydrogen continues to play a role in emerging alternative energy sources, there will be exponential growth in the use and need for more advanced and more robust devices in the future. The types of sensors reviewed in this chapter are (1) room-temperature hydrogen leak sensors; (2) thermometers, particularly useful at low temperature; (3) liquid hydrogen volume and mass gauges; and (4) para/ortho hydrogen ratiometers.

32.1 Room-Temperature Sensors for Hydrogen Leaks: Safety Issues

Hydrogen is used extensively in large-scale industrial plants notably for the processing of fossil fuels (hydrodealkylation, hydrodesulfurization, and hydrocracking) and for the production of ammonia. Large bulk volumes of liquid are used by the National Aeronautics and Space Administration (NASA) for the propulsion of the space shuttle, and recent commercial applications have used hydrogen gas in the food industry as *forming gas*, a mixture of hydrogen and nitrogen, employed as a tracer gas for minute leak detection in the packaging of foodstuffs. Hydrogen is also used in atomic welding and was once widely used as the lifting gas in airships. In all of these applications, it is necessary to have fast, robust, rapid readout detectors of hydrogen for the safe operation of the industrial plants where small amounts of hydrogen can accumulate and lead to explosive mixtures. The sensors need to have high selectivity and in particular discriminate against other reducing gases such as carbon monoxide and methanol. The development of practical safety sensors with sensitivities ranging from tens of parts per million up to the low explosive limit (LEL) of 4% by volume is critical for the use of hydrogen as an energy source and for obtaining public acceptance for the widespread use of hydrogen in fuel cells and industrial applications.

With the proliferation of the use of fuel cells and the associated use of small volumes of high-pressure volumes of hydrogen in confined and harsh environments (e.g., hydrogen-fueled vehicles), the detection of small amounts of hydrogen leaking from working parts of the devices becomes increasingly important. Devices must be robust, cheap, and fast responding.

The first practical hydrogen gas sensors [1–3] were based on thick films of palladium [1–3] that operated on the change in electrical conductivity of Pd as it adsorbed H_2 to form PdHx. The lower electrical conductivity of the hydride enables one to detect H_2 concentrations in the range of a few hundred to a few thousand ppm. The sensor usually consists of a meander path of Pd on a solid substrate and is used in an electrical bridge circuit. Because of the sensitivity of the resistance to temperature, the bridge must be accurately regulated and this is a drawback both in terms of the need for a simple design and for fast readout.

An improvement on the original Pd sensors was made by using the expansion of Pd on the adsorption of H_2 rather than the change in electrical resistance as the working property of the sensor [4,5]. Small nanodroplets of Pd are deposited on a substrate to form a network just below the percolation threshold. The adsorption of H_2 causes the droplets to expand adding more paths for electrical conductivity and thus a very sensitive sensor element. For high reliability, the sensor must be thermally regulated and precautions needed to be taken to prevent poisoning of the sensor element by CO, H_2S, and SO_2 among other gases. The response times vary from about 100 ms [4] to more than 100 s [5] depending on the design geometry.

Most recently microelectromechanical systems (MEMs) have been developed for much faster response times and improved reliability. In one type of microsensor SnO_2-Ag_2O-PtO_X, nanocrystalline powders were fabricated on a silicon-based substrate with a heater [6]. Each SnO_2 particle has a space charge near its surface that results from adsorption of O_2 to form a layer of negatively charged O-ions on the surface and correspondingly, due to depletion of electrons, a layer of positively charged donor atoms just below the surface. The reaction of oxygen species on the semiconductor surface with hydrogen releases electrons to the SnO_2 conduction band that lowers the potential barrier for electrical conduction. This leads to a change in the resistance of the sensor film in the presence of hydrogen. While thick film sensors of this type are well known [7–9], they have low sensitivity and low selectivity. Notwithstanding, the introduction of noble gas metal oxides such as PtO_x increases the sensitivity and selectivity (to distinguish H_2 from H_2S and ethanol) as a result of the noble metal oxide interactions with SnO_2 particles to produce an electron-deficient space charge. Response times of the order of 10 s for H_2 concentrations ranging from 100 to 1000 ppm have been reported for these MEMs devices [6]. Robust thick film tin oxide semiconductor sensors are now available commercially with good sensitivity for hydrogen, covering 10–40,000 ppm [10], and MEMs devices based on catalytic conversion are available with very fast response times (~1 s) and with sensitivities ranging from a few ppm to 100% of the LEL [11].

Electrochemical sensors have also been widely explored [12]. These sensors are based on the changes induced in the current or electrochemical potential of a cell in which

TABLE 32.1

Comparison of Room-Temperature Hydrogen Gas Sensors

Sensor Type	Operating Principle	Low Limit Sensitivity Reported	Upper Limit Sensitivity Reported	Response Time	Selectivity	Reference
Ultrathin Pd film (monolayer)	Lattice expansion	25 ppm	~2%	0.1 s	NR	[4]
Thin Pd film	Lattice expansion	NR	~4%	~1000 s	NR	[5]
MEMS SnO_2–Ag_2O–PtOx	Change in conductivity	100 ppm	1,000 ppm	8–10 s	Very good	[6]
Schottky junctions PtSi–porous Si	Change in breakdown voltage	10 ppm	200 ppm	6–60 s	NR	[14]
ZnO nanorods	Change in conductivity	~200 ppm	~10%	~150 s	High	[15]
Silicon carbide C-SC film	Change in conductivity	300 ppm	>10%	~10 min	Good	[16]
Electrochemical	Change in current or electrochemical potential	13 ppm	14,000 ppm	10–100 s	Good	[12]

NR, not reported.

hydrogen moves through a porous membrane and dissolves in the electrolyte and travels by diffusion to the working electrode surface. Sensitivities from a few tens of ppm [13] up to 14,000 ppm have been realized, depending on the electrolyte and electrode design. Response times tend to be slow (~10–100 min) (Table 32.1).

32.2 Thermometry and Temperature Control

In the pursuit and use of low temperatures and cryogenics, temperature (T) is a significant variable. Science and industry frequently require its careful measurement and regulation. Usually, secondary thermometers are used, which are calibrated against primary standards used in the International Temperature Scale of 1990 (ITS-90) [17]. (A primary thermometer is one for which no calibration is required.) Their ease of installation and use, rapidity of response, precision and repeatability, and cost must be considered for each application. Regulation of temperature begins with a thermometer possessing the necessary characteristics and then, using a feedback system, either regulates cooling power or, more normally, applies the requisite heat. Heat capacities and thermal time constants in the entire regulation loop must be considered. Temperature measurement and control techniques, with their advantages and limitations, will now be listed.

Temperature is measured using a thermometric medium with a measurable property that changes in a regular and reproducible way with temperature. Since the internal forces and thermal energy kT of the medium change with T, various material properties must be chosen for the thermometric medium in different temperature regimes ($k = 1.38 \times 10^{-23}$ J/K is Boltzmann's constant). In fact, the properties of certain materials define the practical

TABLE 32.2

Thermometer Characteristics and Measurement Techniques

Type of Sensor	Primary or Secondary	Useful Temperature Range (K)	Measure Technique	Ease of Use	Ability to Control Temperature	Typical Sensitivity ΔT/T (%) in Range of Operation	Magnetic Field Use	Cost
Constant-volume gas	S	4–400	Pressure	Difficult	Moderate	0.01	Yes	Moderate
Chromel/gold thermocouple	S	1.5–600	Electromotive force	Easy	Good	0.1	No	Low
Resistance:	S		Current/voltage	Easy	Good	0.1 (may be made as small as 0.000001)		Moderate
Silicon diode		1.5–450					No	
Ruthenium oxide		0.05–300					Yes	
Carbon		0.02–450					No	
Gallium arsenide		0.01–100					Yes	
Platinum		60–600					Yes	
Germanium		0.01–100					No	
Vapor pressure	S	0.3–4.2	Pressure	Difficult	Moderate	0.01	Yes	Moderate
Capacitance	S	0.001–300	Capacitive	Moderate	Good	0.0001	Excellent	Moderate
^{3}He melting pressure	P	0.001–1	Pressure	Difficult	Good	0.0001	Yes, $T \gtrsim 0.01$	High
Noise	P	0.001–1	Voltage	Very difficult	Poor	1	Yes	High
CMN/LCMN	S	0.001–0.5	Electronic magnetic susceptibility	Difficult	Good	0.001	No	Moderate
Vibrating wire viscometer	S	0.001–0.2	AC resonance	Difficult	Good	1	No	Moderate
Nuclear magnetic resonance (platinum)	S	0.000001–0.1	Nuclear magnetic susceptibility	Difficult	Poor	0.1	Must adjust frequency	High
Nuclear orientation	P	0.002–0.07	γ-ray detection	Very difficult Difficult	Poor	1	No	High

temperature scale, as related to thermodynamics, and are continually refined. The use of gas pressure, vapor pressure, electrical resistance, thermal power, melting pressure, and magnetism for low temperature measurements will be described here.

Table 32.2 gives the characteristics of some thermometers in common use. Above 0.65 K, the popular thermometric materials and measurements are often those used for the ITS-90, which is the accepted temperature scale. At lower temperature, the melting pressure of ^{3}He is used [18]. The easiest thermometers to use are ones in which the resistance is measured. These must be calibrated against a standard, such as the pressure of a constant-volume container of gas or the vapor pressure above a cryogen. Using platinum wire, ruthenium oxide, carbon films, and doped semiconductors (Ge and GaAs), one can span the temperature range from above room temperature to 0.05 K or lower with one resistance-measurement instrument.

The better thermometers at low temperature are in general more difficult to construct and use than resistance thermometers. In common use are magnetic susceptibility of electronic and nuclear magnetic materials and the melting pressure of the light isotope of helium ^{3}He.

As one goes to lower temperature, lower sensing currents must be used in resistive thermometers to avoid self-heating of the thermometer. Also, all but the platinum sensors have increasing resistance with decreasing temperatures, sometimes reaching 1 MΩ or more. Temperature-dependent thermal conductivities, thermal contact, and heat capacities result in large variations in the thermal response times of thermometers and systems being regulated at different temperatures. To accommodate these variations, either several different detection devices or an extremely versatile instrument must be employed.

Temperature is regulated using feedback to control either cooling power or heating, using gas or an electrical heater, or some combination of these parameters. If automated measurements are desired over any significant temperature range ($\Delta T/T > 2$), electronic signal gains and time constants must be adjusted to compensate for the changes in thermal properties of the system. Besides determining the temperature, a control unit is used to supply the regulation signal required to control the temperature. Typically, these units process the signal from the thermometer by a combination of proportioning, differentiating, or integrating the signal in time, with individual time constants for each operation. These time constants are adjusted to suit the thermal parameters, such as conductivity, contact, or heat capacity of the low-temperature system. With the advent of microprocessor-based electronics, this can be done automatically to accommodate virtually any system and any temperature range [19].

Regulation is possible using other resistance thermometers, with the previous considerations being applied. Worthwhile regulation has been achieved down to 0.001 K [20]. Below this temperature, adiabatic demagnetization refrigeration must be used, which is a cyclic, one-shot process. Here, temperature regulation is difficult and usually is done piecewise, by controlling the magnetic field on the refrigerant.

32.3 Cryogenic Monitoring of Liquid Hydrogen Levels and Mass Gauging

32.3.1 Introduction to Volume Measurements and Mass Gauging

Substances that are gases at room temperature and atmospheric pressure, such as hydrogen, are often stored and transported as a gas, usually at elevated pressure. Chapter 17 in this book deals with the general area of hydrogen storage. Here, we will review ways in which the amount of hydrogen contained in a vessel may be determined. The common situation of gas held under pressure in a vessel will not be reviewed, since it is impractical, because of bulk and weight of the container, for use in hydrogen economy applications, such as motorized vehicles and spacecraft. In these applications, it is important to constantly monitor the amount of hydrogen (or energy content) present, through level detection or mass gauging, and the state of the hydrogen (ortho or para). Following this section, we review techniques for determining the ortho/para ration in a quantity of hydrogen. Here, liquid level detection and mass gauging will be reviewed.

Determining the amount or energy content of a cryogen is special for at least two reasons. First, because of the very low temperatures (20 K for boiling hydrogen at atmospheric pressure), either special or adapted-from-room-temperature devices must be used, and heat input must be minimized. The requirement of low heat input stems from the second circumstance special to cryogens: abnormally low values of many parameters—boiling temperature, density, difference in gas and liquid density, refractive index, thermal and electrical conductivity, latent heat, and heat capacity.

Any measurement produces heat. Besides losing the liquid one is attempting to measure, sensor heat can easily change the phase of the cryogen from liquid to gas, since usually the cryogen is being held at its boiling point and a cryogen's latent heat and density is much lower than that of standard liquids (e.g., water). Often, both liquid and gas phases are present, and their relatively small density difference (2% for hydrogen compared to 9% for water) makes accurate measurement of liquid or gas challenging. Level detection in a vessel containing liquid may only be used when there is sufficient gravity or acceleration present and the vessel is not undergoing erratic motion.

Mass gauging attempts to measure the actual amount (mass) of the cryogen present. It offers different information than a level measurement, because it measures the total amount of the fluid present, regardless of state. But it is also a more complicated measurement. However, mass gauging must be used in low gravity and/or when the vessel is undergoing nonuniform acceleration since, in these situations, the liquid does not necessarily reside in the *bottom* of the vessel or even in a singly connected mass. These conditions require quite sophisticated gauging techniques that will be described.

The dead reckoning method accounts for hydrogen intake and withdrawal from a vessel by keeping a detailed balance sheet, thereby giving a running record of the amount of hydrogen in the vessel. This technique requires accurate monitoring of the liquid flow and vapor present and use of the hydrogen, which is often not possible. Some of the facets of this monitoring technique will be reviewed.

This cursory review is simply a starting point for learning about sensors for the hydrogen economy. A next step would be to consult the reference volume by Flynn [21] and references therein. However, this area is advancing rapidly, and the present work will be obsolete before going to press.

32.3.2 Liquid Level Detection

By far, the most energy-intensive method of storing hydrogen is in liquid form. In Chapter 17, the vessels for storing this cryogenic liquid are described. Here, we review the various forms of liquid level sensing devices that are in use. These devices are useful in earth's surface gravity and in rockets that are accelerating (during launch), which is when most of the hydrogen in a fuel tank is consumed. Level sensing systems may be divided into two classes: (1) discrete or point and (2) continuous. They may also be categorized by the wavelength of the power they use: (1) direct current (DC), (2) alternating current (AC), (3) radio frequency (rf), and (4) optical. Table 32.3 lists some properties of these various systems.

TABLE 32.3

Point Level Measuring Devices

Point Device	Method	Operation Principle	Temperature	Resolution (μm)	Time Response (ms)	Heat Input	Cost	Difficulty of Use
Hot wire	Resistance	Temperature gradient	Resistor specific	279–381	5–100	Low	Low	Sensitivity changes with temperature, error from condensation, needs large temperature gradient.
Capacitive	Capacitance	Dielectric/density gradient	<373 K	254–279	100	Very low	Low	Corrosion, error from surface buildup.
Optical	Light illumination	Index of refraction	$>LN_2$	203–279	100–300	Medium	Medium	Condensation, bubbles, not for liquids that crystallize or leave solid residue.
Acoustic, ultrasonic	Vibrating diaphragm	Wave damping	30 K	381–762	40–200	Medium	Medium–high	Turbulence and foam lead to errors.
Mechanical	Vibrating paddle	Mechanical damping	No H, He	584–1194	189–869	High	Low	Needs viscous and dense liquids
Noncontact ultrasonic	Piezoelectric	Acoustic resonance	LN_2	381–736	37–173	Low	High	Container must be plastic, not thicker than 1/4 in.

32.3.2.1 Point Level Sensors

Discrete or point sensors indicate the presence of liquid or vapor (wet or dry) and must be of sufficient number to yield the desired precision of volume determination. They are usually arranged along a vertical support tube or wall and, with an accurate profile of the storage container, may be used to determine volume of liquid present. Often, besides volume present, it is important to know if certain devices in the storage vessel, such as refrigeration apparatus, syphon tubes, magnets, or stored material, are immersed in the liquid. Point sensors are particular handy in this case, as one mounted at a critical level may be used to sound an alarm or begin a filling operations, for example.

32.3.2.1.1 Float Switches

The simplest and perhaps most reliable of these devices is a float-mounted switch [22]. Usually a microswitch with a foam float attached to its switch lever, this device changes state because the buoyant force on the foam float changes depending on whether it is immersed in liquid or vapor. This device may be run at low DC or AC current and voltage, dissipating negligible heat and having essentially zero failure rate over its lifetime. The mechanical parts of the switch must be clean and free of wet lubricants, and the foam must be closed-cell. Switch position may be conveyed outside the vessel by an rf link, fiber optic, or metallic wires. Float switches are used in the cryogenic tanks of many spacecraft, including the Delta Rockets, built by Boeing.

32.3.2.1.2 Resistive Elements

Most materials dissipate heat faster when immersed in liquid compared to vapor, because of better thermal contact and conduction in the liquid. If it has a characteristic that depends on temperature (which may also be used as a thermometer), it can be used as a point level detector [23]. Usually, small wires or composite materials (containing some conductor, such as carbon) or oxides (RuO_2) are used. The current through the element (AC or DC) is adjusted to the lowest level that results in a measurable difference in voltage drop across the element depending on whether it is wet or dry. This type of device works better as the ambient temperature of the liquid drops (e.g., working better in liquid H_2 compared to LN_2) but, of course, introduces heat into the liquid. Often, the measurements may be made intermittently, so that the duty cycle of the electrical power can be greatly reduced, minimizing the heat introduced to a negligible amount.

32.3.2.1.3 Diodes

The I–V characteristics of doped semiconducting devices, because of the thermal activation of carriers into the conduction band, are very sensitive to temperature [24]. Small, mass-produced silicon diodes, operating near zero current conduction, produce a voltage drop that is very sensitive to temperature. The rate of heat dissipation, and hence temperature, depends on immersion in liquid or vapor. They may also be used as thermometers. Their huge advantage, being mass-produced semiconducting chips, is their extremely small size, low cost, and yet good reproducible characteristics (I–V curve) at low temperature, if obtained from a single processed silicon wafer.

32.3.2.1.4 Capacitive or Dielectric Gauge

Since the dielectric constant of the liquid is significantly different from the vapor in a cryogenic storage vessel, a point measurement detecting the presence of liquid may be made using a capacitor that allows the liquid or vapor to enter between its plates [25]. A continuous detector of this type may also be constructed (see Section 32.3.2.2.4 following). To increase sensitivity, the gap between the plates may be small, but care must be taken that surface tension does not permanently trap liquid between the plates. It is better to construct a capacitor with large plate area.

32.3.2.1.5 Resonant Damping Probes

The viscosity of liquids is higher than that of their vapors, so any oscillating object whose damping may be monitored can act as a point level detector. These are usually composed of wires, quartz tuning forks, or diaphragms and may be made quite sensitive by operating at a resonant frequency of the active element [26]. Wire motion may be detected by observing the electromotive force (emf) induced as the wire, driven by a small AC current, oscillates perpendicular to a small magnetic field. Wires as small as 0.01 mm have been employed. Tuning forks [27] can be part of a resonant electrical circuit driven by an oscillator. These device, found in most watches, are very small (millimeters) and possess quality factors (Q) of 10,000 or more. Thus, they require very little excitation energy and are extremely sensitive to viscous drag damping. A diaphragm, like a microphone, may be part of a parallel plate capacitor whose motion is driven at resonance by an AC voltage.

32.3.2.1.6 Optical Probe

The use of optical probes, while appearing to have advantages such as lack of electrical connections inside the vessel or negligible heat dissipation, suffers from the low refractive index of most cryogens and lack of difference in refractive index between the liquid and gas phases [28]. For these reasons, most optical point sensors employ the drastic change in Brewster's angle [29], which occurs when the fluid surrounding a solid changes from gas to liquid (or the reverse). An optical fiber, which is terminated in a prism cut to the appropriate angle, transmits most of its light into the vessel when it is covered with liquid but undergoes total internal reflection and returns most of the light back up the fiber, when it is covered with gas.

32.3.2.2 Continuous Level Detectors

Continuous level detectors usually have the advantage of higher resolution but are often much more complicated to build, install, and use. One for which this is not the case is the superconducting wire device used with liquid helium. Now that high-temperature, nonconventional superconducting wire is available, such as MgB_2 [30], these devices can be used in liquid hydrogen. As with liquid helium, this will soon be the dominating method for level detection in liquid hydrogen. Other techniques used in liquid hydrogen are weighting the vessel; differential pressure gauging; capacitive, optical, resonant damping; or a simple float device (Table 32.4).

32.3.2.2.1 Superconducting Wire

A straight wire made of superconducting material, which is twisted with a heater wire (an alloy whose thermal coefficient of resistance is low and electrical resistance is moderate-ohms per foot), may be used with a simple control box to continuously determine liquid level in a cryogenic vessel. The controller passes enough current through the heater wire to drive the probe wire into the normal conducting state where it is in the gas, but allow it to remain superconducting where it is immersed in the liquid cryogen. A simple resistance

TABLE 32.4

Continuous Level Measuring Devices

Device	Method	Operation Principle	Temperature	Resolution (μm)	Time Response (ms)	Heat Input	Cost	Difficulty
Optical	Fiber-optical long-period grating	Internal refraction	L Hydrogen	0.0003	10–180	Low	Medium	Very sensitive to temperature, error from boiling bubbles.
Thermistor	Superconducting wire	Temperature gradient	>He II	150–250	257	Low	Medium	Ice and other deposit buildup, needs low vapor density, small temperature range, affected by He II film flow.
Capacitive	Long capacitor plates/tubes	Dielectric/density gradient	All	838–865	50–370	Very low	Medium	Density of vapor must be significantly different from liquid.
NRA	Nuclear radiation attenuation	Gamma transmission	>LN_2	High	Medium	Medium	High	Liquid must absorb radiation (e.g., bad for H), high noise from thick walls, safety.
Load cell	Digital scale	Mass of container	Any	Low	High	None	Low	Requires constant calibration due to fluid density, container weight, and position.

measurement determines the faction of the wire in the vapor above the liquid. The controller can easily use any duty cycle for the measurements, reducing the heat input to almost nil. The information is easily digitized. The controller, using very little power, may be operated on batteries, so this level meter may be small and attached to a transport dewar. A cable may be attached to both charge the battery and connect the device to a computer (or rf may be employed), so that any number of containers may be remotely monitored [31].

32.3.2.2.2 Float Gauge

Cryogenic storage dewars may use a simple Styrofoam float attached to a stick to indicate level. This technique is prone to fail mainly due to sticking of the float/rod mechanism. Its advantage is that it may be part of a mechanical system for initiating refill or other actions and requires no external power.

32.3.2.2.3 Differential Pressure Gauge

A mechanical pressure gauge, sensitive to the pressure of the hydrostatic head of the column of cryogen (inches of water), may be easily configured with one side reading the pressure at the bottom of the cryogen tank and the other in the vapor at the top of the tank and calibrated to read depth (linear) or volume (if the tank is close to a uniform cylinder or other regular geometry). The measurement is less sensitive for lighter (less dense) cryogens, such as H_2 or He, and may also be degraded by boiling, liquid motion, thermal oscillations, or bubbling. This is the most common level gauge seen on storage tanks. The gauge is often produced by Magnehelic Corporation (Ref. [32]).

32.3.2.2.4 Capacitive or Dielectric Gauge

As mentioned earlier, since the dielectric constant of the liquid is different from the vapor in a cryogenic storage vessel, an extended capacitor measurement detecting liquid level may be made. The continuous detector consists of a vertical pair of concentric tubes that act as the plates of a capacitor, the capacitance of which depends on the liquid level. The relation between capacitance and level is not direct because the dense vapor in the upper part of the capacitor contributes significantly to the measured signal. Hence, with the proper algorithm, this device may be used as total content or mass gauger. To increase sensitivity, the gap between the concentric tubes can be made small, introducing a small error due to capillary rise, which may be corrected.

32.3.3 Mass Gauging Techniques

Sometimes, total molecular content of a vessel is desired. This information, coupled with temperature and pressure data, can be used to determine the total energy, heat, or cooling capability of the contents. Also, in cases where the container is not in steady motion and/or in sufficient gravitation force, mass gauging is necessary simply to get an accurate reading of the contents of the container. This is because under such conditions, the contents do not sit in the bottom of the container, there is no single liquid/gas interface (the contents might be dispersed as globules throughout the container), or the contents are simply sloshing around inside the container. It can be written that there is no really good, reliable way to mass gauge, but several attempts have been made. Most methods depend on measuring some global property of the mass, such as heat capacity or photo absorption. These attempts will be briefly mentioned here.

Mass gauging requires that the entire volume of the cryogenic container be sampled by some probe that indicates the presence of the cryogen. This is done either by having a suitable number of probes distributed throughout the volume [33] or sampling the entire volume by a multiply-reflected light beam or beams [34]. No technique developed to date claims a precision of more than 10%. Limitations include incomplete interrogation of the volume, caused by either limitations of the probe system or properties of the cryogens.

One technique uses small thermometers and heaters to measure temperature rise and hence heat capacity of the material present. This method is limited by the poor thermal conductivity of most cryogens, requiring many sensors. It has been suggested that for some applications, such as propulsion using hydrogen, a material may be added to the fuel to enhance its conductivity and its energy content, such as carbon nanotubes [35]. The data may be radio-frequency telemetered from the sensors in the containers, simplifying sensor installation [36].

Another system uses one or more light beams that are absorbed as they traverse the contents of the container many times [37]. This system requires a highly reflective inner container wall (easily achieved for infrared radiation) and calibration for each type container.

Both of these methods have the advantage that the controller can also tell how rapidly the hydrogen is moving or *sloshing* around the container, which is important for the control of the vehicle and the fuel injections. Even in transport, some sloshing occurs for low-viscosity fluids, such as hydrogen, and an array of level detectors for each separate compartment is needed compared to the naïve single floater approach used for fuel levels in an automobile. Mass gauging satisfies this need. Massive amounts of liquid hydrogen are transported annually for use as a rocket propellant by either NASA or the new commercial rocket launchers, since it is difficult to achieve payload-to-propellant mass ratio higher than hydrogen possesses.

32.4 Ortho–Para Hydrogen Detection

The detection of ortho–para hydrogen ratios is important for the commercial production of liquid hydrogen and the transport of large volumes over long distances. There are two independent molecular species of hydrogen, para and ortho. Para is the lowest lying state, 171 K below the energy level of orthohydrogen. At room temperature, the gas has 75% ortho content (this is referred to as normal hydrogen), while at liquid hydrogen temperatures, it is 99.8% parahydrogen. Most manufacturers use a two-stage catalytic converter to convert ortho- to parahydrogen, first at 77 K, and then at 204 K in the liquefying process. Since conversion at 77 K produces 50% parahydrogen, the remaining conversion, if done at 20.4 K, consumes a significant fraction of the liquid hydrogen and is a major factor in the cost of producing and transporting liquid hydrogen [38]. Great care is therefore taken in the production of liquid hydrogen to reduce the ortho content as much as possible [39]. This is ensured by the use of catalytic converters but the converters age and it is important to monitor the para–ortho content continuously. The energy of conversion of ortho–para is comparable to latent heat of evaporation, and a tanker with 50% ortho content will lose a large fraction of its volume to the heat of ortho–para conversion during transport. The latent heat of conversion is 670 J/g (for normal hydrogen) compared to a latent heat of evaporation of 430 J/g.

The ortho–para ratio is best measured using the large difference in thermal conductivity of the gas at liquid nitrogen temperatures [40–42]. The thermal conductivity of a dilute gas is given by [43]

$$\kappa = \frac{1}{3\sqrt{2}}\frac{C_V}{\sigma_0}v \tag{32.1}$$

where
C_V is the heat capacity of the gas
$v = \sqrt{\frac{8k_BT}{\pi m}}$ is the molecular mean speed at temperature T

σ_0 is the collision cross section and m is the molecular mass. Expression 32.1 is valid only for the case when the mean free path λ satisfies $d << \lambda << L$ where d is the molecular dimension and L is the length of a typical dimension of the thermal conductivity gauge. The ability to accurately measure the ortho–para content using a thermal gauge arises from the large variation of the heat capacities of the two species. The energy levels of the two species are given by

$$E_J = B_J J(J+1)$$

where
B_J = 87.57 K [44] is the rotational constant
J is the angular momentum

J is even for parahydrogen (J = 0,2,4,... for parahydrogen and J = 1,3,5.... for orthohydrogen). Each level contributes a Schottky heat capacity peak [45], and the calculated ratio of the heat capacities is shown in Figure 32.1. The large variation above 100 K makes it

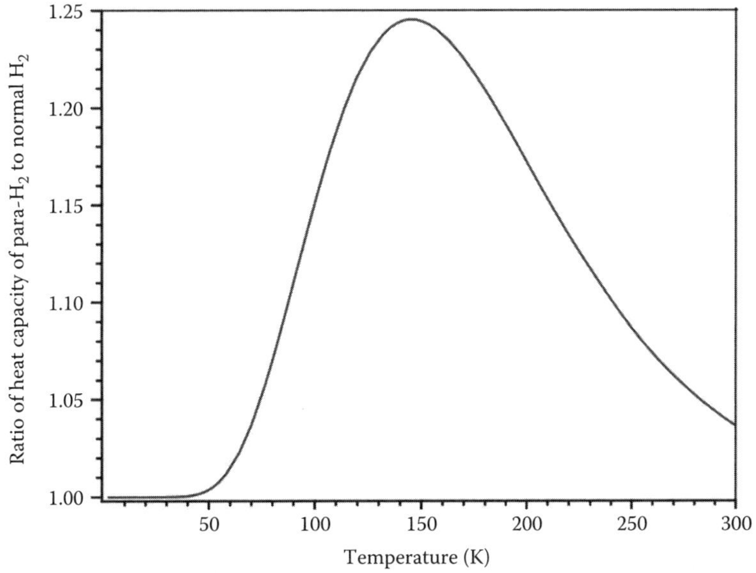

FIGURE 32.1
Temperature dependence of the ratio of the heat capacity of parahydrogen to that of normal hydrogen (75% orthohydrogen).

especially convenient to design thermal conductivity cells that operate with a cold wall at 77 K and with a heated element (wire or plate) that is used to measure the heat flow and thus the conductivity of the gas mixture.

The earliest reported detector [40] used a hot wire element to measure the thermal conductivity. Two Pirani gauges were used in bridge circuit with the sample whose ortho–para ratios were to be determined in one arm and normal hydrogen (75% orthohydrogen) as a reference gas in the other arm. The gauges were immersed in a liquid nitrogen bath

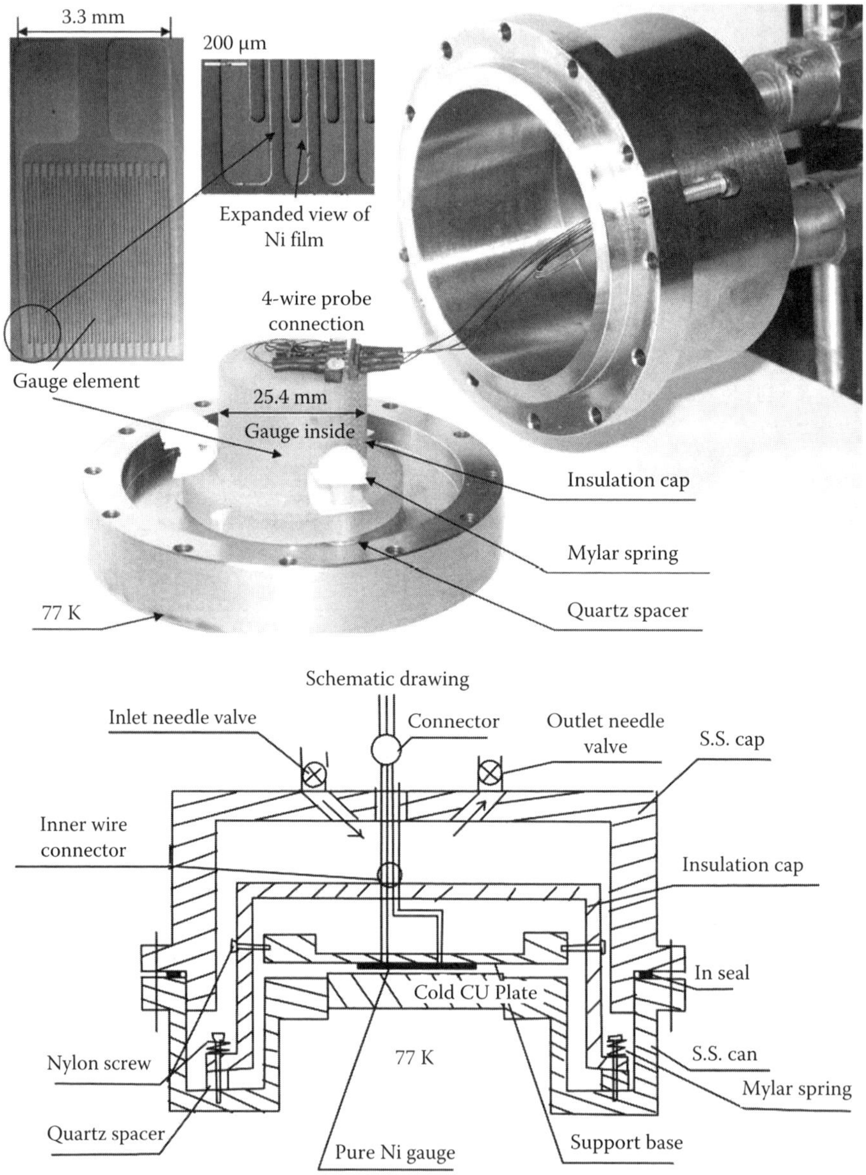

FIGURE 32.2
Flat plate ortho–para ratiometer, showing the gauge element (inset) and the mounting in the low-temperature cell. (Reproduced with permission from Springer Science+Business Media: *J. Low Temp. Phys.*, 134, 2004, 401–405, Zhou, D., Ihas, G.G., and Sullivan, N.S.)

and the pressure was maintained constant in each arm. The unbalance signal provides the measure of the orthohydrogen content and can be calibrated with the use of a sample of parahydrogen prepared by catalytic conversion at liquid helium or liquid hydrogen temperatures [46]. This use of heated wire elements suffered from two problems: (1) the need to have matched Pirani gauges and (2) a relatively strong dependence on the pressure because the size of the heated element was not negligible compared to the mean free path of the gas. These difficulties were overcome with the use of a parallel plate geometry [41,42], which is the preferred method for measuring thermal conductivities of gases.

In the flat plate ortho–para ratiometer shown in Figure 32.2, a commercial microminiature Ni strain gauge (3.3 mm × 4 mm × 10 μm) is used as a low heat capacity hot element located at a short distance (0.75 mm) from a cold wall maintained at 77 K. The low heat capacity and short distances lead to a fast response time (<0.01 s.), which is necessary for practical applications. The response is pressure independent (to better than 0.3%) for $0.10 < P < 0.80$ atm. The electronic readout was obtained by placing the heated element in a feedback circuit that maintained the heated element at constant temperature (typically 90 K), and the power needed to maintain circuit at a constant temperature was used to measure the thermal conductivity of the cell and thus the ortho–para content. As in the case of the heated wire detectors, the flat plate cell is simply calibrated by using reference gases of pure parahydrogen and normal hydrogen (75% orthohydrogen). High purity samples of orthohydrogen can be prepared using a separation column using activated alumina that selectively adsorbs orthohydrogen [47]. The performance of the parallel plate cell can be tested by using the cell to measure the ratio of the thermal conductivity of normal hydrogen to nitrogen that has the value of 7.04.

It is important in the design and operation of ortho–para hydrogen ratiometers to take steps to remove paramagnetic impurities that can lead to rapid ortho–para hydrogen conversion at impurity sites and make the measurements unreliable [48]. Conversion at the walls will convert the hydrogen to the equilibrium value at the wall temperature (~50% at 77 K). For the clean stainless steel construction employed in Ref. [4], this was found to be negligible over the timescale of the measurement. It is also important to remove impurity gases such as nitrogen or helium that would alter the thermal conductivity of the gas and lead to a false measurement.

References

1. Toy, S.M. and Phillips, A., 1973. US Patent 3732076; Toy, S.M., 1973. US Patent 3768975.
2. McDonald, D.D., Mckubre, M.C.H., Scott, A.C., and Wentrek, P.R., 1981. *Ind. Eng. Chem. Fundam.* **20**, 291.
3. Chrstofides, C. and Mandelis, A., 1990. *J. Appl. Phys.* **68**, R1–R30.
4. Xu, T., Zach, M.P., Zhao, Z.L., Rosenbaum, D., Welp, U., Kwok, W.K., and Crabtree, G.W., 2005. *Appl. Phys. Lett.* **86**, 20310.
5. Kiefer, T., Villaneuva, L.G., Fargier, F., Favier, F., and Brugger, J., 2010. *Appl. Phys. Lett.* **97**, 121911.
6. Kim, I.J., Han, S.D., Han, C.H., Gwak, J., Hong, D.U., Jakhar, D., Singh, K.C., and Wang, J.S., 2007. *Sens. Actuat.* **B127**, 441.
7. Gardner, J.W., Varadan, V.K., and Awaldelkarim, O.O., 2001. *Micro Sensors, MEMS and Smart Devices*, John Wiley & Sons, New York.
8. Barsam, N., Stetter, J.R., Findlay, M., and Gopel, W., 2000. *Sens. Actuat.* **B66**, 31.

9. Seal, S. and Shukla, S., 2002. *J. Miner. Metal. Mater.* **54**, 35.
10. Figaro USA, Inc., Arlington Heights, IL 60025.
11. Kebaili Corporation, Irvine, CA 92616–4153.
12. Korotchenko, G., Han, S.D., and Stettler, J.R., 2009. *Chem. Rev.* **109**, 1402.
13. Rasmesh, C., Velayuthan, G., Mwuugesan, N., Ganesan, V., Dhathatheyan, K.S., and Periswami, G., 2003. *J. Solid State Electrochem.* **8**, 511.
14. Raissi, F. and Farivar, R., 2005. *Appl. Phys. Lett.* **87**, 164101.
15. Lupan, O., Chai, G., and Chow, L., 2008. *Microelectron. Eng.* **85**, 2220.
16. Fawcett, T.J., Reues, M., Spetz, A.L., Saddow, S.E., and Wolan, J.T., 2006. *Mater. Res. Soc. Symp. Proc.* **911**, B12.
17. Preston-Thomas, H., 1990. *Metrologia* **27**, 3 and errata in 1990, **27**, 107.
18. Soulen, Jr., R.J. and Fogle, W.E., 1997. *Phys. Today.* (August issue), 36.
19. Scientific Instruments, Inc., 4400 Tiffany Blvd., West Palm Beach, FL 33407.
20. Ni, W., Xia, J.S., Adams, E.D., Haskins, P.S., and McKisson, J.E., 1995. *J. Low Temp. Phys.* **99**, 167 and references therein.
21. Flynn, T.M. 2005. *Cryogenic Engineering*, 2nd edn., Marcel Dekker, New York, 579ff; Casas, J., 1998. *Handbook of Applied Superconductivity*, vol. 2, ed. B. Seeber, Institute of Physics, Dirac House, Temple Back, Bristol, UK, pp. 891–896.
22. Ihas, G.G. and Niemela, J., 2001. *Proceedings of the 13th International Conference on Liquefied Natural Gas*, Chicago, IL, pp. 1–8.
23. Takeda, M., Takeda, M., Matsuno, Y., Kodama, I., Kumakura, H., and Kazama, C. 2009. *IEEE Trans. Appl. Superconduct.* **19**, 764–767; Kajikawa, K., Tomachi, K., Matsuo, M., Sato, S., Funaki, K., Kumakura, H., Tanaka, K. et al., 2008. *J. Phys. Conf. Ser.* **97**, 02140.
24. Rao, M.G., 1988. *CEBAF* **PR-88-022**; Muhlenhaupt, J. and Srnelsey, C.P., 1963. *Carbon Resistors for Cryogenic Liquid Level Measurement*. Cryogenic Engineering Laboratory Division, National Bureau of Standards, Boulder, CO; Juanarena, D.B. and Rao, M.G., 1992. *Cryogenics*, **32**, 39–43.
25. Talpe, J., Stolovitzky, G., and Bekeris, V., 1987. *Cryogenics* **27**, 693–695.
26. Shu, Q.-S., 2000. *Advances in Cryogenic Engineering*, **45B**, 1803ff.
27. Wenger, N.C. and Smetana, J., 1966. *NASA Tech. Note* **TN D-3680**; Ellerbruch, D.A., 1970. *IEEE Trans. Instrum. Meas.* **IM-19**, 412–416.
28. Blažková, M., Človečko, M., Eltsov, V.B., Gažo, E., deGraaf, R., Hosio, J., Krusius, M., Schmoranzer, D., Schoepe, W., and Skrbek, L., 2007. *J. Low Temp. Phys.* **150**, 525–535.
29. Weiss, J.D., 2008. *Opt. Eng.* **39**, 2198–2213.
30. Yang, C., Chen, S., and Yang, G., 2001. *Sens. Actuat.* A **94**, 69–75.
31. Frederick, L., Labbe, G., and Ihas, G.G. 2000. *Physica B* **284–288**, 2028–2029.
32. http://www.dwyer-direct.com/shop/contactus.jsp.
33. Mord, A.J., Snyder, H.A., Kilpatrick, K.A., Hermanson, L.A., Hopkins, R.A., and Vangundy, D.A., 1988. *Ball Aerospace Systems Report* **DRD MA183T**, Contract NAS9–17616; Rogers, A.C., Dodge, F.T., and Behring II, K.A., 1993. *Southwest Research Institute Report* **AIAA 93–1801**; Monti, R., 1994. *ESA J.*, **18**; Van Sciver, S., Adams, T., Caimi, F., Celik, A., Justak, J., and Kocak, D., 2004. Optical mass gauging of solid hydrogen, *Cryogenics*, **44(6–8)**, 501–506.
34. Van Sciver, S.W., Adams, T., Caimi, F., Celik, D., Justak, J., and Kocak, D., 2004. *Cryogenics*, **44**, 501–506; Caimi, F.M., Kocak, D.M., and Justak, J.F., 2006. *AIP Conf. Proc.* **823**, 224–231; Jurns, J.M. and Riogers, A.C., 1995. https//kb.osu.edu/dspace/handle/1811/19230. 2 Huntington Quadrangle Melville, New York 11747, *NASA Contractor Report* **198366**.
35. Ihas, G.G., Mitin, V.F., and Sullivan, N.S., 2003. *J. Low Temp. Phys.* **134**, 437.
36. Ngo, K.D.T., Phipps, A., Nishida, T., Lin, J., and Xu, S., 2006. *ASME Conf. Proc.* **2006**, 513.
37. Van Sciver, S., Adams, T., Caimi, F., Celik, D., Justak, J., and Kocak, D., 2004. *Cryogenics*, **44**, 501.
38. Williamson, Jr., K.D. and Ederskuty, J., 1983. *Liquid Cryogens, VI*, CRC Press Inc., Boca Raton, FL, pp. 96–100.
39. Rogers, J.D., 1953. *Indus. Eng. Chem.* **45**, 1754–1756.
40. Grilly, E.R., 1953. *Rev. Sci. Instrum.* **24**, 72.
41. Devoret, M., Sullivan, N.S., Esteve, D., and Deschamps, P., 1980. *Rev. Sci. Instrum.* **51**, 1220.

42. Zhou, D., Ihas, G.G., and Sullivan, N.S., 2004. *J. Low Temp. Phys.* **134**, 401–405.
43. Reif, F., 1956. *Fundamentals of Statistical and Thermal Physics*, McGraw-Hill Book Company, New York, pp. 478–481.
44. Silvera, I.F., 1980. *Rev. Mod. Phys.* **52**, 393.
45. Kittel, C., 1976. *Introduction to Solid State Physics*, John Wiley & Sons Inc., New York, p. 454.
46. Sullivan, N.S., Zhou, D., and Edwards, C.M., 1990. *Cryogenics* **30**, 734.
47. Sullivan, N.S., Edwards, C.M., Rall, M., and Zhou, D. 1993. *Cryogenics* **33**, 1008.
48. Lu, X., Wu, S., Wang, L., and Su, Z., 2005. *Sens. Actuat.* **B107**, 812.

Appendix A: Thermophysical Properties of Hydrogen

CONTENT

This appendix gives thermodynamic and transport properties of hydrogen as generated from the equations of state presented in Refs. [1–3]. The properties tabulated are density (ρ), energy (*E*), enthalpy (*H*), entropy (*S*), isochoric heat capacity (*C*), isobaric heat capacity (C_p), speed of sound (v_s), and dielectric constant (*D*). All extensive properties are given on a molar basis. The references should be consulted for information on the uncertainties and the reference states for *E*, *H*, and *S*.

T (K)	ρ (mol/L)	*E* (J/mol)	*H* (J/mol)	*S* (J/mol K)	C_v (J/mol K)	C_p (J/mol K)	v_s (m/s)	*D*
***P* = 0.1 MPa (1 bar)**								
15	37.738	−605	−603	11.2	9.7	14.4	1319	1.24827
20	35.278	−524	−521	15.8	11.3	19.1	1111	1.23093
40	0.305	491	818	75.6	12.5	21.3	521	1.00186
60	0.201	748	1,244	84.3	13.1	21.6	636	1.00122
80	0.151	1030	1,694	90.7	15.3	23.7	714	1.00091
100	0.120	1370	2,202	96.4	18.7	27.1	773	1.00073
120	0.100	1777	2,776	101.6	21.8	30.2	827	1.00061
140	0.086	2237	3,401	106.4	23.8	32.2	883	1.00052
160	0.075	2723	4,054	110.8	24.6	33.0	940	1.00046
180	0.067	3216	4,714	114.7	24.6	32.9	998	1.00041
200	0.060	3703	5,367	118.1	24.1	32.4	1054	1.00037
220	0.055	4179	6,009	121.2	23.4	31.8	1110	1.00033
240	0.050	4641	6,638	123.9	22.8	31.2	1163	1.00030
260	0.046	5093	7,256	126.4	22.3	30.6	1214	1.00028
280	0.043	5535	7,865	128.6	21.9	30.2	1263	1.00026
300	0.040	5970	8,466	130.7	21.6	29.9	1310	1.00024
400	0.030	8093	11,421	139.2	21.0	29.3	1518	1.00018
***P* = 1 MPa**								
15	38.109	−609	−583	10.9	10.1	14.1	1315	1.25089
20	35.852	−532	−504	15.5	11.4	18.4	1155	1.23496
40	3.608	399	676	54.1	12.9	28.4	498	1.02209
60	2.098	697	1,173	64.3	13.2	23.5	635	1.01280
80	1.523	994	1,651	71.1	15.4	24.7	719	1.00928
100	1.204	1343	2,174	77.0	18.8	27.7	779	1.00733
120	0.999	1756	2,758	82.3	21.9	30.6	835	1.00608

(continued)

(continued)

T (K)	ρ (mol/L)	E (J/mol)	H (J/mol)	S (J/mol K)	C_v (J/mol K)	C_p (J/mol K)	v_s (m/s)	D
140	0.854	2219	3,390	87.1	23.9	32.5	891	1.00520
160	0.747	2709	4,048	91.5	24.7	33.2	949	1.00454
180	0.663	3204	4,712	95.4	24.6	33.1	1006	1.00404
200	0.597	3693	5,368	98.9	24.1	32.5	1063	1.00363
220	0.543	4170	6,012	102.0	23.5	31.9	1118	1.00330
240	0.498	4634	6,643	104.7	22.9	31.2	1171	1.00303
260	0.460	5087	7,263	107.2	22.3	30.7	1222	1.00279
280	0.427	5530	7,873	109.5	21.9	30.3	1271	1.00259
300	0.399	5966	8,475	111.5	21.6	30.3	1317	1.00242
400	0.299	8091	11,433	120.1	21.0	29.4	1525	1.00182
P = 10 MPa								
20	39.669	−568	−316	13.0	10.9	15.0	1458	1.26198
40	31.344	−209	110	27.3	13.2	27.0	1171	1.20354
60	21.273	255	725	39.7	13.8	32.5	931	1.13527
80	14.830	686	1,360	48.8	15.9	31.1	886	1.09303
100	11.417	1110	1,986	55.8	19.3	31.9	904	1.07109
120	9.357	1571	2,640	61.8	22.4	33.5	941	1.05801
140	7.969	2068	3,323	67.0	24.3	34.6	989	1.04925
160	6.963	2583	4,020	71.7	25.0	34.9	1042	1.04294
180	6.195	3099	4,713	75.7	24.9	34.4	1096	1.03814
200	5.588	3604	5,393	79.3	24.4	33.6	1150	1.03436
220	5.094	4094	6,057	82.5	23.7	32.8	1203	1.03129
240	4.683	4569	6,704	85.3	23.1	32.0	1254	1.02874
260	4.336	5030	7,336	87.8	22.6	31.3	1302	1.02659
280	4.038	5481	7,958	90.1	22.1	30.8	1349	1.02475
300	3.780	5924	8,570	92.3	21.8	30.4	1394	1.02315
400	2.869	8073	11,559	100.9	21.2	29.6	1592	1.01753

Values are given as a function of temperature for several isobars. The phase can be determined by noting the sharp decrease in density between two successive temperature entries; all lines above this point refer of the liquid phase and all lines below refer to the gas phase. If there is no sharp discontinuity in density, all data in the table refer to the supercritical region (i.e., the isobar is above the critical pressure).

References

1. Younglove, B.A., Thermophysical properties of fluids. Part I, *J. Phys. Chem. Ref. Data*, 11, Suppl. 1, 1982.
2. Younglove, B.A. and Ely, J.F., Thermophysical properties of fluids. Part II, *J. Phys. Chem. Ref. Data*, 16, 577, 1987.
3. McCarty, R.D., Thermodynamic properties of helium. *J. Phys. Chem. Ref. Data*, 2, 923, 1973.

Appendix B: Virial Coefficients of Gaseous Hydrogen

Henry V. Kehiaian
University of Paris VII

CONTENT

This appendix gives second virial coefficients of gaseous hydrogen as a function of temperature. Selected data from the literature have been fitted by least squares to the equation

$$B\,(\text{cm}^3\text{mol}^{-1}) = \sum_{i=1}^{n} a(i)[(T_o/T)-1]^{i-1}$$

where T_o = 298.15 K. The table below gives the coefficient $a(i)$ and values of B at fixed temperature increments, as calculated from this smoothing equation.

The equation may be used with the tabulated coefficients for interpolation within the indicated temperature range. It should not be used for extrapolation beyond this range.

A useful compilation of virial coefficient data from the literature may be found in [1].

	T (K)	B (cm^3 mol^{-1})
	15	−230
	20	−151
	25	−108
$a(1)$ = 15.4	30	−82
$a(2)$ = −9.0	35	−64
$a(3)$ = −0.2	40	−52
	45	−42
	50	−35
	60	−24
	70	−16
	80	−11
	90	−7
	100	−3
	200	11
	300	15
	400	18

Reference

1. J. H. Dymond and E. B. Smith, *The Virial Coefficients of Pure Gases and Mixtures: A Critical Compilation*, Oxford University Press, Oxford, U.K., 1980.

Appendix C: van der Waals Constants for Common Gases

CONTENT

The van der Waals equation of state for a real gas is

$$(P + n^2a/V^2)(V - nb) = nRT$$

where
- P is the pressure
- V is the volume
- T is the temperature
- n is the amount of substance (in moles)
- R is the gas constant

The van der Waals constants a and b are characteristic of the substance and are independent of temperature. They are related to the critical temperature and pressure, T_c and P_c, by

$$a = 27R^2T_c^2/64P_c \quad b = RT_c/8P_c$$

The table below gives values of a and b for some common gases. Most of the values have been calculated from the critical temperature and pressure values of the respective gases. Van der Waals constants for other gases may easily be calculated from the critical constants data.

To convert the van der Waals constants to SI units, note that 1 bar L^2/mol^2 = 0.1 Pa m^6/mol^2 and 1 L/mol = 0.001 m^3/mol.

Substance	a (bar L^2/mol^2)	b (L/mol)
Acetic acid	17.71	0.1065
Acetone	16.02	0.1124
Acetylene	4.516	0.0522
Ammonia	4.225	0.0371
Aniline	29.14	0.1486
Argon	1.355	0.0320
Benzene	18.82	0.1193
Bromine	9.75	0.0591
Butane	13.89	0.1164
1-Butanol	20.94	0.1326
2-Butanone	19.97	0.1326
Carbon dioxide	3.658	0.0429

(*continued*)

(continued)

Substance	a (bar L^2/mol^2)	b (L/mol)
Carbon disulfide	11.25	0.0726
Carbon monoxide	1.472	0.0395
Chlorine	6.343	0.0542
Chlorobenzene	25.80	0.1454
Chloroethane	11.66	0.0903
Chloromethane	7.566	0.0648
Cyclohexane	21.92	0.1411
Cyclopropane	8.34	0.0747
Decane	52.74	0.3043
1-Decanol	59.51	0.3086
Diethyl ether	17.46	0.1333
Dimethyl ether	8.690	0.0774
Dodecane	69.38	0.3758
1-Dodecanol	75.70	0.3750
Ethane	5.580	0.0651
Ethanol	12.56	0.0871
Ethylene	4.612	0.0582
Fluorine	1.171	0.0290
Furan	12.74	0.0926
Helium	0.0346	0.0238
Heptane	31.06	0.2049
1-Heptanol	38.17	0.2150
Hexane	24.84	0.1744
1-Hexanol	31.79	0.1856
Hydrazine	8.46	0.0462
Hydrogen	0.2452	0.0265
Hydrogen bromide	4.500	0.0442
Hydrogen chloride	3.700	0.0406
Hydrogen cyanide	11.29	0.0881
Hydrogen fluoride	9.565	0.0739
Hydrogen iodide	6.309	0.0530
Hydrogen sulfide	4.544	0.0434
Isobutane	13.32	0.1164
Krypton	5.193	0.0106
Methane	2.303	0.0431
Methanol	9.476	0.0659
Methylamine	7.106	0.0588
Neon	0.208	0.0167
Neopentane	17.17	0.1411
Nitric oxide	1.46	0.0289
Nitrogen	1.370	0.0387
Nitrogen dioxide	5.36	0.0443
Nitrogen trifluoride	3.58	0.0545
Nitrous oxide	3.852	0.0444
Octane	37.88	0.2374

(continued)

(continued)

Substance	a (bar L^2/mol^2)	b (L/mol)
1-Octanol	44.71	0.2442
Oxygen	1.382	0.0319
Ozone	3.570	0.0487
Pentane	19.09	0.1449
1-Pentanol	25.88	0.1568
Phenol	22.93	0.1177
Propane	9.39	0.0905
1-Propanol	16.26	0.1079
2-Propanol	15.82	0.1109
Propene	8.442	0.0824
Pyridine	19.77	0.1137
Pyrrole	18.82	0.1049
Silane	4.38	0.0579
Sulfur dioxide	6.865	0.0568
Sulfur hexafluoride	7.857	0.0879
Tetrachloromethane	20.01	0.1281
Tetrachlorosilane	20.96	0.1470
Tetrafluoroethylene	6.954	0.0809
Tetrafluoromethane	4.040	0.0633
Tetrafluorosilane	5.259	0.0724
Tetrahydrofuran	16.39	0.1082
Thiophene	17.21	0.1058
Toluene	24.86	0.1497
1,1,1-Trichloroethane	20.15	0.1317
Trichloromethane	15.34	0.1019
Trifluoromethane	5.378	0.0640
Trimethylamine	13.37	0.1101
Water	5.537	0.0305
Xenon	4.192	0.0516

Reference

1. Reid, R. C., Prausnitz, J. M., and Poling, B. E., *The Properties of Gases and Liquids*, 4th edn., McGraw-Hill, New York, 1987.

Appendix D: Mean Free Path and Related Properties of Common Gases

CONTENT

In the simplest version of the kinetic theory of gases, molecules are treated as hard spheres of diameter d, which make binary collisions only. In this approximation, the mean distance traveled by a molecule between successive collisions, the mean free path l, is related to the collision diameter by:

$$l = \frac{kT}{\pi\sqrt{2}Pd^2}$$

where
- P is the pressure
- T is the absolute temperature
- k is the Boltzmann constant

At standard conditions (P = 100,000 Pa and T = 298.15 K) this relation becomes:

$$l = \frac{9.27 \cdot 10^{27}}{d^2}$$

where l and d are in meters.

Using the same model and the same standard pressure, the collision diameter can be calculated from the viscosity η by the kinetic theory relation:

$$\eta = \frac{2.67 \cdot 10^{-20}(MT)^{1/2}}{d^2}$$

where
- η is in units of μPa s
- M is the molar mass in g/mol

Kinetic theory also gives a relation for the mean velocity $\bar{v}$ of molecules of mass m:

$$\bar{v} = \left(\frac{8kT}{\pi m}\right)^{1/2} = 145.5\left(\frac{T}{M}\right)^{1/2} \text{ m/s}$$

Finally, the mean time τ between collisions can be calculated from the relation $\tau\bar{v} = l$.

The table below gives values of l, $\overline{v}$, and τ for some common gases at 25°C and atmospheric pressure, as well as the value of d, all calculated from measured gas viscosities (see Refs. [2,3]). It is seen from the above equations that the mean free path varies directly with T and inversely with P, while the mean velocity varies as the square root of T and, in this approximation, is independent of P.

Gas	d	l	$\overline{v}$	τ
Air	$3.66 \cdot 10^{-10}$ m	$6.91 \cdot 10^{-8}$ m	467 m/s	148 ps
Ar	3.58	7.22	397	182
CO_2	4.53	4.51	379	119
H_2	2.71	12.6	1769	71
He	2.15	20.0	1256	159
Kr	4.08	5.58	274	203
N_2	3.70	6.76	475	142
NH_3	4.32	4.97	609	82
Ne	2.54	14.3	559	256
O_2	3.55	7.36	444	166
Xe	4.78	4.05	219	185

A more accurate model, in which molecular interactions are described by a Lennard-Jones potential, gives mean free path values about 5% lower than this table (see Ref. [4]).

References

1. Reid, R. C., Prausnitz, J. M., and Poling, B. E., *The Properties of Gases and Liquids*, 4th edn., McGraw-Hill, New York, 1987.
2. Lide, D. R. and Kehiaian, H. V., *CRC Handbook of Thermophysical and Thermochemical Data*, CRC Press, Boca Raton, FL, 1994.
3. Vargaftik, N. B., *Tables of Thermophysical Properties of Liquids and Gases*, 2nd edn., John Wiley & Sons, New York, 1975.
4. Kays, G. W. C. and Laby, T. H., *Tables of Physical and Chemical Constants*, 15th edn., Longman, London, U.K., 1986.

Appendix E: Vapor Pressure of Hydrogen, Helium, and Neon below 300 K

CONTENT

This appendix gives vapor pressure of hydrogen, helium, and neon in the temperature range 2 to 44 K. The data have been taken from evaluated sources; references are listed at the end of the table.

Pressures are given in kilopascals (kPa). Note that:

1 kPa = 7.50062 Torr

100 kPa = 1 bar

101.325 kPa = 1 atmos

"s" following an entry indicates that the compound is solid at that temperature.

Helium		**Hydrogen**		**Neon**	
***T* (K)**	***P* (kPa)**	***T* (K)**	***P* (kPa)**	***T* (K)**	***P* (kPa)**
2.2	5.3	14.0	7.90	25.0	51.3
2.3	6.7	14.5	10.38	26.0	71.8
2.4	8.3	15.0	13.43	27.0	98.5
2.5	10.2	15.5	17.12	28.0	132.1
2.6	12.4	16.0	21.53	29.0	173.5
2.7	14.8	16.5	26.74	30.0	223.8
2.8	17.5	17.0	32.84	31.0	284.0
2.9	20.6	17.5	39.92	32.0	355.2
3.0	24.0	18.0	48.08	33.0	438.6
3.1	27.8	18.5	57.39	34.0	535.2
3.2	32.0	19.0	67.96	35.0	646.2
3.3	36.5	19.5	79.89	36.0	772.8
3.4	41.5	20.0	93.26	37.0	916.4
3.5	47.0	20.5	108.2	38.0	1078
3.6	52.9	21.0	124.7	39.0	1260
3.7	59.3	21.5	143.1	40.0	1462
3.8	66.1	22.0	163.2	41.0	1688
3.9	73.5	22.5	185.3	42.0	1939
4.0	81.5	23.0	209.4	43.0	2216
4.1	90.0	23.5	235.7	44.0	2522
4.2	99.0	24.0	264.2	[2]	
4.3	108.7	24.5	295.1		
4.4	119.0	25.0	328.5		

(continued)

(continued)

Helium		Hydrogen		Neon	
T (K)	*P* (kPa)	*T* (K)	*P* (kPa)	*T* (K)	*P* (kPa)
4.5	129.9	25.5	364.3		
4.6	141.6	26.0	402.9		
4.7	153.9	26.5	444.3		
4.8	167.0	27.0	488.5		
4.9	180.8	27.5	535.7		
5.0	195.4	28.0	586.1		
5.1	210.9	28.5	639.7		
		29.0	696.7		
		29.5	757.3		
		30.0	821.4		
		30.5	889.5		
		31.0	961.5		
		31.5	1038.0		
		32.0	1119.0		
		32.5	1204.0		
Refs.	[3,4]		[1]		

References

1. B. A. Younglove, Thermophysical properties of fluids. I. Ethylene, parahydrogen, nitrogen trifluoride, and oxygen, *J. Phys. Chem. Ref. Data*, 11, Supp. 1, 1982.
2. V. A. Rabinovich et al., *Thermophysical Properties of Neon, Argon, Krypton, and Xenon*, Hemisphere Publishing Corp., New York, 1987.
3. R. D. McCarty, *J. Phys. Chem. Ref. Data*, 2, 923, 1973.
4. R. Span and W. Wagner, *J. Phys. Chem. Ref. Data*, 25, 1509, 1996.

Appendix F: Properties of Cryogenic Fluids

CONTENT

This appendix gives physical and thermodynamic properties of eight cryogenic fluids. The properties are:

M	Molar mass in grams per mole
T_t	Triple point temperature in kelvins
P_t	Triple point pressure in kilopascals
ρ_t(l)	Liquid density at the triple point in grams per milliliter
$\Delta_{fus}H@T_t$	Enthalpy of fusion at the triple point in joules per gram
T_b	Normal boiling point in kelvins at a pressure of 101,325 pascals (760 mmHg)
$\Delta_{vap}H@T_b$	Enthalpy of vaporization at the normal boiling point in joules per gram
ρ(l)$@T_b$	Liquid density at the normal boiling point in grams per milliliter
ρ(g)$@T_b$	Vapor density at the normal boiling point in grams per liter
C_p(l)$@T_b$	Liquid heat capacity at constant pressure at the normal boiling point in joules per gram kelvin
C_p(g)$@T_b$	Vapor heat capacity at constant pressure at the normal boiling point in joules per gram kelvin
T_c	Critical temperature in kelvins
P_c	Critical pressure in megapascals
ρ_c	Critical density in grams per milliliter

In the case of air, the value given for the triple point temperature is the incipient solidification temperature, and the normal boiling point value is the incipient boiling (bubble) point. See Ref. [3] for more details.

Property	Units	Air	N_2	O_2	H_2	He	Ne	Ar	Kr	Xe	CH_4
M	g/mol	28.96	28.014	31.999	2.0159	4.0026	20.180	39.948	83.800	131.290	16.043
T_t	K	59.75	63.15	54.3584	13.8		24.5561	83.8058	115.8	161.4	90.694
P_t	kPa		12.463	0.14633	7.042		50	68.95	72.92	81.59	11.696
ρ_t(l)	g/mL	0.959	0.870	1.306	0.0770		1.251	1.417	2.449	2.978	0.4515
$\Delta_{fus}H@T_t$	J/g		25.3	13.7	59.5		16.8	28.0	16.3	13.8	58.41
T_b	K	78.67	77.35	90.188	20.28	4.2221	27.07	87.293	119.92	165.10	111.668
$\Delta_{vap}H@T_b$	J/g	198.7	198.8	213.1	445	20.7	84.8	161.0	108.4	96.1	510.83
ρ(l)$@T_b$	g/mL	0.8754	0.807	1.141	0.0708	0.124901	1.204	1.396	2.418	2.953	0.4224
ρ(g)$@T_b$	g/L	3.199	4.622	4.467	1.3390	16.89	9.51	5.79	8.94		1.816
C_p(l)$@T_b$	J/g K	1.865	2.042	1.699	9.668	4.545	1.877	1.078	0.533	0.340	3.481
C_p(g)$@T_b$	J/g K		1.341	0.980	12.24	9.78		0.570	0.248	0.158	2.218
T_c	K	132.5	126.20	154.581	32.98	5.1953	44.40	150.663	209.40	289.73	190.56
P_c	Mpa	3.766	3.390	5.043	1.293	0.227460	2.760	4.860	5.500	5.840	4.592
ρ_c	g/mL	0.316	0.313	0.436	0.031	0.06964	0.484	0.531	0.919	1.110	0.1627

References

1. Younglove, B. A., *J. Phys. Chem. Ref. Data*, 11, Suppl. 1, 1982.
2. Daubert, T. E., Danner, R. P., Sibul, H. M., and Stebbins, C. C., *Physical and Thermodynamic Properties of Pure Compounds: Data Compilation*, extant 1994 (core with 4 supplements), Taylor & Francis, Bristol, PA (also available as database).
3. Sytchev, V. V. et al., *Thermodynamic Properties of Air*, Hemisphere Publishing, New York, 1987.
4. Jacobsen, R. T., Stewart, R. B., and Jahangiri, M., *J. Phys. Chem. Ref. Data*, 15, 735, 1986. [Nitrogen]
5. Stewart, R. B., Jacobsen, R. T., and Wagner, W., *J. Phys. Chem. Ref. Data*, 20, 917, 1997. [Oxygen]
6. McCarty; R. D., *J. Phys. Chem. Ref. Data*, 2, 923, 1973. [Helium] Also, Donnelly, R. J., private communication.
7. Stewart, R. B. and Jacobsen, R. T., *J. Phys. Chem. Ref. Data*, 18, 639, 1989. [Argon]
8. Setzmann, U. and Wagner, W., *J. Phys. Chem. Ref. Data*, 20, 1061, 1991. [Methane]
9. Vargaftik, N. B., *Thermophysical Properties of Liquids and Gases*, 2nd edn., John Wiley, New York, 1975.

Appendix G: Viscosity of Common Gases

CONTENT

The following appendix gives the viscosity of some common gases as a function of temperature. Unless otherwise noted, the viscosity values refer to a pressure of 100 kPa (1 bar). The notation $P = 0$ indicates the low pressure limiting value is given. The difference between the velocity at 100 kPa and the limiting value is generally less than 1%. Viscosity is given in units of μPa s; note that 1 μPa s = 10^{-5} poise.

		Viscosity in Micropascal Seconds (μPa s)						
		100 K	**200 K**	**300 K**	**400 K**	**500 K**	**600 K**	**Ref.**
	Air	7.1	13.3	18.6	23.1	27.1	30.8	[1]
Ar	Argon	8.0	15.9	22.9	28.8	34.2	39.0	[2,8]
BF_3	Boron trifluoride		12.3	17.1	21.7	26.1	30.2	[13]
ClH	Hydrogen chloride			14.6	19.7	24.3		[13]
F_6S	Sulfur hexafluoride ($P = 0$)			15.3	19.8	23.9	27.7	[10]
H_2	Hydrogen ($P = 0$)	4.2	6.8	9.0	10.9	12.7	14.4	[4]
D_2	Deuterium ($P = 0$)	5.9	9.6	12.6	15.4	17.9	20.3	[11]
H_2O	Water			10.0	13.3	17.3	21.4	[6]
D_2O	Deuterium oxide			11.1	13.7	17.7	22.0	[7]
He	Helium ($P = 0$)	9.7	15.3	20.0	24.4	28.4	32.3	[8]
Kr	Krypton ($P = 0$)	8.8	17.1	25.6	33.1	39.8	45.9	[8]
NO	Nitric oxide		13.8	19.2	23.8	28.0	31.9	[13]
N_2	Nitrogen ($P = 0$)		12.9	17.9	22.2	26.1	29.6	[12]
N_2O	Nitrous oxide		10.0	15.0	19.4	23.6	27.4	[13]
Ne	Neon ($P = 0$)	14.4	24.3	32.1	38.9	45.0	50.8	[8]
O_2	Oxygen ($P = 0$)	7.5	14.6	20.8	26.1	30.8	35.1	[12]
O_2S	Sulfur dioxide		8.6	12.9	17.5	21.7		[13]
Xe	Xenon ($P = 0$)	8.3	15.4	23.2	30.7	37.6	44.0	[8]
CO	Carbon monoxide	6.7	12.9	17.8	22.1	25.8	29.1	[13]
CO_2	Carbon dioxide		10.0	15.0	19.7	24.0	28.0	[9,10]
$CHCl_3$	Chloroform			10.2	13.7	16.9	20.1	[13]
CH_4	Methane		7.7	11.2	14.3	17.0	19.4	[10]
CH_4O	Methanol				13.2	16.5	19.6	[13]
C_2H_2	Acetylene			10.4	13.5	16.5		[13]
C_2H_4	Ethylene		7.0	10.4	13.6	16.5	19.1	[3]
C_2H_6	Ethane		6.4	9.5	12.3	14.9	17.3	[5]
C_2H_6O	Ethanol				11.6	14.5	17.0	[13]

(continued)

(continued)

		Viscosity in Micropascal Seconds (μPa s)						
		100 K	**200 K**	**300 K**	**400 K**	**500 K**	**600 K**	**Ref.**
C_3H_8	Propane			8.3	10.9	13.4	15.8	[5]
C_4H_{10}	Butane			7.5	10.0	12.3	14.6	[5]
C_4H_{10}	Isobutane			7.6	10.0	12.3	14.6	[5]
$C_4H_{10}O$	Diethyl ether			7.6	10.1	12.4		[13]
C_5H_{12}	Pentane			6.7	9.2	11.4	13.4	[13]
C_6H_{14}	Hexane				8.6	10.8	12.8	[13]

References

1. K. Kadoya, N. Matsunaga, and A. Nagashima, Viscosity and thermal conductivity of dry air in the gaseous phase, *J. Phys. Chem. Ref. Data*, 14, 947, 1985.
2. B. A. Younglove and H. J. M. Hanley, The viscosity and thermal conductivity coefficients of gaseous and liquid argon, *J. Phys. Chem. Ref. Data*, 15, 1323, 1986.
3. P. M. Holland, B. E. Eaton, and H. J. M. Hanley, A correlation of the viscosity and thermal conductivity data of gaseous and liquid ethylene, *J. Phys. Chem. Ref. Data*, 12, 917, 1983.
4. M. J. Assael, S. Mixafendi, and W. A. Wakeham, The viscosity and thermal conductivity of normal hydrogen in the limit zero density, *J. Phys. Chem. Ref. Data*, 15, 1315, 1986.
5. B. A. Younglove and J. F. Ely, Thermophysical properties of fluids, II. Methane, ethane, propane, isobutane, and normal butane, *J. Phys. Chem. Ref. Data*, 16, 577, 1987.
6. J. V. Sengers and J. T. R. Watson, Improved international formulations for the viscosity and thermal conductivity of water substance, *J. Phys. Chem. Ref. Data*, 15, 1291, 1986.
7. N. Matsunaga and A. Nagashima, Transport properties of liquid and gaseous D_2O over a wide range of temperature and pressure, *J. Phys. Chem. Ref. Data*, 12, 933, 1983.
8. J. Kestin et al., Equilibrium and transport properties of the noble gases and their mixtures at low density, *J. Phys. Chem. Ref. Data*, 13, 299, 1984.
9. V. Vescovic et al., The transport properties of carbon dioxide, *J. Phys. Chem. Ref. Data*, 19, 1990.
10. R. D. Trengove and W. A. Wakeham, The viscosity of carbon dioxide, methane, and sulfur hexafluoride in the limit of zero density, *J. Phys. Chem. Ref. Data*, 16, 175, 1987.
11. M. J. Assael, S. Mixafendi, and W. A. Wakeham, The viscosity of normal deuterium in the limit of zero density, *J. Phys. Chem. Ref. Data*, 16, 189, 1987.
12. W. A. Cole and W. A. Wakeham, The viscosity of nitrogen, oxygen, and their binary mixtures in the limit of zero density, *J. Phys. Chem. Ref. Data*, 14, 209, 1985.
13. C. Y. Ho, ed., *Properties of Inorganic and Organic Fluids*, CINDAS Data Series on Materials Properties, Vol. V-1, Hemisphere Publishing Corp., New York, 1988.

Appendix H: Binary Diffusion Coefficients of Common Gases

CONTENT

This appendix gives binary diffusion coefficients D_{12} for a number of common gases as a function of temperature. Values refer to atmospheric pressure. The diffusion coefficient is inversely proportional to pressure as long as the gas is in a regime where binary collisions dominate. See Ref. [1] for a discussion of the dependence of D_{12} on temperature and composition.

The first part of the table gives data for several gases in the presence of a large excess of air. The remainder applies to equimolar mixtures of gases. Each gas pair is ordered alphabetically according to the most common way of writing the formula. The listing of pairs then follows alphabetical order by the first constituent.

D_{12}/cm² s⁻¹ for p = 101.325 kPa and the Specified T (K)

System	200	273.15	293.15	373.15	473.15	573.15	673.15
Large excess of air							
Ar–air		0.167	0.189	0.289	0.437	0.612	0.810
CH_4–air			0.210	0.321	0.485	0.678	0.899
CO–air			0.208	0.315	0.475	0.662	0.875
CO_2–air			0.160	0.252	0.390	0.549	0.728
H_2–air		0.668	0.756	1.153	1.747	2.444	3.238
H_2O–air			0.242	0.399	0.638	0.873	1.135
He–air		0.617	0.697	1.057	1.594	2.221	2.933
SF_6–air				0.150	0.233	0.329	0.438
Equimolar mixture							
Ar–CH_4				0.306	0.467	0.657	0.876
Ar–CO		0.168	0.190	0.290	0.439	0.615	0.815
Ar–CO_2		0.129	0.148	0.235	0.365	0.517	0.689
Ar–H_2		0.698	0.794	1.228	1.876	2.634	3.496
Ar–He	0.381	0.645	0.726	1.088	1.617	2.226	2.911
Ar–Kr	0.064	0.117	0.134	0.210	0.323	0.456	0.605
Ar–N_2		0.168	0.190	0.290	0.439	0.615	0.815
Ar–Ne	0.160	0.277	0.313	0.475	0.710	0.979	1.283
Ar–O_2		0.166	0.187	0.285	0.430	0.600	0.793
Ar–SF_6				0.128	0.202	0.290	0.389
Ar–Xe	0.052	0.095	0.108	0.171	0.264	0.374	0.498
CH_4–H_2			0.708	1.084	1.648	2.311	3.070
CH_4–He			0.650	0.992	1.502	2.101	2.784
CH_4–N_2			0.208	0.317	0.480	0.671	0.890
CH_4–O_2			0.220	0.341	0.523	0.736	0.978
CH_4–SF_6				0.167	0.257	0.363	0.482

(*continued*)

(continued)

System	200	273.15	293.15	373.15	473.15	573.15	673.15
CO–CO_2			0.162	0.250	0.384		
CO–H_2	0.408	0.686	0.772	1.162	1.743	2.423	3.196
CO–He	0.365	0.619	0.698	1.052	1.577	2.188	2.882
CO–Kr		0.131	0.149	0.227	0.346	0.485	0.645
CO–N_2	0.133	0.208	0.231	0.336	0.491	0.673	0.878
CO–O_2			0.202	0.307	0.462	0.643	0.849
CO–SF_6				0.144	0.226	0.323	0.432
CO_2–C_3H_8			0.084	0.133	0.209		
CO_2–H_2	0.315	0.552	0.627	0.964	1.470	2.066	2.745
CO_2–H_2O			0.162	0.292	0.496	0.741	1.021
CO_2–He	0.300	0.513	0.580	0.878	1.321		
CO_2–N_2			0.160	0.253	0.392	0.553	0.733
CO_2–N_2O	0.055	0.099	0.113	0.177	0.276		
CO_2–Ne	0.131	0.227	0.258	0.395	0.603	0.847	
CO_2–O_2			0.159	0.248	0 380	0.535	0.710
CO_2–SF_6				0.099	0.155		
D_2–H_2	0.631	1.079	1.219	1.846	2.778	3.866	5.103
H_2–He	0.775	1.320	1.490	2.255	3.394	4.726	6.242
H_2–Kr	0.340	0.601	0.682	1.053	1.607	2.258	2.999
H_2–N_2	0.408	0.686	0.772	1.162	1.743	2.423	3.196
H_2–Ne	0.572	0.982	1.109	1.684	2.541	3.541	4.677
H_2–O_2		0.692	0.782	1.188	1.792	2.497	3.299
H_2–SF_6			0.412	0.649	0.998	1.400	1.851
H_2–Xe		0.513	0.581	0.890	1.349	1.885	2.493
H_2O–N_2			0.242	0.399			
H_2O–O_2			0.244	0.403	0.645	0.882	1.147
He–Kr	0.330	0.559	0.629	0.942	1.404	1.942	2.550
He–N_2	0.365	0.619	0.698	1.052	1.577	2.188	2.882
He–Ne	0.563	0.948	1.066	1.592	2.362	3.254	4.262
He–O_2		0.641	0.723	1.092	1.640	2.276	2.996
He–SF_6			0.400	0.592	0.871	1.190	1.545
He–Xe	0.282	0.478	0.538	0.807	1.201	1.655	2.168
Kr–N_2		0.131	0.149	0.227	0.346	0.485	0.645
Kr–Ne	0.131	0.228	0.258	0.392	0.587	0.812	1.063
Kr–Xe	0.035	0.064	0.073	0.116	0.181	0.257	0.344
N_2–Ne			0.317	0.483	0.731	1.021	1.351
N_2–O_2			0.202	0.307	0.462	0.643	0.849
N_2–SF_6				0.148	0.231	0.328	0.436
N_2–Xe		0.107	0.122	0.188	0.287	0.404	0.539
Ne–Xe	0.111	0.193	0.219	0.332	0.498	0.688	0.901
O_2–SF_6			0.097	0.154	0.238	0.334	0.441

References

1. Marrero, T. R. and Mason, E. A., *J. Phys. Chem. Ref. Data*, 1, 1, 1972.
2. Kestin, J. et al., *J. Phys. Chem. Ref. Data*, 13, 229, 1984.

Appendix I: Diffusion of Common Gases in Water

CONTENT

This appendix gives values of the diffusion coefficient, D, for diffusion of several common gases in water at various temperatures. For simple one-dimensional transport, the diffusion coefficient describes the time-rate of change of concentration, dc/dt, through the equation

$$dc/dt = D\, d^2c/dx^2$$

where x is, for example, the perpendicular distance from a gas–liquid interface. The values below have been selected from the references indicated; in some cases data have been refitted to permit interpolation in temperature.

Gas–liquid diffusion coefficients are difficult to measure, and large differences are found between values obtained by different authors and through different experimental methods. See Refs. [1,2] for a discussion of measurement techniques.

	D (10^{-5} cm^2 s^{-1})						
	10°C	**15°C**	**20°C**	**25°C**	**30°C**	**35°C**	**Ref.**
Ar				2.5			[3,4]
$CHCl_2F$				1.80			[5]
CH_3Br				1.35			[5]
CH_3Cl				1.40			[5]
CH_4	1.24	1.43	1.62	1.84	2.08	2.35	[1]
CO_2	1.26	1.45	1.67	1.91	2.17	2.47	[1]
C_2H_2	1.43	1.59	1.78	1.99	2.23		[2]
Cl_2		1.13	1.5	1.89			[2,6]
HBr				3.15			[6]
HCl				3.07			[6]
H_2	3.62	4.08	4.58	5.11	5.69	6.31	[1]
H_2S				1.36			[2,6]
He	5.67	6.18	6.71	7.28	7.87	8.48	[1,3]
Kr	1.20	1.39	1.60	1.84	2.11	2.40	[1,3]
NH_3		1.3	1.5				[2]
NO_2			1.23	1.4	1.59		[2,6]
N_2				2.0			[2]
N_2O		1.62	2.11	2.57			[2,6]
Ne	2.93	3.27	3.64	4.03	4.45	4.89	[1,3]
O_2		1.67	2.01	2.42			[2,6]
Rn	0.81	0.96	1.13.	1.33	1.55	1.80	[1]
SO_2			1.62	1.83	2.07	2.32	[2]
Xe	0.93	1.08	1.27	1.47	1.70	1.95	[1,3]

References

1. Jähne, B., Heinz, G., and Dietrich, W., *J. Geophys. Res.*, 92, 10767, 1987.
2. Himmelblau, D. M., *Chem. Rev.*, 64, 527, 1964.
3. Boerboom, A. J. H. and Kleyn, G., *J. Chem. Phys.*, 50, 1086, 1969.
4. O'Brien, R. N. and Hyslop, W. F., *Can. J. Chem.*, 55, 1415, 1977.
5. Maharajh, D. M. and Walkley, J., *Can. J. Chem.*, 51, 944, 1973.
6. Landolt-Börnstein, *Numerical Data and Functional Relationships in Science and Technology*, 6th edn., II/5a, *Transport Phenomena I (Viscosity and Diffusion)*, Springer-Verlag, Heidelberg, Germany, 1969.

Index

D

I

N

O

T